DO NOT REMOVE
THIS BOOK FROM
THE LIBRARY

Encyclopedia of Separation Technology

Volume 1

Encyclopedia of Separation Technology

•

Volume 1

A KIRK-OTHMER ENCYCLOPEDIA

Douglas M. Ruthven, Editor

A WILEY-INTERSCIENCE PUBLICATION

JOHN WILEY & SONS

NEW YORK • CHICHESTER • WEINHEIM • BRISBANE • SINGAPORE • TORONTO

This text is printed on acid-free paper.

Copyright © 1997 by John Wiley & Sons, Inc.

All rights reserved. Published simultaneously in Canada.

Reproduction or translation of any part of this work
beyond that permitted by Section 107 or 108 of the
1976 United States Copyright Act without the permission
of the copyright owner is unlawful. Requests for
permission or further information should be addressed to
the Permissions Department, John Wiley & Sons, Inc.,
605 Third Avenue, New York, NY 10158-0012.

Library of Congress Cataloging-in-Publication Data

Encyclopedia of separation technology / editor, Douglas M. Ruthven.
    p. cm.
    "A Wiley-Interscience publication."
    Includes bibliographical references and index.
    ISBN 0-471-16124-1 (cloth : alk. paper)
    1. Separation (Technology)—Encyclopedias. I. Ruthven, Douglas
M. (Douglas Morris), 1938-
    TP156.S45E53   1997
    660'.2842—dc21   96-46795

Printed in the United States of America

10 9 8 7 6 5 4 3 2

CONTENTS

Absorption	1
Introduction	1
Equipment	57
Activated Alumina	60
Activated Carbon	72
Adsorption	94
Adsorption, Gas Separation	129
Adsorption, Liquid Separation	172
Air Pollution Control	199
Air Separation	200
Nitrogen	200
Oxygen	207
Bioseparations	211
BTX Separations	245
Centrifugal Separation	251
Chromatography	299
Chromatography, Chiral Separations	320
Contactors for Chemical Separations	355
Cryogenic Distillation	379
Crystallization	393
Desiccants	440
Desulfurization	465
Cleaning and Desulfurization of Coal	465
Desulfurization of Flue Gases	481
Dewatering	487
Dialysis	514
Diffusion Separation Methods	530
Distillation	584
Distillation, Azeotropic and Extractive	630
Drying	670
Electroseparations	715
Electrodialysis	715
Electrophoresis	728
Energy Conservation in Separation Processes	749
Extraction, Liquid–Liquid	760
Extraction, Liquid–Solid	814
Filtration	Vol. 2
Flocculation	Vol. 2
Flotation	Vol. 2
Gas–Solid Separations	Vol. 2
High Purity Gases	Vol. 2
Hollow-Fiber Membranes	Vol. 2
Inclusion Compounds	Vol. 2
Ion Exchange	Vol. 2
Isotope Separation	Vol. 2
Magnetic Separation	Vol. 2
Mass Transfer	Vol. 2
Membrane Technology	Vol. 2
Metals Separation	Vol. 2
Microbial and Viral Filtration	Vol. 2
Minerals Processing	Vol. 2
Molecular Sieves	Vol. 2
Nuclear Fuel Reprocessing	Vol. 2
Petroleum Refining	Vol. 2
Reverse Osmosis	Vol. 2
Sedimentation	Vol. 2
Separations Process Synthesis	Vol. 2
Silica Gel	Vol. 2
Size Separation	Vol. 2
Stack Gas	Vol. 2
Sulfurous Gases	Vol. 2
Supercritical Fluids	Vol. 2
Ultrafiltration	Vol. 2
Water Treatment	Vol. 2
Desalination	Vol. 2
Municipal water treatment	Vol. 2
Wastewater treatment	Vol. 2
Xylenes and Ethylbenzene	Vol. 2
Zone Refining	Vol. 2

EDITORIAL STAFF

Executive Editor: **Jacqueline I. Kroschwitz**
Editor: **Mary Howe-Grant**
Associate Managing Editor: **Lindy Humphreys**
Copy Editors: **Lawrence Altieri**
Jonathan Lee

CONTRIBUTORS

Rakesh Agrawal, *Air Products and Chemicals Inc., Allentown, Pennsylvania,* Cryogenic distillation

Hazel Aranha-Creado, *Pall Corporation, Port Washington, New York,* Microbial and viral filtration

Malcolm H. I. Baird, *McMaster University, Hamilton, Ontario, Canada,* Extraction, liquid–liquid

Frederick S. Baker, *Westvaco Corporation, Charleston, South Carolina,* Activated carbon

CONTRIBUTORS

Richard W. Baker, *Membrane Technology & Research, Inc., Menlo Park, California,* Membrane technology

Robert Bakish, *Bakish Materials Corporation, Englewood, New Jersey,* Desalination (under Water treatment)

Scott D. Barnicki, *Eastman Chemical Company, Kingsport, Tennessee,* Separations process synthesis

Ulrich Baurmeister, *Akzo Faser AG, Germany,* Dialysis

D. Bhattacharyya, *University of Kentucky, Lexington,* Reverse osmosis

P. F. Bryan, *Chevron Research and Technology Company, Richmond, California,* BTX separations

William Cannella, *Chevron Research and Technology Company, Richmond, California,* Xylenes and ethylbenzene

Michael A. Capone, *Exxon Research and Engineering Company, Florham Park, New Jersey,* Sulfurous gases

Shiao-Hung Chiang, *University of Pittsburgh, Pennsylvania,* Cleaning and desulfurization of coal (under Desulfurization)

James T. Cobb, Jr., *University of Pittsburgh, Pennsylvania,* Cleaning and desulfurization of coal (under Desulfurization)

Alan P. Cohen, *UOP, Des Plaines, Illinois,* Desiccants

Burton B. Crocker, *Consultant, Chesterfield, Missouri,* Equipment (under Absorption); Gas–solid separations; Stack gas

Charles Dickert, *Consultant, Yardley, Pennsylvania,* Ion exchange

David J. Dixon, *South Dakota School of Mines & Technology, Rapid City,* Supercritical fluids

Michael F. Doherty, *University of Massachusetts, Amherst,* Distillation, azeotropic and extractive

Cecil Dybowski, *University of Delaware, Newark,* Chromatography

Matthew Ennis, *Stanford University, Stanford, California,* Nitrogen (under Air separation)

Edward F. Ezell, *The BOC Group, Inc., Murray Hill, New Jersey,* High purity gases

James R. Fair, *The University of Texas at Austin,* Distillation

Stanley A. Gembicki, *UOP, Des Plaines, Illinois,* Adsorption, liquid separation

W. L. Godfrey, *BE Incorporated, Branwell, South Carolina,* Nuclear fuel reprocessing

J. C. Hall, *BE Incorporated, Branwell, South Carolina,* Nuclear fuel reprocessing

Thomas L. Hardenburger, *Air Liquide America Corporation, Tualatin, Oregon,* Nitrogen (under Air separation)

James G. Hansel, *Air Products and Chemicals, Inc., Allentown, Pennsylvania,* Oxygen (under Air separation)

Howard I. Heitner, *Cytec Industries, Stamford, Connecticut,* Flocculation

Shuen-Cheng Hwang, *The BOC Group, Inc., Murray Hill, New Jersey,* High purity gases

James A. Johnson, *UOP, Des Plaines, Illinois,* Adsorption, liquid separation

Keith P. Johnston, *University of Texas at Austin,* Supercritical fluids

Mary A. Kaiser, *E. I. du Pont de Nemours & Co., Inc., Newark, Delaware,* Chromatography

Glenn E. Kinard, *Air Products and Chemicals Inc., Allentown, Pennsylvania,* Cryogenic distillation

P. R. Klinkowski, *Dorr-Oliver, Inc.,* Ultrafiltration

Jeffrey P. Knapp, *E. I. du Pont de Nemours & Co., Inc., Wilmington, Delaware,* Distillation, azeotropic and extractive

Charles T. Kresge, *Mobil Research and Development Corporation, Paulsboro, New Jersey,* Molecular sieves

Günter H. Kühl, *Consultant, Cherry Hill, New Jersey,* Molecular sieves

Michael R. Ladisch, *Purdue University, West Lafayette, Indiana,* Bioseparations

Alan Letki, *Alfa Laval Separation Inc., Warminster, Pennsylvania,* Centrifugal separation

E. J. Lightfoot, *E. I. du Pont de Nemours & Co., Inc., Buffalo, New York,* Mass transfer

Edwin N. Lightfoot, Jr., *University of Wisconsin, Madison,* Mass transfer

Noam Lior, *University of Pennsylvania, Philadelphia,* Desalination (under Water treatment)

Teh C. Lo, *T. C. Lo & Associates, Wayne, New Jersey,* Extraction, liquid–liquid

Peter Luckie, *Penn State University, University Park, Pennsylvania,* Size separation

Michael J. Lysaght, *Cyto Therapeutics, Inc., Providence, Rhode Island,* Dialysis

W. C. Mangum, *University of Kentucky, Lexington,* Reverse osmosis

Kathleen Markey, *Synergen, Inc., Boulder, Colorado,* Electrophoresis (under Electroseparations)

Paul Y. McCormick, *Drying Unincorporated, Newark, Delaware,* Drying

Wayne A. McRae, *Consultant, Mannedork, Switzerland,* Electrodialysis (under Electroseparations)

Charles E. Miller, *Westvaco Corp., Charleston, South Carolina,* Activated carbon

Irving Moch, Jr., *E. I. du Pont de Nemours & Co., Inc., Wilmington, Delaware,* Hollow-fiber membranes

R. T. Moll, *Alfa Laval Separation Inc., Warminster, Pennsylvania,* Centrifugal separation

Booker Morey, *SRI International, Menlo Park, California,* Dewatering

Don Morgan, *Walker Magnetics, Milwaukee, Wisconsin,* Magnetic separation

D. R. Nagaraj, *Cytec Industries, Stamford, Connecticut,* Minerals processing

Anil R. Oroskar, *UOP, Des Plaines, Illinois,* Adsorption, liquid separation

K. Oshima, *New Mexico State University, Las Cruces,* Microbial and viral filtration

Alan Pearson, *Aluminum Company of America, Alcoa Center, Pennsylvania,* Activated alumina

Muthiah Ramanathan, *Roy F. Weston, Inc.,* Wastewater (under Water treatment)

Albert J. Repik, *Westvaco Corporation, Charleston, South Carolina,* Activated carbon

H. Wayne Richardson, *Phibro-Tech, Inc., Sumter, South Carolina,* Metals separation

Ronald W. Rousseau, *Georgia Institute of Technology, Atlanta,* Crystallization

Howard C. Rowles, *Air Products and Chemicals Inc., Allentown, Pennsylvania,* Cryogenic distillation

Scott Rudge, *Synergen, Inc., Boulder, Colorado,* Electrophoresis (under Electroseparations)

CONTRIBUTORS

Douglas M. Ruthven, *University of Maine, Orono,* Adsorption
J. Shacter, *Consultant, Oak Ridge, Tennessee,* Diffusion separation methods
Leonard Shapiro, *Consultant, Warminster, Pennsylvania,* Centrifugal separation
John D. Sherman, *UOP, Des Plaines, Illinois,* Adsorption, gas separation
Jeffrey J. Siirola, *Eastman Chemical Company, Kingsport, Tennessee,* Separations process synthesis
J. E. Singley, *Environmental Science & Engineering, Inc.,* Municipal water treatment (under Water treatment)
James Speight, *Western Research Institute, Laramie, Wyoming,* Petroleum refining
Apryll M. Stalcup, *University of Cincinnati, Ohio,* Chromatography, chiral separations
Dan Steinmeyer, *Monsanto Company, St. Louis, Missouri,* Energy conservation in separation processes
Urs von Stockar, *École Polytechnique Fédérale, Lausanne, Switzerland,* Introduction (under Absorption)
Ladislav Svarovsky, *Consultant Engineers and Fine Particle Software, West Yorkshire, United Kingdom,* Filtration; Sedimentation
W. P. M. van Swaaij, *University of Twente, the Netherlands,* Contactors for chemical separations
W. A. Sweeney, *Chevron Research and Technology Company, Richmond, California,* BTX separations
E. Donald Tolles, *Westvaco Corporation, Charleston, South Carolina,* Activated carbon
G. A. Townes, *BE Incorporated, Branwell, South Carolina,* Nuclear fuel reprocessing
G. F. Versteeg, *University of Twente, the Netherlands,* Contactors for chemical separations
Edward Von Halle, *Consultant, Oak Ridge, Tennessee,* Diffusion separation methods; Isotope separation
Richard J. Wakeman, *University of Exeter, Devon, United Kingdom,* Extraction, liquid–solid
Edwin Weber, *Technische Universität Bergakademie Freiberg, Germany,* Inclusion compounds
William R. Wilcox, *Clarkson College of Technology,* Zone refining
Charles R. Wilke, *University of California, Berkeley,* Introduction (under Absorption)
M. E. Williams, *EET Corporation, Knoxville, Tennessee,* Reverse osmosis
Walter H. Witlock, *The BOC Group, Inc., Murray Hill, New Jersey,* High purity gases
Baki Yarar, *Colorado School of Mines, Golden, Colorado,* Flotation
Carmen M. Yon, *UOP, Des Plaines, Illinois,* Adsorption, gas separation

PREFACE

The *Encyclopedia of Separation Technology* provides A to Z coverage, with authoritative reviews, of the separation processes used in industry and in the laboratory, including both the technology and the underlying science. As a result of increasingly stringent requirements for product purity (for example, the standards for ethical drugs now commonly require enantiomeric purity), separation technology has come to assume a dominant role in the chemical process industries (CPI). Within these industries, as of the mid-1990s, approximately half the capital investment and over half of the operating costs are associated with separation processes.

This two-volume encyclopedia is designed to provide an introduction and review of the various techniques that can be used to effect separations of gases, liquids, and solids for the purification of raw materials, feedstocks, process streams, or finished products. The articles, which have been written and reviewed by experts in the field of separations, are mainly derived from the fourth edition of the well-known *Kirk-Othmer Encyclopedia of Chemical Technology*. Some of the articles have been reprinted without substantial modification, whereas others have been condensed and edited to match the focus of the present work. Several new articles, for example, those on BIOSEPARATIONS, CONTACTORS FOR CHEMICAL SEPARATIONS, and HIGH PURITY GASES, have been written especially for this work in order to cover important separations topics which were not addressed adequately in *Kirk-Othmer*. All articles are self-contained, but extensive cross-references to other relevant entries are included. The contributing authors are acknowledged at the end of each entry. A full list of contributors and their affiliations is also included at the front of each volume.

NOTE ON CHEMICAL ABSTRACTS SERVICE REGISTRY NUMBERS AND NOMENCLATURE

Chemical Abstracts Service (CAS) Registry Numbers are unique numerical identifiers assigned to substances recorded in the CAS Registry System. They appear in brackets in the *Chemical Abstracts* (CA) substance and formula indexes following the names of compounds. A single compound may have synonyms in the chemical literature. A simple compound like phenethylamine can be named β-phenylethylamine or, as in *Chemical Abstracts*, benzeneethanamine. The usefulness of the *Encyclopedia* depends on accessibility through the most common correct name of a substance. Because of this diversity in nomenclature careful attention has been given to the problem in order to assist the reader as much as possible, especially in locating the systematic CA index name by means of the Registry Number. For this purpose, the reader may refer to the CAS Registry Handbook—Number Section which lists in numerical order the Registry Number with the *Chemical Abstracts* index name and the molecular formula; eg, **458-88-8**, Piperidine, 2-propyl-, (*S*)-, $C_8H_{17}N$; in the *Encyclopedia* this compound would be found under its common name, coniine [*458-88-8*]. Alternatively, this information can be retrieved electronically from CAS Online. In many cases molecular formulas have also been provided in the *Encyclopedia* text to facilitate electronic searching. The Registry Number is a valuable link for the reader in retrieving additional published information on substances and also as a point of access for on-line data bases.

In all cases, the CAS Registry Numbers have been given for title compounds in articles and for all compounds in the index. All specific substances indexed in *Chemical Abstracts* since 1965 are included in the CAS Registry System as are a large number of substances derived from a variety of reference works. The CAS Registry System identifies a substance on the basis of an unambiguous computer-language description of its molecular structure including stereochemical detail. The Registry Number is a machine-checkable number (like a Social Security number) assigned in sequential order to each substance as it enters the registry system. The value of the number lies in the fact that it is a concise and unique means of substance identification, which is independent of, and therefore bridges, many systems of chemical nomenclature. For polymers, one Registry Number may be used for the entire family; eg, polyoxyethylene (20) sorbitan monolaurate has the same number as all of its polyoxyethylene homologues.

Cross-references are inserted in the index for many common names and for some systematic names. Trademark names appear in the index. Names that are incorrect, misleading, or ambiguous are avoided. Formulas are given very frequently in the text to help in identifying compounds. The spelling and form used, even for industrial names, follow American chemical usage, but not always the usage of *Chemical Abstracts* (eg, *coniine* is used instead of *(S)-2-propylpiperidine*, *aniline* instead of *benzenamine*, and *acrylic acid* instead of *2-propenoic acid*).

There are variations in representation of rings in different disciplines. The dye industry does not designate aromaticity or double bonds in rings. All double bonds and aromaticity are shown in the *Encyclopedia* as a matter of course. For example, tetralin has an aromatic ring and a saturated ring and its structure

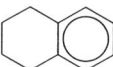

appears in the *Encyclopedia* with its common name, Registry Number enclosed in brackets, and parenthetical CA index name, ie, tetralin [*119-64-2*] (1,2,3,4-tetrahydronaphthalene). With names and structural formulas, and especially with CAS Registry Numbers, the aim is to help the reader have a concise means of substance identification.

Quantity	Unit	Symbol	Acceptable equivalent
*frequency	megahertz	MHz	
	hertz	Hz	1/s
heat capacity, entropy	joule per kelvin	J/K	
heat capacity (specific), specific entropy	joule per kilogram kelvin	J/(kg·K)	
heat-transfer coefficient	watt per square meter kelvin	W/(m^2·K)	
*illuminance	lux	lx	lm/m^2
*inductance	henry	H	Wb/A
linear density	kilogram per meter	kg/m	
luminance	candela per square meter	cd/m^2	
*luminous flux	lumen	lm	cd·sr
magnetic field strength	ampere per meter	A/m	
*magnetic flux	weber	Wb	V·s
*magnetic flux density	tesla	T	Wb/m^2
molar energy	joule per mole	J/mol	
molar entropy, molar heat capacity	joule per mole kelvin	J/(mol·K)	
moment of force, torque	newton meter	N·m	
momentum	kilogram meter per second	kg·m/s	
permeability	henry per meter	H/m	
permittivity	farad per meter	F/m	
*power, heat flow rate, radiant flux	kilowatt	kW	
	watt	W	J/s
power density, heat flux density, irradiance	watt per square meter	W/m^2	
*pressure, stress	megapascal	MPa	
	kilopascal	kPa	
	pascal	Pa	N/m^2
sound level	decibel	dB	
specific energy	joule per kilogram	J/kg	
specific volume	cubic meter per kilogram	m^3/kg	
surface tension	newton per meter	N/m	
thermal conductivity	watt per meter kelvin	W/(m·K)	
velocity	meter per second	m/s	
	kilometer per hour	km/h	
viscosity, dynamic	pascal second	Pa·s	
	millipascal second	mPa·s	
viscosity, kinematic	square meter per second	m^2/s	
	square millimeter per second	mm^2/s	

xviii FACTORS, ABBREVIATIONS, AND SYMBOLS

Quantity	Unit	Symbol	Acceptable equivalent
volume	cubic meter	m^3	
	cubic diameter	dm^3	L (liter) (5)
	cubic centimeter	cm^3	mL
wave number	1 per meter	m^{-1}	
	1 per centimeter	cm^{-1}	

In addition, there are 16 prefixes used to indicate order of magnitude, as follows:

Multiplication factor	Prefix	Symbol	Note
10^{18}	exa	E	
10^{15}	peta	P	
10^{12}	tera	T	
10^{9}	giga	G	
10^{6}	mega	M	
10^{3}	kilo	k	
10^{2}	hecto	h^a	^{a}Although hecto, deka, deci, and centi
10	deka	da^a	are SI prefixes, their use should be
10^{-1}	deci	d^a	avoided except for SI unit-multiples
10^{-2}	centi	c^a	for area and volume and nontech-
10^{-3}	milli	m	nical use of centimeter, as for body
10^{-6}	micro	μ	and clothing measurement.
10^{-9}	nano	n	
10^{-12}	pico	p	
10^{-15}	femto	f	
10^{-18}	atto	a	

For a complete description of SI and its use the reader is referred to ASTM E380 (4) and the article UNITS AND CONVERSION FACTORS which appears in Vol. 24.

A representative list of conversion factors from non-SI to SI units is presented herewith. Factors are given to four significant figures. Exact relationships are followed by a dagger. A more complete list is given in the latest editions of ASTM E380 (4) and ANSI Z210.1 (6).

Conversion Factors to SI Units

To convert from	To	Multiply by
acre	square meter (m^2)	4.047×10^3
angstrom	meter (m)	1.0×10^{-10}†
are	square meter (m^2)	1.0×10^{2}†

†Exact.

FACTORS, ABBREVIATIONS, AND SYMBOLS

To convert from	To	Multiply by
astronomical unit	meter (m)	1.496×10^{11}
atmosphere, standard	pascal (Pa)	1.013×10^{5}
bar	pascal (Pa)	$1.0 \times 10^{5\dagger}$
barn	square meter (m²)	$1.0 \times 10^{-28\dagger}$
barrel (42 U.S. liquid gallons)	cubic meter (m³)	0.1590
Bohr magneton (μ_B)	J/T	9.274×10^{-24}
Btu (International Table)	joule (J)	1.055×10^{3}
Btu (mean)	joule (J)	1.056×10^{3}
Btu (thermochemical)	joule (J)	1.054×10^{3}
bushel	cubic meter (m³)	3.524×10^{-2}
calorie (International Table)	joule (J)	4.187
calorie (mean)	joule (J)	4.190
calorie (thermochemical)	joule (J)	$4.184^{\dagger}$
centipoise	pascal second (Pa·s)	$1.0 \times 10^{-3\dagger}$
centistokes	square millimeter per second (mm²/s)	$1.0^{\dagger}$
cfm (cubic foot per minute)	cubic meter per second (m³/s)	4.72×10^{-4}
cubic inch	cubic meter (m³)	1.639×10^{-5}
cubic foot	cubic meter (m³)	2.832×10^{-2}
cubic yard	cubic meter (m³)	0.7646
curie	becquerel (Bq)	$3.70 \times 10^{10\dagger}$
debye	coulomb meter (C·m)	3.336×10^{-30}
degree (angle)	radian (rad)	1.745×10^{-2}
denier (international)	kilogram per meter (kg/m)	1.111×10^{-7}
	tex‡	0.1111
dram (apothecaries')	kilogram (kg)	3.888×10^{-3}
dram (avoirdupois)	kilogram (kg)	1.772×10^{-3}
dram (U.S. fluid)	cubic meter (m³)	3.697×10^{-6}
dyne	newton (N)	$1.0 \times 10^{-5\dagger}$
dyne/cm	newton per meter (N/m)	$1.0 \times 10^{-3\dagger}$
electronvolt	joule (J)	1.602×10^{-19}
erg	joule (J)	$1.0 \times 10^{-7\dagger}$
fathom	meter (m)	1.829
fluid ounce (U.S.)	cubic meter (m³)	2.957×10^{-5}
foot	meter (m)	$0.3048^{\dagger}$
footcandle	lux (lx)	10.76
furlong	meter (m)	2.012×10^{-2}
gal	meter per second squared (m/s²)	$1.0 \times 10^{-2\dagger}$
gallon (U.S. dry)	cubic meter (m³)	4.405×10^{-3}
gallon (U.S. liquid)	cubic meter (m³)	3.785×10^{-3}
gallon per minute (gpm)	cubic meter per second (m³/s)	6.309×10^{-5}
	cubic meter per hour (m³/h)	0.2271

†Exact.
‡See footnote on p. xiii.

xx FACTORS, ABBREVIATIONS, AND SYMBOLS

To convert from	To	Multiply by
gauss	tesla (T)	1.0×10^{-4}
gilbert	ampere (A)	0.7958
gill (U.S.)	cubic meter (m^3)	1.183×10^{-4}
grade	radian	1.571×10^{-2}
grain	kilogram (kg)	6.480×10^{-5}
gram force per denier	newton per tex (N/tex)	8.826×10^{-2}
hectare	square meter (m^2)	$1.0 \times 10^{4\dagger}$
horsepower (550 ft·lbf/s)	watt (W)	7.457×10^2
horsepower (boiler)	watt (W)	9.810×10^3
horsepower (electric)	watt (W)	$7.46 \times 10^{2\dagger}$
hundredweight (long)	kilogram (kg)	50.80
hundredweight (short)	kilogram (kg)	45.36
inch	meter (m)	$2.54 \times 10^{-2\dagger}$
inch of mercury (32°F)	pascal (Pa)	3.386×10^3
inch of water (39.2°F)	pascal (Pa)	2.491×10^2
kilogram-force	newton (N)	9.807
kilowatt hour	megajoule (MJ)	$3.6^\dagger$
kip	newton (N)	4.448×10^3
knot (international)	meter per second (m/S)	0.5144
lambert	candela per square meter (cd/m^3)	3.183×10^3
league (British nautical)	meter (m)	5.559×10^3
league (statute)	meter (m)	4.828×10^3
light year	meter (m)	9.461×10^{15}
liter (for fluids only)	cubic meter (m^3)	$1.0 \times 10^{-3\dagger}$
maxwell	weber (Wb)	$1.0 \times 10^{-8\dagger}$
micron	meter (m)	$1.0 \times 10^{-6\dagger}$
mil	meter (m)	$2.54 \times 10^{-5\dagger}$
mile (statute)	meter (m)	1.609×10^3
mile (U.S. nautical)	meter (m)	$1.852 \times 10^{3\dagger}$
mile per hour	meter per second (m/s)	0.4470
millibar	pascal (Pa)	1.0×10^2
millimeter of mercury (0°C)	pascal (Pa)	$1.333 \times 10^{2\dagger}$
minute (angular)	radian	2.909×10^{-4}
myriagram	kilogram (kg)	10
myriameter	kilometer (km)	10
oersted	ampere per meter (A/m)	79.58
ounce (avoirdupois)	kilogram (kg)	2.835×10^{-2}
ounce (troy)	kilogram (kg)	3.110×10^{-2}
ounce (U.S. fluid)	cubic meter (m^3)	2.957×10^{-5}
ounce-force	newton (N)	0.2780
peck (U.S.)	cubic meter (m^3)	8.810×10^{-3}
pennyweight	kilogram (kg)	1.555×10^{-3}
pint (U.S. dry)	cubic meter (m^3)	5.506×10^{-4}
pint (U.S. liquid)	cubic meter (m^3)	4.732×10^{-4}

†Exact.

FACTORS, ABBREVIATIONS, AND SYMBOLS

To convert from	To	Multiply by
poise (absolute viscosity)	pascal second (Pa·s)	$0.10^\dagger$
pound (avoirdupois)	kilogram (kg)	0.4536
pound (troy)	kilogram (kg)	0.3732
poundal	newton (N)	0.1383
pound-force	newton (N)	4.448
pound force per square inch (psi)	pascal (Pa)	6.895×10^3
quart (U.S. dry)	cubic meter (m^3)	1.101×10^{-3}
quart (U.S. liquid)	cubic meter (m^3)	9.464×10^{-4}
quintal	kilogram (kg)	$1.0 \times 10^{2\dagger}$
rad	gray (Gy)	$1.0 \times 10^{-2\dagger}$
rod	meter (m)	5.029
roentgen	coulomb per kilogram (C/kg)	2.58×10^{-4}
second (angle)	radian (rad)	$4.848 \times 10^{-6\dagger}$
section	square meter (m^2)	2.590×10^6
slug	kilogram (kg)	14.59
spherical candle power	lumen (lm)	12.57
square inch	square meter (m^2)	6.452×10^{-4}
square foot	square meter (m^2)	9.290×10^{-2}
square mile	square meter (m^2)	2.590×10^6
square yard	square meter (m^2)	0.8361
stere	cubic meter (m^3)	$1.0^\dagger$
stokes (kinematic viscosity)	square meter per second (m^2/s)	$1.0 \times 10^{-4\dagger}$
tex	kilogram per meter (kg/m)	$1.0 \times 10^{-6\dagger}$
ton (long, 2240 pounds)	kilogram (kg)	1.016×10^3
ton (metric) (tonne)	kilogram (kg)	$1.0 \times 10^{3\dagger}$
ton (short, 2000 pounds)	kilogram (kg)	9.072×10^2
torr	pascal (Pa)	1.333×10^2
unit pole	weber (Wb)	1.257×10^{-7}
yard	meter (m)	$0.9144^\dagger$

†Exact.

Abbreviations and Unit Symbols

Following is a list of common abbreviations and unit symbols used in the *Encyclopedia*. In general they agree with those listed in *American National Standard Abbreviations for Use on Drawings and in Text* (*ANSI Y1.1*) (6) and *American National Standard Letter Symbols for Units in Science and Technology* (*ANSI Y10*) (6). Also included is a list of acronyms for a number of private and government organizations as well as common industrial solvents, polymers, and other chemicals.

FACTORS, ABBREVIATIONS, AND SYMBOLS

Rules for Writing Unit Symbols (4):

1. Unit symbols are printed in upright letters (roman) regardless of the type style used in the surrounding text.
2. Unit symbols are unaltered in the plural.
3. Unit symbols are not followed by a period except when used at the end of a sentence.
4. Letter unit symbols are generally printed lower-case (for example, cd for candela) unless the unit name has been derived from a proper name, in which case the first letter of the symbol is capitalized (W, Pa). Prefixes and unit symbols retain their prescribed form regardless of the surrounding typography.
5. In the complete expression for a quantity, a space should be left between the numerical value and the unit symbol. For example, write 2.37 lm, *not* 2.37lm, and 35 mm, *not* 35mm. When the quantity is used in an adjectival sense, a hyphen is often used, for example, 35-mm film. *Exception:* No space is left between the numerical value and the symbols of degree, minute, and second of plane angle, degree Celsius, and the percent sign.
6. No space is used between the prefix and unit symbol (for example, kg).
7. Symbols, not abbreviations, should be used for units. For example, use "A," not "amp," for ampere.
8. When multiplying unit symbols, use a raised dot:

$$N \cdot m \quad \text{for} \quad \text{newton meter}$$

In the case of W·h, the dot may be omitted, thus:

$$Wh$$

An exception to this practice is made for computer printouts, automatic typewriter work, etc, where the raised dot is not possible, and a dot on the line may be used.

9. When dividing unit symbols, use one of the following forms:

$$m/s \quad or \quad m \cdot s^{-1} \quad or \quad \frac{m}{s}$$

In no case should more than one slash be used in the same expression unless parentheses are inserted to avoid ambiguity. For example, write:

$$J/(mol \cdot K) \quad or \quad J \cdot mol^{-1} \cdot K^{-1} \quad or \quad (J/mol)/K$$

but *not*

$$J/mol/K$$

FACTORS, ABBREVIATIONS, AND SYMBOLS

10. Do not mix symbols and unit names in the same expression. Write:

$$\text{joules per kilogram} \quad or \quad \text{J/kg} \quad or \quad \text{J·kg}^{-1}$$

but *not*

$$\text{joules/kilogram} \quad nor \quad \text{joules/kg} \quad nor \quad \text{joules·kg}^{-1}$$

ABBREVIATIONS AND UNITS

A	ampere	AOAC	Association of Official Analytical Chemists
A	anion (eg, HA)		
A	mass number	AOCS	American Oil Chemists' Society
a	atto (prefix for 10^{-18})		
AATCC	American Association of Textile Chemists and Colorists	APHA	American Public Health Association
		API	American Petroleum Institute
ABS	acrylonitrile–butadiene–styrene	aq	aqueous
abs	absolute	Ar	aryl
ac	alternating current, *n.*	*ar*-	aromatic
a-c	alternating current, *adj.*	*as*-	asymmetric(al)
ac-	alicyclic	ASHRAE	American Society of Heating, Refrigerating, and Air Conditioning Engineers
acac	acetylacetonate		
ACGIH	American Conference of Governmental Industrial Hygienists		
		ASM	American Society for Metals
ACS	American Chemical Society	ASME	American Society of Mechanical Engineers
AGA	American Gas Association		
Ah	ampere hour	ASTM	American Society for Testing and Materials
AIChE	American Institute of Chemical Engineers	at no.	atomic number
AIME	American Institute of Mining, Metallurgical, and Petroleum Engineers	at wt	atomic weight
		av(g)	average
		AWS	American Welding Society
		b	bonding orbital
AIP	American Institute of Physics	bbl	barrel
		bcc	body-centered cubic
AISI	American Iron and Steel Institute	BCT	body-centered tetragonal
		Bé	Baumé
alc	alcohol(ic)	BET	Brunauer-Emmett-Teller (adsorption equation)
Alk	alkyl		
alk	alkaline (not alkali)	bid	twice daily
amt	amount	Boc	*t*-butyloxycarbonyl
amu	atomic mass unit	BOD	biochemical (biological) oxygen demand
ANSI	American National Standards Institute		
		bp	boiling point
AO	atomic orbital	Bq	becquerel

C	coulomb	DIN	Deutsche Industrie Normen
°C	degree Celsius		
C-	denoting attachment to carbon	*dl*-; DL-	racemic
		DMA	dimethylacetamide
c	centi (prefix for 10^{-2})	DMF	dimethylformamide
c	critical	DMG	dimethyl glyoxime
ca	circa (approximately)	DMSO	dimethyl sulfoxide
cd	candela; current density; circular dichroism	DOD	Department of Defense
		DOE	Department of Energy
CFR	Code of Federal Regulations	DOT	Department of Transportation
cgs	centimeter-gram-second	DP	degree of polymerization
CI	Color Index	dp	dew point
cis-	isomer in which substituted groups are on same side of double bond between C atoms	DPH	diamond pyramid hardness
		dstl(d)	distill(ed)
		dta	differential thermal analysis
cl	carload		
cm	centimeter	(*E*)-	entgegen; opposed
cmil	circular mil	ϵ	dielectric constant (unitless number)
cmpd	compound		
CNS	central nervous system	*e*	electron
CoA	coenzyme A	ECU	electrochemical unit
COD	chemical oxygen demand	ed.	edited, edition, editor
coml	commercial(ly)	ED	effective dose
cp	chemically pure	EDTA	ethylenediaminetetra- acetic acid
cph	close-packed hexagonal		
CPSC	Consumer Product Safety Commission	emf	electromotive force
		emu	electromagnetic unit
cryst	crystalline	en	ethylene diamine
cub	cubic	eng	engineering
D	debye	EPA	Environmental Protection Agency
D-	denoting configurational relationship		
		epr	electron paramagnetic resonance
d	differential operator		
d	day; deci (prefix for 10^{-1})	eq.	equation
d	density	esca	electron spectroscopy for chemical analysis
d-	*dextro*-, dextrorotatory		
da	deka (prefix for 10^1)	esp	especially
dB	decibel	esr	electron-spin resonance
dc	direct current, *n*.	est(d)	estimate(d)
d-c	direct current, *adj*.	estn	estimation
dec	decompose	esu	electrostatic unit
detd	determined	exp	experiment, experimental
detn	determination	ext(d)	extract(ed)
Di	didymium, a mixture of all lanthanons	F	farad (capacitance)
		F	faraday (96,487 C)
dia	diameter	f	femto (prefix for 10^{-15})
dil	dilute		

FAO	Food and Agriculture Organization (United Nations)	hyd	hydrated, hydrous
		hyg	hygroscopic
		Hz	hertz
fcc	face-centered cubic	^{i} (eg, Pr^i)	iso (eg, isopropyl)
FDA	Food and Drug Administration	i-	inactive (eg, i-methionine)
		IACS	International Annealed Copper Standard
FEA	Federal Energy Administration	ibp	initial boiling point
FHSA	Federal Hazardous Substances Act	IC	integrated circuit
		ICC	Interstate Commerce Commission
fob	free on board		
fp	freezing point	ICT	International Critical Table
FPC	Federal Power Commission	ID	inside diameter; infective dose
FRB	Federal Reserve Board		
frz	freezing	ip	intraperitoneal
G	giga (prefix for 10^9)	IPS	iron pipe size
G	gravitational constant = 6.67×10^{11} N·m^2/kg^2	ir	infrared
		IRLG	Interagency Regulatory Liaison Group
g	gram		
(g)	gas, only as in H$_2$O(g)	ISO	International Organization Standardization
g	gravitational acceleration		
gc	gas chromatography	ITS-90	International Temperature Scale (NIST)
gem-	geminal		
glc	gas–liquid chromatography	IU	International Unit
		IUPAC	International Union of Pure and Applied Chemistry
g-mol wt; gmw	gram-molecular weight		
GNP	gross national product	IV	iodine value
gpc	gel-permeation chromatography	iv	intravenous
		J	joule
GRAS	Generally Recognized as Safe	K	kelvin
		k	kilo (prefix for 10^3)
grd	ground	kg	kilogram
Gy	gray	L	denoting configurational relationship
H	henry		
h	hour; hecto (prefix for 10^2)	L	liter (for fluids only) (5)
ha	hectare	l-	$levo$-, levorotatory
HB	Brinell hardness number	(l)	liquid, only as in NH$_3$(l)
Hb	hemoglobin	LC$_{50}$	conc lethal to 50% of the animals tested
hcp	hexagonal close-packed		
hex	hexagonal	LCAO	linear combination of atomic orbitals
HK	Knoop hardness number		
hplc	high performance liquid chromatography	lc	liquid chromatography
		LCD	liquid crystal display
HRC	Rockwell hardness (C scale)	lcl	less than carload lots
		LD$_{50}$	dose lethal to 50% of the animals tested
HV	Vickers hardness number		

LED	light-emitting diode	N-	denoting attachment to nitrogen
liq	liquid		
lm	lumen	n (as n_D^{20})	index of refraction (for 20°C and sodium light)
ln	logarithm (natural)		
LNG	liquefied natural gas	^{n} (as Bu^n),	
log	logarithm (common)	n-	normal (straight-chain structure)
LOI	limiting oxygen index		
LPG	liquefied petroleum gas	n	neutron
ltl	less than truckload lots	n	nano (prefix for 10^9)
lx	lux	na	not available
M	mega (prefix for 10^6); metal (as in MA)	NAS	National Academy of Sciences
M	molar; actual mass	NASA	National Aeronautics and Space Administration
$\overline{M}_w$	weight-average mol wt		
$\overline{M}_n$	number-average mol wt	nat	natural
m	meter; milli (prefix for 10^{-3})	ndt	nondestructive testing
		neg	negative
m	molal	NF	*National Formulary*
m-	meta	NIH	National Institutes of Health
max	maximum		
MCA	Chemical Manufacturers' Association (was Manufacturing Chemists Association)	NIOSH	National Institute of Occupational Safety and Health
		NIST	National Institute of Standards and Technology (formerly National Bureau of Standards)
MEK	methyl ethyl ketone		
meq	milliequivalent		
mfd	manufactured		
mfg	manufacturing		
mfr	manufacturer	nmr	nuclear magnetic resonance
MIBC	methyl isobutyl carbinol		
MIBK	methyl isobutyl ketone	NND	New and Nonofficial Drugs (AMA)
MIC	minimum inhibiting concentration		
		no.	number
min	minute; minimum	NOI-(BN)	not otherwise indexed (by name)
mL	milliliter		
MLD	minimum lethal dose	NOS	not otherwise specified
MO	molecular orbital	nqr	nuclear quadruple resonance
mo	month		
mol	mole	NRC	Nuclear Regulatory Commission; National Research Council
mol wt	molecular weight		
mp	melting point		
MR	molar refraction	NRI	New Ring Index
ms	mass spectrometry	NSF	National Science Foundation
MSDS	material safety data sheet		
mxt	mixture	NTA	nitrilotriacetic acid
μ	micro (prefix for 10^{-6})	NTP	normal temperature and pressure (25°C and 101.3 kPa or 1 atm)
N	newton (force)		
N	normal (concentration); neutron number		

NTSB	National Transportation Safety Board	qv	quod vide (which see)
O-	denoting attachment to oxygen	R	univalent hydrocarbon radical
o-	ortho	(R)-	rectus (clockwise configuration)
OD	outside diameter	r	precision of data
OPEC	Organization of Petroleum Exporting Countries	rad	radian; radius
		RCRA	Resource Conservation and Recovery Act
o-phen	o-phenanthridine		
OSHA	Occupational Safety and Health Administration	rds	rate-determining step
		ref.	reference
owf	on weight of fiber	rf	radio frequency, n.
Ω	ohm	r-f	radio frequency, adj.
P	peta (prefix for 10^{15})	rh	relative humidity
p	pico (prefix for 10^{-12})	RI	Ring Index
p-	para	rms	root-mean square
p	proton	rpm	rotations per minute
p.	page	rps	revolutions per second
Pa	pascal (pressure)	RT	room temperature
PEL	personal exposure limit based on an 8-h exposure	RTECS	Registry of Toxic Effects of Chemical Substances
		^{s} (eg, Bu^s); sec-	secondary (eg, secondary butyl)
pd	potential difference		
pH	negative logarithm of the effective hydrogen ion concentration	S	siemens
		(S)-	sinister (counterclockwise configuration)
phr	parts per hundred of resin (rubber)	S-	denoting attachment to sulfur
p-i-n	positive-intrinsic-negative	s-	symmetric(al)
pmr	proton magnetic resonance	s	second
p-n	positive-negative	(s)	solid, only as in $H_2O(s)$
po	per os (oral)	SAE	Society of Automotive Engineers
POP	polyoxypropylene		
pos	positive	SAN	styrene-acrylonitrile
pp.	pages	sat(d)	saturate(d)
ppb	parts per billion (10^9)	satn	saturation
ppm	parts per million (10^6)	SBS	styrene–butadiene–styrene
ppmv	parts per million by volume		
ppmwt	parts per million by weight	sc	subcutaneous
PPO	poly(phenyl oxide)	SCF	self-consistent field; standard cubic feet
ppt(d)	precipitate(d)		
pptn	precipitation	Sch	Schultz number
Pr (no.)	foreign prototype (number)	sem	scanning electron microscope(y)
pt	point; part		
PVC	poly(vinyl chloride)	SFs	Saybolt Furol seconds
pwd	powder	sl sol	slightly soluble
py	pyridine	sol	soluble

soln	solution	*trans-*	isomer in which substituted groups are on opposite sides of double bond between C atoms
soly	solubility		
sp	specific; species		
sp gr	specific gravity		
sr	steradian		
std	standard	TSCA	Toxic Substances Control Act
STP	standard temperature and pressure (0°C and 101.3 kPa)	TWA	time-weighted average
		Twad	Twaddell
sub	sublime(s)	UL	Underwriters' Laboratory
SUs	Saybolt Universal seconds	USDA	United States Department of Agriculture
syn	synthetic		
^{t} (eg, Bu^t), *t-*, *tert-*	tertiary (eg, tertiary butyl)	USP	*United States Pharmacopeia*
		uv	ultraviolet
T	tera (prefix for 10^{12}); tesla (magnetic flux density)	V	volt (emf)
		var	variable
t	metric ton (tonne)	*vic-*	vicinal
t	temperature	vol	volume (not volatile)
TAPPI	Technical Association of the Pulp and Paper Industry	vs	versus
		v sol	very soluble
		W	watt
TCC	Tagliabue closed cup	Wb	weber
tex	tex (linear density)	Wh	watt hour
T_g	glass-transition temperature	WHO	World Health Organization (United Nations)
tga	thermogravimetric analysis		
		wk	week
THF	tetrahydrofuran	yr	year
tlc	thin layer chromatography	(Z)-	zusammen; together; atomic number
TLV	threshold limit value		

Non-SI (Unacceptable and Obsolete) Units		Use
Å	angstrom	nm
at	atmosphere, technical	Pa
atm	atmosphere, standard	Pa
b	barn	cm^2
bar†	bar	Pa
bbl	barrel	m^3
bhp	brake horsepower	W
Btu	British thermal unit	J
bu	bushel	m^3; L
cal	calorie	J
cfm	cubic foot per minute	m^3/s
Ci	curie	Bq
cSt	centistokes	mm^2/s
c/s	cycle per second	Hz

†Do not use bar (10^5 Pa) or millibar (10^2 Pa) because they are not SI units, and are accepted internationally only for a limited time in special fields because of existing usage.

Non-SI (Unacceptable and Obsolete) Units		Use
cu	cubic	exponential form
D	debye	C·m
den	denier	tex
dr	dram	kg
dyn	dyne	N
dyn/cm	dyne per centimeter	mN/m
erg	erg	J
eu	entropy unit	J/K
°F	degree Fahrenheit	°C; K
fc	footcandle	lx
fl	footlambert	lx
fl oz	fluid ounce	m^3; L
ft	foot	m
ft·lbf	foot pound-force	J
gf den	gram-force per denier	N/tex
G	gauss	T
Gal	gal	m/s^2
gal	gallon	m^3; L
Gb	gilbert	A
gpm	gallon per minute	(m^3/s); (m^3/h)
gr	grain	kg
hp	horsepower	W
ihp	indicated horsepower	W
in.	inch	m
in. Hg	inch of mercury	Pa
in. H_2O	inch of water	Pa
in.-lbf	inch pound-force	J
kcal	kilo-calorie	J
kgf	kilogram-force	N
kilo	for kilogram	kg
L	lambert	lx
lb	pound	kg
lbf	pound-force	N
mho	mho	S
mi	mile	m
MM	million	M
mm Hg	millimeter of mercury	Pa
mμ	millimicron	nm
mph	miles per hour	km/h
μ	micron	μm
Oe	oersted	A/m
oz	ounce	kg
ozf	ounce-force	N
η	poise	Pa·s
P	poise	Pa·s
ph	phot	lx
psi	pounds-force per square inch	Pa
psia	pounds-force per square inch absolute	Pa
psig	pounds-force per square inch gage	Pa
qt	quart	m^3; L
°R	degree Rankine	K
rd	rad	Gy
sb	stilb	lx
SCF	standard cubic foot	m^3
sq	square	exponential form
thm	therm	J
yd	yard	m

BIBLIOGRAPHY

1. The International Bureau of Weights and Measures, BIPM (Parc de Saint-Cloud, France) is described in Appendix X2 of Ref. 4. This bureau operates under the exclusive supervision of the International Committee for Weights and Measures (CIPM).
2. *Metric Editorial Guide (ANMC-78-1)*, latest ed., American National Metric Council, 5410 Grosvenor Lane, Bethesda, Md. 20814, 1981.
3. *SI Units and Recommendations for the Use of Their Multiples and of Certain Other Units (ISO 1000-1981)*, American National Standards Institute, 1430 Broadway, New York, 10018, 1981.
4. Based on *ASTM E380-89a (Standard Practice for Use of the International System of Units (SI))*, American Society for Testing and Materials, 1916 Race Street, Philadelphia, Pa. 19103, 1989.
5. *Fed. Reg.*, Dec. 10, 1976 (41 FR 36414).
6. For ANSI address, see Ref. 3.

R. P. LUKENS
ASTM Committee E-43 on SI Practice

ABSORPTION

Introduction, 1
Equipment, 57

INTRODUCTION

Absorption, or gas absorption, is a unit operation used in the chemical industry to separate gases by washing or scrubbing a gas mixture with a suitable liquid. One or more of the constituents of the gas mixture dissolves or is absorbed in the liquid and can thus be removed from the mixture. In some systems, this gaseous constituent forms a physical solution with the liquid or the solvent, and in other cases, it reacts with the liquid chemically.

The purpose of such scrubbing operations may be any of the following: gas purification (eg, removal of air pollutants from exhaust gases or contaminants from gases that will be further processed), product recovery, or production of solutions of gases for various purposes. Several examples of applied absorption processes are shown in Table 1.

Gas absorption is usually carried out in vertical countercurrent columns as shown in Figure 1. The solvent is fed at the top of the absorber, whereas the gas mixture enters from the bottom. The absorbed substance is washed out by the solvent and leaves the absorber at the bottom as a liquid solution. The solvent is often recovered in a subsequent stripping or desorption operation. This second step is essentially the reverse of absorption and involves countercurrent contacting of the liquid loaded with solute using an inert gas or water vapor. The absorber may be a packed column, plate tower, or simple spray column, or a bubble column. The packed column is a shell either filled with randomly packed elements or having a regular solid structure designed to disperse the liquid and bring it and the rising gas into close contact. Dumped-type packing elements

2 ABSORPTION

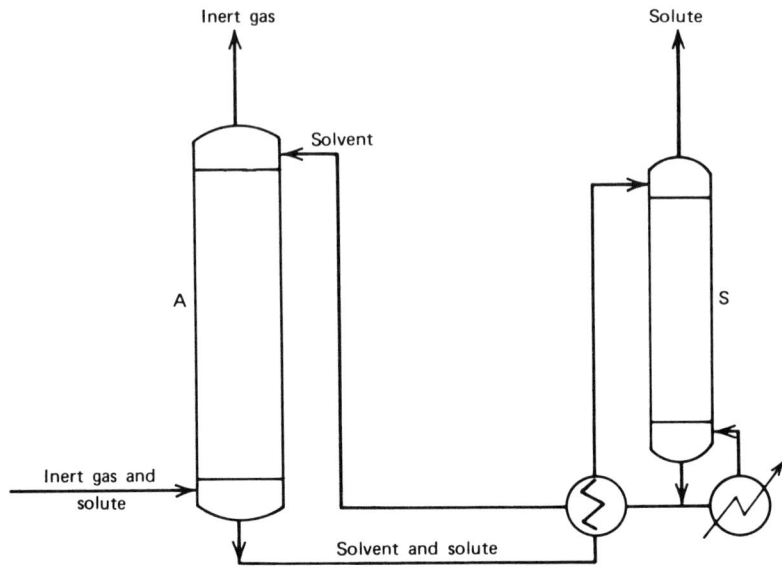

Fig. 1. Absorption column arrangement with a gas absorber A and a stripper S to recover solvent.

come in a great variety of shapes (Fig. 2a–f) and construction materials, which are intended to create a large internal surface but a small pressure drop. Structured, or arranged packings may be made of corrugated metal or plastic sheets providing a large number of regularly arranged channels (Fig. 2g), but a variety of other geometries exists. In plate towers, liquid flows from plate to plate in cascade fashion and gases bubble through the flowing liquid at each plate through a multitude of dispersers (eg, holes in a sieve tray, slits in a bubble-cap tray) or through a cascade of liquid as in a shower deck tray (see DISTILLATION).

The advantages of packed columns include simple and, as long as the tower diameter is not too large, usually relatively cheaper construction. These columns are preferred for corrosive gases because packing, but not plates, can be made from ceramic or plastic materials. Packed columns are also used in vacuum applications because the pressure drop, especially for regularly structured packings, is usually less than through plate columns. Tray absorbers are used in applications where tall columns are required, because tall, random-type packed towers are subject to channeling and maldistribution of the liquid streams. Plate towers can be more easily cleaned. Plates are also preferred in applications having large heat effects since cooling coils are more easily installed in plate towers and liquid can be withdrawn more easily from plates than from packings for external cooling. Bubble trays can also be designed for large liquid holdup.

The fundamental physical principles underlying the process of gas absorption are the solubility of the absorbed gas and the rate of mass transfer. Information on both must be available when sizing equipment for a given application. In addition to the fundamental design concepts based on solubility and mass transfer, many practical details have to be considered during actual plant design

Table 1. Commercial Gas Absorption Processes

Treated gas	Absorbed gas, solute	Solvent	Function
coke oven gas	ammonia	water	by-product recovery
coke oven gas	benzene and toluene	straw oil	by-product recovery
reactor gases in manufacture of formaldehyde from methanol	formaldehyde	water	product recovery
drying gases in cellulose acetate fiber production	acetone	water	solvent recovery
refinery gases	hydrogen sulfide	alkaline solutions	pollutant removal
natural and refinery gases	hydrogen sulfide	solution of sodium 2,6- (and 2,7-) anthraquinonedisulfonate	pollutant removal
products of combustion	sulfur dioxide	water	pollutant removal
	carbon dioxide	ethanolamines	by-product recovery
wet well gas	propane and butane	kerosene	gas separation
ammonia synthesis gas	carbon monoxide	ammoniacal cuprous chloride solution	contaminant removal
roast gases	sulfur dioxide	water	production of calcium sulfite solution for pulping

4   ABSORPTION

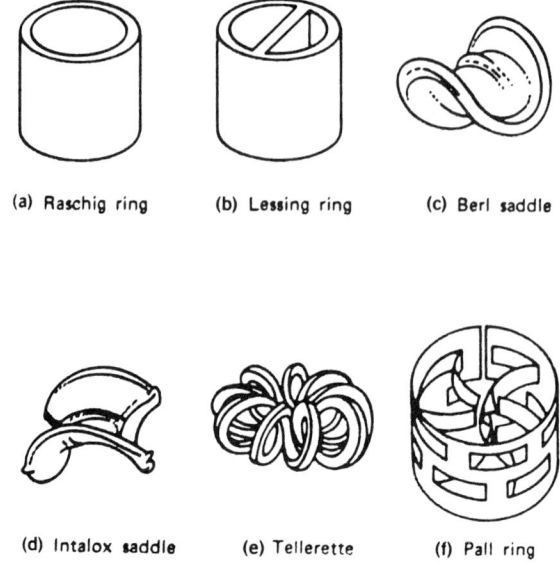

Fig. 2. Packing materials for packed columns. (**a**)–(**f**) Typical packing elements generally used for random packing; (**g**) example of structured packing. (**g**) Courtesy of Sulzer Bros. S.A. Winterthur, Switzerland.

and construction which may affect the performance of the absorber significantly. These details have been described in reviews (1) and in some of the more comprehensive treatments of gas absorption and absorbers (2–5) (see DISTILLATION; MASS TRANSFER).

Gas Solubility

At equilibrium, a component of a gas in contact with a liquid has identical fugacities in both the gas and liquid phase. For ideal solutions Raoult's law applies:

$$y_A = \frac{P_s}{P} x_A \qquad (1)$$

where y_A is the mole fraction of A in the gas phase, P is the total pressure, P_s is the vapor pressure of pure A, and x_A is the mole fraction of A in the liquid. For moderately soluble gases with relatively little interaction between the gas and liquid molecules Henry's law is often applicable:

$$y_A = \frac{H}{P} x_A \qquad (2)$$

where H is Henry's constant. Usually H is dependent upon temperature, but relatively independent of pressure at moderate levels. In solutions containing inorganic salts, H is also a function of the ionic strength. Henry's constants are tabulated for many of the common gases in water (6).

A more general way of expressing solubilities is through the vapor–liquid equilibrium constant m defined by

$$y_A = m x_A \qquad (3)$$

The value of m, also known as equilibrium K value, is widely employed to represent hydrocarbon vapor–liquid equilibria in absorption and distillation calculations. When equation 1 or 2 is applicable at constant pressure and temperature (equivalent to constant m in eq. 3) a plot of y vs x for a given solute is linear from the origin. In other cases, the y–x plot may be approximated by a linear relationship over limited regions. Generally, for nonideal solutions or for nonisothermal conditions, y is a curving function of x and must be determined from experimental data or more rigorous theoretical relationships. The y–x plot, when applied to absorber design, is commonly called the equilibrium line.

Gas solubility has been treated extensively (7). Methods for the prediction of phase equilibria and actual solubility data have been given (8,9) and correlations of the equilibrium K values of hydrocarbons have been developed and compiled (10). Several good sources for experimental information on gas– and vapor–liquid equilibrium data of nonideal systems are also available (6,11,12).

Mass Transfer Concepts

Mass Transfer Coefficients and Driving Forces. In order to determine the size of the equipment necessary to absorb a given amount of solvent per unit time, not only the equilibrium solubility of the solute in the solvent, but also the rate at which the equilibrium is established must be known; ie, the rate at which the solute is transferred from the gas to the liquid phase must be determined. One of the first theoretical models describing the process proposed an essentially stable gas–liquid interface (13). Large fluid motions are presumed to exist at a certain distance from this interface distributing all material rapidly and equally in the bulk of the fluid so that no concentration gradients are developed. Closer to this interface, however, the fluid motions are impaired and the slow process of molecular diffusion becomes more important as a mechanism of mass transfer. The rate-governing step in gas absorption is therefore the transfer of solute through two thin gas and liquid films adjacent to the phase interface. Transfer of materials through the interface itself is normally presumed to take place instantaneously so that equilibrium exists between these two films precisely at the interface. Although this assumption has been confirmed in experiments utilizing many systems and different types of phase interface (14–18), interfacial resistances can develop in some situations (19–24).

The resulting concentration profile is shown in Figure 3. With the passage of time in a nonflowing closed system, the profiles would become straight horizontal lines as the bulk gas and bulk liquid reached equilibrium. In a flowing system, Figure 3 represents conditions at some countercurrent flow point, eg, at a certain height in an absorption tower where, as gas and liquid pass each other, the bulk materials do not have sufficient contact time to attain equilibrium. Solute is continuously transferred from the gas to the liquid and concentration gradients develop when this transfer proceeds at only a finite rate.

The experimentally observed rates of mass transfer (qv) are often proportional to the displacement from equilibrium and the rate equations for the gas and liquid films are

$$N_A = k_G(p_A - p_{Ai}) = k_G P(y_A - y_{Ai}) \tag{4}$$

$$N_A = k_L(c_{Ai} - c_A) = k_L \bar{\rho}(x_{Ai} - x_A) \tag{5}$$

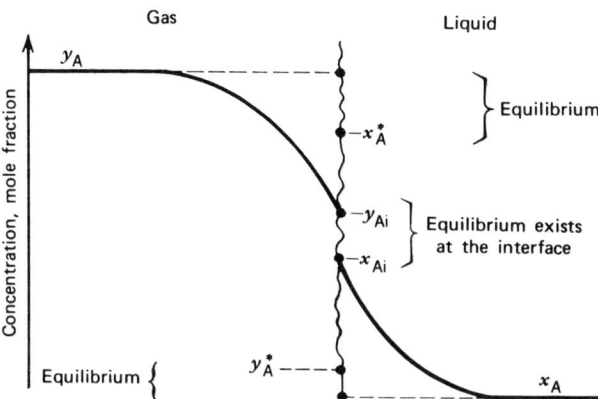

Fig. 3. The two-film concept: y_A and x_A are the concentrations in the bulk of the phases; y_{Ai} and x_{Ai} are the actual interfacial concentrations at equilibrium; y_A^* and x_A^* are the hypothetical equilibrium concentrations which would be in equilibrium with the bulk concentration of the other phase.

where $y_A - y_{Ai}$ and $x_{Ai} - x_A$ are concentration driving forces, k_G is the gas-phase mass transfer coefficient, and k_L is the liquid-phase mass transfer coefficient.

Mass transfer rates may also be expressed in terms of an overall gas-phase driving force by defining a hypothetical equilibrium mole fraction y_A^* as the concentration which would be in equilibrium with the bulk liquid concentration ($y_A^* = mx_A$):

$$N_A = K_{OG} P (y_A - y_A^*) \tag{6}$$

The relationship of the overall gas-phase mass transfer coefficient K_{OG} to the individual film coefficients may be found from equations 4 and 5, assuming a straight equilibrium line:

$$N_A = k_G P (y_A - y_{Ai}) = k_L \bar{\rho} \frac{1}{m} (y_{Ai} - y_A^*)$$

and by comparison with equation 6,

$$\frac{1}{K_{OG}} = \frac{1}{k_G} + \frac{mP}{k_L \bar{\rho}} \tag{7}$$

Expressions similar to equations 6 and 7 may be derived in terms of an overall liquid-phase driving force. Equation 7 represents an addition of the resistances to mass transfer in the gas and liquid films. The analogy of this process to the flow of electrical current through two resistances in series has been analyzed (25).

A representation of the various concentrations and driving forces in a y–x diagram is shown in **Figure 4**. The point representing the interfacial concentrations (y_{Ai}, x_{Ai}) must lie on the equilibrium curve since these concentrations are at equilibrium. The point representing the bulk concentrations (y_A, x_A) may be anywhere above the equilibrium line for absorption or below it for desorption. The slope of the tie line connecting the two points is given by equations 4 and 5:

$$\frac{y_A - y_{Ai}}{x_A - x_{Ai}} = -\frac{k_L \bar{\rho}}{k_G P} \tag{8}$$

In situations where the gas film resistance is predominant (gas film-controlled situation), $k_G P$ is much smaller than $k_L \bar{\rho}$ and the tie line is very steep. y_{Ai} approaches y^* so that the overall gas-phase driving force and the gas-film driving force become approximately equal, whereas the liquid-film driving force becomes negligible. From equation 7 it also follows that in such cases $K_{OG} \approx k_G$. The reverse is true if the liquid film resistance is controlling. Since the example depicted in Figure 4 involves a strongly curved equilibrium line, equation 7 is only valid if the slope of the dashed line between x_A and x_{Ai} is substituted for m. Overall mass transfer coefficients may vary considerably over a certain concentration range as a result of variations in m even if the individual film constants stay essentially constant.

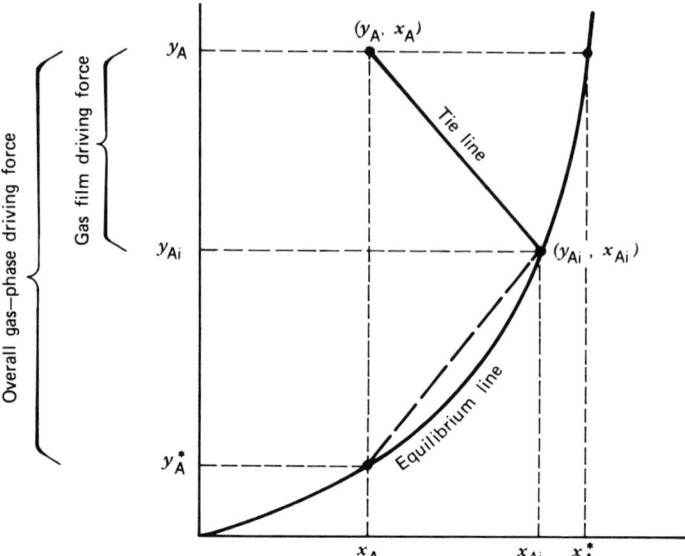

Fig. 4. The driving forces in the y–x diagram.

Film Theory. Many theories have been put forth to explain and correlate experimentally measured mass transfer coefficients. The classical model has been the film theory (13,26) that proposes to approximate the real situation at the interface by hypothetical "effective" gas and liquid films. The fluid is assumed to be essentially stagnant within these effective films making a sharp change to totally turbulent flow where the film is in contact with the bulk of the fluid. As a result, mass is transferred through the effective films only by steady-state molecular diffusion and it is possible to compute the concentration profile through the films by integrating Fick's law:

$$J_A = -D_{AB} \frac{dc_A}{dz} \tag{9}$$

where J_A is the flux of component A relative to the average molar flow of the whole mixture, D_{AB} is the diffusion coefficient of A in B, z is the distance of diffusion, and c_A is the molar concentration of A at a given point in the film. The bulk concentrations are denoted y_{Ab} and x_{Ab} in the following three sections whereas y_A and x_A stand for the concentration at a particular point within the films.

Equimolar Counterdiffusion in Binary Cases. If the flux of A is balanced by an equal flux of B in the opposite direction (frequently encountered in binary distillation columns), there is no net flow through the film and N_A, like J_A, is directly given by Fick's law. In an ideal gas, where the diffusivity can be shown to be independent of concentration, integration of Fick's law leads to a linear concentration profile through the film and to the following expression where $(P/RT)y_A$ is substituted for c_A:

$$N_A = \frac{D_{AB}P}{z_0 RT}(y_{Ab} - y_{Ai}) \qquad (10)$$

thus

$$k_G^0 = \frac{D_{AB}}{z_0 RT} \qquad (11)$$

where k_G is labeled with a zero to indicate equimolar counterdiffusion and z_0 is the effective film thickness. This same treatment is usually adopted for liquids, although the diffusion coefficients are customarily not completely independent of concentration. Substituting $\bar{\rho} x_A$ for c_A, the result is

$$k_L^0 = \frac{D_A}{z_0} \qquad (12)$$

Equations 11 and 12 cannot be used to predict the mass transfer coefficients directly, because z_0 is usually not known. The theory, however, predicts a linear dependence of the mass transfer coefficient on diffusivity.

Unidirectional Diffusion through a Stagnant Medium. An ideally simple gas absorption process involves diffusion of only one component, either the inert gas or the solvent, through a nondiffusing medium. There exists a net flux of material through the film in this case, and therefore the mixture as a whole is not at rest. The total flux of A is now the sum of the flux with respect to the average flow of the mixture, that is still given by Fick's law of diffusion, plus the flux of A caused by the average bulk flow of the mixture itself:

$$N_A = J_A + y_A \sum_j N_j = -\frac{D_{AB}P}{RT}\frac{dy_A}{dz} + y_A N_A \qquad (13)$$

The derivation of this result may be found in various texts (27). Rearranging and integrating equation 13 yields

$$N_A = \frac{D_{AB}P}{z_0 RT}\frac{1}{y_{BM}}(y_{Ab} - y_{Ai}) \qquad (14)$$

where y_{BM} is the logarithmic mean of the stagnant gas concentration through the film:

$$y_{BM} = \frac{(1 - y_{Ab}) - (1 - y_{Ai})}{\ln[(1 - y_{Ab})/(1 - y_{Ai})]} \qquad (15)$$

Therefore

$$k_G' = \frac{D_{AB}}{z_0 RT y_{BM}} \qquad (16)$$

For liquids,

$$k'_L = \frac{D_A}{z_0 x_{BM}} \qquad (17)$$

where

$$x_{BM} = \frac{(1 - x_{Ab}) - (1 - x_{Ai})}{\ln[(1 - x_{Ab})/(1 - x_{Ai})]} \qquad (18)$$

The values of y_{BM} and x_{BM} are near unity for dilute gases.

Multicomponent Diffusion. In multicomponent systems, the binary diffusion coefficient D_{AB} has to be replaced by an effective or mean diffusivity D_{Am}. Although its rigorous computation from the binary coefficients is difficult, it may be estimated by one of several methods (27–29). Any degree of counterdiffusion, including the two special cases "equimolar counterdiffusion" and "no counterdiffusion" treated above, may arise in multicomponent gas absorption. The influence of bulk flow of material through the films is corrected for by the film factor concept (28). It is based on a slightly different form of equation 13:

$$N_A = J_A + y_A t_A N_A \qquad (19)$$

where

$$t_A \equiv \frac{\Sigma_j N_j}{N_A} \qquad (20)$$

Applying the same derivation as for unidirectional diffusion through a stagnant medium, the results turn out to be

$$k_G = \frac{D_{AB}}{z_0 R T Y_f} \qquad (21)$$

$$k_L = \frac{D_{AB}}{z_0 X_f} \qquad (22)$$

where

$$Y_f \equiv \frac{(1 - t_A y_{Ab}) - (1 - t_A y_{Ai})}{\ln(1 - t_A y_{Ab})/(1 - t_A y_{Ai})} \qquad (23)$$

$$X_f \equiv \frac{(1 - t_A x_{Ab}) - (1 - t_A x_{Ai})}{\ln(1 - t_A x_{Ab})/(1 - t_A x_{Ai})} \qquad (24)$$

Y_f and X_f are called the film factors. They are generalized y_{BM} and x_{BM} factors, respectively, and are reduced to them in the case of unidirectional diffusion through a stagnant medium because $t_A = 1$ in this case. The film factors Y_f and X_f correct the mass transfer coefficients for the effect of net flux through the films. In situations having strong counterdiffusion giving rise to a net flow opposed to the diffusion of A, the film factor becomes larger than one and therefore decreases the mass transfer coefficient and the flux of A. For weak or negative counterdiffusion producing a bulk flux parallel to the diffusion of A, the film factor is smaller than unity and thus increases N_A. In extreme situations, counterdiffusion may become large enough to reverse the direction of transport of a given component and force it to diffuse against its own driving force. These situations are characterized by a negative film factor and hence a negative k_G. If equimolar counterdiffusion prevails, t_A becomes zero, the film factor is unity, irrespective of the concentration, and $k_G = k_G^0$.

Rate Equations with Concentration-Independent Mass Transfer Coefficients. Except for equimolar counterdiffusion, the mass transfer coefficients applicable to the various situations apparently depend on concentration through the y_{BM} and Y_f factors. Instead of the classical rate equations 4 and 5, containing variable mass transfer coefficients, the rate of mass transfer can be expressed in terms of the constant coefficients for equimolar counterdiffusion using the relationships

$$k_G Y_f = k_G' y_{BM} = k_G^0 \tag{25}$$

$$k_L X_f = k_L' x_{BM} = k_L^0 \tag{26}$$

This leads to rate equations with constant mass transfer coefficients, whereas the effect of net transport through the film is reflected separately in the y_{BM} and Y_f factors. For unidirectional mass transfer through a stagnant gas the rate equation becomes

$$N_A = k_G^0 P(y_{Ab} - y_{Ai}) \frac{1}{y_{BM}} \tag{27}$$

For multicomponent diffusion,

$$N_A = k_G^0 P(y_{Ab} - y_{Ai}) \frac{1}{Y_f} \tag{28}$$

Equation 28 and its liquid-phase equivalent are very general and valid in all situations. Similarly, the overall mass transfer coefficients may be made independent of the effect of bulk flux through the films and thus nearly concentration independent for straight equilibrium lines:

$$N_A = K_{OGP}^0 (y_{Ab} - y_A^*) \frac{1}{y_{BM}^*} \tag{29}$$

$$= K_{OGP}^0 (y_{Ab} - y_A^*) \frac{1}{Y_f^*} \tag{30}$$

where the logarithmic means in y_{BM}^* and Y_f^* must be taken between y_{Ab} and y_A^*.

Rate equations 28 and 30 combine the advantages of concentration-independent mass transfer coefficients, even in situations of multicomponent diffusion, and a familiar mathematical form involving concentration driving forces. The main inconvenience is the use of an effective diffusivity which may itself depend somewhat on the mixture composition and in certain cases even on the diffusion rates. This advantage can be eliminated by working with a different form of the Maxwell-Stefan equation (30–32). One thus obtains a set of rate equations of an unconventional form having concentration-independent mass transfer coefficients that are defined for each binary pair directly based on the Maxwell-Stefan diffusivities.

Other Models for Mass Transfer. In contrast to the film theory, other approaches assume that transfer of material does not occur by steady-state diffusion. Rather there are large fluid motions which constantly bring fresh masses of bulk material into direct contact with the interface. According to the penetration theory (33), diffusion proceeds from the interface into the particular element of fluid in contact with the interface. This is an unsteady state, transient process where the rate decreases with time. After a while, the element is replaced by a fresh one brought to the interface by the relative movements of gas and liquid, and the process is repeated. In order to evaluate N_A, a constant average contact time τ for the individual fluid elements is assumed (33). This leads to relations such as

$$k_L = 2\left(\frac{D}{\pi \cdot \tau}\right)^{1/2} \tag{31}$$

If, on the other hand, it is assumed that contact times for the individual fluid elements vary at random, an exponential surface age distribution characterized by a fractional rate of renewal s may be used (34). This approach is called surface renewal theory and results in

$$k_L = (Ds)^{1/2} \tag{32}$$

Neither the penetration nor the surface renewal theory can be used to predict mass transfer coefficients directly because τ and s are not normally known. Each suggests, however, that mass transfer coefficients should vary as the square root of the molecular diffusivity, as opposed to the first power suggested by the film theory.

Another concept sometimes used as a basis for comparison and correlation of mass transfer data in columns is the Chilton-Colburn analogy (35). This semi-empirical relationship was developed for correlating mass- and heat-transfer data in pipes and is based on the turbulent boundary layer model and the close analogy between momentum and mass transfer. It must be considerably modified for gas-absorption columns, but it predicts that the mass transfer coefficient varies with D raised to the ⅔ power (4,36). A modern theory for surface rejuvenation has also been published and compared with earlier models (37).

Absorption and Chemical Reaction. In instances where the solute gas is absorbed into a liquid or a solution where it is able to undergo chemical reaction, the driving forces of absorption become far more complex. The solute not only diffuses through the liquid film at a rate determined by the gradient of the concentration, but at the same time also reacts with the liquid at a rate determined by the concentrations of both the solute and the solvent at the point of interest. Inclusion of a term for the chemical reaction in Fick's second law of diffusion, followed by integration of the expression, allows the concentration profiles through the liquid film to be computed based on a particular mass transfer model. The calculations show that these profiles are steeper and the rate of mass transfer higher than without chemical reaction. Thus the results are often expressed as an enhancement factor ϕ, defined as the fractional increase of the liquid film mass transfer coefficient resulting from the chemical reaction (k_L^r/k_L^0). The solutions that have been developed in this manner based on the film, penetration, and surface renewal theories are quite similar for a given type of reaction (38,39). Solutions and estimations of enhancement factors may be found in the literature (4,37–40).

Design of Packed Absorption Columns

Discussion of the concepts and procedures involved in designing packed gas absorption systems shall first be confined to simple gas absorption processes without complications: isothermal absorption of a solute from a mixture containing an inert gas into a nonvolatile solvent without chemical reaction. Gas and liquid are assumed to move through the packing in a plug-flow fashion. Deviations such as nonisothermal operation, multicomponent mass transfer effects, and departure from plug flow are treated in later sections.

Standard Absorber Design Methods. *Operating Line.* As a gas mixture travels up through a gas absorption tower, as shown in Figure 5, the solute A is transferred to the liquid phase and thus gradually removed from the gas. The liquid accumulates solute on its way down through the column so x increases from the top to the bottom of the column. The steady-state concentrations y and x at any given point in the column are interrelated through a mass balance around either the upper or lower part of the column (eq. 34), whereas the four concentrations in the streams entering and leaving the system are interrelated by the overall material balance.

Because the total gas and liquid flow rates per unit cross-sectional area vary throughout the tower (Fig. 5), rigorous material balances should be based on the constant inert gas and solvent flow rates G'_M and L'_M, respectively, and expressed in terms of mole ratios Y' and X'. A balance around the upper part of the tower yields

$$G'_M Y' + L'_M X'_2 = G'_M Y'_2 + L'_M X' \tag{33}$$

which may be rearranged to give

$$Y' = \frac{L'_M}{G'_M}(X' - X'_{A,2}) + Y'_{A,2} \tag{34}$$

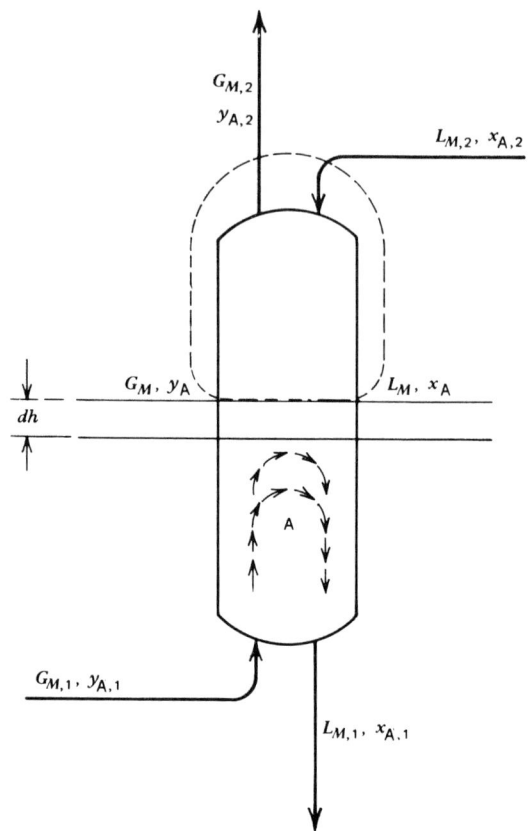

Fig. 5. Mass balance in gas absorption columns. The curved arrows indicate the travel path of the solute A. The upper broken curve delineates the envelope for the material balance of equation 33.

where G'_M, L'_M are in kg·mol/(h·m²) [lb·mol/(h·ft²)] and $Y' \equiv y/(1-y)$ and $X' \equiv x/(1-x)$. The overall material balance is obtained by substituting $Y' = Y'_1$ and $X' = X'_1$. For dilute gases the total molar gas and liquid flows may be assumed constant and a similar mass balance yields

$$y = \frac{L_M}{G_M}(x - x_2) + y_2 \qquad (35)$$

A plot of either equation 34 or 35 is called the operating line of the process as shown in Figure 6. As indicated by equation 35, the line for dilute gases is straight, having a slope given by L_M/G_M. (This line is always straight when plotted in Y'–X' coordinates.) Together with the equilibrium line, the operating line permits the evaluation of the driving forces for gas absorption along the column (Fig. 4). The farther apart the equilibrium and operating lines, the larger the driving forces become and the faster absorption occurs, resulting in the need for a shorter column (Fig. 6).

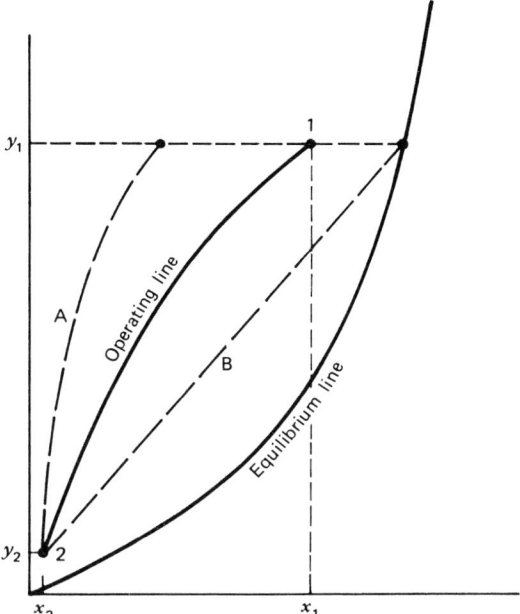

Fig. 6. Operating lines for an absorption system: line A, high L_M/G_M ratio; solid line, medium L_M/G_M ratio; line B, L_M/G_M ratio at theoretical minimum necessary for the removal of the specified quantity of solute. Subscript 1 represents the bottom of tower, 2, the top of tower.

To place the operating line, the flows, composition of the entering gas y_1, entering liquid x_2, and desired degree of absorption y_2, are usually specified. The specification of the actual liquid rate used for a given gas flow (the L_M/G_M ratio) usually depends on an economic optimization because the slope of the operating line may be seen to have a drastic effect on the driving force. For example, use of a very high liquid rate (line A in Fig. 6) results in a short column and a low absorber cost, but at the expense of a high cost for solvent circulation and subsequent recovery of the solute from a relatively dilute solution. On the other hand, a liquid rate near the theoretical minimum, which is the rate at which the operating line just touches the equilibrium line (line B in Fig. 6), requires a very tall tower because the driving force becomes very small at its bottom. Use of a liquid rate on the order of one and one-half times the theoretical minimum is not unusual. In the absence of a detailed cost analysis, the L_M/G_M ratio is often specified at 1.4 times the slope of the equilibrium line (41).

Design Procedure. The packed height of the tower required to reduce the concentration of the solute in the gas stream from $y_{A,1}$ to an acceptable residual level of $y_{A,2}$ may be calculated by combining point values of the mass transfer rate and a differential material balance for the absorbed component. Referring to a slice dh of the absorber (Fig. 5),

$$N_A a \, dh = -d(G_M y) = -G_M \, dy - y \, dG_M \qquad (36)$$

and

$$dG_M = -N_A a\, dh \qquad (37)$$

where a is the interfacial area present per unit volume of packing. Combining equations 36 and 37,

$$-dh = \frac{G_M\, dy}{N_A a(1-y)} \qquad (38)$$

Substituting for N_A from equation 27 and integrating over the tower,

$$h = \int_{y_2}^{y_1} \frac{G_M}{k_G^0 aP} \frac{y_{BM}\, dy}{(1-y)(y-y_i)} \qquad (39)$$

Equation 39 may be integrated numerically or graphically and its component terms evaluated at a series of points on the operating line. y_i is found by placing tie lines from each of these points; the slopes are given by equation 8. Thus equation 39 is a general expression determining the column height required to effect a given reduction in y_A.

Equation 39 can often be simplified by adopting the concept of a mass transfer unit. As explained in the film theory discussion earlier, the purpose of selecting equation 27 as a rate equation is that k_G^0 is independent of concentration. This is also true for the $G_M/k_G^0 aP$ term in equation 39. In many practical instances, this expression is fairly independent of both pressure and G_M: as G_M increases through the tower, k_G^0 increases also, nearly compensating for the variations in G_M. Thus this term is often effectively constant and can be removed from the integral:

$$h = \left(\frac{G_M}{k_G^0 aP}\right) \int_{y_2}^{y_1} \frac{y_{BM}\, dy}{(1-y)(y-y_i)} \qquad (40)$$

$G_M/k_G^0 aP$ has the dimension of length or height and is thus designated the gas-phase height of one transfer unit, H_G. The integral is dimensionless and indicates how many of these transfer units it takes to make up the whole tower. Consequently, it is called the number of gas-phase transfer units, N_G. Equation 40 may therefore be written as

$$h = (H_G)(N_G) \qquad (41)$$

where

$$H_G = \frac{G_M}{k_G^0 aP} \qquad (42)$$

and

$$N_G = \int_{y_2}^{y_1} \frac{y_{BM}\, dy}{(1-y)(y-y_i)} \tag{43}$$

The same treatment for the liquid side yields

$$h = (H_L)(N_L) \tag{44}$$

where

$$H_L = \frac{L_M}{k_L^0 a \bar{\rho}} \tag{45}$$

and

$$N_L = \int_{x_2}^{x_1} \frac{x_{BM}\, dx}{(1-x)(x_i-x)} \tag{46}$$

A similar treatment is possible in terms of an overall gas-phase driving force by substituting equation 29 into equation 38:

$$h = (H_{OG})(N_{OG}) \tag{47}$$

where

$$H_{OG} = \frac{G_M}{K_{OG}^0 a P} \tag{48}$$

and

$$N_{OG} = \int_{y_2}^{y_1} \frac{y_{BM}^*\, dy}{(1-y)(y-y^*)} \tag{49}$$

H_{OG} and N_{OG} are called the overall gas-phase height of a transfer unit and the number of overall gas-phase transfer units, respectively. In the case of a straight equilibrium line, K_{OG}^0 is often nearly concentration-independent as explained earlier. In such cases, use of equation 47 is especially convenient because N_{OG}, as opposed to N_G, can be evaluated without solving for the interfacial concentrations. In all other cases, H_{OG} must be retained under the integral and its value calculated from H_G and H_L at different points of the equilibrium line as

$$H_{OG} = \frac{y_{BM}}{y_{BM}^*} H_G + \frac{m G_M}{L_M} \frac{x_{BM}}{y_{BM}^*} H_L \tag{50}$$

To use all of these equations, the heights of the transfer units or the mass transfer coefficients $k_G^0 a$ and $k_L^0 a$ must be known. Transfer data for packed columns are

18 ABSORPTION

often measured and reported directly in terms of H_G and H_L and correlated in this form against G_M and L_M.

Sometimes the height equivalent to a theoretical plate (HETP) is employed rather than H_G and H_L to characterize the performance of packed towers. The number of heights equivalent to one theoretical plate required for a specified absorption job is equal to the number of theoretical plates, N_{TP}. It follows that

$$h = (\text{HETP})(N_{TP}) \tag{51}$$

which is similar in form to equation 47. HETP is a less fundamental variable than the heights of the transfer units, and it is more difficult to translate HETPs from one situation to another. Only for linear operating and equilibrium lines can they be related analytically to H_{OG} as shown later by equation 81.

Simplified Design Procedures for Linear Operating and Equilibrium Lines. *Logarithmic-Mean Driving Force.* As noted earlier, linear operating lines occur if all concentrations involved stay low. Where it is possible to assume that the equilibrium line is linear, it can be shown that use of the logarithmic mean of the terminal driving forces is theoretically correct. When the overall gas-film coefficient is used to express the rate of absorption, the calculation reduces to solution of the equation

$$L_M(x_1 - y_2) = G_M(y_1 - y_2) = K_{OG}aPh(y - y^*)_{av} \tag{52}$$

where

$$(y - y^*)_{av} = \frac{(y_1 - y_1^*) - (y_2 - y_2^*)}{\ln\left[(y_1 - y_1^*)/(y_2 - y_2^*)\right]} \tag{53}$$

In these cases, a quantitative significance can be given to the concept of a transfer unit. Because $H_{OG} = G_M/K_{OG}aP$, it follows from equations 52 and 47 that

$$N_{OG} = \frac{y_1 - y_2}{(y - y^*)_{av}}$$

Therefore, in this case, one transfer unit corresponds to the height of packing required to effect a composition change just equal to the average driving force.

Number of Transfer Units. For relatively dilute systems the ratios involving y_{BM}, y_{BM}^*, and $1 - y$ approach unity so that the computation of H_{OG} from equation 50 and N_{OG} from equation 49 may be simplified to

$$H_{OG} = H_G + \left(\frac{mG_M}{L_M}\right) H_L \tag{54}$$

$$N_{OG} \approx N_T = \int_{y_2}^{y_1} \frac{dy}{y - y^*} \tag{55}$$

Equation 55 is a rigorous expression for the number of overall transfer units for equimolar counterdiffusion, in distillation columns, for instance.

For cases in which the equilibrium and operating lines may be assumed linear, having slopes L_M/G_M and m, respectively, an algebraic expression for the integral of equation 55 has been developed (41):

$$N_{OG} \approx N_T = \frac{\ln\left[\left(1 - \frac{mG_M}{L_M}\right)\left(\frac{y_1 - mx_2}{y_2 - mx_2}\right) + \frac{mG_M}{L_M}\right]}{1 - \frac{mG_M}{L_M}} \qquad (56)$$

The required tower height may thus be easily calculated using equation 47, where H_{OG} is given by equation 54 and N_{OG} by equation 56.

Rapid Approximate Design Procedure for Curved Operating and Equilibrium Lines. If the operating or the equilibrium line is nonlinear, equation 56 is of little use because mG_M/L_M will assume a range of values over the tower. The substitution of effective average values for m and for L_M/G_M into equations 50 and 56 obviates lengthy graphical or numerical integrations and leads to a quick, approximate solution for the required tower height (4).

The effective average values of m and L_M/G_M were determined in a computational study covering hundreds of hypothetical absorber designs for gas streams containing up to 80 mol % of solute for recoveries from 81 to 99.9%. By numerical integration, precise values were obtained for N_{OG} and N_T. By solving equation 56 numerically for each of the design cases, average values of the slope of the equilibrium line $\overline{m}$ and average flow ratios $L_M/G_M = R_{av}$ were found which gave the same N_T when substituted into equation 56 as the graphical or numerical integration.

It was found that the effective average L_M/G_M ratio, R_{av}, could be correlated satisfactorily as a function of the terminal values R_1 and R_2, of the change in the mole fraction of the absorbed component over the tower, and of the fractional approach to equilibrium y_1^*/y_1 between the concentrated gas entering the tower and the liquid leaving. Figure 7 shows the resulting correlation for cases with $L_M/G_M > 1$. No correlation was obtained when this ratio was less than unity. The effective average slope of the equilibrium line, $\overline{m}$, was correlated as a function of the initial slope m_2, of the slope m_c of the cord connecting y_1^*/x_1 and y_1^*/x_2, and of various other parameters as shown in Figure 8. Figure 8a applies when the equilibrium line is concave upward, ie, $m_c > m_2$; and Figure 8b applies when the curvature is concave downward, $m_c < m_2$.

The recommended design procedure uses the values of $(L_M/G_M)_{av}$ and $\overline{m}$ from Figures 7 and 8 in equation 56 and yields a very good estimation of N_T despite the curvature of the operating and the equilibrium lines. This value differs from N_{OG} obtained by equation 49 because of the $y_{BM}^*/(1-y)$ term in the latter equation. A convenient approach for purposes of approximate design is to define a correction term ΔN_{OG} which can be added to equation 55:

$$N_{OG} = N_T + \Delta N_{OG} \qquad (57)$$

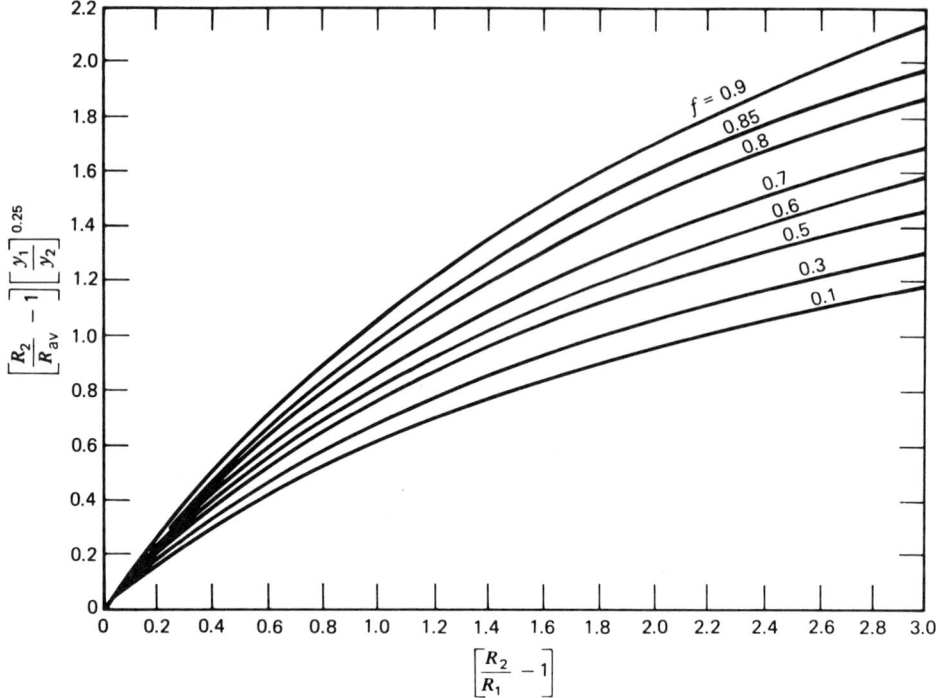

Fig. 7. Design chart for estimation of average-flow ratio in absorption (4). $R_2 = L_M/G_M$ at gas outlet; $R_1 = G_M/L_M$ at gas inlet; y_2 = mole fraction in outlet gas; y_1 = mole fraction in inlet gas; R_{av} = effective average L_M/G_M; $f = y_1^*/y_1$ = fractional approach to equilibrium.

For cases in which y_{BM}^* may be represented by an arithmetic mean (42),

$$\Delta N_{OG} = \frac{1}{2} \ln \frac{1 - y_2}{1 - y_1} \tag{58}$$

Equation 58 is sufficiently accurate for most situations.

The average slopes R_{av} and $\bar{m}$ from Figures 7 and 8 may also be used in equation 54 to compute H_{OG} although equation 50, with some suitable averages of y_{BM}^* and x_{BM}, should be preferred. Use of point values at an effective average liquid concentration given by equation 59 is suggested.

$$\bar{x} = \left(\frac{R_2}{R_{av}} - 1\right) \bigg/ (R_2 - 1) \tag{59}$$

In many situations, however, especially when $m \geq 1$, the results using the simpler equation 54 are virtually the same. The required tower height is finally calculated by means of equation 47.

Multicomponent Mass Transfer Effects. *Equimolar Counterdiffusion.* Just as unidirectional diffusion through stagnant films represents the

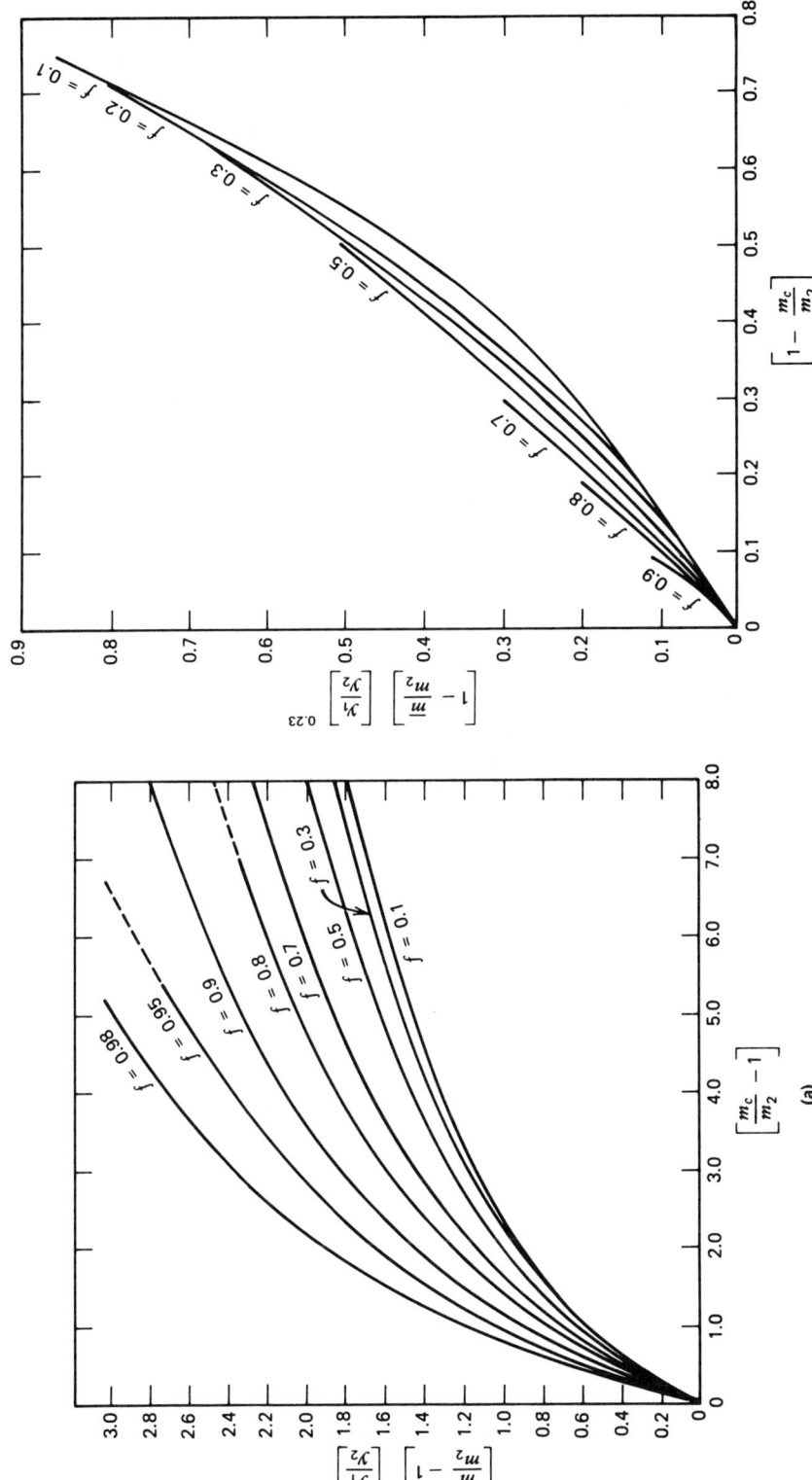

Fig. 8. Correlation of the effective average slopes $\bar{m}$ of the equilibrium line (4). (a) Equilibrium line curved concave upward; (b) equilibrium line curved concave downward.

situation in an ideally simple gas absorption process, equimolar counterdiffusion prevails as another special case in ideal distillation columns. In this case, the total molar flows L_M and G_M are constant, and the mass balance is given by equation 35. As shown earlier, no y_{BM} factors have to be included in the derivation and the height of the packing is

$$h_T = \int_{y_2}^{y_1} H_G \frac{dy}{y - y_i} \tag{60}$$

N_{OG} is given by N_T:

$$N_{OG} = \int_{y_2}^{y_1} \frac{dy}{y - y^*} \tag{61}$$

and H_{OG} is rigorously defined by equation 54. It must, however, be retained under the integral because m usually changes over the tower:

$$h = \int_{y_2}^{y_1} H_{OG} \frac{dy}{y - y^*} \tag{62}$$

General Situation. Both unidirectional diffusion through stagnant media and equimolar diffusion are idealizations that are usually violated in real processes. In gas absorption, slight solvent evaporation may provide some counterdiffusion, and in distillation counterdiffusion may not be equimolar for a number of reasons. This is especially true for multicomponent operation.

A simple treatment is still possible if it may be assumed that the flux of the component of interest A through the interface stays in a constant proportion to the total molar transfer through the interface over the entire tower:

$$\frac{\Sigma N_j}{N_A} = t_A = \text{constant} = \frac{\Sigma \Delta g_j}{\Delta g_A} \tag{63}$$

where Δg_j = total moles of component j absorbed over the tower. It will generally suffice to compute t_A from preliminary estimates of Δg_j and Δg_A, the total mass transfer of each component over the tower.

The mass balance for A is best represented as a straight line in hypothetical coordinates Y^0 and X^0:

$$Y_A^0 = \frac{L_M^0}{G_M^0} (X_A^0 - X_{A,2}^0) + Y_{A,2}^0 \tag{64}$$

where $Y_A^0 = y_A/(1 - t_A y_A)$, $X_A^0 = x_A/(1 - t_A x_A)$, $G_M^0 = G_M(1 - t_A y_A)$, and $L_M^0 = L_M(1 - t_A x_A)$. G_M^0 and L_M^0 are always constant, whereas G_M and L_M are not. For unimolecular diffusion through stagnant gas ($t_A = 1$), Y^0 and X^0 reduce to Y' and X' and G_M^0 and L_M^0 reduce to G_M' and L_M'; equation 64 then becomes equation 34. For equimolar counterdiffusion $t_A = 0$, and the variables reduce to y, x, G_M, and

L_M, respectively, and equation 64 becomes equation 35. Using the film factor concept and rate equation 28, the tower height may be computed by

$$h_T = \int_{y_2}^{y_1} H_G \frac{Y_f}{(1 - t_A y)} \frac{dy}{(y_A - y_{Ai})} \tag{65}$$

y_{Ai} is found as usual through tie lines of the slope

$$\frac{y_A - y_{Ai}}{x_A - x_{Ai}} = -\frac{L_M}{G_M} \frac{H_G}{H_L} \frac{Y_f}{X_f} \tag{66}$$

where

$$\frac{L_M}{G_M} = \frac{L_M^0(1 - t_A y_A)}{G_M^0(1 - t_A x_A)} \tag{67}$$

It may be noted that the above system of equations is very general and encompasses both the usual equations given for gas absorption and distillation as well as situations with any degree of counterdiffusion. The exact derivations may be found elsewhere (43).

Nonisothermal Gas Absorption. *Nonvolatile Solvents.* In practice, some gases tend to liberate such large amounts of heat when they are absorbed into a solvent that the operation cannot be assumed to be isothermal, as has been done thus far. The resulting temperature variations over the tower will displace the equilibrium line on a y–x diagram considerably because the solubility usually depends strongly on temperature. Thus nonisothermal operation affects column performance drastically.

The principles outlined so far may be used to calculate the tower height as long as it is possible to estimate the temperature as a function of liquid concentration. The classical basis for such an estimate is the assumption that the heat of solution manifests itself entirely in the liquid stream. It is possible to relate the temperature increase experienced by the liquid flowing down through the tower to the concentration increase through a simple enthalpy balance, equation 68, and thus correct the equilibrium line in a y–x diagram for the heat of solution as shown in Figure 9.

$$T_L \approx T_{L2} + \frac{(x_A - x_{A2})H_{OS}}{x_A C_{qA} + (1 - x_A)C_{qB}} \tag{68}$$

where T_L is the liquid temperature, °C; T_{L2} is the temperature of the entering liquid, °C; C_{qj} is the molar heat capacity of component j; H_{OS} is the integral mean heat of solution of solute. For each pair of values for x_A and T_L obtained from equation 68 it is possible to evaluate $y^*(x, T_L)$, the concentration in equilibrium with the bulk of the liquid phase, and to place the equilibrium line for the overall driving force (line A in Fig. 9). The line connecting the actual interfacial concentrations (y_i, x_i), line B on Figure 9, does not coincide with line A unless there is no

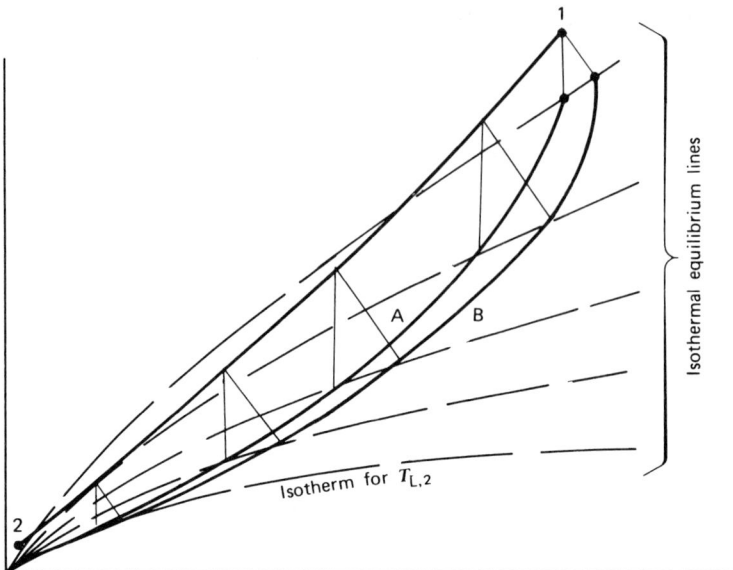

Fig. 9. Simple model of adiabatic gas absorption. A, nonisothermal equilibrium line for overall gas-phase driving force: $y^* = f(x, T_L)$; B, nonisothermal equilibrium line for individual gas-film driving force: $y_i = f(x_i)$.

liquid mass transfer resistance. However, because the interfacial temperature T_i and the bulk liquid temperature T_L usually are virtually equal, the equilibrium concentration y^* and the actual interfacial concentration y_i are connected by an isotherm. Line B on Figure 9 may therefore be constructed as shown on the basis of line A, tie lines, and isothermal equilibrium lines. Line B may be used in conjunction with equation 39 to compute the required depth of packing.

General Case. The simple adiabatic model just discussed often represents an oversimplification, since the real situation implies a multitude of heat effects: (*1*) The heat of solution tends to increase the temperature and thus to reduce the solubility. (*2*) In the case of a volatile solvent, partial solvent evaporation absorbs some of the heat. (This effect is particularly important when using water, the cheapest solvent.) (*3*) Heat is transferred from the liquid to the gas phase and vice versa. (*4*) Heat is transferred from both phase streams to the shell of the column and from the shell to the outside or to cooling coils.

In the general case, the temperature profile is determined simultaneously by all of the four heat effects. The temperature influences the transfer of mass and heat to a large extent by changing the solubilities. This turns the simple gas absorption process into a very complex one and all factors exhibit a high degree of interaction. Computer algorithms for solving the problem rigorously have been developed (43,44). Figure 10 depicts typical profiles through an adiabatic packed gas absorber from one of these algorithms (43,45). The calculations were carried out to solve a design example calling for the removal of 90% of the acetone vapors present in an air stream by absorption into water at an L_M/G_M ratio of 2.5. The air stream contained 6 mol % acetone and was saturated with water; the ambient temperature was 15°C.

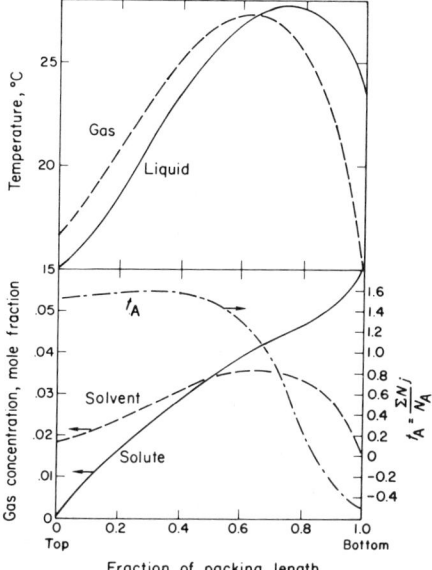

Fig. 10. Computed rigorous profiles through an adiabatic packed absorber during the absorption of acetone into water (43).

It is a typical feature of such calculations that the shapes of the liquid temperature profiles are highly irregular and often exhibit maxima within the column. Such internal temperature maxima have been observed experimentally in plate and packed absorbers (44,46,47), and the measured temperature profiles can be shown to agree closely with rigorous computations. The temperature maximum occurs in part because the heat of solution causes the entering liquid stream to be heated. In the lower part of the tower, however, the heat of absorption is smaller than the opposite heat effects of solvent evaporation and heat transfer to the cold entering gas, so that the net effect is a cooling of the liquid phase. These transfers are reversed in the upper part of the column, as is obvious from Figure 10: the gas gives up heat to the liquid, is cooled, and some of the solvent condenses from the gas stream into the liquid stream, which is heated much faster in this part of the column than would be the case with the absorption alone.

Figure 11 shows the rigorously computed y–x diagram for the same example. The temperature maximum within the column produces a region of reduced solubility reflecting itself in the typical bulge in the middle of the rigorous equilibrium line D. Since less acetone is absorbed in this part of the equipment, the gas concentration curve exhibits a slight plateau (Fig. 10). This example may also serve as a demonstration of the difficulty in estimating the required depth of packing using simplifying assumptions (Table 2). The isothermal approximation failed completely in this case and yielded 1.95 m of required packing as opposed to the rigorously determined value of 3.63 m. Neglecting the temperature increase completely, this model assumes a solubility which is much too large, reflected by equilibrium line A, and thus underestimates the rigorous result by 90%.

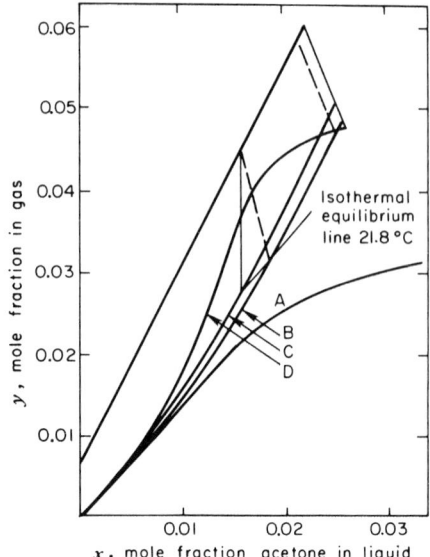

Fig. 11. y–x diagram for adiabatic absorption of acetone into water. A, isothermal equilibrium line at T_{L2}; B, equilibrium line for simple model of adiabatic gas absorption, gas-film driving force; C, equilibrium line for simple model of adiabatic gas absorption, overall gas-phase driving force; D, rigorously computed equilibrium line, gas-film driving force (43).

The standard way to correct for the heat of solution approximately is the simple adiabatic model described on the preceding pages, which yields equilibrium line B if the gas-phase driving force is used and line C on the basis of the overall driving force. This model, however, is a poor representation of the conditions prevailing in the absorber, as demonstrated by the deviation of its equilibrium line B from the rigorous line D. The approximation underestimates the true packing depth value of 3.63 m by more than one-third yielding 2.4 m (Table 2).

Rapid Approximate Design Procedure. For purposes of quick, approximate design of adiabatic packed gas absorbers an empirical correlation of liquid temperature profiles has been developed (45) which may be used in a way similar to equation 68 to compute the necessary tower height. The liquid temperature profiles were rigorously computed for over 90 hypothetical design cases covering wide variations in system properties and operating conditions. The mG_M/L_M factor was varied from maximum values given by pinchpoints between the equilibrium and the operating lines down to values where heat effects were subdued by the heat capacity of the high solvent flow rate. The apparent Henry's law constants were varied from 0.0 to 186.6 kPa (1400 mm Hg) and heats of solution up to 58.6 kJ/mol (14.0 kcal/mol) were taken into account. The investigation was limited to gas concentrations below 15 mol % and to recovery fractions ranging from 90 to 99%. Water was considered to be the most important solvent, but variations in solvent properties have been included to a certain extent.

It was found that the rigorously computed liquid temperature profiles could

Table 2. Comparison of Results of Different Design Calculations[a]

Method used	N_{OG}	Required depth of packing, m
rigorous calculation	5.56	3.63
isothermal approximation	3.30	1.96
simple adiabatic model	4.01	2.38
suggested short-cut design procedure		
graphical integration	5.51	3.60
approximate analytical integration	5.56	3.73

[a]Ref. 45.

be satisfactorily represented as a function of liquid concentration by the empirical equation

$$T_L = T_{L,2} + (T_{L,1} - T_{L,2})X_N + 74.34(X_N^{1.074} - X_N^{1.114}) \Delta T_{max} \qquad (69)$$

where

$$X_N = \frac{x_A - x_{A,2}}{x_{A,1} - x_{A,2}} \qquad (70)$$

$T_{L,1}$ is the temperature of the liquid leaving the tower and may be calculated from an enthalpy balance around the whole tower based on an estimate of the temperature of the leaving gas, $T_{G,2}$. It was found that the rigorous values of $T_{L,1}$ could be predicted rapidly within a standard deviation of 1.2°C if the following semiempirical equations were used for estimating $T_{G,2}$:

$$T_{G,2} = T_{L,2} + \left(\frac{dT_L}{dx_A}\right)_2 \left(\frac{G_M}{L_M}\right)_2 \left(\frac{H_{G,Q}}{H_{OG,A}}\right)(y_{A,2} - y_{A,2}^*) \qquad (71)$$

$$\left(\frac{dT_L}{dx_A}\right)_2 = \frac{L_{M,2}H_{OS} - G_{M,2}H_v m_{B,2}}{L_{M,2}C_{q,2} - G_{M,2}C_{P,2} - G_{M,2}H_v(1 - x_{A,2})(dm_B/dT_L)_2} \qquad (72)$$

where $H_{G,Q} = G_M C_P/h_G a$, the gas-phase height of a heat transfer unit. This quantity may be estimated from H_G for the solute using the Chilton-Colburn analogy; $H_{OG,A} = H_{OG}$ for the solute; H_{OS} is the integral heat of solution for solute; H_v is the latent heat of pure solvent; $m_{B,2}$ is the "solubility" of solvent, y_B^*/x_B at the top of the tower; $(dm_B/dT_L)_2$ is the temperature coefficient of m_B at the top of the tower; and $C_{P,2}$ and $C_{q,2}$ are mean molar heat capacities of gas and liquid at the top of the towers, respectively.

ΔT_{max} in equation 69 is related to the internal temperature maximum and may be estimated from specified variables using the correlations shown in Figure 12. All other quantities in equations 69 and 70 are usually known from design specifications. If applied to values of the liquid concentration, equation 69 yields corresponding temperature values which may serve to evaluate the equilibrium gas concentration y^* as a function of the liquid concentration. This procedure

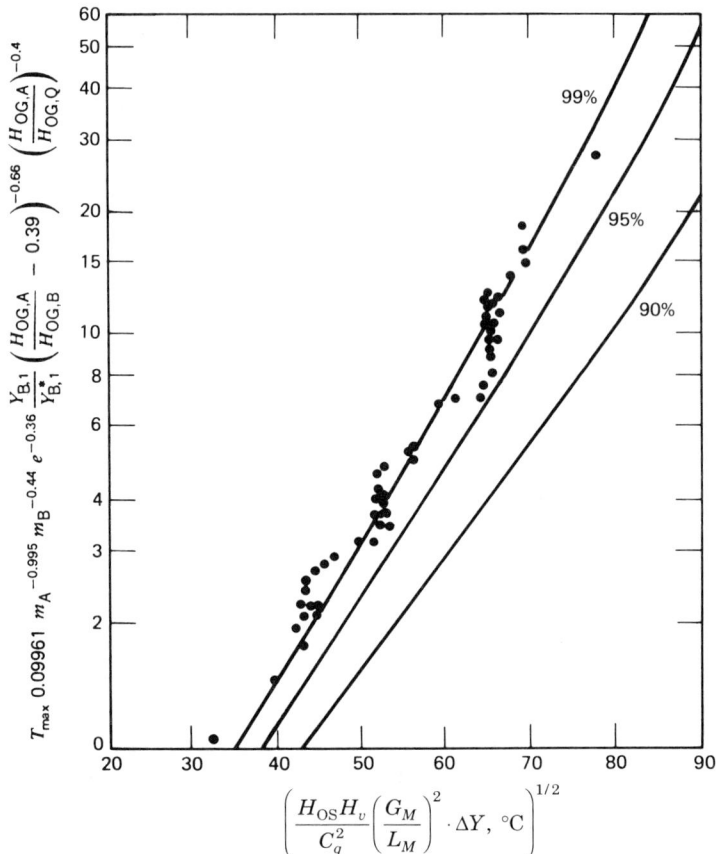

Fig. 12. Correlation of ΔT_{max}. The three lines represent the best fit of a mathematical expression obtained by multidimensional nonlinear regression techniques for 99, 95, and 90% recovery; the points are for 99% recovery. C_q = mean molar heat capacity of liquid mixture, averaged over tower; $\Delta Y = y_{A,1} - y_{A,2}$; m_A = slope of equilibrium line for solute, to be taken at liquid feed temperature; m_B = slope of equilibrium line for solvent, $y_B^*/(1 - x_A)$, to be taken at liquid feed temperature; y_B/y_B^* = saturation of solvent in feed gas; $H_{OG,Q} \approx H_{G,Q}$, the height of a gas phase heat transfer unit. (The H_{OG}s are computed from equation 54; the individual heights of a transfer unit for the solvent and for heat may be estimated from $H_{G,A}$ and $H_{L,A}$ using the Chilton-Colburn analogy. The m in equation 54 is to be evaluated at the temperature of the liquid feed.) Reprinted by permission (45).

yields a good approximation of the rigorous equilibrium line accounting for all heat effects. The required number of transfer units and the tower height may then be computed on the basis of a conventional x–y diagram by graphical integration.

To obviate the tedious graphical integration, a simplified design procedure was developed on the basis of Colburn's analytical solution, equation 56. Substitution of the L_M/G_M ratio presents no problem because this ratio stays fairly constant in the tower at the low concentrations for which Figure 12 is valid. The difficulty arises with m because of the unexpected shapes of nonisothermal equilibrium lines.

The equilibrium line, as approximated by the synthetically correlated temperature profile, is cut into two parts at its inflection point as shown in Figure 13.

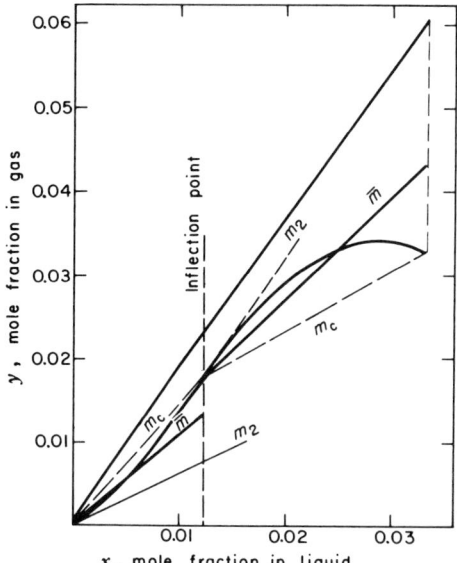

Fig. 13. Definition of effective average slopes of equilibrium line (45).

In each part the equilibrium curve is replaced by a straight line with a slope $\overline{m}$ which, upon substitution into equation 56, yields the same number of transfer units as the graphical integration based on the curved equilibrium line. These effective average slopes $\overline{m}$ were calculated for an expanded set of well over 100 hypothetical design cases by performing the integration numerically and solving for m. They were correlated using a multidimensional, nonlinear regression analysis and the initial slopes and the slopes of the cord of the equilibrium line in the respective section (Fig. 13). The resulting correlation for the dilute section of the y–x diagram, where the equilibrium is curved concave upward, is shown in Figure 14, whereas Figure 15 represents the correlation for the concentrated section where the curvature is concave downward.

The recommended rapid approximate design procedure for adiabatic packed gas absorbers consists of the following steps: (*1*) After the problem is fully specified, the temperature of the leaving product is estimated (eqs. 71, 72) using an enthalpy balance around the tower. (*2*) The maximum temperature $\Delta T_{\max}$ of the convex portion of the liquid temperature profile is estimated on the basis of Figure 12. (Steps 1 and 2 determine the estimated temperature profile completely (eq. 69).) (*3*) A special correlation (Fig. 16) may then be used to find the locus of the inflection point on the equilibrium curve on the y–x diagram without trial and error. (*4*) The equation for the estimated temperature profile (eq. 69) is then used to find the temperature as well as the slope of the equilibrium line at the inflection point. Based on this information, the effective average slopes for the two sections may be read from Figures 14 and 15, and used in equation 56 to determine the necessary number of transfer units. The overall height of a transfer unit is evaluated for each section substituting the values for $\overline{m}$ from Figures 14 and 15 into equation 54.

The total number of transfer units and the total required depth of packing

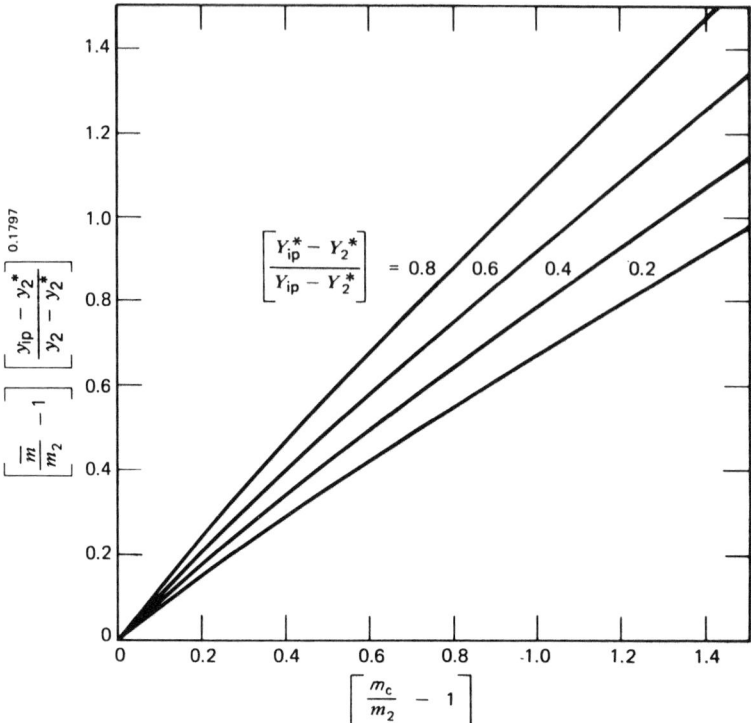

Fig. 14. Correlation of effective average slope $\overline{m}$ of equilibrium line (dilute part of absorber); equilibrium line concave upward. Y_{ip}: gas mole fraction at inflection point (45).

are obtained by adding the respective values for the two sections. Table 2 compares this design procedure, the isothermal and the simple adiabatic approximations, and the rigorous result with respect to the design example, discussed previously, and clearly demonstrates the superior accuracy of the procedure. No serious temperature bulge develops if $(T_{L,2} - T_{L,1})/\Delta T_{max} \geq 4.3$. The simple model of adiabatic gas absorption may then be used to determine the required tower height. Figure 14 may, of course, be used also in this case together with equations 47, 54, and 56 to circumvent the graphical integration and obtain the result rapidly without significant loss of accuracy.

Axial Dispersion Effects. *Effect of Axial Dispersion on Column Performance.* Another assumption underlying standard design methods is that the gas and the liquid phases move in plug-flow fashion through the column. In reality, considerable departure from this ideal flow assumption exists (4) and different fluid particles travel through the packing at varying velocities. The impact of this effect, which is usually referred to as axial dispersion, on the concentration profiles is demonstrated in Figure 17. The effect counteracts the countercurrent contacting scheme for which the column is designed and thus lowers the driving forces throughout the packed bed. Neglect of axial dispersion results in an overestimation of the driving forces and in an underestimation of the number of transfer units needed. It may therefore lead to an unsafe design.

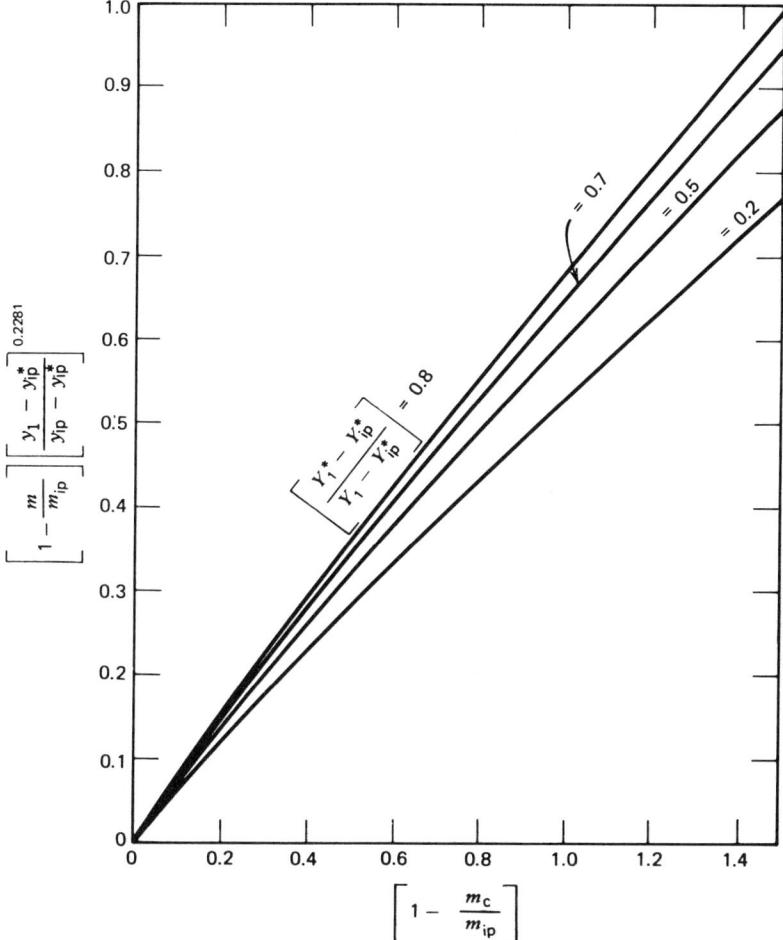

Fig. 15. Correlation of effective average slope $\overline{m}$ of nonisothermal equilibrium line (concentrated part of absorber); equilibrium line concave downward. The subscript ip indicates the inflection point (45).

Determination of separation efficiencies from pilot-plant data is also affected by axial dispersion. Neglecting it yields high H_G or H_L values. Literature data for this parameter have usually not been corrected for this effect.

The extent of axial dispersion occurring in a gas absorber can be determined by measuring the residence time distribution of both gas and liquid. Based on classical flow models of axial dispersion, the result is usually expressed in terms of two Peclet numbers (Pe), one for each phase. Peclet numbers tending towards infinity indicate near-ideal plug flow, whereas vanishing values of Pe indicate axial dispersion to such an extent that the phase begins to become well backmixed. When designing packed columns, the Peclet numbers are usually estimated from literature correlations (48–50). Correlations for predicting Peclet numbers in large scale gas–liquid contactors are quite scarce, but contributions

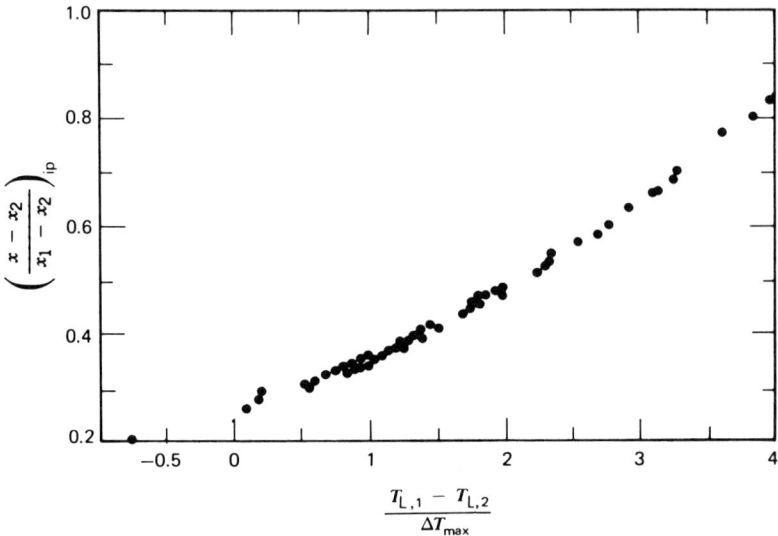

Fig. 16. Correlation of the liquid concentration at which the inflection point of the nonisothermal equilibrium occurs (45).

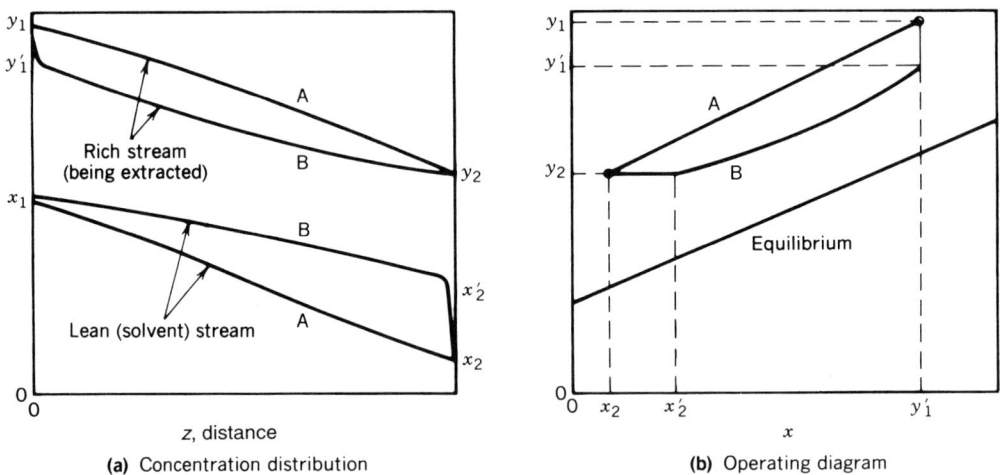

Fig. 17. Effect of axial dispersion in both phases on solute distribution through countercurrent mass transfer equipment. A, piston or plug flow; B, axial dispersion in both streams (diagrammatic). Reprinted with permission (4).

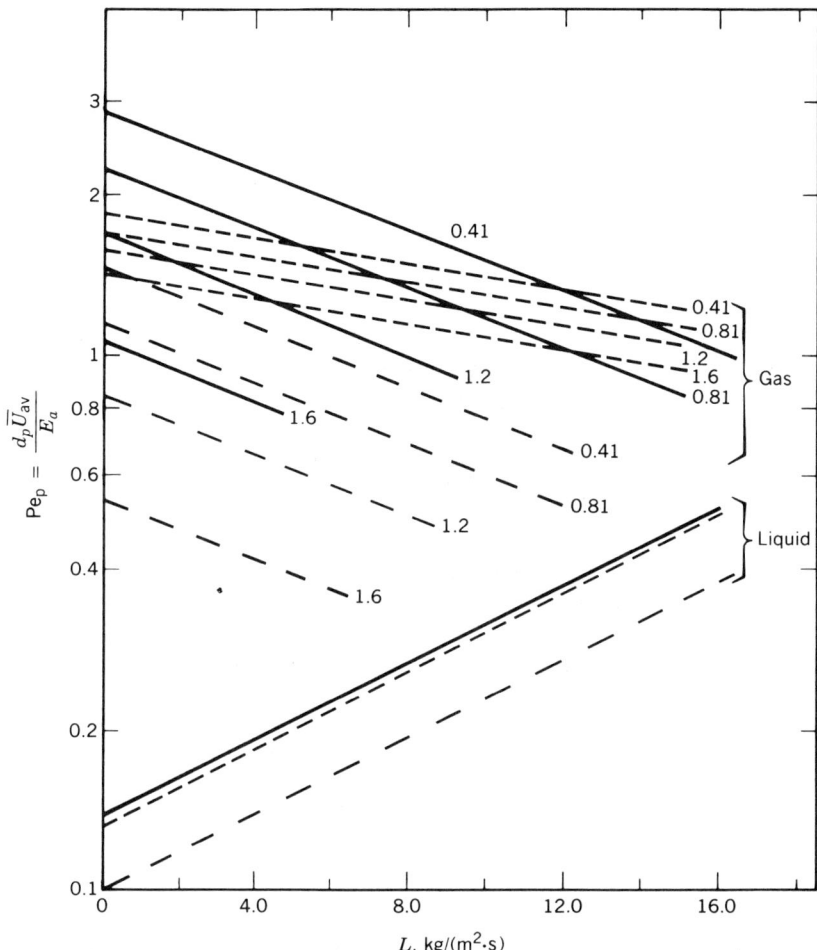

Fig. 18. Peclet numbers in large scale gas–liquid contactors using 2.54-cm Berl saddles (———) or 2.54 cm (— —) or 5.08 cm (- - -) Raschig rings (51). Numbers on lines represent G values = gas flow in kg/(m²·s); d_p = nominal packing size; U_{av} = superficial velocity. To convert kg/(m²·s) to lb/(h·ft²) multiply by 737.5. Reprinted with permission (4).

have been made (51–55). Some of the available data have been described (4) and a review of published correlations for liquid-phase Peclet numbers is also available (55). Figure 18 reproduces some data (51).

When designing packed towers, axial dispersion can be accounted for by incorporating terms for axial dispersion of the solute into the differential mass balance equation 36. The integration of the resulting differential equations is best effected by computer. Analytical solutions for cases having linear equilibrium and operating lines have been developed (56,57). They are, however, not explicit for the design case and are of such enormous complexity that application for design also requires a computer.

Rapid Approximate Design Procedure. Several simplified approximations to the rigorous solutions have been developed over the years (57–60), but

34 ABSORPTION

they all remain too complicated for practical use. A simple method proposed in 1989 (61,62) uses a correction factor accounting for the effect of axial dispersion, which is defined as (57)

$$\text{correction factor} = \frac{\text{NTU}_{ap}}{\text{NTU}} \qquad (73)$$

NTU_{ap} is the "exterior apparent" overall gas-phase number of transfer units calculated neglecting axial dispersion simply on the basis of equation 56, whereas NTU stands for the higher real number of transfer units (N_{OG}) which is actually required under the influence of axial dispersion. The correction factor ratio can be represented as a function of those parameters that are actually known at the outset of the calculation

$$\frac{\text{NTU}_{ap}}{\text{NTU}} = f\left(\text{NTU}_{ap}, \left(\frac{mG}{L}\right), \text{Pe}_x, \text{Pe}_y\right) \qquad (74)$$

Equation 74 is shown graphically in Figure 19**a** for a given set of conditions. Curves such as these cannot be directly used for design, however, because the Peclet number contains the tower height h as a characteristic dimension. Therefore, new Peclet numbers are defined containing H_{OG} as the characteristic length. These relate to the conventional Pe as

$$\text{Pe}_{HTU} = \frac{uH_{OG}}{D_{ax}} = \frac{uh}{D_{ax}}\frac{H_{OG}}{h} \qquad (75a)$$

$$= \text{Pe}\,\frac{1}{N_{OG}} \qquad (75b)$$

The correction factor $(\text{NTU})_{ap}/\text{NTU}$ as a function of Pe_{HTU} rather than Pe is shown in Figure 19**b**. The correction factors given in Figures 19**a** and 19**b** can roughly be estimated as

$$\frac{\text{NTU}_{ap}}{\text{NTU}} \approx 1 - \frac{\text{NTU}_{ap}}{\dfrac{\ln S}{S-1} + \dfrac{\text{Pe}_x\text{Pe}_y}{\text{Pe}_y + S\text{Pe}_x}} \qquad (76)$$

$$\frac{\text{NTU}_{ap}}{\text{NTU}} \approx \frac{\text{Pe}_{HTU,y}\text{Pe}_{HTU,x}}{\text{Pe}_{HTU,y}\,\text{Pe}_{HTU,x} + \text{Pe}_{HTU,y} + S\text{Pe}_{HTU,x}} \qquad (77)$$

In these equations S denotes the stripping factor, mG_M/L_M. Equation 77 is only valid for a sufficiently high number of transfer units so that the correction factor becomes independent of NTU_{ap}.

In the original study (61), $\text{NTU}_{ap}/\text{NTU}$ was calculated for thousands of hypothetical design cases as a function of both Pe and Pe_{HTU}. The results were correlated and empirical expressions were given that can be evaluated on a

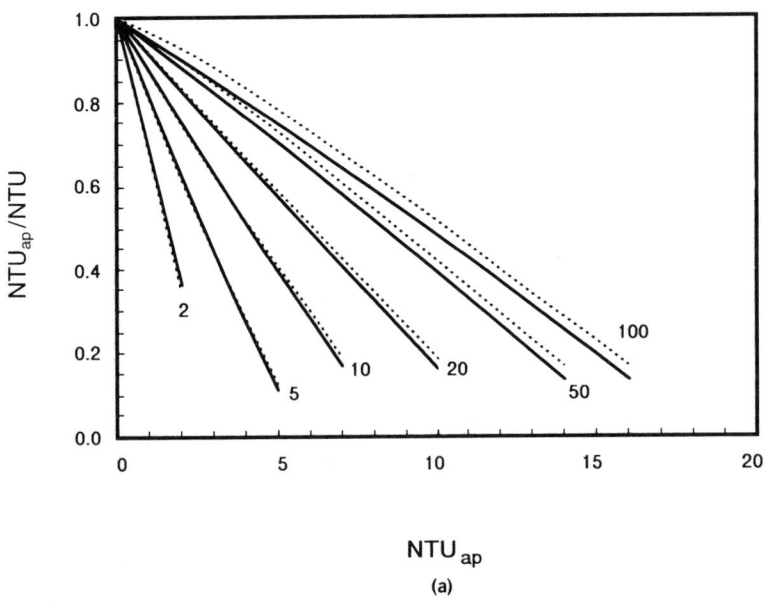

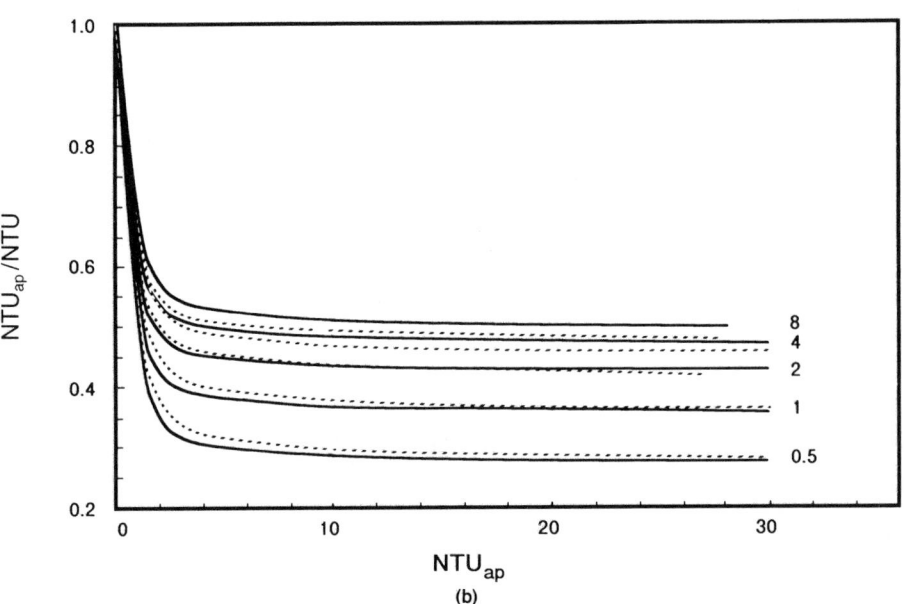

Fig. 19. Correction factor for axial dispersion as a function of NTU_{ap}. Solid lines are rigorous calculations; broken lines, approximate formulas according to literature (61). (**a**) Numbers on lines represent Pe_x values; $Pe_y = 20$; $mG_M/L_M = 0.8$. (**b**) For design calculations. Numbers on lines represent $Pe_{HTU,x}$ values; $Pe_{HTU,y} = 1$; $mG_M/L_M = 0.8$.

handheld calculator, just as equations 76 and 77, but which approximate the computer calculation much better, to within about ±5%.

The recommended rapid design procedure consists of the following steps: (*1*) The apparent N_{OG} is calculated using equation 56. (*2*) The extent of axial dispersion is estimated from literature correlations for each phase in terms of Pe numbers and transformed into Pe_{HTU} values. (*3*) The correction factor NTU_{ap}/NTU is estimated on the basis of the correlation given in the literature (61). A reasonable, conservative estimate may also be obtained using equation 77, provided $NTU_{ap} > 5$. When the apparent number of transfer units is divided by this correction factor, the value of N_{OG} actually required under the influence of axial dispersion is obtained. (*4*) The packed tower height is found by multiplying N_{OG} by the true H_{OG}. In order to obtain values for the latter, pilot-plant data has to be corrected for the influence of axial dispersion. This correction may be made in a manner similar to that described above, but equation 76 would be used to estimate the correction factor rather than equation 77.

Experimental Mass Transfer Coefficients. Hundreds of papers have been published reporting mass transfer coefficients in packed columns. For some simple systems which have been studied quite extensively, mass transfer data may be obtained directly from the literature (6). The situation with respect to the prediction of mass transfer coefficients for new systems is still poor. Despite the wealth of experimental and theoretical studies, no comprehensive theory has been developed, and most generalizations are based on empirical or semiempirical equations.

Liquid-Phase Transfer. It is difficult to measure transfer coefficients separately from the effective interfacial area; thus data is usually correlated in a lumped form, eg, as $k_L a$ or as H_L. These parameters are measured for the liquid film by absorption or desorption of sparingly soluble gases such as O_2 or CO_2 in water. The liquid film resistance is completely controlling in such cases, and $k_L a$ may be estimated as $K_{OL}a$ since $x_i \approx x^*$ (Fig. 4). This is a prerequisite because the interfacial concentrations would not be known otherwise and hence the driving force through the liquid film could not be evaluated.

The resulting correlations fall into several categories. Some are essentially empirical in nature. Examples include (63)

$$H_L = \frac{1}{\alpha}\left(\frac{L}{\mu}\right)^n \left(\frac{\mu_L}{\rho_L D_L}\right)^{0.5} \tag{78}$$

The values of α and n are given in Table 3; typical values for D_L can be found in Table 4. The exponent of 0.5 on the Schmidt number $(\mu_L/\rho_L D_L)$ supports the penetration theory. Further examples of empirical correlations provide partial experimental confirmation of equation 78 (3,64–68). The correlation reflecting what is probably the most comprehensive experimental basis, the Monsanto Model, also falls in this category (68,69). It is based on 545 observations from 13 different sources and may be summarized as

$$H_L = \phi\, C_{fl} \left(\frac{h}{3.05}\right)^{0.15} Sc_L^{0.5} \tag{79}$$

Table 3. Values of Constants for Equation 78

Packing	Size, cm	α^a	α^b	n
Raschig rings	0.95	3120	550	0.46
	1.3	1390	280	0.35
	2.5	430	100	0.22
	3.8	380	90	0.22
	5.1	340	80	0.22
Berl saddles	1.3	685	150	0.28
	2.5	780	170	0.28
	3.8	730	160	0.28

^{a}Valid for units kg, s, m.
^{b}Valid for units lb, h, ft.

Table 4. Diffusion Coefficients for Dilute Solutions of Gases in Liquids at 20°C

Gas	Liquid	D_L, m^2/s × 10^{-9}
CO_2	water	1.78
Cl_2	water	1.61
H_2	water	5.22
HCl	water	0.61
H_2S	water	1.64
N_2	water	1.92
N_2O	water	1.75
NH_3	water	1.83
O_2	water	2.08
acetone	water	1.61
benzene	kerosene	1.41

The packing parameter $\phi(m)$ reflects the influence of the liquid flow rate as shown in Figure 20. C_{fl} reflects the influence of the gas flow rate, staying at unity below 50% of the flooding rate but beginning to decrease above this point. At 75% of the flooding velocity, $C_{fl} = 0.6$. Sc_L is the Schmidt number of the liquid.

Other correlations based partially on theoretical considerations but made to fit existing data also exist (71–75). A number of researchers have also attempted to separate k_L from a by measuring the latter, sometimes in terms of the wetted area (76–78). Finally, a number of correlations for the mass transfer coefficient k_L itself exist. These are based on a more fundamental theory of mass transfer in packed columns (79–82). Although certain predictions were verified by experimental evidence, these models often cannot serve as design basis because the equations contain the interfacial area a as an independent variable.

Gas-Phase Transfer. The height of a gas-phase mass transfer unit, or $k_G a$, is normally measured either by vaporization experiments of pure liquids, in which no liquid mass transfer resistance exists, or using extremely soluble gases. In the latter case, m is so small that the liquid-film resistance in equation 7 is negligible and the gas-film mass transfer coefficient can be observed as $k_G a \approx K_{CG} a$ and

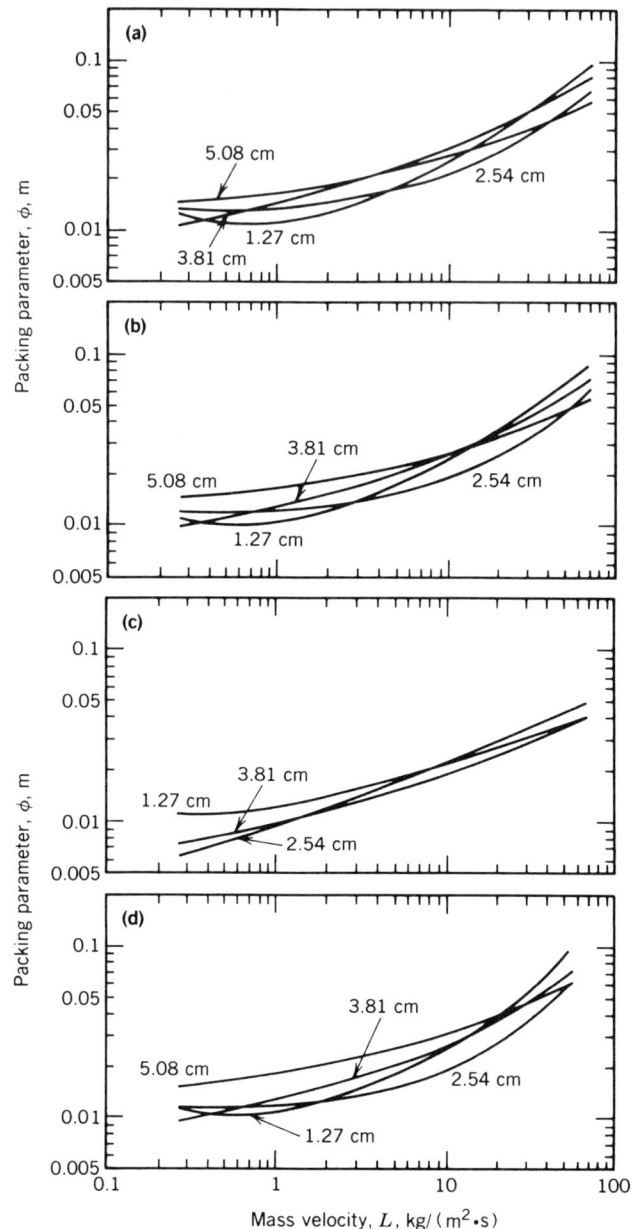

Fig. 20. Improved packing parameters ϕ for liquid mass transfer: (**a**) ceramic Raschig rings; (**b**) metal Raschig rings; (**c**) ceramic Berl saddles; (**d**) metal Pall rings (69). Reprinted with permission (70).

$y_i \approx y^*$. The experiments are difficult because they have to be carried out in very shallow beds. Otherwise, all of the highly soluble gas is absorbed and the driving force cannot be evaluated. The resulting end effects are probably the main reason for the substantial disagreement of the published data, which have been reported to vary some threefold for the same packing and flow rates (83). Furthermore, the

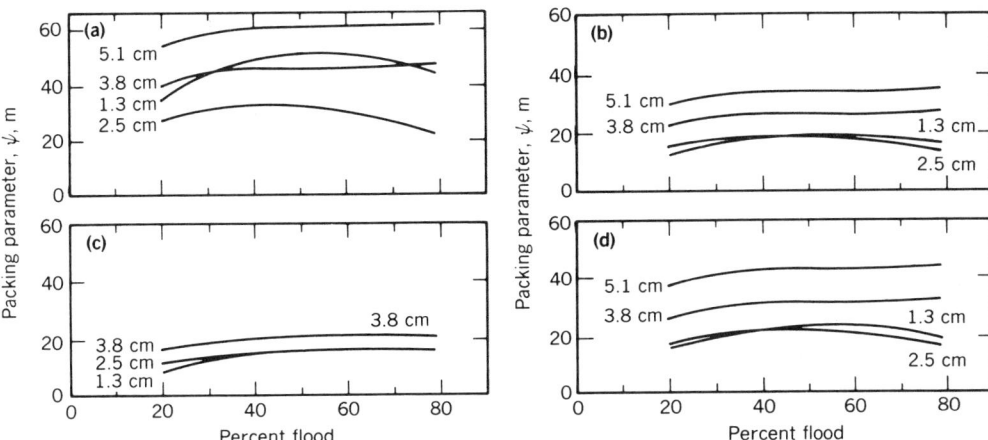

Fig. 21. Improved packing parameters ψ for gas mass transfer: (**a**) ceramic Raschig rings; (**b**) metal Rashig rings; (**c**) ceramic Berl saddles; (**d**) metal Pall rings (69). Reprinted with permission.

effective interfacial areas seem to differ for absorption and vaporization (84). During the absorption experiments, the many stagnant or semistagnant pockets of liquid which exist in a packing tend to become saturated and thus ineffective. This is not the case in vaporization experiments where the total effective interfacial area consists of the surface area of moving liquid plus the semistagnant liquid pockets.

The correlation of H_G based on the most extensive experimental basis is again the Monsanto Model (69):

$$H_G = \psi \frac{(3.28\ d_c')^m (h/3.05)^{1.3}}{(737 L f_\mu f_\rho f_\sigma)^n} \mathrm{Sc}_G^{0.5} \qquad (80)$$

where d_c' is the lesser of 0.61 or column diameter, m; $f_\mu = (\mu_L/\mu_w)^{0.16}$; $f_\rho = (\rho_L/\rho_w)^{-1.25}$; $f_\sigma = (\sigma_L/\sigma_w)^{-0.8}$; Sc_G = Schmidt number of the gas. The packing parameter ψ, in m, depends on the gas flow rate as shown in Figure 21. Values of the diffusivity of various solutes in air and typical Schmidt numbers for use in equation 80 are found in Table 5. The exponents m and n adopt the values of 1.24 and 0.6, respectively, for rings as packing materials, and 1.11 and 0.5, for saddles.

Another type of experiment to measure k_G separately from other factors consists of saturating packings made from porous materials using a volatile liquid and subsequently drying it by passing a stream of inert gas through the packing (85–88). Since the surface of the packing is normally known in these experiments, k_G can be computed. Application of this kind of data to gas absorption design is difficult, however, because of the different, unknown effective interfacial areas when two phases are flowing through the packing. A similar approach was used by evaporating naphthalene from a packing made from this material (78,84). The mass transfer coefficient k_G measured in this manner was then combined with $k_G a$ data (89) to determine the effective area, which was found to be fairly independent

Table 5. Values of the Diffusion Coefficient D_A and of $\mu_G/D_A\rho_G$ for Various Gases in Air at 0°C and at Atmospheric Pressure[a]

Gas	CAS Registry Number	D_A, m²/s × 10⁻⁵		$\dfrac{\mu_G}{\rho_G D_A}$
		Calculated	Experiment	
acetic acid	[64-19-7]		1.05	1.26
acetone	[67-64-1]	0.83		1.60
ammonia	[7664-41-7]	1.62	2.17	0.61
benzene	[71-43-2]	0.72	0.78	1.71
bromobenzene	[108-86-1]	0.67		1.71
butane	[106-97-8]	0.75		1.77
n-butyl alcohol	[71-36-3]		0.69	1.88
carbon dioxide	[124-38-9]	1.19	1.39	0.96
carbon disulfide	[75-15-0]		278.	1.48
carbon tetrachloride	[56-23-5]	0.61		2.13
chlorine	[7782-50-5]	0.92		1.42
chlorobenzene	[108-90-7]	0.61		2.13
chloropicrin	[76-06-2]	0.61		2.13
2,2'-dichloroethyl sulfide (mustard gas)	[505-60-2]	0.56		2.44
ethane	[74-84-0]	1.08		1.22
ethyl acetate	[141-78-6]	0.67	0.72	1.84
ethyl alcohol	[64-17-5]	0.94	1.03	1.30
ethyl ether	[60-29-7]	0.69	0.78	1.70
ethylene dibromide	[106-93-4]	0.67		1.97
hydrogen	[1333-74-0]	5.61		0.22
methane	[74-82-8]	1.58		0.84
methyl acetate	[79-20-9]		0.94	1.57
methyl alcohol	[67-56-1]	1.22	1.33	1.00
naphthalene	[91-20-3]		0.50	2.57
nitrogen	[7727-37-9]	1.33		0.98
n-octane	[111-65-9]		0.50	2.57
oxygen	[7782-44-7]	1.64	1.78	0.74
pentane	[109-66-0]	0.67		1.97
phosgene	[75-44-5]	0.81		1.65
propane	[74-98-6]	0.89		1.51
n-propyl acetate	[109-60-4]	0.67	0.67	1.97
n-propyl alcohol	[71-23-8]	0.81	0.86	1.55
sulfur dioxide	[7446-09-5]	1.03		1.28
toluene	[108-88-3]	0.64	0.72	1.86
water	[7732-18-5]	1.89	2.19	0.60

[a] The value of μ_G/ρ_G is that for pure air, 1.33×10^{-5} m²/s. Diffusion coefficients may be corrected for other conditions by assuming them proportional to $T^{2/3}$ and inversely proportional to P. The Schmidt numbers depend only weakly on temperature (113).

of gas rate up to the loading point. The data bank underlying the Monsanto Model and literature correlations for k_G and k_L (78,84) have also been used (90) to develop a new correlation for packed distillation columns.

Height Equivalent to a Theoretical Plate. Provided both the equilibrium and operating lines are straight, HETP values may be estimated by combining the

H_G and H_L values predicted by the above correlations and by translating the resulting H_{OG} into HETP by combining equations 47, 51, and 56 with equation 85, which is discussed under bubble tray absorption columns:

$$\text{HETP} = \frac{\ln(mG_M/L_M)}{(mG_M/L_M) - 1} H_{OG} \tag{81}$$

HETP values obtained in this way have been compared to measured values in data banks (69) and statistical analysis reveals that the agreement is better when equations 79 and 80 are used to predict H_G and H_L than with the other models tested. Even so, a design at 95% confidence level would require a safety factor of 1.7 to account for scatter.

The situation is very much poorer for structured rather than random packings, in that hardly any data on H_G and H_L have been published. Based on a mechanistic model for mass transfer, a way to estimate HETP values for structured packings in distillation columns has been proposed (91), yet there is a clear need for more experimental data in this area.

Capacity Limitations. Thus far the discussion has been confined to factors affecting the tower height required to perform a specific absorption job. The necessary tower diameter, on the other hand, depends primarily on the total amount of gas and liquid that must be handled. At a given set of flow rates, the diameter of the packing can only be decreased at the expense of a large pressure drop, which in turn generates higher operating costs because more power is needed to blow the gas through the packing. The reason for this is the fact that handling a given total gas flow rate in a smaller tower diameter increases the superficial velocity at which the gas has to be pushed through the packing.

The relationship between the pressure drop per unit of packed height and the superficial gas velocity given in terms of the gas flow rate is shown schematically in Figure 22. In a dry packing, ΔP increases almost as the square of the gas velocity, which is in accord with the turbulent nature of the flow. At low

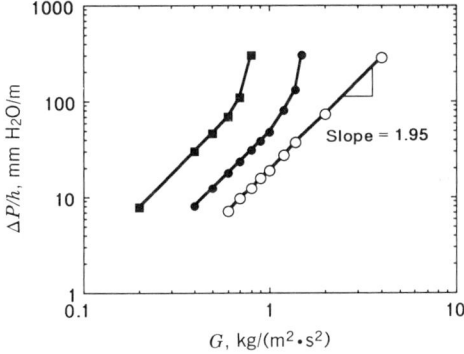

Fig. 22. Schematic representation of typical pressure drop as a function of superficial gas velocity, expressed in terms of $G = \rho_G u_G$, in packed columns. ○, Dry packing; ●, low liquid flow rate; ■, higher liquid flow rate. The points do not correspond to actual experimental data, but represent examples.

liquid flow rates, the curves are somewhat shifted upwards because the presence of liquid films restricts the free section available for gas flow and thus increases the linear gas velocity somewhat. Because the liquid hold-up remains independent of G, the slope of the curve in this log–log plot remains close to 2. At higher pressure drops, however, the upflowing gas impairs the downflow of liquid and excess liquid starts to accumulate in the packing, thereby increasing the hold-up. In this operating region, called the loading zone, the increasing liquid hold-up restricts the free section available for the gas flow further as G becomes larger. Hence the linear gas velocities increase faster than G and the power dependence of ΔP on G starts to rise above 2. The G value at which the curve begins to deviate from the straight line has been defined as the loading point.

If the tower diameter is made too small for a given total gas flow rate, that is, if u_G and G are increased above a certain critical value, ΔP becomes so great that the liquid cannot flow downward anymore over the packing, but is blown out the top of the packing. The vertical asymptotes on Figure 22 indicate the gas rates at which this condition, called flooding, occurs. This gas flow rate at flooding, G_F, determines the theoretical minimum diameter at which the tower is operable, and knowledge of it is therefore very important. Flooding rates have been correlated (92–94).

Both the pressure drop per unit length of packed tower and the gas flooding rate have been correlated for random packings as shown in Figure 23 (95). Such correlations enable predicting the gas flow rate G that will flood the packing at a given L/G ratio. In practice, the tower has to be operated at flow rates considerably less than the flooding rates for safety reasons. It is generally accepted that 50 to 80% of the flooding flow rates can be permitted. The curves plotted on Figure 23 thus also enable prediction of the pressure drop at any chosen operating value of G. The diameter of the column must then be evaluated by comparing this value to the total quantity of gas that the tower is supposed to handle. The correlation shown in Figure 23 can be applied to predict the hydraulic performance of many different packings owing to an adjustable parameter known as the packing factor F_P. Values for F_P have been compiled (96); a few examples are listed in Table 6. Similar flooding rate correlations are available for arranged packings. Figure 24 reports results for four examples of Mellapak types (97). Comparison of Figures 24 and 23 reveals higher capacity limits in structured than in many of the dumped packings. At similar loads, structured packings tend generally to give rise to less pressure drop.

Bubble Tray Absorption Columns

General Design Procedure. Bubble tray absorbers may be designed graphically based on a so-called McCabe-Thiele diagram. An operating line and an equilibrium line are plotted in y–x, Y'–X', or Y^0–X^0 coordinates using the principles for packed adsorbers outlined above (see Fig. 25). The minimum number of plates required for a specified recovery may be computed by assuming that equilibrium is reached between the two phases on each bubble tray. Thus the gas and the liquid leaving a tray are at equilibrium and a hypothetical tray capable of equilibrating the phase streams is termed a theoretical plate. Starting the calcula-

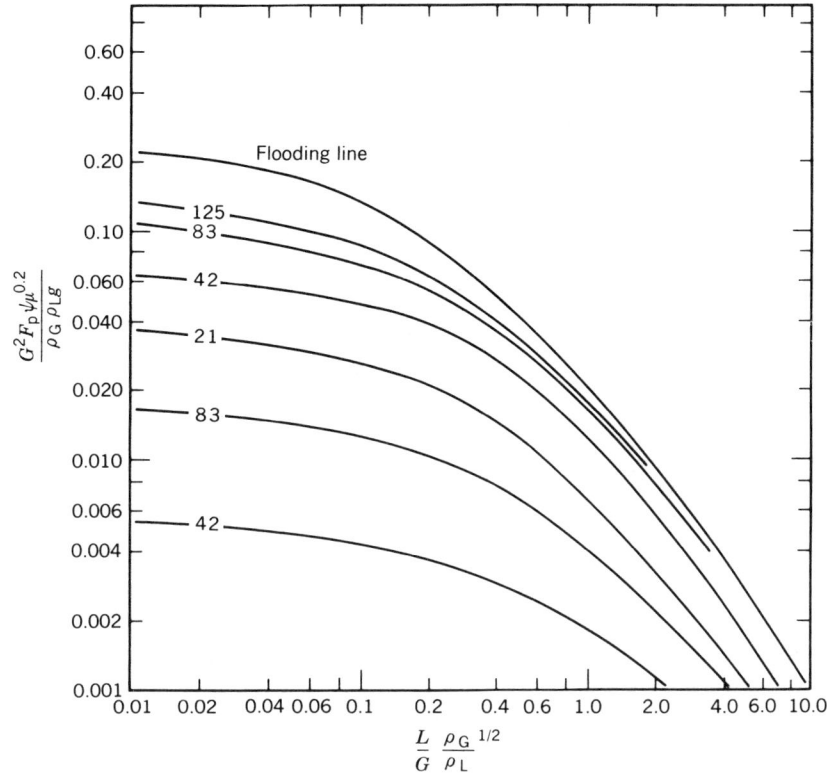

Fig. 23. Pressure drop and flooding correlation for various random packings (95). $\psi = \rho_{H_2O}/\rho_L$, g (standard acceleration of free fall) = 9.81 m/s^2, μ = liquid viscosity in mPa·s; numbers on lines represent pressure drop, mm H$_2$O/m of packed height; to convert to in. H$_2$O/ft multiply by 0.012. Packing factors for various packings have been published (96) and are reproduced in part in Table 6.

Table 6. Characteristics of Dumped Tower Packings[a]

Packing type	Nominal size, mm	Surface area, m^2/m^3	Packing factor, F_P, m^{-1}
Raschig rings, ceramic	6	710	5250
	13	370	2000
	25	190	510
	50	92	215
Raschig rings, steel	25	185	450
	50	95	187
Berl saddles, ceramic	6	900	2950
	13	465	790
	25	250	360
	50	105	150
Pall rings, metal	25	205	157
	50	115	66
Pall rings, polypropylene	25	205	170
	50	100	82

[a] Ref. 96.

44   ABSORPTION

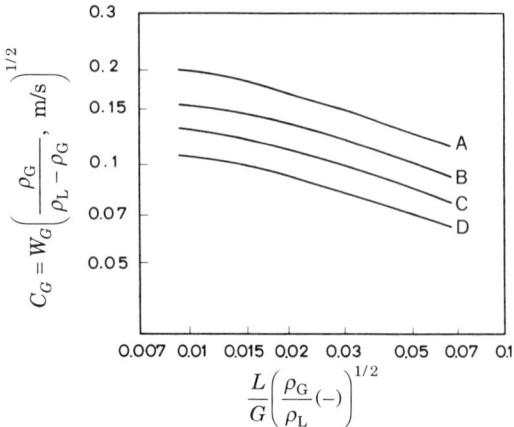

Fig. 24. Souders load diagram for capacity limit determination for four structured packings of the Sulzer-Mellapak type. The solid lines represent the capacity limits of the respective packings as defined by a pressure drop of 1.2 kPa/m: (A) 125 Y; (B) 250 Y; (C) 350 Y; (D) 500 Y. Flooding rates are about 5% higher. Reprinted with permission (97).

tion at the bottom of the tower, where the concentrations are y_{N+1} and x_N (see Fig. 26), the concentration leaving the lowest theoretical plate y_N may be found on the design diagram (Fig. 25**a**) by moving from the operating line vertically to the equilibrium line, because y_N is at equilibrium with x_N. The concentrations between two plates are always related by the operating line. Thus x_{N-1} may be found from y_N by moving horizontally to the operating line. By repeating this sequence of steps until the desired residual gas concentration y_1 is reached, the number of theoretical plates can be counted.

The required number of actual plates, N_P, is larger than the number of theoretical plates, N_{TP}, because it would take an infinite contacting time at each stage to establish equilibrium. The ratio $N_{TP}:N_P$ is called the overall column efficiency. This parameter is difficult to predict from theoretical considerations, however, or to correct for new systems and operating conditions. It is therefore customary to characterize the single plate by the so-called Murphree vapor plate efficiency, E_{MV} (98):

$$E_{MV} \equiv \frac{y_n - y_{n+1}}{y_n^* - y_{n+1}} \tag{82}$$

which indicates the fractional approach to equilibrium achieved by the plate. An efficiency of 80% means that the reduction in solute gas concentration effected by the plate is 80% of the reduction obtained from a theoretical plate. Corresponding actual plates may therefore be stepped off by moving from the operating line vertically only 80% of the distance between operating and equilibrium line (Fig. 25**b**). In some special cases having negligible resistance in the gas phase, E_{MV} values may become unreasonably small. It is then more logical to define a Murphree liquid plate efficiency, E_{ML}, simply by reversing the role of liquid and gas and by focusing on the change in liquid composition across the plate with respect to an equilibrium given by the leaving vapor.

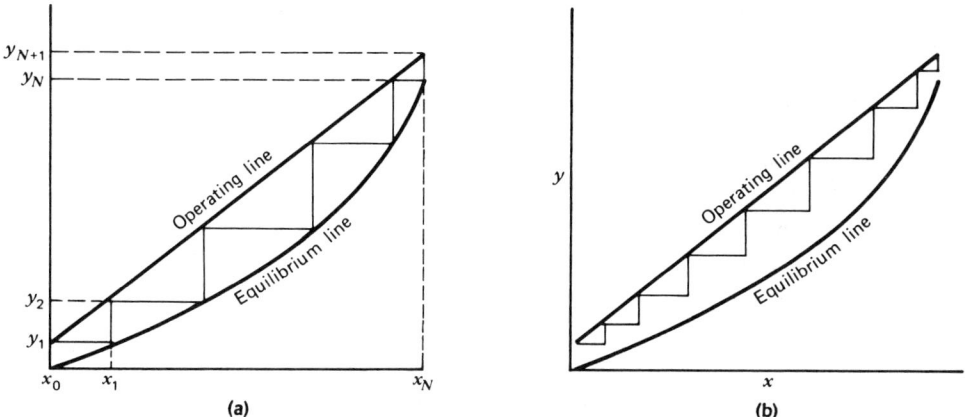

Fig. 25. McCabe-Thiele diagram. (**a**) Number of theoretical plates, 5; (**b**) number of actual plates, 8.

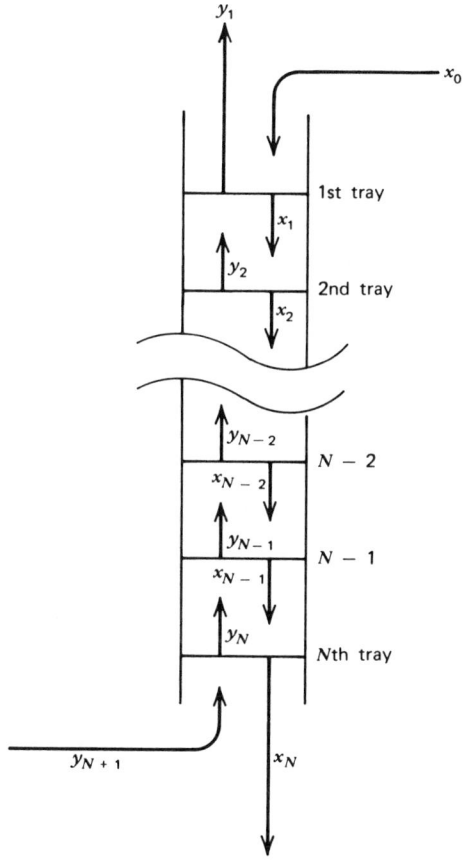

Fig. 26. Numbering plates and concentrations in a bubble tray column.

46 ABSORPTION

Simplified Design Procedure for Linear Equilibrium and Operating Lines. A straight operating line occurs when the concentrations are low such that L_M and G_M remain essentially constant. (The material balance is obtained from equation 35.) In cases where the equilibrium K value does not depend too much on concentration, the use of absorption and stripping factors (99–101) allows rapid calculations for absorption design. One of the simplifying assumptions made in the development of this so-called Kremser-Brown method involves the use of the absorption factor A. The following algebraic expression describes the liquid and vapor flows from the plate (102):

$$A_n = \frac{L_n}{K_n G_n} \quad \text{or} \quad A_{\text{av}} = \frac{L_{\text{av}}}{K_{\text{av}} G_{\text{av}}} \tag{83}$$

where A_n is the absorption factor for each plate n and L_n and G_n are the liquid and vapor flows from the plate. The fractional absorption of any component by an absorber of N plates is expressed in a form similar to the Kremser equation (100,101,103).

$$\frac{Y_{N+1} - Y_1}{Y_{N+1} - Kx_0} = \frac{A^{N+1} - A}{A^{N+1} - 1} \tag{84}$$

where Y_{N+1} = moles of absorbed component entering the column per mole of entering vapor and Y_1 = moles of absorbed component leaving the column per mole of entering vapor. The calculation of plate efficiency (100) is quite sensitive to the choice of equilibrium constants (4,104).

For linear equilibrium and operating lines, an explicit expression for the number of theoretical plates required for reducing the solute mole fraction from y_{N+1} to y_1 has been derived (41):

$$N_{\text{TP}} = \frac{\ln\left[\left(1 - \frac{mG_M}{L_M}\right)\left(\frac{y_{N+1} - mx_0}{y_1 - mx_0}\right) + \frac{mG_M}{L_M}\right]}{\ln \frac{L_M}{mG_M}} \tag{85}$$

This is the one case where the overall column efficiency can be related analytically to the Murphree plate efficiency, so that the actual number of plates is calculable by dividing the number of theoretical plates through equation 86:

$$\frac{N_{\text{TP}}}{N_{\text{P}}} = \frac{\ln\left[1 + E_{\text{MV}}\left(\frac{mG_M}{L_M} - 1\right)\right]}{\ln \frac{mG_M}{L_M}} \tag{86}$$

Nonisothermal Gas Absorption. The computation of nonisothermal gas absorption processes is difficult because of all the interactions involved as described for packed columns. A computer is normally required for the enormous number of plate calculations necessary to establish the correct concentration and temperature profiles through the tower. Suitable algorithms have been developed (46,105) and nonisothermal gas absorption in plate columns has been studied experimentally and the measured profiles compared to the calculated results (47,106). Figure 27 shows a typical liquid temperature profile observed in an adiabatic bubble plate absorber (107). The close agreement between the calculated and observed profiles was obtained without adjusting parameters. The plate efficiencies required for the calculations were measured independently on a single exact copy of the bubble cap plates installed in the five-tray absorber.

A general, approximate, short-cut design procedure for adiabatic bubble tray absorbers has not been developed, although work has been done in the field of nonisothermal and multicomponent hydrocarbon absorbers. An analytical expression which predicts the recovery of each component provided the stripping factor, ie, the group mG_M/L_M, is known for each component on each tray of the column has been developed (102). This requires knowledge of the temperature and total flow (G_M and L_M) profiles through the tower. There are many suggestions about how to estimate these profiles (102, 103, 108). A realistic estimate of the temperature profile for theoretical plates can probably be obtained by the short-cut method developed on the basis of rigorous computer solutions for about 40 different hypothetical designs (108) which closely resemble those of Figure 27.

Plate Efficiency Estimation. *Rate of Mass Transfer in Bubble Plates.* The Murphree vapor efficiency, much like the height of a transfer unit in packed absorbers, characterizes the rate of mass transfer in the equipment. The value of

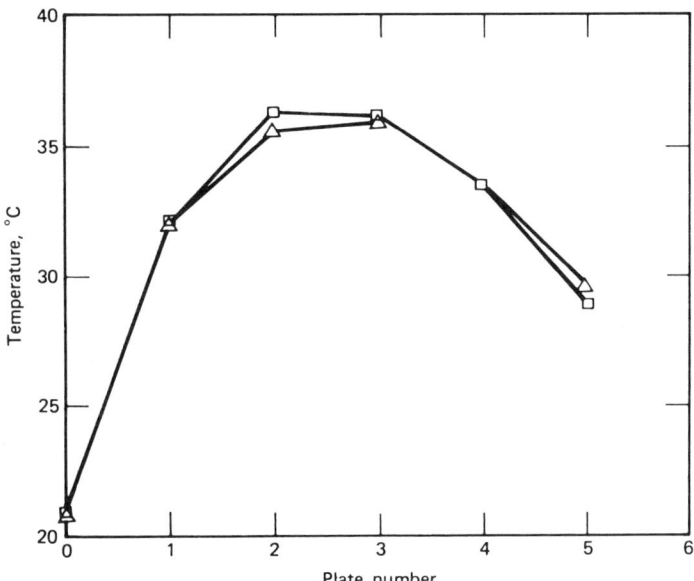

Fig. 27. Computed and experimental liquid temperature profiles in an ammonia absorber with 5 bubble cap trays (107). Water was used as a solvent. $y_{N+1} = 0.123$; $y_1 = 0.0242$; $L'_M/G'_M = 1.757$; $\triangle$, measured; $\square$, calculated.

the efficiency depends on a large number of parameters not normally known, and its prediction is therefore difficult and involved. Correlations have led to widely used empirical relationships, which can be used for rough estimates (109,110). The most fundamental approach for tray efficiency estimation, however, summarizing intensive research on this topic, may be found in reference 111.

In large plates 0.61 m or more in diameter, the efficiency of the tray as a whole may differ from the efficiency observed at some particular point of the tray because the liquid is not uniformly mixed in the direction of the flow on the whole tray. The point value of the efficiency called E_{OG} is thus more closely related to interphase diffusion than E_{MV}. As the gas passes upward through the liquid covering a small area of the plate, mass transfer from gas to liquid occurs in a manner similar to a packed tower of height h_B, the depth of the bubbling area. Under the assumption that the liquid is completely mixed in the vertical direction, and that the gas travels through that minicolumn in a plug-flow-like fashion, the number of transfer units of the bubbling area may be calculated in terms of the gas concentrations above and below the area under consideration by applying the definition of N_{OG}, equation 55. This equation may be integrated by taking y^* as constant and equal to y_n^* because of the well-mixed nature of the liquid phase. By comparing the result with the definition of the plate efficiency, equation 82, formulated for a single point on the plate, the following relationship between the point efficiency and the number of transfer units arises:

$$E_{OG} = 1 - e^{-N_{OG}} \qquad (87)$$

If resistance to transfer is present in both phases, N_{OG} may be expressed as an addition of resistances using equations 54, 47, 41, and 44:

$$\frac{1}{N_{OG}} = \frac{1}{N_G} + \frac{mG_M}{L_M}\frac{1}{N_L} \qquad (88)$$

Hence the point efficiency E_{OG} may be computed if both N_G and N_L in the bubbling area are known. These parameters are determined by the prevailing transfer coefficients, the interfacial area, and by h_B: $N_G = k_G a P h_B / G_M$ and $N_L = k_L a \rho h_B / L_M$.

To estimate the number of transfer units for design, the following empirical correlations which were derived from efficiency measurements employing a variety of trays and operating conditions under the aforementioned assumptions are recommended (111):

$$N_G = (0.776 + 4.63 h_w - 0.238 F + 0.0712 L') \text{Sc}^{-0.5} \qquad (89)$$

where h_w is the weir height in m; $F = u_G \rho_G^{1/2}$ in m/s [(kg/m^3)$^{1/2}$]; L' is the liquid flow rate per average width of stream in m^3/(s·m); and

$$N_L = 3050 D_L^{1/2} (68 h_w + 1) t_L \qquad (90)$$

where $t_L = h_L z_L/L'$ is the liquid residence time in s. The recommended correlation for h_L is

$$h_L = 0.0419 + 0.19 h_w - 0.0135 F + 2.46 L' \qquad (91)$$

Effect of Different Degrees of Mixing. Once E_{OG} is evaluated on the basis of N_G and N_L using equation 87, it has to be translated into E_{MV} by considering the degree of mixing on the tray. It is obvious that for a small plate with completely backmixed liquid,

$$E_{MV} = E_{OG} \qquad (92)$$

If, on the other hand, the liquid flows in a plug-flow-like manner over the tray, but the vapor may be assumed to mix between the trays so that it enters each tray in uniform composition, the result may be calculated according to (112).

$$E_{MV} = \frac{L_M}{mG_M} [e^{(mG_M/L_M)E_{OG}} - 1] \qquad (93)$$

In the case of unmixed vapors between the plates, the equations, being implicit in E_{MV}, have also been solved numerically (112). The results depend on the arrangement of the downcomers and are not too different numerically from equation 93. In reality, however, the liquid is neither completely backmixed nor can the tray be considered as a plug-flow device.

Many theories have been put forth to handle partial liquid backmixing on plates. Early calculations (113–115) used the so-called tanks-in-series model. Recycle models have been derived by assuming partial or complete backmixing of the liquid, or the vapor, or both (116–118). The suggested procedure (111) is based on the eddy diffusion model when the parameter characterizing the degree of liquid backmixing is the dimensionless group known as the Peclet number:

$$\text{Pe} = \frac{z_L^2}{D_E t_L} \qquad (94)$$

The eddy dispersion coefficient D_E has been measured and correlated empirically as

$$D_E^{1/2} = 0.00378 + 0.0179 u_g + 3.69 L' + 0.18 h_w \qquad (95)$$

The nomenclature is the same as that used in equations 89 and 90.

The relationship between E_{MV}, E_{OG}, and Pe has been calculated according to the recommended formulas (110) and presented in tabular form (4). A plot of this data is shown in Figure 28. Complete backmixing is characterized by Pe = 0 ($D_E = \infty$) where $E_{MV} = E_{OG}$. In larger columns, in which the liquid is not completely mixed horizontally, E_{MV} can be seen from Figure 28 to become larger than the point efficiency and may thus even exceed 100% (119). Entrainment, as

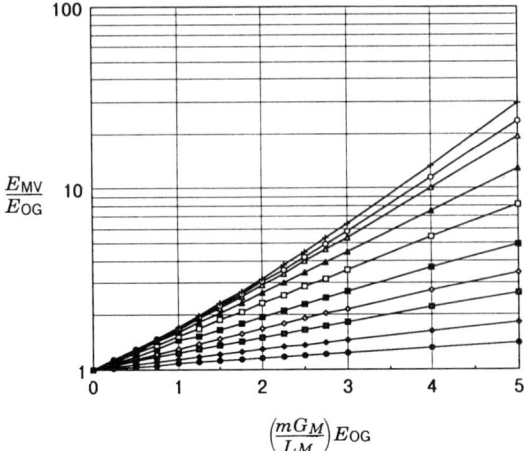

Fig. 28. Relationship between E_{OG} and E_{MV} at different degrees of liquid backmixing (4). Curves represent different Peclet numbers. From top to bottom Pe = ∞, +; Pe = 100, ○; Pe = 50, △; Pe = 20, ▲; Pe = 10, □; Pe = 5, ■; Pe = 3, ◇; Pe = 2, ◪; Pe = 1, ◆; Pe = 0.5, ●.

well as by-passing, are always detrimental to the plate efficiency. Analytical expressions to correct for entrainment and by-passing have been developed (120,121).

Capacity Limitations. The fluid flow capacity of a bubble tray may be limited by any of three principal factors.

1. Flooding, often the most restrictive of the limitations, occurs when the clear liquid height in the downcomer, H_{dc}, exceeds a certain fraction of the tray spacing. During operation, the liquid level in the downcomer builds up as a result of the head necessary to overcome the various resistances to liquid flow, including the friction in the downcomer itself and the hydraulic gradient across the plate. A significant portion in the liquid backup is caused by the need for the liquid to overcome the difference in pressure between the inside and the outside of the downcomer, which in turn is caused by the pressure drop of the vapors through the next higher plate. If the diameter of the column is made too small for a given flow, the vapor pressure drop and thus the liquid backup in the downcomers will increase to the point where some liquid spills onto the next higher tray, and flooding sets in. In principle, the condition may be corrected by increasing the diameter or the tray spacing. A conservative design requires a clear liquid head in the downcomer of no more than half the tray spacing to allow for froth entrapped in the downcomer. The maximum allowable superficial velocity based on the column cross section u_G may be roughly estimated for different tray spacings (6,122,123). A more reliable design would consider each pressure drop contributing to H_{dc} separately.

2. Entrainment occurs when spray or froth formed on one tray enters the gas passages in the tray above. In moderate amounts, entrainment will impair the

countercurrent action and hence drastically decrease the efficiency. If it happens in excessive amounts, the condition is called priming and will eventually flood the downcomers.

3. At high liquid flow, the hydraulic gradient may become so large that the caps near the liquid feed point stops bubbling, and the efficiency suffers.

NOMENCLATURE

Symbol	Definition	Units
A	L_M/mG_M, absorption factor	
a	effective interfacial area per packed volume	m^2/m^3
c_{Ai}	concentration of solute at interface	mol/m^3 or mol/L
C_P	specific heat of gas mixture	$kJ/(mol·K)$
C_q	specific heat of liquid mixture	$kJ/(mol·K)$
D_{ax}	axial dispersion coefficient	m^2/s
D_{AB}	diffusion coefficient of A in B	m^2/s
D_{Am}	effective diffusion coefficient of A in a mixture	m^2/s
D_E	eddy dispersion (or diffusion) coefficient on a tray	m^2/s
D_j	diffusion coefficient of component j	m^2/s
D_L	liquid diffusion coefficient	m^2/s
E_{MV}	Murphree vapor plate efficiency, see equation 82	
E_{OG}	point value of plate efficiency	
F	$u_G \rho_G^{1/2}$, F-factor for bubble tray gas load	$kg^{1/2}/(m^{1/2}·s)$
g_j	moles of component j	mol
G_M	gas flow rate	$kmol/(m^2·s)$
G'_M	flow rate of inert gas	$kmol/(m^2·s)$
G_M^0	generalized molar gas flow rate, see equation 64	
H	Henry's constant	Pa
HETP	height equivalent to a theoretical plate	m
H_G	height of a gas-phase transfer unit, see equation 42	m
H_L	height of a liquid-phase transfer unit, see equation 45	m
H_{OG}	overall gas-phase height of a transfer unit, see equation 48	m
H_{OS}	integral heat of solution for the solute	kJ/mol
H_v	latent heat of pure solvent	kJ/mol
h	total height of packing	m
h_L	clear height of liquid on a tray	m
h_w	weir height	m
h_B	height of bubble layer on a tray	m
J_A	flux of component A due to diffusion	$mol/(m^2·s)$
K	partition coefficient for gas–liquid equilibrium	
K_{OG}	overall gas-phase mass transfer coefficient	$mol/(s·m^2·Pa)$
k_G	gas-phase mass transfer coefficient	$mol/(s·m^2·Pa)$
k_L	liquid-phase mass transfer coefficient	m/s
k_L^r	mass transfer coefficient including effect of chemical reaction	m/s
k^0	mass transfer coefficient for equimolar counterdiffusion	m/s
k'	mass transfer coefficient for unidirectional molecular diffusion through a stagnant gas	m/s

Symbol	Definition	Units
k	mass transfer coefficient for general multicomponent diffusion situation	m/s
L	mass flow rate of liquid	kg/(s·m^2)
L_M	liquid flow rate	kmol/(s·m^2)
L_M^0	generalized liquid flow rate, see equation 64	kmol/(s·m^2)
L_M'	flow rate of solute-free solvent	kmol/(s·m^2)
L'	liquid flow rate on a plate per width of stream	m^3/(s·m)
m	slope of equilibrium line	
$m_{B,2}$	"solubility" of solvent, y_B^*/x_B at the top of the tower	
N_A	flux of solute A through phase interface	mol/(s·m^2)
N_G	number of gas phase transfer units, see equation 43	
N_j	flux of component j through phase interface	mol/(s·m^2)
N_L	number of liquid-phase transfer units, see equation 46	
N_{OG}	number of overall gas-phase transfer units, see equation 49	
N_P	number of actual plates in the column	
N_T	N_{OG} for equimolar counterdiffusion, see equation 57	
N_{TP}	number of theoretical plates	
NTU$_{ap}$	N_{OG} calculated by equation 56, see also equation 73	
NTU	actual value of N_{OG} under the influence of axial dispersion, see equation 74	
P	pressure	Pa
p_A	partial pressure of A	Pa
p_{Ai}	partial pressure of A at the interface	Pa
P_s	vapor pressure of a pure component	Pa
Pe	uh/D_{ax}, dimensionless Peclet number	
Pe$_i$	Peclet number specifically defined for phase i, $i = x$ or y	
Pe$_{HTU}$	uH_{OG}/D_{ax}, dimensionless HTU Peclet number, see equation 75	
Pe$_{HTU,i}$	Peclet number specifically defined for phase i, $i = x$ or y, see equation 77	
R	$L_M:G_M$, flow ratio	
s	surface renewal rate, equation 32	s^{-1}
Sc	$\mu/\rho D$, the dimensionless Schmidt number	
T	temperature	°C or K
t_A	$\Sigma_j N_j/N_A$, parameter indicating the degree of counterdiffusion	
t_L	residence time of liquid on plate, see equation 90	s
u	superficial velocity of gas or liquid phase	m/s
X_f	film factor, see equation 24	
X^0_A	$x_A/(1 - t_A x_A)$, generalized liquid concentration, see equation 64	
X'_A	$x_A/(1 - x_A)$, liquid mole ratio	
x_A	liquid mole fraction of solute	
x_{Ab}	mole fraction in bulk of the liquid phase	
x_{Ai} or x_i	liquid mole fraction of solute at the phase interface	
x_{BM}	logarithmic mean of solvent concentration between the phase interface and the bulk of the liquid, see equation 18	

Symbol	Definition	Units
Y	moles absorbed component per mole of entering vapor, equation 84	
Y'	$y/(1-y)$, mole ratio	
Y_A^0	$Y_A/(1-t_A y_A)$, generalized gas concentration, see equation 64	
Y_f	film factor, see equation 23	
y_A^*	mx_A, concentration in equilibrium with bulk liquid concentration	
y_{Ab}	mole fraction in bulk of gas phase	
y_{Ai} or y_i	gas mole fraction of solute at phase interface	
y_{BM}	logarithmic mean of inert gas concentration between the phase interface and the bulk of the gas phase, see equation 15	
Y_{BM}^*	$[(1-y_A)(1-y_A^*)]/\ln[(1-y_A)/(1-y_{BM}^*)]$, logarithmic mean of inert gas concentration between the equilibrium concentration and the bulk of the gas phase	
Y_f^*	$[(1-t_A y_A)-(1-t_A y_A^*)]/\ln[(1-t_A y_A)/(1-t_A y_A^*)]$, overall gas-phase film factor, see equation 30	
z_L	length of liquid travel on tray	m
Δg_j	total amount of component j absorbed	kmol/(s·m^2)
ΔT_{max}	temperature associated with internal temperature maximum	°C or K
μ	viscosity	mPa·s(=cP)
$\overline{\rho}$	mean density of liquid phase	mol/m^3, mol/L, or g/cm^3
ρ	density of liquid	mol/m^3
σ	surface tension	N/m
ϕ	packing parameter for equation 79	m
ψ	packing parameter for equation 80	m

Subscripts

av	average
A	component A, solute
B	component B, usually nondiffusing
G	gas
L	liquid
i	at gas–liquid interface
b	in the bulk; this symbol is normally omitted
o	liquid feed for plate columns
x	pertaining to liquid phase
y	pertaining to gas phase
M	molar quantity
n	referring to stream leaving the nth tray
1	bottom of a packed column
2	top of a packed column

BIBLIOGRAPHY

"Absorption" in the *Encyclopedia of Chemical Technology*, 1st ed., Vol. 1, pp. 14–32, by E. G. Scheibel, Hoffmann-La Roche, Inc.; in *ECT* 2nd ed., Vol. 1, pp. 44–77, by F. A. Zenz, Squires International, Inc.; in *ECT* 3rd ed., Vol. 1, pp. 53–96, by C. W. Wilke and Urs von Stockar, University of California, Berkeley; *ECT* 4th ed., Vol. 1, pp. 38–93, by U. von Stockar, Ecole Polytechnique Fédérale, and C. R. Wilke, University of California, Berkeley.

1. F. A. Zenz, *Chem. Eng.* **1972,** 120 (Nov. 13, 1972).
2. R. E. Treybal, *Mass Transfer Operations,* 3rd ed., McGraw-Hill Book Co., Inc., New York, 1980.
3. W. S. Norman, *Absorption, Distillation and Cooling Towers,* Longmans, Green & Co., Ltd. (Wiley) New York, 1962.
4. T. K. Sherwood, R. L. Pigford, and C. R. Wilke, *Mass Transfer,* McGraw-Hill Book Co., Inc., New York, 1975.
5. A. Mersmann, H. Hofer, and J. Stichlmair, *Ger. Chem. Eng.* **2,** 249–258 (1979).
6. R. H. Perry and D. Green, eds., *Perry's Chemical Engineer's Handbook,* 6th ed., McGraw-Hill Book Co., Inc., New York, 1984.
7. J. H. Hildebrand and R. L. Scott, *Regular Solutions,* Prentice-Hall, Inc., Englewood Cliffs, N.J., 1962.
8. R. C. Reid, J. M. Prausnitz, and B. E. Poling, *The Properties of Gases and Liquids,* 4th ed., McGraw-Hill Book Co., New York, 1988.
9. A. S. Kertes and co-workers, *Solubility Data Series,* Pergamon Press, Oxford, UK, 1979.
10. W. C. Edmister, *Applied Hydrocarbon Thermodynamics,* Gulf Publishing Co., Houston, Tex., 1961.
11. N. B. Vargaftik, *Tables on the Thermophysical Properties of Liquids and Gases,* John Wiley & Sons, Inc., New York, 1975.
12. J. Gmehling and U. Onken, *Vapor-Liquid Equilibrium Data Collection, Chemistry Data Series,* Dechema, Frankfurt, 1977 ff.
13. W. G. Whitman, *Chem. Metall. Eng.* **29,** 147 (1923).
14. A. F. Ward and L. H. Brooks, *Trans. Faraday Soc.* **48,** 1124 (1952).
15. E. J. Cullen and J. F. Davidson, *Trans. Faraday Soc.* **53,** 113 (1957).
16. W. J. Ward and J. A. Quinn, *AIChE J.* **11,** 1005 (1965).
17. J. L. Duda and J. S. Vrentas, *AIChE J.* **14,** 286 (1968).
18. S. Lynn, J. R. Straatemeier, and H. Kramers, *Chem. Eng. Sci.* **4,** 49, 58, 63 (1955).
19. R. W. Schrage, *A Theoretical Study of Interface Mass Transfer,* Columbia University, New York, 1953.
20. B. Paul, *J. Am. Rocket Soc.* 1321 (Sept. 1962).
21. L. V. Delaney and L. C. Eagleton, *AIChE J.* **8,** 418 (1962).
22. R. Cartier, D. Pindzola, and P. E. Bruins, *Ind. Eng. Chem.* **51,** 1409 (1959).
23. N. A. Clontz, R. T. Johnson, W. L. McCabe, and R. W. Rousseau, *Ind. Eng. Chem. Fundam.* **11,** 368 (1972).
24. J. T. Davies and E. K. Rideal, *Adv. Chem. Eng.* **4,** 1 (1963).
25. C. J. King, *AIChE J.* **10,** 671 (1964).
26. W. K. Lewis and W. G. Whitman, *Ind. Eng. Chem.* **16,** 1215 (1924).
27. R. B. Bird, W. E. Stewart, and E. N. Lightfoot, *Transport Phenomena,* John Wiley & Sons, Inc., New York, 1960.
28. C. R. Wilke, *Chem. Eng. Prog.* **46,** 95 (1950).
29. R. E. Treybal, *Ind. Eng. Chem.* **61,** 36 (1969).
30. R. Krishna and R. Taylor in N. P. Chemerisinoff, ed., *Handbook for Heat and Mass Transfer Operations,* Vol. 2, Gulf Publishing Corporation, Houston, Tex., 1986.

31. J. A. Wesselingh and R. Krishna, *Elements of Mass Transfer,* Technical University Delft, Delft, the Netherlands, 1989.
32. R. Krishna and R. Taylor, *Multicomponent Mass Transfer,* John Wiley & Sons, Inc., New York, 1991.
33. R. Higbie, *Trans. Am Inst. Chem. Eng.* **31,** 365 (1935).
34. P. V. Danckwerts, *Ind. Eng. Chem.* **43,** 1460 (1951).
35. T. H. Chilton and A. P. Colburn, *Ind. Eng. Chem.* **26,** 1183 (1934).
36. G. F. Froment in *Chemical Reaction Engineering* (Advances in Chemistry Series 109), American Chemical Society, Washington, D.C., 1972, p. 19.
37. K. F. Loughlin, M. A. Abul-Hamayel, and L. C. Thomas, *AIChE* **31,** 1614 (1985).
38. P. V. Danckwerts, *Gas-Liquid Reactions,* McGraw-Hill Book Co., Inc., New York, 1970.
39. D. W. van Krevelen and P. J. Hoftijzer, *Rec. Trav. Chim.* **67,** 563 (1948).
40. G. Astarita, *Mass Transfer with Chemical Reaction,* Elsevier, Amsterdam, 1966.
41. A. P. Colburn, *Trans. Am. Inst. Chem. Eng.* **35,** 211 (1939).
42. J. H. Wiegand, *Trans. Am. Inst. Chem. Eng.* **36,** 679 (1940).
43. C. R. Wilke and U. v. Stockar, *Ind. Eng. Chem. Fundam.* **16**(2), 88 (1977).
44. J. D. Raal and M. K. Khurana, *Can. J. Chem. Eng.* **51,** 162 (1973).
45. C. R. Wilke and U. v. Stockar, *Ind. Eng. Chem. Fundam.* **16**(2), 94 (1977).
46. J. Stichlmair, *Chem. Ind. Technol.* **44,** 411 (1972).
47. J. R. Bourne, U. v. Stockar, and G. C. Coggan, *Ind. Eng. Chem. Process Des. Dev.* **13,** 124 (1974).
48. I. A. Furzer, *Ind. Eng. Chem. Fundam.* **23,** 159 (1984).
49. N. Kolev and Kr. Semkov, *Chem. Eng. Prog.* **19,** 175 (1985).
50. N. Kolev and Kr. Semkov, *Vt Verfahrenstechnik* **17,** 474 (1983).
51. W. E. Dunn, T. Vermeulen, C. R. Wilke, and T. T. Word, *Report UCRL 10394 (1962),* Univ. of California Radiation Laboratory as cited in reference 4.
52. W. E. Dunn, T. Vermeulen, C. R. Wilke, and T. T. Word, *Ind. Eng. Chem. Fundam.* **16,** 116 (1977).
53. E. T. Woodburn, *AIChE J.* **20,** 1003 (1974).
54. M. Richter, *Chem. Tech.* **30,** 294 (1978).
55. U. von Stockar and P. F. Cevey, *Ind. Eng. Chem. Process Res. Dev.* **23,** 717 (1984).
56. T. Miyauchi and T. Vermeulen, *Ind. Eng. Chem. Fundam.* **2,** 113 (1963).
57. C. A. Sleicher, *AIChE J.* **5,** 145 (1959).
58. S. Stemerding and F. J. Zuiderweg, *Chem. Engr.* CE 156–CE 159 (May 1963).
59. J. C. Mecklenburgh and S. Hartland, *The Theory of Backmixing,* Wiley-Interscience, New York, 1975, Chapt. 10.
60. J. S. Watson and H. D. Cochran, *Ind. Eng. Chem. Process Res. Dev.* **10,** 83–85 (1971).
61. U. von Stockar and Xiao-Ping Lu, *A Simple and Accurate Short-cut Procedure to Account for Axial Dispersion in Counter-current Separation Columns, Ind. Eng. Chem. Research,* 1991, in press.
62. U. von Stockar, *ACHEMASIA, 1989* (International Meeting on Chemical Engineering & Biotechnology, Beijing, China), Dechema & Ciesc, 1989, p. 43.
63. T. K. Sherwood and F. A. L. Holloway, *Trans. Inst. Am. Chem. Eng.* **36,** 39 (1940).
64. F. F. Rixon, *Trans. Instn. Chem. Engrs.* **26,** 119 (1948).
65. H. A. Koch, L. F. Stutzman, H. A. Blum, and L. E. Hutchings, *Chem. Eng. Prog.* **45,** 677 (1949).
66. E. L. Knoedler and C. F. Bonilla, *Chem. Eng. Prog.* **50,** 125 (1954).
67. J. E. Vivian and C. J. King, *AIChE J.* **120,** 221 (1964).
68. D. Cornell, W. G. Knapp, and J. R. Fair, *Chem. Eng. Prog.* **56**(7), 68 (1960).
69. W. L. Bolles and J. R. Fair, *Chem. Eng.* **89**(July 12), 109 (1982).
70. P. H. Au-Yeung and A. B. Ponter, *Can. J. Chem. Eng.* **61,** 481 (1983).

71. D. M. Mohunta, A. S. Vaidyanathan, and G. S. Laddha, *Ind. Chem. Eng.* **11,** 73 (1969).
72. J. B. Zech and A. B. Mersmann, *I. Chem. E. Symp. Ser.* **56,** 39 (1979).
73. R. Mangers and A. B. Ponter, *Ind. Eng. Chem. Process Des. Dev.* **19,** 530 (1980).
74. Mei Geng Shi and A. B. Mersmann, *Ger. Chem. Eng.* **8,** 87 (1985).
75. R. Billet and M. Schultes, *Paper given at AIChE Annual Meeting,* Washington, D.C. (1988).
76. D. W. van Krevelen and P. J. Hoftijzer, *Rec. Trav. Chim. Pays-Bas* **66,** 49 (1947).
77. K. Onda, H. Takeuchi, and Y. Okumoto, *J. Chem. Eng., Jpn.* **1,** 56 (1968).
78. H. L. Schulman, C. F. Ulrich, A. Z. Proulx, and J. O. Zimmerman, *AIChE J.* **1,** 253 (1955).
79. J. F. Davidson, *Trans. Instn. Chem. Engrs.* **37,** 131 (1959).
80. J. Bridgewater and A. M. Scott, *Trans. Instn. Chem. Engrs.* **52,** 317 (1974).
81. A. B. Ponter and P. H. Au-Yeung, *Can. J. Chem. Eng.* **60,** 94 (1982).
82. R. Echarte, H. Campana, and E. A. Brignole, *Ind. Eng. Chem. Process Des. Dev.* **23,** 349 (1984).
83. E. J. Lynch and C. R. Wilke, *AIChE J.* **1,** 9 (1955).
84. H. L. Shulman, C. F. Ullrich, and N. Wells, *AIChE J.* **1,** 247 (1955).
85. B. W. Gamson, G. Thodos, and O. A. Hougen, *Trans. Am. Inst. Chem. Eng.* **39,** 1 (1943).
86. R. G. Eckert and O. A. Hougen, *Chem. Eng. Prog.* **45,** 188 (1949).
87. C. R. Wilke and O. A. Hougen, *Trans. Am. Inst. Chem. Eng.* **41,** 445 (1945).
88. M. Hobson and G. Thodos, *Chem. Eng. Prog.* **47,** 370 (1951).
89. L. Fellinger, Dissertation, Massachusetts Institute of Technology, 1941; see also reference 113.
90. J. L. Bravo and J. R. Fair, *Ind. Eng. Chem. Process Des. Dev.* **21,** 162 (1982).
91. J. R. Hufton, J. L. Bravo, and J. R. Fair, *Ind. Eng. Chem. Res.* **27,** 2096 (1988).
92. T. K. Sherwood, G. H. Shipley, and F. A. L. Holloway, *Ind. Eng. Chem.* **30,** 765 (1938).
93. W. E. Lobo, L. Friend, F. Hashmall, and F. Zenz, *Trans. Am. Inst. Chem. Eng.* **41,** 693 (1945).
94. F. A. Zenz and R. A. Eckert, *Pet. Refiner.* **40,** 130 (1961).
95. R. A. Eckert, *Chem. Eng. Prog.* **66**(3), 39 (1970).
96. J. R. Fair, D. E. Steinmeyer, W. R. Penney, and B. B. Crocker, in ref. 6, pp. 18–23.
97. L. Spiegel and W. Meier, *Chem. Eng. Symp. Ser.* **104,** A203 (1987).
98. E. V. Murphree, *Ind. Eng. Chem.* **17,** 474 (1925).
99. A. Kremser, *Nat. Pet. News* **22**(21), 42 (1930).
100. G. G. Brown and M. Souders, *Oil Gas J.* **31**(5), 34 (1932).
101. M. Souders and G. G. Brown, *Ind. Eng. Chem.* **24,** 519 (1932).
102. G. Horton and W. B. Franklin, *Ind. Eng. Chem.* **32,** 1384 (1940).
103. W. C. Edmister, *Ind. Eng. Chem.* **35,** 837 (1943).
104. G. G. Brown and M. Souders, *The Science of Petroleum,* Vol. 2, Oxford University Press, New York, 1938, Sect. 25, p. 1557.
105. J. R. Bourne, U. v. Stockar, and G. C. Coggan, *Ind. Eng. Chem. Process Des. Dev.* **13,** 115 (1974).
106. J. Stichlmair and A. Mersmann, *Chem. Ing. Technol.* **43,** 17 (1971).
107. U. von Stockar, *Gasabsorption mit Wärmeeffekten, Diss. Nr 4917,* ETH-Zurich, 1973.
108. W. R. Owens and R. N. Maddox, *Ind. Eng. Chem.* **60**(12), 14 (1968).
109. H. G. Drickamer and J. R. Bradford, *Trans. Am. Inst. Chem. Eng.* **39,** 319 (1943).
110. A. E. O'Connell, *Trans. Am. Inst. Chem. Eng.* **42,** 741 (1946).
111. Research Committee, *Bubble Tray Design Manual,* American Institute of Chemical Engineers, New York, 1958.
112. W. K. Lewis, *Ind. Eng. Chem.* **28,** 399 (1936).
113. T. K. Sherwood and R. L. Pigford, *Absorption and Extraction,* McGraw-Hill Book Co., Inc., New York, 1952.

114. M. Nord, *Trans. Am. Inst. Chem. Eng.* **42,** 863 (1946).
115. M. F. Gautreaux and H. E. O'Connell, *Chem. Eng. Prog.* **51,** 232 (1955).
116. E. D. Oliver and C. C. Watson, *AIChE J.* **2,** 18 (1956).
117. L. A. Warzel, Dissertation, University of Michigan, Ann Arbor, Mich., 1955.
118. V. M. Ramm, *Absorption of Gases,* Israel Program for Scientific Translations, Jerusalem, 1968, Chapt. 3.
119. G. G. Brown, M. Souders, H. V. Nyland, and W. H. Hessler, *Ind. Eng. Chem.* **27,** 383 (1935).
120. A. P. Colburn, *Ind. Eng. Chem.* **28,** 526 (1936).
121. C. P. Strand, *Chem. Eng. Prog.* **59,** 58 (1963).
122. M. Souders and G. G. Brown, *Ind. Eng. Chem.* **26,** 98 (1934).
123. J. R. Fair, *Petro/Chem Eng.* **33**(10), 45 (Sept. 1961).

General References

References 2, 3, 4, and 113 are general references.
A. H. P. Skelland, *Diffusional Mass Transfer,* John Wiley & Sons, Inc., New York, 1974.

URS VON STOCKAR
École Polytechnique Fédérale, Lausanne

CHARLES R. WILKE
University of California, Berkeley

EQUIPMENT

Several types of contacting equipment are used for gas absorption including various combinations of columns and towers. For particulate fuel gases the countercurrent packed tower is the usual choice because this configuration maximizes the average driving force for mass transfer (qv). Plastic packings having extended surface area and high void space with wetted temperatures below 85°C minimize pressure drop and provide high mass transfer and constant liquid film renewal. Insoluble particulates and heavy loads of soluble ones plug counterflow packing rapidly; concurrent flow tends to reduce plugging. Six months of plug-free operation of a parallel-flow bed absorbing SiF_4 in water has been reported (1).

The cross-flow packed scrubber (Fig. 1) (2) is even more plug-resistant and has been used extensively as a pollutant absorber. Typical design parameters for gas absorption only are gas-flow rate, and liquid flow rate, L, of 2.44 and 2.03 kg/(s·m²), respectively. When particulates are present, sprays directed at the bed-retaining grillwork are added upstream. Most of the solids are impacted on the first 150 mm of packing in the gas-flow direction. To remove deposited solids, the liquid rate over the first 300 mm of packing is increased to $L = 13.56$ kg/(s·m²), maintaining a rate of $L = 2.71$ kg/(s·m²) over the remainder of the bed. Solids loading up to 11 g/m³ have been successfully handled. A single transfer unit has been achieved in a gas-flow depth of 200 mm absorbing HF in water. Because scale-up can be a problem, a computer program has been developed for the necessary calculations (3).

Open horizontal spray chambers (1) and vertical spray towers have both

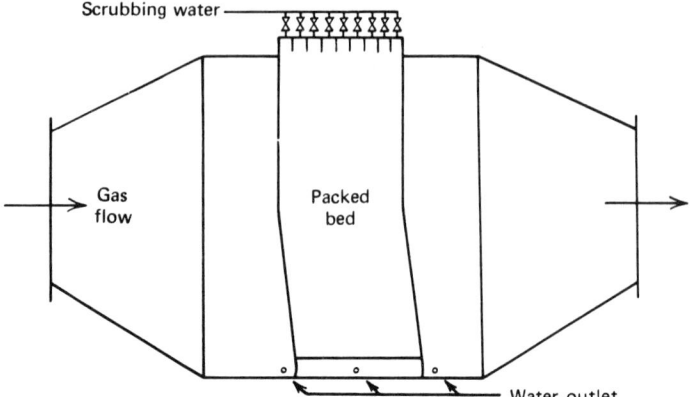

Fig. 1. Cross-flow packed scrubber.

been used when solids are present. Cyclonic spray towers provide slightly better scrubbing when the optimum spray droplet size is used. Figure 2 illustrates various spray chambers. When a large number of transfer units are required, single tower absorption contacting may be unsatisfactory. Loss of countercurrency resulting from spray entrainment limits the number of transfer units achievable in a single tower. Using vertical spray towers (Fig. 2**b**), 5.8 transfer units have been attained (4). Seven transfer units in a commercial cyclonic spray tower have been reported (5) and 3.5 transfer units have been reported in horizontal spray towers. The venturi scrubber is advantageous for particulate collection, especially submicrometer particles, along with gas absorption, but it has been indicated that these scrubbers are limited to 3 transfer units (4).

Water is the most common absorption liquid. It is used for removing highly soluble acidic gases such as HCl [7647-01-0], HF [7664-39-3], and SiF_4 [7783-61-1], especially if the last contact is with water of alkaline pH; NH_3 [7664-41-7] can also be recovered in water if the final contact is acidic. Problems can arise in the initial absorption stages when contacting high concentration gases and volatile neutralizing agents. Vapor phase reactions can produce a submicrometer smoke which is often difficult to wet and collect. These problems can be avoided if initial contact is made at points in the tower where reactant vapor pressures are low. Gases such as SO_2, Cl_2 [7782-50-5], and H_2S [7783-06-4] can also be absorbed more readily in an alkaline solution. Absorption of SO_2 has been practiced in two coal-fired power plants in England using seawater. Tremendous once-through water quantities have been used; thus the discharge water is still unsaturated with SO_2. Scrubbing of SO_2 using alkaline ammonium salt solutions, as in the Cominco (6) process, has also been practiced. Many absorption processes have been commercialized for removing SO_2 from coal-fired power plant flue gas. Organic liquids such as dialkylaniline, the various ethanolamines, and methyldiethanolamine [105-59-9] can also be used for absorption of particulate-free acidic gases. Low volatility oils and solvents such as kerosene [8008-20-6] can be used to absorb organic vapors as long as the scrub liquid volatility is low enough to prevent vapor loss and atmospheric contamination. Control of volatile organic compound (VOC) emissions from small industrial sources by absorption, adsorption (qv), and

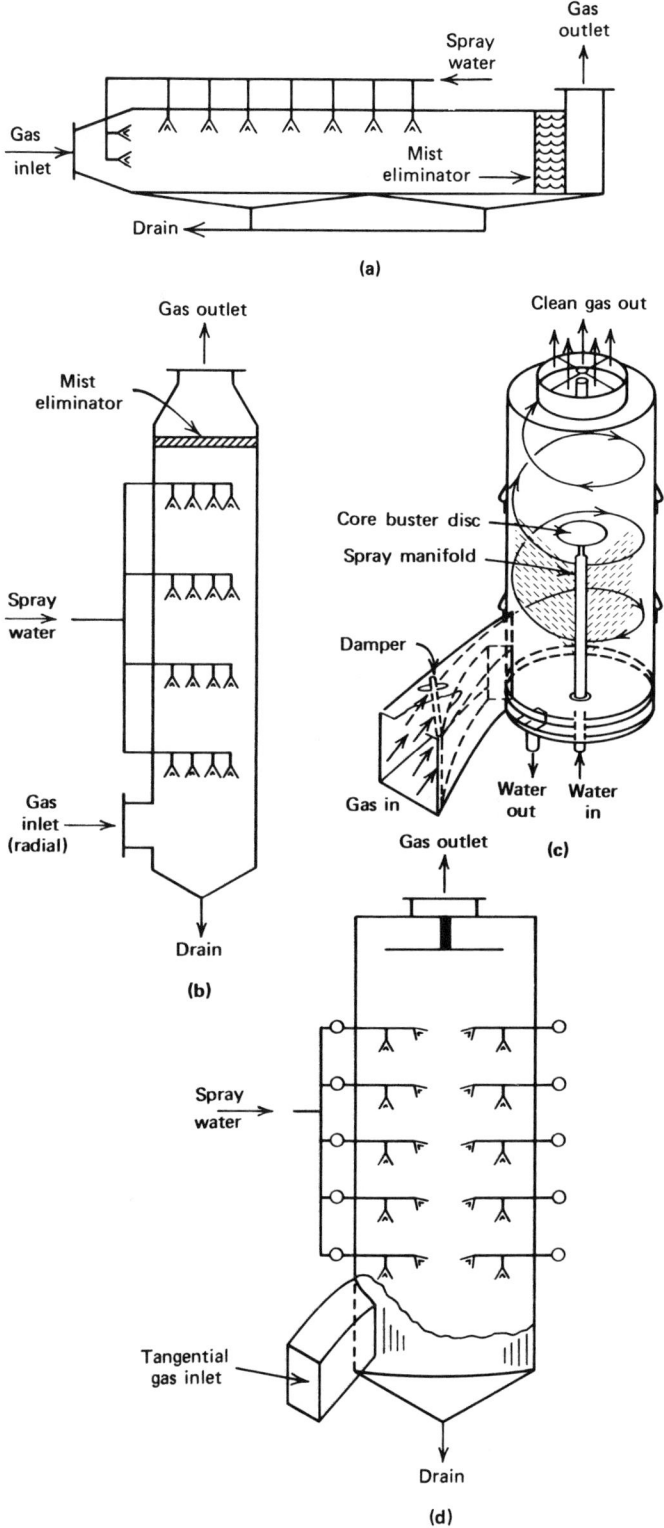

Fig. 2. Types of spray towers: (**a**) horizontal spray chamber; (**b**) simple vertical spray tower; (**c**) cyclonic spray tower, Pease-Anthony type; (**d**) cyclonic spray tower, external sprays.

condensation has been compared (7). Organic compound oxidation and fire are other hazards to be considered.

Disposal of recovered gaseous pollutants can be a problem. Precipitation of the pollutant as an insoluble sludge may be possible through the addition of lime or other reagents. The sludge may be thickened by settling, then dewatered by centrifugation or filtration; however, sludges containing 70% water are not uncommon. Disposal to streams is not feasible and impounding in landfills or tailing ponds is becoming less acceptable. Conversion of the pollutant to a usable form is preferable. Recovered sulfate and sulfite compounds, if ammoniated, may be incorporated into fertilizer or may be used by nearby sulfate–sulfite pulp and paper mills. Recovered halogens can be an even greater problem.

BIBLIOGRAPHY

Material for this article first appeared in "Air Pollution Control Methods" in the *Encyclopedia of Chemical Technology*, 4th ed., Vol. 1, pp. 749–825, by B. B. Crocker, Consultant.

1. H. O. Grant, *Chem. Eng. Prog.* **60,** 53 (Jan. 1964).
2. A. J. Teller, *Chem. Eng. Prog.* **63,** 75 (Mar. 1967).
3. S. V. Cabibbo and A. J. Teller, "The Crossflow Scrubber—A Digital Model for Absorption," *Paper No. 69-186, APCA 62nd Annual Meeting, New York, N.Y., (June 22–26, 1969).*
4. K. E. Lunde, *Ind. Eng. Chem.* **50,** 293 (Mar. 1958).
5. J. P. Jewell and B. B. Crocker, *Proceedings of the 8th Annual Sanitary and Water Resources Engineering Conference,* Vanderbilt University, Nashville, Tenn., 1969, pp. 211–228.
6. W. W. Lehle in W. W. Duecker and J. R. West, eds., *The Manufacture of Sulfuric Acid,* Reinhold Publishing Corp., New York, 1959, pp. 348–352.
7. J. J. Spivey, *Environ. Progress* **7**(1), 31–40 (Feb. 1988).

BURTON B. CROCKER
Consultant

ACTIVATED ALUMINA

The activated aluminas comprise a series of nonequilibrium forms of partially hydroxylated aluminum oxide [*1344-28-1*], Al_2O_3. The chemical composition can be represented by $Al_2O_{(3-x)}(OH)_{2x}$ where x ranges from about 0 to 0.8. They are porous solids made by thermal treatment of aluminum hydroxide [*21645-57-2*] precursors and find application mainly as adsorbents, catalysts, and catalyst supports. Activated alumina, for purposes of this discussion, refers to thermal decomposition products (excluding α-alumina [*12252-63-0*]) of aluminum trihydroxides, oxide hydroxides, and nonstoichiometric gelatinous hydroxides. The term "activation" is used herein to indicate a change in properties resulting from heating (calcining). Other names for these products are active alumina, gamma alumina, catalytic alumina, and transition alumina. Transition alumina is probably the most accurate because the various phases identified by x-ray diffraction are really stages in a continuous transition between the disordered structures immediately following decomposition of the hydrous precursors and the stable α-alumina which is the product of high temperature calcination.

Physical and Chemical Properties

In general, as a hydrous alumina precursor is heated, hydroxyl groups are driven off leaving a porous solid structure of activated alumina. The transformation is topotactic and little change in size or shape of the material is observed at low magnifications. At magnifications higher than about 10,000, changes in texture resulting from recrystallization can be seen. The physical properties of the material are set by the choice of precursor, the forming process, and the activation conditions.

Figure 1 shows the decomposition sequence for several hydrous precursors and indicates approximate temperatures at which the activated forms occur (1). As activation temperature is increased, the crystal structures become more ordered as can be seen by the x-ray diffraction patterns of Figure 2 (2). The similarity of these patterns combined with subtle effects of precursor crystal size, trace impurities, and details of sample preparation have led to some confusion in the literature (3). The crystal structures of the activated aluminas have, however, been well-documented by x-ray diffraction (4) and by nmr techniques (5).

Decomposition of Boehmite. Boehmite [1318-23-6], AlO(OH), can be synthesized having surface areas ranging from about 1 to over 800 m^2/g depending upon the method of preparation (6). The properties of activated boehmite products are strongly influenced by the crystallite size of the precursor material (7). When a crystallized, low surface area boehmite is heated, conversion to gamma alumina occurs at about 725 K yielding a low surface area product having a well-defined x-ray pattern. On further heating, the transition follows the gamma–delta–theta–alpha sequence shown in Figure 1. In contrast, a very high surface area boehmite (also referred to as pseudoboehmite or gelatinous boehmite) decomposes at 575–625 K yielding a high surface area product having a poorly defined gamma alumina x-ray pattern. On further heating, the transformation of this material to delta and theta phases may be retarded or may not occur at all, depending upon trace impurities (8). This sequence is also indicated in Figure 1. Figure 3 shows surface areas of three boehmite samples having different initial surface areas after heating for 16 hours at various temperatures. In the low temperature region, the highest surface area starting material maintains a higher surface area than the others, but in the range where delta–theta phases would be expected, the curves converge. This loss of surface area is a manifestation of coarsening of the activated alumina crystals and is also accompanied by a shift toward larger pore sizes on heating (9). In the high surface area materials, the surface area is mostly attributable to the external surface area of the precursor

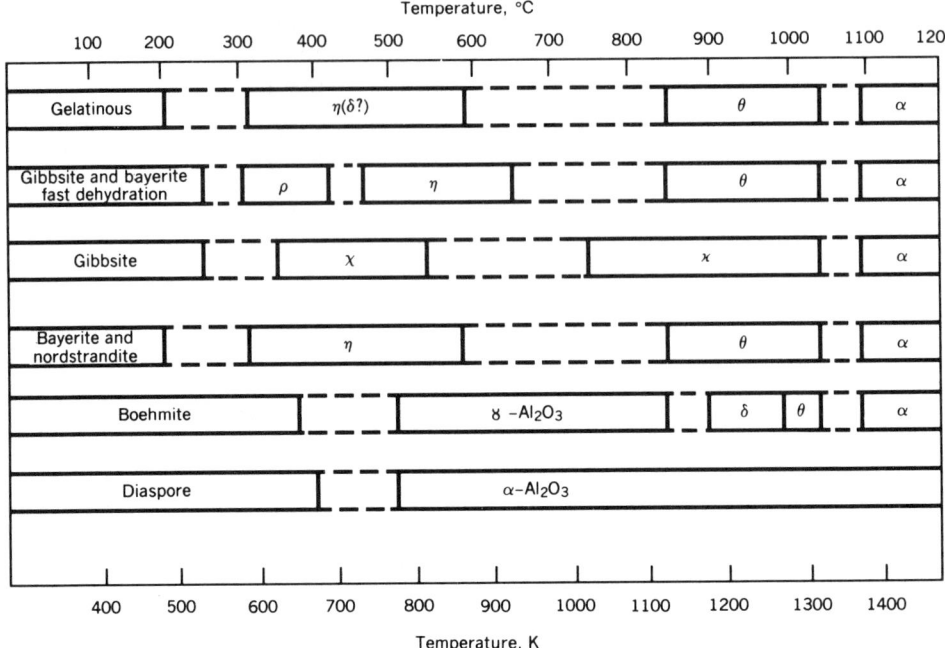

Fig. 1. Decomposition sequence of hydrous aluminas.

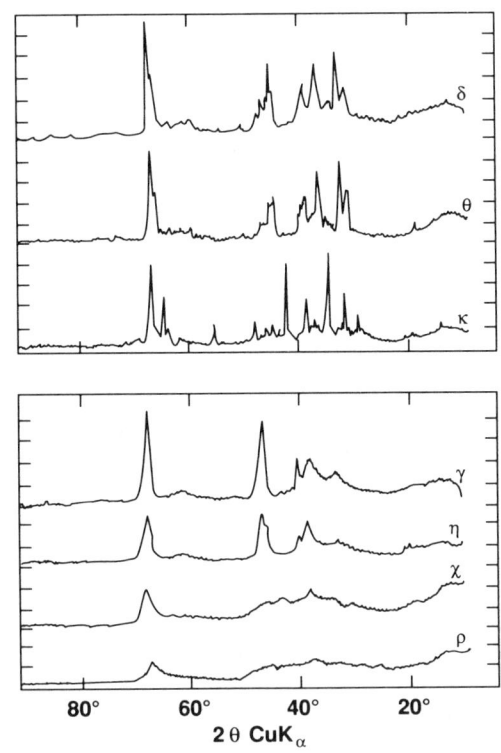

Fig. 2. X-ray diffraction patterns of transition aluminas.

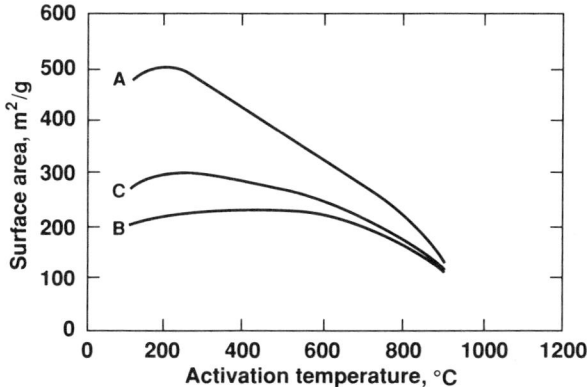

Fig. 3. Surface area of boehmite samples after activation, where A and B are experimental boehmite and C is Catapal SB.

crystallites. Evolution of internal pore structure has been well-documented for activation of well-crystallized boehmite (10), but this is only a minor contribution to surface areas of the commercially important activated boehmites, typically in the 200–350 m^2/g range (11,12). Loss of surface area on heating is inevitable, but can be either retarded or accelerated by various additives (13).

Activation Products of Aluminum Hydroxide. Figure 4 shows the effect of temperature on some properties of activated alumina produced by heating gibbsite [14762-49-3], α-Al(OH)$_3$ (14). As the material is heated, surface area reaches a maximum of 300 m^2/g or more at about 650 K. As temperature is increased further, surface area decreases and the skeletal structure becomes more dense reflecting increased ordering of the crystalline structure during the progression from chi to kappa to alpha. At about 1450 K conversion to alpha alumina occurs with a major rearrangement of crystal structure and corresponding decrease in surface area to about 5 m^2/g. These trends vary somewhat according to precursor crystal size, purity, and the atmosphere of heating.

Unlike the case for boehmite, the surface area of activated gibbsite results mainly from internal porosity rather than external surface of the precursor. There is a minor effect of initial crystal size, however. Some (10% or less) boehmite tends to form when coarse crystals of gibbsite are activated because of hydrothermal conditions which are generated in the particles. On further activation, this boehmite decomposes, but contributes little to the overall surface area. For this reason, activation of fine gibbsite tends to give somewhat higher maximum surface areas than coarser material (15). High humidity in the activating gas can also promote boehmite formation and lower surface area.

Activation of bayerite [20257-20-9] and nordstrandite [13840-05-6] qualitatively follows the pattern shown in Figure 4, but the transition sequence is eta–theta–alpha. The structures of these transition phases are somewhat different from those obtained from gibbsite, reflecting the differences in crystal structure of the hydroxides (16).

If aluminum hydroxide is decomposed by heating at low temperature under vacuum or by rapidly heating at high temperature, a nearly amorphous (x-ray

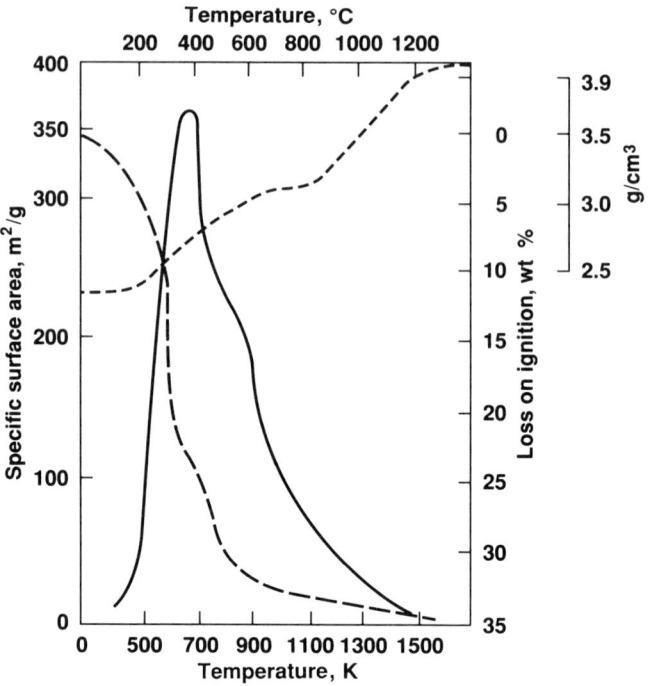

Fig. 4. Properties of activated gibbsite, where (– – –) represents density, (— — —) is the loss on ignition, and (———) is specific surface area.

indifferent) phase known as rho alumina is produced which has the interesting property of recrystallizing (rehydrating) to boehmite or bayerite when mixed with water (17). This behavior is known as rehydration bonding and occurs to a significant degree in hot water at atmospheric pressure (18). Rho alumina can be formed from any of the aluminum hydroxides. The crystal structures are probably somewhat different depending upon which precursor (gibbsite or bayerite) is used, but this cannot be detected by x-ray and rehydration properties are similar. Some recrystallization to boehmite occurs in all of the activated aluminas under severe hydrothermal conditions and generally the degree to which this occurs decreases with increased crystalline order (higher activation temperature). The ease with which rho alumina rehydrates sets it apart from the other activated alumina phases.

Except for rho alumina, the activated aluminas are quite stable when mixed with water in the pH range from about 4 to 10. Below pH 2 and above pH 12 they degrade rapidly.

The surface of activated alumina is a complex mixture of aluminum, oxygen, and hydroxyl ions which combine in specific ways to produce both acid and base sites. These sites are the cause of surface activity and so are important in adsorption, chromatographic, and catalytic applications. Models have been developed to help explain the evolution of these sites on activation (19). Other ions present on the surface can alter the surface chemistry and this approach is commonly used to manipulate properties for various applications.

Manufacturing Processes

The large majority of activated alumina products are derived from activation of aluminum hydroxide, rehydrated alumina, or pseudoboehmite gel. Other commerical methods to produce specialty activated aluminas are roasting of aluminum chloride [7446-70-0], $AlCl_3$, and calcination of precursors such as ammonium alum [7784-25-0], $AlH_7NO_8S_2$. Processing is tailored to optimize one or more of the product properties such as surface area, purity, pore size distribution, particle size, shape, or strength.

Activated Aluminum Hydroxide. The principal precursor for this class of products is gibbsite derived from the Bayer process although a small amount of bayerite is also used for some specialty catalytic applications. Bayer process gibbsite is available in very large tonnages as an intermediate product in aluminum production. It is 99+% pure: the main impurity is sodium oxide [1313-59-3], Na_2O, at 0.2–0.3% on a $Al(OH)_3$ basis. Low cost, relatively high purity, and availability make gibbsite the raw material of choice for many activated alumina products.

Gibbsite is a free-flowing powder having a median particle diameter of about 100 micrometers. Traditionally, activated alumina powders were produced by passing this material through a rotary kiln at an appropriate temperature. However, the aluminum industry has developed fluid bed and flash heating technology for calcination of metallurgical alumina and Bayer alumina-based activated aluminas are often produced in this type of equipment (20,21). The particle sizes of the activated aluminas are essentially the same as that of the precursor powder and in the unground form are amenable to applications involving fluid bed handling. These powders can also be ground to sizes of 10 micrometers or less by standard techniques. Conventional activated aluminas are produced by slow heat treatment whereas "rehydratable" powders are activated within a few seconds.

Larger particle size products can be produced by direct activation of agglomerated gibbsite. The oldest product of this class is made from "scale" which forms on the walls of Bayer process precipitators and is periodically removed in massive chunks, then crushed and activated. This gives a granular product, having particle sizes up to 1 cm or more, which has good mechanical properties and is relatively inexpensive. A similar product is manufactured by high pressure compaction of gibbsite powder. Once the gibbsite agglomerates are formed they are converted to the desired activated alumina product by a combination of crushing, screening, and heat treatment.

Unrefined bauxite [1318-16-7] is also used as a precursor to low cost activated alumina because bauxites can contain as much as 90% gibbsite (dry basis). These materials represent the low cost, low performance end of the activated alumina spectrum of products, but they are still used in significant quantities.

Rehydration Bonded Alumina. Rehydration bonded aluminas are agglomerates of activated alumina, which derive their strength from the rehydration bonding mechanism. Because more processing steps are involved in the manufacture, they are generally more expensive than activated aluminum hydroxides. On the other hand, rehydration bonded aluminas can be produced in a wider range of particle shape, surface area, and pore size distribution.

The generic process used to manufacture these materials begins with an activated powder produced by rapid (flash) activation and grinding of Bayer process gibbsite. This powder is generally composed of a mixture of chi and rho forms, the proportion depending upon the specifics of the activation process. The powder is then mixed with water and formed into a shape by tumbling (22), extrusion (23), or dropping a slurry into an immiscible fluid (oil-drop) (24,25). During or after the forming process, the shapes are heated without drying from several minutes to several hours at temperatures between about 330 and 370 K. This allows partial rehydration to occur, rigidizing the structure. At this point, the alumina is a mixture of pseudoboehmite, bayerite, chi, and amorphous phases (18,22). The shapes are then given a second heat treatment or calcination to establish the desired degree of activation. The surface area, which depends upon the micropores, is largely determined by the second heat treatment whereas the total pore volume is set by specifics of the forming operation. Activated aluminas prepared by this method are mixtures of phases reflecting the composition after rehydration.

Gel-Based Activated Aluminas. Alumina gels can be formed by wet chemical reaction of soluble aluminum compounds. An example is rapid mixing of aluminum sulfate [*17927-65-0*], $Al_2(SO_4)_3 \cdot xH_2O$, and sodium aluminate [*1302-42-7*], $NaAlO_2$, solutions to form pseudoboehmite and a near neutral sodium sulfate [*7757-82-6*] solution (26). After extensive washing to remove the sodium sulfate, the resulting gel can be dried or partially dewatered and formed directly by extrusion or oil drop. If the gel is dried prior to forming, it can be ground to fine particle size, then mixed with water and tumbled or formed by the general processes described above (27). The shapes are then activated to produce relatively pure phase gamma alumina. Gels made from aluminates and aluminum salts must be carefully washed to remove undesirable anions and cations which would be detrimental to the final application.

Hydrolysis of aluminum alkoxides is also used commercially to produce precursor gels. This approach avoids the introduction of undesirable anions or cations so that the need for extensive washing is reduced. Although gels having surface area over 800 m^2/g can be produced by this approach, the commercial products are mostly pseudoboehmite powders in the 200–300 m^2/g range (28). The forming processes already described are used to convert these powders into activated alumina shapes.

"Oil-drop" covers an interesting group of processes to produce small activated alumina spheres or beads by dispersing an aqueous alumina sol or solution in an immiscible liquid. The surface tension effects cause the aqueous droplets to attain a spherical shape and, while in this condition, they gel. Gellation can be accomplished by neutralization with ammonia (26,28), dehydration with alcohol, or through rehydration bonding (24). Beads of very uniform size have been generated by forming droplets of aluminum nitrate [*13473-90-0*], $Al(NO_3)_3$, solution using a specially designed nozzle, allowing them to be partially gelled with ammonia [*7664-41-7*] vapor then falling into an oil–ammonia solution for final gellation (29).

The gel-based products have traditionally been the most expensive and highest performance activated alumina products. They have very good mechanical properties, high surface area, and their purity and gamma-alumina structure make them somewhat resistant to thermal degradation. On the other hand, they are the most difficult to manufacture; and disposal of by-product salts can present an environmental problem.

Economic Aspects

In 1990 U.S. production of activated alumina was about 50,000 t/yr and bulk prices were mostly between $0.60 and $3.00/kg. The least expensive products are those derived directly from Bayer-process gibbsite and powders are generally less expensive than formed products. The soda content (0.2–0.3% Na_2O) of Bayer gibbsite makes it unattractive for many catalytic applications. Gel-based products are normally used where low soda level is required. Soda content of gels prepared from inorganic salts or aluminate solutions is typically about 0.03% whereas soda in alkoxide-based gels is much lower. Specialty activated aluminas having purity as high as 99.99% are also available at a much higher price.

Shaped products used for adsorbent purposes are generally less sophisticated and therefore less expensive than catalytic products. In 1985, it was reported that 10,000 t/yr of activated alumina adsorbents were produced in the United States. North American producers of Bayer process-based activated aluminas include Alcoa, La Roche (formerly Kaiser Chemicals), Discovery, and Alcan. Gel-based activated aluminas are produced by La Roche, Vista, and several of the major catalyst manufacturers. In Europe, principal sources of supply are Rhône-Poulenc and Condea.

Safety and Handling

Activated alumina is a relatively innocuous material from a health and safety standpoint. It is nonflammable and nontoxic. Fine dusts can cause eye irritation and there is some record of lung damage because of inhalation of activated alumina dust mixed with silica [7631-86-9] and iron oxide [1317-61-9] (30). Normal precautions associated with handling of nuisance dusts should be taken. Activated alumina is normally shipped in moisture-proof containers (bags, drums, sling bins) because of its strong desiccating action.

Uses

Catalytic Applications. Activated alumina is used commercially in catalytic processes as a catalyst, catalyst substrate, or as a modifying additive. Activated alumina serves as the catalyst in the Claus process for recovering sulfur from H_2S that originates from natural gas processing or petroleum refinery operations (31). The alumina is generally in the form of spheres, about 5 mm in diameter. This size has evolved as a good compromise between high activity and low pressure drop for fixed bed application (32). Promoting the alumina using a small amount of alkali has been claimed to enhance performance in certain Claus operations (33) (see STACK GAS TREATMENT).

The largest application for activated alumina as a catalyst substrate is in hydrotreating of petroleum feedstocks (qv) (34). The purpose of hydrotreating is threefold: to increase the H/C ratio; to remove O, S, and N impurities; and to remove V, Ni, and other tramp contaminants, especially from residuum of heavier feedstocks. Specialized alumina-based catalysts typically promoted with compounds of Co, Mo, W, and Ni have been developed for these operations (34). The catalysts are usually in the form of extrudates having variously shaped cross sections (35) (circular, lobed, wagon-wheel) about one millimeter in diameter. Spherical catalysts 1 mm or less in diameter have also been described for hydrotreating (24). Much attention has been given to optimizing pore volume and pore size distribution of the activated alumina substrates used in these operations and the "optimum" properties vary with operating conditions and the petroleum feedstock being processed. In the 1990s, worldwide consumption of hydrotreating catalysts was estimated at 30,000–40,000 t/yr, increasing at about 6%/yr (36,37).

Another catalytic application for promoted alumina is in automotive exhaust catalysts which enhance oxidation of hydrocarbons, carbon monoxide, and nitrogen oxide in exhaust gas. There are two general configurations of catalyst used: beads and monoliths. In both systems, the catalytic component is precious metal (platinum, palladium, and rhodium) on alumina (38). The bead system consists of a bed of 3-mm diameter alumina spheres which act as both catalyst substrate and mechanical support. In the monolith system, the mechanical support is provided by a porous ceramic (cordierite [*12182-53-5*]) multichannel "honeycomb" having a thin alumina coating (washcoat) as the catalytic substrate. Automotive exhaust catalyst was the largest volume application (about 18,000 t/yr) for promoted alumina in the mid-1970s and was a significant driving force in improving the technology of low density alumina sphere manufacture. Since that time, however, monoliths have become the system of choice for this application. Beads have maintained only a small share of the market and consequently the volume of activated alumina used in automotive catalysts has dropped significantly (39).

The largest tonnage single application for catalyst particles is in fluid cracking (FCC). 1990 U.S. consumption was projected to be about 180,000 tons (40). These materials are typically made from zeolite having a clay or alumina–silica binder system to provide the necessary mechanical strength for fluid bed handling. Addition of alumina (aluminum hydroxide, pseudoboehmite, or rehydratable alumina) to these formulations has been reported to improve various properties (41,42). Any alumina powder would be activated under FCC processing conditions, thus the activated alumina is used as a modifying additive in the catalytic process. The actual usage of activated alumina in FCC catalysts is unclear because of the highly competitive and proprietary nature of this market.

A number of smaller but nevertheless important applications in which activated alumina is used as the catalyst substrate include: alcohol dehydration, olefin isomerization, hydrogenation, oxidation, and polymerization (43).

Chromatographic Applications. Activated alumina has been used for many years in the separation of various organic compounds by normal phase chromatography (qv) because of its natural hydrophilic surface characteristics. More recently, stable surface coatings have been developed which impart hydrophobic properties to the particle surface (44). These coatings, coupled with improved technology to produce closely sized particles have allowed alumina to compete in reverse-phase chromatographic markets. Compared to silica, the dominant reverse-phase packing material, alumina has better chemical stability at moderately high pH levels, which gives it a natural advantage for separations in this pH range (45).

Membranes. Membranes comprised of activated alumina films less than 20 μm thick have been reported (46). These films are initially deposited via sol–gel technology from pseudoboehmite sols and are subsequently calcined to produce controlled pore sizes in the 2 to 10-nm range. Inorganic membrane systems based on this type of film and supported in solid porous substrates have been introduced commercially. They are said to have better mechanical and thermal stability than organic membranes (47). The activated alumina film comprises only a miniscule part of the total system (see MEMBRANE TECHNOLOGY).

Adsorbent Applications. One of the earliest uses for activated alumina was removal of water vapor from gases and this remains an important application (see GAS SEPARATION). Under equilibrium conditions alumina adsorbs an increasing amount of water as the relative humidity of the contacting gas increases. At 50% rh, for example, a good quality activated alumina can adsorb water at levels of 15 to 20% or more of its own weight (48). By heating the activated alumina to about 525 K under low rh conditions, essentially all of the adsorbed water is removed and the alumina is returned to its original state. This adsorption–regeneration cycle can be repeated hundreds of times with little deterioration of the adsorbent. Industrial scale countercurrent drying systems use this cyclic approach, but generally are designed for lower water loadings because economic cycle times are too short for equilibrium to be established. Another consideration in the large beds of industrial dryers is the amount of heat generated which amounts to about 46 kJ/mol (11 kcal/mol) of water adsorbed. A common strategy is to operate the drying systems under pressure. This partially reduces the water content in the gas before it enters the bed by condensation and also increases the heat capacity of the gas, facilitating heat removal from the bed (49). Under proper operating conditions, moisture in the dried gas can be as low as 11 ppmv. The usual forms of activated alumina used in desiccant applications are granules or spheres about 3 to 12 mm in size. Besides air, a partial list of gases that can be dried includes Ar, He, H_2, CH_4, C_2H_6, C_3H_8, C_2H_2, C_2H_4, C_3H_6, Cl_2, HCl, SO_2, NH_3, and fluorochloroalkanes. Activated alumina is also used to remove water from organic liquids including gasoline, kerosene, oils, aromatic hydrocarbons, cyclohexane, butane and heavier alkanes, butylenes, and many chlorinated hydrocarbons.

Besides drying applications, activated alumina is used to selectively remove various species from gas and liquid systems. In refining and petrochemical operations, activated alumina is used to remove trace HCl from reformer hydrogen, fluorides from hydrocarbons produced by HF alkylation, and a variety of other stream cleanup applications (48). An important example in air pollution abatement is adsorption of fluoride vapors emanating from the aluminum smelting operation. Fluoride-contaminated air from the smelting cell area is countercurrently passed through fluid beds of alumina cell feed which adsorbs the fluorides for recycle. Fluorides are also effectively removed from potable water by activated alumina (50). Fluoride tends to adsorb on alumina at low pH levels and desorb (regenerate) when the pH is increased so in these systems, regeneration of the alumina adsorbent is accomplished by pH control rather than thermal treatment. Arsenic has also been effectively removed from potable water by activated alumina

(51). Activated alumina is receiving renewed attention as an adsorbent and a wealth of information has been published on adsorption characteristics of many chemical species (52,53).

BIBLIOGRAPHY

"Aluminum Oxide (Alumina)" under "Aluminum Compounds" in the *Encyclopedia of Chemical Technology*, 1st ed., Vol. 1, pp. 640–649, by J. D. Edwards, Aluminum Company of America, and A. J. Abbott, Shawinigan Chemicals Limited; in *ECT* 2nd ed., Vol. 2, pp. 41–58, by D. Papée and R. Tertian, Cie de Produits Chimiques et Electrométallurgiques, Péchiney; in *ECT* 3rd ed., Vol. 2, pp. 218–244, by G. MacZura, K. P. Goodboy, and J. J. Koenig, Aluminum Company of America; in *ECT* 4th ed., Vol. 2, pp. 291–302, by A. Pearson, Aluminum Company of America.

1. C. Misra, *Industrial Alumina Chemicals, ACS Monogr. 184,* American Chemical Society, Washington, D.C., 1986, p. 78.
2. K. Wefers and C. Misra, *Oxides and Hydroxides of Aluminum, Alcoa Technical Paper 19,* revised, Alcoa Laboratories, Aluminum Company of America, Pittsburgh, Pa., 1987, p. 52.
3. B. C. Lippens and J. J. Steggerda in B. G. Linsen, ed., *Physical and Chemical Aspects of Absorbents and Catalysts,* Academic Press, New York, 1970, pp. 188–194.
4. Ref. 2, pp. 51–54.
5. C. S. John, N. C. M. Alma, and G. R. Hays, *Appl. Catal.* **6,** 341–346 (1983).
6. M. Astier and K. S. W. Sing, *J. Chem. Tech. Biotechnol.* **30,** 691–698 (1980).
7. D. Aldroft, G. C. Bye, J. G. Robinson, and K. S. W. Sing, *J. Appl. Chem.* **18,** 301–306 (1968).
8. Ref. 3, pp. 189, 190.
9. R. K. Oberlander in B. E. Leach, ed., *Aluminas for Catalysis: Their Preparation and Properties,* Vol. 3, *Applied Industrial Catalysis,* Academic Press, New York, 1984, pp. 98–102.
10. S. J. Wilson and M. H. Stacey, *J. Colloid and Interface Sci.* **82**(2), 507–517 (1981).
11. *Alumina Products and Technology,* Product Data, Kaiser Chemicals Corporation, Baton Rouge, La., 1984. Note: Kaiser is now owned by La Roche Chemicals Corporation.
12. *Calcination of Catapal Aluminas,* Technical Information, Vista Chemical Company, Houston, Tex., 1986.
13. P. Burtin, J. P. Brunelle, M. Pijolat, and M. Soustelle, *Appl. Catal.* **34,** 225–238 (1987).
14. Ref. 2, p. 49.
15. U.S. Pat. 2,876,068 (Mar. 3, 1959), R. Tertian and D. Papée (to Péchiney Co., France).
16. Ref. 3, p. 186.
17. U.S. Pat. 3,226,191 (Dec. 18, 1965), H. E. Osment and R. L. Jones (to Kaiser Aluminum and Chemicals Corporation).
18. K. Yamada, *Nippon Kagaku Kaishi* **9,** 1486–1492 (1981).
19. J. B. Peri, *J. Phys. Chem.* **69**(1), 220–230 (1965).
20. W. M. Fish, *Light Met. 1974* **3,** 673 (1974).
21. U.S. Pat. 2,915,365 (Dec. 1, 1959), F. Saussol (to Pechiney Company, France).
22. U.S. Pat. 3,392,125 (July 9, 1968), A. C. Kelly, H. J. Ducote, and L. R. Barsotti (to Kaiser Aluminum and Chemicals Corporation).
23. U.S. Pat. 3,856,708 (Dec. 14, 1974), V. G. Carithers (to Reynolds Metals Company).

24. U.S. Pat. 4,411,771 (Oct. 25, 1983), W. E. Bambrick and M. S. Goldstein (to American Cyanamid Company).
25. U.S. Pat. 4,579,839 (Apr. 1, 1986), A. Pearson (to Aluminum Company of America).
26. U.S. Pat. 4,390,456 (June 28, 1983), M. G. Sanchez, M. V. Earnest, and N. R. Laine (to W. R. Grace & Company).
27. U.S. Pat. 3,714,313 (Jan. 30, 1973), W. Belding and co-workers (to Kaiser Aluminum and Chemicals Corporation).
28. U.S. Pat. 2,620,314 (Dec. 2, 1952), J. Hoekstra (to Universal Oil Products Company).
29. U.S. Pat. 3,933,679 (Jan. 20, 1976), W. H. Weitzel and L. D. LaGrange (to General Atomics Corporation).
30. N. I. Sax, *Dangerous Properties of Industrial Materials,* 6th ed., Van Nostrand Reinhold, New York, 1984, p. 178.
31. Z. M. George, *Sulfur Removal and Recovery from Industrial Processes,* American Chemical Society, Washington, D.C., 1975, pp. 75–92.
32. *S-100 Activated Alumina for Claus Catalysis,* Case Histories, Alcoa Chemicals Division, Aluminum Company of America, Pittsburgh, Pa., 1985.
33. *SP-100 Promoted Activated Alumina for Claus Catalysis,* Product Data, Alcoa Chemicals Division, Aluminum Company of America, Pittsburgh, Pa., 1984.
34. B. C. Gates, J. R. Katzer, and G. C. A. Schuit, *Chemistry of Catalytic Processes,* McGraw-Hill, New York, 1979, p. 393.
35. U.S. Pat. 3,674,680 (July 4, 1972), G. B. Hoekstra and R. B. Jacobs (to Standard Oil Company).
36. J. C. Downing and K. P. Goodboy, "Claus Catalysts and Alumina Catalyst Materials and Their Application," in L. D. Hart, ed., *Alumina Chemicals Handbook,* American Ceramic Society, Westerville, Ohio, 1990.
37. *Hydrocarbon Process.,* 19, (Apr. 1985).
38. C. J. Pereira, G. Kim, and L. L. Hegedus, *Catal. Rev. Sci. Eng.* **26,** 503–623 (1984).
39. "Catalysis '85", Special Advertising Supplement, *Chem. Week,* 32 (June 26, 1985).
40. Ref. 39, p. 26.
41. U.S. Pat. 4,010,116 (Mar. 1, 1977), R. B. Secor, R. A. Van Nordstrand, and D. R. Pegg (to Filtrol Corporation).
42. U.S. Pat. 4,606,813 (Aug. 19, 1986), J. W. Byrne and B. K. Speronello (to Engelhard Corporation).
43. Ref. 9, pp. 73–76.
44. U. Bien-Vogelsang and co-workers, *Chromatographia* **19,** 170–199 (1984).
45. P. R. Brown and R. A. Hartwick, eds., *High Performance Liquid Chromatography,* John Wiley & Sons, Inc., New York, 1988, pp. 165–168.
46. A. F. M. Leenars, K. Keizer, and A. J. Burggraaf, *J. Mater. Sci.* **19,** 1077–1088 (1984).
47. *Membralox Ceramic Multichannel Membrane Modules,* Technical Brochure, Alcoa/SCT, Aluminum Company of America, Pittsburgh, Pa., 1987.
48. *F-200 Activated Alumina for Adsorption Applications,* Product Data, Alcoa Chemicals Division, Aluminum Company of America, Pittsburgh, Pa., 1985.
49. R. D. Woosley, "Activated Alumina Desiccants," in L. D. Hart, ed., *Alumina Chemicals: Science and Technology Handbook,* American Ceramic Society, Westerville, Ohio, 1990.
50. R. Rubel, Jr. and R. D. Woosley, *J. Am. Water Works Assoc.* **1,** 24–49 (1979).
51. F. Rubel, Jr. and F. S. Williams, *Pilot Study of Fluoride and Arsenic Removal from Potable Water, EPA-600/2-80-100,* U.S. Environmental Protection Agency, Research Triangle Park, N.C., 1980.
52. J. W. Novak, Jr., R. R. Burr, and R. Bednarik, "Mechanisms of Metal Ion Adsorption of Activated Alumina," Vol. 35, *Proc. Int. Symp. on Metals Speciation, Separation, and*

Recovery, Chicago, Ill., July 27–Aug. 1, 1986, Industrial Waste Elimination Research Center of the Illinois Institute of Technology, Chicago, Ill.
53. C. P. Huang and co-workers, "Chemical Interactions Between Heavy Metal Ions and Hydrous Solids," Vol. 1, in Ref. 52.

ALAN PEARSON
Aluminum Company of America

ACTIVATED CARBON

Activated carbon is a predominantly amorphous solid that has an extraordinarily large internal surface area and pore volume. These unique characteristics are responsible for its adsorptive properties, which are exploited in many different liquid- and gas-phase applications. Activated carbon is an exceptionally versatile adsorbent because the size and distribution of the pores within the carbon matrix can be controlled to meet the needs of current and emerging markets (1). Engineering requirements of specific applications are satisfied by producing activated carbons in the form of powders, granules, and shaped products. Through choice of precursor, method of activation, and control of processing conditions, the adsorptive properties of products are tailored for applications as diverse as the purification of potable water and the control of gasoline emissions from motor vehicles.

In 1900, two very significant processes in the development and manufacture of activated carbon products were patented (2). The first commercial products were produced in Europe under these patents: Eponite, from wood in 1909, and Norit, from peat in 1911. Activated carbon was first produced in the United States in 1913 by Westvaco Corp. under the name Filtchar, using a by-product of the papermaking process (3). Further milestones in development were reached as a result of World War I. In response to the need for protective gas masks, a hard, granular activated carbon was produced from coconut shell in 1915. Following the war, large-scale commercial use of activated carbon was extended to refining of beet sugar and corn syrup and to purification of municipal water supplies (4). The termination of the supply of coconut char from the Philippines and India during World War II forced the domestic development of granular activated carbon products from coal in 1940 (5). More recent innovations in the manufacture and use of activated carbon products have been driven by the need to recycle resources and to prevent environmental pollution.

Physical and Chemical Properties

The structure of activated carbon is best described as a twisted network of defective carbon layer planes, cross-linked by aliphatic bridging groups (6). X-ray diffraction patterns of activated carbon reveal that it is nongraphitic, remaining amorphous because the randomly cross-linked network inhibits reordering of the structure even when heated to 3000°C (7). This property of activated carbon contributes to its most unique feature, namely, the highly developed and accessible internal pore structure. The surface area, dimensions, and distribution of the pores depend on the precursor and on the conditions of carbonization and activation. Pore sizes are classified (8) by the International Union of Pure and Applied Chemistry (IUPAC) as micropores (pore width <2 nm), mesopores (pore width 2–50 nm), and macropores (pore width >50 nm) (see ADSORPTION).

The surface area of activated carbon is usually determined by application of the Brunauer-Emmett-Teller (BET) model of physical adsorption (9,10) using nitrogen as the adsorptive (8). Typical commercial products have specific surface areas in the range 500–2000 m^2/g, but values as high as 3500–5000 m^2/g have been reported for some activated carbons (11,12). In general, however, the effective surface area of a microporous activated carbon is far smaller because the adsorption of nitrogen in micropores does not occur according to the process assumed in the BET model, which results in unrealistically high values for surface area (10,13). Adsorption isotherms are usually determined for the appropriate adsorptives to assess the effective surface area of a product in a specific application. Adsorption capacity and rate of adsorption depend on the internal surface area and distribution of pore size and shape but are also influenced by the surface chemistry of the activated carbon (14). The macroporosity of the carbon is important for the transfer of adsorbate molecules to adsorption sites within the particle.

Functional groups are formed during activation by interaction of free radicals on the carbon surface with atoms such as oxygen and nitrogen, both from within the precursor and from the atmosphere (15). The functional groups render the surface of activated carbon chemically reactive and influence its adsorptive properties (6). Activated carbon is generally considered to exhibit a low affinity for water, which is an important property with respect to the adsorption of gases in the presence of moisture (16). However, the functional groups on the carbon surface can interact with water, rendering the carbon surface more hydrophilic (15). Surface oxidation, which is an inherent feature of activated carbon production, results in hydroxyl, carbonyl, and carboxylic groups that impart an amphoteric character to the carbon, so that it can be either acidic or basic. The electrokinetic properties of an activated carbon product are, therefore, important with respect to its use as a catalyst support (17). As well as influencing the adsorption of many molecules, surface oxide groups contribute to the reactivity of activated carbons toward certain solvents in solvent recovery applications (18).

In addition to surface area, pore size distribution, and surface chemistry, other important properties of commercial activated carbon products include pore volume, particle size distribution, apparent or bulk density, particle density, abrasion resistance, hardness, and ash content. The range of these and other

Table 1. Properties of Selected U.S. Activated Carbon Products[a]

		Gas-phase carbons			Liquid-phase carbons		
Manufacturer		Calgon	Norit	Westvaco	Calgon	Norit	Westvaco
Precursor		Coal	Peat	Wood	Coal	Peat	Wood
Product grade		BPL	B4	WV-A 1100	SGL	SA 3	SA-20
Product form		Granular	Extruded	Granular	Granular	Powdered	Powdered
Property	Typical range						
particle size, U.S. mesh[b,c]	<4	12 × 30	3.8 mm	10 × 25	8 × 30	64% <325	65–85% <325
apparent density, g/cm^3	0.2–0.6	>0.48	0.43	0.27	0.52	0.46	0.34–0.37
particle density, g/cm^3	0.4–0.9	0.80		0.50	0.80		
hardness number	50–100	>90	99		>75		
abrasion number							
ash, wt %	1–20	<8	6		<10	6	3–5
BET surface area, N_2, m^2/g	500–2500	1050–1150	1100–1200	1750	900–1000	750	1400–1800
total pore volume, cm^3/g	0.5–2.5	0.8	0.9	1.2	0.85		2.2–2.5
CCl_4 activity, wt %	35–125	>60					
butane working capacity, g/100 cm^3	4–14			>11.0			
iodine number	500–1200	>1050			>900	800	>1000
decolorizing index							
Westvaco	15–25						>20
molasses number							
Calgon	50–250				>200		
Norit	300–1500					440	
heat capacity at 100°C, J/(g·K)[d]	0.84–1.3	1.05			1.05		
thermal conductivity, W/(m·K)	0.05–0.10						

[a]Specific values shown are those cited in manufacturers' product literature (19). Typical ranges shown are based on values reported in the open literature.
[b]Unless otherwise noted.
[c]Approximate mm corresponding to cited meshes are mesh: mm—4: 4.76; 8: 2.38; 10: 2; 12: 1.68; 25: 0.72; 30: 0.59; 325: 0.04.
[d]To convert J to cal, divide by 4.184.

properties is illustrated in Table 1 together with specific values for selected commercial grades of powdered, granular, and shaped activated carbon products used in liquid- or gas-phase applications (19).

Manufacture and Processing

Commercial activated carbon products are produced from organic materials that are rich in carbon, particularly coal, lignite, wood, nut shells, peat, pitches, and cokes. The choice of precursor is largely dependent on its availability, cost, and purity, but the manufacturing process and intended application of the product are also important considerations. Manufacturing processes fall into two categories, thermal activation and chemical activation. The effective porosity of activated carbon produced by thermal activation is the result of gasification of the carbon at relatively high temperatures (20), but the porosity of chemically activated products is generally created by chemical dehydration reactions occurring at significantly lower temperatures (1,21).

Thermal Activation Processes. Thermal activation occurs in two stages: thermal decomposition or carbonization of the precursor and controlled gasification or activation of the crude char. During carbonization, elements such as hydrogen and oxygen are eliminated from the precursor to produce a carbon skeleton possessing a latent pore structure. During gasification, the char is exposed to an oxidizing atmosphere that greatly increases the pore volume and surface area of the product through elimination of volatile pyrolysis products and from carbon burn-off. Carbonization and activation of the char are generally carried out in direct-fired rotary kilns or multiple hearth furnaces, but fluidized-bed reactors have also been used (22). Materials of construction, notably steel and refractories, are designed to withstand the high temperature conditions, ie, >1000°C, inherent in activation processes. The thermal activation process is illustrated in Figure 1 for the production of activated carbon from bituminous coal (23,24).

Bituminous coal is pulverized and passed to a briquette press. Binders may be added at this stage before compression of the coal into briquettes. The briquetted coal is then crushed and passed through a screen, from which the on-size material passes to an oxidizing kiln. Here, the coking properties of the coal particles are destroyed by oxidation at moderate temperatures in air. The oxidized coal is then devolatilized in a second rotary kiln at higher temperatures under steam. To comply with environmental pollution regulations, the kiln off-gases containing dust and volatile matter pass through an incinerator before discharge to the atmosphere.

The devolatilized coal particles are transported to a direct-fired multihearth furnace where they are activated by holding the temperature of the furnace at about 1000°C. Product quality is maintained by controlling coal feed rate and bed temperature. As before, dust particles in the furnace off-gas are combusted in an afterburner before discharge of the gas to the atmosphere. Finally, the granular product is screened to provide the desired particle size. A typical yield of activated carbon is about 30–35% by weight based on the raw coal.

The process for the thermal activation of other carbonaceous materials is

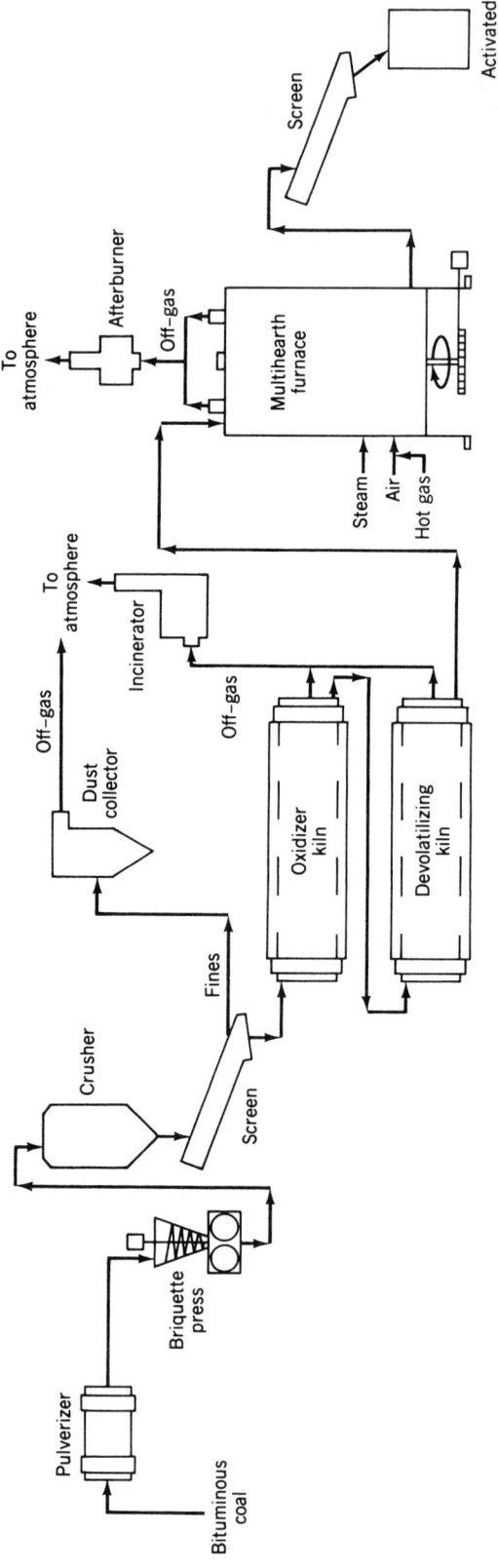

Fig. 1. Thermal activation of bituminous coal.

modified according to the precursor. For example, the production of activated carbon from coconut shell does not require the stages involving briquetting, oxidation, and devolatilization. To obtain a high activity product, however, it is important that the coconut shell is charred slowly prior to activation of the char. In some processes, the precursor or product is acid-washed to obtain a final product with a low ash content (23,25).

Chemical Activation Processes. In contrast to the thermal activation of coal, chemical activation is generally carried out commercially in a single kiln. The precursor, usually wood, is impregnated with a chemical activation agent, typically phosphoric acid, and the blend is heated to a temperature of 450–700°C (26). Chemical activation agents reduce the formation of tar and other by-products, thereby increasing carbon yield. The chemical activation process is illustrated in Figure 2 for the production of granular activated carbon from wood (23,27).

Sawdust is impregnated with concentrated phosphoric acid and fed to a rotary kiln, where it is dried, carbonized, and activated at a moderate temperature. To comply with environmental pollution regulations, the kiln off-gases are treated before discharge to the atmosphere. The char is washed with water to remove the acid from the carbon, and the carbon is separated from the slurry. The filtrate is then passed to an acid recovery unit. Some manufacturing plants do not recycle all the acid but use a part of it to manufacture fertilizer in an allied plant. If necessary, the pH of the activated carbon is adjusted, and the product is dried. The dry product is screened and classified into the size range required for specific granular carbon applications. Carbon yields as high as 50% by weight of the wood precursor have been reported (26).

Novel Manufacturing Processes. Different chemical activation processes have been used to produce carbons with enhanced adsorption characteristics. Activated carbons of exceptionally high surface area (>3000 m^2/g) have been produced by the chemical activation of carbonaceous materials with potassium hydroxide (28,29). Activated carbons are also produced commercially in the form of cloths (30), fibers (31), and foams (32) generally by chemical activation of the precursor with a Lewis acid such as aluminum chloride, ferric chloride, or zinc chloride.

Forms of Activated Carbon Products. To meet the engineering requirements of specific applications, activated carbons are produced and classified as granular, powdered, or shaped products. Granular activated carbons are produced directly from granular precursors, such as sawdust and crushed and sized coconut char or coal. The granular product is screened and sized for specific applications. Powdered activated carbons are obtained by grinding granular products. Shaped activated carbon products are generally produced as cylindrical pellets by extrusion of the precursor with a suitable binder before activation of the precursor.

Shipping and Storage. Activated carbon products are shipped in bags, drums, and boxes in weights ranging from about 10 to 35 kg. Containers can be lined or covered with plastic and should be stored in a protected area both to prevent weather damage and to minimize contact with organic vapors that could reduce the adsorption performance of the product. Bulk quantities of activated carbon products are shipped in metal bins and bulk bags, typically 1–2 m^3 in

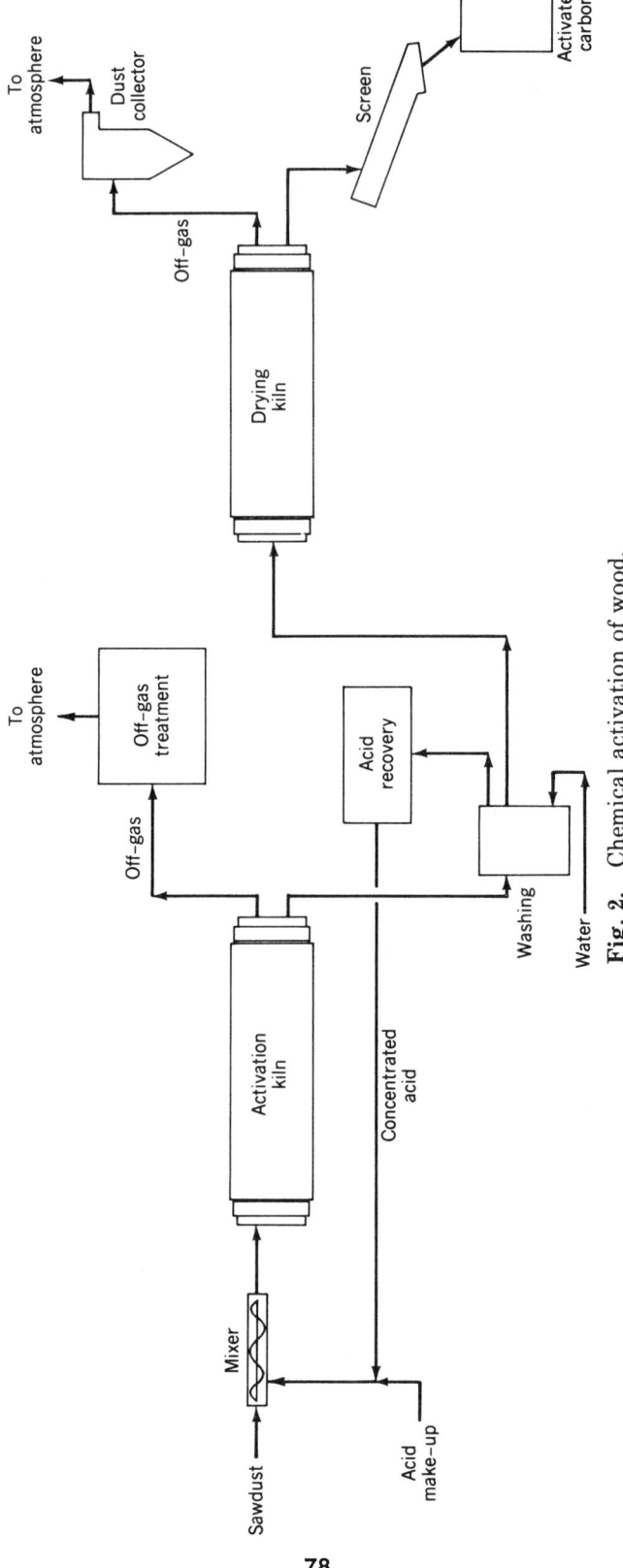

Fig. 2. Chemical activation of wood.

volume, and in railcars and tank trucks. Bulk carbon shipments are generally transferred by pneumatic conveyors and stored in tanks. However, in applications such as water treatment where water adsorption does not impact product performance, bulk carbon may be transferred and stored as a slurry in water.

Specifications. Activated carbon producers furnish product bulletins that list specifications, usually expressed as a maximum or minimum value, and typical properties for each grade produced. Standards helpful in setting purchasing specifications for granular and powdered activated carbon products have been published (33,34).

Economic Aspects

Excluding the former Eastern European bloc countries and China, world production capacity of activated carbon was estimated to be 375,000 metric tons in 1990 (35). The price of most products was 0.70 to 5.50 $/kg, but some specialty carbons were more expensive (36). Forty percent of the production capacity was in the United States, 30% in Western Europe, 20% in Japan, and 10% in other Pacific Rim countries (Table 2).

Table 2. World[a] Production Capacity, Estimated 1990

Country	Capacity, 10^3 t
United States	146
Western Europe	108
Japan	72
Pacific Rim, other	49
Total	375

[a]Excluding Eastern Europe and China.

Production capacity was almost equally split between powdered and nonpowdered activated carbon products. Powdered activated carbon, a less expensive form used in liquid-phase applications, is generally used once and then disposed of. In some cases, however, granular and shaped products are regenerated and reused (35). In 1990 production capacity for granular and shaped products was split with about two-thirds for liquid-phase and one-third for gas-phase applications (37).

Throughout the 1980s production capacity in the United States remained essentially unchanged, but minor fluctuations occurred in response to changes in environmental regulations (38). A similar reaction was noted worldwide (35). The demand for activated carbon has been estimated at 93% of production capacity and near-term growth in demand projected to be approximately 5.5%/yr (39).

In 1970 the U.S. Congress enacted the Clean Air Act, the Clean Water Act, and the Safe Drinking Water Act. Because activated carbon can often be used to help meet Environmental Protection Agency (EPA) regulations, the U.S. activated carbon industry reacted by increasing its production capacity. A proposed amendment to the Safe Drinking Water Act in 1979 required the use of granular activated carbon systems, but the amendment was not enacted. In response to the projected increase in demand for activated carbon, production capacity remained high until the late 1980s, but when the anticipated need did not materialize, some production facilities were shut down. Owing to stricter EPA regulations implementing all three acts in 1990, the industry was expected to increase production capacity by 25% during the early 1990s (35,40).

The estimated production capacity of activated carbon in the United States is shown in Table 3 for seven manufacturers (41). The principal producers are Calgon Carbon (37%), American-Norit (26%), Westvaco (19%), and Atochem (10%). Several other companies purchase activated carbon for resale but do not manufacture products.

Table 3. Production Capacity in the United States, Estimated 1990

Company	Location	Capacity, 10^3 t
Acticarb Division, Royal Oak Enterprises	Romeo, Fla.	6.8
American Norit Co.	Marshall, Tex.	38.6
Barneby and Sutcliffe	Columbus, Ohio	3.0
Calgon Carbon Corp.	Catlettsburg, Ky. and Pittsburgh, Pa.	53.5
Ceca Division, Atochem NA	Pryor, Okla.	15.0
Trans-Pacific Carbon	Blue Lake, Calif.	2.3
Westvaco Corp.	Covington, Va.	27.2
Total		146.4

Western Europe has seven manufacturers of activated carbon. The two largest, Norit and Chemviron (a subsidiary of Calgon), account for 70% of West European production capacity, and Ceca accounts for 13% (42). Japan is the third largest producer of activated carbon, having 18 manufacturers, but four companies share over 50% of the total Japanese capacity (43). Six Pacific Rim countries account for the balance of the world production capacity of activated carbon, 90% of which is in the Philippines and Sri Lanka (42). As is the case with other businesses, regional markets for activated carbon products have become international, leading to consolidation of manufacturers. Calgon, Norit, Ceca, and Sutcliffe-Speakman are examples of multinational companies.

Activated carbon is a recyclable material that can be regenerated. Thus the economics, especially the market growth, of activated carbon, particularly granular and shaped products, is affected by regeneration and industry regeneration capacity. The decision to regenerate an activated carbon product is dependent on the cost, size of the carbon system, type of adsorbate, and the environmental issues involved. Large carbon systems, such as those used in potable and wastewater treatment, generally require a high temperature treatment, which is typically carried out in rotary or multihearth furnaces. During regeneration, carbon losses of 1 to 15% typically occur from the treatment and movement of the carbon (44). However, material loss is compensated for by the addition of new carbon to the adsorber system. In general, regeneration of spent carbon is considerably less expensive than the purchase of new activated carbon. For example, fluidized-bed furnace regeneration of activated carbon used in a 94,600 m^3 per day water treatment system cost only 35% of new material (45). For this system, regeneration using either infrared or multihearth furnaces was estimated to be more expensive but still significantly less so than the cost of new carbon.

Because powdered activated carbon is generally used in relatively small quantities, the spent carbon has often been disposed of in landfills. However, landfill disposal is becoming more restrictive environmentally and more costly. Thus large consumers of powdered carbon find that regeneration is an attractive alternative. Examples of regeneration systems for powdered activated carbon include the Zimpro/Passavant wet air oxidation process (46), the multihearth furnace as used in the DuPont PACT process (47,48), and the Shirco infrared furnace (49,50).

Other types of regenerators designed for specific adsorption systems may use solvents and chemicals to remove susceptible adsorbates (51), steam or heated inert gas to recover volatile organic solvents (52), and biological systems in which organics adsorbed on the activated carbon during water treatment are continuously degraded (53).

Analytical Test Procedures and Standards

Source references for frequently used test procedures for determining properties of activated carbon are shown in Table 4. A primary source is the *Annual Book of American Society for Testing and Materials (ASTM) Standards* (61). Other useful sources of standards and test procedures include manufacturers of activated carbon products, the American Water Works Association (AWWA) (33,34), and the Department of Defense (54).

Health and Safety

Activated carbon generally presents no particular health hazard as defined by NIOSH (62). However, it is a nuisance and mild irritant with respect to inhalation, skin contact, eye exposure, and ingestion. On the other hand, special consideration must be given to the handling of spent carbon that may contain a concentration of toxic compounds.

Activated carbon products used for decolorizing food products in liquid form must meet the requirements of the *Food Chemical Codex* as prepared by the Food & Nutrition Board of the National Research Council (63).

According to the National Board of Fire Underwriters, activated carbons normally used for water treatment pose no dust explosion hazard and are not subject to spontaneous combustion when confined to bags, drums, or storage bins (64). However, activated carbon burns when sufficient heat is applied; the ignition point varies between about 300 and 600°C (65).

Dust-tight electrical systems should be used in areas where activated carbon is present, particularly powdered products (66). When partially wet activated

Table 4. Source References for Activated Carbon Test Procedures and Standards

Title of procedure or standard	Source
Standard Definitions of Terms Relating to Activated Carbon	ASTM D2652
Apparent Density of Activated Carbon	ASTM D2854
Particle Size Distribution of Granular Activated Carbon	ASTM D2862
Total Ash Content of Activated Carbon	ASTM D2866
Moisture in Activated Carbon	ASTM D2867
Ignition Temperature of Granular Activated Carbon	ASTM D3466
Carbon Tetrachloride Activity of Activated Carbon	ASTM D3467
Ball-Pan Hardness of Activated Carbon	ASTM D3802
Radioiodine Testing of Nuclear-Grade Gas-Phase Adsorbents	ASTM D3803
pH of Activated Carbon	ASTM D3838
Determination of Adsorptive Capacity of Carbon by Isotherm Technique	ASTM D3860
Determining Operating Performance of Granular Activated Carbon	ASTM D3922
Impregnated Activated Carbon Used to Remove Gaseous Radio-Iodines from Gas Streams	ASTM D4069
Determination of Iodine Number of Activated Carbon	ASTM D4607
Military Specification, Charcoal, Activated, Impregnated	Ref. 54
Military Specification, Charcoal, Activated, Unimpregnated	Ref. 54
AWWA Standard for Granular Activated Carbon	Ref. 33
AWWA Standard for Powdered Activated Carbon	Ref. 34
BET Surface Area by Nitrogen Adsorption	Refs. 6, 8, 9, 55
Pore Volume by Nitrogen Adsorption or Mercury Penetration	Refs. 10, 56–59
Particle Density	Ref. 60

carbon comes into contact with unprotected metal, galvanic currents can be set up; these result in metal corrosion (67).

Manufacturer material safety data sheets (MSDS) indicate that the oxygen concentration in bulk storage bins or other enclosed vessels can be reduced by wet activated carbon to a level that will not support life. Therefore, self-contained air packs should be used by personnel entering enclosed vessels where activated carbon is present (68).

Liquid-Phase Applications

Activated carbons for use in liquid-phase applications differ from gas-phase carbons primarily in pore size distribution. Liquid-phase carbons have significantly more pore volume in the macropore range, which permits liquids to diffuse more rapidly into the mesopores and micropores (69). The larger pores also promote greater adsorption of large molecules, either impurities or products, in many liquid-phase applications. Specific-grade choice is based on the isotherm (70,71) and, in some cases, bench or pilot scale evaluations of candidate carbons.

Liquid-phase activated carbon can be applied either as a powder, granular, or shaped form. The average size of powdered carbon particles is 15–25 μm (70). Granular or shaped carbon particle size is usually 0.3–3.0 mm. A significant factor in choosing between powdered and nonpowdered carbon is the degree of purifica-

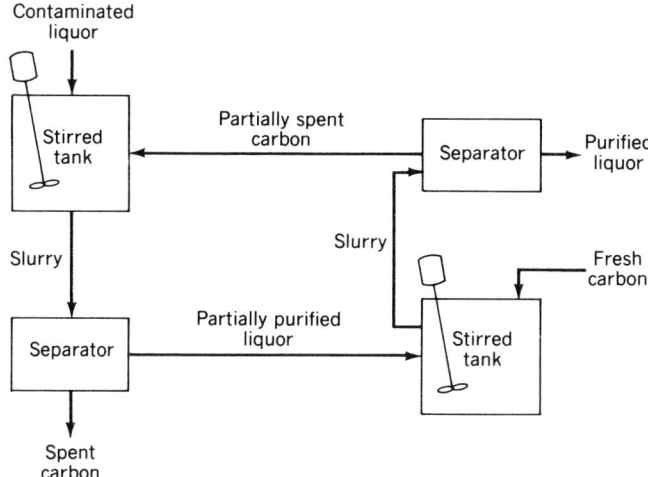

Fig. 3. Multistage countercurrent application of powdered activated carbon.

tion required in the adsorption application. Granular and shaped carbons are usually used in continuous flow through deep beds to remove essentially all contaminants from the liquid being treated. Granular and shaped carbon systems are preferred when a large carbon buffer is needed to withstand significant variations in adsorption conditions, such as in cases where large contaminant spikes may occur. A wider range of impurity removal can be attained by batch application of powdered carbon, and the powdered carbon dose per batch can be controlled to achieve the degree of purification desired (69) (see ADSORPTION, LIQUID SEPARATION).

Batch-stirred vessels are most often used in treating material with powdered activated carbon (72). The type of carbon, contact time, and amount of carbon vary with the desired degree of purification. The efficiency of activated carbon may be improved by applying continuous, countercurrent carbon–liquid flow with multiple stages (Fig. 3). Carbon is separated from the liquid at each stage by settling or filtration. Filter aids such as diatomaceous earth are sometimes used to improve filtration.

Granular and shaped carbons are used generally in continuous systems where the liquid to be treated is passed through a fixed bed (72,73). Compounds are adsorbed by the carbon bed in the adsorption zone (Fig. 4). As carbon in the bed becomes saturated with adsorbates, the adsorption zone moves in the direction of flow, and breakthrough occurs when the leading edge of the adsorption zone reaches the end of the column. Normally at least two columns in series are on line at any given time. When the first column becomes saturated, it is removed from service, and a column containing fresh carbon is added at the discharge end of the series. An alternative approach is the moving bed column (73). In this design the adsorption zone is contained within a single column by passing liquid upward while continuously or intermittently withdrawing spent carbon at the bottom and adding fresh carbon at the top.

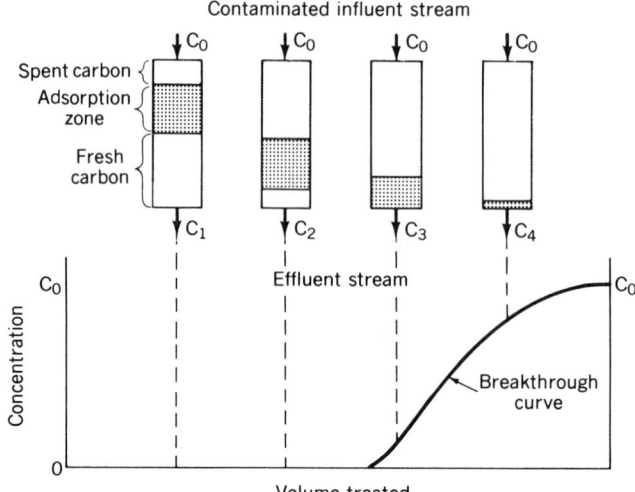

Fig. 4. Adsorption zone and breakthrough curve for fixed bed of granular or shaped activated carbon.

The total activated carbon consumption for liquid-phase applications in the United States in 1987 was estimated to be about 76,700 t, which accounted for nearly 80% of the total activated carbon use. The consumption by application is summarized in Table 5 (74).

Potable Water Treatment. Treatment of drinking water accounts for about 24% of the total activated carbon used in liquid-phase applications (74). Rivers, lakes, and groundwater from wells, the most common drinking water sources, are often contaminated with bacteria, viruses, natural vegetation decay products, halogenated materials, and volatile organic compounds. Normal water disinfection and filtration treatment steps remove or destroy the bulk of these materials (75). However, treatment by activated carbon is an important additional step in many plants to remove toxic and other organic materials (76–78) for safety and palatability.

Groundwater Remediation. Concern over contaminated groundwater sources increased in the 1980s, and in 1984 an Office of Groundwater Protection was created by the EPA (74). This led to an increase in activated carbon consumption in 1987 for groundwater treatment to about 4% of the total liquid-phase usage, and further growth is expected in the 1990s. There are two ways to apply carbon in groundwater cleanup. One is the conventional method of applying powdered, granular, or shaped carbon to adsorb contaminants directly from the water. The other method utilizes air stripping to transfer the volatile compounds from water to air. The compounds are then recovered by passing the contaminated air through a bed of carbon (79,80).

Industrial and Municipal Wastewater Treatment. Wastewater treatment consumes about 21% of the total U.S. liquid-phase activated carbon (74), and governmental regulations are expected to increase demand over the next several years. Wastewater may contain suspended solids, hazardous microorganisms, and toxic organic and inorganic contaminants that must be removed or destroyed

Table 5. Liquid-Phase Activated Carbon Consumption[a], 10^3 t

	Granular-shaped	Powdered	Total
potable water	4.5	13.6	18.1
wastewater			
industrial	6.4	6.6	13.0
municipal	0.9	2.0	2.9
sweetener decolorization	6.8	9.1	15.9
chemical processing and misc.	4.1	2.3	6.4
food, beverage, and oils	0.9	3.9	4.8
pharmaceuticals	2.0	2.3	4.3
mining	1.6	2.5	4.1
groundwater	0.9	2.3	3.2
household uses	1.4	0.9	2.3
dry cleaning	0.7	0.4	1.1
electroplating	0.2	0.4	0.6
Total	*30.4*	*46.3*	*76.7*

[a] In the United States, 1987.

before discharge to the environment. In tertiary treatment systems, powdered, granular, or shaped carbon can be used to remove residual toxic and other organic compounds after the primary filtration and secondary biological treatment (81). Powdered carbon is also used in the PACT process by direct addition of the carbon to the secondary biological treatment step (47) (see WATER, INDUSTRIAL WATER TREATMENT; MUNICIPAL WATER TREATMENT).

Sweetener Decolorization. About 21% of the liquid-phase activated carbon is used for purification of sugar and corn syrup (74). White sucrose sugar is made from raw juice squeezed from sugar cane or sugar beets. The clarified liquor is decolorized using activated carbon, or ion-exchange resins (82). High fructose corn sweeteners (HFCS) are produced by hydrolysis of corn starch and are then treated with activated carbon to remove undesirable taste and odor compounds and to improve storage life. The demands for HFCS rose sharply in the 1980s primarily because of the switch by soft drink producers away from sucrose (83).

Chemical Processing. Activated carbon consumption in a variety of chemical processing applications is about 8% of the total (74). The activated carbon removes impurities to achieve high quality. For example, organic contaminants are removed from solution in the production of alum, soda ash, and potassium hydroxide (82). Other applications include the manufacture of dyestuffs, glycols, amines, organic acids, urea, hydrochloric acid, and phosphoric acid (83).

Food, Beverage, and Cooking Oil. Approximately 6% of the liquid-phase activated carbon is used in food, beverage, and cooking oil production (74). Before being incorporated into edible products, vegetable oils and animal fats are refined to remove particulates, inorganics, and organic contaminants. Activated carbon is one of several agents used in food purification processes. In the production of alcoholic beverages, activated carbon removes haze-causing compounds from beer, taste and odor from vodka, and fusel oil from whiskey (82). The feed water for soft drink production is often treated with carbon to capture undesirable taste and odor compounds and to remove free chlorine remaining from disinfection treatment. Caffeine is removed from coffee beans by extraction with organic solvents, water, or supercritical carbon dioxide prior to roasting. Activated carbon is used to remove the caffeine from the recovered solvents (83).

Pharmaceuticals. Pharmaceuticals account for 6% of the liquid-phase activated carbon consumption (74). Many antibiotics, vitamins, and steroids are isolated from fermentation broths by adsorption onto carbon followed by solvent extraction and distillation (82). Other uses in pharmaceutical production include process water purification and removal of impurities from intravenous solutions prior to packaging (83).

Mining. The mining industry accounts for only 4% of liquid-phase activated carbon use, but this figure may grow as low-grade ores become more common (74). Gold, for example, is recovered on activated carbon as a cyanide complex in the carbon-in-pulp extraction process (82). Activated carbon serves as a catalyst in the detoxification of cyanides contained in wastewater from cyanide stripping operations (73). Problems caused by excess flotation agent concentrations in flotation baths are commonly cured by adding powdered activated carbon (82).

Miscellaneous Uses. Several relatively low volume activated carbon uses comprise the remaining 6% of liquid-phase carbon consumption (74). Small carbon filters are used in households for purification of tap water. Oils, dyes, and other organics are adsorbed on activated carbon in dry cleaning recovery and recycling systems. Electroplating solutions are treated with carbon to remove organics that can produce imperfections when the thin metal layer is deposited on the substrate (82). Medical applications include removal of toxins from the blood of patients with artificial kidneys (83) and oral ingestion into the stomach to recover poisons or toxic materials (82,84). Activated carbon also is used as a support for metal catalysts in low volume production of high value specialty products such as pharmaceuticals, fragrance chemicals, and pesticides (85).

Gas-Phase Applications

Gas-phase applications of activated carbon include separation, gas storage, and catalysis. Although only 20% of activated carbon production is used for gas-phase applications, these products are generally more expensive than liquid-phase carbons and account for about 40% of the total dollar value of shipments. Most of the activated carbon used in gas-phase applications is granular or shaped. Activated carbon use by application is shown in Table 6 (86). Separation processes comprise the main gas-phase applications of activated carbon. These usually exploit the differences in the adsorptive behavior of gases and vapors on activated carbon on the basis of molecular weight and size. For example, organic molecules with a molecular weight greater than about 40 are readily removed from air by activated carbon (see ADSORPTION, GAS SEPARATION).

Solvent Recovery. Most of the activated carbon used in gas-phase applications is employed to prevent the release of volatile organic compounds into the atmosphere. Much of this use has been in response to environmental regulations,

Table 6. Gas-Phase Activated Carbon Uses[a]

Application	Consumption, 10^3 t
solvent recovery	4.5
automotive/gasoline recovery	4.1
industrial off-gas control	3.2
catalysis	2.7
pressure swing separation	1.1
air conditioning	0.5
gas mask	0.5
cigarette filters	0.5
nuclear	0.3
Total	17.4

[a]In the United States, 1987.

but recovery and recycling of solvents from a range of industrial processes such as printing, coating, and extrusion of fibers also provides substantial economic benefits.

The structure of activated carbons used for solvent recovery has been predominantly microporous. Micropores provide the strong adsorption forces needed to capture small vapor molecules such as acetone at low concentrations in process air (87). In recent years, however, more mesoporous carbons, specifically made for solvent recovery, have become available and are giving good service, especially for the adsorption of heavier vapors such as cumene and cyclohexanone that are difficult to remove from micropores during regeneration (87). Regeneration of the carbon is performed on a cyclic basis by purging it with steam or heated nitrogen.

Gasoline Emission Control. A principal application of activated carbon is in the capture of gasoline vapors that escape from vents in automotive fuel systems (88). Under EPA regulations, all U.S. motor vehicles produced since the early 1970s have been equipped with evaporative emission control systems. Most other auto producing countries now have similar controls. Fuel vapors vented when the fuel tank or carburetor are heated are captured in a canister containing 0.5 to 2 L of activated carbon. Regeneration of the carbon is then accomplished by using intake manifold vacuum to draw air through the canister. The air carries desorbed vapor into the engine where it is burned during normal operation. Activated carbon systems have also been proposed for capturing vapors emitted during vehicle refueling, and activated carbon is used at many gasoline terminals to capture vapor displaced when tank trucks are filled (89). Typically, the adsorption vessels contain around 15 m^3 of activated carbon and are regenerated by application of a vacuum. The vapor that is pumped off is recovered in an absorber by contact with liquid gasoline. Similar equipment is used in the transfer of fuel from barges (90). The type of carbon pore structure required for these applications is substantially different from that used in solvent recovery. Because the regeneration conditions are very mild, only the weaker adsorption forces can be overcome, and therefore the most effective pores are in the mesopore size range (91). A large adsorption capacity in these pores is possible because vapor concentrations are high, typically 10–60%.

Adsorption of Radionuclides. Other applications that depend on physical adsorption include the control of krypton and xenon radionuclides from nuclear power plants (92). The gases are not captured entirely, but their passage is delayed long enough to allow radioactive decay of the short-lived species. Highly microporous coconut-based activated carbon is used for this service.

Control by Chemical Reaction. Pick-up of gases to prevent emissions can also depend on the chemical properties of activated carbon or of impregnants. Emergency protection against radioiodine emissions from nuclear power reactors is provided by isotope exchange over activated carbon impregnated with potassium iodide (93). Oxidation reactions catalyzed by the carbon surface are the basis for several emission control strategies. Sulfur dioxide can be removed from industrial off-gases and power plant flue gas because it is oxidized to sulfur trioxide, which reacts with water to form nonvolatile sulfuric acid (94,95). Hydrogen sulfide can be removed from such sources as Claus plant tail gas because it is converted to sulfur in the presence of oxygen (96). Nitric oxide can be removed from flue gas because it is oxidized to nitrogen dioxide. Ammonia is added and reacts catalytically on the carbon surface with the nitrogen dioxide to form nitrogen (97).

Protection Against Atmospheric Contaminants. Activated carbon is widely used to filter breathing air to protect against a variety of toxic or noxious vapors, including war gases, industrial chemicals, solvents, and odorous compounds. Activated carbons for this purpose are highly microporous and thus maximize the adsorption forces that hold adsorbate molecules on the surface. Although activated carbon can give protection against most organic gases, it is especially effective against high molecular weight vapors, including chemical warfare agents such as mustard gas or the nerve agents that are toxic at parts per million concentrations. The activated carbon is employed in individual canisters or pads, as in gas masks, or in large filters in forced air ventilation systems. In air-conditioning systems, adsorption on activated carbon can be used to control the buildup of odors or toxic gases like radon in recirculated air (98).

Inorganic vapors are usually not strongly adsorbed on activated carbon by physical forces, but protection against many toxic agents is achieved by using activated carbon impregnated with specific reactants or decomposition catalysts. For example, a combination of chromium and copper impregnants is used against hydrogen cyanide, cyanogen, and cyanogen chloride, whereas silver assists in the removal of arsine. All of these are potential chemical warfare agents; the Whetlerite carbon, which was developed in the early 1940s and is still used in military protective filters, contains these impregnants (99). Recent work has shown that chromium, which loses effectiveness with age and is itself toxic, can be replaced with a combination of molybdenum and triethylenediamine (100). Oxides of iron and zinc on activated carbon have been used in cigarette filters to absorb hydrogen cyanide and hydrogen sulfide (101). Mercury vapor in air can be removed by activated carbon impregnated with sulfur (102). Activated carbon impregnated with sodium or potassium hydroxide has long been used to control odors of hydrogen sulfide and organic mercaptans in sewage treatment plants (103). Alkali-impregnated carbon is also effective against sulfur dioxide, hydrogen sulfide, and chlorine at low concentrations. Such impregnated carbon is used extensively to protect sensitive electronic equipment against corrosion by these gases in industrial environments (104).

Process Stream Separations. Differences in adsorptivity between gases provides a means for separating components in industrial process gas streams. Activated carbon in fixed beds has been used to separate aromatic compounds from lighter vapors in petroleum refining process streams (105) and to recover gasoline components from natural and manufactured gas (106,107).

Molecular sieve activated carbons are specially made with restricted openings leading to micropores. These adsorbents are finding increasing use in separations utilizing pressure swing adsorption, in which adsorption is enhanced by operation at high pressure and desorption occurs upon depressurization (108). Larger molecules are restricted from entrance into the pores of these carbons and, therefore, are not retained as strongly as smaller molecules. The target product can be either the adsorbed or unadsorbed gases. Examples include separation of oxygen from air and recovery of methane from inorganic gases in biogas production. Hydrogen can be removed from gases produced in the catalytic cracking of gasoline, and carbon monoxide can be separated from fuel gases. Use of pressure swing techniques for gas separation is an area of growing interest in engineering research.

The Hypersorption process developed in the late 1940s used a bed of activated carbon moving countercurrent to gas flow to separate light hydrocarbons from each other and from hydrogen in refinery operations. The application is of interest because of its scale, treating up to 20,000 m^3/h of gas, but the plants were shut down within a few years, probably because of problems related to attrition of the rapidly circulating activated carbon (109). It should be noted, however, that more recently moving-bed and fluid-bed adsorption equipment using activated carbon has been successfully employed for solvent recovery (110).

Gas Storage. Adsorption forces acting on gas molecules held in micropores significantly densify the adsorbed material. As a result, activated carbon has long been considered a medium for lowering the pressure required to store weakly adsorbed compressed gases (111). Work with modern high capacity carbons has been directed toward fueling passenger cars with natural gas, but storage volume targets have not yet been attained (112). Natural gas storage on activated carbon is used commercially in portable welding cylinders (113). These can be refilled easily at about 2000 kPa and holds as much gas as a conventional cylinder pressurized to 6000 kPa (59 atm).

Catalysis. Catalytic properties of the activated carbon surface are useful in both inorganic and organic synthesis. For example, the fumigant sulfuryl fluoride is made by reaction of sulfur dioxide with hydrogen fluoride and fluorine over activated carbon (114). Activated carbon also catalyzes the addition of halogens across a carbon–carbon double bond in the production of a variety of organic halides (85) and is used in the production of phosgene from carbon monoxide and chlorine (115,116).

BIBLIOGRAPHY

"Active Carbon" under "Carbon" in the *Encyclopedia of Chemical Technology*, 1st ed., Vol. 2, pp. 881–889, by J. W. Hassler, Nuchar Active Carbon Division, West Virginia Pulp and Paper Co., and J. W. Goetz, Carbide and Carbon Chemicals Corp.; "Activated Carbon" under "Carbon" in *ECT* 2nd ed., Vol. 4, pp. 149–158, by E. G. Doying, Union Carbide Corp.,

Carbon Products Division; "Activated Carbon" under "Carbon (Carbon and Artificial Graphite)" in *ECT* 3rd ed., Vol. 4, pp. 561–570, by R. W. Soffel, Union Carbide Corp.; in *ECT* 4th ed., Vol. 4, pp. 1015–1037, F. S. Baker and co-workers, Westvaco Corp.

1. H. Jüntgen, *Carbon* **15,** 273–283 (1977).
2. Brit. Pat. 14,224 (1900), R. von Ostrejko; Fr. Pat. 304,867 (1900); Ger. Pat. 136,792 (1901); U.S. Pat. 739,104 (1903).
3. J. W. Hassler, *Forest Products J.* **8,** 25A–27A (1958).
4. J. W. Hassler, *Activated Carbon,* Chemical Publishing Co., Inc., New York, 1963, pp. 1–14. A comprehensive account of the development and use of activated carbon products to about 1960.
5. R. V. Carrubba, J. E. Urbanic, N. J. Wagner, and R. H. Zanitsch, *AIChE Symp. Ser.* **80,** 76–83 (1984).
6. B. McEnaney and T. J. Mays, in H. Marsh, ed., *Introduction to Carbon Science,* Butterworths, London, 1989, pp. 153–196. A good introduction to carbon science in general.
7. H. Marsh and J. Butler, in K. K. Unger, J. Rouquerol, K. S. W. Sing, and H. Kral, eds., *Characterization of Porous Solids, Proceedings of the IUPAC Symposium (COPS I),* Bad Soden a.Ts., FRG, Apr. 26–29, 1987, Elsevier, Amsterdam, The Netherlands, 1988, pp. 139–149.
8. K. S. W. Sing and co-workers, *Pure Appl. Chem.* **57,** 603–619 (1985).
9. S. Brunauer, P. H. Emmett, and E. Teller, *J. Am. Chem. Soc.* **60,** 309–319 (1938).
10. S. J. Gregg and K. S. W. Sing, *Adsorption, Surface Area, and Porosity,* 2nd ed., Academic Press, London, 1982, 303 pp. An indispensable text on the interpretation and significance of adsorption data.
11. H. Marsh, D. Crawford, T. M. O'Grady, and A. Wennerberg, *Carbon* **20,** 419–426 (1982).
12. *Jpn. Chem. Week* **30,** 5 (Mar. 16, 1989).
13. M. M. Dubinin, *J. Colloid Interface Sci.* **46,** 351–356 (1974).
14. K. S. W. Sing, *Carbon* **27,** 5–11 (1989).
15. J. Zawadzki, in P. A. Thrower, ed., *Chemistry and Physics of Carbon,* Vol. 21, Marcel Dekker, Inc., New York, 1989, pp. 147–380. *Chemistry and Physics of Carbon,* published in 23 volumes through 1991, is a primary source of excellent review articles on carbon, many relevant to activated carbon.
16. D. Atkinson, A. I. McLeod, K. S. W. Sing, and A. Capon, *Carbon* **20,** 339–343 (1982).
17. J. M. Solar, C. A. Leon y Leon, K. Osseo-Asare, and L. R. Radovic, *Carbon* **28,** 369–375 (1990).
18. K.-D. Henning, W. Bongartz, and J. Degel, *19th Biennial Conference on Carbon,* Penn State University, Pa., June 25–30, 1990, extended abstracts, pp. 94, 95.
19. Product data bulletins from activated carbon manufacturers, Calgon Carbon Corp., 1990, American Norit Co., 1990, and Westvaco Corp., 1988.
20. T. Wigmans, *Carbon* **27,** 13–22 (1989).
21. F. Derbyshire and M. Thwaites, *Proceedings of the 4th Australian Coal Science Conference,* Brisbane, Australia, Dec. 3–5, 1990, pp. 372–379.
22. U.S. Pat. 3,976,597 (Aug. 24, 1976), A. J. Repik, C. E. Miller, and H. R. Johnson (to Westvaco Corp.).
23. W. Gerhartz, Y. S. Yamamoto, and F. Thomas Campbell, eds., *Ullmann's Encyclopedia of Industrial Chemistry,* 5th ed., Vol. A5, VCH Publishers, New York, 1986, pp. 124–140. Good descriptions of activation processes.
24. Product literature on Pittsburgh activated carbon, Pittsburgh Coke & Chemical Co. (now Calgon Carbon Corp.), Pittsburgh, Pa., ca 1960.
25. U.S. Pat. 4,014,817 (Mar. 29, 1977), B. C. Johnson, R. K. Sinha, and J. E. Urbanic (to Calgon Corp.).

26. A. Cameron and J. D. MacDowall, in J. M. Haynes and P. Rossi-Doria, eds., *Principles and Applications of Pore Structural Characterization, Proceedings of the RILEM/CNR International Symposium,* Milan, Italy, Apr. 26–29, 1983, J. W. Arrowsmith, Ltd., Bristol, UK, 1985, pp. 251–275.
27. R. C. Bansal, J.-B. Donnet, and F. Stoeckli, *Active Carbon,* Marcel Dekker, Inc., New York, 1988, p. 8. A modern treatise on activated carbon based on a comprehensive review of the literature.
28. U.S. Pat. 4,082,694 (Apr. 4, 1978), A. N. Wennerberg and T. M. O'Grady (to Standard Oil Co.).
29. T. Kasuh, D. A. Scott, and M. Mori, *Proceedings of an International Conference on Carbon,* The University of Newcastle upon Tyne, UK, Sept. 18–23, 1988, pp. 146–148.
30. Product literature on activated carbon cloth, Charcoal Cloth Ltd., UK, 1985, and on C-tex products, Siebe Gorman & Co., Ltd., UK, 1985.
31. Product literature on KYNOL activated carbon fibers and cloths, GUN EI Chemical Industry Co., Ltd., Japan, 1987; Product literature on AD'ALL activated carbon fibers, Unitika, Ltd., Japan, 1989.
32. Product literature on KURASHEET activated carbon foam sheets, Kuraray Chemical Co., Ltd., Japan, 1987.
33. *AWWA Standard for Granular Activated Carbon,* ANSI/AWWA B604, American Water Works Association, Denver, Colo., 1991, 32 pp.
34. *AWWA Standard for Powdered Activated Carbon,* ANSI/AWWA B600, American Water Works Association, Denver, Colo., 1990, 32 pp.
35. *The Economics of Activated Carbon,* 3rd ed., Roskill Information Services Ltd., London, 1990, pp. 8, 9.
36. J. Goin, V. von Schuller-Goetzburg, and Y. Sakuma, *Chemical Economics Handbook—SRI International,* Menlo Park, Calif., 1989, pp. 731.2001P, 731.2001Q.
37. Wertheim Schroder, *Calgon Carbon Corp.,* Company Report No. 955346, Mar. 19, 1990.
38. Ref. 36, pp. 731.2000S–731.2000Y.
39. Ref. 37, pp. 4, 5.
40. S. Irving-Monshaw, *Chem. Eng.* **97**(2), 43–46 (1990).
41. Ref. 35, pp. 54–65.
42. Ref. 35, p. 13.
43. Ref. 35, pp. 25–32.
44. W. G. P. Schuliger, *Waterworld News* **4**(1), 15–17 (1988).
45. R. M. Clark and B. W. Lykins, Jr., *Granular Activated Carbon—Design, Operation, and Cost,* Lewis Publishers, Inc., Chelsea, Mich., 1989, pp. 295–338.
46. P. N. Cheremisinoff and F. Ellerbusch, *Carbon Adsorption Handbook,* Ann Arbor Science Publishers, Inc., Ann Arbor, Mich., 1978, pp. 539–626. An excellent reference book on activated carbon, ranging from theoretical to applied aspects.
47. Ref. 46, pp. 389–447.
48. Product literature on PACT systems, Zimpro/Passavant, Inc., Rothschild, Wis., 1990.
49. Ref. 45, p. 51.
50. W. E. Koffskey and B. W. Lykins, Jr., *J. Am. Water Works Assoc.* **82**(1), 48–56 (1990).
51. A. Yehaskel, *Activated Carbon—Manufacture and Regeneration,* Noyes Data Corporation, Park Ridge, N.J., 1978, pp. 202–217. A dated, but still useful summary of key patent literature.
52. P. N. Cheremisinoff, *Pollut. Eng.* **17**(3), 29–38 (1985).
53. R. G. Rice and C. M. Robson, *Biological Activated Carbon—Enhanced Aerobic Biological Activity in GAC Systems,* Ann Arbor Science Publishers, Ann Arbor, Mich., 1982, 611 pp.
54. *Department of Defense Military Specifications,* MIL-C-0013724D(EA), Sept. 22, 1983; MIL-C-0013724D(EA) Amendment 1, Mar. 5, 1986; and MIL-C-17605C(SH), Mar. 22, 1989.

55. S. J. Gregg and K. S. W. Sing, *Adsorption, Surface Area, and Porosity,* 1st ed., Academic Press Inc. (London) Ltd., London, 1967, pp. 308–355.
56. H. M. Rootare, *Advanced Experimental Techniques in Powder Metallurgy,* Plenum Press, New York, 1970, pp. 225–252. A comprehensive review of the use of mercury penetration to measure porosity.
57. G. Horvath and K. Kawazoe, *J. Chem. Eng. Jpn.* **16,** 470–475 (1983).
58. D. Dollimore and G. R. Heal, *J. Appl. Chem.* **14,** 109–114 (1964).
59. M. M. Dubinin and H. F. Stoeckli, *J. Colloid Interface Sci.* **56,** 34–42 (1980).
60. C. Orr, Jr., *Powder Technol.* **3,** 117–123 (1970).
61. *Annual Book of ASTM Standards,* 15.01, Section 15, American Society for Testing and Materials, Philadelphia, Pa., 1989.
62. *1985–1986 Registry of Toxic Effects of Chemical Substances,* Vol. 2, National Institute for Occupational Safety and Health, U.S. Department of Health and Human Services, Washington, D.C., 1987, p. 1475.
63. National Research Council, Assembly of Life Sciences, Division of Biological Sciences, Food and Nutrition Board, and Committee on Codex Specifications, *Food Chemicals Codex,* 3rd ed., National Academy Press, Washington, D.C., 1981, pp. 70, 71.
64. American Society of Civil Engineers, American Water Works Association, and Conference of State Sanitary Engineers, *Water Treatment Plant Design,* American Water Works Association, Inc., New York, 1969, p. 297.
65. J. W. Hassler, *Purification with Activated Carbon,* 3rd ed., Chemical Publishing Co., Inc., New York, 1974, p. 353. Contains much of the information given in reference 4 but with more emphasis on the commercial uses of activated carbon.
66. Ref. 65, pp. 84, 85.
67. U.S. Environmental Protection Agency, *Process Design Manual for Carbon Adsorption,* Swindell-Dressler Co., Pittsburgh, Pa., 1971, pp. 3–68.
68. Material safety data sheets on activated carbon products, available from the manufacturers, 1991.
69. R. A. Hutchins, *Chem. Eng.* **87**(2), 101–110 (1980). A particularly useful paper on liquid-phase adsorption.
70. M. Suzuki, *Adsorption Engineering,* Kodansha Ltd., Tokyo and Elsevier Science Publishers B.V., Amsterdam, The Netherlands, 1990, pp. 11, 35–62.
71. T. F. Speth and R. J. Miltner, *J. Am. Water Works Assoc.* **82**(2), 72–75 (1990).
72. F. L. Slejko, ed., *Adsorption Technology,* Marcel Dekker, Inc., New York, 1985, pp. 23–32. A good account of the theory, design, and application of adsorption systems.
73. Ref. 46, pp. 8–19.
74. Ref. 36, pp. 731.2000V–731.2001L.
75. American Water Works Association, *Water Quality and Treatment,* 3rd ed., McGraw-Hill Book Co., New York, 1971, pp. 1–216.
76. W. J. Weber, Jr. and B. M. Van Vliet, in I. H. Suffet and M. J. McGuire, eds., *Activated Carbon Adsorption of Organics from the Aqueous Phase,* Vol. 1, Ann Arbor Science Publishers, Inc., Ann Arbor, Mich., 1980, pp. 15–41. A comprehensive, two volume treatise with many key references.
77. J. L. Oxenford and B. W. Lykins, Jr., *J. Am. Water Works Assoc.* **83**(1), 58–64 (1991).
78. I. N. Najm and co-workers, *J. Am. Water Works Assoc.* **83**(1), 65–76 (1991).
79. L. W. Canter and R. C. Knox, *Ground Water Pollution Control,* Lewis Publishers, Inc., Chelsea, Mich., 1985, pp. 89–125.
80. Environmental Science and Engineering, Inc., *Removal of Volatile Organic Chemicals from Potable Water—Technologies and Costs,* Noyes Data Corp., Park Ridge, N.J., 1986, pp. 23–40.
81. G. Culp, G. Wesner, R. Williams, and M. V. Hughes, *Wastewater Reuse and Recycling Technology,* Noyes Data Corp., Park Ridge, N.J., 1980, pp. 343–432. A useful review of wastewater treatment with activated carbon.
82. Ref. 65, pp. 87–125, 274–292.

83. Ref. 35, pp. 92–135.
84. M. Smisek and S. Cerny, *Active Carbon—Manufacture, Properties, and Applications,* Elsevier Publishing Co., New York, 1970, pp. 290–294.
85. A. J. Bird, in A. B. Stiles, ed., *Catalyst Supports and Supported Catalysts,* Butterworths, Stoneham, Mass., 1987, pp. 107–137.
86. Ref. 36, pp. 731.2000W, 731.2001M–731.2001P.
87. P. J. Luft and P. C. Speers, Paper 52c, *AIChE Summer National Meeting,* Aug. 19–22, 1990.
88. P. J. Clarke and co-workers, *SAE Trans.* **76,** 824–837 (1967).
89. Product literature on hydrocarbon vapor recovery systems, John Zink Co., Tulsa, Okla., 1990.
90. J. Hill, *Chem. Eng.* **97,** 133–143 (1990).
91. H. R. Johnson and R. S. Williams, *S.A.E. Technical Paper No. 902119,* International Fuels and Lubricants Exposition, Tulsa, Okla., Oct. 23, 1990.
92. D. W. Moeller and D. W. Underhill, *Nucl. Saf.* **22,** 599–611 (1981).
93. M. L. Hyder, *Comm. Eur. Communities [Rep.] EUR 1986, EUR 10580, Gaseous Effluent Treat. Nucl. Install.,* 451–462 (1986).
94. F. J. Ball, S. L. Torrence, and A. J. Repik, *APCA J.* **22,** 20–26 (1972).
95. P. Ellwood, *Chem. Eng.* **76,** 62–64 (1969).
96. J. Klein and K.-D. Henning, *Fuel* **63,** 1064–1067 (1984).
97. E. Richter, *Catal. Today* **7,** 93–112 (1990).
98. M. A. Brisk and A. Turk, *Proc. APCA Ann. Meet.,* 77th **2,** 84–93 (1984).
99. U.S. Pat. 2,920,050 (Jan. 5, 1960), R. J. Grabenstetter and F. E. Blacet (to U.S. Dept. of Army).
100. U.S. Pat. 4,801,311 (Jan. 31, 1989), E. D. Tolles (to Westvaco Corp.).
101. U.S. Pat. 3,460,543 (Aug. 12, 1969), C. H. Kieth, V. Norman, and W. W. Bates, Jr. (to Ligget & Meyers Corp.).
102. R. K. Sinah and P. L. Walker, *Carbon* **10,** 754–756 (1972).
103. W. D. Lovett and R. L. Poltorak, *Water and Sewage Works* **121,** 74–75 (1974).
104. G. N. Brown, M. A. Lunn, C. E. Miller, and C. D. Shelor, *Tappi J.* **66,** 33–36 (1983).
105. S. Dunlop and R. Banks, *Hydrocarbon Process.* **56,** 147–152 (1977).
106. G. F. Russell, *Petrol. Refiner* **40,** 103–106 (1961).
107. T. Scott, *Gas. J.* **303,** 300–307 (1960).
108. E. Richter, *Erdol Kohle, Erdgas, Petrochem.* **40,** 432–438 (1987).
109. C. Berg, *Chem. Eng. Prog.* **47,** 585–590 (1951).
110. *Gastak Solvent Recovery System,* product literature, Kureha Chemical Industry Co., Ltd., New York, 1990.
111. H. Briggs and W. Cooper, *Proc. Roy. Soc. Edinburgh* **41,** 119–127 (1920–1921).
112. J. Braslaw, J. Nasea, and A. Golovoy, *Alternative Energy Sources: Proceedings of the Miami Int. Conf. on Alternative Energy Sources,* 4th ed., Ann Arbor Science Publishers, Ann Arbor, Mich., pp. 261–270, 1980.
113. U.S. Pat. 4,817,684 (Apr. 4, 1989), J. W. Turko and K. S. Czerwinski (to Michigan Consolidated Gas Co.).
114. U.S. Pat. 4,102,987 (July 25, 1978), D. M. Cook and D. C. Gustafson (to The Dow Chemical Company).
115. H. Jüntgen, *Fuel* **65,** 1436–1446 (1986).
116. H. Jüntgen, *Erdol Kohle, Erdgas, Petrochem.* **39**(12), 546–551 (1986).

FREDERICK S. BAKER
CHARLES E. MILLER
ALBERT J. REPIK
E. DONALD TOLLES
Westvaco Corporation

ADSORPTION

Adsorption is the term used to describe the tendency of molecules from an ambient fluid phase to adhere to the surface of a solid. This is a fundamental property of matter, having its origin in the attractive forces between molecules. The force field creates a region of low potential energy near the solid surface and, as a result, the molecular density close to the surface is generally greater than in the bulk gas. Furthermore, and perhaps more importantly, in a multicomponent system the composition of this surface layer generally differs from that of the bulk gas since the surface adsorbs the various components with different affinities. Adsorption may also occur from the liquid phase and is accompanied by a similar change in composition, although, in this case, there is generally little difference in molecular density between the adsorbed and fluid phases.

The enhanced concentration at the surface accounts, in part, for the catalytic activity shown by many solid surfaces, and it is also the basis of the application of adsorbents for low pressure storage of permanent gases such as methane. However, most of the important applications of adsorption depend on the selectivity, ie, the difference in the affinity of the surface for different components. As a result of this selectivity, adsorption offers, at least in principle, a relatively straightforward means of purification (removal of an undesirable trace component from a fluid mixture) and a potentially useful means of bulk separation.

Fundamental Principles

Forces of Adsorption. Adsorption may be classified as chemisorption or physical adsorption, depending on the nature of the surface forces. In physical adsorption the forces are relatively weak, involving mainly van der Waals (induced dipole–induced dipole) interactions, supplemented in many cases by electrostatic contributions from field gradient–dipole or –quadrupole interactions. By contrast, in chemisorption there is significant electron transfer, equivalent to the formation of a chemical bond between the sorbate and the solid surface. Such interactions are both stronger and more specific than the forces of physical adsorption and are obviously limited to monolayer coverage. The differences in the general features of physical and chemisorption systems (Table 1) can be understood on the basis of this difference in the nature of the surface forces.

Heterogeneous catalysis generally involves chemisorption of the reactants, but most applications of adsorption in separation and purification processes depend on physical adsorption. Chemisorption is sometimes used in trace impurity removal since very high selectivities can be achieved. However, in most situations the low capacity imposed by the monolayer limit and the difficulty of regenerating the spent adsorbent more than outweigh this advantage. The higher capacities achievable in physical adsorption result from multilayer formation and this is obviously critical in such applications as gas storage, but it is also an important consideration in most adsorption separation processes since the process cost is directly related to the adsorbent capacity.

Table 1. Parameters of Physical Adsorption and Chemisorption

Parameter	Physical adsorption	Chemisorption
heat of adsorption (ΔH)	low, > 1–5 times latent heat of evaporation	high, > 1–5 times latent heat of evaporation
specificity	nonspecific	highly specific
nature of adsorbed phase	monolayer or multilayer, no dissociation of adsorbed species	monolayer only may involve dissociation
temperature range	only significant at relatively low temperatures	possible over a wide range of temperature
forces of adsorption	no electron transfer, although polarization of sorbate may occur	electron transfer leading to bond formation between sorbate and surface
reversibility	rapid, nonactivated, reversible	activated, may be slow and irreversible

In very small pores the molecules never escape from the force field of the pore wall even at the center of the pore. In this situation the concepts of monolayer and multilayer sorption become blurred and it is more useful to consider adsorption simply as pore filling. The molecular volume in the adsorbed phase is similar to that of the saturated liquid sorbate, so a rough estimate of the saturation capacity can be obtained simply from the quotient of the specific micropore volume and the molar volume of the saturated liquid.

Selectivity. Selectivity in a physical adsorption system may depend on differences in either equilibrium or kinetics, but the great majority of adsorption separation processes depend on equilibrium-based selectivity. Significant kinetic selectivity is in general restricted to molecular sieve adsorbents, eg, carbon molecular sieves, zeolites, or zeolite analogues. In these materials the pore size is of molecular dimensions, so that diffusion is sterically restricted. In this regime small differences in the size or shape of the diffusing molecule can lead to very large differences in diffusivity. In the extreme limit one species (or one class of compounds) may be completely excluded from the micropores, thus giving a highly selective molecular sieve separation. The most important example of such a process is the separation of linear hydrocarbons from their branched and cyclic isomers using a 5A zeolite adsorbent. A second example, where the difference in diffusivities is less extreme but still large enough to produce an efficient separation, is air separation (qv) over carbon molecular sieve or 4A zeolite, in which oxygen, the faster diffusing component, is preferentially adsorbed (see MOLECULAR SIEVES).

A degree of control over the kinetic selectivity of molecular sieve adsorbents can be achieved by controlled adjustment of the pore size. In a carbon sieve this may be accomplished by adjusting the burn-out conditions or by controlled deposition of an easily crackable hydrocarbon. In a zeolite, ion exchange (qv) offers the simplest possibility but controlled silanation or boration has also been shown to be effective in certain cases (1).

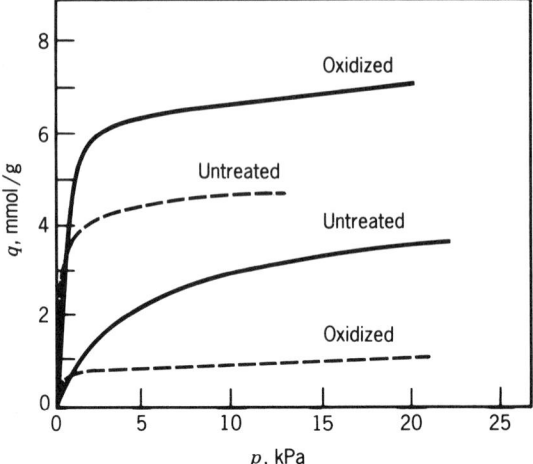

Fig. 1. Equilibrium isotherms for adsorption on activated carbon at 298 K showing the effect of surface modification (2): (———), SO_2; (- - -), n-hexane. To convert kPa to torr multiply by 7.5.

Control of equilibrium selectivity is generally achieved by adjusting the balance between electrostatic and van der Waals forces. This may be accomplished by changing the chemical nature of the surface and also, to a lesser extent, by adjusting the pore size. In carbon adsorbents surface oxidation offers a simple and effective way of introducing surface polarity and thus modifying the selectivity. One example is shown in Figure 1. On an untreated carbon adsorbent n-hexane is adsorbed more strongly than sulfur dioxide, whereas on an oxidized surface the relative affinities are reversed. With zeolite adsorbents, changing the nature of the exchangeable cation by ion exchange or adjusting the silicon–aluminum ratio of the framework, which determines the cation density, are the most common approaches. In some instances the aluminum-free zeolite analogue (a porous crystalline silicate) may be prepared with the same channel geometry but with a nonpolar surface.

Adsorption on a nonpolar surface such as pure silica or an unoxidized carbon is dominated by van der Waals forces. The affinity sequence on such a surface generally follows the sequence of molecular weights since the polarizability, which is the main factor governing the magnitude of the van der Waals interaction energy, is itself roughly proportional to the molecular weight.

Hydrophilic and Hydrophobic Surfaces. Water is a small, highly polar molecular and it is therefore strongly adsorbed on a polar surface as a result of the large contribution from the electrostatic forces. Polar adsorbents such as most zeolites, silica gel, or activated alumina therefore adsorb water more strongly than they adsorb organic species, and, as a result, such adsorbents are commonly called hydrophilic. In contrast, on a nonpolar surface where there is no electrostatic interaction water is held only very weakly and is easily displaced by organics. Such adsorbents, which are the only practical choice for adsorption of organics from aqueous solutions, are termed hydrophobic.

The most common hydrophobic adsorbents are activated carbon and silicalite. The latter is of particular interest since the affinity for water is very low indeed; the heat of adsorption is even smaller than the latent heat of vaporization (3). It seems clear that the channel structure of silicalite must inhibit the hydrogen bonding between occluded water molecules, thus enhancing the hydrophobic nature of the adsorbent. As a result, silicalite has some potential as a selective adsorbent for the separation of alcohols and other organics from dilute aqueous solutions (4).

Capillary Condensation. The equilibrium vapor pressure in a pore or capillary is reduced by the effect of surface tension. As a result, liquid sorbate condenses in a small pore at a vapor pressure that is somewhat lower than the saturation vapor pressure. In a porous adsorbent the region of multilayer physical adsorption merges gradually with the capillary condensation regime, leading to upward curvature of the equilibrium isotherm at higher relative pressure. In the capillary condensation region the intrinsic selectivity of the adsorbent is lost, so in separation processes it is generally advisable to avoid these conditions. However, this effect is largely responsible for the enhanced capacity of macroporous desiccants (qv) such as silica gel or a alumina at higher humidities.

Practical Adsorbents

To achieve a significant adsorptive capacity an adsorbent must have a high specific area, which implies a highly porous structure with very small micropores. Such microporous solids can be produced in several different ways. Adsorbents such as silica gel and activated alumina are made by precipitation of colloidal particles, followed by dehydration (see ALUMINA; SILICA, AMORPHOUS SILICA). Carbon adsorbents are prepared by controlled burn-out of carbonaceous materials such as coal, lignite, and coconut shells (see ACTIVATED CARBON). These procedures generally yield a fairly wide distribution of pore size (Fig. 2). The crystalline adsorbents (zeolite and zeolite analogues) are different in that the dimensions of the micropores are determined by the crystal structure and there is therefore virtually no distribution of micropore size (see MOLECULAR SIEVES). Although structurally very different from the crystalline adsorbents, carbon molecular sieves also have a very narrow distribution of pore size. The adsorptive properties depend on the pore size and the pore size distribution as well as the nature of the solid surface. A simple classification of some of the common adsorbents according to these features is as follows:

Surface polarity	Pore size distribution	
	Narrow	Broad
polar	zeolites (Al rich)	activated alumina silica gel
nonpolar	carbon molecular sieves silicalite	activated carbon

Despite the difference in the nature of the surface, the adsorptive behavior of the

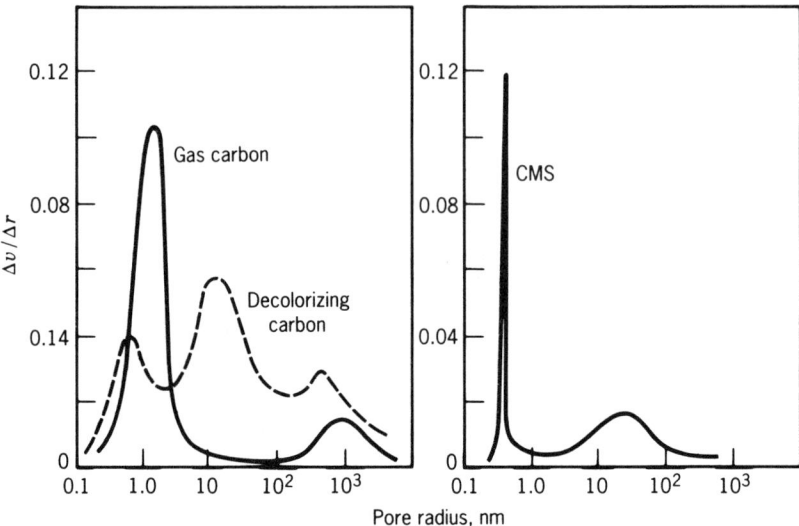

Fig. 2. Pore size distribution of typical samples of activated carbon (small pore gas carbon and large pore decolorizing carbon) and carbon molecular sieve (CMS). $\Delta v / \Delta r$ represents the increment of specific micropore volume for an increment of pore radius.

molecular sieve carbons resembles that of the small pore zeolites. As their name implies, molecular sieve separations are possible on these adsorbents based on the differences in adsorption rate, which, in the extreme limit, may involve complete exclusion of the larger molecules from the micropores.

Important properties and a number of applications of several commercial adsorbents are summarized in Tables 2–4.

Amorphous Adsorbents. The amorphous adsorbents (silica gel, activated alumina, and activated carbon) typically have specific areas in the 200–1000-m^2/g range, but for some activated carbons much higher values have been achieved ($\sim$1500 m^2/g). The difficulty is that these very high area carbons tend to lack physical strength and this limits their usefulness in many practical applications. The high area materials also contain a large proportion of very small pores, which renders them unsuitable for applications involving adsorption of large molecules. The distinction between gas carbons, used for adsorption of low molecular weight permanent gases, and liquid carbons, which are used for adsorption of larger molecules such as color bodies from the liquid phase, is thus primarily a matter of pore size.

In a typical amorphous adsorbent the distribution of pore size may be very wide, spanning the range from a few nanometers to perhaps one micrometer. Since different phenomena dominate the adsorptive behavior in different pore size ranges, IUPAC has suggested the following classification:

> micropores, <2-nm diameter
> mesopores, 2–50-nm diameter
> macropores, >50-nm diameter

Table 2. Properties and Applications of Amorphous Adsorbents

Adsorbent	Pore diameter, nm	Particle density, g/cm^3	Specific area, m^2/g	Applications
activated carbon				
large pore	1–10^3	0.6–0.8	200–600	water purification, sugar decolorizing
small pore	1–10	0.5–0.9	400–1200	removal of light organics
carbon molecular sieve	0.4–0.5, 10–10^2 (bimodal)	0.9–1.0	100–300	air separation (N$_2$ production)
silica gel				
high area	2–10	1.09	800	general purpose desiccants
low area	10–50	0.62	300	
activated alumina	2–10	1.2–1.3	300–400	

Table 3. Properties and Application of Polymeric Adsorbents

Type of polymer	Representative commercial product	Properties	Applications
sulfonated styrene–divinylbenzene copolymers[a] with various degrees of cross-linking	Dowex-50 Amberlite IR120B	pore diameter, porosity, density, etc vary with degree of hydration or dehydration	sugar separations[b], eg, various fructose, glucose
macroreticular sulfonated styrene–divinylbenzene	Diaion HPK-25	porosity ~0.33, microparticles ~80-μm diameter	removal of NH$_3$ or light amines[c]

[a]Ion exchanged to Ca^{2+} form.
[b]Liquid-phase operation.
[c]Gas or liquid phase.

Table 4. Properties and Applications of Crystalline Adsorbents[a]

Structure	Cation	Typical formula of unit cell or pseudocell	Window Obstructed	Window Free	Effective channel diameter, nm	Applications
4A	Na^+	$Na_{12}[(AlO_2)_{12}(SiO_2)_{12}]$	8-ring		0.38	desiccant; CO_2 removal; air separation (N_2)
5A	Ca^{2+}	$Ca_5Na_2[(AlO_2)_{12}(SiO_2)_{12}]$		8-ring	0.44	linear paraffin separation; air separation (O_2)
3A	K^+	$K_{12}[(AlO_2)_{12}(SiO_2)_{12}]$	8-ring		0.29	drying of reactive gases
13X	Na^+	$Na_{86}[(AlO_2)_{86}(SiO_2)_{106}]$		12-ring	0.84	air separation (O_2), removal of mercaptans
10X	Ca^{2+}	$Ca_{43}[(AlO_2)_{86}(SiO_2)_{106}]$	12-ring		0.80	
SrBaX	Sr^{2+}, Ba^{2+}	$Sr_{21}Ba_{22}[(AlO_2)_{86}(SiO_2)_{106}]$	12-ring		0.80	separation of C_8 aromatics
KY	K^+	$K_{56}[(AlO_2)_{56}(SiO_2)_{136}]$		12-ring	0.80	
Mordenite	H^+	$H_8[(AlO_2)_8(SiO_2)_{40}]$		12-ring	0.70	trapping of Kr from nuclear off-gas
AgX	Ag^+	$Ag_8[(AlO_2)_8(SiO_2)_{40}]$		12-ring	0.70	trapping of CH_3I from nuclear off-gas
silicalite/HZSM5	Ag^+	$Ag_{86}[(AlO_2)_{86}(SiO_2)_{106}](SiO_2)_{96}$		12-ring 10-ring	0.84 0.60	removal of organics in aqueous systems

[a] Structural details can be found in refs. 5 and 6. A simplified description is given in Ref. 7.

This division is somewhat arbitrary since it is really the pore size relative to the size of the sorbate molecule rather than the absolute pore size that governs the behavior. Nevertheless, the general concept is useful. In micropores (pores which are only slightly larger than the sorbate molecule) the molecule never escapes from the force field of the pore wall, even when in the center of the pore. Such pores generally make a dominant contribution to the adsorptive capacity for molecules small enough to penetrate. Transport within these pores can be severely limited by steric effects, leading to molecular sieve behavior.

The mesopores make some contribution to the adsorptive capacity, but their main role is as conduits to provide access to the smaller micropores. Diffusion in the mesopores may occur by several different mechanisms, as discussed below. The macropores make very little contribution to the adsorptive capacity, but they commonly provide a major contribution to the kinetics. Their role is thus analogous to that of a super highway, allowing the adsorbate molecules to diffuse far into a particle with a minimum of diffusional resistance.

Crystalline Adsorbents. In the crystalline adsorbents, zeolites and zeolite analogues such as silicalite and the microporous aluminum phosphates, the dimensions of the micropores are determined by the crystal framework and there is therefore virtually no distribution of pore size. However, a degree of control can sometimes be exerted by ion exchange, since, in some zeolites, the exchangeable cations occupy sites within the structure which partially obstruct the pores. The crystals of these materials are generally quite small (1–5 μm) and they are aggregated with a suitable binder (generally a clay) and formed into macroporous particles having dimensions large enough to pack directly into an adsorber vessel. Such materials therefore have a well-defined bimodal pore size distribution with the intracrystalline micropores (a few tenths of a nanometer) linked together through a network of macropores having a diameter of the same order as the crystal size (~ 1 μm).

Desiccants. A solid desiccant is simply an adsorbent which has a high affinity and capacity for adsorption of moisture so that it can be used for selective adsorption of moisture from a gas (or liquid) stream. The main requirements for an efficient desiccant are therefore a highly polar surface and a high specific area (small pores). The most widely used desiccants (qv) are silica gel, activated alumina, and the aluminum rich zeolites (4A or 13X). The equilibrium adsorption isotherms for moisture on these materials have characteristically different shapes (Fig. 3), making them suitable for different applications.

The zeolites have high affinity and high capacity at low partial pressures, shown by the nearly rectangular form of the isotherm. This makes them useful desiccants where a very low humidity or dew point is required. The 3A zeolite is a molecular sieve desiccant, since its micropores are small enough to exclude most molecules other than water. It is therefore useful for drying reactive gases. The major disadvantage of zeolite desiccants is that a high temperature is required for regeneration ($>300°C$), which makes their use uneconomic when only a moderately low dew point is required.

Considerable variation in the moisture isotherm for alumina can be obtained by different preparation and pretreatment. However, in general, the initial slope of the isotherm is not as steep as that of a zeolite, indicating a lower moisture affinity at low partial pressure, but the capacity at high humidities is

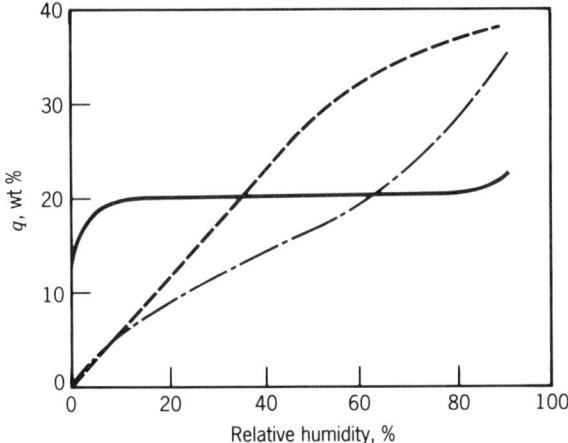

Fig. 3. Adsorption equilibrium isotherms for moisture on three commercial adsorbents: pelletized 4A zeolite (———), silica gel (- - -), and a typical activated alumina (– · –).

often higher than that of a zeolitic adsorbent. Regeneration temperatures are typically in the 250–350°C range. Alumina adsorbents are also more robust than zeolites and less sensitive to deactivation by organics, but they are generally less suitable than the zeolites where very low humidity is the primary requirement.

The isotherm for silica gel is more nearly linear over a wide range of partial pressure, although the affinity for moisture is lower than that of either alumina or the zeolites. However, a correspondingly lower regeneration temperature is also required. This can be as low as 120°C, making silica gel the most suitable candidate for pressure swing driers, desiccant cooling systems (8), and other applications where low grade heat is used for regeneration of the adsorbent.

Loaded Adsorbents. Where highly efficient removal of a trace impurity is required it is sometimes effective to use an adsorbent preloaded with a reactant rather than rely on the forces of adsorption. Examples include the use of zeolites preloaded with bromine to trap traces of olefins as their more easily condensible bromides; zeolites preloaded with iodine to trap mercury vapor, and activated carbon loaded with cupric chloride for removal of mercaptans.

Adsorption Equilibrium

Henry's Law. Like any other phase equilibrium, the distribution of a sorbate between fluid and adsorbed phases is governed by the principles of thermodynamics. Equilibrium data are commonly reported in the form of an isotherm, which is a diagram showing the variation of the equilibrium adsorbed-phase concentration or loading with the fluid-phase concentration or partial pressure at a fixed temperature. In general, for physical adsorption on a homogeneous surface at sufficiently low concentrations, the isotherm should approach a linear form, and the limiting slope in the low concentration region is commonly known as the Henry's law constant. The Henry constant is simply a thermodynamic equilibrium constant and the temperature dependence therefore follows the usual van't Hoff

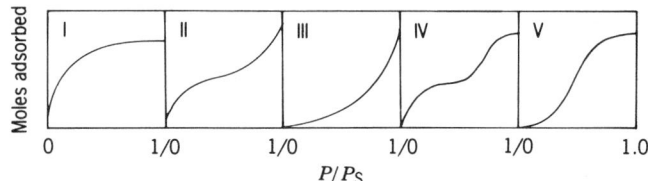

Fig. 4. The Brunaner classification of isotherms (I–V).

equation:

$$\lim_{p \to 0} \left(\frac{\partial q}{\partial p}\right)_T \equiv K' = K'_0 e^{-\Delta H_0/RT} \tag{1}$$

in which $-\Delta H_0$ is the limiting heat of adsorption at zero coverage. Since adsorption, particularly from the vapor phase, is usually exothermic, $-\Delta H_0$ is a positive quantity and K' therefore decreases with increasing temperature. A corresponding dimensionless Henry constant (K) may also be defined, based on the ratio of adsorbed and fluid-phase concentrations:

$$\lim_{c \to 0} \left(\frac{\partial q}{\partial c}\right)_T \equiv K = K_0 e^{-\Delta U_0/RT} \tag{2}$$

Since, for an ideal vapor phase, $p = cRT$, these quantities are related by

$$K = RTK'; \quad -\Delta H_0 = -\Delta U_0 + RT \tag{3}$$

Henry's law corresponds physically to the situation in which the adsorbed phase is so dilute that there is neither competition for surface sites nor any significant interaction between adsorbed molecules. At higher concentrations both of these effects become important and the form of the isotherm becomes more complex. The isotherms have been classified into five different types (9) (Fig. 4). Isotherms for a microporous adsorbent are generally of type I; the more complex forms are associated with multilayer adsorption and capillary condensation.

Langmuir Isotherm. Type I isotherms are commonly represented by the ideal Langmuir model:

$$\frac{q}{q_s} = \frac{bp}{1 + bp} \tag{4}$$

where q_s is the saturation limit and b is an equilibrium constant which is directly related to the Henry constant ($K' = bq_s$). The Langmuir model was originally developed to represent monolayer adsorption on an ideal surface, for which q_s corresponds to the monolayer coverage. However, in applying this model to physical adsorption on a microporous solid, the saturation limit becomes the quantity of sorbate required to fill the micropore volume. This expression is of the correct form to represent a type I isotherm, since at low pressure it approaches

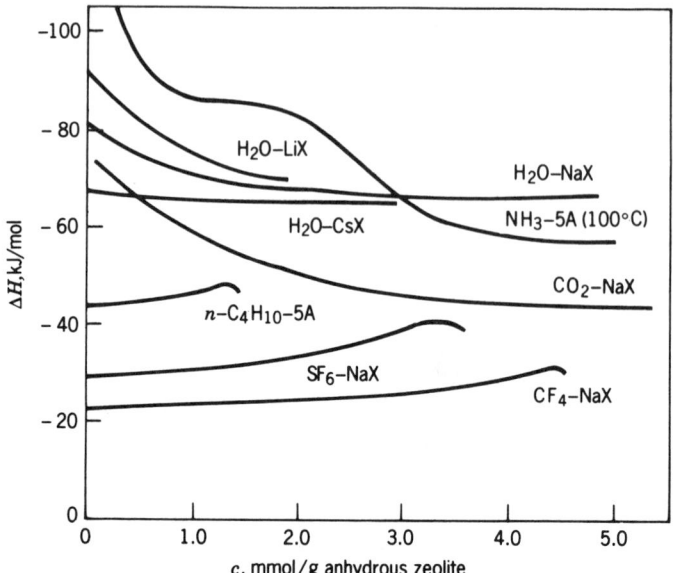

Fig. 5. Variation of isosteric heat of adsorption with adsorbed phase concentration. Reprinted from ref. 10, courtesy of Marcel Dekker, Inc. To convert kJ to kcal divide by 4.184.

Henry's law while at high pressure it tends asymptotically to the saturation limit. The equilibrium constant $b \; (= K'/q_s)$ decreases with increasing temperature (eq. 1); therefore, for a given pressure range, the isotherm approaches rectangular or irreversible form at low temperatures (large b) and linear form at high temperatures (small b).

Although very few systems conform accurately to the Langmuir model, this model provides a simple qualitative representation of the behavior of many systems and it is therefore widely used, particularly for adsorption from the vapor phase. According to the Langmuir model the heat of adsorption should be independent of loading, but this requirement is seldom fulfilled in practice. Both increasing and decreasing trends are commonly observed (Fig. 5). For a polar sorbate on a polar adsorbent (ie, a system in which electrostatic forces are dominant) a decreasing trend is normally observed, since the relative importance of the electrostatic contribution declines at high loadings as a result of preferential occupation of the most favorable sites and consequent screening of cations. In contrast, where van der Waals forces are dominant (nonpolar sorbates), a rising trend of heat of adsorption with loading is generally observed. This is commonly attributed to the effect of intermolecular attractive forces, but other explanations are also possible (11). In homologous series such as the linear paraffins the heat of adsorption increases linearly with carbon number (12).

Freundlich Isotherm. The isotherms for some systems, notably hydrocarbons on activated carbon, conform more closely to the Freundlich equation:

$$q = bp^{1/n} \qquad (n > 1.0) \tag{5}$$

Although the Freundlich expression does not reduce to Henry's law at low concentrations, it often provides a good approximation over a wide range of conditions. This form of equation can be explained as resulting from energetic heterogeneity of the surface. Superposition of a set of Langmuir isotherms with different b values (corresponding to sites of different energy) yields an expression of this form.

Adsorption of Mixtures. The Langmuir model can be easily extended to binary or multicomponent systems:

$$\frac{q_1}{q_{s1}} = \frac{b_1 p_1}{1 + b_1 p_1 + b_2 p_2 + \cdots}; \quad \frac{q_2}{q_{s2}} = \frac{b_2 p_2}{1 + b_1 p_1 + b_2 p_2; + \cdots} \quad (6)$$

Thermodynamic consistency requires $q_{s1} = q_{s2}$, but this requirement can cause difficulties when attempts are made to correlate data for sorbates of very different molecular size. For such systems it is common practice to ignore this requirement, thereby introducing an additional model parameter. This facilitates data fitting but it must be recognized that the equations are then being used purely as a convenient empirical form with no theoretical foundation.

Equation 6 shows that the adsorption of component 1 at a partial pressure p_1 is reduced in the presence of component 2 as a result of competition for the available surface sites. There are only a few systems for which this expression (with $q_{s1} = q_{s2} = q_s$) provides an accurate quantitative representation, but it provides useful qualitative or semiquantitative guidance for many systems. In particular, it has the correct asymptotic behavior and provides explicit recognition of the effect of competitive adsorption. For example, if component 2 is either strongly adsorbed or present at much higher concentration than component 1, the isotherm for component 1 is reduced to a simple linear form in which the apparent Henry's law constant depends on p_2:

$$q_1 \simeq \left(\frac{b_1 q_{s1}}{1 + b_2 p_2}\right) p_1 \quad (7)$$

For an equilibrium-based separation, a convenient measure of the intrinsic selectivity of the adsorbent is provided by the separation factor (α_{12}), which is defined by analogy with the relative volatility as

$$\alpha_{12} = \frac{(X_1/Y_1)}{(X_2/Y_2)} \quad (8)$$

where X and Y refer to the mole fractions in the adsorbed and fluid phases, respectively, at equilibrium. For a system that obeys the Langmuir model (eq. 6) it is evident that $\alpha_{12} = b_1/b_2$ and is thus independent of concentration. The Langmuir isotherm is therefore sometimes referred to as the constant separation factor model.

The assumption of a constant separation factor is often a reasonable approximation for preliminary process design but this assumption is often violated in real systems, where some variation of the separation factor with composition is common and more extreme variations involving azeotrope formation ($\alpha = 1.0$ at a particular composition) and selectivity reversal (α varying from greater than 1.0 to less than 1.0 with changing composition) are not uncommon. There have been many attempts to improve the correlation of equilibrium data by using more complex expressions, one of the more widely used being the Langmuir-Freundlich or loading ratio correlation (13):

$$\frac{q_1}{q_s} = \frac{b_1 p_1^{n_1}}{1 + b_1 p_1^{n_1} + b_2 p_2^{n_2} + \cdots}; \quad \frac{q_2}{q_s} = \frac{b_2 p_2^{n_2}}{1 + b_1 p_1^{n_1} + b_2 p_2^{n_2} + \cdots} \quad (9)$$

This has the advantage that the expressions for the adsorbed-phase concentration are simple and explicit, and, as in the Langmuir expression, the effect of competition between sorbates is accounted for. However, the expression does not reduce to Henry's law in the low concentration limit and therefore violates the requirements of thermodynamic consistency. Whereas it may be useful as a basis for the correlation of experimental data, it should be treated with caution and should not be used as a basis for extrapolation beyond the experimental range.

Ideal Adsorbed Solution Theory. Perhaps the most successful approach to the prediction of multicomponent equilibria from single-component isotherm data is ideal adsorbed solution theory (14). In essence, the theory is based on the assumption that the adsorbed phase is thermodynamically ideal in the sense that the equilibrium pressure for each component is simply the product of its mole fraction in the adsorbed phase and the equilibrium pressure for the pure component at the same spreading pressure. The theoretical basis for this assumption and the details of the calculations required to predict the mixture isotherm are given in standard texts on adsorption (7) as well as in the original paper (14). Whereas the theory has been shown to work well for several systems, notably for mixtures of hydrocarbons on carbon adsorbents, there are a number of systems which do not obey this model. Azeotrope formation and selectivity reversal, which are observed quite commonly in real systems, are not consistent with an ideal adsorbed phase and there is no way of knowing a priori whether or not a given system will show ideal behavior.

Adsorption Kinetics

Intrinsic Kinetics. Chemisorption may be regarded as a chemical reaction between the sorbate and the solid surface, and, as such, it is an activated process for which the rate constant (k) follows the familiar Arrhenius rate law:

$$k = k_0 e^{-E/RT} \quad (10)$$

Depending on the temperature and the activation energy (E), the rate constant may vary over many orders of magnitude.

In practice the kinetics are usually more complex than might be expected on this basis, since the activation energy generally varies with surface coverage as a result of energetic heterogeneity and/or sorbate–sorbate interaction. As a result, the adsorption rate is commonly given by the Elovich equation (15):

$$q = \frac{1}{k'} \ln(1 + k''t) \qquad (11)$$

where k' and k'' are temperature-dependent constants.

In contrast, physical adsorption is a very rapid process, so the rate is always controlled by mass transfer resistance rather than by the intrinsic adsorption kinetics. However, under certain conditions the combination of a diffusion-controlled process with an adsorption equilibrium constant that varies according to equation 1 can give the appearance of activated adsorption.

As illustrated in Figure 6, a porous adsorbent in contact with a fluid phase offers at least two and often three distinct resistances to mass transfer: external film resistance and intraparticle diffusional resistance. When the pore size distribution has a well-defined bimodal form, the latter may be divided into macropore and micropore diffusional resistances. Depending on the particular system and the conditions, any one of these resistances may be dominant or the overall rate of

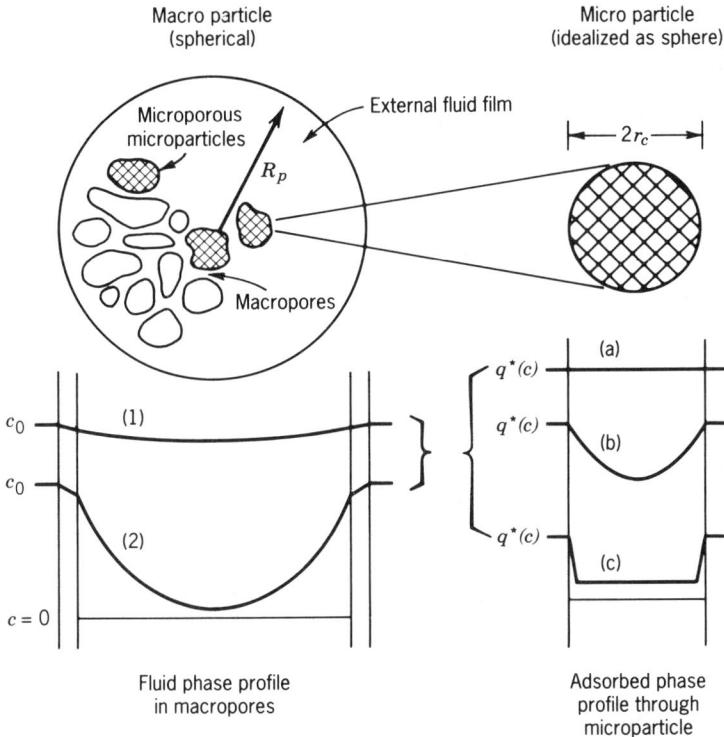

Fig. 6. Concentration profiles through an idealized biporous adsorbent particle showing some of the possible regimes. (1) + (a) rapid mass transfer, equilibrium throughout particle; (1) + (b) micropore diffusion control with no significant macropore or external resistance; (1) + (c) controlling resistance at the surface of the microparticles; (2) + (a) macropore diffusion control with some external resistance and no resistance within the microparticle; (2) + (b) all three resistances (micropore, macropore, and film) significant; (2) + (c) diffusional resistance within the macroparticle and resistance at the surface of the microparticle with some external film resistance.

mass transfer may be determined by the combined effects of more than one resistance.

External Fluid Film Resistance. A particle immersed in a fluid is always surrounded by a laminar fluid film or boundary layer through which an adsorbing or desorbing molecule must diffuse. The thickness of this layer, and therefore the mass transfer resistance, depends on the hydrodynamic conditions. Mass transfer (qv) in packed beds and other common contacting devices has been widely studied. The rate data are normally expressed in terms of a simple linear rate expression of the form

$$\frac{\partial q}{\partial t} = k_f a(c - c^*) \tag{12}$$

and the variation of the mass transfer coefficient (k_f) with the hydrodynamic conditions is generally accounted for in terms of empirical correlations of the general form

$$\text{Sh} \equiv \frac{2k_f R}{D_m} = f(\text{Re}, \text{Sc}) \tag{13}$$

where Re and Sc are the (particle-based) Reynolds and Schmidt numbers. One of the most widely used correlations, applicable to both gas and liquid systems over a wide range of conditions, is (16)

$$\text{Sh} = \frac{2k_f R}{D_m} = 2.0 + 1.1\, \text{Sc}^{1/3} \text{Re}^{0.6} \tag{14}$$

Macropore Diffusion. Transport in a macropore can occur by several different mechanisms, the most important of which are bulk molecular diffusion, Knudsen diffusion, surface diffusion, and Poiseuille flow. In liquid systems bulk molecular diffusion is generally dominant, but in the vapor phase the contributions from Knudsen and surface diffusion may be large or even dominant. The contribution from Poiseuille flow, ie, forced flow through the pore under the influence of the pressure gradient, is generally relatively minor since pressure gradients are usually kept small. However, this is not true in the pressurization and blowdown steps of a pressure swing process, where the contribution from Poiseuille flow can be dominant.

A molecule colliding with the pore wall is reflected in a specular manner so that the direction of the molecule leaving the surface has no correlation with that of the incident molecule. This leads to a Fickian mechanism, known as Knudsen diffusion, in which the flux is proportional to the gradient of concentration or partial pressure. The Knudsen diffusivity (D_K) is independent of pressure and varies only weakly with temperature:

$$D_K = 9700\rho\, (T/M)^{1/2}\ (\text{cm}^2/\text{s}) \tag{15}$$

where ρ is the pore radius (cm) and M the molecular weight. Knudsen diffusion becomes dominant when collisions with the pore wall occur more frequently than collisions between diffusing molecules, ie, when the pore diameter is smaller than the mean free path. Since the mean free path varies inversely with pressure there is a gradual transition from the molecular to the Knudsen regime as the pressure is reduced, but the pressure at which this occurs depends on the pore size. At atmospheric temperature and pressure Knudsen diffusion is dominant in pores of less than about 10 nm diameter. In the intermediate region, which spans the range of the macropores in many commercial adsorbents, both mechanisms are of comparable significance.

The combined effects of Knudsen and molecular diffusion may be estimated approximately from the reciprocal addition rule:

$$\frac{1}{\epsilon_p D_p} = \frac{\tau}{\epsilon_p}\left(\frac{1}{D_K} + \frac{1}{D_m}\right) \tag{16}$$

The factor ϵ_p takes account of the fact that diffusion occurs only through the pore and not through the matrix; τ is a tortuosity factor which accounts for the increased path length and reduced concentration gradient arising from the random orientation of the pores as well as any other geometric effects. In a typical adsorbent $\tau \sim 3.0$ and $\epsilon_p \sim 0.3$, so the effect of these two factors is to reduce the diffusivity by about one order of magnitude relative to the value for a straight cylindrical capillary.

Micropore Diffusion. In very small pores in which the pore diameter is not much greater than the molecular diameter the diffusing molecule never escapes from the force field of the pore wall. Under these conditions steric effects and the effects of nonuniformity in the potential field become dominant and the Knudsen mechanism no longer applies. Diffusion occurs by an activated process involving jumps from site to site, just as in surface diffusion, and the diffusivity becomes strongly dependent on both temperature and concentration.

The true driving force for any diffusive transport process is the gradient of chemical potential rather than the gradient of concentration. This distinction is not important in dilute systems where thermodynamically ideal behavior is approached. However, it becomes important at higher concentration levels and in micropore and surface diffusion. To a first approximation the expression for the diffusive flux may be written

$$J = -Bq\, \partial\mu/\partial z \tag{17}$$

where q is the concentration in the adsorbed phase. Assuming an ideal vapor phase, the expression for the chemical potential is

$$\mu = \mu^\circ + RT \ln p \tag{18}$$

where

$$\frac{\partial \mu}{\partial z} = RT \frac{d \ln p}{dq} \frac{\partial q}{\partial z} \tag{19}$$

110 ADSORPTION

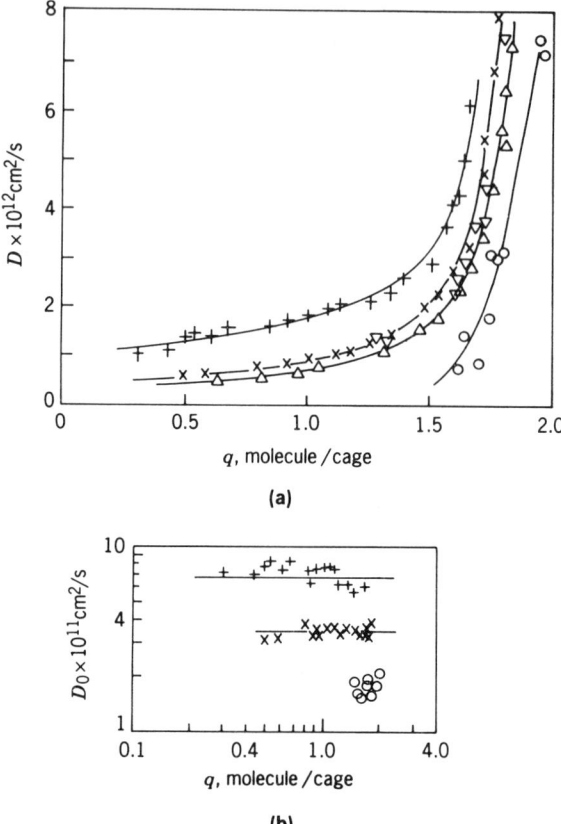

Fig. 7. Variation of (**a**) intracrystalline diffusivity and (**b**) corrected diffusivity (D_0) with sorbate concentration for *n*-heptane in a commercial sample of 5A zeolite crystals: ○, 409 K; △,▽, 439 K (ads, des); ×, 462 K; +, 491 K. Reproduced by permission of National Research Council of Canada from ref. 17.

Combining equations 17 and 18 yields, for the Fickian diffusivity (defined by $J = -D\,\partial q/\partial z$),

$$D = D_0 \frac{d \ln p}{d \ln q} \qquad (20)$$

where $D_0 = BRT$. If the equilibrium relation is linear, $d \ln p/d \ln q = 1.0$ and $D \to D_0$. At higher concentrations the equilibrium relationship is nonlinear, and as a result the diffusivity is generally concentration dependent. For the special case of a Langmuir isotherm (eq. 4), $d \ln p/d \ln q = (1 - q/q_s)^{-1}$ so

$$D = \frac{D_0}{1 - q/q_s} \qquad (21)$$

A rapid increase in diffusivity in the saturation region is therefore to be expected,

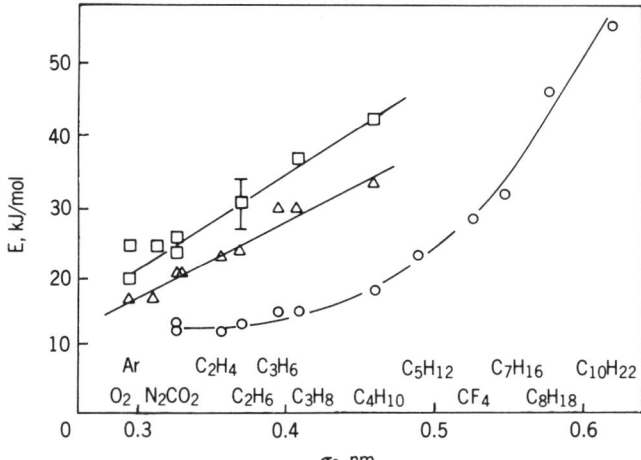

Fig. 8. Variation of activation energy with kinetic molecular diameter for diffusion in 4A zeolite (□), 5A zeolite (○), and carbon molecular sieve (MSC-5A) (△). Kinetic diameters are estimated from the van der Waals co-volumes. From ref. 7. To convert kJ to kcal divide by 4.184.

as illustrated in Figure 7 (17). Although the corrected diffusivity (D_0) is, in principle, concentration dependent, the concentration dependence of this quantity is generally much weaker than that of the thermodynamic correction factor ($d \ln p/d \ln q$). The assumption of a constant corrected diffusivity is therefore an acceptable approximation for many systems. More detailed analysis shows that the corrected diffusivity is closely related to the self-diffusivity or tracer diffusivity, and at low sorbate concentrations these quantities become identical.

The temperature dependence of the corrected diffusivity follows the usual Eyring expression

$$D_0 = D_\infty e^{-E/RT} \qquad (22)$$

in which E is the activation energy or the energy barrier between adjacent sites. In small pore zeolites and carbon molecular sieves the dominant contribution to this energy is the repulsive interaction encountered by the molecule in penetrating the concentration in the pore. As a result, the activation energy shows a well-defined correlation with the molecule diameter and the size of the micropore, as illustrated in Figure 8.

Sorption Rates in Batch Systems. Direct measurement of the uptake rate by gravimetric, volumetric, or piezometric methods is widely used as a means of measuring intraparticle diffusivities. Diffusive transport within a particle may be represented by the Fickian diffusion equation, which, in spherical coordinates, takes the form

$$\frac{\partial q}{\partial t} = D \left(\frac{\partial^2 q}{\partial r^2} + \frac{2}{r} \frac{\partial q}{\partial r} \right) \qquad (23)$$

For a step change in sorbate concentration at the particle surface ($r = R$) at time zero, assuming isothermal conditions and diffusion control, the expression for the uptake curve may be derived from the appropriate solution of this differential equation:

$$\frac{m_t}{m_\infty} = 1 - \frac{6}{\pi^2} \sum_{n=1}^{\infty} \frac{1}{n^2} e^{-n^2 \pi^2 Dt/R^2} \quad (24)$$

or, in the initial region,

$$\frac{m_r}{m_\infty} = \frac{6}{\sqrt{\pi}} \left(\frac{Dt}{R^2}\right)^{1/2} \quad (25)$$

The time constant R^2/D, and hence the diffusivity, may thus be found directly from the uptake curve. However, it is important to confirm by experiment that the basic assumptions of the model are fulfilled, since intrusions of thermal effects or extraparticle resistance to mass transfer may easily occur, leading to erroneously low apparent diffusivity values.

In certain adsorbents, notably partially coked zeolites and some carbon molecular sieves, the resistance to mass transfer may be concentrated at the surface of the particle, leading to an uptake expression of the form

$$\frac{m_t}{m_\infty} = 1 - e^{-k_s t} \quad (26)$$

in place of equation 24. The difference between surface resistance and intraparticle diffusion control is easily apparent from the form of the uptake curves (see Fig. 9). Since both D and k_s are generally concentration dependent, it is preferable to make differential measurements over small concentration steps in order to simplify the interpretation of the experimental data.

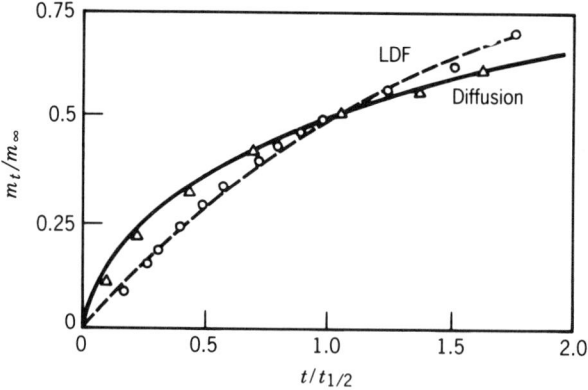

Fig. 9. Uptake curves for N_2 in two samples of carbon molecular sieve showing conformity with diffusion model (eq. 24) for sample 1 ($\triangle$), and with surface resistance model (eq. 26) for sample 2 ($\bigcirc$); LDF = linear driving force. Data from ref. 18.

For a macroporous sorbent the situation is slightly more complex. A differential balance on a shell element, assuming diffusivity transport through the macropores with rapid adsorption at the surface (or in the micropores), yields

$$\epsilon_p \frac{\partial c}{\partial t} + (1 - \epsilon_p) \frac{\partial q}{\partial t} = \epsilon_p D_p \left(\frac{\partial^2 c}{\partial r^2} + \frac{2}{r} \frac{\partial c}{\partial r} \right) \quad (27)$$

Assuming a linear equilibrium relationship (over the range of the small concentration step, $q^* = Kc$), this becomes

$$\frac{\partial c}{\partial t} = \frac{\epsilon_p D_p}{1 + (1 - \epsilon_p)K} \left(\frac{\partial^2 c}{\partial r^2} + \frac{2}{r} \frac{\partial c}{\partial r} \right) \quad (28)$$

which is of the same form as equation 23 but with an effective diffusivity given by

$$D = \frac{\epsilon_p D_p}{1 + (1 - \epsilon_p)K} \quad (29)$$

For adsorption from the vapor phase, K may be very large (sometimes as high as 10^7) and then clearly the effective diffusivity is very much smaller than the pore diffusivity. Furthermore, the temperature dependence of K follows equation 2, giving the appearance of an activated diffusion process with $E \approx (-\Delta U)$.

As a result of these difficulties the reported diffusivity data show many apparent anomalies and inconsistencies, particularly for zeolites and other microporous adsorbents. Discrepancies of several orders of magnitude in the diffusivity values reported for a given system under apparently similar conditions are not uncommon (18). Since most of the intrusive effects lead to erroneously low values, the higher values are probably the more reliable.

Adsorption Column Dynamics

In most adsorption processes the adsorbent is contacted with fluid in a packed bed. An understanding of the dynamic behavior of such systems is therefore needed for rational process design and optimization. What is required is a mathematical model which allows the effluent concentration to be predicted for any defined change in the feed concentration or flow rate to the bed. The flow pattern can generally be represented adequately by the axial dispersed plug-flow model, according to which a mass balance for an element of the column yields, for the basic differential equation governing the dynamic behavior,

$$-D_L \frac{\partial^2 c_i}{\partial z^2} + \frac{\partial}{\partial z}(vc_i) + \frac{\partial c_i}{\partial t} + \left(\frac{1 - \epsilon}{\epsilon} \right) \frac{\partial \bar{q}_i}{\partial t} = 0 \quad (30)$$

The term $\partial \bar{q}_i/\partial t$ represents the overall rate of mass transfer for component i (at time t and distance z) averaged over a particle. This is governed by a mass transfer rate expression which may be thought of as a general functional relationship of the form

$$\frac{\partial \bar{q}}{\partial t} = f(c_i, c_j, \ldots, q_i, q_j, \ldots) \tag{31}$$

This rate equation must satisfy the boundary conditions imposed by the equilibrium isotherm and it must be thermodynamically consistent so that the mass transfer rate falls to zero at equilibrium. It may be a linear driving force expression of the form

$$\frac{\partial \bar{q}_i}{\partial t} = k_s(q_i^* - \bar{q}_i) \tag{32}$$

where $q_i^*(c_i, c_j, \ldots)$ represents the equilibrium adsorbed phase concentration of component i, or it may be a set of diffusion equations with their associated boundary conditions.

For an isothermal system the simultaneous solution of equations 30 and 31, subject to the boundary conditions imposed on the column, provides the expressions for the concentration profiles $c_i(z,t)$, $\bar{q}_i(z,t)$ in both phases. If the system is nonisothermal, an energy balance is also required and since, in general, both the equilibrium concentration and the rate coefficients are temperature dependent, all equations are coupled. Analytical solutions are possible only for the simpler cases: single-component isothermal systems with linear or rectangular equilibrium isotherms. In the general case of a multicomponent nonisothermal system, numerical solutions offer the only practical approach.

The form of the response for an adiabatic three-component system (two adsorbable components in an inert carrier) is illustrated in Figure 10. In general, if there are n components (counting both heat and nonadsorbing species as components), the response contains $(n-1)$ transitions or mass transfer zones, separated by $(n-2)$ plateaus between the initial and final states. When the change imposed at the column inlet involves an increase in the concentration of the more strongly adsorbed species, the concentration at the intermediate plateau will exceed both its initial concentration and its final steady-state concentration. This phenomenon, known as roll-up, results from displacement by the more strongly adsorbed species, which travels more slowly through the column.

Equilibrium Theory. The general features of the dynamic behavior may be understood without recourse to detailed calculations since the overall pattern of the response is governed by the form of the equilibrium relationship rather than by kinetics. Kinetic limitations may modify the form of the concentration profile but they do not change the general pattern. To illustrate the different types of transition, consider the simplest case: an isothermal system with plug flow involving a single adsorbable species present at low concentration in an inert carrier, for which equation 30 reduces to

$$v\frac{\partial c}{\partial z} + \frac{\partial c}{\partial t} + \left(\frac{1-\epsilon}{\epsilon}\right)\frac{\partial \bar{q}}{\partial t} = 0 \tag{33}$$

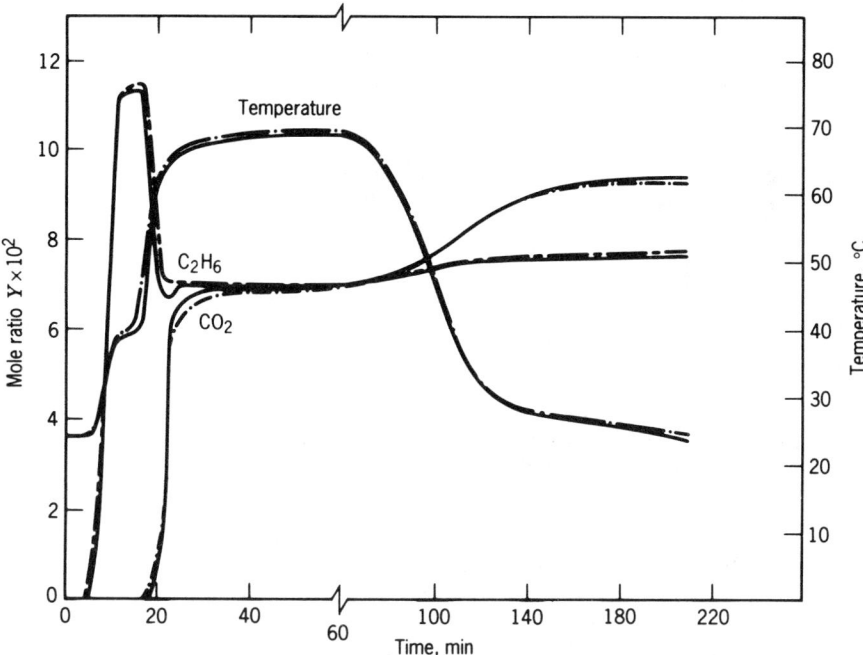

Fig. 10. Comparison of theoretical (———) and experimental (—·—·) concentration and temperature breakthrough curves for sorption of C_2H_6–CO_2 mixtures from a N_2 carrier on 5A molecular sieve. Feed: 10.5% CO_2, 7.03% C_2H_6 (molar basis) at 24°C, 116.5 kPa (1.15 atm). Column length, 48 cm. Theoretical curves were calculated numerically using the linear driving force model with a Langmuir equilibrium isotherm. Experimental data are from ref. 19. From ref. 20, courtesy of Pergamon Press.

Assuming local equilibrium, $\bar{q} = f(c)$ where this function represents the isotherm equation, this becomes

$$\frac{v}{1 + ((1 - \epsilon)/\epsilon)f'(c)} \frac{\partial c}{\partial z} + \frac{\partial c}{\partial t} = 0 \tag{34}$$

where $f'(c) = dq^*/dc$ is simply the slope of the equilibrium isotherm at concentration c.

Equation 34 has the form of the kinematic wave equation and represents a transition traveling with the wave velocity w, given by

$$w = \left(\frac{\partial z}{\partial t}\right)_c = \frac{v}{1 + ((1 - \epsilon)/\epsilon)f'(c)} \tag{35}$$

For a linear system $f'(c) = K$, so the wave velocity becomes independent of concentration and, in the absence of dispersive effects such as mass transfer resistance or axial mixing, a concentration perturbation propagates without changing its shape. The propagation velocity is inversely dependent on the adsorption equilibrium constant.

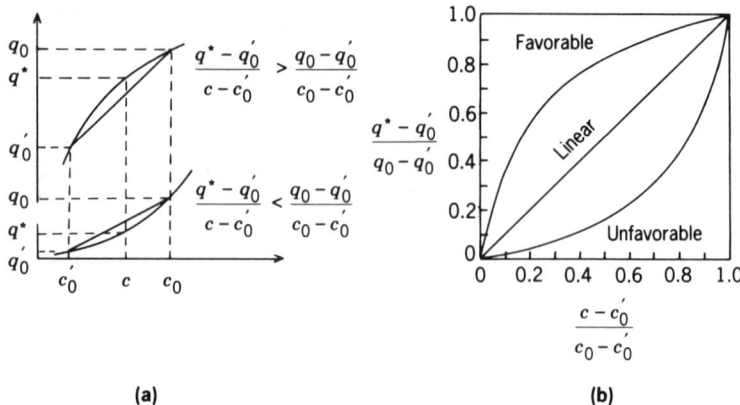

Fig. 11. (a) Equilibrium isotherm and (b) dimensionless equilibrium diagram showing favorable, linear, and unfavorable isotherms.

For a nonlinear system the behavior depends on the shape of the isotherm. If the isotherm is unfavorable (Fig. 11), $f'(c)$ increases with concentration so that w decreases with concentration. This leads to a spreading profile, as illustrated in Fig. 12b. However, if the isotherm is favorable (in the direction of the concentration change), an entirely different situation arises. Then $f'(c)$ decreases with concentration so that w increases with concentration. This would lead to the

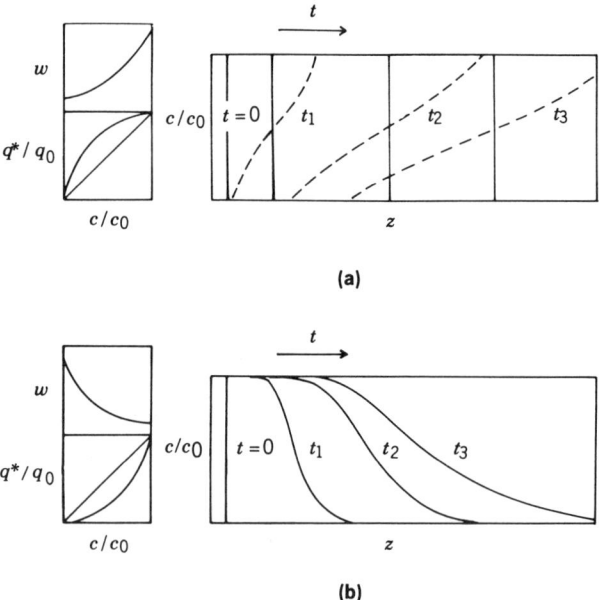

Fig. 12. (a) Development of the physically unreasonable overhanging concentration profile and the corresponding shock profile for adsorption with a favorable isotherm and (b) development of the dispersive (proportionate pattern) concentration profile for adsorption with an unfavorable isotherm (or for desorption with a favorable isotherm). From ref. 7.

physically unreasonable overhanging profiles shown in Figure 12a. In fact, what happens is that the continuous solution is replaced by the equivalent shock transition so that response becomes a shock wave which propagates at a steady velocity w' given by

$$w' = \frac{v}{1 + ((1 - \epsilon)/\epsilon)(\Delta q/\Delta c)} \qquad (36)$$

where $\Delta q/\Delta c$ represents the ratio of the concentration changes in the adsorbed and fluid phases.

Constant Pattern Behavior. In a real system the finite resistance to mass transfer and axial mixing in the column lead to departures from the idealized response predicted by equilibrium theory. In the case of a favorable isotherm the shock wave solution is replaced by a constant pattern solution. The concentration profile spreads in the initial region until a stable situation is reached in which the mass transfer rate is the same at all points along the wave front and exactly matches the shock velocity. In this situation the fluid-phase and adsorbed-phase profiles become coincident, as illustrated in Figure 13. This represents a stable situation and the profile propagates without further change in shape—hence the term constant pattern. The form of the concentration profile under constant pattern conditions may be easily deduced by integrating the mass transfer rate expression subject to the condition $c/c_0 = q/q_0$, where q_0 is the adsorbed phase concentration in equilibrium with c_0.

The distance required to approach the constant pattern limit decreases as the mass transfer resistance decreases and the nonlinearity of the equilibrium isotherm increases. However, when the isotherm is highly favorable, as in many

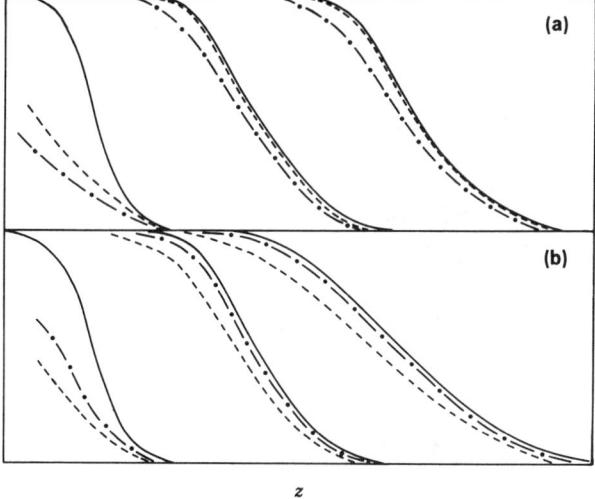

Fig. 13. Schematic diagram showing (**a**) approach to constant pattern behavior for a system with a favorable isotherm and (**b**) approach to proportionate pattern behavior for a system with an unfavorable isotherm. y axis: c/c_0, (——); $\bar{q}/q_0$, (- - -); c^*/c_0, (—·—). From ref. 7.

adsorption processes, this distance may be very small, a few centimeters to perhaps a meter.

Length of Unused Bed. The constant pattern approximation provides the basis for a very useful and widely used design method based on the concept of the length of unused bed (LUB). In the design of a typical adsorption process the basic problem is to estimate the size of the adsorber bed needed to remove a certain quantity of the adsorbable species from the feed stream, subject to a specified limit (c') on the effluent concentration. The length of unused bed, which measures the capacity of the adsorber which is lost as a result of the spread of the concentration profile, is defined by

$$\text{LUB} = (1 - q'/q_0)L = (1 - t'/\bar{t})L \qquad (37)$$

where q' is the capacity at the break time t' and $\bar{t}$ is the stoichiometric time (see Fig. 14). The values of t', $\bar{t}$, and hence the LUB are easily determined from an experimental breakthrough curve because, by overall mass balance,

$$\bar{t} = \frac{L}{v}\left[1 + \left(\frac{1-\epsilon}{\epsilon}\right)\left(\frac{q_0}{c_0}\right)\right] = \int_0^\infty \left(1 - \frac{c}{c_0}\right) dt \qquad (38)$$

$$t' = \frac{L}{v}\left[1 + \left(\frac{1-\epsilon}{\epsilon}\right)\left(\frac{q'}{c_0}\right)\right] = \int_0^{t'} \left(1 - \frac{c}{c_0}\right) dt \qquad (39)$$

Under constant pattern conditions the LUB is independent of column length although, of course, it depends on other process variables. The procedure is therefore to determine the LUB in a small laboratory or pilot-scale column packed with the same adsorbent and operated under the same flow conditions. The length of column needed can then be found simply by adding the LUB to the length calculated from equilibrium considerations, assuming a shock concentration front.

One potential problem with this approach is that heat loss from a small scale column is much greater than from a larger diameter column. As a result, small columns tend to operate almost isothermally whereas in a large column the system is almost adiabatic. Since the temperature profile in general affects the concentration profile, the LUB may be underestimated unless great care is taken to ensure adiabatic operation of the experimental column.

Proportionate Pattern Behavior. If the isotherm is unfavorable, the stable dynamic situation leading to constant pattern behavior can never be achieved.

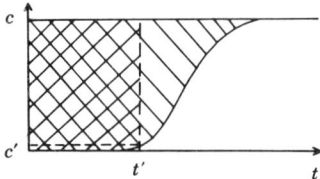

Fig. 14. Sketch of breakthrough curve showing break time t' and the method of calculation of the stoichiometric time $\bar{t}$ and LUB. From ref. 7. ▨ = the integral of equation 38; ▩ = the integral of equation 39.

The situation is shown in Figure 13**b**. The equilibrium adsorbed-phase concentration lies above rather than below the actual adsorbed-phase profile. As the mass transfer zone progresses through the column it broadens, but the limiting situation, which is approached in a long column, is simply local equilibrium at all points ($c = c^*$) and the profile therefore continues to spread in proportion to the length of the column. This difference in behavior is important since the LUB approach to design is clearly inapplicable under these conditions.

Favorable and unfavorable equilibrium isotherms are normally defined, as in Figure 11, with respect to an increase in sorbate concentration. This is, of course, appropriate for an adsorption process, but if one is considering regeneration of a saturated column (desorption), the situation is reversed. An isotherm which is favorable for adsorption is unfavorable for desorption and vice versa. In most adsorption processes the adsorbent is selected to provide a favorable adsorption isotherm, so the adsorption step shows constant pattern behavior and proportionate pattern behavior is encountered in the desorption step.

Detailed Modelling Results. The results of a series of detailed calculations for an ideal isothermal plug-flow Langmuir system are summarized in Figure 15. The solid lines show the form of the theoretical breakthrough curves for adsorption and desorption, calculated from the following set of model equations and expressed in terms of the dimensionless variables ζ, τ, and β:

Differential Balance for Column

$$v \frac{\partial c}{\partial z} + \frac{\partial c}{\partial t} + \left(\frac{1-\epsilon}{\epsilon}\right) \frac{\partial \bar{q}}{\partial t} = 0 \qquad (40)$$

Rate Equation

$$\frac{\partial \bar{q}}{\partial t} = k(q^* - \bar{q}) \qquad (41)$$

Equilibrium Isotherm

$$\frac{q^*}{q_s} = \frac{bc}{1 + bc} \qquad (42)$$

Initial Conditions

$$\bar{q}(z, 0) = c(z, 0) = 0 \quad \text{(adsorption)}$$
$$\bar{q}(z, 0) = q_0, \, c(z, 0) = c_0 \quad \text{(desorption)} \qquad (43)$$

Boundary Conditions

$$c(0, t) = c_0 \quad \text{(adsorption)}$$
$$c(0, t) = 0 \quad \text{(desorption)} \qquad (44)$$

Bed Length Parameter

$$\zeta = \frac{z}{v}\left(\frac{q_0}{c_0}\right)\left(\frac{1-\epsilon}{\epsilon}\right)$$

Dimensionless Time

$$\tau = k\left(t - \frac{z}{v}\right) \qquad (45)$$

Nonlinearity Parameter

$$\beta = 1 - \frac{q_0}{q_s} = \frac{1}{1+bc_0}$$

Also shown are the corresponding curves calculated for the same system assuming a diffusion model in place of the linear rate expression. For intracrystalline diffusion $k = 15D_0/r_c^2$, whereas for macropore diffusion $k = (15\epsilon_p D_p/R_p^2)(c_0/q_0)$, in accordance with the Glueckauf approximation (21).

For linear or moderately nonlinear systems ($\beta \to 1.0$) there is little difference in the response curves for all three models, thus verifying the validity of the Glueckauf approximation. Differences between the models, however, become more significant for a highly nonlinear isotherm ($\beta \to 0$). For linear or near linear systems the adsorption and desorption curves are mirror images, but as the isotherm becomes more nonlinear the adsorption and desorption curves become increasingly asymmetric. The adsorption curve approaches its limiting constant pattern form whereas the desorption curve approaches the limiting proportionate pattern form. In the long-time region the desorption curve is governed entirely by equilibrium, so that the curves for all three rate models again become coincident.

The main conclusion to be drawn from these studies is that for most practical purposes the linear rate model provides an adequate approximation and the use of the more cumbersome and computationally time consuming diffusing models is generally not necessary. The Glueckauf approximation provides the required estimate of the effective mass transfer coefficient for a diffusion controlled system. More detailed analysis shows that when more than one mass transfer resistance is significant the overall rate coefficient may be estimated simply from the sum of the resistances (7):

$$\frac{1}{kK} = \frac{R}{3k_f} + \frac{R^2}{15KD_c} + \frac{R^2}{15\epsilon_p D_p} \qquad (46)$$

Adsorption Chromatography. The principle of gas–solid or liquid–solid chromatography (qv) may be easily understood from equation 35. In a linear multicomponent system (several sorbates at low concentration in an inert carrier) the wave velocity for each component depends on its adsorption equilibrium constant. Thus, if a pulse of the mixed sorbate is injected at the column inlet, the

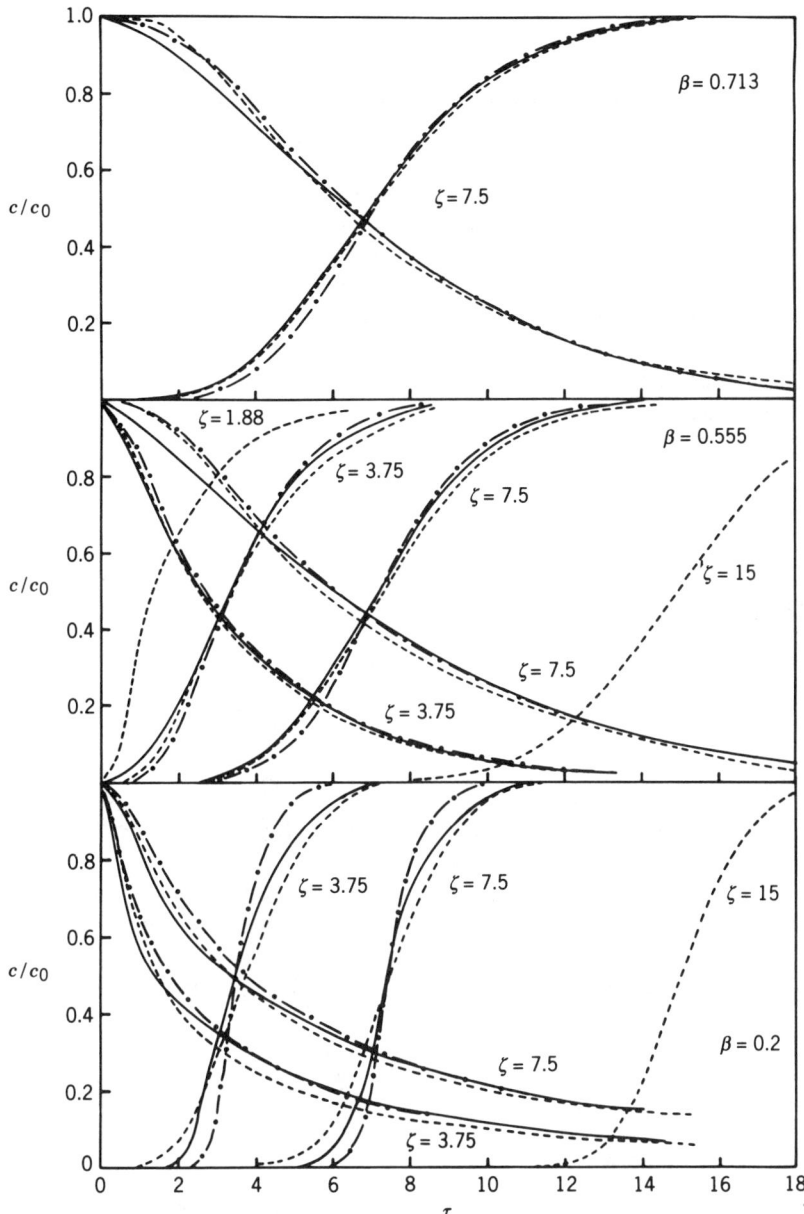

Fig. 15. Theoretical breakthrough curves for a nonlinear (Langmuir) system showing the comparison between the linear driving force (——), pore diffusion (- - -), and intracrystalline diffusion (—·—·) models based on the Glueckauf approximation (eqs. 40–45). From ref. 7.

different species separate into bands which travel through the column at their characteristic velocities, and at the outlet of the column a sequence of peaks corresponding to the different species is detected. Measurement of the retention time ($\bar{t}$) under known flow conditions thus provides a simple means of determining

ADSORPTION

the equilibrium constant (Henry constant):

$$\int_0^\infty \frac{ct\,dt}{\int_0^\infty c\,dt} = \bar{t} = \frac{L}{v}\left[1 + \left(\frac{1-\epsilon}{\epsilon}\right)K\right] \quad (47)$$

In an ideal system with no axial mixing or mass transfer resistance the peaks for the various components propagate without spreading. However, in any real system the peak broadens as it propagates and the extent of this broadening is directly related to the mass transfer and axial dispersion characteristics of the column. Measurement of the peak broadening therefore provides a convenient way of measuring mass transfer coefficients and intraparticle diffusivities. The simplest approach is to measure the second moments of the response peak over a range of flow rates:

$$\sigma^2 \equiv \int_0^\infty \frac{c(t-\bar{t})^2\,dt}{\int_0^\infty c\,dt} \quad (48)$$

Solution of the model equations shows that, for a linear isothermal system and a pulse injection, the height equivalent to a theoretical plate (HETP) is given by

$$H = \frac{\sigma^2}{\bar{t}^2}L = \frac{2D_L}{v} + \frac{2v}{kK}\left(\frac{\epsilon}{1-\epsilon}\right)\left[1 + \frac{\epsilon}{(1-\epsilon)K}\right]^{-2} \quad (49)$$

where $1/kK$ is the overall mass transfer resistance defined by equation 46.

For liquid systems D_L/v is approximately independent of velocity, so that a plot of H versus v provides a convenient method of determining both the axial dispersion and mass transfer resistance. For vapor-phase systems at low Reynolds numbers D_L is approximately constant since dispersion is determined mainly by molecular diffusion. It is therefore more convenient to plot H/v versus $1/v^2$, which yields D_L as the slope and the mass transfer resistance as the intercept. Examples of such plots are shown in Figure 16.

Applications

The applications of adsorbents are many and varied and may be classified as nonregenerative uses, in which the adsorbent is used once and discarded, and regenerative applications, in which the adsorbent is used repeatedly in a cyclic manner involving sequential adsorption and regeneration steps.

Nonregenerative Uses

Desiccant in dual pane windows
Odor removal in health care products
Desiccant in refrigeration and air conditioning systems
Cigarette filters

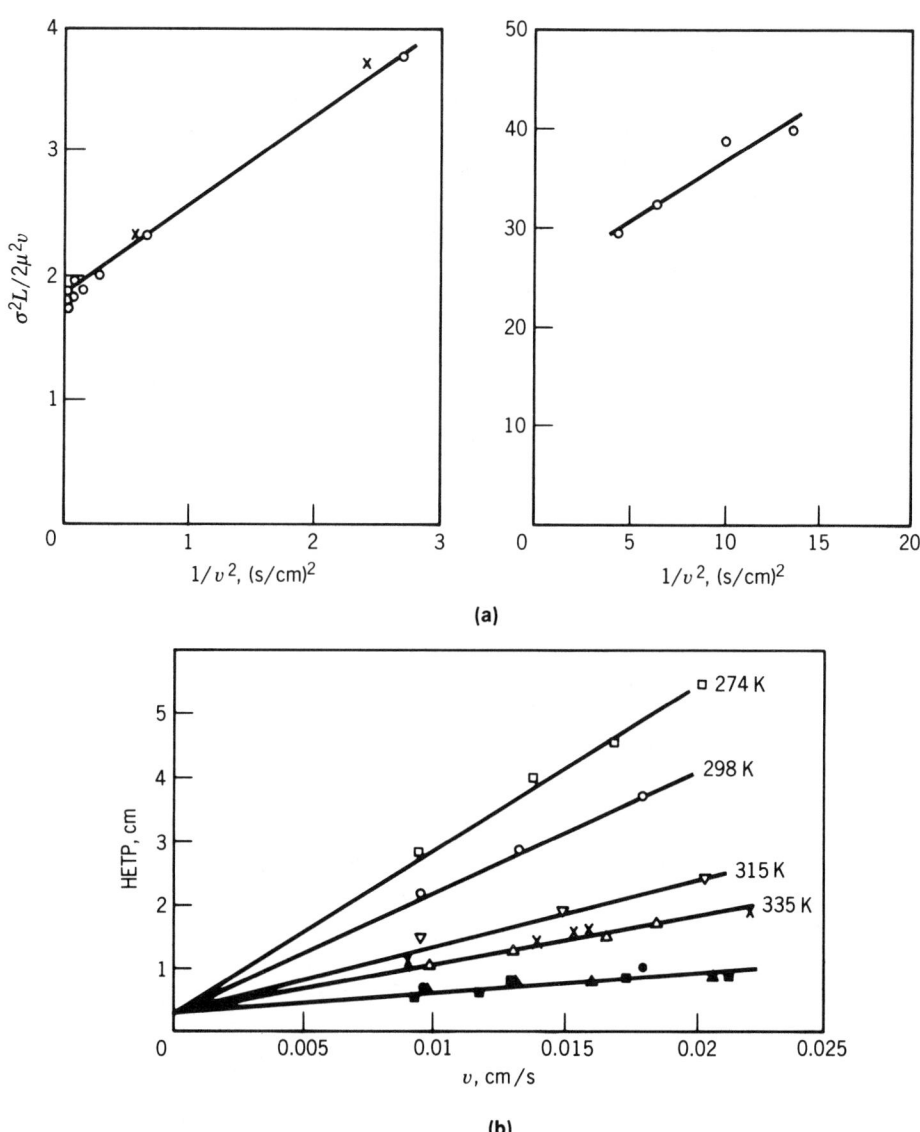

Fig. 16. Plots showing (a) variation of $(\sigma^2 L/2\mu^2 v)$ with $1/v^2$ for O_2 (left plot, ×, 0.84–0.72 mm = 20–25 mesh; ○, 0.42–0.29 mm = 40–50 mesh) and N_2 (right plot, on 3.2-mm pellets) in Bergbau-Forschung carbon molecular sieve and (b) variation of HETP with liquid velocity (interstitial) for fructose (solid symbols), and glucose (open symbols) in a column packed with KX zeolite crystals. From refs. 22 and 23.

Regenerative Uses

Water purification (some systems)

Removal of trace impurities from gases or liquid streams

Bulk separations (gas or liquid)

Low pressure storage of methane

Desiccant cooling (open-cycle air conditioning)

In terms of tonnage, some of the nonregenerative applications, notably as desiccants in dual pane windows, in cigarette filters, and in water purification, are surprisingly important, but most of the important chemical engineering applications are regenerative since, with a few notable exceptions, the cost of the adsorbent is too great to allow nonregenerative use.

The application of adsorbents (generally high area activated carbon) as a means of storing methane fuel (natural gas) at relatively high density under moderate pressure is relatively new. Whereas, capacities of about 180 m^3 STP/m^3 at 3 MPa (30 atm) pressure are achievable (24), somewhat higher capacities are needed to compete with liquid fuels for motor vehicles. However, depending on the cost of crude oil and the potential for improvement of the adsorbent, this technology could become important in the future.

Open-cycle desiccant cooling is another area of emerging technology (8). Rather than cooling and dehumidifying by mechanical work, as in a conventional air conditioning system, in an open-cycle desiccant system dehumidification is achieved directly, while cooling is achieved by controlled evaporation. The energy input is in the form of the heat required to regenerate the desiccant. A significant advantage of this system is that it can be designed to operate with a low regeneration temperature, thus making it possible to utilize low grade heat or even solar heat to drive the system.

Adsorption Separation and Purification Processes. The main area of application of absorption is in separation and purification processes. Many different ways of operating such processes have been devised and it is helpful to consider the various systems according to the mode of fluid–solid contact (see Fig. 17). In a cyclic batch process at least two beds are employed and each bed is successively saturated with the preferentially adsorbed species (or class of species) during the adsorption step and then regenerated during a desorption step in which the direction of mass transfer is reversed to remove the adsorbed species from the bed. In the continuous countercurrent process the adsorbent can be regarded as circulating continuously between the adsorption and desorption beds, in both of which fluid and solid contact in countercurrent flow. More commonly, as in the Sorbex type of process, the adsorbent is not physically circulated but the same effect is achieved in a fixed adsorbent bed equipped with multiple inlet and outlet ports to which the fluid streams are directed in sequence. Such systems can achieve a close approximation to countercurrent flow without the problems inhernet in circulating the solid adsorbent. However, the system is relatively expensive, so it is generally used only for difficult separations (low separation factor) which cannot be carried out efficiently in a simple batch process.

The other major difference between adsorption processes lies in the method by which the adsorbent bed is regenerated. The advantages and disadvantages of three different methods—temperature swing, pressure swing, and displacement—are summarized in Table 5. For efficient removal of trace impurities it is normally essential to use a highly selective adsorbent on which the sorbate is strongly held. Temperature swing regeneration is therefore generally used in such applications. However, in bulk separations all three regeneration methods are widely used.

Process Design. As with any chemical engineering process, the choice of process type and the details of the design are dictated primarily by economic considerations, subject to the overriding requirements of safety and reliability.

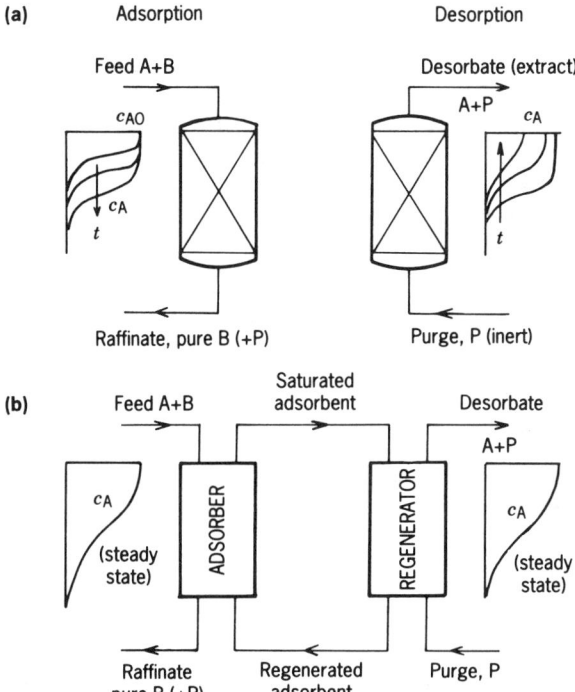

Fig. 17. The two basic modes of operation for an adorption process; (**a**) cyclic batch system; (**b**) continous countercurrent system with adsorbent recirculation. From Ref. 7.

Although the principles of adsorption processes are well understood, most practical designs still rely on a good deal of empiricism since factors such as the aging and deterioration of an adsorbent under practical operating conditions are difficult to predict except from experience.

In general the sophistication of the design procedures is closely related to the level of sophistication of the process. In the simple thermal swing, batch type processes for removal of trace impurities the beds are generally sized on the basis of equilibrium capacity and LUB with a suitable allowance for aging of the adsorbent. The desorption or regeneration temperature is generally selected on the basis of equilibrium data as the minimum temperature which will allow the required specification of the purity of the raffinate product to be easily met. Use of a higher regeneration temperature generally gives a purer product but only at the cost of increased energy consumption and reduction of the service life of the adsorbent. The quantity of purge gas is estimated in two ways: from the overall heat balance with the required temperature rise and from the mass balance with the assumption of equilibrium at the bed outlet. Depending on the particular system and the process conditions, either of these considerations may be limiting. In general, when regeneration is carried out by purging at atmospheric pressure, the purge requirement is determined by the heat balance, but when the regenera-

Table 5. Factors Governing Choice of Regeneration Method[a]

Method	Advantages	Disadvantages
thermal swing	good for strongly adsorbed species; small change in T gives large change in q^*	thermal aging of adsorbent
	desorbate may be recovered at high concentration	heat loss means inefficiency in energy usage
		unsuitable for rapid cycling, so adsorbent cannot be used with maximum efficiency
	gases and liquids	in liquid systems the latent heat of the interstitial liquid must be added
pressure swing	good where weakly adsorbed species is required at high purity	very low P may be required
		mechanical energy more expensive than heat
	rapid cycling—efficient use of adsorbent	desorbate recovered at low purity
displacement desorption	good for strongly held species	product separation and recovery needed (choice of desorbent is crucial)
	avoids risk of cracking reactions during regeneration	
	avoids thermal aging of adsorbent	

[a]Ref. 7.

tion is carried out at elevated pressure, the mass balance is often the major constraint.

The design of pressure swing systems is also largely dependent on the scale-up of pilot-plant units, although, with such systems, the use of a detailed numerical simulation to guide the optimization of the operating conditions is more common. This is true also of countercurrent and simulated countercurrent processes where the initial design is commonly based on a simple McCabe-Thiele diagram (25) (see ADSORPTION, LIQUID SEPARATION).

NOTATION

a	ratio of external surface area to particle volume
b	Langmuir equilibrium constant

B	mobility
c	sorbate concentration in fluid phase
c_0	initial value of c
D	diffusivity
D_e	effective diffusivity
D_m	molecular diffusivity
D_K	Knudsen diffusivity
D_L	axial dispersion coefficient
D_0	corrected diffusivity
D_∞	preexponential factor
D_p	pore diffusivity
E	activation energy
$-\Delta H_0$	limiting heat of adsorption
K	dimensionless equilibrium constant
K_0	preexponential factor
K'	Henry's law constant
K'_0	preexponential factor
k	rate constant
k_0	preexponential factor
k', k''	constants in Elovich equation
k_f	fluid film mass transfer coefficient
k_s	surface mass transfer coefficient
L	bed length
M	molecular weight
m_t	mass adsorbed (or desorbed) at time t
m_∞	mass adsorbed (or desorbed) at $t \to \infty$
p	partial pressure of sorbate
q	adsorbed phase concentration
q_0	initial value of q
$\bar{q}$	adsorbed phase concentration averaged over a particle
q^*	equilibrium value of q
q_s	saturation limit in Langmuir expression
r	radial coordinate
R	radius of adsorbent particle
R	gas constant
t	time
$\bar{t}$	mean residence time or stoichiometric time
t'	break time
T	temperature
$-\Delta U$	change of interval energy on adsorption
v	interstitial fluid velocity
w	wave velocity
w'	shock velocity
z	distance
α	separation factor
ϵ	voidage of adsorbent bed
ϵ_p	porosity of particle
μ	chemical potential
ρ	mean pore radius
σ^2	variance of pulse response
τ	tortuosity factor

BIBLIOGRAPHY

"Adsorptive separation, introduction" in the *Encyclopedia of Chemical Technology*, 3rd ed., Vol. 1, pp. 531–544, by Theodore Vermeulen, University of California, Berkeley; "Adsorption, theoretical" in *ECT* 2nd ed., Vol. 1, pp. 421–459, by Sydney Ross, Rensselaer Polytechnic Institute; "Adsorption, theoretical" in *ECT* 1st ed., Vol. 1, pp. 206–222, by P.H. Emmett, Mellon Institution of Industrial Research; in *ECT* 4th ed., Vol. 1, pp. 493–528, by D.M. Ruthven, University of New Brunswick, Canada.

1. A. Thijs, G. Peters, E. F. Vansant, I. Verhaert, and P. deBievre, *J. Chem. Soc. Faraday Trans. 1* **79**, 2821 (1983).
2. Y. Matsumura, *Proceedings of the First Indian Carbon Conference, New Delhi,* December 1982, pp. 99–106.
3. E. M. Flanigen and co-workers, *Nature* **271**, 512 (1978).
4. S. M. Klein and W. H. Abraham, *AIChE Symp. Ser.* **79**(230), 53 (1984).
5. D. W. Breck, *Zeolite Molecular Sieves,* John Wiley & Sons, Inc., New York, 1974.
6. W. M. Meier and D. H. Olson, *Atlas of Zeolite Structure Types,* Juris Druck and Verlag AG, Zurich, 1978.
7. D. M. Ruthven, *Principles of Adsorption and Adsorption Processes,* Wiley-Interscience, New York, 1984.
8. T. R. Penney and I. Maclaine-Cross, *Proceedings, Desiccant Cooling and Dehumidification Workshop, June 10–11, 1986, Chattanooga, Tenn.,* Sponsored by Electric Power Research Institute, Gas Research Institute, and Tennessee Valley Authority.
9. S. Brunauer, L. S. Deming, W. E. Deming, and J. E. Teller, *J. Am. Chem. Soc.* **62**, 1723 (1940).
10. D. M. Ruthven, *Sep. Purif. Methods* **5**(2), 189 (1976).
11. D. M. Ruthven and K. F. Loughlin, *J. Chem. Soc. Faraday Trans. 1* **68**, 696 (1972).
12. A. V. Kiselev and K. D. Shcherbakova in "Molecular Sieves," *Proceedings 1st International Zeolite Conference, London, 1967,* Society of Chemical Industry, London, 1968.
13. R. Sips, *J. Chem. Phys.* **16**, 490 (1948).
14. A. L. Myers and J. M. Prausnitz, *AIChE J.* **11**, 121 (1965).
15. P. G. Ashmore, *Catalysis and Inhibition of Chemical Reactions,* Butterworths, London, 1963, p. 164.
16. N. Wakao and T. Funazkri, *Chem. Eng. Sci.* **33**, 1375 (1978).
17. I. H. Doetsch, D. M. Ruthven, and K. F. Loughlin, *Can. J. Chem.* **52**, 2717 (1974).
18. J. A. Dominguez, D. Psaris, and A. I. La Cava, *AIChE Symp. Ser.* **84**(264), 73 (1988).
19. D. Basmadjian and D. W. Wright, *Chem. Eng. Sci.* **36**, 937 (1981).
20. A. I. Liapis and O. K. Crosser, *Chem. Eng. Sci.* **37**, 958 (1982).
21. E. Glueckauf, *Trans. Faraday Soc.* **51**, 1540 (1955).
22. D. M. Ruthven, N. S. Raghavan, and M. M. Hassan, *Chem. Eng. Sci.* **41**, 1325 (1986).
23. C. B. Ching and D. M. Ruthven, *Zeolites* **8**, 68 (1988).
24. S. S. Barton, J. A. Holland, and D. F. Quinn, *Proceedings of the 2nd International Conference on Adsorption, Santa Barbara, May 1986,* Engineering Foundation, New York, 1987, p. 99.
25. D. M. Ruthven and C. B. Ching, *Chem. Eng. Sci.* **44**, 1011 (1989).

General References

D. M. Ruthven, *Principles of Adsorption and Adsorption Processes,* Wiley-Interscience, New York, 1984.
M. Suzuki, *Adsorption Engineering,* Kodansba-Elsevier, Tokyo, 1990.
R. T. Yang, *Gas Separation by Adsorption Processes,* Butterworths, Stoneham, Mass., 1987.
P. Wankat, *Large Scale Adsorption and Chromatography,* CRC Press, Boca Raton, Fla., 1986.

D. Basmadjian, *The Little Absorption Book*, CRC Press, Boca Raton, Fla., 1996.

D. M. Ruthven, S. Farooq, and K. Knaebel, *Pressure Swing Adsorption*, VCH Publishers, New York, 1995.

A. E. Rodrigues and D. Tondeur, eds., *Percolation Processes*, NATO ASI No. 33, Sijthoff & Noordhoff, Alpen aan den Rijn, 1980.

A. E. Rodrigues, M. D. Le van, and D. Tondeur, *Adsorption: Science and Technology,* NATO ASI E158, Kluwer, Amsterdam, 1989.

N. Wakao, *Heat and Mass Transfer in Packed Beds,* Gordon & Breach, New York, 1982.

M. Smisek and S. Cerny, *Active Carbon,* Elsevier, Amsterdam, 1970.

T. Vermeulen, M. D. LeVan, N. K. Hiester, and G. Klein, "Adsorption and Ion Exchange," Section 16 of *Perry's Chemical Engineers' Handbook,* 6th ed., McGraw-Hill Book Co., New York, 1984.

DOUGLAS M. RUTHVEN
University of Maine

ADSORPTION, GAS SEPARATION

Gas-phase adsorption is widely employed for the large-scale purification or bulk separation of air, natural gas, chemicals, and petrochemicals (Table 1). In these uses it is often a preferred alternative to the older unit operations of distillation (qv) and absorption (qv).

An adsorbent attracts molecules from a gas stream, the molecules become concentrated on the surface of the adsorbent, and are removed from the gas phase. Many process concepts have been developed to allow the efficient contact of feed gas mixtures with adsorbents to carry out desired separations and to allow efficient regeneration of the adsorbent for subsequent reuse. In nonregenerative applications, the adsorbent is used only once and is not regenerated.

Most commercial adsorbents for gas-phase applications are employed in the form of pellets, beads, or other granular shapes, typically about 1.5 to 3.2 mm in diameter. Most commonly, these adsorbents are packed into fixed beds through which the gaseous feed mixtures are passed. Normally, the process is conducted in a cyclic manner. When the capacity of the bed is exhausted, the feed flow is stopped to terminate the loading step of the process, the bed is treated to remove the adsorbed molecules in a separate regeneration step, and the cycle is then repeated.

The growth in both variety and scale of gas-phase adsorption separation processes, particularly since 1970, is due in part to continuing discoveries of new, porous, high-surface-area adsorbent materials (particularly molecular sieve zeolites) and, especially, to improvements in the design and modification of adsorbents. Increasingly, the development of new applications requires close cooperation in adsorbent design and process cycle development and optimization.

Table 1. Commercial Adsorption Separations

Separation[a]	Adsorbent
Gas bulk separations	
normal paraffins, isoparaffins, aromatics	zeolite
N_2/O_2	zeolite
O_2/N_2	carbon molecular sieve
CO, CH_4, CO_2, N_2, Ar, NH_3/H_2	zeolite, activated carbon
acetone/vent streams	activated carbon
C_2H_4/vent streams	activated carbon
H_2O/ethanol	zeolite
Gas purifications	
H_2O/olefin-containing cracked gas, natural gas, air, synthesis gas, etc	silica, alumina, zeolite
CO_2/C_2H_4, natural gas, etc	zeolite
organics/vent streams	activated carbon, others
sulfur compounds/natural gas, hydrogen, liquified petroleum gas (LPG), etc	zeolite
solvents/air	activated carbon
odors/air	activated carbon
NO_x/N_2	zeolite
SO_2/vent streams	zeolite
Hg/chlor–alkali cell gas effluent	zeolite

[a]Ref. 1.

Adsorption Principles

The design and manufacture of adsorbents for specific applications involves manipulation of the structure and chemistry of the adsorbent to provide greater attractive forces for one molecule compared to another, or, by adjusting the size of the pores, to control access to the adsorbent surface on the basis of molecular size. Adsorbent manufacturers have developed many technologies for these manipulations. Whereas the technologies are considered proprietary, the broad principles are well known.

The attention of this article is focused on physical adsorption, which involves relatively weak intermolecular forces, because most commercial applications of adsorption rely on this phenomenon alone. Chemisorption is discussed only briefly herein relating to specific applications.

Adsorption Forces. Coulomb's Law allows calculations of the electrostatic potential resulting from a charge distribution, and of the potential energy of interaction between different charge distributions. Various elaborate computations are possible to calculate the potential energy of interaction between point charges, distributed charges, etc. See reference 2 for a detailed introduction.

An electric dipole consists of two equal and opposite charges separated by a distance. All molecules contain atoms composed of positively charged nuclei and negatively charged electrons. When a molecule is placed in an electric field between two charged plates, the field attracts the positive nuclei toward the negative plate and the electrons toward the positive plate. This electrical distortion, or polarization of the molecule, creates an electric dipole. When the field is removed, the distortion disappears, and the molecule reverts to its original condition. This electrical distortion of the molecule is called induced polarization; the dipole formed is an induced dipole.

The magnitude of the induced dipole moment depends on the electric field strength in accord with the relationship $\mu_i = \alpha F$, where μ_i is the induced dipole moment, F is the electric field strength, and the constant α is called the polarizability of the molecule. The polarizability is related to the dielectric constant of the substance. Group-contribution methods (2) can be used to estimate the polarizability from knowledge of the number of each type of bond within the molecule, eg, the polarizability of an unsaturated bond is greater than that of a saturated bond.

The total potential energy of adsorption interaction may be subdivided into parts representing contributions of the different types of interactions between adsorbed molecules and adsorbents. Adopting the terminology of Barrer (3), the total energy Φ_{Total} of interaction is the sum of contributions resulting from dispersion energy Φ_D, close-range repulsion Φ_R, polarization energy Φ_P, field–dipole interaction $\Phi_{F-\mu}$, field gradient–quadrupole interaction $\Phi_{\delta F-Q}$, and adsorbate–adsorbate interactions, denoted self-potential Φ_{SP}:

$$\Phi_{Total} = \underbrace{\Phi_D + \Phi_R + \Phi_P}_{\text{nonspecific}} + \underbrace{\Phi_{F-\mu} + \Phi_{\delta F-Q}}_{\text{specific}} + \underbrace{\Phi_{SP}}_{\text{adsorbate–adsorbate}}$$

The Φ_D and Φ_R terms always contribute, regardless of the specific electric charge distributions in the adsorbate molecules, which is why they are called nonspecific. The third nonspecific Φ_P term also always contributes, whether or not the adsorbate molecules have permanent dipoles or quadrupoles; however, for adsorbent surfaces which are relatively nonpolar, the polarization energy Φ_P is small.

The $\Phi_{F-\mu}$ and $\Phi_{\delta F-Q}$ terms are specific contributions, which are significant when adsorbate molecules possess permanent dipole and quadrupole moments. In the absence of these moments, these terms are zero, as is true also if the adsorbent surface has no electric fields, a completely nonpolar adsorbent.

Finally, the Φ_{SP} term is the contribution resulting from interactions between adsorbate molecules. At low coverages of the adsorbent by adsorbate molecules, this contribution approaches zero, and at high coverage it often causes a noticeable increase in the heat of adsorption.

The $\Phi_D + \Phi_R$ (dispersion plus repulsion) terms are known as the London or van der Waals forces. Spherical, nonpolar molecules are well described by the familiar Lennard-Jones 6–12 potential equation:

$$\Phi_D + \Phi_R = 4\epsilon[-(\sigma/r)^6 + (\sigma/r)^{12}]$$

where r is the intermolecular separation distance, and σ (length units) and ϵ (energy units) are constants characteristic of the colliding molecules. Values of force constants σ and ϵ have been compiled (2).

These forces arise from the fact that each molecule contains atoms having a nucleus and surrounded by a cloud of electrons. The electron cloud fluctuates and is nonsymmetrical at various instants in time. Although a nonpolar neutral molecule has no net permanent charge or dipole, these fluctuating electron distributions provide fluctuating dipoles in each molecule. These fluctuating dipoles interact to generate forces between molecules or between adsorbed molecules and adsorbent surfaces. These contributions to the potential energy of adsorption are present even if the adsorbed molecules are nonpolar and even if the adsorbent structure contains no strong electrostatic fields.

The contribution Φ_P is due to the polarization of the molecules by electric fields on the adsorbent surface, eg, electric fields between positively charged cations and the negatively charged framework of a zeolite adsorbent. The attractive interaction between the induced dipole and the electric field is called the polarization contribution. Its magnitude is dependent upon the polarizability α of the molecule and the strength of the electric field F of the adsorbent (4): $\Phi_P = -\frac{1}{2}\alpha F^2$.

The first of the two specific interaction terms $\Phi_{F-\mu}$ is due to the attractive interaction between the permanent dipole moment μ of a molecule and the electric field on the adsorbent surface (4):

$$\Phi_{F-\mu} = -F\mu \cos \Theta$$

where Θ is the dipole–axis/field angle.

The other specific interaction term $\Phi_{\delta F-Q}$ is due to the attractive interaction between the permanent quadrupole moment Q of the molecule and the electric field gradient on the adsorbent surface (4):

$$\Phi_{\delta F-Q} = \frac{1}{2}Q \, dF/dr$$

The final contribution, the self-potential term, Φ_{SP}, is the sum of all the above interactions of adsorbed molecules with each other.

Finally, an analysis of the energies of adsorption on many practical polar and nonpolar adsorbents has shown not only that the magnitude of the Φ_P term depends directly upon the polarizability α, but also that the sum of all of the nonspecific terms taken together, ie, $\Phi_D + \Phi_R + \Phi_P$, increases monotonically, with increasing α (4).

Adsorption Selectivities. For a given adsorbent, the relative strength of adsorption of different adsorbate molecules depends on the relative magnitudes of the polarizability α, dipole moment μ, and quadrupole moment Q of each. These properties of some common molecules are given in Table 2. Often, just the consideration of the values of α, μ, and Q allows accurate qualitative predictions to be made of the relative strengths of adsorption of given molecules on an adsorbent or of the best adsorbent type (polar or nonpolar) for a particular separation.

For example, the strength of the electric field F and field gradient ($\delta F = dF/dr$) of the highly polar cationic zeolites is strong. For this reason, nitrogen is more strongly adsorbed than is oxygen on such adsorbents, primarily because of the stronger quadrupole of N_2 compared to O_2.

In contrast, nonpolar activated carbon adsorbents lack strong electric fields

Table 2. Electrostatic Properties of Common Gases

Molecule	Polarizability $\alpha \times 10^{40}$, $C^2 \cdot m^2/J$ [a]	Dipole moment $\mu \times 10^{30}$, $C \cdot m$ [b]	Quadrupole moment $Q \times 10^{40}$, $C \cdot m^2$ [c]
Ar	1.83	0.00	0.00
H_2	0.90	0.00	2.09
N_2	0.78	0.00	−4.91
O_2	1.77	0.00	−1.33
CO	2.19	0.37	−6.92
CO_2	3.02	0.00	−13.71
CS_2	9.41	0.00	12.73
N_2O	3.32	0.54	−12.02
NH_3	2.67	5.10	−7.39
C_2H_6	4.97	0.00	−3.32
C_6H_6	11.49	0.00	−30.7
HCl	2.94	3.57	13.28

[a] To convert $C^2 \cdot m^2/J$ to cm^3, divide by 1.113×10^{-16}.
[b] To convert $C \cdot m$ to debyes, divide by 3.336×10^{-30}.
[c] To convert $C \cdot m^2$ to Buckinghams, divide by 3.336×10^{-40}.

and field gradients. Such adsorbents adsorb O_2 slightly more strongly than N_2, because of the slightly higher polarizability of O_2. Relative selectivities on nonpolar adsorbents often parallel the relative volatilities of the same compounds. Compounds with higher boiling points are more strongly adsorbed. In this case, the higher boiling O_2 (bp ~90 K) is more strongly adsorbed than is N_2 (bp ~77 K).

The polarizabilities of molecules in a homologous series increase steadily with increasing numbers of atoms. Therefore, the relative strengths of adsorption also increase (along with the boiling points).

For a given adsorbate molecule, the relative strength of adsorption on different adsorbents depends largely on the relative polarizability and electric field strengths of adsorbent surfaces. On the one hand, water molecules, with relatively low polarizability but a strong dipole and moderately strong quadrupole moment, are strongly adsorbed by polar adsorbents (eg, cationic zeolites), but only weakly adsorbed by nonpolar adsorbents (eg, silicalite or nonoxidized forms of activated carbon). On the other hand, saturated hydrocarbons with low molecular weight have greater polarizabilities than does water, but no dipoles and only weak quadrupoles. These molecules are adsorbed less strongly than water on polar adsorbents, but more strongly than water on nonpolar adsorbents. Therefore, polar adsorbents are often called hydrophilic adsorbents and nonpolar adsorbents are called hydrophobic adsorbents.

Isotherms and Isobars. The graphical presentation of the equilibrium adsorbate loading vs adsorbate pressure (or concentration) at constant temperature (Fig. 1) is an adsorption isotherm (1). A graph of the adsorbate loading vs temperature at constant adsorbate pressure (Fig. 2) is an adsorption isobar (1). The greater the strength of adsorption, the greater is the adsorbate loading at a given temperature and partial pressure of the adsorbate up to the point where the maximum adsorption capacity of the adsorbent has been attained.

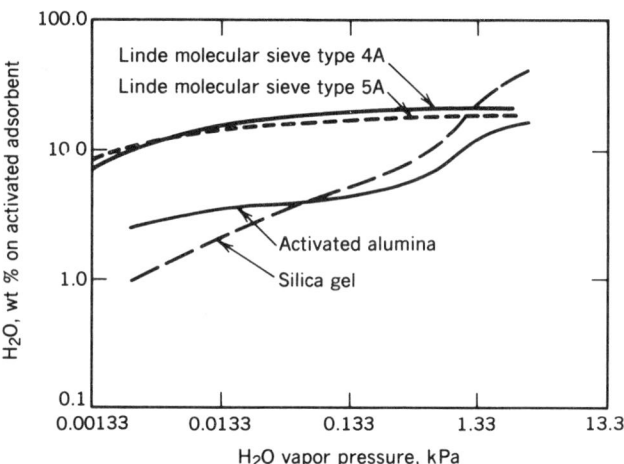

Fig. 1. Water isotherms for various adsorbents (1). Activation conditions: Linde molecular sieves, 350°C and <1.33 Pa; activated alumina, 350°C and <1.33 Pa; silica gel, 175°C and <1.33 Pa. To convert kPa to mm Hg, divide by 0.133.

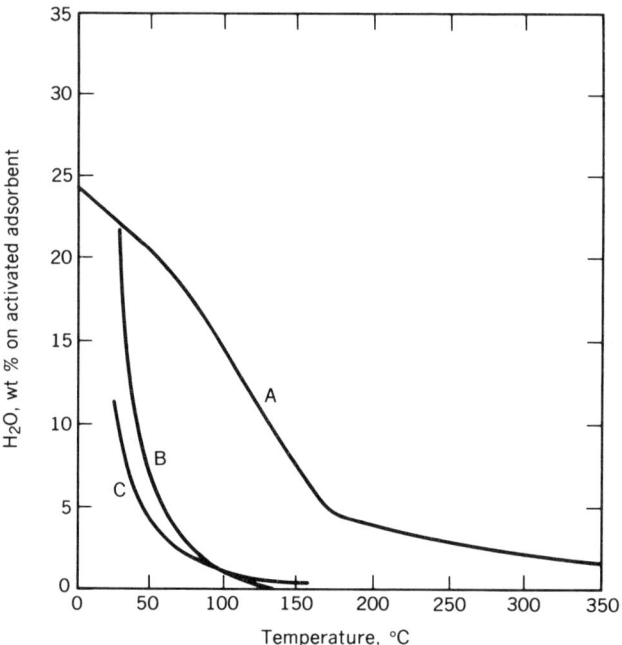

Fig. 2. Water isobars for various adsorbents: Equilibrium H_2O capacity vs temperature for three adsorbents (1); $p_{H_2O} = 1.33$ kPa (10 mm Hg) at 12°C and 101.3 kPa (1 atm). Activation conditions: A, Linde molecular sieve, 350°C and <1.33 Pa; B, activated alumina, 350°C and <1.33 Pa; C, silica gel, 175°C and <1.33 Pa. To convert Pa to mm Hg, multiply by 0.0075.

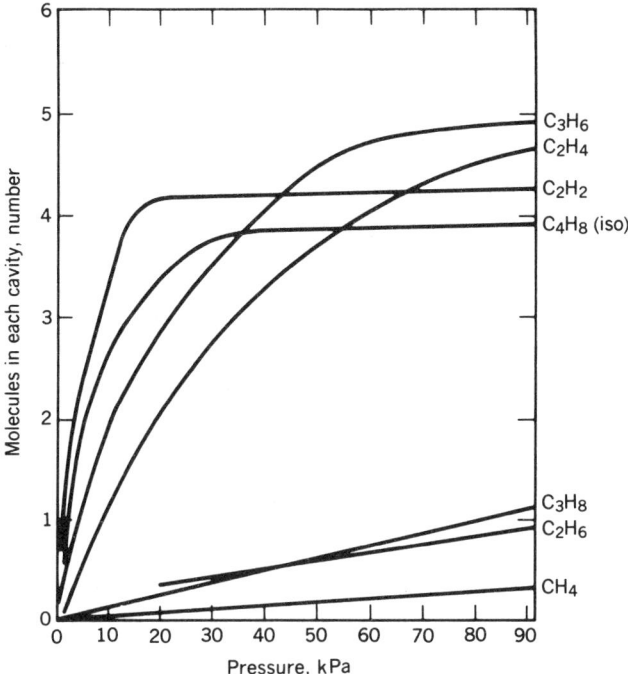

Fig. 3. Adsorption of hydrocarbons by zeolites is much greater for unsaturated hydrocarbons having double or triple bonds. From top to bottom, the curves show adsorption at 150°C of propylene, ethylene, acetylene, and isobutylene (unsaturated) and propane, ethane, and methane (saturated) (5). To convert kPa to mm Hg, multiply by 7.5. Courtesy of *Scientific American*.

The strength of adsorption of unsaturated hydrocarbons by a polar adsorbent (zeolite) is much greater than for saturated hydrocarbons, and increases with increasing carbon number (Fig. 3) (5). This observation may be understood as a consequence of the increasing polarizability of molecules with increasing numbers of bonds and the presence of dipole and stronger quadrupole moments in the unsaturated hydrocarbons compared to the saturated hydrocarbons.

Heats of Adsorption. Physical adsorption processes are exothermic, ie, they release heat. Because the entropy change ΔS on adsorption is negative (adsorbed molecules are more ordered than in the gas phase) and the free energy change ΔG must be negative for adsorption to be favored, thermodynamics ($\Delta G = \Delta H - T \Delta S$) requires the enthalpy change ΔH on adsorption (heat of adsorption) to be negative (exothermic). Adsorption strengths thus decrease with increasing temperature.

The integral heat of adsorption is the total heat released when the adsorbate loading is increased from zero to some final value at isothermal conditions. The differential heat of adsorption δH_{iso} is the incremental change in heat of adsorption with a differential change in adsorbate loading. This heat of adsorption δH_{iso} may be determined from the slopes of adsorption isosteres (lines of constant adsorbate loading) on graphs of $\ln P$ vs $1/T$ (Fig. 4) (6) through the Clausius-

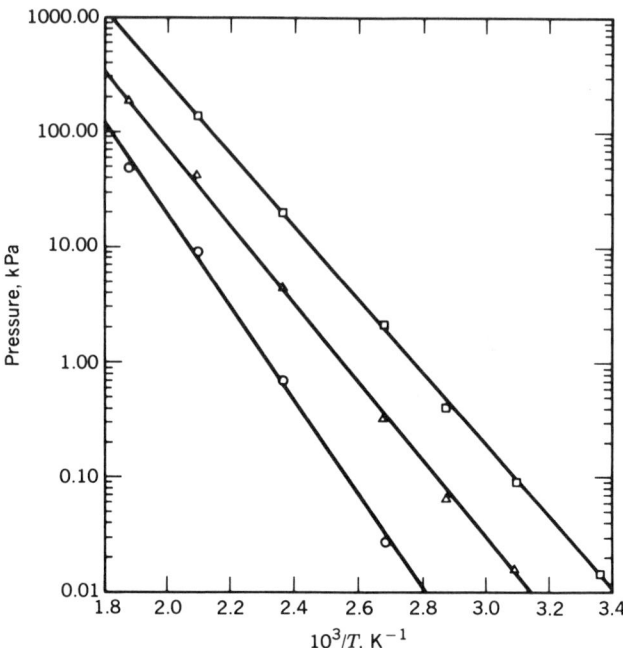

Fig. 4. Adsorption isosteres, water vapor on 4A (NaA) zeolite pellets (6). H_2O loading: (□) 15 kg/100 kg zeolite; (△), 10 kg/100 kg; (○), 5 kg/100 kg. To convert kPa to mm Hg, multiply by 7.5. Courtesy of Union Carbide.

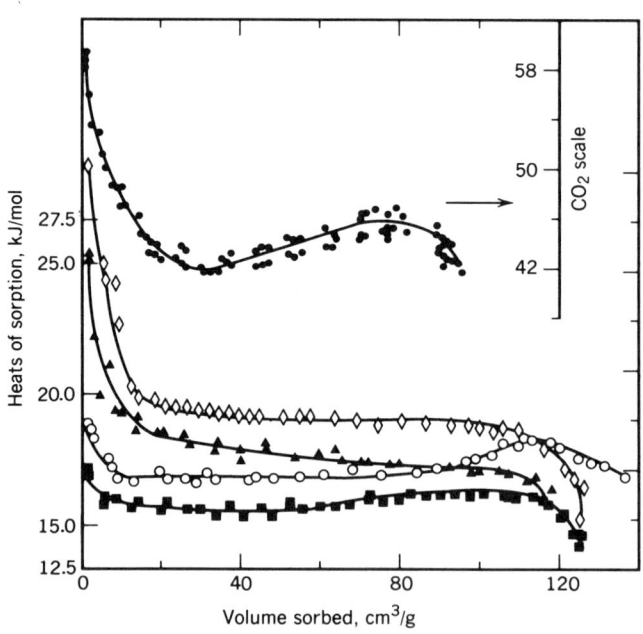

Fig. 5. Differential heats of sorption in natural chabazite (4). (▲) = N_2; (■) = Ar; (○) = O_2; (◇) = CO; (●) = CO_2. See Table 2 for polarizability, dipole moment, and quadrupole moment values for the gases. Volume adsorbed is expressed as cm^3 of adsorbate as liquid. To convert kJ to kcal, divide by 4.184. Courtesy of Academic Press.

Clapeyron relationship:

$$\frac{d \ln P}{d(1/T)} = -\frac{\delta H_{iso}}{R}$$

where R is the gas constant, P the adsorbate absolute pressure, and T the absolute temperature.

Differential heats of adsorption for several gases on a sample of a polar adsorbent (natural zeolite chabazite) are shown as a function of the quantities adsorbed in Figure 5 (4). Consideration of the electrical properties of the adsorbates, included in Table 2, allows the correct prediction of the relative order of adsorption selectivity:

$$Ar < O_2 < N_2 < CO << CO_2$$

At low adsorbate loadings, the differential heat of adsorption decreases with increasing adsorbate loadings. This is direct evidence that the adsorbent surface is energetically heterogeneous, ie, some adsorption sites interact more strongly with the adsorbate molecules. These sites are filled first so that adsorption of additional molecules involves progressively lower heats of adsorption.

All practical adsorbents have surfaces that are heterogeneous, both energetically and geometrically (not all pores are of uniform and constant dimensions). The degree of heterogeneity differs substantially from one adsorbent type to another. These heterogeneities are responsible for many nonlinearities, both in single component isotherms and in multicomponent adsorption selectivities.

In Figure 5, the heat of adsorption of CO_2 increases slightly at the higher adsorbate loadings. This increase is due to the increasing self-potential contribution at the higher loadings.

Isotherm Models. Many efforts have been made over the years to develop isotherm models for data correlation and design predictions for both single component and multicomponent adsorption. Unfortunately, no single model is accurate over broad ranges of adsorbent and adsorbate types, pressures, temperatures, and loadings, especially for multicomponent systems. This is probably due to deficiencies in the models in adequately describing both the heterogeneities of the surface and the effects of the adsorbate on the properties of the adsorbent itself. Most models assume the adsorbent is inert, ie, not altered by the presence of the adsorbate molecules; however, partial changes in some adsorbent properties are commonly observed.

Each of the more popular isotherm models have been found useful for modeling adsorption behavior in particular circumstances. Detailed discussions of derivations, assumptions, strengths, and weaknesses of isotherm models are available (4 and 7–16). Not all of the isotherm models discussed herein are rigorous in the sense of being thermodynamically consistent. For example, specific deficiencies in the Freundlich, Sips, Dubinin-Radushkevich, Toth, and vacancy solution models have been identified (14).

The Sips and related loading ratio correlation (LRC) models fail to properly predict Henry's law behavior, as required for thermodynamic consistency, at the zero pressure limit (8). Thermodynamic inconsistency of the LRC model was noted by the original authors (17); nevertheless, the model is useful in predicting multicomponent performance from single component data and correlating multicomponent data (18). Users of models lacking thermodynamic consistency must take care, however, particularly to avoid extrapolation beyond the range of experimental data.

Thermodynamically Consistent Isotherm Models. These models include both the statistical thermodynamic models and the models that can be derived from an assumed equation of state for the adsorbed phase plus the thermodynamics of the adsorbed phase, ie, the Gibbs adsorption isotherm,

$$\left(\frac{d\Phi}{dP}\right)_T = \frac{qRT}{P}$$

where Φ is the spreading pressure, P the partial pressure of adsorbate, q the adsorbate loading x per quantity w of adsorbent $= x/w$, T the temperature, and R the gas constant. In the following models, $\Theta = q/q_{max}$ is the fractional surface coverage, where q_{max} is the maximum loading. Constants are q_{max}, all K's, all k's, all A's, b, c, n, s, t, β, and τ. The vapor pressure of pure adsorbate is P_0.

Law or model	Equation	Reference
Henry's law	$q = KP_0$	1
Langmuir	$K'P_0 = \Theta/(1 - \Theta)$	19
Volmer	$bP_0 = [\Theta/(1 - \Theta)] \exp[\Theta/(1 - \Theta)]$	20
van der Waals	$K''P_0 = [\Theta/(1 - \Theta)] \exp[\Theta/(1 - \Theta)] \exp[\beta/RT]$	8
Virial	$K'''P_0/x = \exp[2A_1 x + (3/2)A_2 x^2 + \cdots]$	8

Statistical Thermodynamic Isotherm Models. The use of statistical thermodynamic isotherms was begun in the 1930s (21) and expanded in 1960 (22). Examples of the application of this approach to modeling of adsorption in microporous adsorbents are given in References 3, 23–27. Excellent reviews have been written (4,28).

Semiempirical Isotherm Models. Some of the semiempirical isotherm models have been shown to have some thermodynamic inconsistencies. Nevertheless, each has been found to be useful for data correlation and interpolation, as well as for the calculation of some thermodynamic properties.

Models Based on the Polanyi Adsorption Potential:

$$A = RT \ln(P_0/P)$$

Dubinin-Radushkevich. This model (29) is the same as the more general Dubinin-Astakhov equation (30) when $n = 2$.

Dubinin-Astakhov:

$$\Theta = \exp[-(A/E)^n]$$

where n is generally between 1 and 3.

Radke-Prausnitz. This model (31) is also known as the Langmuir-Freundlich model:

$$\Theta = \frac{k'P}{[1 + (k'P)]^\tau} \quad \text{for} \quad 0 < \tau \le 1$$

Toth. This model (32) is represented as:

$$\Theta = \frac{kP}{[1 + (kP^t)^{1/t}]}$$

UNILAN. The uniform distribution, Langmuir local isotherm model (9,12):

$$\Theta = \frac{1}{2s} \ln\left[\frac{(c + Pe^s)}{(c + Pe^{-s})}\right]$$

where Pe is the Peclet number.

BET. The Brunanauer-Emmett-Teller model (33) estimates the coverage corresponding to one monolayer of adsorbate and is used to measure the surface area of solids:

$$\Theta = \frac{b(P/P_0)}{[(1 - P/P_0)(1 - P/P_0 + bP/P_0)]}$$

Isotherm Models for Adsorption of Mixtures. Of the following models, all but the ideal adsorbed solution theory (IAST) and the related heterogeneous ideal adsorbed solution theory (HIAST) have been shown to contain some thermodynamic inconsistencies. References to the limited available literature data on the adsorption of gas mixtures on activated carbons and zeolites have been compiled, along with a brief summary of approximate percentage differences between data and theory for the various theoretical models (16). In the following the subscripts i and j refer to different adsorbates.

Markham and Benton. This model (34) is known as the extended Langmuir isotherm equation for two components, i and j:

$$\Theta_i = K_i P_i / (1 + K_i P_i + K_j P_j)$$
$$\Theta_j = K_j P_j / (1 + K_i P_i + K_j P_j)$$

Leavitt Loading Ratio Correlation (LRC) Method. The LRC model (17) for a single component i parallels Sips model (35):

$$\Theta_i = (K_i P_i)^{1/ni} / [1 + (K_i P_i)^{1/ni}]$$

but with

$$-\ln K_i = A_{1i} + A_{2i}/T$$

For the binary system of components i and j, the LRC model (17) is

$$\Theta_i = (K_i P_i)^{1/ni} / [1 + (K_i P_i)^{1/ni} + (K_j P_j)^{1/nj}]$$

Ideal Adsorbed Solution (IAS) Model. For components i and j, assuming ideal gas behavior, this model (36) is

$$\frac{\Phi A}{RT} = \int_0^{P_i^\circ} [q_i^\circ(P)] \, d(\ln P) = \int_0^{P_j^\circ} [q_j^\circ(P)] \, d(\ln P)$$

$$PY_i = P_i^\circ X_i$$

$$PY_j = P_j^\circ X_j = P_j^\circ (1 - X_i)$$

where P_i° is the vapor pressure of component i, $q_i^\circ(P)$ the equilibrium loading of pure i at pressure P, Y_i the vapor phase mole fraction of component i, and X_i the adsorbed phase mole fraction of component i. These equations are solved simultaneously to determine P_i°, P_j°, and X_i, and the following equations are used to calculate q_i, q_j, and q_{total}:

$$\{1/q_{\text{total}}\} = X_i/[(q_i)(P_i^\circ)] + X_j/[(q_j)(P_j^\circ)]$$

$$q_i = q_{\text{total}} X_i$$

$$q_j = q_{\text{total}} X_j$$

Heterogeneous Ideal Adsorbed Solution Theory (HIAST). This IAS theory has been extended to the case of adsorbent surface energetic heterogeneity and is shown to provide improved predictions over IAST (12).

Vacancy Solution Model. The initial model (37) considered the adsorbed phase to be a mixture of adsorbed molecules and vacancies (a vacancy solution) and assumed that nonidealities of the solution can be described by the two-parameter Wilson activity coefficient equation. Subsequently, it was found that the use of the three-parameter Flory-Huggins activity coefficient equation provided improved prediction of binary isotherms (38).

Adsorption Dynamics. An outline of approaches that have been taken to model mass-transfer rates in adsorbents has been given (see ADSORPTION). Detailed reviews of the extensive literature on the interrelated topics of modeling of mass-transfer rate processes in fixed-bed adsorbers, bed concentration profiles, and breakthrough curves include references 16 and 26. The related simple design concepts of WES, WUB, and LUB for constant-pattern adsorption are discussed later.

Reactions on Adsorbents. To permit the recovery of pure products and to extend the adsorbent's useful life, adsorbents should generally be inert and not react with or catalyze reactions of adsorbate molecules. These considerations often affect adsorbent selection and/or require limits be placed upon the severity of operating conditions to minimize reactions of the adsorbate molecules or damage to the adsorbents.

However, even then, gradual reactions of trace impurities in a feed stream or slowly occurring reactions that modify the adsorbent may still cause a gradual decline in the adsorbent performance, as illustrated in Figure 6 (39). To compensate, adsorbent beds are sized to account for the gradual loss in capacity and to allow use for a given period of time. Most commonly, at the end of its useful life,

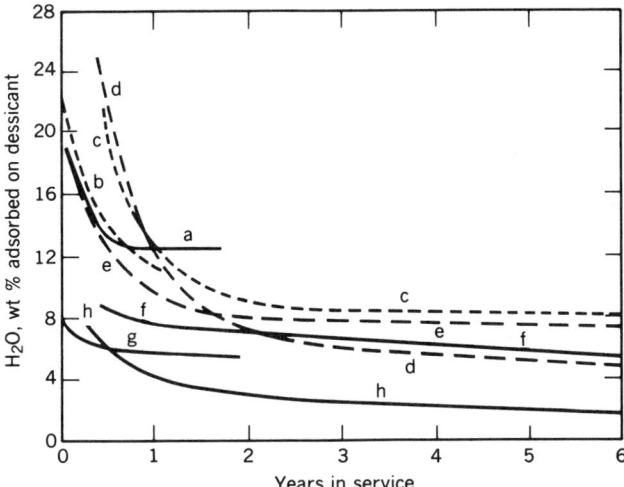

Fig. 6. Adsorption capacity of various dessicants vs years of service in dehydrating high pressure natural gas (39). a, Alumina H-151, gas ~27°C and 123 kPa, from oil and water separators; b, silica gel, gas ~38°C and 145 kPa, from oil absorption plant; c, sorbead, 136-kPa gas from absorption plant; regeneration gas inlet temperature 243°C (maximum allowable dew point −6.7°C); d, sorbead, 40-kPa gas containing propane; regeneration gas temperature 177°C (maximum allowable dew point −34°C); e, sorbead, 1950–1956 data; f, activated alumina, same gas as for Curve d; g, activated bauxite(florite), residue gas from gasoline absorption plant; h, activated alumina, same gas for for Curve c. Courtesy of Gulf Publishing Company.

the adsorbent is dumped from the beds and replaced with fresh adsorbent. However, in some cases, the process equipment is designed to allow periodic *in situ* rejuvenation of the adsorbent, eg, a periodic burning-off of coke accumulated on the adsorbent.

Adsorbent Principles

Principal Adsorbent Types. Commercially useful adsorbents can be classified by the nature of their structure (amorphous or crystalline), by the sizes of their pores (micropores, mesopores, and macropores), by the nature of their surfaces (polar, nonpolar, or intermediate), or by their chemical composition. All of these characteristics are important in the selection of the best adsorbent for any particular application.

However, the size of the pores is the most important initial consideration because, if a molecule is to be adsorbed, it must not be larger than the pores of the adsorbent. Conversely, by selecting an adsorbent with a particular pore diameter, molecules larger than the pores may be selectively excluded, and smaller molecules can be allowed to adsorb.

Pore size is also related to surface area and thus to adsorbent capacity, particularly for gas-phase adsorption. Because the total surface area of a given mass of adsorbent increases with decreasing pore size, only materials containing micropores and small mesopores (nanometer diameters) have sufficient capacity to be useful as practical adsorbents for gas-phase applications. Micropore diameters are less than 2 nm; mesopore diameters are between 2 and 50 nm; and macropores diameters are greater than 50 nm, by IUPAC classification (40).

The practical adsorbents used in most gas phase applications are limited to the following types, classified by their amorphous or crystalline nature.

Amorphous: silica gel, activated alumina, activated carbon, and molecular sieve carbons.

Crystalline: molecular sieve zeolites, and related molecular sieve materials that are not technically zeolites, eg, silicalite, $AlPO_4$s, silicon aluminum phosphates (SAPOs), etc.

Typical pore size distributions for these adsorbents have been given (see ADSORPTION). Only molecular sieve carbons and crystalline molecular sieves have large pore volumes in pores smaller than 1 nm. Only the crystalline molecular sieves have monodisperse pore diameters because of the regularity of their crystalline structures (41).

Activated carbons are made by first preparing a carbonaceous char with low surface area followed by controlled oxidation in air, carbon dioxide, or steam. The pore-size distributions of the resulting products are highly dependent on both the raw materials and the conditions used in their manufacture, as may be seen in Figure 7 (42).

Assuming the pores are large enough to admit the molecules of interest, the most important consideration is the nature of the adsorbent surface, because this characteristic controls adsorption selectivity. Practical adsorbents may also be classified according to the nature of their surfaces.

Highly polar: molecular sieve zeolites with high aluminum and cation contents.

Moderately polar: crystalline molecular sieves with low aluminum and low cation contents, silica gel, activated alumina, activated carbons with highly oxidized surfaces, crystalline molecular sieve $AlPO_4$s.

Nonpolar: silicalite, F-silicalite, other high silica content crystalline molecular sieves, activated carbons with reduced surfaces.

Adsorption Properties. Adsorption isotherms for water on various adsorbents are given in Figure 1, and the corresponding isobars in Figure 2. Not only do the more highly polar molecular sieve zeolites adsorb more water at lower pressures than do the moderately polar silica gel and alumina gel, but these also hold onto the water more strongly at higher temperatures. For the same reason, temperatures required for thermal regeneration of water-loaded zeolites is higher than for less highly polar adsorbents.

Isotherms for H_2O and n-hexane adsorption at room temperature and for O_2 adsorption at liquid oxygen temperature on 13X (NaX) zeolite and on the crystalline SiO_2 molecular sieve silicalite are shown in Figure 8 (43). Silicalite

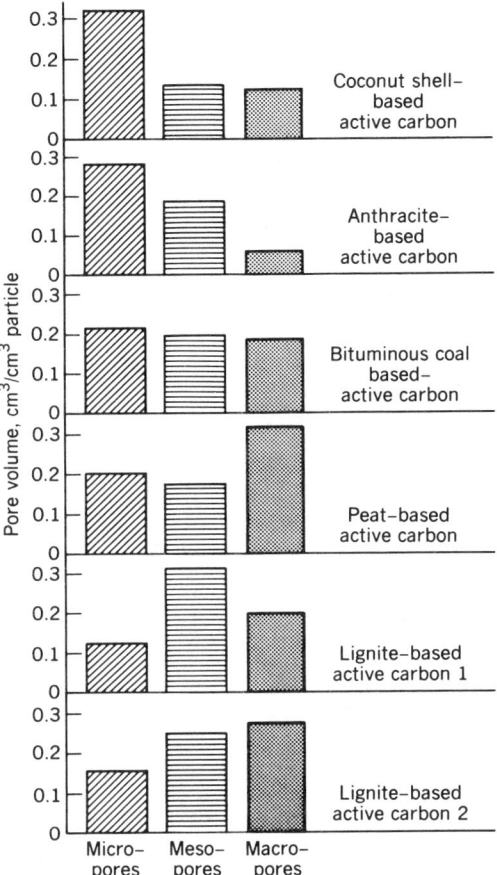

Fig. 7. Pore size distribution in some active carbons obtained using different precursors (42). Courtesy of Marcel Dekker Publishing Company.

adsorbs water very weakly. Further modification of silicalite by fluoride incorporation provides an extremely hydrophobic adsorbent, shown in Figure 9 (44). These examples illustrate the broad range of properties of crystalline molecular sieves.

Activated carbons contain chemisorbed oxygen in varying amounts unless special care is taken to eliminate it. Desired adsorption properties often depend upon the amount and type of chemisorbed oxygen species on the surface. Therefore, the adsorption properties of an activated carbon adsorbent depend on its prior temperature and oxygen-exposure history. In contrast, molecular sieve zeolites and other oxide adsorbents are not affected by oxidizing or reducing conditions (see Fig. 10 (45)).

Water adsorption at low pressures is markedly reduced on a poly(vinylidene chloride)-based activated carbon after removal of surface oxygenated groups by degassing at 1000°C. Following this treatment, water adsorption is dominated by capillary condensation in mesopores, and the size of the adsorption–desorption hysteresis loop increases, because the pore volume previously occupied by water at

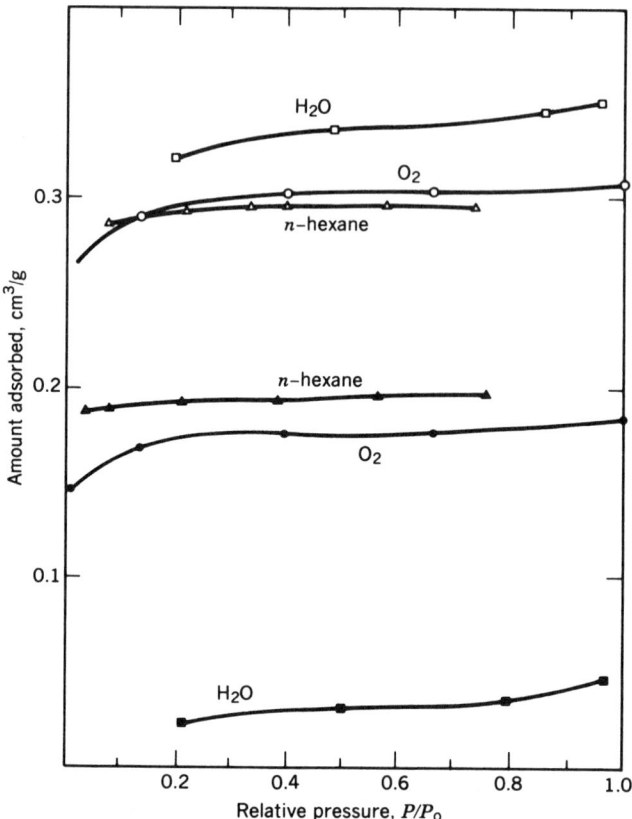

Fig. 8. Adsorption isotherms of H_2O, O_2, and n-hexane on zeolite NaX (open symbols) and silicalite (filled symbols). Oxygen is at $-183°C$ and water and n-hexane (C_6H_{14}) at RT. Volume adsorbed is expressed as cm^3 of adsorbate as liquid. Courtesy of *Nature, London* (43).

the lower pressures now remains empty until the water pressure reaches pressures (~ 0.3 to 0.4 times the vapor pressure) at which capillary condensation can occur.

Typical adsorption isotherms for light hydrocarbons on activated carbon prepared from coconut shells are shown in Figure 11 (46). The polarizabilities and boiling points of these compounds increase in the order

$$CH_4 < C_2H_4 < C_2H_6 < C_3H_6 < C_3H_8$$

The relative strengths of adsorption of these compounds follow the same order, as expected for a nonpolar adsorbent, except that C_3H_6 was adsorbed more strongly than C_3H_8. This result indicates that the surface is weakly polar and that specific (dipole–field and quadrupole–field gradient) contributions to the adsorption potential alter the expected order slightly. This situation may also result from chemisorbed oxygen species on the surface.

Physical Properties. Physical properties of importance include particle size, density, volume fraction of intraparticle and extraparticle voids when

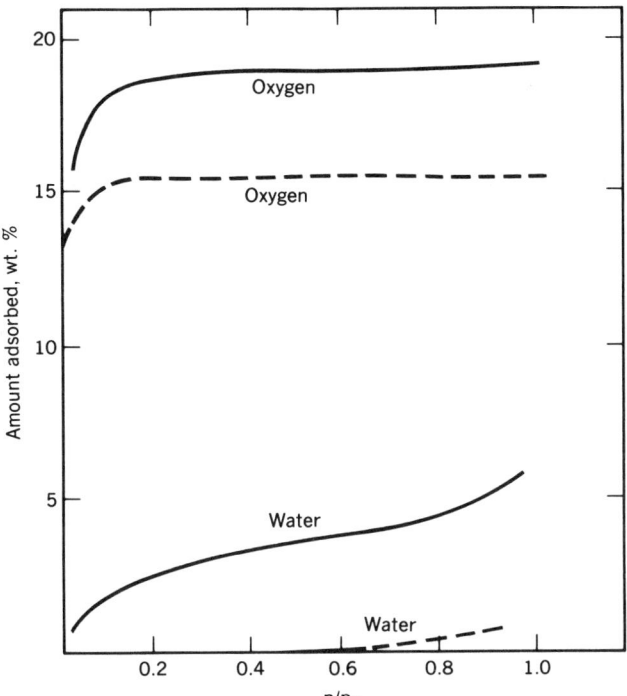

Fig. 9. Adsorption of oxygen (90 K) and water (RT) on silicalite (——) and F-silicalite (---) (44).

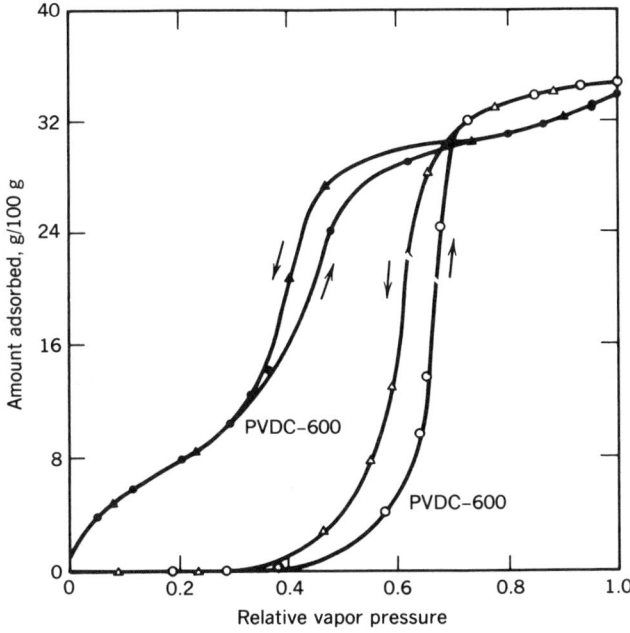

Fig. 10. Adsorption (●, ○)–desorption (▲, △) isotherms of water vapor on poly(vinylidene chloride) (PVDC) carbon before (filled symbols) and after (open symbols) outgassing at 1000°C (46). Courtesy of *Carbon*.

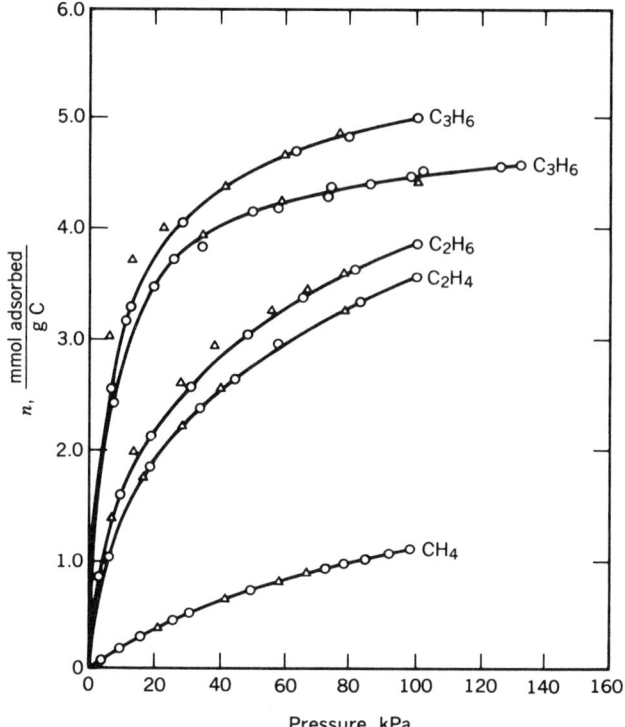

Fig. 11. Adsorption isotherms for hydrocarbons on activated coconut-shell carbon at 25°C (46). ○, Adsorption; △, desorption. To convert kPa to mm Hg, multiply by 7.5. Courtesy of *Industrial and Engineering Chemistry*.

packed into adsorbent beds, strength, attrition resistance, and dustiness. These properties can be varied intentionally to tailor adsorbents to specific applications (see ADSORPTION, LIQUID SEPARATION; ADSORPTION, GAS SEPARATION; ALUMINUM COMPOUNDS, ALUMINUM OXIDE; CARBON, ACTIVATED CARBON; ION EXCHANGE; MOLECULAR SIEVES).

Deactivation. Adsorbent degradation by chemical attack or physical damage is not reversible. Acids or acid gases can react with adsorbents with alkaline surface chemistry, eg, some zeolites, and cause loss of adsorption capacity. Other adsorbents, such as silica gel, are sensitive to alkalies. The constant thermal expansion and contraction in temperature swing adsorption (TSA) processes can cause damage to the internal pore and/or crystal structure. Activated alumina and silica gel can be dehydrated by excessive temperatures. When water is present, hydrothermal cycling can cause explosive steam release that physically damages some adsorbents. Some types of silica gel are susceptible to breakup caused by water droplets; special decrepitation-resistant grades are available.

Adsorption Processes

Adsorption processes are often identified by their method of regeneration. Temperature-swing adsorption (TSA) and pressure-swing adsorption (PSA) are the most frequently applied process cycles for gas separation. Purge-swing cycles and nonregenerative approaches are also applied to the separation of gases. Special applications exist in the nuclear industry. Others take advantage of reactive sorption. Most adsorption processes use fixed beds, but some use moving or fluidized beds.

TEMPERATURE SWING

A temperature-swing or thermal-swing adsorption (TSA) cycle is one in which desorption takes place at a temperature much higher than adsorption. The principal application is for separations in which contaminants are present at low concentration, ie, for purification. The TSA cycles are characterized by low residual loadings and high operating loadings. Figure 12 depicts the isotherms for the two temperatures of a TSA cycle. The available operating capacity is the difference between the loadings X_1 and X_2. These high adsorption capacities for low concentrations mean that cycle times are long, hours to days, for reasonably sized beds. This long cycle time is fortunate, because packed beds of adsorbent respond slowly to changes in gas temperature. A purge and/or vacuum removes the thermally desorbed components from the bed, and cooling returns the bed to

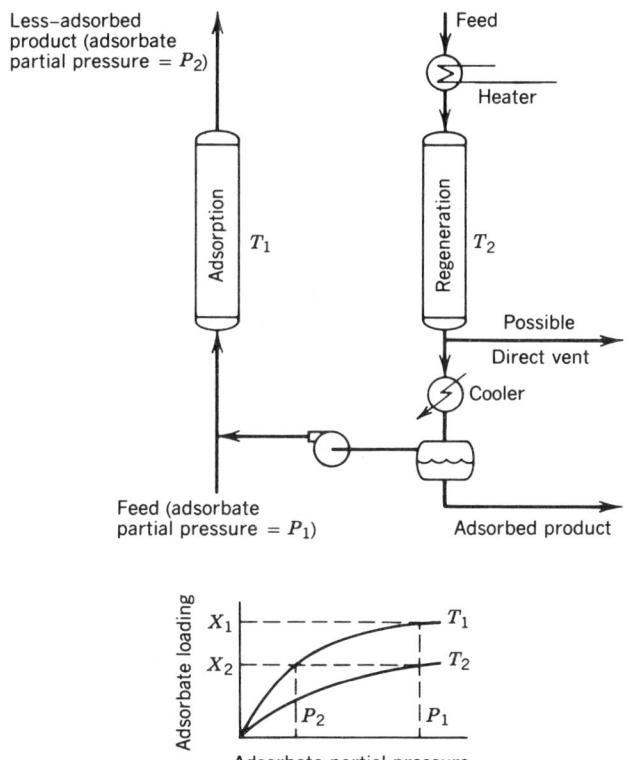

Fig. 12. Temperature-swing cycle (4). Loading X_1 at T_1 and feed partial pressure P_1; X_2 at the higher T_2 and the lower P_2 needed in the product (1).

adsorption condition. Systems in which species are strongly adsorbed are especially suited to TSA. Such applications include drying, sweetening, CO_2 removal, and pollution control.

Principles. In a TSA cycle, two processes occur during regeneration, heating and purging. Heating must provide adequate thermal energy to raise the adsorbate, adsorbent and adsorber temperature, desorb the adsorbate, and make up for heat losses. Regeneration is heating-limited (or stoichiometric-limited) when transfer of energy to the system is limiting. Equilibrium determines the maximum capacity of the purge gas to transfer the desorbed material away. Regeneration is stripping-limited (or equilibrium-limited) when transferring adsorbate away is limiting.

Heating occurs by either direct (external heat exchange to the purge gas) or, less commonly, indirect (heating elements, coils or panels, inside the adsorber) contact of the adsorbent by the heating medium. Direct heating is simpler and is invariably used for stripping-limited regeneration. Microwave fields (47) and dielectric fields (48) are alternative methods for supplying indirect heating. The complexity of indirect heating limits its use to heating-limited regeneration where purge gas is in short supply. Coils or panels can supply indirect cooling as well. The use of steam for the regeneration of activated carbon is a combination of thermal desorption and purge displacement; direct heating is supplied by water adsorption.

Another process for thermal cycling is parametric pumping. In the direct mode of parametric pumping, a single adsorbent bed is indirectly heated and cooled while the fluid feed is pumped forward and backward through the column from reservoirs at each end. For a binary fluid, one component concentrates in one reservoir and one in the other. In the recuperative mode of parametric pumping, the heating and cooling takes place outside the adsorbent column.

Steps. Thermal-swing cycles have at least two steps, adsorption and heating. A cooling step is also normally used after the heating step. A portion of the feed or product stream can be utilized for heating, or an independent fluid can be used. Easily condensable contaminants may be regenerated with noncondensable gases and recovered by condensation. Water-immiscible solvents are stripped with steam, which may be condensed and separated from the solvent by decantation. Fuel and/or air may be used when the impurities are to be burned or incinerated.

The highest regeneration temperatures are the most efficient for desorption. However, heater cost, metallurgy, and the thermal stability of the adsorbent and the fluids must be considered. Silica gel requires the lowest temperatures and the lowest amount of heat of any commercial adsorbent. Activated carbons and aluminas can tolerate the highest temperatures. Although thermal-swing regeneration can be done at the same pressure as adsorption, lowering the pressure can achieve better desorption and is often used; such cycles are actually a hybrid of PSA and TSA. The heating gas is normally used for the cooling step. Rather than cooling the bed, adsorption can sometimes be started on a hot bed. If certain criteria are met (49), the dynamic adsorption performance does not depend on cooling.

Flow Sheet. The most common processing scheme is a pair of fixed-bed adsorbers alternating between the adsorption step and the regeneration steps

(Fig. 12). However, the variations possible to achieve special needs are endless. Flow directions can be varied. Single beds provide interrupted flow, but multiple beds can ensure constant flow. Beds can be configured in lead–trim, parallel trains, series cool–heat, or closed-loop (1). Regeneration may even be *ex situ* rather than *in situ*.

The normal flow direction through a fixed bed is usually in a vertical direction. The mechanical complexities required for horizontal- or annular-flow beds often outweigh the decrease in pressure drop achieved. Because allowable velocities for crushing exceed those for lifting, the cycle step with the highest pressure drop should be downward. All other flows can then be in the same direction as the limiting flow (cocurrent) or in the opposite direction (countercurrent). Each combination of flow directions for heating and cooling produces a different residual of adsorbate (Fig. 13).

Although most applications of fixed bed have multiple adsorber beds to treat continuous streams, batch operation using a single adsorber bed is an alternative. For purification applications, where one vessel can contain enough adsorbent to provide treatment for days, weeks, or even months, the cost savings and simplicity often justify the inconvenience of stopping adsorption treatment periodically for a short regeneration.

When the mass transfer zone is a major portion of an adsorbent bed, the equilibrium capacity is poorly utilized. A lead–trim configuration uses the adsorbent more fully. The feed flows successively through a lead bed and then a trim bed. The lead bed is nearly exhausted before it is taken out of service to be regenerated. When a lead bed is removed from adsorption, the trim bed becomes the lead, and a fully regenerated bed becomes the new trim bed.

When large flows are to be treated, designing and building a single adsorber vessel large enough to treat the entire stream is not practical. Instead, the feed flow is split equally between parallel beds and/or trains of adsorbers. This provides the additional advantage of a convenient method of turning down the process to save on utilities.

At the start of the cooling step, the adsorber vessel is a large heat sink containing valuable energy: the sum of all of the sensible heats of the adsorbent, the vessel, and any internals. Using three adsorber beds—one on adsorption, one on heating, and one on cooling—the purge gas flows in series first to cool a hot bed and then to heat a spent bed. Thus all of the heat from the bed being cooled is recovered.

Thermal energy can be conserved by using a thermal-pulse cycle. When desorption is heat-limited, only a short soak time at temperature completes regeneration. The entire adsorbent bed need not be at desorption temperature before beginning the cooling step. Only a pulse of heating gas that contains the heat of desorption is required to move through the bed, desorbing the adsorbate until it exhausts its thermal energy as it reaches the outlet. Because temperature fronts spread as they move through packed beds, a small excess of heat is added to the stoichiometric quantity to ensure that the outlet reaches the desired level before being cooled.

When the gas available for regeneration is in short supply, the regeneration steps are often carried out in a closed loop. This recycle of the bed effluent back to the inlet has the advantage of concentrating the impurity and making it easier to

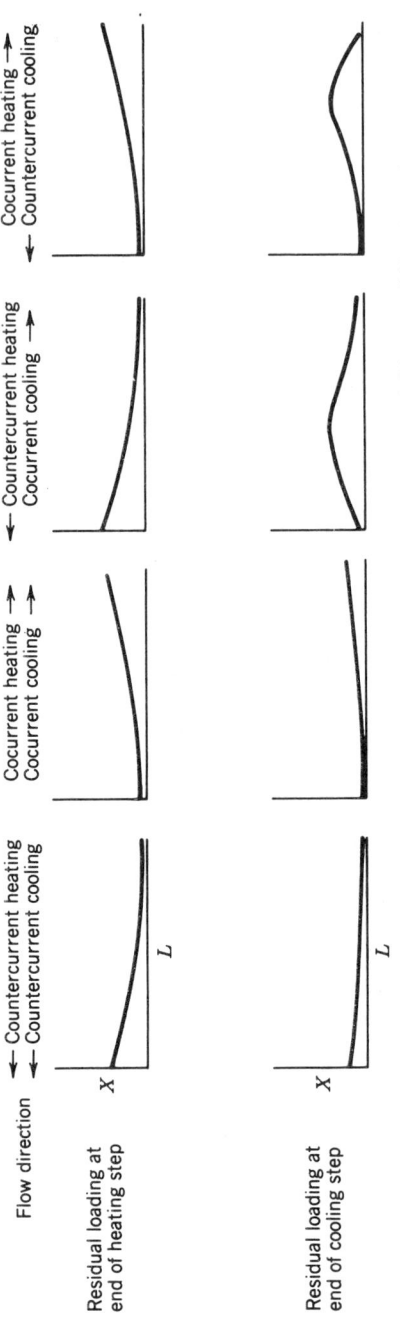

Fig. 13. Shape of residual loading profile (124). L is bed length. Courtesy of *Chemical Engineering*.

separate by condensation or other recovery means. Heating is usually accomplished with a semiclosed loop which has a constant fresh gas makeup and a bleed to draw off the desorbed material. However, contaminant is at a higher level than in an open loop and product purity is harder to achieve.

Drying. The single most common gas phase application for TSA is drying. The natural gas, chemical, and cryogenics industries all use zeolites, silica gel, and activated alumina to dry streams. Adsorbents are even found in mufflers.

Zeolites, activated alumina, and silica gel have all been used for drying of pipeline natural gas. Alumina and silica gel have the advantage of having higher equilibrium capacity and of being more easily regenerated with waste level heat (50–52). However, the much lower dewpoint and longer life attainable with 4A makes zeolites the predominant adsorbent. Special acid-resistant zeolites are used for natural gas containing large amounts of acid gases, such as CO_2 and H_2S.

The low dewpoint that can be achieved with zeolites is especially important when drying feed streams to cryogenic processes to prevent freeze-up at process temperatures. Natural gas is dried before liquefaction to liquefied natural gas (LNG), both in peak demand and in base load facilities (53,54). Zeolites have largely replaced silica gel and activated alumina in drying natural gas for ethane recovery utilizing the cryogenic turboexpander process, and for helium recovery (54). The air to be cryogenically distilled into N_2, O_2 and argon must be purified of both water and CO_2. This purification is accomplished with 13X zeolites (53).

The 4A zeolite, silica gel, and activated alumina all find applications drying synthesis gas, inert gas, hydrocracker gas, rare gases, and reformer recycle H_2. Cracked gas before low temperature distillation for olefin production is a reactive stream. The 3A or pore-closed 4A zeolite size selectively adsorbs water but excludes the hydrocarbons, thus preventing coking (52). This molecular sieving also prevents coadsorption of hydrocarbons which would otherwise be lost during desorption with the water. Small pore zeolites are also applied to the drying of ethylene, propylene, and acetylene as they are drawn from salt cavern, or conventional, storage (54). When industrial gases containing Cl_2, SO_2 and HCl are dried, acid-resistant zeolites are used.

A more recently developed drying application for zeolites is the prevention of corrosion in automobile mufflers (52,55). Internal corrosion in mufflers is caused primarily by the condensation of water and acid as the system cools. A unique UOP zeolite adsorption system takes advantage of the natural thermal cycling of an automobile exhaust system to desorb the water and acid precursors.

Sweetening. Another significant purification application area for adsorption is sweetening. Hydrogen sulfide, mercaptans, organic sulfides and disulfides, and COS need to be removed to prevent corrosion and catalyst poisoning. These are found in H_2, natural gas, deethanizer overhead, and biogas. Often adsorption is attractive because it dries the stream as it sweetens.

In the sweetening of wellhead natural gas to prevent pipeline corrosion, 4A zeolites allow sulfur compound removal without CO_2 removal (to reduce shrinkage), or the removal of both to upgrade low thermal content gas. When minimizing the formation of COS during desulfurization is desirable, calcium-exchanged zeolites are commonly used because these are less catalytically active for the reaction of CO_2 with H_2S to form COS and water. Natural gas for steam–methane reforming in ammonia production must be sweetened to protect the sulfur-sensitive, low temperature shift catalyst. Zeolites are better than activated carbon because mercaptans, COS, and organic sulfides are also removed (54). Many refinery H_2 streams require H_2S and water removal by 4A and 5A zeolites to prevent poisoning of catalysts such as those in catalytic reformers.

Other Separations. Other TSA applications range from CO_2 removal to hydrocarbon separations, and include removal of air pollutants and odors, and purification of streams containing HCl and boron compounds. Because of the high selectivity for CO_2 and the ability to dry concurrently, 4A, 5A, and 13X zeolites are the predominant adsorbents for CO_2 removal by temperature-swing processes. The air fed to an air separation plant must be H_2O- and CO_2-free to prevent fouling of heat exchangers at cryogenic temperatures; 13X is typically used here. Another application for 4A-type zeolite is for CO_2 removal from baseload and peak-shaving natural gas liquefaction facilities.

The removal of volatile organic compounds (VOC) from air is most often accomplished by TSA. Air streams needing treatment can be found in most chemical and manufacturing plants, especially those using solvents. At concentrations from 500 to 15,000 ppm, recovery of the VOC from steam used to regenerate activated carbon adsorbent thermally is economically justified. Concentrations above 15,000 ppm are typically in the explosive range and require the use of inert gas rather than air for regeneration. Below about 500 ppm, recovery is not economically justifiable, but environmental concerns often dictate adsorptive recovery followed by destruction. Activated carbon is the traditional adsorbent for these applications, which represent the second largest use for gas-phase carbons. New forms of activated carbon, such as carbon fabrics (56) and adsorbent wheels (57), have been introduced to reduce the airflow pressure drop, which can result in large utility consumptions (see ACTIVATED CARBON).

A number of inorganic pollutants are removable by TSA processes. One of the principal pollutants requiring removal is SO_2 from flue gases and from sulfuric acid plant tail gases. The Sulfacid and Hitachi fixed-bed processes, the Sumitomo and BF moving-bed processes, and the Westvaco fluidized-bed process all use activated carbon adsorbents for proven SO_2 removal (58). Zeolites with high acid resistance, such as mordenite and clinoptilolite, have proven to be effective adsorbents for dry SO_2 removal from sulfuric acid tail gas (59); special zeolite adsorbents have been incorporated into the UOP PURASIV S process for this application (54).

Zeolites have also proven applicable for removal of nitrogen oxides, NO_x, from wet nitric acid plant tail gas (59) by the UOP PURASIV N process (54). The removal of NO_x from flue gases can also be accomplished by adsorption. The Unitaka process utilizes activated carbon with a catalyst for reaction of NO_x with ammonia, and activated carbon has been used to convert NO to NO_2, which is removed by scrubbing (58). Mercury is another pollutant that can be removed and recovered by TSA. Activated carbon impregnated with elemental sulfur is effective for removing Hg vapor from air and other gas streams; the Hg can be recovered by *ex situ* thermal oxidation in a retort (60). The UOP PURASIV Hg process recovers Hg from chlor–alkali plant vent streams using more conventional TSA regeneration (54). Mordenite and clinoptilolite zeolites are used to remove HCl from Cl_2, chlorinated hydrocarbons, and reformer catalyst gas streams (61). Activated aluminas are also used for such applications, and for the adsorption of fluorine and boron–fluorine compounds from alkylation (qv) processes (50).

PRESSURE SWING

A pressure-swing adsorption (PSA) cycle is one in which desorption takes place at a pressure much lower than adsorption. Its principal application is for bulk separations where contaminants are present at high concentration. The PSA cycles are characterized by high residual loadings and low operating loadings. Figure 14 shows the operating loading $(X_1–X_2)$ that derives from the partial pressure at feed conditions and the lower pressure P_2 at the end of desorption. These low adsorption capacities for high concentrations mean that cycle times must be short, seconds to minutes, for reasonably sized beds. Fortunately, packed beds of adsorbent respond rapidly to changes in pressure. A purge usually re-

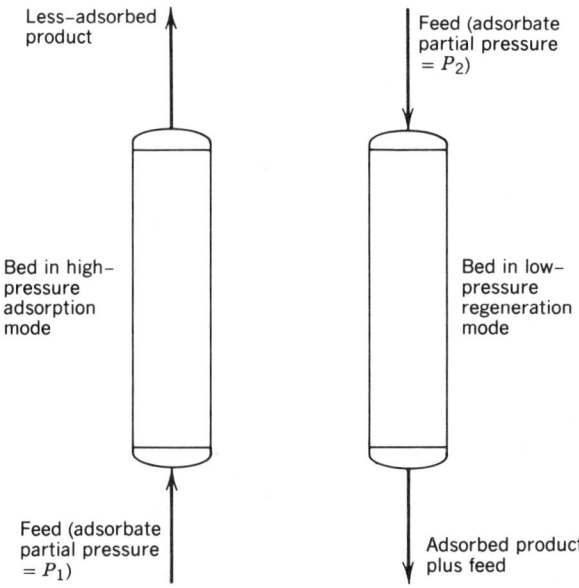

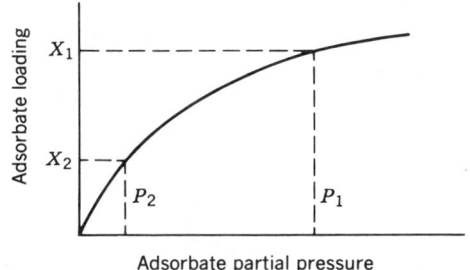

Fig. 14. Pressure-swing cycle (1).

moves the desorbed components from the bed, and the bed is returned to adsorption condition by repressurization. Applications may require additional steps. Systems with weakly adsorbed species are especially suited to PSA adsorption. The applications of PSA include drying, upgrading of H_2 and fuel gases, and air separation. Several broad reviews of PSA have been written (62–64).

Principles. In a PSA cycle two processes occur during regeneration, depressurizing, and purging. Depressurization must provide adequate reduction in the partial pressure of the adsorbates to allow desorption. Enough purge gas must flow through the adsorbent to transfer the desorbed material away. Equilibrium determines the maximum capacity of the gas to accomplish this. These cycles operate at nearly constant temperature and require no heating or cooling steps. They utilize the exothermic heat of adsorption remaining in the adsorbent to supply the endothermic heat of desorption. Pressure-swing cycles are classified as PSA, vacuum-swing adsorption (VSA), pressure-swing parametric pumping (PSPP), or rapid pressure-swing adsorption (RPSA). PSA swings between a high superatmospheric and a low superatmospheric pressure; VSA swings from a superatmospheric pressure to a subatmospheric pressure. Otherwise, the principles involved are the same.

The other means of accomplishing pressure cycling of an adsorbent is parametric pumping, in which a single adsorbent bed is alternately pressurized with forward flow and depressurized with backward flow through the column from reservoirs at each end. Like TSA parametric pumping, one component concentrates in one reservoir and one in the other. As the name implies, pressure-swing parametric pumping embodies pressure changes that are more than pressuring and depressurizing a bed of adsorbent. Significant pressure gradients occur in the bed, much as temperature gradients are imposed in TSA parametric pumping. These gradients are especially critical to the way that RPSA cycles operate and result in much smaller adsorbent beds and simpler processes (57).

In most applications of adsorption, the separation is carried out by adsorbing the more strongly adsorbed species from the less strongly adsorbed. These separations are thus equilibrium-limited. However, an adsorptive separation can also be based on a rate- or kinetically-limited system (65). Slightly larger molecules diffuse more slowly through a microporous adsorbent with properly selected pore diameter. Therefore, in a rapidly cycling process such as PSA, smaller molecules can be preferentially adsorbed even in the absence of any equilibrium selectivity. Indeed rate-limited PSA has preferentially adsorbed oxygen from air on 4A zeolite when the equilibrium selectivity favors N_2 adsorption (52).

Steps. A pressure-swing cycle has at least three steps: adsorption, blowdown, and repressurization. Although not always necessary, a purge step is normally used. In finely tuned processes, cocurrent depressurization and pressure-equalization steps are frequently added.

At the completion of adsorption, the less selectively adsorbed components have been recovered as product. However, a significant quantity of the weakly adsorbed species are held up in the bed, especially in the void spaces. A cocurrent depressurization step reduces the bed pressure by allowing flow out of the bed cocurrently to feed flow and thus reduces the amount of product retained in the voids (holdup), improving product recovery, and increases the concentration of the more strongly adsorbed components in the bed. The purity of the more selectively adsorbed species has been shown to depend strongly on the cocurrent depressurization step for some applications (66). A cocurrent depressurization step is optional because a countercurrent one always exists. Criteria have been developed to indicate when the use of both is justified (67).

None of the selectively adsorbed components is removed from the adsorption vessel until the countercurrent depressurization (blowdown) step. During this step, the strongly adsorbed species are desorbed and recovered at the adsorption inlet of the bed. The reduction in pressure also reduces the amount of gas in the bed. By extending the blowdown with a vacuum (ie, VSA), the productivity of the cycle can be greatly increased.

Additional stripping of the adsorbates from the adsorbent and purging of them from the voids is accomplished by the purge step. This step can occur concurrently with the end of the blowdown or be carried out afterward. This step is accomplished with a flow of product into the product end to provide a low residual of the selectively adsorbed components at the effluent end of the bed.

The repressurization step returns the adsorber to feed pressure and completes the steps of a PSA cycle. Pressurization is carried out with product and/or feed. Pressurizing with product is done countercurrent to adsorption so that purging of the product end continues; indeed it may be merely a continuation of the purge step but with the bed exit valve closed. Pressurizing with feed cocurrent to adsorption in effect begins adsorption without producing any product.

Pressure equalization steps are used to conserve gas and compression energy. They are applied to reduce the quantity of feed or product gas needed to pressurize the beds. Portions of the effluent gas during depressurization, blowdown, and purge can be used for repressurization.

Flow Sheet. The most common processing scheme has two or three fixed-bed adsorbers alternating between the adsorption step and the desorption steps. The simplest two-bed configuration is illustrated in Figure 14. However, the variations possible to achieve special separations are endless. Single beds with external surge vessels provide continuous flow; multiple beds are used to accommodate additional steps. An example of the bed sequencing needed for multiple steps in a four-bed PSA is shown in Figure 15. Beds can be configured in series or parallel to accomplish coproduction.

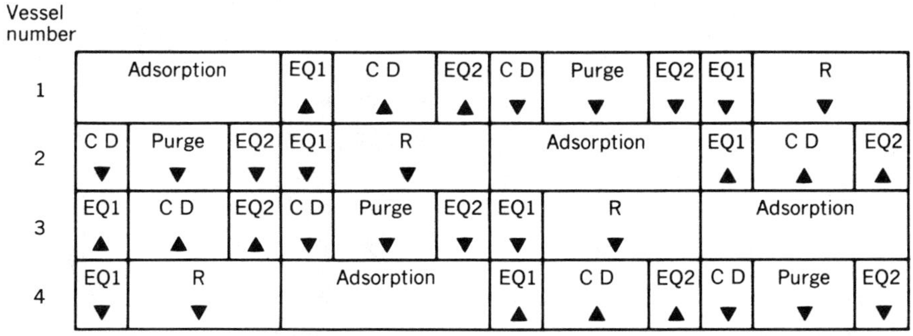

Fig. 15. Four-bed PSA system cycle sequence chart (64). EQ, equalization; C D ▲, cocurrent depressurization; C D ▼, countercurrent depressurization; R, repressurization; ▲, cocurrent flow; ▼, countercurrent flow. Courtesy of American Institute of Chemical Engineers.

The flow directions in a PSA process are fixed by the composition of the stream. The most common configuration is for adsorption to take place up-flow. All gases with compositions rich in adsorbate are introduced into the adsorption inlet end, and so effluent streams from the inlet end are rich in adsorbate. Similarly, adsorbate-lean streams to be used for purging or repressurizing must flow into the product end.

Because RPSA is applied to gain maximum product rate from minimum adsorbent, single beds are the norm. In such cycles where the steps take only a few seconds, flows to and from the bed are discontinuous. Therefore, surge vessels are usually used on feed and product streams to provide uninterrupted flow. Some RPSA cycles incorporate delay steps unique to these processes. During these steps, the adsorbent bed is completely isolated; and any pressure gradient is allowed to dissipate (68). The UOP Polybed PSA system uses five to ten beds to maximize the recovery of the less selectively adsorbed component and to extend the process to larger capacities (69).

Purifications. The major purification applications for PSA are for hydrogen, methane, and drying (qv). One of the first commercial uses was for gas drying in which the original two-bed Skarstrom cycle was used. This cycle uses adsorption, countercurrent blowdown, countercurrent purge, and cocurrent repressurization to produce a dry air stream with less than 1 ppm H_2O (70). About half of all dryers of instrument air use a PSA cycle similar to this one, most commonly using activated alumina or silica gel (71). Zeolites are used to obtain the lowest possible dewpoints. Some applications for drying air do not require a low level of H_2O, but only a significant lowering of the dew point. The pneumatic compressor systems used in vehicle air-brakes are an example; when a 10–30-K dew point depression is needed for higher discharge air temperatures in the presence of compressor oil, zeolites have been demonstrated to have an advantage over activated alumina and silica gel (72).

High purity H_2 is needed for applications such as hydrogenation, hydrocracking, and ammonia and methanol production (see HIGH PURITY GASES). As a significant source of such gas, PSA is able to produce purities as high as 99.9999% using technologies such as the UOP Polybed approach (69). Most H_2 purification by PSA is associated with stream reforming of natural gas and with ethylene-plant and refinery off-gas streams (62). Hydrogen is also available in coke-oven gas, cracked ammonia, and coal-gasification gas. The contaminants that have to be removed by PSA include carbon oxides, N_2, O_2, NH_3, CH_4, and heavier hydrocarbons. To remove these components, adsorbent beds are compounded of activated carbon, zeolites, and carbon molecular sieves.

Bulk Separations. Air separation, methane enrichment, and iso-/normal separations are the principal bulk separations for PSA. Others are the recovery of CO and CO_2.

The PSA process is used to separate air into N_2 and O_2. Many companies market systems for PSA O_2. Zeolites 5A, 13X, clinoptilolite and mordenite, and carbon molecular sieves are commonly used in PSA, VSA, and RPSA cycles. The product purity ranges from 85 to 95% (limited by the argon, which remains with the O_2). About two-thirds of the O_2 produced is employed for electric furnace steel, with lesser amounts for waste water treating and solid waste and kilns (62). Smaller production units are used for patients requiring respiratory inhalation therapy in the hospital and at home (64) and for pilots on board aircraft (73). Enriched air, 25 to 55% O_2, used to enhance combustion, chemical reactions, and ozone production can be produced by tuning PSA processes (63). High purity, up to 99.99%, N_2 is produced by PSA and VSA cycles with zeolites and carbon molecular sieves (74). The major use for the N_2 is inert blanketing, such as in metal heat-treating furnaces. Small units are used to purge aircraft fuel tanks (52) and in the food and beverage industry.

The upgrading of methane to natural gas pipeline quality is another significant PSA separation area. Methane is recovered from fermentation gases of landfills and wastewater purification plants and from poor-quality natural gas wells and tertiary oil recovery when CO_2 is the major bulk contaminant. Fermentation gases are saturated with water and contain "garbage" components such as sulfur and halogen compounds, alkanes, and aromatics (75). These impurities must first be removed by TSA using activated carbon or carbon molecular sieves. The CO_2 is then selectively adsorbed in a PSA cycle using either zeolites or silica gel in an equilibrium separation, or carbon molecular sieve in a kinetic-assisted equilibrium separation (76,77).

One version of the UOP IsoSiv process uses PSA to separate normal paraffins from branched and cyclic hydrocarbons in the C_5 to C_9 range. Zeolite 5A is used because its pores can size-selectively adsorb straight-chain molecules while excluding branched and cyclic species. The normal hydrocarbon fraction has better than 95% purity, and the higher octane isomer fraction contains less than 2% normal hydrocarbons (64).

PURGE SWING

A purge-swing adsorption cycle is one in which desorption takes place at the same temperature and total pressure as adsorption. Regeneration is accomplished either by partial-pressure reduction by an inert gas purge or by adsorbate displacement by an adsorbable gas. Its major application is for bulk separations when contaminants are at high concentration. Like PSA, purge cycles are characterized by high residual loadings, low operating loadings, and short cycle times (minutes). Mixtures of weakly adsorbed components are especially suited to purge-swing adsorption. Applications include the separation of normal from branched and cyclic hydrocarbons, gasoline vapor recovery, and bulk drying of organics.

Principles. Purging must provide adequate reduction in the partial pressure of the adsorbates to allow desorption. With enough purge volume, loadings as high as the loading X_1 in equilibrium with the feed partial pressure P_1 can be achieved, as shown in Figure 16. Reduction in partial pressure operates analogously to the reduction in system pressure in PSA cycles. Equilibrium determines the maximum capacity of the gas to purge the adsorbate. These cycles operate adiabatically at nearly constant inlet temperature and require no heating or cooling steps. As with PSA, purge processes utilize the exothermic heat of adsorption remaining in the adsorbent to supply the endothermic heat of desorption. Purge cycles are divided into two categories, inert purge and displacement purge. In inert-purge stripping, inert refers to the fact that the purge gas is not appre-

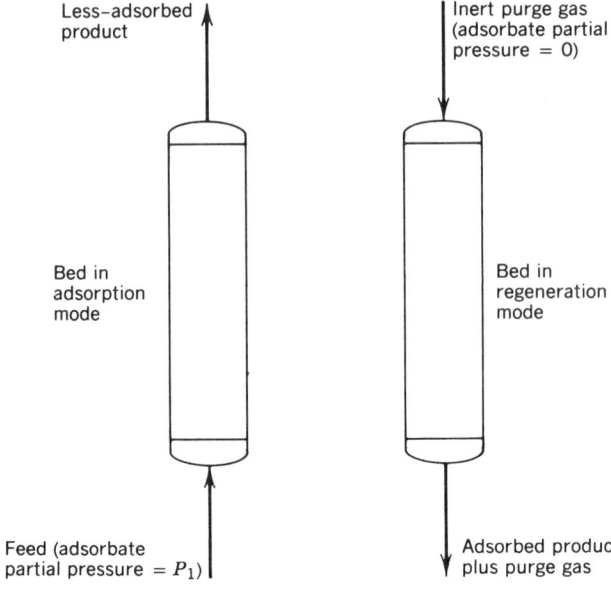

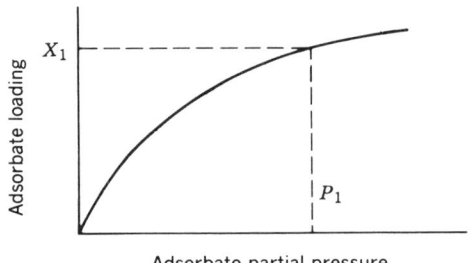

Fig. 16. Inert-purge cycle (1).

ciably adsorbable at the cycle conditions. Inert purging desorbs the adsorbate solely by partial pressure reduction.

In displacement-purge stripping, displacement refers to the displacing action of the purge gas caused by its ability to adsorb at the cycle conditions. This competitive adsorption tends to desorb the adsorbate in addition to the partial pressure reduction of dilution. Displacement purging is not as dependent on the heat of adsorption remaining on the adsorbent, because the adsorption of purge gas can release much or all of the energy needed to desorb the adsorbate. The adsorbate must be more selectively adsorbed than the displacement purge so that it can desorb purge fluid during the adsorption step. The displacement purge gas composition must be carefully selected, because it contaminates both the product stream and the recovered adsorbate and requires distillation as illustrated in Figure 17. The displacement purge is more efficient for less selective adsor-

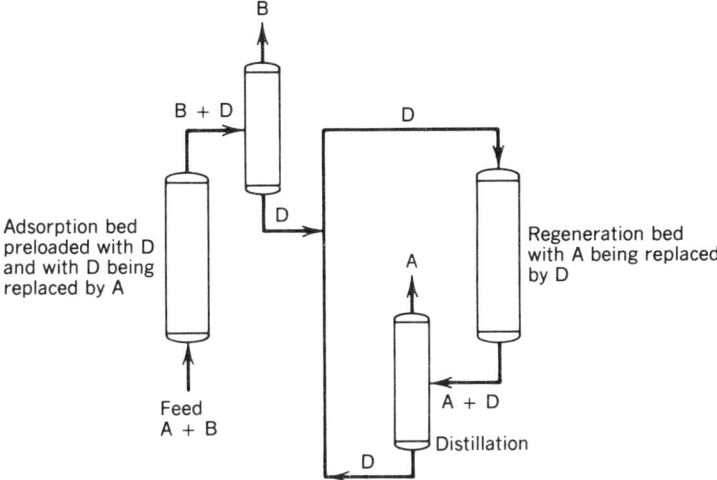

Fig. 17. Displacement-purge cycle (1).

bate–adsorbent systems; systems with high equilibrium loading of adsorbate require more purging (78).

Steps. A purge-swing cycle usually has two steps, adsorption and purge. Sometimes, a cocurrent purge is added. After the adsorption step has been completed and the less selectively adsorbed components have been recovered, an appreciable amount of product is still stored in the bed. A purge cocurrent to feed can increase recovery by displacing the fluid held in the voids.

The more selectively adsorbed components are stripped from the adsorbent bed during the countercurrent purge step. By purging into the product end of the bed, a lower residual loading of the selectively adsorbed species can be achieved in the portion of the adsorber that determines product quality.

Flow Sheet. Most purge-swing applications use two fixed-bed adsorbers to provide a continuous flow of feed and product (Fig. 16). Single beds are used when the flow to be treated is intermittent or cyclic. Because the purge flow is invariably greater than that of adsorption, purge is carried out in the down-flow direction to prevent bed lifting, and adsorption is up-flow.

Applications. Several purge-swing processes for the separation of C_{10}–C_{18} iso- from normal paraffins have been commercialized: Exxon's Ensorb, UOP's IsoSiv, Texaco Selective Finishing (TSF), Leuna Werke's Parex, and the Shell Process (52). All of these processes take advantage of the molecular size selectivity of 5A zeolite, but vary in the purge fluid. Ammonia is used in a displacement-purge cycle in Ensorb. Normal paraffins or light naphtha with a carbon number of two to four less than the feed stream are used for the displacement purge for TSF, Parex, and the Shell Process (79). One version of UOP purge IsoSiv for C_5 to C_9 naphtha employs H_2 in an inert-purge cycle (64). UOP also developed a similar process, OlefinSiv, for the separation of isobutylene from normal butenes using a size-selective zeolite with displacement purge (80).

Since 1971 all U.S. automobile models must have canisters of activated carbon to control gasoline vapors. Any gasoline vapors from the carburetor or the gas tank during running, from the tank during diurnal cycling, and from carburetor hot-soak losses are adsorbed by the carbon and held until they can be regenerated. The vapors are desorbed by an inert purge of air and are drawn into the carburetor as fuel when the engine is running (81). This gas-phase use for activated carbon is the third largest after solvent recovery and air purification.

Another use of inert-purge regeneration is UOP's Adsorptive Heat Recovery (AHR) drying system (82). The AHR system has been commercialized for drying azeotropic ethanol to be blended with gasoline into "gasohol." The process uses a closed loop of N_2 as the inert purge to desorb the water. The AHR process can be economically superior to azeotropic distillation for drying feeds with as much as 20% H_2O.

NONREGENERATIVE PROCESSES

Gas-phase adsorption can also be used when regenerating the adsorbent is not practical. Most of these applications are used where the facilities to effect a regeneration are not justified by the small amount of adsorbent in a single unit. Nonregenerative adsorbents are used in packaging, dual-pane windows, odor removal, and toxic chemical protection.

Applications. Silica gel is the adsorbent most commonly used as a desiccant in packaging. Activated carbon (qv) is used in packaging and storage to adsorb other chemicals for preventing the tarnishing of silver, retarding the ripening or spoiling of fruits, gettering, ie, scavenging, outgassed solvents from electronic components, and removing odors.

Adsorbents are used in dual-pane windows to prevent fogging between the sealed panes that could result from the condensation of water or the solvents used in the sealants (83). Synthetic zeolites (3A, 4A, 13X) or, less frequently, blends of zeolites with silica gel are installed in the spacing strips in double-glazed windows to adsorb water during initial dry-down and any in-leakage and to adsorb organic solvents emitted from the sealants during their cure. The adsorbent or mix of adsorbents applied depends on the sealant system and filling gas used.

The largest use of activated carbon is for the purification of air streams. Much of this carbon is used to treat recirculated air in large occupied enclosures, such as office buildings, apartments, and plants. The carbon is incorporated into thin filterlike frames to treat the large volumes of air with low pressure drop. Odors are also removed from smaller areas by activated carbon filters in kitchen hoods, air conditioners, and electronic air purifiers. On a smaller scale, gas masks containing carbon or carbon impregnated with promoters are used to protect wearers from odors and toxic chemicals. The smallest scale carbon filters are those used in cigarettes. Activated carbon fibers have been formed into fabrics for clothing to protect against vesicant and percutaneous chemical vapors (84).

REACTIVE ADSORPTION

Although chemisorbents are not used as extensively as physical adsorbents, a number of commercially significant processes employ chemisorption for gas purification.

Iron Sponge. An old method for removal of sulfur compounds involves contacting gases containing H_2S and H_2O with α- or γ-ferric oxide monohydrates at approximately 38°C to adsorb the sulfur in the form of ferric sulfide, followed by periodic reoxidation of the surface to form elemental sulfur and to revivify the ferric oxides (85). The iron sponge is reused in this cycle until buildup of sulfur in its pores reduces its effectiveness. The sponge is then replaced. The process is most efficient when the treated gases contain oxygen to allow continuous revivification. Spent adsorbents may be regenerated, eg, by oxidation of the sulfur to SO_2 to be fed to a sulfuric acid plant or by solvent extraction with carbon disulfide, and reused.

Mercury Removal. Trace amounts of mercury found in natural gas in some parts of the world are known to cause significant pinhole corrosion damage to aluminum heat exchanger surfaces in cryogenic coldboxes upstream of liquefied natural gas (LNG) plants (86). Mercury can be removed from such streams, and from other industrial gases by treatment in an *ex situ* TSA regenerative process using an activated carbon adsorbent containing sulfur. Reactions involving the formation of the less volatile mercuric sulfide occur. Alternatively, a newer developed adsorbent that may be employed in either nonregenerative or TSA regenerative process cycles may be used for mercury removal (87).

Nuclear Waste Management. Separation of radioactive wastes provides a number of relatively small scale but vitally important uses of gas-phase purification applications of adsorption. Such applications often require extremely high degrees of purification because of the high toxicity of many radioactive elements.

Delay for Decay. Nuclear power plants generate radioactive xenon and krypton as products of the fission reactions. Although these products are trapped inside the fuel elements, portions can leak out into the coolant through fuel cladding defects and thus can be released to the atmosphere with other gases through an air ejector at the main condenser.

To prevent such release, off-gases are treated in Charcoal Delay Systems, which delay the release of xenon and krypton, and other radioactive gases, such as iodine and methyl iodide, until sufficient time has elapsed for the short-lived radioactivity to decay. The delay time is increased by increasing the mass of adsorbent and by lowering the temperature and humidity. For a boiling water reactor (BWR), a typical system containing 21 t of activated carbon operated at 255 K, at 500 K dewpoint, and 101 kPa (15 psia) would provide about 42 days holdup for xenon and 1.8 days holdup for krypton (88). Humidity reduction is typically provided by a combination of a cooler–condenser and a molecular sieve adsorbent bed.

If the spent fuel is processed in a nuclear fuel reprocessing plant, the radioactive iodine species trapped in the spent fuel elements are ultimately released into dissolver off-gases. The radioactive iodine may then be captured by chemisorption on molecular sieve zeolites containing silver (89).

Other Applications. Many applications of adsorption involving radioactive compounds simply parallel similar applications involving the same compounds in nonradioactive forms, eg, radioactive carbon-14, or deuterium- or tritium-containing versions of CO_2, H_2O, hydrocarbons. For example, molecular sieve zeolites are commonly employed for these separations, just as for the corresponding nonradioactive uses.

MOVING AND FLUIDIZED BEDS

Most adsorption systems use stationary-bed adsorbers. However, efforts have been made over the years to develop moving-bed adsorption processes in which the adsorbent is moved from an adsorption chamber to another chamber for regeneration, with countercurrent contacting of gases with the adsorbents in each chamber. Union Oil's Hypersorption Process (90) is an example. However, this process proved uneconomical, primarily because of excessive losses resulting from adsorbent attrition.

The commercialization by Kureha Chemical Co. of Japan of a highly attrition-resistant, activated-carbon adsorbent as Beaded Activated Carbon (BAC) allowed development of a process employing fluidized-bed adsorption and moving-bed desorption for removal of volatile organic carbon compounds from air. The process has been marketed as GASTAK in Japan and as PURASIV HR (91) in the United States. It is marketed as SOLDACS by Daikin Industries, Ltd.

The discovery (92) that the graphite coating of molecular sieves can dramatically improve attrition resistance without significantly impairing adsorption performance is expected to allow the extension of moving-bed technology to bulk gas separations (93).

Design Methods

Design techniques for gas-phase adsorption range from empirical to theoretical. Methods have been developed for equilibrium, for mass transfer, and for combined dynamic performance. Approaches are available for the regeneration methods of heating, purging, steaming, and pressure swing. Several broad reviews have been published on analytical equations describing adsorption (94,95), of experimental adsorption equilibrium and kinetic data (96), on theoretical models for adsorption processes (97,98), and on adsorption design considerations (1).

Adsorption. In the design of the adsorption step of gas-phase processes, two phenomena must be considered, equilibrium and mass transfer. Sometimes adsorption equilibrium can be regarded as that of a single component, but more often several components and their interactions must be accounted for. Design techniques for each phenomenon exist as well as some combined models for dynamic performance.

Equilibrium. Among the aspects of adsorption, equilibrium is the most studied and published. Many different adsorption equilibrium equations are used for the gas phase. Equally important is the adsorbed phase mixing rule used to predict multicomponent behavior.

Many simple systems that could be expected to form ideal liquid mixtures are reasonably predicted by extending pure-species adsorption equilibrium data to a multicomponent equation. The potential theory has been extended to binary mixtures of several hydrocarbons on activated carbon by assuming an ideal mixture (99) and to hydrocarbons on activated carbon and carbon molecular sieves, and to O_2 and N_2 on 5A and 10X zeolites (100). Mixture isotherms predicted by ideal adsorbed solvent theory (IAST) agree with experimental data for methane + ethane and for ethylene + CO_2 on activated carbon, and for CO + O_2 and for propane + propylene on silica gel (36). A statistical thermodynamic model has been successfully applied to equilibrium isotherms of several nonpolar species on 5A zeolite, to predict multicomponent sorption equilibria from the Henry constants for the pure components (26). A set of equations that incorporate surface heterogeneity into the IAST model provides a means for predicting multicomponent equilibria, but the agreement is only good up to 50% surface saturation (9).

For most models of adsorptive equilibrium, however, the coefficients derived from pure species are not adequate to predict multicomponent equilibrium for nonideal mixtures. Fitting the systems ethane + ethylene + propane on 5A zeolite and H_2S + CO_2 on H-mordenite required using binary parameters with the IAST or the real adsorbed solution theory models (101). A coalescing factor applied to the potential theory did collapse all isotherms to a single curve for activated carbon, zeolites, and silica gel. A binary interaction parameter that is a function of the coalescing factor was needed to gain agreement with binary data (102). For the multicomponent system of H_2, CO, CH_4, CO_2, and H_2S on activated carbon, an interaction parameter was required in the extended Langmuir equation to predict multicomponent equilibrium (103). Cross-correlation coefficients were necessary to apply a statistical model to three nonideal ternary zeolite systems (104). The composition dependence of an activity coefficient, described by the Wilson equation, has been added to the vacancy solution model (VSM) to fit data for hydrocarbons on activated carbon and O_2 + N_2 on 10X zeolite (37). Activity coefficients of the adsorbate–adsorbate interactions or treatment of the surface as heterogeneous are correlative methods that allow extension of the IAST to binary adsorption (10).

Mass Transfer. The degree of approach to equilibrium that can be achieved in adsorption is determined by the mass-transfer rates. One useful design concept is the mass-transfer zone (MTZ), an extension of the ion-exchange zone method (105). Figure 18**b** is a depiction of the adsorbate loading in a fixed bed during adsorption. The ordinate is loading (X) and the abscissa is distance (L) from the inlet of the bed. Between the inlet and the exhaustion point (L_e), the loading is in equilibrium with the feed gas, and this section is called the equilibrium section. From the breakthrough point (L_b) to the outlet of the bed, the adsorbate loading is still at the residual loading level and is unused bed. Mass transfer between the gas and the adsorbent is occurring between the breakthrough and exhaustion points, and so this zone is called the mass-transfer zone (MTZ). The length of the bed, L_b to L_e, is called the mass-transfer zone length (MTZL). The MTZL is usually correlated to flow rate or flow velocity (106).

Most dynamic adsorption data are obtained in the form of outlet concentrations as a function of time as shown in Figure 18**a**. The area iebai measures the removal of the adsorbate, as would the stoichiometric area idcai, and is used to calculate equilibrium loading. For constant pattern adsorption, the breakthrough time (Θ_b), and the stoichiometric time (Θ_s), are used to calculate LUB as $(1 - \Theta_b/\Theta_s)L_{bed}$ (107). This LUB concept is commonly used for drying and desulfurization design in the natural gas industry and for air prepurification before cryogenic distillation.

Another way of subdividing the bed is illustrated in Figure 18**b**. If the mass-transfer resistance were negligible, the MTZ would become a square or stoichiometric front along the line dsc. The area febgf represents used adsorbent loading, while the area ehbe between the potential loading and the actual loading curve eb

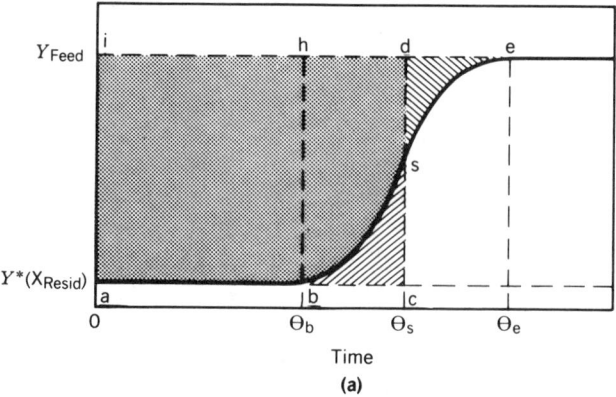

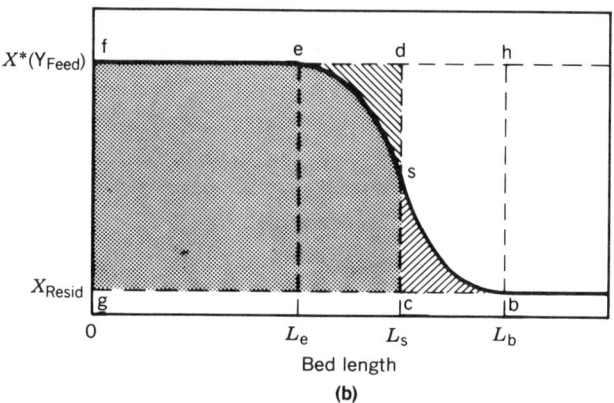

Fig. 18. (a) Time trace of adsorbate composition in an adsorber effluent during adsorption. (b) Adsorbate loading along the flow axis of an adsorber during adsorption (1).

is unused. By material balance, the area fdcgf up to the stoichiometric front would also represent the used capacity. Therefore, areas febgf and fdegf are equal. This portion of the bed up to the stoichiometric point, s, is called the weight of equivalent equilibrium section (WES). The rest of the adsorbent from the stoichiometric point to the breakthrough point is termed the weight of unused bed (WUB), because it is equivalent to a bed with no usable capacity in the stoichiometric interpretation. Adsorption beds can thus be sized by combining a WES calculated from equilibrium data and a WUB derived from kinetic data.

Dynamic Performance. Most models do not attempt to separate the equilibrium behavior from the mass-transfer behavior. Rather they treat adsorption as one dynamic process with an overall dynamic response of the adsorbent bed to the feed stream. Although numerical solutions can be attempted for the rigorous partial differential equations, simplifying assumptions are often made to yield more manageable calculating techniques.

The J-function is a definite integral of an expression including I_0, the modified Bessel function of the first kind. J-function curves use stoichiometric time and the number of theoretical stages as the two parameters to fit breakthrough curves and extend to other conditions. These curves have been approximated for use on microcomputers (108). A phenomenological model requires the determination of two parameters, a transfer coefficient and a linear isotherm constant, from a complete breakthrough curve. The solution to the model is in an infinite series form, which is calculable by a hand-held calculator or personal computer (109). Another method separates the equilibrium from the kinetic effects by constructing effective equilibrium curves. Because the solution to the model involves nonlinear algebraic or differential equations, graphs called solution charts are used to predict breakthrough fronts (110). Theoretical stages form the essence of the discrete cell model graphical procedures, which are applied to flat isotherms and incorporate pore diffusivity and axial dispersion (98). Another solution technique is the use of fast Fourier transforms. Linear isotherms are required, but their applicability for predicting breakthrough curves has been demonstrated for isothermal and nonisothermal adsorbers (111). Another model, with a solution in infinite series form, incorporates separate mass-transfer coefficients for external film, macropore, and micropore resistances (112). Techniques have also been developed to predict breakthrough from fluidized beds. The behavior of organic solvents adsorbed from air on activated carbon was shown to exhibit breakthrough times that can be correlated to the adsorption capacity and the amount of bed expansion (113).

Some specific design methods have been developed for particular applications. Several procedures have been published for the design of gas dryers. The J-function has been applied to silica gel dryers after a correlated correction factor accounted for the nonisothermality (114). In other work on drying with activated alumina and silica gel, constant-pattern LUBs were shown applicable for designs at H_2O contents of less than 0.003 kg/kg air (115). Equilibrium and kinetic parameters for H_2O on activated alumina were determined for a more rigorous nonisothermal model that predicted adiabatic behavior (116). Breakthrough times for several organic vapors on activated carbon respirator cartridges have been found to be predictable by using a theory of statistical moments. In one study, equilibrium data was correlated by the Potential Theory and breakthrough was calculated using the normal probability distribution curve (117). In another, the equilibrium data was represented by a Freundlich equation (118). For heavy hydrocarbon recovery from natural gas on silica gel, the equilibrium data were fit to a Freundlich isotherm and the breakthrough composition was found to have a power dependence on the extent of adsorptive saturation of the adsorbent (111). A WUB design approach was found to predict breakthrough for several organics on activated carbon using a Potential Theory equilibrium curve (119). Correlations of equilibrium capacity and WUB were also found applicable in the removal of H_2S from natural gas using 5A zeolite (120).

Regeneration. In recent years, considerable effort has been expended to better understand and quantify the process of regeneration. Methods are available to predict thermal, purge, and steaming requirements. Models are available to simulate all of the regeneration types, temperature, pressure, and purge swings.

Thermal Requirements. When a temperature-swing cycle is heating limited, the regeneration design is only concerned with transferring energy to the system. Charts for isothermal, linear-isotherm adsorption that were derived by Hougen and Marshall (121) from earlier work on heat transfer from a gas to fixed beds (122) can be reapplied to heat transfer for heating-limited regeneration when the heat of adsorption is negligible (123). This approach has been expanded to include both a correction of one dimensionless time per dimensionless bed length for heat losses and another correction particular to the H_2O–4A zeolite system studied (124). Because cooling is a heating-limited step, it can be calculated by the modified Hougen and Marshall method. As mentioned in the discussion of thermal swing, the cooling step can be performed under some process conditions by starting adsorption with a hot adsorbent; performance is not affected.

Purge Requirements. The amount of purge gas needed in stripping-limited regeneration is similar to that for purge regeneration, but it differs primarily in the temperature at which the isothermal desorption occurs. For a pressure-swing process, the theoretical minimum volumetric purge-to-feed ratio is the ratio of the purge pressure to the feed pressure (123), and one model shows the optimum ratio to be the minimum purge volume that can be used with the given cyclic steady-state conditions (125). For a thermal-swing process, the minimum purge-to-adsorbent ratio is the ratio of the heat capacity of the solid to that of the gas (126). Specific design purge data have been published for purge-swing activated carbon automotive evaporative emissions control (81) and for pressure-swing drying of pneumatic system air (72). The pneumatic system process exhibits an optimum purge ratio for maximizing the attainable dewpoint depression. An isothermal purge-swing model that uses Langmuir equilibrium to simulate adsorbent performance has been presented (127).

Steaming Requirements. The steaming of fixed beds of activated carbon is a combination of thermal swing and displacement purge swing. The exothermic heat released when the water adsorbs from the vapor phase is much higher than is possible with heated gas purging. This cycle has been successfully modeled by equilibrium theory (128).

Temperature Swing. Several fairly comprehensive reviews of thermal-swing adsorption models are found in the literature (96,97,129,130). Many of these models have been used to carry out parametric analyses with a goal of energy minimization. A nonisothermal model for single components using equilibrium theory demonstrated that efficiency improves with increased purge contact time and high heat capacity purge gas but is minimally affected by initial bed loading (131), and the model defined conditions under which the desorption can be continued with a cold stream without additional overall purge gas. That work also introduced the concept that minimum thermal energy is required at the characteristic temperature, that temperature at which the slope of the equilibrium isotherm is equal to the ratio of the heat capacity at the adsorbent to that of the purge gas. A nonequilibrium, mechanistic model with a multicomponent VSM model of equilibrium (132) and a nonequilibrium nonadiabatic computer model with a Langmuir-like isotherm (130) reached similar conclusions on optimization. Another nonequilibrium nonadiabatic computer model with an Antoine equation isotherm was used with supporting data to demonstrate that significant energy can be saved by proper timing of the cooling step (133). Most modeling has assumed that the purge gas is clean even though solvent recovery processes use a recirculated stream when an inert purge gas is employed. Using the method of characteristics and a Freundlich isotherm, an equilibrium theory modeled the incorporation of closed-loop heating and cooling steps (134).

Pressure Swing. Design equations have been developed to predict temperature rise, minimum bed length to retain the heat front, minimum purge rate, and effluent composition (135). A nonequilibrium, nonisothermal simulation program with a Freundlich isotherm equation was found to agree with data for drying with silica gel (136). A somewhat simpler isothermal model using an isotherm approximated by two straight lines successfully calculated the volumetric purge-to-feed ratio needed to achieve varying product dryness using silica gel (137). An adiabatic equilibrium model with a Langmuir isotherm was used to study the blowdown step of a cycle removing CO_2 on activated carbon and 5A zeolite (138). Changing to an isothermal assumption introduced significant errors into the results. The countercurrent pressurization step was investigated with an isothermal equilibrium model using a Langmuir isotherm for O_2 production from air with 5A zeolite (139). The model predicted the dependence of O_2 concentration on countercurrent pressure and was used to study other parameters. An isothermal model with linear isotherms and component-specific pore diffusivity was used and compared to data for the kinetic-limited separation of air by RS-10 zeolite (140). The simulations agreed well with the experimental parametric studies of time and pressure of feed, blowdown, purge, and pressurization. An equilibrium model was formulated to simulate RPSA using a Freundlich isotherm for separation of N_2 and CH_4 (141). Pressure responses, flow rates, and compositions compared favorably as a function of feed pressure, cycle frequency, and product rate. A nonequilibrium, nonisothermal model for RPSA was developed using a linear isotherm and Darcy's law for pressure drop (142). The model predicted performance in agreement with previous data (57) for air separation on 5A zeolite.

Pressure Drop. The prediction of pressure drop in fixed beds of adsorbent particles is important. When the pressure loss is too high, costly compression may be increased, adsorbent may be fluidized and subject to attrition, or the excessive force may crush the particles. As discussed previously, RPSA relies on pressure drop for separation. Because of the cyclic nature of adsorption processes, pressure drop must be calculated for each of the steps of the cycle. The most commonly used pressure drop equations for fixed beds of adsorbent (143,144,145) use a particle Reynolds number ($Re = D_p G/\mu$) and friction factor, f, to calculate the pressure drop, ΔP, per unit length, L, by the equation

$$\frac{\Delta P}{L} = \frac{fG^2}{2g_c D_p \rho}$$

where D_p is the particle diameter, G the mass flux, μ the gas viscosity, and ρ the gas density. Various methods differ in their definition of D_p and f. For up-flow in fixed-bed adsorbers, fluidization occurs when the pressure drop just balances the weight, corrected by any buoyancy:

$$\frac{\Delta P}{L} = \frac{(1 - \epsilon)(\rho_s - \rho)g}{g_c}$$

where ρ_s is the density of the solid. For down-flow in packed beds, the potential for crushing the adsorbent must be checked. Two forces act to crush the particles, pressure drop and the weight of the bed. The sum of these two ($\Delta P + (1 - \epsilon)\rho_s L$) should be kept less than that which is known to cause adsorbent damage.

BIBLIOGRAPHY

"Adsorptive Separation, Gases" in the *Encyclopedia of Chemical Technology*, 3rd ed., Vol. 1, pp. 544–581, by D. B. Broughton, UOP Process Division, UOP Inc.; in *ECT* 4th ed., Vol. 1, pp. 529–573, by J. D. Sherman and C. M. Yon, UOP.

1. G. E. Keller, II, R. A. Anderson, and C. M. Yon, in R. W. Rousseau, ed., *Handbook of Separation Process Technology*, John Wiley & Sons, Inc., New York, 1987, 644–696.
2. J. O. Hirschfelder, C. F. Curtiss, and R. B. Bird, *Molecular Theory of Gases and Liquids*, John Wiley & Sons, Inc., New York, 1954, p. 215, 949, 1110.
3. R. M. Barrer and D. E. W. Vaughan, *J. Phys. Chem. Solids* **32**, 731 (1971).
4. R. M. Barrer, *Zeolites and Clay Minerals as Adsorbents and Catalysts*, Academic Press, London, 1978, p. 164, 174, 185.
5. D. W. Breck and J. V. Smith, *Sci. Am.*, 8 (Jan. 1959).
6. Data from Union Carbide Molecular Sieves, UOP, Tarrytown, N.Y.
7. W. A. Steele, "The Physical Adsorption of Gases on Solids," *Adv. Colloid Interface Sci.* **1**, 3–78 (1967). (Review with 360 refs.).
8. D. M. Ruthven, *Principles of Adsorption and Adsorption Processes*, John Wiley & Sons, Inc., New York, 1984, Chapt. 3, 4, p. 108.
9. A. L. Myers in A. L. Myers and G. Belfort, eds., *Fundamentals of Adsorption*, Engineering Foundation, New York, 1984, pp. 365–381.
10. A. L. Myers in A. I. Liapis, ed., *Fundamentals of Adsorption*, Engineering Foundation, New York, 1987, pp. 3–25.
11. A. L. Myers, *NATO ASI Ser., Ser. E*, **158** (Adsorpt. Sci. Technol.), 15–36 (1989).
12. D. P. Valenzuela and A. L. Myers, *Adsorption Equilibrium Data Handbook*, Prentice Hall, Engelwood Cliffs, N.J., 1989.
13. D. P. Valenzuela and A. L. Myers, *Sep. Purif. Methods* **13**(2), 153–183 (1984).
14. O. Talu and A. L. Myers, *AIChE J.* **34**, 1887–1893 (1988).
15. *Ibid.*, 1931–1932 (1988).
16. R. T. Yang, *Gas Separation by Adsorption Processes*, Butterworths, Boston, 1987, p. 86.
17. C. M. Yon and P. H. Turnock, *AIChE Symp. Ser.* **67**(117), 75 (1971).
18. R. T. Maurer in J. R. Katzer, ed., *Molecular Sieves—II* (ACS Symp. Ser. 40) American Chemical Society, Washington, D.C., 1977, p. 379.
19. I. Langmuir, *J. Am. Chem. Soc.* **40**, 1361 (1918).
20. M. Volmer, *Z. Phys. Chem.* **115**, 253 (1925).
21. R. H. Fowler and E. A. Guggenheim, *Statistical Thermodynamics*, Cambridge University Press, Cambridge, 1939.
22. T. L. Hill, *Introduction to Statistical Thermodynamics*, Addison-Wesley, Reading, Mass., 1960.
23. V. A. Bakaev, *Dokl. Akad. Nauk SSSR* **167**, 369 (1967).
24. L. Riekert, *Adv. Catal.* **21**, 287 (1970).
25. P. Brauer, A. Lopatkin, and G. Ph. Stepanez in E. M. Flanigen and L. B. Sand, eds., *Molecular Sieve Zeolites, Adv. in Chem 102*, American Chemical Society, Washington, D.C., 1971, p. 97.
26. D. M. Ruthven, K. F. Loughlin, and K. A. Holborrow, *Chem. Eng. Sci.* **28**, 701 (1973).
27. D. M. Ruthven, *Nat. Phys. Sci.* **232**(29), 10 (1971).

28. Ref. 8, p. 75ff.
29. M. M. Dubinin and L. V. Radushkevich, *Dokl. Akad. Nauk SSSR, Ser. Khim.* **55,** 331 (1947).
30. M. M. Dubinin and V. A. Astakhov, *Izv. Akad. Nauk. SSSR, Ser. Khim.* **71,** 5 (1971).
31. C. J. Radke and J. M. Prausnitz, *Ind. Eng. Chem. Fundam.* **11,** 445 (1972); *AIChE J.* **18,** 761 (1972).
32. J. Toth, *Acta. Chim. Acad. Sci. Hung.* **69,** 311 (1971).
33. S. Brunauer, P. H. Emmett, and E. Teller, *J. Am. Chem. Soc.* **60,** 309 (1938).
34. E. C. Markham and A. F. Benton, *J. Am. Chem. Soc.* **53,** 497 (1931).
35. R. Sips, *J. Chem. Phys.* **16,** 490 (1948).
36. A. L. Myers and J. M. Prausnitz, *AIChE J.* **11,** 121 (1965).
37. S. Suwanayuen and R. P. Danner, *AIChE J.* **26,** 68, 76 (1980).
38. T. W. Cochran, R. L. Kabel, and R. P. Danner, *AIChE J.* **31,** 268 (1985).
39. A. Kohl and F. Riesenfeld, *Gas Purification,* 4th ed., Gulf Publishing Co., Houston, Tex., 1985, p. 651.
40. K. S. W. Sing and co-workers, *Pure Appl. Chem.* **57,** 603 (1985).
41. D. W. Breck, *Zeolite Molecular Sieves—Structure, Chemistry, and Use,* John Wiley & Sons, Inc., New York, 1974.
42. R. C. Bansal, J.-B. Donnet, and F. Stoeckli, *Active Carbon,* Marcel Dekker, New York, 1988, p. ix.
43. E. M. Flanigen and co-workers, *Nature (London)* **271,** 512 (1978).
44. E. M. Flanigen and R. L. Patton, UOP, Tarrytown, N.Y., private communication.
45. R. C. Bansal, T. L. Dhami, and S. Parkash, *Carbon* **16,** 389 (1978).
46. W. K. Lewis, E. R. Gilliland, B. Chertow, and W. P. Cadogan, *Ind. Eng. Chem.* **42,** 1326 (1950).
47. M. Benchanaa, M. Lallemant, M. H. Simonet-Grange, and G. Bertrand, *Thermochim. Acta.* **152,** 43–51 (1989).
48. H. R. Burkholder, G. E. Fanslow, and D. D. Bluhm, *Ind. Eng. Chem. Fundam.* **25,** 414–416 (1986).
49. D. Basmadjian, *Can. J. Chem. Eng.* **53,** 234–238 (1975).
50. B. Crittenden, *Chem. Eng.* **452,** 21–24 (1988).
51. K. P. Goodboy and H. L. Fleming, *Chem. Eng. Progr.* **80,** 63–68 (1984).
52. D. M. Ruthven, *Chem. Eng. Progr.* **84,** 42–50 (1988).
53. H. L. Brooking and D. C. Walton, *Chem. Eng.* **257,** 13–17 (1972).
54. R. A. Anderson in Ref. 18, pp. 637–649.
55. S. R. Dunne, *Automotive Corrosion and Prevention Conf. Proc.* Society of Automotive Engineers, Warrendale, Pa., 1989, 165–173.
56. R. E. Kenson and J. F. Jackson, Prepared Paper, Air Pollution Control Association, Annual Mtg., 1988.
57. G. E. Keller, II, and R. J. Jones in W. H. Flank, ed., *Adsorption and Ion Exchange with Synthetic Zeolites, Am. Chem. Soc. Symp. Ser. 135,* American Chemical Society, Washington, D.C., 1980, pp. 275–286.
58. H. Juentgen, *Carbon* **15,** 273–283 (1977).
59. J. R. Kiovsky, P. B. Koradia, and D. S. Hook, *Chem. Eng. Progr.* **72,** 98–103 (1976).
60. W. D. Lovett and F. T. Cunniff, *Chem. Eng. Progr.* **70,** 43–47 (1974).
61. A. Dyer, *An Introduction to Zeolite Molecular Sieves,* John Wiley & Sons, Inc., New York, 1988, 102–105.
62. J. R. Martin, C. F. Gotzmann, F. Notaro, and H. A. Stewart, *Adv. Cryog. Eng.* **31,** 1071–1086 (1986).
63. S. Sircar in A. E. Rodrigues, M. D. LeVan, and D. Tondeur, eds., *Adsorption: Science and Technology,* Kluwer Academic Publishers, Dordrecht, Netherlands, 1988, 285–321.

64. R. T. Cassidy and E. S. Holmes, *AIChE Symp. Ser.* **80**, 68–75 (1984).
65. Z. J. Pan, R. T. Yang, and J. A. Ritter in G. E. Keller, II, and R. T. Yang, eds., *New Directions in Sorption Technology,* Butterworths, Boston, 1988.
66. P. Cen and R. T. Yang, *Ind. Eng. Chem. Fundam.* **25**, 758–767 (1986).
67. S. S. Suh and P. C. Wankat, *AIChE J.* **35**, 523–526 (1989).
68. G. E. Keller, II, in T. E. White, Jr., C. M. Yon, and E. H. Wagener, eds., *Industrial Gas Separations, Am. Chem. Soc. Symp. Ser. 223,* American Chemical Society, Washington, D.C., 1983, pp. 145–169.
69. R. T. Cassidy in ref. 57, pp. 248–259.
70. C. W. Skarstrom in N. N. Li, ed., *Recent Developments in Separation Science,* Vol. 2, CRC Press, Boca Raton, Fla., 1975, pp. 95–106.
71. J. W. Armond in R. P. Townsend, ed., *The Properties and Applications of Zeolites,* The Chemical Society, London, 1980, pp. 92–102.
72. J. P. Ausikaitis in ref. 18, pp. 681–695.
73. J. B. Tedor, T. C. Horch, and T. J. Dangieri, *SAFE J.* **12**, 4–9 (1982).
74. M. Kawai and T. Kaneko, *Gas Sep. Purif.* **3**, 2–6 (1989).
75. R. Kumar and J. K. VanSloun, *Chem. Eng. Progr.* **85**, 34–40 (1989).
76. E. Richter, *Erdoel Kohle, Erdgas, Petrochem.* **40**, 432–438 (1987).
77. A. Kapoor and R. T. Yang, *Chem. Eng. Sci.* **44**, 1723–1733 (1989).
78. S. Sircar and R. Kumar, *Ind. Eng. Chem. Proc. Des. Dev.* **24**, 358–364 (1985).
79. R. T. Yang, *Gas Separation by Adsorption Processes,* Butterworths, Stoneham, Mass., 1987.
80. M. S. Adler and D. R. Johnson, *Chem. Eng. Progr.* **75**, 77–79 (1979).
81. P. J. Clarke, J. E. Gerrard, C. W. Skarstrom, J. Vardi, and D. T. Wade, *SAE Trans.* **76**, 824–842 (1968).
82. D. R. Garg and C. M. Yon, *Chem. Eng. Progr.* **82**, 54–60 (1986).
83. J. P. Ausikaitis, *Glass Dig.* **61**, 69–78 (1982).
84. R. N. Macnair and G. N. Arons in P. N. Cheremisinoff and F. Eleerbusch, eds., *Carbon Adsorption Handbook,* Ann Arbor Science, Ann Arbor, Mich., 1978, 819–859.
85. Ref. 39, p. 421.
86. M. D. Bingham, *Field Detection and Implications of Mercury in Natural Gas, Soc. Petrol. Engrs. Production Engrg.,* 120–124 (May, 1990).
87. J. Markovs, UOP, Tarrytown, N.Y., private communication; U.S. Pat. 4,874,525 (Oct. 17, 1989).
88. J. T. Collins, M. J. Bell, and W. M. Hewitt in A. A. Moghissi and co-workers, eds., *Nuclear Power Waste Technology,* American Society of Mechanical Engineers, New York, 1978, Chapt. 4.
89. D. W. Holladay, *A Literature Survey: Methods for the Removal of Iodine Species from Off-Gases and Liquid Waste Streams of Nuclear Power and Nuclear Fuel Reprocessing Plants, with Emphasis on Solid Sorbents,* Report ORNL/TM-6350 (January, 1979), p. 46, available from National Technical Information Service, Springfield, Va.
90. C. Berg, *Pet. Refiner* **30**(9), 241 (Sept. 1951).
91. "Beaded Carbon Ups Solvent Recovery," *Chem. Eng.* **84**(18) (Aug. 29, 1977).
92. U.S. Pat. 4,526,877 (July 2, 1985), A. Acharya and W. E. BeVier.
93. G. E. Keller, III, *Separations: New Directions for an Old Field* (AIChE Monogr. Ser. 17) American Institute of Chemical Engineers, 1987, p. 83.
94. C. Huang and J. R. Fair, *AIChE J.* **34**, 1861–1877 (1988).
95. S. Sircar and A. L. Myers, *Ads. Sci. Technol.* **2**, 69–87 (1985).
96. M. S. Ray, *Sep. Sci. Tech.* **18**, 95–120 (1983).
97. J. W. Carter in R. P. Townsend, ed., *The Properties and Applications of Zeolites,* The Chemical Society, London, 1980, pp. 76–91.

98. D. D. Do, *AIChE J.* **31,** 1328–1337 (1985).
99. R. J. Grant and M. Manes, *Ind. Eng. Chem. Fundam.* **5,** 490–498 (1966).
100. S. J. Doong and R. T. Yang, *Ind. Eng. Chem. Res.* **27,** 630–635 (1988).
101. G. Gamba, R. Rota, G. Storti, S. Carra, and M. Morbidelli, *AIChE J.* **35,** 959–966 (1989).
102. S. D. Mehta and R. P. Danner, *Ind. End. Chem. Fundam.* **24,** 325–330 (1985).
103. J. A. Ritter and R. T. Yang, *Ind. Eng. Chem. Res.* **26,** 1679–1686 (1987).
104. R. Rota, G. Gamba, R. Paludetto, S. Carra, and M. Morbidelli, *Ind. Eng. Chem. Res.* **27,** 848–851 (1988).
105. A. S. Michaels, *Ind. Eng. Chem.* **44,** 1922–1930 (1952).
106. H. M. Barry, *Chem. Eng.* **67,** 105–120 (1960).
107. J. J. Collins, *Chem. Eng. Progr. Symp. Ser.* **63,** 31–35 (1967).
108. S. L. Forbes and D. W. Underhill, *JAPCA* **36,** 61–64 (1986).
109. R. Mohilla, J. Argelan, and R. Szolcsanyi, *Int. Chem. Eng.* **27,** 723–729 (1987).
110. D. Basmadjian and C. Karayannopoulos, *Ind. Eng. Chem. Proc. Des. Dev.* **24,** 140–149 (1985).
111. C. L. Humphries, *Hydrocarbon Process.* **45,** 88–95 (1966).
112. P. I. Cen and R. T. Yang, *AIChE J.* **32,** 1635–1641 (1986).
113. H. Hori, I. Tanaka, and T. Akiyama, *JAPCA* **38,** 269–271 (1988).
114. H. Lee and W. P. Cummings, *Chem. Eng. Progr. Symp. Series* **63**(74), 42–49 (1967).
115. L. C. Eagleton and H. Bliss, *Chem. Eng. Progr.* **49,** 543–548 (1953).
116. J. W. Carter, *Br. Chem. Eng.* **14,** 303–306 (1969).
117. O. Grubner and W. A. Burgess, *Environ Sci. Technol.* **15,** 1346–1351 (1981).
118. Y. E. Yoon and J. H. Nelson, *Am. Ind. Hyg. Assoc. J.* **45,** 517–524 (1984).
119. L. A. Jonas and J. A. Rehrmann, *Carbon* **11,** 59–64 (1973).
120. C. W. Chi and H. Lee, *AIChE Symp. Ser.* **69,** 95–101 (1973).
121. O. A. Hougen and W. K. Marshall, *Chem. Eng. Progr.* **43,** 197–208 (1947).
122. C. C. Furnas, *Trans. Am. Inst. Chem. Eng.* **24,** 142–193 (1930).
123. G. M. Lukchis, *Chem. Eng.* **80,** (13), 111–116; (16), 83–87; (18), 83–90 (1973).
124. C. W. Chi, *AIChE Symp. Ser.* **74,** 42–46 (1977).
125. R. P. Underwood, *Chem. Eng. Sci.* **41,** 409–411 (1986).
126. R. Kumar and G. L. Dissinger, *Ind. Eng. Chem. Proc. Des. Dev.* **25,** 456–464 (1986).
127. I. Zwiebel, R. L. Gariepy, and J. J. Schnitzer, *AIChE J.* **18,** 1139–1147 (1972); **20,** 915–923 (1974).
128. A. Jedrzejak and M. Paderewski, *Int. Chem. Eng.* **28,** 707–712 (1988).
129. J. L. Bravo, Report of DOE Contract No. DE-AS07-831D12473, 1984, 150–181.
130. J. M. Schork and J. R. Fair, *Ind. Eng. Chem. Res.* **27,** 457–469 (1988).
131. D. Basmadjian, K. D. Ha, and C. Y. Pan, *Ind. Eng. Chem. Proc. Des. Dev.* **14,** 328–340 (1975).
132. C. Huang and J. R. Fair, *AIChE J.* **35,** 1667–1677 (1989).
133. M. M. Davis and M. D. LeVan, *Ind. Eng. Chem. Res.* **28,** 778–785 (1989).
134. A. Jedrzejak, *Chem. Eng. Technol.* **11,** 352–358 (1988).
135. D. H. White, Jr., and P. G. Barkley, *Chem. Eng. Progr.* **85,** 25–33 (1989).
136. K. Chihara and M. Suzuki, *J. Chem. Eng. Japan* **16,** 293–299 (1983).
137. J. W. Carter and M. L. Wyszynski, *Chem. Eng. Sci.* **38,** 1093–1099 (1983).
138. R. Kumar, *Ind. Eng. Chem. Res.* **28,** 1677–1683 (1989).
139. J. L. Liow and C. N. Kenney, *AIChE J.* **36,** 53–65 (1990).
140. H. Shin and K. S. Knaebel, *AIChE J.* **34,** 1409–1416 (1988).
141. P. H. Turnock and R. H. Kadlec, *AIChE J.* **17,** 335–342 (1971).
142. S. J. Doong and R. T. Yang, *AIChE Symp. Ser.* **84,** 145–154 (1988).
143. S. Ergun, *Chem. Eng. Progr.* **48,** 89–94 (1952).

144. M. Leva, *Chem. Eng.* **56,** 115–117 (1949).
145. L. E. Brownell, H. S. Dombrowski, and C. A. Dickey, *Chem. Eng. Progr.* **46,** 415–422 (1950).

JOHN D. SHERMAN
CARMEN M. YON
UOP

ADSORPTION, LIQUID SEPARATION

Nearly every chemical manufacturing operation requires the use of separation processes to recover and purify the desired product. In most circumstances, the efficiency of the separation process has a significant impact on both the quality and the cost of the product (1). Liquid-phase adsorption has long been used for the removal of contaminants present at low concentrations in process streams. In most cases, the objective is to remove a specific feed component. Alternatively, when the contaminants are not well defined, the objective is the improvement of feed quality defined by color, taste, odor, and storage stability (2–5).

In contrast to trace impurity removal, the use of adsorption for bulk separation in the liquid phase on a commercial scale is a relatively recent development. The first commercial operation occurred in 1964 with the advent of the UOP Molex process for recovery of high purity *n*-paraffins (6–8). Since that time, bulk adsorptive separation of liquids has been used to solve a broad range of problems, including individual isomer separations and class separations. The commercial availability of synthetic molecular sieves (qv) and ion-exchange (qv) resins and the development of novel process concepts have been the two significant factors in the success of these processes (see CONTACTORS FOR CHEMICAL SEPARATIONS).

Adsorbate–Adsorbent Interactions

An adsorbent can be visualized as a porous solid having certain characteristics. When the solid is immersed in a liquid mixture, the pores fill with liquid, which at equilibrium differs in composition from that of the liquid surrounding the particles. These compositions can then be related to each other by enrichment factors that are analogous to relative volatility in distillation. The adsorbent is selective for the component that is more concentrated in the pores than in the surrounding liquid.

The choice of separation method to be applied to a particular system depends largely on the phase relations that can be developed by using various separative agents. Adsorption is usually considered to be a more complex operation than is the use of selective solvents in liquid–liquid extraction (see EXTRACTION, LIQUID–LIQUID), extractive distillation, or azeotropic distillation (see DISTILLATION, AZEOTROPIC AND EXTRACTIVE). Consequently, adsorption is employed when higher selectivities than those obtained with solvents can be achieved.

A significant advantage of adsorbents over other separative agents lies in the fact that favorable equilibrium-phase relations can be developed for particular separations. Adsorbents can be produced that are much more selective in their affinity for various substances than are any known solvents. This selectivity is particularly true of the synthetic crystalline zeolites containing exchangeable cations. These zeolites became available in the early 1960s under the name of molecular sieves (qv) (9).

An example of unique selectivity is provided by the use of 5A molecular sieves for the separation of linear hydrocarbons from branched and cyclic types. In this system only the linear molecules can enter the pores. The others are completely excluded because of larger cross sections. Thus the selectivity for linear molecules with respect to other types is infinite. In the more usual case, all the feed components access the selective pores, but some components of the mixture are adsorbed more strongly than others. A selectivity between the different components that can be used to accomplish separation is thus established.

Another example of unique selectivities is the separation of olefins from paraffins in feed mixtures containing about five successive molecular sizes, eg, C_{10} to C_{14}. Liquid–liquid extraction might be considered for this separation. However, polar solvents give solubility patterns of the type shown in Figure 1. Each olefin is more soluble than the paraffin of the same chain length, but the solubility of both species declines as chain length increases. Thus, in a broad-boiling mixture, solubilities of paraffins and olefins overlap and separation becomes impossible. In contrast, the relative adsorption of olefins and paraffins from the liquid phase on the adsorbent used commercially for this operation is shown

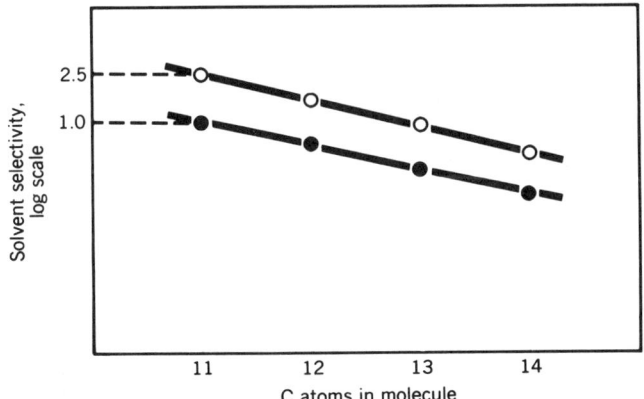

Fig. 1. Liquid–liquid extraction selectivity: ○, olefins; ●, paraffins.

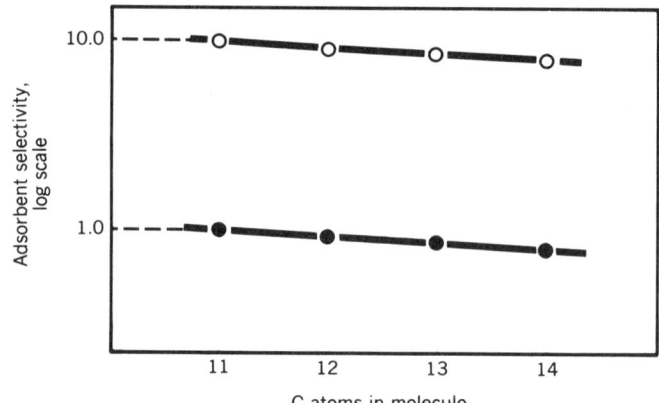

Fig. 2. Liquid-phase selectivity of UOP Olex adsorbent: ○, olefins; ●, paraffins.

in Figure 2. Not only is there selectivity between an olefin and paraffin of the same chain length, but also chain length has little effect on selectivity. Consequently, the complete separation of olefins from paraffins becomes possible.

Unique adsorption selectivities are employed in the separation of C_8 aromatic isomers, a classical problem that cannot be easily solved by distillation, crystallization, or solvent extraction (10). Although p-xylene [106-42-3] can be separated by crystallization, its recovery is limited because of the formation of eutectic with m-xylene [108-58-3]. However, either p-xylene, m-xylene, o-xylene [95-47-6], or ethylbenzene [100-41-4] can be extracted selectively by suitable modification of zeolitic adsorbents.

Literature dealing with adsorbent–adsorbate interactions in liquid phase is largely confined to patents (11–43). Although theoretical consistency tests exist for such data (44), the search for an adsorbent of suitable selectivity remains an art.

Practical Adsorbents

The search for a suitable adsorbent is generally the first step in the development of an adsorption process. A practical adsorbent has four primary requirements: selectivity, capacity, mass transfer rate, and long-term stability. The requirement for adequate adsorptive capacity restricts the choice of adsorbents to microporous solids with pore diameters ranging from a few tenths to a few tens of nanometers.

Traditional adsorbents such as silica [7631-86-9], SiO_2; activated alumina [1318-23-6], Al_2O_3; and activated carbon [7440-44-0], C, exhibit large surface areas and micropore volumes. The surface chemical properties of these adsorbents make them potentially useful for separations by molecular class. However, the micropore size distribution is fairly broad for these materials (45). This characteristic makes them unsuitable for use in separations in which steric hindrance can potentially be exploited (see ACTIVATED ALUMINA).

Typical nonsieve, polar adsorbents are silica gel and activated alumina. Equilibrium data have been published on many systems (11–16,46,47). The order of affinity for various chemical species is: saturated hydrocarbons < aromatic hydrocarbons = halogenated hydrocarbons < ethers = esters = ketones < amines = alcohols < carboxylic acids. In general, the selectivities are parallel to those obtained by the use of selective polar solvents; in hydrocarbon systems, even the magnitudes are similar. Consequently, the commercial use of these adsorbents must compete with solvent-extraction techniques.

The principal nonpolar-type adsorbent is activated carbon. Equilibrium data have been reported on hydrocarbon systems, various organic compounds in water, and mixtures of organic compounds (11,15,16,46,47). With some exceptions, the least polar component of a mixture is selectively adsorbed; eg, paraffins are adsorbed selectively relative to olefins of the same carbon number, but dicyclic aromatics are adsorbed selectively relative to monocyclic aromatics of the same carbon number (see ACTIVATED CARBON).

Polymeric resins [81133-25-7], $-(C_{10}H_{10})_n-$, are widely used in the food and pharmaceutical industries as cation–anion exchangers for the removal of trace components and for some bulk separations, such as fructose from glucose (48). These resins are primarily attractive for aqueous-phase separations and offer a fairly wide potential range of surface chemistries to fit a number of separation needs. For example, polymeric resins are effective in partitioning by size and molecular weight and may also be effective in ion exclusion (see ION EXCHANGE).

In contrast to these adsorbents, zeolites offer increased possibilities for exploiting molecular-level differences among adsorbates. Zeolites are crystalline aluminosilicates containing an assemblage of SiO_4 and AlO_4 tetrahedra joined together by oxygen atoms to form a microporous solid, which has a precise pore structure (49). Nearly 40 distinct framework structures have been identified. Table 1 and Figure 3 summarize some of those structures that have been widely used in the chemical industry. The versatility of zeolites lies in the fact that widely different adsorptive properties may be realized by the appropriate control of the framework structure, the silica-to-alumina ratio (Si/Al), and the cation form. For example, zeolite A, shown in Figure 4, has a three-dimensional isotrophic channel structure constricted by an eight-membered oxygen ring. Its effective pore size can be controlled at about 0.3, 0.4, and 0.45 nm by exchanging with potassium, sodium, and calcium, respectively. The potassium form, with 0.3-nm pores, is used for removing water from olefinic hydrocarbons. The sodium form can be used to efficiently remove water from nonreactive hydrocarbons, such as alkanes. The substitution of calcium can provide a pore size that admits n-paraffins and excludes other hydrocarbons.

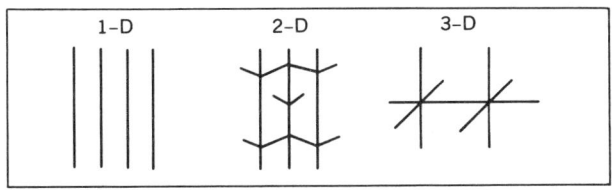

Fig. 3. Schematic diagram of molecular sieve pore structure. See Table 1.

Table 1. Molecular Sieve Pore Structures

Common name	Ring size, number of atoms	Free aperture, nm	Pore structure[a]	CAS Registry Number	Formula
faujasite	12	0.74	3-D	[12173-28-3]	$(Ca, Mg, Na_2, K_2)_{29.5}[(AlO_2)_{59}(SiO_2)_{133}] \cdot 235 H_2O$
mordenite	8	0.29 × 0.57	1-D	[12173-98-7]	$Na_8[(AlO_2)_8(SiO_2)_{40}] \cdot 24 H_2O$
	12	0.67 × 0.7	1-D		
L	12	0.71	1-D		$K_9[(AlO_2)_9(SiO_2)_{27}] \cdot 22 H_2O$
ZSM-5	10	0.54 × 0.56	1-D	[58339-99-4]	$(Na, TPA^b)_3 [(AlO_2)_3(SiO_2)_{93} \cdot 16 H_2O]$
	10	0.51 × 0.56	1-D		
Erionite	8	0.36 × 0.52	2-D	[12150-42-8]	$(Ca, Mg, Na_2, K_2)_{4.5}[(AlO_2)_9(SiO_2)_{27}] \cdot 27 H_2O$
A	8	0.42	3-D		$Na_{12}[(AlO_2)_{12}(SiO_2)_{12}] \cdot 27 H_2O$

[a]See Figure 3.
[b]TPA = tetrapropylammonium.

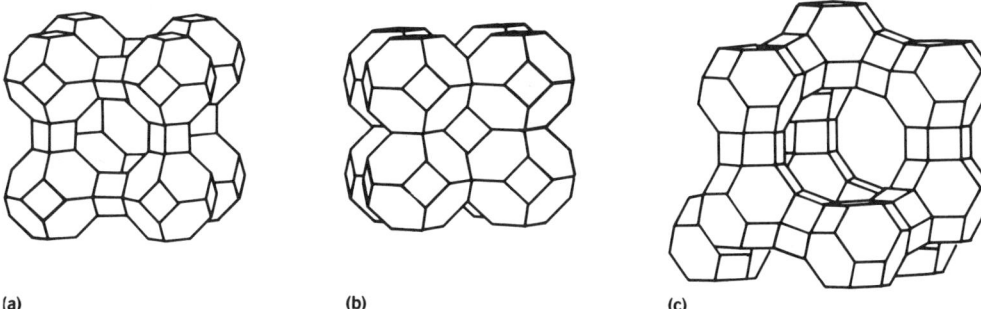

Fig. 4. Three zeolites with the same structural polyhedron, cubo-octahedrons. (a) Type A, $Na_{12}[(AlO_2)_{12}(SiO_2)_{12}]\cdot 27H_2O$; (b) sodalite [1302-90-5]; (c) faujasite (Type X, Y), where $X = Na_{86}[(AlO_2)_{86}(SiO_2)_{106}]\cdot 264H_2O$; $Y = Na_{56}[(AlO_2)_{56}(SiO_2)_{136}]\cdot 250H_2O$

Large-pore zeolites, X, Y, and mordenites, have pores defined by 12-membered oxygen rings with a free diameter of 0.74 nm. The framework structure of X and Y faujasites sketched in Figure 4 consists of a total of 192 SiO_2 and AlO_2 units. The Si/Al (atomic) ratio for X is generally between 1.0 and 1.5, whereas for Y it is between 1.5 and 3.0. With suitable procedures, Y can be dealuminated to make ultrastable Y where Si/Al ratios exceed 100. Adsorption properties of faujasites are stongly dependent on not only the cation form, but also the Si/Al ratio. The flexibility provided by faujasites in the adsorption of C_8 aromatics is shown in Table 2. The selectivity order, from the most selectivity adsorbed to the least selectively adsorbed, can be changed significantly by the choice of zeolite properties.

The separation of fructose from glucose illustrates the interaction between the framework structure and the cation (Fig. 5) (50). Ca^{2+} is known to form complexes with sugar molecules such as fructose. Thus, Ca–Y shows a high selectivity for fructose over glucose. However, Ca–X does not exhibit high selectivity. On the other hand, K–X shows selectivity for glucose over fructose. This polar nature of faujasites and their unique shape-selective properties, more than the molecular-sieving properties, make them most useful as practical adsorbents.

Polymeric cation-exchange resins are also used in the separation of fructose from glucose. The UOP Sarex process has employed both zeolitic and polymeric resin adsorbents for the production of high fructose corn syrup (HFCS). The

Table 2. Selectivity of Zeolites in C_8H_{10} Aromatic Systems

	Adsorbent[a]			
	No. 1	No. 2	No. 3	No. 4
p-xylene	1	2	3	4
ethylbenzene	2	1	4	3
m-xylene	3	3	1	2
o-xylene	4	4	2	1

[a]Key: 1 = most selectively adsorbed, 4 = least selectively adsorbed.

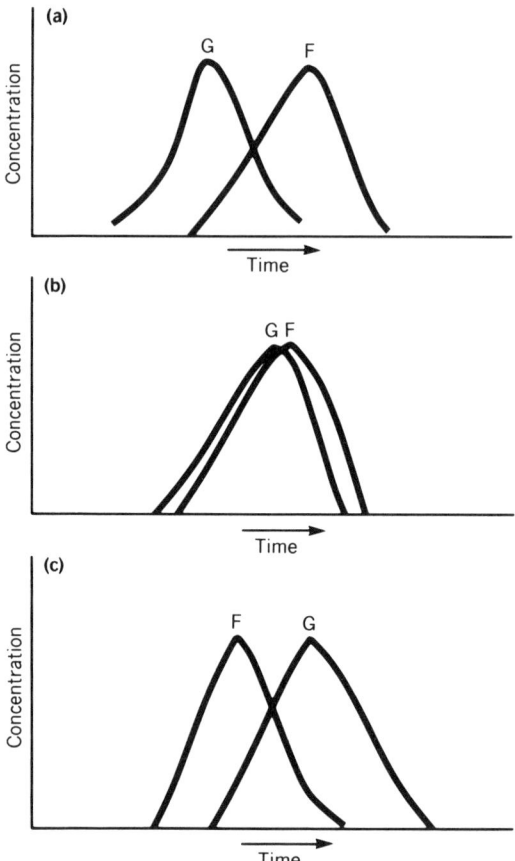

Fig. 5. Fructose(F)–glucose(G) separation on faujasite adsorbents. (**a**) Ca–Y adsorbent; (**b**) Ca–X adsorbent; (**c**) K–X adsorbent.

operating characteristics of these two adsorbents are substantially different and have been compared in terms of fundamental characteristics such as capacity, selectivity, and adsorption kinetics (51).

The zeolite and the resin adsorbents show different adsorption isotherm characteristics, particularly at higher concentration (51). The resin adsorbent isotherm is slightly concave upward, whereas the zeolite isotherm is linear, or even slightly concave downward. Resins, therefore, have an advantage in a UOP Sarex operation that involves high feed-solids concentration.

In addition to the fundamental parameters of selectivity, capacity, and mass-transfer rate, other more practical factors, namely, pressure drop characteristics and adsorbent life, play an important part in the commercial viability of a practical adsorbent.

Pressure Drop Characteristics. Ion-exchange resins are compressible and exhibit a characteristic stress–strain relationship, as shown in Figure 6. In addition, they undergo shrinking and swelling as a result of osmotic pressure variation resulting from concentration changes. Zeolite adsorbents are rigid and do not exhibit much strain with pressure. When resins are used as adsorbents, the

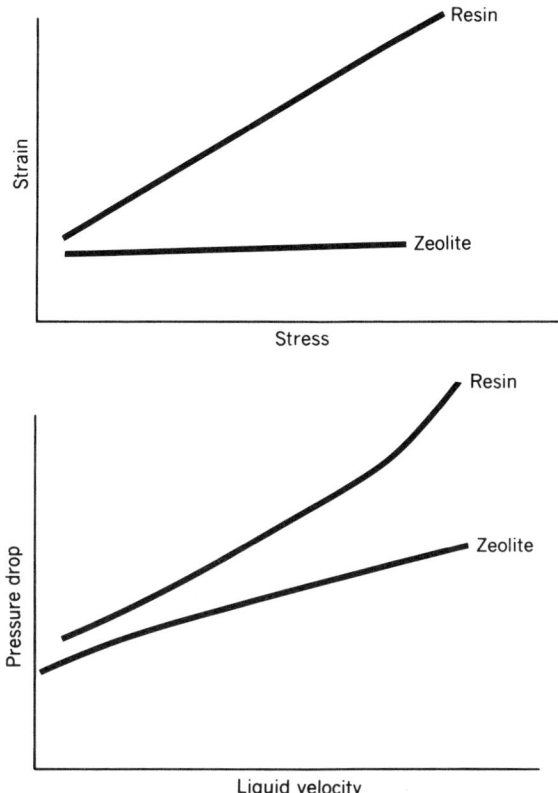

Fig. 6. Compressibility characteristics of adsorbents.

implications of shrinking and swelling and compressibility must be considered to ensure safe operation below the design pressure drop. The pressure drop can often become a major factor in the determination of maximum throughput.

Adsorbent Life. Long term stability under rugged operating conditions is an important characteristic of an adsorbent. By their nature zeolites are not stable in an aqueous environment and must be specially formulated to enhance their stability in order to obtain several years of service. Polymeric resins do not suffer from dissolution problems. However, they are prone to chemical attack (52).

Commercial Processes

Industrial-scale adsorption processes can be classified as batch or continuous (53,54). In a batch process, the adsorbent bed is saturated and regenerated in a cyclic operation. In a continuous process, a countercurrent staged contact between the adsorbent and the feed and desorbent is established by either a true or a simulated recirculation of the adsorbent.

The efficiency of an adsorption process is significantly higher in a continuous mode of operation than in a cyclic batch mode (55). In a batch chromatographic operation, the liquid composition at a given level in the bed undergoes a cyclic change with time, and large portions of the bed do not perform any useful function at a given time. In continuous operation, the composition at a given level is invariant with time, and every part of the bed performs a useful function at all times. The height equivalent of a theoretical plate (HETP) in a batch operation is roughly three times that in a continuous mode. For difficult separations, batch operation may require 25 times more adsorbent inventory and twice the desorbent circulation rate than does a continuous operation. In addition, in a batch mode, the four functions of adsorption, purification, desorption, and displacement of the desorbent from the adsorbent are inflexibly linked, whereas a continuous mode allows more degrees of freedom with respect to these functions, and thus a better overall operation.

Continuous Countercurrent Processes

The need for a continuous countercurrent process arises because the selectivity of available adsorbents in a number of commercially important separations is not high. In the p-xylene system, for instance, if the liquid around the adsorbent particles contains 1% p-xylene, the liquid in the pores contains about 2% p-xylene at equilibrium. Therefore, one stage of contacting cannot provide a good separation, and multistage contacting must be provided in the same way that multiple trays are required in fractionating materials with relatively low volatilities.

The hypersorption process (56) developed by Union Oil Company in the early 1950s for the recovery of propane and heavier components from natural gas is the earliest example of large-scale countercurrent adsorption processes. This process used an activated carbon adsorbent flowing as a dense bed continuously downward through a rising gas stream. A unit built for the Dow Chemical Company had a gas capacity of more than 500,000 m^3/day. However, this process proved to be less economical than cryogenic distillation and is no longer in operation. A number of commercial moving-bed designs exist mainly for ion exchange. A good review of these designs can be found in reference 57.

Since the 1960s the commercial development of continuous countercurrent processes has been almost entirely accomplished by using a flow scheme that simulates the continuous countercurrent flow of adsorbent and process liquid without the actual movement of the adsorbent. The idea of a simulated moving bed (SMB) can be traced back to the Shanks system for leaching soda ash (58).

Such a concept was originally used in a process developed and licensed by UOP under the name UOP Sorbex (59,60). Other versions of the SMB system are also used commercially (61). Toray Industries built the Aromax process for the production of p-xylene (20,62,63). Illinois Water Treatment and Mitsubishi have commercialized SMB processes for the separation of fructose from dextrose (64–66). The following discussion is based on the UOP Sorbex process.

MOVING-BED OPERATION

A hypothetical moving-bed system and a liquid-phase composition profile are shown in Figure 7. The adsorbent circulates continuously as a dense bed in a closed cycle and moves up the adsorbent chamber from bottom to top. Liquid

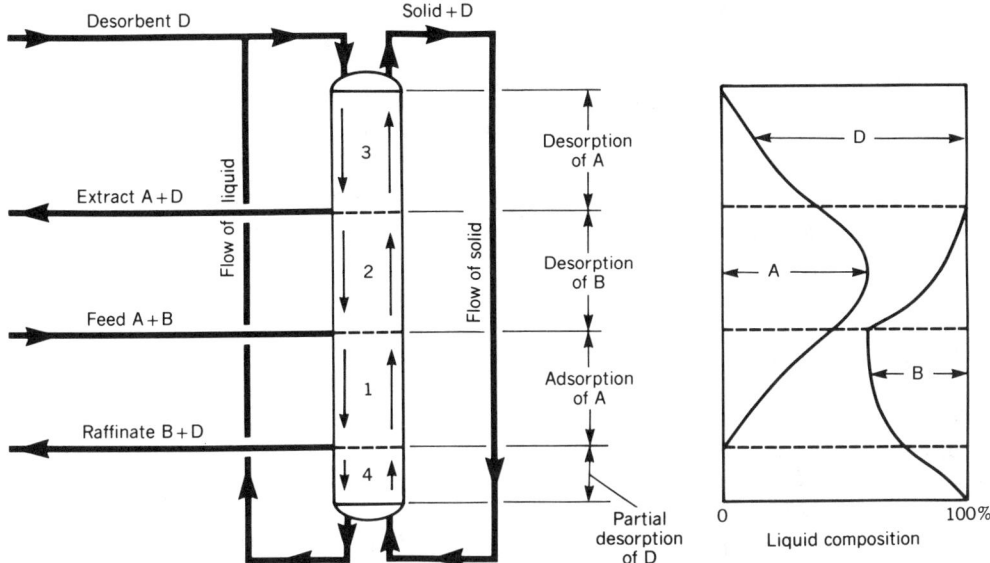

Fig. 7. Adsorptive separation with moving bed.

streams flow down through the bed countercurrently to the solid. The feed is assumed to be a binary mixture of A and B, with component A being adsorbed selectively. Feed is introduced to the bed as shown.

Desorbent D is introduced to the bed at a higher level. This desorbent is a liquid of different boiling point from the feed components and can displace feed components from the pores. Conversely, feed components can displace desorbent from the pores with proper adjustment of relative flow rates of solid and liquid.

Raffinate product, consisting of the less strongly adsorbed component B mixed with desorbent, is withdrawn from a position below the feed entry. Only a portion of the liquid flowing in the bed is withdrawn at this point; the remainder continues to flow into the next section of the bed. Extract product, consisting of the more strongly adsorbed component A mixed with desorbent, is withdrawn from the bed; again, only a portion of the flowing liquid in the bed is withdrawn, and the remainder continues to flow into the next bed section.

The positions of introduction and withdrawal of net streams divide the bed into four zones, each of which performs a different function as described below.

Zone 1. The primary function of this zone is to adsorb A from the liquid. The solid entering at the bottom carries only B and D in its pores. As the liquid stream flows downward, countercurrent to this solid, component A is transferred from the liquid stream into the pores of the solid. At the same time, component D is desorbed (transferred from the pores to the liquid stream) to make room for A.

Zone 2. The primary function of this zone is to remove B from the pores of the solid. When the solid arrives at the fresh feed point, the pores contain the quantity of A that was adsorbed in Zone 1. However, the pores also contain a large quantity of B, because the solid has just been in contact with fresh feed. The liquid entering the top of Zone 2 contains no B, only A and D. As the solid moves upward, countercurrent to this stream, B is gradually displaced from the pores and is replaced by A and D. Thus, when the solid arrives at the top of Zone 2, the pores contain only A and D. By proper regulation of the liquid rate in Zone 2, B can be desorbed completely from the pores. This B desorption can be accomplished without simultaneously desorbing all of A, because A is more strongly adsorbed than B.

Zone 3. The function of this zone is to desorb A from the pores. The solid entering the zone carries A and D in the pores; the liquid entering the top of the zone consists of pure D. As the solid rises, A in the pores is displaced by D.

Zone 4. The purpose of this zone is to act as a buffer to prevent component B, which is at the bottom of Zone 1, from passing into Zone 3, where it would contaminate extracted component A. When the adsorbent leaves Zone 3, the pores are completely filled with desorbent. The liquid entering the top of Zone 4 is of raffinate composition and contains B and D. If the flow rate in Zone 4 is properly regulated, component B will be readsorbed completely from the liquid, preventing its entry into Zone 3, where it would contaminate the product A.

Difficulties of Moving-Bed Operation. The use of a moving bed introduces the problem of mechanical erosion of the adsorbent. Obtaining uniform flow of both solid and liquid in beds of large diameter is also difficult. The performance of this type of operation can be greatly impaired by nonuniform flow of either phase.

The use of a series of fluidized beds may be considered when solid overflows from each bed to the next. However, this arrangement involves a sacrifice in mass-transfer efficiency because the number of theoretical equilibrium trays cannot exceed the number of physical beds. In contrast, the flow through dense and fixed beds of adsorbent, as practiced in chromatography, can provide hundreds of theoretical trays in beds of modest length. Another disadvantage of a fluidized-bed operation is the large-sized equipment required to contain a given inventory of adsorbent (see MASS TRANSFER). In view of these difficulties, only a few fluidized-bed operations are practiced commercially. The Purasiv HR system using beaded activated carbon for the recovery of solvent is one example. This process uses a staged fluidized bed for adsorption and a moving bed for regeneration (67).

SIMULATED MOVING-BED OPERATION

In the moving-bed system of Figure 7, solid is moving continuously in a closed circuit past fixed points of introduction and withdrawal of liquid. The same results can be obtained by holding the bed stationary and periodically moving the positions at which the various streams enter and leave. A shift in the positions of the introduction of the liquid feed and the withdrawal in the direction of fluid flow through the bed simulates the movement of solid in the opposite direction.

Of course, moving the liquid feed and withdrawal positions continuously is impractical. However, approximately the same effect can be produced by providing multiple liquid-access lines to the bed and periodically switching each stream to the adjacent line. Functionally, the adsorbent bed has no top or bottom and is equivalent to an annular bed. Therefore, the four liquid-access positions can be moved around the bed continually, always maintaining the same distance between the various streams.

The commercial application of this concept (68) is portrayed in Figure 8,

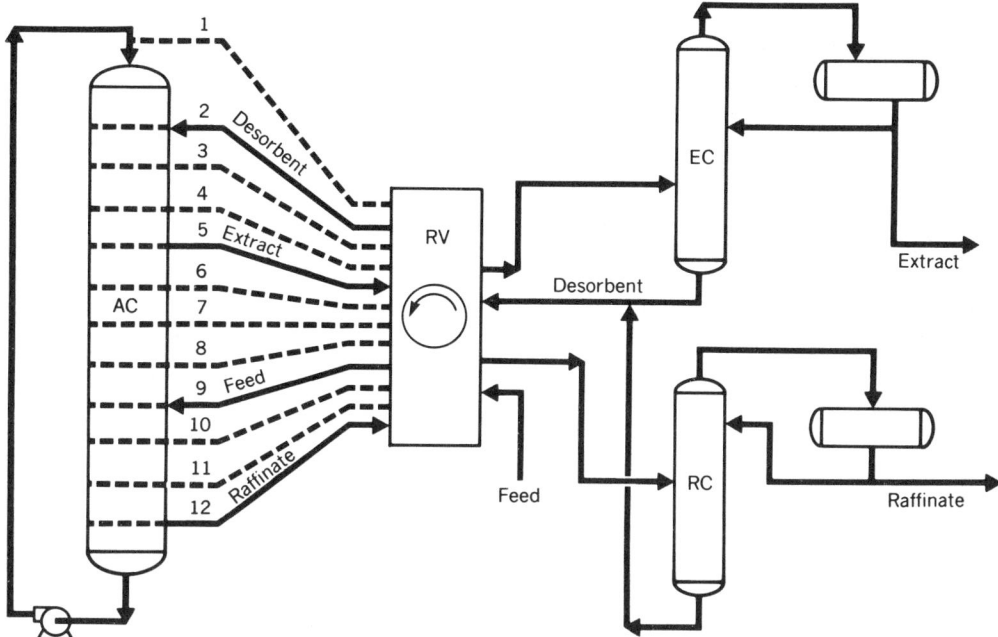

Fig. 8. UOP Sorbex simulated moving bed for adsorptive separation. AC = adsorbent chamber, RV = rotary valve, EC = extract column, RC = raffinate column.

which shows the adsorbent as a stationary bed. A liquid circulating pump is provided to pump liquid from the bottom outlet to the top inlet of the adsorbent chamber. A fluid-directing device known as a rotary valve (69,70) is provided. The rotary valve functions on the same principle as a multiport stopcock in directing each of several streams to different lines. At the right-hand face of the valve, the four streams to and from the process are continuously fed and withdrawn. At the left-hand face of the valve, a number of lines are connected that terminate in distributors within the adsorbent bed.

At any particular moment, only four lines from the rotary valve to the adsorbent chamber are active. Figure 8 shows the flows at a time when lines 2, 5, 9, and 12 are active. When the rotating element of the rotary valve is moved to its next position, each net flow is transferred to the adjacent line; thus, desorbent enters line 3 instead of line 2, extract is drawn from 6 instead of 5, feed enters 10 instead of 9, and raffinate is drawn from 1 instead of 12.

Figure 7 shows that in the moving-bed operation, the liquid flow rate in each of the four zones is different because of the addition or withdrawal of the various steams. In the simulated moving-bed of Figure 8, the liquid flow rate is controlled by the circulating pump. At the position shown in Figure 8, the pump is between the raffinate and desorbent ports, and therefore should be pumping at a rate appropriate for Zone 4. However, after the next switch in position of the rotary valve, the pump is between the feed and raffinate ports, and should therefore be pumping at a rate appropriate for Zone 1. Stated briefly, the circulating pump must be programmed to pump at four different rates. The control point is altered each time an external stream is transferred from line 12 to line 1.

To complete the simulation, the liquid-flow rate relative to the solid must be the same in both the moving-bed and simulated moving-bed operations. Because the solid is physically stationary in the simulated moving-bed operation, the liquid velocity relative to the vessel wall must be higher than in an actual moving-bed operation.

The primary control variables at a fixed feed rate, as in the operation pictured in Figure 8, are the cycle time, which is measured by the time required for one complete rotation of the rotary valve (this rotation is the analog of adsorbent circulation rate in an actual moving-bed system), and the liquid flow rate in Zones 2, 3, and 4. When these control variables are specified, all other net rates to and from the bed and the sequence of rates required at the liquid circulating pump are fixed. An analysis of sequential samples taken at the liquid circulating pump can trace the composition profile in the entire bed. This profile provides a guide to any changes in flow rates required to maintain proper performance before any significant effect on composition of the products has appeared. Various aspects of process control are described in the patent literature (71–73).

Temperature and pressure are not considered as primary operating variables: temperature is set sufficiently high to achieve rapid mass-transfer rates, and pressure is sufficiently high to avoid vaporization. In liquid-phase operation, as contrasted to vapor-phase operation, the required bed temperature bears no relation to the boiling range of the feed, an advantage when heat-sensitive stocks are being treated.

MODELING OF UOP SORBEX SYSTEMS

The theoretical performance of the commercial simulated moving-bed operation is practically identical to that of a system in which solids flow continuously as a dense bed countercurrent to liquid. A model in which the flows of solid and liquid are continuous, as shown in Figure 7, is therefore adequate.

The operation is modeled in terms of theoretical equilibrium trays having the same significance as in fractionating columns (see DISTILLATION). Solid and liquid are assumed to flow continuously through hypothetical well-mixed theoretical trays in which equilibrium is attained. The number of theoretical trays has no relation to the number of bed segments in the actual operation. Each segment can be equivalent to many theoretical trays. The number is determined by bed height, mass-transfer coefficient, and flow rates. Axial mixing is generally of much greater significance in liquid than in vapor systems because of the greater mass of process fluid in the voids relative to that in the pores. To allow for the effect of axial mixing in the liquid phase, a second parameter is introduced into the model by assuming that the solid entrains a certain quantity of liquid from deck to deck.

The relationship of this type of model to a true differential analysis has been discussed for the case of linear equilibrium and first-order kinetics (74,75). A minor extension of this work leads to the following relations for a bed section in which flow rates of solid and liquid are constant. For the number of theoretical trays, n,

$$n = KkH/Lz \tag{1}$$

where K is the adsorption equilibrium constant; k the mass-transfer coefficient; H the bed height; L the net liquid-flow rate; and $z = j\ln[j/(j-1)]$, where j is an integer. The entrainment factor per unit liquid flow rate is given by

$$e = KkD°/L^2 \tag{2}$$

where $D°$ is the axial diffusion coefficient. Equations 1 and 2 are approximations that apply for efficient beds at reasonable operating conditions. These equations appear to be ambiguous because the values of n and e are different, depending on the component being considered. However, this ambiguity is not of great practical significance because each zone of the system is performing a function that is critical with respect to only one component. Values of n and e corresponding to the equilibrium properties of the critical component in each zone should be used.

The model of theoretical equilibrium trays with entrainment is readily treated by computer with methods analogous to those used for the design of fractionating columns.

McCABE-THIELE DIAGRAM

The McCabe-Thiele approach has been developed to describe the Sorbex process (76). Two feed components, A and B, with a suitable adsorbent and a desorbent, C, are separated in an isothermal continuous countercurrent operation. If A is the more strongly adsorbed component and the system is linear and noninteracting, the flows in each section of the process must satisfy the following constraints for complete separation of A from B:

Section	Condition
IV	$S/(D + F - E - R) > K_{CB}$
I	$S/(D + F - E) > K_{BA}$
II	$S/(D - E) < K_{AB}$
III	$S/D < K_{CA}$

The required direction of the net flow of each component is illustrated in Figure 9. The UOP Sorbex process has four flow-rate variables (S/F, D/F, E/F, and R/F) and four inequality constraints, one for each section of the bed. Once the equilibrium is fixed, the only remaining degree of freedom is the margin by which the inequality constraints are fulfilled. Once that is decided, the inequality constraints become four equations that define all flow-rate ratios for the system. Once the flow rates are fixed, a preliminary estimate of the number of theoretical stages in each section may be obtained by a McCabe-Thiele diagram, shown in Figure 10.

McCabe-Thiele diagrams for nonlinear and more practical systems with pertinent inequality constraints are illustrated in Figures 11 and 12. The convex isotherms are generally observed for zeolitic adsorbents, particularly in hydrocarbon separation systems, whereas the concave isotherms are observed for ion-exchange resins used in sugar separations.

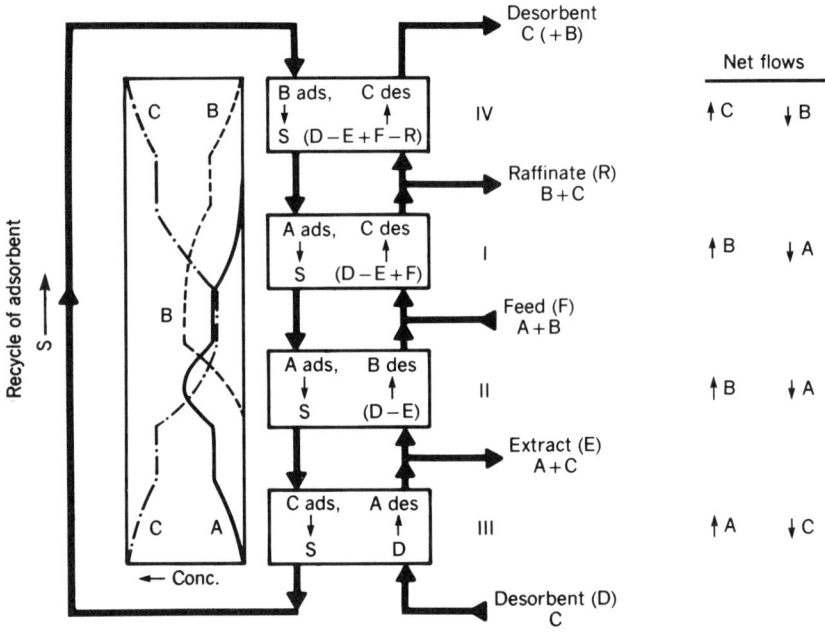

Fig. 9. Schematic diagram of a UOP Sorbex process. D, E, F, R, and S represent flow rates for desorbent, extract, feed, raffinate, and net solids, respectively.

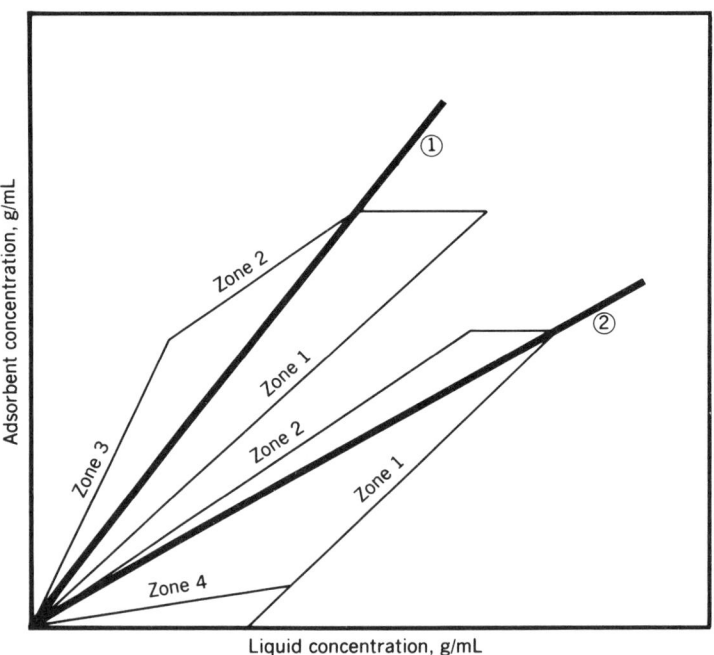

Fig. 10. UOP Sorbex operation with linear isotherms. Slope of ① $= K_1$; slope of ② $= K_2$. Conditions for separation: $K_1 > K_2$, $L_3/S \geq K_1$, $K_2 \leq L_2/S \leq K_1$, $K_2 \leq L_1/S \leq K_1$, $L_4/S \leq K_2$, $L_1 - L_2 = F$; where $K =$ adsorption coefficient, $L =$ net liquid flow rate, $S =$ net solids flow rate, and $F =$ feed flow rate.

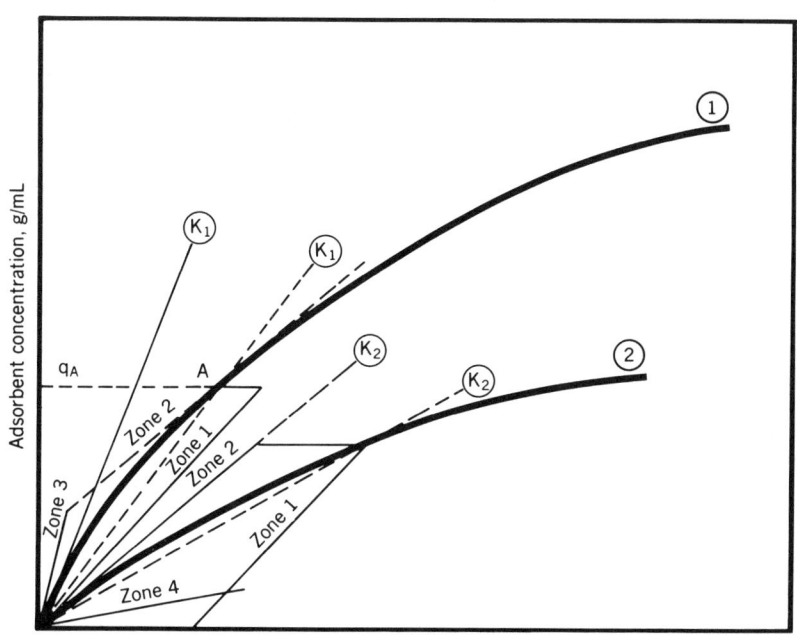

Fig. 11. UOP Sorbex operation with convex isotherms. Conditions for separation: at point A, slope of ① $= K_2$, $K_1 > K_2$, $L_3/S \geq K_1$, $K_2 \leq L_2/S \leq K_T \leq K_1$, $K_2' \leq L_1/S \leq K_1'$, $L_4/S \leq K_2$, $L_1 - L_2 = F$; where K, L, S, and F are as defined in Figure 10.

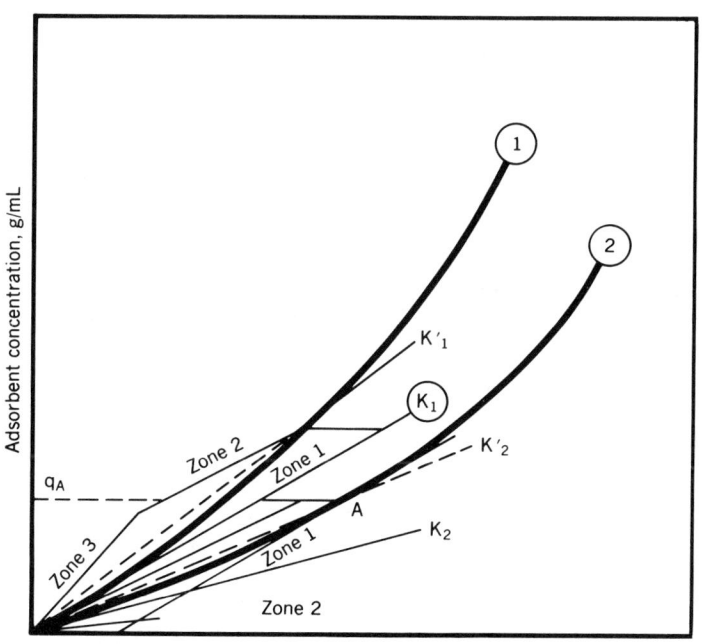

Fig. 12. Conditions for separation: At point A, slope of ② $= K_1$, $K_1 > K_2$, $L_3/S > K_1'$, $K_2 \leq K_T \leq L_1/S \leq K_1$, $K_2' \leq L_2/S \leq K_1'$, $L_4/S \leq K_2$; where K, L, and S are as defined in Figure 10. Adsorbent concentration of component 2 less than q_A.

UOP Sorbex Applications

The first UOP Sorbex process was licensed in 1962 as a UOP Molex process to separate n-paraffins from branched paraffins, cyclic paraffins, and aromatics. This plant started up in 1964. Products were used for the manufacture of biodegradable detergents. Since then, about 80 units have been put on-stream in a variety of applications that produce in excess of 8×10^6 t/yr of products. The extent of commercialization, as of July 1989, is shown in Table 3. The problems associated with large-scale commercial operation have been solved as is evidenced by the satisfactory operation of a single p-xylene unit with a capacity of nearly 400,000 t/yr. The performance of rotary valves has been excellent. The valves have required only routine maintenance. The largest adsorbent bed in commercial use as of 1989 had a diameter of 6.7 m and was performing as well as a pilot plant with a bed diameter of 0.1 m. This performance was achieved by paying particular attention to the internal design of large adsorbent chambers to ensure good distribution and uniform flow of liquids (77–9). Evolutionary improvements in the UOP Sorbex process design have reduced product impurities from tenths of a percent to a few hundred parts per million.

Table 3. UOP Sorbex Processes for Commodity Chemicals

UOP Processes	Separation	Licensed units
Parex	p-xylene from C_8 aromatics	44
Molex	n-paraffins from branched and cyclic hydrocarbons	21
Olex	olefins from paraffins	6
Cymex	p- or m-cymene from cymene isomers	1
Cresex	p- or m-cresol from cresol isomers	1
Sarex	fructose from dextrose plus polysaccharides	5
total		78

***n*-Paraffin Separation.** The UOP Molex process is used to separate n-paraffins from branched paraffins and aromatics in a variety of applications. In the C_5–C_6 range, the Molex process is used to enhance the octane number of gasoline. As illustrated in Figure 13, the UOP Molex process is integrated with a catalytic isomerization process, UOP Penex, to achieve maximum isomerization octane (80). This octane upgrade is a result of the separation of the low octane n-paraffins from the high octane branched paraffins and the recycle of the n-paraffins to UOP Penex for further conversion. The magnitude of the octane increase for the UOP Penex-Molex process over the once-through UOP Penex process depends on the C_5–C_6 content of the feed. Typically, the product octanes are 4 to 5 numbers higher, for instance, 83.5 RON (research octane number) for the UOP Penex process and 88.5 RON for the UOP Penex-Molex process.

Linear paraffins in the C_{10} to C_{15} range are used for the production of alcohols and plasticizers and biodegradable detergents of the linear alkylbenzene sulfonate and nonionic types. The UOP Molex process is used to extract n-paraffins from a hydrotreated kerosine (6–8).

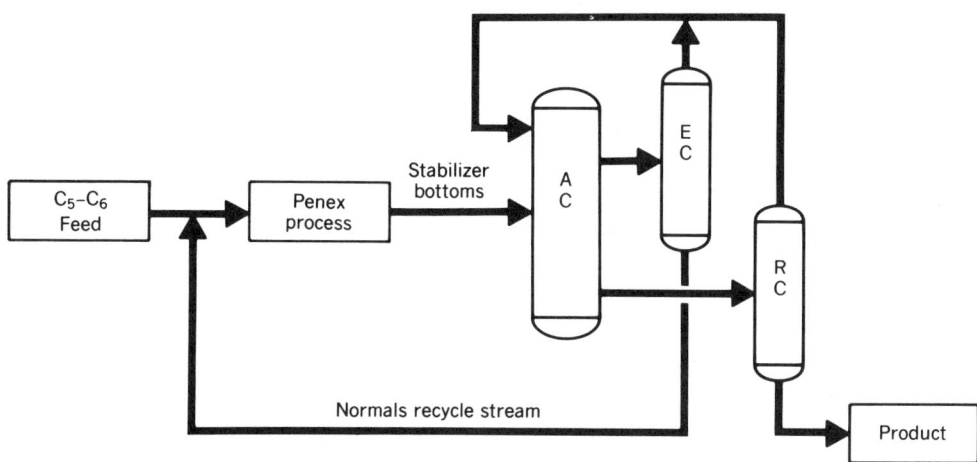

Fig. 13. UOP Penex-Molex process. AC = adsorbent chamber, EC = extract column, RC = raffinate column.

The fermentation of n-paraffins in the C_{10} to C_{23} range for protein production has provided another outlet for these hydrocarbons. Because it operates in liquid phase, the UOP Molex process can readily accomplish the separation of n-paraffins from such a wide boiling feedstock.

Olefin–Paraffin Separation. The catalytic dehydrogenation of n-paraffins offers a route to the commercial production of linear olefins. Because of limitations imposed by equilibrium and side reactions, conversion is incomplete. Therefore, to obtain a concentrated olefin product, the olefins must be separated from the reactor effluent (81–85), and the unreacted n-paraffins must be recycled to the catalytic reactor for further conversion.

The performance of the adsorptive section of such a combination in a commercial installation is shown in Table 4. The feedstock includes C_{11}–C_{14} components, and olefins are recovered at about 94% efficiency.

The olefin product contains 1.1% of residual n-paraffins. Essentially similar results have been obtained in commercial operations on C_8–C_{10} and C_{15}–C_{18} feedstocks. The desorbents used are generally hydrocarbon mixtures of lower boiling range than the feed components. The concentrated olefin stream may then be used for production of detergent alcohols.

p-Xylene Separation. p-Xylene finds wide use as a precursor for the production of polyester fibers and plastics. Before the advent of adsorptive tech-

Table 4. Commercial Operation of Linear C_{11}–C_{14} Olefin Extraction

Component	Feed, wt %	Extract, wt %	Raffinate, wt %
n-olefins	9.0	96.2	0.6
n-paraffins	90.1	1.1	98.5
other components[a]	0.9	2.7	0.9
total	*100.0*	*100.0*	*100.0*

[a]Including aromatics and branched-chain aliphatics.

niques, the *p*-xylene was commonly separated from hydrocarbon mixtures by crystallization. However, recovery was limited to 55 to 60% because of the formation of eutectic mixtures.

Since 1971 mainly adsorptive separation processes are used to obtain high purity *p*-xylene (55,84–86). A typical commercial process for the separation of *p*-xylene from other C_8 aromatics produces about 99.8% purity *p*-xylene at greater than 95% recovery.

Ethylbenzene Separation. Ethylbenzene [*100-41-4*], which is primarily used in the production of styrene, is difficult to separate from mixed C_8 aromatics by fractionation. A column of about 350 trays operated at a reflux:feed ratio of 20 is required. The adsorptive operation has been performed successfully in pilot plants (see Table 5). About 99% of the ethylbenzene in the feed was recovered at a purity of 99.7%. This operation, the UOP Ebex process, requires only about 40% of the energy required by fractional distillation.

Table 5. Ethylbenzene Separation, Pilot-Plant Scale

Component	Feed, wt %	Ethylbenzene, wt %	Residue, wt %
ethylbenzene	30.5	99.7	0.4
p-xylene	12.9	0.1	18.4
m-xylene	35.8	0.1	51.3
o-xylene	20.8	0.1	29.9
total	*100.0*	*100.0*	*100.0*

Fructose–Dextrose Separation. Fructose–dextrose separation is an example of the application of adsorption to nonhydrocarbon systems. An aqueous solution of the isomeric monosaccharide sugars, $C_6H_{12}O_6$, fructose and dextrose (glucose), accompanied by minor quantities of polysaccharides, is produced commercially under the designation of "high" fructose corn syrup (HFCS) by the enzymatic conversion of cornstarch. Because fructose has about double the sweetness index of dextrose, the separation of fructose from this mixture and the recycling of dextrose for further enzymatic conversion to fructose is of commercial interest.

The UOP Sarex process has been used since 1978 for the separation of high purity fructose from a mixture of fructose, glucose, and polysaccharides (87,88). The pilot-plant performance of fructose–glucose separation is given in Table 6.

Table 6. Separation of Fructose from High Fructose Corn Syrup, Pilot-Plant Scale

Component	Feed, wt %[a]	Extract, wt %[a]	Raffinate, wt %[a]
fructose	41.9	95.0	7.0
dextrose	53.1	5.0	84.7
other saccharides	5.0	0.0	8.3
total	*100.0*	*100.0*	*100.0*

[a]Dry basis.

Aromatic and Nonaromatic Hydrocarbon Separation. Aromatics are partially removed from kerosines and jet fuels to improve smoke point and burning characteristics. This removal is commonly accomplished by hydroprocessing, but can also be achieved by liquid–liquid extraction with solvents, such as furfural, or by adsorptive separation. Table 7 shows the results of a simulated moving-bed pilot-plant test using silica gel adsorbent and feedstock components mainly in the C_{10}–C_{15} range. The extent of extraction does not vary greatly for each of the various species of aromatics present. Silica gel tends to extract all aromatics from nonaromatics (89).

Table 7. Separation of Aromatics from Nonaromatics in C_{10}–C_{15} Light Cycle Oil

Components	Feed, vol %	Extract, vol %	Raffinate, vol %	Extraction, %
aromatics				
alkylbenzenes	17.9	34.3	7.5	76
tetralins, indanes	8.3	14.8	4.2	71
indenes	1.7	3.4	0.6	80
naphthalenes	13.6	30.0	1.6	86
acenaphthenes, biphenyls	4.4	9.1	0.8	89
tricyclic aromatics	1.7	3.3	0.2	92
total aromatics	*47.6*	*95.0*	*14.9*	*82*
nonaromatics	52.4	5.0	85.1	4
total	*100.0*	*100.0*	*100.0*	*41*

Citric Acid Separation. Citric acid [77-92-9] and other organic acids can be recovered from fermentation broths using the UOP Sorbex technology (90–92). The conventional means of recovering citric acid is by a lime and sulfuric acid process in which the citric acid is first precipitated as a calcium salt and then reacidulated with sulfuric acid. However, this process generates significant byproducts and thus can become inefficient.

UOP has developed a UOP Sorbex process for the recovery and purification of citric acid from fermentation broths. The process provides technical-grade citric acid, $C_6H_8O_7$, which can be further recrystallized to obtain food-grade citric acid (qv).

Separation of Fatty Acids. Tall oil is a by-product of the pulp and paper manufacturing process and contains a spectrum of fatty acids, such as palmitic, stearic, oleic, and linoleic acids, and rosin acids, such as abietic acid. The conventional refining process to recover these fatty acids involves intensive distillation under vacuum. This process does not yield high purity fatty acids, and moreover, a significant degradation of fatty acids occurs because of the high process temperatures. These fatty and rosin acids can be separated using a UOP Sorbex process (93–99) (Tables 8 and 9).

Table 8. UOP Sorbex Separation of Fatty Acids from Rosin Acids in Distilled Tall Oil

Composition	Feed, wt %	Extract, wt %	Raffinate, wt %
rosin acid	33.0	1.50	94.50
linoleic acid	32.7	48.50	2.00
oleic acid	32.75	48.508	2.00
neutrals	1.50	1.50	1.50

Table 9. UOP Sorbex Separation of Saturated and Unsaturated Tall Oil Fatty Acids

Composition	Feed, wt %	Extract, wt %	Raffinate, wt %
oleic acid	48.55	92.00	5.00
linoleic acid	48.55	5.00	92.00
rosin acid	1.50	1.50	1.50
neutrals	1.50	1.50	1.50

Cyclic-Batch Processes

Continuous processes have wide application in different areas of the chemical industry. The separation efficiency of a continuous process is generally higher than that of a batch or cyclic-batch process. However, in some applications the cyclic-batch process may be preferred because of the complexity of design and the difficulty of controlling the continuous processes. Examples of commercial cyclic-batch adsorption processes operating in liquid phase include the UOP methanol recovery (UOP MRU) and oxygenate removal (UOP ORU) processes, which separate oxygenates from C_4 hydrocarbons; the UOP Cyclesorb process, which separates fructose from glucose; and ion-exclusion processes for recovering sucrose from molasses.

UOP MRU–ORU Processes. Methyl *tert*-butyl ether [1634-04-4] (MTBE) is an additive that is used to increase the octane value of gasoline. In environmentally sensitive locations where a minimum oxygen content for gasoline is specified, MTBE can provide the desired oxygen content while increasing octane and maintaining compatibility with the hydrocarbon components of gasoline. The MTBE synthesis typically requires an excess of methanol [67-56-1] relative to isobutylene [115-11-7] to achieve as high a conversion as possible. Downstream of the reactor, this excess methanol must be separated from the MTBE and unreacted C_4 hydrocarbons so that it can be recycled back to the reactor. A two-stage water wash is conventionally used for methanol recovery. The UOP MRU process offers an alternative means of recovering the 3 to 4% methanol contained in the C_4 stream (100).

In most cases, the linear olefins in the C_4 stream are subsequently sent to alkylation or polymerization processes, both of which are sensitive to trace oxygenates. The UOP ORU process can be used to remove trace dimethyl ether, *tert*-butyl alcohol, and water down to concentrations of a few parts per million. Figure 14 shows how the UOP MRU and UOP ORU processes are integrated with the MTBE and downstream plants. The UOP MRU process is a multibed thermal swing adsorption process that operates in a cyclic-batch adsorption mode. Methanol is preferentially adsorbed over the less polar hydrocarbons. The desorption step is accomplished by a temperature swing using the reactor feed. The UOP ORU process involves liquid-phase adsorption of the remaining trace oxygenates from the effluent C_4 stream followed by vapor-phase desorption using an external regenerant. Both processes are in commercial use.

UOP Cyclesorb Process. The UOP Cyclesorb process was developed by UOP in the mid-1980s as an alternative to the UOP Sarex process. The UOP Cyclesorb process is simpler in design and yet attains reasonable efficiency of

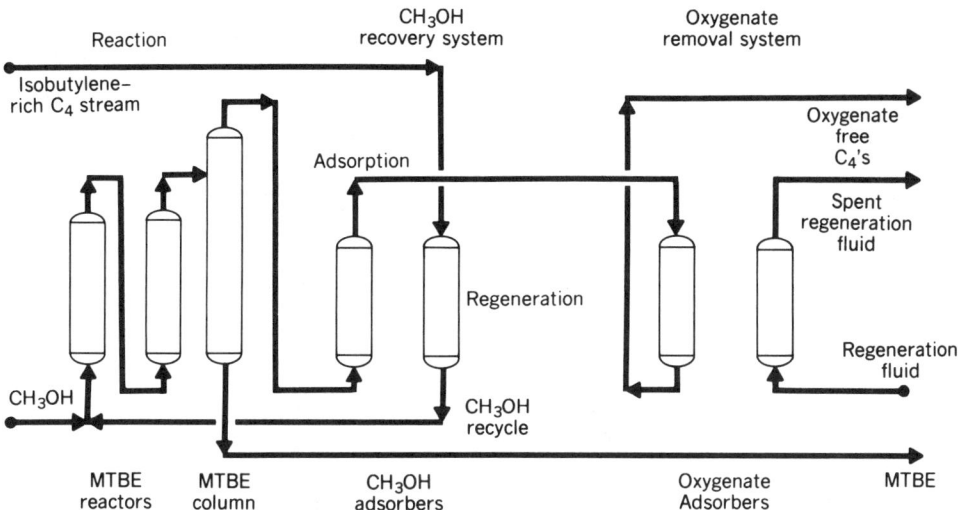

Fig. 14. UOP raffinate process treatment process. MTBE = methyl *tert*-butyl ether.

separation. The UOP Cyclesorb is a continuous recycle chromatographic process, wherein a series of chromatographic columns are used to develop the separation of fructose from glucose, and a series of internal recycle streams of impure and dilute portions of the chromatograph are used to improve the efficiency (101,102).

A schematic diagram of a six-vessel UOP Cyclesorb process is shown in Figure 15. The UOP Cyclesorb process has four external streams: feed and desorbent enter the process, and extract and raffinate leave the process. In addition, the process has four internal recycles: dilute raffinate, impure raffinate, impure extract, and dilute extract. Feed and desorbent are fed to the top of each column, and the extract and raffinate are withdrawn from the bottom of each column in a predetermined sequence established by a switching device, the UOP rotary valve. The flow of the internal recycle streams is from the bottom of a column to the top of the same column in the case of dilute extract and impure raffinate and to the top of the next column in the case of dilute raffinate and impure extract.

Such a flow establishes a chromatographic profile in each column that is moving from top to bottom. Also, the profile is staggered from one column to the other as each column is in a different stage of development of the chromatogram. The concentration at any given point in a column varies with time, but the variation is more gradual than can be expected in a batch chromatographic process. Therefore, all portions of the column perform a useful function at any given time.

Ion-Exclusion Processes for Sucrose. Molasses, which is a by-product of raw cane or beet sugar manufacturing processes, is a heavy, viscous liquid that is separated from the final low grade massecuite from which no more sugar can be crystallized by the usual methods. Molasses has a reasonably high sugar content. The recovery of sucrose from molasses has been the object of intense investigation for more than 50 years. In 1953, the ion-exclusion process was introduced by the

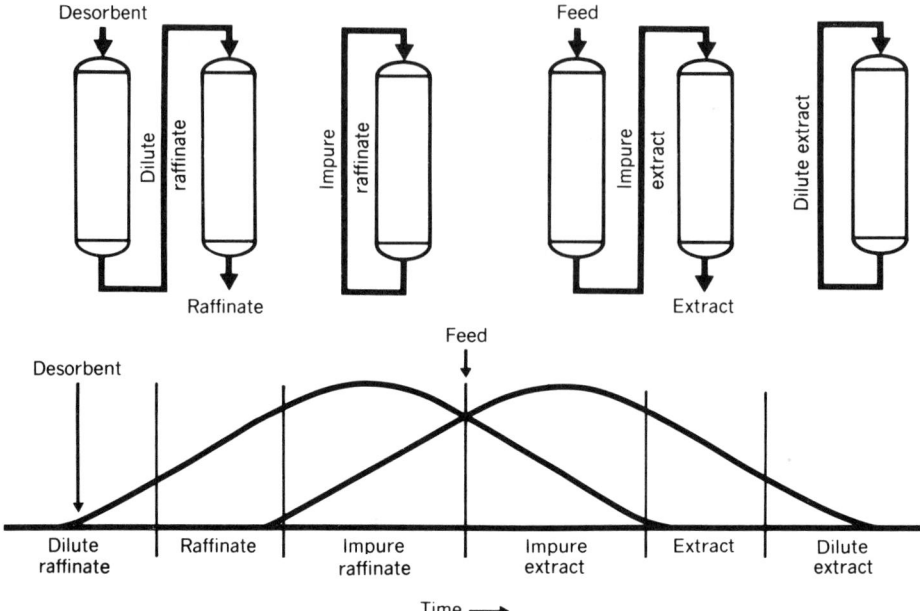

Fig. 15. UOP Cyclesorb process flow.

Dow Chemical Company (103). This process, which was developed to separate the ionic from the nonionic constituents of molasses, was based on certain ion-exchange resins having a different affinity for nonionic species than for ionic species under equilibrium conditions.

The ion-exclusion process for sucrose purification has been practiced commercially by Finn Sugar (104). This process operates in a cyclic-batch mode and provides a sucrose product that does not contain the highly molassogenic salt impurities and thus can be recycled to the crystallizers for additional sucrose recovery.

Liquid Chromatography

Conventional liquid chromatography (qv) has not attained great commercial significance in the area of large-scale bulk separations from the liquid phase. In 1952 the Sun Oil Company announced such a process, designated as the Arosorb process, for separating aromatic hydrocarbons from naphthas with a silica gel (105). However, the Arosorb process did not attain wide commercial usage, largely because of the simultaneous development of efficient liquid–liquid extraction processes for the same application.

However, chromatographic processes still have a considerable applicability (106). For instance, in small-scale operations, the greater simplicity of the chromatograph may more than compensate economically for the larger adsorbent inventory and desorbent usage. Chromatography may also be advantageous when it is required to separate several pure products from a single feed stream. A simulated moving-bed system can yield only two well-separated fractions from a single feed stream.

The separating power of a chromatographic process arises from the development of many theoretical plates to achieve adsorption equilibrium within a column of moderate length. Even though the separation factor between two components may be small, any desired resolution may be achieved with sufficient theoretical plates.

The height equivalent to a theoretical plate (HETP) for a chromatographic column is approximated by the van Deemter equation (107):

$$\text{HETP} = A + B/v + Cv$$

where v is the interstitial velocity, and A, B, and C are related to such factors as particle size, bed porosity, and adsorption equilibrium constant. To obtain a small HETP, and therefore an efficient separation process, uniform packing is essential. Such packing requires careful loading procedures to eliminate variation in voidage and stratification of particle sizes and shapes across the cross section of the bed.

In analytical chromatography, the primary objective is to maximize the resolution between two components subject to some restrictions on the maximum time of elution. As a result, the feed pulse loading is minimized, and the number of theoretical plates is maximized. In preparative chromatography, the objective is to maximize production rate as well as reduce capital and operating costs at a given separation efficiency. The adsorption column is therefore commonly run under overload conditions with a finite feed pulse width. The choice of operating conditions for preparative chromatography has been discussed (53). In production chromatography, the optimal pulse sequence occurs when the successive pulses of feed are introduced at intervals such that the feed components are just resolved both within a given sample and between adjacent samples.

Gas versus Liquid Adsorption

The question of whether adsorption should be done in the gas or liquid phase is an interesting one. Often the choice is clear. For example, in the separation of nitrogen from oxygen, liquid-phase separation is not practical because of low temperature requirements. In C_{10}–C_{14} olefin separation, a gas-phase operation is not feasible because of reactivity of feed components at high temperatures. Also, in the case of substituted aromatics separation, such as p-xylene from other C_8 aromatics, the inherent selectivities of individual components are so close to one another that a simulated moving-bed operation in liquid phase is the only practical choice.

However, in some cases, the answer is not clear. A variety of factors need to be taken into consideration before a clear choice emerges. For example, UOP's Molex and IsoSiv processes are used to separate normal paraffins from non-normals and aromatics in feedstocks containing C_5–C_{20} hydrocarbons, and both processes use molecular sieve adsorbents. However, Molex operates in simulated moving-bed mode in liquid phase, and IsoSiv operates in gas phase, with temperature swing desorption by a displacement fluid. The following comparison of UOP's

Molex and IsoSiv processes indicates some of the primary factors that are often used in decision making:

Factor	Molex	IsoSiv
adsorbent/extract	1	2
desorbent/extract	1	3
tolerance to feed contaminants	low	high

Outlook

Liquid adsorption processes hold a prominent position in several applications for the production of high purity chemicals on a commodity scale. Many of these processes were attractive when they were first introduced to the industry and continue to increase in value as improvements in adsorbents, desorbents, and process designs are made. The UOP Parex process has seen three generations of adsorbent and four generations of desorbent. Similarly, liquid adsorption processes can be applied to a much more diverse range of problems than those presented in Table 3.

A surprisingly large number of important industrial-scale separations can be accomplished with the relatively small number of zeolites that are commercially available. The discovery, characterization, and commercial availability of new zeolites and molecular sieves are likely to multiply the number of potential solutions to separation problems. A wider variety of pore diameters, pore geometries, and hydrophobicity in zeolites and molecular sieves as well as more precise control of composition and crystallinity in existing zeolites are expected to help to broaden the applications for adsorptive separations.

BIBLIOGRAPHY

"Adsorptive separation, liquids," in the *Encyclopedia of Chemical Technology,* 3rd ed., Vol. 1, pp. 563–581, D. B. Broughton, UOP Process Division, UOP, Inc.; in *ECT* 4th ed., Vol. 1, pp. 573–598, by S. A. Gembicki and co-workers, UOP.

1. *Separation and Purification: Critical Needs and Opportunities,* National Research Council Report, National Academy Press, 1987.
2. C. L. Mantell, *Adsorption,* 2nd ed., McGraw-Hill, Inc., New York, 1951.
3. V. R. Deitz, *Bibliography of Solid Adsorbents,* N.B.S. Circular 566, National Bureau of Standards, Washington, D.C., 1956.
4. T. Vermeulen, N. K. Heister, and G. Klein, *Perry's Chemical Engineers Handbook,* 4th ed., McGraw-Hill, Inc., New York, 1963, Sect. 16.
5. *Adsorption Handbook,* Pittsburgh Chemical Company, Activated Carbon Division, 1961.
6. D. B. Broughton, *Chem. Eng. Prog.* **64,** 60 (1968).
7. D. B. Broughton and A. G. Lickus, *Pet. Refiner* **40**(5), 173 (1961).
8. D. B. Carson and D. B. Broughton, *Pet. Refiner* **38**(4), 130 (1959).
9. D. W. Breck, *Zeolite Molecular Sieves,* John Wiley & Sons, Inc., New York, 1974.
10. L. Berg, *Chem. Eng. Prog.* **65**(9), 52 (1969).
11. A. E. Herschler and T. S. Mertes, *Ind. Eng. Chem.* **47,** 193 (1955).

12. D. Haresnape, F. A. Fidler, and R. A. Lowry, *Ind. Eng. Chem.* **41,** 2691 (1949).
13. B. J. Mair and M. Shamaiengar, *Anal. Chem.* **30,** 276 (Feb. 1958).
14. B. J. Mair, A. L. Gaboriault, and F. D. Rossini, *Ind. Eng. Chem.* **39,** 1072 (1947).
15. S. Eagle and J. W. Scott, *Ind. Eng. Chem.* **42,** 1287 (1950).
16. A. E. Hirschler and S. Amon, *Ind. Eng. Chem.* **30,** 276 (Feb. 1958).
17. U.S. Pat. 3,133,126 (May 12, 1964), R. N. Fleck and C. G. Wright (to Union Oil Company).
18. Brit. Pat. 1,108,305 (Apr. 3, 1968), D. W. Peck, R. R. Gentry, and H. E. Frite (to Union Carbide Corp.).
19. U.S. Pat. 3,843,518 (Oct. 22, 1974), E. M. Magee and F. J. Healy (to Esso Research & Engineering Company).
20. U.S. Pat. 3,761,533 (Sept. 25, 1973), S. Otani and co-workers (to Toray Industries Inc.).
21. U.S. Pat. 3,686,343 (Aug. 22, 1972), R. Bearden and R. J. De Feo, Jr. (to Esso Research & Engineering Company).
22. U.S. Pat. 3,724,170 (Apr. 3, 1973), P. T. Allen, B. M. Drinkard, and E. H. Vager (to Mobil Oil Corp.).
23. U.S. Pat. 3,626,020 (Dec. 7, 1971), R. W. Neuzil (to Universal Oil Products Co.).
24. U.S. Pat. 3,558,730 (Jan. 26, 1971), R. W. Neuzil (to Universal Oil Products Co.).
25. U.S. Pat. 3,558,732 (Jan. 26, 1971), R. W. Neuzil (to Universal Oil Products Co.).
26. U.S. Pat. 3,663,638 (May 16, 1972), R. W. Neuzil (to Universal Oil Products Co.).
27. U.S. Pat. 3,686,342 (Aug. 22, 1972), R. W. Neuzil (to Universal Oil Products Co.).
28. U.S. Pat. 3,734,974 (May 22, 1973), R. W. Neuzil (to Universal Oil Products Co.).
29. U.S. Pat. 3,706,813 (Dec. 19, 1972), R. W. Neuzil (to Universal Oil Products Co.).
30. U.S. Pat. 3,851,006 (Nov. 26, 1974), A. J. de Rosset and R. W. Neuzil (to Universal Oil Products Co.).
31. U.S. Pat. 3,698,157 (Oct. 17, 1972), P. T. Allen and B. M. Drinkard (to Mobil Oil Corp.).
32. U.S. Pat. 3,917,734 (Nov. 4, 1975), A. J. de Rosset (to Universal Oil Products Co.).
33. U.S. Pat. 3,665,046 (May 23, 1972), A. J. de Rosset (to Universal Oil Products Co.).
34. U.S. Pat. 3,510,423 (May 5, 1973), R. W. Neuzil (to Universal Oil Products Co.).
35. U.S. Pat. 3,723,561 (Mar. 27, 1973), J. W. Priegnitz (to Universal Oil Products Co.).
36. U.S. Pat. 3,851,006 (Nov. 26, 1974), A. J. de Rosset (to Universal Oil Products Co.).
37. F. Wolf and K. Pilchowski, *Chem. Technol.* **23**(11), (1971) (in Ger.).
38. R. M. Moore and J. R. Katzer, *AIChE J.* **18,** 816 (1972).
39. C. N. Satterfield and C. S. Cheng, *AIChE J.* **18,** 710 (1972).
40. F. Wolf, K. Pilchowski, K. H. Mohrmann, and E. Hause, *Chem. Technol.* **27**(12), (1975) (in Ger.).
41. S. K. Suri and V. Ramkrishna, *Trans. Faraday Soc.* **65**(6), 1960 (1969).
42. J. F. Walter and E. B. Stuart, *AIChE J.* **10,** 889 (1964).
43. U.S. Pat. 3,929,669 (Dec. 30, 1975), D. H. Rosback and R. W. Neuzil (to Universal Oil Products Co.).
44. S. Sircar and A. L. Meyers, *AIChE J.* **17,** 186 (1971).
45. R. T. Yang, *Gas Separation by Adsorption Processes,* Butterworth, London, 1986.
46. E. Heftmann, ed., *Chromatography,* Van Nostrand-Reinhold, New York, 1975.
47. J. J. Kipling, *Adsorption from Solutions of Non-Electrolytes,* Academic Press, Inc., New York, 1965.
48. F. C. Nachod and J. Schubert, *Ion-Exchange Technology,* Academic Press, Inc., New York, 1956.
49. R. M. Barrer, *Zeolites and Clay Minerals as Sorbents and Molecular Sieves,* Academic Press, Inc., London, 1978.
50. J. A. Johnson and A. R. Oroskar, "Sorbex Technology for Industrial Scale Separation," in H. G. Karge and J. Weitkamp, eds., *Zeolites as Catalysts, Sorbents, and Detergent Builders,* Elsevier Science Publishers BV, Amsterdam, 1989.

51. C. Ho, C. B. Ching, and D. M. Ruthven, *Ind. Eng. Chem. Res.* **26,** 1407 (1987).
52. S. A. Fisher and G. Otten, *Proceedings of the 42nd International Water Conference,* Engineering Society of Western Pennsylvania, Pittsburgh, 1981.
53. D. M. Ruthven, *Principles of Adsorption and Adsorption Processes,* John Wiley & Sons, Inc., New York, 1984.
54. G. E. Keller II, in T. E. Whyte and co-workers, eds., *Industrial Gas Separation* (ACS Symposium Series No. 223), American Chemical Society, Washington, D.C., 1983.
55. D. B. Broughton, R. W. Neuzil, J. M. Pharis, and C. S. Brearly, *Chem. Eng. Prog.* **66**(9), 70 (1970).
56. C. Berg, *Trans. Am. Inst. Chem. Eng.* **42,** 665 (1946).
57. P. C. Wankat, *Large Scale Adsorption and Chromatography,* CRC Press, Boca Raton, Fla., 1986.
58. R. E. Treybal, *Mass Transfer Operations,* 3rd ed. McGraw Hill, New York, 1980.
59. D. B. Broughton, H. J. Bieser, and M. C. Anderson, *Pet. Int. (Milan),* **23**(3), 91 (1976) (in English).
60. D. B. Broughton, H. J. Bieser, and R. A. Persak, *Pet. Int. (Milan),* **23**(5), 36 (1976) (in English).
61. P. E. Barker and G. Gavelson, *Separation and Purification Methods* **17,** 1 (1988).
62. S. Otani and co-workers, *Chem. Econ. Eng. Rev.* **3**(6), 56 (1971).
63. S. Otani, *Chem. Eng.* **80**(9), 106 (1973).
64. *Making Waves in Liquid Processing,* Illinois Water Treatment Company, IWT Adsep System, Rockford, Ill., 1984, **VI** (1).
65. Tetsua Hirota, *Sugar Azucar* (Jan. 1980).
66. Advertisement, *Sugar Azucar* (March 1980).
67. *Chem. Eng.* 39, (Aug. 29, 1977).
68. U.S. Pat. 2,985,589 (May 23, 1961), D. B. Broughton and C. G. Gerhold (to Universal Oil Products Co.).
69. U.S. Pat. 3,040,777 (June 26, 1962), D. B. Carson (to Universal Oil Products Co.).
70. U.S. Pat. 3,192,954 (July 6, 1965), C. G. Gerhold and D. B. Broughton (to Universal Oil Products Co.).
71. U.S. Pat. 3,268,604 (Aug. 23, 1966), D. M. Boyd (to Universal Oil Products Co.).
72. U.S. Pat. 3,268,603 (Aug. 23, 1966), D. M. Boyd (to Universal Oil Products Co.).
73. U.S. Pat. 3,131,232 (Apr. 28, 1964), D. B. Broughton (to Universal Oil Products Co.).
74. T. Miyauchi and T. Vermeulen, *Ind. Eng. Chem. Fundam.* **2,** 304 (1963).
75. S. Hartland and J. C. Mecklenburgh, *Chem. Eng. Sci.* **21,** 1209 (1966).
76. D. M. Ruthven and C. B. Ching, *Chem. Eng. Sci.* **44,** 1011 (1989).
77. U.S. Pat. 3,208,833 (Sept. 28, 1964), D. B. Carson (to Universal Oil Products Co.).
78. U.S. Pat. 3,214,247 (Oct. 26, 1964), D. B. Broughton (to Universal Oil Products Co.).
79. U.S. Pat. 3,523,762 (Aug. 11, 1970), D. B. Broughton (to Universal Oil Products Co.).
80. R. J. Schmidt, B. H. Johnson, and J. A. Weiszmann, 1987 National Petroleum Refiners Association Annual Meeting, San Antonio, Tex., Mar. 1987.
81. D. B. Broughton and R. C. Berg, *Hydrocarbon Process.* **48**(6), 115 (1969).
82. D. B. Broughton and R. C. Berg, National Petroleum Refiners Association 1969 Annual Meeting, Mar. 23, 1969, technical paper AM-69-38.
83. J. A. Johnson, S. R. Raghuram, and P. R. Pujado, "Olex: A Process for Producing High Purity Olefins," presented at the AIChE Summer National Meeting, Minneapolis, Minn., Aug. 1987.
84. G. Koenig, *Erdoel Kohle* **26,** 323 (1973) (in German).
85. D. P. Thornton, *Hydrocarbon Process.* **49**(11), 151 (1970).
86. F. H. Adams, *Eur. Chem. News* **62** (Oct. 13, 1972).
87. R. W. Neuzil and R. A. Jensen, 85th National Meeting of the AIChE, Philadelphia, Pa., June, 1978.

88. A. J. de Rosset, R. W. Neuzil, and D. J. Korous, *Ind. Eng. Chem. Proc. Des. Dev.* **15,** 261 (1978).
89. D. B. Broughton and L. C. Hardison, "Hydrocarbon-Type Separation by Unisorb," Paper presented at the 27th Midyear Meeting of the American Petroleum Institute, Division of Refining, San Francisco, Calif., May 15, 1962.
90. U.S. Pat. 4,720,579 (Jan. 19, 1988), S. Kulprathipanja (to UOP, Inc.).
91. U.S. Pat. 4,851,573 (June 25, 1989), S. Kulprathipanja, J. Priegnitz, and A. R. Oroskar (to UOP, Inc.).
92. U.S. Pat. 4,851,574 (July 25, 1989), S. Kulprathipanja (to UOP, Inc.).
93. U.S. Pat. 4,529,551 (1985), M. T. Cleary, S. Kulprathipanja, and R. W. Neuzil (to UOP, Inc.).
94. U.S. Pat. 4,534,900 (1985), M. T. Cleary (to UOP, Inc.).
95. U.S. Pat. 4,495,094 (1985), M. T. Cleary (to UOP, Inc.).
96. U.S. Pat. 4,495,106 (1985), M. T. Cleary and W. C. Laughlin (to UOP, Inc.).
97. U.S. Pat. 4,521,343 (1985), T. H. Chao and M. T. Cleary (to UOP, Inc.).
98. S. A. Gembicki, S. M. Shah, and M. T. Cleary, *Pulp Chemicals Association,* Pine Mountain, Ga., Mar. 10, 1983.
99. R. W. Johnson and E. Fritz, *Fatty Acids in Industry,* Marcel Dekker, New York, 1989.
100. A. Benchikha and D. R. Garg, "The C4 Raffinate Treatment Process; Methanol Recovery/Oxygenate Removal," Presentation at HUELS MTBE Symposium in Marl, West Germany, Sept. 6, 1988.
101. U.S. Pat. 4,402,832 (Sept. 6, 1983), C. G. Gerhold (to UOP, Inc.).
102. U.S. Pat. 4,478,721 (Oct. 23, 1984), C. G. Gerhold (to UOP, Inc.).
103. M. Wheaton and W. C. Bauman, *I&EC Eng. and Proc. Dev.* **45,** 228 (1953).
104. H. Hongisto and H. Heikkila, *Sugar Azucar* **56,** 60 (Mar. 1978).
105. W. H. Davis, J. I. Harper, and E. R. Weatherley, *Pet. Refiner* **31,** 109 (1952).
106. C. J. King, *Separation Processes,* McGraw Hill, Inc., New York, 1971.
107. J. J. van Deemter, F. J. Zuiderweg, and A. Klinberg, *Chem. Eng. Sci.* **5,** 271 (1956).

STANLEY A. GEMBICKI
ANIL R. OROSKAR
JAMES A. JOHNSON
UOP

AIR POLLUTION CONTROL

The traditional method of handling gaseous waste by atmospheric dispersion from a tall stack is generally no longer acceptable and some form of purification of the stack gas is generally mandatory. The separation problems associated with the control of gaseous emissions may be divided into two classes: removal of gaseous pollutants (see STACK GAS; SULFUROUS GASES) and removal of particulate matter (see GAS–SOLID SEPARATIONS).

AIR SEPARATION

Nitrogen, **200**
Oxygen, **207**

NITROGEN

Manufacture and Processing

Atmospheric air is the feedstock for all commercial nitrogen production processes. Nitrogen [7727-37-9] is separated from air commercially by cryogenic distillation (qv), pressure swing adsorption (see ADSORPTION, GAS SEPARATION), membrane permeation, or hydrocarbon combustion processes. Cryogenic distillation is the most cost-effective technology for production of large quantities of relatively pure nitrogen and is the most commonly used. Pressure swing adsorption (PSA) and membrane permeation are the most economical processes for production of lower purity nitrogen in low to moderate volume ranges (25–500 m^3/h (1000–20,000 SCFH)). All statements of volume (m^3) are at normal conditions (t = 25°C, pressure = 101.3 kPa (1 atm)). Both PSA and membrane permeation are rapidly growing technologies. Industry estimates indicate that noncryogenic separation should eventually account for greater than 30% of all commercial nitrogen production (1). Combustion-based processes are in decline in most applications owing to displacement by noncryogenic processes but are still widely used in heat treatment where residual contaminants play an active process role. The choice of the most economic technology is principally driven by required nitrogen purity and flow rate as shown in Figure 1. Liquid nitrogen is produced exclusively from cryogenic processes.

Cryogenic Air Separation. Cryogenic air separation has been in commercial use for the production of nitrogen since the beginning of the twentieth century. Most nitrogen is produced in large tonnage cryogenic distillation plants. Oxygen

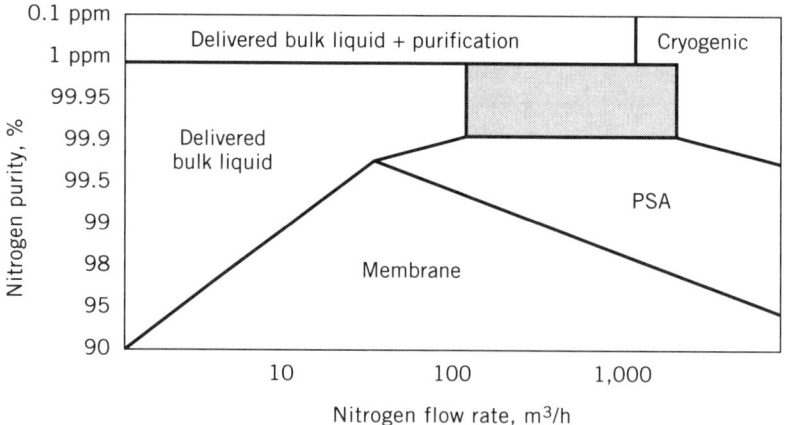

Fig. 1. Approximate economic range of nitrogen supply technologies (at median site conditions). Shaded area represents bulk liquid or PSA membrane plus deoxo or liquid assist cryogenic. To convert m^3/h to SCFH, multiply by 40.

and argon are coproducts. The nitrogen and oxygen are either utilized directly in gaseous form at adjacent industrial facilities and distribution is effected by pipeline, or some or all is liquified to enable distribution and storage in vacuum-insulated vessels. The delivered liquid nitrogen is then used directly or vaporized as needed, or is vaporized and stored under pressure in cylinders.

If the required gaseous nitrogen flow rate is relatively constant and is greater than about 400 m^3/h (15,000 SCFH), an on-site cryogenic production plant may be the most economical mode of supply. Figure 2 shows a typical cryogenic nitrogen separation process. Air is compressed, cooled to remove excess water vapor, and purified to remove residual carbon dioxide, water vapor, and other contaminants which could freeze in the process by either reversing heat exchangers or molecular sieve pressure swing adsorption (Fig. 3). In the reversing heat-exchanger process, the incoming compressed air is cooled by countercurrent heat exchange with cold oxygen-rich waste gas and nitrogen product exiting the process. Residual contaminants in the air are frozen out onto the heat-exchanger surfaces. At periodic intervals, the flow through the heat exchangers is reversed to remove the contaminants. In the molecular sieve process, two beds of activated alumina and activated carbon are used to remove the contaminants by reversible adsorption. The incoming air flow alternates between beds to allow regeneration of the depleted bed. The molecular sieve process is advantageous in nitrogen plants that may have insufficient waste gas available to clean out the reversing heat exchangers or that must be started up and shut down frequently. Most nitrogen plants use the molecular sieve process owing to increased reliability and lower cost (see MOLECULAR SIEVES).

The clean compressed air is then further cooled to near the dew point of air (ca $-176°C$ at 450 kPa) through countercurrent heat exchange with outflowing streams of waste gas and cold nitrogen product and is separated into its constituents in a single distillation column. Nitrogen product emerges from the top of the column and is available at ca 140–450 kPa (20–65 psig). Additional refrigeration is required to accommodate heat losses in the system from process efficiencies and imperfect thermal insulation. This is provided by either expanding a portion of the incoming compressed air through an expansion turbine coupled to an external brake to provide isoenthalpic cooling, or through the vaporization of a small quantity of delivered liquid nitrogen. In larger plants with an expansion turbine, liquid nitrogen in volumes between 1 to 5% of the capacity of the plant can be produced for use as backup. If more liquid nitrogen production ($\leq 10\%$) or higher nitrogen product pressures (ca 450–1000 kPa (65–145 psig)) are required, an alternative cycle can be used which expands a portion of the oxygen-rich waste gas rather than the incoming air through the expansion turbine.

Capacities of on-site cryogenic nitrogen plants range from 250 to over 50,000 m^3/h (10,000 to over 2,000,000 SCFH). A level of 1 ppm residual O_2 is readily achievable using 1990s technology. Through the addition of more separation trays and at the expense of reduced nitrogen recovery, levels of less than 1 ppb O_2 are possible for demanding semiconductor manufacturing applications. In 1993, the approximate fully installed capital cost of a standard 1300 m^{3}h (50,000 SCFH) plant was $2 million, and that of 5250 m^3/h (200,000 SCFH) plant was $4.6 million. Electrical power requirements range from 0.20–0.30 kWh/m^3 (0.53–0.79 kWh/100 SCF).

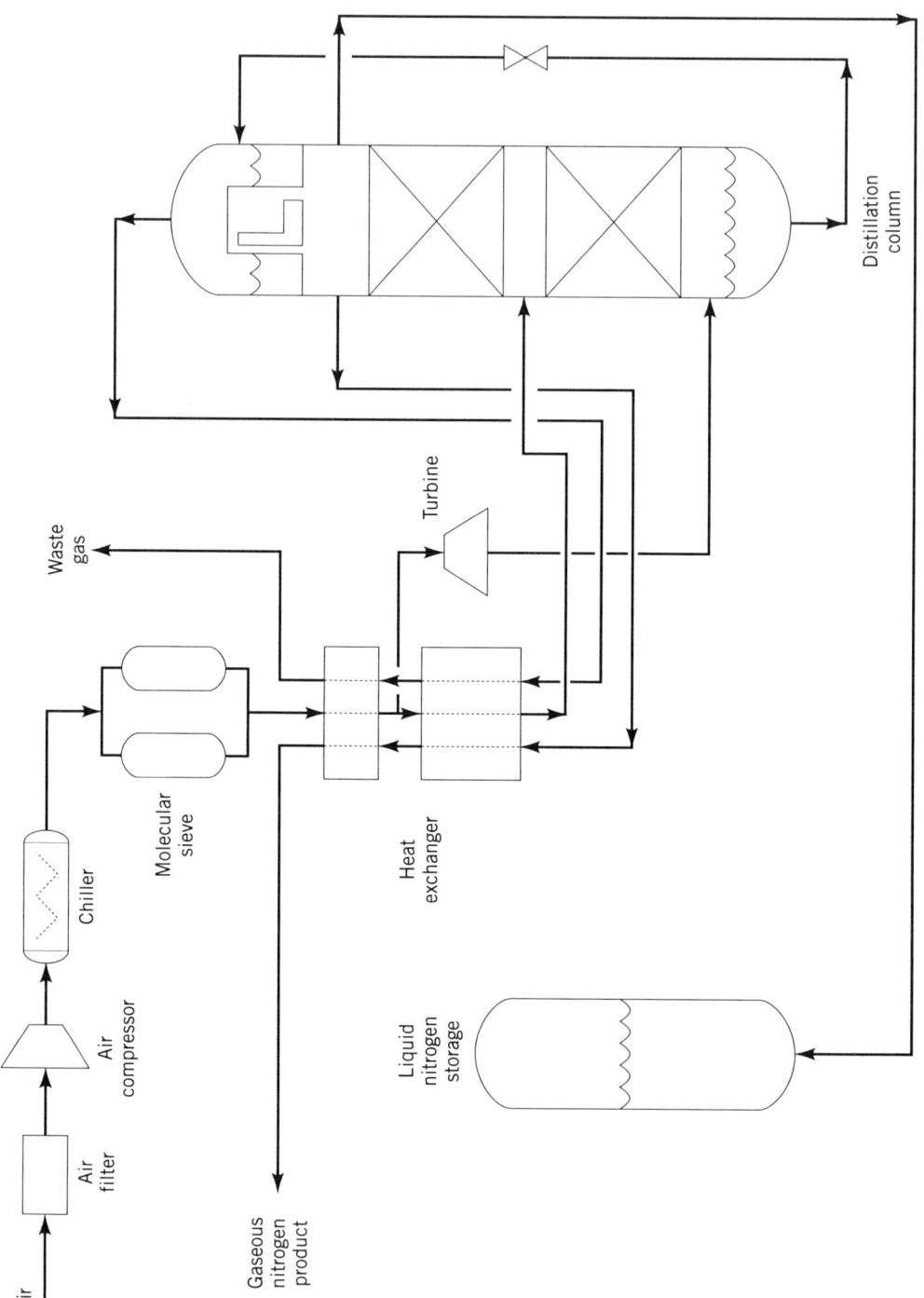

Fig. 2. Cryogenic nitrogen generation process.

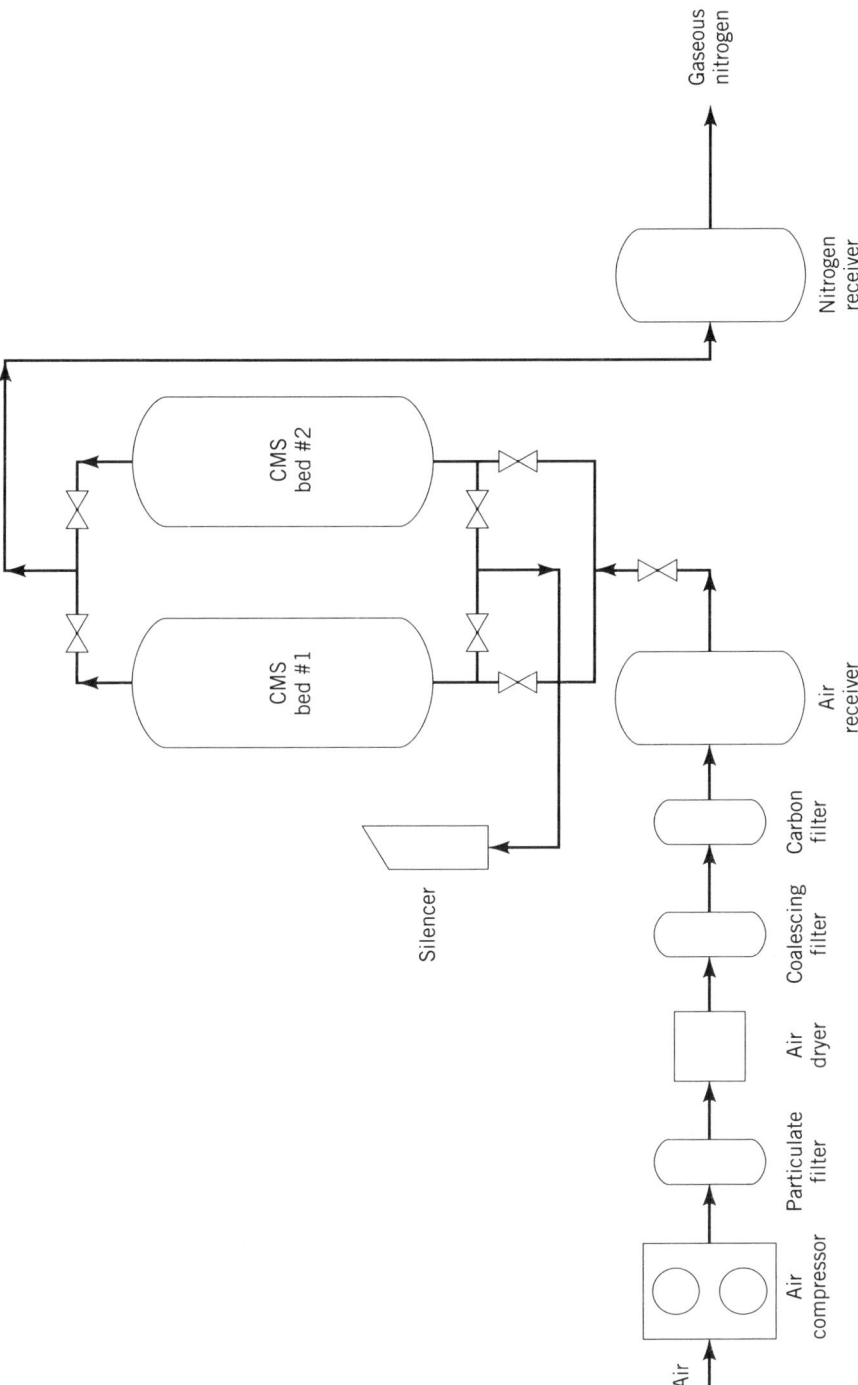

Fig. 3. Pressure swing adsorption nitrogen generation system. CMS = carbon molecular sieve.

Pressure Swing Adsorption. PSA systems operate on the principle of reversible selective adsorption of oxygen on a carbon molecular sieve (CMS). This process was invented in 1965 and commercialized in the 1970s by the German research institute on coal, Bergbau-Forschung GmbH. Modern CMS is manufactured from bituminous coal which is oxidized in air at a temperature below the ignition point and mixed with an organic binder to form pellets which are carbonized and processed. These pellets have pores with diameters similar in size to oxygen and nitrogen molecules. Oxygen is adsorbed by the CMS at a faster rate than nitrogen due to polarity and molecular size effects. Oxygen is a smaller molecule than nitrogen.

A typical nitrogen PSA system is shown in Figure 3. Air is compressed, filtered, cooled to remove excess moisture, and passed alternately through two beds of CMS. Oxygen, carbon dioxide, and water vapor are selectively adsorbed in the CMS matrix. When one bed is saturated with oxygen, the air flow is diverted to the second. The first bed is depressurized to atmosphere, releasing the adsorbed oxygen, carbon dioxide, and water vapor and the process is repeated. Product nitrogen is collected in a nitrogen receiver and is available at ca 690 kPa (100 psig). Atmospheric argon is concentrated in the product nitrogen.

Nitrogen PSA systems are economical for producing gaseous nitrogen on-site at flow rates ranging from 25 to 800 m^3/h (1,000 to 30,000 SCFH) at purities up to 99.8% (~0.2% residual oxygen). More production can be obtained from a given system at lower purities, provided sufficient air is available. For higher purities, the residual oxygen can be combined with hydrogen in a supplemental deoxo system. The water reaction product is removed in a regenerative dryer. However, residual hydrogen may be present in the product nitrogen and there are significant additional costs of hydrogen supply and the deoxo system.

In 1993, the approximate installed capital cost for a 4000-SCFH PSA plant was $150,000, and $350,000 for a 20,000-SCFH system, both producing nitrogen at 99.5% purity. Electrical power requirements ranged from 0.40 to 0.48 kWh/m^3 (1.05–1.25 kWh/100 SCF) for 99.5% nitrogen.

Membrane Permeation. The use of hollow-fiber polymeric membranes for air separation is a new technology, commercially viable since the early 1980s (see HOLLOW-FIBER MEMBRANES). Membrane systems are displacing PSA systems in the lower purity, lower flow rate range. Membrane nitrogen generation systems operate on the principle of selective gaseous permeation through a membrane. Commercially available systems use hollow-fiber membranes (qv) fabricated from polymers such as polysulfones, polyimides, and polycarbonates which permeate oxygen faster than nitrogen through a solution–diffusion mechanism. A typical membrane fiber outer diameter is 100–200 μm with a wall thickness of 30–50 μm. In the membrane process shown in Figure 4, air is compressed and passed through a series of filters to remove any residual oil, which can be detrimental to membrane longevity, from the compressor and excess water vapor. The air is then heated to the optimum process temperature for the given polymer (usually 40–60°C) and is fed axially into the center of thousands of the hollow fibers packed in a tube-and-shell configuration. The nitrogen is concentrated during its passage down the fibers and is collected as the nitrogen product. Atmospheric argon is also concentrated in the product stream (see MEMBRANE TECHNOLOGY).

Nitrogen membrane systems are economical for producing gaseous nitrogen

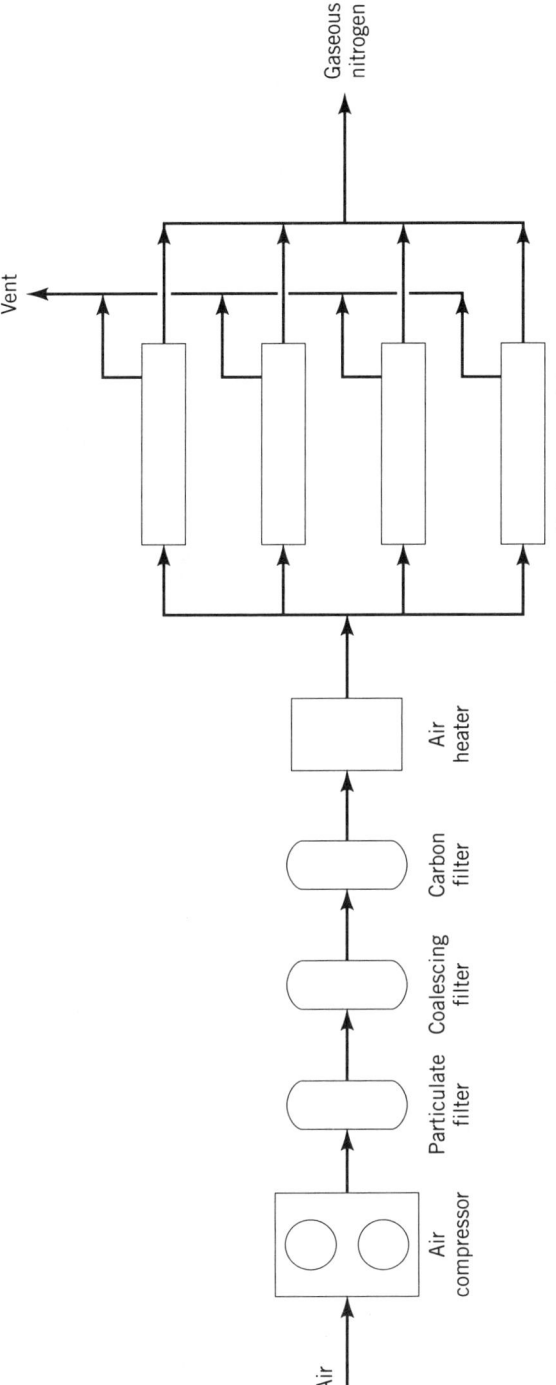

Fig. 4. Membrane permeation nitrogen generation system.

at flow rates ranging from 3 to 3000 m³/h (100–100,000 SCFH), depending on purity produced. Purities up to 99.5% are economical, depending on flow rate. As with PSA systems, the residual oxygen can be removed by use of a supplemental deoxo system. Membrane technology has proven versatile because of the simplicity of the process and the light weight of the membranes.

In 1993, the approximate installed capital cost of a 26 m³/h (1000 SCFH) membrane system was $25,000, and $100,000 for a 260 m³/h (10,000 SCFH) system at 98% purity (balance oxygen). Electrical power requirements ranged from 0.30 to 0.60 kWh/m³ (0.80–1.50 kWh/100 SCF).

Inert Gas Generators. Inert gas generators remove atmospheric oxygen by combustion with natural gas or propane. These systems find wide application in metal heat treating but have been largely displaced by other more cost-effective and reliable nitrogen generation technologies in other applications. In metal heat treating, inert gas generators are commonly called exothermic generators. Combustion products include controllable amounts of CO, H_2, CO_2, and water vapor, depending on the specific air–fuel gas input mixture. Excess water vapor and CO_2 can be removed through a refrigerant dryer and a molecular sieve adsorber, respectively. Two types of exothermic atmospheres, rich and lean, are commonly specified in heat treating applications. Rich exothermic gas contains 10–21% CO and H_2, and 5% CO_2, produced with a 6.5 to 1 air–fuel gas input ratio. Lean exothermic gas contains 1–4% CO and H_2, and 11% CO_2, produced with a 9.0 to 1 air–fuel gas input ratio (2). The balance of both atmospheres is nitrogen with a dew point of 40°C. These gas mixtures are commonly used for annealing ferrous and nonferrous materials where decarburization and brightness are not factors. Exothermic generators produce surplus heat which can be used to generate steam.

A significant application for liquid nitrogen is the cryogenic grinding of plastic or heat-sensitive materials. Plastic materials become brittle at cryogenic temperatures enabling easier grinding or deflashing operations. Cryogenic grinding is performed on both thermoplastic and thermosetting resins, old rubber tires for recycling, spices, coffee, coloring agents and pigments, and wax. In the United States, 40×10^6 m³/yr (1.5 billion SCF/yr) of liquid nitrogen is consumed in these applications (3).

Other liquid nitrogen applications include freezing biological specimens such as livestock semen and whole blood, cryosurgery, shrink fitting metal parts, and paint removal. Liquid nitrogen can be used for ground freezing to allow excavation in unstable wet soils (4), freeze plugging pipe sections to allow repairs while the rest of the pipeline remains pressurized (5), and cooling concrete in hot weather (6).

BIBLIOGRAPHY

"Nitrogen" in the *Encyclopedia of Chemical Technology,* 1st ed., Vol. 9, pp. 404–406, by E. S. Gould, Polytechnic Institute of Brooklyn; in *ECT* 2nd ed., Vol. 13, pp. 857–863, by J. W. Hall, Union Carbide Corp.; in *ECT* 3rd ed., Vol. 15, pp. 932–941, by R. W. Schroeder, Union Carbide Corp.; in *ECT* 4th ed., Vol. 17, pp. 156–171, by T. L. Hardenburger, Air Liquide America Corp., and M. Ennis, Stanford University.

1. *Chem. Eng.* **97,** 37 (1990).

2. *Plant Eng.* **46,** A4–A7 (1992).
3. R. G. Scurlock, ed., *History and Origins of Cryogenics,* Oxford University Press, New York, 1992, p. 243.
4. *Cryogenics* **26,** 572 (Oct. 1986).
5. C. Schelling, *Plant Eng.* **40,** 55–57 (July 24, 1986).
6. C. Ibbetson, *Civil Eng.,* 24–26 (June 1987).

General References

K. D. Timmerhaus and T. M. Flynn, *Cryogenic Process Engineering*, Plenum Press, New York, 1989.
R. G. Scurlock, ed., *History and Origins of Cryogenics,* Oxford University Press, New York, 1992.
Handbook of Compressed Gases, 3rd ed., Compressed Gas Association, Arlington, Va., 1990.

THOMAS L. HARDENBURGER
Air Liquide America Corporation

MATTHEW ENNIS
Stanford University

OXYGEN

Manufacture

Commercial oxygen [7782-44-7], both gaseous and liquid, at about 99.5% purity is produced by cryogenic distillation in air separation plants. In these plants the air is cleaned, dried, compressed, and refrigerated until it partially liquefies at about 80 K (see CRYOGENIC DISTILLATION). The air is then distilled into its components (Table 1). Commercial gaseous oxygen at about 90–93% purity is produced from air by vacuum swing adsorption (VSA) processes (see ADSORPTION, GAS SEPARATION). The VSA method is the fastest growth portion of oxygen production.

Air was first liquefied and oxygen subsequently separated about 1900. Technologies introduced after World War II substantially improved the air separation process for high purity oxygen production. In the 1960s the production of high volume, low cost oxygen for the manufacture of chemicals and petrochemicals, and especially for the basic oxygen process (BOP) for the manufacture of steel, provided a distinct advantage. Since about 1980 a number of further improvements have occurred in the cryogenic air separation process resulting in significant efficiency, productivity, and reliability gains. The most modern cryogenic plants operate at thermodynamic efficiencies in the vicinity of 35%, a number significantly higher than for other industrial processes, such as petroleum refining.

Process improvements include (*1*) improved energy transfer in heat exchangers via improved thermal design of the main heat exchangers; (*2*) more complex heat integration yielding more optimal distillation column performance; (*3*) improved expander–compressor efficiencies; (*4*) reduced pressure drops

Table 1. Gaseous Composition of Air[a]

Constituent[b]	Vol %	Boiling point, K
Fixed components		
nitrogen	78.084 ± 0.004	77.36
oxygen	20.946 ± 0.002	90.18
argon	0.934 ± 0.001	87.28
carbon dioxide	0.033 ± 0.003	194.68[c]
neon	$(1.821 ± 0.004) \times 10^{-3}$	27.09
helium	$(5.239 ± 0.05) \times 10^{-4}$	4.215
krypton	$(1.14 ± 0.01) \times 10^{-4}$	119.81
xenon	$(8.7 ± 0.1) \times 10^{-6}$	165.04
hydrogen	ca 5×10^{-5}	20.27
Impurities[d]		
water	0.1–2.8	
methane	1.5×10^{-4}	
carbon monoxide	$(6-100) \times 10^{-6}$	
sulfur dioxide	0.1 to 1.0	
nitrous oxide	5×10^{-5}	
ozone	$(1-10) \times 10^{-6}$	
nitrogen dioxide	$(5-200) \times 10^{-8}$	
radon	6×10^{-18}	
nitric oxide	[e]	

[a]Refs. 1–3.
[b]Composition of dry air is constant to an altitude of 20 km.
[c]Sublimation temperature. Liquid CO_2 does not exist at 101 kPa (1 atm).
[d]In ambient air, including dusts, pollen, and local pollutants.
[e]Trace amounts.

within distillation columns via improved distillation tray design and the use of packing in place of trays; (5) use of down-flow reboilers; (6) use of high (up to ca 3100 kPa (450 psi)) pressure operating cycles; and (7) improved front-end purification equipment requiring less energy to reactivate (4,5).

Process improvements have been combined into a number of operating cycles. The choice of which cycle to use depends on a number of factors, including the ratio of oxygen:nitrogen produced; the ratio of liquid oxygen:nitrogen vs gaseous oxygen:nitrogen produced; and the final pressure required of the gaseous oxygen and nitrogen.

Improvements in instrumentation and automatic controls, especially the introduction of digital, ie, computerized, control systems, have provided significant gains in personnel productivity as well as in the ability to optimize process parameters to maintain peak efficiencies. The standardization of (especially) smaller cryogenic air separation plants has further reduced design and construction costs. These types of plants often operate unattended. Even moderately large plants may be unattended during certain periods of operation.

The principal impurity in 99.5% oxygen is argon because of the closeness of the boiling points. There has been a high demand for argon in the 1990s so that the oxygen is further refined (via distillation) to recover the argon. As a result, oxygen purity often rises to 99.8%. This additional refining has also influenced the selection of operating cycles.

Cryogenic Separation. In the cryogenic air separation process, the ambient intake air is compressed and the moisture and carbon dioxide removed, either by cooling the air or by adsorption systems. In the former, the moisture and carbon dioxide are deposited as both liquid and ice, including solid carbon dioxide. These deposits are sublimed or removed in a second step after switching (reversing) the flows of air and waste nitrogen, thus warming the deposits. In the adsorption method, water, carbon dioxide, and trace impurities, eg, hydrocarbons, are removed by zeolite and silica gel-type adsorbents. The adsorbent bed is regenerated on a periodic basis, typically by flushing with a hot waste oxygen–nitrogen stream while the main air flow is diverted to a second bed. This adsorption process for removal of moisture and carbon dioxide has similarities to the vacuum swing adsorption process for the production of oxygen. The choice of processing technique, ie, reversing exchangers or adsorption, depends on the total output of pure oxygen and nitrogen expected from the unit as well as the size of the unit, but adsorption (qv) has largely become the method of choice in new plants.

Many of the trace impurities within the incoming air are either frozen out and trapped in the reversing exchangers or are removed together with the moisture and carbon dioxide by adsorption. All modern plants also include silica-gel adsorbers in the oxygen-rich liquid circuit as well as in the guard circuit used to control any accumulation of undesirable materials in the oxygen pool of the oxygen distillation column (see ADSORPTION, LIQUID SEPARATION). Silica gel is particularly effective in removing acetylene, which in early plants was responsible for initiating explosions of itself or other hydrocarbons concentrated in the liquid oxygen.

After compression and removal of impurities, the air is cooled in heat exchangers and expanded to low pressure through a turbine, to recover energy, or through a valve. Liquid air, which forms at about 80 K, is separated via a distillation column. The column as well as the heat exchangers and the associated piping are placed within a cold box, which is packed with insulation to minimize heat transfer (qv) between streams and to protect the system from the ambient air external to the cold box.

Adsorptive Separation. A noncryogenic air separation process, which is increasingly employed for small- to moderate-scale oxygen production units, is based on the adsorption of nitrogen (but not oxygen) onto zeolites (see ADSORPTION, GAS SEPARATION) (6). This batch process, known as vacuum swing adsorption (VSA), typically uses two identical switching beds, each containing two strata. The first stratum removes water and carbon dioxide; the second adsorbs nitrogen from the flowing air. In the two-bed system, while unit one is on-stream adsorbing first water and carbon dioxide and then nitrogen from the air, unit two is being evacuated to remove the previously adsorbed nitrogen. The product oxygen is substantially unaffected. After a certain period, the second bed is brought into sequential use, while the first is evacuated, etc. Depending on the operating cycle chosen, the product may be up to about 93% oxygen. The balance is nitrogen and argon. Moisture and carbon dioxide residuals are in the low ppm range. The oxygen is produced at about 24 kPa (3.5 psig) and must be compressed if the oxygen is required at higher pressures. The flow of the oxygen is unsteady but the use of a surge tank or a compressor can even out the oxygen flow.

VSA plants range from small hospital units to very large units producing as much as 229,000 m^3/d (8.0 × 10^6 ft^3/d). Fully assembled units in moderate sizes can be transported readily. Energy costs of larger units approach the equivalent cryogenic unit. The VSA plant can be started quickly and readily shut down. The largest use for the 90–93% purity, low pressure product is for oxygen enrichment in combustion furnaces. Oxygen may also be produced by pressure swing adsorption (PSA) units. Sales of PSA units, however, are normally for very small applications.

Another noncryogenic technique for oxygen production involves the electrolysis of water. This technique is used only to an insignificant extent. However, if the use of hydrogen energy were to grow significantly, hydrogen production via electrolysis could also grow; concomitantly, relatively large quantities of oxygen would be coproduced. The preferred method for hydrogen production as of 1995 was chiefly from natural gas.

BIBLIOGRAPHY

"Oxygen" in the *Encyclopedia of Chemical Technology*, 1st ed., Vol. 9, pp. 713–734, by R. F. Benenati, Polytechnic Institute of Brooklyn; in *ECT* 2nd ed., Vol. 14, pp. 390–409, by A. H. Taylor, Airco Industrial Gases Division, Air Reduction Co., Inc.; in *ECT* 3rd ed., Vol. 16, pp. 653–673, by A. H. Taylor, Airco, Inc.; in *ECT* 4th ed., Vol. 17, pp. 924–940, by J. G. Hansel, Air Products and Chemicals, Inc.

1. B. A. Mirtov, *Gaseous Composition of the Atmoshere and Its Analysis*, trans., NASA TTF-145, OTS 64-11023, U.S. Dept. of Commerce, Washington, D.C., 1964, p. 22.
2. Handbook of Compressed Gases, 3rd ed., Compressed Gas Association, Inc., Van Nostrand Reinhold Co., Inc., 1990.
3. W. Braker and A. Mossman, eds., *Matheson Gas Data Book*, 6th ed., Matheson, Inc., Lyndhurst, N.J., 1980.
4. D. R. Bennett and co-workers, *Cryogenic Air Separation Equipment Design*, AIChE Tutorial on Cryogenic Technology, Houston, Tex., 1993.
5. R. Agrawal and co-workers, "Impact of Low Pressure Drop Structure Packing on Air Distillation," Institute of Conference on Distillation, Birmingham, UK, 1992.
6. S. Sircar, "Air Fractionation by Adsorption," *Separ. Sci. Technol. Chem. Eng.* **23**, 2379 (1988).

JAMES G. HANSEL
Air Products and Chemicals, Inc.

BIOSEPARATIONS

The large-scale purification of proteins and other bioproducts is the final production step, prior to product packaging, in the manufacture of therapeutic proteins, specialty enzymes, diagnostic products, and value-added products from agriculture. These separation steps, taken to purify biological molecules or compounds obtained from biological sources, are referred to as bioseparations. Large-scale bioseparations combine art and science. Bioseparations often evolve from laboratory-scale techniques, adapted and scaled up to satisfy the need for larger amounts of extremely pure test quantities of the product. Uncompromising standards for product quality, driven by commercial competition, applications, and regulatory oversight, provide many challenges to the scale-up of protein purification. The rigorous quality control embodied in current good manufacturing practices, and the complexity and lability of the macromolecules being processed provide other practical issues to address (1).

Recovery and purification of new biotechnology products is the fastest growing area of bioseparations. Biotechnology was broadly defined in 1991 by the U.S. Office of Technology Assessment as "any technique that uses living organisms (or parts of organisms) to make or modify products, to improve plants or animals, or to develop microorganisms for specific uses." The new biotechnology, introduced in 1970, involves directed manipulation of the cell's genetic machinery through recombinant deoxyribonucleic acid (DNA) techniques and cell fusion. The new biotechnology was first applied on an industrial scale in 1979. Since then it has fundamentally expanded the utility of biological systems, so that biological molecules for which there is no other means of industrial production can be generated. Substantial manufacturing capability is expected to be needed to bring about the full application of this biotechnology (2). The recovery, purification, and packaging of biotechnological products for delivery to the consumer is undergoing unprecedented growth.

Manufacturing approaches for selected bioproducts of the new biotechnology impact product recovery and purification. The most prevalent bioseparations method is chromatography (qv). Thus the practical tools used to initiate scaleup of process liquid chromatographic separations starting from a minimum amount of laboratory data are given.

Economic Aspects

The development of biotechnology processes in the biopharmaceutical and bioproduct industries is driven by the precept of being first to market while achieving a defined product purity, and developing a reliable process to meet validation requirements. The economics of bioseparations are important, but are likely to be secondary to the goal of being first to market. The cost of a lost opportunity in a tightly focused market where there is room for only a few manufacturers can be devastating for products which take 5 to 10 years and $100 to 200×10^6 to develop. After process and product are validated, the cost of change in any portion of the procedure can also be great, if only to satisfy regulatory constraints. Hence, once the manufacturing process is in place, changes are likely to be considered only if significant improvements result.

The three main sources of competitive advantage in the manufacture of high value protein products are first to market, high quality product, and low cost (3). The first company to market a new protein biopharmaceutical, and the first to gain patent protection, enjoys a substantial advantage. The second company to enter the market may find itself enjoying only one-tenth of the sales. In the absence of patent protection, product differentiation becomes very important. Differentiation reflects a product that is purer, more active, or has a greater lot-to-lot consistency.

Biopharmaceuticals and Protein Products. Purification of proteins is a critical and expensive part of the production process, often accounting for ≥50% of total production costs (2). Hence, bioseparation processes have a significant impact on manufacturing costs. For small-volume, very high value biotherapeutics (Table 1), however, these costs may be considered secondary to the first to

Table 1. Unit Values and Relative Production Quantities for Selected Approved Biopharmaceuticals, 1990–1991[a]

Product	Year approved	Selling price, $/g[b]	Quantity for 200×10^6 in sales, kg
human insulin	1982	375	530.0
tissue plasminogen activator	1987	23,000	8.7
human growth hormone	1985	35,000	5.7
erythropoetin (Epogen)	1989	840,000	0.24
GM CSF	1991	384,000	0.52
G-CSF	1991	450,000	0.44

[a] Adapted from Ref. 2 with additional data from Ref. 4.

[b] Values are approximate and are likely to decrease.

market principle unless a lower cost competitor surfaces. Annual 1995 sales were $700 million for human insulin (5), $300 million for tissue plasminogen activator, and $220 million for human growth hormone (6). The most successful bioproduct in biotechnology history, recombinant erythropoetin (EPO), had worldwide sales estimated at $1.6 to $2.6 billion in 1995 (5, 7, 8). Epogen is a genetically engineered version of erythropoetin [11096-26-7], which is produced by the kidneys and stimulates blood stem cells to mature into red blood cells. Epogen can reverse the severe anemia often caused by kidney disease. Amgen's sales of this product, together with Neupogen (a recombinant protein that directs blood stem cells to become bacteria-fighting neutrophils), was about $1.8 billion in 1995 (9).

Bioproduct Separations

The task of quickly specifying, designing, and scaling-up a bioproduct separation is not simple. These separations are carried out in a liquid phase using macromolecules which are labile, and where conformation and heterogeneous chemical structure undergoing even subtle change during purification may result in an unacceptable product. A typical purification scheme for biopharmaceutical proteins involves the harvesting of protein-containing material or cells, concentration of protein using ultrafiltration (qv), initial chromatographic steps, viral clearance steps, additional chromatographic steps, again concentration of protein using ultrafiltration, and finally formulation (10).

Biosynthetic Human Insulin from *E. coli*. Insulin [9004-10-8], a polypeptide hormone, stimulates anabolic reactions for carbohydrates, proteins, and fats thereby producing a lowered blood glucose level. Porcine insulin [12584-58-6] and bovine insulin [11070-73-8] were used to treat diabetes prior to the availability of human insulin [11061-68-0]. All three insulins are similar in amino acid sequence, although the sequence variation of amino acid residues could lead to immunogenic responses. Eli Lilly's human insulin was approved for testing in humans in 1980 by the U.S. FDA and was placed on the market by 1982 (11,12).

Human insulin was the first animal protein to be made in bacteria in a sequence identical to the human pancreatic peptide. Expression of separate insulin A and B chains were achieved in *Escherichia coli* K-12 using genes for the insulin A and B chains synthesized and cloned in frame with the β-galactosidase gene of plasmid pBR322 (13,14). Insulin's small size, 21 amino acids for the A-chain, mol wt = 2300; and 30 for the B-chain, mol wt = 3400, together with the absence of methionine (Met) and tryptophan (Trp) residues, were critical elements both in the decision to undertake cloning of this peptide hormone and in the rapid development of the manufacturing process. The Met and Trp residues, produced as a consequence of engineering and expression in *E. coli*, are hydrolyzed by reagents used during the recovery process. The presence of these amino acids in insulin would have resulted in the hydrolysis and destruction of the product (12).

Recovery and Purification. The production of Eli Lilly's human insulin requires 31 principal processing steps of which 27 are associated with product recovery and purification (13). The production process for human insulin, based on a fermentation which yields proinsulin, provides an instructive case study on the range of unit operations which must be considered in the recovery and purification of a recombinant product from a bacterial fermentation. Whereas the exact sequence has not been published, the principal steps in the purification scheme are outlined in Figure 1a.

The fermentation product is a fusion protein where a portion of the Trp enzyme is connected to proinsulin through a Met residue (Fig. 1b). The *E. coli* contains a plasmid having the proinsulin gene connected to the Trp promoter. The Trp operon is turned on when the fermentation media is allowed to become depleted of tryptophan and the production of a fusion protein of proinsulin occurs. An inclusion body, ie, a large body of aggregated protein and nucleic acids occupying about half of the cell volume is formed (see Fig. 1). Because formation of inclusion bodies causes cell growth to stop, premature formation of inclusion bodies results in lower productivity. Hence, the Trp switch is an important practical tool in maximizing productivity. At this point the fermentation is complete, and protein recovery, dissolution, protein refolding, and purification is carried out (12). Following inclusion body recovery, CNBr, a hydrolytic agent which specifically attacks Met and Trp linkages, cleaves away the fusion protein from the proinsulin (see Fig. 1b). No Met or Trp occurs in proinsulin, so the proinsulin molecule is left intact. The proinsulin is then subjected to oxidative sulfitolysis, refolding to its proper conformation, purification, and enzyme treatment to remove the peptide connecting the insulin A- and B-chains. The crude insulin consisting of A- and B-chains in their proper conformation is then further purified using a sequence of ion-exchange (qv), reversed-phase, and size-exclusion chromatography (3,12,14).

Desamidation of asparagine or glutamine residues can occur readily in either acidic or neutral solutions; disulfide exchange reactions can occur at alkaline pH and cause formation of isomeric monomers or aggregated forms (multimers) of the protein (13). Deamidation products of insulin, referred to as desamido insulin, can form during processing. These insulin variants require high resolution chromatography techniques to remove. Therefore, a multimodal sequence of chromatographic separations for the crude recombinant insulin is required.

Ion-exchange chromatography removes most of the impurities and is followed by reversed-phase chromatography which separates insulin from structurally similar insulin-like components. Then size-exclusion (gel-permeation) chromatography is introduced to remove salts and other small molecules from the insulin. The best pH range for the acetonitrile mobile phase for reversed-phase chromatography is reported to be 3.0–4.0. This is well below the isoelectric pH of 5.4, gives excellent resolution, and minimizes deamidation of the insulin if the residence time in the reversed-phase column is less than several hours (12,14). This sequence (14) follows the principle of orthogonality of separation sequence, ie, each step is based on a different property, in this case charge, solubility, and size, respectively (1). Near the end of the chromatography sequence, the insulin may be concentrated by precipitation to form insulin zinc crystals.

Process Equipment Volumes. Product recovery involves cell lysis, centrifugation, refolding, buffer exchange, chromatography, precipitation, and filtration. Some of these steps are repeated. The volumes of the individual chromatography columns are estimated to range from 50 to 1000 L. These volumes are small com-

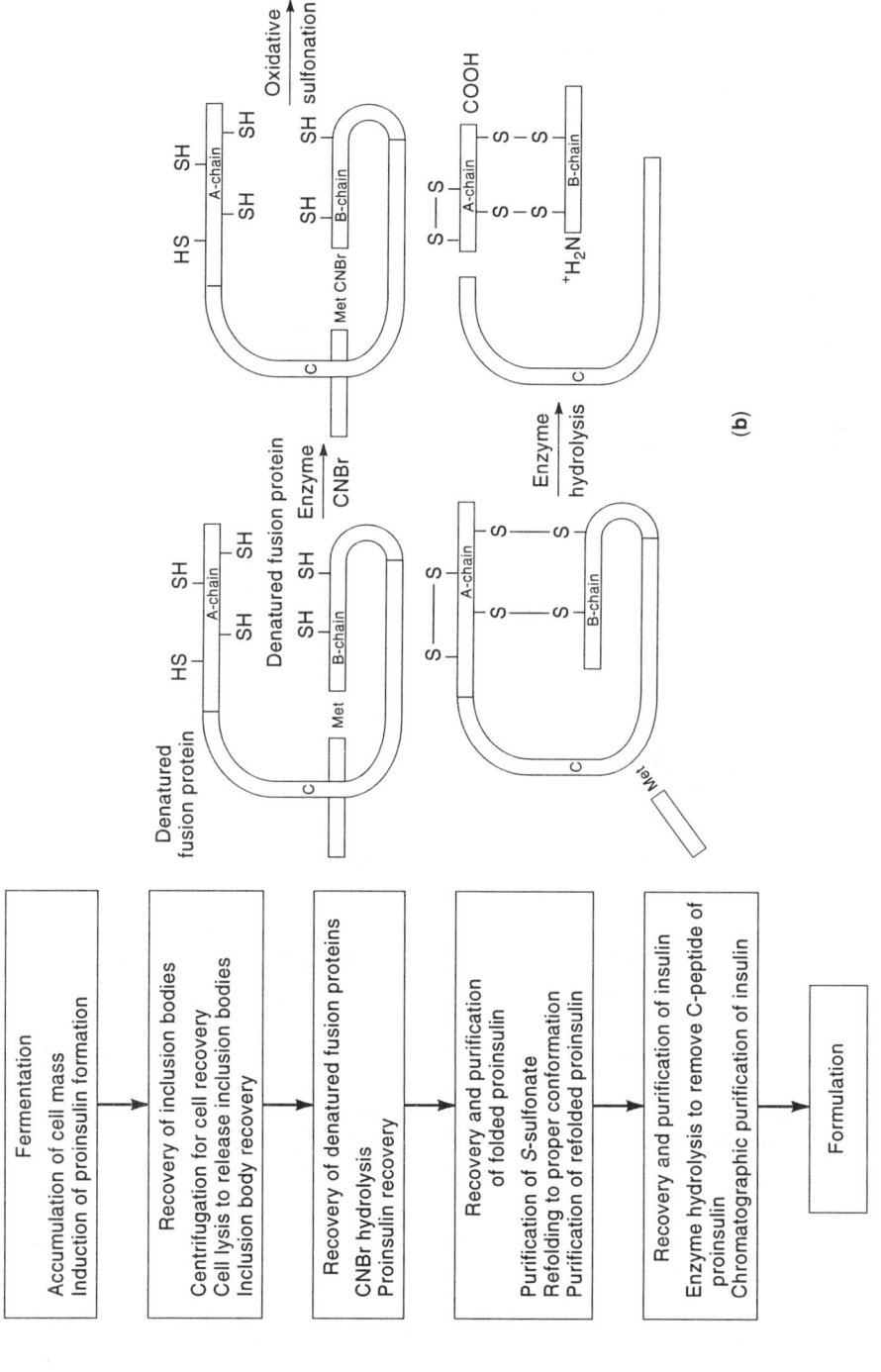

Fig. 1. (a) Process flow sheet for human insulin production, recovery, and purification (12); (b) corresponding steps in recovery of biosynthetic human insulin.

pared to other types of chemical recovery processes, but are large in the context of biotechnology manufacturing. The largest column for the reversed-phase step is 80 L for an insulin loading of 1.2 kg per run (3). Assuming the total amount of recombinant insulin produced annually in a typical plant is on the order of 1000 to 2000 kg, this size column would enable processing as much as 30% (>300 kg) of the annual output of insulin.

Yield Losses. The numerous steps incur a built-in yield loss. For example, if only 2% yield loss were to be associated with each step, the overall yield for a purification sequence of 10 steps would be as in equation 1:

$$\eta = 100\,(1 - L/100)^n = 100\,(1 - 0.02)^{10} = 81.7\% \qquad (1)$$

where η denotes yield, L the percent yield loss, and n the number of steps. If the yield loss at each step were 5%, the overall yield would only be 60%. Maximizing recovery at each step is important.

The purification of human insulin involves five separate alterations in the molecular structure, and hence, changes in physicochemical properties during its recovery and purification. The various forms are fusion protein, denatured aggregate, denatured monomers, properly folded proinsulin, and finally insulin. Whereas various purification procedures are used repeatedly, thus introducing more steps in the process, the chance of removing contaminants is maximized because the contaminants are not as likely to change chemically in the same way as the insulin molecule. The final purification steps rely on multiple properties of the insulin, such as size, hydrophobicity, ionic charge, and crystallizability (13). The final purity level is reported to be >99.99% (15).

Tissue Plasminogen Activator from Mammalian Cell Culture. Tissue-type plasminogen activator or tissue plasminogen activator [*105857-23-6*] (t-PA) was originally identified in tissue extracts in the late 1940s (15). Other known plasminogen activators include streptokinase from bacteria, urokinase from urine, and prourokinase from plasma (16). In 1981 the Bowes melanoma cell line was found to secrete t-PA (known as mt-PA) at 100 × higher concentrations, making possible the isolation and purification of this enzyme in sufficient quantities that antibodies could be generated and assays developed to lead to cloning of the gene for this enzyme and subsequent expression of the enzyme in both *E. coli* and a chinese hamster ovary (CHO) cell line (15,17).

Comparison of the melanoma and recombinant forms of the enzymes showed the same activity toward dissolution of blood clots. Comparison to urokinase, another thrombolytic agent, served as the basis for introducing recombinant t-PA into clinical trials in 1984 (17). Two pilot studies demonstrated that mt-PA resulted in thrombolysis without significant fibrinogenolysis. Fibrinogen, the precursor to fibrin, is important to the clotting of blood. Because mt-PA was available in limited quantities, recombinant t-PA (rt-PA) was used to carry out the first significant clinical trial. Doses of 0.375–0.75 mg rt-PA/kg body weight was found to be effective in humans for achieving 70% recannalization. Another pilot study confirmed that a dose of 80 mg over three hours gave the same results (17). The comparison of rt-PA (injected intravenously) to streptokinase IV (injected intracoronary) produced sufficiently favorable results to end the trial early, and make the results public in 1985, resulting in the use of t-PA for heart attacks

(15). A trial completed in 1996 showed that t-PA administered within three hours of a stroke caused by a clot in the brain facilitated full recovery of 31% of stroke patients. Hence, another use of rt-PA is likely to develop (18).

Approximately 500,000 Americans suffer strokes each year. Many of the 80% that survive suffer paralysis and impaired vision and speech, often needing rehabilitation and/or long-term care. Hence, whereas treatment using rt-PA is likely to be expensive (costs are $2200/dose for treating heart attacks), the benefits of rt-PA could outweigh costs. In the case of heart attacks, the 10 times less expensive microbially derived streptokinase can be used. There is no competing pharmaceutical for treatment of strokes (18,19). Consequently, the cost of manufacture of rt-PA may not be as dominant an issue as would be the case for other types of bioproducts.

Characteristics of t-PA. Tissue plasminogen activator, a proteolytic, hydrophobic enzyme, has a molecular weight of 66,000, 12 disulfide bonds, 4 possible glycosylation sites, and a bridge of 6 amino acids connecting the principal protein structures (17,20,21). Only three of the sites (Asn-117–184, -448) are actually glycosylated (16). When administered to heart attack victims it dissolves clots consisting of platelets in a fibrin protein matrix and acts by clipping plasminogen, an active precursor protein found in the blood, to form plasmin, a potent protease that degrades fibrin (17,21). Whereas plasminogen activator is found in blood and tissues, concentrations are low (17).

The concentration of t-PA in human blood is 2–5 ng/mL, ie, 2–5 ppb. Plasminogen activation is accelerated in the presence of a clot, but the rate is slow. The dissolution of a clot requires a week or more during normal repair of vascular damage (17). Prevention of irreversible tissue damage during a heart attack requires that a clot, formed by rupture of an atherosclerotic plaque, be dissolved in a matter of hours. This rapid thrombolysis (dissolution of the clot) must be achieved without significant fibrinogenolysis elsewhere in the patient.

rt-PA is derived from a biological source, transformed CHO cells, and by definition is a biologic, not a drug. It is generally not possible to define biologics as discrete chemical entities or demonstrate a unique composition. Other biologics include blood fractionation products such as albumin and Factor VIII, and both live and killed viral vaccines.

The process used to make a biologic is closely monitored and regulated by regulatory agencies, because a significant change in the process may result in a product which is different from that previously reviewed and regulated, and hence may require a new license. Process changes made during the investigational new drug (IND) development stage, and before the license is approved, are more easily incorporated into a new product (from a regulatory point of view) than after the license is generated (15).

t-PA Production. Recombinant technology provides the only practical means of rt-PA production. The amount of t-PA required per dose is on the order of 100 mg. Cell lines of transformed CHO cells, selected for high levels of rt-PA expression using methotrexate, are grown in large fermenters (21). The purification steps for rt-PA must therefore separate out cells, virus, and DNA. The literature on the industrial practice of recovery and purification of rt-PA generated by suspension culture of chinese hamster ovary cells is limited (15). Recovering a protein derived from mammalian cells involves a number of steps (15). One pos-

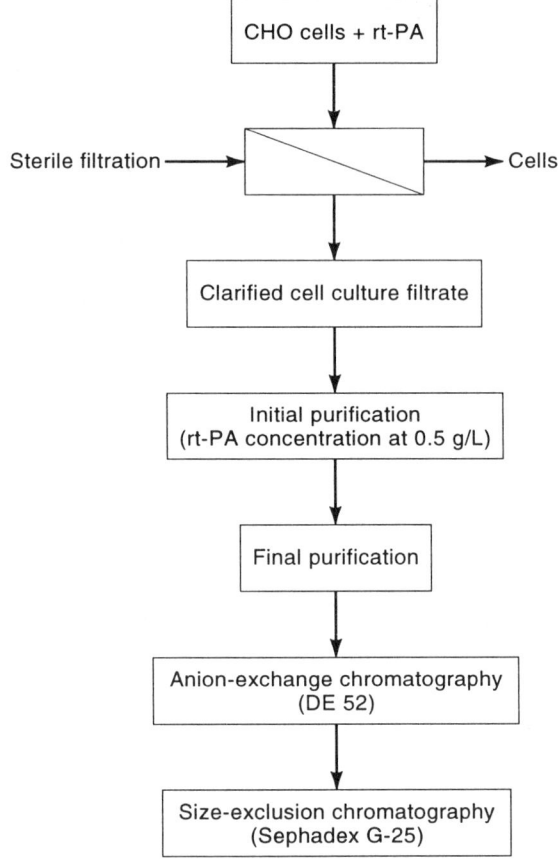

Fig. 2. Outline of possible steps in the recovery and purification sequence for recombinant tissue plasminogen activator derived from recombinant CHO cells.

sible scheme is shown in Figure 2. The culture medium is separated from the cell by sterile filtration (see MICROBIAL AND VIRAL FILTRATION). This is followed by additional removal by cross-flow filtration, ultrafiltration (qv), and chromatography to remove DNA and remaining viruses. The product protein then undergoes purification by chromatography.

The separation of cells from the culture media or fermentation broth is the first step in a bioproduct recovery sequence. Whereas centrifugation is common for recombinant bacterial cells (see CENTRIFUGAL SEPARATION), the final removal of CHO cells utilizes sterile-filtration techniques. Saftey concerns with respect to contamination of the product with CHO cells are addressed by confirming the absence of cells in the product, and their relative noninfectivity with respect to immune competent rodents injected with a large number of CHO cells.

The possibility that DNA from recombinant immortal cell lines such as CHO cells could cause oncogenic (gene altering) events resulting in cancer (22) was a primary concern during development of the rt-PA purification sequence.

Data suggest that DNA, by itself, is inactive *in vivo*; removal of the DNA, however, is still a concern. The goal for rt-PA purification is to reduce the DNA to <10 pg/dose (10^{-11} g/dose). A level of 0.1 pg has been achieved, representing a ~9 log reduction in the DNA, and requiring special assays to detect and quantify these very low levels of DNA in the final product (15).

Sensitive, specific, and if possible, rapid assays for product and potential contaminants are an essential part of separation methods selected for the purification sequence. Ultrafiltration (qv) followed by ion-exchange (qv) chromatography (qv) and then a final round of ultrafiltration concentrate the dilute protein while purifying it (23). Precipitation prior to chromatography could remove unwanted proteins before the sample is injected into the first liquid chromatography column. At the initial purification stage the rt-PA concentration is 0.5 g/L; the DNA is at 0.11 ng/mL. The use of anion-exchange chromatography (DE step) appears to be particularly effective in removing the DNA. Studies using another product, IgM, derived from cell culture showed DNA clearance may be enhanced by predigestion (hydrolysis) of the DNA using nucleases prior to the anionic-exchange chromatography step (24).

Independent Assays for Proving Virus Removal. Retroviruses and viruses can also be present in culture fluids of mammalian cell lines (15,24). Certainly the absence of virus can be difficult to prove. Model viruses, eg, NIH Rausher leukemia virus and NZB Xenotropic virus, were spiked into fluids being purified, and their removal subsequently validated when subjected to the same purification sequence as used for the product.

Viral clearance can be achieved by use of chaotropes such as urea or guanidine, pH extremes, detergents, heat, formaldehyde, proteases or DNA uses, organic solvents such as formaldehyde, or ion-exchange or size-exclusion chromatography. The protein product must be stable at the conditions used to deactivate or remove the virus. Because only the inactivation or removal which can be measured counts as validation, a sequence of orthoganol removal/fractionation steps must be used (1, 15, 24). For a fluid spiked with 10^6 virus particles/mL, if the sensitivity of detection after each treatment is 10^2 particles/mL, the analytical technique could only show a removal of 10^4 particles/mL. Hence to achieve evidence of 12 logs of clearance (10^6–10^{-6}), three different mechanisms of analysis would need to be used assuming each gives 4 logs of clearance (20). It is not valid to use the same approach, ie, the same step repeated three times, to achieve the 12 log reduction. Total clearance, based on the product of three separation steps ($10^4 \times 10^4 \times 10^4 = 10^{12}$), requires that the three steps be totally independent or orthogonal.

Another example of virus clearance is for IgM human antibodies derived from human B lymphocyte cell lines where the steps are precipitation, size exclusion using nucleases, and anion-exchange chromatography (24). A second sequence consists of cation-exchange, hydroxylapatite, and immunoaffinity chromatographies. Each three-step sequence utilizes steps based on different properties. The first sequence employs solubility, size, and anion selectivity; the second sequence is based on cation selectivity, adsorption, and selective recognition based on an anti-u chain IgG (24).

Purification of Human Cell-Line t-PA. The sequence of steps making up the initial and final purifications of recombinant t-PA from CHO cells is propri-

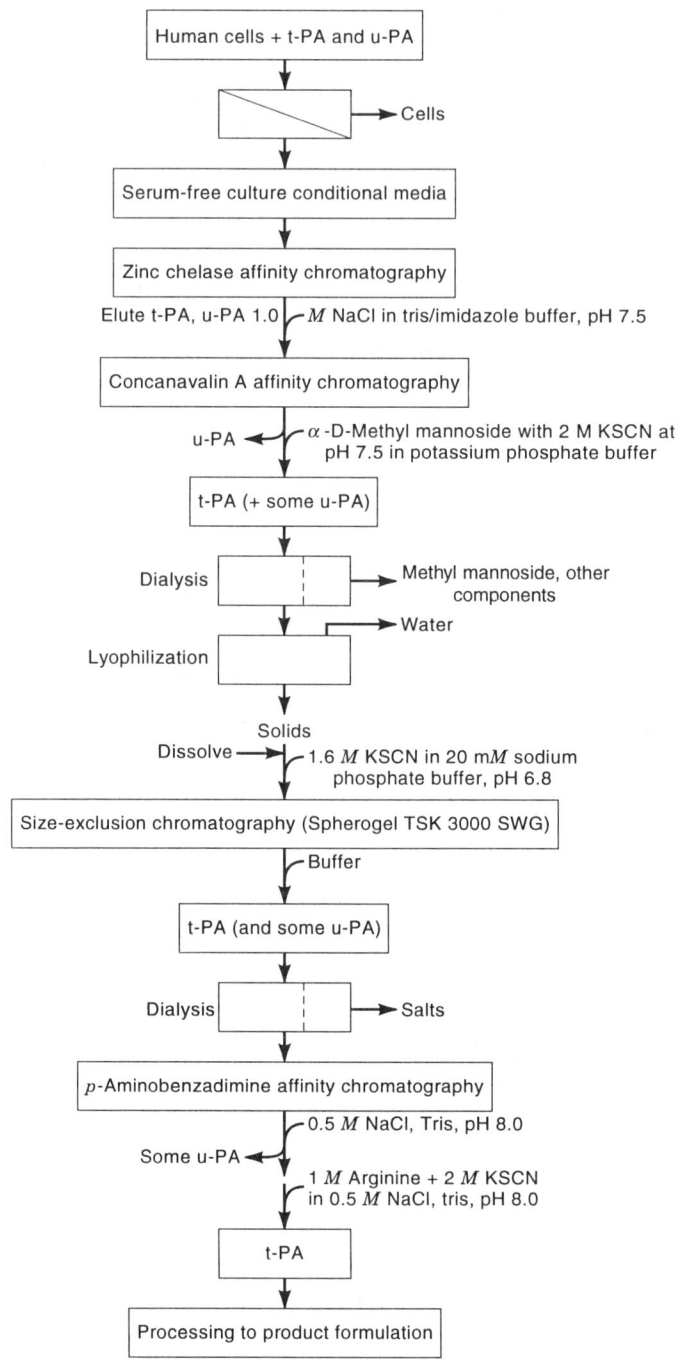

Fig. 3. Overview of purification sequence for the nonrecombinant tissue plasminogen activator (t-PA) which also contains urokinase plasminogen activator (u-PA). Serum-free culture conditional media is from normal human cell line. The temperature for all steps, except for size-exclusion chromatography (22°C), was 4°C. Adapted from Ref. 16.

etary. A detailed experimental protocol for t-PA derived from normal, nonrecombinant human cultured cells (ATCC CRL-1459, American Type Culture Collection, Rockville, Maryland), however, is available (16). This provides insights into the types of chromatography steps which might be employed for purification of rt-PA. Human cell t-PA also contains urokinase plasminogen activator (u-PA) which, except for a single glycosylation (at Asn-302), is structurally similar to t-PA and tends to co-purify. The sequence in Figure 3 shows the steps for fractionating the two proteins. The yield is only 20 mg from 1400 L, illustrating the critical role of a recombinant cell line in obtaining both high yields and higher selectivity in producing a specific type of protein.

Because the culture media contained both t-PA and u-PA, this separation required several extra affinity chromatography steps, as well as dialysis/buffer exchange between the different chromatography columns (see Fig. 3). The salts and buffers added during the purification sequence must also be removed from the product at various points, adding significant complexity to the purification sequence. Desalting and buffer exchange constitute significant separation steps in the production of almost all biotechnology protein products.

Adsorption of t-PA to process equipment surfaces consisting of either stainless steel or glass was minimized by adding the detergent polyoxyethylene sorbitan monooleate (Tween 80) to the serum-free culture conditioned media at 0.01% (vol/vol). The equipment was also rinsed, before use, with phosphate buffered saline (PBS) containing 0.01% Tween 80. Hydrophilic, plastic equipment was used whenever possible. All buffers were sterile filtered. Sterile filtration of liquids and gases is usually carried out using 0.2 or 0.45 µm filters.

Manufacture of Biologics and Government Regulation

The difference between biologics and drugs is not only a matter of definition, it is also a process design issue. To compensate for the incomplete analytical capability to define biologics, regulatory agencies include parameters of the process used to make biologics in the control and monitoring. Changes in the process may yield a different product from that previously reviewed and approved. A different product requires a new license (15). Thus substantial barriers exist in terms of effort, money, and time to making significant changes in processes used to produce licensed biologics. Process changes are to be expected during the investigational new drugs (IND) phase and before the license is approved, but significant changes are rarely made after licensing. The time which can elapse between conception of an idea for a process change and granting of a new license can be as much as two years and cost several million dollars.

The definition of biologics versus drugs continues to evolve. Assignment is made on a case by case basis (25). Section 351 of the Public Health Service Act defines a biologic product as "any virus, therapeutic serum, toxin, antitoxin, vaccine, blood, blood component or derivative, allergenic product, or analogous product . . . applicable to the prevention, treatment, or cure of diseases or injuries in man." Biologics are subject to licensing provisions that require that both the manufacturing facility and the product be approved. All licensed products are subject to specific requirements for lot release by the FDA. In comparison, drugs are approved under section 505 of the FD & C act (21 USC 301-392), where there is not lot release by the FDA except for insulin products. Insulin, growth hormone, and many other hormones have been treated as drugs, whereas erythropoietin (EPO), which also fulfills the criteria of a hormone, was reviewed in the biologic division of the FDA. Insulin is derived from a bacterial fermentation; EPO is obtained from mammalian cell culture. Hormones, for the most part, are expected to be reviewed as drugs.

The design of bioseparation unit operations is greatly influenced by these governmental regulations. The constraints on process development grow as a recovery and purification scheme undergo licensing for commercial manufacture.

Protein Chromatography. Proteins and nucleotides are macromolecular biomolecules. Mixtures of biomolecules are fractionated based on differences in charge; molecular weight, shape, and size; solubility in organic solvents; surface hydrophobic character; and types of active sites using ion-exchange, size-exclusion (gel-permeation), and reversed-phase, hydrophobic interaction (surface hydrophobicity), and affinity chromatographies, respectively. The appropriate separation may be selected from these five basic classes of chromatography. More than 30% of the purification steps for laboratory-scale protein purification procedures use ion-exchange and/or gel filtration, and at least 20% use affinity chromatography (23). This pattern is likely to be consistent with industrial practice. The following represent some chromatography column options for biopharmaceutical proteins (10). Sepharose, Sephadex, and Sephacryl resins are supplied by Pharmacia; Spherodex, Spherosil, and Trisacryl resins are supplied by Sepracor, Inc.; Toyopearl resins are supplied by TosoHaas; Fractogel resins are supplied by E. Merck Separations; Bakerbond resins and silicas are supplied by J. T. Baker. For adsorption, silica may be used.

Ion exchange	*Gel permeation*	*Hydrophobicity*
DEAE Sepharose Fast Flow LC	Agarose, 16%	Toyopearl Bulyl-650 M
DEAE Sepharose Fast Flow HC	Sephadex G25 Medium	Bulyl Spherodex M
DEAE Spherodex M	Sephadex G75	Toyopearl Ether-650 M
DEAE Spherosil M	Trisacryl Plus GF 03 M	Octyl Spherodex M
DEAE Trisacryl Plus M	Trisacryl Plus GF 10 M	Phenyl Spherodex M
Toyopearl DEAE-650 (M)	Trisacryl Plus GF 20 M	Toyopearl Phenyl-650 M
Fractogel EMD DMAE-650 (M)	Sephacryl S-100HR	Bakerbond Hi-Propyl
Fractogel EMD DEAE-650 (M)	Sephacryl S-200HR	
Q Sepharose Fast Flow	Toyopearl HW-50F	
QMA Spherodex M	Toyopearl HW-55F	
QMA Spherosil M		*Affinity*
QMA Trisacryl Plus M		Blue Sepharose CL-6B
Toyopearl QAE-550 C		Red Sepharose CL-6B
SP Sepharose Fast Flow		Blue Spherodex M
SP Sepharose High Performance		Baseline Blue Trisacryl M
SP Sepharose Big Bead		Heparin Sepharose CL-6B
SP Trisacryl Plus M		Heparin Spherodex M
Toyopearl SP-650 M		Toyopearl AF-Chelate-650 M (Copper)
Fractogel EMD SO_3-650 M		Toyopearl AF-Chelate-650 M (Zinc)

Reversed-phase chromatography is widely used as an analytical tool for protein chromatography, but it is not as commonly found on a process scale for protein purification because the solvents which make up the mobile phase, ie, acetonitrile, isopropanol, methanol, and ethanol, reversibly or irreversibly denature proteins. Hydrophobic interaction chromatography appears to be the least common process chromatography tool, possibly owing to the relatively high costs of the salts used to make up the mobile phases.

Liquid Chromatographs. The basic equipment for liquid chromatography is shown in Figure 4. The column is packed with an adsorbent, ie, the stationary phase. The mixture to be separated is pushed through the column by the eluent or mobile phase. Isocratic chromatography, carried out at a constant flow rate, buffer composition, and temperature, is usually associated with size-exclusion separations. Gradient chromatography typically uses a constant flow rate and temperature, but the composition of the element is altered by mixing two or more buffer reservoirs to achieve a steadily changing salt concentration or changes in pH. The gradients formed are reported in terms of concentration at the inlet of

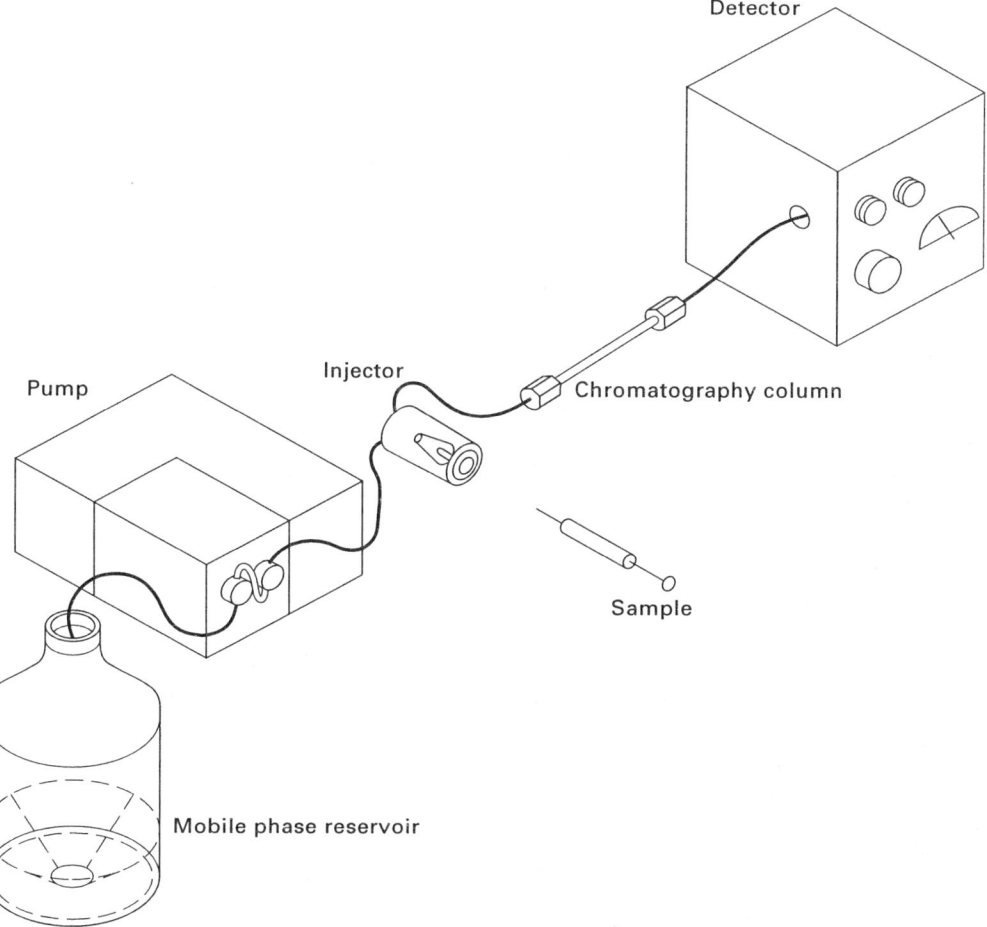

Fig. 4. Schematic of a process liquid chromatography system (16).

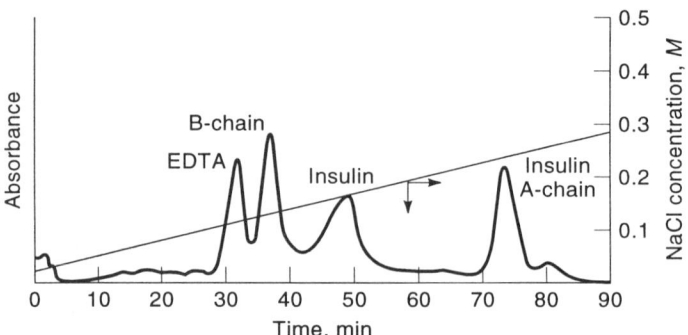

Fig. 5. Anion-exchange separation of insulin, and insulin A- and B-chains, over diethylaminoethyl (DEAE) in a 10.9 × 200 mm column having a volume of 18.7 mL. Sample volume is 0.5 mL and protein concentration in 16.7 mM Tris buffer at pH 7.3 is 1 mg/mL for each component in the presence of EDTA. Eluent (also 16.7 mM Tris buffer, pH 7.3) flow rate is 1.27 mL/min, and protein detection is by uv absorbance at 280 nm. The straight line depicts the salt gradient. Courtesy of the American Chemical Society (48).

the chromatography column; protein is detected at the column outlet. A chromatogram of the type illustrated in Figure 5 results.

The concentration profile of the gradient at the outlet of the column can be significantly different from the profile at the inlet when a component (referred to as a modulator) in the eluting buffer adsorbs onto the stationary phase in a nonideal manner causing the gradient to deform as it passes through the column. What appears to be anomalous peak behavior can result including self-concentration of a peak and the appearance of shoulders or multiple peaks for single sample components known to be homogeneous. This can occur for gradients used in reversed-phase, ion-exchange, or affinity chromatography (27–29).

Ion-Exchange Chromatography. Ion-exchange chromatography is initiated by eluting an injected sample through a column using a buffer but no NaCl or other displacing salt. The protein, which has charged sites spread over its surface, displaces anions or cations previously equilibrated on the stationary phase, ie, the protein sites exchange with the salt counterions associated with the ion-exchange stationary phase. A protein having a greater number and/or density of charged sites displaces or exchanges more ions and hence binds more strongly than a protein having a lower charge number or charge density.

Proteins deform and change shape in response to the environment. Hence, a protein left on the surface of an ion-exchange resin for a day or longer may slowly start to unfold exposing an increasing number of charged sites to bind with the ion-exchange resin. It is possible that this process can continue until the protein binds so strongly that it is impossible to desorb the protein without dissolving it, in NaOH, for example, and destroying it. To prevent such a situation, ion-exchange chromatography must be completed in a matter of hours or less.

After the column is completed, proteins of similar size and shape are separated by differential desorption from the ion exchanger by using an increasing salt gradient of the mobile phase. The more weakly bound macromolecules elute first; the most tightly bound elute last, at the highest salt concentration. Figure

5 is an example of an anion-exchange separation. Prior to injection of the sample, the column was equilibrated with the 16.7 mM Tris buffer; the EDTA stabilized the solubility of the insulin sample injected onto the column. All of the proteins are initially retained on the anion-exchange stationary phase (DEAE 650M) during loading of the sample onto the column. Subsequent application of the NaCl gradient, formed by the controlled mixing of buffers from two reservoirs of mobile phase, elutes the proteins. One reservoir contains only the 16.7 mM Tris buffer; the second contains 0.5 M NaCl in the same buffer. Following the elution of the last peak, the column may be flushed using a buffer at a high (2.5 M NaCl) salt concentration to verify that all proteins are desorbed. In some cases, a cleaning procedure is performed by passing methanolic NaOH through the column. The column is then re-equilibrated using the salt-free buffer, by pumping approximately 10 column volumes of the buffer through the stationary phase, or until the pH of the effluent and influent are the same to prepare the column for another injection.

Amphoteric Properties Determine Conditions for Ion-Exchange Chromatography. Proteins, amphoteric polymers of acidic, basic, and neutral amino acid residues, carry both negatively and positively charged groups on the surface, the ratio depending on pH (30). The isoelectric point (pI) is the pH at which a protein has a equal number of positive and negative charges. Proteins in solution at a pH > pI have a net negative charge. Below the pI, proteins have a net positive charge. Many proteins have a pI < 7 and are processed using buffers having a pH of 7 to 8. Thus anion exchangers (positively charged stationary phases) are popular for protein chromatography. Ion-exchange matrices derivatized having negatively charged groups are cation exchangers. These bind positively charged proteins, ie, cations, when the mobile phase pH is < pI.

The selection of the pH of the buffer, as well as the type of ion-exchange (anion or cation) stationary phase is a function of the amphoteric nature of the protein and protein stability as a function of pH. For example, for a protein stable at pH > 6.5, having pI = 5.5, an anion exchanger is appropriate when the separation is run in a buffer of pH > 6.5. If this protein were stable at pH = 5, < pI = 5.5, a cation exchanger and buffer of pH < 5.0 would be appropriate.

The ion-exchange (qv) groups used to derivatize stationary phases for the purification of proteins are summarized in Table 2. Corresponding buffers are given in Table 3. Strong anion and cation exchangers are almost fully ionized at pH = 3–11 and coincide with the pH range of protein purification. Weak anion and cation exchangers have a narrower pH range over which they are ionized. Anion exchangers are preferred because desorption of the protein is more readily accomplished at lower salt concentrations.

Size Exclusion (Gel-Permeation) Chromatography. Size-exclusion chromatography is often referred to as gel-permeation chromatography because the stationary phases are usually made up of soft spherical particles which resemble gels. Separation occurs by a molecular sieving effect (see MOLECULAR SIEVES; SIZE SEPARATION). The larger molecules, which explore less of the intraparticle void fraction (ie, pores) than smaller molecules, elute first because the former spend less time inside the stationary phase than the latter. Separation can be achieved if the porosity of the stationary phase is properly selected and there is a significant difference in the sizes of the molecules to be separated. This size difference

Table 2. Ion-Exchange Groups Used in Protein Purification[a]

Name	Abbreviation	Formula
Weak anion		
aminoethyl	AE	$-C_2H_4NH_3^+$
diethylaminoethyl	DEAE	$-C_2H_4NH(C_2H_5)_2^+$
Weak cation		
carboxy	C	$-COO^-$
carboxymethyl	CM	$-CH_2COO^-$
Strong anion		
trimethylaminoethyl	TAM	$-CH_2N(CH_3)_3^+$
triethylaminoethyl	TEAE	$-C_2H_4N(C_2H_5)_3^+$
diethyl-2-hydroxypropylaminoethyl	QAE	$-C_2H_4N^+(C_2H_5)_2CH_2CH(OH)CH_3$
Strong cation		
sulfo	S	$-SO_3^-$
sulfomethyl	SM	$-CH_2SO_3^-$
sulfopropyl	SP	$-C_3H_6SO_3^-$

[a] Courtesy of IRL Press (30).

Table 3. Buffers for Ion-Exchange Chromatography[a]

Buffer	pK	Buffering range
Anion exchange		
L-histadine	6.15	5.5–6.0
imidazole	7.00	6.6–7.1
triethanolamine	7.77	7.3–7.7
Tris	8.16	7.5–8.0
diethanolamine	8.80	8.4–8.8
Cation exchange		
acetic acid	4.76	4.8–5.2
citric acid	4.76	4.2–5.2
Mes	6.15	5.5–6.7
phosphate	7.20	6.7–7.6
hepes	7.55	7.6–8.2

[a] Courtesy of IRL Press (38).

is measured in terms of hydrodynamic ratio. To select the stationary phase having the appropriate pore size requires that the size of the proteins be known.

The apparatus utilized to carry out size-exclusion (gel-permeation) chromatography is analogous to that used for isocratic operating conditions (see Fig. 4). The column is packed with a gel-filtration stationary phase, selected according to the molecular weight of the protein of interest (31). A variety of commercially available gel-filtration matrices facilitate separations ranging from molecular weights of 50 to 10^8 (Fig. 6). However, a single gel having a porosity which is capable of sieving molecules over the entire separation range does not exist.

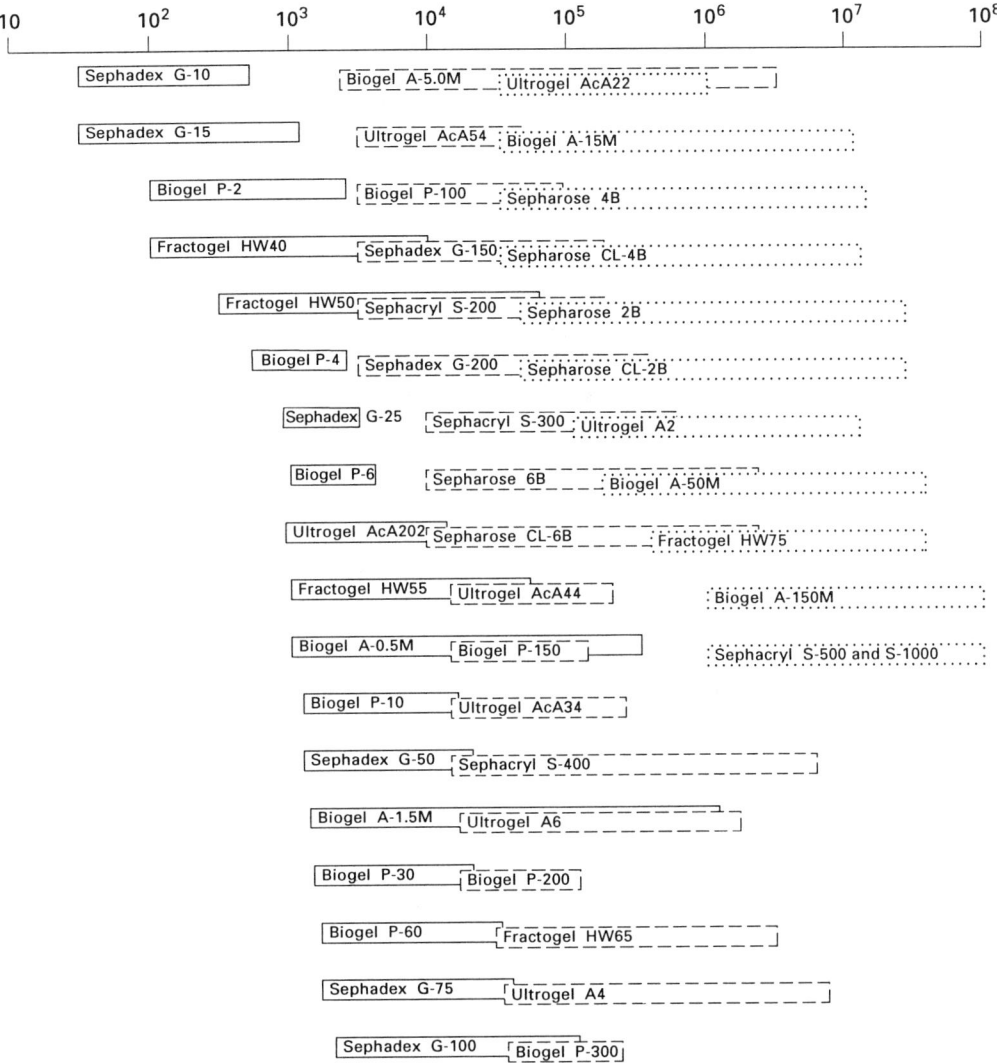

Fig. 6. Fractionation ranges of commercially available gel-filtration matrices: (□) small, (▫) medium, and (▣) large (31).

An example of a size-exclusion chromatogram is given in Figure 7 for both a bench-scale (23.5 mL column) separation and a large-scale (86,000 mL column) run. The stationary phase is Sepharose CL-6B, a cross-linked agarose with a nominal molecular weight range of ~5000–2 × 10^6 (see Fig. 6) (31).

Buffer Exchange and Desalting. A primary use of size-exclusion chromatography (sec) is for removal of salt or buffer from the protein, ie, desalting and buffer exchange (32). The difference in molecular weights is large; salts generally have a mol wt < 200, whereas mol wts of proteins are between 10,000 and 60,000.

Alternative methods of desalting and buffer exchange include continuous diafiltration, countercurrent dialysis (ccd), a membrane separation technique,

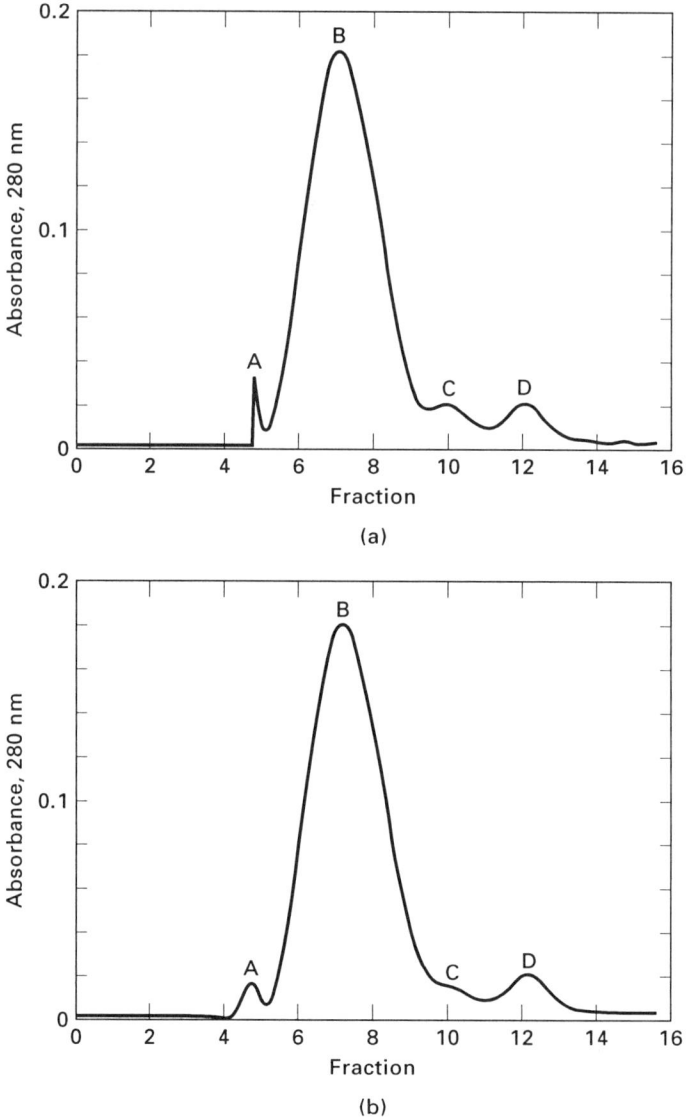

Fig. 7. Chromatograms of size-exclusion separation of IgM (mol wt = 800,000) from albumin (69,000) where A–D correspond to IgM aggregates, IgM, monomer units, and albumin, respectively, using (**a**) FPLC Superose 6 in a 1 × 30-cm long column, and (**b**) Sepharose CL-6B in a 37-cm column. Courtesy of American Chemical Society (24).

and cross-flow filtration, which uses membranes (see MICROBIAL AND VIRAL FILTRATION). Both of those methods rely on filtration at a molecular scale, using membranes having porosities which reject proteins but allow passage of salts. Membrane methods are often preferred for an unspecified protein because these procedures are less costly and enable higher throughput than size-exclusion chromatography. Buffer exchange, used to remove denaturing agents in order to induce refolding of proteins, to remove buffers between purification steps, or to remove buffers and other reagents from the final product, is usually carried out at later steps in a recovery sequence (see Figs. 2 and 3). Equations for calculating separation efficiencies and recovery yields for all three methods for a specific case study using a recombinant protein of 160,000 mol wt are available (32). Size-exclusion chromatography using Sephadex G-25M gel-filtration media in this case had disadvantages compared to the membrane filtration techniques giving 100 × less complete ion removal, 130–1200% greater dilution, 30% higher cost, 66% higher (eluting) buffer requirement, 50% higher space on the plant floor, 50% higher operating time, and half the throughput. However, cross- or tangential-flow filtration (tff) required up to 90 passes through a pump whereas sec and ccd were single pass. Other disadvantages of tff include frequent changeout of membranes and relatively large volumes of protein feed being required for laboratory-scale tests, a particular disadvantage for recombinant products. The more recent testing of a novel size-exclusion stationary phase, however, which facilitates rapid preparative (process-scale) separation of salts from protein in less than seven minutes, shows that process size-exclusion chromatography is capable of high throughput while reducing the volume needed to obtain proper refolding of the protein. Salts causing the protein to be denatured were rapidly removed and reduced to a level where protein refolding occurred (33). The use of sec is likely to continue to be a widely practiced technology in industry. Rapid size-exclusion columns for the purpose of buffer exchange have been developed which enable desalting to be achieved at linear velocities of 500–600 cm/h (33), significantly increasing throughput and reducing operating time and plant floor space. Further, sec using gel-filtration media on cellulosic-based material has a special niche for the partial and controlled separation of denaturing salts from recombinant proteins for purposes of refolding. The development of such rapid desalting techniques is important because of the larger volumes of proteins needing to be processed in industry.

Column Size. Size-exclusion chromatography columns are generally the largest column on a process scale. Separation is based strictly on diffusion rates of the molecules inside the gel particles. No proteins or other solutes are adsorbed or otherwise retained owing to adsorption, thus, significant dilution of the sample volume can occur, particularly for small sample volumes. The volumetric capacity of this type of chromatography is determined by the concentration of the proteins for a given volume of the feed placed on the column.

The volume of the solvent between the point of injection and the peak maximum of the eluting protein is defined as the elution volume, V_e (Fig. 8). The fluid volume between the particles of the stationary phase is the extraparticulate void volume or exclusion volume, V_o. The porosity of the stationary phase, determined by the extent of cross-linking of the polymers which make up the particles of a gel-permeation matrix, determines the extent to which a protein or other solute can explore the intraparticulate void volume. The higher the cross-linking, the smaller the effective pore size and the lower the molecular weight (or size) of the molecule which is excluded from the gel. Hence, the apparent porosities of a gel-permeation column are a function of the molecular probes used to measure it. For a large molecule that is completely excluded, the void volume is equivalent to the extraparticulate void volume, V_o. Molecules that are small enough to penetrate the gel have an elution volume $> V_o$. A small molecule, such as a salt, can potentially explore almost all of the bed volume and has the following elution volume:

$$V_e = V_o + K_d V_s \qquad (2)$$

where K_d represents the fraction of the volume of the mobile phase inside the particle, V_s, which can be explored by the molecular probe. For a probe small enough to explore all of the intraparticle void volume, K_d is 1 and the elution volume is $V_o + V_s$. Because the combined volume of the fluid between the particles, V_o, and inside the particles, V_s, cannot exceed the total volume of the column, V_t, V_e must be less than column volume V_t. All components injected into a size-exclusion column thus elute in one column volume, and the total stationary-phase volume required to process a given feed stream is proportional to the inlet concentration and volume of the feed. For example, for a typical inlet concentration of protein of 10 g/L, in a 100 L volume of feed, a column volume of at least 100 L is needed for size-exclusion chromatography. In comparison, an ion-exchange column having an adsorption capacity of 50 g/L would only require 20 L of column volume for the same feed.

Elution and Sample Volumes. The elution volume, V_e, is measured from the beginning of the injection to the center of the peak maximum for a Gaussian peak, if the sample volume is negligible relative to the elution volume (see Fig. 8a). A negligible injection volume is defined as being ≤ 2% of the elution volume. The elution volume for samples larger than this are measured from the halfway point of the volume injected to the center of the eluting peak (see Fig. 8b). In samples which are so large that a plateau region is obtained having the same concentration as the sample (see Fig. 8c), the elution volume is measured from the start of the sample injection to the inflection point of the peak.

Sample dilution in gel permeation can be 10-fold or more for small volumes. Hence, proper representation of the inlet concentration (eg, see Fig. 8a), would require that the injection pulse be much higher because the area under the eluting peak and under the injected sample should be the same. Similarly, dilution of the injected sample would occur in the cases represented by Figures 8b and 8c, although the difference in heights between injected pulse and the maximum of the eluting peak would be smaller, because diffusion of the solute away from the center of mass occurs at the leading (left side) and tailing (right side) edges of the peak. If the peak is broad enough, ie, the sample volume is large enough, the solute at the center of the peak is not diluted owing to diffusion and the peak maximum is therefore equal to that of the injected sample.

Distribution Coefficients. Gel-permeation stationary-phase chromatography normally exhibits symmetrical (Gaussian) peaks because the partitioning of the solute between mobile and stationary phases is linear. Criteria more sophisticated than those represented in Figure 8 are seldom used (34).

The elution volume, V_e, and therefore the partition coefficient, K_D, is a function of the size of solute molecule, ie, hydrodynamic radius, and the porosity characteristic of the size-exclusion media. A protein of higher molecular weight is not necessarily larger than one of lower molecular weight. The hydrodynamic radii can be similar, as shown in Table 4 for ovalbumin and α-lactal-

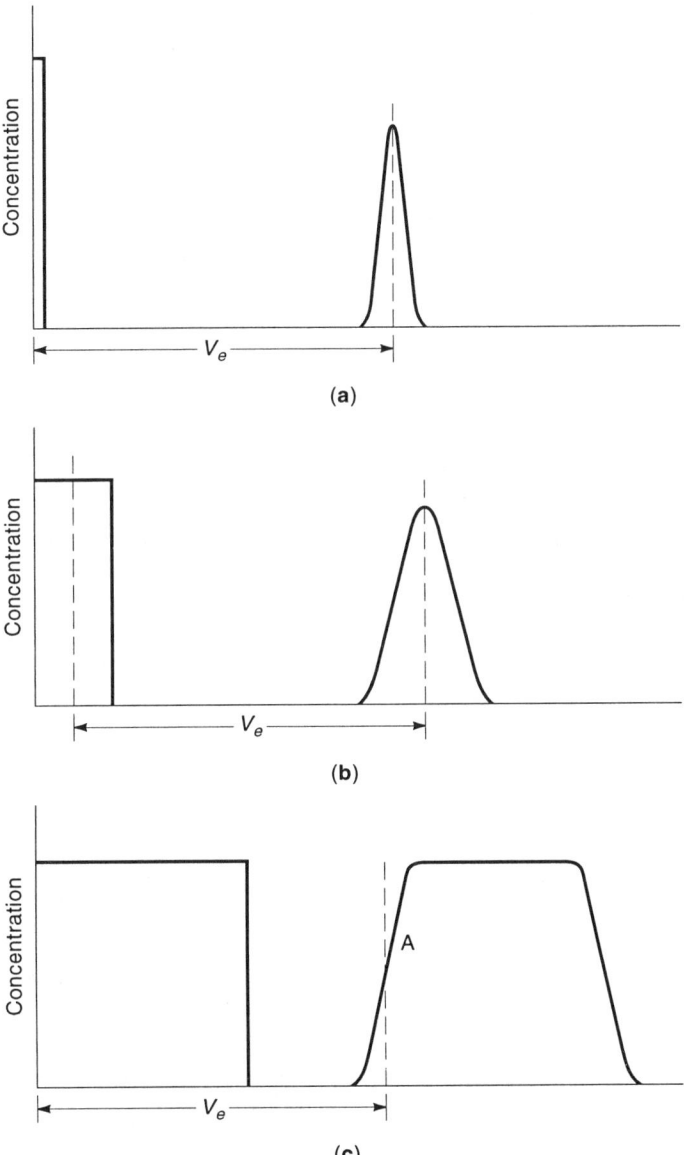

Fig. 8. Representation of measurement of elution volume, V_e, as a function of sample volume: (**a**) <2% of bed volume, (**b**) >2%, and (**c**) >2% and giving a plateau region which has the same concentration as the injected sample; A represents the inflection point. See text. Courtesy of Pharmacia (34).

bumin. The molecular weights of these proteins differ by 317%; their radii differ by only 121%.

Some types of size-exclusion phases, based on silica or macroporous polymeric materials, have rigid pores and defined pore size distributions. The dominant types of gels used in industry are dextran cross-linked with epichlorophy-

Table 4. Properties of Standard Proteins[a]

Protein	Mol wt	pI	Radius, nm	Asymmetry, f/fo[b]
bovine serum albumin	66,000	5.74	3.50	1.29
ovalbumin	45,000	5.08	2.78	1.16
α-lactalbumin	14,200	4.57	2.30	1.18
myoglobin	16,900	7.10	2.40	1.18

[a]Adapted from Ref. 5.
[b]Frictional coefficients of f, solvated protein, and fo, nonsolvated sphere.

dran (Sephadex), agarose prepared from agar (Sepharose), or allyl dextran cross-linked with N,N'-methylenebisacrylamide (Sephacryl). These materials imbibe significant quantities of water and have bed volumes ranging from 2 to 3 mL/g dry weight of stationary phase for Sephadex G-10 (nominal molecular weight cutoff of 10,000) to 20–25 mL/g dry weight of stationary-phase Sephadex G-200 (nominal molecular weight cutoff of 200,000). Their structures resemble a cross-linked spider web, where the extent of cross-linking or association between hydrated polymer chains, rather than specific pore sizes, determine the apparent pore size distribution. The hydrophilic character of the polymers which make up these gels require cross-linking to prevent dissolution. The hydrophilic character is compatible with the majority of industrially relevant proteins, most of which can be denatured by hydrophobic surfaces but preserved in active confirmation at hydrophilic conditions. This property can be offset by poor flow properties, however, particularly for lightly cross-linked gels, because these gels are soft and have a tendency to compress when flow rates exceed a threshold which decreases with decreasing extents of cross-linking. Hence, Sephadex G-10 has the highest cross-linking and flow stability, and the lowest specific bed volume, but also has the lowest effective pore size or porosity, limiting its sieving capabilities. Sephadex G-200 has the lowest cross-linking and flow stability and the highest specific bed volume and effective pore size. Sephadex 200 is useful for separating high molecular weight proteins, but at relatively low flow rates (Table 5).

The distribution coefficient, K_d, represents the fractional volume of a specific stationary phase explored by a given solute, represented by equation 3:

$$K_d = \frac{V_e - V_o}{V_s} \quad (3)$$

where V_o is the void volume, V_s is the volume of the solvent (usually aqueous buffer) inside the gel which is available to very small molecules, and V_e is the elution volume of a small volume of injected molecular probe. The measurement of V_s is difficult, requiring use of an ion or small molecule which freely diffuses into all of the fluid volume inside the gel particles and then is readily detected at the outlet of the column. Radioactive ^{23}Na and D_2O have been used. The latter is detected by a refractive index detector. An indirect measurement of V_s is more convenient and adequate. The column void volume (Fig. 9) may be measured

Table 5. Comparison of Gel-Permeation Stationary Phase[a]

Sephadex G-X[b]	Specific volume water mL/g dry gel	Permeability, K_o	Operating pressure[c], kPa[d]	Flow rate[c] water, mL/(cm²·h)
10	2–3	19	f	f
15	2.5–3.5	18	f	f
25	4–6	9–290[e]	f	f
50	9–11	13.5–400[e]	f	f
75	12–15		160	77
100	15–20		96	50
150	20–30		36	23
200	30–40		16	12

[a] Adapted from Ref. 34.
[b] Corresponds to the nominal cutoff value for mol wt × 10^3, eg, G-10 has a mol wt cutoff value of ~10,000.
[c] Value is maximum unless otherwise noted.
[d] To convert kPa to cm H_2O, multiply by 10.2.
[e] Depends on particle size (dp); as dp increases, K_o increases.
[f] May be calculated using Darcy's law: $U = K_o (\Delta P/L)$, where U is linear flow as mL/(cm²·h), L is bed length in cm, ΔP is pressure drop over gel bed in kPa[d], and the maximum pressure is 30.4 kPa[d] (310 cm H_2O).

using a soluble, high molecular weight target molecule which, because it does not explore any of the internal fluid volume of the stationary phase, is only distributed in the mobile phase. Blue dextran, a water-soluble, sulfonated, blue-colored dextran having mol wt > 669,000, manufactured by Pharmacia, and DNA (Type III from salmon tests) have been employed (26). The total column volume, V_t, can be calculated from the dimensions of the bed, although the direct measurement of column volume using water displacement before packing is more accurate. The difference, $V_t - V_o$, is then taken as an approximation of V_s. On this basis, K_{av},

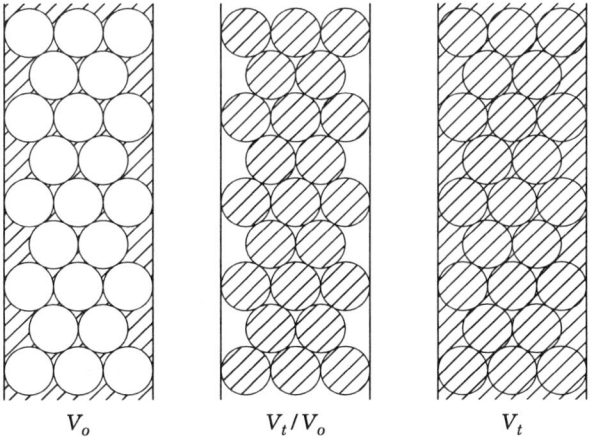

Fig. 9. Diagrammatic representation of V_t and V_o. V_t/V_o includes the volume of the solid material forming the matrix of each bead. Courtesy of Pharmacia (34).

the fraction of stationary phase volume available for a given solute species, is defined as in equation 4:

$$K_{av} = \frac{V_e - V_o}{V_t - V_o} \qquad (4)$$

The constant K_{av}, is not a true partition coefficient because the difference, $V_t - V_o$, includes the solids and the fluid associated with the gel or stationary phase. By definition, V_s represents only the fluid inside the stationary-phase particles and does not include the volume occupied by the solids which make up the gel. Thus K_{av} is a property of the gel, and like K_d it defines solute behavior independently of the bed dimensions. The ratio of K_{av} to K_d should be a constant for a given gel packed in a specific column. Other definitions which have been used for quantitating elution behavior in gel-permeation columns include V_e/V_t, V_e/V_o, and the retention coefficient, $R = V_o/V_e$ (34).

Selectivity curves result from measured values of K_{av} plotted vs log (mol wt) enabling molecular weight determination of globular proteins having similar asymmetry factors. A sphere has an asymmetry factor of 1; an ellipsoid has a factor > 1. Such curves are linear (Fig. 10), the intercept increasing with increasing porosity (decreasing cross-linking). Extrapolation of these curves through the x-axis yields the molecular weights of probes which should be completely excluded from the gels because these target molecules are larger than the largest pores. Theoretically, the K_{av} for a given molecular probe should have a value between 0 and 1. A completely excluded molecule has $K_{av} = 0$; molecules able to completely explore the fluid inside the stationary-phase particle have $K_{av} = 1$. If K_{av} is less than zero, then channeling owing to a poorly packed bed is a probable cause. If the K_{av} is greater than 1, an interaction (adsorption) of the molecular probe with the stationary phase is a likely explanation.

Gel-permeation media are extremely versatile and may be used for separation of particles such as viruses (Fig. 11) as well as proteins (34). Separations of proteins and other particles having sizes equivalent to a molecular weight of 40 $\times 10^6$ are possible using the agar-based Sepharose-type gel. This particular gel

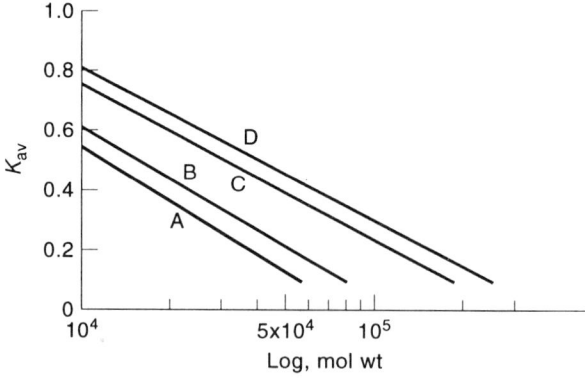

Fig. 10. Selectivity curves A–D for Sephadex G-75, G-100, G-150, and G-200, respectively, for globular proteins. Courtesy of Pharmacia (34).

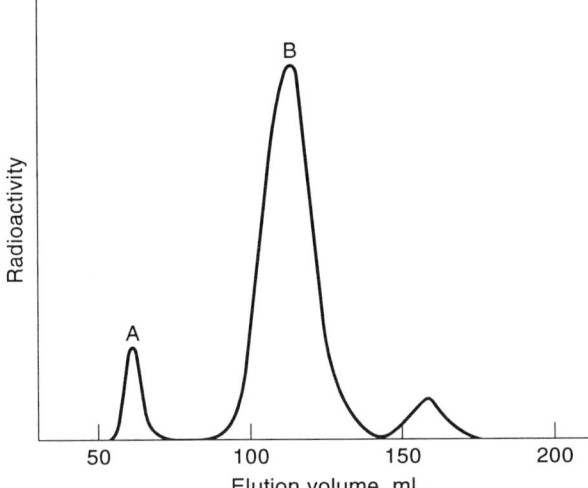

Fig. 11. Separation of ^{32}P-labeled A, adenovirus Type 5, and B, poliovirus Type 1 on Sepharose 2B where the column is 2.1×56 cm; eluent, $0.002\ M$ sodium phosphate, pH 7.2, containing $0.15\ M$ NaCl; and flow rate, $2\ \text{mL/(cm}^2\cdot\text{h})$. Courtesy of Pharmacia (34).

has a limited temperature range for operation, however. It melts upon heating to 40°C (34).

The gels having a larger extent of cross-linking, particularly Sephadex G-10, G-15, and G-25, may retain through weak adsorption some types of aromatic molecules, and consequently impart reversed-phase properties to the gel. This may be the result of the weakly hydrophobic character of the cross-linking agents used in the synthesis of these gels.

Reversed-Phase Chromatography. Stationary phases for reversed-phase chromatography consist principally of silica particles, silica supports having a hydrocarbon bonded phase, or polymeric materials based on either vinyl or styrene–divinylbenzene copolymers (35–41). Mobile phases commonly used in reversed-phase chromatography are aqueous methanol, 2-propanol, and acetonitrile. These solvents are often mixed with acidic buffers containing small amounts of acids such as trifluoroacetic acid or hexafluorobutyric acid. These acids reduce the pH of the mobile phase to 3 and give sharper peaks by suppressing ionization of silanol groups of silica-based reversed-phase supports, and minimizing ionic effects (42).

The most prevalent type of stationary phase is made of silica or another type of inorganic support derivatized and bonded with an octadecyl (C_{18}) or octyl (C_8) coating. Nonderivatized silicas are sometimes utilized for process chromatography. Much like ion-exchange chromatography, the organic component, sometimes referred to as a modifier, is mixed with water or buffer to form an increasing gradient of the modifier. This gradient serves to elute the components, which are initially adsorbed onto the stationary phase in order of increasing hydrophobic character. Methanol is used as the modifier for eluting weakly adsorbed, hydrophilic peptides; 2-propanol is used to elute strongly adsorbed, hydrophobic peptides. Acetonitrile is widely used for separations of proteins and many other types of molecules because this solvent exhibits favorable mass-transfer properties, lower viscosity (and back-pressure) than the other solvents, and good eluting strength. Methanol has a higher aqueous heat of mixing than the other solvents, and this can lead to solvent degassing and bubble formation in the column. Bubbles interfere both with the operation of the column, causing peak dispersion, and detection of the peaks at the column outlet. Bubbles give anomalous peaks in spectrophotometric detectors and refractive index detection (42).

Product Monitoring and Peptide Mapping. Reversed-phase chromatography is widely used for analysis of proteins. Historically, the principal use of reversed-phase chromatography has been in the analysis and process separations of peptides, amino acids, and organic compounds which are characterized by lower molecular weight and solubility in acetonitrile or alcohol gradients. These mobile phases denature proteins and some polypeptides. Hence, reversed-phase chromatography is infrequently used for process-scale purification of proteins. Rather the excellent protein resolving power of this type of chromatography is employed on an analytical scale using columns packed with 2–10 mL of stationary phases having 1–5 µm particle sizes and for sample volumes which typically range from 1 to 10 µL.

One purpose in monitoring a protein product is to detect the presence of a change in which as little as one amino acid has been chemically or biologically altered or replaced during the manufacturing process. Variant amino acid(s) in a protein may not affect protein retention during reversed-phase chromatography if the three-dimensional structure of the polypeptide shields the variant residue from the surface of the reversed-phase support (20). Reversed-phase chromatography discriminates between different molecules on the basis of hydrophobicity. Because large proteins may contain only small patches of hydrophobic residues, these patches may not correlate to the molecular modifications which a reversed-phase analytical method seeks to detect. The reversed-phase method must therefore be completely validated, and preferably combined with controlled chemical and/or proteolytic hydrolysis followed by chromatography or electrophoresis (see ELECTROSEPARATIONS) of the cleared protein to give a map of the resulting peptide fragments (20, 43).

A peptide map is generated by cleaving a previously purified protein using chemicals or enzymes. Hydrolytic agents having known specificity are used to perform limited proteolysis followed by resolution and identification of all the peptide fragments formed. Identification of changes, and reconstruction of the protein's primary structure, is then possible. Reagents and enzymes which cleave specific bonds are discussed in the literature (44).

An example of a peptide map generated by trypsin hydrolysis of recombinant tissue plasminogen activator (rt-PA) is shown in Figure 12. The chromatogram shows the resolving power of reversed-phase high performance liquid chromatography in separating peptides obtained from t-PA in which the disulfide bonds had been reduced and alkylated prior to enzyme hydrolysis. The small peptides formed have little or no three-dimensional structure. Hence, measurable shifts in elution profiles occur when there are variant amino acids because a single amino acid change in a peptide has a larger effect on its solubility and retention than the same change has in a protein. The replacement of arginine at position 275 in a

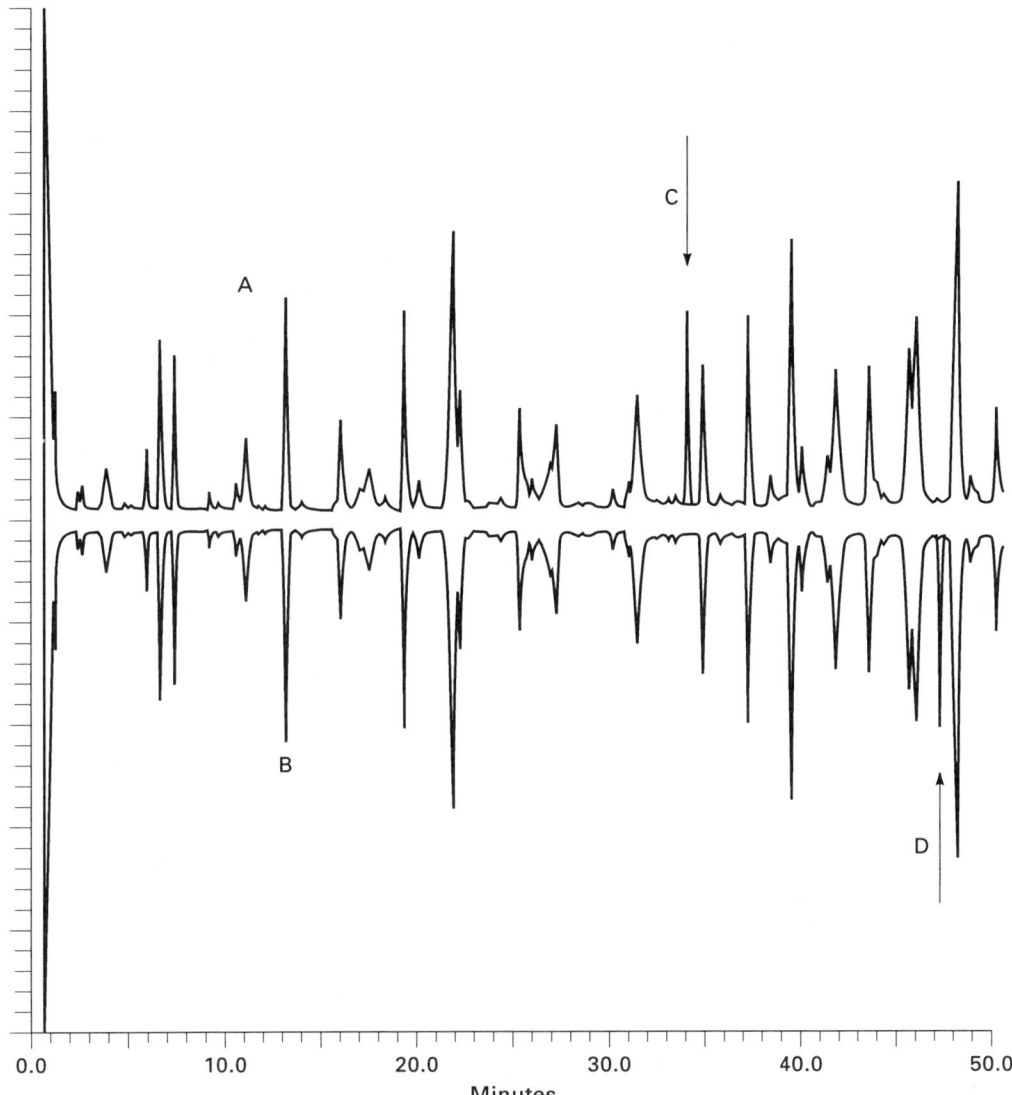

Fig. 12. Tryptic map of rt-PA (mol wt = 66,000) showing peptides formed from hydrolysis of reduced, alkylated rt-PA. Separation by reversed-phase octadecyl (C_{18}) column using aqueous acetonitrile with an added acidic agent to the mobile phase. Arrows show the difference between A, normal, and B, mutant rt-PA where the glutamic acid residue, D, has replaced the normal arginine residue, C, at position 275. Courtesy of Marcel Dekker (43).

normal t-PA molecule with glutamic acid results in a significant peak shift (see Fig. 12) (43), showing how tryptic mapping can be a suitable method for monitoring lot-to-lot consistency of this particular recombinant product (20).

Reversed-phase high performance liquid chromatography has come into use for estimating the purity of proteins and peptides as well. However, before employed, a high performance liquid chromatographic (hplc) profile of a given protein must be completely validated (43).

Insulin Purification. An example of the purification of recombinant product by reversed-phase chromatography is recombinant insulin, a polypeptide hormone. Insulin consists of 51 amino acid residues in two chains and is relatively small. Reversed-phase chromatography is used after most of the other impurities have been removed by a prior ion-exchange step (see Fig. 1) (12). The method utilizes a process-grade C_8 reversed-phase support (Zorbex) having a particle size of 10 μm (14). Partially purified insulin crystals, dissolved in a water-rich mobile phase, are applied to the column and then eluted in a linear gradient generated by mixing 0.25 M aqueous acetic acid to 60% acetonitrile. The acidic mobile phase gives excellent resolution of insulin from structurally similar insulin-like components. The ideal pH is from 3.0 to 4.0, below insulin's isoelectric point, pI = 5.4. Under mildly acidic conditions insulin may deamidate to monodesamido insulin, but if the reversed-phase separation is done within a matter of hours, the deamidation can be minimized.

This reversed-phase chromatography method was successfully used in a production-scale system to purify recombinant insulin. The insulin purified by reversed-phase chromatography has a biological potency equal to that obtained from a conventional system employing ion-exchange and size-exclusion chromatographies (14). The reversed-phase separation was, however, followed by a size-exclusion step to remove the acetonitrile eluent from the final product (12,14).

Whereas recombinant proteins produced as inclusion bodies in bacterial fermentations may be amenable to reversed-phase chromatography (42), the use of reversed-phase process chromatography does not appear to be widespread for higher molecular weight proteins.

Reversed-Phase Process Chromatography. Polypeptides, peptides, antibiotics, alkaloids, and other low molecular weight compounds are amenable to process chromatography by reversed-phase methods. There are numerous examples of bioproducts which have been purified using reversed-phase chromatography. The manufacture of salmon calcitonin, a 32-residue peptide used for treatment of post-menopausal osteoporosis, hypercalcemia, and Paget's disease of the bone, includes reversed-phase chromatography. This peptide, commercially prepared on a kilogram scale by a solid-phase synthesis, is then purified by a multimodal purification train. Reversed-phase chromatography is the dominant technique used by Rhône-Poulenc Rorer (45).

Another example is the purification of a ß-lactam antibiotic, where process-scale reversed-phase separations began to be used around 1983 when suitable, high pressure process-scale equipment became available. A reversed-phase microparticulate (55–105 μm particle size) C_{18} silica column, with a mobile phase of aqueous methanol having 0.1 M ammonium phosphate at pH 5.3, was able to fractionate out impurities not readily removed by liquid–liquid extraction (37). Optimization of the separation resulted in recovery of product at 93% purity and 95% yield. This type of separation differs markedly from protein purification in feed concentration (50–200 g/L for cefonicid vs 1 to 10 g/L for protein), molecular weight of impurities (<5000 compared to 10,000–100,000 for proteins), and throughputs (1–2 mg/(g stationary phase·min) compared to 0.01–0.1 mg/(g·min) for proteins).

Reversed-phase separation was also found to purify diastereomer precursors used in the chemical synthesis of the insect sex phermone of *Limantria dispar*, a pest which attacks oak trees. The liquid chromatography columns tested had dimensions of up to 15 cm id by 130 cm long, and were able to purify up to 708 g of starting material in 4.1 L sample using a column having 23 L of stationary phase. The throughput is estimated to have been on the order of 0.2–0.4 mg/(g·min), where separation was obtained using a gradient of hexane and diethyl ether.

Small Particle Silica Columns. Process-scale reversed-phase supports can have particle sizes as small as 5–25 µm. Unlike polymeric reversed-phase sorbents, these small-particle silica-based reversed-phase supports require high pressure equipment to be properly packed and operated. The introduction of axial compression columns has helped promote the use of high performance silica supports on a process scale. Resolution approaching that of an analytical-scale separation can be achieved using these columns that can also be quickly packed. These columns consist of a plunger fitted into a stainless steel column. The particles are placed into the column in a slurry. The plunger then squeezes or compacts the bed in an axial direction to give a stable, tightly packed bed. This type of column must be operated at pressures of up to 10 MPa (100 bar), but also gives excellent resolution in run times of an hour or less (36).

Hydrophobic Interaction Chromatography. Hydrophobic interactions of solutes with a stationary phase result in their adsorption on neutral or mildly hydrophobic stationary phases. The solutes are adsorbed at a high salt concentration, and then desorbed in order of increasing surface hydrophobicity, in a decreasing kosmotrope gradient. This characteristic follows the order of the lyotropic series for the anions: phosphates > sulfates > acetates > chlorides > nitrates > thiocyanates. Anions which precipitate proteins less effectively than chloride (nitrates and thiocyanates) are chaotropes or water structure breakers, and have a randomizing effect on water's structure; the anions preceding chlorides, ie, phosphates, sulfates, and acetates, are polar kosmotropes or water structure makers. These promote precipitation of proteins. Kosmotropes also promote adsorption of proteins and other solutes onto a hydrophobic stationary phase (46). These kosmotropes have other beneficial characteristics which include increasing the thermal stability of enzymes, decreasing enzyme inactivation, protecting against proteolysis, increasing the association of protein subunits, and increasing the refolding rate of denatured proteins. Hence, utilization of hydrophobic interaction chromatography is attractive for purification of proteins where recovery of a purified protein in an active and stable conformation is desired (46,47).

Salt Effects. The definition of a capacity factor k' in hydrophobic interaction chromatography is analogous to the distribution coefficient, K_{av}, in gel-permeation chromatography:

$$k' = \frac{V_e - V_o}{V_o} \qquad (5)$$

However, because protein retention owing to adsorption can occur, the value of k' can be greater than one, ie, elution of the most retained peak need not occur after

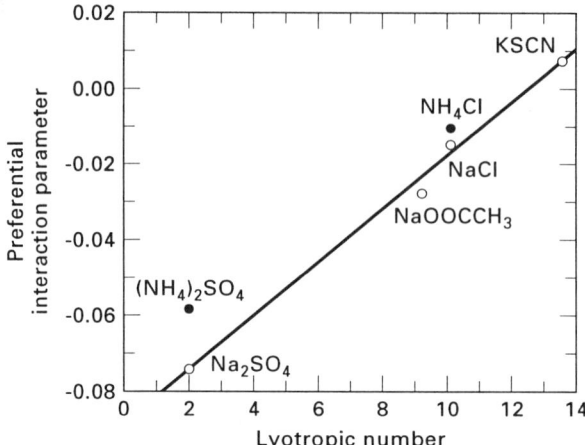

Fig. 13. Preferential interaction parameter vs lyotropic number for lysozyme on (○) bovine serum albumin and (●) Toyopearl. Courtesy of American Chemical Society (47).

one column volume of mobile phase has passed through the column. The retention behavior of lysozyme on a polymeric hydrophobic interaction support follows the preferential interaction parameter of the lyotropic series of anions (Fig. 13). The preferential interaction parameter is a measure of the net salt inclusion or exclusion in the hydration layer. The higher the value, the larger the disrupting effect of the salt. This analysis led to derivation and experimental validation of the capacity factor for lysozyme with respect the lyotropic number of the anion for a hydrophilic vinyl polymer support having an average particle diameter of 30 μm, and average pore size of 100 nm. This capacity factor has the following form:

$$k' = a[C]^d[N_x - b] + h \tag{6}$$

where a, b, d, and h are protein specific parameters, N_x is the lyotropic number, and C is salt concentration in M. Hydrophobic interaction parameters can then be estimated from experimental peak retention data as changes in retention time upon a change in salt or salt concentration. An example of how salt type and concentration affect retention of lysozyme is illustrated in Figure 14. A similar functional relation was found for myoglobin with respect to a hydrophilic vinyl polymer derivatized using butyl groups (47).

Various types of proteins have been purified using hydrophobic interaction chromatography including alkaline phophatase, estrogen receptors, isolectins, strepavidin, calmodulin, epoxide hydrolase, proteoglycans, hemoglobins, and snake venom toxins (46). In the case of cobra venom toxins, the order of elution of the six cardiotoxins supports the hypothesis that the mechanism of action is related to hydrophobic interactions with the phospholipids in the membrane.

The recovery of recombinant chymosin from a yeast fermentation broth showed that large-scale hydrophobic interaction chromatography could produce an acceptable product in one step. Chymosin, which used to be obtained from the

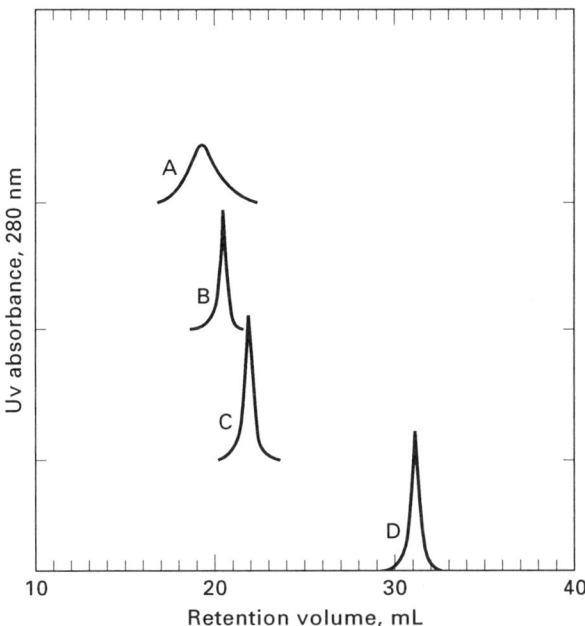

Fig. 14. Chromatographic retention of 20 µL of a 3 mg/mL solution of lysozyme on Toyopearl HW-65S using a 50 cm × 8 min ID column in 1.3 M ammonium salt, 20 mM Tris mobile phase at 1 mL/min for A, NH_4I; B, NH_4Cl; C, NH_4OOCCH_3; and D, $(NH_4)_2SO_4$. Courtesy of American Chemical Society (47).

lining of the stomachs of calves, is used in cheesemaking, and its cost is an issue. Because the capacity of the hydrophobic interaction stationary-phase is limited, an alternative method has been developed in which the enzyme is extracted into a two-phase polyethylene glycol (PEG) salt system. The partition for the chymosin into PEG coefficient is 100, and hence enables efficient recovery of this in one step. Together with a subsequent ion-exchange step, this method gives a suitably purified chymosin. The use of hydrophobic interaction chromatography may have helped to indicate that two-phase extraction is a viable approach (10).

Affinity Chromatography. The concept of affinity chromatography, credited to the discovery of biospecific adsorption in 1910, was reintroduced as a means to purify enzymes in 1968 (49). Substrates and substrate inhibitors diffuse into the active sites of enzymes irreversibly or reversibly binding there. Conversely, if the substrate or substrate inhibitor is immobilized through a covalent bond to a solid particle of stationary phase having large pores, the enzyme should be able to diffuse into the stationary phase and bind with the substrate or inhibitor. Because the substrate is small (mol wt < 500) and the enzyme large (> 15,000), the diffusion of the enzyme to its binding partner at a solid surface can be sterically hindered. The placement of the substrate at the end of an alkyl or glycol chain tethered to the stationary phase's surface reduces hindrance and forms the basis of affinity chromatography. This concept has also been applied to ion-exchange chromatography under the names of tentacle or fimbriated stationary phases.

The realization that enzymes could be selectively retained in a chromatography column packed with particles of immobilized substrates or substrate analogues led to experiments with other pairs of binding partners. Numerous applications of affinity chromatography developed, given the specific and reversible yet strong affinity of biological macromolecules for numerous specific ligands or effectors. These interactions have been exploited for purposes of highly selective, but often expensive protein purifications, recovery of messenger ribonucleic acid (mRNA) in some recombinant DNA applications, and study of mechanisms of protein binding with effector molecules (49).

Minimization of Nonspecific Binding. The purpose of affinity chromatography is the highly selective adsorption and subsequent recovery of the target biomolecule. Loss of specificity occurs when macromolecules, other than the targeted materials, adsorb onto the stationary phase owing to hydrophobic or ionic interactions. For example, a spacer arm, which allows the binding ligand on the column to be located away from the matrix surface, can improve accessibility and reduce steric hindrance to the immobilized ligand, often decreasing selectivity. Hexamethylenediamine, a common spacer arm used initially in affinity chromatography, has been the source of strong, nonspecific binding. This hydrophobic character has been decreased by interposing an ether or secondary amine, such as 3,3-diaminodipropylamine, in the middle of the spacer arm.

The ideal matrices for anchoring binding ligands are nonionic, hydrophilic, chemically stable, and physically robust. The most popular matrices are polysaccharide based, principly owing to their hydrophilic character and history of use as size-exclusion or gel-permeation gels (see Fig. 6), although glass beads, polyacrylamide gels, cross-linked dextrans (Sephadex), and agarose synthesized into a bead form have all been used (49). In particular, agaroses such as Biogel A (Bio-Rad), Ultragel A (IBF), and Sepharose (Pharmacia) are popular (49,50). Cross-linked agaroses (Sepharose CL and Sepharose FF by Pharmacia) are physically more stable than Sepharose and are suitable for attaching affinity ligands. Both forms of the agaroses have an open porosity which allows proteins to readily diffuse inside. Affinity chromatography results are usually reported in terms of specific activity of the final product (activity per mg of material) and the amount of biomolecule recovered (% yield).

Activation of the Stationary-Phase Surface. Activation of polysaccharide, silica, or polyacrylamide stationary phases involve the formation of a reactive intermediate, covalently attached to the surface, to which a difunctional alkyl-, aryl-, or glycol spacer is subsequently joined. The other end of the spacer is subsequently reacted with the ligand. Cyanogen bromide, CNBr, has been widely used to activate agarose and dextran gels (49). The attachment of ligands, and sometimes activation of supports, is generally carried out in the laboratories of the chromatography process developers because fully prepared affinity stationary phases are not as widely available as stationary phases for the other types of chromatography.

Multistep Processes. An excellent synopsis and industrial viewpoint of affinity chromatography is available (51). Ligands range from the low molecular weight components, eg, arginine and benzamidine, which both bind trypsin-like proteases, triazine dyes, and metal chelates; to high molecular weight ligands, eg, protein A, immunoglobins, and monoclonal antibodies. The blood factor VIII, purified by monoclonal affinity techniques, was approved by the U.S. FDA in

1988. Key limitations of affinity chromatography as an industrial separation technique are leaching of bound ligands from the column into the product at ppm levels, nonspecific interactions resulting in contamination of the target molecule, and failure of the affinity ligand to differentiate all variant forms of a protein or polypeptide (51). For example, polyclonal antibodies do not distinguish desamido-insulin, which contains a deamidated asparagine, from insulin. Because many antibody preparations cannot differentiate between minor structural changes in proteins, affinity chromatography must be followed by other separation steps, and does not provide a one-step purification.

Receptor Affinity Chromatography. Protein or polypeptide ligands used in preparing receptor affinity supports are themselves products of fermentation of recombinant microorganisms and are subjected to a separate sequence of purification steps, prior to being reacted with a functionalized stationary phase to form the affinity support. The resulting affinity chromatography columns are expensive when viewed on the basis of cost of support/unit volume of stationary phase. The cost/benefit ratio would still be attractive because process-scale columns can be small (volumes on the order of 1–10 L). Moreover, as with other types of affinity chromatography, purification of dilute but highly active protein is possible.

Receptor affinity chromatography is a selective form of immunoaffinity chromatography which is based on antigen–antibody interactions. Its growth as a separations tool was promoted by the development of monoclonal antibody-producing hybrid cell lines in 1975. Monoclonal antibodies are dilute (at 50 µg/mL) when produced in tissue culture supernatant. The supernatant contains significant amounts of other proteins used to grow the B-cell clone which generates the monoclonal antibody. The monoclonal antibodies are used principally as diagnostic tools, and are recovered using protein A columns (52).

BIBLIOGRAPHY

1. R. C. Willson and M. R. Ladisch, in M. R. Ladisch, R. C. Willson, C-C. Painton, and S. E. Builder, eds., *ACS Symp. Ser.* **427**, 1–13 (1990).
2. Committee on Bioprocess Engineering, National Research Council, *Putting Biotechnology to Work: Bioprocess Engineering*, National Academy of Sciences, Washington, D.C., 1992, pp. 2–22.
3. S. M. Wheelwright, *Protein Purification: Design and Scale-up of Downstream Processing*, Hanser Publishers, Munich, Germany, 1991, pp. 1–9, 61, 213–217.
4. C. A. Bisbee, *GEN* **13**(14), 8–9 (1993).
5. A. M. Thayer, *C&EN* **74**(33), 13–21 (1996).
6. A. M. Thayer, *C&EN* **74**(1), 22–23 (1996).
7. R. Rawls, *C&EN* **74**(32), 31 (1996).
8. J. A. Wells, *Science* **273**(5274), 449–450 (1996).
9. W. Roush, *Science* **273**(5273), 300–301 (1996).
10. V. B. Lawlis and H. Heinsohn, *LC-GC* **11**(10), 720–729 (1993).
11. M. Bernon and J. Bodelle, in Y-Y. H. Chien and J. L. Gueriguian, eds., *Drug Biotechnol. Reg.* **13**, xv-xxiii (1991).
12. M. R. Ladisch and K. L. Kohlmann, *Biotechnol. Prog.* **8**(6), 469–478 (1992).
13. W. F. Prouty, in Ref. 11, pp. 221–262.
14. E. P. Kroeff, R. A. Owens, E. L. Campbell, R. D. Johnson, and H. I. Marks, *J. Chromatog.* **461**, 45–61 (1989).

15. S. E. Builder, R. van Reis, N. Paoni, and J. Ogez, "Process Development and Regulatory Approval of Tissue-Type Plasminogen Activator," in *Proceedings of the 8th International Biotechnology Symposium*, Paris, July 17–22, 1989.
16. N. K. Harakas, J. P. Schaumann, B. T. Connolly, A. J. Wittwer, J. V. Orlander, and J. Feder, *Biotechnol. Prog.* **4**(3), 149–158 (1988).
17. S. E. Builder and E. Grossbard, *Transfus. Med. Rec. Technol. Adv.*, 303–313 (1986).
18. M. Baringa, *Science* **272**(5262), 664–666 (1996).
19. J. O'C. Hamilton, *Business Week*, **3478**, 118, 122 (1996).
20. S. E. Builder and W. S. Hancock, *Chem. Eng. Prog.* **84**(8), 42–46 (1988).
21. J. D. Watson, M. Gilman, J. Witkowski, and M. Zoller, *Recombinant DNA*, 2nd ed., W. H. Freeman and Co., New York, 1992, pp. 458–460.
22. B. Alberts, D. Bray, J. Lewis, M. Raff, K. Roberts, and J. D. Watson, *Molecular Biology of the Cell*, 2nd ed., Garland Publishing, New York, 1989, pp. 1203–1212.
23. S. V. Ho, in Ref. 1, pp. 14–34.
24. G. B. Dove, G. Mitra, G. Roldan, M. A. Shearer, and M-S. Cho, in Ref. 1, pp. 194–209.
25. S. Sobel, in Ref. 11, pp. 499–511.
26. J. K. Lin, B. J. Jacobsen, A. N. Pereira, and M. R. Ladisch, in W. A. Wood and S. T. Kellog, eds., *Methods in Enzymology*, Vol. 160, Academic Press, Inc., San Diego, Calif., 1988, pp. 145–159.
27. A. Velayudhan and M. R. Ladisch, *Anal. Chem.* **63**(18), 2028–2032 (1991).
28. A. Velayudhan, R. L. Hendrickson, and M. R. Ladisch, *AIChE J.* **41**(5), 1184–1193 (1995).
29. A. Velayudhan and M. R. Ladisch, *Ind. Eng. Chem. Res.* **34**(8), 2805–2810 (1995).
30. S. Roe, in E. L. V. Harris and S. Angal, eds., *Protein Purification Methods—A Practical Approach*, IRL Press, Oxford, U.K., 1989, pp. 200–216.
31. A. Z. Preneta, in Ref. 30, pp. 293–306.
32. R. T. Kurnik, A. W. Yu, G. S. Blank, A. R. Burton, D. Smith, A. M. Athalye, and R. van Reis, *Biotechnol. Bioeng.* **45**, 149–157 (1995).
33. K. H. Hamaker, J. Liu, R. J. Seely, C. M. Ladisch, and M. R. Ladisch, *Biotechnol. Prog.* **12**, 184–189 (1996).
34. *Gel Filtration—Theory and Practice*, Pharmacia Fine Chemicals, Uppsala, Sweden, 1979, pp. 1–19, 30–35, 46, 50, and 57.
35. P. C. Sadek, P. W. Carr, R. M. Doherty, M. J. Kamlet, R. W. Taft, and M. H. Abraham, *Anal. Chem.* **57**, 2971–2978 (1985).
36. H. Colin, P. Hilaireau, and J. De Tournemire, *LC-GC* **8**(4), 302–312 (1990).
37. A. M. Cantwell, R. Calderone, and M. Sienko, *J. Chromatog.* **316**, 133–149 (1984).
38. R. L. Gustafson, R. L. Albright, I. Heisler, J. A. Lirio, and O. T. Reid, *Ind. Eng. Chem. Prod. Res. Dev.* **7**(2), 107–115)1968).
39. D. J. Pietrzyk and J. D. Stodola, *Anal. Chem.* **53**(12), 1822–1828 (1981).
40. D. J. Pietrzyk and C-H. Chu, *Anal. Chem.* **49**(6), 757–764 (1977).
41. J. Morris and J. S. Fritz, *LC-GC* **11**(7), 513–517 (1993).
42. G. K. Sofer and L. E. Nystrom, *Process Chromatography: A Practical Guide*, Academic Press, Ltd., London, 1989, pp. 128–129.
43. R. L. Garnick, M. J. Ross, and R. A. Baffi, in Ref. 11, pp. 263–313.
44. E. A. Carrey, in T. E. Creighton, ed., *Peptide Mapping in Protein Structure: A Practical Approach*, IRL Press, Oxford, U.K., 1990, pp. 117–144.
45. E. Flanigan, *High Performance Liquid Chromatography in the Production and Quality Control of Salmon Calcitonin*, Purdue University, West Lafayette, Ind., 1991, p. 207.
46. B. F. Roettger and M. R. Ladisch, *Biotechnol. Adv.* **7**, 15–29 (1989).
47. B. R. Roettger, J. A. Myers, M. R. Ladisch, and F. E. Regnier, in Ref. 1, pp. 80–92.
48. M. R. Ladisch, R. L. Hendrickson, and K. L. Kohlmann, in Ref. 1, pp. 93–103.
49. I. Parikh and P. Cuatrecases, *CEN* **63**, 17–32 (1985).

50. J. A. Asenjo and I. Patrick, in E. L. V. Harris and S. Angal, eds., *Protein Purification Applications—A Practical Approach*, IRL Press, Oxford, U.K., 1990, pp. 1–28.
51. S. K. Basak and M. R. Ladisch, *Anal. Biochem.* **226**, 51–58 (1995).
52. P. Bailon and D. V. Weber, *Nature* **335**(6193), 839–840 (1988).

MICHAEL R. LADISCH
Purdue University

BTX SEPARATIONS

Benzene [71-43-2], toluene [108-88-3], and the isomeric xylenes are the lowest molecular weight aromatics. All are important petrochemical feedstocks. Considerable quantities of these components are commonly obtained directly from the pipe still of a typical refinery and as by-products of ethylene [74-85-1]. Depending on demand, these sources are often augmented by additional production through various reforming and dehydrogenation processes.

The complexity of the separations involved in recovering pure benzene, toluene, and xylenes from typical feedstocks depends on many factors including the nature and quantity of the nonaromatic components. Where nonaromatics make up only a relatively small fraction of the feedstock, simple distillation (qv) processes can generally be used to separate relatively pure streams of benzene, toluene, and mixed xylenes. Where nonaromatics comprise a relatively large proportion of the feed, an extraction or extractive distillation process is generally needed (see XYLENES AND ETHYLBENZENE).

The separation of the C_8-aromatics stream (mixed xylenes plus ethylbenzene) into its isomeric components cannot be accomplished by distillation. Either adsorption (qv), eg, the PAREX process, crystallization (qv), or some combination of these processes is generally employed.

An option for avoiding the cost of extraction is to increase the severity of the BTX formation step. This reduces the quantity of residual paraffins, and, depending on the BTX formation process, may leave the BTX clean enough to purify by distillation. The final impurity concentrations may still be too high for merchant sale, but may be acceptable in some downstream operations. For example, xylenes going into an isomerization/separation loop for *p*-xylene production can contain some paraffins if the isomerization catalyst is capable of decomposing them. Disadvantages of high severity processing are the increased catalyst fouling rate and the potential increase in undesirable olefin impurities.

Extraction and Extractive Distillation

The choice of an extraction or extractive distillation solvent depends on boiling point, polarity, thermal stability, selectivity, aromatics capacity, and upon the feed

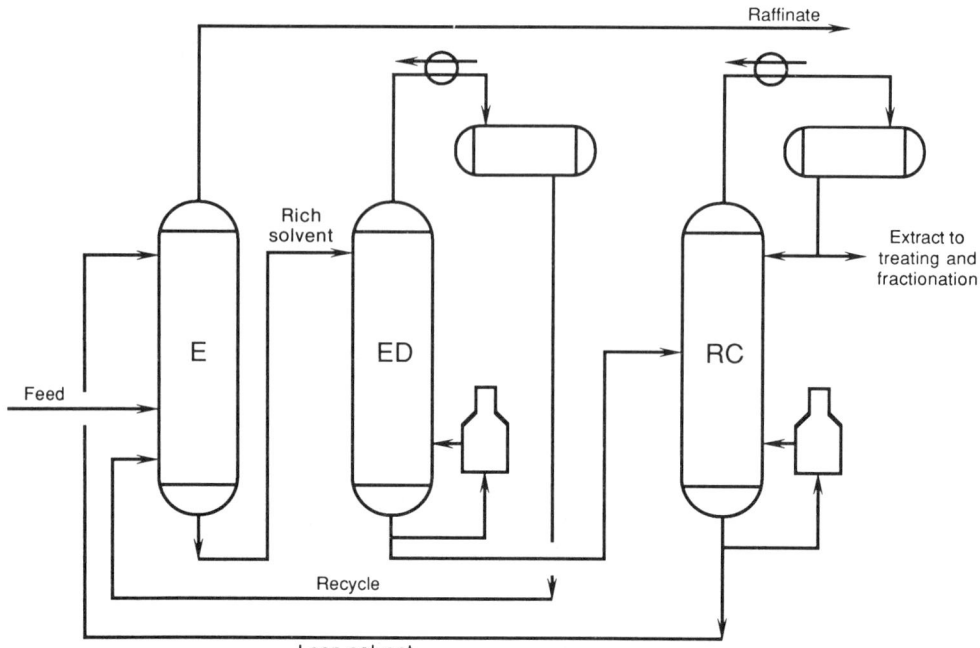

Fig. 1. Shell sulfolane extraction process. E, extraction; ED, extractive distillation; RC, recovery column. Courtesy of UOP, Inc.

aromatic content (see EXTRACTION). Capacity, defined as the quantity of material that is extracted from the feed by a given quantity of solvent, must be balanced against selectivity, defined as the degree to which the solvent extracts the aromatics in the feed in preference to paraffins and other materials. Most high capacity solvents have low selectivity. The ultimate choice of solvent is determined by economics. The most important extraction processes use either sulfolane or glycols as the polar extraction solvent.

Sulfolane [126-33-0], used in UOP and Shell processes (1,2) offers good thermal and hydrolytic stability, high density and boiling point, and a good balance of solvent properties. Its high density and boiling point make it easy to separate from the hydrocarbon streams. A diagram of a sulfolane extraction unit is shown in Figure 1. Fresh feed enters the extractor and flows countercurrent to the down flowing solvent. The raffinate is withdrawn at the top of the extractor and leaves the system after water washing. The solvent, now rich in aromatics, is sent to the top of the extractive stripper where the nonaromatic hydrocarbons are removed. Aromatics and sulfolane are separated in the recovery column. The lean solvent is recycled to the extractor and the aromatics are washed with water and removed. The recovery of benzene and toluene is usually 99+%; of C_8-aromatics 97%; and of C_9^+-aromatics 75–90%.

The widely employed UOP Udex process uses a glycol solvent (3). Diethylene glycol was used in early versions of the process; however, increased capacity was obtained by adding dipropylene glycol or, in some cases, a change was made to triethylene glycol. Further improvement was made by using tetraethylene glycol (4).

Table 1. Extractive Processes for BTX Recovery

Company process	Solvent	CAS Registry Number	Reference
Extraction			
Shell Process	sulfolane	[126-33-0]	1,2
UOP Udex Process	diethylene glycol	[111-46-6]	3
	triethylene glycol	[112-27-6]	
	tetraethylene glycol	[112-60-7]	4
Union Carbide Tetra Process	tetraethylene glycol	[112-60-7]	5
Lurgi Arosolvan	N-methyl-2-pyrrolidinone and monoethylene glycol	[872-50-4] [107-21-1]	6,7
Institut Français du Petrôle	dimethyl sulfoxide (DMSO)	[67-68-5]	8
SNAM Progetti Formex	N-formylomorpholine	[4394-85-8]	9
Howe-Baker Aromex	diglycolamine	[929-06-6]	10
Krupp-Koppers Morphylex	N-formylmorpholine	[4394-85-8]	11
Extractive distillation			
Institut Français du Petrôle DMF	dimethylformamide (DMF)	[68-12-2]	12
Krupp-Koppers Octenar	N-formylmorpholine	[4394-85-8]	12
Lurgi Distapex	N-methyl-2-pyrrolidinone	[872-50-4]	12
UOP Sulfolane	sulfolane	[126-33-0]	12

The Union Carbide Tetra process also employs tetraethylene glycol (5). Other extraction processes are included in Table 1.

Extractive distillation, using similar solvents to those used in extraction, may be employed to recover aromatics from reformates which have been prefractionated to a narrow boiling range. Extractive distillation is also used to recover a mixed benzene–toluene stream from which high quality benzene can be produced by postfractionation; in this case, the toluene product is less pure, but is still acceptable as a feedstock for dealkylation or gasoline blending. Extractive distillation processes for aromatics recovery include those listed in Table 1.

Downstream Processing. In addition to extraction, various downstream operations are often carried out on the BTX product to produce products in proportions to fit the market demand. A typical aromatics processing scheme is shown in Figure 2 in which benzene, *p*-xylene, and *o*-xylene are the products.

After the crude BTX is formed, by reforming in this case, a heart cut is sent to extraction. Actually, the xylenes and heavier components are often sent to downstream processes without extraction. The toluene produced is converted to

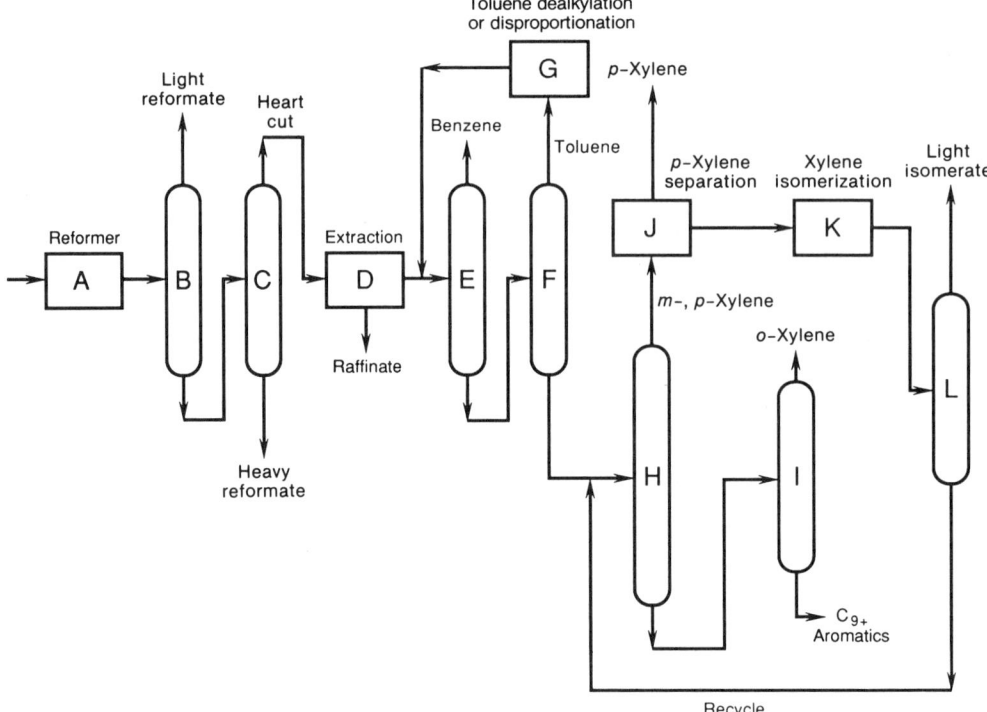

Fig. 2. General BTX processing sequence.

benzene, a more valuable petrochemical, by running the toluene through a hydrodealkylation unit. This catalytic unit operates at 540–810°C with an excess of hydrogen. Another option is to disproportionate toluene or toluene plus C_9 aromatics to a mixture of benzene and xylenes using a process such as UOP's Tatoray or Mobil's Selective Toluene Disproportionation process (STDP) (13).

The o-xylene [95-47-6] in Figure 2 is recovered by a two-stage distillation. First o-xylene is separated (split) from m-xylene [108-38-3] and the other C_8-aromatics in a superfractionating column, the xylene splitter (Unit H). The bottoms, a mixture of o-xylene and C_9^+-aromatics, is redistilled (rerun) in Unit I to recover o-xylene of 96+% purity.

The distillate (overhead) from Unit H, containing mostly ethylbenzene [100-41-4], p-xylene, and m-xylene, and some o-xylene becomes the feed for the p-xylene separation process (Unit J).

p-Xylene [106-42-3] can be purified by crystallization (qv), or adsorption (qv). When a typical reformate-derived C_8-aromatic mixture is cooled, p-xylene crystallizes first. Most plants employing crystallization operate at –60 to –75°C, depending on feed composition (14). The process is limited by a eutectic temperature below which o- or m-xylene also crystallize. The solubility of p-xylene in the remaining C_8-aromatic mixture over the range of –60 to –75°C is 9.6 to 6.2%.

UOP's Parex process can be used to purify p-xylene by adsorption (15). Toray has a similar process. These processes take advantage of the fact that p-xylene is adsorbed more easily than the other C_8 aromatics by a suitable molecular sieve.

The *p*-xylene is desorbed by either a lighter or heavier hydrocarbon which is subsequently removed by distillation. *p*-Xylene is recovered in about 97% yield (see ADSORPTION).

The mother liquor obtained from the crystallization, or the raffinate after removal by adsorption, is isomerized using an acidic catalyst to convert *m*-xylene to the *o*- and *p*-isomers (Unit K, Fig. 2).

In the isomerization step, light and heavy impurities are generated, including saturated hydrocarbons, benzene, toluene, and C_9^+-aromatics. These are removed by recycling the isomerate through the xylene splitter and the light isomerate column (Unit L). If the light isomerate contains much benzene or toluene, it can be recycled to Unit G. *m*-Xylene may be recovered between the xylene splitter and the *p*-xylene separation plant by selective sulfonation followed by regeneration or via complex formation with a mixture of HF and BF_3 (16).

To this point the presence of ethylbenzene in the mixed xylenes has been ignored. The amount can vary widely, but normally about 15% is present. The isomerization process must remove the ethylbenzene in some way to ensure that it does not build up in the isomerization loop of Figure 2. The ethylbenzene may be selectively cracked (17) or isomerized to xylenes (18) using a platinum catalyst. In rare cases the ethylbenzene is recovered in high purity by superfractionation.

There are many variations of the basic processing loop shown in Figure 2. Processing to produce only BT is common, often in conjunction with a toluene-to-benzene dealkylation unit. If benzene and toluene are not to be recovered, Column B may be used to remove toluene and lighter components. In that case, Units E, F, and G would be eliminated. To recover *p*-xylene only, the xylene splitter is reduced in size and is used to split between *o*-xylene and the C_9^+ aromatics. The *o*-xylene rerun still is then eliminated.

Environmental Considerations

BTX processing has come under steadily increasing pressure to reduce emissions and workplace exposures. Reductions in the permissible levels of both benzene and total aromatics (BTX) in gasoline have been legislated. Whereas all BTX components are to be controlled, the main focus is on benzene which is considerably more toxic than the others and is classified as a known carcinogen (19).

Workplace exposure limits for benzene have been regulated to levels as low as 0.5 ppm (20). Industrial emissions affecting the public are low enough that the EPA considers mostly indoor sources, such as smoking, automobile exhausts, and consumer products, as the greater hazard (21).

The stringent controls over manufacturing and handling benzene may dissuade refiners from installing new equipment to extract benzene from gasoline to satisfy gasoline requirements. Instead modification of reforming operations to produce less benzene or to convert the remaining benzene to other compounds is expected. Overall, this is likely to result in further segregation of petrochemicals BTX production from gasoline production.

BIBLIOGRAPHY

"BTX Processing" in the *Encyclopedia of Chemical Technology*, 3rd ed., Vol. 4, pp. 264–277, by D. L. Ransley, Chevron Research Co.; in *ECT* 4th ed., Vol. 4, pp. 590–605, by W. A. Sweeney and P. F. Bryan, Chevron Research and Technology Co.

1. C. E. Beecher, R. E. Mann, and M. L. Renquist, *Oil Gas J.* **63,** 80 (Nov. 22, 1965).
2. *Hydrocarbon Process.*, 194 (Sept. 1982).
3. *Hydrocarbon Process.*, 248 (Sept. 1970).
4. G. S. Somekh and B. I. Friedlander, *Hydrocarbon Process.*, 127 (Dec. 1969).
5. *Hydrocarbon Process.*, 195 (Sept. 1982).
6. E. Muller, *Chem. Ind. (London)*, 518 (June 2, 1973).
7. E. Muller and G. Hochfield, *7th World Petrol. Congr.* **4,** 13 (1967).
8. B. Choffe and co-workers, *Hydrocarbon Process.*, 188 (May 1966).
9. E. Cinelli and co-workers, *Hydrocarbon Process.*, 141 (Apr. 1971).
10. W. T. Jones and V. Payne, *Hydrocarbon Process.*, 91 (Mar. 1973).
11. M. Stein, *Hydrocarbon Process.*, 139 (Apr. 1973).
12. K. J. Day and T. M. Snow, in E. G. Hancock, ed., *Toluene, the Xylenes, and Their Industrial Derivatives*, Elsevier, Amsterdam, the Netherlands, 1982, Chapter 3.
13. *Hydrocarbon Process.*, 93 (Nov. 1989).
14. D. L. McKay, G. H. Dale, and D. C. Tabler, *Chem. Eng. Progress* **62,** 104 (1966).
15. M. Seko, T. Miyake, and K. Inada, *Hydrocarbon Process.*, 133 (Jan. 1980).
16. T. Ueno, in T. C. Lo and co-eds., *Handbook of Solvent Extraction*, John Wiley & Sons, Inc., New York, 1983, p. 575.
17. D. B. Broughton, *186th ACS National Meeting, Am. Chem. Soc. Div. Pet. Chem. Prepr.* **28**(4), 1072 (Aug. 1983).
18. C. V. Berger, *Hydrocarbon Process.*, 173 (Sept. 1973).
19. K. B. Clansky, ed., *A Guide to Industrial Chemicals Covered under Major Federal Regulatory and Advisory Programs,* Roytech, Burlingame, Calif., 1990.
20. *California OSHA Regulations,* Sec. 5218, 1990.
21. J. Raloff, *Sci. News Lett.* **136,** 245 (Dec. 1989).

W. A. SWEENEY
P. F. BRYAN
Chevron Research and Technology Company

CENTRIFUGAL SEPARATION

Centrifugal separation is a mechanical means of separating the components of a mixture of liquids or of liquids and solid particles. The material is accelerated in a centrifugal field which acts upon the mixture in the same manner as a gravitational field. The centrifugal field can, however, be varied by changes in rotational speed and equipment dimensions, whereas gravity is essentially constant. Commercial centrifugal equipment can reach an acceleration of 20,000 times gravity (20,000 G); laboratory equipment can reach up to 360,000 G. Most centrifugation equipment is intended to separate immiscible or insoluble components from a liquid medium. The ultracentrifuge and gas centrifuge represent special cases that establish separation gradients on a molecular scale. The usual gravitational operations, such as sedimentation (qv) or flotation (qv) of solids in liquids, drainage or squeezing of liquids from solid particles, and stratification of liquids according to density, are accomplished more effectively in a centrifugal field (see SIZE SEPARATION).

The development of theory for centrifugation equipment has been slow. Flow patterns in the centrifuge bowls are complex and difficult to model mathematically. The concept of a theoretical capacity factor for sedimentation depends only on equipment characteristics and is independent of the system (1,2). The theoretical effect of particle size distribution in single and multistage centrifugation has been demonstrated (3). Extensive application of centrifugation to dewatering (qv) and thickening of relatively soft solids and hydrogels associated with industrial and municipal waste treatment has resulted in changes in centrifuge design. A sound theoretical basis for centrifugal drainage or squeezing of liquids from solids has been only partially developed for compressible solids. Many aspects are in need of amplification.

Herein centrifugal separation in a liquid medium is discussed. A brief presentation of centrifugal separation in a gaseous medium is also offered.

Theory

Separation by Density Difference. A single solid particle or discrete liquid drop settling under the acceleration of gravity in a continuous liquid phase accelerates until a constant terminal velocity is reached. At this point the force resulting from gravitational acceleration and the opposing force resulting from frictional drag of the surrounding medium are equal in magnitude. The terminal velocity largely determines what is commonly known as the settling velocity of the particle, or drop under free-fall, or unhindered conditions. For a small spherical particle, it is given by Stokes' law:

$$v_g = \frac{\Delta \delta d^2 g}{18 \mu} \tag{1}$$

where v_g = the settling velocity of a particle or drop in a gravitational field; $\Delta \delta = \delta_S - \delta_L$ = the difference between true mass density of the solid particle or liquid drop, and that of the surrounding liquid medium; d = the diameter of the solid particle or liquid drop; g = the acceleration of gravity; and μ = the absolute viscosity of the surrounding medium.

Stokes' law can be readily extended to a centrifugal field:

$$v_s = \frac{\Delta \delta d^2 \omega^2 r}{18 \mu} = v_g \left(\frac{\omega^2 r}{g} \right) \tag{2}$$

where v_s = the settling velocity of a particle or drop in a centrifugal field; ω = the angular velocity of the particle in the settling zone; and r = the radius at which settling velocity is determined. Analogous equations describe the terminal velocity of a light particle or drop rising in a heavier continuous medium.

The settling velocity, v_s, is relative to the continuous liquid phase where the particle or drop is suspended. If the liquid medium exhibits a motion other than the rotational velocity, ω, the vector representing the liquid-phase velocity should be combined with the settling velocity (eq. 2) to obtain a complete description of the motion of the particle (or drop).

These concepts are used to analyze separations in the bottle centrifuge, the imperforate bowl centrifuge, and the disk centrifuge. Separation by density difference in other types of centrifuges can be analyzed by analogy.

The Bottle Centrifuge. Analysis of the performance of a bottle centrifuge is based on the model shown in Figure 1. A solid or liquid particle is considered in an initial position, X, at a radius, r, from the axis of rotation. If equation 2 is applied to this specific particle, assuming that $v_s = dr/dt$, then

$$\int_r^{r_C} \frac{dr}{r} = \int_0^t v_g \left(\frac{\omega^2}{g} \right) dt \tag{3}$$

where r_C = the radius of the sedimented cake, and t = the time during which the particle is subjected to centrifugal acceleration. Integration of equation 3 leads to the following:

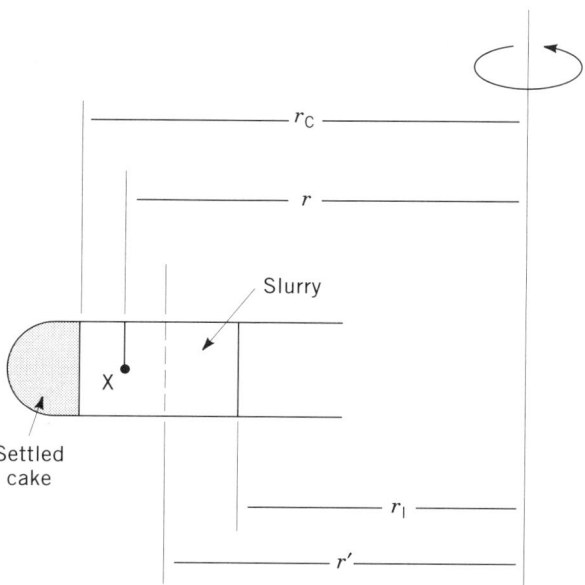

Fig. 1. Separation in a bottle centrifuge, where X is the initial position of a particle (drop). See text for definition of terms.

$$\ln \frac{r_C}{r} = v_g \left(\frac{\omega^2}{g}\right) t \qquad (4)$$

A radius, r, that divides the volume of supernatant into two equal parts can be defined as follows:

$$(r - r_1) = (r_C - r) \text{ or } r' = \frac{r_C + r_1}{2} \qquad (5)$$

where r_1 is the radius of free surface of liquid. Assuming the presence of more than one particle in the feed as well as a uniform initial particle distribution, then each of the two volumes defined by r' initially contains the same number of particles. By making the further assumption that the particles in the suspension are identical, a settling time, t, is chosen so that those particles starting from radius r' all reach the cake r_C after time t. Under these conditions, one-half of the particles that were in suspension at $t = 0$ are sedimented after the time, t, has elapsed. The other half, initially located above the level defined by the radius r', remain in suspension. If r is replaced (eq. 4) by r', a sedimentation condition is established that is referred to as 50% cutoff. An effective capacity, Q_0, for the bottle centrifuge is determined by the ratio between the volume, V, occupied by the slurry in the bottle, and the spinning time, t, calculated from equation 4:

$$Q_0 = \frac{V}{t} 2 v_g \left(\frac{\omega^2}{g} \frac{V}{2 \ln 2r_C/(r_C + r_1)}\right) \qquad (6)$$

In equation 6, v_g characterizes the settling behavior of the solid particles or liquid drops in the suspension, whereas the second part of the right-hand side refers to speed and size of the centrifuge and is expressed by the capacity factor Σ_B. For a bottle centrifuge, it takes the following form:

$$\Sigma_B = \frac{\omega^2 V}{2\,g(2r_C/(r_C + r_l))} \qquad (7)$$

This capacity factor has the dimension of an area and represents the area of a static gravity settling tank having a separation performance equal to that of the rotating bottle centrifuge handling the same particles. By combining equations 6 and 7, and by eliminating volume, V, the following relation is obtained:

$$\frac{Q_0}{\Sigma_B} = \frac{2g}{(\omega^2 t)} \cdot \ln\left(\frac{2r_C}{r_C + r_l}\right) \qquad (8)$$

Equation 8 provides the basis of comparison for the performance of various bottle centrifuges having the same material, and also, under certain circumstances, of other types of sedimentation centrifuges, if geometric dissimilarities are also considered.

The capacity factor, Σ_B, defined by equation 7, is derived from a set of assumptions. An additional assumption is specific to the bottle centrifuge. Namely, a particle is considered sedimented when it reaches the surface of the cake without contacting the tube wall.

The Imperforate Bowl Centrifuge. In an imperforate bowl centrifuge the flow of the continuous liquid phase is nominally axial, except for areas immediately adjacent to the feed inlet and effluent outlet. Tubular solid-bowl basket and imperforate bowl conveyor-discharge centrifuges satisfy this definition.

The mathematical model chosen for this analysis is that of a cylinder rotating about its axis (Fig. 2). Suitable end caps are assumed. The liquid phase is introduced continuously at one end so that its angular velocity is identical everywhere with that of the cylinder. The flow is assumed to be uniform in the axial direction, forming a layer bound outwardly by the cylinder and inwardly by a free air–liquid surface. Initially the continuous liquid phase contains uniformly distributed spherical particles of a given size. The concentration of these particles is sufficiently low that their interaction during sedimentation is neglected.

Under these circumstances, the settling motion of the particles and the axial motion of the liquid phase are combined to determine the settling trajectory of these particles. The trajectory of particles just reaching the bowl wall near the point of liquid discharge defines a minimum particle size that starts from an initial radial location and is separated in the centrifuge. A radius r is chosen to divide the liquid annulus in the bowl into two equal volumes initially containing the same number of particles. Half the particles of size d present in the suspension are separated; the other half escape. This is referred to as a 50% cutoff.

The feed rate corresponding to this condition is related to the bowl geometry

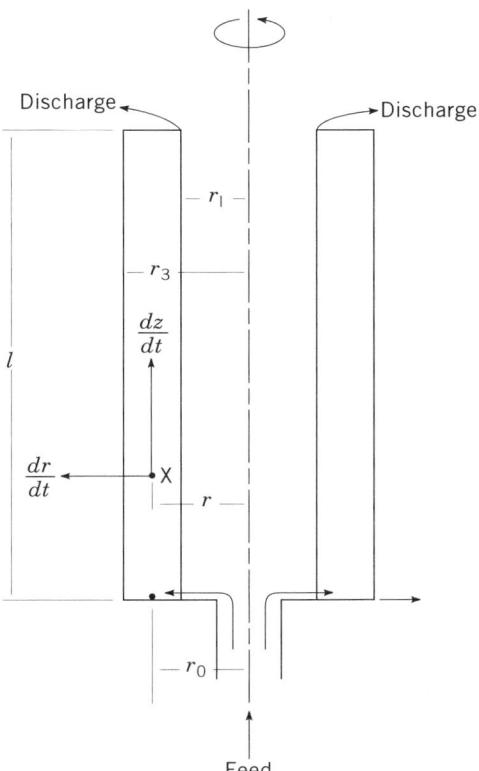

Fig. 2. Separation in a basket or tubular centrifuge, where X is the initial position. Other terms are defined in the text.

r_1, r_3, and l; the bowl angular speed, ω; and the Stokes' settling velocity, v_g (eq. 2).

$$Q_0 = 2v_g\left(\frac{\omega^2}{g}\right)2\pi l\left(\frac{3}{4}r_3^2 + \frac{1}{4}r_1^2\right) \tag{9}$$

where v_g = the Stokes' settling velocity (see eq. 1); ω = the angular velocity of the centrifuge; g = the gravitational acceleration; l = the length of settling zone; r_3 = the radius of the inside wall of the cylinder; and r_1 = the radius of the free surface of the liquid layer in the cylinder. Equation 9 can be rewritten as equations 10 and 11:

$$Q_0 = 2\, v_g \Sigma_T \tag{10}$$

where

$$\Sigma_T = 2\pi l\left(\frac{\omega^2}{g}\right)\left(\frac{3}{4}r_3^2 + \frac{1}{4}r_1^2\right) \tag{11}$$

Equation 10 estimates the flow or throughput rate, above which particles of size d are less than 50% sedimented, and below which over 50% are mostly collected. Equations 10 and 11 are also applicable to the light particles rising in a heavy phase liquid, provided that r_3 and r_1 are interchanged in equation 11.

The theoretical capacity factor, Σ_T, defined by equation 11 has the dimension of an area and can be interpreted as the area of a gravity settling tank where the separation performance is equal to the centrifuge provided that the factor v_g is the same for both. This restriction is required because the particles suspended in the continuous phase can be deaggregated and further dispersed by the vigorous shearing to which the feed is subjected during acceleration in the centrifuge. If this effect is not considered in comparison with that of the settling tank, centrifugal sedimentation might be less favorable in practice than is anticipated by theory.

Equation 11 can be further reduced to facilitate understanding of its use and application. If, instead of the radii of pond surface, r_1, and bowl wall, r_3, a mean radius, r_m, is introduced, equation 11 can be rewritten as follows:

$$\Sigma_T = 2\pi r_m l \left(\frac{\omega^2 r_m}{g} \right) \quad (12)$$

Equation 12 shows that Σ_T can be expressed as the product of a mean sedimentation area ($2\pi r_m l$) and the G level ($\omega^2 r_m/g$), and therefore reflects the increased sedimentation rate expected through a defined area having centrifugal acceleration instead of gravity.

Disk Centrifuge. The separation of particles inside a disk stack is illustrated in Figure 3. The continuous liquid phase, containing solid or liquid particles to be separated, flows from the outside of the disk stack, having radius

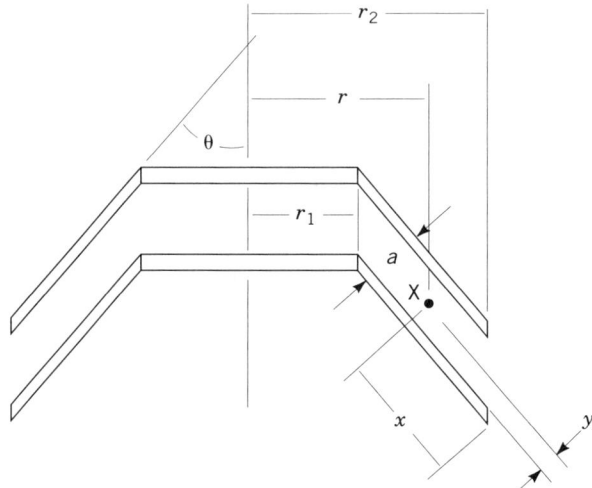

Fig. 3. Separation in a disk centrifuge, where X is the initial position of the particle. Other terms are defined in the text.

r_2, to the inside discharge opening, having radius r_1. Assuming that the liquid phase is evenly divided between the spaces formed by the disks, the flow in each disk space is Q_0/n, where Q_0 is the total flow through the entire disk stack and n is the number of spaces (disks). The flow of the continuous liquid phase is also assumed to be in a radial plane and parallel to the surfaces of the disks, or as having the same angular velocity as the stack.

Here again an equation is established (2) to describe the trajectory of a particle under the combined effect of the liquid transport velocity acting in the x-direction and the centrifugal settling velocity in the y-direction. Equation 13 determines the minimum particle size which originates from a position on the outer radius, r_2, and the midpoint of the space, a, between two adjacent disks, and just reaches the upper disk at the inner radius, r_1. Particles of this size initially located above the midpoint of space a are all collected on the underside of the upper disk; those particles initially located below the midpoint escape capture. This condition defines the throughput, Q_0, for which a 50% recovery of the entering particles is achieved. That is,

$$Q_0 = 2v_g \left(\frac{2\pi n \omega^2}{3\ g} \cot\ \theta \left(r_2^3 - r_1^3\right) \right) \qquad (13)$$

where v_g = the Stokes' settling velocity (see eq. 1); n = the number of spaces between disks in a stack; ω = the angular velocity of the centrifuge; θ = half the included angle of the disks; r_2 = the outer radius of the disks; and r_1 = the inner radius of the disks (see Fig. 3). If v_g, which describes the settling characteristics of the particles, is separated from parameters relating to the geometry and rotational speed of the disk stack, then,

$$Q_0 = 2v_g \Sigma_D \qquad (14)$$

where

$$\Sigma_D = \frac{2\pi n \omega^2}{3\ g} \left(\cot\ \theta \left(r_2^3 - r_1^3\right) \right) \qquad (15)$$

Again, Σ_D corresponds to the area of a gravity settling tank capable of the same separation performance as the disk stack defined by the parameters included in equation 15.

Separation by Drainage. The theory covering drainage in a packed bed or particles is incomplete, and requires more development for a centrifugal field. Liquid is held within the bed by various forces. Removal involves several flow mechanisms. In addition, the centrifugal acceleration changes with radius in the bed, causing changes in packing tendencies of particles and accelerating forces on the residual liquid.

There are three types of liquid content in a packed bed: (*1*) in a submerged bed, there is liquid filling the larger channels, pores, and interstitial spaces; (*2*) in a drained bed, there is liquid held by capillary action and surface tension at points of particle contact, or near-contact, as well as a zone saturated with liquid corresponding to a capillary height in the bed at the liquid discharge face of the cake; and (3) essentially undrainable liquid exists within the body of each particle or in fine, deep pores without free access to the surface except perhaps by diffusion or compaction.

The last type of liquid can be removed by evaporation (qv) or displacement by another liquid but cannot be removed by simple flow in either a gravitational or centrifugal field. There is no sharp distinction between the first two types. The rate and extent of liquid removal from a submerged bed during drainage depends on the physical characteristics of the components, the force of the centrifugal field, and the time of exposure to the field. The residual liquid content of a drained cake consists largely of the capillary and irremovable types at the time of discharge from the separation equipment.

During cake formation and drainage the liquid moves into and through the bed in three different ways. During cake deposition, a continuous head of liquid ranging in composition from feed to clarified supernate may exist over the deposited cake. After feed is stopped, a layer of essentially clarified liquid may still exist over slow-draining cakes. Wash liquor, if it is used, may also create a liquid layer over the deposited cake. Drainage under these conditions requires continuous flow through the cake. The interstitial spaces are assumed to be full. When the free liquid layer no longer exists above the cake, the free liquid surface moves through the cake to an equilibrium position at the capillary height, leaving behind the larger voids filled with gas or vapor. Then, after bulk drainage of the larger voids, liquid still exists in the cake's upper zone in a film covering the surfaces of the solids and in partially filled voids having very restricted outlets. In time, some of this liquid flows as a film to the continuous liquid layer at the capillary height.

If a cake is sufficiently impermeable to permit the buildup of a feed or wash liquid head, flow through the cake approaches steady-state conditions except for changes in compaction or cake thickness as more feed is added, or in the liquid head if the drainage rate differs from the liquid rate addition. An equation for full-pore flow in a centrifugal field has been developed. The hypotheses set forth and for the most part proved are as follows: flow radiates out from the rotation axis and the effect of the gravitational field is negligible; voids at all points are filled with liquid that moves in laminar flow through the cake; kinetic energy changes of the liquid in the cake may be neglected, ie, the filter medium is sufficiently permeable so that it does not run full of liquid; and ambient pressure exists at the outer face of the cake. Essentially incompressible solids produce very similar cake permeabilities in a vacuum, under pressure, and during centrifugal filtration, although significant local variations in permeability may occur because of irregularities in the feed and its distribution. When pores are filled, the flow rate of the supernatant liquid through the cake (4,5) is as follows:

$$Q = \left(\frac{\pi K \omega^2 h}{\mu_L g}\right)\left(\frac{r_m^2 - r_L^2}{\ln(r_m/r_C)}\right) \qquad (16)$$

where K is the permeability; h is the basket height; r_C and r_m are radii from the axis of rotation to the inner and outer faces of cake, respectively; ω is the angular velocity; and r_L is the radius from the axis to the inner face of the liquid layer. Comparison to the usual term α for cake resistance in pressure filtration shows that $K = \delta_L g/\alpha$. The functions related to cake thickness, $\ln(r_m/r_C)$; liquid layer thickness, $(r_m^2 - r_L^2)$; angular velocity, ω^2; and viscosity, μ_L, have been verified by experiment. The exponent on the angular velocity varies for materials that exhibit cake compression as speed is increased, but compression effects are minor on starch, chalk, and kieselguhr, ie, loose or porous diatomite (qv). In practice, the exponent of ω can probably be assumed to have a value of two and experimental variations in permeability can be absorbed by changes in the permeability coefficient. The real value of equation 16 lies in its use for estimating the effect of changes in operating variables for a material where the characteristics are already known from experimental data or plant operation.

Low permeability cakes draining under the conditions of equation 16 are usually handled in perforate basket centrifuges that have relatively long (20 min to several hours) cycles. Following elimination of the supernatant liquid, unsteady-state drainage of the cake may often be neglected. These slow-draining cakes often support such a large capillary height, eg, 90–99% of void volume for chalk (5), that little additional dewatering (qv) is obtained after completion of free liquid drainage. Solids of larger particle size, or freer drainage characteristics, are handled in automatic perforate baskets or continuous screen centrifuges. Low final moistures are usually achieved. For cakes of high permeability, the period when a liquid layer exists above the cake is either short or nonexistent. The time cycle depends chiefly on the rate of film drainage at the completion of bulk liquid flow. Under these conditions film drainage and permanent residual moisture are most important.

The quantity of undrainable residual moisture cannot be predicted without the benefit of experimental data. Equation 17 (6) indicates the important parameters where the exponents were determined using limited experimentation. Introducing the approximation that $s\delta_s$ is proportional to $1/\bar{d}$, where s is the specific surface area per weight of solid, the modified equation for undrainable liquid becomes

$$S^\infty = k\left(\frac{1-\epsilon}{\epsilon}\right)\left(\frac{1}{\bar{d}^2 G}\right)^{1/4}\left(\frac{\sigma \cos \Psi}{\delta_L}\right)^{1/4} \qquad (17)$$

where S^∞ is the fraction of void volume occupied by liquid after infinite drainage time, k is an experimental coefficient, ϵ is the void fraction, $\bar{d}$ is the mean particle diameter, G is $\omega^2 r/g$, σ is the surface tension, and Ψ is the wetting angle. Appreciable internal porosity of the particles can badly distort an experimental value of S.

For cakes of high permeability, the capillary drain height may be an insignificant fraction of cake thickness, and film drainage becomes the controlling factor in a centrifugal field (7). Under unsteady-state conditions, equation 18 represents the drainable liquid left in the cake as a function of the centrifugal filtration parameters:

$$S - S^\infty = \frac{3}{\pi} \frac{s'}{\epsilon} \omega \left(\frac{\mu_2}{\delta_L}\right) \left(\frac{h}{2 r_m - h}\right) \left(\frac{1}{t - t'}\right)^{1/2} \quad (18)$$

where S is the fraction of void volume occupied by liquid at time t, s' is surface area/volume of cake, h is cake thickness, t' is time at which free liquid surface enters the cake, and μ_2 is the viscosity of the surrounding medium.

It is difficult to obtain void volume data for a cake under drainage conditions in a centrifuge. Prediction of these values from filter cake data is uncertain because the compressive force in centrifugation increases with radius throughout the cake depth, and the effective mass of a particle, proportional to $(\delta_S - \delta_L)$ when the cake is submerged, becomes essentially δ_S after bulk drainage is completed. For engineering purposes, it is simpler to approximate the ratio of the volume of liquid to the volume of solid. Assuming that $G \approx \omega^2(2r_m - h)/g$, and $s' \approx 6(1 - E)/\bar{d}$ as for spheres, equation 18 becomes

$$q - q^\infty = \left(\frac{k'}{\bar{d}}\right)\left(\frac{\mu_L}{\delta_2}\right)^{1/2}\left(\frac{h}{Gt}\right)^{1/2} \cdot 100 \quad (19)$$

where q is the ratio of the volume of liquid to the volume of solid. This measurement is readily obtained from the volume percentage of liquid in cake and is also easily converted by a density ratio to a weight ratio of liquid to solid. The value for q^∞, the ratio of the liquid volume to the volume of solid at infinite time, may also be applied when known.

Reasonable experimental agreement was obtained (7) using the exponents given in equation 19 for relatively slow-draining chalk and kieselguhr. Figure 4

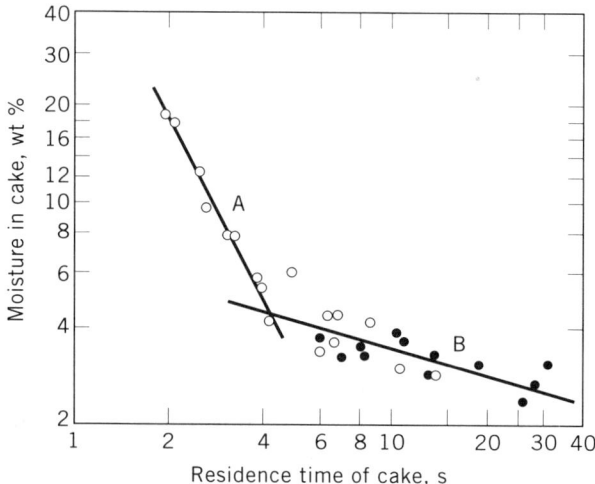

Fig. 4. Drainage of salt crystals in a cylindrical screen pusher-discharge centrifuge (8), where the cake thickness is 3.3 cm, the centrifugal field = 320 G, and the crystals 14 wt % <250 μm. (●) Represents moisture in the discharge cake, and (○) moisture in the cake by material balance with drainage flows; line A has slope = −1.9 and B, −0.3.

(8) shows the effect of drain time after the disappearance of a free liquid head above the cake. The sharp break probably indicates completion of bulk drainage and start of drainage by film flow only. Drainage before the break is rapid and proportional to $t^{-1.9}$. After the break in the curve, the value $t^{-0.3}$ indicates that film drainage becomes controlling if low residual moisture is required. This exponent, -0.3, is appreciably lower than the -0.5 in theory but is close to -0.25, the value previously obtained (9). Considerable data indicate the validity of $G^{-1/2}$; limited data corroborate the theoretical exponent of 1/2 for kinematic viscosity.

Experimental exponents for cake thickness vary from 0.5 to as much as 3.0. The theoretical value of 1/2 may be approached only by incompressible cakes of a narrow range of sizes. The proper and characteristic value for the mean particle size, $\bar{d}$, is difficult to ascertain. In practice, the most finely divided particles, eg, 10–15 wt % of solids, almost wholly determine the liquid content of a cake, regardless of the rest of the size distribution. It seems reasonable to use a $\bar{d}$ closely related to liquid content, eg, the 10% point on a cumulative weight-distribution curve.

Σ-Concept and Its Application. The assumptions and conditions for deriving equations 7, 11, and 15 impose limitations on the application of the Σ-concept and fall into two groups. The first concerns the particulate material. Particles (or drops) are assumed to be spherical in shape and uniform in size. These should not deaggregate, deflocculate, coalesce, or flocculate during passage through the zone in which separation occurs. Initially, particles are evenly distributed in the continuous liquid phase, where their concentration is low enough for them to settle as individual particles without interaction. The settling velocity, v_s, of the particles is such that the Reynolds number does not exceed 1.0, thus ensuring that the deviation from Stokes' law does not exceed about 10%. The settling velocity, v_g, in a gravity field, or v_s in a centrifugal field, is theoretically never reached because the accelerating time required for a particle to reach its terminal velocity is infinite. However, a particle of up to 100 μm approaches 90% of its terminal velocity in milliseconds. The time available for the particles to settle in a centrifuge is enough for each individual particle to be almost at its theoretical settling velocity, despite variation in G.

The second group of assumptions and conditions concerns flow conditions. Flow is assumed to be streamlined. Fresh feed is introduced uniformly into the full space available for its flow. In an imperforate bowl centrifuge, this condition requires that the continuous liquid phase immediately occupy the full liquid layer thickness between the free surface and the inside radius of the cylindrical wall. In a disk centrifuge, the continuous phase is assumed to divide evenly between all the disk spaces axially as well as circumferentially. In the disk stack, flow lines of the continuous phase are directed radially everywhere. In any radial plane, the velocity profile normal to the disk surface is a symmetrical plug flow. In any imperforate bowl centrifuge, the continuous phase rotates everywhere at the same angular velocity as the bowl, ie, there is no forward or back swirl. The displacement of the flow pattern of the continuous phase by the layer of deposited material may be neglected. Remixing at the interface of the separated material is negligible. Finally, the detrimental effect resulting from heavy separated material crossing the fresh feed stream outside of the disk stack may be neglected.

Few of the assumed conditions are fully satisfied in practice. The last three items relate to potential interference between separated phases. Such interference can occur and leads to poor sedimentation performance if an excessive volume of the sedimented phase is retained in the centrifuge.

Excessive volume of solids may be retained in the bowl of conveyor centrifuges if (1) the conveyor volumetric displacement is not sufficient to handle the sedimentation rate of solids; (2) the sedimented solids cannot be successfully conveyed and discharged over the solids port until a sufficient layer has been built up inside the bowl; and (3) solids do not easily slide outwardly on the underside of the disk of a disk centrifuge.

In the case of the nozzle disk centrifuge, the flow of the solids phase through the discharge nozzles may be so restricted that an excessive layer can accumulate inside the bowl shell. When this layer reaches the zone utilized by the fresh feed stream entering the disk stack, reentrainment of the sedimented solids by the fresh feed may lead to poor sedimentation performance.

The sedimentation phenomenon that the Σ-concept attempts to describe quantitatively is only part of the total task that the centrifuge has to accomplish. Thus, attempts to predict separation performance solely on the basis of Σ-concepts have sometimes given disappointing results.

Nevertheless, the Σ-concept is a valuable tool, allowing in theory a comparison between geometrically and hydrodynamically similar centrifuges operating on the same feed material. Equations 7, 11, and 15 show that the sedimentation performance of any two similar centrifuges having the same feed suspension is the same if the quantity Q_0/Σ is the same for each. In practice, an efficiency factor, e, is often introduced to extend the use of Σ so as to compare dissimilar centrifuges. This factor takes into consideration differences in feeding, discharging, flow, turbulence, and remixing that exist in different types of centrifuges operating on the same feed material. The flow rate, Q_{0_2}, of a No. 2 centrifuge can thus be compared to the rate, Q_{0_1}, of a No. 1 centrifuge operating on the same feed. For equal sedimentation performance,

$$\frac{Q_{0_2}}{\Sigma_2 e_2} = \frac{Q_{0_1}}{\Sigma_1 e_1} \quad \text{or} \quad Q_{0_2} = Q_{0_1}\left(\frac{e_2}{e_1}\right)\left(\frac{\Sigma_2}{\Sigma_1}\right) \tag{20}$$

If the two centrifuges are geometrically and hydrodynamically similar, then $e_1 = e_2$ and equation 20 can be simplified to

$$Q_{0_2} = Q_{0_1} \frac{\Sigma_2}{\Sigma_1} \tag{21}$$

The relation of equation 21 for similar centrifuges requires identical sedimentation performance characteristics when operating on the same material.

The Σ-concept permits scale-up between similar centrifuges solely on the basis of sedimentation performance. Other criteria and limitations, however, should also be investigated. Scale-up analysis for a specified solids concentration, for instance, requires knowledge of solids residence time, permissible accumulation of solids in the bowl, G level, solids conveyability, flowability, compressibility, limitations of torque, and solids loading. Extrapolation of data from one size centrifuge to another calls for the application of specific scale-up mechanisms for the particular type of centrifuge and performance requirement.

Other Sedimentation Scale-Up Equations. Some centrifuge suppliers use an area-equivalent, Ae, description instead of Σ; others use KQ or Lf_2 values. All of these are in units of area. For a disk centrifuge,

$$\Sigma_D = \frac{2\pi n \omega^2}{3\,g} \cot\theta \left(r_2^3 - r_1^3\right) \tag{15}$$

$$KQ = \frac{2\pi n \omega^{1.5}}{3\,g} \cot\theta \left(r_2^{2.75} - r_1^{2.75}\right) \tag{22}$$

$$Lf_2 = \frac{2\pi n \omega^2}{3\,g} \cot\theta \left(r_2^3 \frac{r_1}{r_2} - \left(\frac{r_1}{r_2}\right)^2 - \left(\frac{r_1}{r_2}\right)^3\right) \tag{23}$$

For an imperforate tubular centrifuge,

$$Ae_{3/4} = \frac{2\pi l \omega^2}{g}(0.75\ r_{0.75}^2) \tag{24}$$

$$\Sigma = \frac{2\pi l \omega^2}{g}(0.75\ r_3^2 + 0.25\ r_1^2) \tag{25}$$

All of these equations work in the scale-up of geometrically similar centrifuges. The KQ reduces the effect of rotational speed from ω^2 to $\omega^{1.5}$ and the disk radius from r^3 to $r^{2.75}$, based on empirical experience. The Lf_2 uses the projected cylinder area of the disks at their mean radius (see eq. 12 of Σ_T for imperforate bowls). $Ae_{3/4}$ is the cylindrical area at 3/4 of the bowl wall radius.

When testing a new material for centrifugal separation, a bottle centrifuge is usually used to obtain the general G range needed, and to choose the centrifuge type and size. To estimate size, the Q_0/Σ_B must be determined for the bottle centrifuge (eq. 7) and then used in equation 21 to determine the Σ value of the centrifuge to be used. The efficiency factor assumed for disk and imperforate centrifuge has been found to be $\sim e = 0.5$, compared to the bottle centrifuge.

Factors Influencing Centrifugal Sedimentation. The sedimentation velocity of a particle is defined by equations 1 and 2. Each of the terms therein effects separation.

Viscosity. Sedimentation rate increases with decreased viscosity, μ, and viscosity is dependent on temperature. Often mineral oils, which are highly viscous at room temperature, have a viscosity that is reduced by a factor of 10 at 70–80°C. Tar, solid at room temperature, is a low viscosity liquid at 150–200°C and can be clarified of inorganic solids at high flow rates. Even the viscosity of water changes significantly when the temperature changes between 10 and 35°C (10).

Density Difference Between Particle and Liquid. Separation cannot take place if $\Delta \delta = 0$. Some mineral oils have the same density as water at room temperature. If it is heated to 80°C, the reduction of the density of water is less than that of the mineral oil, resulting in the water becoming heavier. Therefore separation is possible. Dilution of a liquid by a solvent, eg, molasses by water or heavy oil by naphtha, results in lower density and lower viscosity of the liquid. Solvent stripping takes place at a later stage.

Particle Size. Doubling particle size, d, increases sedimentation by a factor of 4. Thus, methods to increase size are important. Additives are commonly used to flocculate many fines into an agglomerate that acts as a single large particle. Many chemical companies offer a wide range of organic and inorganic flocculating products, in dry or liquid form. Bench tests are usually required to determine best type and dose, ie, to optimize the flocculent choice. pH control, electrostatic devices, and mechanical coalascers are used to combine fine liquid drops in emulsions to produce larger particles. Special care must be exercised to pump, pipe, and feed mixtures of these easily breakable particle agglomerates, thus preventing the large particles from becoming fine before sedimentation.

Particle Shape. Whereas the Stokes' particle is assumed to be a sphere, very few real solids are actually spherical. Flat and elongated particles sediment slower than spheres. For maximum sedimentation rate, the particle should be as spherical as possible.

Particle Size Distribution. Almost every feed slurry is a mixture of fine and coarse particles. Performance depends on the frequency of distribution of particle size in the feed. Figure 5 shows that whereas all of the coarse particles having a diameter greater than some $d_{\lim}$ are separated, fewer of the very fine particles are, at any given feed rate. The size distribution frequency of particles in feed and centrate for a fine and coarse feed are quite different. More coarse particles separate out than fine ones. Classification of solids by size is often done by centrifugal sedimentation.

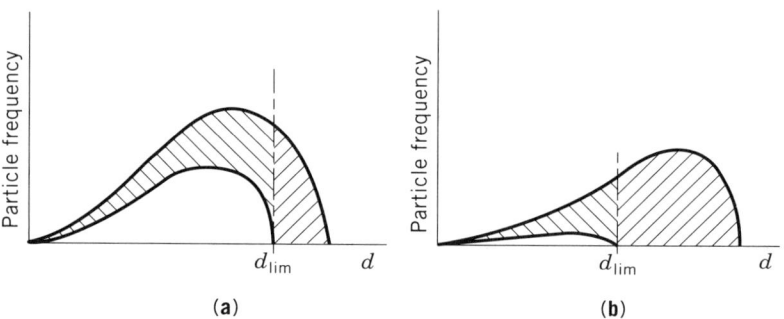

Fig. 5. Particle distribution (upper line) before and (lower line) after action of the separator where the cross-hatched areas represent the particles separated out. By definition, all particles of $d > d_{\lim}$ are separated out. A number of particles having $d < d_{\lim}$ are also separated. (**a**) Fine and (**b**) coarse particle dispersion (10).

Sedimentation Velocity. Velocity of sedimentation, v_s, is equal to throughput Q when divided by Σ. If the concentration in the centrate, c, is divided by

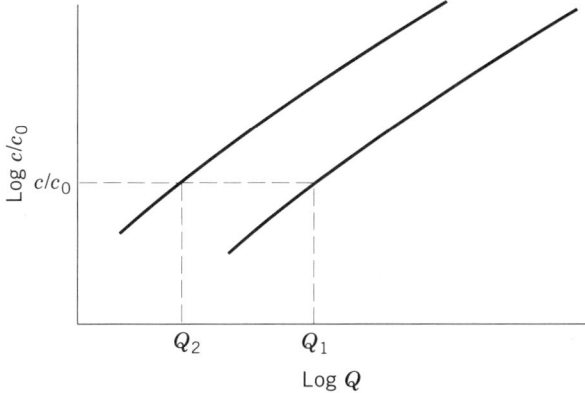

Fig. 6. Relative performance, E_{rel}, where $E_{rel} = Q_1/Q_2$.

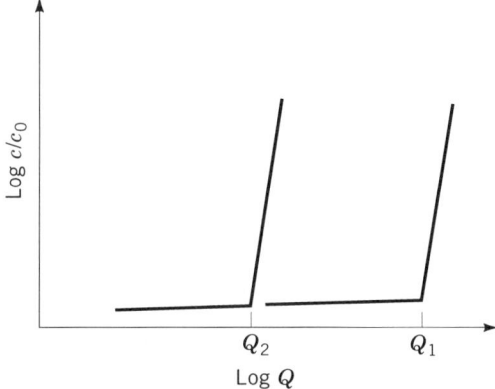

Fig. 7. Determination of E_{rel} for particle distribution within narrow limits. See Fig. 6.

that of the feed, c_0, then c/c_0 gives the unsedimented percentage. A log–log plot of c/c_0 against the centrate rate, Q, usually results in a near-straight line for normal particle size distributions (Fig. 6) (10). In comparing the level of performance at the same percent unsedimented ratio for the same feed material and conditions, the ratio Q_1/Q_2 should equal Σ_1/Σ_2. For materials where the size distribution is very narrow, however, the performance curve is not a straight line (Fig. 7) (10). Almost all of the particles are sedimented at rates below the break in the curve.

Liquid–Liquid-Phase Behavior

Liquid drops, suspended in a continuous liquid medium, separate according to the same laws as solid particles. After reaching a boundary, these drops coalesce to form a second continuous phase separated from the medium by an interface that may be well- or ill-defined. The discharge of these separated layers is controlled by the presence of dams in the flow paths of the phases. The relative radii of these dams can be shown by simple hydrostatic considerations to determine the radius of the interface between the two separated layers. The radius is defined by

$$r_i^2 - r_h^2 = \frac{\delta_l}{\delta_h}\left(r_i^2 - r_l^2\right) \quad (26)$$

where r_i, r_h, and r_l are the radii of the interface, the liquid surface at the heavy discharge dam, and the liquid surface at the light discharge dam, respectively, and δ_h and δ_l are the densities of the heavy and light phases, respectively. Control of the interface radius, achieved by varying r_h or r_l for the desired ratio, is an important factor in liquid–liquid separation, as it determines whether the heavy or light phase is exposed to the greater clarifying effect (10).

Equation 26 is accurate only when the liquids rotate at the same angular velocity as the bowl. As the liquids move radially inward or outward these must be accelerated or decelerated as needed to maintain solid-body rotation. The radius of the interface, r_i, is also affected by the radial height of the liquid crest as it passes over the discharge dams, and these crests must be considered at higher flow rates.

Centrifuge Components

Power, Energy, and Drives. Centrifuges accomplish their function by subjecting fluids and solids to centrifugal fields produced by rotation. Electric motors are the drive device most frequently used; however, hydraulic motors, internal combustion engines, and steam or air turbines are also used. One power equation applies to all types of centrifuges and drive devices.

The total power, P_T, needed to run a centrifuge, ie, delivered by the drive device, is equal to sum of all losses:

$$P_T = P_P + P_S + P_F + P_W + P_{BD} + P_{CP} \quad (27)$$

where P_P, the process power, $= Q\delta\omega^2 r^2$; Q is the flow rate of liquid or solids; δ = mass density; and r = discharge radius of material being discharged. Power for each liquid and the solid phase must be added to get P_P. P_S, the solids process power, $= T_C \cdot \Delta N$ for scroll decanters, where T_C = conveyor torque and ΔN = differential speed between bowl and conveyor. P_F is the friction power, ie, loss in bearings, seals, gears, belts, and fluid couplings. P_W, the windage power, $= K\mu^{0.2}\rho^{0.8}N^3 D^{4.5}$ and μ = viscosity of surrounding gas; ρ = density of gas; D = rotor outside diameter; N = rpm; and K = shape constant. Increased density owing to gas pressure increases the windage power, and this may be very significant for high pressure applications. Also, many hydrocarbon gases are heaver than air, resulting in high windage power. Very high G centrifuges often operate in a vacuum to avoid excessive windage power. For constant G, the scale-up windage power $\propto D^3$; for constant rotor stress, the scale-up of windage power $\propto D^{1.5}$.

Windage power is a very important loss for large machines and must be determined. Whereas windage power can be calculated from drawing dimensions (11), it is preferable to measure the windage power for an actual rotor, and then extrapolate using the formula given for windage power for a geometrically similar (larger or smaller) size. Doubling the size of a rotor while maintaining the g level results in eight times the windage power loss.

P_{CP} is the power absorbed by the centripetal pump. The centrate kinetic energy is partially recovered by the pump, which delivers the centrate flow at a positive pressure. The added power must be supplied by the centrifuge main drive, but use of a centripetal pump avoids the need for a separate centrate pump. The power required to bring the feed to the centrifuge is supplied by a feed pump at the feed tank, not by the centrifuge driver. This power may be significant where the feed pressure required at high flow rates is high.

For scroll centrifuges having back-drives, P_{BD} is the back-drive power:

$$P_{BD} = K \frac{T_C}{R} N_{BD} \qquad (28)$$

where T_C = conveyor torque; R = gear box ratio; and N_{BD} = back-drive speed. This power, provided by the centrifuge drive device, is absorbed by the brake as heat, or if a regenerative device is used, recovered by the electric power supply. Regenerative back-drives reduce total power consumption. A high gear box ratio results in lower back-drive power. Direct hydraulic motor conveyor drive devices get their power from an external hydraulic power supply, not from the main drive motor, and must meet the direct conveyor torque demands. The equation for direct hydraulic conveyor power is $P_{BD} = KT_C\Delta N$; however, hydraulic losses in the power supply, rotary seals, and the hydraulic motor must be added. Eddy current brakes, d-c motors, and variable-frequency a-c motors are the most common variable-speed back-drive devices.

The choice of the main drive, usually an a-c motor, must include starting specifications. Various methods are used to start centrifuges. These include mechanical, fluid, and hydraulic couplings, as well as WYE-DELTA electric motor starters and electronic motor controls. Centrifuges are high inertia rotating devices, often taking 1–12 min to accelerate to operating speed. Details of the mass moment of inertia, friction, and windage losses must be considered to specify a drive device. The inertia seen by the drive device, when comparing centrifuges operating at constant G, is proportional to D^4. Small rotors are easy to start, but large rotors must be carefully reviewed.

Disk machines having nozzles to discharge the solids phase through small backward-pointing nozzles must have this power included in the calculations. Some thickening scroll centrifuges also use such nozzles, usually with an intermittent flow. P_N is the nozzle power:

$$P_N = Q_N \delta \omega r_N (\omega r_N - v_N \cos \phi) \qquad (29)$$

where Q_N = the volume rate of the material discharged through the nozzles, δ_N = the mass density of the material discharged through the nozzles, ω = the angular velocity of the centrifuge, r_N = the radius at which the nozzles discharge, v_N = the linear discharge velocity out of the nozzles, and ϕ = the angle measured between the direction of the nozzle and the tangent to the circle of radius, r.

The discharge velocity out of the nozzles is given by equation 30:

$$v_N = C(2p_N/\delta_N)^{1/2} \qquad (30)$$

where C = a discharge coefficient, and p_N = the hydraulic pressure at radius r_N resulting from the rotation of the bowl. The pressure p_N is given by equation 31:

$$p_N = \frac{\delta_n \omega^2}{2}\left(r_N^2 - r^2\right) \qquad (31)$$

where r is either the free-surface radius of the heavy liquid phase in the case of liquid–liquid and liquid–liquid–solids separation, or the free-surface radius of the liquid phase in the case of liquid–solids separation; and δ is intermediate between the mass densities of the heavy phase and that of the nozzle stream, in the case of liquid–liquid–solids separation, or the liquid phase and that of the nozzle stream, in the case of liquid–solid separation.

The coefficient C in equation 30 is a function of two phenomena. First, the presence of viscous friction accounts for a small loss of energy. Nozzle orifice contraction results in discharge coefficients which range from 0.5 to 0.85. The coefficient falls in the upper portion of this range when the length of the nozzle is two or three times its diameter, and in the lower end of the range where the nozzle diameter is more than five times its length.

Special vortex nozzle designs which deliver lower flows using a large opening have been used to reduce plugging problems. Also, viscosity-sensitive nozzles, where flow is increased as viscosity is increased, are used to control variation in solids concentrations.

Filtering centrifuges must consider the power needed to bring the solids to a final radius, r_S. The radius used to determine the centrate power in equation 27 is the outside radius of the rotor supporting the filtering screen. Many filtering centrifuges discharge solids at a reduced bowl speed to avoid particle breakage. The main drive is often designed to recover the rotor kinetic energy during deceleration.

The energy absorbed by the liquid and solids stream, P_P, is transferred in the feed zone of the rotor or conveyor. One-half of the total energy is converted to the kinetic energy associated with the tangential velocity of the pond surface. Because the pond surface is at or near the discharge dam radius, this kinetic energy is dissipated as heat when the centrate is discharged to the stationary casing. The other half of the total energy is lost as turbulence in or near the feed zone. The intensity of this turbulence can break friable particles, or create tight emulsions of two immiscible liquids, both making the separation which follows more difficult. Reducing the pond radius reduces the total process power and particle degrading and thus improves total separation performance. Reducing the pond inside radius by 25% reduces total power by 44%.

Materials of Construction and Operational Stress. Before a centrifugal separation device is chosen, the corrosive characteristics of the liquid and solids as well as the cleaning and sanitizing solutions must be determined. A wide variety of materials may be used. Most centrifuges are austenitic stainless steels;

however, many are made of ordinary steel, rubber or plastic coated steel, Monel, Hastelloy, titanium, duplex stainless steel, and others. The solvents present and of course the temperature environment must be considered in elastomers and plastics, including composites.

Once the material choice based on corrosion is made, a careful analysis of the stresses produced by rotation for the particular type of centrifuge is required, so that for the given liquid and solids specific gravities a maximum operating speed can be determined. In general, the metals used are ductile and elastic in the operating speed range. The usual limits are the ultimate strength and yield strength (0.2% offset). The stresses of the centrifuge bowl are primarily tangential stress and axial stress. These are the result of the weight of the bowl material itself, and the need to contain the rotating process solids and liquids within the bowl.

The geometry of the bowl parts is important, and for intermittently discharging centrifuges, fatigue strength must be considered. In general the tangential stress in a bowl wall is σ_T.

$$\sigma_T = \sigma_{self} + \sigma_p \tag{32}$$

where

$$\sigma_{self} = K \delta_M \omega^2 r^2 \tag{33}$$

$$\sigma_p = K \delta_p \omega^2 (r^2 - r_s^2) \tag{34}$$

where δ_M = mass density of bowl material; r = bowl radius; δ_p = mass density of process (δ_p is usually the wet solids, but can include heavy liquid phase); r_s = inside radius of the pond; and K is a function of bowl geometry and includes stress concentrations owing to changes in section, holes, slots, fillets, etc. The total stress increases with the square of bowl speed, ω^2, and the pond radius, and is directly proportional to the liquid and solid density.

Within the bowl, the pressure, p, developed also exerts axial separation forces. The internal pressure in a bowl is shown in equation 35:

$$p = \delta_p \omega^2 (r^2 - r_0^2)/2 \tag{35}$$

The axial projected area $A = \pi(r^2 - r_0^2)$ and the average pressure is $1/2\ p_{max}$. The axial force $F_A = p_{max} A/2$, where

$$F_A = \delta_p \omega^2 \pi (r^2 - r_0^2)^2 / 4 \tag{36}$$

The stress resulting from the axial forces must be considered in analyzing the parts which resist axial separation, such as bolts, nuts, rings, etc. On scroll centrifuges the axial force owing to the axial component of the conveyor torque must also be considered.

Typically the total stress on the bowl is 50–65% self-stress, and 35–50%

process stress. The axial stress is usually 5–10% of the tangential stress. When the bowl material density is low (such as titanium), or where solids are heavy (as is coal), or when the pond surface is close to the inside wall (as when using shallow pond scroll centrifuges), these proportions differ. All centrifuges must have a factor of safety so that the maximum stress at any point is less than the yield strength. The ratio of the material strength to the actual stress is the factor of safety. Factor of safety is set by the manufacturer, and sometimes, especially in Europe, by government regulation. Every centrifuge supplier sets the limits of bowl speed, temperature, and liquid and solids density. On some very high G preparative or zonal centrifuges, the number of cycles of bowl use must be recorded and the rotor retired after a given number, because fatigue determines the bowl life. Most disk and scroll centrifuges are made from forgings, centrifugal castings, or fabrications. Each manufacturer must carefully monitor the strength and ductile properties of the basic bowl material to ensure that the required factor of safety is maintained. A rotor failure during operation is very serious, and can not only destroy the centrifuge, but also damage nearby equipment and injure operators.

Many of the centrifugal separation applications are abrasive and erosive to centrifuge parts because of the high relative velocity or high contact pressure between the particles and part. There are many cases of erosion and corrosion wear. Areas of wear must be protected using materials which resist both mechanical and chemical attack. In general the main areas of abrasive wear are the feed zone surfaces, where nonrotating process slurries are accelerated to speed; the tips and faces of the conveyor flights on scroll centrifuges; solids discharge openings; screens, where solids slide across the screen; and stationary collection surfaces that receive the impact of solids being discharged from the bowl. Many materials are used. Examples are sintered tungsten carbide bound with cobalt or nickel; sintered aluminum oxide ceramic; sprayed and fused or weld-deposited hard-surfacing alloys; and elastomeric coatings or inserts. There is great variation in the design and application of these materials which effect the actual working life in service and the cost to rebuild the worn areas or replace the worn insert.

Maintenance operating costs are lower if replaceable wear protection is used, compared to the requirement to rebuild (and thus rebalance) the bowl materials. In very severe applications, such as coal (qv) or coal refuse dewatering that use screens, wedge wire screens made from ceramic or carbide materials are required. The use of carbide or ceramic flight-tip protection on continuous scroll centrifuges has permitted economic use in applications such as dewatering (qv) tar sands as well as coal and coal refuse, and dewatering and thickening mixed primary and secondary sewage sludges. Continuous nozzle discharge disk and scroll centrifuges would not be feasible without replaceable ceramic or carbide nozzle inserts. The use of relatively soft abrasion-resistant elastomers, such as urethanes, has been successfully applied in the solids receiver housing and the feed zone targets of scroll centrifuges. Hard chrome plating has been used on the first few disks of disk centrifuges. Stellite or carbide inserts are often used at the solids discharge opening.

Noise. Centrifuges, as do any rotating equipment, create noise. When the motion of air or gas entrained by a rotating bowl shell is deflected or otherwise disturbed, its energy is transferred to the environment through the casing or chutes. This mechanism suggests that the noise level created by the bowl is related to the surface linear speed of the rotor. High linear speed is important for maintaining separating capacity, so the noise level should be reduced without reducing the rotational speed, if possible.

In addition to surface speed, rotor imbalance, surface irregularities, clearance between covers and rotor, resonance in the supporting structure, conditions of installation, and particularly the drive motor contribute to centrifuge noise. An inadequate supporting platform can amplify centrifuge vibration. Open piping and venting and discharge connections allow noise generated inside the unit to escape. Discharge connections should be tight, yet flexible enough to prevent transfer of vibrational energy to plant piping. Size, spacing, materials of construction, and other properties of the centrifuge room also affect noise level. Sound-absorbing materials or enclosures should be provided when other means are inadequate.

The electric motors are often the noisiest component of the centrifuge assembly. Most standard motors in the 75–250 kW range develop noise levels of 85 dbA (weighted sound pressure level using filter A, per the ANSI standard). A quiet motor can reduce this level by 5 dbA and should be used whenever noise is of concern.

Equipment. Centrifugation equipment that separates by density difference is available in a variety of sizes and types and can be categorized by capacity range and the theoretical settling velocities of the particles normally handled. Centrifuges that separate by filtration produce drained solids and can be categorized by final moisture, drainage time, G, and physical characteristics of the system, such as particle size and liquid viscosity.

For optimum results, a combination of several types of equipment may be used, eg, a gravity separator for oil recovery from sludge at a petroleum refinery. The sludge, an aqueous suspension of 1–5% oil and 5–30% solids below 50 μm, is screened to remove trash, and degritted in a cyclone to eliminate the coarse solids that would cause excessive abrasion. An imperforate bowl conveyor-discharge centrifuge then removes 60–70% of the solids in oil-free condition. The resulting oil-in-water emulsion, stabilized by residual fine silt, is passed through a 0.25-mm (60-mesh) screen, and sent to a disk centrifuge that discharges an oil stream at 0.5–2% bottom sediment and water, and an oil-free peripheral nozzle discharge containing the remaining solids in water.

Equipment Materials and Abrasion Resistance. Stainless steel, especially Type 316, is the construction material of choice and can resist a variety of corrosive conditions and temperatures. Carbon steels are occasionally used. Rusting may, however, cause time-consuming maintenance and can damage mating locating surfaces, which increases the vibration and noise level. Titanium, Hastelloy, or high nickel alloys are used in special instances, at a considerable increase in capital cost.

Abrasion, a serious problem in some applications, requires the addition of hard-surfacing materials to points exposed to abrasive wear (12). The severity of wear depends on the nature, size, hardness, and shape of particles as well as the frequency of contact, the force exerted against the wearing parts, and solids loading as related to feed rate and solids concentration.

A wide range of abrasion-resistant materials is available. Nickel–chrome–boron and cobalt–chrome–tungsten hard-surfacing alloys have been used for many years. Composite coatings of nickel-base alloys containing crushed tungsten carbide particles, applied by flame spraying and fusing, are also used. Solid tungsten carbide, pressed to shape and sintered at high temperatures, provides the best protection. Tungsten carbide plates brazed or bonded to stainless steel supports, which in turn can be easily welded to portions of a centrifuge such as conveyor flights, have been very successful. Ceramics have been used where minor impact and abrasive particle pressures are involved.

Centrifuges

Manufacturers of sedimentation and filtration centrifuges can be found in Reference 13.

Sedimentation Equipment. Centrifugal sedimentation equipment is usually characterized by limiting flow rates and theoretical settling capabilities. Feed rates in industrial applications may be dictated by liquid handling capacities, separating capacities, or physical characteristics of the solids. Sedimentation equipment performance is illustrated in Figure 8 on the basis of nominal clarified effluent flow rates and the applicable Q_0/Σ values. The latter are equivalent to twice the theoretical gravity settling velocities. In liquid–solid separations, the effluent rate represents the clarified stream of the liquid medium and does not include the volume of solids discharged or the volume of medium discharged with the solids. The effluent rate of liquid–liquid separation refers to the clarified, heavy, or light continuous phase that usually occupies the greater volume within the separating equipment. The flow range for a particular piece of equipment does not represent its absolute limitations, but the normal flows for good clarification in standard applications. Similarly, large particles can always be sedimented.

As an additional guide, the Q_0/Σ values are correlated with the equivalent spherical particle diameter by Stokes' law, as in equation 1. A density difference $\Delta\delta$ of 1.0 g/cm^3 and a viscosity of 1 mPa·s(=cP) are assumed, thus conversion to other physical characteristics of the system requires that the particle size scale be adjusted to equate a particle of 1.0 μm diameter to its Q_0/Σ in cm/s, according to the relationship $Q_0/\Sigma = 10^{-7} \times 1.09\,\Delta\delta/\mu$, for $\Delta\delta$ in g/cm^3, and viscosity μ in Pa·s. For interpretation of the particle sizes, the scale refers to the 50% cutoff particle size, and under actual centrifugation conditions the value of Σ, determined from Figure 8, must be increased by efficiency factors to give the theoretical value of Σ.

Figure 8 serves as a guide to the types of equipment that can handle a given separation. Other characteristics further narrow selection. For example, for the separation at $Q_0 = 3.5 \times 10^{-3}$ m^3/s (50 gpm) of kaolin clay solids from an aqueous suspension, where the particle density is 2.55 g/cm^3 and the size ranges from 0.25 to 30 μm with 55% <2 μm, the 1.0-μm point on the particle size scale would be equivalent to $Q_0/\Sigma = 1.69 \times 10^{-4}$ cm/s. Assuming that a high recovery is desired, a disk centrifuge is required, and recovery of most particles greater than 0.4 μm is satisfactory, the Q_0/Σ equivalent to 0.4 μm on

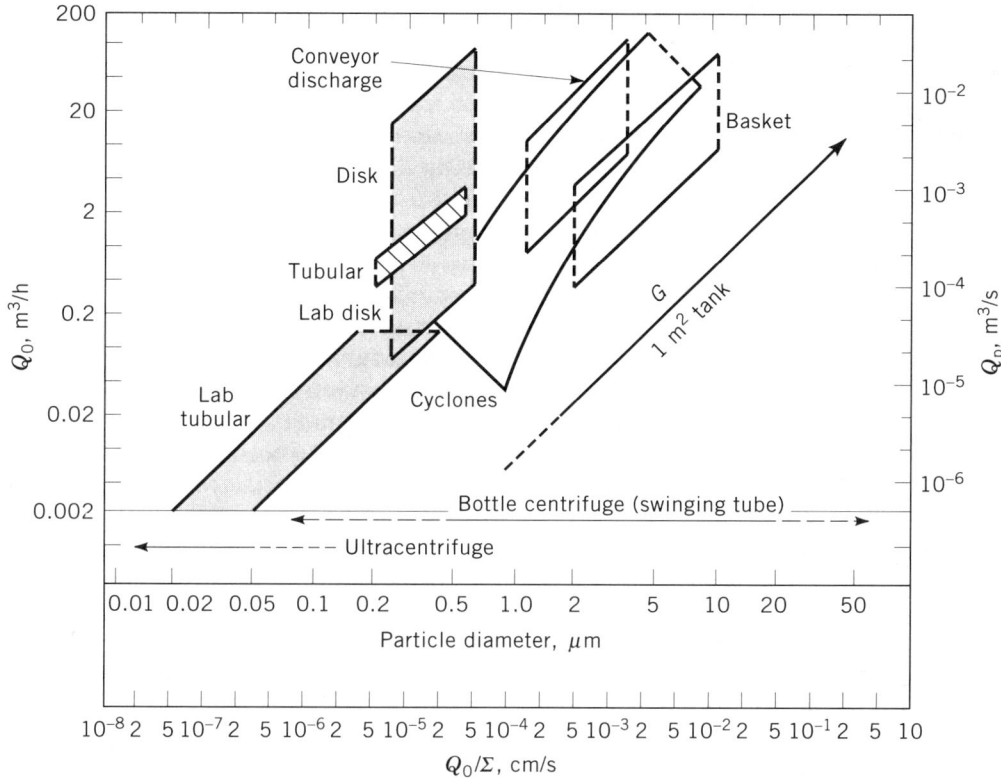

Fig. 8. Sedimentation equipment performance where the particles have a $\Delta\delta$ value of 1.0 g/cm³ and a viscosity, μ, value of 1 mPa·s(=cP). The value of Q_0/Σ is twice the settling velocity at $G = 1$, and Q = overflow discharge rate in measurements given.

the adjusted scale is about 2.3×10^{-5} cm/s. Using a disk machine efficiency of 40%, the centrifuge needed would have a Σ value of

$$\Sigma = \frac{3.15 \times 10^{-3} \times 10^6}{0.40 \times 2.3 \times 10^{-5}} = 34.3 \times 10^7 \text{ cm}^2$$

Because clay tends to pack hard, only the continuous nozzle discharge bowl would be satisfactory. Intermittently discharging disk centrifuges could not be used.

If classification of solids were desired, several other types of centrifuges could be used as well, assuming that only particles over about 2 μm were to be removed from the suspension and that the oversize stream should be highly concentrated for disposal. Figure 8 shows that a conveyor-discharge or basket centrifuge or standard cyclone could theoretically be used. Because cyclones cannot concentrate the oversize as much as the centrifuges, these are less satisfactory for this example. If the feed concentration were low, eg, less than a few percent solids, a basket centrifuge would be used with intermittent discharge of solids. A conveyor-discharge centrifuge gives almost as good a concentration of oversize as the basket and is a more efficient classifier. The conveyor-discharge centrifuge would thus be the better choice and is actually used in the kaolin industry.

In general, solids-retaining batch and batch automatic machines are limited to low feed concentrations to minimize the time required to unload the solids. Continuous disk centrifuges can have higher feed concentration. The limit is the underflow concentration. Conveyor discharge centrifuges can handle high feed concentration and are limited only by the volume of solids displacement, or torque capacity.

Using flocculent to create aggregated solids, varying degrees of deflocculation of the solid particles may occur during acceleration in the centrifugal field. An additional problem is the removal of the resulting soft, slimy cakes under centrifugal force. Scroll centrifuges have been successfully used to discharge soft, slippery solids continuously. Specially modified, automated basket centrifuges can handle a broad range of soft sludges, often without polymer addition. Disk centrifuges are particularly well-suited for clarification of streams containing solids such as aluminum hydroxide or waste-activated sludge.

The disk centrifuge having high capacity and G level is normally used for separating a liquid–liquid mixture or for clarification of such a mixture containing fine solids (14). Specially modified conveyor centrifuges are also used for three-phase separations. Settling velocity is a criterion of selection, but the actual separating of emulsified liquid–liquid mixtures may not strictly follow a settling theory. To break an emulsion, a threshold level of centrifugal force may have to be exceeded. In addition, drainage of liquid from the continuous-phase film of the emulsion has a time factor. Centrifugal force and time cannot be calculated interchangeably as Σ-theory would indicate. In centrifugation equipment, coalescence of the dispersed liquid occurs coincidentally with its separation because there is neither time nor space for appreciable interfacial retention of unbroken emulsion.

Batch equipment, such as the bottle centrifuge or ultracentrifuge, does not have a real throughput capacity. By increasing the time of operation, according to Figure 8, the smallest particle size of solids usually sedimented in the bottle centrifuge may be 0.1 μm, at which size Brownian movement controls. In the ultracentrifuge, separation can be achieved down to molecular size, perhaps 0.005 μm, where Brownian movement is controlling. There is clearly no limitation to the larger particles that may be settled in bottle centrifuges so that an arbitrary upper limit is indicated for practical minimum conditions of 1000 rpm and 10 s. Similarly, for the ultracentrifuge the upper limit for Q_0/Σ was estimated for minimum conditions of one hour and 5000 rpm.

Centrifugal Sedimentation Equipment. Commercial sedimentation centrifuges are characterized principally by how solids are discharged, and the general dryness of these solids. There are batch and automatic batch solid bowl machines which collect the solids at the bowl wall. Solids are removed very dry. Almost any solid is collectable, even those that are very soft and compressible.

Disk-type solid bowl machines are batch, batch automatic, and continuous. The solids are removed in many different ways, but are usually wet. Scroll centrifuges discharge solids continuously and usually drier than disk and imperforate batch types. Generally disk centrifuges have the highest values of Σ

or KQ for a given size and therefore the best ability to collect fine particles at a high rate.

Bottle Centrifuge. A bottle centrifuge is designed to handle small batches of material for laboratory separations, testing, and control. The basic structure is usually a motor-driven vertical spindle supporting various heads or rotors. A surrounding cover reduces windage, facilitates temperature control, and provides a safety shield. Accessories include timer, tachometer, and manual or automatic braking. Bench-top bottle centrifuges operate at 500–5000 rpm, producing centrifugal fields up to 3000 G in the lower speed range, and operate up to 20,000 rpm with 34,000 G in the high speed units. Larger models operate up to 6000 rpm and develop 8000 G, using special attachments that permit 40,000 G. These models may also be equipped with automatic temperature control down to $-10°C$ and other programmable controls to manage the cycle.

There are three types of rotors: swinging bucket, fixed-angle head, or small perforate or imperforate baskets for larger quantities of material. In the swinging bucket type, the bottles are vertical at rest but swing to a horizontal radial position during acceleration so that solids are deposited in a pellet at the bottom of the tube. Although sedimenting particles must travel up to the full depth of the liquid layer, which requires appreciable time, the long path of travel and the perpendicularity of the sedimenting boundary to the axis of the tube are distinct advantages in effecting fractional sedimentation. Heads carrying fixed tubes at a 35–50° angle reduce centrifuging time because the maximum distance traveled by a particle is the secant of the tube angle times the diameter of the tube. Particles strike the wall and slide down the tube to collect near the bottom, but the angle makes it difficult to measure relative volumes of supernatant liquid and sedimented solids. Rotors carry 2 to 16 metal containers having tubes and bottles of various sizes and shapes. Containers range in capacity from capillaries for microanalysis to a 1-L maximum, limiting the batch capacity of this type of centrifuge to 4 L. Although glass bottles and tubes are generally used, plastic and metal containers are available for high speed operation or corrosive liquids. Tubes are usually cylindrical, tapered, and graduated; special shapes for analytical work are available, including pear-shaped tubes having capillary tips for measuring small quantities of solids.

The bottle centrifuge is primarily used in the laboratory to separate small quantities of material. It is also used for standard analyses including many ASTM methods and in preliminary testing for scale-up to commercial centrifuges. The Σ_B value for free-swinging tubes is determined by equation 7 and the Q_0/Σ_B value by equation 8; Q_0/Σ_B data can be prepared by bottle centrifuge (5,15). The bottle centrifuge has been used to study waste treatment sludges for estimation of cake concentrations and feasibility of handling in a conveyor centrifuge (16). Because compaction of solids is largely a function of the centrifugal field force and the exposure time, the bottle centrifuge can also be used to study these parameters, although not always in the range applicable to industrial equipment. Similarly, drainage of packed solids can be studied by using tubes with fretted glass or perforated metal bottoms. Closed containers should be used to prevent drying by windage.

Specialty rotors permit ordinary bottle centrifuges to achieve some of the results previously considered possible only in ultracentrifuges. A modified zonal

276 CENTRIFUGAL SEPARATION

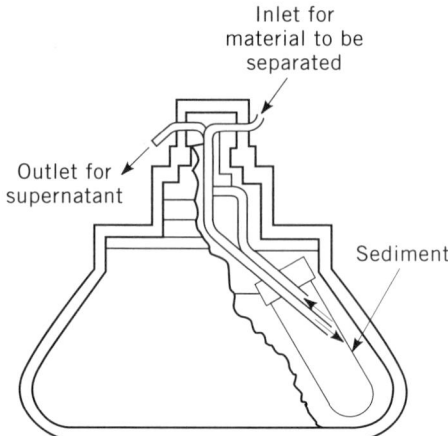

Fig. 9. Tube-type continuous-flow rotor. Courtesy of Sorvall.

rotor, shown in Figure 9, permits collection of sediment using continuous addition of feed and discharge of centrate.

Preparation Ultracentrifuge. Preparation ultracentrifuges are suitable for a range of applications, such as processing quantities of subcellular particles, viruses, and proteins (qv). Many design variations are available and only the common features are considered here. Preparation ultracentrifuges range in operating speed from 20,000 rpm, generating about 40,000 G, to 75,000 rpm and about 500,000 G. The rotor is surrounded by a high strength cylindrical casing and underdriven by an electric motor. To avoid overheating of the rotor by air friction at these speeds, the pressure in the casing is reduced to about 0.13 Pa (1 μm Hg). Sensors (qv) monitor the temperature and a cooling system controls the temperature in the range of −15 to 30°C within ±1°C. Electronic controls maintain the rotor speed within a required narrow range and may be automatically programmed for sequential changes in speed, including control of the acceleration and deceleration (17).

Preparation ultracentrifuges are guaranteed for several billion revolutions and can be rebuilt using relatively few parts. Among the great number of rotors available are batch rotors and those accepting feed and discharging centrate continuously during rotation. Batch rotors include angle and swinging-bucket types as well as those having vertical tubes parallel to the axis of rotation, which present a very short sedimenting distance and time requirement. Swinging-bucket rotors are also used for density-gradient separations or volume evaluation of the settled cake.

Separation by selective sedimentation on the basis of size and density of the particles may be satisfactory for polydisperse particle systems. However, the cake contains a range of material depending on its starting position in the container. Selectivity of separation can be improved by introducing the sample near the surface after the container is up to speed. Reslurrying and recentrifuging may be necessary to achieve purer fractions. Isopycnic separation improves initial separation efficiency where particles differ in density. If the density of the medium is intermediate to the range of densities of particles, higher density particles settle, whereas others remain suspended or rise regardless of size.

Zonal Centrifuge. The use of density gradients in centrifuge rotors greatly increases the sharpness of separations and the quantities of material that can be handled. In principle, the density gradient is established normal to the axis of rotation of the rotor and the highest density is located at the outer radius of the rotor. Low molecular weight solutes such as cesium chloride, sucrose, or potassium citrate, which are compatible with many systems in solution, are frequently used. A natural gradient may be formed by introducing a homogeneous solution and centrifuging for long periods of time. Continuous or step gradients may also be formed by introducing successive layers of solution, the composition of which varies continuously or stepwise from low to high density, where the latter displaces the former toward the center of the rotor (17–28) (Fig. 10).

In the simpler rotors using batch containers having swinging, angle, or vertical tubes, the gradient is introduced while the rotor is at rest and then accelerated to speed. The gradient shows relatively little mixing. Slowing the rotor gradually at the end of the run allows retention of the gradient and permits collection of the material banded isopycnically (Fig. 11).

More sophisticated rotors can be loaded with gradient and sample while rotating. When the batch is finished or the bands are sufficiently loaded with material, the bowl may be stopped slowly and the reoriented layers displaced under static conditions. Rotors may also be designed to establish gradients and isopycnic bands of sample and then be unloaded dynamically by introducing a dense solution near the edge of the rotor as shown in Figure 12.

Particles in the gradient may be separated on the basis of sedimentation rate; a sample introduced at the top of the preformed gradient settles according to density and size of particles, but the run is terminated before the heaviest particles reach the bottom of the tube. If the density of all the particles lies within

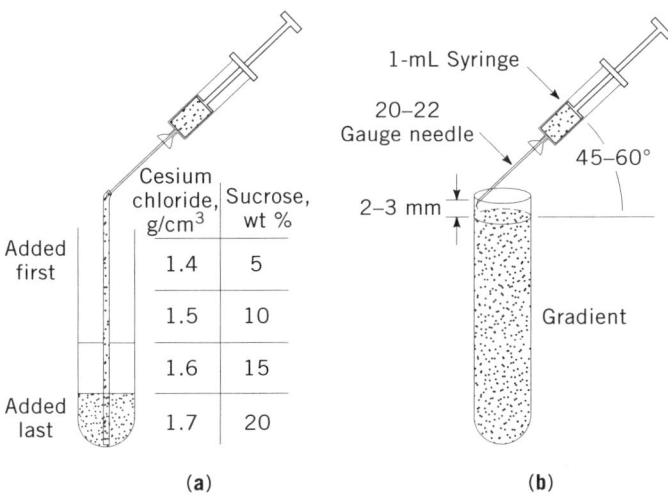

Fig. 10. (a) Forming a gradient and (b) applying the sample to the gradient before inserting tubes in a centrifuge rotor.

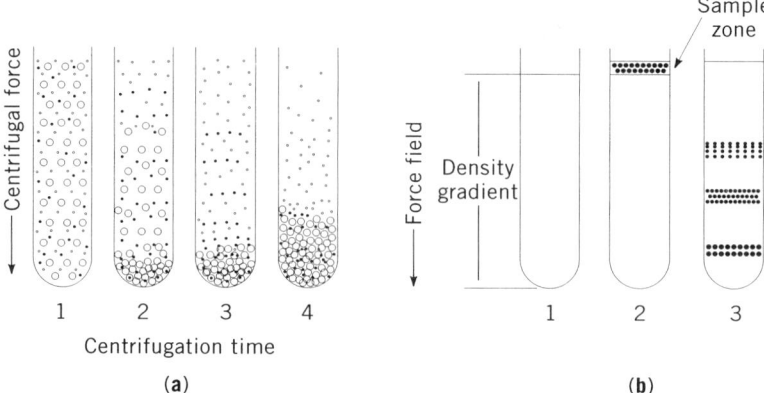

Fig. 11. (**a**) Differential centrifugation (pelleting), where time 1 < time 2 < time 3 < time 4. (**b**) Rate zonal separation in a swinging-bucket rotor, where tube 1 represents the density gradient solution, tube 2 the sample plus the gradient, and tube 3 the separation of sample particles under a centrifugal force, where the particles move at differing rates, depending on their mass.

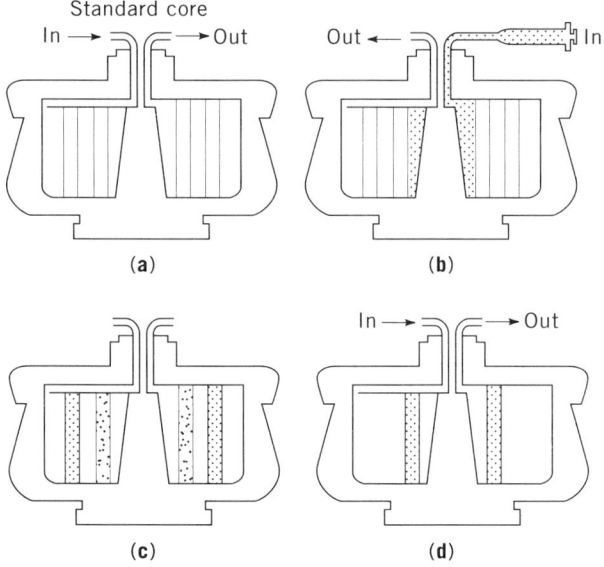

Fig. 12. Dynamic loading and unloading of a zonal rotor. (**a**) Gradient is loaded while rotor is spinning at 2000 rpm; (**b**) a sample is injected at 2000 rpm, followed by injection of overlay; (**c**) particles separated when the rotor is running at speed; and (**d**) contents are unloaded by introducing a dense solution at the rotor edge, displacing fractions at the center.

the range of the density limits of the gradient, and the run is not terminated until all particles have reached an equilibrium position in the density field, equilibrium separation takes place. The steepness of the gradient can be varied to match the breadth of particle densities in the sample.

Rotors are made of titanium or aluminum and may be cylindrical or bowl-shaped (see Fig. 12). Larger bowls reach 100,000 G; smaller units reach

250,000 G. The tubular rotors permit feed rates up to 60 L/h at 150,000 G or 120 L/h in a larger unit at 90,000 G. Such centrifuges may be used to separate relatively large quantities of viral material from larger quantities of cellular and subcellular matter, as, for example, in the production of vaccines (see SIZE SEPARATION).

Tubular Centrifuges. Tubular centrifuges (Fig. 13) separate liquid–liquid mixtures or clarify liquid–solid mixtures having less than 1% solids content and fine particles. Liquid is discharged continuously, whereas solids are removed manually when sufficient bowl cake has accumulated. For industrial use, the cylindrical bowls are 100–180 mm in diameter with length-to-diameter ratios ranging from 4–8. Bowl speeds up to 17,000 rpm generate centrifugal accelerations up to 20,000 G at the bowl wall (14). Because of the small bowl diameter, however, Σ_T values according to equation 11 result in flow rates in the range of 0.1–4 m^3/h (0.5–16 gpm). The tubular centrifuge handles low to medium flows and theoretical particle settling velocities in the range of 5×10^{-6} to 5×10^{-5} cm/s. Designs for up to 101.3 kPa (1 atm) and 300°C are available. Clean in place (CIP) and sterilize in place (SIP) are available. Drive motors of 1.5–7.5 kW are used.

The laboratory tubular centrifuge is similar to the industrial model. It operates with a motor or turbine drive at speeds to 50,000 rpm, generating 65,000 G at the latter speed in the 4.5 cm diameter bowl. The nominal capacity range is 30–2400 cm^3/min. This centrifuge is uniquely capable of separating far finer particles than any other production centrifugation equipment except the bottle centrifuge. It is widely used in the production of flu virus.

A long, hollow, cylindrical bowl is suspended by a flexible spindle and driven from the top as shown in Figure 13. Axial ribs in the bowl ensure full acceleration of the liquid during its short time in the bowl. Feed is jetted into the bottom of the bowl and clarified liquid overflows at the top, leaving deposited solids as compacted cake on the bowl wall. The clarifying performance of the bowl is reduced as the deposited cake decreases the effective outer radius of the bowl in accordance with equation 11. Consequently, cake capacity of the industrial model is limited to 0.1–10 L. For liquid–liquid separation, the interface position (eq. 26) is determined by selection of ring-dam diameter or by the length of a hollow nozzle-type screw dam.

The tubular centrifuge was long used for the purification of contaminated lubricating oils because of the high centrifugal force developed and the simplicity of its operation. Colloidal carbon and moisture are removed from transformer oils to maintain dielectric strength; carbon and acid sludges are removed from diesel engine lubricating oils; and water and solid contaminants are removed from steam turbine lubricating oil. Polishing operations include the removal of small quantities of solids in the clarification of varnish, cider, fruit juices (qv), and even highly viscous chicle. In vegetable oil refining, oil losses in the semisolid soap stock are kept low by compaction of the soap phase under high centrifugal force. Automatic disk centrifuges which do not require manual solids unloading have largely replaced the batch-operating tubular. The laboratory tubular centrifuge is used to recover fine solids in batch preparations too large for bottle centrifuge separation, to estimate scale-up rates in larger centrifuges, and to analyze particle size distributions involving settling rates too low for

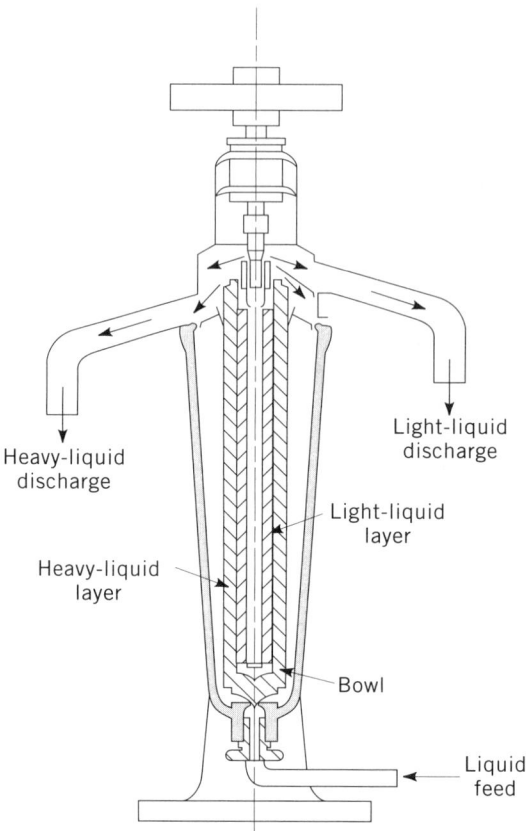

Fig. 13. A tubular centrifuge. Courtesy of Alfa Laval.

feasible gravity sedimentation (19). Modern machines having variable-frequency drives are available (14).

Disk Centrifuge. Centrifuges that channel feed through a large number of conical disks to facilitate separation combine high flow rates with high theoretical capacity factors (see Fig. 8). For industrial units flow rates up to 250 m^3/h (1100 gpm) can be obtained on easy separations, and theoretical settling velocities may range from 8×10^{-6} to about 5×10^{-5} cm/s. Both liquid–liquid and liquid–solid separations are performed using feed solids concentration below 15% and small particle sizes. As seen from equation 15, the theoretical capacity factor depends on the number of disks, which is limited by the height of the disk stack. The performance is proportional to the cube of the disk diameter.

Several of the assumptions in the development of Σ_D do not apply in practice (20), and mathematical representation of the actual flow pattern has been difficult to achieve. Computer studies of the flow to and between the disks, as well as experimental analysis of flow patterns within the disk stack, have improved the effectiveness of disks. Not all disk machine designs can be compared using Σ_D alone. Details such as type and size disk spacer, number of spacers, and location of the feed holes with respect to the spacers and solids discharging ports are very important. Also the method of feed acceleration is especially important in liquid–liquid separations and in the presence of fragile solids. If the disk centrifuge design is geometrically similar, scaling up by Σ_D from one speed and size of stack to another is reasonably accurate.

The outstanding feature of the disk bowl design is a stack of thin cones, commonly referred to as disks, which are separated by thin spacers. These are so arranged that the mixture to be clarified must pass through the disk stack before discharge. The resulting stratification of the liquid medium greatly reduces the sedimenting distance required before a particle reaches a solid surface and can be considered removed from the process stream. The angle of the cones to the axis of rotation is great enough to ensure that solid particles deposited on the surfaces slide, either individually, or as a concentrated phase according to the difference between their density and that of the medium.

The general flow patterns for a liquid–liquid separator and a recycle clarifier, respectively, are illustrated in Figure 14. Feed enters near the center of the bowl from either the top or the bottom, depending on the support, and is accelerated by vanes to the radius at which it enters the disk stack. When the disk stack is used for one phase, as in clarification or classification, the feed is distributed to the stock through the zone between the outer edges of the disks and the bowl wall. The clarified medium is discharged at a relatively small radius, generally at the top of the bowl. For a liquid–liquid separation, with or without solids, feed is distributed by a number of feed channels. The interface of the two coalesced liquid phases is located at the disk feed holes, by appropriate selection (eq. 26) of the heavy- and light-phase discharge radii. During handling of the two liquids, the heavier moves toward the edge of the disks and the lighter moves inward. Separate channels in the bowl and separate cover compartments segregate the discharges.

Solids in either phase are sedimented to the underside of the disks and slide outward along the surfaces because of their density. The aggregated solids must move by free settling from the outer edges of the disks to the bowl wall; some may be reentrained into new feed material, and carried into the disk stack, which accounts in part for actual performance falling short of theoretical prediction.

Commonly used with disk centrifuges are centripetal pumps (Fig. 15) that discharge the clarified liquid phases under pressures up to 0.7 MPa (100 psi) at reduced aeration, and scoop the rotating liquid out by using a stationary impeller. Interface location can be altered by varying the centripetal pump-back pressure, allowing interface control without shutdown to change a discharge weir. Centripetal pumps are capable of discharging at rates to 250 m^3/h (1100 gpm), and often eliminate the need for a tank and conventional pump.

To maximize cake capacity, the simplest disk centrifuge bowl (Fig. 15a) is designed having a nonperforate bowl wall parallel to the axis. Feed solids should not exceed 0.5%. Bowl diameters of industrial units range from 180–600 mm with operating speeds from 8000 to 4500 rpm; the disks, between 30 and ≥200, are stacked at spacings of 0.4–6 mm; the half-angle is frequently 35–40° because solids handling is not critical. This type of centrifuge was originally employed for the separation of cream from milk and is still used widely in this field. Other uses include purification of fuel and lubricating oils having a low percentage of solids, separation of wash water from fats and vegetable or fish oils, and removal of moisture and solids from jet fuel. Solids which move readily in plastic flow can

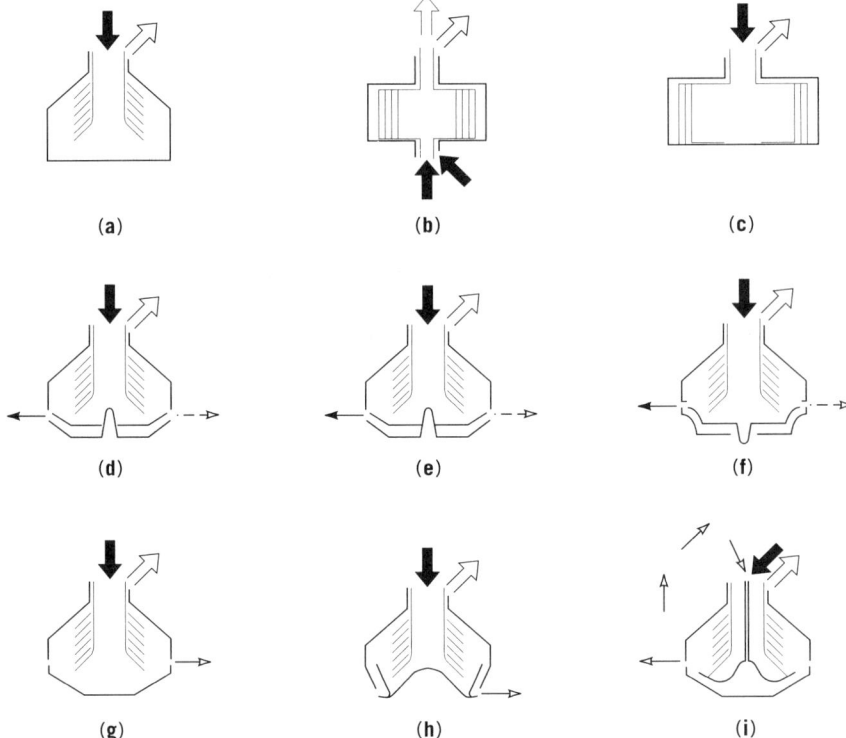

Fig. 14. Disk centrifuge bowls, where bowl diameters range from 10–90 cm; feed flow rates from 0.06×10^{-3} to 38×10^{-3} m^3/s (1–600 gpm); and operating temperatures from 10–90°C, and (➡) represents the main inlet, (⇒) the main outlet, (→) continuous flow solids, (⋯→) intermittent flow solids, and (→) auxiliary liquid. (**a–c**) Solid wall bowl: base design, spiral cylinder inserts, and cylinder inserts, respectively; (**d–f**) bowl for intermittent solids discharges: base design, radial peripheral parts, and shoe; peripheral parts and shoe with nozzles for both continual and intermittent solids discharge; and axial peripheral parts and shoe, respectively; and (**g–i**) bowl with nozzles: base design and peripheral nozzles, nozzles at reduced diameter, and peripheral nozzles and solids circulation, respectively.

be continuously discharged as the heavy phase, for example, in the separation of soap stock from oil in vegetable oil refining.

Continuous discharge of solids, as a slurry, is achieved by sloping the inner walls of the bowl toward a peripheral zone containing between 8–24 orifices, commonly called nozzles, as shown in Figure 14**g–i**. The nozzles must be spaced closely enough so that the natural angle of repose of the solids deposited between the nozzles does not cause a buildup of cake to reach into the disk stack and interfere with clarification. The size of the nozzles is limited because the fluid pressure at the wall, which can be 6.9–13.8 MPa (1000–2000 psi), produces high nozzle velocities (see eq. 30). On the other hand, the nozzles must be large enough to prevent obstruction by individual particles; nozzle diameters at least four times the size of the largest particle are satisfactory. Replaceable, wear-resistant bushings in the nozzles provide orifices in the range of 0.8–2.5 mm. From 5–50% of the feed may be discharged with the solids

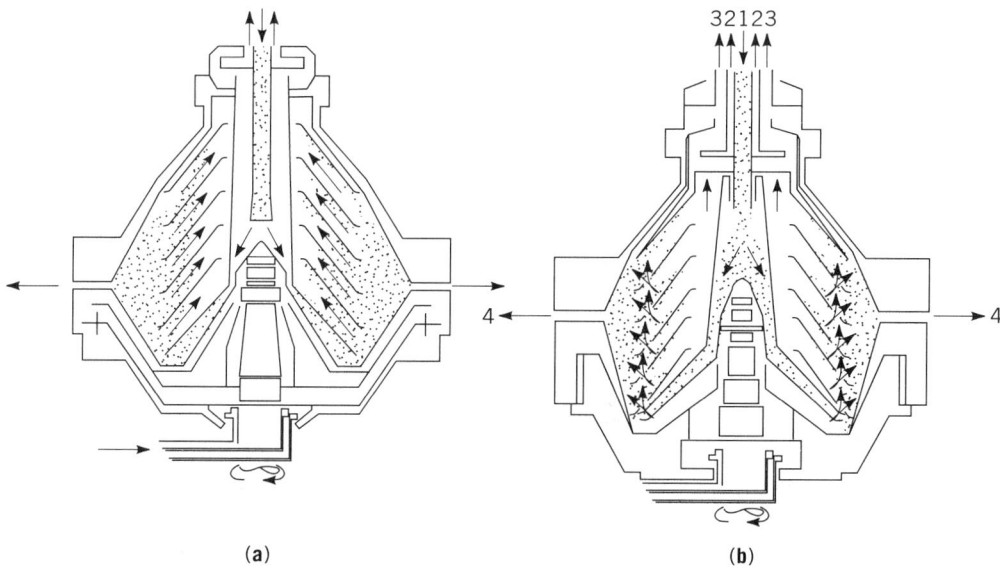

Fig. 15. (a) A clarifier and (b) a purifier of the paring disk-type design, where intermittent discharges of solids are designated by →, and 1 represents feed; 2, light phase; 3, heavy phase; and 4, solids.

through the nozzles. The upper limit of solids concentration depends on the particle packing characteristics but seldom exceeds 20 times that in the feed. The nozzle flow is directed backward with respect to rotation to recover much of the pressure energy and to reduce centrifuge power (see eq. 29). Bowl diameters range from 100 mm in laboratory units to 900 mm in industrial units and speeds from 12,000–3000 rpm, respectively; power requirements are between 10–150 kW. Centrifugal accelerations to 12,000 G are obtained at the wall in the smaller bowls. Feed rates range from 1–136 m^3/h (4–600 gpm).

Applications include kaolin clay dewatering, separation of fish oils from press liquor, starch and gluten concentration, clarification of wet-process phosphoric acid, tar sands, and concentrations of yeast, bacteria, and fungi from growth media in protein synthesis (14).

A variant of the continuous-discharge disk centrifuge provides for introduction of a recycle stream. Restrictions on number and size of nozzles sometimes prevent adequate concentration of the discharged solids to satisfy further process requirements. To increase this concentration, a portion of the discharged slurry is returned to the feed, thereby increasing the overall loading of solids in the bowl. A more efficient method, shown in Figure 14**i**, is to return some discharged slurry through a recycle system to the region of the nozzles where the higher density recycle stream preferentially joins the nozzle flow.

A further modification of the disk centrifuge provides peripheral ports that are opened only intermittently to discharge sludge (see Fig. 14**d**–**f**, and Fig. 15). This type is employed for medium quantities of solids (1–4%) for which neither continuous discharge nor batch operation is suitable, and for solids that break down under the shear forces of nozzle discharge and are therefore not suitable for nozzle recycle. Intermittent discharge provides longer holding time and better concentration of solids, often at the expense of decreased disk size and reduced bowl throughput. The frequency and duration of opening can be controlled to discharge very high concentrations of sludge by partial emptying of the bowl.

These centrifuges are available in 180-mm bowl laboratory and 460–600-mm bowl industrial units. Disks are mostly spaced at 0.6–1.0 mm and the half-angle is 40–45°. In the industrial units, bowls with 60–200 or more disks operate at maximum speeds of 4400–6500 rpm and require 10–50 kW. Feed flow rates range from 2.5–114 m³/h (10–500 gpm); temperatures from 30–90°C. Theoretical capacity factors are generally lower than in continuous discharge bowls of the same size. The disk outside diameter is smaller in order to provide solids-holding space of 4–20 L. Applications are limited to free-flowing solids that do not pack, and include recovery of wool grease from wool scouring liquor, orange juice clarification, recovery of soya protein, clarification of animal fats and food extracts, and purification of marine and jet engine fuels and lube oils.

Other modifications have special but more limited applications. A centrifugal bowl may contain, instead of disks, several annular baffles that take the liquid through a labyrinth path before discharge. The multiple cylinders increase cake capacity to as much as 70 L for easily sedimented solids. This centrifuge is used for clarification of food syrups and antibiotics, and for recovery of heavy metallic salts and catalysts (see Fig. 14c).

Continuous Conveyor Discharge Centrifuges. Imperforate bowl conveyor-discharge centrifuges collect solids by sedimentation and continuously discharge both liquid and solid material. These centrifuges have bowl diameters of 150–1400 mm and are essentially tubular shells with a length-to-diameter ratio of 1.5–5.2, as shown in Figure 16. Deposited solids are moved by a helical screw conveyor operating at a differential speed of 0.5–100 rpm with respect to the bowl. Centrifugal fields are lower than in disk or tubular centrifuges because of the conveyor and its associated mechanism. Maximum speeds range from 300–9000 rpm. Figure 8 shows that particles of intermediate settling velocities, such as 1.5 to about 15 × 10⁻⁴ cm/s, are handled at medium to large flow rates. For clarification, this type of centrifuge recovers medium and coarse particles from feeds at high or low solids concentration. Particle sizes less than about 2 μm are normally not collected without the addition of flocculating agents. For classification of solids, the flow rates are higher than for clarification and the overflow usually contains most of the finer solids. Feed flow rates range from 1 to 136 m³/h (4–600 gpm).

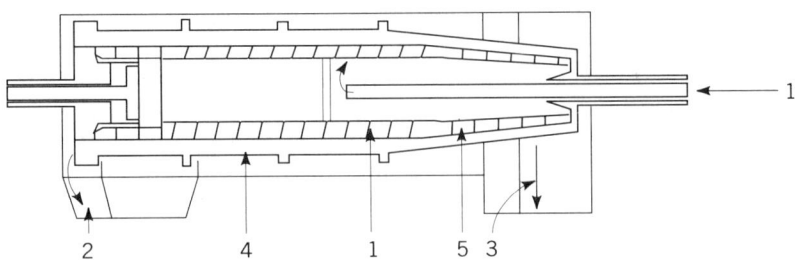

Fig. 16. An imperforate bowl conveyor-discharge centrifuge, where 1 corresponds to feed suspension; 2, to liquid phase; 3, to solid phase; 4, to liquid pool; and 5, to dry beach.

Discharged solids are not as dry as those obtained by centrifugal filtration. Coarse crystals may discharge at 2–10% moisture, ground limestone at 15–20%, and kaolin clay in the filler range (1–10 μm) at 30–35%. Compressible, amorphous, and fibrous materials, such as sewage sludges, can be dewatered to 60–75% moisture, and meat rendering solids at 60–70% moisture plus 6–8% liquid fat. Operating pressures up to 1.03 MPa (150 psi) and temperatures up to 200°C are standard and 300°C is available.

Feed is introduced through a stationary axial tube. Solids adhering as sediment to the bowl wall are removed from the liquid by a conveyor, moved up a sloping beach, and usually discharged at a radius smaller than that of the liquid discharge. Fine and flocculent solids compact under the liquid and show relatively little drainage on the beach. Coarse crystals and fibers do drain on the beach to a low residual moisture. The liquid level in the bowl is maintained by ports adjustable to the desired overflow radius. Considerable variation in design is available for the bowl shell, flight angle and pitch, beach angle and length, conveyor speed, feed position and type, and patterns of liquid and solids movement through the bowl. Bowl shapes having a high l/d exhibit high clarification capacity but may have wetter solids owing to the greater amount of fine particles in the solids discharge.

Bowl designs include countercurrent or concurrent movement of the phases. In countercurrent flow, feed enters near the conical–cylindrical intersection; liquid flows toward the far end of the bowl to discharge over dams; and deposited solids move toward the cake end with the conveyor. In concurrent flow, feed enters at the end away from the cake discharge, liquid and settled solids move in the same axial direction toward the cake end, axial conduits or a skimming device remove centrate from the pond surface, and solids move up the beach. The actual liquid flow between the helical flights is more tangential than axial so the concurrent/countercurrent description is not particularly significant. More important is the problem of introducing the feed stream exactly where the deposited solids are deepest, and the resulting reslurrying in countercurrent designs. Also there is the problem of discharging clarified centrate just where the solids are deepest, and the resulting reslurrying in concurrent designs. Care of feed introduction in countercurrent, and centrate discharge in concurrent, are important factors which determine separation efficiency. Plugging of return tubes and maintenance of the feed/centrate seal are problems unique to concurrent designs. In either type, solutions of flocculating agents may be introduced in the piping before entering the centrifuge, either in the feed tube, the feed acceleration zone, or in the pond after feed acceleration. Owing to the complexity of the chemistry of polyelectrolyte flocculation rate and efficiency, and floc damage owing to turbulence, the best solution for optimum polyelectrolyte use must be determined experimentally. A wash can be applied to the solids on the beach, but efficient rinsing depends on careful design to direct the rinse liquid to the solids. The wash is not collected separately but is discharged with the mother liquor.

A vertical decanter design using a bowl elastically suspended from a spindle with no bottom bearing is often used for high pressure and high temperature applications instead of horizontal axis decanters. Vertical units are easier to seal for pressure operation, and are well suited to accommodate bowl expansion at high temperatures. A vertical, elastically supported rotor as compared to a horizontal design does not offer separation advantages. There are mechanical advantages, however: the vertical design uses only one seal to three in the horizontal design. Moreover, the process connections are rigid (vertical) rather than flexible (horizontal), thermal expansion is not critical, and noise is lower.

Although Σ_T (eq. 11) indicates a reduced sedimentation performance level at increased liquid depth, this occurs only for coarse solids. The optimum pond depth varies with feed zone design, tendency of deposited solids to redisperse, conveyor differential speed, and particularly the depth of the cake layer required to produce a given solids concentration. Deep ponds are generally more effective for soft, slimy solids because conveying problems may reduce performance level and prevent complete clarification, even at low flow rates. The difficulty with which a scroll centrifuge is able to discharge soft, slippery solids has limited use in the past. To move the solids up the beach the solids must remain at bowl speed, with the helical flights moving the solids inward, against the high G. Often the soft solids slide back into the pond, building up in the bowl and ruining sedimentation performance. Setting the pond level inward of the solids discharge radius and separating the feed portion of the pond from the cone end using a deeply immersed baffle permits the centrifugal pressure head developed to assist the conveyor in discharging the solids. This method is used to thicken secondary sewage sludge from feed, ie, solids concentrations of 0.5–1% to solids concentrations of 5–10%. This thickening reduces total flow to the next stage of treatment by an order of magnitude. Flow rates up to 100 m³/h (440 gpm) have been achieved without the use of polyelectrolytes. A small amount of polyelectrolytes, however, can double this rate.

Dewatering sewage sludge to high solids levels has been achieved by the use of high (5–25 kg/t) polyelectrolyte dosage, increased (2000–4000 G) G level, and longer solids residence time in the bowl. Longer residence has been achieved by mechanically restricting the solids near or at the beach and reducing the differential speed to 0.5–4 rpm. The torque between the bowl and conveyor is a good measure of solids dryness, and use of this factor has permitted automatic differential speed control to maintain a constant solid dryness at varying feed concentration and rate. Differential speed control can be obtained by driving the conveyor using a hydraulic motor mounted on the bowl and using an external hydraulic power supply to deliver variable rates of high pressure oil to the rotor by means of a rotary union. Control is also achieved by use of a planetary gearbox mounted on the bowl, the gear box receiving controlled torque, and speed from a variable speed motor or electrical brake. All methods are programmable using microprocessors. Higher torque differential speed controllers have permitted centrifuge capacity and solids dryness to achieve levels of 100 m³/h (440 gpm) at 35–40 wt % solids.

Final dewatering of sewage, especially anaerobically digested sludge, results from the solids almost filling the interior space between the conveyor hub and bowl shell interior. The use of polymers as flocculating agents results in easy separation in spite of a greatly reduced residence time for the centrate. The polymer also conditions the solids so that the pressure resulting from the deep layer of solids and the force needed to move the solids axially and inwardly toward discharge compresses and squeezes additional water from the space between and within the flocs. A drier discharged cake results. Compatibility studies (29) can be made to characterize particular solids. Improvements in dryness is proportional to $G^{1.5}t$, where G is the level at the mean solids radius and t is the solids residence time within the centrifuge. On a scroll decanter having an $l/d = 4.2$, the solids have been found to occupy 75–80% of the bowl. The centrate clarification occurs just before discharging.

Theoretical studies (30) comparing the ability to dewater compressible solids by sedimenting and filtering centrifuges to pressure filters, have shown that at high G levels, scroll decanters produce drier cakes than pressure filtration.

Scroll centrifuges, which include perforated screens, are used to dewater coarser solids to a very high degree. Feed slurry is introduced conventionally into an imperforate scroll machine. The dewatered solids are deposited onto a cylindrical screen with an inside diameter equal to the solids discharge diameter. The conveyor moves the solids axially over the bar screen area where further dewatering and washing takes place. The centrate from the screen is recycled to the feed to further recover fines while the solids pass out of the solids discharge. Typical screen decanter applications are fine coal dewatering and production of purified parazylene (12).

The capacity of decanters can be limited by any one of several factors (31): centrate clarity, usually a function of Σ; solids dryness, imposed by requirements of the next process step; conveyor torque, limited by rating of gear box, hydraulic motor, or back-drive capacity; chatter, by torsional instability resulting from stick-slip action of conveyor flights on fusible solids; swallowing capacity, by the ability of conveyor to accept feed without rejection; power, by the rating of drive motor; solids purity, where purity is obtained by rinsing; erosion rate, where abrasive particles are present; size purity, for particle classifications; solids volume, where solids feed concentration is high; fused solids, rate at which pasties are formed; and control, the maximum rate at which process control can be maintained.

Special designs have been offered which include one or more of the following features in vertical or horizontal construction: vapor-tight and pressure-tight enclosures; clean in place (CIP) and sterilize in place (SIP) rotors; three-phase separations having two immiscible liquids and solids; sanitary construction; a wide range of helices, pitches and leads, beach angles, compound beach angles, l/d ratios, pond depth/d ratios, and ribbed or grooved beaches or bowls; centrate centripetal pump devices; and abrasion protection systems, including complex flight tip geometries of ceramic, carbide, and conventional hard-surfacing.

In clarifying operations, the conveyor discharge centrifuge recovers many types of crystals, meal from fish press liquor, and polymers, such as poly(vinyl chloride) and polyolefins. It is also used to dewater coal (24) and to concentrate solids from flue gas desulfurization sludges. Vertical designs, vapor-tight or under pressure, are applied to terephthalic acid, polypropylene, and catalyst recovery. Classification includes separation of particles over 2 μm from kaolin coating clay, and of particles over about 5 μm in the mill discharge of ground TiO_2; selective recovery of calcium carbonate from lime-treated waste sludges permits calcining and recycling of the lime without an overwhelming recycle load of inert material. The conveyor centrifuge is frequently used to rough out medium and coarse solids before a second separation stage such as a disk centrifuge handling refinery sludges.

Centrifugal Filtration Equipment. The important parameters of centrifugal filtration equipment (4,32) are screen area, level of centrifugal acceleration in the final drainage zone, and cake thickness. The latter affects both residence time and volumetric throughput rate. As indicated by equation 19, the particle size of the solids and the kinematic viscosity of the mother liquor also strongly affect the final moisture content. A limited correlation has been developed for the performance of perforate basket and conveyor discharge conical screen bowls, but the range of materials handled and the complexity of the drainage and washing operations do not lend themselves to broad correlations. The variables of correlation may be useful in a particular study, especially if more than one type of centrifuge is involved. An example of centrifugal filtration is the recovery of salt crystals from a mother liquor of about 3×10^{-6} m²/s (3 cSt) viscosity; the crystal size is 170 μm at the 15% level and drained cake bulk density about 1.5×10^3 kg/m³ (95 lb/ft³). The cone screen is 25.4 cm at the larger diameter and the automatic cycle basket centrifuge is 68.6 cm in diameter; other parameters for final values of $q = 7.6\%$ in both centrifuges are given in Table 1.

A conveniently expressed coordinate for plotting filtration performance is the drainage number, $\overline{d}(G)^{1/2}/\nu$, where $\overline{d}$ is the mean particle diameter in micrometers, ν is the kinematic viscosity of the mother liquor in m²/s (Stokes $\times 10^{-4}$) at the drainage temperature, and G is $\omega^2 r/g$; r is the largest screen radius in a conical bowl. Because the final moisture content of a cake is closely related to the finest 10–15% fraction of the solids and is almost independent

Table 1. Centrifuge Parameter Values for $q = 7.6\%$

Properties	Centrifuge	
	Cone screen	Automatic basket
G	1440	865
bowl		
diameter, cm	25.4	68.6
speed, rpm	3180	1500
$\overline{d}(G)^{1/2}/\nu$	3.750×10^6	2.880×10^6
time, s		
feed and spin		12
spin after rinse		27
drain time	0.6[a]	39
rinse[b]		10
unload and screen rinse		4
cycle time		53
$qt^{1/2}$, %·s$^{1/2}$	5.89	47.46
h, cm	0.51	3.81
screen area, cm²	930	7675
approximate cake rate, kg/s	1.19	0.84

[a]Differential speed = 60 rpm, 5/8 turn per helical flight.
[b]Rinse is shown only in the basket centrifuge because rinse efficiency in the cone screen is fairly low compared to the basket. The latter may be selected for the application if good rinsing is needed, and the conveyor-discharge conical screen centrifuge selected if no rinsing is necessary.

of the coarser material, it is suggested that $\overline{d}$ be used at the 15% cumulative weight level of the particle size distribution instead of the usual 50% point.

The other coordinate is $qt^{1/2}$, where q is the percentage of final moisture on the discharged cake, as the volume of mother liquor per unit volume of solids, and t is the drainage time in seconds. A weight ratio may be used for q, but the volume ratio makes the function more universal by eliminating densities. For a conveyor discharge conical screen bowl, the helical conveyor is assumed to control the residence time of the solids, so that time becomes the number of turns of helical flight around the conveyor hub divided by the differential speed between the conveyor and the bowl. For a pusher centrifuge, t is the retention time on the screen as controlled by length and diameter of screen, thickness of cake, and frequency and length of stroke of the cake. In a basket centrifuge, dead and unload times should not be included in the calculation of drain time. Bulk drainage is completed so quickly that film drainage is usually controlling. Thus, drain time t is approximated by the sum of feeding time, spin time prior to rinsing, and spin time after rinsing but prior to unloading according to the theory of cyclical centrifuges (33). Filtration correlations generally show a spread as a function of cake thickness. Conical screen bowls characteristically have short residence times and achieve good drainage by maintaining thin layers of cake. Smaller perforate baskets operate having cakes 5–10 cm in thickness, whereas larger baskets may carry cakes up to 15-cm thick. Pusher centrifuges and high speed peeler baskets may handle cakes ranging in thickness from 5–20 cm.

Perforate Basket Centrifuges. The simplest and most common form of centrifugal filter is a perforate-wall basket centrifuge, consisting of a cylindrical bowl having a diameter ranging from about 100–2400 mm and a diameter-to-height ratio ranging from 1–3. The wall is perforated with a large number of holes, more than adequate for the drainage of most liquid loads, and is lined with a filter medium. In the simplest case, the medium is a single layer of fabric or metal cloth or screen. In high speed basket centrifuges, one or more backup screens of relatively large mesh support a finer mesh filter surface. The method of discharging accumulated solids distinguishes three types of basket centrifuge: those that are stopped for discharge, those that are decelerated to a very low speed for discharge, and those that discharge at full speed (34).

Basket centrifuges that must be stopped for discharge are available in many sizes. The bowls are usually supported on a vertical spindle. Designs vary from a 300-mm diameter basket of 30-L cake capacity (0.4-kW motor, 2100 rpm, and about 800 G) to a 1500-mm diameter basket having 0.5-m³ cake capacity (14-kW motor, 600 rpm, and 300 G). Basket cake volumes are always nominal and must be modified according to cake density. Construction materials include carbon or stainless steel, Monel, Inconel, titanium, and a variety of rubber and plastic coatings. Normal operation includes pressures up to 35 kPa (5 psi) and temperatures up to 180°C. This type of centrifuge is used if a variety of materials must be filtered in small batches, if equipment must be sterilized between batches, or if the production rate is too low to warrant more automation. These centrifuges are also used in removing liquid from crystalline materials, in drying bulk materials such as raw leafy vegetables, and in clarifying process liquids and waste streams. Cycle times are above 10 min. Large (dia = 1300–1500 mm) baskets operating at 700–800 rpm may use an inner perforated container which mounts in the bowl but is removable for bulk loading outside the centrifuge. Such units are well suited to laundry and dyeing purposes and for dewatering of textiles, yarn, raw stock, feathers, and hair. Particle sizes range from very fine to 500 μm. Feed flow rates vary from 1–20 m^3/h (4.5–90 gpm). A syphon to control the rate of filtrate removal improves dewatering (35).

Improved control systems and rising labor costs have led to manually or automatically controlled cycles with mechanical unloading at reduced speed. Baskets typically load at low to medium speed, accelerate to 900–1800 rpm for drainage and washing, decelerate to 35–75 rpm for mechanical unloading, and then start the cycle again. Cycle times range from 2–6 min. Cycles of 30 min for slow drainage or multiple rinses are common. Both a top-driven suspended bowl and underdriven bowl with three-point casing suspension designs are available. The cake is discharged in 20–120 s by means of a single or multiple plow that leaves a heel of cake on the filter medium. The heel can be completely removed with the help of a plastic-tipped plow and a perforated protecting plate. Filter media vary from perforated plates having 3-mm holes to 37-μm (400-mesh) Dutch twill. Baskets are always bottom discharge; a valve mechanism may be used to seal the bottom if the basket is fed with the whole charge at one time, so that an appreciable liquid layer develops.

The most fully automated basket or peeler centrifuge feeds and discharges at spin speed, and normally operates at cycle times of less than 3 min at pressures up to 1.03 MPa (150 psi) and temperatures from −70 to 120°C, and in special cases to 350°C. It is primarily used for materials draining freely in a high centrifugal field, for medium tonnages, and for multiple rinses where nearly complete segregation of the rinses and motor liquor is desired. Ideally, the feed should have a constant composition and high concentration. For this purpose, a gravity slurry concentrating tank or cyclone is often installed in front of the centrifuge. The bowl rotates on a horizontal axis with a metal screen as the filter medium and discharges solids by cutting them out with a hydraulically operated knife. Bowls are 300–1200 mm in diameter and handle loads of 28–170 L (1–6 ft^3) of cake at bowl speeds of 1000–2500 rpm, producing maximum centrifugal fields of about 1250 G. The solids discharge through a front chute, which simplifies sealing against pressure. A distributor riding on the cake during feeding maintains uniform cake thickness, gives better bowl balance, and improves washing efficiency. Power requirements range from 15–150 kW. Materials of construction include carbon or stainless steel and Monel. The high speed during feeding and discharge may cause deformation of particles and breakage of crystals, whereas the heel of cake left on the screen may lose permeability through glazing or plugging with fractured fines. The heel can be conditioned by suitable washes after one or more cycles. Table 2 lists two process cycles.

This type of centrifuge is also used on borax and boric acid, p-xylene, sodium bicarbonate, and sodium chloride from glycerol or electrolytic caustic, in addition to dewatering of various products, eg, potato starch, or dewatering and washing slurries, eg, calcium hypochlorite.

Inverting Filter Centrifuge. Another batch automatic horizontal perforated bowl centrifuge inverts the flexible filter to discharge the solids. Feed slurry may be deposited on the inside surface of a cloth filter, with the bucket end completely closed. When the interior is full of dewatered solids, the bowl is decelerated to a

Table 2. Basket Centrifuge Operation on Free- and Slow-Draining Particles

Operation	Type of crystal	
	Free-draining[a]	Slow-draining[b]
solids handling rate, t/h	20–24	1.5–2
conditioning rinse, s	1	0
feeding time, s	7	25
wash time, s	5	25
drain time, s	12	30
unloading time, s	2	6
total cycle, s	27	86

[a] For example, ammonium sulfate.
[b] Requiring several rinses, eg, polyolefins.

slow speed and piston and closure plates move axially, inverting the filter cloth so that the solids reside on the outside diameter of the cloth. Very little residual material remains on the filter cloth surface for the next cycle. Centrifuges are available with diameters of 300–1300 mm, and operate at 720–1920 g. Filter cake rates of 75–250 kg/h are achieved at excellent rinsing efficiencies and low crystal breakage. Drives are electric or hydraulic, with the axial movement hydraulically actuated via a rotary seal. Power demands are low, from 3 kW for the smallest to 100 kW for the largest.

Continuous Cylindrical Screen Centrifuges. Continuous filtering centrifuges are used for very fast draining that do not require extremely dry final products. Rinsing efficiency varies considerably; power requirements are usually low; initial slurry concentrations can be somewhat more variable and not as high as for the high speed basket. Continuous centrifugal filters are equipped with either a cylindrical or a conical screen. Both types are made without a retaining lip on the solids discharge end of the bowl and employ various methods to move the solids through the bowl.

The cylindrical screen centrifuge deposits solids at one end of the bowl in a layer 6–80 mm thick and pushes the annular ring of cake axially through the bowl by means of a reciprocating piston (36,37). Washes are collected in separate sections of the casing but are not as distinctly separated as with basket centrifuge sequential operation. Drained solids at the end of the bowl are thrown into a casing that is separated by baffles from the liquid discharge zones. Bowls rotate on a horizontal axis, range in diameter from 200–1200 mm, and have capacities of 1–25 t of solids/h. Centrifugal fields of 300–600 G are common. To reduce the fines loss and to facilitate movement of the cake on the screen, maximum speeds are rarely used. Power requirements range from 4–60 kW. The cylindrical screens are generally of the bar screen type, with 0.1–0.5-mm spacing. The reciprocating piston operates at 20–100 strokes/min, with stroke lengths up to 80 mm. The thickness of the cake (max about 80 mm) depends on the packing and draining characteristics, the bulking tendency of the layer in front of the pusher, and the frictional resistance of the cake on the screen. Friction also depends on the construction of the bar screen.

The feed slurry is introduced by gravity flow of 20–200 t/h or screw conveyor to an imperforate distributing cone that deposits the slurry at its original

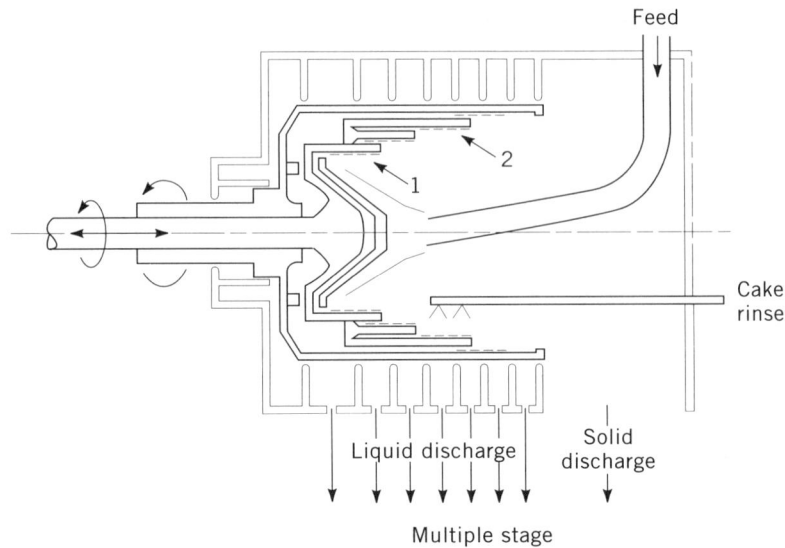

Fig. 17. Multiple-stage pusher centrifuge. Screens 1 and 2 reciprocate with pusher.

concentration immediately in front of the pusher, as shown in Figure 17. The cake must not buckle at this point, so slurries of 40% concentration are generally necessary. Because fast drainage is required, feed particle size should exceed 150 μm; medium and coarse crystals and granules or fibrous solids can be handled in this type of centrifuge. Crystal breakage is low within the basket but some breakage does occur during discharge.

To handle materials that form a soft or plastic cake or have a high frictional resistance, the cylindrical screen may consist of two to six steps with successively larger diameters, as shown in Figure 17. Alternate steps reciprocate with the piston. Thus, the cake is pushed across only a short length of screen before redistribution on the next step at a slightly larger diameter. Drainage and washing efficiency are increased by redistribution of the cake under these conditions. This type of centrifuge is used on sugar, where the high viscosity of the mother liquor causes slow drainage, a high degree of plasticity in the partially drained cake, and poor penetration of wash liquor.

Another type of continuous, cylindrical screen centrifuge discharges the cake by moving it axially through the bowl with a helical conveyor. Crystal breakage through conveyor action is greater than with the pusher-type mechanism. Applications include the handling of copper as trisodium phosphate and purification of paraxylene.

Conical Screen Centrifuge. In conical screen centrifuges the angle of the bowl causes or assists the cake to move axially and redistributes it in an increasingly thin layer which improves drainage characteristics. The feed slurry is deposited at the small end of the cone, where most drainage occurs. The drained solids are discharged from the large end, which has no retaining ring. Screens generally have 0.08–1.5-mm slots or perforated plate holes. Screens less than about 0.25-mm thick require a backup screen to extend screen life. There are three types of conical screen centrifuges: those that are self-discharging,

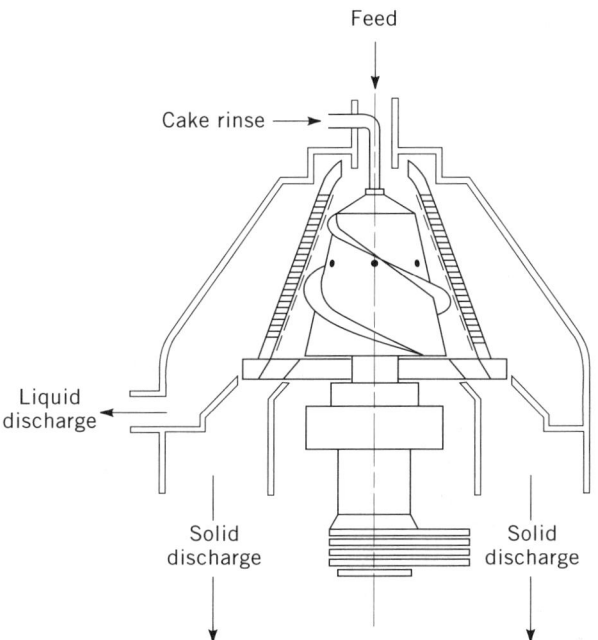

Fig. 18. Conical screen conveyor discharge centrifuge.

those that discharge by means of a helical conveyor (Fig. 18), and those that apply an axial vibration or oscillation to the bowl or the bowl and casing.

In its simplest form, the cone angle is slightly larger than the angles of repose of the solids at any stage in their drying cycle. Some bowls are made with two angles, such that the more shallow is at the small end, and the steeper, where solids concentrate, is higher and at the large end. Horizontal or vertical bowls of 500–1000-mm large-end diameter operate at speeds up to 2600 rpm, producing up to 2500 G; cone angles vary from 20–35° and selection of the proper cone angle and suitable screen surface is critical for each application. Temperature control of viscous feeds is necessary to maintain proper distribution and drainage. Feed slurry, usually under gravity flow, is introduced at the small end of the basket, where it is accelerated and spread evenly over the periphery of the screen. Viscous sugar massecuites are successfully handled at capacities of 2–7 t/h on a 750-mm diameter basket. Loss of fines is greater than on the automatic basket centrifuges, but improved rinse performance increases sugar purity. Rinsing efficiency is not generally high on this type of cone and segregation of rinse liquor is incomplete. This centrifuge is also used for the drying of crystalline materials such as ammonium sulfate and separation or dewatering of fine fiber in wet corn milling.

Conical screens with a helical conveyor turning slightly faster than the bowl can handle a variety of materials. The vertical baskets are underdriven and the larger diameter is downward, as shown in Figure 18. Horizontal axis designs are also available. Feed is introduced at the small end and the rate at which solids move through the drainage zone is controlled to some degree by the differential speed of the conveyor. Bowls have large-end diameters of 250–500 mm, and lengths about two-thirds of the large diameter; bowl angles range from 10–20°, and operating speeds are 2500–3800 rpm, giving up to 3500 G. Solids capacities range from 1–30 t/h, with feed capacities of 1–15 m^3/h (260–4000 gph). Centrifuge casings may be vapor-tight but are not intended for pressure operation; maximum temperature is about 150°C. Power requirements, low for the tonnages handled, range from 7–30 kW. Applications include dewatering of medium and coarse crystals, deoiling of proteinaceous solids, and removal of solids from fruit and vegetable pulps and other food slurries.

The third type of cone screen centrifuge operates with bowl angles of 13–18° and assists solids discharge by a vibratory motion of the bowl or bowl and casing. These units usually have underdriven bowls having 500–1100-mm larger diameter at the upper end; horizontal axis designs are also available where diameter-to-length ratios range from 1–2. Operating speeds are normally 300–500 rpm, and solids capacities range from 25–150 t/h. Pressurized units are not available, and operating temperatures range to 100°C. Power requirements are 15–35 kW. Bar screens are frequently used, and applications are largely on coal dewatering; particle sizes from 30 mm down to 0.25 mm (60 mesh) are easily handled. These centrifuges are also used in dewatering of potash and other crystalline solids.

Gas Centrifugal Separation

Highly developed centrifuges are used to enrich uranium for nuclear application. Gaseous uranium hexafluoride, UF$_6$, is introduced into a very high speed tubular rotor, causing the lighter ^{235}U-fraction to separate from that of the heavier ^{285}U. The enrichment that is achieved using one centrifuge is not large enough, so in a commercial plant, the centrifuges are arranged in cascades, where the enriched fraction passes up the cascade for additional enrichment stages, while the depleted fraction passes down the cascade. Each stage of enrichment requires many centrifuges to multiply the total capacity, and many stages to result in an adequately enriched product. The feed has 0.7% ^{235}U; the product has 3–4% ^{235}U. This ultimately results in thousands of individual centrifuges, interconnected to each other in the cascade. Extensive successful development has taken place to improve the performance of each centrifuge and reduce the total number of gas centrifuges needed to achieve a satisfactory product at an economic cost per unit of separation power, ie, kg SW/a, where SW = separative work in kg/yr.

A gas centrifuge separative power, U, can be expressed as

$$U = \frac{Dp}{RT}\left(\frac{\Delta M \omega^2 r^2}{2\,RT}\right)^2 \frac{\pi l}{2} f \qquad (37)$$

where D, p, and T are the gas properties, density, pressure, and temperature, respectively; R is the ideal gas constant; ΔM is the difference in the molecular weight of the gaseous isotopes; $\omega^4 r^4$ = surface velocity v^4 of the rotor; l = length of the rotor; and f = flow factor, which is a function for the actual gas flow pattern within the centrifuge rotor.

In order to increase the separative power of a gas centrifuge, the surface velocity must be as high as possible, the length of the rotor as long as possible, and the gas flow within the rotor introduced and withdrawn from optimum locations, to result in maximum enrichment. Figure 19 is a schematic of a gas centrifuge which operates at surface speeds of 600 m/s, has a diameter of 50 cm (19.7 in.) (147,000 G), a length of 500 cm (16.4 ft), and rotates in a vacuum contained within a stationary housing. Owing to the slender shape ($1/d = 10$), the required high speed is well above the first natural frequency of the rotor (even above the third critical). Upper and lower magnetic bearings locate and damp vibration during operation and when passing through the critical speeds. The rotor hub is driven by a rotating magnetic field generated by the stationary stator. The weight is magnetically reduced, with only a needle bearing in the bottom to support and locate the rotor.

The feed gas is introduced near the rotor axis. Enriched and depleted gases are extracted by stationary pitot-like scoops. The location and shape of these tubes, and the baffles within the rotor, greatly effect the gas flow which

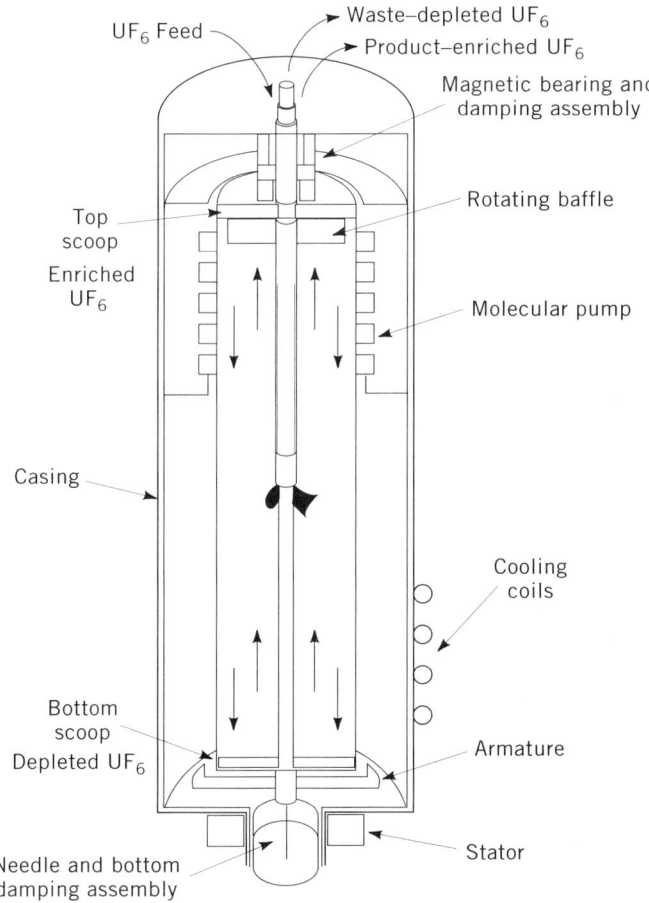

Fig. 19. The modern gas centrifuge and its main components.

recirculates within the rotor, reaching enrichment equilibrium at a given feed rate. A vacuum is maintained around most of the rotor. The UF_6 leakage around the stationary axial post is confined to the top of the case by the use of a molecular pump.

Gas centrifuge effectiveness has increased 10–15 times from the 1975 pilot plant owing to increases in speed from stronger bowl (38–41) materials and dynamic bearing and dampening systems. A better understanding of the gas flow within the bowl has also contributed. It is expected that the next generation of gas centrifuges, installed in the same footprint, should have two to three times the separation power by using carbon fiber composites (surface speed >700 m/s) and further refinement of the feed and extraction devices. Commercial gas centrifuge enrichment plants exist in England, Holland, Germany, and Japan.

NOMENCLATURE

Symbol	Definition
$Ae_{3/4}$	the cylindrical area at 3/4 of the bowl wall radius
a	distance; space between adjacent disks
C	discharge coefficient
c	concentration
d	diameter of particle; particle size
$\overline{d}$	mean particle diameter
e	efficiency factor
E	energy
E_{rel}	relative performance
G	level at the mean solids radius
G	ratio of centrifugal acceleration to acceleration gravity
g	acceleration of gravity
h	cake thickness; height
K	permeability
k	experimental coefficient in eq. 17
k'	experimental coefficient in eq. 19
l	length, particularly from feed zone to centrate discharge
m	torque
n	number of disks
P	power
p	pressure
q	volume of undrained liquid (mother liquor)/unit volume of solid, %
q_∞	volume of undrained liquid (mother liquor)/unit volume of solid at infinite time, %
Q	flow rate, volumetric
r	radius
s	external surface area/weight of solid
s'	surface area/volume of cake
S	fraction of void volume occupied by liquid at time, t
S^∞	fraction of void volume occupied by liquid at infinite time
t	time during which material is exposed to separation effect
t'	time at which free liquid enters the cake in eq. 18
T	torque

Symbol	Definition
v	velocity
V	volume
α	cake resistance constant in pressure filtration, length/volume
δ	mass density = mass/unit volume
Δ	difference, particularly of density
ϵ	void fraction
μ	viscosity
ω	angular velocity of rotation motion
Ψ	wetting angle
ϕ	angle between direction of the nozzle and tangent to circle intersecting nozzle axis at discharge section
σ	surface tension or material stress
Σ	theoretical capacity factor
θ	half-included angle of disks
ν	kinematic viscosity

Subscripts

B	bottle centrifuge
BD	back-drive
C	cake, or conveyor torque
D	disk centrifuge
f	film flow
F	friction power
g	settling velocity of particle in gravity field
h	heavy phase
i	interface
l	free surface of liquid or light phase
L	liquid medium
m	mean value
n	nozzle
M	filter medium
N	nozzle
P	process power
p	process
S	solid medium
s	settling velocity of particle in centrifugal field
W	windage power

BIBLIOGRAPHY

"Centrifugal Separation" in the *Encyclopedia of Chemical Technology,* 1st ed., Vol. 3, pp. 501–521, by M. H. Hebb and F. M. Smith, The Sharples Corp.; in *ECT* 2nd ed., Vol. 4, pp. 710–758, by A. C. Lavanchy and F. W. Keith, Jr., The Sharples Co., and J. W. Beams (Gas Centrifugal Separation), University of Virginia; in *ECT* 3rd ed., Vol. 5, pp. 194–233, by A. C. Lavanchy and F. W. Keith, Jr., Pennwalt Corp.; in *ECT* 4th ed., Vol. 21, pp. 828–875, by A. Letki and R. T. Moll, Alfa Laval Separation Inc., and L. Shapiro, Consultant.

1. C. M. Ambler, *Chem. Eng. Prog.* **48**, 150 (1952).
2. F. W. Keith, Jr. and R. T. Moll in R. A. Young and P. Cheremisinoff, eds., *Wastewa-*

ter Physical Treatment Processes, Ann Arbor Science Publishers, Ann Arbor, Mich., scheduled 1978.
3. J. Murkes, *Brit. Chem. Eng.* **14**(12), 636 (1969).
4. H. P. Grace, *Chem. Eng. Prog.* **49**, 427 (1953).
5. J. A. Storrow, *AIChE J.* **3**, 528 (1957).
6. W. Batel, *Chem. Ing. Tech.* **33**, 541 (1961).
7. E. Nenninger, Jr. and J. A. Storrow, *AIChE J.* **4**, 305 (1958).
8. Technical data, Sharples Research Laboratory, 1961.
9. J. O. Maloney, *Ind. Eng. Chem.* **48**, 482 (1956).
10. Technical data, Alfa Laval Separation, Warminster, Pa., May 1993.
11. T. Theodorsen and A. Regier, *Experiments on Drag of Revolving Disk, Cylinders and Streamline Rods at High Speeds*, Report No. 793 National Advisory Committee for Areonautics 11944, pp. 367–384.
12. Technical data, Alfa Laval Separation, Warminster, Pa., 1984.
13. *Chem. Eng.*, 353–356 (Aug. 1993).
14. H. Axelsson, *Centrifugal Separations—Principles and Techniques*, Alfa Laval Separation, Tumba, Sweden, presented at the Bioprocess Technology Program, University of Virginia, Charlottesville, Va., Oct. 17–25 1991.
15. C. M. Ambler and F. W. Keith, Jr., in A. Weissberger, ed., *Techniques of Chemistry*, 3rd ed., Vol. XII, Wiley-Interscience, New York, 1978, Chapt. VI.
16. P. A. Vesilind, *Treatment and Disposal of Waste Water Sludges*, Ann Arbor Science Publishers, Ann Arbor, Mich., 1974.
17. O. M. Griffith, *Techniques of Preparative, Zonal and Continuous Flow Ultracentrifugation*, Spinco Division of Beckman Instruments, Inc., Palo Alto, Calif., 1975.
18. G. B. Cline in E. S. Perry and C. F. van Oss, eds., *Progress in Separation and Purification*, Wiley-Interscience, New York, 1971, pp. 299–306.
19. T. Lee and C. W. Weber, *Anal. Chem.* **39**, 620 (1967).
20. C. A. Willus and B. Fitch, *Chem. Eng. Prog.* **69**, 73 (Sept. 1973).
21. D. E. Albertson and E. E. Guidi, Jr., *J. WPCF* **41**, 607 (Apr. 1969).
22. F. A. Records, *Chem. Eng.*, 81 (Jan. 1974).
23. R. J. Woolcock, *Filtr. Sep.* **12**, 174 (Mar./Apr. 1975).
24. J. J. Halloran, No. TIS-5039, *Annual Meeting, Slurry Transport Association*, Houston, Tex., Aug. 24–25, 1976.
25. R. Day, *Chem. Eng. Prog.* **69**, 67 (Sept. 1973).
26. L. F. Kelsall, *Trans. Inst. Chem. Eng. London* **30**, 87 (1952).
27. D. F. Kelsall, *Chem. Eng. Sci.* **2**, 254 (1953).
28. D. Bradley, *The Hydrocyclone*, Pergammon Press, London, 1965.
29. P. A. Vesilund, B. Zhang; *J. WPCF*, **56**(12), 1231–1237 (Dec. 1984).
30. F. M. Tiller, N. B. Hsyung, *Chem. Eng. Prog.*, 20–28 (Aug. 1993).
31. W. Gosile, *German Chem. Eng.* 3 (1980).
32. R. Day, *Chem. Eng.*, 81 (May 13, 1974).
33. F. A. Records, *Chem. Proc. Eng.* **52**, 47 (Nov. 1971).
34. Technical data, Alfa Laval Separation, Warminster, Pa., Sept. 1991.
35. K. Lilley and G. Huhtsch, *Filtr. Sep.* **12**, 70 (Jan.–Feb. 1975).
36. D. K. Baumann and D. B. Todd, *Chem. Eng. Prog.* **69**, 62 (Sept. 1973).
37. P. M. T. Brown, *Chem. Proc. Eng.* **52**, 65 (Nov. 1971).
38. T. T. Edwards and M. D. Holmes, *Nucl. Eng. U.K.* **28**(6), 174–177 (Nov.–Dec. 1987).
39. I. D. Heriot, *Uranium Enrichment by Gas Centrifuge*, EUR 11486EN, for the Commission of the European Communities, Urenco Ltd., Luxembourg, Belgium, 1988.
40. *Fundamentals of Uranium Enrichment, Pt. 2. Methods: Coaseous Diffusion and Gas Centrifuges*, Nuclear Fuel Cycle Training Course, X/ITP—269/P2; DE89 011789 U.S.

Air Force Technical Application Center Project T/9427/NP/DOE; The International Technology Programs Div., Martin Marietta Energy Systems Inc., Oak Ridge, Tenn., May 1989.
41. K. Cohen, *The Theory of Isotope Separation as Applied to Large Scale Production of 235V*, McGraw-Hill Book Co., Inc., New York, 1951.

ALAN LETKI
R. T. MOLL
Alfa Laval Separation Inc.

LEONARD SHAPIRO
Consultant

CHROMATOGRAPHY

Chromatography is a technique for separating and quantifying the constituents of a mixture. Separation techniques are essential for the characterization of the mixtures that result from most chemical processes. Chromatographic analysis is used in many areas of science and engineering: in environmental studies, in the analysis of art objects, in industrial quality control, in analysis of biological materials, and in forensics. Most chemical laboratories employ one or more chromatographs for routine analysis (1).

The first scientific reports demonstrating chromatographic phenomena appeared in the 1890s. However, the era of analytical chromatography began in 1903 when a paper was published describing the separation and identification of the components of a mixture of structurally similar yellow and green chloroplast pigments in leaf extracts. A solution of these extracts in carbon disulfide was passed through a column packed with chalk (2). The application of a pure solvent to the development of a chromatogram was significant, as was the explanation of adsorption (qv) as the mechanism by which separation occurred and the realization that this technique was potentially valuable as a means for identifying compounds other than by color. In 1906 the term chromatography was coined for these processes, coming from the combination of two Greek roots "chroma," meaning color, and "graphe," meaning writing (3).

Chromatographic separations rely on fundamental differences in the affinity of the components of a mixture for the phases of a chromatographic system. Thus chromatographic parameters contain information on the fundamental quantities describing these interactions and these parameters may be used to deduce stability constants, vapor pressures, and other thermodynamic data appropriate to the processes occurring in the chromatograph.

The importance of chromatographic processes to science can be gauged in many ways. For example, the 1952 Nobel Prize in chemistry was awarded for the development of liquid–liquid partitioning, or liquid–liquid chromatography (4). A key element in the development of this technique was the realization that both liquid phases need not move simultaneously to effect separation. In addition to liquid–liquid chromatography, this led to the development of gas–liquid chromatographic techniques. From a practical standpoint the importance of chromatography is evidenced by the number of chromatographic instruments that exist in laboratories (1).

Chromatography is not restricted to the analytical laboratory. Preparative chromatography is often the preferred tool for process separations. In addition, on-line chromatographic devices are used as detectors in many chemical processes such as reaction kinetics. Preparative chromatography is performed primarily to purify materials for subsequent use, eg, for additional analyses or for creating high purity materials required by some processes. Whereas all forms of chromatography are used for preparative purposes, liquid chromatography has been of highest value, especially in biological and pharmaceutical applications. Usually the column capacity is stretched to the limit to allow the greatest quantity of sample to be added to the mobile phase. The detector for preparative chromatography should be nondestructive and allow adequate discrimination between the product and its impurities (see BIOSEPARATIONS).

Principles

The principle of chromatographic separation is quite straightforward. A mixture is allowed to come into contact with two phases, one referred to as the stationary phase and the other as the mobile phase. The stationary phase is contained in a column or sheet through which the mobile phase moves in a controlled manner relative to the stationary phase, carrying with it any material that may prefer to mix with it. For a comparison with other methods, see Table 1 which lists various separation techniques, the phenomena upon which they are based, and the corresponding methods of separation. In addition, there are separation techniques based on molecular geometry, which exploit differences within a single phase, for example, molecular sieving, field flow fractionation, gel-filtration chromatography, size exclusion chromatography, gas diffusion, inclusion complexes, ultrafiltration, and dialysis (qv) (see MOLECULAR SIEVES).

Because of differences in affinity of the mixture's constituents for the mobile phase, as compared to the stationary phase, these constituents tend to be swept along with the mobile phase at different rates. This selective interaction is known as partitioning. In adsorption chromatography the constituents in the dissolved sample compete with the mobile phase for the active sites on the stationary phase. To remove constituents adsorbed on the stationary phase, the mobile-phase chromatographic strength is increased by modifying the mobile phase to have a greater affinity for the stationary phase than the adsorbed sample constituents. To determine the effects of adsorption or partitioning on retention of substances on the column, a detector measures either the time required to travel a given distance or the distance traveled in a fixed time. The detector may be as simple

Table 1. Separation Techniques

	Methods		
Phenomenon	Heat[a]	Physicochemical[a] interaction	Nonuniformities of concentration[b]
adsorption, surface activity		gas–solid chromatography, liquid–solid chromatography, thin-layer chromatography, supercritical fluid chromatography	foam fractionation
electromigration			electrophoresis electrodecantation
exchange partition coefficient	zone refining zone melting	ion exchange gas–liquid chromatography, extraction, liquid–liquid chromatography, paper chromatography	
volatility	distillation		

[a] These methods create a second phase having a different concentration.
[b] These methods exploit differences within a single phase.

as the human nose or the human eye or as complex as a microsensor. A plot of detector response versus time of travel for a fixed distance is called a chromatogram. In preparative chromatography a device may be attached to the end of a column to collect the separated components of a mixture.

The nature of the stationary and mobile phases in a particular chromatographic experiment determines the efficacy of component separation in a particular mixture. A wide variety of combinations of stationary and mobile phases is used as is shown in Figure 1. The stationary phase may be a solid or a liquid supported on a solid. The mobile phase may be a gas, a liquid, or a material such as a supercritical fluid. A particular chromatographic technique is specified by naming the mobile phase, followed by the stationary. Thus, gas–liquid chromatography (glc) is a system that uses a gaseous mobile phase in contact with a film of liquid stationary phase.

In the analytical chromatographic process, mixtures are separated either as individual components or as classes of similar materials. The mixture to be separated is first placed in solution, then transferred to the mobile phase to move through the chromatographic system. In some cases, irreversible interaction with the column leaves material permanently attached to the stationary phase. This process has two effects: because the material is permanently attached to the stationary phase, it is never detected as leaving the column and the analysis of the mixture is incomplete; additionally, the adsorption of material on the stationary phase alters the ability of that phase to be used in future experiments. Thus it is extremely important to determine the ultimate fate of known materials when used in a chromatographic system and to develop a feeling for the kinds of materials in an unknown mixture before use of a chromatograph.

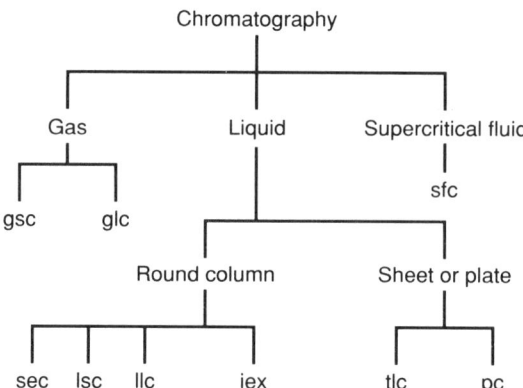

Fig. 1. Classification of chromatographic systems where gsc is gas–solid chromatography; glc, gas–liquid chromatography; sec, size-exclusion chromatography; lsc, liquid–solid chromatography; llc, liquid–liquid chromatography; iec, ion-exchange chromatography; tlc, thin-layer chromatography; pc, paper chromatography; and sfc, supercritical-fluid chromatography.

Development of the Chromatogram. The term development describes the process of performing a chromatographic separation. There are several ways in which separation may be made to occur, eg, frontal, displacement, and elution chromatography. Frontal chromatography uses a large quantity of sample and is usually unsuited to analytical procedures. In displacement and elution chromatography, much smaller amounts of material are used.

Passing impure material through a packed bed is frequently used for purifying large quantities of fluids such as gases and solvents. In such techniques the primary concern is removal of impurities through retention on the column packing or stationary phase. To carry out this procedure, a stationary phase that selectively retains the impurities, usually via adsorption, is sought. A significant factor is the amount of impurity the stationary phase can contain before it is saturated. The higher the capacity, the better the material for purification. The two principal applications of this technique are the purification of gas or solvent and selective sorption of trace materials from a fluid.

The three principal types of chromatography are frontal, displacement, and elution. Elution is by far the most common. Frontal chromatography is a technique in which the sample is introduced onto a column continuously. In essence the sample that is collected at the end of the column is the mobile phase, free of materials that adsorb/absorb on the stationary phase. Once the bed, ie, the stationary phase, is saturated and can no longer remove the impurities, the material coming off the column contains these materials. Using an appropriate detector, the condition at which this transition occurs can be determined, thereby determining the capacity of the column. This technique is called frontal chromatography or frontal analysis. Figure 2a shows an example of an integral chromatogram from a frontal analysis.

In displacement chromatography a small sample is displaced by the much more strongly held mobile phase, so that the sample is gradually pushed through the column as the mobile phase advances. As this happens, the components are

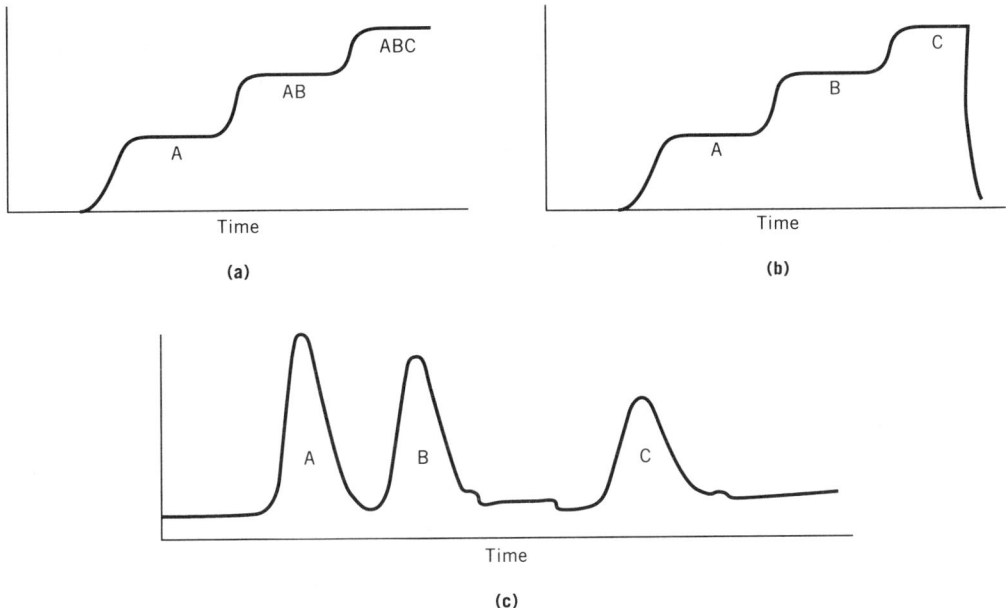

Fig. 2. Chromatograms of a mixture containing three components A, B, and C, where A is less sorbed than B, and B is less sorbed than C; (**a**) frontal analysis; (**b**) displacement analysis; and (**c**) differential elution chromatogram.

dispersed into bands that can then either be excised to obtain the pure material or displaced from the column. Such techniques are useful for the generation of quantities of pure material, particularly in application to problems in environmental analyses. Figure 2**b** shows an integral chromatogram obtained by displacement analysis.

Both frontal and displacement chromatographies suffer a significant disadvantage in that once a column has been used, part of the sample remains on the column. The column must be regenerated before reuse. In elution chromatography all of the sample material is usually removed from the column during the chromatographic process, allowing reuse of the column without regeneration. Most analytical applications of chromatography employ elution methods where a small sample is put onto the column, at the column head as a plug or a band. The sample is applied, sometimes by injection, while a stream of eluent, the mobile phase, is moving through the column. Because of the difference in affinities of the sample's components for the stationary phase, constituents travel through the column at different rates and elute at different times. Figure 2**c** shows a typical differential elution chromatogram.

Gas Chromatography

The most frequently used chromatographic technique is gas chromatography (gc) for which instrumentation was first offered commercially in 1955 by Burrell Corp.,

Perkin-Elmer, and Podbielniak. Five additional companies offered instrumentation in 1956. Gas chromatographs were the most frequently mentioned analytical instrumentation planned for purchase in surveys in 1990, and growth in sales was projected to remain around 6% through 1995 (1,5).

Gas chromatography, depending on the stationary phase, can be either gas–liquid chromatography (glc) or gas–solid chromatography (gsc). The former is the most commonly used. Separation in a gas–liquid chromatograph arises from differential partitioning of the sample's components between the stationary liquid phase adsorbed on a porous solid, and the gas phase. Separation in a gas–solid chromatograph is the result of preferential adsorption on the solid or exclusion of materials by size.

A second way of classifying gas chromatographic separations is by use. If the desired result is an analysis of the sample, the technique is analytical gas chromatography. If the desired result is the production of purer materials through fractionation in the chromatographic column, the technique is known as preparative gas chromatography. If the analytical chromatography is performed as part of the control of a manufacturing process, the technique is known as process gas chromatography.

Columns. The chromatographic column is often described as the heart of the chromatographic system, because it is the single part of the system that must be present to effect a separation. Columns come in a wide variety of sizes and shapes. They are frequently tubes made of various materials coated on or bonded to stationary phase. In general, the larger the diameter of the column, the poorer the separation of the components. Large (>2 mm internal diameter) columns are often used in preparative chromatography, whereas medium (diameter between 1 and 2 mm) columns are used for analytical chromatography. Typically such columns are coated with a liquid phase, which is either chemically bonded or coated on porous particles that are presumed to be inert, ie, the solid supports do not affect the separation process. A schematic of a packed column is shown in Figure 3.

Capillary Chromatography. Capillary, or wall-coated open tubular (WCOT), columns are fine tubes having internal diameters in the range of 0.20–0.75 mm. Use of these columns affects high resolution gas chromatography, ie, remarkably efficient separations of components. Columns having diameters in the range of 0.20–0.35 mm generally give the best separations. Capillary columns require the use of a sample splitter, a device to allow a representative aliquot of

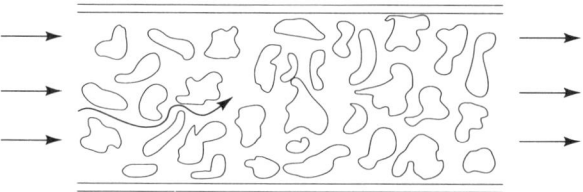

Fig. 3. A packed chromatographic column showing the thin column walls and the irregularly shaped solid support coated with a liquid phase. The arrows indicate the movement of the mobile phase.

sample onto the column, because the sample capacity of a capillary column is not very great. Capillary columns are small and thus have a small (ca 100 ng/component) capacity. Sample size is therefore a critical factor in achieving the high resolutions of which such columns are capable. Columns of 0.32 mm internal diameter are popular because such columns retain the high resolution feature, but accept a larger (ca 500 ng/component) capacity. A 0.53-mm diameter column can usually be installed on a classical gas chromatograph without any modification of the system and accepts samples having up to ca 2000 ng/component.

Columns having internal diameters in the range of 0.53–0.75 mm are known as wide bore open tubular columns. The 0.75-mm diameter column is particularly useful for such coupled tandem techniques as gas chromatography–mass spectrometry (gc–ms), gas chromatography–Fourier transform infrared spectroscopy (gc–ftir), and gas chromatography–nuclear magnetic resonance spectroscopy (gc–nmr), in which a greater quantity of material is needed for detection. These columns are also used with detectors of low sensitivity such as the thermal conductivity detectors. Capacities up to 15,000 ng can be achieved using 0.75-mm diameter columns.

The Liquid Phase. The stationary phase in an open tubular column is generally coated or chemically bonded to the wall of the capillary column in the same way the phase is attached to the support of a packed column. These are called nonbonded and bonded phases, respectively. In capillary columns there is no support material or column packing.

The greater the thickness of the liquid-phase (film) coating, the more coating per unit length and the greater the retarding potential of the column for components attracted to the stationary phase. Thick films (0.5–5.0 μm) are usually used to separate low boiling materials. For most other applications, columns having coatings of 0.2–0.4 μm are employed.

Chemically bonded phases are usually more resilient than nonbonded phases, tending not to wash out as large amounts of solvent pass through the column, and having much better thermal stability than do the nonbonded phases. Frequently a chemically bonded phase can be identified to effect a given separation at the same efficiency as a nonbonded one, thus the bonded phases are generally preferred

In some cases, increased stability of the stationary phase can be achieved by acting on it chemically. The phase can be cross-linked to give more mechanical, chemical, and thermal stability. Free-radical initiators such as peroxides, ozone, and azo-compounds are used to initiate cross-linking. Gamma-irradiation is also sometimes used. Cross-linked phases are very stable and can withstand solvent washing to clean the interior of the chromatographic column. Not all phases are capable of being cross-linked, however, and care should be taken to determine if the phase is bonded or cross-linked, or nonbonded, before attempting to clean the column.

Several hundred types of liquid phases are commercially available. These have been used individually or in combination with other liquid phases, inorganic salts, acids, or bases. The selection of stationary phases for a particular application is beyond the scope of this article, however, it is one of the most important chromatographic tasks. Stationary phase selection is discussed at length in books, journal articles, and catalogs from vendors. See general references for examples.

The Support Material. The support is the inert frame onto which the liquid phase is applied. The most common support materials for packed-column gas chromatography are the diatomaceous earths. These are the remains of diatoms, single-cell algae. The porous siliceous material has pores approximately 1 µm in diameter. These materials are typically treated with sodium carbonate to approximately 900°C. Following this treatment, they are sieved to obtain material of reasonably uniform dimensions (see SIZE SEPARATIONS). Supports are classified by the particle size or the screen mesh through which the particles pass. Most chromatographic packings are 80/100 mesh (177–149 µm) where the first number indicates the grid through which the material passes and the second number is the grid which stops it.

Once separated, the supports are washed using acid (HCl) and silanized, ie, treated with dimethyldichlorosilane [75-78-5], (DMCS), to reduce the polarity. Silanizing replaces adjacent SiOH groups with nonpolar CH_3 caps. In addition to diatomaceous earths, supports of carbon, halocarbons, eg, Teflon, and glass beads are in use by various chromatographers.

$$\begin{array}{c} \text{OH} \quad \text{OH} \\ | \quad\quad | \\ -\text{Si}-\text{O}-\text{Si}- \\ | \quad\quad | \\ \text{surface} \end{array} + (CH_3)_2SiCl_2 \longrightarrow \begin{array}{c} CH_3 \; CH_3 \\ \diagdown \; \diagup \\ \text{Si} \\ \diagup \; \diagdown \\ \text{O} \quad\quad \text{O} \\ | \quad\quad | \\ -\text{Si}-\text{O}-\text{Si}- \\ | \quad\quad | \\ \text{surface} \end{array} + 2\;HCl$$

The Stationary Phase in Gas–Solid Chromatography. In gas–solid chromatography the packing itself is responsible for separation, because it interacts directly with the sample through its surface or its pore structure. Gas–solid chromatography has some advantages: (1) there is no liquid phase that can bleed, ie, evaporate or degrade, at higher temperatures that may interfere with detection of the sample, particularly at lower levels of sample; (2) the column usually has greater thermal stability than gas–liquid columns; and (3) there are certain unique properties to this system. The disadvantages of gas–solid chromatography mainly have to do with (1) the strong interactions between the sample and the surface of many packings, especially for polar samples; (2) the inhomogeneous surface of the solid; (3) the small sample capacity relative to liquid-phase systems; and (4) limitations on kinds of samples that may be analyzed.

Separation by Molecular Size. The most common type of solid-phase packing in gas–solid chromatography is a molecular sieve. Molecular sieves (qv) are zeolites or carbon sieves that have a regular pore structure and are used almost exclusively for the separation of small molecules such as permanent gases, eg, oxygen, nitrogen, carbon monoxide, argon, and nitric oxide, or low carbon-number hydrocarbons (qv).

Porous organic polymers have been used for separating low molecular-weight mixtures containing halogenated or sulfur-containing compounds, water, alcohols, glycols, free fatty acids, ketones, esters, and aldehydes. Such porous polymers usually have a maximum operating temperature lower than many common liquid phases.

Silica, alumina, and other metal oxides and salts have been used as the stationary phase in gas–solid chromatographic systems. The applicability of these materials is limited by the difficulty of producing a consistent, resilient, reproducible material.

Column Tubing. The chromatographic column is contained in a tubing, the composition of which may have a dramatic effect on the separation process, because the sample components may also interact with the walls of the tube. Some of the materials used for columns are

Material	Uses/Comments
stainless steel	adequate for packed columns for hydrocarbons, permanent gases, and nonpolar materials
glass-lined stainless steel	combines innate inertness of glass with the durability of stainless steel, however, cannot easily determine if lining is broken
glass	inert, but inflexible; for packed columns, must be custom fitted to instrument
copper	good heat-transfer properties, but not as inert as stainless steel or glass
nickel	more inert than copper
Teflon	good for reactive systems such as sulfur gas analyses; columns do not pack well
aluminum	more active than stainless steel; used for high temperature applications
fused silica	most inert; commonly used for capillary columns; fused silica is usually protected by a silicon nitride and polyimide coating to prevent breakage

For sensitive compounds such as certain pharmaceuticals, steroids, and pesticides, the standard practice is to use columns packed in glass tubes. The surface of glass is more nearly inert than are the surfaces of metal tubes. Glass columns also have the advantage of being transparent, giving a visible means of examination for column degradation or contamination. Packing efficiency is also easy to monitor in glass columns. Column voids may be visually detected. Glass, however, is more fragile than metal. Packing and chromatograph assembly of glass containing columns must be carried out with the utmost care to ensure that the column is not damaged in either of these operations.

Tubes made of metal such as stainless steel, nickel, copper, or aluminum are much more resistant than glass to damage in handling. Stainless steel is often preferred because it is less active than other metals. If corrosive samples are used, however, it is sometimes necessary to contain the column in a tubing made of a material such as Teflon. Columns made of Teflon generally are difficult to pack. Additionally, connections to the tubing that are free of leaks, which may degrade the performance of the chromatographic system, are difficult to make.

For capillary columns fused silica is the material of choice for the column container. It has virtually no impurities (<1 ppm metal oxides) and tends to be quite inert. In addition, fused silica is relatively easily processed and manufacture of columns from this material is reproducible. In trace analysis, inertness of tubing is an important consideration to prevent all of the tiny amounts of sample from becoming lost through interaction with the wall during an analysis.

Fused-silica columns are externally coated using a protective polyimide layer to improve strength and durability, and to provide a measure of protection against reaction of the silica with water in the environment. The fused-silica column is an inherently straight wire of material, ie, its resting state is straight, not coiled. To use the material in a chromatographic oven, it must be wound onto a frame that secures it in the coiled configuration, after which it is inserted into the oven. The process of creating a fused-silica column is complex and requires sophisticated, expensive equipment, a high temperature (2000°C) furnace, and a laser-based system for determining the trueness of diameter of the ultimate product. On the other hand, capillary columns made of glass are easily made on an inexpensive drawing machine in the chromatography laboratory.

In addition to fused-silica and glass capillary columns, there are several designs for glass capillary columns, although few are widely used. One system is the so-called porous layer open tubular (PLOT) column. These are made from glass that has been pulled to a capillary with an internal layer of solid packing, often similar to the material used in gas–solid chromatography. Subsequently this material is removed by etching to produce a column with many pores at the surface of the capillary. Support-coated open tubular (SCOT) columns are glass capillary columns having a coating on the column wall. Because this phase, frequently a liquid one, is different from the wall material, additional mechanisms for separation are possible. Micropacked columns are glass capillaries packed using small-mesh particles similar to those in liquid-chromatographic columns.

The Mobile Phase. The purpose of the mobile phase, also called the carrier gas, is to transport the sample through the chromatographic column. The selection of carrier gas is often dictated by the type of detector attached to the chromatographic system. To achieve the best performance, gases of relatively higher molecular weight such as nitrogen, carbon dioxide, or argon are used as a carrier. If these gases do not permit sufficiently high gas velocities, then a lower molecular-weight gas such as helium or hydrogen should be used. The purity of the carrier gas is an important consideration because gas passes through the column and impurities could interfere with chromatographic separation. Gases used are generally of 99.995% purity or better. Two particularly troubling contaminants are water and air, which can affect the stability of the liquid phase in a packed column. The best compromise for column performance and safety is helium for capillary columns and either helium or nitrogen for packed columns.

Detectors. The function of the gc detector is to sense the presence of a constituent of the sample at the outlet of the column. Selectivity is the property that allows the detector to discriminate between constituents. Thus a detector selective to a particular compound type responds especially well to compounds of that type, but not to other chemical species. The response is the signal strength generated by a given quantity of material. Sensitivity is a measure of the ability of the detector to register the presence of the component of interest. It is usually given as the quantity of material that can be detected having a response at twice the noise level of the detector.

By far the most used detector is the thermal conductivity detector (TCD). Detectors like the TCD are called bulk-property detectors, in that the response is to a property of the overall material flowing through the detector, in this case the thermal conductivity of the stream, which includes the carrier gas (mobile phase) and any material that may be traveling with it. The principle behind a TCD is that a hot body loses heat at a rate that depends on the composition of the material. Most materials have lower thermal conductivities than helium, the typical carrier gas, and most organic materials have similar thermal conductivities, thus the TCD is often used for quantitative analysis of separation of organics. Of course, the thermal conductivities of organic compounds are not exactly the same, so very accurate results require an evaluation of the response factor for each material, essentially an evaluation of the thermal conductivity of each material.

Another type of detector, one most extensively used in capillary chromatography, is the flame-ionization detector (FID). The principle behind its operation is the detection of a current from ions formed when organic materials are burned in a small hydrogen–oxygen flame at the end of the column. Typically a voltage is applied across this region and the small current carried by the ions is detected using a sensitive electrometer. One of the primary advantages of the FID is that it is, in general, more sensitive than the TCD. In addition, it does not respond to materials such as water, carbon dioxide, carbon monoxide, and most simple sulfur-containing gases. This is an advantage in the analysis of certain samples, for example traces of organics in water. The principal disadvantage of the FID is that it destroys the separated material in the process of detection.

A third detector type is the electron-capture detector (ECD) which is very selective for the detection of highly electronegative compounds in the effluent such as chlorinated hydrocarbons, many pesticides, and polychlorinated biphenyls (see ELECTROSEPARATIONS). Its principle of action is the interaction of such compounds with electrons emitted from a radioactive source. A detector sensitive to these beta emissions that is positioned across the stream from a source senses a drop in emissions when the stream contains compounds that capture the emitted electrons. Thus the response to the passage of materials is a loss of signal. The sources generally used in these detectors are nickel-63 or tritium. One disadvantage is the radioactive source, which requires special handling. Also, the linear response range is not very great and the detector is subject to high background noise, unless care is taken to eliminate column, carrier-gas, or sample contaminants.

The flame-photometric detector (FPD) is selective for organic compounds containing phosphorus and sulfur, detecting chemiluminescent species formed in a flame from these materials. The chemiluminescence is detected through a filter by a photomultiplier. The photometric response is linear in concentration for phosphorus, but it is second order in concentration for sulfur. The minimum detectable level for phosphorus is about 10^{-12} g/s; for sulfur it is about 5×10^{-11} g/s.

The alkali flame-ionization detector (AFID), sometimes called a thermionic (TID) or nitrogen–phosphorus detector (NPD), has as its basis the fact that a phosphorus- or nitrogen-containing organic material, when placed in contact with an alkali salt above a flame, forms ions in excess of thermal ionic formation, which can then be detected as a current. Such a detector at the end of a column then reports on the elution of these compounds. The mechanism of the process is not clearly understood, but the enhanced current makes this type of detector popular for trace analysis of materials such as phosphorus-containing pesticides.

The mass spectrometer (ms) is a common adjunct to a chromatographic system. The combination of a gas chromatograph for component separation and a mass spectrometer (gc/ms) for detection and identification of the separated components is a powerful tool, particularly when the data are collected using an on-line data-handling system. Qualitative information inherent in the separation can be coupled with the identification of structure and relatively straightforward quantification of a mixture's components.

Infrared (ir) spectrometers are gaining popularity as detectors for gas chromatographic systems, particularly because the Fourier transform infrared (ftir) spectrometer allows spectra of the eluting stream to be gathered quickly. Gc/ir data are valuable alone and as an adjunct to gc/ms experiments. Gc/ir is a definitive tool for identification of isomers.

Plasma atomic emission spectrometry is also employed as a detection method for gc. By monitoring selected emission lines a kind of selective detection based on elemental composition can be achieved.

Theory. Most theoretical models of gas chromatographic processes are based on analogy to processes such as distillation (qv) or countercurrent extraction experiments (6). The separation process is viewed as a type of successive partitioning of the components of a mixture between the stationary and mobile phases similar to the partitioning that occurs in distillation columns. In those experiments an important parameter is the number of theoretical plates of which the column may be considered to be composed; the greater the number of theoretical plates, the greater the efficiency of the column for achieving separations of similar components. In gas chromatography, the equivalent measure of efficacy is the height equivalent to theoretical plates (HETP), which measures the ultimate ability of the column to separate like components. This quantity depends on many instrumental parameters such as wall or particle diameter, type of carrier gas, flow rate, liquid-phase thickness, etc. The theoretical expression relating these various parameters is called the van Deemter equation, which relates the change in efficiency to the flow.

$$\text{HETP} = A + B/\mu + C\mu$$

where A is the eddy diffusion or multipath effect term, B is the molecular diffusion term, μ is the linear gas velocity or flow rate, and C is the resistance to mass transfer term.

The expanded version of the van Deemter equation is used to help understand the relationships between the packing parameters and the gas flow.

$$\text{HETP} = 2\lambda d_p + \frac{2\gamma D_{(g)}}{\mu} + \frac{8k'd_f^2}{\pi^2(l + k')^2 D_{(\text{liq})}}\mu$$

where λ is a packing constant, d_p is the average particle diameter of the solid support, γ is the tortuosity factor used to account for the tortuosity of the gas channels in the column, $D_{(g)}$ is the diffusivity of the solute in the gas phase, k' is the capacity factor, d_f is the effective thickness of the liquid film coated on the support, and $D_{(liq)}$ is the diffusivity of the solute in the liquid phase. A plot of HETP versus µ gives a hyperbola with a minimum HETP. This minimum is the optimal flow rate where the column operates most efficiently.

Inlet Systems. The inlet or injector is the means by which the sample is introduced onto the chromatographic column. The process of sample introduction requires one to create a representative aliquot of the sample at the beginning of the column without degradation or without discrimination among the components of the sample. Most inlets operate on the principle that a sample can be vaporized quickly, assuming it is not already a gaseous material, after being squirted out of a microliter syringe into a small, heated volume, usually at about 50°C hotter than the maximum temperature of the column during the experiment. This vaporized material is quickly swept as a narrow band onto the column by a flow of carrier gas. Once on the column, the components interact with the stationary phase and begin to travel along with the carrier gas at differing rates, depending on the strengths of interaction with the stationary phase. Sample introduction is critical to the proper operation of a gas chromatograph. In order that reliable data be obtained, a representative sample must reach the column in a narrow band so that all components begin the process at the same time. Selective interaction of the components with the stationary phase then allows different components to reach the detection region at different times. Whereas direct injection from a syringe is the most widely used technique for sample introduction in gc, other means such as injection valves, pyrolyzers, headspace samplers, thermal desorbers, and purge-and-trap samplers are found in various applications.

Inlets for syringe sampling are divided into two main categories: one for packed-column and the other for capillary-column devices. For packed columns, all material injected is carried by the mobile phase onto the column. The inlet is usually an open tube, but sometimes, albeit rarely, the inlet itself may be packed, eg, to assure that the first centimeters of the column do not become contaminated with degradation products or nonvolatile materials that may affect the efficacy of the column.

Capillary columns require increased care in injection because of the much smaller capacity of the column. There are four different inlet designs: direct, split/splitless, programmed-temperature vaporization, and cool on-column injectors. Direct inlets are generally used with capillary columns of larger diameter and work much as direct injectors for packed columns. Split/splitless injectors operate in two modes. In the split mode, most of the sample introduced into the inlet goes out a vent that has less resistance to flow than the column. Thus, in the split mode, a smaller amount of sample actually enters the column than was introduced into the inlet. Splitters usually remove 90–99% of the volume of the material injected. A primary problem with this injection mode is that the material finding its way onto the column is not always representative of the sample. Discrimination resulting from differential boiling points of the components is such that certain components of the mixture are more likely to be vented than introduced onto the column. Such a problem is not related to the reproducibility of the injection itself.

In the splitless mode, the vent is turned off and everything injected goes onto the column. After a short period, the vent is opened and any residual solvent is vented. The splitless mode is found particularly in trace analytical schemes. Splitless sample injection is an art, and it requires practice to ensure reproducible introduction of sample onto the column. This type of injection is usually used for qualitative analysis.

Programmed-temperature vaporizers are flexible sample-introduction devices offering a variety of modes of operation such as split/splitless, cool-sample introduction, and solvent elimination. Usually the sample is introduced onto a cool injection port liner so that no sample discrimination occurs as in hot injections. After injection, the temperature is increased to vaporize the sample.

Cool on-column injection is used for trace analysis. All of the sample is introduced without vaporization by inserting the needle of the syringe at a place where the column has been previously stripped of liquid phase. The injection temperature must be at or below the boiling point of the solvent carrying the sample. Injection must be rapid and no more than a very few, usually no more than two, microliters may be injected. Cool on-column injection is the most accurate and reproducible injection technique for capillary chromatography, but it is the most difficult to automate.

Temperature Considerations. The inlet, detector, and the oven compartment where the column is kept, are usually controlled at different temperatures, because each part serves a different function that is best performed in a specified temperature range. In practice, the maximum oven temperature expected to be reached in the course of an analysis that is high enough to achieve the desired result in minimum time is chosen. This temperature should also be low enough to minimize the probability of column liquid-phase degradation. Generally, retention time is halved for every 30°C the temperature is increased. The injection port's temperature is usually slightly higher than the maximum oven temperature, but low enough to minimize thermal degradation or thermal rearrangement of sample components. Ideally, the thermal energy in the injection port will cause instantaneous vaporization without causing a loss of separation efficiency by spreading the sample over a large volume. The detector temperature is usually 10–30°C higher than the injector, but low enough to avoid thermal degradation of the column's liquid phase in that part of the column near the detector.

For materials with a wide boiling range, temperature programming is often used. The initial temperature, if possible, should be near the boiling point of the most volatile component; the final temperature should be near or, if possible, slightly higher than the least volatile component. The heating rate from low temperature to high temperature is usually empirically determined to obtain the most efficient separation in the shortest possible analysis time. In the consideration of heating rate, cool-down time must be counted as part of the analysis time when doing multiple-sample analyses.

Liquid Chromatography

Liquid chromatography (lc) refers to any chromatographic process in which the mobile phase is a liquid. Traditional column chromatography, thin-layer chromatography (tlc), paper chromatography (pc), and high performance liquid chromatography (hplc) are all members of this class of processes. Modern liquid chromatographic techniques originated in the late 1960s and early 1970s. Developments in hplc were driven by improvements in instrumentation, column packings, and theoretical understanding of the various separation processes involved. For example, use of pressurized mobile phases in place of gravity-driven ones in chromatography greatly shortened the time necessary for a separation. Other terms used for hplc are high speed or high pressure chromatography.

Liquid chromatography is complementary to gas chromatography because samples that cannot be easily handled in the gas phase, such as nonvolatile compounds or thermally unstable ones, eg, many natural products, pharmaceuticals, and biomacromolecules, are separable by partitioning between a liquid mobile phase and a stationary phase, often at ambient temperature. Developments in the technology of lc have led to many separations, done by gc in the past, to be carried out by liquid chromatography (see BIOSEPARATIONS).

An advantage of liquid chromatography is that the composition of the mobile phase, and perhaps of the stationary phase, can be varied during the experiment to provide greater efficacy of the separation. There are many more combinations of mobile and stationary phases to effect a separation in lc than one would have in a similar gas chromatographic experiment, where the gaseous mobile phase often serves as little more than a convenient carrier for the components of the sample.

In classical column chromatography the usual system consisted of a polar adsorbent, or stationary phase, and a nonpolar solvent, mobile phase, such as a hydrocarbon. In practice, the situation is often reversed, in which case the technique is known as reversed-phase lc.

Paper chromatography originated in the 1940s and tlc in the 1950s. In these techniques a chamber is usually used to isolate the column, which is a piece of filter paper in pc and a glass plate coated with an adsorbent such as silica gel in tlc, from the laboratory environment. The chromatogram is developed by allowing a mobile phase to creep through the column, carrying with it materials soluble to various extents. After this process has proceeded sufficiently, the column is removed from the solvent tank and the mobile phase evaporated. The separated components are visualized elsewhere, for example under an ultraviolet lamp in which various fluorescent bands indicate how fluorescent materials are separated by the movement of the solvent. Paper chromatography is sometimes described as a type of liquid–liquid chromatography, because the paper inherently contains bound water that acts as a stationary liquid phase. In tlc the usual mechanism for separation is partitioning resulting from adsorption on the stationary phase.

Paper and thin-layer chromatography may be further classified as either one- or two-dimensional, by direction, eg, ascending, descending, ascending/descending, and by capacity as either analytical or preparative. In one-dimensional pc or tlc, a small spot of sample is applied, usually from a micropipet, at a point near the edge from which solvent is to enter the paper or plate. In traditional ascending chromatography, the plate or paper is dipped in the mobile phase in the bottom of the chamber and the mobile phase is allowed to rise through the column by capillary action. When the mobile phase nears the top, the plate or paper is removed from the chamber and allowed to dry. Sometimes at this point a visualizing agent is sprayed or dipped to allow detection of the components. The distance from the origin to the migrated spot is used to calculate a retardation factor (r_f), the ratio of the distance a component travels to the distance the solvent front has traveled. Retardation factors are always less than or equal to one and are used to characterize the partitioning of a component for a particular solvent and stationary phase. In two-dimensional chromatography, after the plate or paper is removed from one solvent and allowed to dry, it may be placed in another tank with a different solvent entering the paper or plate at 90° to the direction of travel of the first solvent. The result is a further resolution of components that may have had similar partitioning in the first solvent. One-dimensional tlc plates have a very high capacity, eg, up to 20 samples and standards applied, versus one-at-a-time single spots in two-dimensional tlc, or in column chromatography (hplc or gc). A critical advantage of tlc is that a fresh plate is used for each analysis. Another important advantage is that many visualizing techniques are available that are not available using lc detectors. In addition, the spots or, more frequently, the band in prep tlc, can be physically cut from the plate.

Columns. As for gc, the column is the heart of the chromatographic system. Columns for modern lc can be packed with a variety of materials: inert particles bonded to a liquid phase (liquid–liquid chromatography); a porous gel as for size-exclusion chromatography (sec) or gel-permeation chromatography (gpc); an ion-exchange resin as for ion-exchange chromatography (iec); or an affinity adsorbent as for affinity chromatography (ac). Most column tubing is about 4.5 mm in internal diameter having a 0.16 cm wall; the columns are generally 10 to 15 cm in length. Tubing materials for lc are

Material	Uses/Comments
316 stainless steel	general-utility material; good for high pressure systems
poly(ether ether ketone) (PEEK)	inert to almost all organic solvents except methylene chloride, thetrahydrofuran, dimethylsulfoxide, and concentrated nitric and sulfuric acids; holds to 34 MPa (5000 psi); good for metal-free biological systems
tefzel	common for metal-free applications; inert
titanium	withstands pressures to 34 MPa (5000 psi); corrosion-resistant; expensive
fused silica	used for capillary lc
glass	limited pressure range
glass-lined stainless steel	difficult to know when glass is broken

The interior surface of the stainless steel tubing is usually polished to allow the particles to pack efficiently. An important consideration is uniformity of the particles, which are generally about 5 μm in diameter for typical liquid chromatographic experiments.

Packings. Most packings for lc are made of chemically modified silica gel having functional groups covalently attached to the surface of the particles.

Reversed-phase packings have covalently bonded octadecyl groups (a C_{18}-phase) or octyl groups (a C_8-phase) at the surface to provide a nonpolar environment. Sometimes the further reaction of these materials with other reagents to attach trimethylsilyl groups (endcapping) is attempted. This treatment is generally supposed to cover the regions of the surface that are not covered by the first treatment, eliminating interactions that may degrade the efficiency of the column. In addition to the C_{18} and C_8 column packings, other species used for chemically binding to the support particles include phenyl, nitro, and amino groups. For size-exclusion chromatography, porous polymers such as polystyrene–divinylbenzene are sometimes used, as are the usual treated and untreated silicas. Silica columns are limited to pH below about 8.

The Mobile Phase. The great power of lc to separate the components of a mixture lies in the differential solubility of the components in the mobile liquid phase and the stationary phase. In isocratic lc, the composition of the mobile phase remains constant throughout the course of the experiment. However, the effective separating power of lc can often be enhanced by changing the composition of the mobile phase during the course of the experiment. This process, known as gradient lc, is analogous to programming the column temperature in gc. In gradient lc, the composition is deliberately altered over the course of the experiment to achieve more rapid separation. Frequently the switch is from a weak solvent for a given material to a strong one. The change can be made in a single step or by slowly varying the composition of the mobile phase with time during the separation process. Most processes involve two solvents or solvent mixtures, although there are some cases in which three solvents are used. Obviously, the more solvents used, the more complex the program of mixing. An important consideration in gradient lc is the selection of a detector. The detector must be compatible with all the solvents used in the separation process.

Solvent-delivery systems ensure uniform transfer of the mobile phase to the column. These devices must give reproducibly uniform flow without pulsations in order to ensure reproducible retention times and peak areas for analyses. Because of the small diameter of the particles used in modern hplc, there is a high resistance to flow through the column, and high pressure pumps are required. To obtain the best signal from a detector, it is important that the detector be insensitive to pump strokes at all flows. The pump materials must not only be able to withstand such pressures, but must not be affected by the solvents in the system. For gradient elution lc, the pump should have a small mixing (hold-up) volume. This minimizes memory effects from solvent changes. Pump parameters for liquid chromatography are

Property	Comment
noise	pulsation from piston movement
drift	flow change as a function of time
accuracy	ability to deliver a set flow
precision (short-term)	constancy of volume output over a short time period, eg, 15 min
resettability	ability to match exactly the flow parameters for run to run

There are two approaches to carrying out a gradient elution: high pressure mixing and low pressure mixing. In the high pressure mode two or more pumps are programmed to pump mobile phase components into a mixing chamber before the mixed mobile phase enters the head of the liquid chromatographic column. The output of each is independently controlled, allowing almost any kind of gradient to be formed at the outlet of the mixing chamber. In the low pressure approach, the gradient is formed by mixing two or more solvents at atmospheric pressure, then pumping the mixture onto the column using a single high pressure pump. A primary advantage of the low pressure approach is lower cost because only a single pump is used instead of the two or more required in the high pressure mode.

Reciprocating pumps are those most commonly used in high performance lc. The single-piston type usually has inlet and outlet check valves with some mechanism such as variable stroke frequency to minimize the effect of pump pulsations. Dual-piston pumps operate with the pistons 180° out of phase to minimize pulsations. For this system to work optimally, the piston units must be identical.

Sample introduction onto liquid chromatographic columns is usually accomplished with a sampling valve. A sample loop of volume between 5 and 50 µL attached to the sampling valve is filled or partly filled with sample solution. At the time of introduction of the sample, the sample valve is either manually or pneumatically actuated so flow of solvent through the sample loop moves sample onto the column. Because the sample loops themselves are sometimes made of small-diameter tubing, the sample is often prefiltered to remove particles that may clog these loops. Prefilling also assures that the inlet frits and columns are not clogged by the sample.

Detectors. Lc detectors must be compatible with the solvent system (mobile phase) and are optimized for sensitivity, stability, and speed of response. They are designed to retain the quality of the separation. No versatile, universal detector is in use for lc, as, eg, the flame-ionization or thermal-conductivity detectors are for gc. Instead, the most common detector found in lc is the ultraviolet (uv) detector, a selective detector that measures the absorption of radiation at a specified wavelength. These devices may be set at a fixed wavelength or the wavelength may be variable and only sensitive to materials that absorb radiation in the range of the detector. Selective detectors are also known as being solute-property detectors. Each measures some property of only the sample component. Uv detectors are relatively insensitive to temperature or flow changes, but the response can be sensitive to solvent composition, which can effect sample absorption characteristics, as in gradient–elution chromatography.

The fluorescence detector, perhaps the most sensitive of the commonly used detectors in lc, is limited in its utility to the detection of materials that fluoresce or have derivatives that fluoresce. These detectors find particular use in analysis of environmental and food samples, where measurements of trace quantities are required.

Electrochemical detectors sense electroreducible and electrooxidizable compounds at low concentrations. For these detectors to work efficiently, the mobile phase (solvent) must be conductive and not subject to electrochemical decomposition.

Another classification of detector is the bulk-property detector, one that measures a change in some overall property of the system of mobile phase plus sample. The most commonly used bulk-property detector is the refractive-index (RI) detector. The RI detector, the closest thing to a universal detector in lc, monitors the difference between the refractive index of the effluent from the column and pure solvent. These detectors are not very good for detection of materials at low concentrations. Moreover, they are sensitive to fluctuations in temperature.

Conductivity detectors, commonly employed in ion chromatography, can be used to determine ionic materials at levels of parts per million (ppm) or parts per billion (ppb) in aqueous mobile phases. The infrared (ir) detector is one that may be used in either nonselective or selective detection. Its most common use has been as a detector in size-exclusion chromatography, although it is not limited to sec. The detector is limited to use in systems in which the mobile phase is transparent to the ir wavelength being monitored. It is possible to obtain complete spectra, much as in some gc–ir experiments, if the flow is not very high or can be stopped momentarily.

Affinity Chromatography. This technique involves the use of a bioselective stationary phase placed in contact with the material to be purified, the ligate. Because of its rather selective interaction, sometimes called a lock-and-key mechanism, this method is more selective than other lc systems based on differential solubility. Affinity chromatography is sometimes called bioselective adsorption.

Chiral Chromatography. Chiral chromatography is used for the analysis of enantiomers, most useful for separations of pharmaceuticals and biochemical compounds (see CHROMATOGRAPHY, CHIRAL). There are several types of chiral stationary phases: those that use attractive interactions, metal ligands, inclusion complexes, and protein complexes. The separation of optical isomers has important ramifications, especially in biochemistry and pharmaceutical chemistry, where one form of a compound may be bioactive and the other inactive, inhibitory, or toxic.

Ion-Exchange Chromatography. In iec, the column contains a stationary phase having ionic groups such as a sulfonate or carboxylate. The charge of these groups is compensated by counterions such as sodium or potassium. The mobile phase is usually an ionic solution, eg, sodium chloride, having pH and salt concentrations that act as the separation variables, having ions similar to the counterions. Ionic samples are introduced into the mobile phase, and retardation in movement results from ion exchange with the stationary phase of the form:

$$M^+ + SO_3Na \rightleftarrows Na^+ + SO_3M$$

where M^+ is the ion in the sample to be analyzed and the sulfonate groups are assumed to be a part of the stationary phase. The more the ion interacts with the exchanger, the more strongly it is retained. For cation-exchange chromatography, positively charged ions are separated, as shown in the equation. In anion-exchange chromatography, negatively charged ions in the sample interact with and bind to cationic stationary phases (see ION EXCHANGE).

Ion chromatography (ic), a novel form of ion-exchange chromatography, is a technique in which a weak ion-exchange column is used for separation. Detection is usually done conductimetrically. After passing through the weak ion-exchange column, the eluent passes through a subsequent column called a stripper column, in which the stream, usually made acidic or basic in the ion-exchange column, is neutralized. The stripper column may be a hollow fiber suppressor as well as a column, or the suppressor may be absent, with high conductivity being suppressed electronically. This stream then gives no conductimetric response in this condition; however, when added ions are present, such as happens when sample is passing through the stripper column, the conductivity of the solution changes and a signal is detected. Ic is a powerful technique for examining low concentrations of anions and cations. It has the advantage over selective ion-electrode analysis in that it simultaneously gives information on many ions in a single experiment.

Ion-pair chromatography (ipc), a variant of iec, is also sometimes called paired-ion chromatography (pic), soap chromatography, extraction chromatography, or chromatography with a liquid ion exchanger. In this technique the mobile phase consists of a solution of an aqueous buffer and an organic cosolvent containing an ion of charge opposite to the charge on the sample ion. The sample ion and the solvated ion form an ionic pair that is soluble in the stationary phase. Thus retention is determined by the ability to form the ion pair as well as the solubility of the complex in the stationary phase.

In size-exclusion chromatography (sec) or gel-permeation chromatography (gpc), the material with which the column is packed has pores in a certain range of sizes. Molecules or solvent-molecule complexes too large to pass through these pores pass rapidly through the column, whereas molecules or complexes of suffiently small size are retained and are the last to exit the column. Molecules of intermediate size are partially retained and elute from the chromatographic column at intermediate times. SEC is extremely useful as a tool for characterization of polymer materials because the retention mechanism is reproducible enough to give good comparative data. SEC can also give valuable information about the distribution of sizes of molecules in a sample.

Supercritical-Fluid Chromatography

Supercritical-fluid chromatography (sfc), developed in the late 1960s, was not used extensively until the early 1980s. This technique is the link between gc and lc, because its mobile phase, a supercritical fluid, has physicochemical properties intermediate between a gas and a liquid. The physiochemical properties of the mobile phase are strong factors determining the selectivity, sensitivity toward a component, and efficiency of separation in the chromatographic process. Supercritical fluids (q.v.), for example, can be viewed as dense gases that cannot become liquid. The density of a supercritical material increases continously with pressure at constant temperature and its solvating power increases with pressure, because the solubility of materials in a solvent usually increases with density. Hence this can be used as a powerful means of changing retention. Carbon dioxide is the mobile phase most often used in sfc.

Sfc can be performed with either capillary or lc-like packed columns. Carbon dioxide is compatible with chromatographic hardware, is readily available, and is noncorrosive. The most important detector for sfc is the flame-ionization detector because the mobile phase does not give a significant background signal. Most early applications of sfc were in the separation of petroleum products, however, as of the 1950s many of those separations are carried out using hplc. More recent applications of sfc include separations in fields as diverse as natural products, drugs, foods, pesticides, herbicides, surfactants, and polymers. These are a direct result of the advantages that sfc has over other forms of chromatography because of low operating temperature, selective detection, and sensitivity to molecular weight.

BIBLIOGRAPHY

"Chromatography, Affinity" in the *Encyclopedia of Chemical Technology,* 3rd ed., Vol. 6, pp. 35–54, by A. H. Nishikawa, Hoffmann-LaRoche Inc., in *ECT* 4th ed., Vol. 6, pp. 207–228, by M. A. Kaiser, Dupont, and C. Dybowski, University of Delaware.

1. G. Wilkinson, *Today's Chemist* **29** (Dec. 1990).
2. M. S. Tswett, *Proc. Warsaw Soc. Nat. Sci.* (1903).
3. M. S. Tswett, *Ber. Deut. Bot. Ges.* **24,** 384 (1906).
4. A. J. P. Martin and R. L. M. Synge, *Biochem. J. (London)* **35,** 1358 (1941).
5. *Res. Dev.*, 19 (Jan. 1991).
6. R. L. Grob, *Modern Practice of Gas Chromatography*, Wiley-Interscience, John Wiley & Sons, Inc., New York, 1985.

General References

L. R. Snyder, J. L. Glajsch, and J. J. Kirkland, *Practical HPLC Method Development*, Wiley-Interscience, John Wiley & Sons, Inc., New York, 1988.

L. R. Snyder and J. J. Kirkland, *Introduction to Modern Liquid Chromatography*, Wiley-Interscience, John Wiley & Sons, Inc., New York, 1989.

M. S. Klee, *GC Inlets—An Introduction*, Hewlett Packard, Avondale, Pa., 1990.

J. M. Miller, *Chromatography: Concepts and Contrasts*, Wiley-Interscience, John Wiley & Sons, Inc., New York, 1988.

W. Jennings, *Gas Chromatography with Glass Capillary Columns*, Academic Press, Inc., New York, 1980.

M. Lee, F. Yang, and K. D. Bartle, *Open Tubular Column Gas Chromatography*, Wiley-Interscience, John Wiley & Sons, Inc., New York, 1984.

R. L. Grob, *Chromatographic Analysis of the Environment*, Marcel Dekker, New York, 1983.

R. P. W. Scott, *Small Bore Liquid Chromatography Columns—Their Properties and Uses*, Wiley-Interscience, John Wiley & Sons, Inc., New York, 1984.

N. A. Parris, *Instrumental Liquid Chromatography*, Elsevier, Amsterdam, The Netherlands, 1984.

R. L. Grob and M. A. Kaiser, *Environmental Problem Solving Using Gas and Liquid Chromatography*, Elsevier, Amsterdam, The Netherlands, 1982.

B. Fried and J. Sherma, *Thin-Layer Chromatography, Techniques and Applications*, Marcel Dekker, New York, 1986.

C. Horvath and J. Nikelly, *Analytical Biotechnology: Capillary Electrophoresis and Chromatography*, ACS Books, Washington, D.C., 1990.

N. Grinberg, *Modern Thin-Layer Chromatography*, Marcel Dekker, New York, 1990.

S. Ahuja, *Chiral Separations by Liquid Chromatography*, ACS Books, Washington, D.C., 1991.

H. F. Walton and R. D. Rocklin, *Ion Exchange in Analytical Chemistry*, CRC Press, Boca Raton, Fla., 1990.

B. J. Hunt and S. R. Holding, *Size Exclusion Chromatography*, Blackie, Glasgow, Scotland, 1989.

H. J. Cortes, *Multidimensional Chromatography: Techniques and Applications*, Marcel Dekker, New York, 1990.

P. R. Haddad and P. E. Jackson, *Ion Chromatography: Principles and Applications*, Elsevier, Amsterdam, The Netherlands, 1990.

MARY A. KAISER
E. I. du Pont de Nemours & Co., Inc.

CECIL DYBOWSKI
University of Delaware

CHROMATOGRAPHY, CHIRAL SEPARATIONS

Chiral separations are concerned with separating the pairs of molecules, called enantiomers or optical isomers, which can exist as nonsuperimposable mirror images. Enantiomers possess identical physical and chemical properties within an achiral environment. Stereoisomers other than enantiomers, ie, diastereomers, are identified by distinct physical and chemical properties including melting points, spectral characteristics, and rates of reaction with both chiral and achiral reactants. Enantiomers, however, are only distinguished when in the presence of a homochiral environment such as polarized light, chiral solvents, chiral reagents, or chiral molecules such as biomolecules, eg, nucleic acids, proteins, and carbohydrates.

Although chirality is often associated with compounds containing a tetrahedral carbon having four different substituents, other atoms, such as phosphorus or sulfur, may also be chiral. In addition, molecules containing a center of asymmetry, such as hexahelicene, tetrasubstituted adamantanes, and substituted allenes, or molecules having hindered rotation, such as some 2,2'-disubstituted binaphthyls, may also be chiral. Compounds exhibiting a center of asymmetry are called atropisomers.

Scientists have known since the time of Louis Pasteur (1) that optical isomers can behave differently in a chiral environment. The two chiral molecules in a pair of enantiomers rotate a plane of polarized light with equal intensities, but in opposite directions. The dextrorotatory isomer (+ or *d*) rotates the plane of polarized light clockwise; the levorotatory isomer (– or *l*) rotates the plane of polarized light counterclockwise. An equal mixture of (+) and (–)-enantiomers is a racemic mixture or racemic compound and does not rotate a plane of polarized light.

Since about 1980 there has been a growing awareness of the implications arising from the fact that many drugs are chiral. Absorption, metabolism, and biological activities of organic compounds are influenced by molecular interactions with asymmetric biomolecules. These interactions, which involve hydrophobic, electrostatic, inductive, dipole–dipole, hydrogen bonding, van de Waals forces, steric hindrance, and inclusion complex formation give rise to enantioselective differentiation (2). Within a series of similar structures, substantial differences in biological effects, molecular mechanism of action, distribution, or metabolic events may be observed. For example, (R)-carvone [6485-40-1] (**1**) has the odor of spearmint, whereas (S)-carvone [2244-16-8] (**2**) has the odor of caraway (3).

The amino acids L-leucine, L-phenylalanine, L-tyrosine, and L-tryptophan all taste bitter, whereas their D-enantiomers taste sweet (4). D-Penicillamine [52-67-5], a chelating agent used to remove heavy metals from the body, is a relatively nontoxic drug effective in the treatment of rheumatoid arthritis, but L-penicillamine [1113-41-3] produces optic atrophy and subsequent blindness (5). L-Penicillamine is roughly eight times more mutagenic than its enantiomer. Such enantioselective mutagenicity is likely due to differences in renal metabolism (6). (R)-Thalidomide (**3**) is a sedative–hypnotic; (S)-thalidomide (**4**) is a teratogen (7).

Unfortunately, the (R)-thalidomide, containing some of the (S)-enantiomer, was given to pregnant women resulting in a large number of fetal deaths and congenital malformations. It is still unclear whether the administration of pure (R)-thalidomide (**3**) would have prevented the observed teratogenicity. The stereocenter of thalidomide is labile and may racemize under physiological conditions (8).

Enantiomers differ in absorption rates across membranes, especially where active transport mechanisms are involved (9), binding with different affinities to plasma proteins (10) and undergoing alternative metabolic and detoxification processes (11). Ibuprofen is an example of a chiral drug which undergoes rapid inversion *in vivo* (12). In addition, there are several examples of achiral (or prochiral) drugs being biotransformed into chiral entities. In some cases, the enantiomeric ratios produced *in vivo* by laboratory animals may differ from that produced in humans. For example, cimetidine [51484-61-9] (**5**), used to treat peptic ulcers and marketed as Tagamet, is achiral. However, cimetidine sulfoxide (**6**), a principal metabolite, is chiral by virtue of the oxidation of the sulfur to the sulfoxide. The lone pair of electrons on the sulfur constitutes the fourth group. In humans, the (+)-enantiomer predominates (2.4:1); in rats, although (+)-cimetidine sulfoxide is produced in excess, the enantiomeric ratio approaches racemic (1.3:1) (13).

Whereas it might seem that administration of enantiomerically pure substances would be preferred, the diuretic indacrinone (**7**), is an example of a drug for which one enantiomer mediates the harmful effects of the other (14). (+)-Indacrinone, the diuretically active enantiomer or eutomer, causes uric acid retention; the other enantiomer (distomer) causes uric acid elimination. Thus, administration of a mixture of the two enantiomers, although not necessarily racemic, may have therapeutic value.

Although a great deal of the work in chiral separations is related to pharmaceuticals, the agricultural and the food and beverage industries are affected as well. For instance, several chiral pesticides are used commercially. It is possible that the enantiomers may differ in their persistence in the environment and their effectiveness against specific pests. For example, the neurotoxic action of the pesticide ethyl-4-nitrophenyl phenylphosphonothionate (EPN) resides almost entirely in the (*S*)-enantiomer; the desired insecticidal activity resides entirely in the (*R*)-enantiomer (15). This raises the question of whether the pesticide may be safer and more effective if applied as an enantiomerically pure formulation. In the food and beverage industry, many of the constituents that confer flavor or aroma are chiral. For instance, the configuration of the 4-alkyl-substituted γ-lactones responsible for much of the flavor in fruits is almost exclusively (*R*) (16). Often, the two enantiomers have very different aromas or flavors and the presence of any of the unnaturally occurring enantiomer may confer an off-flavor to the substance. This may be indicative of racemization under adverse storage conditions, adulteration, or formulation from nonnatural sources.

The growing awareness of the implications of chirality to the pharmaceutical industry has spurred tremendous effort toward stereoselective synthetic strategies and the development of new chiral catalysts. Dramatically improved techniques in asymmetric syntheses (17), chiral separations (18), analytical techniques, and stereochemical characterization have led to the widespread production and biological evaluation of numerous homochiral molecules. The appearance in the late 1980s of the journals *Chirality* and *Tetrahedron Asymmetry* provided testament to this vastly expanding research area (19–21). The U.S. FDA requires that both enantiomers of a drug be individually tested when associated toxicities occur near the effective dose of the racemic substance (22,23). The enantiomeric purity of these substances or their chiral precursors needs to be determined. Also, there are many chiral compounds for which no stereospecific synthetic pathways have been devised. Thus, there is also a tremendous need not only for analytical scale (<5–10 mg), but also for bulk-scale chiral separations. Whether analyzing drugs or synthetic precursors for enantiomeric purity, monitoring biological or environmental samples for chiral discrimination, or trying to resolve kilogram quantities of a racemic drug, there are a variety of reasons for performing chiral separations. The purpose of the separation dictates, to some extent, the method employed (see also BIOSEPARATIONS).

Traditionally, chiral separations have been considered among the most difficult of all separations. Conventional separation techniques, such as distillation (qv), liquid–liquid extraction (see EXTRACTION, LIQUID–LIQUID), crystallization (qv), or even some forms of chromatography (qv), are usually based on differences in analyte solubilities or vapor pressures. However, in an achiral environment, enantiomers or optical isomers have identical physical and chemical properties. The general approach, then, is to create a chiral environment to achieve the desired chiral separation. This requires chiral analyte–chiral selector interactions having more specificity than is obtainable using conventional techniques.

A variety of strategies have been devised to obtain chiral separations. The focus herein is on chromatographically based methodologies. Other methods include crystallization and stereospecific enzymatic-catalyzed synthesis or degradation. In crystallization methods, racemic chiral ions are typically revolved by the addition of an optically pure counterion, thus forming diastereomeric complexes. Product yields of the desired enantiomer are at most 50%.

Enzymatically based methods depend on the stereospecificity of an enzyme-catalyzed reaction, such as lipase-catalyzed esterification, to enantioselectively degrade the unwanted enantiomer or to produce the desired enantiomer. Because only one enantiomer undergoes the reaction, the subsequent separation is reduced to separating two different species. For example, in the case of enzyme-catalyzed esterification, the originally difficult enantiomeric separation is reduced to the separation of the ester of one optical isomer from the alcohol or acid of the other optical isomer of the original starting material and may be accomplished using a variety of conventional separation methodologies. One disadvantage of enzymatically based methods is that only one enantiomer is obtained and there is usually no analogous method for producing the opposite enantiomer.

An alternative method of creating a chiral environment is to derivatize a chiral analyte using an optically pure reagent thus producing diastereomers. The resultant distereomers, containing more than one chiral center, have slightly different melting and boiling points and can often be separated using conventional methods. A number of chiral derivatizing agents, as well as the types of compounds for which they are useful, have been developed and are listed in Table 1. Limitations of this approach include lack of suitable functionality in the analyte that can be derivatized with an appropriate enantiomerically pure derivatizing agent, unavailability of a derivatizing agent of sufficiently high or at least known optical purity, removal of the derivatizing group, enantiodiscrimination during derivatization, potential racemization either during derivatization or removal of the chiral derivatizing group (not always possible), and the additional validation required to confirm that the enantiomeric ratio of the final product corresponds to the original enantiomeric ratio.

Thin-Layer Liquid Chromatography

One method for creating a chiral environment is to add an optically pure chiral selector to a liquid mobile phase. Many of the chiral selectors available as chiral stationary phases for high performance liquid chromatography (hplc) originated as chiral mobile-phase additives, particularly in thin-layer chromatography (tlc). Chiral mobile-phase additives have several advantages over chiral stationary phases and continue to be the predominant mode for chiral separations by tlc (24). First of all, the chiral selector added to a mobile phase can be readily changed. Then the use of chiral mobile-phase additives allows chiral separations to be accomplished using less expensive, conventional stationary phases. A wider variety of chiral selectors are available to be used as chiral mobile-phase additives than are available as chiral stationary phases, thus providing considerably increased flexibility. Finally, the use of chiral mobile-phase additives may provide valuable insight into the chromatographic conditions and/or likelihood of success for a potential chiral stationary-phase chiral selector. This is particularly important for the development of new chiral stationary phases because of the difficulty, time, and cost involved.

Chiral mobile additives, however, do pose some unique problems. Many chiral agents are expensive or are not commercially available and therefore must be synthesized. The presence of the chiral additive in the mobile phase may also interfere with detection or recovery of the analytes. Finally, enantiomeric impurity of the chiral mobile-phase additive may add analytical complications.

Tlc offers several advantages. Besides being inexpensive, tlc can be used to rapidly screen mobile-phase conditions, eg, organic modifier content, pH, etc, chiral selectors, and analytes. Several different analytes may be run simultaneously on the same plate. Usually, no pre-equilibration of the mobile and stationary phases is required. In addition, only small amounts of mobile phase, and therefore chiral mobile-phase additive, are required. Another significant advantage is that the analyte can always be unambiguously found on the tlc plate.

Two principle mechanisms for chiral separations using chiral mobile-phase additives, analogous to models developed for ion-pair chromatography, have been proposed to explain the chiral selectivity obtained using these additives. In one

Table 1. Analyte Functional Groups and Chiral Derivatizing Reagents

Analyte functional group	Class	Derivatizing reagent Example	Product
carboxylic acid[a]	alcohol	(−)-menthol	ester
	amine	1-phenylethylamine	amide
		1-(1-naphthyl)ethylamine	
amine			
primary	aldehyde	o-phthaldialdehyde–2-mercaptoethanol	isoindole
primary and secondary	anhydrides	γ-butyloxycarbonyl-L-leucine anhydride	amide
		O,O-dibenzoyltartaric anhydride	
	acyl halides	(R)-(−)-methylmandelic acid chloride	amide
		α-methoxy-α-trifluoromethylphenylacetyl chloride	
	isocyanates	α-methylbenzyl isocyanate	urea
		1-(1-naphthyl)ethyl isocyanate	
	isothiocyanate	2,3,4,6-tetra-O-acetyl-β-D-glucopyranosyl isothiocyanate	thiourea
		α-methylbenzyl isothiocyanate	
	chloroformates[b]	(−)-menthyl chloroformate	carbamate
		(+)-1-(9-fluorenyl)ethyl chloroformate	
alcohols	acyl halides	(−)-menthoxy acid chloride	ester
		(S)-O-propionylmandelyl chloride	
	anhydrides	(S,S)-tartaric anhydride	ester
	chloroformate	(−)-menthyl chloroformate	carbonate
	isocyanate	α-methylbenzyl isocyanate	carbamate

[a] Acid- or base-catalyzed.
[b] Can N-dealkylate tertiary amines.

model, the chiral mobile-phase additive and analyte enantiomers form diastereomeric complexes in solution. The diastereomers may have slightly different physical properties such as mobile-phase solubilities or slightly different affinities for the stationary phase. Thus, the chiral separation can be achieved using conventional columns.

An alternative model has been proposed in which the chiral mobile-phase additive is thought to modify the conventional, achiral stationary phase *in situ*, thus dynamically generating a chiral stationary phase. In this case, the enantioseparation is governed by the differences in the association between the enantiomers and the chiral selector in the stationary phase.

Several different types of chiral additives have been used including chiral counterions such as $(1R)$-$(-)$-ammonium-10-camphorsulfonic acid, cyclodextrins, proteins, and various amino acid derivatives such as N-benzoxycarbonyl-glycyl-L-proline. Both $(1R)$-$(-)$-ammonium-10-camphorsulfonic acid and N-benzoxycarbonyl-glycyl-L-proline have been used under normal phase conditions, promoting ion-pair associations. In contrast, the cyclodextrins, proteins, and amino acid derivatives have been used exclusively under aqueous mobile-phase conditions.

Chiral stationary phases have been used less extensively in tlc than in high performance liquid chromatography (hplc), in large part perhaps owing to lack of availability. The cost of many chiral selectors as well as the accessibility and success of chiral additives may have inhibited widespread commercialization. Chiral stationary phases in tlc have been limited to phases based on normal or microcrystalline cellulose or triacetylcellulose sorbents or silica-based sorbents that have been chemically modified or physically coated to incorporate chiral selectors into the stationary phase. Of the silica-based materials, only the ligand-exchange phases are commercially available. Typically, the ligand-exchange selector is comprised of an amino acid residue to which a long hydrocarbon chain has been attached. The hydrocarbon chain of the functionalized amino acid is either chemically bonded to the substrate or intercalates in between the chains of a reversed-phase stationary phase thus immobilizing the chiral selector. The bidentate amino acid chiral selector is thought to reside close to the surface of the stationary phase and participates as a ligand in the formation of a biligand complex with a divalent metal ion and the chiral bidentate analyte (Fig. 1). Differences in the stabilities of the diastereomeric complexes thus formed give rise to the chiral separation.

Chiral Separation Validation. Chiral separation in tlc may be accomplished by recovering the individual analyte spots from the plate and subjecting this material to some type of chiroptical spectroscopy. Alternatively, the plates may be analyzed using a scanning densitometer. Scanning densitometers irradiate the surface of the plate at a specified wavelength in the ultraviolet or visible regions and can measure the intensity of the reflected beam. A trace of the reflected beam vs distance has the general appearance of a chromatogram (Fig. 2) (25). The relative peak heights or areas of the two enantiomers obtained at two or more different wavelengths should remain constant because the extinction coefficients of the enantiomers are identical at every wavelength.

Although chiral mobile-phase additives have been used in hplc, the large amounts of solvent, and thus chiral mobile-phase additive, required to pre-equilibrate the stationary phase renders this approach much less attractive than for tlc.

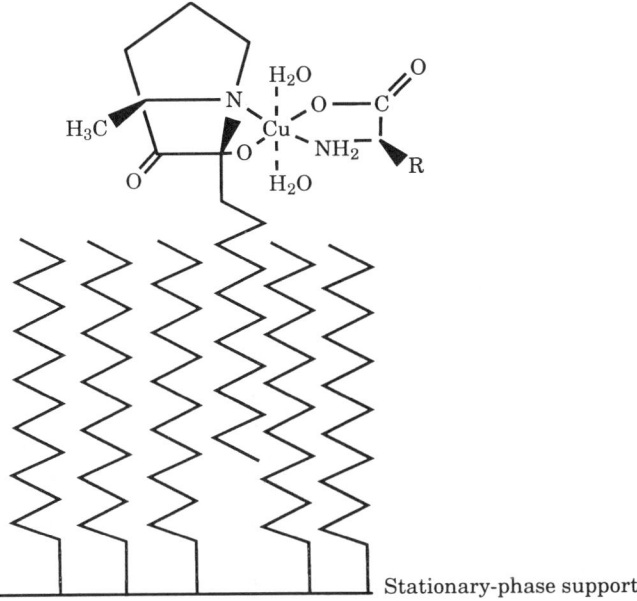

Fig. 1. Schematic of a ligand-exchange chiral selector complexed with a chiral analyte.

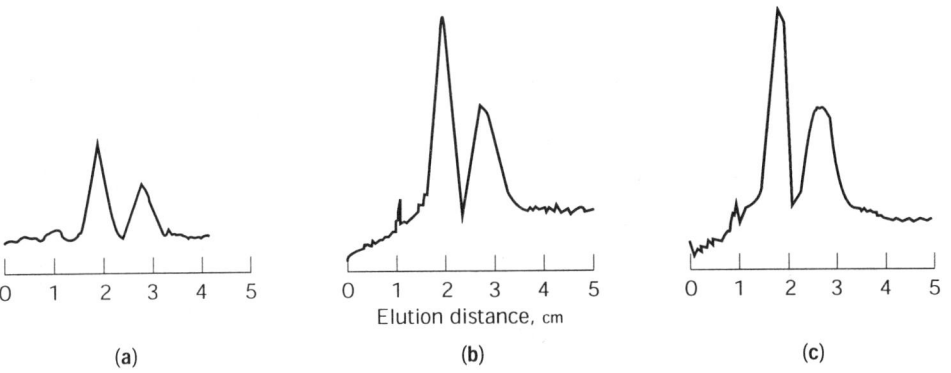

Fig. 2. Tlc densitometer scans showing the resolution of isoproterenol on a high performance tlc silica gel plate obtained using a mobile phase consisting of 6.8 mM (1R)-(−)-ammonium-10-camphorsulfonic acid in 75:25 (vol/vol) methylene chloride:methanol. Detector absorbance was (**a**) 254 nm, (**b**) 275 nm, and (**c**) 300 nm.

High-Performance Liquid Chromatography

Chiral Stationary Phases. Most chiral chromatographic separations are accomplished using chromatographic stationary phases which incorporate a chiral selector. The chiral separation mechanisms are generally thought to involve the formation of transient diastereomeric complexes between the enantiomers and the stationary-phase chiral ligand. Differences in the stabilities of these complexes account for the differences in the retention observed for the two enantiomers. Often, the use of a chiral stationary phase allows for the direct separation of the enantiomers without the need for derivatization. One advantage offered by the use of chiral stationary phases is that the chiral selector need not be enantiomerically pure, only enriched. In addition, for chiral stationary phases having a well-understood chiral recognition mechanism, assignment of configuration, eg, (R) or (S), may be possible even in the absence of optically pure standards. However, chiral stationary phases have some limitations. The specificity required for chiral discrimination limits the broad applicability of most chiral stationary phases, thus there is no universal chiral stationary phase. The cost of most chiral columns are typically much higher (ca three times) than for conventional columns. In contrast to conventional chromatographic columns, chiral stationary phases are generally not as robust, require more careful handling than conventional columns, and usually, once column performance has begun to deteriorate, cannot be returned to their original performance levels. In many cases, chromatographic column choice or mobile-phase optimization for chiral stationary phases is not as straightforward as with conventional stationary phases. For many of the chiral stationary phases, adequate chiral recognition models, used to guide selection of the appropriate column for a given separation, have yet to be developed. Thus, column selection is often reduced to identifying structurally similar analytes for which chiral resolution methods have been reported in the scientific literature or chromatographic supply catalogs and adapting a reported method for the chiral pair to be resolved.

An additional complication, sometimes arising with the use of chiral stationary phases, may occur when the analytes either exist as conformers or can undergo inversion during the chromatographic analysis. Figure 3 illustrates a typical chromatogram obtained for oxazepam, one of the chiral benzodiazepines which can undergo ring opening and inversion at the chiral center (26). The peaks appear to have a plateau between them sometimes referred to as Batman peaks. Although the appearance of Batman peaks is not unique to chiral separations, the specificity of chiral analyte–chiral selector interactions may increase the frequency of their occurrence. This effect can often be suppressed by lowering column temperature.

A large number of different types of chiral stationary phases have been commercialized including the cyclodextrin phases (27), the chirobiotic phases (28), the π–π interaction phases (29,30), and the protein phases (31–35), as well as the cellulosic and amylosic phases (36,37) and chiral crown ether phases (38,39). As of 1996 there were over 50 different chiral columns commercially available for hplc. Table 2 summarizes the types of columns available as well as typical applications and mobile-phase conditions. Each of these chiral stationary phases are successful at separating large numbers of enantiomers which, in many cases, are unresolvable using any of the other chiral stationary phases. Despite the large number and variety of chiral stationary phases, there remains a large number of enantiomeric compounds which are unresolvable. In addition, incomplete understanding of the chiral recognition mechanisms of many of these chiral stationary phases limits the realization of the full potential of the existing chiral stationary phases and hampers development of new chiral stationary phases.

Ligand-Exchange Phases. Among the earliest reports of chiral separations by liquid chromatography (lc) were those based on ligand exchange (40). Chiral separations by ligand exchange in hplc is accomplished using bidentate amino acid

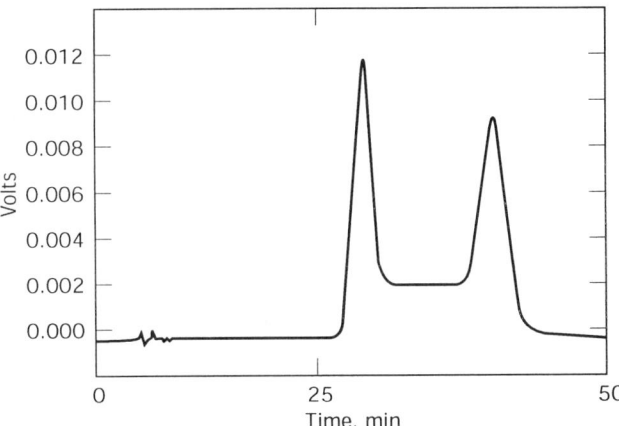

Fig. 3. Chromatogram illustrating the chiral separation obtained from oxazepam on a sulfated cyclodextrin hplc column (4.6 mm ID × 25 cm) using a 10% acetonitrile–buffer (25 mM ammonium acetate, pH 7).

ligands, immobilized on a chromatographic substrate bound to a divalent metal cation (see Fig. 1). This participates in the formation of a diastereomeric complex with a bidentate chiral analyte and the ligand. Although almost any amino acid can form the basis for the chiral selector, proline and hydroxyproline exhibit the most widespread utility. Also, although other metals can be used, copper(II) is usually the metal of choice and is added to the aqueous buffer mobile phase.

Table 2. Classes of Hplc Chiral Stationary Phases

Column chiral selector	Mobil-phase conditions	Analyte features required
Pirkle	nonpolar organic	π-acid or π-basic moieties for charge-transfer complex; hydrogen-bonding or dipole stacking capability near chiral center
protein	phosphate buffers	aromatic near chiral center; organic acids or bases; cationic drugs
cyclodextrin	aqueous buffers; polar organic	good fit between chiral cavity or chiral mouth of cyclodextrin and hydrophobic moiety; hydrogen-bonding capability near chiral center
ligand exchange	aqueous buffers	α-hydroxy or α-amino acids near chiral center; can do nonaromatic
chiral crown ether	0.01 N perchloric acid	primary amines near chiral center; can do nonaromatic
macrocyclic antibiotics	aqueous buffers; nonpolar and polar organic	amines, amides, acids, esters; aromatic; hydrophobic moiety
cellulosic and amylosic	nonpolar organic	aromatic

The dependence of chiral recognition on the formation of the diastereomeric complex imposes constraints on the proximity of the metal binding sites in the analyte. These sites are usually either a hydroxy or an amine α to a carboxylic acid. Principle advantages of this technique include the ability to assign configuration in the absence of standards and to enantioselectively resolve nonaromatic analytes. Other advantages are the compatability of the stationary phase with aqueous mobile phases and the availability of a stationary phase having the opposite enantioselectivity. Moreover, there is predictability in regard to the likelihood of successful chiral resolution for a given analyte based on a well-understood chiral recognition mechanism.

Pirkle Phases. The first commercially available chiral column for liquid chromatography was introduced in 1980. This first-generation column, named a Pirkle phase after its originator, was based on N-(3,5-dinitrobenzoyl)phenylglycine which was immobilized on a silica support. Of all of the commercially available chiral stationary phases for liquid chromatography, the chiral recognition mechanism for the Pirkle phases are probably the best understood.

Chiral recognition on Pirkle phases is thought to depend on complimentary interactions between the analyte and the selector. These interactions, which may be π–π, steric, hydrogen-bonding, or dipole–dipole interactions, contribute to the overall stability of the diastereomeric association complexes which form between the individual enantiomers and the chiral selector in the stationary phase. The π–π interactions arise through association of aromatic systems having complementary electron-withdrawing (nitro) and electron-donating (alkyl) substituents. The electron-deficient aromatic system may be referred to as π-acidic; the electron-rich system, π-basic. Three unique interactions emanating from the chiral centers of the analyte and the chiral ligand in the stationary phase seem to be required for successful chiral recognition. Thus a model invoking three unique points of interaction is sometimes referred to as the three-point interaction model (41). To promote analyte–selector interactions, functional groups are often introduced into the analyte through achiral derivatization. For example, amines may be derivatized using 3,5-dinitrobenzoyl chloride to introduce a π-acid aromatic group to promote diastereomeric complexation with a π-basic (R)-(N)-(2-naphthyl)alanine chiral selector in the stationary phase. Derivatization often has the additional benefit of enhancing solute solubility.

Reciprocity, an important concept introduced by Pirkle, exploited the notion that analytes which were well resolved using a particular chiral selector would likely be good candidates for chiral selectors to enantioselectively resolve analytes similar to the particular chiral selector. For instance, the first generation Pirkle phase incorporating N-(3,5-dinitrobenzoyl)phenylglycine was successful at resolving compounds containing naphthyl moieties near the stereogenic center. This spawned a second generation of Pirkle phases based on β-naphthylvaline which were successful at resolving analytes containing a 3,5-dinitrobenzoyl group.

Nonpolar organic mobile phases, such as hexane using ethanol or 2-propanol as typical polar modifiers, are most commonly used with these types of phases. Under these conditions, retention seems to follow normal phase-type behavior, eg, increased mobile-phase polarity produces decreased retention. The normal mobile-phase components only weakly interact with the stationary phase and are easily displaced by the chiral analytes thereby promoting enantiospecific interactions. Some of the Pirkle-type phases have also been used, to a lesser extent, in the reversed-phase mode.

The Whelk-O-1 (**8**) phase is the most recent addition to this type of chiral stationary phase as of 1996. Chiral centers are designated by asterisks. This selector has a wedge-like chiral surface where one edge offers the π-basic tetrahydrophenanthrene ring system, and the other edge is comprised of a 3,5-dinitrobenzoyl π-acidic moiety. The amide linkage between the two ring systems presents dipole stacking and hydrogen-bonding interaction sites. The presence of both π-acid and π-base features, as well as the inherent rigidity of the chiral selector, confers greater versatility than any of the previous Pirkle-type phases, imposing fewer constraints on both analyte structural features required for successful enantioresolution and mobile-phase conditions. Indeed, this chiral stationary phase has demonstrated considerable chiral selectivity for naproxen [22204-53-1], warfarin [81-81-2], and the p-chloro analogue of warfarin under reversed-phase conditions.

(**8**)

Cyclodextrin Phases. Among the most successful of the liquid chromatographic reversed-phase chiral stationary phases have been the cyclodextrin-based phases (42,43). Cyclodextrins (CDs) are macrocyclic compounds comprised of D-glucose bonded through 1,4-α-linkages and produced enzymatically from starch. The greek letter which precedes the name indicates the number of glucose units incorporated in the CD, eg, α = 6, β = 7, γ = 8, etc. Cyclodextrins are toroidal-shaped molecules having a relatively hydrophobic internal cavity. The exterior is relatively hydrophilic because of the presence of the primary and secondary hydroxyls. The primary C6 hydroxyls are free to rotate and can partially block the CD cavity from one end. The mouth of the opposite end of the CD cavity is encircled by the C2 and C3 secondary hydroxyls. The restricted conformational freedom and orientation of these secondary hydroxyls is thought to be responsible for the chiral recognition inherent in these molecules (44). The most commonly used cyclodextrin in hplc is the β-cyclodextrin [7585-39-9], having seven glucose units (Fig. 4). The mechanism thought to be responsible for the chiral selectivity observed with the cyclodextrin phases is based on the formation of an inclusion complex between the hydrophobic moiety of the chiral analyte and the hydrophobic interior of the cyclodextrin cavity, thought to be shaped like a truncated cone (Fig. 5). Preferential complexation between one optical isomer and the cyclodextrin through stereospecific interactions with the secondary hydroxyls which line the mouth of the cyclodextrin cavity results in the enantiomeric separation. Unlike the Pirkle-

332 CHROMATOGRAPHY, CHIRAL

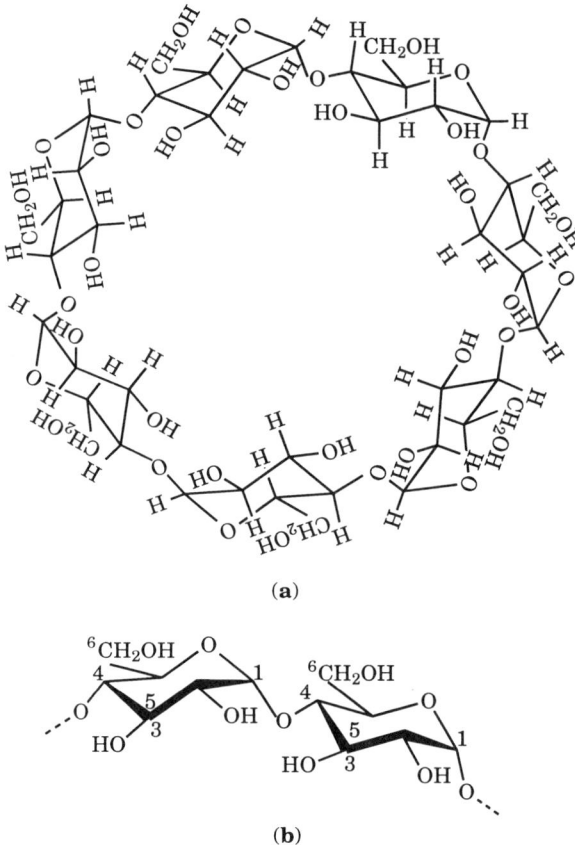

Fig. 4. Structural diagram of (**a**) β-cyclodextrin and (**b**) two of the glucopyranose units illustrating details of the α-(1,4) glycosidic linkage and C-1 (D) chair conformation.

type phases, enantiospecific interactions between the analyte and the cyclodextrin are not the result of a single, well-defined association. For cyclodextrins there is more of a statistical averaging of all the potential interaction with each interaction weighted by its energy or strength of interaction.

Vast amounts of empirical data suggest that chiral recognition on cyclodextrin phases in the reversed-phase mode require the presence of an aromatic moiety which can fit into the cyclodextrin cavity, that there be hydrogen bonding groups in the molecule, and that the hydrophobic and hydrogen-bonding moieties should be in close proximity to the stereogenic center. Chiral recognition seems to be enhanced if the stereogenic center is positioned between two π-systems or incorporated in a ring.

Most of the chiral separations reported using the native cyclodextrin-based phases have been accomplished in the reversed-phase mode using aqueous buffers containing small amounts of organic modifiers. Polar organic mobile phases, however, have gained in popularity because of their ease of removal from the sample and reduced tendency to accelerate column degradation relative to the hydroorganic mobile phases (45). In these cases, because the more nonpolar

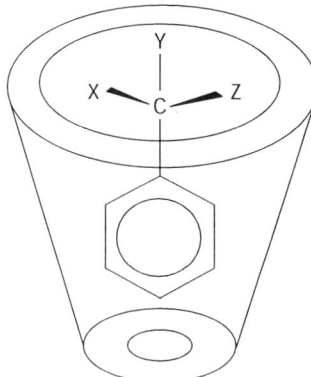

Fig. 5. Schematic of a hydrophobic inclusion complex between a chiral analyte, C_6H_5CXYZ, and a cyclodextrin.

component of the mobile phase is thought to occupy the cyclodextrin cavity, the analyte is thought to sit atop the mouth of the cyclodextrin much like a lid.

Limitations of the chiral selectivity of the native cyclodextrins fostered the development of various functionalized cyclodextrin-based chiral stationary phases, including acetylated, sulfated, 2-hydroxypropyl, 3,5-dimethylphenylcarbamoylated, and 1-naphthylethylcarbamoylated cyclodextrin. Each of the glucose residues contribute three hydroxyl groups to which a substituent may be appended, thus each cyclodextrin contributes multiple sites for derivatization. In the bonded phases, the cyclodextrins are thought to be tethered in the silica substrate through one or two spacer ligands. Typical degrees of substitution per β-cyclodextrin, which has 21 hydroxyls, range from three to ten. Hence, there are many residual hydroxyls on each cyclodextrin.

The substituents of these functionalized cyclodextrins seem to play a variety of roles in enhancing chiral recognition. In some cases, the substituent may only serve to enlarge the chiral cavity or may provide alternative interaction sites. For instance, in the case of the naphthylethylcarbamoylated cyclodextrin, the naphthyl ring provides a π-basic site and the carbamate linkage provides additional hydrogen bonding and dipole interaction sites not available with native cyclodextrin. On the sulfated cyclodextrin phase, the sulfate group presents the potential for ion-pair formation unavailable with the native cyclodextrin. The introduction of 2-hydroxypropyl- and 1-napthylethylcarbamoyl substituents incorporates additional stereogenic centers onto the cyclodextrin. In some cases, the configuration of the substituent was found to dominate the enantiomeric elution order. However, in other cases, the enantiomeric elution order was found to be independent of the configuration of the substituent.

In some cases the chiral selectivity of the cyclodextrin seemed to synergize the chiral selectivity of one configuration of the substituent while antagonizing the chiral selectivity of the oppositely configured substituent. One feature of these functionalized cyclodextrins that is particularly attractive is that many of them exhibit enantioselectivity under hydroorganic reversed-phase as well as normal-phase and polar organic mobile-phase conditions and that each set of conditions can provide chiral separations for analytes which are not resolved under any of the other type of mobile-phase conditions. Further, the chromatographic mode, eg, reversed phase to normal phase, can be readily changed with no deleterious impact on chiral recognition as long as routine care is taken to avoid problems with solvent immiscibility. The naphthylethylcarbamoylated cyclodextrin phase was considered to be one of the first multimodal chiral stationary phases (46).

Immobilization. Cyclodextrin stationary phases utilize cyclodextrins bound to a solid support in such as way that the cyclodextrin is free to interact with solutes in solution. These bonded phases consist of cyclodextrin molecules linked to silica gel by specific nonhydrolytic silane linkages (47). The stable cyclodextrin-bonded phase is sold commercially under the trade name Cyclobond (Advanced Separation Technologies, Whippany, New Jersey). The vast majority of all reported hplc separations on CD-bonded phases utilize this media which was also the first chiral stationary phase developed for use in the reversed-phase mode.

Applications. The first widely applicable lc separation of enantiomeric metallocene compounds was demonstrated on β-CD bonded-phase columns. Thirteen enantiomeric derivatives of ferrocene, ruthenocene, and osmocene were resolved (48). Retention data for several of these compounds are listed in Table 3, and Figure 6a shows the lc separation of three metallocene enantiomeric pairs. β-Cyclodextrin bonded phases were used to resolve several racemic and diastereomeric, 2,2-binaphthyldiyl crown ethers (51). These compounds do not contain a chiral carbon but still exist as enantiomers because of the staggered position of adjacent naphthyl rings, and a high degree of chiral recognition was attained for most of these compounds (51).

The β-CD column exhibits excellent selectivity for enantiomers of certain amino acid derivatives. Underivatized amino acids are apparently too small to bind tightly to the β-CD cavity and show no enantiomeric resolution. When a substituent such as a dansyl group is present on the amino acid, strong inclusion complexes with β-CD are formed and baseline separation is achieved (see Table 3) (52). Either the amino or the carboxylate group of the amino acid can be derivatized to obtain chiral recognition. Derivatization of both groups, however, tends to reduce chiral recognition. It is possible to detect as little as 0.2% of one enantiomer in a racemic mixture as shown in Figure 6b, thus providing an extremely sensitive test of optical purity (47).

Table 3 gives chromatographic data for different classes of enantiomeric drugs resolved by β-CD bonded phases (50). Drugs for which resolution factors, R_s, greater than 1.0 were obtained include mephenytoin, ketoprofen, chlorpheniramine, and the barbiturates mephobarbital and hexobarbital. Cyclodextrin-bonded phases provide a rapid and specific technique for the pharmacological evaluation of racemic drugs.

Many diastereomers, geometric isomers, and epimers can be successfully resolved using cyclodextrin phases (47,52). For example, the four epimers of estroil [50-27-1], $C_{18}H_{24}O_3$, were separated using β-CD, and *cis*-benzo(a)pyrene and *trans*-benzo(a)pyrene were completely resolved on a γ-CD column. Diastereomeric drugs such as the cinchona alkaloids and antiestrogens have also been separated. Cyclodextrin columns are also of great utility in separating

Table 3. Retention Data for Racemic Compounds Separated on a β-Cyclodextrin Stationary Phase[a]

Compound	CAS Registry Number	Molecular formula	k'[b]	α[c]	R_s[d]
Complexes					
1-ferrocenyl-1-methoxyethane		$C_{13}H_{17}FeO$	3.8	1.12	1.58
(S)-(1-ferrocenylethyl)-thioethanol		$C_{14}H_{17}FeSO$	3.0	1.23	2.13
1-ruthenocenylethanol		$C_{12}H_{14}RuO$	4.6	1.11	1.56
alanine β-naphthylamide	[74144-49-3]	$C_{13}H_{14}N_2O \cdot HCl$	5.1	1.20	2.0
dansyl-leucine	[102783-70-0]	$C_{18}H_{24}N_2O_4S \cdot C_6H_{13}N$	3.0	1.40	2.4
dansylphenylalanine	[42808-06-0]	$C_{21}H_{22}N_2O_4S \cdot C_6H_{13}N$	3.1	1.23	1.11
β-Adrenergic blocker					
propranolol hydrochloride	[3506-09-0]	$C_{16}H_{21}NO_2 \cdot HCl$	2.78	1.04	1.40
Antihistamine					
chlorpheniramine	[132-22-9]	$C_{16}H_{19}ClH_2$	5.86	1.07	1.51
Sedative–anticonvulsants					
mephenytoin	[50-12-4]	$C_{12}H_{14}N_2O_3$	0.48	1.33	1.83
mephobarbital	[115-38-8]	$C_{13}H_{14}N_2O_3$	14.80	1.14	1.60
Nonsteroidal antiinflammatory					
ketoprofen	[22071-15-4]	$C_{16}H_{14}O_3$	7.67	1.06	1.24

[a] Refs. 47, 49, and 50.
[b] k' = capacity factor for the first eluting isomer.
[c] α = selectivity factor and ratio of the capacity factor of the last to the first eluting isomer.
[d] R_s = resolution factor = 2 (distance between peaks)/(sum of the bandwidths of the two peaks).

structural isomers such as the ortho-, meta-, and para-isomers of nitroaniline, xylene, cresols, nitrophenols, and substituted benzoic acids (53). Cyclodextrin-bonded phases are used as nonconventional reversed phases for routine analyses as a result of the unusual selectivity of the cyclodextrin columns. Uses in

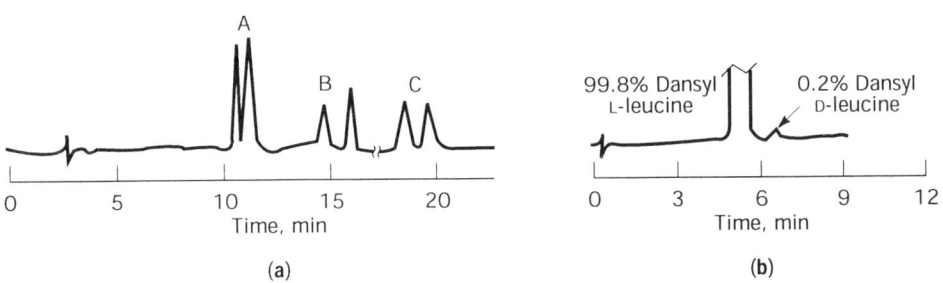

Fig. 6. Chromatogram showing (**a**) the lc separation of A, (±)-(S)-(1-ferrocenylethyl)thioethanol; B, (±)-1-ferrocenyl-1-methoxyethane; and C, (±)-1-rutheno-cenylethanol, on a 25-cm β-cyclodextrin column (see Table 3), and (**b**) the potential use of a β-cyclodextrin column to determine optical purity when one of the enantiomers if present at very low concentration (47,48).

routine analyses include the separation of a series of barbiturates, mycotoxins, polycyclic aromatic hydrocarbons, vitamins and selected dipeptides (54). Cyclodextrin stationary phases can be used in the normal-phase mode with hexane–2-propanol mobile phases. Solutes adsorb to the hydroxyl groups on the outside of the cyclodextrin, rather than forming inclusion complexes. Separations in the normal-phase mode tend to be analogous to those of diol columns.

Mechanism of Separation. There are several requirements for chiral recognition. *(1)* Formation of an inclusion complex between the solute and the cyclodextrin cavity is needed (49,52). This has been demonstrated by performing a normal-phase separation, eg, using hexane–2-propanol mobile phase, on a β-CD column. The enantiomeric solute is then restricted to the outside surface of the cyclodextrin cavity because the hydrophobic solvent occupies the interior of the cyclodextrin. *(2)* The inclusion complex formed should provide a relatively tight fit between the hydrophobic species and the cyclodextrin cavity. This is evident by the fact that β-CD exhibits better enantioselectivity for molecules the size of biphenyl or naphthalene than it does for smaller molecules. Smaller compounds are not as rigidly held and appear to be able to move in such a manner that they experience the same average environment. *(3)* The chiral center, or a substituent attached to the chiral center, must be near to and interact with the mouth of the cyclodextrin cavity. When these three requirements are fulfilled the possibility of chiral recognition is favorable.

The unidirectional 2- and 3-hydroxyl groups located at the mouth of the cyclodextrin cavity appear to be of particular importance in chiral recognition. This is seen in Figure 7, which shows computer-generated projections of the lowest free-energy inclusion complexes of (R)- and (S)-propranolol with β-CD (50). The (R)- and (S)-propranolol are placed identically inside the cyclodextrin cavity and the hydroxyl groups attached to the chiral carbon for the enantiomers are

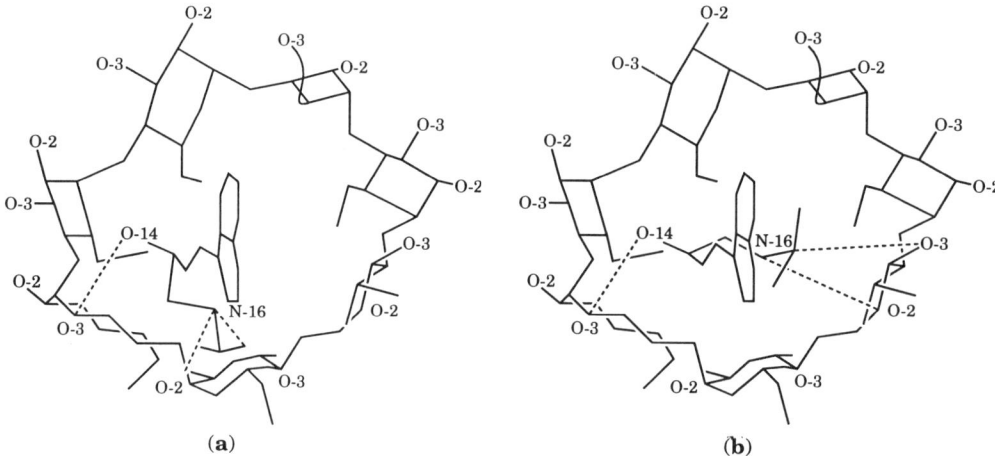

Fig. 7. Computer projections of β-cyclodextrin inclusion complexes of (**a**) (R)-propranolol and (**b**) (S)-propranolol from x-ray crystallographic data. Dotted lines represent potential hydrogen bonds (see text). The configurations shown represent the optimal orientation of each isomer on the basis of the highest degree of hydrogen bonding and complexation (50).

placed in the same position for ideal hydrogen bonding to the 3-hydroxyl group of the cyclodextrin. Important differences exist between the complexes with respect to the secondary amine group. In the (R)-complex the respective bond distances between the nitrogen and the cyclodextrin 2- and 3-hydroxyl groups are 0.33 and 0.28 nm, respectively. This allows for two reasonable hydrogen bond interactions. The same amine group in the (S)-propranolol complex is positioned less favorably for hydrogen bonding: closest bond distances are 0.38 and 0.45 nm, respectively. These models suggest that (R)-propranolol can preferentially interact with β-cyclodextrin in a way that the (S)-isomer cannot, resulting in chiral recognition by the cyclodextrin molecule.

α_1-*Acid Glycoprotein Phases.* *Properties and Structure.* α_1-Acid glycoprotein (α_1-AGP) has a molecular mass of about 41,000 and consists of a peptide chain having 181 amino acid residues and five carbohydrate units (55,56). Two cystine disulfide cross-linkages connect residues 5 and 147 and residues 72 and 164. The carbohydrate units comprise 45% of the molecule and contain sialic acid, hexosamine, and neutral hexoses. In phosphate buffer the isoelectric point of the protein is 2.7. AGP is a very stable protein and tolerates organic solvents as well as high temperatures. AGP columns may be used over a wide pH range without being denatured. Denaturation, as measured by changes in the optical rotation of the molecule, may be caused by boiling in distilled water or adding 10 M LiBr, 10 M urea, or 5 M HCl (55).

Immobilization. The solid-phase support used for bonding AGP is silica gel. The general preparation is based on bonding the protein by charge forces to diethylaminoethylsilica followed by a cross-linking procedure (57). A final reduction to secondary amines is performed using cyanoborohydride. The silica has a particle diameter of 10 µm, a large pore volume, and a large surface area. The commercially available EnantioPac AGP column (LKB Products, Bromma, Sweden) is manufactured using this procedure. A second-generation AGP column, Chiral-AGP (ChromTech AB, Norsburg, Sweden), uses covalent linkage of the protein onto silica with cross-linking of adjacent protein molecules (58,59). This latter packing contains 5 µm spherical silica particles having a smaller pore volume and a smaller surface area giving a stationary phase that is more mechanically stable and resistant to hydrolytic attack. Bonding capacity of the chiral phase and loading capacity of solute are directly dependent on the amount of protein that is bound. The amount of AGP bound on the silica depends on the bonding technique and on the accessible surface area of the silica.

Applications. Retention and selectivity of solutes on AGP is regulated by the concentration and properties of modifiers in the aqueous-based mobile phase. Retention and enantioselectivity usually decrease with increasing concentration of uncharged organic modifiers such as methanol, ethanol, 2-propanol, and acetonitrile in the mobile phase. However, in some cases, enantioselectivity can be improved by adding uncharged modifiers to the mobile phase (60). For example, addition of 2-propanol as an organic modifier gave improved chiral resolution for mephenytoin [50-12-4], $C_{12}H_{14}N_2O_2$, and methylphenobarbital. On the other hand, the enantioselectivity for the tertiary amines mepivacaine and bupivacaine is unaffected by additions of up to 8% of propanol to the mobile phase, despite the fact that the retention of these solutes decreases markedly under these conditions (59).

Column retention is generally increased if the modifier has a charge opposite to that of the solute; a modifier having the same charge as the solute generally decreases retention. For some compounds the addition of charged modifiers is essential to achieve chiral recognition. The fenthiazin derivative, propiomazine [362-29-8], $C_{20}H_{24}N_2OS$, was not resolved on an AGP column when phosphate buffer, pH 7.55, was used as the mobile phase. After addition of 1 mM of the tertiary amine N,N-dimethyloctylamine (DMOA) to the mobile phase, however, the enantiomers of propiomazine were baseline resolved (61) and longer retention times were obtained. Increase in retention is believed to be caused by competition between the solute and DMOA in binding to the AGP. The changes in selectivity are believed to result from reversible changes in the conformation of AGP brought about by changes in pH. Similar effects on selectivity and retention have also been observed by addition of tetrapropylammonium bromide or tetrabutylammonium bromide (56,57,62,63).

Changes in the pH and temperature of the mobile phase can have a significant effect on both retention and selectivity. Changing the pH of the mobile phase has a profound effect on retention and selectivity for basic, acidic, and nonprotolytic compounds. For example, the separation factors for hexobarbital [56-29-1], $C_{12}H_{16}N_2O_3$ (weakly acidic) and metoprolol [37350-58-6], $C_{15}H_{25}NO_3$ (basic) increase with increasing pH, but the enantioselectivity for stronger acids such as 2-phenoxypropionic acid (pK_a = 4.6) generally decreases (62). Column temperature also strongly influences retention and enantioselectivity. It has been generally reported that retention, resolution, and separation factors decrease with increasing temperature, whereas efficiency increases.

AGP columns have wide application for the direct separation of enantiomers of many different classes of drugs, amines, acids, and nonprotolytic compounds (59,64). Acidic drugs resolved include ibuprofen [15687-27-1], $C_{13}H_{18}O_2$, ketoprofen [22071-15-4], $C_{16}H_{14}O_3$, and naproxen [22204-53-1], $C_{14}H_{14}O_3$, and basic drugs such as disopyramide [3737-09-5], $C_{21}H_{29}N_3O$, tropicamide [1508-75-4], $C_{17}H_{20}N_2O_2$, atropine [51-55-8], $C_{17}H_{23}NO_3$, and homatropine [87-00-3], $C_{16}H_{21}NO_3$, have also been separated (62,65). Table 4 lists some racemic compounds that have been completely resolved using a Chiral-AGP column (59). The AGP columns are also commonly used in the determination of enantiomers present at low concentrations in biological fluids such as plasma and urine (60,66). Metoprolol was extracted from plasma and injected on a Chiral-AGP column for separation (66). It was possible to measure as little as 2 nmol/L plasma using fluorescence detection after separation.

Mechanism of Separation. Whereas the scientific basis of separation is well documented, ie, differences in binding between drug enantiomers and proteins, the mechanism for chiral recognition is not clearly understood. It is well known that the conformation of a native protein in solution can be altered by addition of organic modifiers and changes in pH. It is assumed that the AGP molecule has a high degree of flexibility even after bonding to the silica surface. Therefore adding modifiers to the mobile phase can alter the AGP molecule so that new chiral phases having different binding properties are induced (59,62).

Bovine Serum Albumin Chromatographic Phases. Properties and Structure. Bovine serum albumin (BSA) is a globular protein having a molecular mass of 66,210. It consists of 581 amino acids in a single chain, and 17 intrachain disulfide

Table 4. Baseline-Resolved Racemic Compounds Using Chiral-AGP

Compound	CAS Registry Number	Molecular formula	k'^a	α^b
alprenolol	[13655-52-2]	$C_{15}H_{23}NO_2$	17.9	1.19
atenolol	[29122-68-7]	$C_{14}H_{22}N_2O_3$	2.91	1.36
bupivacaine	[2180-92-9]	$C_{18}H_{28}N_2O$	6.66	1.29
cyamenazine		$C_{19}H_{21}N_3S$	5.95	1.55
ephedrine	[90-81-3]	$C_{10}H_{15}NO$	3.86	1.34
ketamine	[299-42-3]	$C_{13}H_{16}ClNO$	7.00	1.26
metoprolol	[37350-58-6]	$C_{15}H_{25}NO_3$	4.72	1.26
oxprenolol	[6452-71-7]	$C_{15}H_{23}NO_3$	16.1	1.29
pheniramine	[86-21-5]	$C_{16}H_{20}N_2$	11.3	1.33
pindolol	[13523-86-9]	$C_{14}H_{20}N_2O_2$	13.9	1.21
verapamil	[52-53-9]	$C_{27}H_{38}N_2O_4$	15.6	1.32
warfarin	[81-81-2]	$C_{19}H_{16}O_4$	8.18	1.39

^{a} k' = capacity factor for the first eluting isomer.
^{b} α = selectivity factor (see Table 3).

bridges form nine double loops (67). Having an isoelectric point of 4.7, BSA is a relatively acidic protein. It is highly soluble in water, but like most globular proteins it precipitates from solution at high salt concentrations. BSA exhibits hydrophobic character and numerous examples of organic compounds binding to albumins have been reported (68). Whereas hydrophobic interactions contribute greatly to the total affinity of organic ligands for BSA, there are other contributions to consider, mainly electrostatic interactions, hydrogen bonding, and charge-transfer processes.

The first observation of the enantioselective properties of an albumin was made in 1958 (69) when it was discovered that the affinity for L-tryptophan exceeded that of the DL-tryptophan [54-12-6], $C_{11}H_{12}N_2O_2$, on BSA immobilized to Sepharose (70). After extensive investigation of the chromatographic behavior of numerous racemic compounds under different mobile-phase conditions, a BSA-SILICA hplc column (Resolvosil-R-BSA, Macherey-Nagel GmvH, Duren, Germany) was introduced in 1983.

Retention and stereoselectivity on the BSA columns can be changed by the use of additives to the aqueous mobile phase (71). Hydrophobic compounds generally are highly retained on the BSA, and a mobile-phase modifier such as 1-propanol can be added to obtain reasonable retention times. The retention and optical resolution of charged solutes such as carboxylic acids or amines can be controlled by pH and ionic strength of the mobile phase.

Applications. Various *N*-derivatives of amino acids are resolvable on BSA columns. These *N*-amino acid derivatives include benzenesulfonyl-, phthalimido-, 5-dimethylamino-1-naphthalenesulfonyl- (DANSYL-), 2,4-dinitrophenyl- (DNP-), and 2,3,6-trinitrophenyl- (TNP-) derivatives (71). Amines such as Prilocain, (±)-2-(propylamino)-*o*-propiono-toluidide, a local anesthetic (Astra Pharmaceutical Company), are also resolved on BSA. The aromatic amino acids DL-tryptophan, 5-hydroxy-DL-tryptophan, DL-kynurenine [343-65-7], $C_{10}H_{12}N_2O_3$, and 3-hydroxy-DL-kynurenine [484-78-6], and drugs such as warfarin, phenprocoumon, and benzodiazepine derivatives can be separated on BSA as well.

Other Protein-Based Phases. Other proteins used as chiral stationary phases for hplc include bovine and human serum albumin, ovomucoid, avidin, and cellobiohydrolase. In most cases, the protein is immobilized onto γ-aminopropyl silica and covalently attached using a cross-linking reagent such as N,N'-carbonyldiimidazole. The tertiary structure or three-dimensional organization of proteins is thought to be important for activity and chiral recognition. Therefore, mobile-phase conditions which cause protein denaturation or loss of tertiary structure must be avoided.

Typically, the mobile phases used with the protein-based chiral stationary phases consist of aqueous phosphate buffers. Often small amounts of organic modifiers, such as methanol, ethanol, propanol, or acetonitrile, are added to reduce hydrophobic interactions with the analyte and to improve enantioselectivity. In some cases, it is thought that the presence of organic modifiers, such as N,N-dimethyloctylamine or octanoic acid, in the mobile phase may be playing an active role in enhancing chiral recognition through allosteric effects. In these cases, the absorption of the organic modifier onto the protein is thought to induce conformational changes in the overall tertiary structure of the protein which may enhance chiral recognition by a variety of mechanisms including changing the accessibility of various sites on the protein.

The chiral recognition mechanism for these protein-based phases is not well understood. In some cases, it is thought that analytes may form inclusion complexes with hydrophobic pockets within the polymer matrix. These hydrophobic interactions may couple with hydrogen bonding, electrostatic interactions, and π–π or dipole stacking to individual amino acid residues thus contributing to stereospecific orientational constraints within the hydrophobic pockets. Optimization of chromatographic conditions and selection of analytes that can be successfully resolved on these phases is usually done empirically. In addition, the large molecular weight of the biopolymers dictates that the amount of chiral selector that can be immobilized on the column packing material is very small. Although the protein is large, relative to the analytes, the actual region of the protein which effects the chiral separation may be very small. Thus, the capacity, ie, the amount of material resolvable during a single chromatographic run, of these columns is generally fairly small (<~0.1 mg) and the columns are easily overloaded.

An interesting application of the protein-based phases is various protein binding and displacement experiments which can be done fairly routinely. For instance, chiral stationary phases derived from the serum albumin, one of the most abundant blood proteins which functions as a transport protein, from several different animal species including rabbit, mouse, and human have been used. Differences in the enantioselectivity toward a particular drug of a serum albumin column derived from human and a column derived from some other animal species might indicate that species might not be a good animal model during drug development thus obviating the need for animal testing (72).

Chiral separations on protein-based phases may also provide useful information on drug interactions. For instance, the effect of the individual enantiomers of warfarin on the enantioselectivity of human serum albumin toward benzodiazepinones has been studied using a human serum column with warfarin as a mobile-phase additive (72).

Cellulose Triacetate and Cellulose Derivatives. Properties and Structure. Cellulose [9004-34-6] and other polysaccharides have long been known to have chiral recognition properties. Cellulose is readily available, inexpensive, and has good chemical stability. Microcrystalline cellulose triacetate [9012-90-3] (MCA), which is commercially available, is the product of the heterogeneous acetylation of microcrystalline cellulose (73). Various cellulose ester derivatives supported on macroscopic silica gel are available as hplc columns from Daicel (Daicel Chemical Industries, Ltd., Tokyo, Japan). These columns have good mechanical stability and mobile phases such as hexane–2-propanol or alcohols are used. The mechanism of chiral recognition on cellulose-based phases is unknown, although hydrogen bonding and ligand inclusion play a part in resolution (74,75).

Applications. MCA is used for the resolution of many classes of chiral drugs. Polar compounds such as amines, amides, imides, esters, and ketones can be resolved (76). A phenyl or cycloalkyl group near the chiral center seems to improve chiral selectivity. Nonpolar racemates have also been resolved, but charged or dissociating compounds are not retained on MCA. Mobile phases used with MCA columns include ethanol and methanol.

The Daicel columns can be used for the separation of a wide variety of compounds, including aromatic hydrocarbons having hydroxyl groups, carbonyls and sulfoxides, barbiturates, and β-blockers (77,78). There are presently nine different cellulose derivative-based columns produced by Daicel Chemical Industries. The different columns each demonstrate unique selectivities so that a choice of stationary phases is available to accomplish a separation.

Cellulosic and Amylosic Phases. Cellulose and amylose are comprised of the same glucose subunits as the cyclodextrins. In the case of cellulose, the glucose units are attached through 1,4-β-linkages resulting in a linear polymer. In the case of amylose, the 1,4-α-linkages, as are found in the cyclodextrins, are thought to confer helicity to the polmeric chain.

Cellulosic phases as well as amylosic phases have also been used extensively for enantiomeric separations (79,80). Most of the work has been with various derivatives of the native carbohydrate. The enantioresolving abilities of the derivatized cellulosic and amylosic phases are reported to be dependent on the types of substituents on the aromatic moieties which are appended onto the native carbohydrate (81). Some of the derivatives include the following:

Cellulosic	Amylosic
triacetate	
tribenzoate	
tribenzylether	
tricinnamate	
triphenylcarbamate	triphenylcarbamate
tris-3,5-dichlorophenylcarbamate	
tris-3,5-dimethylphenylcarbamate	tris-3,5-dimethylphenylcarbamate
tris-1-phenylethylcarbamate	tris-1-phenylethylcarbamate

With the exception of the microcrystalline cellulose(I) triacetate and tribenzoate materials, which are sufficiently robust to be used directly as packing material, most of the commercially available cellulosic and amylosic phases are comprised of mixtures of exhaustively derivatized polymers which are coated onto large pore γ-aminopropyl silica. Even though these coated polymeric phases exhibit admirable enantioselectivities, they have some potential disadvantages. The large polymer size requires the use of fragile, large pore silica. The fact that the chiral selector for these phases is coated onto the silica restricts the types of mobile phases that can be used. In addition, the secondary structure of the polymer seems to be important in the chiral recognition mechanism. The secondary structure of the coated amylose or cellulose may be altered irreversibly by storing the columns in polar solvents leading to disastrous consequences for chiral separations. The polymeric nature of the chiral selectors for these phases and the importance of the secondary structure also hamper the development of adequate models for the chiral recognition mechanism for these phases (82). Despite these limitations, the cellulosic and amylosic phases have enjoyed tremendous success at resolving structurally diverse compounds.

The chiral recognition sites on these polymeric carbohydrate phases are thought to be channels or grooves in the polymer matrix and that analytes are included into these channels. Evidence for inclusion is provided by the enhanced chiral recognition observed for many analytes as the steric bulk of the alcohol mobile-phase modifier increases. Chiral recognition seems to require the presence of an aromatic ring, for π–π interactions, and polar sites of unsaturation or hydrogen-bonding functionalities. As in the case of the Pirkle-type phases, these chiral stationary phases are usually used in the normal-phase mode and mobile phases typically consist of hexane and 2-propanol although there have been some reports of these phases being used in the reversed-phase mode.

Chirobiotic Phases. The chirobiotic chiral stationary phases are based on macrocyclic antibiotics such as vancomycin [1404-93-9] (**9**) and teicoplanin (**10**).

(9)

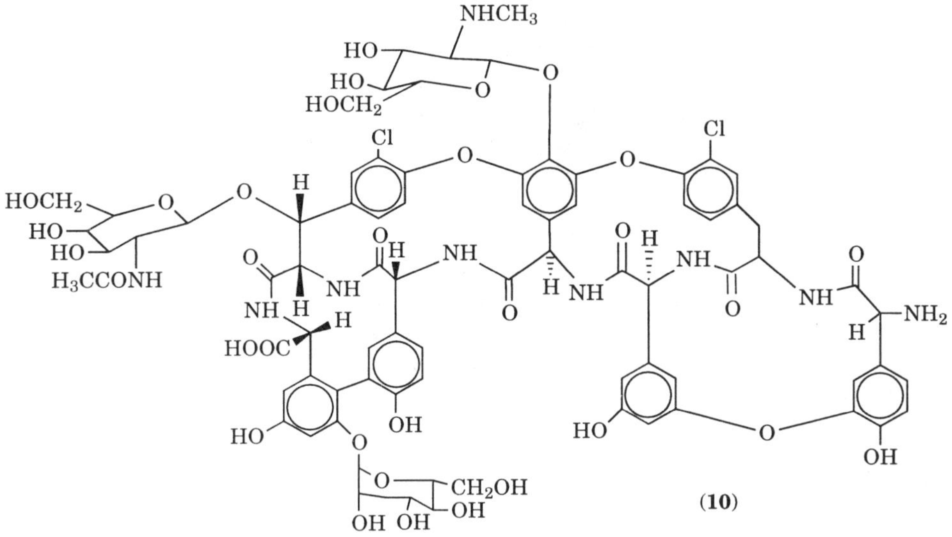

(10)

These chiral selectors, originally used as chiral additives in capillary zone electrophoresis, incorporate aromatic and carbohydrate as well as peptide and ionizable moieties. The presence of aromatic groups, allowing for π–π interactions, and the macrocyclic rings, offering potential inclusion complexation, give these phases some of the advantages of the protein-based phases, eg, peptide and hydrogen-bonding sites, and the carbohydrate-based phases but with greater sample capacity and greater mobile-phase flexibility. Inded, these phases seem to be truly multimodal in that they have demonstrated chiral selectivity in the normal, polar organic and reversed-phase modes. In the normal and polar organic-phase modes, π–π interactions and dipole stacking are thought to play a predominant role in chiral selector–analyte interactions. In the reversed-phase mode, hydrogen bonding, inclusion complexation and, for charged analytes, electrostatic interactions are thought to dominate the interactions. In addition, the use of such well-defined chiral selectors facilitates method development and optimization.

Chiral Crown Ether Phases. Chiral crown ethers based on 18-crown-6 (**11**) can form inclusion complexes with ammonium ions and protonated primary amines.

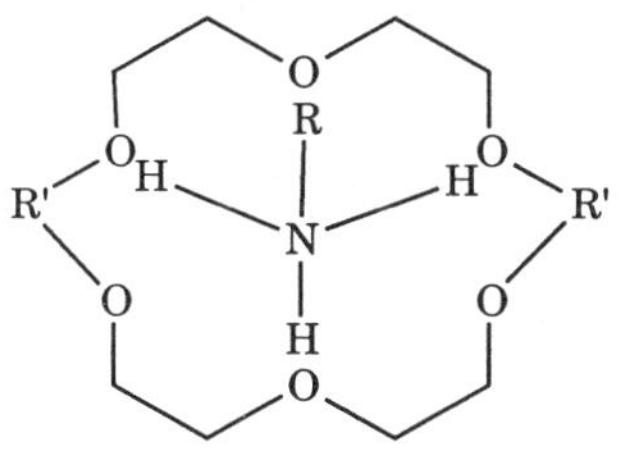

(11)

Immobilization of these chiral crown ethers on a chromatographic support provides a chiral stationary phase which can resolve most primary amino acids, amines, and amino alcohols. Significantly, the chiral crown ether phase is unique in that it is one of the few liquid chromatographic chiral stationary phases which does not require the presence of an aromatic ring to achieve chiral separations. Mobile phases used with this stationary phase are typically 0.01 N perchloric acid containing small amounts of methanol or acetronitrile. One advantage of these phases is that both configurations of the chiral stationary phase are commercially available.

Chiral Synthetic Polymer Phases. Chiral synthetic polymer phases can be classified into three types. In one type, a polymer matrix is formed in the presence of an optically pure compound to molecularly imprint the polymer matrix. Subsequent to the polymerization, the chiral template is removed, leaving the polymer matrix with chiral cavities. The degree of cross-linking in the polymer matrix and degree of association between the template molecule and the monomer, governed by the type and concentration of the monomer, the concentration of the template, and the solvent and temperature or pressure under which polymerization takes place, all play a role in selectivity. The selectivities achieved from these phases are generally excellent, thus facilitating semipreparative separations. However, the applicability of these chiral stationary phases are generally limited to the analyte used for the imprint and a limited number of analogues. In addition, these types of phases generally exhibit poor efficiency in large part because the polymeric matrix contributes to nonstereospecific binding. Advantages of this approach include the ability to prepare reciprocal phases and the predictability of the enantiomeric elution order.

Another type of synthetic polymer-based chiral stationary phase is formed when chiral catalysts are used to initiate the polymerization. In the case of poly(methylmethacrylate) (PMMA) polymers, the chirality of the polymer arises from the helicity of the polymer and not from any inherent chirality of the individual monomeric subunits (83).

A third type of synthetic polymer-based chiral stationary phase is produced when a chiral selector is either incorporated within the polymer network or attached as pendent groups onto the polymer matrix, analogous to methods used to produce polymeric chiral stationary phases for gas chromatography (gc) (84). The polymers can be either coated onto a silica substrate and comonomers bearing silane functional groups may be added for subsequent reaction with the silica, or the silica may be chemically modified to incorporate monomer-bearing silanes. Chemical bonding of the polymer to the substrate eases the mobile-phase restrictions imposed on the coated chiral polymer stationary phases.

In general, the synthetic polymeric phases seem to have polarities analogous to diol-type phases and a wide range of mobile-phase conditions have been used including hexane, various alcohols, acetronitrile, tetrahydrofuran, dichloromethane and their mixtures, as well as aqueous buffers. In contrast to conventional stationary phases and as is the case with the multimodal cyclodextrin phases, capacity factors vs mobile-phase composition often exhibit a minima.

Validation for Hplc. Chiral separations present special problems for hplc validation. Typically, in the absence of spectroscopic confirmation, eg, mass spectral or infrared data, conventional separations are validated by analyzing pure samples under identical chromatographic conditions. Often, two or more chromatographic stationary phases, known to interact with the analyte through different retention mechanisms, are used. If the pure sample and the unknown have identical retention times under each set of conditions, the identity of the unknown is assumed to be the same as the pure sample. However, often the chiral separation obtained from one type of column may not be achievable by any other type of chiral stationary phase. In addition, pure enantiomers are generally not available.

Most commonly, uv or uv–vis spectroscopy is used as the basis for detection in hplc. When using a chiral stationary phase, confirmation of a chiral separation may be obtained by running the sample more than once, using the same mobile-phase conditions but monitoring at different wavelengths. Because enantiomers have identical spectra in an achiral environment, the ratio of the peaks for the two enantiomers should be independent of wavelength. Although not absolute proof of a chiral separation, this approach does provide strong supporting evidence.

As in tlc, another method to validate a chiral separation is to collect the individual peaks and subject each to some type of optical spectroscopy such as circular dichroism or optical rotary dispersion. Enantiomers have mirror image spectra in these spectroscopic techniques, ie, the negative maximum for one enantiomer corresponds to the positive maximum for the other. One problem with this approach is that the analytes are diluted in the mobile phase. Thus, the sample must be injected several times and the individual peaks collected and subsequently concentrated to obtain adequate concentrations for spectral analysis.

Alternatively, a chiroptical spectroscopy can be used as the basis for detection on-line using commercially available optical rotary disperson or circular dichroism-based detectors. Optical rotary disperson instruments are analogous to refractive index-based detectors for conventional chromatography in that they are universal, do not require the presence of a chromophore in the analyte, and have the least sensitivity of the optical detectors. Circular dichroic detection is more sensitive than optical rotary dispersion-based detection, but requires not only the presence of a uv-chromophore in the analyte but also that the chromophore not be too distant from the asymmetric center of the analyte. Both types of detectors produce positive and negative peaks for the two enantiomers. However, neither chiroptical detector can distinguish a fair separation from a poor separation in which there is considerable overlap of the two peaks (Fig. 8**d** and 8**f**). This is because the signals generated by the two enantiomers have opposite signs and thus any overlap causes cancellation of signal. Further, peak overlap results in nonlinear detector response with regard to concentration. Therefore, some other detection method must be used in conjunction with either of the chiroptical detection methods. Nevertheless, chiroptical detection can be advantageous when there is considerable overlap of two peaks. In this case, chiroptical detection (Fig. 8**f**) reveals that the leading and tailing edges of the peak are enantiomerically enriched. This is not apparent from the chromatogram obtained using nonchiroptical detection (Fig. 8**e**).

Another method for validating lc chiral separations is to couple the chromatographic system to a mass spectrometer. In mass spectrometry, high energy ions are used to bombard molecules exiting the column. The impact of the high energy ions causes the molecules to fragment into various ions which are then

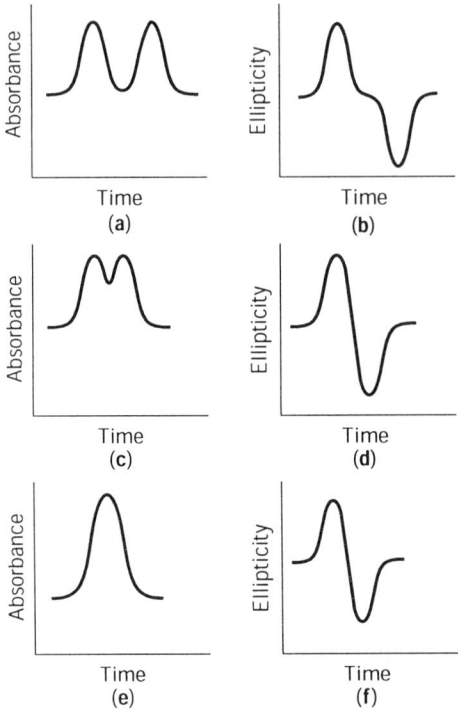

Fig. 8. Simulated chromatograms of chiral separations obtained using nonchiroptical detection (**a, c, e**) and chiroptical detection (**b, d, f**) illustrating the effect of peak overlap on the resultant chromatogram, where (**a**) and (**b**), (**c**) and (**d**), and (**e**) and (**f**) represent three separate samples having increasing overlap.

sent to a mass discriminator. The ion fragments are detected and a fragmentation pattern or mass spectrum is reconstructed. Enantiomers have identical fragmentation patterns. Hence, identical fragmentation patterns for two peaks in the chromatogram confirms a chiral separation.

Gas Chromatography

Although chiral stationary phases for gas chromatography (gc) were introduced before liquid chromatographic chiral stationary phases, development of gc chiral stationary phases has lagged behind for a variety of reasons. First of all, analysis by gc requires that the analyte be volatile and thermally stable. This condition often requires that the analyte be derivatized using an achiral reagent prior to chromatographic analysis to enhance sample volatility. In some cases, derivatization may actually enhance detector response or chiral interactions. For example, trifluoroacetylation amplifies electron capture detection. However, the presence of more than one type of functionality, eg, both an amine and an alcohol in the analyte, each having differing reactivities toward the derivatizing agent, may add additional complications. Typical achiral derivatizing reagents

Table 5. Analyte Functional Groups and Typical Achiral Derivatizing Reagents

Derivatizing agent		Analyte
Type	Example	functional group
alkylsilyl	N-trimethylsilylimidazole	alcohols
		thiols
		carboxylic acid
	N,O-bis(trimethylsilyl)trifluroracetamide	amines
acyl, haloacyl,	acetic acid	alcohols
or anhydride	heptafluorobutyryl chloride	amines
	trifluoroacetyl chloride	amides
		oximes
		thiols
		ketones
alcohol	methanol	carboxylic acids
alkyl halides	methyl bromide	carboxylic acids
diazoalkyl	diazomethane	carboxylic acids
		sulfonic acids
		phenols
isocyanate	isopropylisocyanate	alcohols
		amines
		hydroxy acids
phosgene		β-amino alcohols
		β-amino thiols
		diols
		N-methylamino acids
alkyl hydroxylamine	methylhydroxylamine	ketones

as well as the appropriate functionality required in the analyte are listed in Table 5.

The utility of gas chromatography for chiral separations was also hampered because the high column temperatures typically used in gc tend to accelerate racemization of the stationary phase, thus decreasing column longevity. These temperatures also tend to accelerate racemization of the analyte. In addition, the differences in the stabilities of the diastereomeric complexes formed between the enantiomers and the stationary phase tends to be overcome at the gc column temperatures. Finally, preparative scale separations are generally harder to implement in gc than in hplc. However, the inability of most liquid chromatographic methods to chirally resolve small nonaromatic compounds which are frequently used as chiral synthetic building blocks as well as improvements in gc column technology has led to renewed interest in chiral stationary phases for gc.

Gc chiral stationary phases can be broadly classified into three categories: diamide, cyclodextrin, and metal complex.

Diamide Chiral Separations. The first chiral stationary phase for gas chromatography, reported in 1966 (85), was based on N-trifluoroacetyl (N-TFA) L-isoleucine lauryl ester coated on an inert packing material. It was used to resolve the trifluoroacetylated derivatives of amino acids. Related chiral selectors used by other workers included *n*-dodecanoyl-L-valine-*t*-butylamide and *n*-dodecanoyl-(*S*)-α-(1 naphthyl)ethylamide. The presence of the long alkyl groups lowered chiral selector volatility, reducing but not entirely eliminating, column bleed and improving column longevity.

The first commercially available chiral column was the Chiralsil-val (Fig. 9), introduced in 1976 (86) for the separation of amino acid-type compounds by gas chromatography. It is based on a polysiloxane polymer containing chiral side chains incorporating L-valine-*t*-butylamide. The polysiloxane backbone improved the thermal stability of these chiral stationary phases relative to the original coated columns and extended the operating temperatures up to 220°C. The column is effective for the separation of perfluoroacylated and esterified amino acids, amino alcohols, and some chiral sulfoxides. Another polysiloxane-based chiral stationary phase incorporating L-valine-(*R*)-α-phenylethylamide appended onto hydrolyzed XE-60 was found to be particularly successful at resolving perfluoroacetylated amino alcohol derivatives (87). Chiral separations for a wider range of compounds, including amino alcohols, α-hydroxy acids, diols, and ketones, than had previously been obtainable using these types of stationary phases were achieved (88).

The chiral recognition mechanism for these types of phases has been attributed primarily to hydrogen bonding and dipole–dipole interactions between the analyte and the chiral selector in the stationary phase. Chiral recognition has been postulated to be involved in the formation of transient five- and seven-membered association complexes between the analyte and the chiral selector (89).

On each of these amino acid-based chiral stationary phases, the configuration of the most retained enantiomer corresponds to the configuration of the chiral selector in the stationary phase. For example, L-amino acids were retained longer on the *N*-trifluoroacetyl (N-TFA) L-isoleucine lauryl ester column. Thus, configuration of the analytes can be assigned even if no optically pure material is available as long as optically pure standard materials are available for structurally related compounds. Another advantage of these types of phases is that stationary phases incorporating the chiral selector with either configuration may be readily prepared or are commercially available. Thus, the elution order may readily be reversed by using a column containing the chiral selector with the

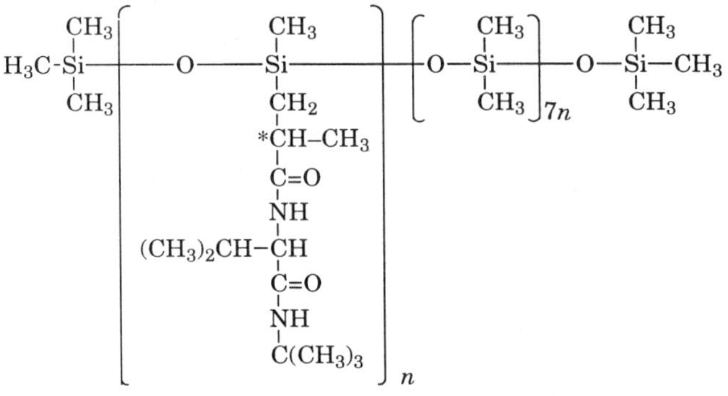

Fig. 9. Structure of Chiralsil-val.

opposite configuration, providing another tool for chiral separation validation. Also, quantitation of a very small peak in the presence of a very large peak is generally easier if the smaller peak elutes first.

Metal Complexes. Complexation gas chromatography, introduced in 1980 (90), employs transition metals, eg, nickel, cobalt, manganese, or rhodium, complexed with chiral terpenoid ketoenolate ligands such as 3-trifluoroacetyl-1(*R*)-camphorate (**12**), 1(*R*)-3-pentafluorobenzoylcamphorate, or 3-heptafluorobutanoyl-(1(*R*),2(*S*))-pinanone-4-ate.

(12)

In most cases, the chiral selector is dissolved in a polymer matrix coated on the interior walls of a capillary column. This class of chiral columns is particularly adept at enantioresolving some olefins and oxygen-containing compounds such as ketones, ethers, alcohols, spiroacetals, oxiranes, and esters. Many of these compounds lack suitable functionality for derivatization with chiral reagents and thus are not amenable to diastereomer formation. As is the case with many of the chiral stationary phases, the chiral recognition mechanism is not sufficiently refined to allow for prediction of analyte absolute configuration on the basis of retention times except for a limited number of cases. Nevertheless, these columns allow for the direct chiral separation of compounds which are important synthetic precursors.

Cyclodextrins. Whereas the native cyclodextrins, which are thermally stable, have been used extensively in liquid chromatographic chiral separations, their utility in gc applications was hampered because high crystallinity and insolubility in most organic solvents made them difficult to formulate into a gc stationary phase. Some functionalized cyclodextrins form viscous oils suitable for gc stationary-phase coatings, however, and these have been used either neat or diluted in a polysiloxane polymer as chiral stationary phases for gc (91). Some of the derivatized cyclodextrins which have been adapted to gc phases are 3-*O*-acetyl- and 3-*O*-butyryl-2,6-di-*O*-pentyl, 2,6-di-*O*-methyl-3-*O*-trifluoroacetyl, 2,6-dipentyl, 2-*O*-methyl-3,6-di-*O*-pentyl, permethyl, permethylhydroxypropyl, perpentyl, and propionyl. Although these derivatized cyclodextrins are often coated, neat, onto the capillary column inner walls, some work has been done to tether the cyclodextrins to a polysiloxane backbone to enhance the thermal stability of the resultant phase (92). Some of the separations obtained using these materials are quite remarkable and include compounds such as halogenated alkanes (Fig. 10), alcohols, alkenes, bicyclic compounds, and simple alkanes (93).

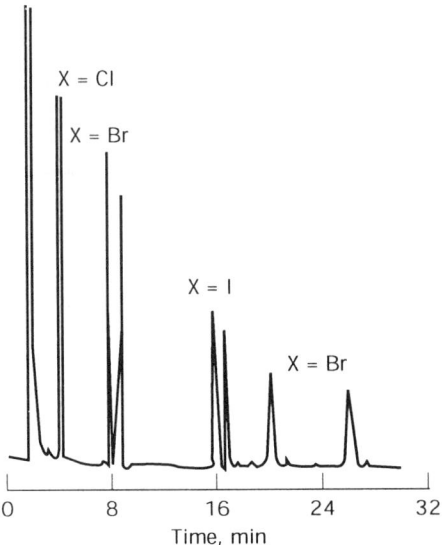

Fig. 10. Enantiomeric separations of the 2-monohalohydrocarbons on a 2,6-O-dipentyl-3-O-trifluoroacetyl-γ-cyclodextrin-coated capillary column (10 m, 0.25 mm ID). Column temperature, 30°C; nitrogen carrier gas, 20.7 kPa (3 psi).

The chiral recognition mechanism of these cyclodextrin-based phases is not entirely understood. Thermodynamic and column capacity studies, however, indicate that the analytes may interact with the functionalized cyclodextrins by either associating with the outside or mouth of the cyclodextrin or by forming a more traditional inclusion complex with the cyclodextrin (see Fig. 5). As in the case of the metal complex chiral stationary phases, configuration assignment is generally not possible in the absence of pure chiral standards.

Validation. The special problems for validation presented by chiral separations can be even more burdensome for gc because most methods of detection in gc destroy the sample. Even when nondestructive detection, eg, thermal conductivity, is used, individual peak collection is generally more difficult than in other forms of chromatography. Thus, off-line chiroptical analysis is not usually an option. However, gc can be readily coupled to a mass spectrometer and this is routinely used to validate a chiral separation.

Capillary Zone Electrophoresis

Capillary electrophoresis (ce) or capillary zone electrophoresis (cze) is a relatively recent addition to the arsenal of analytical techniques (94,95) and has been demonstrated as a powerful chiral separation method. Its high resolution capability and lower sample loading relative to hplc makes it ideal for the separation of minute amounts of components in complex biological mixtures (96,97).

In a ce experiment, a thin capillary is filled with a run buffer and a voltage is applied across the capillary (see ELECTROSEPARATIONS). The underlying impetus for separations in ce is, in general, derived from the fact that charged species migrate in response to an applied electric field proportionately to their charge and inversely proportionately to their size. Thus, given equivalent charges, lighter analytes have higher electrophoretic mobilities than heavier analytes; given equivalent sizes, more highly charged species have higher mobilities than lesser charged or neutral species. In fact, neutral species have no intrinsic electrophoretic mobility. Species having opposite charges have electrophoretic mobilities in opposing directions.

Chiral separations by ce have been performed almost exclusively using chiral additives to the run buffer (98). The advantages of this approach are identical to those for chiral mobile-phase additives in tlc. Many of the chiral selectors used as mobile-phase additives in tlc and as immobilized ligands in hplc have been used successfully in ce including proteins (99), functionalized cyclodextrins (100), various carbohydrates (101,102), assorted functionalized amino acids (103), chiral ion-pairing agents (104), and some macrocyclic antibiotics (105). Other ce chiral selectors that have not been used as immobilized chiral selectors in hplc include bile salts (106) and dextran sulfate (107).

Theoretical models (108,109) (eq. 1), where μ represents the mobility of the analyte in the free and complexed states, K represents the binding constants of enantiomers 1 and 2, and [CA] is the molar concentration of the additive, reveal that in general two conditions must be met to achieve chiral separations by ce.

$$\mu_1 - \mu_2 = \frac{[\mu_{1c} - \mu_{1f}][K_1 - K_2][CA]}{[1 + K_1[CA]][1 + K_2[CA]]} \qquad (1)$$

First, there must be differences in the binding constants of the two enantiomers to the chiral selector. Secondly, because the intrinsic electrophoretic mobilities of the enantiomers are identical in the free state, there must be a significant difference in the mobilities of the analyte in the complexed and free state. Chiral selectors have generally been shown to be more effective when the intrinsic electrophoretic mobility of the additive is in the opposite direction of the intrinsic electrophoretic mobility of the analyte.

Chiral separations by ce is a rapidly growing field offering tremendous flexibility with regard to chiral selector choice. In addition, because the additive is typically in the run buffer, there is virtually no column pre-equilibration. However, ce instruments tend to cost more than most chromatographic systems, sample capacity is much smaller for ce than for an analogous hplc, sample recovery is not a trivial problem in ce, and run-to-run reproducibility for ce tends to be much worse than for most chromatographic methods. Nevertheless, the flexibility, minimal sample, and/or chiral selector required and the extremely high resolving power of ce ensure that this technique will continue to play an important role in future chiral separations.

BIBLIOGRAPHY

Materials for this article concerning cyclodextrins and protein phases first published in "Biopolymers, Analytical Techniques," in the *Encyclopedia of Chemical Technology*, 4th ed., Vol. 4, pp. 187–208, by T. Ward, Millsaps College, and concerning definitions of chiral compounds in "Pharmaceuticals, Chiral" in *ECT* 4th ed., Vol. 18, pp. 511–553, by D. T. Witiak and A. T. Hopper, University of Wisconsin-Madison.

1. L. Pasteur, *Comptes Rendus de l'Academie des Sciences*, **26**, 535 (1848).
2. J. Snopek and I. Jelinek, *J. Chromatograph.* **609**, 1–17 (1992); P. R. Andrews, in H. Kubinyi, ed., *3D QSAR in Drug Design: Theory, Methods and Applications*, ESCOM Science Publishers, the Netherlands, 1993, pp. 13–21.
3. G. F. Russel and J. I. Hills, *Science* **172**, 1043–1044 (1971); L. Freeman and J. G. Miller, *Science* **172**, 1044–1046 (1971).
4. J. Solms, L. Vuataz, and R. H. Egli, *Experientia* **21**, 692–694 (1965).
5. T. Z. Csaky, *Cutting's Handbook of Pharmacology*, 6th ed., Appleton-Century-Crofts, New York, 1979, p. 161.
6. H. Glatt and F. Oesch, *Biochem. Pharm.* **34**, 3725–3728 (1985).
7. G. von Blaschke, H. P. Kraft, K. Finkentscher, and F. Kohler, *Arzneim. Forsch./Drug Res.* **29**, 1640 (1979).
8. K. Eger, M. Jalalian, E. J. Verspohl, and N.-P. Luke, *Arzneim. Forsch./Drug Res.* **40**, 1073–1075) 1990).
9. R. J. Ott and K. M. Giacomini, in I. W. Wainer, ed., *Drug Stereochemistry Analytical Methods and Pharmacology Second Edition, Revised and Expanded*, Marcel Dekker, Inc., New York, 1993, pp. 281–314.
10. T. A. G. Noctor, in Ref. 9, pp. 337–364.
11. N. P. E. Vermeulen and J. M. Koppele, in Ref. 9.
12. W. J. Wechter, D. G. Loughhead, R. J. Reischer, G.J. Van Giessen, and D. G. Kaiser, *Biochem. Biophys. Res. Comm.* **61**, 883 (1974).
13. R. A. Kuzel, S. K. Bhasin, H. G. Oldham, L. A. Damani, J. Murphy, P. Camilleri, and A. J. Hutt, *Chirality* **6**, 607 (1994).
14. S. A. Tobert, *Clin. Pharmacol. Ther.* **29**, 344 (1981).
15. H. Ohkawa, *Bull. Environ. Contamin. Toxicol.* **18**, 534 (1977).
16. H. G. Schmarr, A. Mosandl, and K. Grob, *Chromatographia* **29**, 125 (1990).
17. S. Kotha, *Tetrahedron* **50**, 3639–3662 (1994).
18. S. Levin and S. Abu-Lafi, in P. R. Brown and E. Grushka, eds., *Advances in Chromatography*, Vol. 33, Marcel Dekker, Inc., New York, 1993, pp. 233–266.
19. E. J. Ariëns, *TiPS* **14**, 68–75 (1993).
20. E. J. Ariëns, E. W. Wuis, and E. J. Veringa, *Biochem. Pharmacol.* **37**, 9–18 (1988).
21. B. Holmstedt, H. Frank, and B. Testa, *Chirality and Biological Activity*, Alan R. Liss, Inc., New York, 1990.
22. U.S. Food and Drug Administration, *Chirality* **4**, 338–340 (1992).
23. W. H. De Camp, *Chirality* **1**, 2–6 (1989).
24. D. W. Armstrong, J. R. Faulkner, Jr., and S. M. Han, *J. Chromatogr.* **452**, 323 (1988).
25. J. D. Duncan, D. W. Armstrong, and A. M. Stalcup, *J. Liq. Chromatogr.* **13**, 1091 (1990).
26. A. M. Stalcup, S. Gratz, and Y. Jin, unpublished results, University of Cincinnati, Ohio, 1996.
27. D. W. Armstrong and W. DeMond, *J. Chromatogr. Sci.* **22**, 411 (1984).
28. D. W. Armstrong, Y. Tang, S. Chen, Y. Zhou, C. Bagwill, and J. R. Chen, *Anal. Chem.* **66**, 1473 (1994).
29. W. H. Pirkle, J. M. Finn, B. C. Hamper, J. L. Schreiner, and J. R. Pribish, *Am. Chem. Soc. Symp. Ser.* (185) (1982).
30. W. H. Pirkle and P. G. Murray, *J. Liq. Chromatogr.* **13**, 2123 (1990).
31. J. Hermansson and M. Eriksson, *J. Liq. Chromatogr.* **9**, 621 (1986).
32. G. Schill, I. W. Wainer, and S. A. Barkin, *J. Liq. Chromatogr.* **9**, 641 (1986).
33. S. Allenmark, *J. Liq. Chromatogr.* **9**, 425 (1986).

34. M. Okamoto and H. Nakazawa, *J. Chromatogr.* **508**, 217 (1990).
35. T. Miwa, H. Kuoda, S. Sakashita, N. Asakawa, and Y. Miyake, *J. Chromatogr.* **511**, 89 (1990).
36. R. Isaksson, P. Erlandsson, L. Hansson, A. Holmberg, and S. Berner, *J. Chromatogr.* **498**, 257 (1990).
37. Y. Okamoto, R. Aburatani, K. Hatano, and K. Hatada, *J. Liq. Chromatogr.* **11**, 2147 (1988).
38. T. Shinbo, T. Yamaguchi, K. Nishimura, and M. Sugiura, *J. Chromatogr.* **405**, 145 (1987).
39. M. Hilton and D. W. Armstrong, *J. Liq. Chromatogr.* **14**, 9 (1991).
40. V. A. Davankov, A. A. Kurganov, and A. S. Bochov, *Adv. Chromatogr.* **22**, 71 (1983).
41. C. E. Dalgliesh, *J. Chem. Soc.*, 3940 (1952).
42. S. M. Han and D. W. Armstrong, in A. M. Krstulovic, ed., *Chiral Separations by HPLC*, John Wiley & Sons, Inc., New York, 1989, pp. 208–287.
43. D. W. Armstrong, T. J. Ward, R. D. Armstrong, and T. E. Beesley, *Science* **232**, 1132 (1986).
44. T. J. Ward and D. W. Armstrong, in M. Zief and L. J. Crane, eds., *Chromatographic Chiral Separations*, Marcel Dekker, Inc., New York, p. 131.
45. D. W. Armstrong, S. Chen, C. Chang, and S. Chang, *J. Liq. Chromatogr.* **15**, 545 (1992).
46. D. W. Armstrong, M. Hilton, and L. Coffin, *LC-GC* **9**, 647 (1992).
47. D. W. Armstrong and W. DeMond, *J. Chromatogr. Sci.* **22**, 411 (1984); U.S. Pat. 4,539,399 (1985), D. W. Armstrong.
48. D. W. Armstrong, W. DeMond, and B. P. Czech, *Anal. Chem.* **57**, 481 (1985).
49. T. J. Ward and D. W. Armstrong, *J. Liq. Chromatogr.* **9**(2,3), 407 (1986).
50. D. W. Armstrong, T. J. Ward, R. D. Armstrong, and T. E. Beesley, *Science* **232**, 1132 (1986).
51. D. W. Armstrong, T. J. Ward, A. Czech, B. P. Czech, and R. A. Bartsch, *J. Org. Chem.* **50**, 5556 (1985).
52. D. W. Armstrong and co-workers, *Anal. Chem.* **57**, 234 (1985).
53. C. A. Chang, Q. Wu, and L. Tan, *J. Chromatogr.* **361**, 199 (1986).
54. D. W. Armstrong, W. DeMond, W. L. Hinze, and T. E. Riehl, *J. Liq. Chromatogr.* **8**, 261 (1985).
55. K. Schmid, in F. W. Putman, ed., *The Plasma Proteins*, Academic Press, Inc., New York, 1975, p. 184.
56. G. Schill, I. W. Wainer, and S. A. Barkan, *J. Liq. Chromatogr.* **9**, 641 (1986).
57. J. Hermansson, *J. Chromatogr.* **298**, 67 (1984).
58. Chiral-AGP, *Application Note NO. 1*, Chrom Tech AB, Norsburg, Sweden, 1988.
59. J. Hermansson, *Trends Anal. Chem.* **8**, 251 (1989).
60. M. Enquist and J. Hermansson, *Chirality* **1**(3), 209 (1989).
61. J. Hermansson, *J. Chromatogr.* **316**, 537 (1984).
62. J. Hermansson and M. Eriksson, *J. Liq. Chromatogr.* **9**, 621 (1986).
63. G. Schill, I. W. Wainer, and S. A. Barkan, *J. Liq. Chromatogr.* **365**, 73 (1986).
64. J. Hermansson and G. Schill, in M. Zief and L. J. Crane, eds., *Chromatographic Chiral Separations*, Vol. 40, Marcel Dekker, New York, 1988, p. 245.
65. E. Arvidsson, S. O. Jansson, and G. Schill, *J. Chromatogr.* **506**, 579 (1990).
66. B. A. Persson, K. Balmer, P. O. Lagerstrom, and G. Schill, *J. Chromatogr.* **500**, 629 (1990).
67. Th. Peters, Jr., in Ref. 55, Vol. 1, p. 133.
68. M. C. Meyer and D. E. Guttman, *J. Pharm. Sci.* **57**, 895 (1968).
69. R. H. McMenamy and J. L. Oncley, *J. Biol. Chem.* **223**, 1436 (1958).
70. K. K. Stewart and R. F. Doherty, *Proc. Nat. Acad. Sci.* **70**, 2850 (1973).

71. S. Allenmark, *J. Liq. Chromatogr.* **9**, 425 (1986).
72. E. Domenici, C. Bertucci, P. Salvadori, and I. W. Wainer, *J. Pharm. Sci.* **80**, 164 (1991).
73. A. Ichida and T. Shibata, in Ref. 64, pp. 219–243.
74. Y. Okamoto, M. Kawashima, and K. Hatada, *Chem. Lett.*, 739 (1984).
75. Y. Okamoto, I. Okamoto, and K. Hatada, *J. Am. Chem. Soc.* **106**, 5357 (1984).
76. G. Blaschke, *J. Liq. Chromatogr.* **9**, 341 (1986).
77. M. Zief and L. J. Crane, eds., *Chromatographic Chiral Separations*, Vol. 40, Marcel Dekker, New York, 1988.
78. D. W. Armstrong and S. M. Han, *CRC Crit. Rev. Analyt. Chem.* **19**(3), 175 (1988).
79. H. Hopf, W. Grahn, D. G. Barrett, A. Gerdes, J. Hilmer, J. Hucker, Y. Okamoto, and Y. Kaida, *Chem. Ber.* **123**, 841 (1990).
80. Y. Okamoto, Y. Kaida, R. Aburatani, and K. Hatada, *J. Chromatogr.* **477**, 367 (1989).
81. Y. Okamoto, K. Hatano, R. Aburatani, and K. Hatada, *Chem. Lett.*, 715 (1989).
82. T. Shibata, I. Okamoto, and K. Ishii, *J. Liq. Chromatogr.* **9**, 313 (1986).
83. Y. Okamoto, K. Suzuki, K. Ohta, K. Hatada, and H. Yuki, *J. Am. Chem. Soc.* **101**, 4763 (1979).
84. G. Balschke, W. Bröker, and W. Fraenkel, *Angew. Chem.* **98**, 808 (1986).
85. E. Gil-Av, B. Feibush, and R. Charles-Sigler, *Tetrahedron Lett.* **10**, 1009 (1966).
86. H. Frank, G. J. Nicholson, and E. Bayer, *J. Chromatogr. Sci.* **15**, 174 (1974).
87. W. A. König, I. Benecke, and S. Sievers, *J. Chromatogr.* **217**, 71 (1981).
88. W. A. König, E. Steinbach, and K. Ernst, *J. Chromatogr.* **301**, 129 (1984).
89. B. Feibush, A. Balan, B. Altman, and E. Gil-Av, *J. Chem. Soc. Perkin II*, 1230 (1979).
90. V. Schurig, *Chromatographia* **13**, 263 (1980).
91. H. P. Nowotny, D. Schmalzing, D. Wistuba, and V. Schurig, *J. High Resolut. Chromatogr. Chromatogr. Commun.* **12**, 383 (1989).
92. D. W. Armstrong, Y. Tang, T. Ward, and M. Nichols, *Anal. Chem.* **65**, 1114 (1993).
93. W.-Y. Li, H. L. Jion, and D. W. Armstrong, *J. Chromatogr.* **509**, 303 (1990).
94. W. G. Kuhr, *Anal. Chem.* **62**, 403R (1990).
95. B. L. Karger, *Am. Lab.*, 23 (Oct. 1993).
96. J. W. Jorgenson and K. D. Lukacs, *Anal. Chem.* **53**, 1298 (1981).
97. A. S. Cohen, A. Paulus, and B. L. Karger, *Chromatographia* **24**, 15 (1987).
98. T. J. Ward, *Anal. Chem.* **66**, 632A (1994).
99. P. Sun, N. Wu, G. Barker, and R. A. Hartwick, *J. Chromatogr.* **648**, 475 (1993).
100. Y. Y. Rawjee and G. Vigh, *Anal. Chem.* **66**, 619 (1994).
101. A. M. Stalcup and N. M. Agyei, *Anal. Chem.* **66**, 3054 (1994).
102. H. Soini, M. Stefansson, M.-L. Riekkola, and M. V. Novotny, *Anal. Chem.* **66**, 3477 (1994).
103. P. Gozel, E. Gassman, H. Michelson, and R. N. Zare, *Anal. Chem.* **59**, 44 (1987).
104. A. M. Stalcup and K. H. Gahm, *J. Microcolumn Sep.* **8**, 145 (1996).
105. D. W. Armstrong, K. Rundlett, and G. L. Reid, *Anal. Chem.* **66**, 1690 (1994).
106. T. O. Cole, M. J. Sepaniak, and W. L. Hinze, *J. High Resolut. Chromatogr. Chromatogr. Commun.* **13**, 570 (1990).
107. N. M. Agyei, K. H. Gahm, and A. M. Stalcup, *Anal. Chim. Acta* **307**, 185 (1995).
108. S. A. C. Wren and R. C. Rowe, *J. Chromatogr.* **603**, 235 (1992).
109. A. Guttman, A. Paulus, A. S. Cohen, N. Grinberg, and B. L. Karger, *J. Chromatogr.* **448**, 41 (1988).

General References

C. F. Poole and S. K. Poole, *Chromatography Today*, Elsevier Science Publishers BV, Amsterdam, the Netherlands, 1991.

I. W. Wainer, *Drug Stereochemistry: Analytical Methods and Pharmacology*, Marcel Dekker, Inc., New York, 1993.

W. A. König, *The Practice of Enantiomer Separation by Capillary Gas Chromatography*, Hüthig Verlag, Heidelberg, Germany, 1987.

P. Schreier, A. Bernreuther, and M. Huffer, *Analysis of Chiral Organic Molecules*, Walter de Gruyter & Co., Berlin, 1995.

G. Subramanian, *A Practical Approach to Chiral Separations by Liquid Chromatography*, VCH, Weinheim, Germany, 1994.

APRYLL M. STALCUP
University of Cincinnati

CLATHRATES. See INCLUSION COMPOUNDS.

COAL. See DESULFURIZATION, CLEANING AND DESULFURIZATION OF COAL.

CONTACTORS FOR CHEMICAL SEPARATION

In the chemical process industry, separation of mixtures into streams of high purity compounds is frequently necessary. Mixtures may originate from feedstock impurities, incomplete conversion of the reactants, and/or occurrence of undesirable side reactions. Feedstocks for the process industry are often complex mixtures such as mineral oil, coal, petroleum distillates, or vegetable fats. It is often desirable to separate the components of these mixtures prior to usage.

For the separation of mixtures a large number of processes and techniques have been developed. These can be divided into two classes. The first is composed of separations based on differences in physicochemical properties, eg, boiling point in the case of distillation (qv), and diffusivity in some membrane applications (see MEMBRANE TECHNOLOGY). In the second class of processes the separation is achieved by means of a selective chemical reaction. Examples include the removal of the acidic gases, carbon dioxide [124-38-9], CO_2, and hydrogen sulfide [7783-06-4], H_2S, by aqueous alkanolamine solutions, and selective removal of hydrogen [1333-74-0], H_2, via pressure-swing operation of fixed beds consisting of metal hydride-forming alloys. It is the second class, the chemical separation processes, which are discussed herein. Special emphasis is placed on the contactor selection.

The implementation of a chemical reaction in separation processes is rarely simple and straightforward. Aspects such as reactor selection, regeneration of the intermediate product(s) in order to produce the pure compound, and the possible degradation of the various components used must all be considered. Often an additional complication is the occurrence of two or more physical phases in the separation process, eg, gas–liquid or liquid–liquid, which results in mass-transfer (qv) phenomena accompanied by chemical reactions. Owing to these transport influences, the type of contactor becomes very important because the separation efficiency can be substantially affected by the reactor and the operational mode. Selectivity in separations by chemical reaction can therefore be realized by the reaction itself; by the difference in rate of reactions, ie, kinetic selectivity; or by manipulating mass-transfer rates in combination with chemical reactions, ie, overall kinetic selectivity.

The selection or design of a contactor for the separation by chemical processes is often very complicated, even when a classical contactor is chosen to reduce risks. For a new process the various steps involved in the chemical separation need to be identified, eg, mass transfer, diffusion and chemical reaction, reaction schemes or networks, by-product formation, deactivation, etc. This identification is not only for the separation but also for regeneration. For commercial development the route to a scientifically based design is often too expensive to be followed completely. Thus only a partial knowledge of the elementary processes may be obtained before proceeding.

For example, for a chemical separation process involving a gas and a liquid phase in a tray column, the expected bulk concentrations of the gas and liquid phases in a differential laboratory reactor, operating at values similar to the process conditions, could be simulated for different positions in the column. The gas- and liquid-phase mass-transfer coefficients and interfacial area at which the transport fluxes of the various components involved must be measured. Although this would involve some iteration and detailed information of the tray column envisaged, designs can be made without too much knowledge of the details of the reactive system. Such an approach has often led to useful results.

When more detailed knowledge of the reactions is available, selection or design of a special contactor exclusively devoted to the duty of the chemical separation process is possible. An important problem, however, is that data on full-scale performance for nonstandard contactors are lacking, thus increasing scale-up risks. Indeed, introduction of nonstandard contacting equipment in chemical separation processes is difficult.

Newly developed contactors may play an important role in the decoupling of mass transfer and chemical reaction as these contactors often have special mass-transfer properties. Moreover, it is possible to demonstrate improvements that can be commercially attained in accepted contactors by exaggerating the effects of nonstandard contactors. Simple and low risk/high gain first applications are necessary for acceptance on a commercial scale. Examples of chemical separation processes involving gas–liquid systems, gas–solid systems, and slurry contactors are given herein. For any new contactor, many hurdles have to be taken. Strong advantages should be evident before such a development is applied on a full industrial scale.

Gas–Liquid Reactive System

Removal of the acid components H_2S and CO_2 from sour natural gases by means of reversible absorption (qv) in liquids, commonly referred to as gas treating, is a well-known industrial operation. Alkanolamine solutions are frequently used as solvents for these separation processes (1). Depending on the process requirements, eg, selective removal of H_2S or CO_2 bulk removal, several options for alkanolamine-based solvents and varying compositions of the solution and gas–liquid contactors have been presented. In many situations it is economically attractive to remove only H_2S (2). Thus this selective process is discussed herein.

For alkanolamine-based solvents the reaction between H_2S and the amine can be regarded as instantaneous when compared to the mass-transfer rate. The reaction of CO_2 has a much lower, finite rate and this difference in reactivity can be used to improve on the H_2S selectivity. Owing to the instantaneous reaction between H_2S and the liquid amine the mass-transfer resistance is frequently in the gas phase. The absorption of CO_2 is determined mainly by liquid-phase resistance which can be affected by the reaction. This ultimately leads to enhancement of the mass-transfer rate. From the difference in the location of the mass-transfer resistances it can be concluded that gas–liquid contactors having high ratios of the gas-phase to liquid-phase mass-transfer coefficient are preferred for selective removal of H_2S.

Selection and screening of the various solvents and contactors for selective removal of H_2S may be carried out using an absorber/desorber model (3). In the absorber model, the molar flux of the various compounds on each tray or segment of the packed-bed reactor have to be determined. The molar flux, J_i, of component i at the trays in units of mol/(m²·s) can be calculated according to equation 1:

$$J_i = (C_{G,i} - C_{L,i}/He_i)/(1/k_G + 1/He_i k_L E_i) \quad (1)$$

where $C_{G,i}$ and $C_{L,i}$ are molar concentrations and k_G and k_L are mass-transfer rates of i in the gas and in the liquid, respectively; E_i is the enhancement factor, determined by the rates of the reactions that take place during the absorption process; and He_i is the physical solubility of the gaseous components. Multiplying equation 1 by the interfacial area of the local position of the contactor yields the absorption rate. Thus, absorption rate is determined by (1) the mass-transfer parameters k_G, k_L, and the interfacial area which are strongly dependent on the type of gas–liquid contactor; (2) the solubility of the gaseous components and if a solvent is used in which chemical reactions occur; and (3) the chemical equilibrium of the absorption process which influences the driving

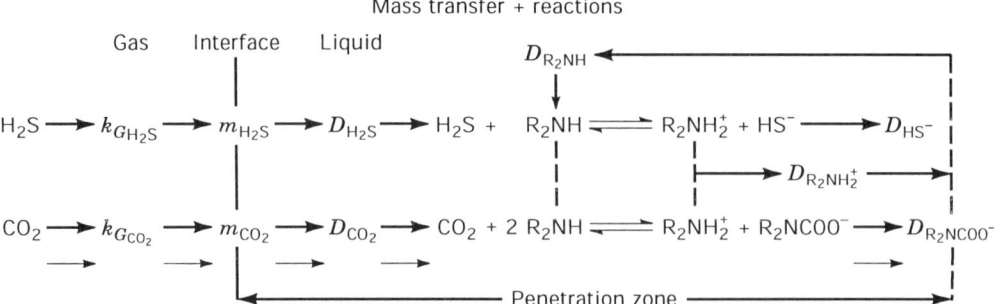

Fig. 1. Schematic of the mass-transfer process and subsequent reactions for the simultaneous absorption of H_2S and CO_2 in aqueous alkanolamine, where D_i corresponds to the diffusion constant of component i and $k_{G,i}$ and m_i are the mass-transfer coefficients of i in the gas phase and at the interface, respectively.

force for absorption. The determination of the equilibrium composition of a partially loaded amine solution can be estimated if sufficient information is available on the equilibrium constants of the occurring reactions and if allowance is made for the nonideal behavior of the liquid phase (4,5) and E_i, the enhancement factor.

For the calculation of the molar fluxes an absorption model is required (6–8). In alkanolamine-based processes a large number of reversible, parallel, and consecutive reactions take place during the absorption process. The mass-transfer and subsequent reactions are presented schematically (Fig. 1) for the simultaneous absorption of H_2S and CO_2 in an aqueous solution of a secondary alkanolamine. Basically this type of absorption model consists of a number of material balances for each participating species in the treating process. Therefore the model results in a set of nonlinear partial differential equations which must be solved numerically (7,9) because an analytical solution method is not available. The local rate of absorption of component i can be calculated using the absorption model if sufficient information is available on the mass-transfer coefficients, reaction rates, equilibrium composition, and physicochemical parameters. Because the application of alkanolamines in gas-treating processes has been extensively studied, a large amount of data are available, permitting a scientifically based approach for process selection.

Selective Removal of H_2S. The local selectivity, σ_{local}, of a gas-treating process may be approximated by equation 2:

$$\sigma_{local} = \text{rate of absorption } H_2S/\text{rate of absorption } CO_2 = J_{H_2S}/J_{CO_2} \quad (2)$$

The values of the local selectivities depend on the magnitudes of the driving forces of CO_2 and H_2S. Therefore a selectivity factor, S, is defined to which these forces contribute (eq. 3).

$$S = \frac{(\text{rate of absorption of } H_2S/\text{driving force of } H_2S)}{(\text{rate of absorption of } CO_2/\text{driving force of } CO_2)} \quad (3)$$

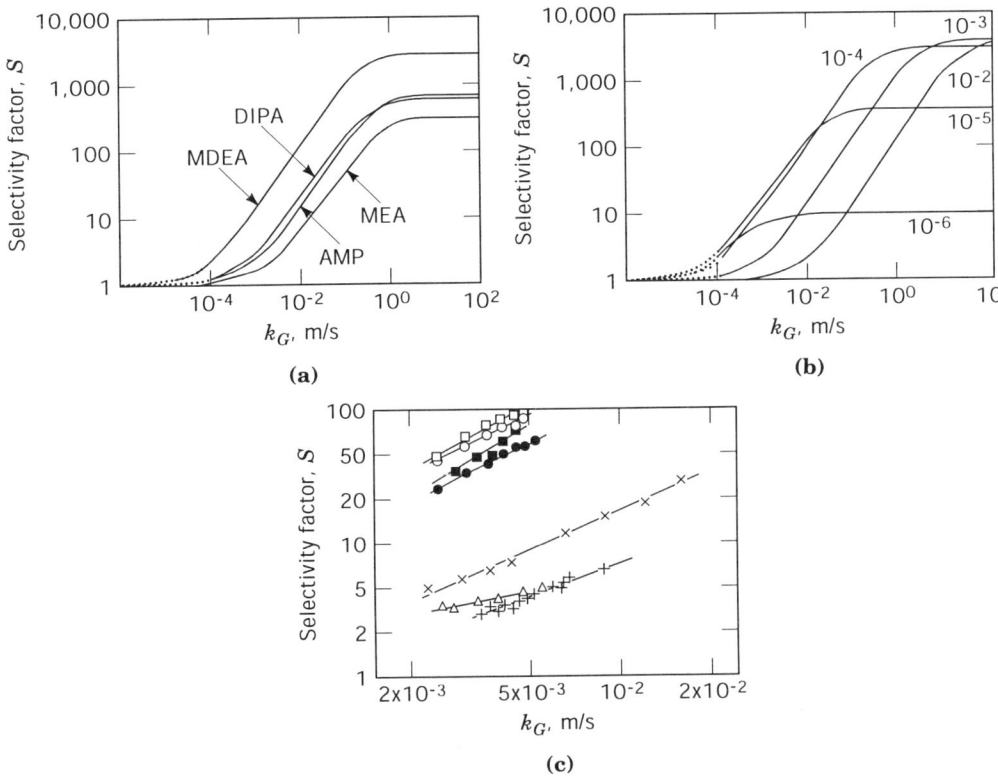

Fig. 2. Selectivity factors for aqueous alkanolamine solutions where AMP is 2-amino-2-methyl-1-propanol; DIPA, diisopropanolamine; MDEA, methyldiethanolamine [105-59-9]; MEA, monoethanolamine [141-43-5]; and DEA, diethanolamine [111-42-2]: (**a**) and (**b**) are calculated values using the absorption model and amine concentrations of 2000 mol/m^3, where in (**a**) k_L is 10^{-4} m/s and in (**b**), for MDEA, the numbers represent various k_L values in m/s; (**c**) experimental values where the concentration of the amine, C_{am} (10^3 mol/m^3) and N_L (s^{-1}), respectively, are (□) MDEA 0.9, 0.33; (○) MDEA 0.9, 0.82; (■) MDEA 1.5, 0.35; (●) MDEA 1.5, 0.75; (×) DIPA (10) 2.0; (△) DIPA 2.4; and (+) DEA 2.0 (8).

Using this model it is possible to simulate the influence of the mass-transfer coefficients, k_G and k_L, on the selectivity factor S.

The results of simulations using the absorption model are given for four aqueous amine solutions in Figure 2. Figure 2a clearly shows that a good selection of the solvent has a pronounced effect on the selectivity of the treating process. Figure 2b indicates that for small values of k_G, all S values become equal to one, ie, the selectivity which can be attained for complete gas-phase limitation for both components. Increasing k_G leads first to higher values of S, then a maximum is reached and S remains constant. For the latter asymptotic situation the absorption of CO_2 and H_2S are both liquid-phase-limited, resulting in considerable interaction. The selectivity factor which can be attained in the region between the asymptotics depends on both k_G and k_L. There appear to be two possible routes to improving the selectivity of a process treating H_2S. First, selection of a proper solvent; second, manipulation of the mass-transfer coefficients, ie, selection of the best gas–liquid contactor. In Figure 2c experimentally observed selectivity factors are shown (8,10). The experimental results agree well with the calculated ones.

Gas–Liquid Contactors for Selective Removal of H_2S. The selectivity of a treatment process can be affected by the ratio k_G/k_L (see Fig. 2c). For the conventional tray contactors and the countercurrent operated packed-bed reactors, however, the improvement of the selectivity which can be achieved is very limited. Thus other gas–liquid contactors have been presented for use.

It has been demonstrated theoretically that in a fixed-bed reactor having cocurrent downflow operation a substantial increase in selectivity may be obtained (3). The main disadvantage is that one contacting stage is, at best, equivalent to one equilibrium stage. This is not, however, a serious problem for most amine absorbers if countercurrent arrangements are used for the cocurrent devices (3). An overview of specifically designed equipment, such as swirl tubes, spray cyclone contactors, and centrifugal reactors, for the selective removal of H_2S is available (11). Experimental data for both the spray cyclone and the centrifugal reactors performance for acid-gas treating applications have been presented. The spray cyclone reactor (Fig. 3) has yielded experimentally observed selectivity ratios of 12 to 33 for 1000 mol/m^3 MEA and 92 to 154 for 1000 mol/m^3 DIPA (13), and in a second fairly similar apparatus, for an aqueous MDEA solution of 1000 mol/m^3, values for S ranged from 13 up to 548 (14).

Absorption in centrifugal reactors has been reviewed (15). A rotating contactor, filled with packing material having a high specific contact area for distillation purposes has been presented (16). The application of this reactor for selective H_2S removal has been discussed (17). A centrifugal reactor (Fig. 4) which can

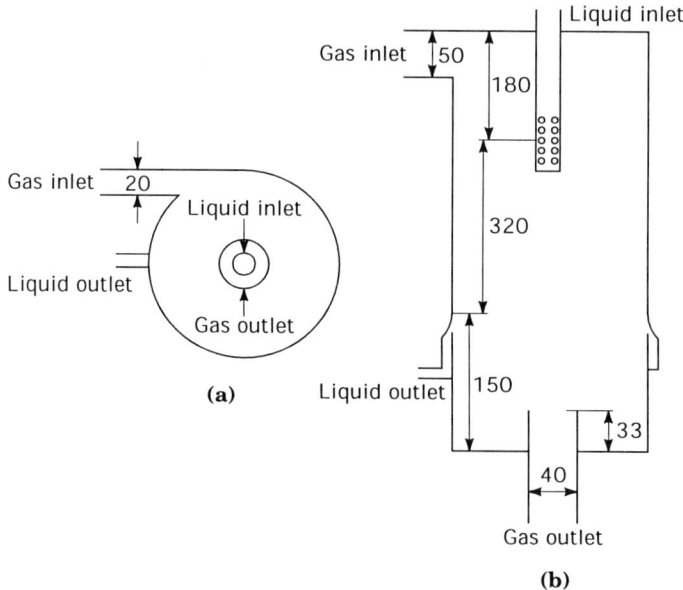

Fig. 3. Two views of a spray cyclone reactor where the numbers represent sizes in mm: (**a**) top and (**b**) side view (12).

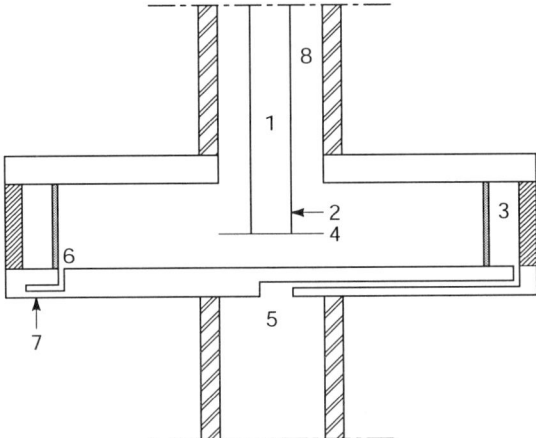

Fig. 4. Schematic of a centrifugal reactor where 1 = the liquid inlet; 2, electrode; 3, gas distribution; 4, nozzle; 5, gas inlet; 6, liquid outlet; 7, electrode; and 8, gas outlet, and (□) represents a lexan; (▨), stainless steel; (■) sintered stainless steel; and (▨) aluminum.

be regarded as a rotating bubble column has been developed (12). However, in a manner similar to the rotating contactor (16), packing material can be incorporated. In this centrifugal reactor values for the volumetric liquid-phase mass-transfer coefficient, $k_L a$, were observed varying between 1 and 20 s^{-1}. Data for H$_2$S absorption (14) show that the reactor can be regarded as one (ideal) equilibrium stage (14). Thus the reactor is suitable for selective H$_2$S removal; however, some kind of staging is required for deep H$_2$S removal. A detailed comparison of the equipment for selective removal of H$_2$S has been presented (14). Results are summarized in Table 1.

Microporous Membranes as Gas–Liquid Contactors. A newer type of gas–liquid contactor has been developed involving the incorporation of membranes in two-phase contacting equipment. This class of equipment can be regarded as dispersion-free contactors that can be applied in gas–liquid and liquid–liquid separation processes (18–21). The membrane acts as a fixed interface and keeps the gas and liquid phases separated while the transport of reactants/products can take place through the membrane. Depending on the membrane material, the liquid, and the pressure of both phases, the pores of the membrane are filled with either liquid or gas. These two types of membranes are often referred to as wetted and nonwetted, respectively.

Hollow-fiber membrane modules (HFMM) offer some significant advantages compared to conventional equipment such as bubble columns, tray columns, or packed beds. In a hollow-fiber membrane module the interfacial area between the gas and liquid phases is formed by the membranes, whereas in conventional absorbers it is mainly determined by the direct interaction between gas and liquid flow. The amount of interfacial area that can be realized in HFMMs is thus much larger than the values encountered in the conventional contactors (up to 1000 m^2/m^3) (22,23). For the hollow-fiber modules, numbers of 10^4 m^2/m^3 have been reported (24).

Table 1. Properties of Contactors for Selective Acid-Gas Treating[a,b]

Equipment	k_G/k_L	Liquid holdup	Complexity Construction	Complexity Cascading	Size	Reliability	Gas-liquid separation	Scale-up	Flexibility	Pressure drop
tray column reactor	−	±	±	−	−	+	−	+	±	+
cocurrent packed bed	+	±	+	−	−	+	−	+	−	−
spray cyclone	+	+	+	−	−	+	+	?	−	−
centrifugal	−	+	−	+	+	?	+	+	+	−

[a] Ref. 14.
[b] A + sign indicates an advantage; −, a disadvantage; and ?, the parameter has yet to be thoroughly tested.

The HFMM has a large operational flexibility because the gas and liquid flows are separated by the membrane. Contrary to conventional absorbers where the operational flexibility is severely limited by phenomena such as flooding, loading, entrainment, etc, no mutual influence is possible. Another attractive feature is that scaling-up of this contactor is straightforward. If the process can be described satisfactorily in one fiber, the design of the full-scale process is simple.

The mass-transfer rates in the module, however, are limited by the mass-transfer resistances in the gas and liquid phase and the additional resistance introduced by the membrane. In conventional contactors the gas and liquid flows are often turbulent. In HFMMs using small fibers and small channels between the fibers, the gas and liquid flows are usually laminar. Mass-transfer resistances are generally lower in turbulent flow than in laminar. Therefore the mass-transfer rate in the gas and liquid on both sides of the membrane in the module is expected to be lower. Also the additional mass-transfer resistance of the membrane, which can be regarded as a stagnant layer, reduces the mass-transfer rate.

Mass-transfer phenomena in HFMMs for gas–liquid systems have been studied (25,26). For processes in which mass transfer was determined by the transport in the liquid phase, the active mass-transfer area is equal to the total membrane area, regardless of the porosity of the fiber. For processes where liquid flows through the fibers, the influence of fiber diameter, d; diffusivity in the liquid, D; liquid viscosity, z; and liquid velocity, v, on mass transfer can be correlated extremely well using the Graetz-Leveque solution (eq. 4), derived for the analogous case of heat transfer (Fig. 5):

$$Sh = 1.62(Re \cdot Sc \cdot d/Z)^{1/3} = 1.62(vd^2/DZ)^{1/3} = 1.62Gz^{1/3} \text{ for } Re \cdot Sc \cdot d/Z > 20 \quad (4a)$$

$$Sh = 3.67 \text{ for } Re \cdot Sc \cdot d/Z < 10 \quad (4b)$$

where Re and Sc are the Reynolds and Schmidt numbers, respectively; Z = heat exchange; and G, gas flux. From equation 4 it can be concluded that the value of the mass-transfer coefficient for a liquid phase flowing through the lumen of the fiber is of the same order of magnitude ($\approx 10^{-4}$ m/s) as in conventional contactors such as a bubble column, for inside diameters, d, of about 0.5 mm. For liquid flowing around regularly packed fibers mass transfer has been described satisfactorily using a correlation derived from a numerical solution for the similar heat-transfer problem (27). Correlating mass transfer in liquid flowing around irregularly packed fibers has not been possible because of the undefined dimensions of the different channels between the fibers.

An important disadvantage of the application of membrane modules is the occurrence of maldistribution on both sides of the membrane owing to the effect of the thickness of the membrane which affects the actual pore diameter in the lumen, and local variations in the stacking arrangement of the membrane fibers resulting in stagnant zones and preferential flow paths owing to different hydrodynamical resistances. The estimation of the latter effect on maldistribution is complicated and requires the availability of three-dimensional hydrodynamical one-phase flow models in combination with a mass-transfer model.

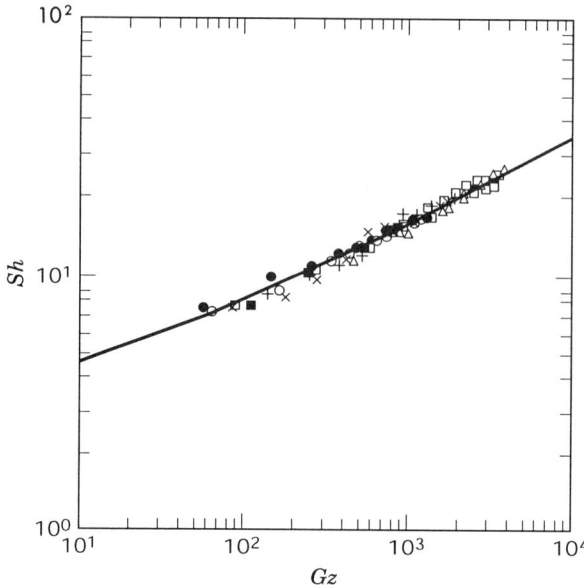

Fig. 5. Absorption data measured in two hollow fiber modules having (○) 0, (●) 15, (□) and (△) 50, (■) 65, (×) 80, and (+) 90 wt % glycerol in water mixtures flowing through the fibers. The solid line represents equation 4.

Absorption processes accompanied by chemical reactions can also be described in HFMMs (28). Chemically enhanced absorption in liquids flowing laminarly through a HFMM was studied for the absorption of carbon dioxide in solutions of sodium hydroxide. This process can be simulated using a numerical model assuming an irreversible second-order reaction. The model calculations agree well with the experiments (28).

Gas–Solids Reactive Systems

An example of a gas–solids reactive system is the regenerative desulfurization of flue gas and fuel gas. The common feature is that in a reactor section an acid gas (SO_2 or H_2S) is removed from a gas mixture usually containing other acidic gases by a suitable solid acceptor. This solid acceptor should be regenerated and must produce a concentrated stream of sulfur components ready for further processing. The processing of flue gas gives rise to requirements that differ from fuel gas processing (see STACK GAS TREATMENT).

Flue Gas Treating. The main reactions for the process discussed herein, SO_x removal using copper on silica or alumina, are given in Table 2. Apart from the regenerative DeSoxing reactions, DeNoxing is carried out simultaneously. The whole system, both adsorption (qv) and regeneration, takes place at ca 400°C. In addition to hydrogen, other reducing agents, eg, CO and CH_4, possibly mixed, can be used for the regeneration.

Whereas selection of the type of adsorber can be based on the usual factors,

Table 2. Reactions in Flue Gas Treatment

Reaction	ΔH at 25°C, kJ/mol[a,b]
Absorption	
1. $2\,Cu + O_2 \rightarrow 2\,CuO$	-155
2. $2\,CuO + 2\,SO_2 + O_2 \rightarrow 2\,CuSO_4$	-318
3. $4\,NO + 4\,NH_3 + O_2 \rightarrow 4\,N_2 + 6\,H_2O$	-407[c]
Regeneration	
4. $CuO + H_2 \rightarrow Cu + H_2O$	-86.6
5. $CuSO_4 + 2\,H_2 \rightarrow Cu + SO_2 + 2\,H_2O$	-10.6

[a] To convert J to cal, divide by 4.184.
[b] Value is per mole of Cu unless otherwise indicated.
[c] Value is per mole of NO.

eg, contacting patterns, contact time required, interfacial area, etc, because of the large amounts of gas to be treated in flue gas reactors, pressure drop and gas velocity are also very important. Moreover, a certain amount of fly ash must be acceptable if the process is to be applicable. The classical packed-bed swing reactor is therefore not suitable. A modified version, the parallel passage reactor, has been developed by Shell and commercial operation has been reported (29,30). A honeycomb system could also be suitable if swing operation is allowed.

The intermittent operation of the CuO process is complicated (30) because the system goes from oxidative to reducing conditions during the cycle (Table 3) creating difficult valve problems. Also problematic are the cyclic production of SO_2 and the replacement of the catalyst or sorbent. Thus a continuously operating adsorber would be preferable. Moving-bed reactors may be a solution if pressure drop can remain acceptably low and particle blockage can be avoided. One attractive feature could be simultaneous particle removal. A system operating along these lines was developed in the United States by Rockwell International and PETC based on cross-current flow (31). Nevertheless, for this type of operation pressure drop remains relatively high even at low gas velocities. A second

Table 3. Properties of Gas–Solids Contactors for the CuO–SO_2 Process[a]

	Reactor[b]			
Parameter	Parallel passage	Moving bed	Fluid bed	Trickle flow
continuous operation	−	±	+	+
countercurrent contacting	+	±	−	+
solids holdup	+	+	+	±
pressure drop, kPa[c]	3	1.5	4.5	1–1.5
superficial gas velocity at STP, m/s[d]	1.9	0.079	0.5	0.5–1

[a] A + indicates an advantage; −, a disadvantage.
[b] Mass transfer occurs fast enough for each of these systems.
[c] To convert kPa to psi, multiply by 0.145.
[d] Values are typical.

possibility for continuous operation is the fluidized bed and related reactors. This option has also been developed by PETC (32–34) and others (35,36). Dense beds are used because the intrinsic chemical reaction probably does not allow the use of, eg, risers.

Table 3 gives an overview of the contactors used for flue gas treatment. Included is the gas–solids trickle flow (GSTF) system used in Twente (37) and at ECN, the Netherlands, for the CuO process.

The basic ideas of the GSTF contactor is not new, but its properties have been studied and industrial application realized, as a heat exchanger, only relatively recently. In the GSTF a dilute flow of solid particles is contacted countercurrently with an upward gas flow over a regularly stacked packing. There are several favorable properties including low pressure drop, limited axial dispersion in the gas and solids phase, and excellent heat and mass transfer between both phases (38–40). When well designed, a regular stacked packing built up from bars or tubes can provide a good radial redistribution of both phases. The pressure drop is relatively low because the particles are only partially supported by the gas phase and owing to frequent collisions with the packing, the gas often does not reach its terminal velocity. Use of millimeter-sized particles (Table 4; Fig. 6) allows high superficial gas velocities suitable for flue gas cleaning.

The trickle flow adsorber was tested for the CuO process at fully integrated conditions, complete with regeneration both in a small unit (Fig. 7) and in a pilot-plant scale unit. Holdup and pressure drop of the bench-scale unit are given in Figure 8. In the small-scale unit over 50% desulfurization could be realized at a column height of only one meter. For the pilot plant at ECN, where the height of the packing element is five meters, conversions of 95% have been observed. Model calculations based on axial dispersed plugflow for both phases, inserting realistic values for the parameters under full-size operation, revealed that owing to the countercurrent operation, the value of the heat extraction factor, and other parameters, an important temperature peak should be present in the adsorption column (Fig. 9). The occurrence of such a peak, which may be advantageous for the process, was indeed observed at the pilot plant operated at ECN. As of late 1996, the trickle flow process is available for demonstration.

Applications for this contactor are expected to be in continuous process

Table 4. Packing Properties of the Pilot Plant

Property	Value
packing elements, m	
diameter	0.01
length	0.06
pitch	
horizontal	0.02
vertical	0.01
packing porosity, $\varepsilon(-)$	0.607
packing length, L, m	0.53[a]
drag constant, K/m	185

[a] Two sections.

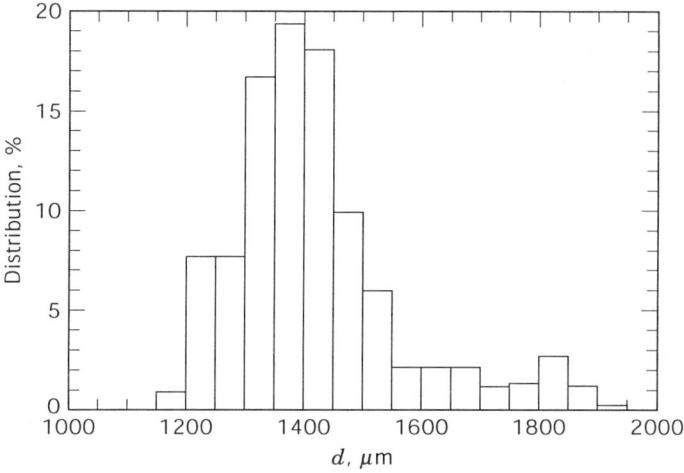

Fig. 6. Particle size distribution of the applied silica support.

where countercurrent flow is needed, relative fast adsorption reactions occur, ie, operation is close to the $k_G a$ limit, low pressure drop is required, low attrition rate of the sorbent is available, high gas-phase conversion occurs, and possible integration with heat exchange may be an advantage.

Fuel Gas Treating. High temperature H_2S removal from coal gas for power generation would increase electricity generation efficiency by eliminating a liquid wash step and, moreover, avoid a wastewater stream of condensed water from the product gas. A process based on Fe_2O_3 as active sorbent deposited by the deposition–precipitation method on a fluid-cracking catalyst (FCC) carrier has been investigated. For full details on this process and the reactor models see Reference 41. A schematic is shown in Figure 10. For conditions of $P = 3.0$ MPa (435 psi) and $T = 400°C$,

Adsorption $\qquad Fe_2O_3 + H_2 + 2\,H_2S \rightarrow 2\,FeS + 3\,H_2O$

Regeneration $\qquad 2\,FeS + 3.5\,O_2 \rightarrow Fe_2O_3 + 2\,SO_2$

The selection of the adsorber contactor is based more or less on the same principles as is the flue gas desulfurization, ie, continuous processes are preferable to avoid switching over from reducing to oxidizing conditions in the same apparatus and to avoid important transients in the removed and concentrated acid gases. There are, however, a few differences of substantial importance. First, because this process must be operated at high pressures (~3 MPa), high gas velocities are dictated in order to avoid large diameters of high pressure vessels. Secondly, the pressure drop is much less important.

The reaction is very rapid compared to the mass-transfer rate. This leads to the selection of very small particles and short contact times. The reaction rates observed using a bench-scale riser plant put in an electrically heated oven (Figs. 11 and 12) were so fast that at nearly all conditions mass-transfer limitation occurred. A typical contactor for this process would be a circulating fluidized-bed or riser reactor.

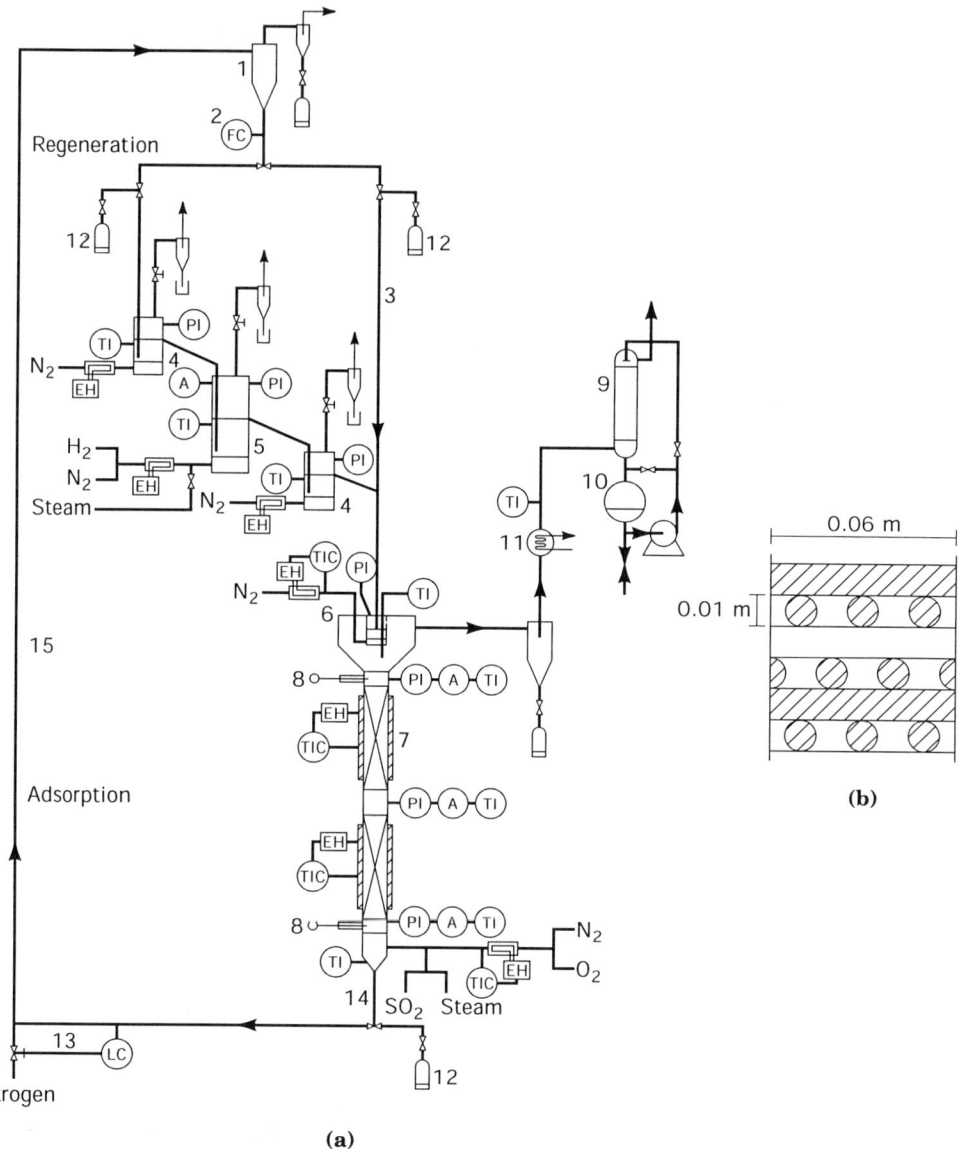

Fig. 7. The CuO process using a gas–solid trickle flow absorber: (**a**) schematic of the University of Twente bench-scale plant, where 1 is the settling chamber; 2, solids feed unit; 3, bypass; 4, purge fluidized bed; 5, regenerator; 6, absorber top section; 7, GSTF section; 8, guillotine valve; 9 wet scrubber; 10, NaOH solution; 11, spiral cooler; 12, sample bottle; 13, solids transport control; 14, downcomer; and 15, riser (TIC = temperature indicator and controller; EH, electrical heater; PI and TI, pressure and temperature indicators; FC, flow controller; and A, adsorbent); and (**b**) the packing configuration where ▨ and ⌀ represent packing elements (see Table 4).

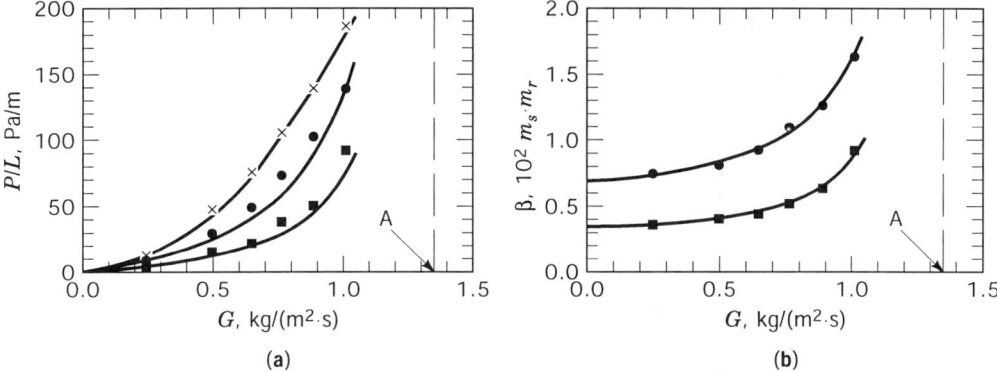

Fig. 8. Data as a function of gas flux, G, for the GSTF reactor where lines were calculated with a hydrodynamical model (37). Also shown, A, is the gas flux at which particle terminal velocity is reached in the smallest cross-sectional area, $G = \rho_g u_t/2$: (●) and (■) correspond to values of $S = 0.80$ and 0.37 kg/(m²·s), respectively, for (**a**), the pressure drop over the reactor for $\Delta P_s/L$ at 75°C, where (×) corresponds to $\Delta P/L$ values, and (**b**) the particle holdup, β, at 375°C (37). To convert Pa to mm Hg, multiply by 145.

Although these systems have been used in several applications such as catalytic cracking, drying, calcining, etc, very little is known about the gas–solids contacting performance, especially at elevated pressures. Contacting is very important if high gas-phase conversions are the goal as is the case in coal-gas desulfurization. In principle, countercurrent operation should also be preferred to obtain high solids conversions in one pass, but this factor has been shown to be less important (41).

It is well known that in riser reactions radial segregation of gas and solids occurs. This is a kind of core-annular flow that adversely influences the gas–solids contacting. Therefore a gas–solids contactor was developed consisting

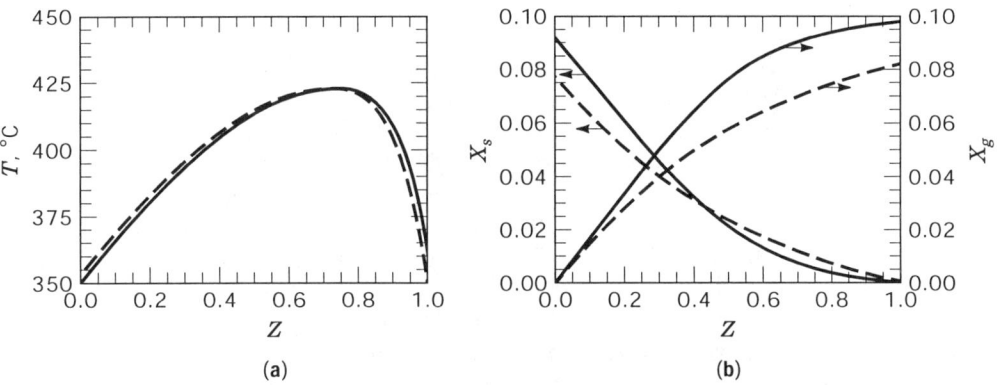

Fig. 9. (**a**) Axial temperature profiles calculated for a 15 m high GSTF DeSO$_x$ plant, showing the temperature, T, peak owing to internal heat exchange, Z, where (—) represents gas-phase, T_g, and (- - -) solids-phase, T_s, temperatures, and (**b**) conversion degree of gas, X_g, and solids, X_s, phases where (—) is the numerical and (- - -) the analytical solution, the latter at isothermal conditions that neglect heat exchange. Such temperature peaks and corresponding conversion increases have been observed in a pilot plant at ECN, the Netherlands (37).

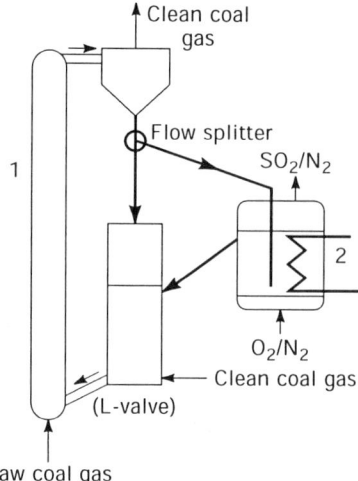

Fig. 10. Reactor setup for continuous coal-gas desulfurization process where 1 corresponds to the absorber, a circulating fluid bed, and 2, to the regenerator, a fluidized bed; (→) corresponds to gas and (➤) solids flow.

of a riser reactor having stacked bar packing (Fig. 13). The function of the bars is to eliminate the radial profiles of the gas velocity and solids density profiles. It was also hoped that the presence of the bars would increase $k_G a$ values at a given solids circulation rate and that the axial profile of the solids holdup would be reduced. The concept was extended to incorporate several stages in one single pressure vessel as shown schematically in Figure 14. Overall countercurrent contacting can be achieved while in each sector cocurrent contacting occurs. The multistage reactor would only be required if the reaction rate of the acceptor solids decreased dramatically upon acid-gas loading of the solids. This was not the case for this reaction as, owing to the packing, the solids holdup was increased and no visual segregation could be detected.

The mass-transfer rate was measured by an adsorption technique and found to be increased by several orders of magnitude in the 0.06×0.06 m² column as a result of the presence of the packing (42). Erosion and attrition are expected to be important factors and both are expected to depend on the design factors of the equipment, the solids, and on gas and solid velocities. The largest-scale tests were performed in a column of 0.125×0.125 m² using a cracking catalyst at ambient conditions, gas fluxes of 1.2–4.5 kg/(m²·s), and solids fluxes up to 40 kg/(m²·s). Neither excessive attrition nor erosion were observed in the short trial.

An open-circulating fluid bed without internals may show sufficiently effective gas–solids contacting for this particular process. The uncertainty in scaling-up needs to be overcome. As of late 1996, however, accurate data on gas–solids contacting under these conditions are lacking.

Reactive Slurry Systems

In industry large amounts of hydrogen are used for various processes, eg, ammonia and methanol synthesis, and hydrocracking. Usually in these processes a hydrogen

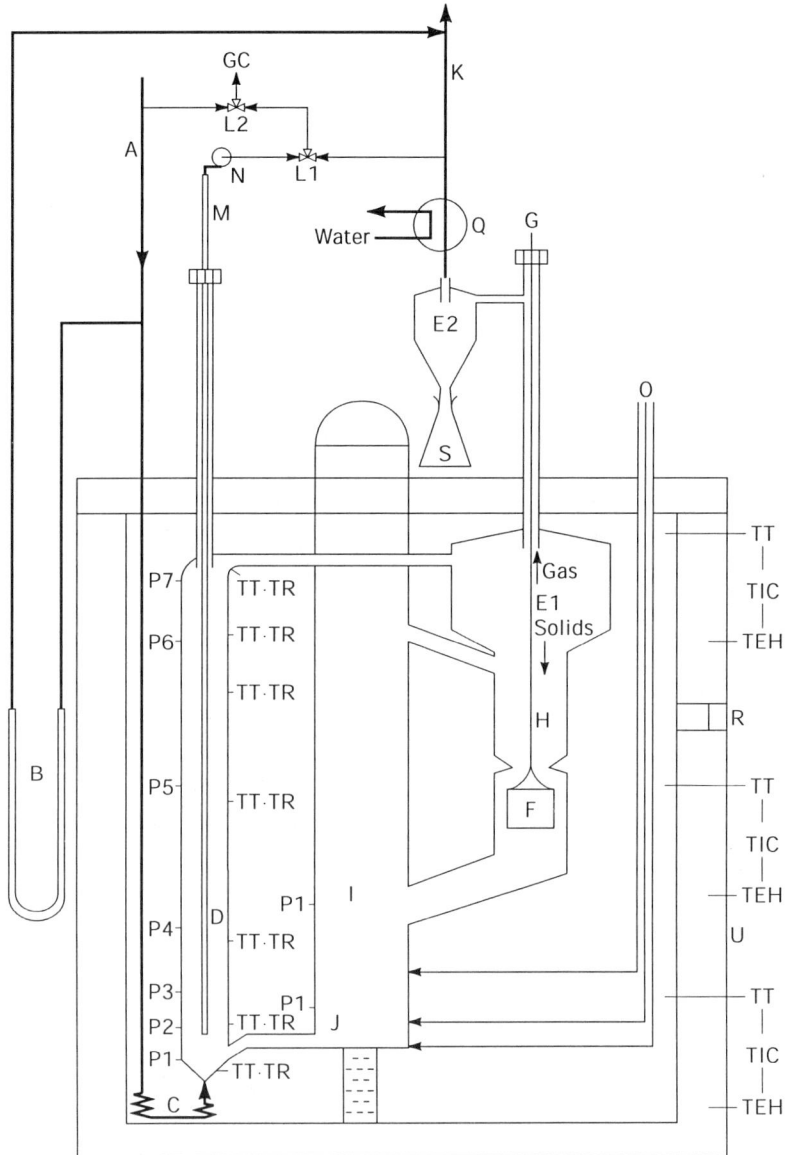

Fig. 11. Lab-scale riser reactor for the hot coal-gas desulfurization process where the length is ~ 1.2 m; diameter riser ≈ 0.015 m. A represents the gas inlet; B, overpressure safety; C, gas heating spiral; D, riser section; E1 and E2, cyclones; F, G, and H, solid flux measuring sections; I, standpipe (30 mm dia, 0.75 length); J, L-valve used for solid flow control; K, gas outlet; L1 and L2, gas sample valves for gas chromatograph; M, gas sampling probe; N, gas pump; O, gas inlet for aeration of standpipe and operation of L-valve; P1–P7, pressure indicators; Q, condenser; R, window in oven wall; S, solid sample flask; and U, column oven. The temperature transmitter (TT), temperature registration (TR), temperature indicator and controller (TIC), and electrical heater (TEH) are also represented.

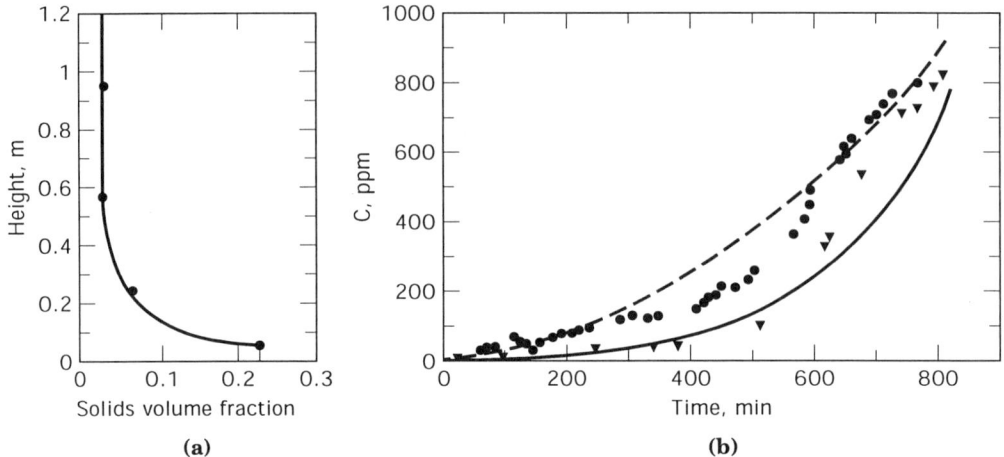

Fig. 12. (a) Axial solids volume fraction profile calculated from the axial pressure profile according to Reference 41; (b) H_2S concentration, C (initially 1070 ppm), at (▼) the outlet and (●) the probe tip located 0.15 m above the inlet as a function of time, where U_g = 1.8 m/s, G_s = 37 kg/(m^2·s), and $\bar{\beta}(0-L)$ = 0.05. The (—) corresponds to mass-transfer coefficients, k_Ga = 275 kg/(m^2·s), and (- - -), k_Ga = 700 kg/(m^2·s).

concentration or recovery from the main gas or off-gas stream, respectively, is an important step. A large number of techniques have been proposed (43), for instance membrane separation, cryogenic separation, and pressure-swing adsorption. Applicability of any of these techniques depends on the process requirements. Owing to substantial progress in the development of hydride-forming metal alloys, processes based on this chemistry have been proposed for hydrogen storage and recovery.

Hydrogen reacts reversibly and generally exothermically with a hydride-forming metal alloy, M:

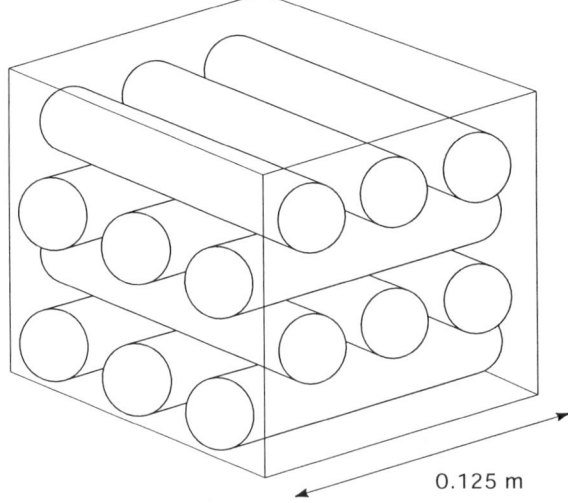

Fig. 13. Stacked packing used to prevent large-scale segregation of solids and gas in the riser reactor.

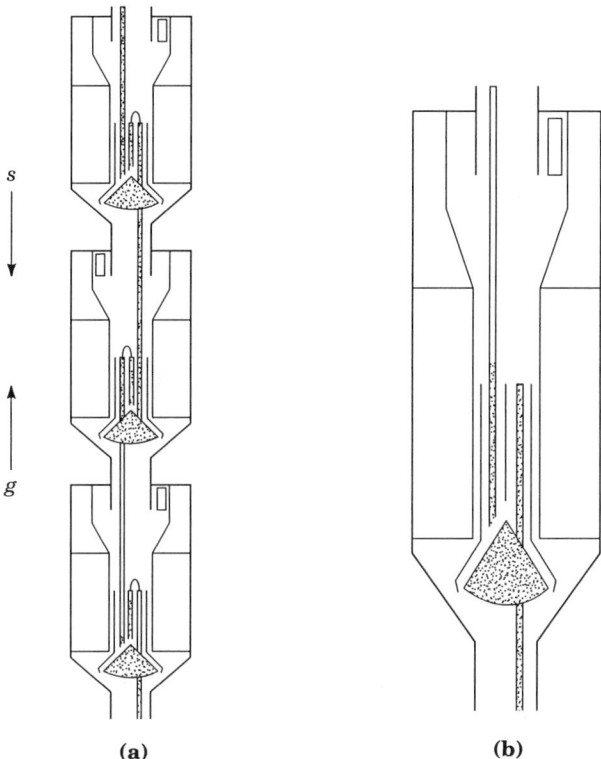

Fig. 14. (a) Overall countercurrent setup of (b) cocurrent packed riser sections, where g is gas; s, solid.

$$M + n/2\, H_2 \to MH_n + \text{heat}$$

The hydrogen absorption capacity in the metal alloys is very high, usually resulting in a density of hydrogen larger than that of liquid hydrogen. This property must be attributed to absorption of hydrogen in the interstitial sites of the crystal structure. A characteristic of the hydride-forming metal alloys is that the amount of hydrogen that has been absorbed is a function of the hydrogen pressure (Fig. 15).

For low values of F and low hydrogen pressures, the hydrogen is dissolved in the octahedral interstices of the metal alloy and a strong correlation exists between hydrogen pressure and F-value for this, the α-phase. For higher values of F a new metal hydride phase, β-phase, appears. Characteristic for the simultaneous occurrence of the α-phase and β-phase is that, as a consequence of the Gibb's phase rule, the hydrogen pressure remains constant for varying values of F. This constant value is called the plateau and is represented in Figure 15 as line AB. At higher values of F the α-phase disappears and hydrogen dissolves in the β-phase.

Usually the proposed hydride-forming metal alloy-based hydrogen recovery techniques are gas–solid processes in which the standard means of operation is pressure–temperature swing in packed beds of hydride-forming metal alloy pellets. For this type of process severe drawbacks exist: discontinuous operation and all the disadvantages of swing techniques, poor heat transfer, decrepitation of the parti-

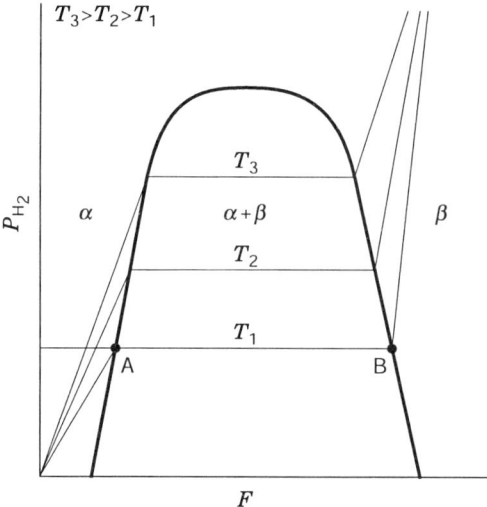

Fig. 15. Pressure composition isotherms for hydrogen absorption by metal hydrides where F = number of hydrogen atoms/number of hydride-forming metal atoms, and $T_3 > T_2 > T_1$ = temperature of P_{eq}. See text.

cles, and undesirable emission of fines. Dispersion of the hydride-forming metal alloys in a chemically inert liquid has been suggested to overcome these disadvantages (44). A continuously operated process would thus be possible in which nearly all the disadvantages would be eliminated. Moreover, because of the possibility of operating countercurrently, an efficient absorption process could be expected (45).

The introduction of a liquid phase in which the hydride-forming metal alloy is suspended means that two additional mass-transfer steps must be considered:

$$H_2 (g) \rightarrow H_2 (l)$$
$$H_2 (l) \rightarrow H_2 (s)$$

that is, mass transfer of hydrogen from gas to liquid, and from liquid to solids. These can substantially affect the hydrogen removal rate. The various steps for both absorption and desorption of hydrogen in or from the slurry are presented in Figure 16. The restricted solubility of the gaseous component, in this situation hydrogen, in the liquid can have a rate-decreasing effect.

The increase in resistance against mass transfer can be regarded as a substantial disadvantage of slurry processes in general. Despite this, however, slurries are frequently applied in industry. The absorption of gases into fine solid particles–liquid slurries can be increased by means of higher solubility or enhancement by the presence of very small particles near the gas–liquid interface. Alternatives for increasing the solubility is very limited because this parameter is completely determined by the physicochemical properties of the system. Enhancement of the absorption rate owing to the presence of small particles, however, may occur if the particles are small enough to be accommodated in the mass-transfer boundary layer. This subject has been extensively reviewed (46,47). The values that can be attained for absorption enhancement are affect-

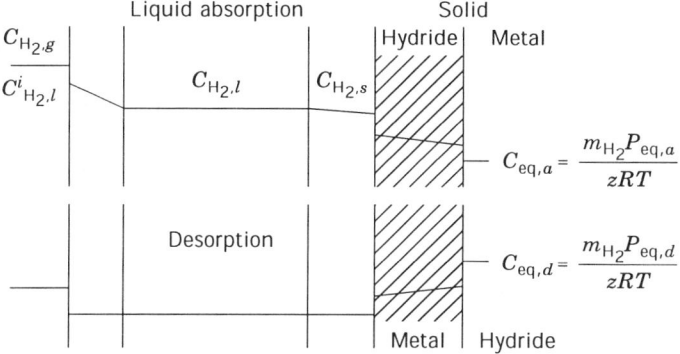

Fig. 16. Film model representation for hydrogen absorption and desorption in a slurry where $C_{H_2,g}$, $C_{H_2,l}$, and $C_{H_2,s}$ represent the concentration of hydrogen in the gas, liquid, and solid phase, respectively. Also shown are the concentrations of hydrogen at equilibrium for the absorption $C_{eq,a}$, and desorption, $C_{eq,d}$ processes.

ed by the type of solid particles. For instance, activated carbon (qv) preferentially adheres to the gas–liquid interface resulting in a locally higher concentration and a more pronounced enhancement (48). This effect has also been demonstrated theoretically (49).

An illustration for a continuous hydrogen separation process using a hydride-forming metal alloy dissolved in a liquid is shown in Figure 17. Also given are the corresponding pressure composition isotherms at absorber and desorber conditions. This type of process has been investigated on a pilot-plant

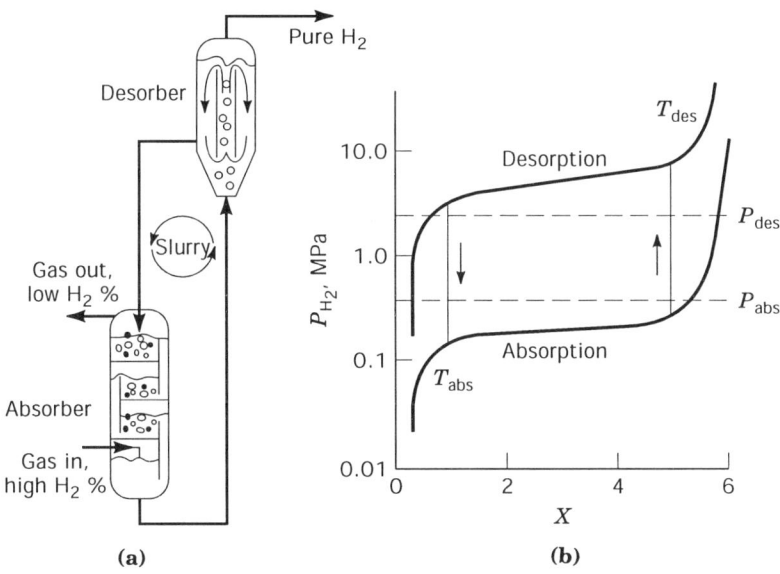

Fig. 17. A continuous H_2 separation process using a slurry of $LaNi_5$ and silicone oil: (**a**) schematic illustration of the process and (**b**) hydride pressure isotherms at absorber (P_{abs}, T_{abs}) and desorber (P_{des}, T_{des}) conditions. To convert MPa to psi, multiply by 145.

scale (50) using LaNi$_5$ silicone oil slurries in a valve tray column. Depending on the solids fraction in the slurry, the absorption rate was determined by reaction or mass transfer for low and high fractions, respectively. Furthermore, it was concluded that the hydride-forming metal alloy preferentially moves away from the interface to the bulk. Therefore no enhancement was observed although this was expected on the basis of the solids fraction (51). The pilot-plant experiments showed that this type of slurry process is technically feasible. On the trays no sedimentation problems occurred and high mass-transfer rates could be achieved. The solubility of hydrogen was, however, very low and combined with the absence of enhancement, the observed overall absorption rates remained low.

NOMENCLATURE

Symbol	Definition	Units
c_i	concentration component i	mol/m^3
d	diameter/particle diameter	m
D_i	diffusivity for component i	m^2
E_i	enhancement factor for component i	
F	constant defined by eq. 6	
G	molar flow gas phase	mol/s
	gas flux	kg/(m^2·s)
Gz	Graetz number	
ΔH^0	reaction enthalpy	kJ/mol
H	column height	m
He_i	Henry coefficient component i	Pa·m^3/mol
J_i	molar flux component i	mol/(m^2·s)
k_G	gas-phase mass-transfer coefficient	m/s
k_L	liquid-phase mass-transfer coefficient	m/s
L	molar flow liquid phase	mol/s
m_i	dimensionless mobility for component i	
N_L	number of rotations of stirrer in model reactor	s^{-1}
ΔP	pressure drop	Pa
R	ideal gas constant	J/(mol·K)
Re	Reynolds number	
S	selectivity factor defined by eq. 3	
Sc	Schmidt number	
Sh	Sherwood number	
T	temperature	K or °C
U	superficial velocity	m/s
v	velocity	m/s
X	indentical to constant F	
x	molar fraction liquid phase	
y	molar fraction gas phase	
Z	length	m
z	compressibility factor	
α	α-phase of metal hydride	
β	holdup	
	β-phase of the metal hydride	

Symbol	Definition	Units
ρ	density	kg/m^3
σ	selectivity defined by eq. 2	

Subscripts / superscripts

a, abs	absorption
d, des	desorption
e, eq	at equilibrium conditions
exp	experimental
G	gas phase
i	component i
L	liquid phase
local	local
s	solid
1	section 1
2	section 2
n	section n
i	interface

BIBLIOGRAPHY

1. A. L. Kohl and F. C. Riesenfeld, *Gas Purification*, Gulf, Houston, Tex., 1985.
2. M. W. McEwan and A. Marmin, *Proceedings LNG-6*, Vol. II, Kyoto, Japan, 1980.
3. P. M. M. Blauwhoff, B. Kamphuis, W. P. M. van Swaaij, and K. R. Westerterp, *Chem. Eng. Proc.* **19**, 1 (1985).
4. P. M. M. Blauwhoff and W. P. M. van Swaaij, *Proceedings of the 2nd International Conference on Phase Equilibria and Fluid Properties in the Chemical Industry*, EFCE Publication 11, Berlin, 1980, p. 78.
5. T. Chakravarty, Ph.D. dissertation, Clarkson University, Potsdam, N.Y., 1985.
6. G. F. Versteeg, J. A. M. Kuipers, F. P. H. van Beckum, and W. P. M. van Swaaij, *Chem. Eng. Sci.* **44**, 2295 (1989).
7. G. F. Versteeg, J. A. M. Kuipers, F. P. H. van Beckum, and W. P. M. van Swaaij, *Chem. Eng. Sci.* **45**, 183 (1990).
8. H. Bosch, J. A. M. Kuipers, W. P. M. van Swaaij, and G. F. Versteeg, *Gas Sep. Pur.* **3**, 75 (1989).
9. R. Cornelisse, A.A.C.M. Beenackers, F. P. H. van Beckum, and W. P. M. van Swaaij, *Chem. Eng. Sci.* **35**, 1245 (1980).
10. P. M. M. Blauwhoff and W. P. M. van Swaaij, *Chem. Eng. Proc.* **19**, 67 (1985).
11. R. C. Darton, P. J. Hoek, J. A. M. Spaninks, M. M. Suenson, and E. F. Wijn, *I. Chem. E. Symp. Ser.* **104**, A323 (1987).
12. G. F. Versteeg and W. P. M. van Swaaij, *I. Chem. E. Symp. Ser.* **104**, B139–B154 (1987).
13. F. J. M. Schrauwen and D. Thoeness, *Chem. Eng. Sci.* **43**, 2189 (1988).
14. H. Bosch, G. F. Versteeg, and W. P. M. van Swaaij, *Gas Sep. Pur.* **8**, 513–520 (1989).
15. F. Rumford and I. J. Rae, *Trans. Instn. Chem. Eng.* **34**, 195 (1956).
16. C. Ramshaw, *Chem. Eng.* **13** (Feb. 1983).
17. P. Carnell and P. Starkey, *Chem. Eng.* **30** (Nov. 1984).
18. Z. Qi and E. L. Cussler, *J. Mem. Sci.* **23**, 321 (1985).
19. *Ibid.*, p. 333.
20. R. Prasad and K. K. Sirkar, *AIChE J.* **34**, 177 (1988).

21. D. O. Cooney and C. C. Jackson, *Chem. Eng. Comm.* **79**, 153 (1989).
22. A. Laurent and J. C. Charpentier, *Chem. Eng. J.* **8**, 85 (1974).
23. H. van Landeghem, *Chem. Eng. Sci.* **35**, 1912 (1980).
24. S. L. Matson, J. Lopez, and J. A. Quinn, *Chem. Eng. Sci.* **38**, 503 (1983).
25. H. Kreulen, G. F. Versteeg, C. A. Smolders, and W. P. M. van Swaaij, *Chem. Eng. Sci.* **48**, 2093 (1993).
26. H. Kreulen, G. F. Versteeg, C. A. Smolders, and W. P. M. van Swaaij, *J. Mem. Sci.* **78**, 197 (1993).
27. O. Miyatake and H. Iwashita, *Int. J. Heat Mass Trans.* **33**, 416 (1990).
28. H. Kreulen, G. F. Versteeg, C. A. Smolders, and W. P. M. van Swaaij, *J. Mem. Sci.* **78**, 217 (1993).
29. J. E. Dautzenberg, J. E. Naber, and A. J. J. van Ginneken, *Chem. Eng. Prog.* **67**, 86 (1971).
30. J. M. Burke, EPA No. 600/7-82-064, U.S. EPA, Washington, D.C., 1982.
31. D. Stelman, J. P. Ampaya, and J. C. Newcomb, DOE contract DE-AC22-83 PC 60262, U.S. DOE, Washington, D.C., 1987.
32. D. H. McCrea, A. J. Forney, and J. G. Meyers, *JAPCA* **20**, 819 (1970).
33. J. T. Yeh, C. J. Drummond, and J. I. Joubert, *Environ. Prog.* **6**, 44 (1987).
34. J. T. Yeh, R. J. Demski, J. P. Strakey, and J. I. Joubert, *Environ. Prog.* **4**, 223 (1985).
35. R. J. Best and J. G. Yates, *Ind. Eng. Chem. Proc. Des. Dev.* **16**, 347 (1977).
36. C. Laguerie and D. Barreteau, *Sadhana* **10**, 49 (1987).
37. J. H. A. Kiel, W. Prins, and W. P. M. van Swaaij, in Ref. 13, pp. 539–548.
38. A. W. M. Roes and W. P. M. van Swaaij, *Chem. Eng. J.* **17**, 81 (1979).
39. A. W. M. Roes and W. P. M. van Swaaij, *Chem. Eng. J.* **18**, 29 (1979).
40. A. B. Verver and W. P. M. van Swaaij, *Chem. Eng. Sci.* **4**, 435 (1987).
41. A. G. J. van der Ham, W. Prins, and W. P. M. van Swaaij, *Proceedings of the CFB-4 Conference*, 1993.
42. A. G. J. van der Ham, W. Prins, and W. P. M. van Swaaij, *Powder Technol.* (1993).
43. "Gas Handbook," *Hydrocarbon Proc.*, 92 (1986).
44. Eur. Pat. Appl. 94,136 (1982), A.A.C.M. Beenackers and W. P. M. van Swaaij.
45. Rep. Eur. 9687EN (1985), K. J. Ptasinski, A.A.C.M. Beenackers, W. P. M. van Swaaij, R. D. Holstvoogd, and G. F. Versteeg (to Commission of European Communities, Brussels).
46. E. Alper and W. D. Deckwer, *Nato-ASI Series E, Vol. 2, No. 73*, M. Nijhoff, The Hague, 1983, p. 199.
47. A.A.C.M. Beenackers and W. P. M. van Swaaij, *Nato Series,* Vol. 110, M. Nijhoff, Boston, Mass., 1986, p. 463.
48. O. J. Wimmers, Ph.D. Thesis, University of Amsterdam, the Netherlands, 1987.
49. R. D. Holstvoogd, W. P. M. van Swaaij, and L. L. van Dierendonck, *Chem. Eng. Sci.* **43**, 2181 (1988).
50. R. D. Holstvoogd, W. P. M. van Swaaij, G. F. Versteeg, and E. D. Snijder, *Zeit. Physikal. Chem. Neue Folge*, **164**, 1429 (1989).
51. J. P. M. H. Knippels, R. D. Holstvoogd, G. J. de Laat, K. J. Ptasinsky, G. F. Versteeg, and W. P. M. van Swaaij, EG-report, EN 3B-0107 NL, 1990.

W. P. M. VAN SWAAIJ
G. F. VERSTEEG
University of Twente, the Netherlands

CROWN ETHERS. See INCLUSION COMPOUNDS.

CRYOGENIC DISTILLATION

Cryogenic distillation is of industrial importance in the production of nitrogen and oxygen from air, purified hydrogen from refinery and petrochemical plant off gases, and the recovery of helium from natural gas.

Air Separation

A considerable number of cryogenic concepts were developed near the end of the 19th and beginning of the 20th century to liquefy and separate air into its main constituents: 78% nitrogen, 21% oxygen and 1% argon by volume. By 1905, Carl von Linde's double distillation column process to produce practically pure oxygen [7782-44-7] and nitrogen [7727-37-9] was already known. Argon [7440-37-1] was industrially produced by 1912 (1). Trace quantities of other rare gases such as neon, krypton, and xenon are present in air and can be recovered by proper modifications to a cryogenic double-distillation column air-separation plant (see AIR SEPARATION) (2).

Figure 1 shows a cryogenic air-separation process using a double-distillation column of the Linde-type. There is a side rectifying column to recover crude argon (3). Feed air is compressed in a multistage compressor to about 650 kPa (94 psi) with interstage cooling. The compressed air is cooled in an aftercooler by cooling water and the condensed water is removed in an aftercooler separator. To avoid freezing of water and carbon dioxide in the cryogenic part of the plant, the feed air is passed through an adsorbent bed of molecular sieves. Along with water and carbon dioxide, several trace impurities in the air such as acetylene, ethylene, butane, and other heavier hydrocarbons are also adsorbed. This alleviates some of the safety problems arising from the formation of hydrocarbon–oxygen mixtures in older plants where molecular-sieve adsorbers are not used.

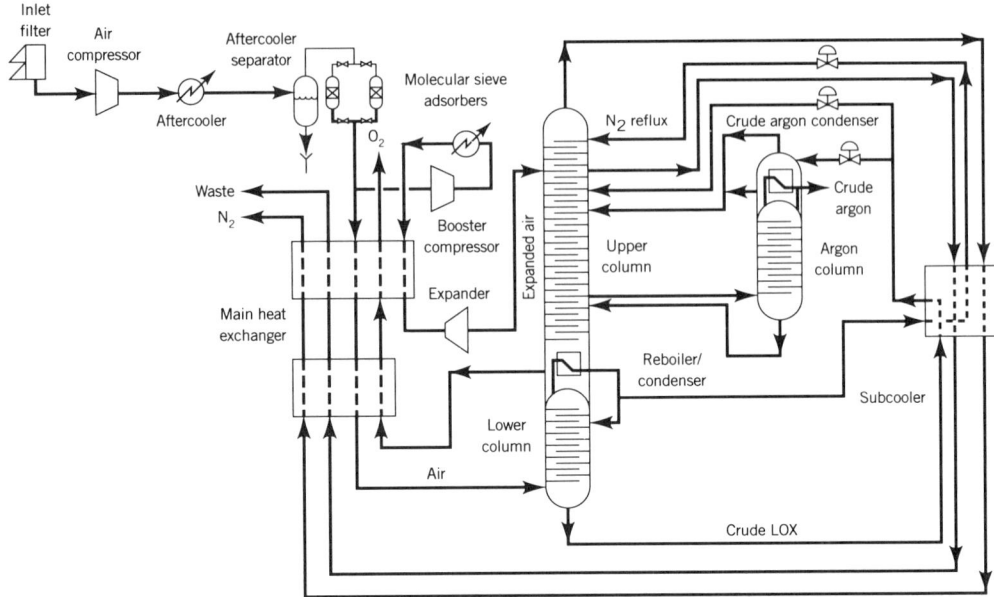

Fig. 1. Cryogenic air separation process. LOX = liquid oxygen.

The purified air stream is split into two streams. One stream of about 7–12% of the total air flow is boosted in pressure by about 60–100 kPa (9–15 psi) in a booster compressor. The boosted air is cooled by cooling water and then in the main heat exchanger against the returning cold product nitrogen and oxygen streams. The cooled air stream is work expanded to near atmospheric pressure and fed to the distillation column. This expander provides the needed refrigeration for the plant. Usually the work extracted from the expander drives the booster compressor such that an additional source of energy is not required.

The principal air stream is cooled in the main heat exchanger to near its dew point. Nitrogen and oxygen are separated by distillation in a two-stage distillation process. The first stage, the lower column, which operates at about 600 kPa (87 psi) separates the feed air into nitrogen and an oxygen-enriched liquid (crude LOX) stream, which is eventually fed to the upper column. The upper column, which operates at close to ambient pressure, produces pure oxygen and nitrogen streams. Nitrogen reflux for both columns is generated at the top of the lower column. Nitrogen vapor at the top of the lower column is condensed against the liquid oxygen at the bottom of the upper column by heat exchange between the two streams in a reboiler-condenser. The pressure difference between the two columns provides the temperature difference between the condensing and boiling fluids.

Argon boils between oxygen and nitrogen, thus a peak in the argon concentration of about 7–15% occurs in the lower section of the upper column. A vapor stream is drawn from the upper column near the location of the peak argon concentration and is distilled in a third (argon) column. Typically, a vapor stream is drawn from the upper column at a point where the concentration of nitrogen is low (25–100 vppm) and the concentration of argon is 7–12%. Reflux for the argon column is provided by heat exchange between the vapor at the top of this column and a portion of the crude LOX stream in a crude argon condenser. The vaporized crude LOX is fed to the upper column and the condensed argon-rich stream is returned as reflux to the argon column. The liquid from the bottom of the argon column is returned to an appropriate location in the upper column. A crude argon stream containing 2–5% O_2 and 0.05–1% N_2 is withdrawn from the top of the argon column. The oxygen produced from the bottom of the upper column can easily have an oxygen purity of 99.5% or greater. The remainder is mainly argon. The nitrogen product from the top of the upper column can be produced with an oxygen concentration below 5 ppm by volume (vppm).

The heat exchangers used in cryogenic air-separation plants are generally brazed aluminum plate-fin heat exchangers. Since the early 1930s, sieve trays have been used predominantly for cryogenic distillation. The use of structured packing with much lower pressure drop per theoretical stage of separation is, however, becoming increasingly common in the upper and argon columns. Use in the upper column decreases the power consumption by the air compressor as the pressure of the feed air to the lower distillation column can be decreased by about 60 kPa (9 psi). The use of structured packing in the argon column provides more theoretical stages of separation for the same permissible pressure drop between the bottom and top of the argon column. This allows argon production from the argon distillation column with oxygen impurity as low as 1 vppm, either eliminating or reducing the size of the expensive downstream processes used historically to produce pure argon product (4).

In the field of semiconductor device manufacturing, the trends toward device miniaturization and the need for high production yield are leading to gas specifications of ultrahigh purities. This requires that concentration of all impurities in each of the nitrogen, oxygen, and argon products utilized be less than 10 vppb. The impetus to decrease the impurity level is expected to continue as devices continue to decrease in size. Gases with parts-per-trillion impurity concentrations are expected to be needed. To meet this challenge, cryogenic processes that modify the scheme shown in Figure 1 by employing additional columns have been developed. For example, use of an additional stripping column in conjunction with the upper column to coproduce a large fraction of oxygen product as ultrahigh purity oxygen has been described (5).

Liquid Nitrogen, Liquid Oxygen, and Liquid Argon. Large quantities of liquid nitrogen are typically produced by liquefying gaseous nitrogen from an air separation plant. For this purpose, a gaseous nitrogen stream is compressed to a fairly high pressure and work expanded at successive lower temperatures to create colder temperatures for liquefying nitrogen. Low pressure gaseous nitrogen from the air-separation unit is combined with a recycle nitrogen stream and compressed to about 600 kPa (87 psi) in a feed compressor. The compressed nitrogen is combined with another recycle nitrogen stream and further compressed to about 3 MPa (435 psi) in a recycle compressor. Both the feed and recycle compressors are high efficiency, multistage centrifugal machines. The nitrogen stream is then further compressed in two booster compressors to about 4.5 MPa (653 psi), which is above the critical pressure of nitrogen. Each of these booster compressors is linked to and driven by an expander providing refrigeration to the process. The final compressed nitrogen stream is then sent to a heat exchanger. Large portions of this stream are expanded in two expanders and recycled, while a smaller fraction is cooled to below the critical temperature of nitrogen by heat exchange with the expanded streams. The inlet temperatures to the expanders are chosen such that the temperature differences between the cooling and warming streams in the heat exchanger are minimized to reduce the liquefaction power requirement. The cooled high pressure nitrogen stream is then isenthalpically let down in pressure across valves in two stages to provide the desired liquid nitrogen product stream near ambient pressure.

Liquid argon can be produced directly from the crude argon condenser shown in Figure 1 by increasing the flow of air through the expander, but expander refrigeration is not economic for production of large quantities of liquid oxygen. Typically, for this purpose, refrigeration is provided by adding liquid nitrogen from a nitrogen liquefier to the top of the upper column while liquid oxygen is withdrawn from the bottom of this column.

Hydrogen Purification

Cryogenic separation is used extensively for recovery and purification of hydrogen [1333-74-0] from refinery and petrochemical plant off-gases. The most common applications are in recovery of hydrogen from catalytic reformer gas, ethylene plant off-gas, hydrodealkylation (HDA) recycle gas, hydrodesulfurization (HDS) off-gas and methanol–ammonia synthesis purge gases (6). Hydrogen can also be coproduced when carbon monoxide is recovered by cryogenic separation of a stream–methane reformer off-gas.

The simplest form of hydrogen purification uses partial condensation to condense light hydrocarbons and trace impurities such as argon, carbon monoxide, and nitrogen from the relatively noncondensable hydrogen. Because of the high relative volatility of hydrogen to light hydrocarbons, in the range of 100 to 300 for hydrogen–methane at separation temperatures of 135–100 K, cryogenic separation can easily attain a hydrogen purity of 90–96% without distillation. Product hydrogen is recovered at close to feed gas pressure. Typical hydrogen recoveries are 90–98%.

At optimum conditions, ie, feed gas rate of at least 5000 STP m^3/h feed pressure of at least 2.8MPa (406 psi), and feed hydrogen content less than 80%, all of the refrigeration requirements for the cryogenic hydrogen purification process can usually be obtained by Joule-Thomson expansion of the condensed hydrocarbons. At less favorable conditions, auxiliary refrigeration is required. This refrigeration can typically be supplied by inexpensive package Freon or propane units providing refrigeration at temperatures of 250 to 210 K.

Compact brazed aluminum plate-fin heat exchangers can be used in most cryogenic hydrogen purification applications. The use of these relatively low cost heat exchangers, combined with low separation energy requirements, results in a highly economical process for hydrogen purification.

One of the most common applications of cryogenic hydrogen purification is to recover hydrogen and reject light hydrocarbons from the recycle loop of an HDA

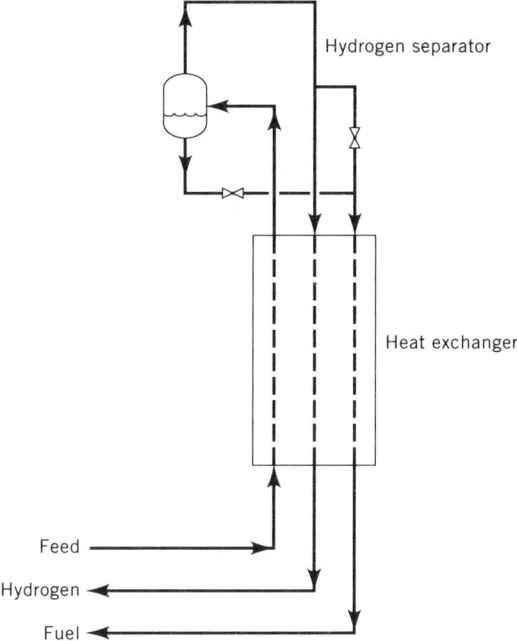

Fig. 2. Hydrogen purification process.

unit that converts toluene or other aromatics to benzene (see BTX SEPARATIONS) (Fig. 2). After pretreatment of the feed gas to remove water and aromatic impurities that would freeze at low temperatures, the feed gas is cooled to about 120 K to condense most of the hydrocarbons and provide a product hydrogen purity of 90–92%. The condensed hydrocarbons are flashed to a low pressure, typically 200–500 kPa (29–73 psi) revaporized and warmed for refrigeration recovery and then sent to the plant fuel system. The product hydrogen is rewarmed at pressure and recycled to the HDA unit.

Light Olefin and Liquefied Petroleum Gas Recovery. The relatively simple cryogenic purification process for hydrogen recovery can easily be adapted (7) to recover a crude light olefin or propane and heavier hydrocarbon (LPG) stream. The pretreated feed gas is cooled to an intermediate temperature, in the range of 240 to 200 K for propylene/LPG recovery or 180 to 150 K for ethylene recovery. The uncondensed vapor is then further cooled to condense the remaining methane and residual heavy hydrocarbons. The two condensed hydrocarbon streams are flashed separately to a lower pressure, revaporized and warmed with the product hydrogen stream for refrigeration recovery. A flash separator can be added to reduce the amount of light impurities in the crude olefin/LPG stream. A demethanizer or de-ethanizer column can also be incorporated to provide a high purity olefin or LPG product, but requires auxiliary refrigeration.

Nitrogen Rejection and Helium Recovery

Cryogenic distillation has been used extensively in the processing of natural gas for nitrogen removal and for helium [7440-59-1] recovery (8,9).

Nitrogen Rejection. Two basic processes are used for nitrogen rejection from natural gas; the single-column heat-pumped process, and the double-column process. Earlier processes utilized multistage flash columns for helium recovery from natural gas (10).

In the single-column heat-pumped process (Fig. 3), feed gas is cooled and fed to a high pressure column which operates at 2.1–2.8 MPa (304–406 psi). Nitrogen vapor is withdrawn from the overhead of the column and natural gas liquids from the bottom. A closed-loop methane heat pump supplies reboiler heat and condenser cooling to effect the separation. The natural gas liquids stream is flashed to a low pressure, revaporized, and warmed with the rejected nitrogen for refrigeration recovery. The nitrogen rejected at high pressure is suitable for reinjection into oil reservoirs to maintain pressure and enhance oil recovery.

The double-column process (Fig. 4) is similar to the process described in

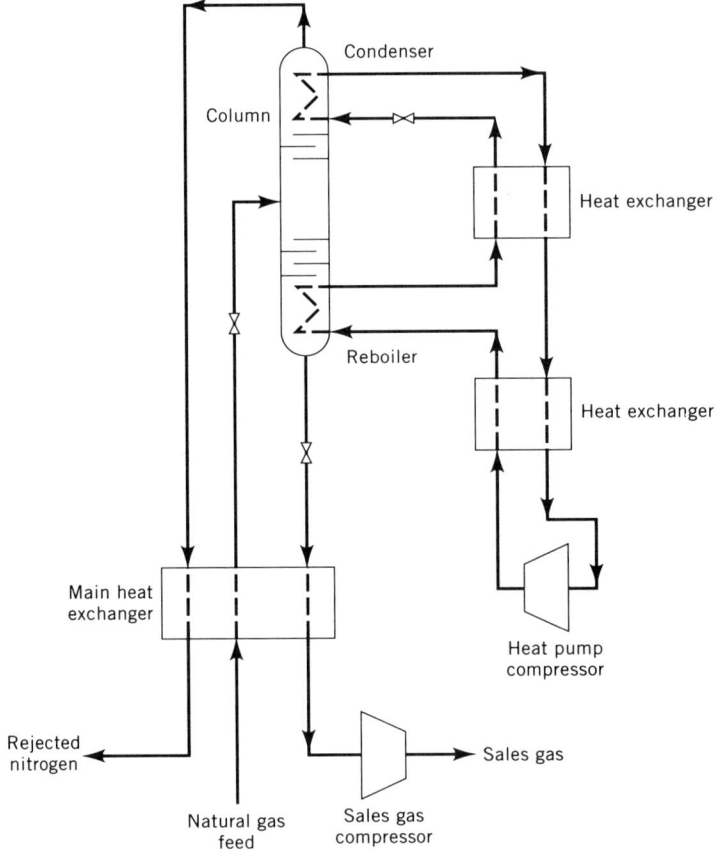

Fig. 3. Single-column heat pumped process for nitrogen rejection.

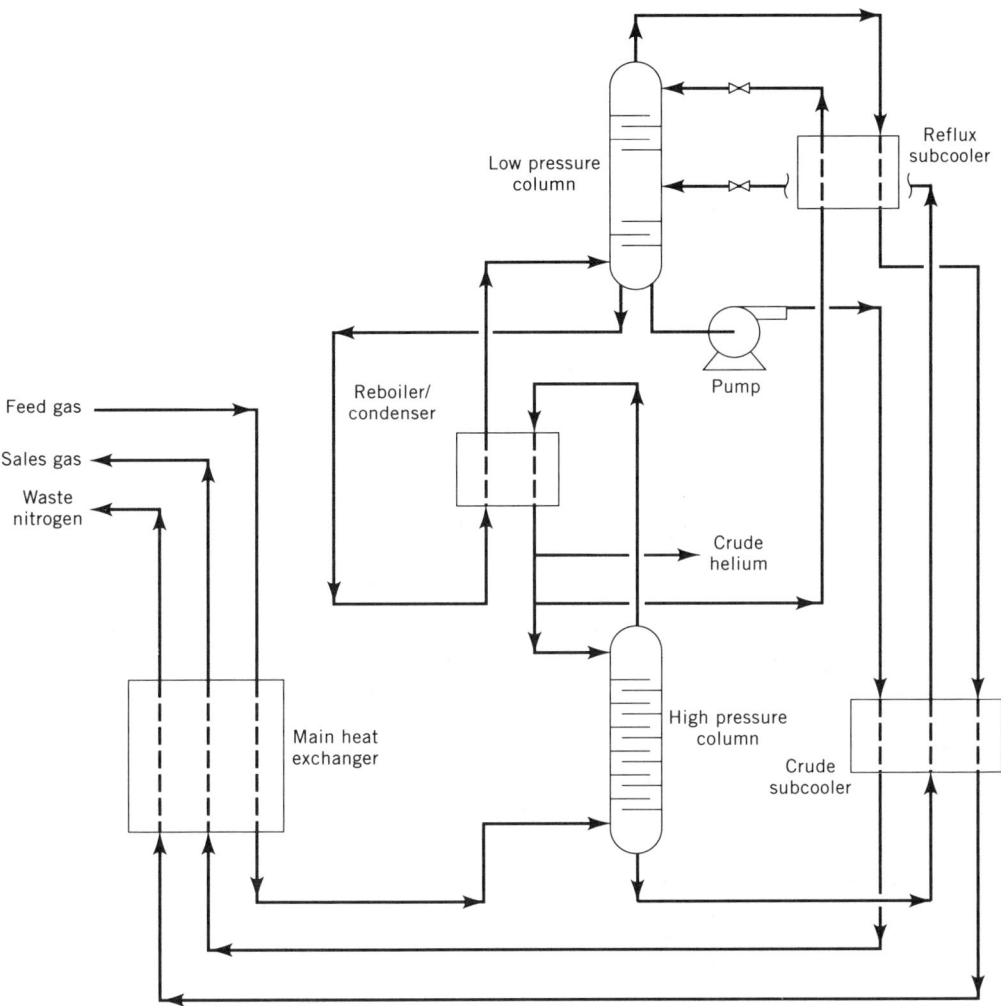

Fig. 4. Double-column process for nitrogen rejection.

Figure 1 for air separation. Feed gas is cooled and fed to a high pressure column operating at 1–2.5 MPa (145–363 psi). Nitrogen from the top of the high pressure column and a crude natural gas liquid stream from the bottom of the high pressure column are further processed in a low pressure column at about 150 kPa (22 psia) to complete the separation. Reboiler heat for the low pressure column is provided by condensing nitrogen at the top of the high pressure column, which is then used to reflux both columns. The natural gas liquids from the bottom of the low pressure column are pumped to a higher pressure, revaporized, and warmed with the low pressure nitrogen from the top of the low pressure column for refrigeration recovery.

Crude helium (containing 50–70% helium, associated hydrogen and neon, 1–3% methane, and the balance nitrogen) can easily be obtained by minor enhancements to the nitrogen rejection unit, particularly with natural gases containing 0.5% or more helium. For example, by operating the double-column condenser in a partial condensation mode, a stream of uncondensed vapor at about 50% helium concentration can be obtained. This crude helium stream can be fed directly to helium purification and liquefaction units.

Natural gas liquids (NGL) recovery can also be integrated into a nitrogen rejection unit to achieve economies in equipment and operating cost (11). Integration of the heat exchange equipment and optimization of the refrigeration in the nitrogen rejection and NGL sections of the plant can provide significant overall savings. Typical ethane recoveries of 70–85% can be obtained.

Helium Purification and Liquefaction. Helium, which is the lowest-boiling gas, has only 1 degree K difference between its normal boiling point (4.2 K) and its critical temperature (5.2 K), and has no classical triple point (12,13). It exhibits a phase transition at its lambda line (running from 2.18 K at 5.03 kPa (0.73 psia) to 1.76 K at 3.01 MPa (437 psia)) below which it exhibits superfluid properties (13).

Helium is commercially recovered from natural gas. If helium is present in a natural gas it usually occurs in concentrations between 0.2% to 2%, although it has been found in concentrations up to 8%. In the United States, significant quantities of helium exist in gas fields of Wyoming and the panhandle regions of Texas and Oklahoma and in southwestern Kansas. Helium is also found in natural gas in Algeria, Canada, Poland, China, and the North Sea. As part of the United States' helium conservation program of the 1960s, several plants were built to recover helium. These plants typically processed 12×10^6 STP m^3/d of gas. A 400 km pipeline to collect the crude helium was also built. Crude helium from this pipeline is stored in the Cliffside Reservoir near Amarillo, Texas for later withdrawal and processing. Helium from Wyoming gas is liquefied without intermediate storage as crude helium.

Helium is normally concentrated in stages, first from the initial field concentration to crude, then to pure helium for liquefaction or sale as pure gas. Crude helium is easily recovered from a plant rejecting nitrogen from natural gas. Final upgrading of helium from crude to pure is typically done by pressure-swing adsorption (PSA) in combination with cryogenic partial condensation (14) (see ADSORPTION, GAS SEPARATION). Helium in natural gas streams is accompanied by neon and hydrogen that cannot be separated from helium in the PSA unit. However, both must be removed to prevent their freezing in the helium liquefaction process.

A process for final upgrading (purifying) of the helium is shown in Figure 5 (14). Crude helium, which may be at low temperature from previous processing, is combined with PSA purge gas recycle, cooled to 80 K and partially condensed to provide a vapor stream of about 90% helium concentration. Liquid from the partial condensation is warmed and separated to provide a waste stream and a nitrogen stream. The waste stream is used to regenerate the driers, and a part of the nitrogen stream may be liquefied. The upgraded 90% helium stream, with air added to provide oxygen for catalytic combustion of the hydrogen, is preheated and passed over a hydrogen removal catalyst, then cooled to condense most of the water (Fig. 6). The upgraded helium stream then flows through the PSA unit and to liquefaction. The PSA unit removes most of the remaining impurities, including water and nitrogen but excluding neon and unreacted hydrogen, to less than 10

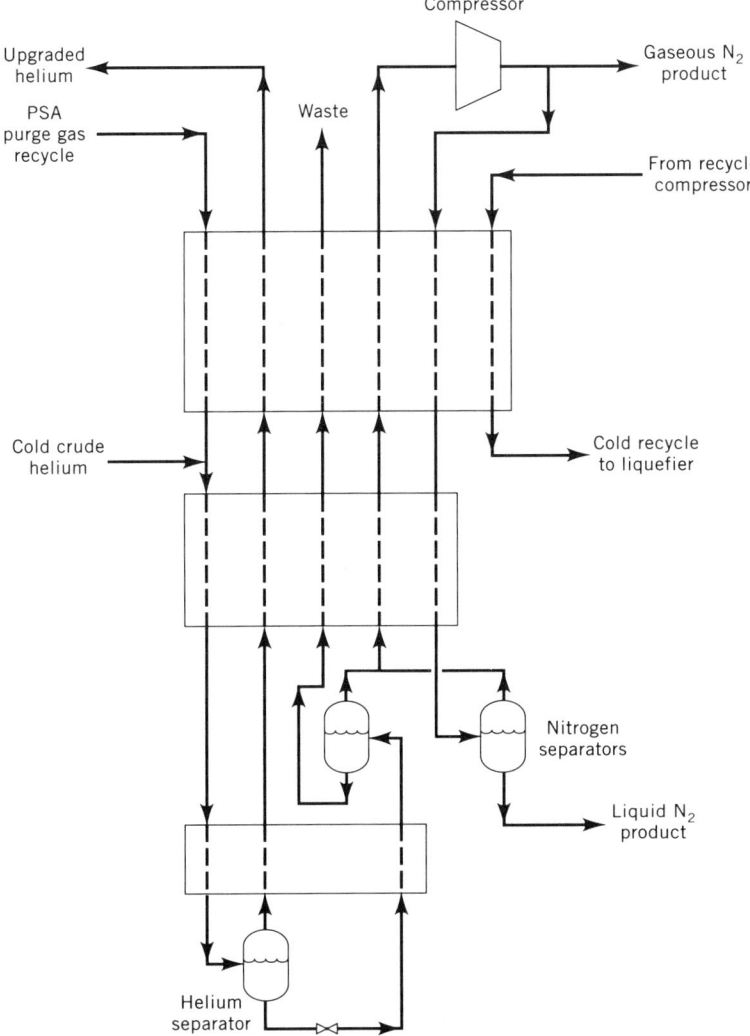

Fig. 5. Crude helium upgrading process.

vppm. The trace impurities are removed in an 80 K adsorber in the liquefier, and neon and any unreacted hydrogen are adsorbed in a 20 K adsorber. A portion of the pure product from the PSA is used to regenerate the PSA beds, then compressed, dried, and recycled to the inlet of the upgrader where it is combined with crude feed for high overall helium recovery.

Helium is liquefied by cooling to 80 K with either liquid nitrogen or the exhaust from a helium expander, then cooled below 80 K with refrigeration supplied by several helium expanders, followed by expansion through a dense-fluid wet expander. The number of expanders used in a liquefaction cycle varies with the plant capacity. Figure 7 shows a liquefier with 5 expanders (14). Helium is usually cooled at 2 MPa pressure. The main flow from the exhaust of the expanders returns to compression at 0.2 MPa (29 psi) and a small stream from

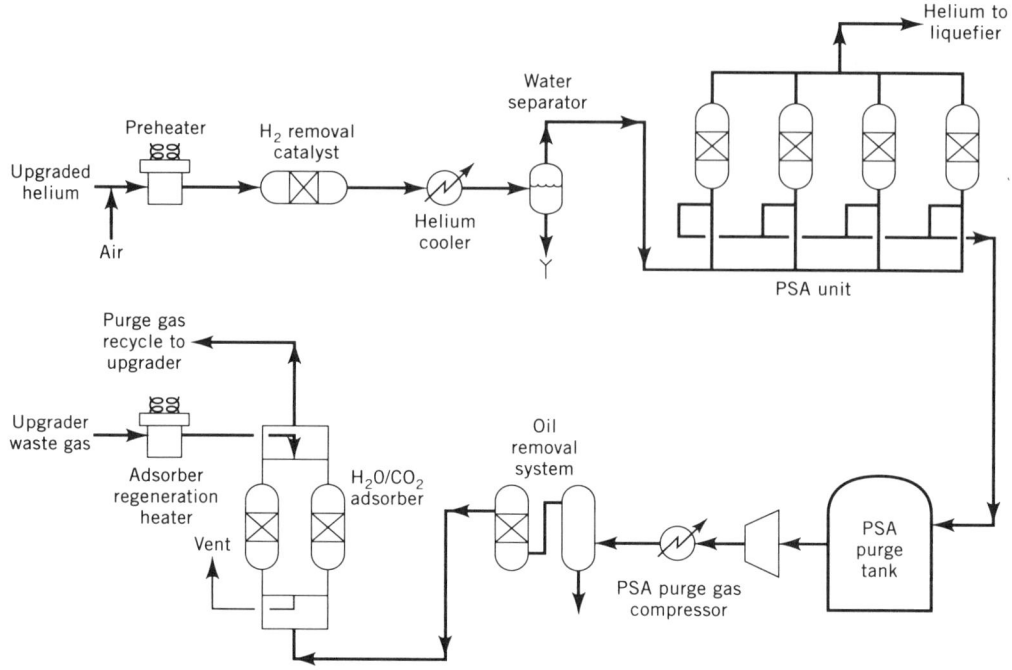

Fig. 6. Helium purification process.

liquid Dewar boil-off and liquid trailer returns at 0.1 MPa. Heat exchangers in the liquefier are brazed aluminum plate-fin type, designed for temperature approaches as tight as 0.1 degree K to minimize energy requirements. The helium expanders are usually centrifugal with either oil- or gas-bearings, and are braked by either an oil cup loader or a closed-loop helium compressor. The main recycle compressors are reciprocating or (more typically) oil-flooded screw machines.

Large helium refrigerators are used to cool superconducting magnets to 4.5 K in accelerators for high-energy physics experiments (15). The process configuration for a helium refrigerator is similar to that for a liquefier, and the refrigerator may produce some liquid. The process flows vary with the ratio of 4 K heat load relative to the liquid production rate, because a refrigerator has balanced flows cooling and warming, with the heat load at the cold end, whereas a liquefier heat load is distributed over the temperature range from 300 to 4 K. Temperatures below 4.2 K are achieved by subatmospheric pressures maintained by cold compressors or by warm vacuum pumps. A refrigerator/liquefier can trade 100 W of refrigeration at 4 K for 1 g/s of liquid helium product. Liquid helium refrigerators are being considered for superconducting generators, transmission lines, and electric storage devices.

Equipment

Machinery. Compressor selection for a cryogenic process plant depends on the fluid, the volumetric flowrate, the pressures involved, the compressor effi-

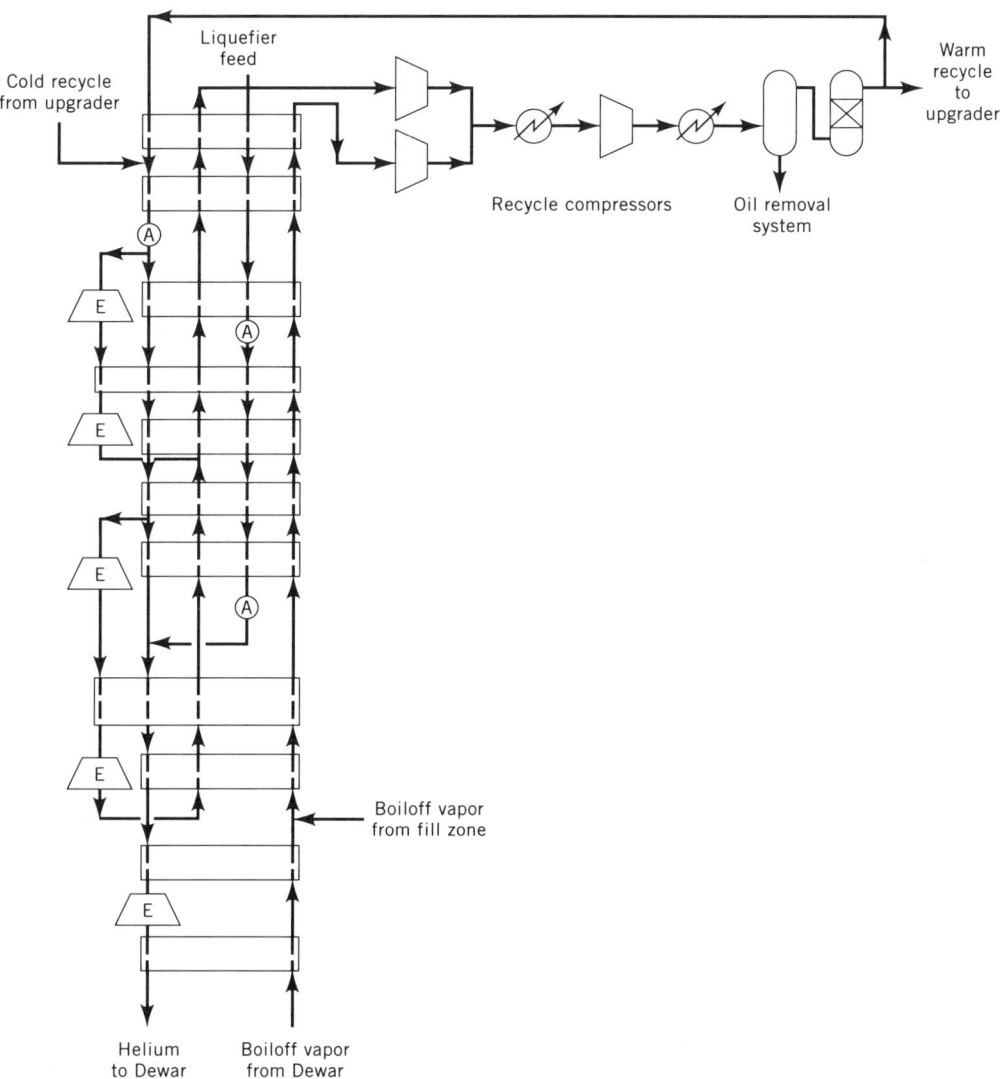

Fig. 7. Helium liquefaction process. A represents adsorber; E, expander.

ciency, and the cost of energy and capital. Centrifugal compressors are lower in installed cost than reciprocating machines, and are preferred if the volumetric flows and pressures of the process allow them to be applied. For large volumetric flows, axial compressors are used. At very high pressures with small volumetric flows, reciprocating machines are required. Sometimes a combination of more than one kind of compression stage is used to yield the most cost effective and efficient system. For example, an oxygen compression system delivering oxygen at about 10 MPa (1450 psi) might employ several centrifugal stages followed by several reciprocating stages.

Gases of low molecular weight are difficult to compress in centrifugal compressors because of the low pressure rise imparted by a wheel that generates a reasonable head rise. This results in a large number of wheels being required to achieve a reasonable total pressure rise. Consequently, hydrogen and helium are seldom compressed using centrifugal compressors. Oil-flooded screw compressors are commonly used for helium processes operating at pressures under 2.4 MPa (348 psi), which is typical for helium liquefiers and refrigerators. Oil removal equipment must be used following lubricated reciprocating compressors or oil-flooded screw compressors.

Axial compressors have been applied to baseload LNG plants, but the majority of those plants were using centrifugal compressors as of the early to mid-1990s. Drivers for cryogenic plant compressors can be electric motors, internal combustion gas engines, gas turbines, or steam turbines. Cold compressors are in use with suction temperatures around 4 K to provide temperatures below 4 K by reducing the pressure under which helium is boiling. These cold compressors use magnetic bearings.

Expanders provide refrigeration by extracting work from a fluid, thereby reducing its temperature. Gas expanders can be either reciprocating or centrifugal, and can be loaded (braked) by electric generators, gas blowers or compressors, oil-film cups, or oil pump brakes. The extracted work can be usefully recovered in an electric generator or compressor. Centrifugal expanders are lower in first cost and in maintenance cost. Reciprocating expanders may be required for low volumetric flows and/or high expansion pressure ratios of low molecular weight gases. Bearings for gas expanders can be oil or gas bearing type (static or dynamic). Reverse-running liquid pumps have been used as liquid expanders. Isentropic efficiencies for gas expanders can be as high as 85–90% for machines with high discharge volumetric flow.

Heat Exchangers. The two most prominent types of heat exchangers used in cryogenic service are the coil wound, tube-in-shell exchanger and the plate and fin (core) exchanger (16). Cryogenic process efficiency is highly dependent on maintaining close temperature approaches between all the cooling and warming streams. These heat exchangers are normally designed to accommodate at least three, and as many as eight or nine, countercurrent streams in a single unit. Accurate thermodynamic and physical property data as well as accurate heat transfer and pressure drop correlations are required for the design of these types of heat exchangers. The design must provide for uniform flow distribution to achieve close temperature approaches and must account for longitudinal heat conduction.

The coil wound heat exchanger consists of multiple tubes helically wound on a mandrel, usually with spacers between each tube layer. The tubes are inserted into tube sheets at both ends of the tube bundle, with separate tube sheets to accommodate each tube circuit. The tube bundle is enclosed in a shell with inlet and outlet nozzles for the shellside fluid. This type of heat exchanger is usually constructed of aluminum or stainless steel. Large aluminum coil-wound heat exchangers for base-load LNG plants may be up to 5 m in diameter, 45 m long, and weigh up to 200 metric tons.

The plate and fin heat exchanger consists of corrugated sheets (fins) stacked between flat sheets (plates). The corrugated sheets are usually perforated or serrated to enhance the heat transfer performance of the exchanger. The stack is brazed in a closely controlled vacuum furnace and headers and nozzles are then welded to the brazed stack to form the separate circuits of the unit. This type of heat exchanger is normally constructed of aluminum, but stainless steel units are also becoming available.

Distillation Columns. In a cryogenic air separation plant, distillation (qv) accounts for the major fraction of the total energy consumption. The low relative volatility characteristic of many cryogenic separations requires the use of many stages. Columns operating at low pressures are consequently designed for a low pressure drop per theoretical stage of separation. Sieve trays with small perforations (~1 mm) provide excellent efficiency at low tray spacing for the clean fluids of cryogenic distillation processes. The minimum pressure drop across distillation sieve trays is set by the requirement to maintain a stable biphase zone of gas bubbling through liquid, corresponding to about 35 mm of liquid per theoretical stage. A double column of the type shown in Figure 1 may use as many as 150 trays. Therefore it is important to keep the spacing between the trays as small as possible to minimize capital cost. Usually tray spacing is in the 10 to 20 cm range. Important manufacturers each have their own design of the tray geometry, including multipass cross flow trays, split cross flow trays, and circular flow trays. Recently, low pressure drop structured packings have been used to replace sieve trays in some sections of the upper column and the argon column of air-separation units.

In some cryogenic hydrocarbon separation plants such as a nitrogen rejection unit, both flow rate and feed composition may change over the life of the plant. This requires the use of valve or bubble cap trays with high turndown capability for distillation. In some applications, the presence of foaming or high froth liquids requires tray spacing as high as 60 cm.

Insulation. Cryogenic insulation should economically reduce heat leak into the system so that its impact on the overall refrigeration requirement is minimized. Insulation can be categorized as unevacuated bulk type (eg, purged rockwool or perlite), rigid foam (eg, foam–glass or urethane), vacuum-jacketed (VJ), evacuated powder (eg, perlite), and multilayer insulation (MLI) (eg, evacuated aluminimized Mylar).

Process equipment for cryogenic plants is often enclosed in a cold box consisting of a steel frame with panels that is filled with perlite or rockwool and purged with a gas that will not condense on the exterior of the equipment and piping. This method of insulation is commonly used on plants for air separation, hydrocarbon recovery, hydrogen purification, and nitrogen refrigeration. Equipment supports or suspension systems are designed for low conduction heat leak using materials with low thermal conductivity, long conduction paths, and small cross-sections. Small refrigerators commonly use evacuated MLI. For the large equipment sizes of baseload LNG plants, individual equipment items and piping are separately insulated with urethane. Lines to transfer liquid cryogens can be insulated with rigid foam, vacuum, evacuated MLI, or evacuated MLI with heat shield, depending on the impact of heat leak and the frequency of cooldown. Bayonet fittings and valves for VJ lines are designed for low heat leak using long conduction paths with small cross sections.

Storage tanks for large volumes (>1000 m^3) of LNG, liquid nitrogen, and liquid oxygen often use unevacuated perlite insulation that is pressurized with boil-off gas from the tank's liquid. Tanks for similar volumes of liquid H_2 are usually insulated with evacuated perlite. Smaller tanks for LNG, liquid nitrogen, and liquid oxygen use evacuated perlite insulation. Tanks for liquid helium and smaller ones for liquid hydrogen are insulated with evacuated MLI, and the liquid helium tanks usually incorporate a heat shield using boiloff vapor or liquid nitrogen to intercept part of the heat leak.

Transport vehicles for cryogenic liquids have an inner vessel holding the liquid and a surrounding vacuum perlite or MLI insulation, all enclosed within an outer vessel. Liquid helium vehicles are insulated between the inner and outer vessels by vacuum MLI and a heat shield cooled by liquid nitrogen which is carried on board. These liquid helium containers can be filled with liquid helium in the U.S. and transported by ship to Europe or Asia without venting helium for 45 days (17). Barges and rail-cars to transport cryogens are in everyday service in the U.S. in support of the space program. LNG ships of 125,000 m^3 cargo capacity use rigid foam insulation.

Safety

The possibility of an uncontrolled release of a cryogenic fluid such as liquid oxygen, liquid methane, or liquid hydrogen from storage and during handling must be carefully considered during the design of a cryogenic facility. The level of risk is often reduced by providing dikes for secondary containment of liquid spills. Procedures to protect personnel from cryogenic burns and asphyxia and to protect the nearby equipment from embrittlement failure must be carefully considered and followed. When trace impurities in the feed streams can lead to the combination of an oxidant with a flammable cryogen (eg, solid oxygen in liquid hydrogen) or a combustible with an oxidant (eg, acetylene in liquid oxygen), special precautions must be taken to eliminate them. Many materials react with pure oxygen, so care must be taken in the selection of materials in contact with oxygen and in the cleaning of oxygen systems prior to use. Potential ignition sources must be minimized, particularly in oxygen compression and in systems for handling oxygen at elevated pressures (18).

BIBLIOGRAPHY

"Cryogenics" in the *Encyclopedia of Chemical Technology*, 1st ed., Suppl. 2, pp. 272–281, by H. R. Morrison, Linde Company, Division of Union Carbide Corporation; in *ECT* 2nd ed., Vol. 6, pp. 471–481, by H. R. Morrison, Linde Company, Division of Union Carbide Corporation; in *ECT* 3rd ed., Vol. 7, pp. 227–242, by C. L. Newton, Air Products and Chemicals, Inc.; in *ECT* 4th ed., Vol. 7, pp. 662–683, by R. Agrawal and co-workers, Air Products and Chemicals, Inc.

1. R. A. Thévenot, *History of Refrigeration Throughout the World*, trans. from French by J. C. Fidler, International Institute of Refrigeration, Paris, 1979.
2. W. H. Isalski, *Separation of Gases*, Clarendon Press, Oxford, 1989.
3. R. E. Latimer, *Chem. Eng. Prog.* **63**(2), 35 (1967).
4. R. Agrawal and D. W. Woodward, "Impact of Structured Packing on Distillation of Argon from Air", *Proceedings of the XVIIIth International Congress of Refrigeration*, Vol. I, Montréal, Canada, 1991, p. 162.

5. U.S. Pat. 5,049,173 (Sept. 17, 1991), T. E. Cormier, R. Agrawal, D. W. Woodward and A. Prentice (to Air Products and Chemicals, Inc.).
6. H. C. Rowles, D. M. Nicholas, H. L. Vines and M. F. Hilton, *Energy Prog.* **6**(1), 25 (1986).
7. W. K. Lam, W. J. Stupin and G. Christensen, "Recover Valuable Off-Gases by the Braun ROE Process," paper presented at *AIChE National Meeting*, New Orleans, La., Apr. 6–10, 1986.
8. H. L. Vines, *Chem. Eng. Prog.*, 46 (Nov. 1986).
9. C. A. Pruitt and J. V. O'Brien, *Oil and Gas J.*, 78 (Oct. 9, 1989).
10. L. S. Gaumer, *Chem. Eng. Prog* **63**(5), 72 (1967).
11. A. Goethe and D. J. Mawer, "A New Integrated Nitrogen Rejection Process with NGL Recovery," *paper presented at Gas Processing Association (GPA) Annual Convention*, Denver, Colo., Mar. 16–18, 1987.
12. R. D. McCarty, *Thermophysical Properties for Helium-4 from 2 to 1500K with Pressures to 1000 Atmospheres*, NBS-TN-631, U.S. Dept. of Commerce, Nov. 1972.
13. S. W. Van Sciver, *Helium Cryogenics*, Plenum Press, New York, 1986.
14. R. R. Olsen and R. F. Pahade, in M. McAshan, ed., *Supercollider I*, Proc. of IISSC Conf., Feb. 8–10, 1989, New Orleans, Plenum Press, New York, 1989.
15. J. R. Sanford and D. M. Matthews, eds., *Site-Specific Conceptual Design of the Superconducting Super Collider, SSCL-SR-1056*, SSC Laboratory, Dallas, Tex., July, 1990.
16. R. E. Lowe, *Chem. Eng.* **17,** 131 (Aug., 1987).
17. Technical literature, Gardner Cryogenics, Bethlehem, Penn., 1992.
18. B. R. Dunbobbin, J. G. Hansel and B. L. Werley, in J. M. Stoltzfus and K. McIlroy, eds., *Flammability and Sensitivity of Materials in Oxygen Enriched Atmospheres*, Vol. 5, ASTM STP 1111, American Society for Testing and Materials, Philadelphia, 1991.

RAKESH AGRAWAL
HOWARD C. ROWLES
GLENN E. KINARD
Air Products and Chemicals, Inc.

CRYSTALLIZATION

Crystallization is one of the oldest unit operations in the portfolio of industrial and/or laboratory separations. Almost all separation techniques involve formation of a second phase from a feed, and processing conditions must be selected that allow relatively easy segregation of the two or more resulting phases. This is a requirement for crystallization also, and there are a variety of other properties of the solid product that must be considered in the design and operation of a crystallizer. Interactions among process, function, product, and phenomena important in crystallization are illustrated in Figure 1.

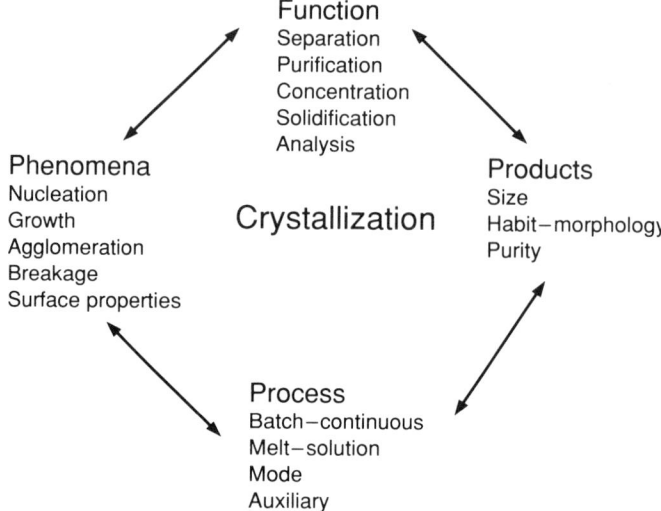

Fig. 1. Crystallization.

Function. There are several possible functions that can be achieved by crystallization (Fig. 1). These include separation, purification, concentration, solidification, and analysis.

Separation. Sodium carbonate [497-19-8] (soda ash) is recovered from a brine by first contacting the brine with carbon dioxide to form sodium bicarbonate. Sodium bicarbonate has a lower solubility than sodium carbonate and can be readily crystallized. The primary function of crystallization in this process is separation. A high percentage of sodium bicarbonate is solidified in a form that makes subsequent separation of the crystals from the mother liquor economical. With the available pressure drop across filters that separate liquid and solid, the capacity of the process is determined by the rate at which liquor flows through the filter cake. That rate is set by the crystal size distribution produced in the crystallizer.

Separation of a chemical species from a mixture of similar compounds can also be achieved by melt crystallization, which is, for example, an important means of separating *para*-xylene [106-42-3] (*p*-xylene) from the ortho and meta isomers. *p*-Xylene is crystallized at the top of a vertical column and crystals are moved downward countercurrently to liquid. The liquid flowing upward is generated by adding heat to melt the crystals at the bottom of the column; a portion of the melt is removed as product and the remainder flows up the column to contact the downward-flowing crystals. Effluent mother liquor, consisting almost entirely of the ortho and meta isomers of xylene, is removed from the top of the column (see BTX SEPARATIONS).

Concentration. The concentration of fruit juice requires removal of solvent (water) from the natural juice. This is commonly done by evaporation, but the derived juices may lose flavor components or undergo thermal degradation during evaporation. In freeze concentration, solvent is crystallized (frozen) in a relatively pure form to leave behind a solution having a solute concentration higher than the original mixture. Significant advantages in product taste have been observed in the application of this process to concentration of certain fruit juices.

Purification. The objective of crystallization also can be purification of a chemical species. For example, L-isoleucine [73-32-5] (an essential amino acid) is separated by crystallization from a fermentation broth that has been filtered and subjected to ion exchange (qv). The recovered crystals contain impurities deleterious to use of the product, and these crystals are, therefore, redissolved and recrystallized to enhance purity.

Solidification. Production of a product in a form suitable for use and acceptable to the consumer also may be an objective of a crystallization process. For example, the appearance of sucrose (sugar) varies with local customs, and deviations from that custom could lead to an unacceptable product. A final crystallization may thus be called for to bring the product appearance into compliance with expectations.

Analysis. Many analytical procedures calling for determination of molecular structure are aided by crystallization or require that the unknown compound be crystalline. Methodologies coupling crystallization and analytical procedures are not covered herein.

Products. In all of the instances in which crystallization is used to carry out a specific function, product requirements are a central component in determining the ultimate success of the process. These requirements grow out of how the product is to be used and the processing steps between crystallization and recovery of the final product. Key determinants of product quality are the size distribution (including mean and spread), the morphology (including habit or shape and form), and purity. Of these, only the last is important with other separation processes.

Crystal size distribution (CSD) determines several important processing and product properties, including crystal appearance, separation of crystals from liquor, reactions, dissolution and other processes and properties involving surface area of the crystalline product, crystal transportation, and crystal storage. In fact, experience indicates that a large fraction of crystallizer troubleshooting cases have been initiated to solve problems associated with inadequate throughput of filters or centrifuges; when solutions are found they generally involve manipulation of CSD.

It is often important to control the CSD of pharmaceutical compounds, eg, in the synthesis of human insulin, which is made by recombinant DNA techniques (1). The most favored size distribution is one that is monodisperse, ie, all crystals are of the same size, so that the rate at which the crystals dissolve and are taken up by the body is known and reproducible. Such uniformity can be achieved by screening or otherwise separating the desired size from a broader distribution or by devising a crystallization process that will produce insulin in the desired form. The latter of these options is preferable, and considerable effort has been expended in that regard.

Process. In each crystallization system there is a need to form crystals, to cause the crystals to grow, and to separate the crystals from residual liquid. There are various ways to accomplish these objectives (Fig. 1) leading to a multitude of processes designed to meet requirements of product yield, purity, and, uniquely, crystal size distribution.

Phenomena. The critical phenomena in crystallization are, as shown in Figure 1, nucleation and growth kinetics, interfacial phenomena, breakage, and agglomeration. Nucleation leads to the formation of crystals, either from a solution or a melt. Growth is the enlargement of crystals caused by deposition of solid material on an existing surface. The relative rates at which nucleation and growth occur determine the crystal size distribution; qualitatively, when the rate of nucleation is high relative to growth rate, crystals formed are small and numerous. Agglomeration is the formation of a larger particle through two or more smaller particles (crystals) sticking together. It is prevalent in many processes, and agglomeration can be essential for solid–liquid separation or it can be undesirable because it may adversely affect crystal quality. Breakage of crystals is almost always undesirable because it is detrimental to crystal appearance and it can lead to excessive fines and have a deleterious effect on crystal purity. Interfacial phenomena influence solid–liquid separation, flow characteristics of slurries, agglomeration, and crystal morphology.

Solid–Liquid Equilibria and Mass and Energy Balances

Solubility. Solid–liquid equilibrium, or the solubility of a chemical compound in a solvent, refers to the amount of solute that can be dissolved at constant temperature, pressure, and system composition; in other words, the maximum concentration of the solute in the solvent at static conditions. In a system consisting of a solute and a solvent, specifying system temperature and pressure fixes all other intensive variables. In particular, the composition of each of the two phases is fixed, and solubility diagrams of the type shown for a hypothetical mixture of R and S in Figure 2 can be constructed. Such a system is said to form an

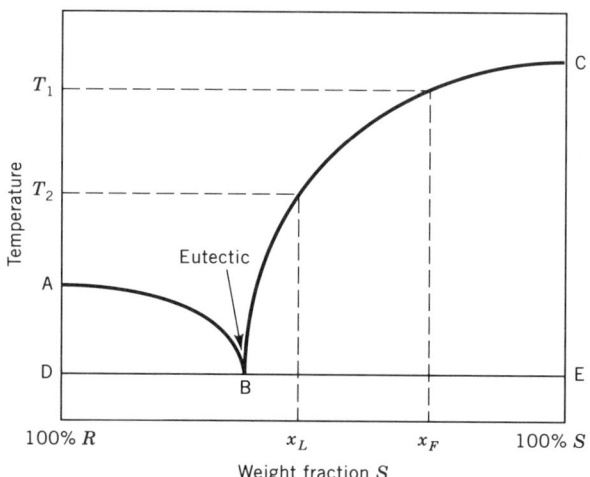

Fig. 2. Solubility diagram for a hypothetical system. The curves AB and BC represent solution compositions that are in equilibrium with solids whose compositions are given by the lines AD and CE. If AD and CE are vertical along the respective axes, the crystals are pure R and S, respectively. Crystallization from any solution whose composition is to the left of the vertical line through point B produes crystals of pure R, whereas solutions to the right of the line produce crystals of pure S. A solution whose composition falls on the line through B produces a solid mixture that has a composition identical to the liquid solution.

eutectic, ie, there is a condition at which both R and S crystallize into a solid phase at a fixed ratio that is identical to their ratio in solution. Consequently, there is no change in the composition of residual liquor as a result of crystallization.

Several features of the hypothetical system in Figure 2 can be used to illustrate proper selection of crystallizer operating conditions and limitations placed on the operation by system properties. Suppose a saturated solution at temperature T_1 is fed to a crystallizer operating at temperature T_2. Because the feed is saturated, the weight fraction of S in the feed is given as shown in Figure 2. The maximum crystal production rate P_{max} from such a process depends on the value of T_2 and is given by

$$P_{max} = Fx_F - Lx_L \tag{1}$$

where F is the feed rate to the crystallizer, and L is the solution flow rate leaving the crystallizer. No other stream is fed to or removed from the crystallizer. Note that the lower limit on T_2 is given by the eutectic point B.

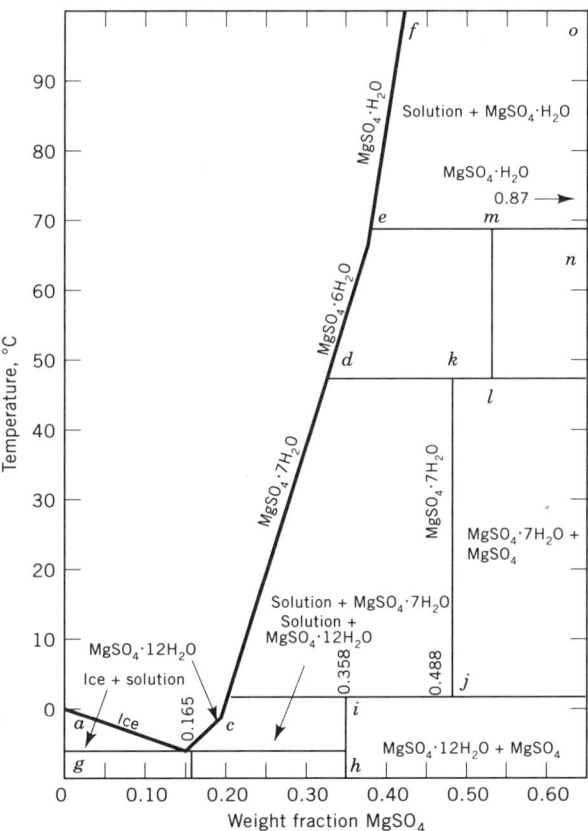

Fig. 3. Solubility diagram for magnesium sulfate in water.

Figure 3, the equilibrium behavior of magnesium sulfate [7487-88-9] in water, is illustrative of systems that form hydrated salts. Equilibrium solution concentrations are plotted as curves *ab, bc, cd, de,* and *ef*; the solid phases that are in equilibrium with these solutions have compositions given by the lines *ag, hi, jk, lm,* and *no,* respectively. Ice is the solid phase whose composition is given by *ag,* and crystals containing differing ratios of water of hydration to magnesium sulfate constitute the solids represented by the other lines. Specifically, the line *no* represents magnesium sulfate monohydrate ($MgSO_4 \cdot H_2O$), which has one water molecule per molecule of magnesium sulfate, whereas the lines *ml, kj,* and *ih* represent the hexahydrate, heptahydrate, and dodecahydrate forms, respectively. The weight fraction of $MgSO_4$ in each of the crystal forms is shown in Figure 3, and as with all crystalline materials having water of hydration, the solute balance of equation 1 must be modified to read

$$x_c P_{max} = F x_F - L x_L \qquad (2)$$

where x_c is the mass fraction of solute in the crystal, eg, x_c is 0.488 when the crystalline substance is magnesium sulfate heptahydrate. Differences in the forms of magnesium sulfate crystals affect the dependence of solubility on temperature, which is reflected by the slopes of the solution composition curves.

The dependence of solubility on temperature affects the mode of crystallization. For example, Figure 4 shows that the solubility of potassium nitrate [7757-79-1] is strongly influenced by the system temperature but that temperature has little influence on the solubility of sodium chloride [7647-14-5]. As a consequence, a reasonable yield of KNO_3 crystals can be obtained by cooling a saturated feed solution; on the other hand, cooling a saturated sodium chloride solution accomplishes little crystallization, and evaporation is required to increase the yield of sodium chloride crystals.

The production of many high value chemicals requires maximizing separation from a relatively dilute solution. It is common in such instances to use a combination of methods to reduce solute solubility in the feed solution. Figure 5,

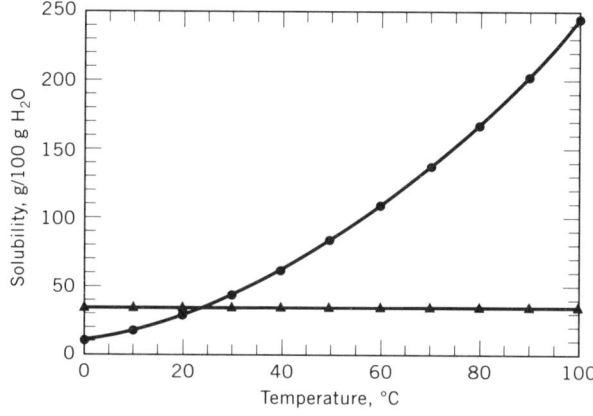

Fig. 4. Solubilities of NaCl ▲, and KNO_3 ●, in water.

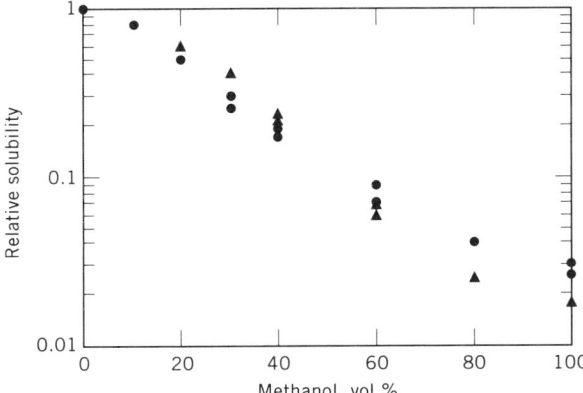

Fig. 5. Solubilities of L-serine in mixtures of methanol and water. Relative solubility = solubility in $CH_3OH-H_2O \div$ solubility in H_2O; ●, 10°C; ▲, 30°C. Reproduced by permission of the American Institute of Chemical Engineers (4).

for example, illustrates that the addition of methanol to a saturated aqueous solution of L-serine [56-45-1] can reduce solubility by more than an order of magnitude.

Solubility data can be found in a variety of units, and conversion from one set of units to another often is required before computation of yield can be performed. Guides to such conversions are available. It is often most convenient, however, to express solubility and compositions in mixed streams in terms of mass ratios, ie, mass of solute per mass of solvent.

Supersaturation. The thermodynamic driving force for both crystal nucleation and growth is the key variable in setting the mechanisms and rates by which these processes occur. Supersaturation is defined rigorously as the deviation from thermodynamic equilibrium, which is the difference between the chemical potential of the solute at the existing conditions of the system μ and the chemical potential of the solute equilibrated at the system conditions μ^*. Less abstract definitions involving measurable system properties such as temperature, concentration, or mass or mole fraction also have been used to express supersaturation. Consider, for example, a system at temperature T with a solute concentration c. A saturated solution having a concentration c would be at temperature T^*, whereas a saturated solution having a temperature T would have a solute concentration c^*.

Expressions of supersaturation can then be formulated as follows: (1) the difference between the chemical potential of the system and the chemical potential at saturation, $\mu - \mu^*$, where the chemical potential is a function of both temperature and concentration; (2) the difference between the solute concentration and the concentration at equilibrium, $c - c^*$; (3) the difference between the system temperature at equilibrium and the actual temperature, $T^* - T$; (4) the ratio of the solute concentration and the equilibrium concentration, c/c^*, which is known as relative saturation; or (5) the ratio of the difference between the solute concentration and the equilibrium concentration to the equilibrium concentration, $s = (c - c^*)/c^*$, which is known as relative supersaturation. This term has often been represented by σ; s is used here because of the frequent use of σ for interfacial energy or surface tension and for variance in distribution functions.

Any of these definitions of supersaturation can be used over a moderate range of system conditions. For example, a difference in chemical potential $\Delta\mu = \mu - \mu^*$ is proportional to both $c - c^*$ and $T^* - T$ over a modest range of conditions. Of the five expressions given, however, the second is most useful in calculating the yield from a crystallizer, the third provides information that may be most useful in the control of a crystallizer, and the fifth is most commonly used in correlating the dependence of nucleation and growth kinetics on supersaturation.

Although the first of these expressions is the most nearly fundamentally correct, it is difficult to evaluate because of inadequate capabilities of determining chemical potential as a function of temperature and composition. The next most appropriate driving force for crystal growth is that given by the last of the possibilities, ie, $s = (c - c^*)/c^*$. This definition of supersaturation will be used throughout the ensuing discussion, but it should be recognized that the validity of doing so is limited to low supersaturations. Solute concentrations used in determining supersaturation should include any water of hydration associated with the solute in the crystal at equilibrium.

Mass and Energy Balances. The formulation of mass and energy balances follows procedures outlined in many basic texts (2). The use of solubilities to calculate crystal production rates from a cooling crystallizer was demonstrated by the discussion of equations 1 and 2. Subsequent to determining the yield, the rate at which heat must be removed from such a crystallizer can be calculated from an energy balance:

$$F\hat{H}_F = P\hat{H}_C + L\hat{H}_L + Q \tag{3}$$

where F, P, and L are feed rate, crystal production rate, and mother liquor flow rate, respectively; $\hat{H}$ is the specific enthalpy of the stream corresponding to the subscript; and Q is the required rate of heat transfer. As F, P, and L are known or can be calculated from a simple mass balance, determination of Q requires methods of estimating specific enthalpies.

If appropriate enthalpy data are unavailable, estimates can be obtained by first defining reference states for both solute and solvent. Often the most convenient reference states are crystalline solute and pure solvent at an arbitrarily chosen reference temperature. The reference temperature selected usually corresponds to that at which the heat of crystallization $\Delta\hat{H}_C$ of the solute is known. The heat of crystallization is approximately equal to the negative of the heat of solution. For example, if the heat of crystallization is known at T_{ref}, then reasonable reference conditions would be the solute as a solid and the solvent as a liquid, both at T_{ref}. The specific enthalpies then could be evaluated as

$$\hat{H}_F = x_F \Delta\hat{H}_C + C_{pF}(T - T_{\text{ref}}) \tag{4}$$

$$\hat{H}_C = C_{pC}(T - T_{\text{ref}}) \tag{5}$$

$$\hat{H}_L = x_L \Delta\hat{H}_C + C_{pL}(T - T_{\text{ref}}) \tag{6}$$

where x_F and x_L are the mass fractions of solute in the feed and mother liquor, respectively. All that is required now to determine the required rate of heat transfer is the indicated heat capacities, which can be estimated or measured experimentally.

Now suppose some of the solvent is evaporated in the crystallizer, as is shown in Figure 6. Independent balances can be written on total mass and solute:

$$F = V + L + P \tag{7}$$

$$x_F F = x_L L + x_C P \tag{8}$$

where F = feed rate, V = vapor withdrawal rate, L = liquid (filtrate) withdrawal rate, and P = crystal production rate, and x_j = solute content of stream j in units consistent with flow rate. There is an equilibrium expression relating x_L, x_C, and the temperature or pressure at which the operation is conducted. In addition, an energy balance must be satisfied:

$$F\hat{H}_F + Q = V\hat{H}_V + L\hat{H}_L + P\hat{H}_C \tag{9}$$

The specific enthalpies ($\hat{H}_j$) in equation 9 can be determined as described earlier, provided the temperatures of the product streams are known. Evaporative cooling crystallizers operate at reduced pressure and may be considered adiabatic (Q = 0). As with of many problems involving equilibrium relationships and mass and energy balances, trial-and-error computations are often involved in solving equations 7 through 9.

The mass balance on a crystallizer is related to the growth kinetics that occur within the unit. This may be illustrated by considering systems in which crystal growth kinetics are sufficiently fast to use essentially all of the supersaturation provided by the crystallizer. Under such conditions (referred to in the crystallization literature as class II or fast-growth behavior), the solute concentration in the mother liquor can be assigned a value corresponding to saturation. Alternatively, should supersaturation in the mother liquor be so great as to affect the

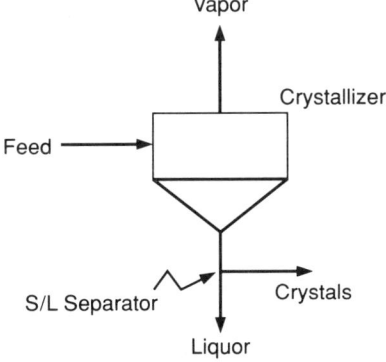

Fig. 6. Schematic diagram of a simple crystallizer.

solute balance, the operation is said to follow class I or slow-growth behavior. An expression coupling the rate of growth to a solute balance must be used to describe such a system.

Crystallization Kinetics

Along with operating variables of the crystallizer, nucleation and growth determine such crystal characteristics as size distribution, purity, and shape or habit.

Nucleation. Crystal nucleation is the formation of an ordered solid phase from a liquid or amorphous phase. Nucleation sets the character of the crystallization process, and it is, therefore, the most critical component in relating crystallizer design and operation to crystal size distributions.

Mechanisms. Classical nucleation theory is based on homogeneous and heterogeneous mechanisms, both of which call for the formation of crystals through a process of sequentially combining the constituent units that form a crystal. Heterogeneous and homogeneous mechanisms are referred to as primary nucleation because existing crystals play no role in the nucleation. Primary nucleation can be illustrated by considering a hypothetical experiment in the context of the solubility data in Figure 7. Assume that a solution is at a concentration and temperature corresponding to point A on the figure. The solution is undersaturated, so any crystals present in the system would dissolve. If the concentration is increased at constant temperature, for example by evaporation, the path followed would cause the solution to reach saturated conditions at point B. Once the concentration becomes greater than that at B, the solution is supersaturated and any crystals present in the system would grow. However, experience shows that nucleation would not occur until the concentration reaches point C, which defines what is called the metastable limit. If this procedure is repeated at various temperatures, a metastable limit curve could be drawn as shown.

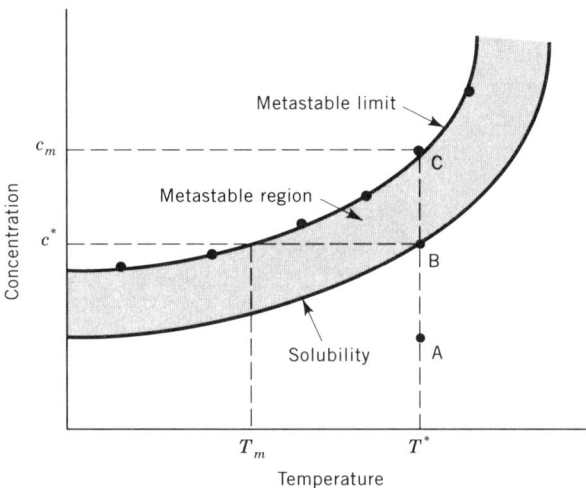

Fig. 7. Solubility and primary nucleation in a hypothetical experiment.

Within the metastable region, primary nucleation will not occur, and the width of the metastable zone, as reflected by $c_m - c^*$ or $T^* - T_m$, varies from one substance to another. Furthermore, the width can vary for the same substance with composition, the presence of impurities that alter interfacial tension, and various other factors.

Both homogeneous and heterogeneous mechanisms require relatively high supersaturation, and they exhibit a high order dependence on supersaturation. These factors often lead to production of excessive fines in systems where primary nucleation mechanisms are important. The classical theoretical treatment of primary nucleation results in the expression (5):

$$B^0 = A \exp\left(-\frac{16\pi\sigma^3 v^2}{3k^3 T^3 [\ln(s+1)]^2}\right) \quad (10)$$

where k is the Boltzmann constant, σ is surface energy per unit area, v is molar volume, and A is a constant. This equation can be simplified by recognizing that $\ln(s+1)$ approaches s as s approaches 0. So for small supersaturations,

$$B^0 = A \exp\left(-\frac{16\pi\sigma^3 v^2}{3k^3 T^3 s^2}\right) \quad (11)$$

The most important variables affecting nucleation rate are shown by equations 10 and 11 to be interfacial energy, temperature, and supersaturation.

The high order dependence of nucleation rate on supersaturation is especially important because a small swing in supersaturation may produce an enormous change in nucleation rate. This gives rise to the often-observed phenomenon of having a clear liquor transformed to a slurry of fine crystals with only a slight increase in supersaturation, for example, by decreasing the solution temperature.

The catalytic effect of solid particles (as in heterogeneous nucleation) is to reduce the energy barrier to formation of a new phase. This, in effect, can reduce the interfacial energy σ significantly.

The metastable limit can provide an empirical approach to modeling primary nucleation. This limit, which was first observed in 1951 (6), must be determined through experimentation, and nucleation rate is correlated with the following equation

$$B^0 = k(c - c_m)^i \quad (12)$$

where the equilibrium concentration c^* is less than the concentration at the metastable limit, c_m. Values of c_m are often very close to c^* for many inorganic systems, and satisfactory correlations have been obtained with c^* substituted for c_m in equation 12 (7); in other words,

$$B^0 = k(c - c^*)^i \quad (13)$$

where the parameters k and i must be evaluated from experimental data.

Secondary nucleation is crystal formation through a mechanism involving the solute crystals; crystals of the solute must be present for secondary nucleation to occur. Thorough reviews have been given (8,9).

Several features of secondary nucleation make it more important than primary nucleation in industrial crystallizers. First, continuous crystallizers and seeded batch crystallizers have crystals in the magma that can participate in secondary nucleation mechanisms. Second, the requirements for the mechanisms of secondary nucleation to be operative are fulfilled easily in most industrial crystallizers. Finally, low supersaturation can support secondary nucleation but not primary nucleation, and most crystallizers are operated in a low supersaturation regime that improves yield and enhances product purity and crystal morphology.

Secondary nucleation can occur as the result of several mechanisms that have been identified in selected systems and include the following. *Initial breeding* results from immersion of seed crystals in a supersaturated solution. It is thought to be caused by dislodging extremely small crystals that were formed on the surface of larger crystals during drying. Although this mechanism is unimportant in continuous and unseeded batch crystallization, it can be significant in the operation of seeded batch crystallizers. Several process variables have been shown to influence nucleation rates caused by initial breeding (10). *Contact nucleation* results from collisions of crystals with one another, and/or crystallizer internals, and/or an impeller in an agitator or circulation pump. The collision energy required for contact nucleation is small and does not result in macroscopic degradation (breakage) of the contacted crystal. *Shear breeding* results when supersaturated solution flows by a crystal surface and carries with it crystal precursors believed formed in the region of the growing crystal surface. In a study of nucleation of $MgSO_4 \cdot 7H_2O$ by this mechanism, it was found (11) that high levels of supersaturation were required for it to produce significant numbers of nuclei.

Process Variables Affecting Contact Nucleation. Pioneering studies elucidated many factors affecting contact nucleation (12–14). The number of crystals produced by a controlled impact of an object with a seed crystal depends on energy of impact, supersaturation at impact, supersaturation at which crystals mature, hardness of the impacting object, area of impact, angle of impact, and system temperature. Although it is impossible to account quantitatively for all of these variables, certain generalizations can be drawn from the research on this nucleation mechanism.

Based on experimental observations, the following expression was proposed (15) for systems at constant supersaturation:

$$B^0 = k_N \exp(E - E_t) \qquad (14)$$

The same researchers proposed that a relationship of impact energy E to crystallizer variables must include the mass of the impacting crystal m_c, the rotational velocity of the impeller providing mixing ω, and the fraction of the available energy actually transmitted to the crystal ϵ:

$$E = f(\omega, \epsilon, m_c) \qquad (15)$$

Correlations of nucleation rates with crystallizer variables have been developed for a variety of systems. Although the correlations are empirical, a mechanistic hypothesis regarding nucleation can be helpful in selecting operating variables for inclusion in the model. Two examples are (1) the effect of slurry circulation rate on nucleation has been used to develop a correlation for nucleation rate based on the tip speed of the impeller (16) and (2) the scaleup of nucleation kinetics for sodium chloride crystallization provided an analysis of the role of mixing and mixer characteristics in contact nucleation (17). Published kinetic correlations have been reviewed through about 1979 (18). In a later section on population balances, simple power-law expressions are used to correlate nucleation rate data and describe the effect of nucleation on crystal size distribution.

Supersaturation has been observed to affect contact nucleation, but the mechanism by which this occurs is not clear. There are data (19) that infer a direct relationship between contact nucleation and crystal growth. This relationship has been explained by showing that the effect of supersaturation on contact nucleation must consider the reduction in interfacial supersaturation due to the resistance to diffusion or convective mass transfer (20).

Still another possible role of supersaturation is that it affects the solution structure and causes the formation of clusters of solute molecules. These clusters may participate in nucleation, although the mechanism by which this would occur is not clear. Evidence of the existence of cluster formation in supersaturated solutions has been presented for citric acid (21); while others have examined the phenomenon in greater detail (22,23).

The ease with which nuclei can be produced by contact nucleation is a clear indication that this mechanism is dominant in many industrial operations. Research on contact nucleation is continuing with the objective of building an understanding of the phenomenon that will allow its successful inclusion in models describing commercial systems.

Crystal Growth. At least two resistances determine growth kinetics, those associated with integration or incorporation of the crystalline unit (for example, solute molecules) into the crystal surface (lattice) and molecular or bulk transport of the unit from the surrounding solution to the crystal face. The primary concern here is with surface incorporation. A simple set of experiments in which the rate of advance of a crystal face is measured can be used to illustrate these two resistances. Data given in Figure 8 show the effect of solution velocity over a crystal face at three different conditions. As the solution velocity increases from low values, the growth rates also increase. At about 24 cm/s, however, the growth rates approach constant values. Such behavior indicates that both bulk mass transfer and surface incorporation are important below 24 cm/s but above this velocity, surface incorporation provides the dominant resistance to growth.

Growth Models. Numerous models have been proposed to describe surface reaction kinetics, including those that assume crystals grow by layers and others that consider growth to occur by the movement of a continuous step. Each model results in a specific relationship between growth rate and supersaturation, but none can be used for a priori predictions of growth kinetics. Insights regarding the roles of certain process variables can be obtained, however, and with additional research predictive capabilities may be achieved. For these reasons and because of the extensive literature on the subject (5,24–26) all that will be pointed

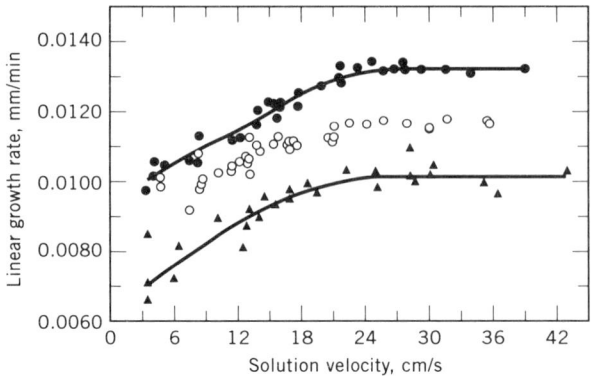

Fig. 8. Effect of solution velocity on the growth rate of the ⟨110⟩ face of $MgSO_4 \cdot 7H_2O$. Concentration of $MgSO_4$: ●, 29.02 wt % at 30.8°C; ○, 28.89 wt % at 30.8°C; ▲, 18.89 wt % at 31.3°C.

out here are the key aspects of the physical models and the resulting relationship between growth and supersaturation predicted by each theory.

Models used to describe the growth of crystals by layers call for a two-step process: (1) formation of a two-dimensional nucleus on the surface and (2) spreading of the solute from the two-dimensional nucleus across the surface. The relative rates at which these two steps occur give rise to the mononuclear two-dimensional nucleation theory and the polynuclear two-dimensional nucleation theory. In the mononuclear two-dimensional nucleation theory, the surface nucleation step occurs at a finite rate, whereas the spreading across the surface is assumed to occur at an infinite rate. The reverse is true for the polynuclear two-dimensional nucleation theory. From the mononuclear two-dimensional nucleation theory, growth is related to supersaturation by the equation.

$$G = C_1 h A [\ln(1+s)]^{1/2} \exp\left[-\frac{C_2}{T^2 \ln(1+s)}\right] \qquad (16)$$

where C_1 and C_2 are system-dependent constants, h is the height of the nucleus, A is surface area, and s and T are as defined earlier. The polynuclear two-dimensional theory produces the equation

$$G = \left(\frac{C_3}{T^2[\ln(1+s)]^{3/2}}\right) \exp\left[-\frac{C_2}{T^2 \ln(1+s)}\right] \qquad (17)$$

where C_3 is a system-dependent constant. Finally, if both formation of the two-dimensional nucleus and spreading of the surface layer are important in determining growth rate, the following equation can be derived:

$$G = C_4 s^{2/3} [\ln(1+s)]^{1/6} \exp\left[-\frac{C_2}{3T^2 \ln(1+s)}\right] \qquad (18)$$

where C_4 is a system-dependent constant.

Equations 16–18 can be simplified considerably by recognizing that in many systems the quantity s is much less than 1. In that case, $\ln(1+s)$ is approximately s. Making this substitution, the growth rate from the mononuclear two-dimensional nucleation theory becomes

$$G = C_1 hAs^{1/2} \exp\left(-\frac{C_2}{T^2 s}\right) \tag{19}$$

For the polynuclear two-dimensional nucleation theory

$$G = \left(\frac{C_3}{T^2 s^{3/2}}\right) \exp\left(-\frac{C_2}{T^2 s}\right) \tag{20}$$

For both steps occurring at similar rates

$$G = C_4 s^{5/6} \exp\left(-\frac{C_2}{T^2 s}\right) \tag{21}$$

The screw dislocation theory (27), often referred to as the BCF theory (after its formulators), shows that the dependence of growth rate on supersaturation can vary from a parabolic relationship at low supersaturation to a linear relationship at high supersaturation. In the BCF theory, growth rate is given by

$$G = C\left(\frac{\epsilon s^2}{\sigma_1'}\right) \tanh\left(\frac{\sigma_1'}{\epsilon s}\right) \tag{22}$$

where ϵ is screw dislocation activity and σ_1' is a system-dependent quantity that is inversely proportional to temperature. The dependence of growth rate on supersaturation is linear if the ratio $\sigma_1'/\epsilon s$ is large, but the dependence becomes parabolic as the ratio becomes small. This is because $\tanh\left(\frac{\sigma_1'}{\epsilon s}\right) \longrightarrow \frac{\sigma_1'}{\epsilon s}$ as $\frac{\sigma_1'}{\epsilon s}$ becomes small (supersaturation becomes large), and $\tanh\left(\frac{\sigma_1'}{\epsilon s}\right) \longrightarrow 1.0$ as $\frac{\sigma_1'}{\epsilon s}$ becomes large (supersaturation becomes small). It is thus possible to observe variations in the dependence of growth rate on supersaturation for a given crystal-solvent system.

An empirical approach can also be used to relate growth kinetics to supersaturation with a power-law function of the form

$$G = k_G s^g \tag{23}$$

where k_G and g are constants determined by fitting the equation to growth-rate data. Such an approach should be valid over small ranges of supersaturation, and analysis of the theories discussed above shows that the more fundamental equations can be fit by equation 23 over limited ranges of supersaturation. For example, using the empirical approach to describe systems in which the screw dislocation model was applicable would limit g to values between 1 and 2, assuming ϵ was independent of supersaturation.

All the models described above indicate the importance of system temperature on growth rate. Dependencies of growth kinetics on temperature are often expressed in terms of an Arrhenius expression:

$$k_G = k_G^0 \exp\left(-\frac{\Delta E_G}{RT}\right) \tag{24}$$

where k_G is a growth rate coefficient of the type required in equation 23, k_G^0 is a constant, and ΔE_G is an activation energy. The magnitude of ΔE_G can be as large as that for many chemical reactions, 42 kJ/mol (>10 kcal/mol).

Both supersaturation and temperature can have different effects on the growth rates of different faces of the same crystal. Such occurrences have implications with respect to crystal habit, and these are dealt with in a later section.

Effects of Impurities and Solvent. The presence of impurities usually decreases the growth rates of crystalline materials, and problems associated with the production of crystals smaller than desired are commonly attributed to contamination of feed solutions. Strict protocols should be followed in operating units upstream from a crystallizer to minimize the possibility of such occurrences. Equally important is monitoring the composition of recycle streams so as to detect possible accumulation of impurities. Furthermore, crystallization kinetics used in scaleup should be obtained from experiments on solutions as similar as possible to those expected in the full-scale process.

The effects of a solvent on growth rates have been attributed to two sets of factors (28): one has to do with the effects of solvent on mass transfer of the solute through adjustments in viscosity, density, and diffusivity; the second is concerned with the structure of the interface between crystal and solvent. The analysis (28) concludes that a solute-solvent system that has a high solubility is likely to produce a rough interface and, concomitantly, large crystal growth rates.

Crystal Growth in Mixed Crystallizers. Multicrystal magma studies usually involve examination of the rate of change of a characteristic crystal dimension or the rate of increase in the mass of crystals. The characteristic dimension depends on the method used in the determination of size; eg, the second-largest dimension is measured by sieve analyses, whereas both electronic-zone-sensing and laser-light-scattering instruments provide estimates of an equivalent spherical diameter. If the rate of change of a crystal mass dM_c/dt is measured, the quantity can be related to the rate of change in the crystal characteristic dimension by the equation

$$\frac{dM_c}{dt} = \frac{d(\rho_c k_v L^3)}{dt} = 3\rho_c k_v L^2 \frac{dL}{dt} \tag{25}$$

where ρ_c is crystal density and k_v is the volume shape factor. Because an area shape factor can be defined by the equation

$$k_a = A_c/L^2 \tag{26}$$

and G is defined as dL/dt,

$$\frac{dM_c}{dt} = 3\rho_c \left(\frac{k_v}{k_a}\right) A_c G \tag{27}$$

The formulation of a population balance requires defining growth rate as the rate of change of the characteristic dimension

$$G = \frac{dL}{dt} \tag{28}$$

and solution of the resulting differential population balance requires a knowledge of the relationship between growth rate and size of the growing crystals. Moreover, this relationship can often be deduced from the form of population density data. A special condition, which simplifies such balances, results when all crystals in the magma grow at the same rate. Crystal-solvent systems that show this behavior are said to follow the ΔL law (29) whereas systems that do not are said to exhibit anomalous growth.

Anomalous growth means that growth rates of crystals in a magma are not identical or that the growth rate of an individual crystal or mass of crystals is not constant. Two theories have been used to explain growth rate anomalies: size-dependent growth and growth rate dispersion. Both alter the form of the population density function obtained from perfectly mixed continuous crystallizers; unfortunately, such behavior cannot be used to distinguish between size-dependent growth and growth rate dispersion, as both have the same qualitative effects on population density.

Size-dependent Crystal Growth. A number of empirical expressions correlate the apparent effect of crystal size on growth rate (30). The most commonly used correlation uses three empirical parameters to correlate growth rate with crystal size:

$$G = G^0(1 + \gamma L)^b; \quad b < 1 \tag{29}$$

where G^0, γ, and b are determined from experimental data. There have been attempts to relate the kinetic parameter b to crystallizer variables; the only success in this regard (31) showed a qualitative dependence on crystallizer volume. Several theories have been proposed to explain size-dependent growth kinetics, but none has been substantiated by direct observation or used to predict the onset of such behavior. One explanation seems particularly appealing: larger crystals impact impellers and other crystallizer internals with higher frequency and energy than smaller crystals; therefore, the larger crystals are recipients of more surface breaks and irregularities that lead to higher growth rates.

Growth Rate Dispersion. This phenomenon is the exhibition of different growth rates by crystals in a magma, even though they may have the same size and are exposed to identical conditions. It is now generally accepted that many observations originally attributed to size-dependent growth were due to growth rate dispersion. Such erroneous interpretations were the result of similarities in the effects of the two types of behavior on crystal size distributions.

The effects of growth rate dispersion on a population of sucrose crystals were first characterized by a linear correlation of the variance of the population about a mean size L with the extent of growth (30). It was demonstrated later (32,33) that growth rate dispersion could account for anomalous characteristics in the population density of crystals obtained from continuous, steady-state crystallizers. Later studies examined batch crystallization data to show that apparent size-dependent growth of potassium alum crystals was a manifestation of growth rate dispersion (34). Crystals of citric acid monohydrate generated by contact nucleation were found to exhibit growth rate dispersion but not size-dependent growth (35).

Two distinctly different mechanisms leading to growth rate dispersion have experimental support. The first assumes that all crystals have the same time-averaged growth rate, but the growth rates of individual crystals fluctuate about some mean value (36). Direct evidence of random fluctuations in growth rates has been reported for magnesium sulfate heptahydrate (37) and potassium alum (38). The second assumes that crystals are formed with a characteristic distribution of growth rates, but individual crystals retain a constant growth rate throughout their residence in a crystallizer. Findings on citric acid (35), potassium nitrate (39), and ammonium dihydrogen phosphate (40) support this mechanism.

Surface integration is thought to be the primary factor in both mechanisms of growth rate dispersion. The BCF theory indicates that the growth rate of a crystal face depends on the number, sign, and location of screw dislocations on the surface of a growing crystal. Collisions of crystals with each other and crystallizer internals result in changes in the dislocation network of a crystal, thereby leading to random fluctuations of growth rates. Changes in the dislocation networks also occur simply due to the imperfect growth of crystal faces. A distribution of growth rates is a result of the varying dislocation networks and densities among nuclei and seed crystals.

Although evidence exists for both mechanisms of growth rate dispersion, separate mathematical models were developed for incorporating the two mechanisms into descriptions of crystal populations: random growth rate fluctuations (36) and growth rate distributions (33,40). Both mechanisms can be included in a population balance to show the relative effects of the two mechanisms on crystal size distributions from batch and continuous crystallizers (41).

Crystal Characteristics

The morphology (including crystal shape or habit), size distribution, and purity of crystalline materials can determine the success in fulfilling the function of a crystallization operation.

Morphology. A crystal is highly organized, and constituent units, which can be atoms, molecules, or ions, are positioned in a three-dimensional periodic pattern called a space lattice. A characteristic crystal shape results from the regular internal structure of the solid with crystal surfaces forming parallel to planes formed by the constituent units. The surfaces (faces) of a crystal may exhibit varying degrees of development, with a concomitant variation in macroscopic appearance.

If atoms, molecules, or ions of a unit cell are treated as points, the lattice structure of the entire crystal can be shown to be a multiplication in three dimensions of the unit cell. Only 14 possible lattices (called Bravais lattices) can be drawn in three dimensions. These can be classified into seven groups based on their elements of symmetry. Moreover, examination of the elements of symmetry (about a point, a line, or a plane) for a crystal shows that there are 32 different combinations (classes) that can be grouped into seven systems. The correspondence of these seven systems to the seven lattice groups is shown in Table 1.

The general shape of a crystal is referred to as its habit. The appearance of the crystalline product and its processing characteristics (such as washing and filtration) are affected by crystal habit. Relative growth rates of the faces of a crystal determine its shape; faster-growing faces become smaller than slower-growing faces and, in the extreme, may disappear from the crystal altogether. Growth rates depend on the presence of impurities, rates of cooling, temperature, solvent, mixing, and supersaturation. Furthermore, the importance of each of these factors may vary from one crystal face to another. For example, consider Figure 9 which shows that the $\langle 111 \rangle$ face grows between 1.6 and 2.2 times as fast as the $\langle 110 \rangle$ face at the conditions examined. These results account for the elongated crystal shape exhibited by magnesium sulfate heptahydrate crystals. In addition, the effects of supersaturation and temperature are different on the growth rates of the two faces studied. Such behavior leads to changes in habit as the temperature and/or supersaturation are changed in a crystallizer.

A number of studies have shown that various additives can be included in a process stream to alter crystal habit (5). Prediction of such behavior is difficult and extensive laboratory or bench-scale experiments may be required to evaluate the effectiveness of habit modifiers. More recently, some measure of success has

Table 1. The 14 Bravais Lattices[a]

Type of symmetry	Lattice	Crystal system
cubic	cube	regular
	body-centered cube	
	face-centered cube	
tetragonal	square prism	tetragonal
	body-centered square prism	
orthorhombic	rectangular prism	orthorhombic
	body-centered rectangular prism	
	rhombic prism	
	body-centered rhombic prism	
monoclinic	monoclinic parallelepiped	monoclinic
	clinorhombic	
triclinic	triclinic parallelepiped	triclinic
rhomboidal	rhombohedron	triclinic
hexagonal	hexagonal prism	hexagonal

[a]Ref. 5.

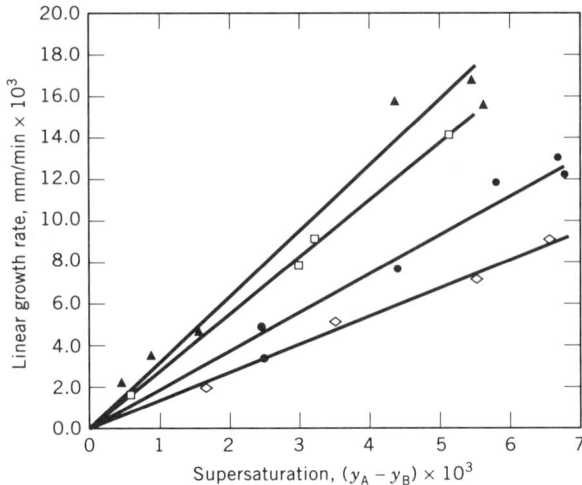

Fig. 9. Growth rate of faces of $MgSO_4 \cdot 7H_2O$. ▲, $\langle 111 \rangle$ face at $T_{sat} = 38.0°C$; ■, $\langle 111 \rangle$ face at $T_{sat} = 33.5°C$; ●, $\langle 110 \rangle$ face at $T_{sat} = 38.0°C$; ◆, $\langle 110 \rangle$ face at $T_{sat} = 33.5°C$; solution velocity = 1.2 cm/s (42). Reprinted with permission of the American Chemical Society.

been achieved with altering the habit of organic crystals based on the molecular structure and characteristics of the crystallizing species. One category of additives affecting crystal habit is surfactants (43). Should an additive enhance the properties of a crystalline material, for example, by making it easier to filter, the expense associated with its use may be warranted. Significant efforts toward tailoring additives so that they have specific effects on crystal habit have been made by a number of research groups (44,45).

Polymorphism is a condition in which a specific chemical substance may crystallize into different forms. For example, ammonium nitrate exhibits four changes in form (5) between $-18°$ and $125°C$:

$$\text{liquid} \underset{169.6°C}{\rightleftharpoons} \text{cubic} \underset{125.2°C}{\rightleftharpoons} \text{trigonal} \underset{84.2°C}{\rightleftharpoons} \text{orthorhombic I} \underset{32.3°C}{\rightleftharpoons} \text{orthorhombic II} \underset{-18°C}{\rightleftharpoons} \text{tetragonal}$$

Transitions from one polymorphic form to another may be accompanied by changes in specific volume, which may lead to destruction of the crystal and containers in which the substance is stored.

A specific polymorph may be absolutely essential for a crystalline product, for example, one polymorph may have a more desirable color or greater hardness or disperse in water more easily than another polymorph. Often, one polymorphic form is more stable than another (for example, at $80°C$ the orthorhombic I form of ammonium nitrate is more stable than the trigonal form) at conditions to which a product is exposed. An interesting approach to keeping a less stable polymorph from transforming to a more stable, but less desireable, polymorphic form uses an additive to block rearrangement of the molecular structure leading to the undesired form (46).

Agglomeration. Many of the analyses of industrial crystallizers require that the particle recovered from the crystallizer consist of a single crystal. It is only with this type of system that single growth rates are likely to be exhibited and, moreover, many of the properties of the crystal are affected deleteriously by agglomeration. Purity, for example, typically is diminished when agglomeration occurs. Countering the negative aspects of agglomeration is recognition that in many systems the single crystals produced by normal crystal growth would be too small to be separable using conventional solid–liquid separation equipment. In such instances, there would be no product without agglomeration.

The needs associated with a greater understanding of agglomeration fall into at least two separate areas: identifying the variables affecting agglomeration and accounting properly for agglomeration in a population balance around a crystallizer. The latter of these has been addressed (47,48), but the subject matter is considered beyond the scope of the present discussion.

Several process variables that influence agglomeration of copper sulfate pentahydrate [7758-99-8] crystals have been identified (49). Particles originally appearing to be single crystals were found to be agglomerates of complex structure. Electron photomicrographs showed that the agglomerates were formed early in the life of crystals comprising the agglomerates and that the growth of these agglomerates had followed a complex and unpredictable pattern. The percentages of agglomerates in the particles recovered from a series of mixed-suspension, mixed-product experiments were found to increase with increasing supersaturation, increasing magma density, and decreasing agitation. The observations fit with a hypothesis that the agglomeration resulted when two or more crystals came together and were bonded through overgrowth of contact areas. Such a hypothesis is inadequate, however, in predicting when agglomeration will occur and the key variables that can be adjusted to control the agglomeration.

Purity. Although crystallization has been employed extensively as a separation process, purification techniques using crystallization have become increasingly important. Mechanisms by which impurities can be incorporated into crystalline products include adsorption of impurities on crystal surfaces (50), solvent entrapment in cracks, crevices and agglomerates, and inclusion of pockets of liquid (51). An impurity having a structure sufficiently similar to the material being crystallized can also be incorporated into the crystal lattice by substitution or entrapment (52,53). Among these mechanisms, inclusion formation has been extensively studied (54–58). It has also been suggested that the purity may be directly linked to size and habit of product crystals, but the interaction does not appear to be simple. It has been noted (59) that the key to producing high purity crystals was to maintain the supersaturation at a low level so that large crystals were obtained. Others have found that reducing the size of ammonium perchlorate crystals resulted in a substantial decrease in moisture due to inclusion (58).

Key factors in solving problems associated with crystal purity are the location, ie, on the surface or incorporated in the crystal, and the nature of the impurity. Impurities are on the exterior of host crystals as a result of adsorption; wetting by a solvent that contains the impurities; or entrapment of impure solvent in cracks, crevices, agglomerates, and aggregates. Incorporation of impurities within crystals comes about through formation of inclusions (also referred to as occlusions) of solvent, lattice substitution, or lattice entrapment. Obviously, the characteristics of an impurity determine whether it is positioned on the surface or the interior of host crystals. Three key impurity types are solutes similar to the product, solutes dissimilar from the product, and the solvent. The effects of process variables on the purity of L-isoleucine crystallized from aqueous solutions containing other amino acids (impurities) have been determined (60). Another study has examined how the methanol content of L-serine crystals could be minimized when the mode of crystallization is addition of methanol to saturated aqueous solutions (4). Mixing and the rate at which supersaturation is generated are important in both of these cases.

Crystal Size Distributions. Particulate matter produced by crystallization has a distribution of sizes that varies in a definite way over a specific size range. A crystal size distribution (CSD) is most commonly expressed as a population (number) distribution relating the number of crystals at each size to the size or as a mass (weight) distribution expressing how mass is distributed over the size range. The two distributions are related and affect many aspects of crystal processing and properties, including appearance, solid–liquid separation, purity, reactions, dissolution, and other properties involving surface area.

Population density (n) has dimensions number/(volume)(length); it is a key quantity in the discussion of CSD, a function of the characteristic crystal dimension L, and it is defined so that it is independent of the magnitude of the system. When a total population density is used, the symbol is $\bar{n}$ and the units are number/length. Population density is defined by letting ΔN be the number of crystals per unit system volume in a size range from L to $L + \Delta L$, so that

$$n = \lim_{\Delta L \to 0} \frac{\Delta N}{\Delta L} \tag{30}$$

The arbitrary system volume on which n is based must be defined before the population density function has meaning. For example, the volume may be that of the slurry or the clear liquor in the system.

The function N in equation 30 is a cumulative number distribution representing the number of crystals per unit volume in the distribution that have a characteristic dimension less than L'. Therefore,

$$N(L') = \int_0^{L'} n \, dL \tag{31}$$

and the fraction of the crystals in the distribution $F(L')$ that have a size less than L' can be calculated as

$$F(L') = \frac{N(L')}{N_{\text{tot}}} \tag{32}$$

A mass (weight) density function, given the symbol m and having dimensions mass/(volume)(length), can be defined analogously to population density; letting ΔM be the mass of crystals per unit system volume in the size range L to $L + \Delta L$,

$$m = \lim_{\Delta L \to 0} \frac{\Delta M}{\Delta L} \tag{33}$$

The two density functions can be related through a simple shape factor as follows. Suppose the mass of a single crystal is M_c and the characteristic dimension of that crystal is L. If the crystal is from a population in which shape is not a function of size, then the mass of any crystal from that population is related to characteristic dimension by a volume shape factor:

$$M_c = k_v \rho L^3 \qquad (34)$$

Recognizing that the mass of crystals in a sample is the product of the number of crystals and the mass of a single crystal, mass and population densities may be related by the expression

$$\Delta M = k_v \rho L^3 \Delta N \qquad (35)$$

where ρ is crystal density. Dividing by ΔL and taking the limit as this quantity approaches zero,

$$m = k_v \rho L^3 n \qquad (36)$$

The function M used in the above equations is a cumulative mass distribution function, representing the mass of crystals having a characteristic dimension less than L'. The total mass of crystals per unit volume is related to population density by the equation

$$M_T = k_v \rho \int_0^\infty L^3 n \, dL \qquad (37)$$

This quantity, which is often referred to as magma density or solids concentration (mass of crystals per unit system volume), is often an important process variable. A cumulative mass fraction of crystals having a size less than L' can also be defined as

$$W(L') = \frac{M(L')}{M_T} = \frac{k_v \rho \int_0^{L'} L^3 n(L) dL}{M_T} \qquad (38)$$

Moments of a distribution often provide information that can be used to characterize particulate matter. The jth moment of the population density function n is defined as

$$m_j = j\text{th moment} = \int_0^\infty L^j n \, dL \qquad (39)$$

It can be demonstrated that the total number of crystals, the total length, the total area, and the total volume of crystals, all in a unit of system volume, can be evaluated from the zero, first, second, and third moments of the population density function.

An average crystal size can be used to characterize a CSD. However, the average can be determined on any of several bases, and the basis selected must be specified for the average to be useful. More than 20 different averaging procedures have been proposed, yet none is generally satisfactory or preferred (5).

The complete characterization of a particulate material requires development of a functional relationship between crystal size and population or mass. The functional relationship may assume an analytical form (7), but more frequently it is necessary to work with data that do not fit such expressions. As such detail may be cumbersome or unavailable for a crystalline product, the material may be more simply (and less completely) described in terms of a single crystal size and a spread of the distribution about that specified dimension.

The dominant crystal size, L_D, is most often used as a representation of the product size, because it represents the size about which most of the mass in the distribution is clustered. If the mass density function defined in equation 33 is plotted for a set of hypothetical data as shown in Figure 10, it would typically be observed to have a maximum at the dominant crystal size. In other words, the dominant crystal size L_D is that characteristic crystal dimension at which $dm/dL = 0$. Also shown in Figure 10 is the theoretical result obtained when the mass density is determined for a perfectly mixed, continuous crystallizer within which invariant crystal growth occurs. That is, mass density is found for such systems to follow a relationship of the form $m = aL^3 \exp(-bL)$, where a and b are system-dependent parameters.

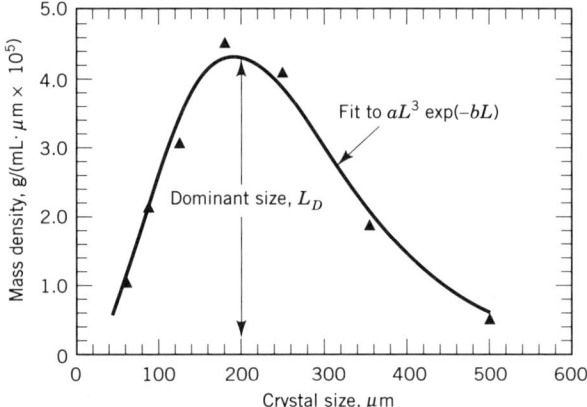

Fig. 10. Determination L_D from plot of mass density function.

The coefficient of variation (cv) of a distribution is a measure of the spread of the distribution about some characteristic size. It is often used in conjunction with dominant size to characterize crystal populations through the equation

$$cv = \frac{\sigma}{L_D} \qquad (40)$$

where σ is the standard deviation of the distribution. The coefficient of variation of the mass density function about the dominant crystal size is given by

$$cv = \left(\frac{m_3 m_5}{m_4^2} - 1\right)^{1/2} \qquad (41)$$

Population Balances and Crystal Size Distributions

Population balances and crystallization kinetics may be used to relate process variables to the crystal size distribution produced by the crystallizer. Such balances are coupled to the more familiar balances on mass and energy. It is assumed that the population distribution is a continuous function and that crystal size, surface area, and volume can be described by a characteristic dimension L. Area and volume shape factors are assumed to be constant, which is to say that the morphology of the crystal does not change with size.

A balance is formulated around a control volume V_T on the number of crystals in any size range, say L_1 to L_2. It must account for crystals that enter and leave the size range by convective flow and crystals that enter and leave the size range by crystal growth. Crystal breakage and agglomeration are assumed to be negligible in the present analysis, and it is assumed that crystals are formed by nucleation at size zero. The rate of crystal growth G is defined as the rate of change of the characteristic crystal dimension L; that is, $G = dL/dt$.

Consider the crystallizer shown in Figure 11. If it is assumed that the crystallizer is well mixed with a constant slurry volume V_T, then, as shown (7), the following partial differential population balance can be derived:

$$\frac{\partial(nG)}{\partial L} + \frac{Q_o n}{V_T} - \frac{Q_i n_i}{V_T} = -\frac{\partial n}{\partial t} \qquad (42)$$

If the crystallizer is now assumed to operate with a clear feed ($n_i = 0$), at steady state ($\partial n/\partial t = 0$), and if the crystal growth rate G is invariant and a mean

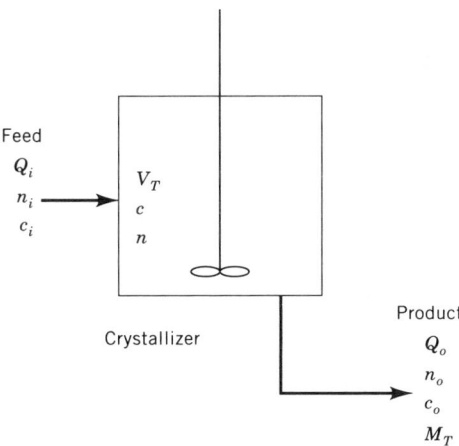

Fig. 11. Schematic diagram of a simple, perfectly mixed crystallizer.

residence time τ is defined as V_T/Q_0, then the population balance can be written as

$$G\frac{dn}{dL} + \frac{n}{\tau} = 0 \qquad (43)$$

τ is often referred to as the drawdown time to reflect that it is the time required to empty the contents from the crystallizer if the feed is set to zero. Equation 43 can be integrated using the boundary condition $n = n^0$ at $L = 0$:

$$n = n^0 \exp\left(-\frac{L}{G\tau}\right) \qquad (44)$$

If the magma volume V_T is allowed to vary in the system on which equation 42 is based, the population balance becomes

$$\frac{\partial n}{\partial t} + \frac{\partial(nG)}{\partial L} + n\frac{\partial(\ln V_T)}{\partial t} + \frac{Q_0 n}{V_T} = 0 \qquad (45)$$

The crystallizer model that led to the development of equations 44 and 45 is referred to as the mixed-suspension, mixed-product removal (MSMPR) crystallizer.

Determination of Crystallization Kinetics. Under steady-state conditions, the total number production rate of crystals in a perfectly mixed crystallizer is identical to the nucleation rate, B^0. Accordingly,

$$B^0 = \frac{1}{\tau}\int_0^\infty n\,dL \qquad (46)$$

For crystallizers following the constraints leading to equation 44,

$$B^0 = n^0 G \qquad (47)$$

Combining equations 45 and 48

$$n = \frac{B^0}{G}\exp\left(-\frac{L}{G\tau}\right) \qquad (48)$$

Analysis of equation 48 shows that a single sample taken either from inside the crystallizer or from the product stream will allow evaluation of nucleation and growth rates at the system conditions. Figure 12 shows a plot of typical population density data obtained from a crystallizer meeting the stated assumptions. The slope of the plot of such data may be used to obtain the growth rate, and the product of the intercept and growth rate gives the nucleation rate.

Many industrial crystallizers operate in a well-mixed or nearly well-mixed manner, and the equations derived above can be used to describe their performance. Furthermore, the simplicity of the equations describing an MSMPR

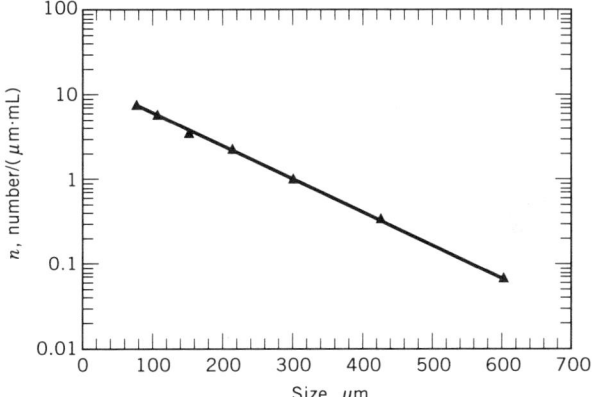

Fig. 12. Plot of population density as a function of size for KNO_3. $\tau = 15$ min. For the line, $n = 16.528 * \exp(-0.0090426L)$; $R = 0.99752$; slope $= -1/G\tau$; intercept $= n^0 = B^0/G$.

crystallizer make experimental equipment configured to meet the assumptions leading to equation 44 useful in determining nucleation and growth kinetics in systems of interest.

From a series of runs at different operating conditions, a correlation of nucleation and growth kinetics with appropriate process variables can be obtained; the resulting correlation can then be used to guide either crystallizer scaleup or the development of an operating strategy for an existing crystallizer. The variables affecting nucleation and growth kinetics include temperature, supersaturation, magma density, and external stimuli, such as agitation or circulation rate of the magma. Empirical power-law functions are used most frequently in correlating nucleation and growth rates, but the choice of the independent variables can be justified from a mechanistic perspective. For example, systems that are believed to follow secondary nucleation mechanisms should include a variable such as magma density that reflects the character of crystals in the crystallizer. The most commonly used power-law functions are

$$B^0 = k_1 s^b M_T^j \tag{49}$$

$$G = k_2 s^g \tag{50}$$

It is often difficult to measure supersaturation, especially in systems that have high growth rates. Even though the supersaturation in such systems is so small that it can be neglected in writing a solute mass balance, it is important in setting nucleation and growth rates. In such instances it is convenient to substitute growth rate for supersaturation by combining equations 49 and 50 to get

$$B^0 = k_n G^i M_T^j \tag{51}$$

The constant k_n may depend on process variables such as temperature, rate of agitation or circulation, presence of impurities, and other variables. If sufficient data are available, such quantities may be separated from the constant by adding more terms in a power-law correlation. The term k_n is specific to the operating equipment and generally is not transferrable from one equipment scale to another. The system-specific constants i and j are obtainable from experimental data and may be used in scaleup, although j may vary considerably with mixing conditions. Illustration of the use of data from a commercial crystallizer to obtain the kinetic parameters k_n, i, and j is available (61).

Mass Balance Constraints. From the schematic diagram of a continuous crystallizer shown in Figure 11, the following mass balance on solute can be constructed:

$$Q_i c_i = Q_0 c_0 + Q_0 M_T \tag{52}$$

c_0 is determined by system kinetics and constrained by a solid–liquid equilibrium (solubility) relationship, which gives the equilibrium concentration c^* at the system conditions. The system (solute–solvent and crystallizer) is characterized by the magnitude of the supersaturation $(c_0 - c^*)$ remaining in the solution exiting the crystallizer. If the mass balance is closed by substituting c^* for c_0 in equation 52, the system is said to be a fast-growth or class II system. If the mass balance is not closed, significant supersaturation remains in the solution and the system is said to be a slow-growth or class I system. In other words, for class I (slow-growth) systems: $c_0 > c^*$ and

$$M_T = \frac{Q_i}{Q_0} c_i - c_0 \tag{53}$$

Values of process variables such as residence time may change the system kinetics that change c_0 and, in turn, M_T.

For class II (fast-growth) systems: $c_0 = c^*$ and

$$M_T = \frac{Q_i}{Q_0} c_i - c^* \tag{54}$$

Process variables do not change c^* and, therefore, M_T is constant over modest ranges of operating conditions.

CSD Characteristics for MSMPR Crystallizers. The perfectly mixed crystallizer described in the preceding discussion is highly constrained and the form of crystal size distributions produced by such systems is fixed. Such distributions have the following characteristics.

(1) Moments of the distribution can be calculated for MSMPR crystallizers by the simple expression

$$m_j = j! n^0 (G\tau)^{j+1} \tag{55}$$

Properties of the distribution such as total number of crystals per unit volume, total length of crystals per unit volume, total area of crystals per unit volume, and total volume of solids (crystals) per unit volume may be explicitly evaluated from the moment equations.

(2) The dominant crystal size L_D is given by $L_D = 3G\tau$. This quantity is also the ratio m_3/m_2, which is often given the symbol $\overline{L}_{3,2}$.

(3) From the definition of the coefficient of variation given by equation 41, $cv = 50\%$ for an MSMPR crystallizer. Such a cv may be too large for certain commercial products, which means either the crystallizer must be altered or the product must be screened to separate the desired fraction.

(4) The magma density M_T (mass of crystals per unit volume of slurry or liquor) may be obtained from the third moment of the population density function and is given by

$$M_T = 6\rho k_v n^0 (G\tau)^4 \tag{56}$$

Although magma density is a function of the kinetic parameters n^0 and G, it often can be measured independently. In such cases, it should be used as a constraint in evaluating nucleation and growth rates from measured crystal size distributions (62), especially if the system of interest exhibits the characteristics of anomalous crystal growth.

(5) A pair of kinetic parameters, one for nucleation rate and another for growth rate, describe the crystal size distribution for a given set of crystallizer operating conditions. Variation in one of the kinetic parameters without changing the other is not possible. Accordingly, the relationship between these parameters determines the ability to alter the characteristic properties (such as dominant size) of the distribution obtained from an MSMPR crystallizer (7).

Preferential Removal of Crystals. Crystal size distributions produced in a perfectly mixed continuous crystallizer are highly constrained; the form of the CSD in such systems is determined entirely by the residence time distribution of a perfectly mixed crystallizer. Greater flexibility can be obtained through introduction of selective removal devices that alter the residence time distribution of materials flowing from the crystallizer. The functions of classified removal are best described in terms of idealized models of clear-liquor advance, classified-fines removal, and classified-product removal.

Clear-liquor advance is simply the removal of mother liquor from the crystallizer without simultaneous removal of crystals. The primary objective of *fines removal* is preferential withdrawal from the crystallizer of crystals whose size is below some specified value. Such crystals may be redissolved and the resulting solution returned to the crystallizer. *Classified-product removal* is carried out to remove preferentially those crystals whose size is larger than some specified value.

The effects of each selective removal function on CSD can be described in terms of the population density function n. It is convenient to define flow rates in terms of clear liquor, which requires the population's density function to be defined on a clear-liquor basis. In the present discussion, only systems exhibiting invariant crystal growth are considered.

Clear-liquor advance reduces the quantity of liquor that must be processed by solid–liquid separation equipment (for example, a filter or a centrifuge). The reduction in liquor flow through the separation equipment may allow use of smaller equipment for a fixed production rate or increased production through fixed equipment.

The function of clear-liquor advance can be illustrated by considering a simple operation, shown in Figure 13, in which Q_i, Q_{CL}, and Q_0 represent volumetric flow rates of clear-liquor fed to the crystallizer, in the clear-liquor advance, and in the output slurry. In such systems the population density function is given by the expression

$$n = n^0 \exp\left(-\frac{L}{G\tau_p}\right) \tag{57}$$

where $\tau_p = V/Q_0$. It is clear that increasing Q_{CL} decreases Q_0 and thereby increases the residence time of the crystals in the crystallizer.

Clearly, the form of the population density function resulting from a clear-liquor advance system is identical to that expected from perfectly mixed systems in which τ_{crystals} is identical to τ_{liquor}. Unless the increase in magma density associated with clear-liquor advance results in significant increases in nucleation, some increase in the dominant crystal size can be expected. It has been observed that the increase in L_D may be greater than predicted from theory. This is caused by the stream being removed as clear liquor containing varying amounts of fines, which means the system characteristics are those of classified-fines removal.

As an idealization of the classified-fines removal operation, assume that two streams are withdrawn from the crystallizer, one corresponding to the product stream and the other a fines removal stream. Such an arrangement is shown schematically in Figure 14. The flow rate of the clear solution in the product stream is designated Q_0 and the flow rate of the clear solution in the fines removal stream is set as $(R - 1)Q_0$. Furthermore, assume that the device used to separate fines from larger crystals functions so that only crystals below an arbitrary size

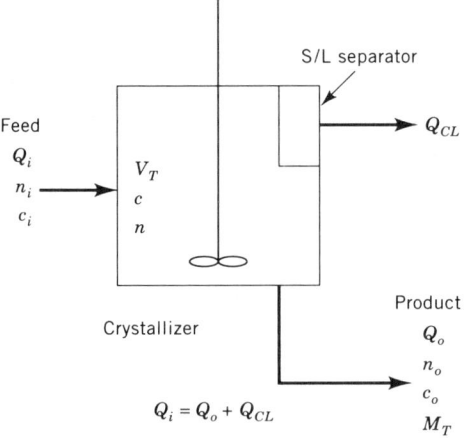

Fig. 13. Simplified schematic diagram of clear-liquor advance or double-draw off (DDO).

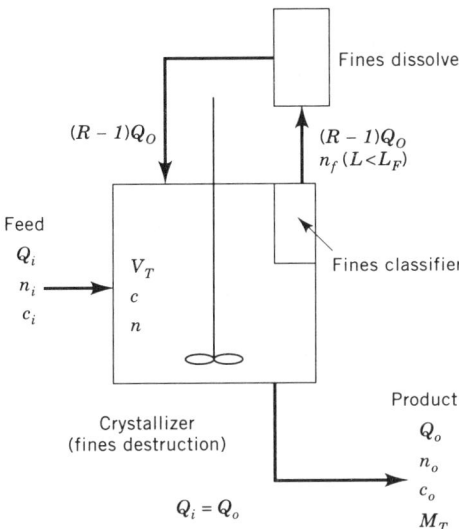

Fig. 14. Simplified schematic diagram of classified-fines removal.

L_F are in the fines removal stream and that all crystals below size L_F have an equal probability of being removed in the fines removal stream. Under these conditions, the crystal size distribution is characterized by two mean residence times, one for the fines and the other for crystals larger than L_F. These quantities are related by the equations

$$\tau_F = \frac{V}{RQ_0} = \frac{\tau}{R} \quad (L < L_F) \tag{58}$$

$$\tau = \frac{V}{Q_0} \quad (L \geq L_F) \tag{59}$$

where V is the volume of clear solution in the crystallizer. Recognize that the ratio of the probability of a crystal smaller than L_F being removed from the crystallizer to that of crystals larger than L_F being removed is $R = \tau/\tau_F$.

For systems following invariant growth the crystal population density in each size range decays exponentially with the inverse of the product of growth rate and residence time. For a continuous distribution, the population densities of the classified fines and the product crystals must be the same at size L_F. Accordingly, the population density for a crystallizer operating with classified-fines removal is given by

$$n = n° \exp\left[-\frac{RL}{G\tau}\right] \quad (L \leq L_F) \tag{60}$$

$$n = n° \exp\left[-\frac{(R-1)L_F}{G\tau}\right] \exp\left[-\frac{L}{G\tau}\right] \quad (L > L_F) \tag{61}$$

Figure 15 shows how the population density function changes with the addition of classified-fines removal. It is apparent from the figure that fines removal increases the dominant crystal size, but it also increases the spread of the distribution.

A simple method for implementation of classified-fines removal is to remove slurry from a settling zone in the crystallizer. The settling zone can be created by constructing a baffle that separates the zone from the well-mixed portion of the vessel, as is the case for a draft-tube-baffle crystallizer, or in small-scale systems, by simply inserting a length of pipe into the crystallizer chamber. The separation of crystals in the settling zone is based on the dependence of settling velocity on crystal size; only those crystals having a settling velocity greater than the upward velocity of the slurry remain in the crystallizer. As the cross-sectional area of a settling zone is invariant, the flow rate of slurry through the zone determines the cut size L_F, and it also determines the parameter R used in equations 59–62.

Classified removal of course material also can be used, as shown in Figure 16. In a crystallizer equipped with idealized classified-product removal, crystals above some size L_C are removed at a rate Z times the removal rate expected for a perfectly mixed crystallizer, and crystals smaller than L_C are not removed at all. Larger crystals can be removed selectively through the use of an elutriation leg, hydrocyclones, or screens. Using the analysis of classified-fines removal systems as a guide, it can be shown that the crystal population density within the crystallizer magma is given by the equations

$$n = n^0 \exp\left[-\frac{L}{G\tau}\right] \quad (L \leq L_C) \tag{62}$$

$$n = n^0 \exp\left[\frac{(Z-1)L_C}{G\tau}\right] \exp\left[-\frac{ZL}{G\tau}\right] \quad (L > L_C) \tag{63}$$

where τ is defined as the residence time V/Q_0. Figure 17 shows the effects of classified-product removal on crystal size distribution. The characteristics of the

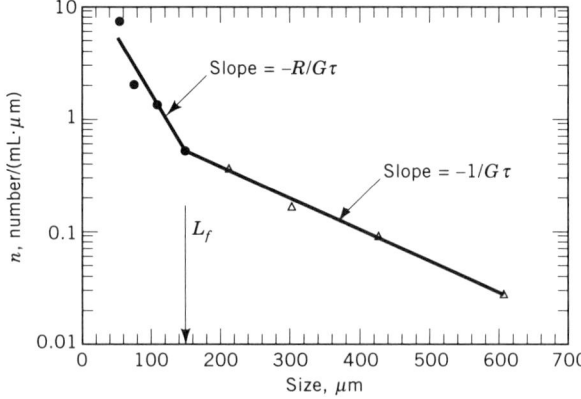

Fig. 15. Population density function for product from crystallizer with classified-fines removal. Cut size $L_F = 150$ μm; $R = 3.7$.

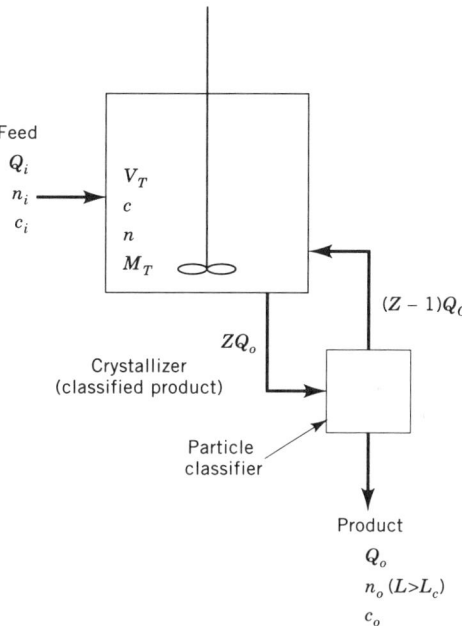

Fig. 16. Classified withdrawal of course crystals.

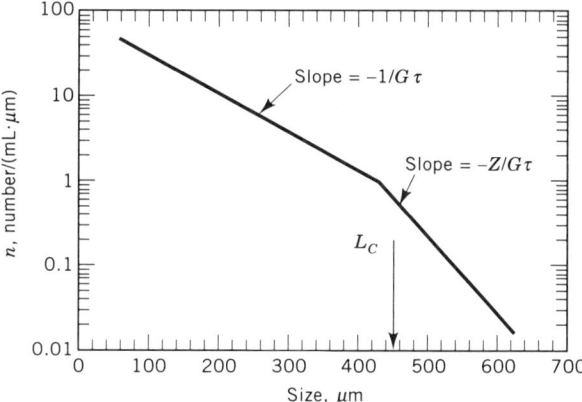

Fig. 17. Effect of classified-product removal on population densities within a crystallizer and in the crystallizer product.

crystal size distribution obtained from a system with classified product removal show a narrower distribution (reduced coefficient of variation) and smaller dominant size. A more complete discussion of the implications of classified-product removal, with particular attention given to the distinction between the crystal population densities within the crystallizer and in the product has been given (7).

It is possible to obtain both a narrowing of the distribution and an increase in dominant size by combining preferential removal of fines and course crystals.

Including the idealized removal functions for fines and course crystals in a population balance and assuming invariant crystal growth will result in a population density function within the crystallizer given by equations 64–66, Figure 18 illustrates the effects of both removal functions on population density. This plot of population density results from sampling the magma within a crystallizer, not from sampling the product stream.

$$n = n^0 \exp\left[-\frac{RL}{G\tau}\right] \quad \text{for } L \leq L_F \quad (64)$$

$$n = n^0 \exp\left[-\frac{(R-1)L_F}{G\tau}\right] \exp\left[-\frac{L}{G\tau}\right] \quad \text{for } L_F < L < L_C \quad (65)$$

$$n = n^0 \exp\left[-\frac{(R-1)L_F}{G\tau}\right] \exp\left[\frac{(Z-1)L_C}{G\tau}\right] \exp\left[-\frac{ZL}{G\tau}\right] \quad \text{for } L \geq L_C \quad (66)$$

The model of the crystallizer and selective removal devices that led to equations 64–66 is referred to as the R-Z crystallizer. It is an obvious idealization of actual crystallizers because of the perfect cuts assumed at L_F and L_C. However, it is a useful approximation to many systems and it allows qualitative analyses of complex operations. The R-Z model may also be representative of inadvertant classification, ie, fines or course crystals may be preferentially removed from a crystallizer without installation of specific hardware to accomplish such an objective.

Although many commercial crystallizers operate with some form of selective crystal removal, such devices can be difficult to operate because of fouling of heat exchanger surfaces or blinding of screens. In addition, several investigations identify interactions between classified fines and course product removal as causes of cycling of a crystal size distribution (7). Often such behavior can be minimized or even eliminated by increasing the fines removal rate (63,64).

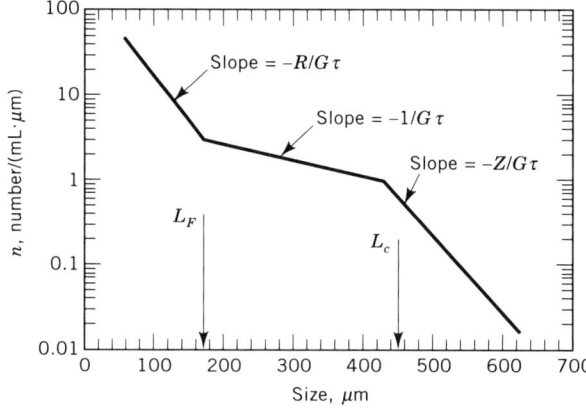

Fig. 18. Population density functions of crystals within a crystallizer, having both classified-fines and classified-product removal and of crystals in the product from such a crystallizer.

Batch Crystallization. Crystal size distributions obtained from batch crystallizers are affected by the mode used to generate supersaturation and the rate at which supersaturation is generated. For example, in a cooling mode there are several avenues that can be followed in reducing the temperature of the batch system, and the same can be said for the generation of supersaturation by evaporation or by addition of a nonsolvent or precipitant. The complexity of a batch operation can be illustrated by considering the summaries of seeded and unseeded operations shown in Figure 19.

Seeded operation	Unseeded operation
Prepare system ↓	Prepare system ↓
Initiate generation of supersaturation ↓	Initiate generation of supersaturation ↓
Add selected quantity of seed crystals having specified CSD. ↓	
Nucleation is initiated by secondary mechanisms involving the seed crystals or low supersaturation and high surface area of seed crystals eliminate or minimize nucleation; seed crystals grow ↓	Supersaturation reaches the metastable limit and nucleation is initiated; supersaturation drops rapidly as crystals formed begin to grow ↓
Nucleation continues by secondary mechanisms and growth continues throughout the run until the batch achieves equilibrium and/or is dumped	No further nucleation occurs until crystals initially formed grow sufficiently large to participate in secondary nucleation or supersaturation again becomes high enough to bring about primary nucleation; growth and perhaps nucleation continue throughout the run until the batch achieves equilibrium and/or is dumped

Fig. 19. Batch crystallizer operation.

The crystal size distributions resulting from the operating strategies outlined in Figure 19 depend greatly on the use of seeding, the rate at which supersaturation is generated, and those variables that are important in the prevailing mechanism of nucleation. Figures 20 and 21 summarize the qualitative variations in CSD that may be observed in batch crystallization and the role of adding seed crystals to such systems.

More quantitative relationships of the CSD obtained from batch operations can be developed through formulation of a population balance. Using a population density defined in terms of the total crystallizer volume rather than on a specific basis ($\bar{n} = nV$), the general population balance given by equation 42 can be modified in recognition of there being no feed or product streams:

$$\frac{\partial(nV)}{\partial t} + \frac{\partial(GnV)}{\partial L} = \frac{\partial \bar{n}}{\partial t} + \frac{\partial(G\bar{n})}{\partial L} = 0 \qquad (67)$$

The solution to this equation requires an initial condition ($\bar{n}$ at $t = 0$) and a boundary condition ($\bar{n}$ at a specific value of L). Assuming that crystals are formed at zero size gives the boundary condition:

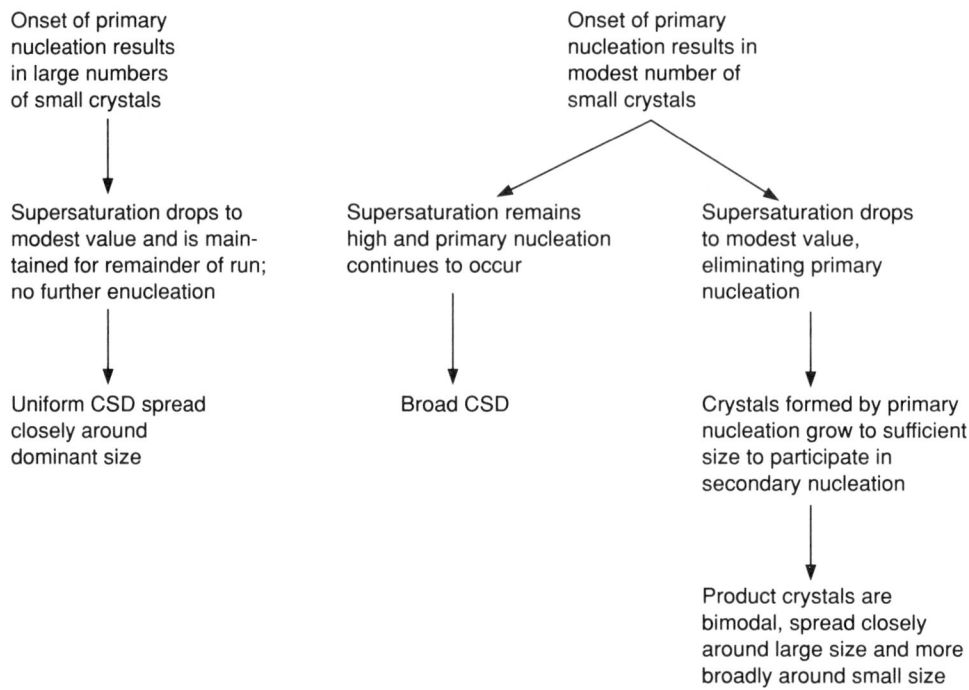

Fig. 20. CSD characteristics from batch crystallization without seeding.

$$\overline{n}(0,t) = \overline{n}^0(t) = \frac{\overline{B}^0(t)}{G(0,t)} \tag{68}$$

Identification of an initial condition is difficult because of the problem of specifying the size distribution at the instant nucleation occurs. The difficulty is mitigated through the use of seeding, which would mean that the initial population density function would correspond to that of the seed crystals:

$$\overline{n}(L,0) = \overline{n}_S(L) \tag{69}$$

where $\overline{n}_S$ is the population density function of the seed crystals.

Moments of the population density function, which are given by

$$\overline{m}_j = \int_0^\infty L^j \overline{n} dL \tag{70}$$

are especially useful in modeling crystal size distributions in batch operations and in the development of equations relating a control variable to time. Recognizing that the zero moment is the total number of crystals in the system it can be shown that

$$\frac{d\overline{m}_0}{dt} = \overline{n}^0 G = \overline{B}^0 = \frac{d\overline{N}_T}{dt} \tag{71}$$

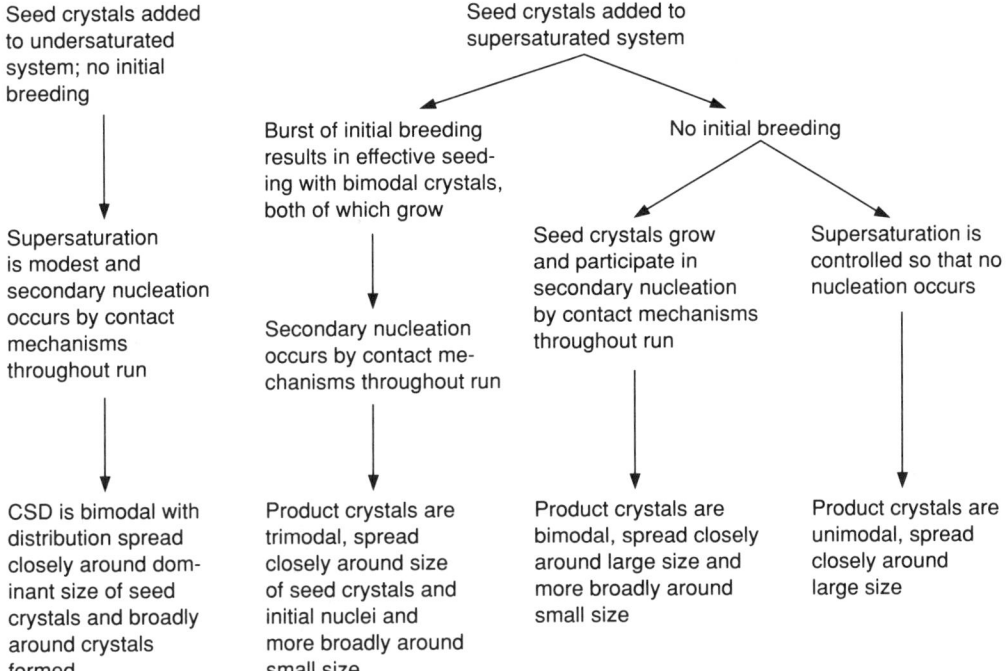

Fig. 21. CSD characteristics from batch crystallization with seeding.

From the relationships of moments to properties of the distribution, the following equations can be derived:

$$\frac{d\overline{m}_1}{dt} = G\overline{m}_0 \Rightarrow \frac{d\overline{L}_T}{dt} = \overline{N}_T G \qquad (72)$$

$$\frac{d\overline{m}_2}{dt} = 2G\overline{m}_1 \Rightarrow \frac{d\overline{A}_T}{dt} = 2k_a \overline{L}_T G \qquad (73)$$

$$\frac{d\overline{m}_3}{dt} = 3G\overline{m}_2 \Rightarrow \frac{d\overline{M}_T}{dt} = 3\left(\frac{k_v}{k_a}\right)\rho \overline{A}_T G \qquad (74)$$

where $\overline{N}_T$ is total number of crystals, $\overline{L}_T$ is total crystal length, $\overline{A}_T$ is total surface area of the crystals, and $\overline{M}_T$ is total mass of crystals in the crystallizer. A solute balance must also be satisfied:

$$\frac{d(Vc)}{dt} + \frac{d\overline{M}_T}{dt} = 0 \qquad (75)$$

where V is the system volume, and c is solute concentration in the solution.

Control of supersaturation is an important factor in obtaining crystal size distributions of desired characteristics, and it would be useful to have a model relating rate of cooling or evaporation or addition of diluent required to maintain a specified supersaturation in the crystallizer. Contrast this to the uncontrolled situation of natural cooling in which the heat transfer rate is given by

$$Q = UA(T - T_c) \qquad (76)$$

where U is a heat-transfer coefficient, A is the area available for heat transfer, T is the temperature of the magma, and T_c is the temperature of the cooling fluid. If U and T_c are constants, the maximum heat transfer rate and the highest rate at which supersaturation is generated are at the beginning of the process when T is highest. These conditions can lead to excessive primary nucleation and the formation of incrustations on the heat-transfer surfaces.

Better product characteristics are obtained through control of the rate at which supersaturation (cooling, evaporation, and addition of a nonsolvent or precipitant) is generated. An objective of the operation may be to maintain the supersaturation at some constant prescribed value, usually below the metastable limit associated with primary nucleation. For example, the batch may be cooled slowly at the beginning of the cycle and more rapidly at the end.

Formulations of population balances on batch crystallizers have been illustrated, and a variety of operating strategies have been considered (65). The results are often complex and present difficult control schemes at best. For example, suppose a model is needed to guide the operation of a batch seeded crystallizer so that isothermal solvent evaporation can be accomplished at a rate that gives a constant crystal growth rate and no nucleation. It is shown that the evaporation rate required is a cubic function of time and the corresponding rate of heat input to the crystallizer must be controlled accordingly. If cooling was to be used rather than evaporation, a similar analysis would show that the dependence of crystallizer temperature on time is highly nonlinear. Although the development of a strategy for generating supersaturation can be aided by such analyses, the initial conditions in the models derived are based on properties of seed crystals added to the crystallizer.

The advantages of selective removal of fines from a batch crystallizer have been demonstrated (66,67). These experimental programs showed narrowing of crystal size distributions and suggest significant reductions in the fraction of a product that would consist of fines or undersize material.

Crystallizers and Crystallization Operations

Crystallization equipment can vary in sophistication from a simple stirred tank to a complicated multiphase column, and the operation can range from allowing a vat of liquor to cool through exchanging heat with the surroundings to the complex control required of batch cyclic operations. In principle, the objectives of these systems are all the same: to produce a pure product at a high yield with an acceptable crystal size distribution. However, the characteristics of the crystallizing system and desired properties of the product often dictate that a specific crystallizer be used in a particular operating mode.

Crystallization from Solution. Crystallization techniques are related to the methods used to induce a driving force for solids formation and to the medium from which crystals are obtained. Several approaches are defined in the following discussion.

Cooling crystallizers use a heat sink to remove both sensible heat from the feed stream and the heat of crystallization released as crystals are formed. The heat sink may be no more than the ambient surroundings of a batch crystallizer, or it may be cooling water or another process stream.

Evaporative crystallizers generate supersaturation by removing solvent, thereby increasing solute concentration. These crystallizers may be operated under vacuum, and, in such circumstances, it is necessary to have a vacuum pump or ejector as a part of the unit. If the boiling point elevation of the system is low (that is, the difference between the boiling point of a solution in the crystallizer and the condensation temperature of pure solvent at the system pressure), mechanical recompression of the vapor obtained from solvent evaporation can be used to produce a heat source to drive the operation.

Evaporative-cooling crystallizers are fed with a liquor that is at a temperature above that in the crystallizer. As the feed is introduced to the crystallizer, which is at reduced pressure, solvent flashes, thereby concentrating the solute in the resulting solution and reducing the temperature of the magma. The mode of this operation can be degenerated to that of a simple cooling crystallizer by returning condensed solvent to the crystallizer body.

Salting-out crystallization operates through the addition of a nonsolvent to the magma in a crystallizer. The selection of the nonsolvent is based on the effect of the solvent on solubility, cost, properties that affect handling, interaction with product requirements, and ease of recovery. The effect of adding a nonsolvent can be quite complex as it increases the volume required for a given residence time and may produce a highly nonideal mixture of solvent, nonsolvent, and solute from which the solvent is difficult to separate.

Reactive crystallization addresses those operations in which a reaction occurs to produce a crystallizing solute. The concentration of the solute formed generally is greater than that corresponding to solubility. In a subset of systems, the solubility is nearly zero and, concomitantly, the supersaturation produced by reaction is large. These are often referred to as *precipitation* operations, and crystal size distributions from them contain a large fraction of fine crystals.

Supercritical fluid solvents are those formed by operating a system above the critical conditions of the solvent. Solubilities of many solutes in such fluids often is much greater than those found for the same solutes but with the fluid at subatmospheric conditions. Recently, there has been considerable interest in using supercritical fluids as solvents in the production of certain crystalline materials because of the special properties of the product crystals. Rapid expansion of a supercritical system rapidly reduces the solubility of a solute throughout the entire mixture. The resulting high supersaturation produces fine crystals of relatively uniform size. Moreover, the solvent poses no purification problems because it simply becomes a gas as the system conditions are reduced below critical.

Crystallizers. The basic requirements of a system involving crystallization from solution are as follows: (*1*) a means of generating supersaturation in a fashion commensurate with the requirements of producing a satisfactory crystal size distribution, (2) a vessel to provide sufficient residence time for crystals to grow to a desired size, and (3) mixing to provide a uniform environment for crystal growth. There are numerous manufacturers of crystallization equipment; in addition, many chemical companies design their own crystallizers based on expertise developed within their organizations. Rather than attempt to describe the variety of special crystallizers that can be found in the marketplace, this section provides a brief general survey of types of crystallizers that use the modes outlined above. Greater detail can be found in the literature (68,69).

The forced-circulation crystallizer is a simple unit designed to provide high heat-transfer coefficients in either an evaporative or a cooling mode. Figure 22 shows a schematic diagram of a forced-circulation crystallizer that withdraws a slurry from the crystallizer body and pumps it through a heat exchanger where heat may be either added to or removed from the slurry. Heat transferred to the circulating magma causes evaporation of solvent as the magma is returned to the crystallizer, whereas heat removal lowers the temperature of the circulating magma. Forced circulation is used to control circulation rates and velocities past the heat-transfer surfaces.

When cooling is the selected mode by which supersaturation is generated, heat can be transferred through an external cooling surface, as shown in Figure 22, or through coils or a jacket internal to the crystallizer body. The higher heat-transfer coefficients that can be achieved with forced circulation allows the temperature difference between heat source and sink to be minimal, thereby reducing formation of encrustations on the heat transfer surface. The operation of cooling crystallizers is limited by the tendency of the solute to form encrustations on the cooling surface, so that the temperature of the cooling fluid and the temperature decrease of the slurry flowing through the heat exchanger may be limited. It is not uncommon to limit the decrease in magma temperature to about 3 to 5°C; therefore, both the circulation rate and heat-transfer surface must be large.

The feed in cooling crystallizers should be rapidly mixed with the magma so as to minimize the occurrence of regions of high supersaturation, which lead to

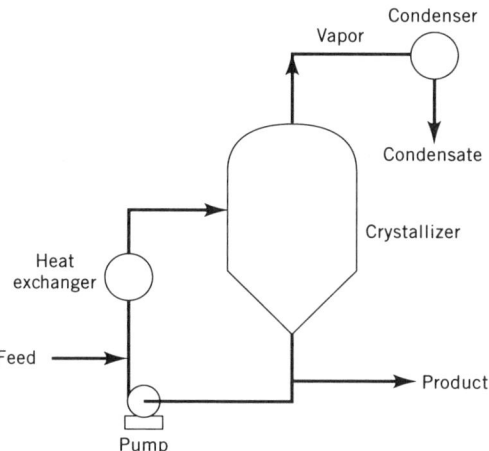

Fig. 22. Schematic diagram of forced circulation crystallizer.

excessive nucleation. Another factor that can lead to degradation of the crystal size distribution is the type of pump used in the circulation loop; an inappropriate pump can cause attrition of the crystals through abrasion, fracture, or shear, and most commercial systems use specially designed axial-flow pumps that provide high flow rates and low heads.

If the characteristics of the system are such that the operating temperature of the crystallizer is low in comparison to the temperature of cooling water, or if there are severe problems with the formation of encrustations, direct-contact refrigeration can be used. A refrigerant is mixed with the crystallizer contents and vaporized at the magma surface. On vaporizing, the refrigerant removes sufficient heat from the magma to cool the feed and remove the heat of crystallization. The refrigerant vapor must be compressed, condensed, and recycled for the process to be economical. Moreover, the refrigerant must be insoluble in the liquor to minimize losses and product contamination.

Scale formation on the heat exchanger surfaces or at the vapor–liquid interface in the crystallizer can cause operational problems with evaporative crystallizers. Such problems can be overcome by not allowing vaporization or excessive temperatures within the exchanger and by proper introduction of the circulating magma into the crystallizer. The latter may be accomplished by introducing the magma below the surface of the magma in the crystallizer, so that all vaporization occurs from a well-mixed zone or by introducing the magma so as to induce a swirling motion that is intended to dislodge encrustations from the wall of the crystallizer at the vapor–liquid interface.

Special devices for classification of crystals may be used in some applications. Figure 23 shows a draft-tube-baffle (DTB) crystallizer designed to provide preferential removal of both fines and classified product. Feed is introduced to the fines circulation line so that nuclei resulting from feed introduction can be dissolved as the stream flows through the fines dissolution exchanger. A quiescent zone is formed between the baffle extending into the chamber and the outside wall of the crystallizer. Flow through the quiescent zone can be adjusted so that crystals below a certain size (determined by settling velocity) are removed in the fines dissolution circuit.

Another type of crystallizer is the Oslo-type unit shown in Figure 24. In units of this type, the object is to form a supersaturated solution in the upper chamber and then relieve the supersaturation through growth in the lower chamber. The use of the downflow pipe in the crystallizer provides good mixing in the growth chamber.

Melt Crystallization. The use of a solvent can be avoided in some systems. In such cases, the system operates with heat as a separating agent, as do several processes involving crystallization from solution, but formation of crystalline material is from a melt of the crystallizing species rather than a solution.

For the following reasons, melt crystallization holds great promise in situations in which it can be substituted for crystallization from solution: (*1*) Without the need to recover and maintain the purity of a solvent, processing costs are reduced substantially. (*2*) Because there is no contaminated solvent to handle, melt crystallization may be more environmentally benign. (*3*) Energy costs found in evaporative crystallization obviously would be reduced if it is possible to produce a desired solid without the need to evaporate solvent. (*4*) Melt crystallization

434 CRYSTALLIZATION

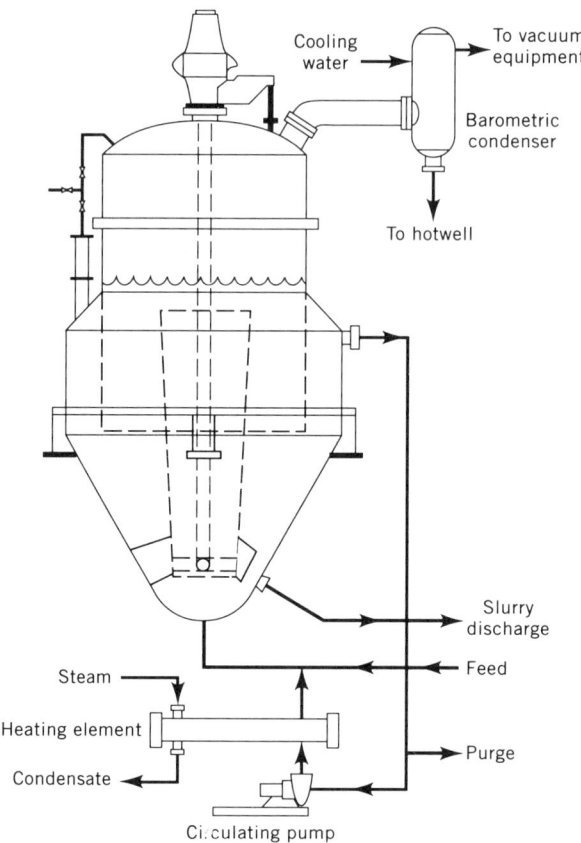

Fig. 23. Schematic diagram of draft-tube-baffle crystallizer.

may be a reasonable alternative to other separation and purification processes, because the heat of vaporization of most volatile organic materials is between two and five times their heat of fusion. An analysis of the energy requirements in melt processes concludes that such processes can compete with other thermal separation techniques only if the plant is well designed and the process precisely controlled (70).

Melt crystallization is carried out either with a suspension of crystals or an advancing front (layer) of solids, although a more complete categorization of melt crystallization is available (71). Following is a brief review of processes in which melt crystallization is used; a more complete review, including a worked out case study for system design, is available (69).

A suspension of crystals formed from the melt may be contacted by well-mixed mother liquor or the crystals may be moved countercurrently to liquor flow in a vertical or horizontal column. In column crystallizers, crystals are moved in a specific direction by gravity or rotating blades. The crystals are melted by the addition of heat when they reach a designated end of the crystallizer; a portion of the melt is removed as product and the remainder is returned to the system to flow countercurrently to and to wash the product crystals.

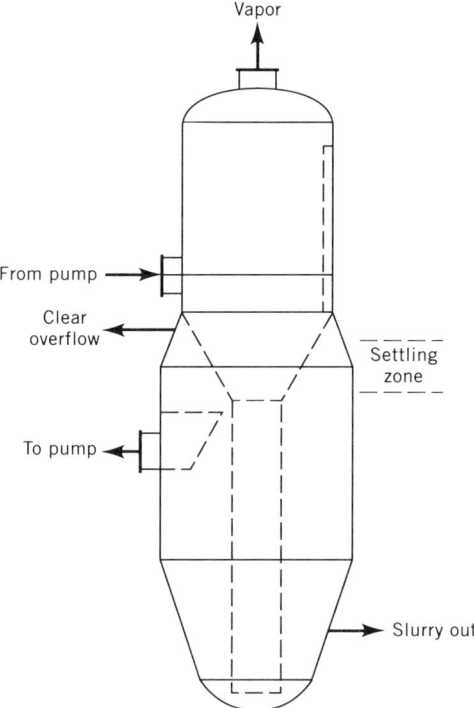

Fig. 24. Schematic diagram of an Oslo crystallizer.

One of the early column crystallizers was that introduced for the separation of xylene isomers (see XYLENES AND ETHYLBENZENE). In this unit, shown schematically in Figure 25, p-xylene crystals are formed in a scraped-surface chiller above the column and fed to the column. The crystals move downward countercurrently to impure liquid in the upper portion of the column and melted p-xylene in the lower part of the column. Impure liquor is withdrawn from an appropriate point near the top of the column of crystals while pure product, p-xylene, is removed from the bottom of the column. The pulse unit drives melt up the column as reflux and into a product receiver.

A horizontal column is typified by the Brodie Purifier, which is shown schematically in Figure 26. Feed enters the column between recovery and refining sections, and crystals exit the refining section and pass through a purifying section. The purifying section is a wash column in which the crystals are contacted with melt generated at the bottom of the column.

In advancing-front or layer melt crystallizations, mother liquor flows over a cooled surface on which material is crystallized. The advancing front of crystals grows in the direction from the cooled surface into the mother liquor. A variety of techniques can be used to take advantage of this type of operation.

Figure 27 is a schematic diagram of the MWB process, which uses an operation in which there are several steps in a batch cycle. Crystal growth is on the inside of a battery of tubes through which melt is flowing, and the melt may flow

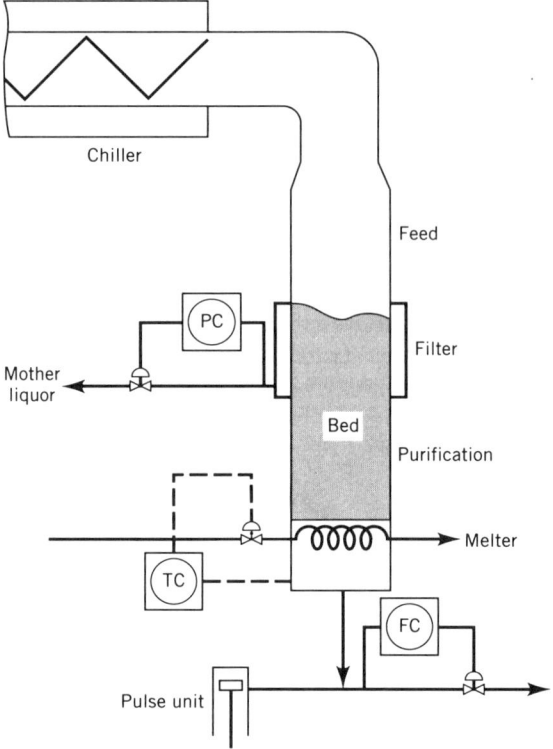

Fig. 25. Schematic diagram of a system used to separate xylene isomers (69). PC = pressure control, TC = temperature control, and FC = flow control.

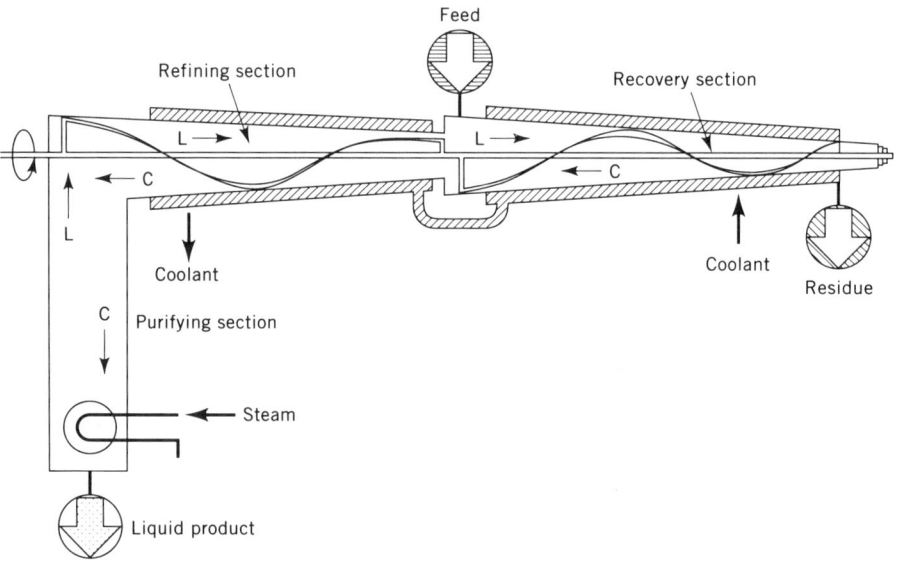

Fig. 26. Schematic diagram of Brodie Purifier. L = liquid; C = crystals (69).

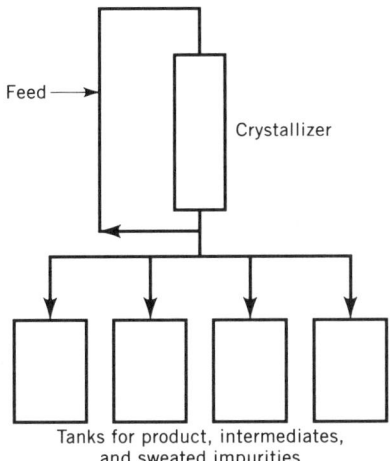

Fig. 27. Schematic diagram of the MWB melt crystallization process.

as a falling film or it can be pumped through the tubes. The process includes the following steps:

(1) Flow of mother liquor through the cooled tubes is initiated, and crystals are grown on the tube surfaces. The heat transfer rate should be controlled so as to moderate crystal growth, thereby producing a relatively uniform layer of high purity solids.

(2) When sufficient crystal mass has been formed, melt flow is stopped and residual mother liquor is drained from the unit.

(3) The solid purity is enhanced by applying heat and causing impurities to flow from the heated solids through a process known as sweating.

(4) After sweating, either the cycle is returned to step 1 and additional solids are deposited or the solids present on the tube wall are melted and recovered as product.

(5) The recovered product melt can be put through the cycle again to increase purity, or fresh feed can be introduced to the cycle.

BIBLIOGRAPHY

"Crystallization," in the *Encyclopedia of Chemical Technology,* 1st ed., Vol. 4, pp. 619–636, by A. Ralph Thompson, University of Pennsylvania; in *ECT* 2nd ed., Vol. 6, pp. 482–515, by J. W. Mullin, University of London; in *ECT* 3rd ed., Vol. 7, pp. 243–285, by J. W. Mullin, University of London; in *ECT* 4th ed., Vol. 7. pp. 683–729, by R. W. Rosseau, Georgia Institute of Technology.

1. National Research Council, *Frontiers in Chemical Engineering: Research Needs and Opportunities*, National Academy Press, Washington, D.C., 1988, Chapt. 3.
2. R. M. Felder and R. W. Rousseau, *Elementary Principles of Chemical Processes*, 2nd ed., John Wiley & Sons Inc., New York, 1986.
3. R. H. Perry and D. W. Green, eds., *Perry's Chemical Engineers' Handbook*, McGraw-Hill Book Co., Inc., New York, 1984, p. 3-99.
4. H. Charmolue and R. W. Rousseau, *AIChE J.* **37,** 1121 (1991).

5. J. W. Mullin, *Crystallization*, 2nd ed., CRC Press, Cleveland, Ohio, 1972.
6. H. A. Miers and F. Issac, in H. E. Buckley, *Crystal Growth*, John Wiley & Sons, Inc., New York, 1951, p. 7.
7. A. D. Randolph and M. A. Larson, *Theory of Particulate Processes*, 2nd ed., Academic Press, Inc., New York, 1988.
8. J. Garside and Roger J. Davey, *Chem. Eng. Commun.* **4,** 393 (1980).
9. M. A. Larson, *AIChE Symp. Ser.*, **80**(240), 39 (1984).
10. M. W. Girolami and R. W. Rousseau, *Ind. Eng. Chem. Process Des. Dev.* **25,** 66 (1986).
11. C. Y. Sung, J. Estrin, and G. R. Youngquist, *AIChE J.* **19,** 957 (1973).
12. R. E. A. Mason and R. F. Strickland-Constable, *Trans. Faraday Soc.*, **62**(pt. 2), 455 (1966).
13. D. P. Lal, R. E. A. Mason, and R. F. Strickland-Constable, *J. Cryst. Growth* **5,** 1 (1969).
14. N. A. Clontz and W. L. McCabe, *Chem. Eng. Prog. Symp. Ser.* **67**(110), 6 (1971).
15. R. W. Rousseau and W. L. McCabe, *Paper presented at the World Congress on Chemical Engineering*, Amsterdam, June 1976.
16. R. C. Bennett, H. Fiedelman, and A. D. Randolph, *Chem. Eng. Prog.* **69** (7), 86(1973).
17. P. A. M. Grootscholten, L. D. M. v.d. Brekel, and E. J. deJong, *Chem. Eng. Res. Des.* **62,** 179 (1984).
18. J. Garside and M. B. Shah, *Ind. Eng. Chem. Process Des. Dev.* **19,** 509 (1980).
19. C. Y. Tai, W. L. McCabe, and R. W. Rousseau, *AIChE J.* **21,** 351 (1975).
20. C. Y. Tai, J.-F. Wu, and R. W. Rousseau, *J. Cryst. Growth* **116,** 294 (1992).
21. J. W. Mullin and C. L. Leci, *Phil. Mag.* **19,** 1075 (1969).
22. P. M. McMahon, K. A. Berglund, and M. A. Larson in S. J. Jančić and E. J. deJong, eds., *Proceedings of the 9th Symposium on Industrial Crystallization*, Elsevier, Amsterdam, The Netherlands, 1984, p. 229.
23. R. M. Ginde and A. S. Myerson, *J. Cryst. Growth* **116,** 41 (1992).
24. J. Garside, *AIChE Symp. Ser. No. 240*, **80,** 23 (1984).
25. R. F. Strickland-Constable, *Kinetics and Mechanism of Crystallization*, Academic Press, Inc., New York, 1968.
26. M. Ohara and R. C. Reid, *Modeling Crystal Growth Rates from Solution*, Prentice-Hall Press, Englewood Cliffs, N.J., 1973.
27. W. K. Burton, N. Cabrera, and F. C. Frank, *Philos. Trans. R. Soc. Lond.* **243**(A866), 299 (1951).
28. J. R. Bourne, *AIChE Symp. Ser. No. 193*, **76,** 59 (1980).
29. W. L. McCabe, *Ind. Eng. Chem.* **21,** 30 (1929).
30. E. T. White and P. G. Wright, *CEP Symp. Ser.* **67**(110), 81 (1971).
31. J. Garside and S. J. Jančić, *Chem. Eng. Sci.* **33,** 1623 (1978).
32. A. H. Janse and E. J. de Jong in E. J. de Jong and S. J. Jančić, eds., *Industrial Crystallization '78*, North-Holland, Amsterdam, The Netherlands, 1979, p. 135.
33. A. H. Janse and E. J. de Jong in J. W. Mullin, ed., *Industrial Crystallization*, Plenum Press, New York, 1976, p. 145.
34. M. W. Girolami and R. W. Rousseau, *AIChE J.* **31,** 1821 (1985).
35. K. A. Berglund and M. A. Larson, *AIChE Symp. Ser. No. 215*, **78,** 9 (1982).
36. A. D. Randolph and E. T. White, *Chem. Eng. Sci.* **32,** 1067 (1977).
37. C. Y. Lui, H. S. Tsuei, and G. R. Youngquist, *Chem. Eng. Prog. Symp. Ser.* **67**(110), 43 (1971).
38. H. J. Human, W. J. P. van Enckevork, and P. Bennema in J. J. Jančić and E. J. de Jong, eds., *Industrial Crystallization 81*, North-Holland, Amsterdam, The Netherlands, 1982, p. 387.
39. K. A. Berglund, E. L. Kaufman, and M. A. Larson, *AIChE J.* **25,** 867 (1983).
40. K. E. Blem and K. A. Ramanarayanan, *AIChE J.* **33,** 677 (1987).
41. R. C. Zumstein and R. W. Rousseau, *AIChE J.* **33,** 121 (1987).

42. N. A. Clontz, R. T. Johnson, W. L. McCabe, and R. W. Rousseau, *Ind. Eng. Chem. Fundam.* **11,** 368 (1972).
43. R. C. Zumstein and R. W. Rousseau, *Ind. Eng. Chem. Res.* **28,** 334 (1989).
44. S. N. Black, R. J. Davey, and M. Halcrow, *J. Cryst. Growth* **79,** 765 (1986).
45. M. Lahav and L. Leiserowitz in A. Mersmann, ed., *Proceedings of the 11th Symposium on Industrial Crystallization*, 1990, p. 609.
46. G. Henning in Ref. 45, p. 113.
47. M. J. Hounslow, R. L. Ryall, and V. R. Marshall, *AIChE J.* **34,** 1821 (1988).
48. P. Marchal, R. David, J. P. Klein, and J. Villermaux, *Chem. Eng. Sci.* **43,** 59 (1988).
49. R. C. Zumstein and R. W. Rousseau, *Chem. Eng. Sci.* **44,** 2149 (1989).
50. M. W. Girolami and R. W. Rousseau, *J. Cryst. Growth* **71,** 220 (1985).
51. N. Hiquily and C. Laguérie in S. J. Jančić and E. J. deJong, eds., *Industrial Crystallization '84*, Elsevier, Amsterdam, The Netherlands, 1984, p. 79.
52. C. Balarew in Ref. 38.
53. S. N. Black and R. J. Davey, *J. Cryst. Growth* **90,** 136 (1988).
54. K. G. Denbigh and E. T. White, *Chem. Eng. Sci.* **21,** 739 (1966).
55. D. D. Edie and D. J. Kirwan, *Ind. Eng. Chem. Fundam.* **12,** 100 (1973).
56. A. S. Myerson and D. J. Kirwan, *Ind. Eng. Chem. Fundam.* **4,** 414 (1977).
57. A. S. Myerson and D. J. Kirwan, *Ind. Eng. Chem. Fundam.* **4,** 420 (1977).
58. N. Hiquily, C. Laguérie, and J. P. Couderc, *Chem. Eng. J.* **30,** 1 (1985).
59. D. H. Jenkins and H. N. Sinha in Ref. 51.
60. R. C. Zumstein, T. Gambrel, and R. W. Rousseau in A. S. Myerson and K. Toyokura, eds., *Crystallization as a Separations Process*, ACS Symp. Ser. No. **438,** 1990, Washington, D.C., p. 85.
61. R. C. Zumstein and R. W. Rousseau, *AIChE Symp. Ser.* **83** (253), 130 (1987).
62. R. W. Rousseau and R. M. Parks, *Ind. Eng. Chem. Fundam.* **20,** 71 (1981).
63. J. R. Beckman and A. D. Randolph, *AIChE J.* **23,** 510 (1977).
64. R. W. Rousseau and T. R. Howell, *Ind. Eng. Chem. Process Des. Dev.* **21,** 606 (1982).
65. N. S. Tavare, J. Garside, and M. R. Chivate, *Ind. Eng. Chem. Process Des. Dev.* **19,** 653 (1980).
66. A. G. Jones, A. Chianese, and J. W. Mullin, in Ref. 51.
67. G. L. Zipp and A. D. Randolph in J. Nývlt and S. Žáček, eds., *Proceedings of the 10th Symposium on Industrial Crystallization*, Elsevier, New York, 1989, p. 469.
68. R. Bennett in Ref. 3, Chapt. 19.
69. R. W. Rousseau and C. G. Moyers in R. W. Rousseau, ed., *Handbook of Separation Process Technology*, John Wiley & Sons, Inc., New York, 1987, Chapt. 11.
70. K. Wintermantel and G. Wellinhoff in Ref. 45, p. 703.
71. J. Ulrich, Y. Özoguz, and K. Wangnick in B. Biscans, N. Gabas, and C. Laguérie, eds., *Proceedings Crystallization Industrielle et Precipitation*, Lavoisier Technique et Documentation, Paris, 1991.

RONALD W. ROUSSEAU
Georgia Institute of Technology

DESICCANTS

Many substances take up (sorb) water from their surroundings by one or more of a number of different physical or chemical mechanisms. Many common materials have an affinity for water, such as wood, paper, natural fibers, polymers, solvents, or salts. Some materials have sufficient capacity for water or efficiency for drying as well as appropriate physical and chemical properties that they are classed as drying agents, or desiccants. These substances are widely used for removing water from gases, liquids, and solids. Desiccants may be liquids or solids. They may be used repetitively by regenerating the desiccant after use to return it to its active state, or they may be used only once. If the desiccant is used only once, it may last the life of the article being dried or may be discarded when spent. Drying agents are used either in a static (batchwise) or dynamic (continuous or semicontinuous) mode. Their use may be further classified as open system, if fluid flows through the system, or closed system if it does not. Examples of the industrial uses of desiccants, designated as dynamic or static and open- or closed-system applications, are given in Table 1. The list is not all inclusive and ignores various laboratory uses (see DRYING).

Desiccants have varied fundamental characteristics in terms of water capacity and the rate of water sorption. The degree of water removal achieved, or efficiency, is usually given in terms of the water content remaining in the substance that has been dried. This water content can be expressed in several ways, such as humidity ratio, relative humidity (at atmospheric pressure only), relative saturation (at elevated pressure or in liquids), dew point (used at any temperature), ice or frost point (used below 0°C), or parts per million (ppm) by weight or volume. The effectiveness of any drying agent can be measured in terms of its water capacity. In static applications, this capacity is usually the true equilibrium capacity. In dynamic systems, the rate of water removal must be taken into account. Usually, to allow for mass-transfer zones, an additional amount of drying agent is used (see MASS TRANSFER). In these instances the dynamic capacity, also termed breakthrough capacity, falls short of the true equilibrium capacity (1).

Table 1. Applications of Desiccants

Industry	Application	Classification
compressed air	prevent freeze-up and corrosion in air-actuated components	dynamic, open
air separation	prevent ice formation in heat exchangers before cryogenic distillation	dynamic, open
natural gas	prevent corrosion and hydrate formation in pipelines, remove water before cryogenic hydrocarbon recovery, dry liquefied petroleum gas (LPG) to prevent freeze-ups during vaporization	dynamic, open
petrochemical	remove moisture before low temperature fractionation, remove moisture during the rejuvenation or burnoff of spent catalysts, prevent side reactions during catalytic refining	dynamic, open
chemical	remove water that is a diluent or contaminant of some finished product, remove water prior to or during polymerization reactions, prevent caking and corrosion	static or dynamic, closed or open
storage and shipping	prevent food deterioration and corrosion of equipment by relative humidity control	static or dynamic, closed or open
moisture vapor control	lower the dew point in sealed spaces where condensation could occur	static, closed
vapor compression refrigeration	remove moisture from circulating refrigerants	dynamic, closed
space cooling	dry air to permit cooling by evaporation of water	dynamic, open
dehumidification	dry ambient air for air-conditioning or for storage, manufacture, or drying of moisture-sensitive parts or materials	dynamic, open
corrosion control	reduce the dew point in automobile exhaust systems during cool down to reduce internal cold condensate corrosion	static, semiclosed
absorption refrigeration	cyclic absorption and stripping of water with liquid desiccants to produce chilled water for air conditioning and process cooling	dynamic, open

Mechanism

The drying mechanisms of desiccants may be classified as follows: Class 1: chemical reaction, which forms either a new compound or a hydrate; Class 2: physical absorption with constant relative humidity or vapor pressure (solid + water + saturated solution); Class 3: physical absorption with variable relative humidity or vapor pressure (solid or liquid + water + diluted solution); and Class 4: physical adsorption.

These mechanisms are characterized by the relative magnitudes of the heats of reaction, solution, or adsorption (qv). All useful drying mechanisms are exothermic. Phosphorus pentoxide is a Class 1 drying agent that reacts with water to form a polyphosphoric acid (2):

$$x\ P_2O_5 + (x+1)\ H_2O(g) \longrightarrow HO\text{-}(HPO_3)_{2x}\text{-}H \quad \Delta H = 109.5\ \text{kJ/mol (26.17 kcal/mol)} \quad (1)$$

Class 1 drying agents (and zeolites, which are Class 4 desiccants) liberate the largest amounts of heat and should be used with appropriate care. Calcium chloride is a Class 1 drying agent that reacts with water to form a hydrate:

$$CaCl_2 + H_2O(g) \longrightarrow CaCl_2 \cdot H_2O \quad \Delta H = 70\ \text{kJ/mol (16.7 kcal/mol)} \quad (2)$$

The more highly complexed hydrates of calcium chloride ($CaCl_2 \cdot nH_2O$ where $n > 2$) may also exhibit the characteristics of a Class 2 drying agent, because the hydrated species can physically absorb additional water to form a saturated solution. The term absorption is used to describe the phenomenon that occurs when a gas or vapor penetrates the solid structure to produce a saturated solution:

$$CaCl_2 \cdot 6H_2O + n\ H_2O(g) \longrightarrow \text{saturated solution} \quad (3)$$

Because calcium chloride has a number of hydrates, the one that is in equilibrium with a saturated solution is a function of the temperature. In this case, the solid is dissolved as it absorbs water to form the saturated solution, and three phases are present: solid, saturated solution, and vapor. Systems having these three phases, or two solids and a vapor phase, have a constant vapor pressure at a given temperature. Therefore, Class 2 drying agents can be used to maintain a constant relative humidity.

Ethylene glycol is an example of a Class 3 drying agent. Because the solution produced is unsaturated only two phases, solution and vapor, exist:

$$\text{ethylene glycol} + n\ H_2O(g) \longrightarrow \text{dilute solution} \quad (4)$$

For this system, the vapor pressure is a function of both temperature and the concentration of water in the dilute solution.

Molecular sieve zeolites are an example of a Class 4 drying agent (see MOLECULAR SIEVES). Water is removed by physical adsorption, but at no time does the adsorbent change phase or dissolve.

$$\text{molecular sieve adsorbent} + H_2O(g) \longrightarrow H_2O\ (\text{adsorbed}) \quad (5)$$

Adsorption (qv) is a phenomenon in which molecules in a fluid phase spontaneously concentrate on a solid surface without any chemical change. The adsorbed molecules are bound to the surface by weak interactions between the solid and gas, similar to condensation (van der Waals) forces. Because adsorption is a surface phenomenon, all practical adsorbents possess large surface areas relative to their mass.

Desiccants can lose water capacity and drying efficiency by taking up moisture during storage. They should therefore be analyzed before use. If necessary, the materials should be reactivated (regenerated) before putting them in service.

Compatibility

Desiccants must be chemically compatible with the material being dried. Ideally, the desiccant and the material should not react because such a reaction may produce harmful or undesirable by-products. For example, in the refrigeration and air-conditioning industry, new ozone-safe refrigerants are being proposed and tested to replace the chlorofluorocarbons in vapor compression refrigeration systems. The desiccants that keep these fluids dry must not react appreciably with the refrigerants or lubricants in the systems. Compatibility tests are conducted in which the desiccant and fluid are contacted under pressure often in the presence of system materials, such as lubricants and metals (3). The mixture is aged for various periods, typically seven days or more, at temperatures above those expected at the desiccant location in the refrigeration system to increase the reaction rates. Following the exposure, the desiccant is analyzed for degradation products and retention of water capacity and physical properties.

Static Drying

Many liquids are dried batchwise rather than continuously. The drying agent is added to the liquid and sufficient time is allowed to dry the product. The liquid is then separated from the drying agent by filtration, decantation, or distillation. Drying agents employing Class 1 or 2 mechanisms are generally used for these applications.

Desiccants Used in Static Drying. The most commonly used desiccants are discussed in this section. Activated alumina (qv), silica gel, and molecular sieves (qv), discussed under dynamic, solid drying agents, are also widely used in static or batch-drying situation.

Barium Oxide. Barium oxide [1304-28-5] (Class 1, nonregenerative) is used primarily as a laboratory drying agent (4). It is the only drying agent that continues to dry even at red heat. Barium oxide is relatively expensive and cannot be regenerated by conventional methods. Therefore, it is not used extensively in commercial applications.

Calcium Chloride. Calcium chloride [10043-52-4], $CaCl_2$ (Class 1 or 2, regenerative), can be either a solid or liquid drying agent (4). Its principal advantage is that its cost is low enough to permit discarding after use in small units. Commercial anhydrous calcium chloride is available in a range of compositions from $CaCl_2 \cdot 0.05H_2O$ to $CaCl_2 \cdot 0.25H_2O$. Figure 1 is a phase diagram for calcium chloride.

Calcium Oxide. Also called lime or quicklime (4,5), calcium oxide [1305-78-8], CaO (Class 1, nonregenerative), is relatively inexpensive. It is prepared by

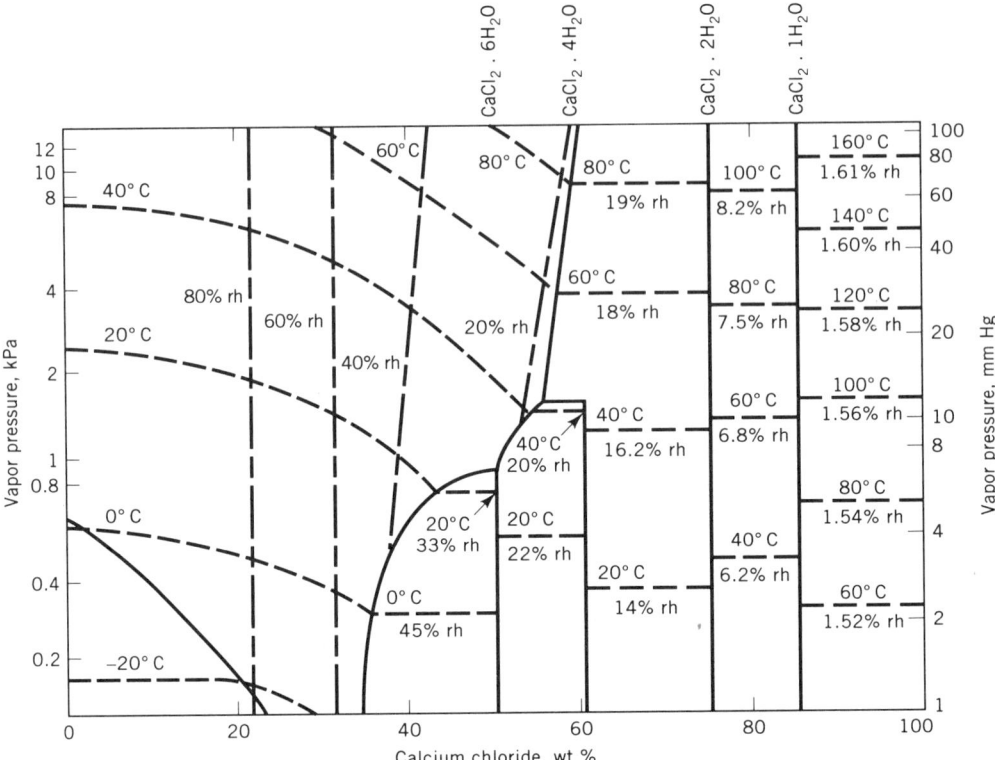

Fig. 1. Vapor pressure and relative humidity over $CaCl_2$ solutions and solids. The straight horizontal lines in the right-hand portion represent two solid phases and a gas phase for vertical line intersections. In addition, a solid phase, saturated solution, and a vapor phase occur in the regions between the vertical lines. The lower left-hand corner shows the ice solution line. The region in between, with skewed isothermal lines, represents unsaturated solutions; the vapor pressure varies as a function of temperature.

roasting calcium carbonate (limestone) and is available in a soft and a hard form according to the way in which it was burned. For desiccant service, soft-burned lime should always be used. Calcium oxide is most commonly used to dehydrate liquids and is most efficient when it can be heated to speed the reaction rate. The reaction product is calcium hydroxide, which crumbles as it picks up moisture.

Calcium Sulfate. Calcium sulfate [7778-18-9] (Class 1, regenerative) is sold under the trade name Drierite (4,6). It occurs in nature in the anhydrous form, $CaSO_4$, and in the hydrated form, $CaSO_4 \cdot 2H_2O$, commonly known as gypsum. When prepared properly, small capillaries form within the granules and increase the somewhat low water capacity of the material by sorbing additional moisture. The first stages of water removal occur by adsorption and a chemical reaction to form a hemihydrate [10034-76-1], $CaSO_4 \cdot \frac{1}{2}H_2O$. The material can be regenerated repeatedly by heating to about 200°C. However, above 300°C, it loses some of its desiccating power. Calcium sulfate is used extensively because it is chemically inert to most materials, reusable, and inexpensive.

Lithium Chloride. Of the metal halides, calcium bromide [7789-41-5], $CaBr_2$, zinc chloride [7646-85-7] $ZnCl_2$, $CaCl_2$, and lithium chloride [7447-41-8] LiCl,

(Class 1, nonregenerative) are the most effective for water removal (4). All are available in the form of deliquescent crystals. The hydrates of LiCl are LiCl·nH_2O, where n = 1, 2, or 3. Lithium chloride solutions are more stable in air and less corrosive than the other metal halides. The high solubility of lithium carbonate [554-13-2], Li_2CO_3, usually eliminates scale formation problems.

Perchlorates. The three common perchlorates (Class 1, nonregenerative) used as drying agents are barium perchlorate [13465-95-7], $Ba(ClO_4)_2$, lithium perchlorate [7791-03-9], $LiClO_4$, and magnesium perchlorate [10034-81-8] $Mg(ClO_4)_2$. The last is the most efficient with drying action above 100°C. Even the higher hydrate form has good drying capacity. Perchlorates are strong oxidizing agents and should never be used in the presence of organic compounds because the mixture is highly explosive. For this reason, perchlorates are usually not regenerated.

Phosphorus Pentoxide. The compound P_2O_5 (Class 1, nonregenerative) is made by burning phosphorus in dry air. It removes water first by adsorption (qv), followed by the formation of several forms of phosphoric acid (2). Phosphorus pentoxide [1314-56-3] has a high vapor pressure and should only be used below 100°C. Its main drawback is that as moisture is taken up, the surface of the granules becomes wetted and further moisture removal is impeded. For this reason, phosphorus petoxide is sometimes mixed with an inert material.

Sodium and Potassium Hydroxides. Sodium hydroxide [1310-73-2] and potassium hydroxide [1310-58-3] (Class 1, nonregenerative) are commonly used when moisture and carbon dioxide or hydrogen sulfide must be removed simultaneously (4). Fused sticks or solutions of the alkali hydroxides are frequently used. These materials must be handled with care to prevent serious skin burns.

Capacity and Efficiency. Figure 2 shows the drying capacity of selected desiccants as a function of relative humidity. The higher capacity desiccants go through the various hydrate levels to yield fairly broad ranges of constant relative humidities as moisture is picked up. However, these compounds do not produce very low relative humidities or dew points. Figure 3 shows the water-vapor pressure over several desiccants as a function of temperature. The addition of more desiccant lowers the vapor pressure. The best performance, or lowest dew point, occurs with excess drying agent. The minimum dew points attainable in air at room temperature and atmospheric pressure are given in Table 2.

The efficiency ranking of desiccants in drying air is not always the same as that observed in drying other materials. Other materials may interact with the desiccants to reduce drying effectiveness. From a study of the efficiency of some 25 desiccants for drying several families of laboratory solvents and reagents it was concluded that molecular sieves are the desiccants of choice in most cases (9–17).

Closed-System Drying. Equilibrium capacity is the principal consideration in the design of closed nonregenerative, relatively static drying systems. The total amount of moisture to be removed must first be calculated from the volume of the system and the initial water concentration. Depending on the final moisture content desired, a drying agent can be selected based on its compatibility with the material to be dried and its ability to produce the final dew point. The equilibrium capacity of the drying agent must be determined at the system temperature and the final water concentration. An amount of drying agent must be used so that

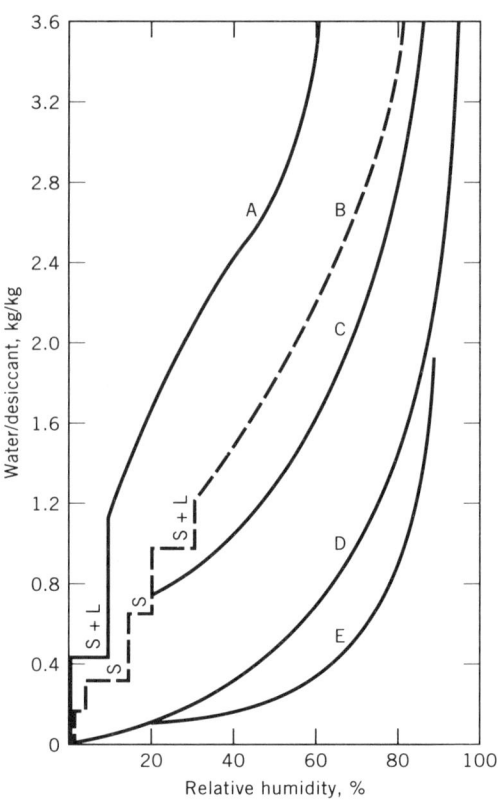

Fig. 2. Drying capacity of selected drying agents in liquid form. A represents lithium chloride; B, calcium chloride; C, sulfuric acid; D, glycerol; E, triethylene glycol. S is solid, L is liquid.

Table 2. Performance of Chemical Desiccants in Drying Air at Room Temperature

Substance	Minimum humidity[a]	Minimum dew point at 101.3 kPa,[b] °C
P_2O_5	0.0193	−98
BaO	0.503	−80.5
KOH fused	1.546	−73
CaO	2.32	−70.5
H_2SO_4	2.32	−70.5
$CaSO_4$ anhyd.	3.87	−67
Al_2O_3	3.87	−67
KOH sticks	10.83	−60
NaOH fused	123.7	−40
$CaBr_2$	139.2	−39
$CaCl_2$ fused	262.9	−33
NaOH sticks	618.6	−25
$Ba(ClO_4)_2$	634.0	−24.5
$ZnCl_2$	657.2	−24
$ZnBr_2$	896.9	−21
$CaCl_2$ granular	1159.8	−18

[a] mg H_2O/kg dry air.
[b] To convert kPa to mm Hg, multiply by 7.5.

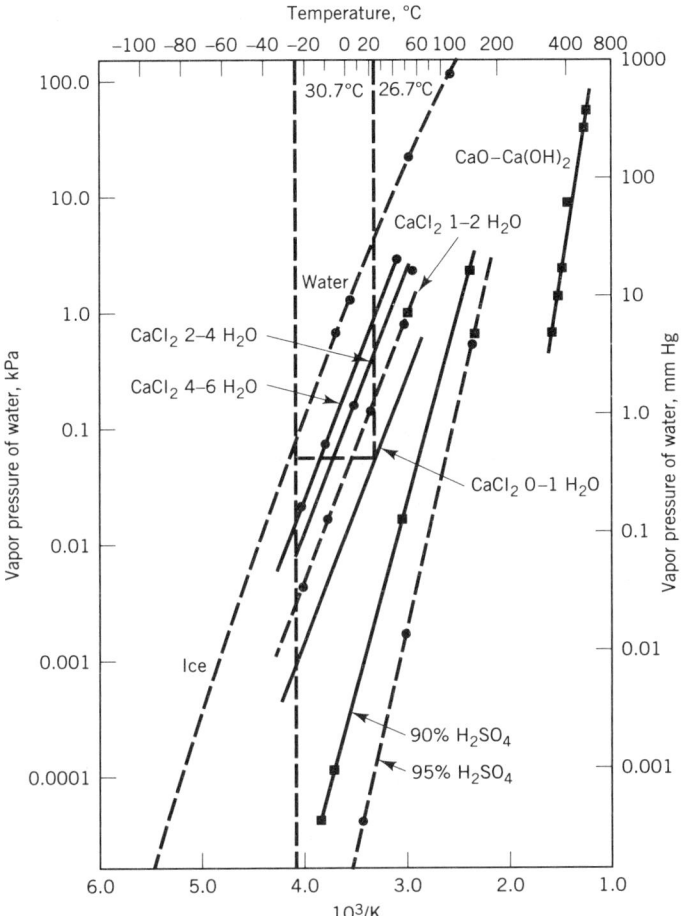

Fig. 3. Water-vapor pressure over ice and several drying agents.

the total amount of water to be removed does not exceed the predetermined equilibrium capacity of the agent. The agent and the material are then placed in intimate contact. After sufficient time, the system comes into equilibrium with the drying agent.

Although equilibrium capacity is the prime concern, few of these closed systems are truly static. Even closed systems have dynamic features or other non-steady-state aspects, such as temperature fluctuations, moisture ingression, and drydown rates.

Closed systems are usually nonregenerative: the desiccant charge is designed for the life of the system or is replaceable. Often in a nominally closed (static) drying system, the total amount of moisture to be removed consists of both the water initially in the system and that which leaks into the system over its lifetime. In multipane (insulating glass) windows, for example, molecular sieve desiccant is used to keep the air (or gas-filled) space dry to prevent condensation in cold weather. The glass panes are typically separated by an aluminum spacer, which is a hollow channel partially filled with the desiccant. The glass and spacer are joined and sealed with organic sealants. Although a small amount of desiccant is required to dry the gas initially sealed into the space, a much larger amount is typically used to adsorb moisture that is slowly transmitted into the space from the outside. The transmission of water vapor through the sealant is driven by the difference in water partial pressure between the desiccated space and the ambient air. Thus to specify the mass of desiccant needed requires knowledge of the moisture vapor transmission rate and specification of the design life of the unit as well as the moisture content of the gas initially sealed into the space.

Because the system likely is nonisothermal, the analysis of a closed-desiccant system requires knowledge of the temperature of the desiccant as well as the dew point (ice point) or water concentration (partial pressure) specification. Indeed, the whole system may undergo periodic temperature transients that may complicate the analysis. For example, in dual-pane windows the desiccant temperature is approximately the average of the indoor and outdoor temperatures after a night of cooling. However, after a day in the sun, the desiccant temperature becomes much warmer than the outdoor temperature. When the sun sets, the outdoor pane cools quickly while the desiccant is still quite warm. The appropriate desiccant for such an application must have sufficient water capacity and produce satisfactory dew points at the highest temperatures experienced by the desiccant.

Another aspect to consider in the design of closed-drying systems is the drydown time. The drydown time is the period required for the system to dry down from its initial water concentration (or partial pressure) to a concentration that approaches equilibrium with the desiccant. During this time, the system is not fully protected from the negative effects of the moisture that the desiccant is designed to remove. In such a system, the instantaneous drying rate is proportional to the water content at any time (18).

The drying rate is represented by differential equation (eq. 6) where h is mass transfer coefficient $1/(h \cdot cm^2)$; A, specific surface area of desiccant beads, cm^2/g; m, mass of desiccant, g; C, concentration by weight of water in the fluid being dried; C', concentration of water at the surface of the desiccant, ie, concentration of water in the fluid that would be in equilibrium with the instantaneous loading on the desiccant, wt·ppm; and t = time, h.

$$\frac{dC}{dt} = -hAm(C - C') \qquad (6)$$

If C' is approximately constant during the initial drydown period, as it is in many closed-system applications, then the water concentration decays exponentially with time. The rate equation can be integrated to give the relationship between water concentration and time:

$$\frac{C}{C_0} = \exp[-hAM(t - t_0)] \qquad (7)$$

where C_0 is the water concentration at the initial time t_0.

If required, the drydown can be hastened by increasing desiccant mass, particle surface area, or mass-transfer coefficient. The mass-transfer coefficient can be altered to some extent by the design of the desiccant container.

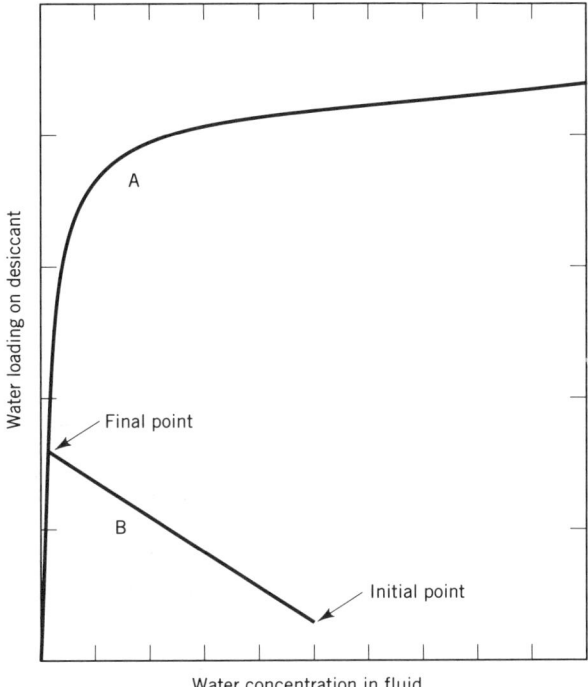

Fig. 4. Drydown path to equilibrium in a closed system. A represents the equilibrium curve and B, the material balance line.

Because a material balance on water must be satisfied during the drydown as well as afterward, the path from the initial concentration to equilibrium can be represented graphically by a material balance line and an equilibrium curve. The coordinates of the starting point on the material balance line are the initial water contents of the fluid to be dried and the desiccant. The slope of the line is the ratio of fluid mass to desiccant mass. The line terminates at its intersection with the equilibrium curve (Fig. 4).

An interesting and novel use of a solid desiccant, the reduction of cold condensate corrosion in automotive exhaust systems, illustrates a hybrid closed–open system. Internal corrosion occurs in mufflers when the water vapor in the exhaust condenses after the engine is turned off and the muffler cools. Carbon dioxide dissolves in the condensate to form an acidic soup. In an essentially closed static drying step, an acid- and heat-resistant desiccant located in the muffler adsorbs water vapor from the exhaust gas as it cools to prevent formation of corrosive acidic condensate. When the engine is restarted, the system becomes open, and the desiccant is regenerated by the hot exhaust gas to be ready for the next cooldown step (19).

Dynamic Desiccants

Continuous drying is employed when drying a volume of gas or liquid in a batchwise fashion is not practical. When a solid, dynamic desiccant is used, the fluid stream is passed over a fixed bed of the drying agent, which must have the physical properties to allow the fluid to pass readily through. When liquid desiccants are used, the drying is usually achieved by countercurrent contact of the gas (flowing up) against the liquid (flowing down). The desiccants are usually regenerable.

Liquid Desiccants. Glycols and sulfuric acid [7664-93-9] are the principal examples.

Sulfuric Acid. This compound, H_2SO_4 (Class 3, regenerative) is used extensively throughout the chemical industry to dry acidic and corrosive gases. It has good capacity and drying capability as illustrated by the vapor pressure curves in Figure 5. At 25°C, the dew point attainable in gases dried with 95% sulfuric acid is less than −75°C.

Sulfuric acid is used in circulating towers like those depicted in Figure 6; the gas flows countercurrent to the acid. The spent acid is removed from the primary contactor at a 50% H_2SO_4 content and may be used for chemical manufacturing or recycled after reconcentration. The makeup, or recycled, acid is usually introduced at 93% H_2SO_4 (sp gr 1.84, 66° Bé or higher) because this moisture content establishes the final moisture level of the dried product gas. Sulfuric acid is highly corrosive, and protective clothing and eye protection must be provided.

Glycerol, Glycol, and Other Polyhydric Alcohols. The alcohols (Class 3, regenerative) widely used to dry gases (20, 21), can only produce dew points in the

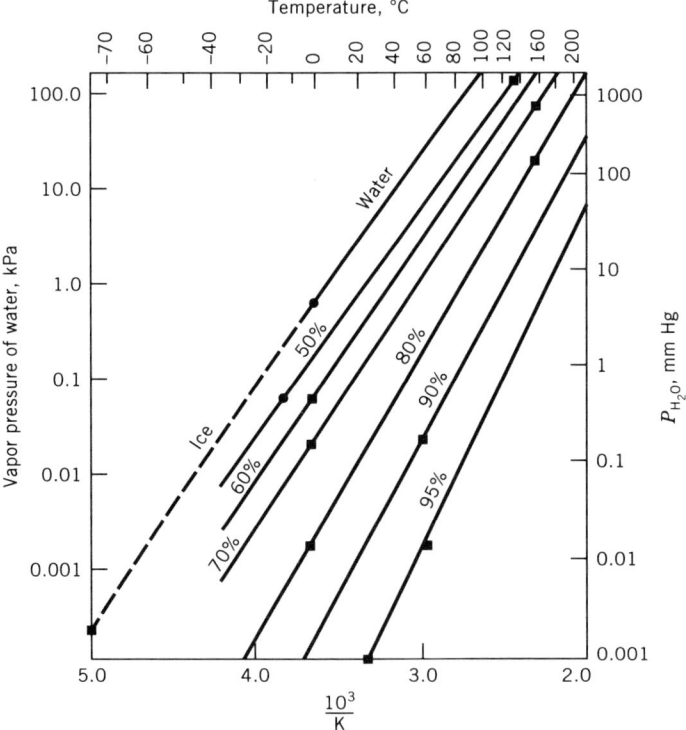

Fig. 5. Vapor pressure of water over sulfuric acid solutions. Percentage of H_2SO_4 noted on each curve.

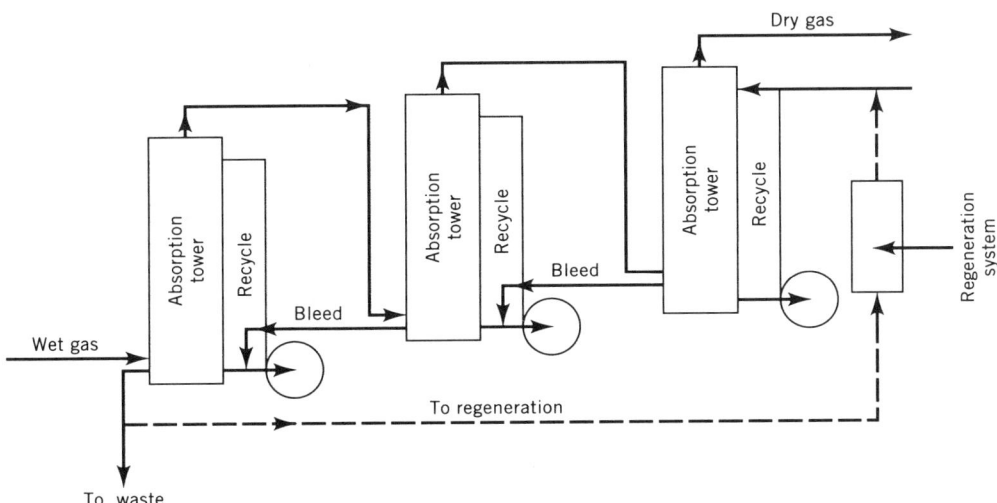

Fig. 6. Continuous sulfuric acid drying system.

range of −15 to 0°C. Whereas these compounds have somewhat lower capacity than sulfuric acid (Fig. 2), they are effective when either injected or employed in a multistage contactor to achieve dew point depression.

Ethylene glycol [107-21-1], diethylene glycol [111-46-6], and triethylene glycol [112-27-6] are used extensively in the natural gas industry to inhibit hydrate formation (22,23). Diethylene glycol (DEG) is the most widely used. The dew point that can be achieved in the treated gas (24,25) is a function of the percentage of glycol present and the contact temperature (Fig. 7). The principal advantages of glycol dehydration are low cost, ease of regeneration, and minimal losses of the drying agent caused by solubility or vapor pressure during subsequent recovery or reclamation.

A typical flow diagram of a glycol dehydration plant is shown in Figure 8. The absorber, or gas–liquid contactor, operates with glycol flowing downward, countercurrent to the gas stream. The regenerator, or stripping column, operates at low pressures to keep the column temperature below the decomposition temperature of the glycol. The regenerated glycol is then recirculated to the absorber.

Solid Desiccants. The solid desiccants used in dynamic applications fall into a class called adsorbents (see ADSORPTION). Because these are used in large packed beds through which the gas or liquid to be treated is passed, the adsorbents are formed into solid shapes that allow them to withstand the static (fluid plus solid head) and dynamic (pressure drop) forces imposed on them. The most common shapes are granules, extruded pellets, and beads.

Activated Alumina. This material (Class 4, regenerative) is made by the calcination of an alumina gel or aluminum oxide trihydrate [21645-51-2], $Al_2O_3 \cdot 3H_2O$, into various crystalline phases of transition aluminas (26). Depending on the manufacturing procedure and starting material, the final product has different degrees of specific surface and pore volume. Table 3 lists the physical properties of a typical Grade A activated alumina.

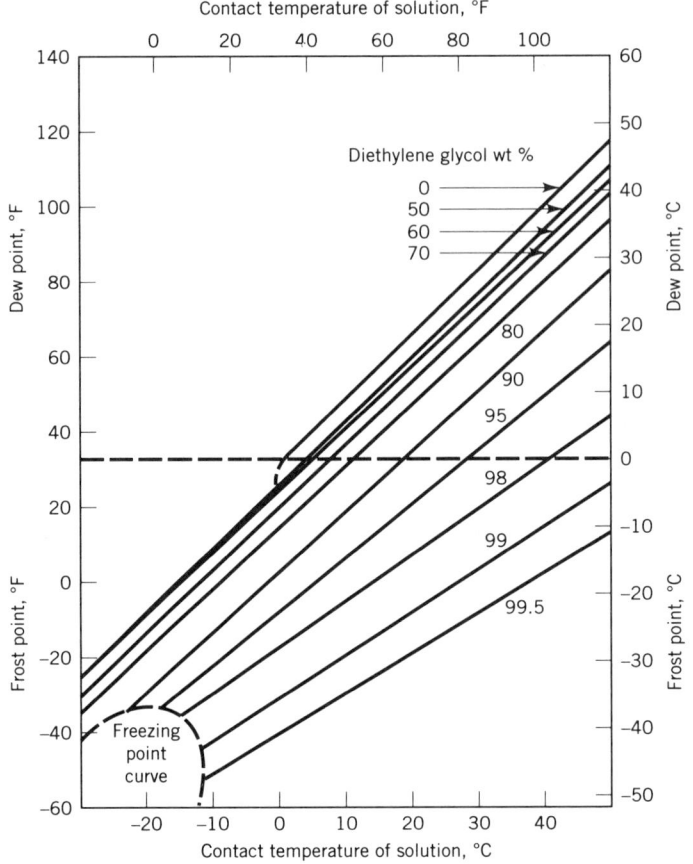

Fig. 7. Dew points of aqueous diethylene glycol solutions.

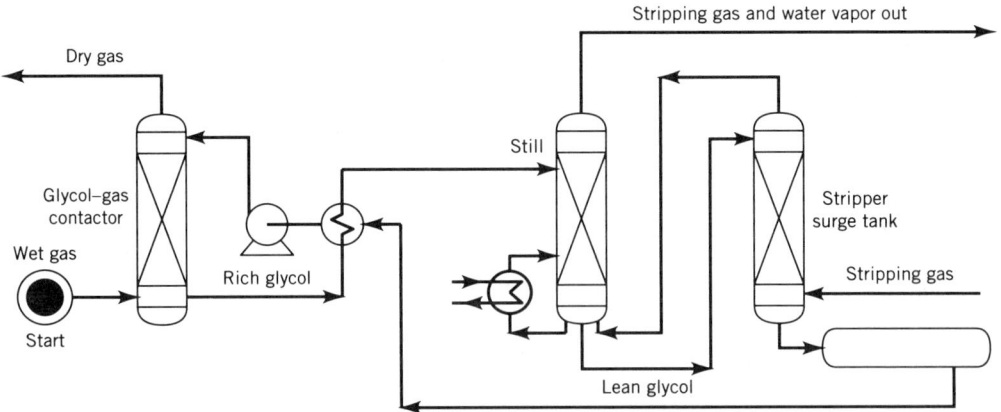

Fig. 8. Natural gas diethylene glycol dehyration system.

Table 3. Physical Properties of Solid Adsorbents

Property	Grade A activated alumina	Silica gel	Type 4A molecular sieve
surface area, m²/g	320	832	750
bulk density, kg/m³	800	720	670
maximum heat of adsorption, J/g H$_2$O[a]	ca 1395	ca 930	ca 4180
specific heat, J/(kg·K)[a]	1005	921	1046
reactivation temperature, °C	150–315	125–275	200–315
pore volume, % of total	ca 50	ca 55	ca 48
pore size, nm	1–7.5	1–40	0.42
pore volume, cm³/100 g	40	43	28.9

[a]To convert J to cal, divide by 4.184.

The water removal mechanism is adsorption, which is the mechanism for all Class 4 drying agents. The capacity of such materials is often shown in the form of adsorption isotherms as depicted in Figures 9a and 9b. The initial adsorption mechanism at low concentrations of water is believed to occur by monolayer coverage of water on the adsorption sites. As more water is adsorbed, successive layers are added until condensation or capillary action takes place at water sat-

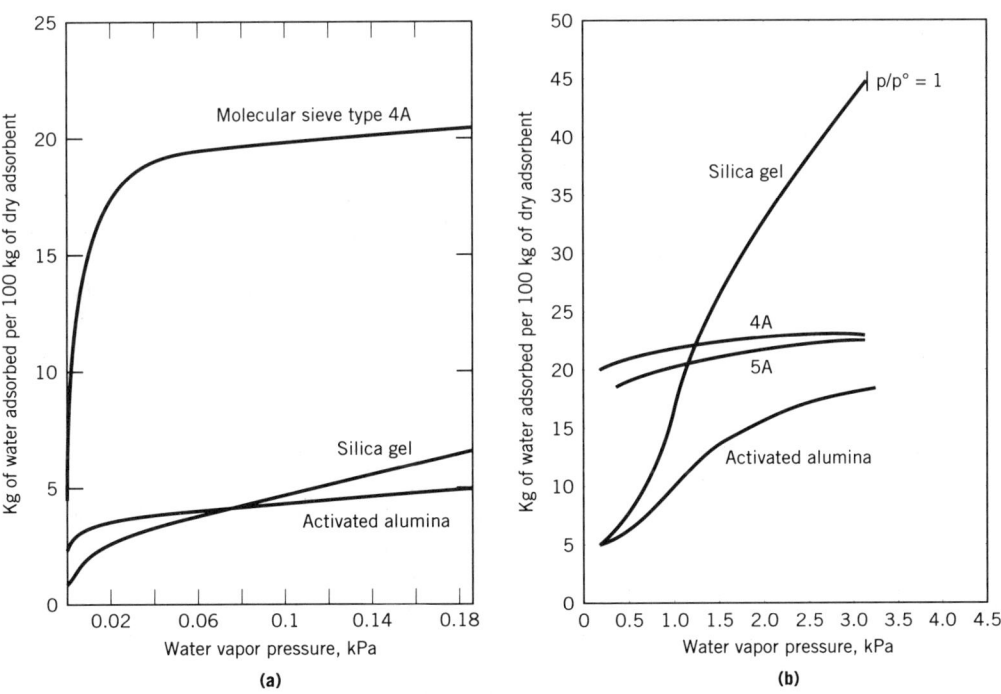

Fig. 9. Water adsorption isotherms at 25°C (to convert kPa to mm Hg, multiply by 7.5). 4A and 5A designate molecular sieves.

uration levels greater than about 70% relative humidity. At saturation, all the pores are filled; and the total amount of water adsorbed, expressed as a liquid, represents the pore volume of the adsorbent.

In regenerating activated alumina, heating to a temperature of 150 to 200°C is sufficient to recover nearly all of the initial water capacity. This regeneration is usually accomplished by passing a heated gas through the adsorbent bed. Unless the surface is fouled by the gas or liquid being dried, the life of activated alumina (qv) is good and usually depends on the number of regeneration cycles.

Silica Gel. Silica gel [7631-86-9] (Class 4, regenerative) is made by dehydration of high purity silica hydrogel (27). The final product is high in purity (99.7% SiO_2), which contributes to its chemical inertness. The typical physical properties are listed in Table 3. The pore size of silica gel has the broadest range of the three primary solid desiccants. Therefore, it can adsorb larger molecules in addition to water (critical diameter = 0.265 nm). The average pore diameter is about 2.5 nm.

The capacity of silica gel is shown in Figures 9**a** and 9**b**, and the shape of the isotherm is similar to activated alumina. At saturation ($p/p° = 1.0$), silica gel, which takes up 40 kg H_2O/100 kg of adsorbent, has the highest capacity of the desiccants shown. However, some high capacity silica gels tend to shatter in the presence of liquid water. When liquid water may be present, a lower capacity, water-resistant silica gel must be used.

The normal regeneration temperature for silica gel is 175°C. In hydrocarbon service, higher temperatures (225–275°C) are recommended to desorb heavy hydrocarbons, which tend to foul the adsorbent during prolonged use.

Molecular Sieves. Molecular sieve desiccants (Class 4, regenerative) are members of a class of materials called zeolites (28). Although some occur in nature, commercial molecular sieve zeolites are usually synthetic (see MOLECULAR SIEVES). They are crystalline framework aluminosilicates containing alkali metal cations. The structure extends in three dimensions by a network of AlO_4 and SiO_4 tetrahedra linked to one another by the sharing of oxygen atoms. Molecular sieves possess the high porosity that is characteristic of all adsorbents. In addition, the ordered crystalline structure of the molecular sieve provides pores of a constant size. In contrast, the pores in activated alumina and silica gel are nonuniform in size, as shown in Table 3. Silica gel is an amorphous material with no crystal structure. The pore structure in activated alumina is not contained within the alumina crystals but is formed by the spaces between randomly agglomerated crystals.

The pore size of molecular sieves can be enlarged or diminished by appropriate cation exchange. Many commercial types are available with pore openings ranging from 0.3 nm to about 1.0 nm. For example, type 3A (the potassium form of zeolite A) has a nominal pore opening of 0.3 nm (3 Å) and type 4A (sodium form) has a nominal 0.4 nm opening. Type 5A (calcium form) with a 0.5 nm opening and type 13X (sodium form of zeolite X) with an 0.85 nm opening are also available.

All these forms adsorb water molecules. The constant size and adjustable pore opening permit the exclusion of many other gaseous and liquid molecules from the internal pore structure, hence the name molecular sieve. This feature provides unique advantages in certain applications. Fouling of the adsorption surface by compounds with high molecular weights can be prevented by excluding them. Molecular sieves with larger pores can also be used simultaneously to dry and purify process streams, eg, by the adsorption of carbon dioxide or sulfur compounds in addition to water.

Because of their ordered structure, molecular sieves have high capacity at low water concentrations and do not exhibit a capillary condensation pore-filling mechanism at high water concentrations. The desiccating properties of the material are still good at elevated temperatures (Fig. 10). A dew point of $-75°C$ can be obtained in a gas dried at 90°C with a molecular sieve that adsorbs water to the level of 1 wt %. In normal operations at ambient temperature, dew points of $<-100°C$ have been measured.

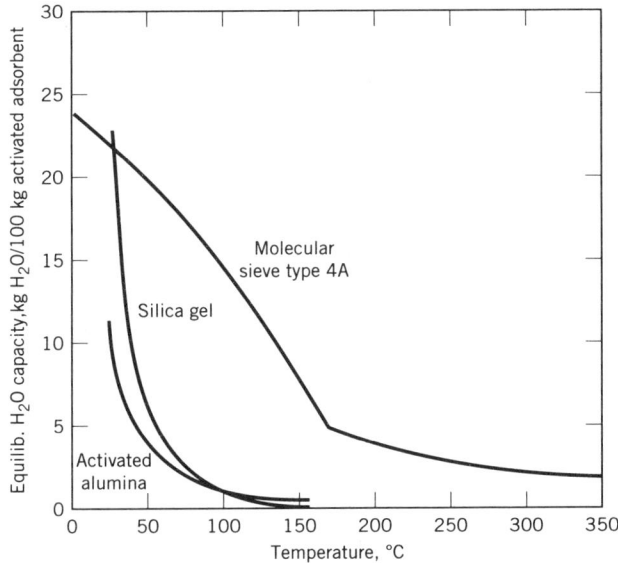

Fig. 10. Adsorbent isobars (p_{H_2O} = 1.38 kPa). To convert kPa to atm, divide by 101.3.

Molecular sieves are also inert to most fluids and are physically stable when wetted with water. Strong inorganic acids or alkalies and temperatures above 700°C should be avoided. Mildly acidic streams can be dried with a molecular sieve of an acid-resistant type (29).

Design of Dynamic Adsorption Drying Systems

Adsorbent drying systems are typically operated in a regenerative mode with an adsorption half-cycle to remove water from the process stream and a desorption half-cycle to remove water from the adsorbent and to prepare it for another adsorption half-cycle (8,30,31). Usually, two beds are employed to allow for continuous processing. In most cases, some residual water remains on the adsorbent after the desorption half-cycle because complete removal is not economically practical. The difference between the amount of water removed during the adsorption and desorption half-cycle is termed the differential loading, which is the working capacity available for dehydration.

The two most common types of drying systems operate on either a pressure-swing cycle or a thermal-swing cycle to take advantage of the difference in water loading on the desiccant with changes in pressure and temperature (Fig. 11) (28). A pressure-swing cycle uses a high pressure adsorption step and a low pressure desorption step and does not require an elevated temperature for regeneration. However, some amount of thermal energy can be added, if desired, during the desorption half-cycle to increase the desorption efficiency. This type of system operates with small differential water loadings. Therefore, short adsorption and desorption times are used (1–60 min). Higher effluent dew points are characteristic of this type of dehydration operation. A thermal-swing cycle requires an elevated temperature during the desorption step. Depressurization can also be used during this step to improve regeneration efficiency. This type of cycle operates with the highest differential loadings, and longer cycle times are normally used (4–24 h). If low dew points are required, a thermal-swing cycle should be employed.

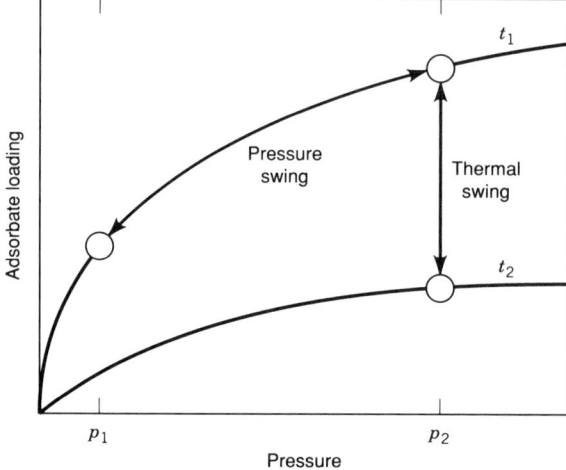

Fig. 11. Schematic illustration of the thermal-swing and pressure-swing cycles. In thermal swing, the differential loading, Δx, is given by: $\Delta x_t = x_{t_1} - x_{t_2}$ at p_2; in pressure swing: $\Delta x_p = x_{p_2} - x_{p_1}$ at t_1.

In many cases a two-bed dehydrator system, with one bed on adsorption and the other on regeneration, can process all the fluid to be dried. Figure 12 is a schematic of a simple two-bed natural gas dehydrator; the same scheme can be used to dry air. Natural gas is first passed through a gas–liquid separator to ensure single-phase operation. The gas passes downflow through a molecular sieve drying tower. The dried product natural gas may then be sent for further processing, eg, cryogenic hydrocarbon recovery of liquefied petroleum gas (LPG) condensates. The residue or lean gas from the cold section of such a plant is commonly used for regeneration. The gas is heated and passed through the exhausted adsorbent bed (Tower B) with the flow countercurrent to the gas drying step. Tower B is then cooled to feed temperature by flowing gas around the heater

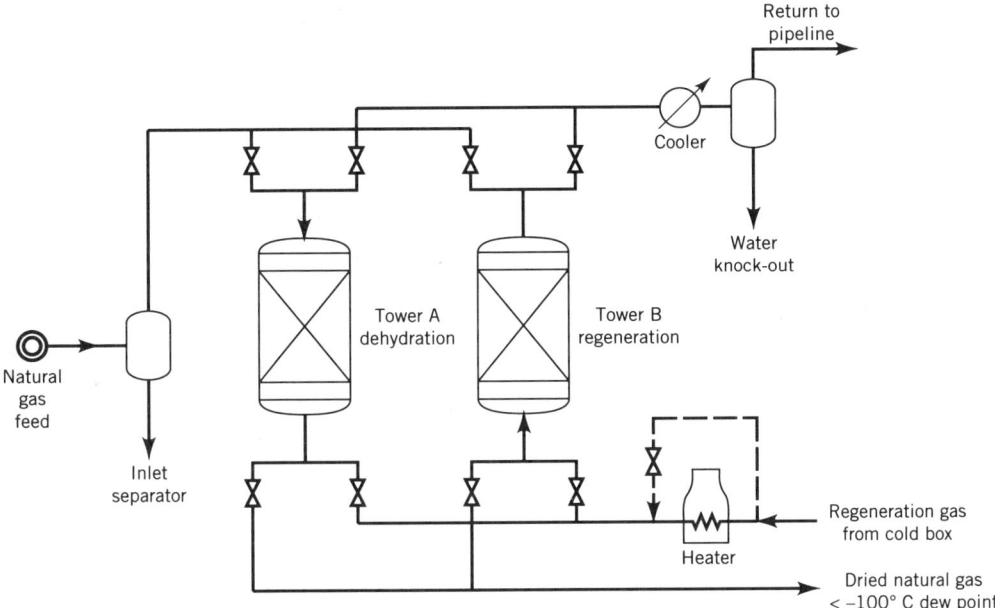

Fig. 12. Molecular sieve natural gas dehydrator.

directly to the bed. The spent regeneration gas then passes through a cooler, and liquid water is removed before it is returned to the pipeline. Depending on the process conditions, the amount of regeneration gas required is about 5–10% of the total amount of gas dried (see CRYOGENIC DISTILLATION).

Single-bed systems can also be used if the demand for drying is intermittent. For example, pressure-swing air dryers installed on heavy trucks usually employ a single desiccant bed to dry compressed air for operating the brakes. When the pressure in the primary air storage reservoir in the system reaches a set point, a blowdown to atmospheric pressure and a purge of the dryer bed using air from a purge reservoir take place (32). A single-bed, thermal-swing system can also be used for drying intermittent flows. Very dry gas is often needed to regenerate or cool adsorbent beds for gas purification (see ADSORPTION, GAS SEPARATION). A single-bed auxiliary dryer can be used for this purpose (33).

Three-bed solid adsorption systems are also used for drying gases. The three beds can be arranged so that two beds are drying and one is in regeneration or one bed is drying and two are in regeneration, one in heating and one in cooling. When two beds are used in drying, they can be in series flow, one bed in the "lead" position and the other in the "lag" or "trim" position, or they can be in parallel flow, where the adsorption fronts are staggered.

In designing a gas drying system, the engineer must often estimate the water content of the gas to be dried. If the gas is air at atmospheric pressure, psychrometric charts are available to give the specific humidity (mass of water per mass of dry air) for given conditions of dry bulb temperature and relative humidity or wet bulb temperature (34,35). Often, however, the gas is under pressure and is known to be saturated with water under given temperature and pressure conditions. In this case, an estimate of the water content can be made by using the vapor pressure of water $p°$ at the given temperature at total pressure π, as reported in the steam tables.

$$y = \frac{p°}{\pi} \qquad (8)$$

Although this estimate is sometimes sufficient, it may well fail at higher pressures (36). Fortunately, several references are available that provide data and estimating procedures for high pressure hydrocarbons (C_1–C_4), natural gas, hydrogen, and nitrogen (37–54).

Adsorption Plots. Isotherm plots are the most common method of presenting adsorption data. An isotherm is a curve of constant temperature: the adsorbed water content of the adsorbent is plotted against the water partial pressure in equilibrium with the adsorbent. An isostere plot shows curves of constant adsorbed water content: the vapor pressure in equilibrium with the adsorbent is plotted against temperature. Figure 13 shows isosteres for the three primary adsorbents described previously. In this case, the dew points for the three adsorbents are plotted at 0.5, 5, and 10 kg H_2O/100 kg of adsorbent. At equilibrium and at a given adsorbed water content, the dew point that can be obtained in the treated fluid is a function only of the adsorbent temperature. The slopes of the isosteres indicate that the capacity of molecular sieves is less temperature sensitive than

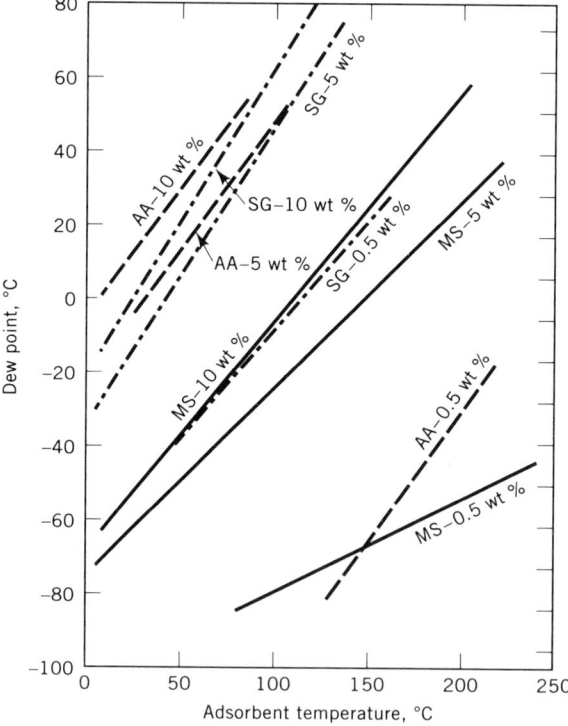

Fig. 13. Adsorbent isosteres for activated alumina (AA), silica gel (SG), and molecular sieves (MS).

that of silica gel or activated alumina. In another type of isostere plot, the natural logarithm of the vapor pressure of water in equilibrium with the desiccant is plotted against the reciprocal of absolute temperature. The slopes of these isosteres are proportional to the isosteric heats of adsorption of water on the desiccant (see ADSORPTION, GAS SEPARATION).

Mass Transfer and Useful Capacity. The term useful capacity, also referred to earlier as breakthrough capacity, differs from the equilibrium capacity shown on Figures 9a and 9b. The useful capacity is a measure of the total moisture taken up by a packed bed of adsorbent at the point where moisture begins to appear in the effluent. Thus the drying process cycle must be stopped before the adsorbent is fully saturated. The portion of the bed that is not saturated to an equilibrium level is called the mass-transfer zone (see MASS TRANSFER).

The parameters affecting the size and shape of a mass-transfer zone are adsorbent type, adsorption isotherm shape, flow rate, packed-bed depth, adsorbent particle size, physical properties of the carrier fluid, temperature, pressure, and the concentration of water in the carrier fluid. For example, as the particle size increases for a given fluid flow rate, the pressure drop decreases. However, as the particle size increases, the mass-transfer resistance increases, and the adsorbent takes longer to achieve its equilibrium water capacity. As a result, larger amounts of desiccant are required. Therefore, the optimal particle size is a compromise between pressure drop (energy consumption) and desiccant utilization. If conditions are chosen that are favorable to mass transfer, eg, small particle size, then the mass-transfer zone is small when compared to the total amount of packed bed employed in drying service. In this case, bed utilization is high and the breakthrough or useful capacity closely approaches the true equilibrium capacity. More often, conditions cannot be optimized on the basis of absorbent needs but are fixed by the needs of the drying process. This situation may dictate unfavorable mass-transfer conditions when pressure drop is limited or practical packed-bed diameters and depths must be employed. Bed utilization is then reduced and the breakthrough capacity falls short of the equilibrium capacity.

Figure 14 depicts the location of the water front in packed beds of adsorbents

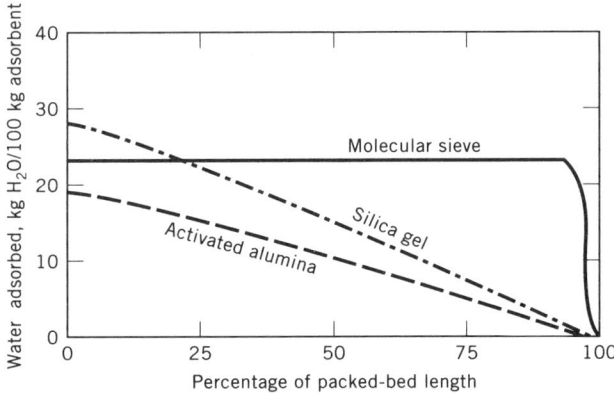

Fig. 14. Position of water front in packed bed of adsorbent during dynamic dehydration. Conditions: 50% rh; 10.2 cm/s air; particle size = ca 0.167 cm; temperature = 25°C; contact time = 1.7 s.

at a short but typical contact time for dehydration. The percentage of useful capacity as compared to equilibrium capacity is about 50% for activated alumina and silica gel and more than 90% for the molecular sieve. The reason for this difference in performance lies in the shape of the isotherm. Under the Brunauer classification, the molecular sieve has a type I isotherm shape, which is favorable for adsorption mass transfer (see ADSORPTION). This isotherm produces a self-sharpening adsorption front, or constant pattern behavior. Silica gel and activated alumina have non-type I isotherms, which are unfavorable for adsorption mass transfer. Their isotherms produce an expanding adsorption front or proportional pattern behavior. Thus the highest equilibrium capacity adsorbent may not always result in the highest useful capacity in a dynamic system. If higher moisture contents can be tolerated in the effluent, then a larger fraction of the mass-transfer zone can be allowed to leak through into the effluent stream. This technique improves bed utilization at the expense of an increase in the effluent dew point.

Economic Aspects

Cost-Effectiveness of Desiccants. The cost-effectiveness of a regenerable Class 4 (physical adsorption) desiccant is a function of the initial investment and the operating cost. The initial investment depends on the desiccant bed weight, which determines the costs of the desiccant and the desiccant vessel. The investment also includes other equipment costs, such as compressors or blowers for gas-phase systems, and regeneration equipment, such as a regeneration compressor and heater. The principal operating costs are the energy costs associated with thermal-swing regeneration and compression costs associated with fluid pressure drop across the bed.

The required desiccant weight is a function of several factors: the water removal requirements (mass/time), the cycle time, the equilibrium loading of water on the desiccant at the feed conditions, the residual water loading on the desiccant after regeneration, and the size of the mass-transfer zone of the desiccant bed. These factors, in turn, depend on the flow rate, temperature, pressure, and water content of both the fluid being dried and the regeneration fluid (see ADSORPTION, GAS SEPARATION).

Shorter cycle times produce smaller bed sizes. The minimum cycle time is usually dictated by the minimum regeneration time required to heat and cool the bed. Systems of greater than two beds provide some flexibility in regeneration time but add to investment costs.

Operating costs consist of compression costs for drying gases and the energy costs for thermally regenerated systems. Labor is also a cost, although most systems are designed to run automatically with a minimum of operator attention. Compression costs are a function of bed size and design and desiccant particle characteristics as well as flow rate, pressure, and temperature. Regeneration costs in thermal-swing systems include the sensible heating of the desiccant and vessel from the feed temperature to the regeneration temperature and the heat of adsorption of water on the desiccant. Some designs with two beds in regeneration (one in heating and one in cooling) allow for recovery of sensible heat.

Desiccant replacement is another operating cost, although desiccant life is

Table 4. Applications of Primary Class 4 Desiccants[a]

	Desiccants, t		
Application	Molecular sieve	Silica gel	Activated alumina
natural gas processing	5500	1200	230
insulating glass	3900	1300	
refrigeration	3900	320	2300
packaging	small	1400	
air and inert gas drying	180	410	3400
alkylation	small		2700

[a]Estimated 1990 U.S. usage.

usually several years in regenerable applications. Desiccant life depends on the stability of the desiccant in the given service, the frequency of regeneration, the presence of reactive contaminants, and the possibility of upsets to normal operation.

Market Data. The largest U.S. manufacturer of molecular sieves for adsorbent and desiccant use is UOP, which has a production capacity of 18–20 million kg/year. W.R. Grace and Zeochem have about 7 and 2 million kg/year capacity, respectively (55). W.R. Grace is the largest producer of silica gel desiccants. Activated alumina for use as adsorbent and desiccant is produced by LaRoche Chemicals (formerly Kaiser) and by Aluminum Company of America. About one-third of the U.S. supply of activated alumina adsorbent and desiccant is imported by Rhône-Poulenc.

The largest users of molecular sieve desiccants are the natural gas processing, insulating glass, and refrigeration industries (Table 4). Much smaller quantities of silica gel are used in these applications. Silica gel dominates the packaging industry, where the material is used to protect electronic equipment and pharmaceuticals, for example, from moisture. Silica gel is also used in dehumidification of buildings. The principal uses of activated alumina are in refrigeration, where its primary function is to adsorb organic acids rather than water, air drying, and alkylation feed stream drying in oil refineries. The data in Table 4 are estimates of 1990 U.S. usage and are considered the best available (56).

Future Applications

Energy Storage. Reactivating a desiccant stores the reactivation energy in the dehydrated desiccant. This energy-storage feature is useful if the energy source is intermittent or seasonal, such as solar energy, or interruptable. Research suggests that this energy-storage feature is especially useful if the desiccant is used to dry air for agricultural applications (57).

Desiccant Cooling. Considerable work is now being done in the field of desiccant-based cooling systems. In these systems, a desiccant is used to produce an extremely dry air stream. The dry air is then cooled in a heat exchanger and humidified with a water spray. The evaporation of the water absorbs heat from the air and produces a cold, almost saturated air stream for air conditioning. The desiccant is thermally regenerated with exhaust air, which is heated with solar energy, natural gas, or waste heat from power generation. Both liquid (58) and solid desiccant systems have been studied. The solid desiccant systems typically employ a wheel, which rotates a bed of desiccant particles through sectors in which drying, thermal regeneration, and cooling take place (59). These systems are extensions of existing dehumidification technology, with a strong emphasis placed on minimizing the thermal regeneration requirements. To this end, extensive basic studies have been carried out of solid desiccants and the properties of the hypothetical desiccant ideally suited to low temperature thermal regeneration. Research shows that the shape of the desiccant isotherm and the thermal properties of the desiccant, such as heat capacity, are equally important in desiccant selection (60). Organic polymers as well as inorganic materials are being considered as desiccant materials (61).

BIBLIOGRAPHY

"Drying Agents" in the *Encyclopedia of Chemical Technology,* 1st ed., Vol. 5, pp. 266–276, by J. F. Skelly, The M.W. Kellogg Co.; in *ECT* 2nd ed., Vol. 7, pp. 378–398, by B. K. Beecher, Wyandotte Chemicals Corp.; in *ECT* 3rd ed., Vol. 8, pp. 114–130, by J. P. Ausikaitis, Union Carbide Corp.; "Desiccants" in *ECT* 4th ed., Vol. 7, pp. 1031–1055, by A. P. Cohen, UOP.

1. J. J. Collins, *Chem. Eng. Prog-Sym. Ser.* **63**(74), 31 (1967).
2. A. D. F. Toy and E. N. Walsh, *Phosphorus Chemistry in Everyday Living*, 2nd ed., American Chemical Society, Washington, D.C., 1987, pp. 233–235.
3. A. P. Cohen, "Test Methods for the Compatibility of Desiccants with Alternative Refrigerants," *ASHRAE Transactions*, Vol. 99, Part 1, American Society of Heating, Refrigerating and Air-Conditioning Engineers, Atlanta, Ga., 1993.
4. O. A. Hougen and F. W. Dodge, *The Drying of Gases*, J. W. Edwards, Ann Arbor, Mich., 1947, p. 30.
5. R. S. Boynton, *Chemistry and Technology of Lime and Limestone*, John Wiley & Sons, Inc., New York, 1966.
6. *Drierite Product Catalog*, W. A. Hammond Drierite Co., Xenia, Ohio.
7. M. Shepherd, "Relative Efficiencies of Drying Agents," *International Critical Tables*, Vol. 3, McGraw-Hill, Inc., New York, 1928.
8. K. G. Davis and K. D. Manchanda, *Chem. Eng.* **81,** 102 (Sept. 16, 1974).
9. D. R. Burfield, K. H. Lee, and R. H. Smithers, *J. Org. Chem.* **42**(18), 3060–3065 (1977). Part 1 of a series of nine papers. Contains data on efficiency of drying benzene, 1,4-dioxane, and acetonitrile with 4A molecular sieve, Al_2O_3, silica gel, CaH_2, $LiAlH_4$, Na, Na–K alloy, P_2O_5, $CaCl_2$, Na_2SO_4, KOH, $CaSO_4$, and $MgSO_4$.
10. D. R. Burfield, G. H. Gan, and R. H. Smithers, *J. Appl. Chem. Biotechnol.* **28,** 23 (1978). Part 2, efficiency of drying 1,4-dioxane, tetrahydrofuran, acetonitrile, and benzene with 3A and 4A molecular sieves.
11. D. R. Burfield and R. H. Smithers, *J. Org. Chem.* **43**(20), 3966–3968 (1978). Part 3, efficiency of drying acetone, dimethylformamide, dimethylsulfoxide, and hexamethylphosphoric triamide with Al_2O_3, BaO, B_2O_3, CaH_2, CaO, $CaSO_4$, $CuSO_4$, KOH, K_2CO_3, molecular sieves 3A and 4A, Na–K alloy, and P_2O_5.
12. D. R. Burfield and R. H. Smithers, *J. Chem. Tech. Biotechnol.* **30,** 491–496 (1980). Part 4, efficiency of drying 1,4-dioxane, toluene, and dichloromethane with cationic exchange resins.

13. D. R. Burfield, R. H. Smithers, and A. S. C. Tan, *J. Org. Chem.* **46,** 629–631 (1981). Part 5, efficiency of drying pyridine, 2-methylpyridine, 2,6-dimethylpyridine, 2,4,6-trimethylpyridine, triethylamine, diisopropylamine, and 1,3-propanediamine with Al_2O_3, BaO, CaC_2, CaH_2, CaO, $CaSO_4$, KOH, molecular sieves 3A and 4A, Na, and silica gel.
14. D. R. Burfield and R. H. Smithers, *J. Chem. Educ.* **59**(8), 703–704 (1982). Part 6, efficiency of drying water-saturated diethylether with $MgSO_4$, $CaSO_4$, $CaCl_2$, cation-exchange resin, molecular sieves 4A and 5A, Na_2SO_4, K_2CO_3, polyacrylamide, and Al_2O_3.
15. D. R. Burfield, R. H. Smithers, and A. S. C. Tan, *J. Org. Chem.* **48,** 2420–2422 (1983). Part 7, efficiency of drying methanol, ethanol, 2-butanol, *t*-butyl alcohol, and 1,2-ethanediol with molecular sieves 3A and 4A, $MgSO_4$, CaC_2, B_2O_3, BaO, CaO, Mg, Al, and benzene azeotrope.
16. D. R. Burfield, G. T. Hefter, and D. S. P. Koh, *J. Chem. Tech. Biotechnol.* **34A,** 187–194 (1984). Part 8, drying with molecular sieve 3A.
17. D. R. Burfield, *J. Org. Chem.* **49,** 3852–3854 (1984). Part 9, efficiency of drying 1,4-dioxane, water-saturated diethylether, and acetonitrile with $CaSO_4$, $CaCl_2$, and molecular sieves 3A and 4A.
18. A. P. Cohen and S. R. Dunne, "Review of Automotive Air-Conditioning Drydown Rate Studies: The Kinetics of Drying Refrigerant 12," *ASHRAE Transactions*, Vol. 93, Part 2, American Society of Heating, Refrigerating, and Air-Conditioning Engineers, Atlanta, Ga., 1987.
19. S. R. Dunne, *Automotive Corrosion and Prevention Conference Proceedings*, Society of Automotive Engineers, Warrendale, Pa., 1989, pp. 165–173.
20. I. Mellan, *Polyhydric Alcohols*, Spartan Books, Washington, D.C., 1962.
21. G. O. Curme, Jr. and F. Johnston, eds., *Glycols*, American Chemical Society Monograph No. 114, Rheinhold Publishing Corp., New York, 1953.
22. L. D. Polderman, "The Glycols as Hydrate Point Depressants in Natural Gas Systems," *Proceedings of the Gas Conditioning Conference*, University of Oklahoma, Norman, 1958.
23. "Triethylene Glycol," Technical Bulletin F-49191A-ICD, Union Carbide Corp., Danbury, Conn., 1989.
24. "Gas Treating Chemicals," Technical Bulletin F-4 1335C, Union Carbide Chemicals and Plastics Co., Danbury, Conn., 1980.
25. "A Guide to Glycols," Technical Bulletin 117-00991-4/91, The Dow Chemical Company, Midland, Mich., 1981.
26. C. Misra, *Industrial Alumina Chemicals*, ACS Monograph 184, American Chemical Society, Washington, D.C., 1986, pp. 107–120.
27. R. K. Iler, *The Chemistry of Silica*, John Wiley & Sons, Inc., New York, 1979, pp. 462–621.
28. D. W. Breck, *Zeolite Molecular Sieves: Structure, Chemistry, and Use*, Wiley-Interscience, New York, 1974, 715 pp.
29. P. N. Kraychy and A. Masuda, *Oil Gas J.* **64** (1966).
30. D. Basmadjian, *Adv. Drying*, **3,** 307–357 (1984). A review of thermal-swing industrial drying applications with 108 references.
31. A. L. Weiner, *Chem. Eng.* **81,** 92 (1974).
32. J. P. Ausikaitis, "Pneumatic System Air Drying by Pressure Swing Adsorption," *ACS Symposium Series, No. 40, Molecular Sieves-II*, American Chemical Society, Washington, D.C., 1977.
33. U.S. Pat. 4,484,933 (Nov. 27, 1984), A. P. Cohen (to Union Carbide Corp.).
34. *ASHRAE Handbook 1989 Fundamentals*, American Society of Heating, Refrigerating, and Air-Conditioning Engineers, Atlanta, Ga., 1989, Chapt. 6.

35. *Perry's Chemical Engineers' Handbook*, 6th ed., McGraw-Hill, New York, 1989, p. 12-4.
36. S. B. Adler, H. Ozkardesh, and C. F. Spencer, *Oil Gas J.*, 107–113 (Nov. 17, 1980).
37. J. M. Campbell, *Gas Conditioning and Processing*, 3rd ed., Campbell Petroleum Series, Norman, Olka., 1974, 266 pp.
38. S. C. Sharma, "Equilibrium Water Content of Gaseous Mixtures," Ph.D. dissertation, University of Oklahoma, Norman, 1969. Available from University Microfilms, Ann Arbor, Mich., order #69-8601.
39. R. G. Anthony and J. J. McKetta, *J. Chem. Eng. Data* **12**(1), 17 (1967).
40. Ref. 39, p. 21.
41. R. G. Anthony and J. J. McKetta, *Hydrocarbon Process.* **47**(6), 131 (1968).
42. E. P. Bartlett, *J. Am. Chem. Soc.* **49**, 65 (1927).
43. R. Kobayashi and D. L. Katz, *Ind. Eng. Chem.* **45**(2), 440 (1953).
44. C. C. Li and J. J. McKetta, *J. Chem. Eng. Data.* **8**(2), 247 (1963).
45. J. J. McKetta and D. L. Katz, *Ind. Eng. Chem.* **40**(5), 853 (1948).
46. R. H. Olds, B. H. Sage, and W. N. Lacey, *Ind. Eng. Chem.* **34**(10), 1223 (1942).
47. F. Pollitzer and E. Strebel, *Z. Phys. Chem.* **110,** 768 (1924).
48. H. H. Reamer, R. H. Olds, B. H. Sage, and W. N. Lacey, *Ind. Eng. Chem.* **35**(7), 790 (1943).
49. H. H. Reamer, R. H. Olds, B. H. Sage, and W. N. Lacey, *Ind. Eng. Chem.* **36**(4), 381 (1944).
50. M. Rigby and J. M. Prausnitz, *J. Phys. Chem.* **72**(1), 330 (1968).
51. J. N. Robinson, "Estimation of the Water Content of Sour Natural Gas," presented at *Canadian Natural Gas Processing Association*, Calgary, Alberta, Mar. 9, 1978.
52. A. W. Saddington and N. W. Krase, *J. Am. Chem. Soc.* **56,** 353 (1934).
53. A. H. Wehe and J. J. McKetta, *J. Chem. Eng. Data* **6**(2), 167 (1961).
54. J. J. McKetta and A. H. Wehe, *Petrol. Ref.* **37**(8), (1958). Data and estimating procedures for saturated water content of natural gas.
55. M. Smart, "Zeolites," *Chemical Economics Handbook*, SRI International, Menlo Park, Calif., 599.1000, 1992.
56. *Adsorbents & Desiccants, Markets and Technology, 1983–1990*, Oxenham Technology Associates, Inc., New York, 1985.
57. W. M. Miller, *Ener. Ag.* **2**, 341–354 (1983).
58. J. R. Howell, *NATO ASI Ser. Ser E* **129,** 374–387 (1987).
59. L. G. Harriman, III, ed., *The Dehumidification Handbook*, 2nd ed., Munters Cargocaire, Amesbury, Mass., 1990, pp. 3-14 to 3-18.
60. R. K. Collier, Jr., *Advanced Desiccant Materials Assessment, Phase II, Final Report*, Gas Research Institute, GRI 88/0125, Chicago, Jan. 1988.
61. A. W. Czanderna, *Polymers as Advanced Materials for Desiccant Applications: 1987*, Solar Energy Research Institute, Golden, Colo., 1988.

ALAN P. COHEN
UOP

DESULFURIZATION

Cleaning and desulfurization of coal, **465**
Desulfurization of flue gases, **481**

CLEANING AND DESULFURIZATION OF COAL

Coal is a primary source of energy for the United States and is expected to continue to be so into the twenty-first century. However, combustion of raw coal directly in the furnace of an electric power generating plant yields flue gases containing sulfur oxides, SO_x; nitrogen oxides NO_x; and compounds of toxic metals. These materials, which are principally derived from impurities in the coal feed, may be reduced below permissible emission levels by various processes (1). The coal can be cleaned before it is fed to the furnace; the contaminants can be captured in a solid sorbent during, or immediately following, combustion as the hot product gases pass through the boiler; or the cool flue gases can be cleaned after leaving the heat exchange region.

Coal Cleaning

In 1990 coal production in the United States reached 0.9 billion metric tons (2) and worldwide production was estimated to be over four billion metric tons. As early as 1982 it was estimated that at least 50% of the world coal production was cleaned in some manner before use (3). As higher quality coal reserves continue to be depleted and more stringent environmental regulations on pollutants, particularly sulfur oxides, are enacted, this percentage should approach 90–100%.

Impurities. The three categories of potential pollutants in coal are sulfur, nitrogen, and ash. Sulfur and ash are associated with both the mineral and organic portions of coal, whereas nitrogen is mainly associated with the organic matter (4).

Most commercial coals of the eastern United States contain 0.5–4.0 wt % sulfur; most western coals contain less than 1 wt % sulfur. Sulfur is present in coal as sulfate, pyrite, and organic sulfur. Sulfate sulfur is of minor concern as its concentration in coal is much less than 1 wt %. Furthermore, sulfate compounds can be easily removed by washing because of their high solubility in water. No definite relationship between the organic and pyritic sulfur coal contents has been established. In the United States, both the organic and pyritic sulfur content in raw coal may vary from 20–80% of the total sulfur. Theoretically, organic sulfur cannot be removed from coal unless chemical bonds are broken or the organic sulfur compound is extracted. Thus the amount of organic sulfur present sets the lowest limit to which a coal can be cleaned by physical methods. The pyritic particles may be macroscopic or microscopic in size. For some coals, pyritic sulfur is finely dispersed in the coal matrix and removal of this form of sulfur by physical cleaning methods can be difficult.

Nitrogen, unlike pyritic sulfur, is mostly chemically bound in organic molecules in the coal and therefore not removable by physical cleaning methods. The nitrogen content in most U.S. coals ranges from 0.5–2.0 wt %.

Coal ash is derived from the mineral content of coal upon combustion or utilization. The minerals are present as discrete particles, cavity fillings, and aggregates of sulfides, sulfates, chlorides, carbonates, hydrates, and/or oxides. The key ash-forming elements and compounds are (4,5):

Minor elements	Trace elements
Pollutant	*Named as hazardous*
sulfur	
nitrogen	beryllium
	fluorine
Ash-forming	arsenic
sodium	selenium
potassium	cadmium
iron	mercury
calcium	lead
magnesium	manganese
silica	copper
alumina	chromium
titania	

Minor elements contribute ≥ 1 wt % to the ash; trace elements contribute ≤0.1 wt %. The degree of de-ashing achievable by physical cleaning depends on the distribution of mineral matter in the coal. In some cases, a considerable amount of the mineral matter can be removed; in other cases, especially where the mineral matter is distributed throughout the coal as microscopic particles, de-ashing by physical cleaning is not practical.

Concern over the release of hazardous trace elements from the burning of coal has been highlighted by the 1990 Clean Air Act Amendments. Most toxic elements are associated with ash-forming minerals in coal (5). As shown in Table 1, levels of many of these toxic metals can be significantly reduced by physical coal cleaning (6).

Conventional Coal Preparation Plants. Coal cleaning (preparation) is based principally on size and density differences, with the exception of flotation

Table 1. Effect of Coal Cleaning on Trace Elements[a]

Coal	Trace element content, ppm							
	Cd	Cr	Cu	F	Hg	Mn	Ni	Pb
feed	3.15	55	25	156	0.20	53	26	18
product[b]	0.05	28	10	71	0.09	7.9	11	3.0
reduction, %	98	49	60	54	55	85	58	83

[a]Upper Freeport Coal, W.Va., 0.075 mm top size, ie, all particles ≤ 0.075 mm.
[b]Float at 1.40 specific gravity.

(qv). In this manner physical impurities, ie, ash and pyrite, may be removed from coal. Four general categories of coal preparation plants can be defined based on levels or degrees of cleaning (7): level 1 involves crushing and screening only; level 2, coarse coal cleaning only; level 3, coarse coal and partial fine coal cleaning; and level 4, total cleaning, ie, all size fractions are cleaned. At each successive level, the process design becomes increasingly more sophisticated.

In a typical modern coal preparation plant, shown in Figure 1, coal is subjected to (1) size reduction and screening; (2) separation of impurities; and (3) dewatering (qv) and drying (qv). Size reduction is accomplished in rotary or roll crushers. More impurities are liberated as the coal size is reduced. Then, coal is screened, either wet or dry, to separate the various size fractions. Before treatment, the crushed raw coal is divided into coarse (>10 mm), intermediate (0.6–10 mm), and fine (<0.6 mm) sizes. Coarse coal is cleaned using one or more pieces of equipment based on gravity separation, such as jigs, or dense-medium baths. The intermediate size coals are usually cleaned using dense-medium baths/cyclones, jigs, concentrating tables, or spirals. Fine size coals can only be effectively treated by nongravimetric washing methods, such as froth flotation.

The product of any wet-separation process must be dewatered or dried depending on the mode of transportation and use. Coarse coal can be easily dewatered by natural drainage using screens. Intermediate size coal is dewatered using sieve bends or centrifuges. For fine coal, dewatering may require not only more complicated mechanical devices, such as centrifuges and vacuum filters, but also thermal drying, to achieve an acceptable moisture content.

In 1992 there were more than 400 physical coal cleaning plants throughout the United States having a total capacity of over 400 million metric tons of raw coal per year. Table 2 shows the types of coal cleaning equipment used in these plants. Historically, jigs are the equipment of choice and remain the most popular device for cleaning coal. The use of dense-medium baths and cyclones has been increasing steadily however, particularly where difficult-to-clean coals are involved and where the relative density differences between coal and refuse are small. Concentrating tables, eg, Deister shaking tables, have been employed by many plants. These provide good separation, especially in removing pyrite from coal. Pneumatic-dry-separation processes are less likely to be used because of the inability to achieve sharp separations, partly because of changes in mining laws requiring large quantities of water to be sprayed on the coal during mining and handling to suppress dust. The efficiency of pneumatic processes is severely impeded by added moisture.

Conventional coal cleaning processes can remove about 50% of pyritic sulfur and 30% of total sulfur. For northern Appalachian region coals it has been shown that a greater sulfur reduction can be achieved by applying physical coal cleaning to finer size coals (Table 3) (8).

Advanced Coal Cleaning Technologies. As the easy-to-remove relatively clean coals are gradually mined out and as fuel specifications become more stringent in order to meet environmental regulations, the need for advanced fine coal cleaning processes has grown. For any fine coal cleaning process, two characteristics tend to dominate. As coal particles are crushed into finer size, the specific surface area increases and the mass of each particle becomes smaller. This leads to the development of surface force-controlled processes or advanced density-

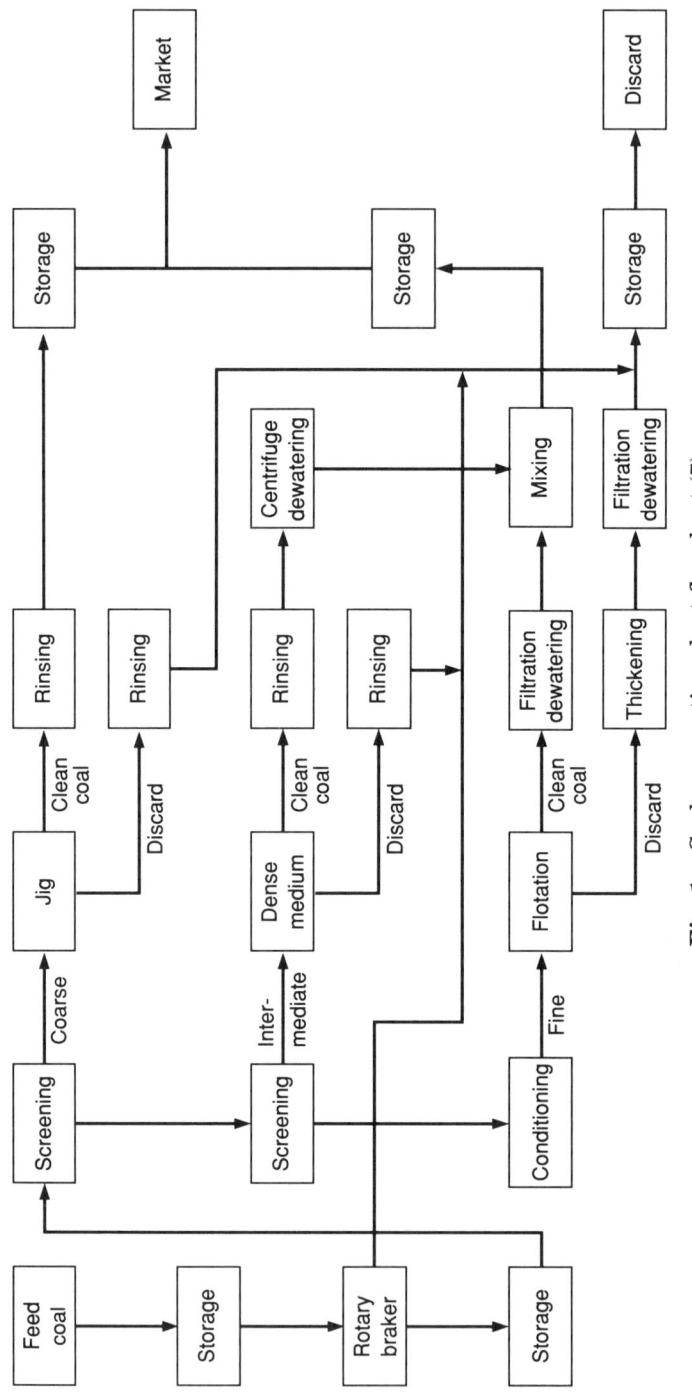

Fig. 1. Coal preparation plant flowsheet (7).

Table 2. Mechanical Cleaning of U.S. Coal by Equipment Type

Equipment	Annual percentage						
	1940	1950	1960	1970	1975	1978	1983[a]
jigs	46.0	47.4	50.0	43.0	46.6	46.6	39.3
dense medium vessels	6.5	14.6	24.3	31.4	32.6	33.2	37.6
concentrating tables	2.3	2.4	11.3	13.6	10.7	10.5	10.5
classifiers[b]	7.6	9.1	4.0	1.1	2.3	2.7	5.7
launders	15.9	5.8	2.8	1.6	1.0	0.6	0.8
others[c]	21.7	20.7	7.0	5.5	2.5	1.9	1.3
flotation			0.6	3.3	4.3	4.5	4.8
Total cleaned, %[d]	*22.2*	*38.5*	*65.7*	*53.6*	*41.2*	*33.8*	*42.6*

[a]1983 is the last year for which these data were collected.
[b]Includes cyclones.
[c]Includes pneumatic separation.
[d]As percentage of total production.

Table 3. Effects of Physical Cleaning on Sulfur Reduction in Coal[a]

Cleaning level	Products, 10^6 t[b]		Sulfur reduction	
	Coal	Sulfur	10^6 t	%
no cleaning	95	2.69		
nominal cleaning, 100% particle top size	89	1.82	0.87	32
3.8 cm[c]	88	1.73	0.96	36
1.0 cm[c]	87	1.50	1.19	44
0.14 cm[d]	85	1.15	1.54	57

[a]Northern Appalachian Region Coals.
[b]Tonnage necessary to produce a heating value of 2.87×10^{12} MJ.
[c]At specific gravity = 1.40.
[d]At specific gravity = 1.3.

based processes that are quite different from the specific gravity-controlled processes found in coarse coal cleaning. In general, advanced processes are capable of producing a deep-cleaned coal product having low ash and low sulfur content. However, as of this writing, most of these processes are in the small-scale demonstration stage and have not been tested on a commercial scale.

Flotation. The application of flotation (qv) to coal cleaning is a relatively new development in the United States. In 1960, only 0.6% of the clean coal produced came from flotation. However, by 1983 flotation accounted for about 5% of the clean coal production (Table 2). Utilization of the flotation process is expected to grow rapidly because more fine size coal is produced as a result of beneficiation schemes that require significant size reduction of the raw coal prior to cleaning to enhance the liberation of pyrite and ash minerals.

The flotation process usually involves three steps: (*1*) the conditioning of the coal surface in a slurry with reagents, (*2*) adhesion of hydrophobic coal particles to gas bubbles, and (3) the separation of the coal-laden bubbles from the slurry. In the conventional flotation process, when the coal particles become attached to air bubbles, the particles are allowed to rise to the top of the flotation cell and form a stable froth layer (9). A mechanical scraper is used to remove the froth layer and separate the clean coal product from the refuse-laden slurry.

Reverse flotation is a two-stage process (10). The first stage is a conventional froth flotation in which most of the high ash refuse and some of the coarser or liberated pyrite are rejected as tailings. The coal froth concentrate and some dilution water is fed to a second stage flotation where a hydrophilic colloid, such as starch or dextrine, is added to depress the coal, followed by a xanthate collector to float the pyrite. This process has been successfully tested at 90 kg/h. The process is specifically designed for pyritic sulfur removal. When tested on a Ohio No. 9 seam coal sample, this process achieved a 93.8% reduction of pyritic sulfur.

Although froth flotation is recognized as the best available fine coal cleaning technique, it becomes ineffective when the particle size is much smaller than 0.1 mm or when the feed contains a large amount of clay, resulting in low coal recovery or poor selectivity. A solution to these problems is the use of modified flotation devices.

The KEN-FLOTE column (11) is one of several column flotation processes based on a countercurrent principle. The feed slurry containing reagents is introduced into the column just below the froth zone. Air is injected at the bottom of the column via an air sparger. Wash water is sprayed within the froth zone to reject the entrained impurities from the froth. Test results on this column indicate that a 6% ash product coal having a combustible-recovery of 75–80% can be obtained. A 70–80% pyrite reduction is also claimed. Figure 2 shows the operation of such a column.

The packed-bed flotation column (12) utilizes a stack of corrugated plates as the packing elements arranged in blocks positioned at right angles to each other. These stacked corrugated plates provide a tortuous flow path to attain intimate particles–bubbles contact and limit impurity entrainment. It also features countercurrent flow of air and pulp in the column. It is reported that less than 1% ash in product coal has been attained for a two-stage cleaning.

Another modification is the use of microbubble column flotation (13). In this process, smaller bubbles are generated to enhance the recovery of micrometer-sized particles. A countercurrent flow of feed slurry is also used to further enhance the bubble–particle attachment. The process is capable of producing ultraclean coals containing less than 0.8% ash.

Similarly, small (0.2–0.6 mm) air bubbles are introduced into a 2.6-m Deister Flotaire column at an intermediate level allowing rapid flotation of readily floatable material in the upper recovery zone. The bottom air permits longer retention time of the harder-to-float particles in the presence of micrometer-sized bubbles at a reduced downward velocity. The first commercial unit went on stream in 1986. It was used to improve the recovery of <0.6 mm (−28 mesh) coal in the plant's tailings. An average of 5.5% increase in coal recovery resulted from its use (14). The second commercial use processed <0.15 mm (−100 mesh) coal feed.

A new flotation cell developed by AFT, Inc. (15) is designed to promote a "skin" flotation process for deep cleaning of fine size coal. In this process plant fines are slurried and then treated with flotation reagents to enhance the natural

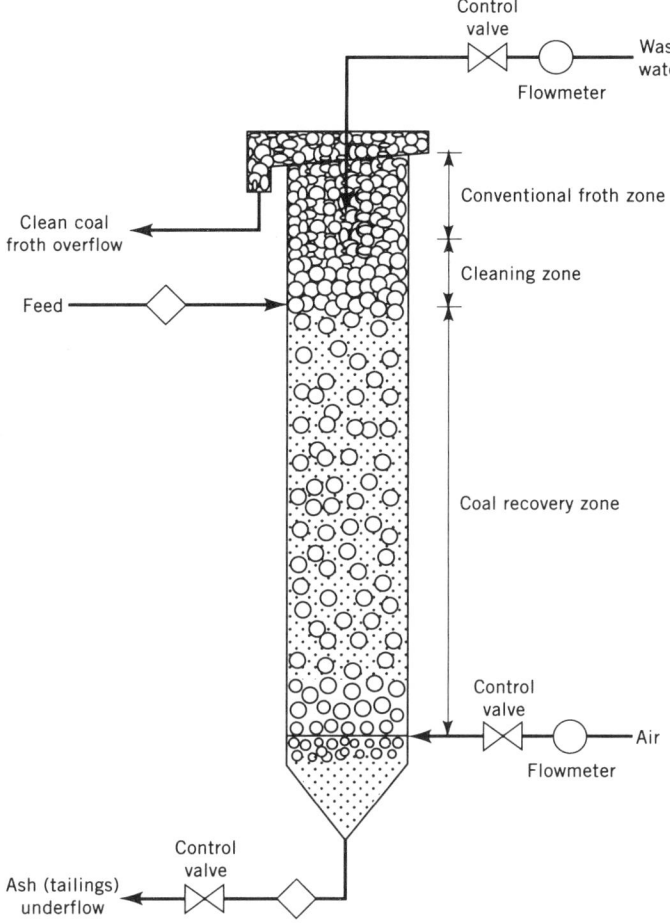

Fig. 2. Diagram of KEN-FLOTE column flotation cell. Courtesy of CAER, University of Kentucky (14).

hydrophobicity of the coal. The conditioned feed slurry is sprayed through nozzles onto the surface of the Sprayflot flotation cell. This operation provides excellent opportunity for air bubble attachment by the air-avid coal particles, which remain on the surface of the cell. The associated hydrophilic impurities remain dispersed in the slurry throughout the cell. The feed to flotation cell contains flocs that form during the conditioning of the feed slurry prior to beneficiation. Such flocs are normally contaminated with entrained mineral matter. The shearing action of the slurry being induced through the spray nozzles is designed to eliminate such flocs and thus provide an improved clean coal product. The froth generated is shown to be drier than a normal flotation froth, thus providing a significant cost advantage during the clean coal dewatering operation.

In 1981, a novel flotation device known as the air-sparged hydrocyclone, shown in Figure 3, was developed (16). In this equipment, a thin film and swirl flotation is accomplished in a centrifugal field, where air sparges through a porous

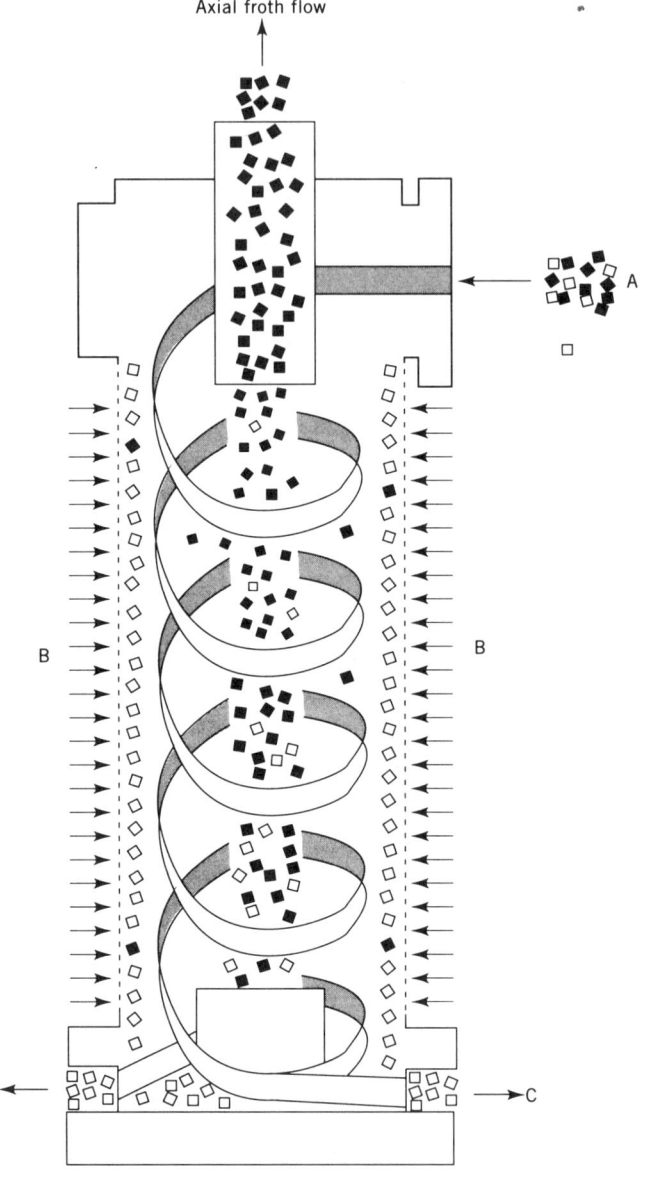

Fig. 3. Air-sparged hydrocyclone, where A represents the tangential feed that establishes swirl flow; B, the area of small bubbles formed by high shear at the porous wall; and C, the outlet for the (□) hydrophilic particles rejected by the swirl flow. The (■) hydrophobic particles are in the axial froth flow. Courtesy of Professor J. D. Miller, University of Utah (16).

wall. Because of the enhanced hydrodynamic condition, separation of fine hydrophobic particles can be readily accomplished. Also, retention times can be reduced to a matter of seconds. Thus, this device provides up to 200 times the throughput of conventional flotation cells at similar yields and product qualities.

Agglomeration-Based Fine Coal Cleaning. Most recently a search for nonaqueous collectors or reagents for fine coal cleaning has been undertaken. A number of liquids have been tested and found to be suitable as agglomeration agents. These include heavy oil, Freon, pentane, hexane, heptane, 2-methylbutane, methyl chloride, and liquid carbon dioxide.

The use of a water-immiscible liquid to separate coal from impurities is based on the principle that the coal surface is hydrophobic and preferentially wetted by the nonaqueous medium whereas the minerals, being hydrophilic, remain suspended in water. Hence, separation of two phases produces a clean coal containing a small amount of a nonaqueous liquid, eg, oil, and an aqueous suspension of the refuse. This process is generally referred to as selective agglomeration.

One of the best known examples is the spherical agglomeration process, developed by the National Research Council of Canada (17,18). This is a two-stage fine coal cleaning process using No. 2 fuel oil as the agglomerant at a rate of about 6% on a dry solid basis. A typical flow diagram is shown in Figure 4. In the first stage, fine coal–oil agglomerates are formed under high shear agitation condition. The formed agglomerates are then fed to the second-stage low shear contactor where the agglomerate size increases. The agglomerated product is in turn separated from fine clay and other high ash material on a Vor-Siv screen. The screen oversize material is further dewatered in high speed screen-bowl centrifuges to about 12% moisture. The product coal is pelletized using a lignin binder prior to storage.

The Otisca-T process (19) is a three-step selective agglomeration process. First, the coal is ground to minus 0.002 mm in a controlled environment. Then, the finely ground coal is agglomerated using a low molecular-weight hydrocarbon such as pentane, leaving the associated pyrite and mineral matter in the water phase. Finally, the agglomerant (pentane) is recovered for reuse. In this process, heating value recovery has been achieved in the range of 93–98% with pyritic sulfur reductions up to 90%. Product coal ash contents less than 1% are routinely obtained.

The LICADO process based on *li*quid *ca*rbon *dio*xide (20,21), is a novel nonaqueous process for fine coal cleaning. Liquid CO_2 is used both as an agglomerant and a transport agent. The process relies on selective agglomeration of coal particles and transport of coal–liquid CO_2 agglomerates from the aqueous phase to the CO_2 phase to achieve the desired separation. It was reported that >90% pyritic sulfur rejection and over 85% clean coal recovery have been achieved. A principal advantage of the LICADO process is that the product coal contains very little moisture and requires no further dewatering. A continuous research unit of the LICADO process is shown in Figure 5.

Heavy-Medium and Heavy-Liquid Cycloning Processes. Heavy-medium cyclones are widely used for cleaning intermediate size coals and application has been extended to fine coal cleaning by using very fine magnetite particles to provide the desired specific gravity of the processing medium. For example, the Micro-

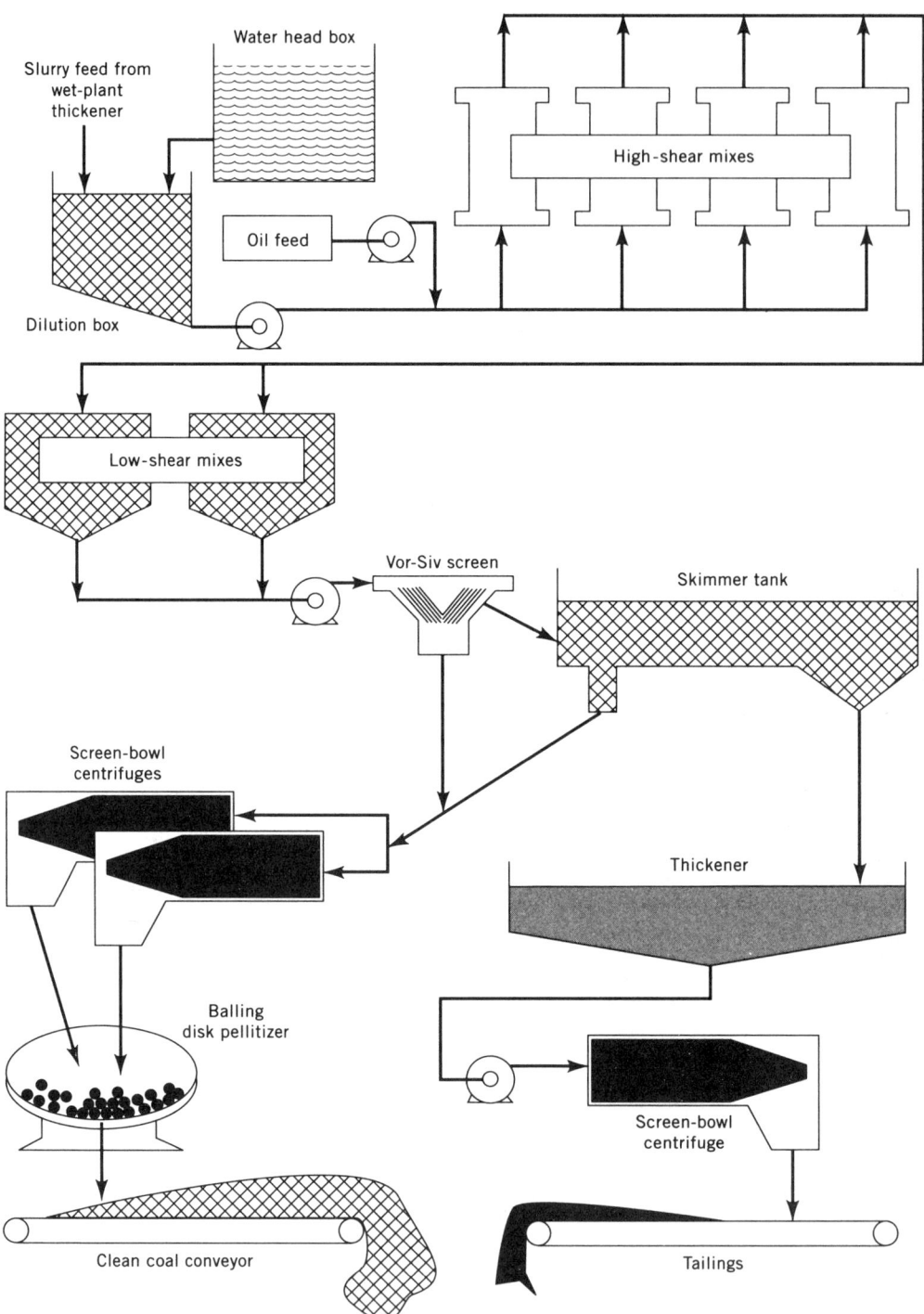

Fig. 4. Flowsheet of Florence Mining Co. oil agglomeration process (7).

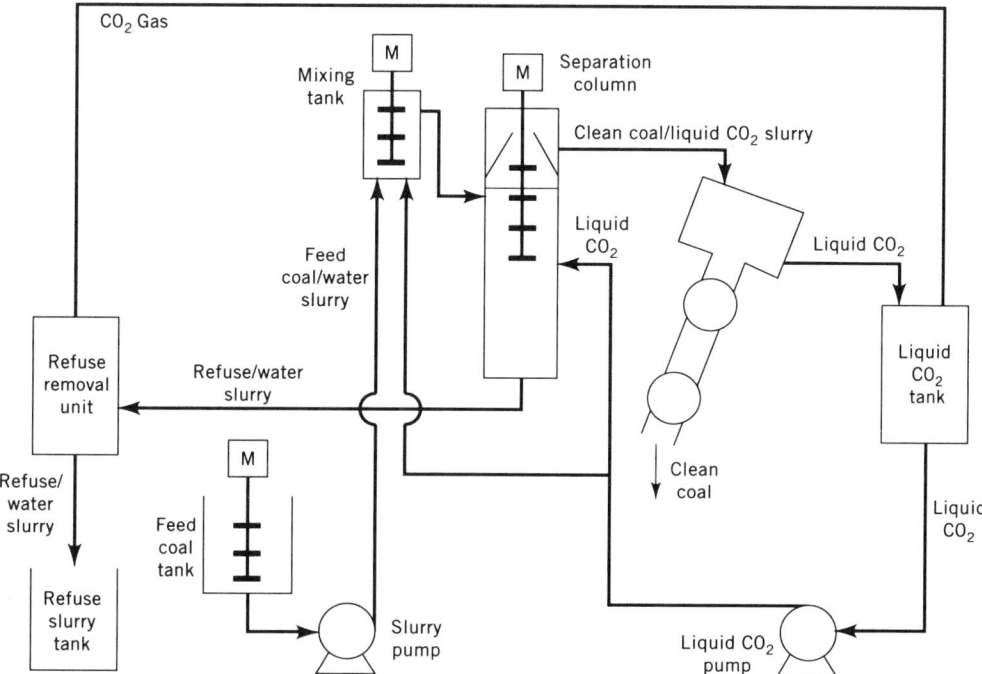

Fig. 5. LICADO continuous research unit where M represents motor-driven mixer (21).

Mag process (22) uses a magnetite medium ground to less than 0.01 mm top size. An alternative method to handle very fine size coal particles is to use heavy-liquids as media for the cyclone separators. Examples of these heavy-liquids include: Freon-113, methylene chloride, sulfuric acid, zinc chloride, calcium chloride, and sugar solutions. All heavy-liquid cycloning processes are capable of effecting separation down to about 0.04 mm (325 mesh) coal particles. Unfortunately, most of these liquids cannot be used in commercial applications because of the potential hazards to workers or the environment and high cost.

Dry Coal Cleaning. Developments in the areas of magnetic and electrostatic separation as a means of cleaning coals in the dry state include high gradient magnetic separation (HGMS), triboelectrostatic separation (TESS), and dry coal purifier (D-CoP).

Several coal cleaning processes have been developed based on the differences in magnetic properties of coal, which is diamagnetic, and pyrite, which is paramagnetic, and some of the ash-forming minerals which are weakly magnetic. The best known example is the HGMS technique which can be used for both dry and wet coal cleaning (23). A magnetic field as strong as 2 Tesla (20 kG) has been employed to capture fine-sized feebly magnetic particles more efficiently. The separation efficiency can be further increased by using a matrix column filled with fine filaments made of ferromagnetic materials, such as stainless steel wool. The HGMS process is usually operated in a cyclic mode. In the first cycle, pyrite and other impurities are captured in the filament matrix. The trapped particles are then flushed out by high pressure air in the next cycle before returning to the first cycle. This technique has been tested on a variety of coals having ash contents in the range of 10–28% and sulfur contents up to 6.5%. The pyrite rejection ranged from 14–94%. The high rejections were usually associated with low clean coal recoveries (see MAGNETIC SEPARATION).

In the TESS process, dry pulverized coal is blown rapidly past a copper baffling device that imparts a positive charge on the coal particles and a negative charge on the pyrite and mineral matter particles (24). The charged particles are immediately introduced into an electrostatic separator where negatively charged plates attract the positively charged coal and positively charged plates attract the pyrite and mineral matter. Ash reductions of up to 93% and pyritic sulfur reductions greater than 95% were obtained in processing minus 0.037 mm coal in a two-stage configuration. However, the yield was as low as 40%.

Another dry coal cleaning process is the dry coal purifier (D-CoP) (25). This technique processes crushed coal in an air-fluidized bed with magnetite particles. When a fluidized bed contains particles of different densities and sizes, there is a tendency at near minimum bubbling conditions for the solids to stratify in the vertical direction according to density, and to a lesser extent, size. In a case of a bed consisting of magnetite particles and crushed coal, the clean fraction of the coal segregates at the top of the bed, the liberated minerals settling toward the bottom. As a consequence, the ash content of the coal at the top of the bed is lowered, thereby permitting recovery of coal having significantly reduced amounts of pyrite and other materials. Because it is a dry process and involves fine coal particles, D-CoP can be integrated directly into a pulverized coal power plant.

Coal Desulfurization

Chemical and Biological Coal Cleaning. Whereas physical coal cleaning is capable of removing most of the ash (mineral matters) and inorganic (pyritic) sulfur, it cannot be used to remove organic sulfur. For most bituminous coals in the United States, 40 to 60% of the sulfur content is bound in organic matter. Organic sulfur can be classified into four types (26): thiols, sulfides, disulfides, and heterocyclic thiophenes, such as dibenzothiophene [*132-65-0*], $C_{12}H_8S$. The sulfur from these compounds can only be removed using chemical or biological methods. Some of these processes have progressed to the miniplant stage. However, most are still at laboratory scale.

Oxidative Desulfurization Process. Oxidative desulfurization of finely ground coal, originally developed by The Chemical Construction Co. (27,28), is achieved by converting the sulfur to a water-soluble form with air oxidation at 150–220°C under 1.5–10.3 MPa (220–1500 psi) pressure. More than 95% of the pyritic sulfur and up to 40% of the organic sulfur can be removed by this process.

The applicability of a nitrogen dioxide oxidative cleaning process for 10 coals on the laboratory scale has been examined via a pilot-plant program (29). In this process, dry pulverized coal is treated with gaseous NO_2, to convert the sulfur to an alkali-soluble form at about 120°C and 342 kPa (496 psi). The coal is then washed using aqueous sodium hydroxide solution to dissolve the sulfur compounds. The waste effluent is treated with lime to regenerate the caustics for reuse and the gypsum formed in the process goes to disposal. The generated gaseous sulfur dioxide and sulfur trioxide are removed by a gas scrubbing system. The process is capable of removing essentially all of the pyrite sulfur and up to 40% of the organic sulfur from a Lower Kittanning coal (29).

Coal Cleaning by Reactive Leaching. During World War II, Germany developed the first chemical coal cleaning technique to produce ultraclean coal (30). A run-of-mine bituminous coal (26 t/h) was cleaned by multistage flotation to less than 1% ash, and then leached using a mixture of 0.4% hydrofluoric acid and 1.4% hydrochloric acid to obtain 0.5% ash coal. This development reached the pilot-plant stage. A flotation product assaying 0.8% ash was cleaned to 0.28% ash and 10% moisture.

More recently, the molten caustic leaching (MCL) process developed by TRW, Inc. has received attention (28,31,32). This process is illustrated in Figure 6. A coal is fed to a rotary kiln to convert both the mineral matter and the sulfur into water- or acid-soluble compounds. The coal cake discharged from the kiln is washed first with water and then with dilute sulfuric acid solution countercurrently. The effluent is treated with lime to precipitate out calcium sulfate, iron

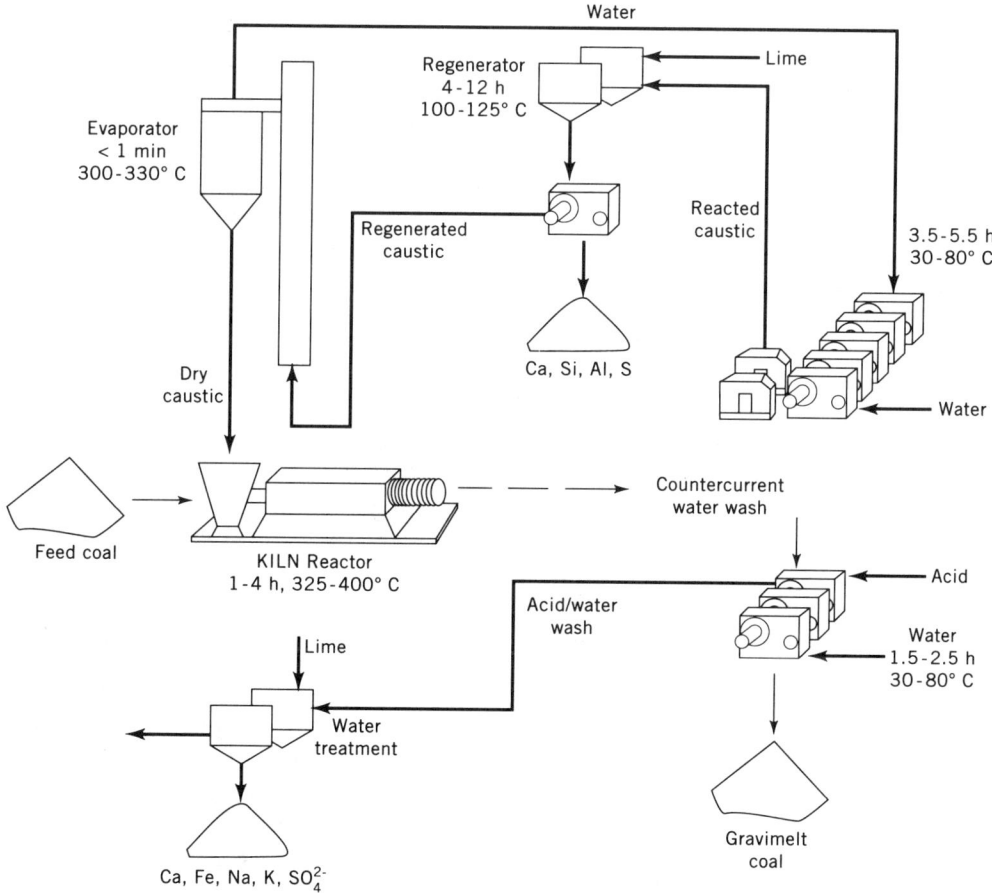

Fig. 6. Flow sheet for TRW's molten caustic leaching process (7).

hydroxide, and sodium–iron hydroxy sulfate. The MCL process can typically produce ultraclean coal having 0.4 to 0.7% sulfur, 0.1 to 0.65% ash, and 25.5 to 14.8 MJ/kg (6100–3500 kcal/kg) from a high sulfur, ie, 4 wt % sulfur and ca 11 wt % ash, coal. The moisture content of the product coal varies from 10 to 50%.

Based on the same principle, several other chemical leaching processes, including the promoted oxidative leaching, wet-oxygen leaching, hydrothermal leaching, and hydrogen peroxide–sulfuric acid leaching processes have been developed and exhibited promising desulfurization characteristics (28).

Microwave Desulfurization. Microwave desulfurization of coal is another modification of the alkali leaching method. A mixture of coal and sodium hydroxide is heated at about 250°C to promote reaction of the sodium hydroxide with the sulfur contained in the coal. The wet-coal cake is then irradiated with microwaves for 25 to 45 seconds under an inert atmosphere. The coal is washed with water and acid to remove soluble sulfide, usually Na_2S, and other solubilized mineral matter. The entire process can be repeated several times to obtain a desired level of sulfur and ash removal. In tests using an Illinois No. 6 coal containing 15.4 wt % ash and 3.4 wt % sulfur, nearly all of the pyritic sulfur, 75% of the organic sulfur, and 87% of the ash were removed, resulting in a clean coal product having less than 2% ash and 0.7% sulfur by weight (33).

Chlorinalysis. Chlorine can be used to remove the sulfur from coal. The coal is contacted with chlorine gas in methylchloroform at about 75°C (34). After separation of the coal from the slurry, the solvent is recovered by distillation. The chlorinated coal is washed with water and finally dechlorinated by heating at about 300–350°C. The process is capable of extracting about 90% of the pyritic sulfur and up to 70% of the organic sulfur from some coals. This process could thus expand the availability of low sulfur solid fuels by making a greater portion of the high organic sulfur coals available as a clean source of energy. However, much research and development is needed to overcome several technical problems, including chlorine retention by the coal, chlorine regeneration, and recycling.

Self-Scrubbing Coal. A novel coal cleaning process marketed by Custom Coal International (35) is designed to produce a self-scrubbing coal. In this process, crushed run-of-mine coal is first cleaned by using a heavy-medium bath to remove noncombustible material, including 90% of the pyritic sulfur content of the coal. Limestone-based additives then are mixed with the beneficiated coal to produce a clean coal product. These additives react with the remaining organic sulfur, which is released during combustion, to remove an additional 70–80% of resulting SO_2 from the coal, effecting a reduction of 80–90% of total sulfur.

Microbial Coal Cleaning. Some of the organic sulfur compounds, such as thiophene and dibenzothiophene (DBT), can be degraded by a variety of microorganisms that often proliferate in petroleum-saturated soil. A strain of *Pseudomonas,* named CB-1, that can convert thiophenic sulfur to sulfate has been isolated (36), as have a number of other microbes including CB-2, which is an *Acinetobacter.* Both CB-1 and CB-2 have been tested in a continuous bench-scale unit. The coal slurry can contain as high as 26 wt % solids. The organisms are grown in continuous fermenters at 25 to 35°C and then fed to the reactor as a thick broth. A process flow sheet is depicted in Figure 7. Organic sulfur removal is in the neighborhood of 25% and the combined use of microbes, either simulta-

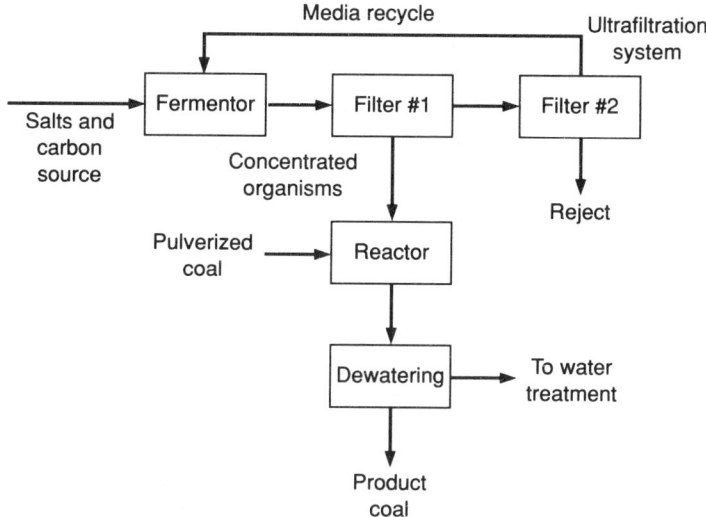

Fig. 7. Flow sheet for ARC's microbial coal cleaning process (7).

neously or sequentially, could potentially improve organic sulfur rejection. The limiting factors appear to be those of accessibility and residence time. Therefore, finer size coal should be used not only to improve accessibility of microbes to coal particle surfaces but also to reduce the overall retention time in the bioreactor. Additional information and a detailed review of conventional and advanced coal cleaning technologies may be found in Chapters 7 and 14 of Reference 7.

BIBLIOGRAPHY

"Coal Conversion Processes, Desulfurization" in the *Encyclopedia of Chemical Technology*, 3rd ed., Vol. 6, pp. 306–324, by E. Stambaugh, Battelle Memorial Institute; in *ECT* 4th ed., Vol. 6, pp. 511–540, by S. Chiang and J. T. Cobb, Jr., University of Pittsburgh.

1. F. T. Princiotta, *Chem. Eng. Prog.* **74**(2), 58–64 (Feb. 1978).
2. *Coal*, National Coal Association, Mar. 1992, p. 9.
3. E. Zimmerman, *9th International Coal Preparation Congress*, New Delhi, India, Nov. 29–Dec. 4, 1982, pp. 32–44.
4. H. H. Lowery, *Chemistry of Coal Utilization*, Supplementary Vol. I, pp. 119–149, 1963.
5. "Trace Elements in Coal—An Interlaboratory Study of Analytical Techniques," *Technical Bulletin*, Consolidation Coal Co., Oct. 1987.
6. C. T. Ford and A. A. Price, "Evaluation of the Effect of Coal Cleaning on Fugitive Elements," Phase II, Part I, *Bituminous Coal Research*, BCR Report L-1082, 1980.
7. J. W. Leonard and B. C. Hardinge, ed., *Coal Preparation*, 5th ed., Society for Mining, Metallurgy and Exploration, Inc., Littleton, Colo., Chapts. 7, pp. 271–496, and 14, pp. 966–1005, 1991.
8. J. A. Cavallaro and co-workers, *Sulfur and Ash Reduction Potential and Selected Chemical and Physical Properties of United States Coals*, DOE/PETC-91/2, Jan. 1990.
9. D. W. Fuerstenau, in P. Somasundaran, ed., *Fine Particles Processing*, Vol. 1, AIME, New York, 1980, p. 669.

10. K. J. Miller and A. W. Deurbrouck, *Physical Cleaning of Coal: Present and Developing Methods*, Marcel Dekker, Inc., 1982, pp. 255–291.
11. B. K. Parekh and co-workers, *Column Flotation '88*, AIME, pp. 227–233, 1980.
12. A. F. Taggart, *Elements of Ore Dressing*, John Wiley & Sons, Inc., New York, 1951.
13. R. H. Yoon and co-workers, *2nd Int. Conf. on Processing and Utilization of High Sulfur Coals*, Carbondale, Ill., Sept. 27–Oct. 1, 1987, pp. 533–543.
14. D. E. Zipperian and U. Svensson, *Column Flotation '88*, 43–54, 1988.
15. L. E. Burgess and co-workers, *5th International Symposium on Coal Slurry Combustion and Technology*, Tampa, Fla., 1983, pp. 255–268.
16. J. D. Miller and M. C. Van Camp, *Tran. AIME* **272**, 1575 (1982).
17. C. E. Capes and K. Darcovich, *Powder Technol.* **40**, 43–52 (1984).
18. J. C. Knight, *Coal*, 58–60 (May 1989).
19. D. V. Keller, Jr. and W. M. Burry, *Coal Prep.* **8**, 1 (1990).
20. U.S. Pat. 4,613,429 (1986), S. H. Chiang and G. E. Klinzing (to University of Pittsburgh).
21. G. Araujo and co-workers, *Energy Prog.* **7**(2), 72 (1987).
22. R. D. Stoessner and co-workers, *Coal Prep '86*, 1986, pp. 5–34.
23. R. E. Hucko and C. P. Maronde, *9th International Coal Preparation Congress*, F2, 1982, pp. F2-1–F2-16.
24. R. E. Hucko and co-workers, "Status of DOE-sponsored Advanced Coal Cleaning Processes," in R. R. Klimpel and P. T. Luckie, eds., *Industrial Practice of Fine Coal Processing*, SME, Inc., Littleton, Colo., 1989.
25. E. Levy and co-workers, *8th Pittsburgh Coal Conference*, Pittsburgh, Pa., Oct. 14–18, 1991, pp. 1021–1026.
26. R. Markuszewski and co-workers, *Div. Fuel Chem.* **49**, 187–194 (1980).
27. S. Friedman, R. B. LaCount, and R. P. Warzinski, *173rd National Meeting of ACS*, New Orleans, La., Mar. 21–25, 1977, pp. 100–105.
28. R. R. Oder and co-workers, *Coal Conference*, American Mining Congress, Pittsburgh, Pa., May 1977.
29. U.S. Pat. 3,909,211 (1975), A. F. Diaz and E. D. Guth (to KVB Engineering Inc.).
30. A. Crawford, *Trans.* **3**(4), 204–219 (Jan. 1952).
31. J. L. Anastasi and co-workers, "Molten-Caustic-Leaching (Gravimelt Process) Integrated Test Circuit Operation Results," Report to the Gravimelt Process Advisory Board, Summer 1989.
32. J. L. Anastasi and co-workers, *5th Annual Coal Preparation, Utilization and Environmental Control Contractors Conference*, U.S. Department of Energy, Pittsburgh, Pa., July 31–Aug. 3, 1989.
33. C. K. Richardson and co-workers, *3rd Annual Pittsburgh Coal Conference*, Pittsburgh, Pa., Sept. 8–12, 1986, pp. 130–141.
34. P. S. Ganguli and co-workers, *Div. Fuel Chem. Prepr., ACS*, **21**(7), 118 (1976).
35. J. K. Kindig, *8th Pittsburgh Coal Conference*, Pittsburgh, Pa., Oct. 14–18, 1991, pp. 231–236.
36. U.S. Pat. 4,562,156 (1985), J.D. Isbister (to Atlantic Research Corp.).

SHIAO-HUNG CHIANG
JAMES T. COBB, JR.
University of Pittsburgh

DESULFURIZATION OF FLUE GASES

In 1983 there were 116 flue-gas desulfurization (FGD) systems in service, representing 47 gigawats-electric of power generation capacity (1). As of 1992, more than 150 coal-fired boilers in the United States operated with FGD systems. The total electrical generating capacity of these plants has risen to 72 gigawatts (2). FGD processes are classified into (1) wet-throwaway, (2) dry-throwaway, (3) wet-regenerative, and (4) dry-regenerative processes (3).

Flue Gas Desulfurization

Wet-Throwaway Processes. By 1978, wet-throwaway systems were in commercial operation: lime scrubbing, limestone slurry scrubbing, and dual alkali (4). Lime/limestone wet scrubbing (Fig. 1) remains the most common postcombustion control technique applied to utility boilers (2). The waste product from the scrubbers can either be sent to a landfill or be upgraded by oxidation to become saleable gypsum.

Whereas it is not precisely a lime/limestone wet scrubber because it uses alkali already present in the coal ash, the Colstrip FGD system has been most expansively described (5). Three scrubber modules, operating in parallel, are used on two 360 MWe coal-fired boilers at the Colstrip Station of the Montana Power Co. and the Puget Sound Power and Light Co. Each module consists of a downflow venturi scrubber, centered within an upflow spray tower contactor, and is designed to clean 120 MWe of equivalent gas flow under normal conditions and 144 MWe under emergency conditions. Thus, when one module is off line, the boilers can still operate at 80% capacity, using the remaining two modules to clean the generated flue gas. Test data show that the levels of pollutants in the plant emissions are well below the vendor guarantee and the applicable federal standards. Scrubber availability and plant load for a 22-month period shortly after startup and a number of operating details for the FGD system may also be found in Reference 5.

In the design of a lime/limestone scrubber, there are numerous considerations to be evaluated, including particulate removal (if any), ash removal, scrubber type and configuration, scaling prevention, absorbent feed control, water balance, mist elimination, reheat, and sludge disposal (6). In addition, the size of the boiler is a factor. FGD systems for smaller industrial coal-fired boilers can be purchased as packages, as opposed to the large specially engineered systems for field construction at utility stations (7). These latter systems are typically designed as vertical towers and use either lime or limestone as the sorbing agent. If the sulfur content is high or liquid waste is not permitted, the double alkali process may be used, where the sodium ion is replaced by a calcium ion to form an insoluble precipitate of the sulfur compounds. The precipitate is then filtered from the liquid stream and the regenerated liquid, containing the original sodium ions, is returned to the scrubber.

Dry-Throwaway Processes. Dry-throwaway systems were the precursor of processes that removed SO_2 in the ductwork, eg, the BCZ and IDS processes. Here, however, the device is a spray chamber similar to the wet scrubbers such as the

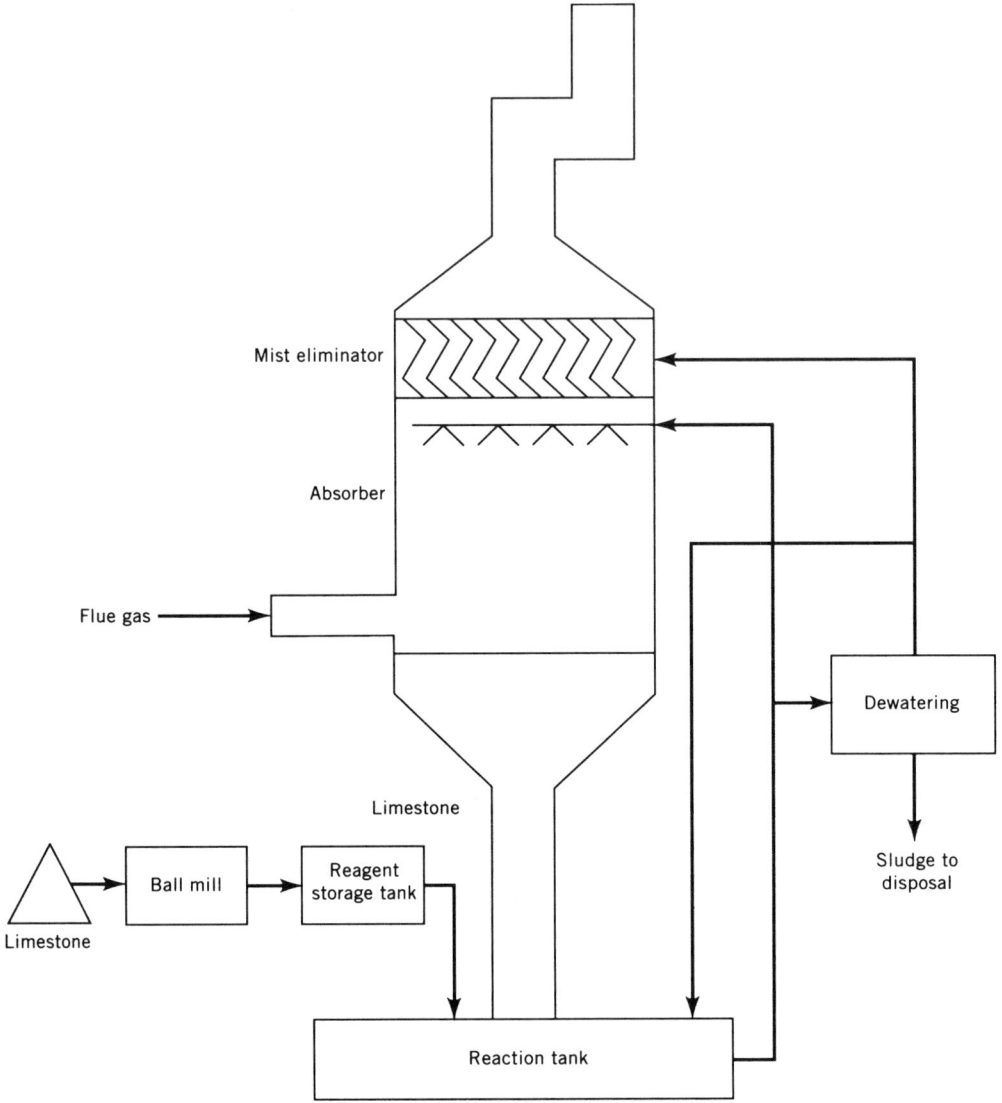

Fig. 1. Limestone FGD system. Reproduced by permission of the American Institute of Chemical Engineers (2).

three modules of the Colstrip installation (Fig. 2). Into the upper portion of the chamber a slurry or clear solution containing sorbent is sprayed. Water evaporates from the droplets, the sorbent reacts with SO_2 both before and after drying, and the dry product is removed in a downstream baghouse or ESP (8). Unfortunately, dry scrubbing is much less efficient than wet scrubbing and lime, instead of the much less expensive limestone, is required to remove SO_2 effectively. Consequently, a search has been conducted for more reactive sorbents (8–11).

One commercial dry-scrubbing process is the system designed and installed by MikroPul Corp. at Strathmore Paper Company's 3.2 MWe PC cogeneration

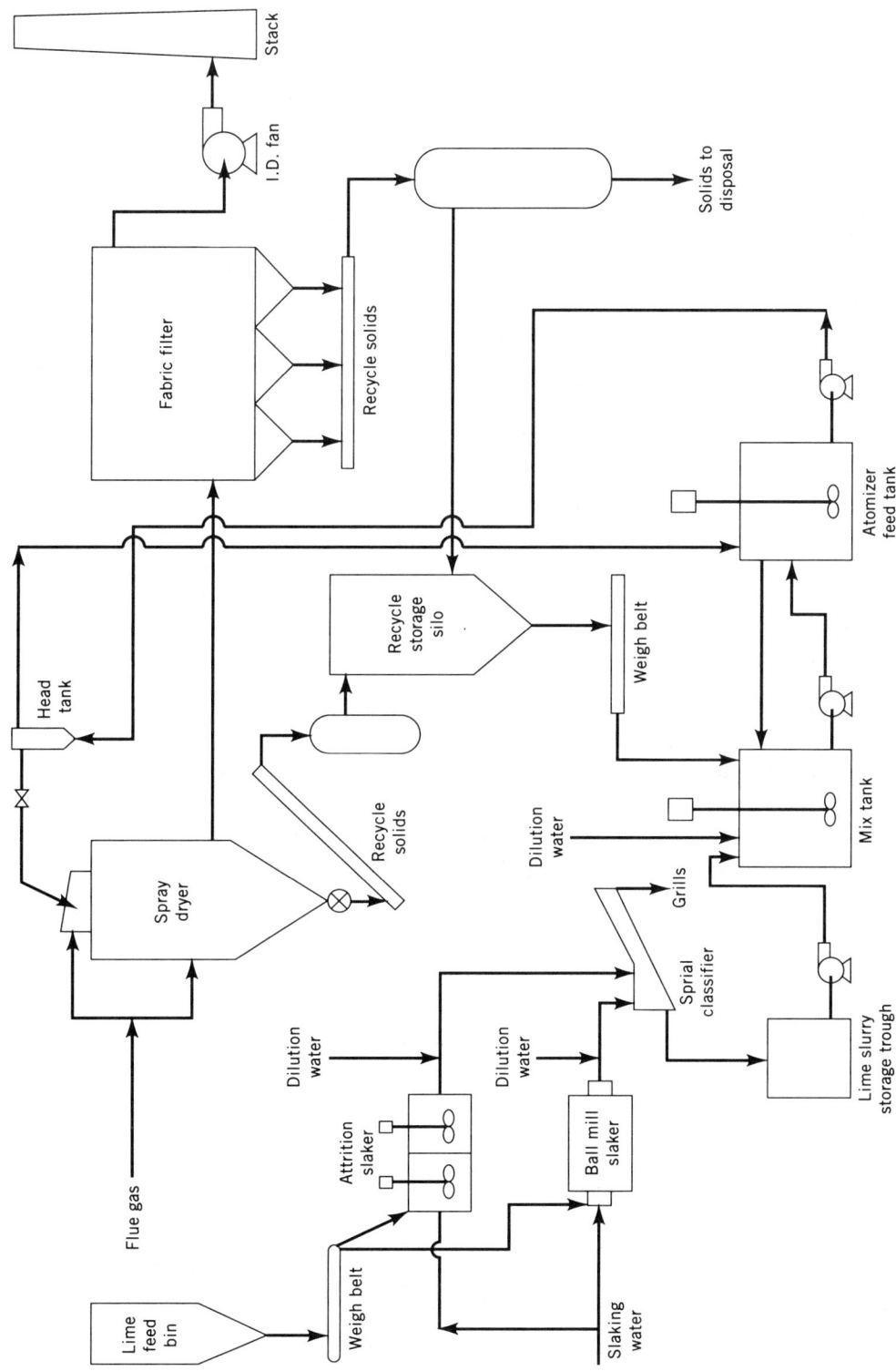

Fig. 2. Lime spray dryer process flow diagram. Reproduced by permission of the American Institute of Chemical Engineers, 1991 (2).

boiler in Woronoco, Massachusetts (12). The system consists of a slaked-lime spray drier reactor and a fabric filter. The stainless-steel spray drier is 4.4 m in diameter and 8.5 m tall. Four flue gas/sorbent slurry diffusers, containing two-fluid, external-mix-type nozzles, are installed at the top of the chamber. Early modifications to the chamber included changes in its aerodynamics to improve mixing and changes in the slurry feed and distribution system. The unit generally removes 80% SO_2 at a Ca:S ratio of 1.8.

Regenerative Processes. By 1984 two commercial regenerative processes, the Davy McKee Wellman-Lord system and the United Engineers & Constructors magnesium oxide system were in operation in five generating stations, for a total of eleven units, in the United States (13). The Wellman-Lord process, of which only one unit is still in operation and which is no longer offered commercially, uses an aqueous solution of sodium sulfite to absorb SO_2 and form sodium bisulfite. The sodium bisulfite is thermally decomposed in the regenerator to re-form sodium sulfite and release SO_2 and the sulfite is returned to the absorber. The SO_2 is converted to either sulfur or sulfuric acid.

The MgO system shown in Fig. 3 uses an aqueous solution of $Mg(OH)_2$ to absorb SO_2, forming crystalline $MgSO_3 \cdot 3H_2O$. After removal by centrifugation the $MgSO_3 \cdot 3H_2O$ is dried in a direct-fired rotary kiln to produce anhydrous magnesium sulfite [7757-88-2], $MgSO_3$, which may then be shipped to a fluidized-bed calciner. In the calciner $MgSO_3$ is decomposed to magnesium oxide [1309-48-4], MgO, and SO_2. MgO absorbers were installed at Units 1 and 2 of the Eddystone Station and Unit 1 of the Cromby Station of Philadelphia Electric Co., and regenerators were built at sulfuric acid plants in Delaware and New Jersey by Allied Corp. and Essex Chemical, respectively. By February 1984 about 23,800 t of sulfuric acid had been made from by-product SO_2 and the scrubbers had consistently reached 96–98% removal of sulfur dioxide. The MgO system continues to operate well, but the regenerative process market has moved toward other systems, particularly the NOXSO process, which provides combined NO_x/SO_2 removal.

The NOXSO process has many of the elements of a traditional dry scrubber (14–16). The sorbent for both NO_x and SO_2 is sodium carbonate impregnated on a high surface-area gamma alumina. From a fluidized-bed absorber the sorbent is first heated from 120 to 600°C, driving off the NO_x and loosely-bound SO_2. The sorbent then passes to the first chamber of a regenerator where it is contacted with natural gas to remove most of the sulfur (about 70%), which it carries into that chamber. In the second chamber of the regenerator steam drives off the remaining sulfur. The sulfur-laden gases from both chambers are combined and processed to yield products for sale. A contract to construct a 5 MWe proof-of-concept plant for the NOXSO process at either Boiler 10 or Boiler 11 of Ohio Edison's Toronto Station was signed in 1989. A full-scale demonstration is in operation at Ohio Edison's Niles Station (see CONTACTORS FOR CHEMICAL SEPARATION PROCESSES).

Development efforts regarding regenerative processes have also focused on higher temperature sorption using fluidized beds of metals supported on porous solids. The most advanced of these developments is the WSA-SNOX cleaning technology, offered by Haldor Topsoe, Inc. (17). After NO_x removal by selective catalyst reduction (SCR), the SO_2 in the flue gas is oxidized to SO_3 over a conventional sulfuric acid catalyst. Upon cooling, the SO_3 is hydrated to sulfuric acid, which is

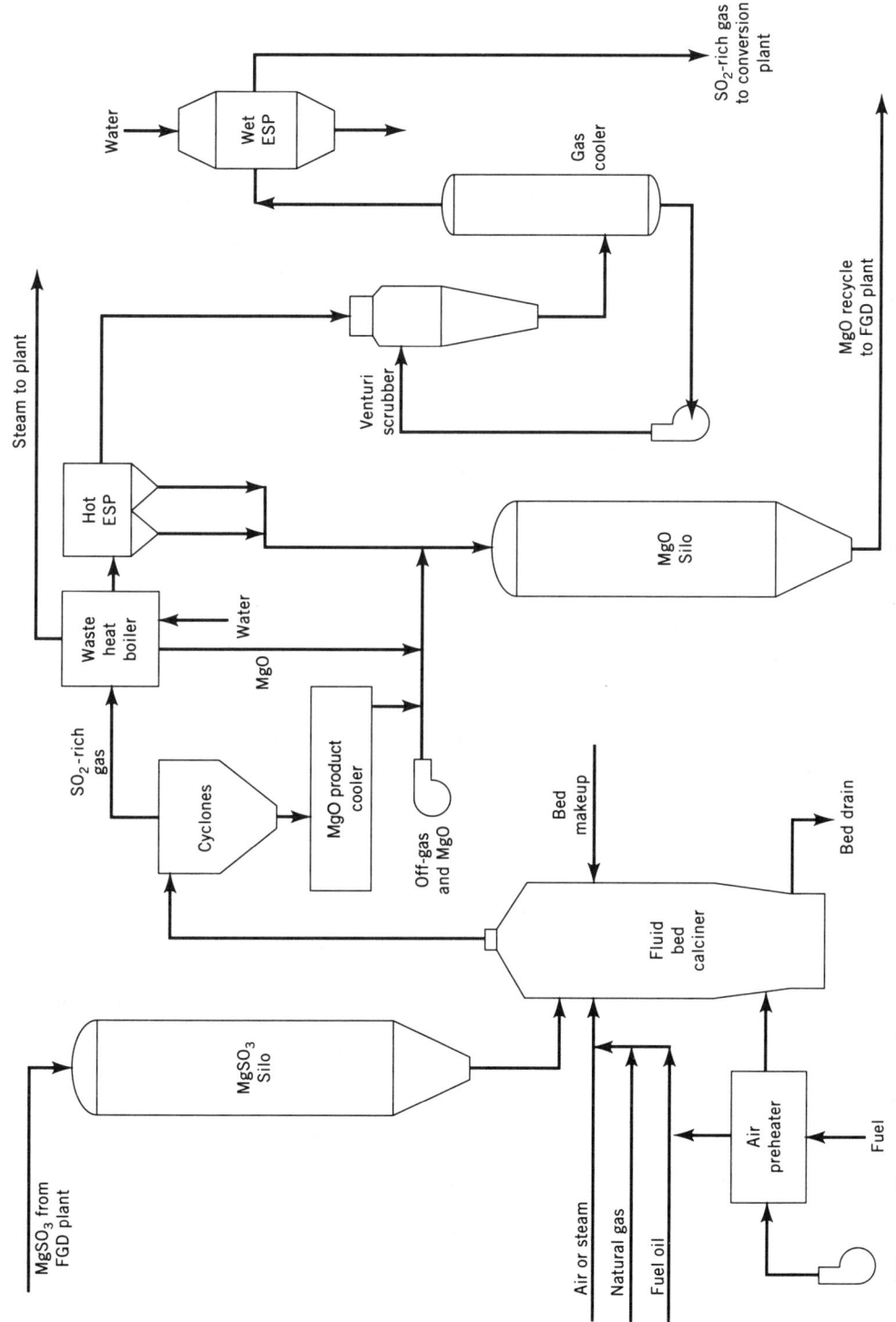

Fig. 3. The MgO desulfurization system: regeneration. Reproduced by permission of the American Institute of Chemical Engineers, 1984 (13).

condensed, concentrated, and stored for sale. In 1988 several WSA-SNOX units were in operation in Europe and the first one in the United States was being planned by Ohio Edison for Boiler No. 2 of the Niles Station.

At a much earlier stage in the research and development cycle, fluidized-bed processes use porous sorbents containing copper oxide (18), cerium oxide (19), and other metal oxides (20).

When the Clean Air Act of 1990 was signed into law, electric utilities were required to establish plans and initiate projects to comply with that Act's Title IV. Each utility had to evaluate how the various commercial and emerging clean coal systems fit into the utility's technical and business environment resulting in strategies to utilize fuel switching and wet throwaway FGD processes almost exclusively (21–23).

BIBLIOGRAPHY

"Coal Conversion Processes, Desulfurization" in the *Encyclopedia of Chemical Technology*, 3rd ed., Vol. 6. pp. 306–324, by E. Stambaugh, Battelle Memorial Institute; in *ECT* 4th ed., Vol. 6, pp. 511–540, by S. Chiang and J. T. Cobb, Jr., University of Pittsburgh.

1. K. E. Yeager, *Pollut. Eng.*, 24–28 (June 1984).
2. M. Maibodi, *Environ. Prog.* **10**(4), 307–313 (Nov. 1991).
3. R. McInnes and R. Van Royen, *Chem. Eng.* **97**(9), 124–127 (Sept. 1990).
4. F. T. Princiotta, *Chem. Eng. Prog.* **74**(2), 58–64 (Feb. 1978).
5. C. Grimm and co-workers, *Chem. Eng. Prog.* **74**(2), 51–57 (Feb. 1978).
6. A. V. Slack, *Chem. Eng. Prog.* **74**(2), 71–75 (Feb. 1978).
7. J. D. Brady, *Chem. Eng. Prog.* **80**(9), 59–62 (Sept. 1984).
8. H. T. Karlsson and co-workers, *J. Air Pollut. Control Assoc.* **33**(1), 23–28 (Jan. 1983).
9. C. Jorgensen and co-workers, *Environ. Prog.* **6**(2), 26–32 (Feb. 1987).
10. J. M. Markussen and H. W. Pennline, *6th Pittsburgh Coal Conference*, Pittsburgh, Pa., Sept. 25–29, 1989, pp. 299–308.
11. A. Pakrasi and co-workers, *J. Air Waste Manage. Assoc.* **40**(7), 987–992 (July 1990).
12. T. V. Reinauer and co-workers, *Chem. Eng. Prog.* **79**(3), 74–81 (Mar. 1983).
13. C. Murawczyk and J. S. MacKenzie, *Chem. Eng. Prog.* **80**(9), 62–68 (Sept. 1984).
14. J. L. Haslbeck and co-workers, *6th Pittsburgh Coal Conference*, Pittsburgh, Pa., Sept. 25–29, 1989, pp. 319–329.
15. J. L. Haslbeck and co-workers, *7th Pittsburgh Coal Conference*, Pittsburgh, Pa., Sept. 10–14, 1990, pp. 330–339.
16. J. L. Haslbeck and co-workers, *8th Pittsburgh Coal Conference*, Pittsburgh, Pa., Oct. 14–18, 1991, pp. 479–484.
17. D. R. Juist and co-workers, *5th Pittsburgh Coal Conference*, Pittsburgh, Pa., Sept. 12–16, 1988, pp. 1165–1176.
18. J. T. Yeh and co-workers, *Environ. Prog.* **4**(4), 223–228 (Nov. 1985).
19. W. G. Wilson and co-workers, *8th Pittsburgh Coal Conference*, Pittsburgh, Pa., Oct. 14–18, 1991, pp. 457–463.
20. O. Faltsi-Saravelou and J. A. Vasalos, *Ind. Eng. Chem. Res.* **29**(2), 251–258 (Feb. 1990).
21. C. W. Schwartz, *Energy Prog.* **2**(4), 207–212 (Dec. 1982).
22. M. M. Peplowski, *Coal* **97**(2), 39–41 (Feb. 1992).
23. S. A. Mitnick, *Electr. J.*, 44–49 (Jan./Feb. 1992).

SHIAO-HUNG CHIANG
JAMES T. COBB, JR.
University of Pittsburgh

DEWATERING

Dewatering is the last process applied to separate water from a solid, unless thermal drying (qv) is used. Dewatering is usually a mechanical process that presses residual water from solids or displaces the water with a gas, and the energy required is negligible compared with the heat required for drying (1). Thus there is a significant incentive for adding a dewatering step to a process. Whereas the broader term deliquoring includes the separation of nonaqueous liquids from solids, the discussion herein is specific to water removal.

In municipal wastewater treatment, dewatering is regarded as the final process used to achieve a water content of ≤85% in the sludge (2,3) or to change the behavior of sludge to that of a solid (4). Many other industries regard 85% water content as *feed* to a primary liquid–solid separation device, such as a thickener, and consider 15% moisture (85% solids) an appropriately dewatered product. Dewatering, then, is not defined by moisture content or by the use of a type of equipment (5).

Maximizing mechanical dewatering is a matter of selecting the most appropriate equipment, optimizing the variables that affect dewatering on the equipment, and optimizing the feed. Optimizing the feed is usually most important. However, if the feed contains a product the properties of which should not be changed or if dewatering is easy, then equipment selection is the most important variable. Selection of equipment is dependent on the particular feed stream, for example, cotton (qv), fruit juices (qv), or sewage, and the field is in constant development. Aids to selecting equipment are found in the literature (see General References). Once the choice of equipment has been narrowed, eg, centrifuges, belt presses, or the like, testing on the equipment is extremely important.

Interaction of Water and Solids

Water associates with solids in a range of energies (6). The energy needed for water removal is indicated in Figure 1. The highest dewatering energies are associated with a monolayer or less of water, which is not generally considered moisture content.

Dewatering processes, normally concerned with water bound in capillaries, affect the water and solids by changing the size distribution of capillary radii, reducing adhesion of water to the solids, displacing water from the capillaries, and reducing the energy required to cause flow in the capillaries. These effects are achieved by three methods (1). (*1*) The particulate matrix may be compacted by applying stress, that is, forces can be induced by frictional drag of the liquid as it flows through the pores. Body forces can arise from gravity or centrifugal motion, and boundary stresses can be applied with rolls, membranes, pistons or screw presses, or acoustic energy (see MEMBRANE TECHNOLOGY). (*2*) The water may be displaced, usually with a gas, by application of vacuum or pressure. Centrifugal force in a pusher centrifuge also results in the water being displaced by a gas. (*3*) An electrical field may be applied to a slurry of charged particles (6).

Although all the techniques are effective, in industrial applications there is rarely time to achieve an equilibrium reduced saturation state (see FILTRATION),

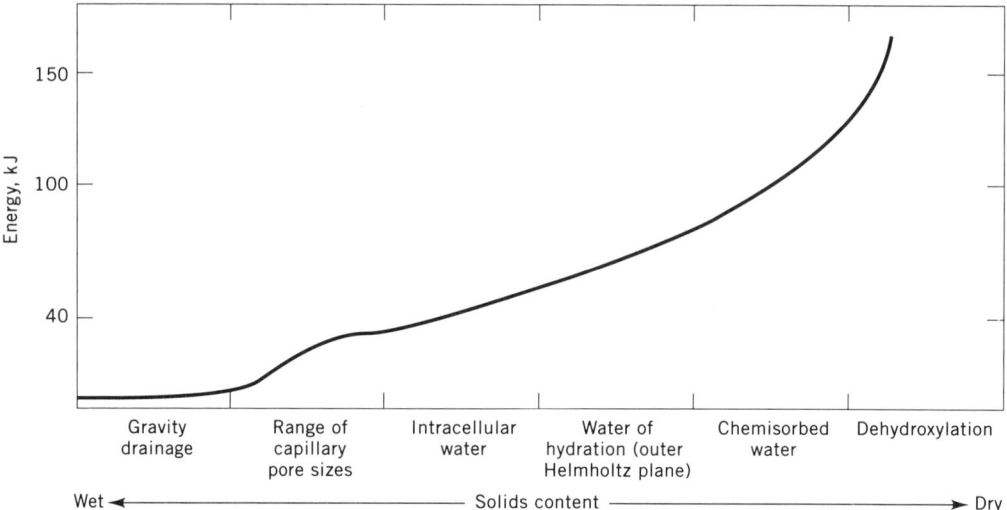

Fig. 1. Energy required for removing 1 mol of water from solids. To convert kJ to kcal, divide by 4.184.

so variables that affect only the kinetics of dewatering and not the equilibrium and residual moisture are also very important. The most important kinetic variables in displacing the liquid from the solid are increases in pressure differentials and viscosity reduction.

Cake Dewatering

The most important function of filtration (qv) is the formation of a filter or centrifuge cake from a slurry (see also CENTRIFUGAL SEPARATION). The most important function of dewatering is removal of the liquid from the resulting cake. Theories of cake dewatering are covered elsewhere (7–12). Although theory lags behind practice, the principal variables controlling dewatering have been identified and both dewatering rate as well as equilibrium and residual moisture in fully defined cakes can be predicted. The behavior of relatively incompressible cakes can be modeled using changes in porosity with pressure. For more compressible cakes, empirical models have been developed (13). Theoretical models of expression dewatering of compressible cakes exist (12). Very few industrial applications of dewatering have consistent or fully characterized cakes, and no process in the design stage can fully characterize a filter cake.

Ways to reduce the final moisture content of a centrifuge cake include the use of steam, surfactants, or flocculants (see FLOCCULATION), as well as pretreatments by pelletizing, oil agglomeration, thermal treatment, and freeze-thaw processes. The main dewatering variable in the centrifuge itself is centrifugal force sufficient to expel the liquid from the pores in the cake through the filter medium. In some newer centrifuges the pitch of the screw is shortened, applying compression to the cake to further reduce cake pore volume and, with it, entrained water (14). Other basket centrifuge designs incorporate baffles to prevent materials from slipping out of the basket before dewatering (35).

Cake dewatering is related to cake formation; in a filter cake, the final moisture content is dependent on many variables that also control cake formation. Pretreatment processes, for example, affect both cake formation and cake dewatering. Cake dewatering is achieved by compacting the solids; displacing the residual liquid in the cake with another phase, usually a gas; or applying an electrical current to remove the liquid. Each process relies on different properties of the cake.

Compaction, Compression, and Expression. Compaction is a newer term for compression and is used to describe the movement of particles relative to one another within a device until the matrix of particles gains enough strength to resist further consolidation (16). Compaction occurs in a plate and frame filter both while the chamber is filling and at the end of the cycle when the chamber is nearly full and the pressure rises steeply. Compactibility (or compressibility) describes the reduction in volume of the particle matrix. Compaction also takes place in the bed of a thickener as the solids continuously deposit on the top of the bed and a thickened slurry is withdrawn from the bottom.

Compression virtually always reduces the permeability of the filter cake. The reduced permeability in highly compressible cakes is greatest at the surface of the filter medium, where the pressure drop is greatest. Further dewatering of these cakes in a filter press can be achieved by reversing the flow on a plate-and-frame filter and using the compressed "skin" of filter cake as a membrane to squeeze water from the center of the cake, where pressure differentials are lower (17). Another method of applying compression is by delayed cake formation using continuous pressure filters. These filters easily remove large quantities of water and produce a highly thickened discharge, which sometimes forms a rope as it discharges. Continuous pressure filters can be very effective for thickening dilute slurries and for washing in slurry form (18).

Expression is the application of mechanical stress to a matrix of particles in the fully formed cake (12). Expression, which refers to squeezing the solids rather than the slurry (19), is used to further reduce void space in order to separate more liquid from the solids. Expression dewatering is most effective with compressible cakes (12,20) of particle sizes below 50 μm (7), superflocculated cakes or cakes otherwise containing loosely entrapped liquid, and organic solids containing liquid-filled cells that must be ruptured for effective dewatering. Filters that use expression include variable-chamber plate-and-frame filters, belt-filter presses, very high pressure belt presses (21,22), tube presses, and screw presses. Tube presses can exert pressures up to 14 MPa (2000 psi) (23), the highest pressure of any of these expression devices. After pressing, an air-blow step can be added to further reduce the entrained moisture (23). Screw presses are used primarily with fibrous or polymeric materials containing entrapped liquid. Pressures up to 110 MPa (16,000 psi) have been measured in these devices at the points between the screw and screen when solids are present (24).

Significant improvements were made in the 1980s and early 1990s in high capacity, automated variable volume filters that incorporate automatic pressure filtration, expression, washing, and air displacement. Some of the large plate-and-frame automatic presses can operate at up to 2 MPa (ca 285 psig), with up to 100 chambers (25,26).

If expression is effective, it reduces the permeability of the cake being compacted and, as a consequence, the resistance to flow of the liquid increases considerably (27). The effectiveness of expression is governed by cake thickness, specific resistance, consolidation properties, and shear forces.

Displacement Dewatering. Replacement of the liquid in the voids of a cake by another liquid or a gas is termed displacement dewatering. Air displacement can be accomplished by using a pressure difference to force the liquid from the pores in the cake. The types of filters that provide displacement dewatering include virtually all vacuum filters, rotary pressure disk and drum filters, the Lasta (25), Larox (28), and Vertipress (29) automatic filter presses, hyperbaric filters, tube press filters (23), and a hybrid continuous pressure filter–expression press (30). In a centrifuge, displacement dewatering is accomplished by applying a body force directly to the liquid by the spinning motion. The factors that control displacement dewatering are

Property	Variable
cake	particle size and size distribution, shape, packing, dimensions of the cake
fluid	density, viscosity
interfacial forces	surface tension (gas–liquid), interfacial tension (solid–liquid, gas–solid)
other	temperature, pressure gradient, rate of displacement

In a study on dewatering methods for peat, displacement dewatering was done using acetone, a polar solvent having a lower heat of vaporization than water. Dewatering was improved in terms of both the pressure filtering step and the quantity of heat required. Less heat was required to dry the cake and recover the acetone from the filtrate by distillation (31).

The rate of displacement dewatering increases by increasing the driving force, bed permeability, and filter area, and decreasing viscosity or cake thickness. The dewatered cake becomes drier by increasing the driving force, bed permeability, and the contact angle or decreasing the surface tension. There are a number of techniques to achieve these results.

Expression Dewatering of Fibrous Materials. Fibrous materials are frequently dewatered in belt-filter, screw, disk, and roll presses and in batch pot and cage presses. Table 1 lists applications of screw, roll, and pot presses. Screw and high pressure belt presses are continuous and have replaced batch pot and cage presses in most applications. Traditionally, however, batch presses have been used for squeezing cocoa butter from cocoa beans, which require pressures up to 41 MPa (6000 psi) (39). A description of many types of batch presses is included in Reference 40.

Screw presses (Fig. 2) do not produce a clear liquid product. Frequently, the product is further filtered in a filter press to give a clear liquid product. Press aids are added to feed materials containing fine particles or particles that can deform and plug the slots in the cage of a screw press. Typical press aids include sawdust, rice hulls, perlite, and diatomaceous earth. A vertical screw press is a continuous press that has been used for dewatering sewage sludge (2).

Table 1. Applications of Screw, Roll, and Pot Presses[a]

Material	Liquid, %	
	In feed	In product
paper pulp	97	50
	90	65
wood chips	85	50
sugar cane	68	43
oilseeds		
high oil content[b]	>30	3–7
low oil content[c]	<30	3–6
cocoa (separation of cocoa butter)	53	12
food[d]	60–90	10–30
polymers, elastomeric and thermoplastic[e]	60	5–8
rendered tissue	20[f]	6–10
sewage sludge[g]	98	85

[a]Refs. 32–37.
[b]Includes copra, cottonseed, corn germ, peanuts, flax, safflower, sunflower, sesame, palm kernels, and linseed.
[c]Includes soybean, rice bran, and dry-process corn germ.
[d]Includes apples, carrots, coffee grounds, fish, grapes, pineapples, and tomatoes.
[e]Includes ABS, nitriles, styrene–butadiene rubber (SBR), natural rubber, and ethylene–propylene-diene rubber (EPDM).
[f]Fat.
[g]The sludge was steam-heated in the press and treated with CaO and polymer. About 95% of the solids were retained in the cake (38).

A disk press, shown in Figure 3, can achieve a compression ratio of about 4:1 and produce a paper pulp having a consistency of 45–50% solids. It has also been used on brewers' spent grains and coffee grounds. The two surfaces of the rotating, converging press disks have a screen backing that retains and presses the solids while letting liquid pass through the screen. Slurry is fed at the wide part of the space between the disks, and the slurry is carried through the maximum compression zone before being released as the disks diverge (41). A similar device, called the shoe rotary press, has been tested on both fine coal and fine coal refuse (42).

A different type of press is the Vari-Nip shown in Figure 4 (43). A slurry is forced into the vat at up to 240 kPa (20 psig). Two porous rolls rotate in a pressurized vat of pulp slurry. As the rolls rotate together into the slurry, the differential pressure across the face of the rolls forces dewatering of the pulp and deposits fibers on the roll faces, forming fiber mats. The pressate, which has passed through the roll perforations, drains to the discharge ports. As the rotation continues, the mat enters the nip areas, and the rolls press the mat together, forcing dewatering to a high consistency. The resulting mat is then guided to a breaker conveyer (43). This press is somewhat similar to drilled press rolls used in paper making.

Improving Cake Dewatering. *Viscosity Reduction.* Equations relating the rate of liquid flow through a filter cake can be simplified to

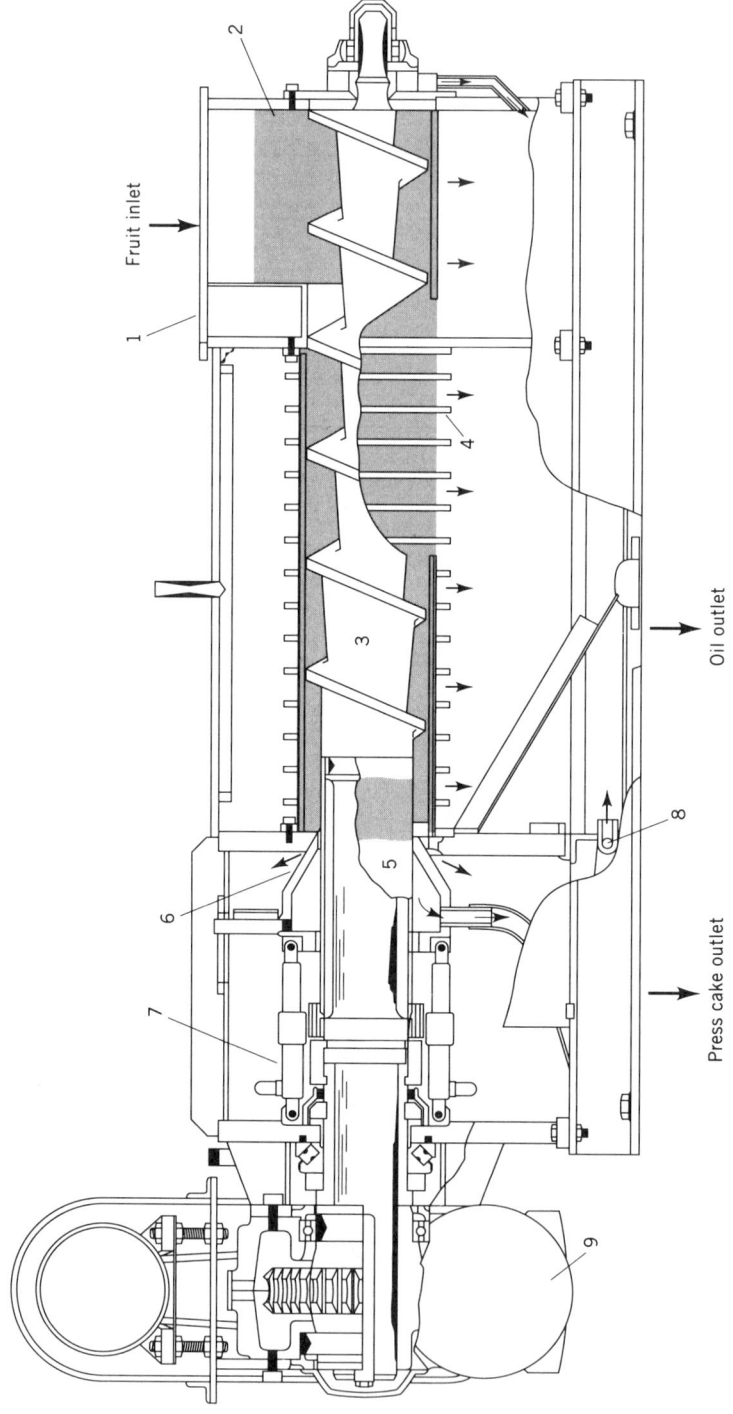

Fig. 2. Cross section of screw press used for fruit juice (32). 1, Hopper; 2, perforated sheets; 3, main shaft; 4, perforated cage; 5, draining cylinder; 6, cone; 7, hydraulic cylinder; 8, draining cylinder oil; 9, gear box. Courtesy of the French Oil Mill Machinery Co.

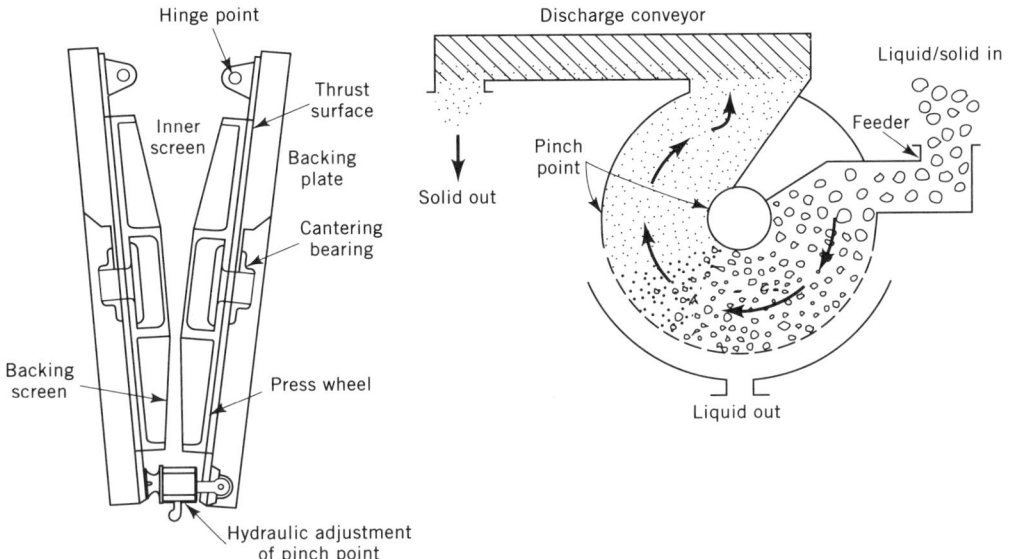

Fig. 3. Schematic of a disk press (41). Courtesy of Bepex Corp., a subsidiary of Berwind Corp.

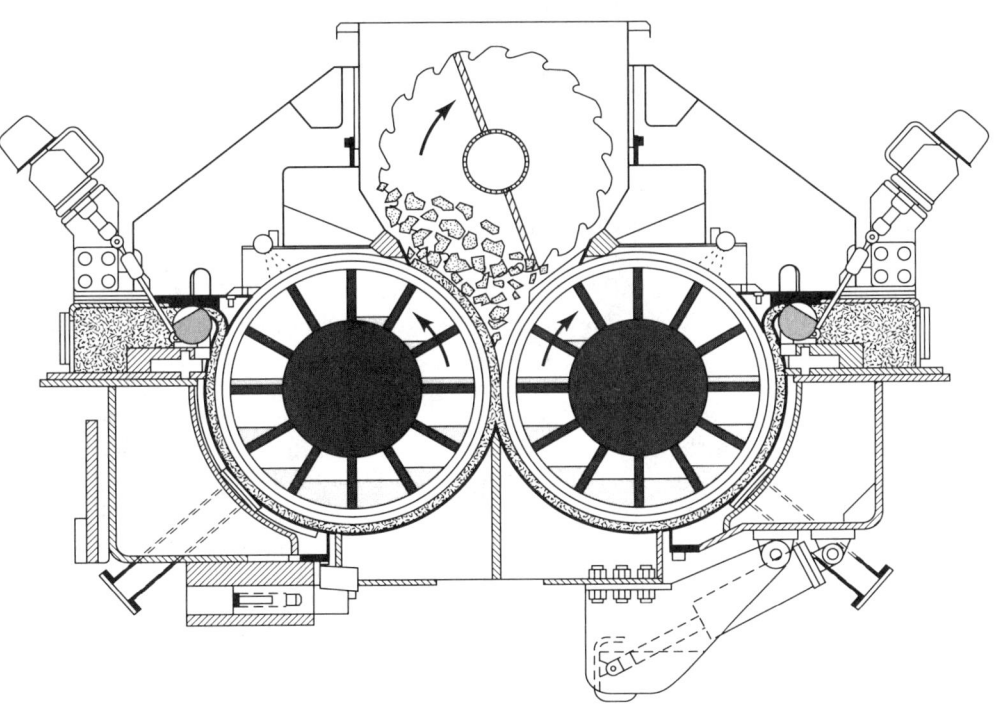

Fig. 4. Schematic of the Vari-Nip press (43). Courtesy of Ingersoll-Rand Co.

$$\frac{V}{A} = \frac{K\Delta P}{\mu l}$$

where V in units of m³/s is the flow rate through the cake, A in m², is the area of the cake, K, m², is the cake permeability, ΔP, Pa, is the pressure drop across the cake, l in m, is the thickness of the cake, and μ is the cake viscosity (Darcy's law). Viscosity, a kinetic variable, does not appear in equations describing reduced saturation levels in a cake (44). Because most practical filtration is limited by the time available for the steps of cake forming and dewatering, an increase in the flow rate of filtrate during dewatering translates directly to lower cake moisture levels. If the liquid is water, the viscosity can drop by a factor of three as the temperature rises from 15°C to 80°C, and consequently the flow rate of water through the cake is tripled. The addition of moderate quantities of salts, polymers, or small amounts of less viscous miscible liquids, such as alcohols, has very little effect on the viscosity of water. Temperature is the most important control of this variable.

The usual method of heating to improve dewatering is to apply low pressure steam to the filter cake as it is in the dewatering phase of the filter cycle. Steam has been used on rotary vacuum filters to dewater fine coal and on horizontal belt filters to dewater pipeline coal. Steaming typically removes an additional 0.7–1.5 kg of water per kilogram of steam applied and reduces moisture levels in coal filter cakes by about 5% (45–47). The effectiveness of steam depends directly on the permeability of the cake. In highly permeable cakes, up to 90% of the contained moisture can be removed. Generally, steam is ineffective on filter cakes in which particles smaller than 10 μm predominate (48). However, permeability is the critical factor: one of the largest installations of steam-assisted filtration is on <20 μm nonmagnetic taconites (49).

Use of Surfactants. Although the use of steam to improve dewatering is consistently beneficial, the effects of surfactants on residual moisture are highly inconsistent. Additions of anionic, nonionic, or sometimes cationic surfactants of a few hundredths weight percent of the slurry, 0.02–0.5 kg/t of solids (50), are as effective as viscosity reduction in removing water from a number of filter cakes, including froth-floated coal, metal sulfide concentrates, and fine iron ores (Table 2). A few studies have used both steam and a surfactant on coal and iron ore and found that the effects are additive, giving twice the moisture reduction of either treatment alone (44–46,49).

Surfactants aid dewatering of filter cakes after the cakes have formed and have very little observed effect on the rate of cake formation. Equations describing the effect of a surfactant show that dewatering is enhanced by lowering the capillary pressure of water in the cake rather than by a kinetic effect. The amount of residual water in a filter cake is related to the capillary forces holding the liquids in the cake. Laplace's equation relates the capillary pressure (P_c) to surface tension (σ), contact angle of air and liquid on the solid (θ) which is a measure of wettability, and capillary radius (r_c), or a similar measure applicable to filter cakes.

$$P_c = \frac{2\sigma \cos \theta}{r_c}$$

Table 2. Effect of Surfactants on Residual Moisture in Filter Cakes

Material	Moisture content, %	
	Without surfactant	With surfactant
sulfide flotation concentrates[a]	15	12
	12	9
iron ore	17	15
	21	17
coal	20–22	16
	36–40	30–34
	6	9
silica sand	12	8

[a]$CuFeS_2$, MoS_2, and ZnS.

Surfactants lower the surface tension of water, typically from 72 to ca 30–35 mN/m (= dyn/cm), and many surfactants have a strong effect on the contact angle when used at low concentrations. Both changes help dewatering. Too much surfactant, near or above the critical micelle concentration (CMC), reverses the effect that the surfactant has on contact angle at lower concentrations, and at or above the CMC there is no further lowering of surface tension. At the higher concentrations, the surfactant loses some of its beneficial effect on dewatering, as shown in Figure 5. The beneficial effects of surfactants on dewatering are most pronounced in cakes that have been partially deslimed or in cakes of partially hydrophobic particles (eg, flotation concentrates) that are adsorbed onto each other. Surfactants at or above CMC have little practical effect on extremely fine cakes, where pores are small and the cake has no further opportunity to consolidate. A number of filter cakes do not respond to surfactant addition at any level.

pH Adjustment. Virtually all solids become charged in water, either by reaction with the water to form surface hydroxyl groups that can ionize or by adsorption of ions from the water. The charges on the particles affect how much water is bound to the particle. Reducing the charge on the particle increases the

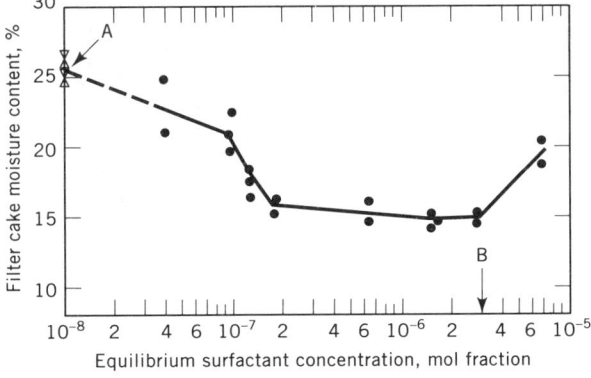

Fig. 5. Effect of surfactant concentration on moisture content of <500 μm coal filter cake (51). Point A represents zero surfactant concentration; Point B, the critical micelle concentration (CMC).

amount of dewatering possible in conventional dewatering equipment. For example, adjusting the pH of peat to approximately 3 increases dewatering, and reducing the charge on coal using pH and metal ions improves the results of pressure and vacuum filtration (2,52,53). When an electrical field is externally applied for dewatering, the effect of the charge changes.

Use of Flocculants. In the minerals industry and in water treatment, the primary purpose of flocculants is to improve sedimentation rates and overflow clarity in thickening operations. Flocculants can also have a beneficial effect on dewatering in a filter or centrifuge. Generally, the flocculants that work best on sedimentation are not the best for filtration. For example, large, loose flocs are effective in causing rapid settling but trap water in the filter cake and can deform easily and block the filter medium. Any excess polymer may stick to the filter medium, causing further blinding (54).

Other flocculants are capable of improving filtration rates up to 100 times, especially of fine clay, sludges, or tailings (55). One of the main uses of flocculants in filtration is to make extremely slow filtering slurries filterable at reasonable rates. In addition, flocculants are critical in making belt-filter presses and dewatering centrifuges effective. There are extensive and helpful reviews on the selection of flocculants and the effects on the performance of dewatering devices (1,56–59). In municipal sludge processing, where often no flocculant is added to the primary thickening devices, flocculants are subsequently chosen to improve dewatering rates.

Although filtration rates can be much faster with flocculants, the final cake moisture is often higher in a flocculated cake (60–63). In contrast, using flocculants optimized for filtration, coal, and other mineral slurries can be dewatered to moisture contents significantly lower than the untreated cake (64–68). The advantages of rapid filtration rates can also be preserved. Flocculants that provide better filtration tend to form flocs having the following characteristics (65):

Floc characteristics	Beneficial effects
small	reduces intrafloccular water in the late stages of filtration; reduces pickup problems resulting from gravity settling in the rotary filter chamber
strong	prevents floc breakdown owing to suspending agitation in the filter tank; resists collapse and premature loss of cake permeability in the early interfloccular stage of filtration
equisized	prevents localized breakthrough of air, cake shrinkage, cracking, and early loss of vacuum
good fines capture (into floc structure)	provides good filtrate clarity; prevents cloth binding and poor discharge

Pumping a polymer-flocculated slurry to a filter degrades or destroys the floccules. To repair the damage, other flocculants, chosen for their optimum filtration characteristics (68), can be added. For example, to filter froth-floated coal (nominally <0.5-mm particle size), a medium molecular-weight anionic flocculant (average molecular weight of 10^7) is used. For sedimentation, much higher molecular-weight polymers are more effective (64,65).

In addition to specifying molecular weight, the chemical structure of polyacrylamide flocculants has a significant effect on final moisture content. Two references (59,65) show the marked effects of both chemical structure and molecular weight on filtration rates and are useful guides to flocculant selection for coal- and clay-containing fine slurries (Fig. 6). Further dewatering of a flocculated filter cake can be achieved by using a surfactant dewatering aid, as described. The effectiveness of surfactants as dewatering aids seems neither to impair nor be impaired by the flocculant (65). Additions of 1–5 kg/t of an insoluble but highly water-absorbable polyacrylamide superabsorbent to a very sticky coal fines filter cake convert the material from a glue to a friable material that can be handled in normal material handling equipment (66).

Sludge Conditioning. Sludge conditioning is the chemical, physical, or heat treatment of wastewater sludges to improve dewatering (2,3,69–71). Because sludge handling costs can be 25–50% of the total cost of wastewater treatment, dewatering is critical to cost control. A number of substances have been added to thickened sludge to increase the permeability of the filter or centrifuge cake. In addition to polymer flocculants, coagulants such as ferric and ferrous chloride, alum, and lime have been added, which chemically react in the sludge and improve dewatering. Diatomaceous earth, fly ash, and ash derived from incinerating dewatered sludge or bark have been added as body feeds to improve sludge dewatering.

Biological processing of sludge reduces the amount of sludge needing dewatering and changes the dewatering properties. The changes are not always helpful. A comparison of high biological oxidation of wastewater sludges, to minimize the amount of sludge that needed disposal, showed higher overall disposal costs compared with the costs of less oxidation and twice as much "wasting" (removal and disposal) of sludge (72). In an unusual application of biological conditioning, enzymes have been used to aid in the dewatering of phosphatic clay ponds (73).

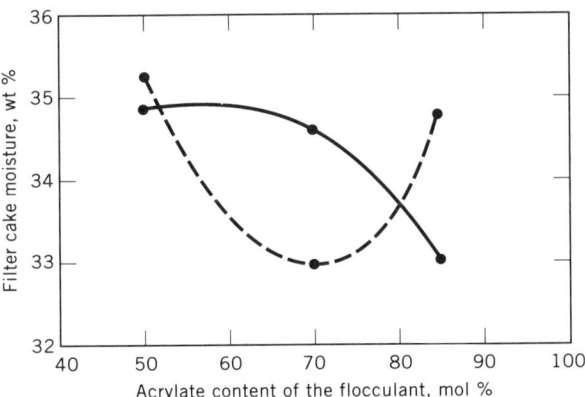

Fig. 6. Effect of acrylate content on coal refuse filter-cake moisture (59): (——), the effect on a longer chain flocculant (mol wt of 8×10^5); (---), shorter chain flocculant (mol wt of 2.5×10^6). Concentration of flocculant is 150 g/t.

498   DEWATERING

The Hi-Compact mechanical dewatering process of Humboldt Wedag takes advantage of sludge conditioning to achieve a high degree of secondary dewatering of highly compressible sludges. The sludge is first flocculated and settled. The thickened sludge is then chopped into pellets that are coated with an incompressible filter aid, such as ash, fine coal, or another material. The coated pellets are then compressed in a batch press at about 5000 kPa (50 atm) to remove 60% of the remaining water. The incompressible coating on the pellets provides a network of channels for the expressed water to follow out of the pressed cake (74).

Comparisons are available on the relative performance and costs for dewatering municipal sludges (2). The relative performance of different filters and conditioners on waste sludges is shown in Table 3. The same sludge was treated on two belt-filter presses, two different centrifuges, and rotary vacuum filter (75). In another study, a variable chamber filter press, fixed-volume filter press, continuous belt-filter press, and rotary vacuum filter were compared for performance, capacity, and capital and operating costs (69).

Figure 7 shows the ranges of solids content achieved by various dewatering methods. The high solids centrifuges can achieve the same or higher solids content achieved by belt presses on municipal sludges. For recommended test procedures and expected results, consult References 2, 3, and 69–71. Particle size is the most important variable in dewatering municipal sludges (25) and is directly related to many of the other variables correlated with moisture content (76). The amount of <5 μ particles is the largest variable, other than the nature of the sludge itself, in determining dewatering behavior (2).

Use of Mechanical Vibration. Vibration of the sludge or cake can further release entrapped moisture. Pretreatment of organic sludges and highly hydrated inorganic sludges using ultrasonic energy reduced by half the amount of polyelectrolyte polymer needed to achieve the same moisture content of the dewatered cake (77). Vibration of thixotropic slurries improved dewatering on a vacuum filter by breaking up trapped air and improving capillary channels (78). Other devices use ultrasonics to break up loose agglomerates, often of cosmetics (qv), so that these pass through a strainer (79).

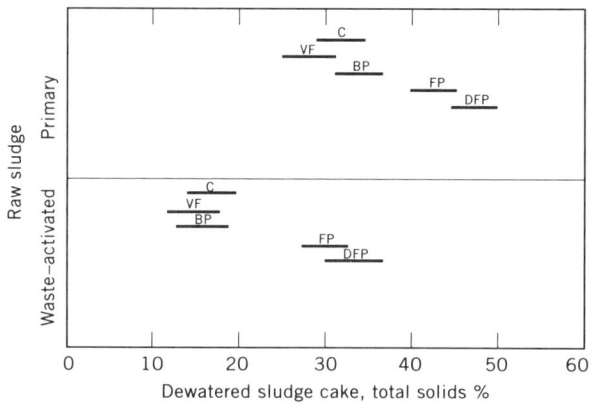

Fig. 7. Solids content of sludges dewatered by C, solid bowl centrifuge; VF, vacuum filter; BP, belt press; FP, filter press, and DFP, diaphragm filter press (2).

Table 3. Comparison of Filters on Aerobically Digested Sludges[a]

Property	Belt-filter press		Filter press		Rotary vacuum filter	Centrifuge
	Standard	Vacuum-assisted	Plates[b]	Cloth[b]		
cake produced, % solids	12–20	14–25	35–50	35–50	3–6	
operating pressure, MPa[e]	[c]	[c]	0.7–1.4	0.7–1.4	0.1[d]	2000–3000 g[f]
relative space requirements	2.6	2.6	3.0	3.0	1.0	
relative cost, $[d]	2.2	2.8	3.0	5.2	1.0	
maximum size[d], m²	60	60	400	60	100	
usual sludge conditioning	polymer	polymer	chemical and filter aid	chemical and filter aid	chemical and filter aid	polymer, 0–10 kg/t

[a]Ref. 69.
[b]Movable parts.
[c]Gravity plus pressure rolls.
[d]Values given are approximate.
[e]MPa unless otherwise noted. To convert MPa to psi, multiply by 145.
[f]Values are in units of gravitational acceleration.

Less Common Commercial Dewatering Processes

When solids dewatering is known to be a problem early in the process-design stage of a plant or is serious enough to warrant consideration of a range of dewatering alternatives, two approaches are available, and both can be used together. First, processes that begin the dewatering process while in the original suspension may be used. A second approach is to extract water or apply unusual desaturating forces to water present in sludges and cakes.

Agglomeration of Suspended Solids. *Pelletizing Precipitation.* Typical, cold, soda lime water softening generates a 5–15% soupy calcium carbonate sludge, which has been dumped into lagoons for disposal. By controlling reaction conditions of lime and hard water and providing for recirculating seeds of sand or calcium carbonate, precipitation of the carbonate can be controlled to form on the sand and to grow to 1.5-mm tight pellets. The pellets dewater by draining to less than 10% moisture (80,81). Control of crystal growth and crystal habit is used to improve dewatering in the production of phosphoric acid and in scrubbing of flue gas with lime (82,83). In each process, a precipitate of gypsum is formed. The most frequent applications of crystal growth regulation are to prevent scaling and to control freezing, for example, of trainloads of coal. The chemical principles are similar.

Additions of new flocculants after conventional thickening produce further dewatering of mineral slimes. A clay flocculated with polyacrylamides and rotated in a drum can produce a growth of compact kaolin pellets (84), which can easily be wet-screened and dewatered. A device called a Dehydrum, which flocculates and pelletizes thickened sludges into round, 3-mm pellets, was developed for this purpose. Several units reported in commercial operation in Japan thicken fine refuse from coal-preparation plants. The product contains 50% moisture, compared with 3% solids fed into the Dehydrum from the thickener underflow (85). In Poland, commercial use of the process to treat coal fines has been reported (86), and is said to compare favorably both economically and technically to thickening and vacuum filtration.

The U.S. Bureau of Mines has run large-scale tests on a similar process for treating <600-μm coal tailings. The tailings of a flotation cell are treated at a concentration of 3–7% solids. After mixing the slurry with 0.14 kg of polyethylene oxide flocculant per metric ton of dry solids and working for less than 30 s, the solids dewater to 50–70% moisture (87,88). The process has been demonstrated at a rate of 2300 L/min of waste slurry. Waste phosphate slimes have been similarly successfully consolidated from 4% to 40% solids (87,89). Mechanisms of flocculation in concentrated suspensions have been studied relatively little; however, conditions favoring pelletizing flocculation are described for colloidal latex suspensions (90).

Other applications include dewatering extremely fine (0.1 μm) laterite leach tailings (91). These pelletizing processes should be compared in flocculant consumption and operating and capital costs with belt-filter presses.

Oil Agglomeration. Pelletizing can also be accomplished using chemicals which make hydrophilic particles hydrophobic and thus agglomerate the particles into tight hydrophobic clusters (92,93). These processes, eg, Cattermole and Murex, were first used to make selective separations of a relatively minor amount of sulfide minerals (1–2%) in a slurry containing mostly silicates and carbonates. A more recent agglomeration process that was selective for coal added Freon to a slurry of <30 μm coal and mineral matter. The Freon caused the fine coal to agglomerate into large pellets that were recovered from the slurry on coarse static screens. The only remaining water was occluded in the pellets. The Freon, with its low heat capacity and boiling point (24°C), was recovered in a dryer (94).

A successful variation of oil agglomeration was used for removal and dewatering of soot from a 1–3% solids suspension consisting of <5-μm particles in refinery process waters (Fig. 8). Heavy oil was added to the dilute slurry and intensely agitated in a multistage mixer. The soot agglomerated with the oil to form 3–5 mm pellets that were easily screened from the water (95). The pellets contained only 5–10% water. The process was modified to recover very fine clean coal, and it produced highly uniform, hard, spherical pellets 1–2 mm in diameter.

This process has been applied to certain mineral oxides. Examples include recovery of cassiterite, SnO_2, from silicates (96), gold from ores (97), and ilmenite, $FeTiO_2$, from silicates (98). In each case, the normally hydrophilic mineral was treated with a surfactant (often a fatty acid) under conditions that would selectively coat the desired mineral. Additional oil is then added to agglomerate the treated minerals. Because of the cost of added reagent, the particles should have some intrinsic value. If no excess oil is added, but enough reagent is added and highly agitated, the pelletizing process is known as shear flocculation (99–101).

Another modification of dewatering methods using oil-agglomeration techniques combines the water exclusion of agglomeration with the ability of froth flotation to thicken. Dissolved air flotation (DAF) or induced air flotation (102) is used in over 300 municipalities for thickening wastewater solids and is also used at a much larger number of oil-well sites, refineries, and food-processing and rendering plants for removal of oil from wastewater. For oily wastes, or for selective removal of oily solids, DAF works very effectively. Typically, however, the thickened product contains only 2–4% solids. Because of poor performance, high energy consumption, and problems with controlling volatile organic gases and odors in the excess air, use as a thickening device is declining in favor of centrifuges and

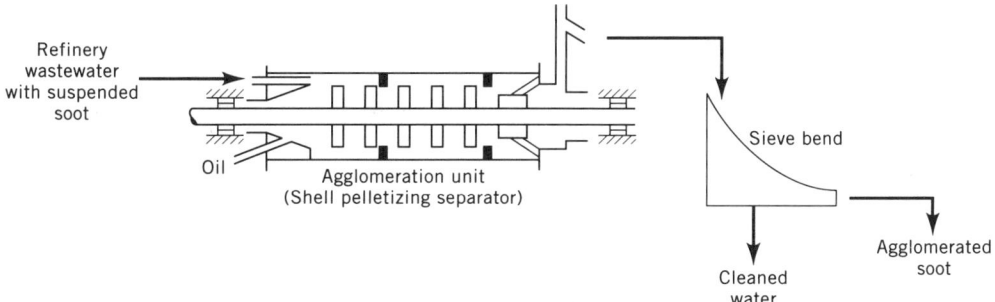

Fig. 8. Shell pelletizing separator (92).

belt presses. On food and industrial wastes, as a separator of oily wastes, DAF usage is strong. Preceding the operation with oil agglomeration or shear flocculation is particularly useful for inorganic particles finer than 15 µm.

Extraction Processes. *Oil-Phase Extraction.* In processes of dewatering by wetting a solid with an immiscible phase, another step in water displacement is possible. The use of very large quantities of an immiscible liquid allows extraction or transfer of the particles from one phase into the other, and the particles remain in a dispersed state in the new phase. To extract solids (as small as 0.1 µm) from water, a hydrophobic surface on the solid is needed. This surface is usually provided by using flotation reagents such as long-chain fatty acids, alkyl sulfonates, amines, or xanthates. Conversely, water has been used to agglomerate and extract hydrophilic solids that are dispersed in inorganic liquids (103). Crud, a concentration of solids at the liquid–liquid interface in normal solvent extraction, represents partial extraction (104). Pigment flushing is a technique used in paint (qv) and ink manufacture for transferring paint particles from an aqueous solution, where formation takes place, to a dispersed state in a nonaqueous carrier (105,106). Because the pigments are first filtered to remove soluble salts but are not dried before dispersion in the oil phase, the process function includes dewatering. Similarly, a process for manufacturing ferrofluids (stable dispersions of submicrometer magnetite in kerosene) consists of precipitating magnetite from water, adding oleic acid to coat the fine precipitate, and then contacting the wet filter cake with kerosene. The magnetite transfers into the kerosene and forms a stable suspension (107).

Solvent Extraction of the Liquid. Water contained in a cake or slurry can be extracted from the solids by dissolving the water in a solvent that is less expensive to evaporate than the contained water alone. The Institute of Gas Technology (IGT) has developed a laboratory process called solvent dewatering based on the principle that the solubility of water in selected solvents changes significantly with a change in temperature. In one example, hot solvent is mixed with wet peat and the water–solvent solution is then decanted. Upon cooling, the water precipitates as a separate phase (108). As of this writing, the process has not proved economical. The Resource Conservation Company has used chilled triethylamine (TEA) to dissolve water from sludge, and then warmed the extracted liquid to form an immiscible water phase. The effect of temperature on water solubility is opposite to the IGT solvent. TEA requires only 309 kJ/kg (129 Btu/lb) to evaporate, less than water (93,109). Similar processes have been considered for desalination (see WATER TREATMENT, DESALINATION).

Thermal Processes. *Thermal Drying.* The solvent-extraction processes discussed have progressively included evaporation. In this aspect, the Carver-Greenfield process of multiple-effect evaporation of water from sludges is an important alternative to very late stage dewatering. Organic sludges of about 20% solids are fed to the unit, along with enough oil to keep the sludge moving in the processing equipment. Using three effects in the evaporators, only 700–900 kJ/kg (300–387 Btu/lb) of evaporated water are required, compared with the 2.3 MJ/kg (1000 Btu/lb) needed in a single-effect evaporator (38 + 110) (see DRYING). A plant using this process to treat municipal wastewater sludges began operations in 1991 in Ocean County, New Jersey.

Thermal Treatment. A number of dewatering processes alter the interaction of solid with liquid. Most depend on making a hydrophilic solid hydrophobic by adding small quantities of surfactants and oils. Many biological sludges cannot be economically treated with reagents to provide hydrophobicity, and such treatments would have no effect on water bound inside the mycelium. Thermal treatment is intended for these organic sludges. Partial wet-air oxidation lowers the specific cake resistance of many biological gels and colloidal sludges by 50–100 times. For example, if a municipal sludge that normally thickens to 5% solids is successfully thermally treated, it thickens to 10–15% solids. On filtering the thickened sludge, the cake formed from the untreated sludge has about 15% solids. The thermally treated sludge can be filtered to 30–50% solids (2,111).

Thermal treatment consists of heating the sludge under a pressure of about 2.4 MPa (350 psi) and to temperatures of 150–225°C (111) for 15–40 min either with (low pressure oxidation) or without (heat treatment) additional air (2). Reactions, including partial oxidation, occur that change the nature and consume 1–5% of the solids. Unfortunately, this process produces significant quantities of acetic acid and other short-chain, soluble organics. Whereas in 1979 there were over 100 installations operating on sewage sludge, in 1992, because of improvements in mechanical dewatering processes, only 30 to 35 remain.

At least five related dewatering processes have been applied to peat and lignite. Peat and lignite have a high absorbed-moisture content (90% in peat and 40–50% water in lignite) and have a tendency to break down to undesirable fines and to become pyrophoric when dried. Steam drying comprises a family of processes very different from steam dewatering of filter cakes. These processes involve heating the peat or lignite containing the initial water content in an autoclave to temperatures of 150–200°C under pressures of about 1.3 MPa (189 psi) for about 15 min. This treatment causes the solids to shrink, eliminating water from pores and removing carboxylic acid and its salts from the surface. The steam treatment itself, considered separately from subsequent evaporative flashing, allows 30–50% of the initial water to drain or be pressure filtered from the product (112). Higher temperature and pressure lead to greater dewatering, and pressures up to 10 MPa (1500 psi) have been tested (113). These high pressures produce a completely dewatered lignite having high stability and little tendency to reabsorb moisture.

The only commercially used process in this group is the Fleissner process, developed in 1927 for drying lignite. One plant, operating in Austria between 1927 and 1960, achieved a capacity of 1700 t/d. In 1982, a number of other plants licensed by Fleissner were operating (114). There were also related processes in the pilot-plant stage (95,113–116), including one for dewatering peat with a capacity of 50,000 metric tons per year (117).

By raising the pressure, temperature, and available oxygen, virtually all the organic solids can be oxidized to CO_2 and H_2O. Ignition occurs at 200–225°C and wet-air oxidation is then autogenous. At 250–300°C, reactions occur rapidly. The vapor pressure of water at 300°C is about 9 MPa (1300 psi). Rather than needing 25–30% solids to achieve autogenous combustion of a sludge in air, only 0.5–1% organics in water is needed for autogenous wet-air oxidation (118). In supercritical water oxidation (SCWO), the temperature and pressure are raised still further to the critical point of water, 374°C and 22 MPa (3190 psi), which completely burns sewage sludge and toxic organics (119–121).

Freeze–Thaw Dewatering. Slow freezing of hydroxide, clay, and municipal sludges affects the water-retention properties of the solids when the frozen slurry is thawed. Two plants used this process on water-treatment sludges. The sludge is gelatinous aluminum hydroxide with organic and inorganic matter that typically thickens to 1–3% solids. In tests in the United Kingdom, the sludge, after a cycle of freezing and thawing, became a sandy, granular material that drained without needing a filter. A few small plants use freeze–thaw dewatering for municipal wastewater sludges (3–7% solids) and achieve a solids content of 25% with natural drainage after thawing the frozen sludge (2). The probable mechanisms of dewatering by freezing have been described (12). The ability to withstand freeze–thaw cycling is an important consideration in latex-paint manufacture (106). Similar considerations must be important in some frozen-food formulations.

Freeze Crystallization. Freezing may be used to form pure ice crystals, which are then removed from the slurry by screens sized to pass the fine solids but to catch the crystals and leave behind a more concentrated slurry. The process has been considered mostly for solutions, not suspensions. However, freeze crystallization has been tested for concentrating orange juice where solids are present. Commercial applications include fruit juices, coffee, beer, wine, and vinegar. A test on milk was begun in 1989 (123). Freeze crystallization has concentrated pulp and paper black liquor from 6% to 30% dissolved solids and showed energy savings of over 75% compared with multiple-effect evaporation. Only 35–46 kJ/kg (15–20 Btu/lb) of water removed was consumed in the process (124).

Clathrate Freezing. Clathrate freezing uses methane or ethane under pressure, where 1 mol of ethane traps 18 mol of water. Methane clathrates apparently can form in natural gas pipelines at room temperature. The process has been studied for dewatering wastewater treatment sludge, using Freon 11 (125). It has also been considered for removing water from the black liquor derived in the kraft pulping process for making paper fiber.

Capillary Suction Processes. The force needed to remove water from capillaries increases proportionately with a decrease in capillary radius, exceeding 1400 kPa (200 psi) in a 1-μm-diameter capillary. Some attempts have been made to use this force as a way to dewater sludges and cakes by providing smaller dry capillaries to suck up the water (27). Sectors of a vacuum filter have been made of microporous ceramic, which conducts the moisture from the cake into the sector and removes the water on the inside by vacuum. Pore size is sufficiently small that the difference in pressure during vacuum is insufficient to displace water from the sector material, thus allowing a smaller vacuum pump to be effective (126).

Electromagnetic Processes. *Electrical Enhancement of Dewatering.* Electrophoresis can be used to prevent a filter cake from forming on a filter medium while allowing water to pass through the medium from the slurry. Electrophoresis is used to move the particles upstream, opposite to the liquid movement, in order to prevent blinding of the medium (see ELECTROSEPARATIONS).

Once a matrix of particles is formed, whether filter cake, thickened underflow, or soil, applying a current to the fluid causes a movement of ions in the water and, with the ions, water of hydration. The phenomenon is called electroosmosis. The pressure generated on the fluid is given by (127):

$$P = \frac{2 \zeta E D}{\pi r^2}$$

where P = pressure in Pa; ζ = zeta potential in V; E = electric field in V/m; D = dielectric constant; r = radius of capillary in m. The amount of water moved is proportional to the intensity and time that power is applied, proportional to the zeta potential of the solid, and inversely proportional to the conductivity of the fluid (128). Results are often measured in kWh/t of dewatered product or in kWh/t of water removed.

High pressures are generated in the small capillary openings. Unlike pressure generated on a fluid by an externally applied force, however, the largest forces are generated at the shear plane of the liquid and the solid in the pores. The effect of a particle size on capillary retention force is shown in Table 4 (26). To calculate the pore radius used in the table, it is assumed that the pore is a cylinder that can just pass between three monosized spheres. The entries following P_E are the pressures developed by electroosmosis in those same pores, assuming a field of 3000 V/m. To generate that field in an electrolyte, a current of 600 A/m² must flow, and, in this case, it is assumed to be created in a 10^{-4} M, 1:1 electrolyte (0.2 S/m conductivity) (129).

In most applications, far less current and lower voltages are used. For example, in dewatering clay soils to stabilize dams, foundations, or dredged spoil, 20–100 V/m are commonly applied (130,131). In soil stabilization power is applied for weeks to months.

The effectiveness and costs of electroosmotic dewatering on a large number of clay-containing tailings from metallic, nonmetallic, and coal mines has been shown (132,133). The process can be used *in situ* or in a batch dewatering cell. One large test dewatered a very old, stable, 50% solid slime generated by a coal-washing plant. Applying 37 kWh/t of the final product, moisture was reduced to 19% in 24 h. Using a batch dewatering cell, 1100 t/week of slimes were dewatered (132). There is interest in using the technique to dewater hazardous waste sludge ponds before excavating. In the Netherlands, electroosmosis is used for *in situ* washing of metals from the soil.

Table 4. Effect of Particle Size on Capillary Retention Force

Parameter	Diameter, μm				
	50	20	15	12	1
$r_c = 0.165\ d/2$, μm	4.125	1.65	1.24	0.99	0.0825
P_c^a, kPa[b]	35.5	88.2	117	147	1760
P_E^c, kPa[b]				17.2	2430

[a] $P_c = 2 \sigma \cos \theta / r_c$, capillary pressure.
[b] To convert kPa to psi, multiply by 0.145.
[c] P_E = electrical pressure.

506 DEWATERING

Electroosmosis with Vibration. A commercially available electroacoustic dewatering (EAD) filter combines ultrasonic vibration of the cake with electroosmosis of water in the cake to achieve greater dewatering. Figure 9 shows a 2-m wide commercial machine developed for processing the cakes produced by conventional dewatering methods. The electrical current provides the force to move the liquid; an appropriate level of ultrasonic agitation helps to consolidate the cake, releases trapped gases and liquids, and maintains a liquid continuum for current to low. Typically, half the remaining water is removed from the dewatered cake fed to the EAD filter (135).

Magnetic Enhancement of Dewatering. *Liquid–Solid Separation.* When magnetic forces are considered for liquid–solid separations, it is usually for thickening and filtration rather than for dewatering. The Frantz Ferrofilter, commonly used to remove suspended ferromagnetic impurities from liquids, is somewhat analogous to a depth filter in its use of multiple collection sites and lack of a definable porous filter medium surface (135). The Ferrofilter principle has been extended to very large magnetized volumes of 1.7 m^3 and at high fields, from 0.15 T (1500 gauss) for the Ferrofilter to 2.0 T (2×10^4 gauss) for the new machines. The large, high gradient magnetic filters use a depth filter medium of 430 stainless-steel wool to form high gradients in the high field. Commercially, these are widely used for removing paramagnetic impurities from kaolin (136), and new superconducting magnets are now also commercially used (137). Less intense versions are used to separate and dewater magnetic iron ores and nonferromagnetic iron ores, for example, itabirite, and other ores containing specular hematite (138,139).

Fig. 9. Two-meter wide electroacoustic dewatering press (134). Courtesy of Ashbrook-Simon-Hartley (134).

Large-volume magnetic separators have been used to remove 90% of the suspended mill scale solids in steel rolling-mill wastewater, and for cleanup of steam-boiler water (see MAGNETIC SEPARATION). A much wider range of potential uses of magnetic separation for clarifying, filtering, and dewatering is possible when nonmagnetic impurities are made magnetic. In Japan, plating wastes with dissolved Cr, Mn, Cu, Zn, Cd, and so on, have been precipitated as magnetic ferrites (qv) and recovered from the wastewater with a simple magnetic separator. Details of this commercial process have been reported (140,141).

A simple separator used to recover the magnetic particles consists of a series of disks mounted on a shaft. Each disk has a number of permanent magnets mounted flush with the surface at its perimeter. The disks rotate into and out of the liquid containing the suspended magnetic material and lift the magnetic particles out of the stream. The magnets are then scraped clean (Fig. 10). Very low residence times are needed for removal of the particles compared to settling or flotation (142).

Adsorption of nonmagnetic suspended materials onto magnetic seeds has been proposed and tested for removal of suspended solids from drinking water (143) and of bacteria from municipal wastewater effluent. Using aluminum sulfate as a coagulant and 100–1000 ppm of magnetite, 80–90% of the suspended solids were removed (144). In laboratory experiments, 0.1% of magnetite was added to sewage, and the magnetite quickly became coated with biological flocs. These settled under the influence of a fairly weak magnetic field of 0.04 T in 4% of the time required for settling the untreated feed. The authors estimated that this would allow a fourfold increase in flow rate, a fivefold decrease in sludge volume, and an effluent with half the normal BOD content over normal wastewater treatment (146).

At least one study has specifically evaluated the use of a high gradient magnetic separator for dewatering a paramagnetic mineral slurry, malachite, $CuCO_3 \cdot Cu(OH)_2$, of an average particle size of 4 µm. Initial slurries of 2–12% solids were passed through the magnetic matrix and allowed to drain. When the field was turned off, the mineral could be successfully washed off the matrix to give a 40% solid slurry (146).

Finally, selective separation and dewatering of one suspended substance in a slurry containing different minerals or precipitates is possible by selectively adsorbing a magnetic material (usually hydrophobic) onto a solid that is also naturally or chemically conditioned to a hydrophobic state. This process (Murex) was

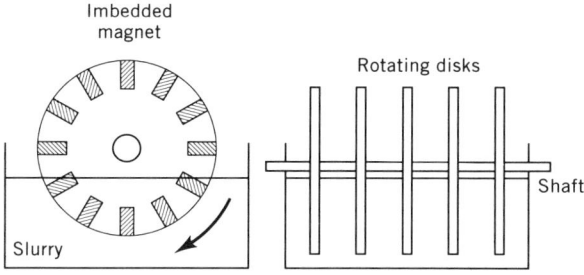

Fig. 10. Rotating magnetic-disk separator (142).

used on both sulfide ores and some oxides (145). More recently, hydrocarbon-based ferrofluids were tested and shown to selectively adsorb on coal from slurries of coal and mineral matter, allowing magnetic recovery (147). Copper and zinc sulfides were similarly recoverable as a dewatered product from waste-rock slurries (148).

Economic Aspects

Dewatering, a part of the liquid–solid separation equipment and supplies market, is not well segmented. The same equipment is often used for both separation and dewatering. The larger market for U.S. industrial and municipal liquid–solid separation equipment, not including disposable cartridges or membranes or high purity electronic or pharmaceuticals filtration, has been estimated at $1 billion for 1991 (149). Earlier estimates of the European market were $1.2 billion in 1988. Municipal water and wastewater treatment accounts for about $300 million, the largest segment, followed by the chemicals and allied industries segment, at $160 million. Other market segments are pulp and paper, general manufacturing, food and beverages, mineral processing, and oil and gas exploration and production, and electric utilities.

The market for filter equipment alone was about $580 million in 1991 (149), and pressure filters are on the order of $115–$160 million of the total filter market. About 500 U.S. companies manufacture liquid–solid separation equipment. Some of the manufacturers in this worldwide business include Ametek, Anderson International Corp., Arus-Andritz Inc., Ashbrook-Simon-Hartley, Bird Machine Co., Inc., Bepex Corp., Black Clawson, Centrifugal and Mechanical Industries, Denver Process Equipment Co., Dorr-Oliver Inc., Ebara Infilco, Envirex Inc., Eimco Process Equipment Co., Hitachi Plant Engineering and Construction Co., Ltd., Infilco-Degremont Inc., Ingersoll Rand Co., JWI Inc., KHD Humboldt Wedag, Krauss-Maffei Corp., Komline-Sanderson Engineering Corp., Kurita Machinery Manufacturing Co., Ltd., Larox Oy, Mitsubishi Kakoki Kaisha, Ltd., Rosenmund Inc., Sharples Division, Alfa-Laval Separation Inc., Sparkler Filters Inc., U.S. Filter, Wemco, and Zimpro-Passavant Environmental Systems Inc.

Flocculants and coagulants are sometimes used as pretreatments before dewatering. The market for flocculants and coagulants for water and wastewater treatment in the United States for 1989 was about $250 million, or 68,000 t, provided by 18 companies. In Europe, 22,000 t of flocculants and coagulants, made by 15 companies, had a market value of $115 million. In Japan, 12 companies made 23,000 t, valued at $184 million (150).

BIBLIOGRAPHY

"Dewatering" in the *Encyclopedia of Chemical Technology*, 3rd ed., Supplement, pp. 310–339, by B. Morey, Telic Technical Services; in *ECT* 4th ed., Vol. 8, pp. 30–58, by B. Morey, SRI International.

1. F. M. Tiller and C. S. Yeh, *Filtr. Sep.* **27,** 129 (1990).

2. O. E. Albertson and co-workers, *Dewatering Municipal Wastewater Sludges Design Manual*, EPA/625/1-87/014, Sept. 1987.
3. W. E. Stanley, *Sludge Dewatering, Manual of Practice 20*, Water Pollution Control Federation, Washington, D.C., 1969, p. 101.
4. P. A. Vesilind, *Treatment and Disposal of Wastewater Sludges*, Ann Arbor Science Publishers, Ann Arbor, Mich., 1974, Chapt. 6.
5. P. W. Thrush, ed., *Dictionary of Mining, Metallurgical and Related Terms*, U.S. Government Printing Office, Washington, D.C., 1968, p. 319; T. C. Collocott, ed., *Chambers Dictionary of Science and Technology*, Barnes and Noble, New York, 1971; D. N. Lapedes, *McGraw-Hill Dictionary of Scientific and Technical Terms*, 2nd ed., McGraw-Hill Book Co., New York, 1978.
6. H. Sato and co-workers, *Filtr. Sep.* **19**, 492 (1982).
7. H. B. Gala and S. H. Chiang, *Filtration and Dewatering; Review of Literature*, Report #DOE/ET/14291-1, U.S. Department of Energy, Washington, D.C., 1980.
8. L. Svarofsky, *Solid–Liquid Separations*, 3d ed., Butterworths, London, 1990.
9. R. J. Wakeman, *Filtration Post-Treatment Processes*, Elsevier Publishing Co., Amsterdam, 1975.
10. D. B. Purchas, *Solid/Liquid Separation*, Uplands Press, Croyden, UK, 1981.
11. F. M. Tiller, C. S. Yeh, and W. F. Leu, *Sep. Sci. Tech.* **22,** 1037 (1987).
12. F. M. Tiller and C. S. Yeh, *AIChE J.* **33**(8), 1241 (Aug. 1987).
13. A. Rushton and M. A. A. Arab, *Filtr. Sep.* **26,** 181 (1989).
14. *Humboldt Centripress ADS (Advanced Dewatering System)*, Humboldt Decanter, Inc., Norcross, Ga., 1990.
15. R. J. Wakeman and Fan Deshun, *Chemical Engineering Design Research* **69**(A5), 403 (1991).
16. F. M. Tiller, in B. M. Moudgil and B. J. Scheiner, eds., *Flocculation and Dewatering*, Engineering Foundation, New York, 1989, p. 89.
17. F. M. Tiller and L.-L. Horng, *AIChE J.* **29**(2), 297 (Mar. 1983).
18. *Continuous Pressure Filter for the Process Industries*, Ingersoll-Rand, Nashua, N.H., 1992.
19. M. Shirato and co-workers, *Filtr. Sep.* **7,** 277 (1970).
20. H. G. Schwartzberg, J. R. Rosenau, and G. Richardson, *AIChE Symp. Ser.* **73**(163), 177 (1977).
21. *Eimco Expressor Press*, Eimco Process Equipment Co., Salt Lake City, Utah, 1990.
22. *Magnum Press*, Bulletin MP-201, Parkson Corp., Fort Lauderdale, Fla., 1987.
23. *High Pressure Filtration Using the Tube Filter Press: A Technical and Economic Review*, Alfa-Dyne Inc., Cleveland, Ohio, 1991; J. Quilter, *Indust. Miner. Mag., Energy Suppl.*, 29 (Mar. 1983).
24. D. K. Bredeson, *J. Am. Oil Chem. Soc.* **60,** 163A (1983).
25. *Lasta Automatic Filterpress*, Ingersoll-Rand, Nashua, N.H., 1991.
26. A. F. Westergard, *Eng. Min. J.*, 60 (June 1983).
27. R. J. Wakeman and A. Rushton, *Filtr. Sep.* **13,** 450 (1976).
28. *Filtr. News*, 10 (Nov. 1989).
29. S. A. Bratten and S. V. Tracy, "Improved Concentrate Dewatering Utilizing Variable Volume Pressure Filters," paper presented at *Annual Meeting Society for Mining, Metallurgy and Exploration*, Denver, Colo., 1991.
30. *Filtr. News*, 32 (Jan. 1992).
31. M. Münter and U. Grén, *Filtr. Sep.* **27,** 264 (1990).
32. *Elastomer and Polymer Processing Systems*, Bulletin MPR 76, 1976, *Pre-Press*, Bulletin FO 175; and *French Dual Cage Screw Press*, Bulletin OP8130, 1981, The French Oil Mill Machinery Co., Piqua, Ohio.
33. L. H. Tindale and S. R. Hill-Haas, *J. Am. Oil Chem. Soc.* **53,** 265 (1976).

34. J. A. Ward, *J. Am. Oil Chem. Soc.* **53,** 261 (1976).
35. D. K. Bredeson, *J. Am. Oil Chem. Soc.* **55,** 762 (1978).
36. *Anderson Duo Crackling Expeller Presses*, Bulletin Duo 375-2, 1981; *Anderson Rubber and Plastic Polymer Dewatering and Drying Equipment*, Bulletin RDD 73, 1980; and *Anderson Expeller Presses*, Bulletin 359 R, 1980, Anderson International Corp., Cleveland, Ohio.
37. *Pressmaster Press*, Bulletin SB82-002B, Beloit Corp., Jones Division, Dalton, Mass., 1982.
38. K. Ohmiya and S. Takahashi, *J. Water Pollut. Control Fed.* **52,** 943 (1980).
39. *Carver Cocoa Presses*, Bulletins HV-A and FP-1, Fred S. Carver, Inc., Menomonee Falls, Wisc., 1981.
40. R. H. Perry and C. H. Chilton, *Chemical Engineers Handbook*, 5th ed., McGraw-Hill Book Co., New York, 1973, pp. 19-101–19-104.
41. *V-Press*, Bulletin 64-5, Bepex Corp., Rietz Division, Santa Rosa, Calif., 1982; Ref. 14, p. 516.
42. B. K. Parekh and J. P. Matoney, in J. W. Leonard, III, ed., *Coal Preparation*, 5th ed., Society for Mining, Metallurgy and Exploration, Inc., Littleton, Colo., 1991, Chapt. 8.
43. *Technical Bulletin 2-2-16/1-B* and *Vari-Nip Technical Discussion*, Ingersoll-Rand, Nashua, N.H.
44. C. E. Silverblatt and D. A. Dahlstrom, *Ind. Eng. Chem.* **46,** 1201 (1954).
45. C. S. Simons and D. A. Dahlstrom, *Chem. Eng. Prog.* **62**(1), 75 (1966).
46. A. F. Baker and A. W. Duerbrouck, in A. C. Partridge, ed., *Proceedings of the International Coal Preparation Congress*, 1977.
47. J. H. Brown, *Can. Min. Metall. Bull.* **58,** 315 (1965); *Transactions Can. Inst. Min. Met.* **68,** 105 (1965).
48. F. M. Tiller and J. R. Crump, *Chem. Eng. Prog.* **74,** 65 (Oct. 1977).
49. U.S. Pat. 4,107,028 (Aug. 15, 1978), R. K. Emmett, S. D. Heden, and R. A. Summerhays (to Envirotech Corp.).
50. S. M. Moos and R. E. Dugger, *Min. Eng.* **31,** 1479 (1979).
51. H. B. Gala, S. H. Chiang, and W. W. Wen, *Proceedings World Filtration Congress III*, The Filtration Society, Downington, Pa., 1982.
52. B. Herath, P. Geladi, and C. Albano, *Filtr. Sep.* **26,** 53 (1989).
53. J. G. Groppo and B. K. Parekh, *Effect of Metal Ions on Vacuum Filtration of Coal*, Society for Mining, Metallurgy and Exploration, Annual Meeting, Denver, Colo., 1991.
54. A. Rushton, *Filtr. Sep.* **13,** 573 (1976).
55. P. J. Lafforgue and co-workers, paper presented at *Society of SME-AIME Annual Meeting*, Feb. 1982, preprint 82-22, available from the United Engineering Society Library, New York.
56. *Proceedings of the Consolidation and Dewatering of Fine Particles Conference*, University of Alabama, Aug. 1982, available from U.S. Bureau of Mines, Tuscaloosa, Ala.
57. *Proceedings of the Progress in the Dewatering of Fine Particles Conference*, University of Alabama, Apr. 1981, available from the U.S. Bureau of Mines, Tuscaloosa, Ala.
58. F. N. Kemmer and J. McCallion, eds., *Nalco Water Handbook*, McGraw-Hill Book Co., New York, 1979, Chapts. 8–9.
59. M. E. Lewellyn and S. S. Wang, in R. B. Seymour and G. A. Stahl, eds., *Macromolecular Solutions Solvent-Property Relationships in Polymers*, Pergamon Press, New York, 1982, pp. 134–150.
60. S. K. Nicol, *Proc. Australas Inst. Min. Metall.*, 37 (Dec. 1976).
61. M. J. Pearse and T. Barnett, *Filtr. Sep.* **17,** 460 (1980).
62. Ref. 14, p. 53.

63. R. Leutz and M. Clement, *Filtr. Sep.* **7,** 193 (1970).
64. S. K. Mishra, in B. M. Moudgil and B. J. Scheiner, eds., *Flocculation and Dewatering*, 1989, Engineering Foundation, New York, p. 89.
65. M. J. Pearse, in Ref. 56, pp. 41–89.
66. G. M. Moody, *Trans. Inst. Min. Metall.* **99,** C137 (1990).
67. V. P. Mehrotra and co-workers, *Filtr. Sep.* **19,** 197, (1982).
68. R. J. Schwartz, *Sludge Dewatering, Manual of Practice 20*, Water Pollution Control Federation, Washington, D.C., 1969, pp. 13–40.
69. A. F. Cassel and B. P. Johnson, *Evaluation of Devices for Producing High Solids Sludge Cake*, NTIS Report No. PB80-111503, National Technical Information Service, Washington, D.C., 1980.
70. D. DiGregorio and J. F. Zievers, in W. W. Eckenfelder, Jr., and C. J. Santhanam, eds., *Sludge Treatment*, Marcel Dekker, Inc., New York, 1981, Chapt. 6, pp. 142–207.
71. *Wastewater Engineering*, 2nd ed., McGraw-Hill Book Co., New York, 1979, Chapt. 11.
72. G. Smith, "Optimizing Operation of Low-Load Aeration Systems: Wasting More ... and Paying Less", presented at *15th Annual Conference of the Alabama Association of Water Pollution Control*, Orange Beach, Ala., Nov. 1991, available from Envirex Inc.
73. M. Anazia, in *Mining Eng.*, 485 (May 1990).
74. *Hi-Compact Method—A Purely Mechanical Process for Maximum Secondary Dewatering of Sludges*, Bulletin 5-400e, KHD Humboldt Wedag AG, Cologne, Germany, 1988.
75. B. Sawyer, R. Watkins, and C. Lue-Hing, *Proceedings 31st Industrial Waste Conference*, Purdue University, Lafayette, Ind., 1976, p. 537.
76. P. R. Karr and T. M. Keinath, *J. Water Pollut. Control Fed.* **50,** 1911 (1978).
77. J. Bien, *Filtr. Sep.* **25,** 425 (1988).
78. *Filtr. Sep.* **27,** 163 (1990).
79. *Fuji Micro-Sonic Filter*, Fuji Filter Manuf. Co., Tokyo, 1988.
80. *Spiractor*, Bulletin 5852, Permutit Co., Inc., Paramus, N.J., 1979.
81. *SWA/KW Reactor*, Esmil Water Systems Ltd., Buckinghamshire, UK, 1976, 1987.
82. D. A. Dahlstrom, in M. P. Freeman and J. A. FitzPatrick, eds., *Theory, Practice and Process Principles for Physical Separations*, Engineering Foundation, New York, 1977, pp. 261–273; *EPRI Report F.P. 937*, Electric Power Research Institute, Palo Alto, Calif., 1979.
83. A. D. Randolph and D. Etherton, *Study of Gypsum Crystal Nucleation and Growth Rates in Simulated Flue Gas Desulfurization Liquors*, EPRI Report CS1885, Electric Power Research Institute, Palo Alto, Calif., 1981.
84. M. Yusa and A. M. Gaudin, *Am. Ceram. Soc. Bull.* **43,** 402, (1964).
85. M. Yusa and co-workers, in A. C. Partridge, ed., *Proceedings of the 7th International Coal Preparation Congress*, Australian National Committee, Sydney, 1976.
86. J. Szczpya, in P. Somasundaran, ed., *Fine Particles Processing*, Society of Mining Engineers of AIME, Littleton, Colo., 1980, p. 1676.
87. B. J. Scheiner and A. G. Smelley, *Dewatering of Thickened Phosphate Clay Waste from Disposal Ponds*, Paper A81-6, The Metallurgical Society of AIME, Warrendale, Pa., 1981; J. R. Pederson, ed., *U.S. Bureau of Mines Research 81*, U.S. Government Printing Office, Washington, D.C., 1981, p. 83.
88. B. J. Scheiner and co-workers, "New Dewatering Techniques for Fine Particle Waste," *Proc. 16 Int. Min. Proc. Cong.*, 1951, Elsevier Publishing Co., Amsterdam, 1988.
89. B. J. Scheiner and M. M. Ragin, *Society for Mining Metallurgy and Exploration Transactions* **284,** 1801 (1988).
90. K. Higashitani and T. Kubota, *Powder Technol.* **51,** 61 (1987).

91. R. M. Hoover and P. V. Avotins, *Development of Polymer Pelletization for Enhancing Solid Liquid Separation of Leached Laterite Residue*, Paper A78-13, The Metallurgical Society of AIME, Warrendale, Pa., 1978.
92. V. P. Mehrotra and co-workers, *Int. J. Miner. Process.* **11,** 175 (1983).
93. V. P. Mehrotra and co-workers, *Min. Eng. (NY)* **32,** 1230 (1980).
94. D. V. Keller, Jr., in Ref. 56, pp. 152–171.
95. F. J. Zuiderweg and co-workers, *Chem. Engineer (London)*, 223 (July 1968).
96. F. W. Meadus and co-workers, *Can. Min. Metall. Bull.*, 968 (1966).
97. F. W. Meadus and co-workers, *Can. Min. Metall. Bull.*, 1326 (1969).
98. I. E. Puddington and B. D. Sparks, *Miner. Sci. Eng.* **7,** 282 (1975).
99. P. T. L. Koh and L. T. Warren, *13th International Mineral Processing Congress*, Warsaw, Poland, 1979.
100. A. M. Gaudin and P. Malozemoff, *J. Phys. Chem.* **37,** 599 (1933).
101. A. M. Gaudin and P. Malozemoff, *Trans. Am. Inst. Min. Metall. Engrs.* **112,** 303 (1934).
102. O. E. Albertson, *Sludge Thickening, Manual of Practice FD1*, Task Force on Sludge Thickening, Water Pollution Control Federation, Washington, D.C., 1980, p. 33.
103. H. M. Smith and I. E. Puddington, *Can. J. Chem.* **38,** 1911 (1960).
104. G. M. Ritcey and A. W. Ashbrook, *Solvent Extraction Principles and Applications to Process Metallurgy*, Elsevier Publishing Co., Amsterdam, 1979, Part II, p. 669.
105. R. Stratton Crawley, in P. Somasundaran and M. Arbiter, eds., *Beneficiation of Mineral Fines*, National Science Foundation, Society of Mining Engineers, AIME, Littleton, Colo., 1979, p. 317.
106. D. Bass, *Paint Manuf.*, 5 (Jan. 1957).
107. G. W. Reimers and S. E. Khalafalla, *Preparing Magnetic Fluids by a Peptizing Method*, U.S. Bureau of Mines Technical Progress Report 59, U.S. Bureau of Mines, Washington, D.C., Sept. 1972; U.S. Pat. 3,843,540 (Oct. 22, 1974), G. W. Reimers and S. E. Khalafalla (to U.S. Department of the Interior).
108. C. L. Tsaros, in J. W. White and B. F. Feingold, eds., *Peat Energy Alternatives*, Institute of Gas Technology, Chicago, 1980.
109. *Chem. Eng.*, 82 (June 4, 1979).
110. S. A. Raksit, *Carver-Greenfield Pilot Demonstration*, LA-OMA Project Los Angeles Department of Public Works, Los Angeles, 1978.
111. J. Jacknow, *Sludge* **2**(4), 26, (July 1979).
112. Can. Pat. 1,010,477 (Nov. 8, 1977), E. J. Wasp (to Bechtel International Corp.).
113. W. H. Oppelt and co-workers, *Drying North Dakota Lignite to 1500 Psi by the Fleissner Process*, Report of Investigations 5527, U.S. Bureau of Mines, Washington, D.C., 1959.
114. B. Stanmore, D. N. Boria, and L. E. Paulson, *Steam Drying of Lignite: A Review of Processes and Performance*, DOE/GFETC/R1-82/1 (DE82007849), U.S. Department of Energy, available from National Technical Information Service, Washington, D.C., 1982.
115. J. B. Murray and D. G. Evans, *Fuel* **51,** 290 (1972).
116. U.S. Pats. 4,052,168; 4,129,420 (1977), E. Koppelman; G. Parkinson, *Chem. Eng.*, 77 (Mar. 27, 1978).
117. J. Rohr, Wheelabrator-Frye, Hampton, N.H., personal communication, 1992.
118. D. F. Othmer, *Mech. Eng.*, 30 (Dec. 1979).
119. J. Josephson, *Environ. Sci. Technol* **16,** 548A (1982).
120. R. W. Shaw and co-workers, *Chem. Eng. News* **69**(51), 26 (Dec. 1991).
121. *Supercritical Water Oxidation Engineering Bulletin*, U.S. Environmental Protection Agency, EPA 540/S-92/006, 1992.
122. G. S. Logsdon and E. Edgerley, Jr., *J. Am. Water Works Assoc.* **63,** 734 (Nov. 1971).

123. J. Douglas and A. Amarnath, *Freeze Concentration: an Energy-Efficient Separation Process*, EPRI Journal, p. 17, 1989.
124. H. E. Davis and C. J. Egan, *AIChE Symp. Ser. 207* **77,** 50 (1981).
125. B. Molayem and T. Bardakci, *Dewatering Wastewater Treatment Sludge by Clathrate Freezing: a Bench-Scale Study*, EPA, NTIS PB 86-239779/AS, 1986.
126. *Filtr. Sep.* **28,** 238 (1991).
127. A. W. Adamson, *Physical Chemistry of Surfaces*, 3rd ed., John Wiley & Sons, Inc., New York, 1974, p. 212.
128. N. C. Lockhart, in Ref. 51, pp. 325–332.
129. M. P. Freeman, in G. Hetsrom, ed., *Handbook of Multiphase Systems*, Hemisphere, New York, 1982, Chapt. 9.3, pp. 9-9–9-115.
130. B. A. Segall and co-workers, *ASCE Geotech Engineering Division J. GT* **106,** 1148 (1980).
131. C. A. Fetzer, *Proceedings ASCE, Journal of Soil Mechanics and Foundations Division* **93 SM4,** 85 (1967).
132. R. H. Sprute and D. J. Kelsh, "Dewatering Fine Particle Waste Suspensions with Direct Current," *Encyclopedia of Fluid Mechanics*, Gulf Publishing, Houston, Tex., 1986, Chapt. 27.
133. R. H. Sprute and D. J. Kelsh, *Electrokinetic Densification of Solids in a Coal Mine Sediment Pond—A Feasibility Study*, Bureau of Mines Report of Investigations 9137, 1988.
134. T. Schiene, *ElectroAcoustic Dewatering*, Ashbrook-Simon-Hartley, Houston, Tex., 1990.
135. *Frantz Ferrofilter Magnetic and Electromagnetic Separators*, Bulletins EM and PM, S. G. Frantz Co., Inc., Trenton, N.J., 1980.
136. C. Mills, *Ind. Miner. (London)*, 41 (Aug. 1977).
137. *Superconducting High Gradient Magnetic Separator System*, Eriez Magnetics, Erie, Pa., 1991.
138. J. E. Lawver and D. M. Hopstock, *Miner. Sci. Eng.* **6,** 154 (July 1974).
139. D. M. Thayer and P. B. Linkson, *Trans. AIME* **270,** 1897 (1981).
140. N. Nojiri and co-workers, *J. Water Pollut. Control Fed.*, **52,** 1898 (1980).
141. Y. Tamaura and co-workers, *Water Res.* **13,** 21 (1979).
142. M. Miura and T. M. Williams, *Chem. Eng. Prog.*, 66 (Apr. 1978).
143. B. A. Bolto and co-workers, *J. Polym. Sci. Polym. Symp.*, 211 (1975).
144. R. R. Oder and B. I. Horst, *Filtr. Sep.* **13,** 363 (1976).
145. A. Faseur and co-workers, *Filtr. Sep.* **25,** 344 (1988).
146. P. Chakrabarti and co-workers, *Filtr. Sep.* **19,** 105 (1982).
147. T. A. Sladek and C. H. Cox, *Coal Preparation Using Magnetic Separation*, Vol. 4, *Evaluation of Magnetic Fluids for Coal Beneficiation*, EPRI Report CS1517, Energy Electric Power Research Institute, Palo Alto, Calif., July 1980.
148. U.S. Pats. 1,043,851 (Nov. 1912); 1,043,850 (Nov. 1912); 996,491 (Aug. 1911); 993,717 (June 1911), A. A. Lockwood.
149. *Filtr. Sep.* **29,** 278 (1992).
150. *SRI International Estimates*, SRI International, Menlo Park, Calif., 1991.

General References

O. E. Albertson and co-workers, *Dewatering Municipal Wastewater, Sludges Design Manual*, EPA/625/1-87/014, Sept. 1987.

H. B. Gala and S. H. Chiang, *Filtration and Dewatering: Review of Literature*, DOE/ET/14291-1, 1980.

Liquid Filtration Manual, Sedimentation and Centrifugation Manual, McIlvaine Co., Northbrook, Ill.

D. B. Purchas, *Solid-Liquid Separation Technology*, Uplands Press, 1981.
L. Svarovsky, *Solid–Liquid Separation*, 3rd ed., Butterworths, 1990.

Equipment Selection

Liquid Filtration Newsletter, Sedimentation and Centrifugation Newsletter, McIlvaine Co., Northbrook, Il.
E. Mayer, *Filtration News*, 24 (May 1988).
A. Ruston, *Selection and Use of Liquid/Solid Separation Equipment*, Institution of Chemical Engineers, 1982.

Solid/Liquid Separation Scaleup

O. E. Albertson and co-workers, *Dewatering Municipal Wastewater, Sludges Design Manual*, EPA/625/1-87/014, Sept. 1987.
D. B. Purchas and R. J. Wakeman, Uplands Press, 1986.
F. M. Tiller and C. S. Yeh, *Filtr. Sep.* **27**, 129 (1990).

Equipment Optimization

D. A. Dahlstrom, *Coal* **95** (4), 52 (1990).
B. J. Scheiner, *Fluid/Particle Separation Journal* **1,** 46 (1990).

Pretreatment

B. M. Moudgil and B. J. Scheiner, eds., *Flocculation and Dewatering*, Engineering Foundation, New York, 1989.
R. J. Wakeman, *Filtration Post Treatment Processes*, Elsevier, Amsterdam, 1975.

BOOKER MOREY
SRI International

DIALYSIS

Dialysis is a membrane separation process in which one or more dissolved species flow across a selective barrier in response to a difference in concentration. It is the earliest molecularly separative membrane process to be identified and described (1) (see MEMBRANE TECHNOLOGY). The mode of transport is diffusion, and separation occurs because small molecules diffuse more rapidly than larger ones, and also because the degree to which membranes restrict solute transport usually increases with permeant size. The basic principles are illustrated in Figure 1. Solute c is present at concentrations c' and c'' on opposite sides of a membrane. In the absence of differences in pressure, temperature, or electrical potential, Fick's phenomenological first-order description of diffusion, published in 1855 (2), states that solute moves from a region of greater to lesser concentration and at a rate

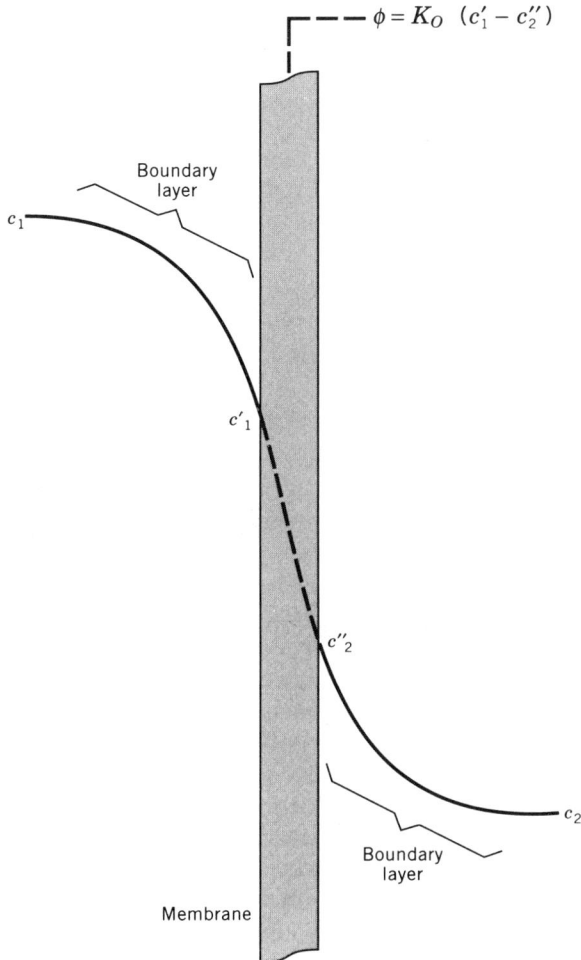

Fig. 1. General dialysis is a process by which dissolved solutes move through a membrane in response to a difference in concentration and in the absence of differences in pressure, temperature, and electrical potential. The rate of mass transport or solute flux, ϕ, is directly proportional to the difference in concentration at the membrane surfaces (eq. 1). Boundary layer effects, the difference between local and wall concentrations, are important in most practical applications.

proportional to the difference. In equation 1, ϕ = unit solute flux in g/cm²s; D = diffusion coefficient, cm²/s; c = concentration in g/cm³; x = distance in cm; and the minus sign accounts for the convention that flux is considered positive in the direction of decreasing concentration.

$$\phi = -D\frac{\partial c}{\partial x} \qquad (1)$$

Diffusion coefficients decrease roughly in proportion to the square root of molecular weight, are widely tabulated for aqueous solutions, or may be estimated from the Stokes Einstein equation (3). Ignoring boundary layer effects for the moment, and by assuming that diffusion within the membrane is analogous to that in free solution, equation 1 can be integrated across a homogeneous membrane of thickness d to yield the following equation, where S represents the dimensionless solute partition coefficient, ie, the ratio of solute concentration in external solution to that at the membrane surface, and D_M represents solute diffusion within the membrane and is assumed independent of solute concentration.

$$\phi = \frac{SD_M \, \Delta c}{d} \qquad (2)$$

The product SD_M is often termed permeability; if two or more solutes are dialysing at the same time, the degree of separation or enrichment is proportional to the ratio of their permeabilities. The closer the permeability of a membrane is to that of an equivalent thickness of free solution, the more rapid is the resultant dialytic transport. Considerable effort has been devoted to understanding how the physical and chemical properties of a membrane determine its permeability. The simplest approaches are geometric and consider the membrane to comprise a series of parallel pores that provide a topographic obstacle to hard noninteracting permeant molecules (4); far more complex analyses are also available (5,6). As a general rule, permeability for a particular species increases with porosity (solute content) of the membrane and with the diameter of its pores. Equation 2 also states that the mass flow rate of solute is inversely proportional to membrane thickness, but the degree of separation (selectivity) is independent of thickness. For this reason, membranes are always made as thin as possible consistent with the requirements of mechanical strength and reliability. Equation 2 is often further simplified to the following expression for flux per unit of membrane area, where thickness is incorporated into an overall membrane mass-transfer coefficient, K_M, with units of cm/s.

$$\phi = K_M \, \Delta_C \qquad (3)$$

Dialysis transport relations need not start with Fickian diffusion; they may also be derived by integration of the basic transport equation (7) or from the phenomenological relationships of irreversible thermodynamics (8,9).

Solutions adjacent to the membranes are rarely well mixed, and the resistance to transport resides not just in the membrane but also in the fluid regions, termed boundary layers, on both the dialysate and feed side. Boundary layer effects typically account for from 25 to 75% of overall resistance. They are minimized by rapid convective flow tangential to the surface of the dialysing membrane. When fluid pathways are thin, juxtamembrane flow is laminar, and boundary layer resistance decreases with increasing wall shear rate. Where geometry permits higher Reynolds numbers, flow becomes turbulent and resistance varies with net tangential velocity. Geometric turbulence promoters are often employed. All tactics to reduce boundary layer result in higher energy utilization. Quantitatively, the membrane resistance becomes part of an overall coefficient K_O which, for conceptual purposes, is broken down into three independent and reciprocally additive components:

$$\frac{1}{K_O} = \frac{1}{K_B} + \frac{1}{K_M} + \frac{1}{K_D} \qquad (4)$$

$$R_O = R_B + R_M + R_D \qquad (5)$$

where K is device-averaged mass-transfer coefficient (or permeability) in cm/s, R is device-averaged resistance in s/cm, and the subscripts B, M, and D respectively denote the feedstream, membrane, and dialysate. Note that K_M in equation 4 is identical to that in equation 3. K_B can be estimated for many relevant conditions of geometry and flow using mass transport analysis based on wall Sherwood numbers (10). K_M is best obtained by measurements employing special test fixtures in which boundary layer resistances are negligible or known (11,12). K_D is more problematic, and is usually obtained by extrapolations based on Wilson plots (13). Boundary layer theory, as well as techniques for correlation, estimation, and prediction of the constituent mass-transfer coefficients have been reviewed in two particularly lucid monographs (14,15). Overall solute transport is obtained from local flux by mass balance and integration; for the most common case of countercurrent flow:

$$\phi = (c'_i - c''_i)\frac{Q_B}{A} \frac{\exp\left[\frac{K_O A}{Q_B}\left(1 - \frac{Q_B}{Q_D}\right)\right] - 1}{\exp\left[\frac{K_O A}{Q_B}\left(1 - \frac{Q_B}{Q_D}\right)\right] - \frac{Q_B}{Q_D}} \qquad (6)$$

where c'_i and c''_i represent inlet concentrations in the feed and dialysate streams in g/cm^3, A represents membrane surface area in cm^2, Q_B and Q_D are feed and dialysate flow rates in cm^3/min, and ϕ and K_O are as defined in equations 3 and 4. Derivations of this relationship and similar expressions for cocurrent or crossflow geometries can be found in the literature (14,16,17) (see MASS TRANSFER).

Dialysis is a highly constrained process. Molecular diffusion is slow in the context of industrial dimensions. The driving force is set by the system itself, decreases in the course of purification, and is not amenable to extrinsic augmentation. The permeant species is not recovered in pure form, and is necessarily more dilute in the dialysate than in the starting stream. Low energy utilization is offset by high capital costs. For these reasons, dialysis has been largely limited to laboratory separations or specialized *in vivo* pharmacological investigations, and has enjoyed very limited success as a broad-based commercial unit operation. But the slow and gentle nature of dialysis has a special appeal for biologic applications, particularly when partial purification of the feed stream, rather than recovery of a product, is intended. Commercially significant examples include the adjustment of alcohol content of beverages and the removal of salts from solutions of proteins or other biologic macromolecules. However, the most successful and widespread application of dialysis—or for that matter of any membrane process—is the support of patients with kidney failure by repeated intermittent blood cleansing. In 1992 nearly half a million patients were maintained on dialysis, and the worldwide commercial aspects of this enterprise exceeded 15 billion U.S. dollars. Dialysis is closely related to membrane gas separation (qv), pervaporation, ultrafiltration, and controlled release of pharmaceuticals. Particularly common is diafiltration, combined simultaneous dialysis, and ultrafiltration (qv).

Industrial Dialysis

The recovery of caustic from hemicellulose in the rayon process was well established in the 1930s (18), and is still used in modern times (19) (see PULP). Very few new industrial applications of dialysis emerged during the 1940–1980 period. More recently, interest has reawakened in isobaric dialysis as a unit operation for the removal of alcohol from beverages (20,21) and in the production of products derived from biotechnology (22,23).

Alcohol-free beer has grown in popularity since the early 1980s in response to changing life-styles and legislative restraints on alcohol consumption. Markets are also developing for alcohol-free wine. By the end of 1992, 40 key beer breweries worldwide had installed dialysis plants with an annual capacity of more than 189,000 m^3 (5 × 10^7 gal) of beer. The process is illustrated in Figure 2. Alcohol is removed from beer by dialysis, the dialysate is distilled to remove alcohol, and the raffinate is recycled as a dialysate stream. The combination of dialysis and distillation preserves the flavor of the product (24). Dialysis is isothermal so the beer need not be heated. Higher boiling alcohols, esters, and carbohydrates that impart the special flavor to the beverage are already present in the dialysate and thus are not removed from the feed stream. A typical commercial installation is shown in Figure 3.

Dialysis plays an important role in the expanding biotechnology industry, but rarely as a stand-alone unit operation. It is applicable to the removal of salts from heat-sensitive or mechanically labile compounds such as vaccines, hormones, enzymes, and other bioactive cell secretions. In these instances, process efficiency is almost always increased by combining dialysis with ultrafiltration in the proc-

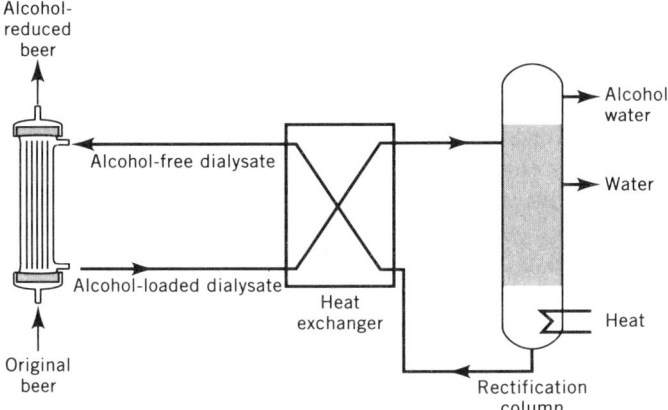

Fig. 2. Schematic of alcohol reduction in beverages. Countercurrent dialysis is combined with distillation. The separation process is isothermal, and high boiling ingredients, present in the dialysate, are preserved. In this fashion, alcohol removal is accomplished with minimal perturbation in flavor.

Fig. 3. A commercial dialysis facility showing the dialysis section of a German brewery where alcohol is removed from beer. Technical dialysis modules contain up to 50,000 capillaries and around 23 m^2 (250 ft^2) of membrane surface area. Typical plants might contain between 50 and 100 modules. Courtesy of Holstein and Kappert Processtechnik GmBH, Dortmund, Germany.

ess known as diafiltration. Dialysis provides a simple means to control media and extracellular environment in bioreactors. Dialyzers can also offer the basis for a novel bioreactor design: the extraluminal region of a hollow fiber dialyzer provides an excellent growth environment for mammalian cells when the lumen is perfused with oxygen and nutrients (25). In the production of monoclonal antibodies, for example, a small benchtop bioreactor can readily equal the antibody production of several thousand mice. This technology, in its early stages of development, is far from mature.

Laboratory Dialysis

Until the early 1960s, laboratory investigators relied on dialysis for the separation, concentration, and purification of a wide variety of biologic fluids. Examples include removal of a buffer from a protein solution or concentrating a polypeptide with hyperosmotic dialysate. Specialized fixtures were sometimes employed; alternatively, dialysis tubes, ie, cylinders of membrane about the size of a test tube and sealed at both ends, were simply suspended in a dialysate bath. In recent years, dialysis as a laboratory operation has been replaced largely by ultrafiltration and diafiltration (see MICROBIAL AND VIRAL FILTRATION).

Microdialysis is a highly specialized application of the technique (26–28). In its simplest form, a U-shaped dialysis capillary is surgically implanted into the tissue of a living animal. Isotonic dialysate is pumped through the tubing at a flow rate low enough to allow equilibration with small solutes in the host's extracellular fluid. Concentration of solutes in the exiting fluid thus approaches those in the extracellular portion of the tissue. This technique is extremely useful because it permits uninterrupted sampling of the chemistry of individual tissues or body compartments without drawing blood. Once the implant is established, a microdialysis probe is capable of sampling continuously for days or even weeks. The procedure is most widely used in rodent studies, and is most popular for direct implantation into the brain by standard animal neurosurgical techniques. Microprobe designs range from straightforward U tubes to complex concentric capillaries. Perfusate flow rate is extremely low, around 10^{-6} mL/min. Microdialysis is performed on anesthetized animals usually under microprocessor control with on-line analysis of the eluate. Between 500 and 1000 articles have appeared in the literature describing microdialysis experiments in animals. The technique is likely to increase in popularity in the future, though therapeutic application seems a remote possibility.

Hemodialysis

Serious kidney disease, surprisingly uncommon in relation to the complexity of the organ, strikes between 1 in 5,000 and 1 in 10,000 of the population per year. The origin of kidney disease may be genetic, traumatic, metabolic, vascular, or immunologic, and the response of the kidney, although essentially sclerotic, may be reversible or permanent, local or systemic, rapid or slow, or any combination thereof (29,30). Kidney failure, as distinct from kidney disease, occurs when renal function has declined to the point that the kidney can no longer satisfactorily perform its homeostatic and excretory functions. Kidneys have an abundance of overcapacity, thus patients become overtly symptomatic and identifiably diseased only after about 90% of function has been lost. When kidneys decline further and loss of capacity exceeds about 95%, some form of renal replacement therapy is required. The alternatives include kidney transplantation and dialysis.

Despite widespread consensus that a successful transplant is the most satisfactory form of therapy for end stage renal disease, a chronic shortage of donor organs limits the number of patients receiving transplantation to about 18,000 per year (31,32). The remainder of renal failure patients require maintenance dialysis. About 12% elect continuous ambulatory peritoneal dialysis (CAPD); the remaining 88% elect hemodialysis. In CAPD, approximately 2 liters of a sterile, nonpyrogenic and hypertonic solution of glucose and electrolyte are instilled via gravity flow into the peritoneal cavity through an indwelling catheter four times per day. Intraperitoneal fluid partially equilibrates with solutes in the plasma, and plasma water is ultrafiltered due to osmotic gradients. After 4–5 h, except at night when the exchange is lengthened to 9–11 h to accommodate sleep, the peritoneal fluid is drained, and the process repeated. Patients perform the exchanges themselves in 20–30 min, at home or in the work environment, after a training cycle which lasts only 1–2 weeks. The literature on CAPD is abundant, but is well summarized in reference texts (33,34) and review articles (35,36).

The remaining 88% of untransplanted patients with kidney failure receive hemodialysis. This is an intermittent therapy with patients typically having thrice-weekly treatments of from 2.5 to 4 hours. Although most hemodialysis is performed in free-standing treatment centers, it may also be provided in a hospital or performed by the patient at home. The hemodialysis circuit consists of two fluid pathways. The blood side is entirely disposable, though many centers re-use some or all circuit components in order to reduce costs. It comprises a 16-gauge needle for access to the circulation (usually through a fistula created in the patient's forearm), lengths of dioctyl phthalate plasticized poly(vinyl chloride) tubing including a special tubing segment adapted to fit into a peristaltic blood pump, the hemodialyzer itself, a venous bubble trap and an open mesh screen filter, various ports for samples and gauge connections, and a return cannula. Components of the blood-side circuit are supplied in sterile and nonpyrogenic condition; ethylene oxide is the most common sterilant, although both radiation and steam sterilization are rapidly gaining favor. The dialysate side is essentially a machine capable of proportioning out glucose and electrolyte concentrates with water to provide dialysate of appropriate composition, pumping dialysate past a restrictor valve and through the hemodialyzer at subatmospheric pressure, and monitoring temperature, circuit pressures, and flow rates. During treatment the patient's blood is anticoagulated with heparin. Typical blood flow rates are 200–350 mL/min; dialysate flow rates are usually 500 mL/min. Straightforward techniques have been developed to prime the blood side with sterile saline prior to use and to rinse back nearly all the formed elements after treatment. Although most mass transport occurs by diffusion, circuits are operated with pressure on the blood side controlled to 13.3 to 66.7 kPa (100 to 500 mm Hg) higher than on the dialysate side. This provides an opportunity to remove 2 to 4 liters of fluid along with the solute; higher rates of fluid removal are technically possible but physiologically unacceptable. Hemodialyzers must be designed with high enough hydraulic permeabilities to provide adequate fluid removal at the upper pressure range, but not so high that excessive dewatering will occur at the lower pressure ranges.

Figure 4 is a schematic of a typical hemodialyzer. Although other geometries are still employed, the preferred format is a hollow fiber hemodialyzer about 25 cm

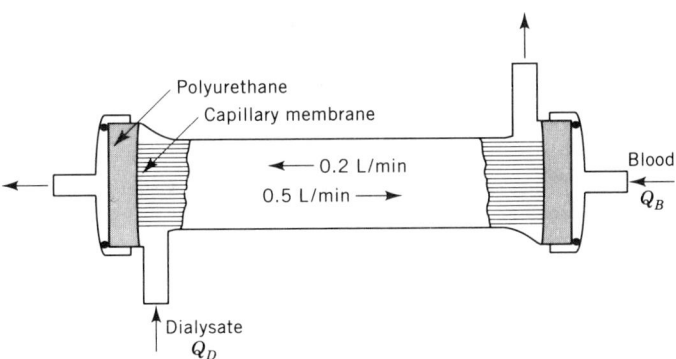

Fig. 4. Schematic of a hemodialyzer. The design of a dialyzer is close to that of a shell and tube heat exchanger. Blood enters through an inlet manifold, is distributed to a parallel bundle of fibers, and exits into a collection manifold. Dialysate flows countercurrent in an external chamber; the blood and dialysate are separated from the fibers by a polyurethane potting material. Housings are typically prepared from acrylate or polycarbonate. Production volume is greater than 50 million units per year and cost is very low, around $10 U.S. in 1992.

in length and 5 cm in diameter. Devices typically contain 6,000 to 10,000 capillaries, each with an inner diameter of 200 µm and a wall thickness of around 10 µm. Mean total membrane surface area is 1.1 ± 0.4 m^2. Well over 60 million hemodialyzers were produced in 1992. Because of economies of scale, unit price was on the order of $10 per unit, much lower than would be anticipated from the complexity of the device or by comparison with other membrane products. The hemodialyzer rarely represents more than 10% of the cost of a treatment session. This therapy is extensively described in the literature. By mid-1992, Med Line contained over 29,000 citations on hemodialysis. Several excellent reference texts provide concise and comprehensive coverage of all aspects of hemodialysis (37–40).

Engineering Aspects of Hemodialysis. Engineering interest in hemodialysis is concentrated on the optimization of the hemodialysis membrane (4,41), the dependency of solute removal on membrane and device characteristics (14,15), and quantitation of hemodialysis therapy through urea pharmacokinetics (42–44).

Hemodialysis membranes vary from one another in chemical composition, transport properties, and in their biocompatibility, defined here as the capacity of a material to avoid recognition and response by various host defense mechanisms (Table 1). Table 2 divides hemodialysis membranes into three classes: cellulosics, modified cellulosics, and synthetics. Cellulosics are prepared from regenerated cellulose by the cupramonium process; these extremely hydrophilic structures sorb water, bind it tightly, and form true hydrogels as is illustrated in the left hand panels of Figure 5. Their principle advantage is low unit cost; this is complemented by the strength of the highly crystalline cellulose which allows membranes to be made very thin and thus provides effective small-solute transport in relatively small hemodialyzers (see HOLLOW-FIBER MEMBRANES). The vulnerabilities of regenerated cellulose are its limited permeability to larger molecules, and the presence of labile nucleophilic groups that trigger complement activation and transient leukopenia during the first hour of a hemodialysis session. The advantages appear to outweigh the disadvantages: over 70% of the hemodialyzers were prepared from cellulosics in the early 1990s. Most were supplied by Akzo Faser AG under the trade name Cuprophan. At the opposite end of the spectrum are membranes prepared from synthetic, engineering thermoplastics, such as polysulfone and polyamide. These materials form anisotrophic membranes with foamlike or trebacular cross sections (see right hand panel in Fig. 5).

Table 1. Contemporary Hemodialysis Membrane Characteristics

Hydraulic permeability[a]	Solute clearance		Market share[b]	Absolute growth
	mol wt 250	mol wt >1000		
low-flux				
KUFR = 2–6	high	low	70%	steady
middle-flux				
KUFR = 5–12	high	medium	20%	growing
high-flux/high-performance				
KUFR = 10–200	high	high	10%	growing

[a]KUFR = ultrafiltration coefficient in mL/h·m^2·mm Hg.
[b]Estimated 1992.

Table 2. Polymeric Materials for Dialysis Membranes

Material	Manufacturer
Regenerated cellulosics	
Cuprophan	Akzo
cuprammonium cellulose	Asahi
	Terumo
SCE[a]	Teijin
	Althin
Synthetically modified cellulose	
Hemophan	Akzo
cellulose acetate	Akzo
	Toyobo
	Althin
	Teijin
cellulose triacetate	Toyobo
SMC[b]	Akzo
Synthetics	
polysulfone	Akzo
	Fresenius
	NMC
	Kurary
	Kawasumi
polycarbonate	Gambro
polyamide	Gambro
polyacrylonitrile	Hospal
	Asahi
SPAN[c]	Akzo
EVAL[d]	Kawasumi/Kuraray
PMMA[e]	Torray

[a] SCE = saponified cellulose ester.
[b] SMC = specially modified cellulose.
[c] SPAN = sulfonated polyacrylonitrile.
[d] EVAL is a poly(vinyl alcohol), a copolymer of ethylene and vinyl alcohol.
[e] PMMA = poly(methyl methacrylate).

They appear less active to the complement cascade and other physiologic identifiable defense mechanisms. In addition to this improved biocompatibility, these membranes are the least restrictive in transport to larger molecules. Drawbacks are increased cost and sufficiently high hydraulic permeability to require specialty control mechanisms and to raise concerns over the biologic quality of dialysate fluid. Roughly 10% of hemodialyzers are produced from such hydrophobic membranes. A middle group, also accounting for 10–15% of total hemodialyzer production, comprises both derivatized cellulosics, eg, cellulose diacetate, and synthetic hydrophilic polymers. Because of regulatory vigilence, all hemodialysis membranes in use are both safe and effective; there is no sound epidemiologic evi-

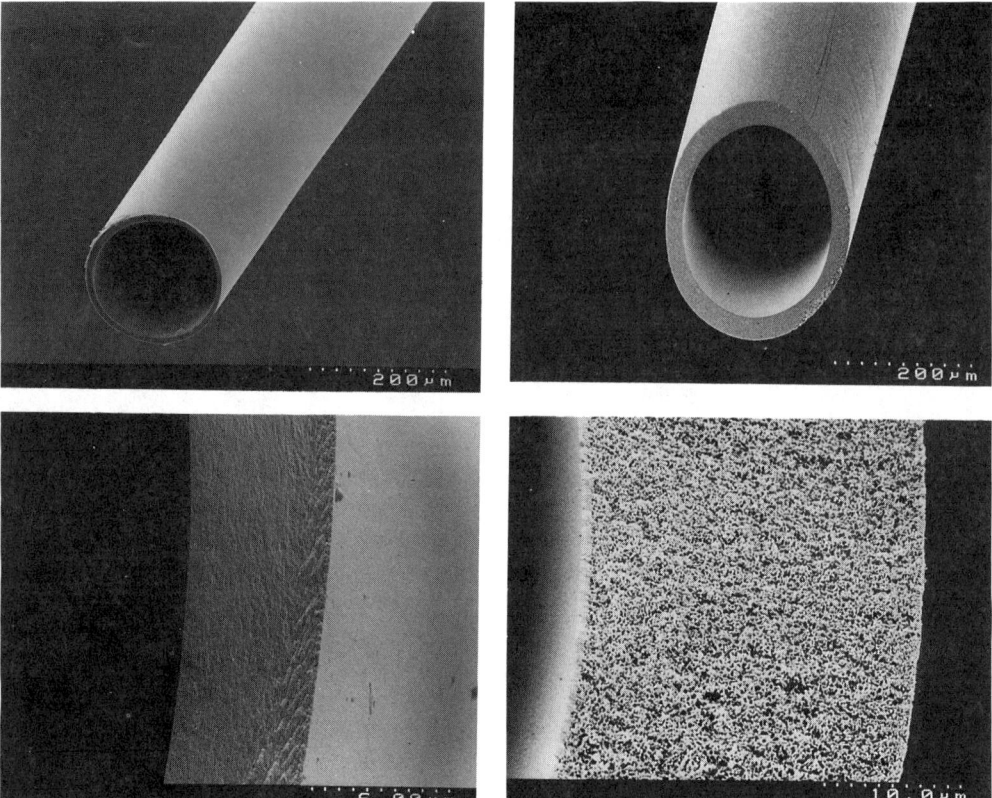

Fig. 5. Scanning electron micrographs of hollow fiber dialysis membranes. Membranes in left panels are prepared from regenerated cellulose (Cuprophan) and those on the right from a copolymer of polyacrylonitrile. The cellulosic materials are hydrogels and the synthetic thermoplastic forms a microreticulated open cell foam with a tight skin on the inner wall. Pictures at top are membrane cross sections; those below are of the wall region. Dimensions as indicated.

dence that selection of one membrane over another alters a patient's morbidity, mortality, or quality of life.

The clinical performance of a hemodialyzer is usually described in terms of clearance, a term having its roots in renal physiology, which is defined as the rate of solute removal divided by the inlet flow concentration as shown in equation 7, where Cl is clearance in mL/min and all other terms are as defined previously except that, in deference to convention, flow rates are now expressed in minutes rather than seconds and feed side (c') is now synonymous with blood flow on the luminal side.

$$Cl = \frac{\phi A}{c'_i} = \frac{Q_B(c'_i - c'_o)}{c'_i} \qquad (7)$$

Note that the numerator in each of the ratios in equation 7 represents the rate of solute removal from the patient. By mass balance, clearance is related to mass-transfer coefficient K_O as defined earlier in equations 3, 4, and 5, and where each of the three expressions equal rate of mass removal in g/s.

$$K_O A \, \Delta c = \phi A = Cl c_i' \tag{8}$$

For consistency, clearance here is expressed in cm^3/s although the more common clinical units, and those used later in this chapter, are mL/min. Combination and rearrangement of equations 6–8 allows clearance to be estimated from mass-transfer coefficient and vice versa; the conditions of countercurrent flow with no dialysate recycling are shown below.

$$Cl = Q_B \frac{\exp\left[\frac{K_O A}{Q_B}\left(1 - \frac{Q_B}{Q_D}\right)\right] - 1}{\exp\left[\frac{K_O A}{Q_B}\left(1 - \frac{Q_B}{Q_D}\right)\right] - \frac{Q_B}{Q_D}} \tag{9}$$

$$K_O = \frac{Q_B}{A\left(1 - \frac{Q_B}{Q_D}\right)} \ln\left[\frac{1 - \frac{Cl}{Q_D}}{1 - \frac{Cl}{Q_B}}\right] \tag{10}$$

Similar expressions for other conditions of geometry and flow are found in References 14 and 15.

Clearance decreases with increasing permeant molecular weight and depends in complex fashion upon blood and dialysate flow rate and upon device geometry. Detailed engineering analyses are available in References 14–17. As a general rule in most contemporary dialyzers, the clearance of small solutes such as urea (mol wt = 58) and creatinine (mol wt = 113), has either approached a maximum (clearance can never exceed blood flow rate) or is limited by boundary layers adjacent to the membrane. For these solutes changes in membrane permeability or membrane surface area does not significantly affect clearance whereas increases in blood flow leads to increased clearance. In contrast, larger solutes such as inulin (mol wt ~ 5200) or beta-2-microglobulin (mol wt = 11,118), are membrane limited. Their clearance increases, often linearly, with increasing membrance surface area, but is largely unaffected by changes in blood or dialysate flow rate. These relationships are illustrated in Figure 6 and summarized in Table 3.

Urea Pharmacokinetics. Pharmacokinetics summarizes the relationships between solute generation, solute removal, and concentration in a patient's blood stream. In the context of hemodialysis, this analysis is most readily applied to urea, which has, as a consequence, become a surrogate for other uremic toxins in the quantitation of therapy and in attempts to describe its adequacy. In the simplest case, a patient is assumed to have no residual renal function. Urea is generated from the breakdown of dietary protein, accumulates in a single pool equivalent to the patient's fluid volume, and is removed uniformly from that pool during hemodialysis. A mass balance around the patient yields the following differential equation:

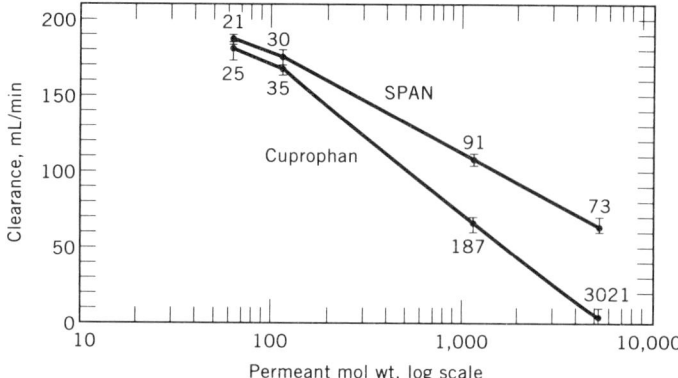

Fig. 6. Solute transport in hemodialysis. Clearance vs solute mol wt for dialyzers prepared from the two different membranes illustrated in Figure 5. Numbers next to points represent R_O in min/cm calculated from equations 10 and 5. Data is *in vitro* at 37°C with saline as the perfusion fluid. Lumen flow, dialysate flow, and transmembrane pressure were 200 mL/min, 500 mL/min, and 13.3 kPa (100 mm Hg); area = 1.6 m². Inulin clearance of the SPAN fiber was elevated by inulin transported by the filtering fluid.

$$\frac{d(cV)}{dt} = G - Clc \qquad (11)$$

where c = whole blood urea concentration normally expressed as mg % (mg/100 mL), V = urea distribution volume in the patient in mL, G = urea generation rate in mg/min, t = time from onset of hemodialysis in minutes, and Cl = urea clearance in mL/min.

Urea concentration in the United States medical literature is often reported as BUN (blood urea nitrogen), which is urea concentration, usually in mg/dL, multiplied by a factor of 0.47. V, in equation 11, can be measured by dilution studies, but is often estimated in kinetic modeling studies as 58% of patient weight. Generation is calculated from a knowledge or an estimate of patient protein intake (each gram of protein consumed produces about 250 mg of urea; see References 43 and 44 for more exact correlations based on metabolic studies of uremic patients). Thus a 70 kg patient, consuming a typical 1.0 g of protein per

Table 3. Effects of Changes in Conditions of Geometry and Flow on Hemodialyser Clearance

	Effect on clearance of	
Parameter increased	Low mol wt solutes[a]	High mol wt solutes[b]
blood flow rate	increases	little or no effect
dialysate flow rate	little effect	little or no effect
membrane surface area	little effect	almost linear increase
membrane permeability	little effect	almost linear increase

[a]Mol wts of <200, eg, urea (60), creatinine (113), or uric acid (158).
[b]Mol wts of >1000, vitamin B_{12} (1355) or inulin (~5200).

kg of body weight per day, would produce 28 g of urea distributed over a fluid volume of 40.6 L and, in the absence of any clearance, urea concentration would increase by 70 mg % (mg/100 mL) every 24 hours. The reduction of urea concentration during hemodialysis is readily obtained from equation 11 by neglecting intradialytic generation and changes in volume where c^o and c^t represent the urea concentrations in blood at the beginning and during the course of treatment.

$$c^t = c^o \exp\left(-\frac{Cl\,t}{V}\right) \tag{12}$$

A 3.5 h treatment of a 70 kg patient (V = 40.6 liters) with a urea clearance of 200 mL/min should result in a 64% reduction in urea concentration or a value of 0.36 for the ratio c^t/c^o; this parameter almost always falls between 0.30 and 0.45. The increase in urea concentration between hemodialysis treatments is obtained from equation 13, again assuming a constant V, where c^o is the urea concentration in the patient's blood at the end of the hemodialysis, and c^t the concentration at time t during the intradialytic interval.

$$c^t = c^o + \frac{G}{V}t \tag{13}$$

Urea concentration typically increases by about 50 to 100 mg/100 mL/24 h. Even a small residual clearance will prove numerically significant and, for oliguric patients, the slightly more complex formulas given in References 43 and 44 should be employed. The exponential decay constant in equation 12, $Cl\,t/V$, is the net normalized quantity of hemodialysis therapy. It is calculated simply by multiplying the urea clearance in mL/min by the duration of hemodialysis, also in minutes, and dividing by the distribution volume in mL, which, in the absence of a better estimate, is taken as 0.58 × the patient weight. This parameter provides an index of the adequacy of hemodialysis (45) and based on retrospective analysis of various therapy formats, a value of 1.0 or greater for urea proposedly provides an adequate amount of hemodialysis for most patients. Although not without its critics, this approach has found nearly universal clinical acceptance, and represents the current prescriptive norm to hemodialysis therapy.

Maintenance hemodialysis has grown and expanded beyond the expectations of even the most enthusiastic of its earliest proponents. Figure 7 is a plot of the overall estimated dialysis population by year since 1970. The population at the end of 1992 exceeded 475,000; another 500,000 patients or so have received therapy at one time but have since died or had transplants. Maintenance dialysis is now available to some extent in all but the poorest nations. In economically advanced countries, excepting the United Kingdom, it is rendered as a virtual entitlement. The worldwide mean cost of a single dialysis patient is about $30,000 per year (47). The aggregate economic magnitude of the medical application of hemodialysis thus approaches $15 billion.

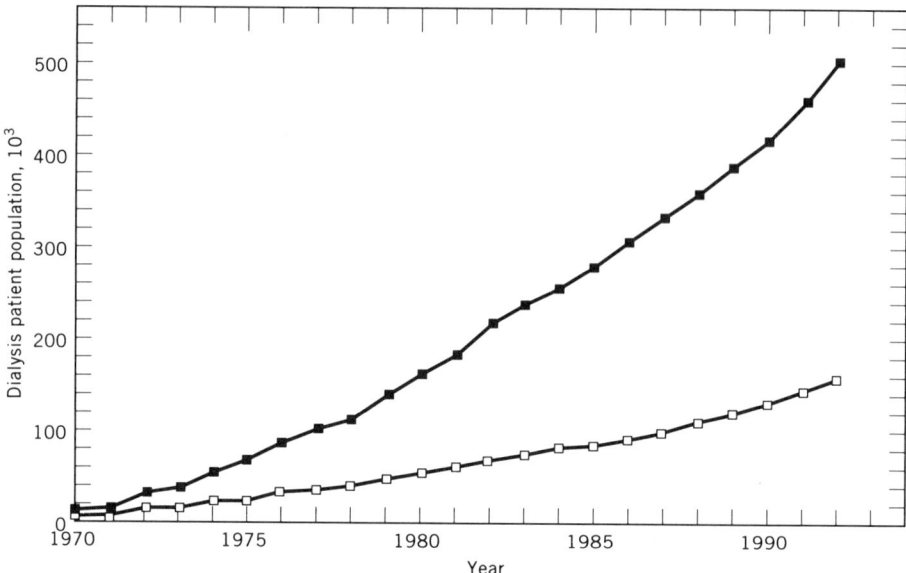

Fig. 7. Estimate of the total number of patients receiving maintenance dialysis over the past 20 years. Totals include both hemodialysis and peritoneal dialysis, but exclude transplant recipients. The fraction of patients receiving peritoneal dialysis has grown steadily from 0% in 1978 to about 12% in 1992. These data were combined from various regional registries and industry sources; demographic estimates of this ilk are accurate to within 5% (46). □, United States; ■, worldwide.

BIBLIOGRAPHY

"Dialysis and Electrodialysis" in the *Encyclopedia of Chemical Technology,* 1st ed., Vol. 5, pp. 1–20, by F. K. Daniel, Chemical Consultant; "Dialysis" in *ECT* 2nd ed., Vol. 7, pp. 1–21, by E. F. Leonard, Columbia University; in *ECT* 3rd ed., Vol. 7, pp. 564–579, by E. F. Leonard, Columbia University; in *ECT* 4th ed., Vol. 8, pp. 58–74, by M. J. Lysaght, Cytotherapeutics Inc., and U. Baurmeister, Akzo Faser HG.

1. T. Graham, *Phil. Mag.* **49,** 337 (1866).
2. A. Fick, *Pogg. Ann.* **94,** 59–86 (1855).
3. A. Einstein, *Ann. Physik.* **19,** 289 (1906).
4. M. J. Lysaght, *Contrib. Nephrol.* **61,** 1–17 (1988).
5. W. Pusch, *Desalination* **59,** 105–198 (1986).
6. H. Strathmann, *Trennung von Molekularan Mischungen mit Hilfe Synthetischer Membranen*, Steinkopff, Darmstadt, FRG, 1959, pp. 15–55.
7. R. Schloegl, *Stofftransport durch Membranen*, Steinkopff, Darmstadt, FRG, 1966.
8. O. Kedem and A. Katachalsky, *Biochem. Biophys. Acta* **27,** 229–246 (1961).
9. A. Katachalsky and P. F. Curran, *Nonequilibrium Thermodynamics in Biophysics*, Harvard University Press, Cambridge, Mass., 1967.
10. C. K. Colton and co-workers, *AIChE J.* **17,** 772–780 (1971).
11. C. K. Colton and co-workers, *J. Biomed. Mater. Res.* **5,** 459–488 (1971).
12. E. Klein and co-workers, *J. Membr. Sci.* **2,** 349–364 (1977).
13. E. F. Leonard and W. Bluemle, *Trans. NY Acad. Sci.*, 585–598 (1959).

14. C. K. Colton and E. G. Lowrie, in B. M. Brenner and F. C. Rector, eds., *The Kidney*, 2nd ed., Saunders Publishing Co., Philadelphia, 1981, pp. 2425–2489.
15. C. K. Colton, *Blood Purif.* **5,** 202–251 (1987).
16. A. S. Michaels, *Trans. Am. Soc. Artif. Intern. Organs* **12,** 387–392 (1966).
17. F. A. Gotch and co-workers, *The Evaluation of Hemodialyzers*, DHEW publication NIH 72-103, Washington, D.C. 1972.
18. H. B. Volrath, *Chem. Met. Eng.* **43,** 303 (1936).
19. T. Nishiwaki and S. Itoi, *Jpn. Chem. Q.* **41,** 36 (1982).
20. H. Moonen and H. J. Niefind, *Desalination* **41,** 327–335 (1982).
21. F. Jonaaon, in P. M. Bungay, H. K. Lonsdale, and M. N. de Pinho, eds., *Synthetic Membranes: Science, Engineering, Applications*, Reidel, Dordrecht, 327–335, 1982.
22. C. Heath and G. Belfort, *Int. J. Biochem.* **22,** 823–835 (1990).
23. J. M. Piret and C. L. Cooney, *Biotechnol. Adv.* **8,** 763–783 (1990).
24. H. G. Tilgner and F. J. Schmitz, *Eur. Pat* 36,175 (1981).
25. P. M. Knazek and co-workers, *Science* **178,** 65–67 (1974).
26. U. Ungerstedt, *J. Int. Med.* **230,** 365–373 (1991).
27. N. Lindefors, G. Amberg, and U. Ungerstedt, *Pharm. Methods.* **22,** 141–156 (1989).
28. G. Amberg and N. Lindefors, *Pharm. Methods* **22,** 157–183 (1989).
29. H. Smith, *From Fish to Philosopher*, Doubleday, Garden City, N.J., 1961.
30. B. M. Brenner and F. C. Rector, eds., *The Kidney*, 3rd ed., Saunders Publishing Co., Philadelphia, 1986.
31. P. A. Keown and C. R. Stiller, *Kidney Int. Suppl* **24,** S145-9 (1988).
32. T. B. Strom and N. L. Tilney, in Ref. 30, pp. 1941–1985.
33. R. Gokal, *Continuous Ambulatory Peritoneal Dialysis*, Churchill Livingston, Edinburgh, 1987.
34. K. D. Nolph, *Peritoneal Dialysis*, 3rd ed., Martinus Nihjoff, The Hague, 1988.
35. K. D. Nolph, A. S. Lindblad, and W. J. Novak, *N. Engl. J. Med.* **318,** 1595–1600 (1988).
36. M. J. Lysaght and P. C. Farrell, *J. Membr. Sci.* **44,** 5–33 (1989).
37. J. F. Maher, *Replacement of Renal Function by Dialysis*, 3rd ed., Kluwer, Boston, 1989.
38. H. J. Gurland, *Uremia Therapy*, Springer-Verlag, Berlin, 1987.
39. T. H. Frost, *Technical Aspects of Renal Dialysis*, Pittman Press, Kent, UK, 1977.
40. A. R. Nissenson, R. Fine, and D. Gentile, *Clinical Dialysis*, Appleton & Lange Century Crofts, Nowalk, Conn., 1984.
41. H. Strathmann and H. Goehl, *Contrib. Nephrol.* **78,** 119–141 (1990).
42. E. G. Lowrie, ed., *Kidney Int.* **23,** suppl. 13 (1983).
43. J. A. Sargeant and F. Gotch, in Ref. 37, pp. 87–143.
44. P. C. Farrell, *Dialysis Kinetics: ASAIO Primers in Artificial Organs 4*, J. B. Lippincott, Philadelphia, 1988.
45. F. Gotch and J. A. Sargent, *Kidney Int.* **23,** S103-6 (1983).
46. M. J. Lysaght, Ph.D. dissertation, University of New South Wales, Australia, 1989, Chapt. 2.
47. P. W. Eggers, *N. Engl. J. Med.* **318,** 223–229 (1988).

MICHAEL J. LYSAGHT
CytoTherapeutics, Inc.

ULRICH BAURMEISTER
Akzo Faser AG

DIFFUSION SEPARATION METHODS

Ordinary diffusion involves molecular mixing caused by the random motion of molecules. It is much more pronounced in gases and liquids than in solids. The effects of diffusion in fluids are also greatly affected by convection or turbulence. These phenomena are involved in mass-transfer processes, and therefore in separation processes (see MASS TRANSFER; SEPARATION PROCESS SYNTHESIS). In chemical engineering, the term diffusional unit operations normally refers to the separation processes in which mass is transferred from one phase to another, often across a fluid interface, and in which diffusion is considered to be the rate-controlling mechanism. Thus, the standard unit operations such as distillation (qv), drying (qv), and the sorption processes, as well as the less conventional separation processes, are usually classified under this heading (see ABSORPTION; ADSORPTION; ADSORPTION, GAS SEPARATION; ADSORPTION, LIQUID SEPARATION).

A number of special processes have been developed for difficult separations, such as the separation of the stable isotopes of uranium and those of other elements (see ISOTOPE SEPARATION). Two of these processes, gaseous diffusion and gas centrifugation, are used by several nations on a multibillion dollar scale to separate partially the uranium isotopes and to produce a much more valuable fuel for nuclear power reactors. Because separation in these special processes depends upon the different rates of diffusion of the components, the processes are often referred to collectively as diffusion separation methods. There is also a thermal diffusion process used on a modest scale for the separation of helium-group gases and on a laboratory scale for the separation of various other materials. Thermal diffusion is not discussed herein.

The most important industrial application of the diffusion separation methods has been for the enrichment of uranium-235 [5117-96-1], ^{235}U. Natural uranium consists mostly of ^{238}U and 0.711 wt % ^{235}U plus an inconsequential amount of ^{234}U. The United States was the first country to employ the gaseous diffusion process for the enrichment of ^{235}U, the fissionable natural uranium isotope, during the 1940s and 1950s.

As of the early 1990s, diffusion separation methods were being employed or developed internationally in (1) Argentina, which had a gaseous diffusion project; (2) Brazil, where work was ongoing on gas centrifuges; (3) China, which had gaseous diffusion and gas centrifuges under development; (4) France, including Eurodif, owned by France, Italy, Spain and Belgium, which had a gaseous diffusion plant at Tricastin, plus a topping plant at Pierrelatte; (5) Germany, which had a large-scale centrifuge plant (Urenco); (6) India, which had gas centrifuges; (7) Japan, which had gas centrifuges; (8) the Netherlands, where there was a large-scale Urenco gas centrifuge plant; (9) Pakistan, which had gas centrifuges; (10) South Africa, which had a version of an advanced vortex tube process and had been working on the gas centrifuge process; (11) the CIS, which had large-scale gaseous diffusion and gas centrifuge plants; and (12) the U.K., which had gaseous diffusion and Urenco gas centrifuge plants.

In the United States, a group of domestic investors, including Duke Power, and Urenco, had applied for permission to construct a new gas centrifuge plant in Louisiana.

General Process and Design Selection

For difficult separations, such as isotope separations that involve the separation of molecules having very similar physical and chemical properties, the enrichment that can be obtained in a single equilibrium stage or transfer unit of the process is quite small. Hence, an extremely large number of these elementary separating units must be connected to form a separation cascade in order to achieve most desired separations. Consequently, very large separation systems requiring large amounts of energy are needed, and the total energy requirement is one of the most important cost considerations.

The energy or power required by any separation process is related more or less directly to its thermodynamic classification. There are, broadly speaking, three general types of continuous separation processes: reversible, partially reversible, and irreversible.

Reversible Processes. Distillation (qv) is an example of a theoretically reversible separation process. In fractional distillation, heat is introduced at the bottom stillpot to produce the column upflow in the form of vapor which is then condensed and turned back down as liquid reflux or column downflow. This system is fed at some intermediate point, and product and waste are withdrawn at the ends. Except for losses through the column wall, etc, the heat energy spent at the bottom vaporizer can be recovered at the top condenser, but at a lower temperature. Ideally, the energy input of such a process is dependent only on the properties of feed, product, and waste. Among the diffusion separation methods discussed herein, the centrifuge process (pressure diffusion) constitutes a theoretically reversible separation process.

Partially Reversible Processes. In a partially reversible type of process, exemplified by chemical exchange, the reflux system is generally derived from a chemical process and involves the consumption of chemicals needed to transfer the components from the upflow into the downflow at the top of the cascade, and to accomplish the reverse at the bottom. Therefore, although the separation process itself may be reversible, the entire process is not, if the reflux is not accomplished reversibly.

Insofar as the consumption of chemicals is concerned, it is obvious that the total consumption of reflux-producing chemicals is proportional to the interstage flows, or width of the cascade, but independent of the number of stages in series, or length of the system.

Irreversible Processes. Irreversible processes are among the most expensive continuous processes. These are used only in special situations, such as when the separation factors of more efficient processes (that is, processes that are theoretically more efficient from an energy point of view) are found to be uneconomically small. Except for pressure diffusion, the diffusion methods discussed herein are essentially irreversible processes. Thus, gaseous diffusion, in which gas expands from a region of high pressure to one of low pressure, mass diffusion, in which a vapor flows from a region of high partial pressure to one of low partial pressure, and thermal diffusion, in which heat flows from a high temperature source to a low temperature sink, are all irreversible processes. In contrast with reversible and partially reversible processes, the energy demand in an irreversible process is distributed over the whole cascade in direct proportion to the distribution flow.

DIFFUSION SEPARATION METHODS

In gaseous diffusion, the cascade consists of individual stages that are connected in series. In each stage part of the gaseous feed is forced through a diffusion membrane or barrier with holes smaller than the mean free path of the gas (see MEMBRANE TECHNOLOGY). Because of slightly greater mobility, the lighter components flow preferentially through the barrier. This enriched portion of the feed is transported to a neighboring stage, up the cascade, where the lighter components tend to concentrate. The other portion of the gas that does not pass through the barrier is rejected to a neighboring stage, down the cascade, where the heavier components tend to concentrate. The feed to each stage is thus composed of combined upflow and downflow from neighboring stages.

In pressure diffusion, a pressure gradient is established by gravity or in a centrifugal field. The lighter components tend to concentrate in the low pressure (center) portion of the fluid. Countercurrent flow and cascading extend the separation effect.

Irreversible processes are mainly applied for the separation of heavy stable isotopes, where the separation factors of the more reversible methods, eg, distillation, absorption, or chemical exchange, are so low that the diffusion separation methods become economically more attractive. Although application of these processes is presented in terms of isotope separation, the results are equally valid for the description of separation processes for any ideal mixture of very similar constituents such as close-cut petroleum fractions, members of a homologous series of organic compounds, isomeric chemical compounds, or biological materials.

Cascade Design

Less conventional diffusional separation operations are characterized by the relatively small separations that can be obtained by the elementary separation mechanism. That is, the changes in fluid composition attained in gaseous diffusion across the barrier, in thermal diffusion between the hot and cold walls, in mass diffusion between the inlet and the condensing surface for the sweep vapor, and in the centrifuge between the axis of the rotor and its periphery, are all quite small. Thus, a large number of separating units must be employed. Cascade is the term given to the aggregation of separating units that have been interconnected so as to be able to produce the desired material. The optimum arrangement of the separating units in a separation cascade generally minimizes the unit cost of product, and its design is a problem common to all separation processes. In a stagewise separation process such as gaseous diffusion, each unit of equipment consists of one separation stage.

The Separation Stage. A fundamental quantity, α, exists in all stochastic separation processes, and is an index of the steady-state separation that can be attained in an element of the process equipment. The numerical value of α is developed for each process under consideration in the subsequent sections. The separation stage, which in a continuous separation process is called the transfer unit or equivalent theoretical plate, may be considered as a device separating a feed stream, or streams, into two product streams, often called heads and tails, or product and waste, such that the concentrations of the components in the two effluent streams are related by the quantity, α. For the case of the separation of a binary mixture this relationship is

$$\left(\frac{y}{1-y}\right)\Big/\left(\frac{x}{1-x}\right) = \alpha \qquad (1)$$

where y is the mol fraction of the desired component in the upflowing (heads) stream from the stage and x the mol fraction of the same component in the downflowing (tails) stream from the stage. The quantity α is usually called the stage separation factor.

For the case of separating a binary mixture, the following conventions are used. The concentrations of the streams are specified by the mol fraction of the desired component. The purpose of the separation process is usually to obtain one component of the mixture in an enriched form. If both components are desired, the choice of the desired component is an arbitrary one. The upflowing stream from the separation stage is the one in which the desired component is enriched, and by virtue of this convention, α is defined as a quantity the value of which is greater than unity. However, for the processes considered here, α exceeds unity by only a very small fraction, and the relationship between the concentrations leaving the stage can be written, without appreciable error, in the form

$$y - x = (\alpha - 1)x(1 - x) \qquad (2)$$

A separation stage or transfer unit operating on a binary mixture is shown schematically in Figure 1. In a cascade of separating stages, the feed stream can be formed by mixing the downflowing stream from the stage above and the upflowing stream from the stage below. The quantity, θ, ie, the fraction of the combined stage feed that goes into the stage upflow stream, is termed the cut of the stage. In cascades ordinarily designed for difficult separations, the stage cut is normally very nearly equal to one-half. In the case of a theoretical plate in a continuous process, the feed consists of two separate streams, one from above and one from below. In cascades for either stagewise or continuous processes the upflow rate L and the downflow rate L' (or $L(1 - \theta)/\theta$) are very nearly equal. For continuous process units, the length S is the length of equipment necessary to satisfy the requirement of equation 1, that the streams leaving the unit be related by α; it is usually called the height of a transfer unit (HTU) or the height equivalent to a theoretical plate (HETP). Although the HTU and HETP are defined differently and are not precisely equivalent to each other, the difference between them becomes negligible when the value of the quantity $\alpha - 1$ is small.

The Separative Capacity. The separation stage is characterized not only by the separation factor α but also by its capacity or throughput of which the upflow L is a measure, and in the case of the continuous process, also by the length S. It is therefore desirable to define and determine a quantity indicative of the amount of useful separative work that can be done per unit time by a single stage. Such a quantity is called the separative capacity of the stage. It is postulated that the separation stage does useful work on the streams it processes, hence increasing their net value. The value of a stream must be a function of its concentration; let this value function be designated by $v(x)$. Then the separative capacity of the

534 DIFFUSION SEPARATION METHODS

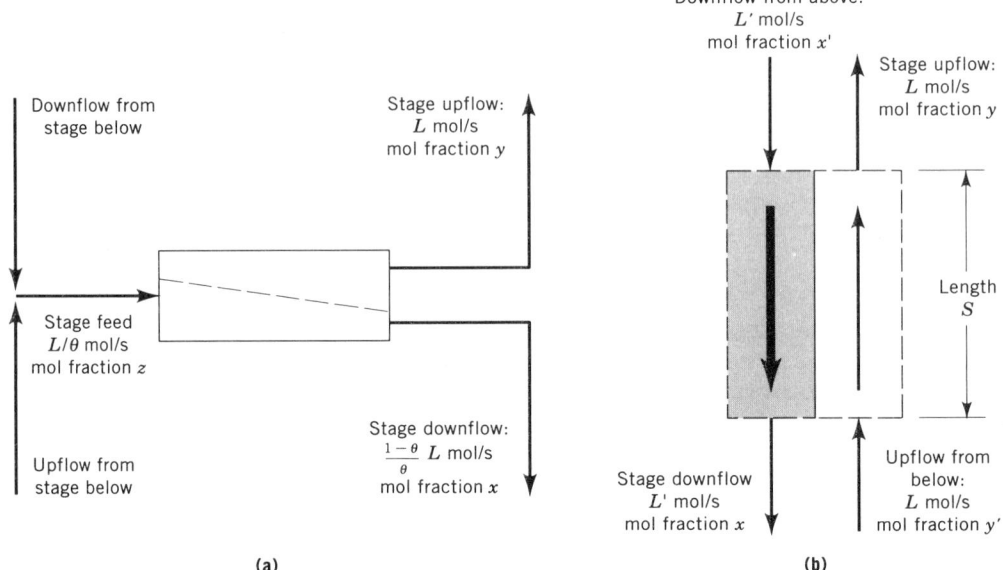

Fig. 1. The analogy between the separation stage and the transfer unit or equivalent theoretical plate: (**a**) in a stagewise process; (**b**) in a continuous-separation process. The terms are defined in text.

stage by definition is set equal to the increase in value it creates. The separative capacity of a unit is a very useful concept and permits comparisons to be made between different separation processes.

The separative capacity of the stage, termed δU, is set equal to the net increase in value of the streams it processes (see Fig. 1):

$$\delta U = Lv(y) + \frac{1-\theta}{\theta} Lv(x) - \frac{1}{\theta} Lv(z) \qquad (3)$$

The value functions appearing in equation 3 may be expanded in Taylor series about x and, because the concentration changes effected by a single stage are relatively small, only the first nonvanishing term is retained. When the value of z is replaced by its material balance equivalent, ie, equation 4:

$$z = (1-\theta)x + \theta y \qquad (4)$$

the separative capacity of the stage is given by:

$$\delta U = \tfrac{1}{2} L (1-\theta)(y-x)^2 v''(x) \qquad (5)$$

where $v''(x)$ is the second derivative of the value function. The concentrations y and x are related by equation 2; thus the separative capacity can also be written as:

$$\delta U = \tfrac{1}{2} L (1 - \theta)(\alpha - 1)^2 x^2 (1 - x)^2 v''(x) \tag{6}$$

As it is desirable that the separative capacity of the stage be independent of the concentration of the material with which it is operating, the terms in the equation involving the concentration are set equal to a constant, taken for convenience to be unity, and the separative capacity of a single stage operating with a cut of one-half is seen to be:

$$\delta U = \tfrac{1}{4} L (\alpha - 1)^2 \tag{7}$$

Thus, the separative capacity of a stage is directly proportional to the stage upflow as well as to the square of the separation effected.

Equivalent Theoretical Plate. The separative capacity of a theoretical plate in a continuous process can be obtained in the same manner. By equating the separative capacity of the unit to the net increase in value of the four streams handled (eq. 8):

$$\delta U = Lv(y) + L'v(x) - Lv(y') - L'v(x') \tag{8}$$

After expansion in Taylor series about the concentration x and replacing the concentration y' by its material balance equivalent:

$$\delta U = L(y - x')(x' - x)v''(x) \tag{9}$$

The separative capacity of the equivalent theoretical stage in the continuous process is seen to depend on the concentration difference between the countercurrent streams as well as on the concentration difference between the top and bottom of the stage. The separative capacity is zero when x' is equal to y or x' is equal to x; inspection shows that it attains a maximum value when x' is equal to the arithmetic average of x and y and that this maximum value is:

$$(\delta U)_{\max} = \tfrac{1}{4} L(y - x)^2 v''(x) = \tfrac{1}{4} L (\alpha - 1)^2 \tag{10}$$

Thus, the maximum value of the separative capacity of a theoretical plate in a continuous process is equal to that of a single separation stage when both units have the same value of the $L(\alpha - 1)^2$ product. When the continuous process is operated so as to yield its maximum separative capacity, the concentrations y' and x' of the streams entering the unit are equal and the similarity between the separation stage and the theoretical plate is accentuated because, for this case, both may be considered to separate a single feed stream into two product streams having concentrations related by α. The definition of a theoretical plate in the continuous process is essentially arbitrary and not required; however, it is a useful concept, permitting both the stagewise and continuous processes to be treated with the same set of cascade equations.

The Value Function. The value function itself is defined, as has been indicated above, by the second-order differential:

$$v''(x) = 1/[x^2(1-x)^2] \qquad (11)$$

In the design of cascades, a tabulation of $v(x)$ and of $v'(x)$ is useful. The solution of the above differential equation contains two arbitrary constants. A simple form of this solution results when the constants are evaluated from the boundary conditions $v(0.5) = v'(0.5) = 0$. The expression for the value function is then:

$$v(x) = (2x - 1)\ln[x/(1-x)] \qquad (12)$$

and for the derivative of the value function (eq. 13):

$$v'(x) = \frac{2x-1}{x(1-x)} + 2 \ln \frac{x}{1-x} \qquad (13)$$

Therefore, $v(1-x) = v(x)$ and $v'(1-x) = -v'(x)$.

Application. In addition to providing a relatively simple means for estimating the production of separation cascades, the separative capacity is useful for solving some basic cascade design problems; for example, the problem of determining the optimum size of the stripping section.

It can be assumed that P, y_P, and x_F for the cascade have been specified, and that the cost of feed and the cost per unit of separative work, the product of separative capacity and time, are known. The basic assumption is that the unit cost of separative work remains essentially constant for small changes in the total plant size. The cost of the operation can then be expressed as the sum of the feed cost and cost of separative work:

$$C_{\text{total}} = (C_F)(F)(\Delta t) + C_{\Delta U}[Pv(y_P) + Wv(x_W) - Fv(x_F)]\Delta t \qquad (14)$$

where C_{total} is the total cost of operation for the period of time Δt, and C_F and $C_{\Delta U}$ are the cost per unit of feed and the cost per unit of separative work, respectively. The optimum value of x_W is that which minimizes the total cost and can be found by differentiating the total cost with respect to x_W under the restrictions that P, y_P, and x_F remain constant, and setting the result equal to zero. The result of this procedure is that the optimum x_W is the solution to equation 15:

$$v(x_F) - v(x_W) - (x_F - x_W)v'(x_W) = C_F/C_{\Delta U} \qquad (15)$$

When the cascade is operated using the optimum x_W, the cost of producing material at any other concentration, y_P, is given by:

$$C_{\text{total}} = C_{\Delta U} P[v(y_P) - v(x_W) - (y_P - x_w)v'(x_W)]\Delta t \qquad (16)$$

obtained by combining equations 14 and 15. An equation of this form can be used to establish the value of material of different concentrations from separation cascades.

Cascade Gradient Equations. An arrangement of separation stages to form a simple cascade is shown in Figure 2. A simple cascade is one that divides a

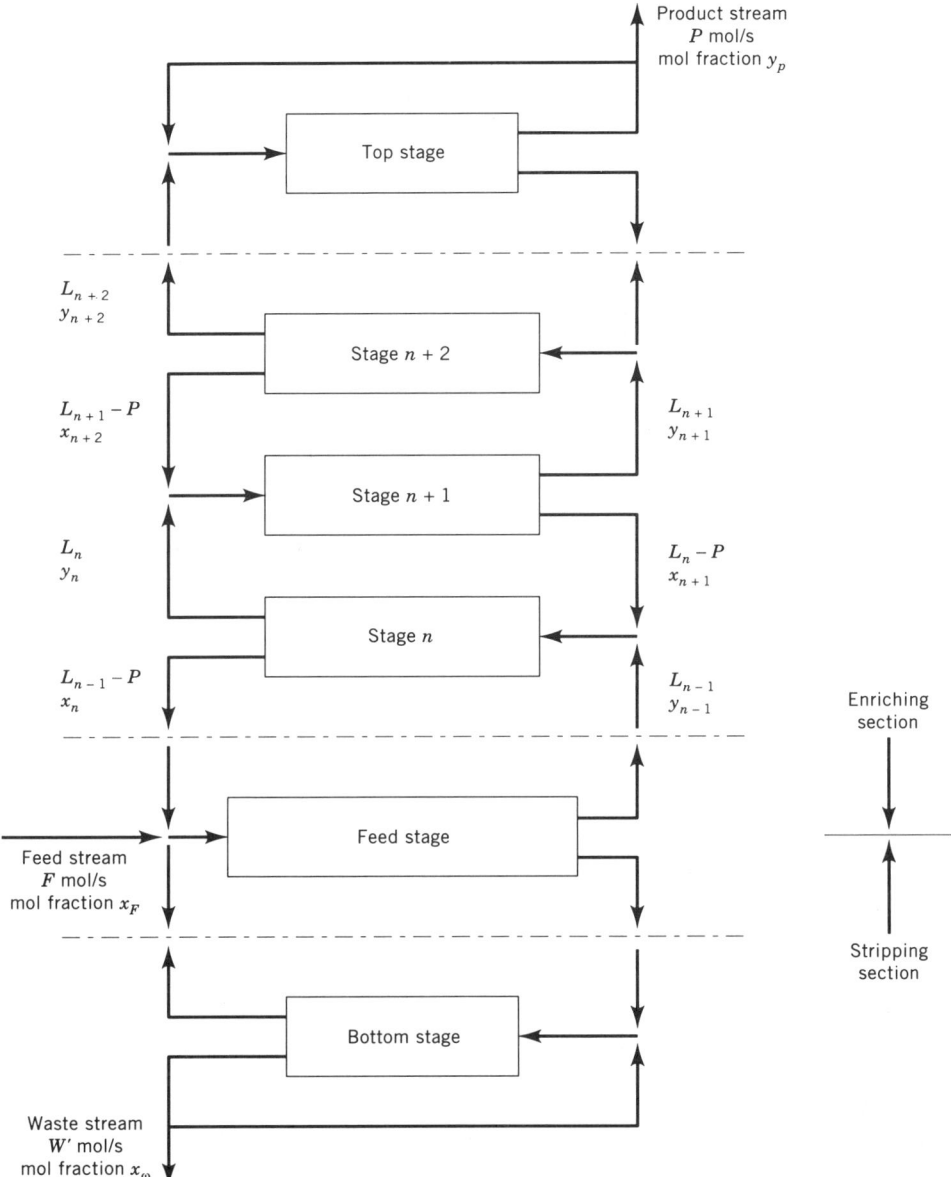

Fig. 2. Separation stages arranged to form a simple cascade. Terms are defined in text.

single cascade feed stream into a product stream and a waste stream. Additional side streams, however, could easily be handled. To be consistent with the conventions given for the single stage, the desired component is assumed to be enriched in the product stream at the top of the cascade. The cascade feed is introduced at some intermediate stage between the top and bottom of the cascade. The portion of the cascade that lies above the feed point is termed the enriching section; that which lies below the feed point is termed the stripping section. The gradient equations for the cascade are obtained from a combination of the material balance equations, frequently called the operating-line equations, and the α relationship, usually called the equilibrium-line equation. From a material balance around the top of the cascade down to, but not including, stage n of the enriching section, is obtained the operating-line equation:

$$L_n y_n = (L_n - P)x_{n+1} + P y_P \tag{17}$$

which can be combined with the equilibrium-line (eq. 2) to give:

$$x_{n+1} - x_n = \frac{L_n}{L_n - P}[(\alpha - 1)x_n(1 - x_n) - (P/L_n)(y_P - x_n)] \tag{18}$$

For the case under consideration, where the value of $\alpha - 1$ is quite small, it follows that everywhere in the cascade, except possibly at the extreme ends, the stage upflow is many times greater than the product withdrawal rate. Thus $L/(L - P)$ can be set equal to unity. Furthermore when the value of $\alpha - 1$ is small, the stage enrichment $x_{n+1} - x_n$ can be approximated by the differential ratio dx/dn without appreciable error. The gradient equation for the enriching section of a simple cascade therefore takes the form

$$dx/dn = (\alpha - 1)x(1 - x) - (P/L)(y_P - x) \tag{19}$$

Similarly, one obtains a gradient equation for the stripping section that has the form

$$dx/dn = (\alpha - 1)x(1 - x) - (W/L)(x - x_W) \tag{20}$$

Equations 19 and 20 are the basic equations for cascade design. Although these equations were derived from a consideration of a cascade composed of discrete separation stages, equations of the same form are also obtained for cascade designs based on continuous or differential separation processes. For use in the case of continuous separation processes, however, the term dx/dn, which is the enrichment per stage, is usually replaced by the equivalent terms $S\, dx/dz$, where S is the stage length and dx/dz the enrichment per unit length of process equipment. These equations may then be used to calculate the output from a given cascade configuration, that is, from a cascade for which the variation of $\alpha - 1$ and L is known as a function of the stage number.

Minimum Length or Minimum Number of Stages. It is evident from the gradient equations that the enrichment per stage decreases as the withdrawal rate increases. Thus the minimum number of stages required to span a given concentration difference is obtained when no material is withdrawn from the cascade. This mode of operation ($P = W = F = 0$) is frequently called total reflux operation. Integration of the gradient equation for this case with $\alpha - 1$ taken to be constant gives:

$$N_{\min} = \frac{1}{\alpha - 1} \ln\left(\frac{x_T}{1 - x_T} \bigg/ \frac{x_B}{1 - x_B}\right) \tag{21}$$

where the concentration range to be spanned is from the concentration x_B at the bottom to concentration x_T at the top. As an example of the magnitudes involved, consider the enrichment of ^{235}U by gaseous diffusion from $x_B = 0.005$ to $x_T = 0.90$. For a value of α equal to 1.0043 the minimum number of diffusion stages required is 1742.

Minimum Width or Minimum Stage Upflow. It also follows directly from the gradient equations that if the withdrawal rates from the cascade are nonzero, it is necessary that the stage upflow from the stage at which the cascade concentration is x must exceed some critical value in order that there be any enrichment at that point in the cascade. This critical value is called the minimum stage upflow and is obtained by setting dx/dn equal to zero in the gradient equation. Thus, for any point in the enriching section the minimum stage upflow is given by

$$L_{min} = P(y_P - x)/[(\alpha - 1)x(1 - x)] \tag{22}$$

For the case of enriching ^{235}U to 90 mol % product concentration, the stage upflow at the feed point ($x_F = 0.0072$) must therefore exceed 29,046 times the product withdrawal rate. It can now be seen from a consideration of the minimum stage upflow that the approximation made in deriving the gradient equation, ie, taking the quantity $(1 - P/L)$ equal to unity, introduces negligible error except possibly in the immediate vicinity of the withdrawal points. The condition that arises in a cascade at points where the stage upflow approaches the value L_{min} is commonly called pinching.

Gradient Equations for a Square Section. A section of a cascade composed of identical stages, that is, a number of stages having the same separation factor and the same stage upflow, is called a square section. For sections of this type the gradient equations are readily integrable. For a section in the enricher of a cascade producing material at rate P and concentration y_P, the solution can be written:

$$N_{sect} = [(\alpha - 1)(X_1 - X_0)]^{-1} \ln \frac{(X_1 - x_B)(x_T - X_0)}{(X_1 - x_T)(x_B - X_0)} \tag{23}$$

where X_1 and X_0 are the roots of a quadratic equation and are given by:

$$X_1, X_0 = \frac{L(\alpha - 1) + P \pm \{[L(\alpha - 1) + P]^2 - 4L(\alpha - 1)Py_P\}^{1/2}}{[2L(\alpha - 1)]} \tag{24}$$

and N_{sect} is the number of stages necessary to span the concentration difference from x_B at the bottom of the section to x_T at the top of the section. Equation 23 is also obtained for a square section in the stripping section, but in the case of the stripper the value of X_1 and X_0 are given by:

$$X_1, X_0 = \frac{\{L(\alpha - 1) - W \pm \{[L(\alpha - 1) - W]^2 + 4L(\alpha - 1)Wx_W\}^{1/2}}{[2L(\alpha - 1)]} \tag{25}$$

Graphical Solution. Some of the preceding concepts can be illustrated graphically by means of a McCabe-Thiele diagram, shown in Figure 3. In such a diagram

540 DIFFUSION SEPARATION METHODS

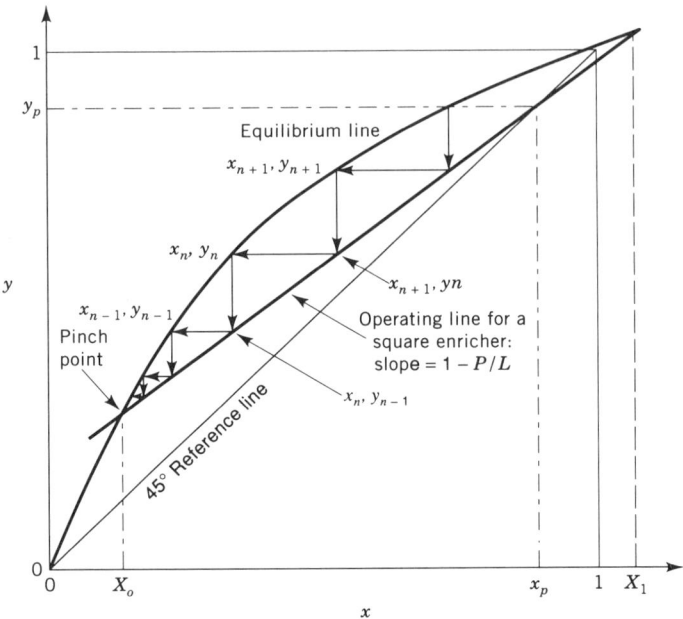

Fig. 3. A McCabe-Thiele diagram for a hypothetical square cascade section illustrating pinching. Terms are defined in text.

the equilibrium line, equations 1 or 2, and the operating line, equation 17, are plotted on a set of x, y coordinates. For the case of a square cascade section, the operating line is straight and has the slope $(1 - P/L)$; if the section lies at the product withdrawal end of the cascade, its opening line passes through the point $x = y = y_P$, as shown. The number of stages required to span a given concentration difference, or conversely the concentration difference obtained across a square section with a given number of stages can be illustrated in such a figure.

It is evident from the construction that the closer the operating line lies to the 45-degree reference line, the fewer the number of stages required to span a given concentration difference. The minimum number of stages is therefore required when the operating line coincides with the 45-degree line, that is, when P/L is equal to zero. It is also evident that in the neighborhood of a point of intersection of the operating and equilibrium lines the enrichment per stage becomes quite small and is equal to zero at the point of intersection itself. At such a point pinching is said to occur, and the origin of the term is made clear by the diagram. In order for a cascade section to span a concentration difference from x_B to y_P, it follows that the operating line may not intersect the equilibrium line before the concentration x_B is reached; the value of L for which the two curves intersect at x_B is the minimum upflow corresponding to the cascade concentration x_B. It is also noteworthy that the values X_0 and X_1 appearing in equations 23 through 25 are the x coordinates of the two points of intersection of the operating line of a square section with the equilibrium line. Although the graphical solution of the cascade gradient equation is simple in principle and exact in theory, it becomes quite cumbersome in practice when processes having separation factors close to unity and hence cascades with thousands of stages are under consideration. For this reason analytic solutions to the gradient equation are usually preferred.

The Ideal Cascade. A cascade of particular interest to design engineers is the ideal cascade: a continuously tapered cascade (ie, L is a continuously varying function of x or n) that has the property of minimizing the sum of the stage upflows of all the stages required to achieve a given separation task. Because, in general, the total volume of the equipment required and the total power requirement of the cascade are directly proportional to the sum of the stage upflows, a consideration of the ideal plant requirements often permits a good economic estimate of the unit cost of product to be made without having to resort to the much more painstaking labor of designing a real (as opposed to ideal) cascade to accomplish the separation job. A simple, intuitive approach to the ideal cascade concept in the case of a cascade composed of discrete stages follows. Again, the resulting equations are also valid for a cascade based on a continuous or differential separation process.

For the case of a stagewise enrichment process the ideal cascade may be defined as one in which there is no mixing of streams of unequal concentrations. Clearly, the mixing of streams of unequal concentrations in the cascade to form the feed to a separation stage constitutes an inefficiency because it is precisely the reverse of the process taking place in the stage itself. Figure 2 shows that the no-mixing condition at the entrance to stage $n + 1$ requires that $y_n = x_{n+2}$. If the enrichment per stage is essentially constant, x_{n+2} may be written as $x_n + 2(dx/dn)$. The concentration y_n is related to x_n by the α-relationship (eq. 2). Thus the no-mixing condition leads directly to the gradient equation for the ideal plant:

$$\frac{dx}{dn} = \frac{\alpha - 1}{2} x(1 - x) \qquad (26)$$

The number of stages required to span a given concentration difference in an ideal plant in which all stages have the same separation factor is therefore

$$N_{\text{ideal}} = \frac{2}{\alpha - 1} \ln \left(\frac{x_T}{1 - x_T} \bigg/ \frac{x_R}{1 - x_R} \right) = 2N_{\text{min}} \qquad (27)$$

and is twice the minimum number of stages required. The combination of equations 19 and 26 gives the equation for the stage upflow at any point in the enricher of an ideal cascade that is twice the minimum upflow

$$L_{\text{ideal}} = \frac{2P(y_P - x)}{(\alpha - 1)x(1 - x)} = 2L_{\text{min}} \qquad (28)$$

Equations 27 and 28 can be used in conjunction, along with the corresponding equations for the stripping section, to produce an ideal plant profile such as is shown in Figure 4 where L_{ideal} is plotted against N_{ideal} for the example of an ideal

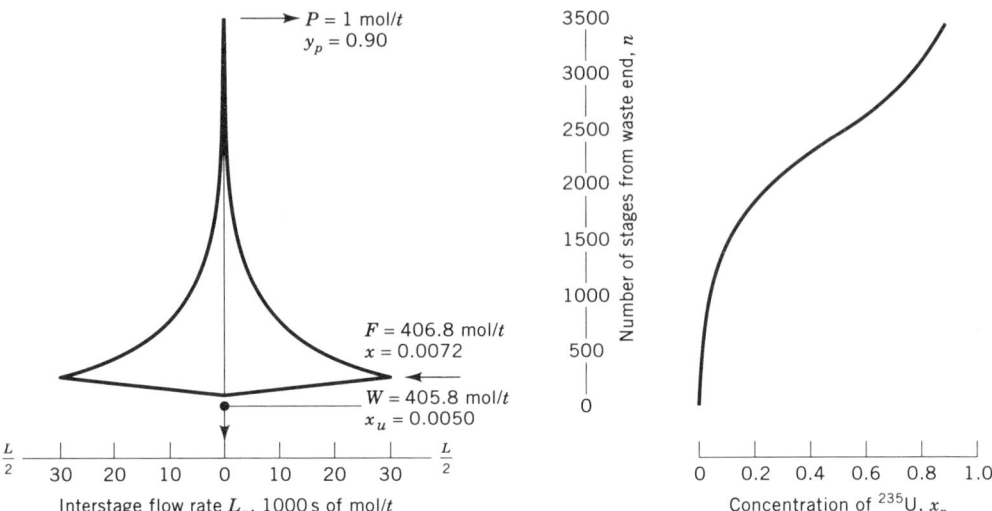

Fig. 4. Characteristics of an ideal separation cascade for uranium isotope separations. For this cascade: $\alpha = 1.0043$; $N_T = 3484$ stages; $\Delta U = 153.08$ mol/t; $\Sigma L_T = 33.116 \times 10^2$ mol/t, where t represents unit time.

cascade to produce one mol of uranium per unit time enriched to 90 mol % in ^{235}U from natural feed containing 0.72 mol % ^{235}U, with a waste stream rejected at a concentration of 0.5% ^{235}U. The characteristic lozenge shape of the ideal cascade is evident: no two stages in either the enricher or stripper are the same size. It can be deduced from the above statements that, except at the terminals, the operating line for an ideal cascade on a McCabe-Thiele diagram is a curved line lying midway between the equilibrium line and the 45° reference line.

Total Upflow in an Ideal Plant. The sum of the upflows from all of the stages in the ideal plant, or more simply, the total upflow, is the area enclosed by the cascade shown in Figure 4. An analytical expression for this quantity is obtained as the summation of all the stage upflows in the enriching section expressed as an integral:

$$\sum^{\text{enr}} L_n = \int L\, dn = \int_{x_F}^{y_P} L\, \frac{dn}{dx}\, dx \qquad (29)$$

For the ideal cascade, L is given by equation 28 and dx/dn by equation 26. Making these substitutions:

$$\sum^{\text{enr}} L_n = \int_{x_F}^{y_P} \frac{4\,P}{(\alpha-1)^2} \frac{(y_P - x)}{x^2(1-x)^2}\, dx \qquad (30)$$

However, recalling the definition of the value function, equation 11, and assuming that the value of α is the same for all stages, the integral may be written in the form:

$$\sum_{}^{enr} L_n = \frac{4P}{(\alpha - 1)^2} \int_{x_F}^{y_P} (y_P - x) v''(x) dx \quad (31)$$

which is readily integrated by parts to give:

$$\text{(total upflow)}_{enr} = [4P/(\alpha - 1)^2][v(y_P) - v(x_F) - (y_P - x_F)v'(x_F)] \quad (32)$$

The equation for the total flow in the stripping section is obtained in the same manner:

$$\text{(total upflow)}_{str} = [4W/(\alpha - 1)^2][v(x_W) - v(x_F) - (x_W - x_F)v'(x_F)] \quad (33)$$

The total flow in the cascade is then given by the sum of equations 32 and 33, which can be simplified with the use of the cascade material balances:

$$P + W = F \quad (34)$$

and

$$Py_P + Wx_W = Fx_F \quad (35)$$

to give the convenient form

$$\text{(total upflow)}_{cascade} = [4/(\alpha - 1)^2][Pv(y_P) + Wv(x_W) - Fv(x_F)] \quad (36)$$

For the example considered above, the total cascade upflow is found to be 33×10^6 mols per unit time.

The second term in brackets in equation 36 is the separative work produced per unit time, called the separative capacity of the cascade. It is a function only of the rates and concentrations of the separation task being performed, and its value can be calculated quite easily from a value balance about the cascade. The separative capacity, sometimes called the separative power, is a defined mathematical quantity. Its usefulness arises from the fact that it is directly proportional to the total flow in the cascade and, therefore, directly proportional to the amount of equipment required for the cascade, the power requirement of the cascade, and the cost of the cascade. The separative capacity can be calculated using either molar flows and mol fractions or mass flows and weight fractions. The common unit for measuring separative work is the separative work unit (SWU) which is obtained when the flows are measured in kilograms of uranium and the concentrations in weight fractions.

The great utility of the separative capacity concept lies in the fact that if the separative capacity of a single separation element can be determined, perhaps from equations 7 or 10, then the total number of such identical elements required in an ideal cascade to perform a desired separation job is simply the ratio of the separative capacity of the cascade to that of the element. The concept of an ideal plant is useful because moderate departures from ideality do not appreciably affect the results. For example, if the upflow in a cascade is everywhere a factor of m times the ideal upflow, the actual total upflow required to perform a separative task is $m^2/(2m - 1)$ times the ideal cascade total upflow. Thus, if the upflow is 20% greater than ideal at every point in that cascade ($m = 1.2$), the number of separation elements would be only 2.86% greater than that calculated from ideal cascade considerations.

Equations for Large Stage Separation Factors. The preceding results have been obtained with the use of equation 2 and by replacing the finite difference, $x_{n+1} - x_n$, by the differential, dx/dn, both of which are valid only when the quantity $(\alpha - 1)$ is very small compared with unity. However, there has been renewed interest, partly because of the development of the gas centrifuge process to commercial status, in the design of cascades composed of stages with large stage separation factors. When the stage separation factor is large, the number of stages required in an ideal cascade in which all stages have the same separation factor is given by

$$N_{ideal} = \frac{2}{\ln\alpha} \ln\left(\frac{y_P}{1 - y_P} \bigg/ \frac{x_W}{1 - x_W}\right) - 1 \tag{37}$$

instead of equation 27. When dealing with cascades composed of stages having large separation factors, it is somewhat more convenient to calculate the sum of all the stage feed flows in the cascade rather than the sum of all the stage upflows as was done in the case when $(\alpha - 1)$ is small. When $(\alpha - 1)$ is small with respect to unity, the stage feed flow is essentially just twice the stage upflow rate, and the stage feed flow rate in an ideal cascade (see eq. 28) is:

$$(L/\theta)_{ideal} = \frac{4P(y_P - x)}{(\alpha - 1)x(1 - x)} \tag{38}$$

However, when α is large, the corresponding equation for the stage feed rate takes the form

$$(L/\theta)_{ideal} = \frac{(\alpha)^{1/2} + 1}{(\alpha)^{1/2} - 1} \frac{P(y_P - z)}{z(1 - z)} \tag{39}$$

The sum of the stage feed flow rates of all of the stages in an ideal cascade is just twice the total cascade upflow rate when $(\alpha - 1)$ is small with respect to unity, or

$$(\text{total stage feed})_{cascade} = \frac{8}{(\alpha - 1)^2} [Pv(y_P) + Wv(x_W) - F_V(x_F)] \tag{40}$$

but is given by

$$(\text{total stage feed})_{cascade} = \frac{2}{\ln\alpha} \frac{(\alpha)^{1/2} + 1}{(\alpha)^{1/2} - 1} [Pv(y_P) + WV(x_W) - FV(x_F)] \tag{41}$$

when α is larger. It can be seen that when α is close to unity, equation 41 gives the same result as equation 40. However, if α is equal to 1.1, the total stage feed would be underestimated by 9.2%, and if α is equal to 2.0, the total stage feed would be underestimated by 42.4%, using equation 40 instead of equation 41. Some of the work in this area of cascade theory has dealt with the design of cascades using asymmetric isotope separation stages (1,2) with the analysis of two-up, one-down cascades, that is, cascades in which the upflow from the nth stage in the cascade bypasses the $(n + 1)^{th}$ stage and is reintroduced at the $(n + 2)^{th}$ stage.

Real Cascades. Although the ideal cascade minimizes the volume of equipment and the energy requirements, the cost of the cascade is generally not minimized because production economies are realized in the manufacture of the process equipment when a large number of identical units are produced. Thus, a minimum-cost cascade consists of a number of square cascade sections rather than uniformly tapered nonidentical stages. A first approximation to the optimum practical cascade, once the size (length and width) of the individual separating units available is known, is obtained by fitting the ideal plant shape with square sections in some intuitively appealing manner, as illustrated in Figure 5.

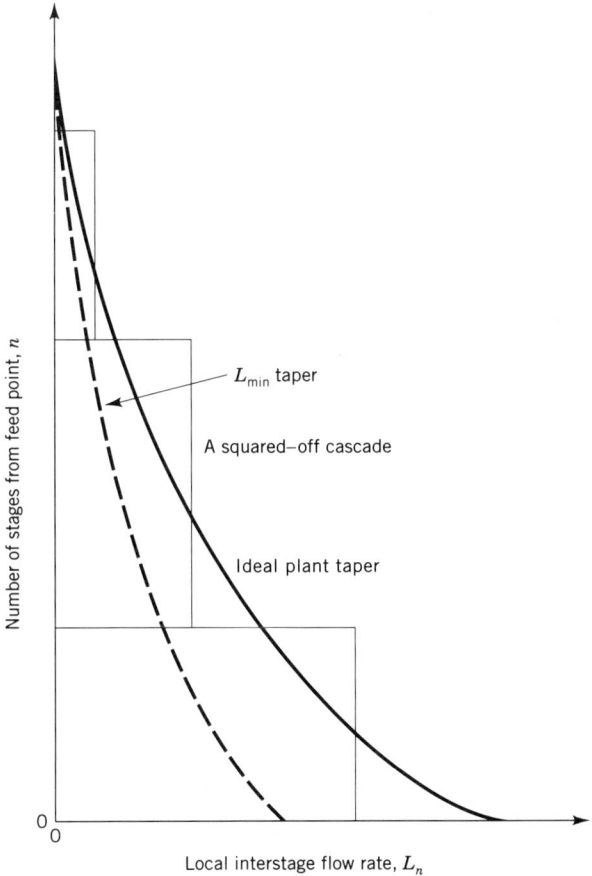

Fig. 5. Design of a real cascade obtained by squaring off an ideal enriching section.

Figure 5 shows an ideally tapered enricher that has been replaced by three square cascade sections, a process called squaring-off the cascade (3–6). During the squaring-off process, two essential requirements must be kept in mind: The interstage flow in all-square sections must always exceed the local value of L_{min} at all points in the cascade, and the squared-off cascade must contain a total number of stages which exceeds N_{min}. In order for the squared-off cascade to give a performance closely resembling that of the ideal cascade, the shape of the squared-off cascade should approximate the shape of the ideal cascade.

In the final analysis the problem of determining the optimum practical cascade is rather complex. Equipment performance and costs need to be related to the selected independent process variables. The main process equipment usually consists of a large number of separating units, pumping and heat exchange equipment, control devices, and connecting piping. The whole process is provided with services, and auxiliary systems and feed and withdrawal facilities. It is usually enclosed by a building and surrounded by land of the proper type. Sizes and cost of these important items must be related to the process variables. The details of procedures used for the optimization of real gaseous diffusion cascades are presented in Reference 7, and the problems of optimization of real gas centrifuge cascades are discussed in Reference 8.

Time-Dependent Cascade Behavior. The period of time during which a cascade must be operated from start-up until the desired product material can be withdrawn is called the equilibrium time of the cascade. The equilibrium time of cascades utilizing processes having small values of $\alpha - 1$ is a very important quantity. Often a cascade may prove to be quite impractical because of an excessively long equilibrium time. An estimate of the equilibrium time of a cascade can be obtained from the ratio of the enriched inventory of desired component at steady state, H, to the average net upward transport of desired component over the entire transient period from start-up to steady state, $\bar{\tau}$. In equation form this definition can be written as

$$T_{eq} = H/\bar{\tau} = \frac{1}{\bar{\tau}} \int_0^N h_n(x_n - x_F) dn \qquad (42)$$

where h_n is the holdup of the nth stage. The average net upward transport for the entire transient period is not usually known; the initial and final values of the net transport, however, are known. At start-up the concentration gradient is flat, because the column is filled with material at feed concentration, and the transport is a maximum. Using this transport in equation 42 gives a lower limit for the equilibrium time

$$(T_{eq})_{min} = H/[L(\alpha - 1)x_F(1 - x_F)] \qquad (43)$$

At steady state, with a fully developed gradient, the net transport is $P(y_P - x_F)$, which is a lower limit for net upward transport. Substituting this into equation 42, leads to an expression for an upper limit for the equilibrium time:

$$(T_{eq})_{max} = H/[P(y_P - x_F)] \qquad (44)$$

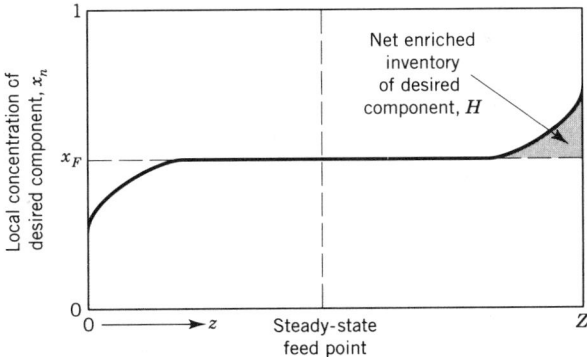

Fig. 6. Concentration gradient in a column or cascade a short time after start-up.

Equations 43 and 44 thus yield a lower and upper limit, respectively, and used together usually give a satisfactory estimate for the equilibrium time of a cascade.

Examination of equation 42 shows that T_{eq} is directly proportional to the average stage holdup of process material. Thus, in conjunction with the fact that liquid densities are on the order of a thousand times larger than gas densities at normal conditions, the reason for the widespread use of gas-phase processes in preference to liquid-phase processes in cascades for achieving difficult separations becomes clear.

The unsteady-state behavior of a separation unit is, furthermore, of interest because it can be used for the experimental determination of the separation parameters of the unit. If the holdup of the separating unit is known, the separation factor, $\alpha - 1$, can be obtained from a knowledge of the transient behavior of the unit during start-up. Figure 6 shows the concentration gradient in a square column shortly after start-up. As long as the gradient is flat at the feed point as shown, an equation much like equation 44 for the maximum equilibrium time can be used to relate the enriched holdup to the elapsed time. From a knowledge of the gradient, the enriched holdup H can be computed, and with L known the separation factor $\alpha - 1$ can be computed from

$$(\alpha - 1) = H/[Lx_F(1 - x_F)\,\Delta t] \qquad (45)$$

where Δt is the elapsed time since start-up when the column was filled with material at feed concentration.

The Gaseous Diffusion Process

The gaseous diffusion separation process depends on the separation effect arising from the phenomenon of molecular effusion (that is, the flow of gas through small orifices). When a mixture of two gases is confined in a vessel and is in thermal equilibrium with its surroundings, the molecules of the lighter gas strike the walls of the vessel more frequently, relative to its concentration, than the molecules of the heavier gas. This is caused by the greater average thermal velocity of the lighter molecules. If the walls of the container are porous with holes large enough to permit the escape of individual molecules, but sufficiently small so that bulk flow of the gas as a whole is prevented (that is, with pore diameters approaching mean-free-path dimensions of the gas), then the lighter molecules escape more readily than the heavier ones, and the escaping gas is enriched with respect to the lighter component of the mixture. The equation for the separation factor, α, for this process reflects the relative ease of light versus heavy molecules in escaping through the pores. Indeed, α^*, the ideal separation factor, is the ratio of the two molecular velocities. Because the kinetic energies, $\frac{1}{2} mv^2$, of the two species are the same, α^*, the ratio of the two velocities is equal also to the square root of the inverse ratio of the two molecular weights. In 1895 Rayleigh and Ramsey used this method to separate argon from nitrogen, and in 1920 it was employed to slightly enrich the concentration of the neon-22 isotope.

A primary improvement in diffusion separation technology was the development in 1932 of a cascade of diffusion stages for isotope separation by an arrangement similar to that shown in Figure 7. Using a 24-stage cascade an appreciable enrichment in the isotopes of neon was obtained. Subsequently (9), almost pure deuterium was obtained from a cascade of 48 stages, and the isotopes of nitrogen and of carbon were enriched (10) using a 34-stage cascade.

The large plants built for the separation of uranium isotopes following World War II are outstanding applications of the gaseous diffusion process. In the United States this work culminated in the construction of the gaseous diffusion cascade called the K-25 plant at Oak Ridge, Tennessee. Other plants were built at Oak Ridge, Paducah, Kentucky, and Portsmouth, Ohio, in the United States (11–13) and at Capenhurst, England (14), and at Pierrelatte and Tricastin, France (15). Gaseous diffusion plants have also been reported to be in operation in the former USSR and in the People's Republic of China.

Process Description. The basic unit of a gaseous diffusion cascade is the gaseous diffusion stage. The main components are the converter holding the barrier in tubular form, motors, and compressors moving the gas between stages, a heat exchanger removing the heat of compression introduced by the stage compressors, the interstage piping, and special instruments and controls to maintain the desired pressures and temperatures.

Figure 7 is a schematic representation of a section of a cascade. The feed stream to a stage consists of the depleted stream from the stage above and the enriched stream from the stage below. This mixture is first compressed and then cooled so that it enters the diffusion chamber at some predetermined optimum temperature and pressure. In the case of uranium isotope separation the process gas is uranium hexafluoride [7783-81-5], UF_6. Within the diffusion chamber the gas flows along a porous membrane or diffusion barrier. Approximately one-half of the gas passes through the barrier into a region of lower pressure. This gas is enriched in the component of lower molecular weight, ^{235}U. The enriched fraction, upon leaving the diffusion chamber, is directed to the stage above where it is recompressed to the barrier high-side pressure. The gas that does not pass through the barrier is depleted with respect to the light component. This depleted fraction, upon leaving the chamber, passes through a control valve, and is directed to the stage below where it too is recompressed to the barrier high-side pressure.

DIFFUSION SEPARATION METHODS 549

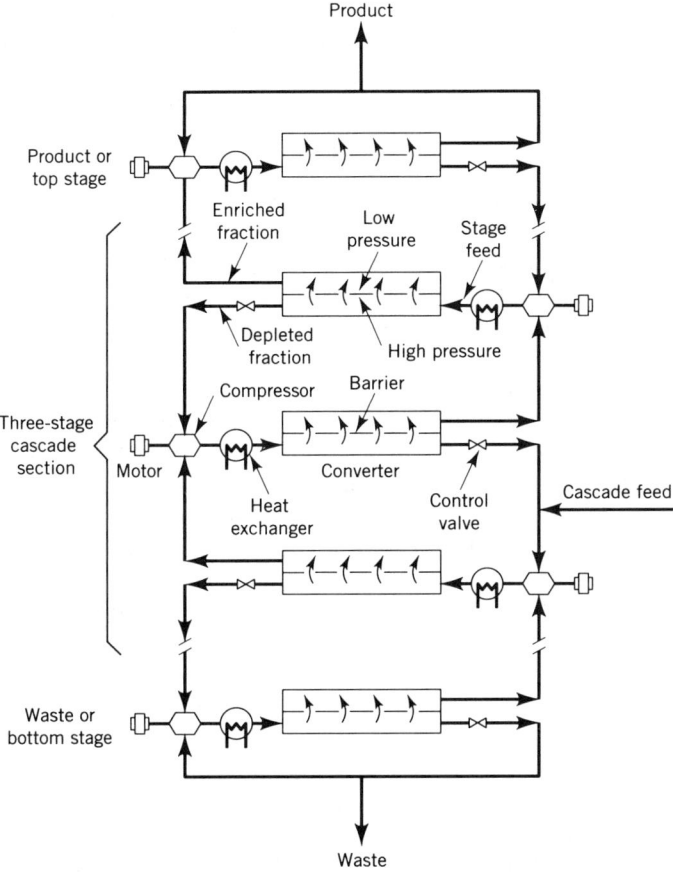

Fig. 7. A cascade of gaseous diffusion stages.

However, because it is necessary in this case to compensate only for the frictional losses and the control valve pressure drop, the compression ratio may not need to be as high as that for the enriched fraction. Thus there is some freedom in the design of the stage; that is, two compressors might be used, a larger one for the enriched fraction and a smaller one for the depleted fraction; or the pressure drop across the control valve may be made equal to the pressure drop across the barrier so that both streams can be recompressed by the same compressor (this scheme, although wasteful of power, might make up for it in savings in equipment costs), or some other compromise mode of operation might be used.

For operational efficiency a number of gaseous diffusion stages are operated together in units referred to as cells and buildings. Cells and buildings can be removed from operation for routine maintenance and bypassed without disturbing the diffusion cascade.

Successful operation of the gaseous diffusion process requires a special, fine-pored diffusion barrier, mechanically reliable and chemically resistant to corrosive attack by the process gas. For an effective separating barrier, the diameter of the pores must approach the range of the mean free path of the gas molecules, and in order to keep the total barrier area required as small as possible, the number of pores per unit area must be large. Seals are needed on the compressors to prevent both the escape of process gas and the inflow of harmful impurities. Some of the problems of cascade operation are discussed in Reference 16.

The need for a large number of stages and for the special equipment makes gaseous diffusion an expensive process. The three United States gaseous diffusion plants represent a capital expenditure of close to 2.5×10^9 dollars (17). However, the gaseous diffusion process is one of the more economical processes yet devised for the separation of uranium isotopes on a large scale.

Stage Design. The important parameters of a separation cascade employing gaseous diffusion stages are the stage separation factor and the size of a stage required to handle the desired stage flows. Both of these parameters depend on the characteristics of the barrier.

Barrier Characteristics. The barrier material must be fine-pored and have many pores per unit area. Preparation and characterization of such a material presents a difficult technological problem. The characteristics of a barrier suitable for the separation of isotopes by gaseous diffusion are discussed in Reference 18, including various effective pore sizes and pore size distributions. Experimental techniques used to evaluate barrier characteristics are presented in Reference 19. These techniques include adsorption methods, electron microscopy, x-ray analysis, porosity measurements with mercury, permeability measurements with liquids and gases, and measurements of separation effectiveness. Barrier materials have pore sizes in the range of 10–30 nm (20). One electrolytic technique leads to a thin sheet material having about 10^{10} pores per square centimeter and radii on the order of 15 nm (21). The separating performance of a barrier has been evaluated by means of a 12-stage pilot plant (22).

Barrier Flow. An ideal separation barrier is one that permits flow only by effusion, as is the case when the diameter of the pores in the barrier is sufficiently small compared to the mean free path of the gas molecules. If the pores in the barrier are treated as a collection of straight circular capillaries, the rate of effusion through the barrier is governed by Knudsen's law (eq. 46):

$$N = \frac{4}{3}(2\pi MRT)^{-1/2} \frac{\phi d}{l}(p_f - p_b) \tag{46}$$

where N is the molar flow of gas per unit area through the barrier, M is its molecular weight, R is the gas constant, T is the absolute temperature, ϕ is the fraction of the barrier area open to flow, d is the effective pore diameter, l is the pore length or thickness of the barrier, and p_f and p_b are the high- and low-side pressures of the barrier, respectively. In practice not all of the flow through the barrier is effusive flow. Through those pores where the diameters are of the order of the mean free path or greater, a nonseparative Poiseuille flow occurs. The two types of flow are additive and the total flow can be represented by

$$N = \frac{a}{(M)^{1/2}}(p_f - p_b) + \frac{b}{\mu}(p_f^2 - p_b^2) \tag{47}$$

where a and b are functions of temperature for a particular barrier and μ is the viscosity of the gas. The first term on the right is the contribution to the total flow of the effusive flow, the second that of the nonseparative Poiseuille flow. Because the pressure dependence of each type of flow is different, the constants a and b can be evaluated from a series of measurements at different pressures (24).

The Fundamental Separation Effect. An ideal-point separation factor can be defined on the basis of the separation obtained when a binary mixture flows through an ideal barrier into a region of zero back pressure. For this case an expression of the form of equation 46 can be written for each component. The flow of light component through the barrier is proportional to $p_f x'/(M_A)^{1/2}$, and the flow of heavy component is proportional to $p_f(1 - x')/(M_B)^{1/2}$ where x' is the mol fraction of the light component on the high-pressure side of the barrier and M_A and M_B are the mol wts of the light and heavy components, respectively. The concentration, y', of the effusing gas is therefore

$$y' = \frac{x'/(M_A)^{1/2}}{x'/(M_A)^{1/2} + (1 - x')/(M_B)^{1/2}} \qquad (48)$$

and from the definition of α (eq. 1), it follows that the ideal-point separation factor is equal to

$$\alpha^* = \frac{y'/(1-y')}{x'/(1-x')} = (M_B/M_A)^{1/2} \qquad (49)$$

which, for the case of uranium isotope separation using UF_6, is equal to 1.00429. As has been pointed out, this is also the expression for the ratio of the velocity of the light molecules to that of the heavy molecules.

The Stage Separation Factor. The stage separation factor, in all probability, is appreciably different from the ideal-point separation factor because of the existence of four efficiency terms:

(1) A Barrier Efficiency Factor. In practice, diffusion plant barriers do not behave ideally; that is, a portion of the flow through the barrier is bulk or Poiseuille flow which is of a nonseparative nature. In addition, at finite pressure the Knudsen flow (25) is not separative to the ideal extent, that is, $(M_A/M_B)^{1/2}$. Instead, the degree of separation associated with the Knudsen flow is less separative by an amount that depends on the pressure of operation. To a first approximation, the barrier efficiency is equal to the Knudsen flow multiplied by a pressure-dependent term associated with its degree of separation, divided by the total flow.

(2) A Back-Pressure Efficiency Factor. Because a gaseous diffusion stage operates with a low-side pressure p_b which is not negligible with respect to p_f, there is also some tendency for the lighter component to effuse preferentially back through the barrier. To a first approximation the back-pressure efficiency factor is equal to $(1 - r)$, where r is the pressure ratio p_b/p_f.

(3) A Mixing Efficiency Factor. As the gas flows along the high-pressure side of the diffusion barrier, it becomes, as a result of the effusion process, preferentially depleted with respect to the lighter component in the neighborhood imme-

diately adjacent to the barrier. As a result a concentration gradient perpendicular to the barrier is set up on the high-pressure side, and the average concentration x' of the light component in the bulk of the gas flowing past a point is greater than x'', the concentration of the light component at the surface of the barrier at that point. The mixing efficiency factor is equal to the ratio $(y' - x')/(y' - x'')$ as indicated in Figure 8. A value for the point mixing efficiency factor can be calculated from a consideration of diffusion through an effective film representing the resistance to diffusion. It is given by an expression of the form: $\exp(-Nl_f/\rho D)$, where l_f is the thickness of the effective film.

(4) A Cut-Correction Factor. The stage separation factor has been defined as relating the concentrations in the streams leaving the stage. Because the concentrations, x' and y', on each side of the barrier are changing continuously as the gas flows through the diffusion stage, the relationship between the concentrations of the streams differs from the point relationship. This difference is taken into account with the cut correction factor. If the gas on the high-pressure side flows through the stage with no appreciable mixing taking place in the direction

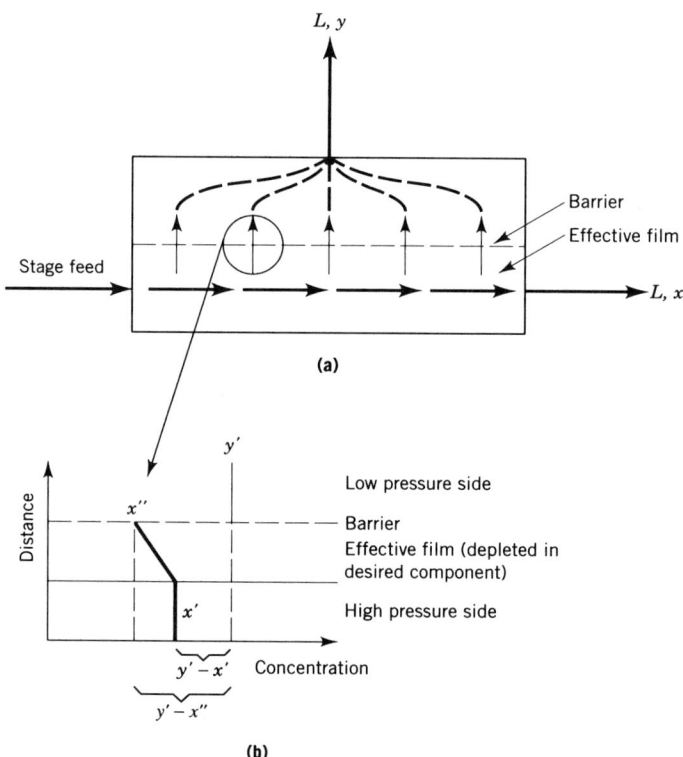

Fig. 8. Flow of process gas through a gaseous diffusion stage. (**a**) Gaseous diffusion stage; (**b**) local concentration profiles near the diffusion barrier; y' = point concentration of light component on low pressure side; x' = average concentration of light component in bulk of gas on high pressure side of barrier flowing past the specified point; x'' = point concentration of light component at surface of barrier on the high pressure side; $y' - x''$ = separation that would be obtained across barrier in the absence of the effective film; and $y' - x'$ = actual separation obtained across barrier, taking a film into account.

of flow, and if the effused fraction is withdrawn from the stage directly upon passing through the barrier, the cut correction can be calculated from material balance considerations. For this case, the exit or stream concentration of the stage upflow is equal to the average concentration of the effused gas, whereas the exit or stream concentration of the downflow is equal to the terminal concentration of the uneffused gas which is, of course, at maximum and not average depletion. Consequently, the stage separation factor relating to exit concentration is greater than the point separation factor, and the cut correction factor exceeds unity. (This phenomenon is analogous to cross-flow in a plate distillation column.)

The stage separation factor can therefore be related to the ideal-point separation factor by an equation of the form

$$(\alpha - 1) = (E_b)(E_p)(E_M)(E_c)(\alpha^* - 1) \tag{50}$$

where

$$E_b = \text{barrier efficiency} = \frac{\text{separative flow through barrier}}{\text{total flow through barrier}} \tag{51}$$

$$E_p = \text{back-pressure efficiency} = 1 - r \tag{52}$$

$$E_M = \text{mixing efficiency} = \exp^{-Nlf/\rho D} \tag{53}$$

and

$$E_c = \text{cut correction} = \frac{1}{\theta} \ln \frac{1}{1 - \theta} \tag{54}$$

For the usual case where approximately one-half of the gas entering the stage passes through the barrier, the value of the cut, θ, is equal to 0.5 and the cut correction takes on the value 1.386.

These efficiency factors are discussed in more detail in References 26–28. Actually, the barrier and back-pressure efficiencies are interrelated and cannot be formulated independently, except only as an approximation. A better formation that has been found to fit the experimental results is:

$$(E_b)(E_p) = (1 - r)/[1 + (1 - r)(p_f/p^*)] \tag{55}$$

where p^* is a constant, the value of which must be determined experimentally. It may be noted that for an ideal barrier, $E_b = 1$, p^* is equal to ∞, and the back-pressure efficiency is given by equation 52.

Separative Capacity. An expression for the separative capacity of a single gaseous diffusion stage where the upflow rate is L mols per unit time, given in equation 7, can be written as

$$\delta U = 1/4 \, L \left(\frac{1 - r}{1 + (1 - r)(p_f/p^*)} \right)^2 E_M^2 \left(\frac{1}{\theta} \ln \frac{1}{1 - \theta} \right)^2 (\alpha^* - 1)^2 \tag{56}$$

554   DIFFUSION SEPARATION METHODS

For a very high quality barrier ($p^* \longrightarrow \infty$), the separative capacity of a stage having a mixing efficiency of 100% and operating at a cut of one-half would be:

$$\delta U = \frac{1.921}{4} L(1 - r)^2(\alpha^* - 1)^2 \tag{57}$$

If the power requirement of the gaseous diffusion process were no greater than the power required to recompress the stage upflow from the pressure on the low-pressure side of the barrier to that on the high-pressure side, then the power requirement of the stage would be $LRT \ln (1/r)$ for the case where the compression is performed isothermally. The power requirement per unit of separative capacity would then be given simply by the ratio

$$\frac{\text{power requirement}}{\text{unit separative capacity}} = \frac{2.082 \, RT \ln (1/r)}{(1 - r)^2(\alpha^* - 1)^2} \tag{58}$$

This quantity is minimized when the stage is operated at a pressure ratio across the barrier corresponding to $r = 0.285$. Furthermore, if power were the only economic consideration, the stage would be operated at this pressure ratio. However, as the value of r is decreased from this optimum, although the cost of power is increased, the number of stages required and hence the capital cost of the plant is decreased. Thus, in practice a compromise between these factors is made.

The optimum pressure level for gaseous diffusion operation is also determined by comparison; at some pressure level the decrease in equipment size and volume to be expected from increasing the pressure and density is outweighed by the losses that occur in the barrier efficiency. Nevertheless, because it is well known that the cost of power constitutes a large part of the total cost of operation of gaseous diffusion plants, it can perhaps be assumed that a practical value of r does not differ greatly from the above optimum. Inclusion of this value in the preceding equations yields

$$\delta U = 0.246 \, L(\alpha^* - 1)^2 \tag{59}$$

and

$$\frac{\text{power requirement}}{\text{unit separative capacity}} = \frac{5.11 \, RT}{(\alpha^* - 1)^2} \tag{60}$$

The actual power requirement is greater than that given by equation 58 or 60 because of the occurrence of frictional losses in the cascade piping, compressor inefficiencies, and losses in the power distribution system.

Plant Operation and Costs. The operation and economics of the three United States gaseous diffusion plants running in 1972 is discussed in References 29 and 30. These plants were operated as a single gaseous diffusion complex such that interplant shipments occurred so as to optimize the overall system. Independent operation of the plants would have resulted in about a 1% loss in separative work.

In 1985, owing to the declining demand by the nuclear power industry for enriched uranium, the Oak Ridge gaseous diffusion plant was taken out of operation and, subsequently, was shut down. The U.S. gaseous diffusion plants at Portsmouth, Ohio and Paducah, Kentucky remain in operation and have a separative capacity of 19.6 million SWU (separative work unit) per year which as of this writing is not fully utilized.

Information on the design of new gaseous diffusion plants is available (30). The shape of a gaseous diffusion cascade, based on 1970 U.S. technology, having 8.75 million SWU/yr separative capacity, designed to produce uranium containing 4% ^{235}U from natural feed having tails at 0.25% ^{235}U is shown in Figure 9.

In 1973, Eurodif, a multinational consortium organized under French leadership, decided to build a large gaseous diffusion plant at Tricastin in France. This plant was completed in 1982, has a separative capacity of 10.8 million SWU/yr, and is based on French gaseous diffusion technology. Some design and progress information regarding this cascade is available (31–33). The engineering design of the French gaseous diffusion stage, although functionally the same as the United States stage, differs appreciably in appearance, and the motor, compressor, and diffuser are arranged vertically and are contained in a single housing in the French plant. Some of the main features of the Tricastin plant are UF$_6$ rates in kg/h of product, feed, and waste are 600, 3025, and 2425, respectively; percentage ^{235}U in the product, feed, and waste is 1.35–3.15, 0.72, and 0.25, respectively; stages in the enricher are 220 small sizes at 0.6 MW, 280 medium sizes at 1.6 MW, and 320 large sizes at 3.3 MW; stages in the stripper are 60 small sizes, 120 medium, and 400 large at 0.6, 1.6, and 3.3 MW, respectively; the total cascade power is 3100 MW, and the total cascade separative capacity is 10.7 × 10^6 SWU/yr; and the specific power requirement is 2538 kWh/SWU.

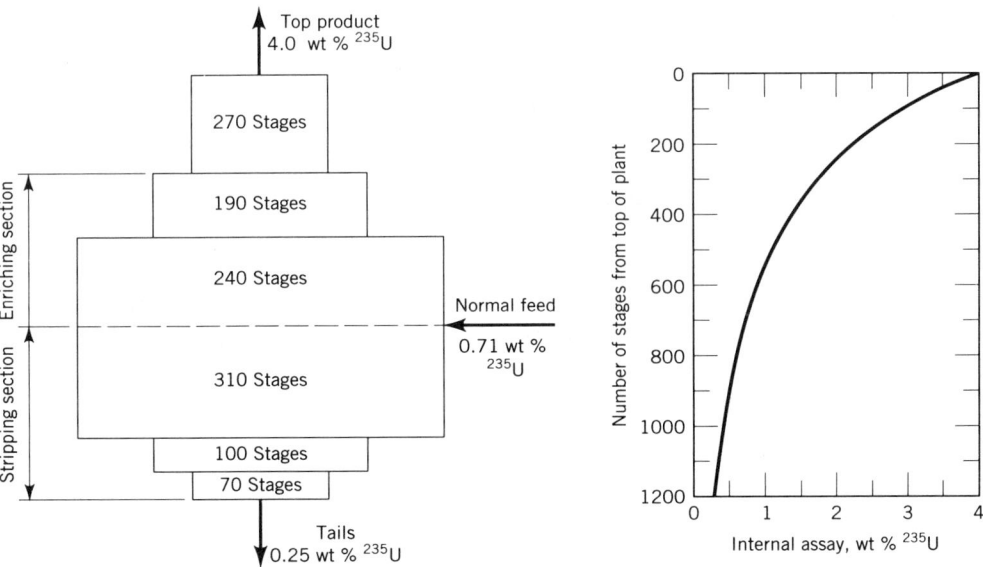

Fig. 9. Schematic diagram and internal gradient for an 8.75 million-SWU/yr plant.

From equation 60 one can obtain a theoretical power requirement of about 900 kWh/SWU for uranium isotope separation assuming a reasonable operating temperature. A comparison of this number with the specific power requirements of the United States (2433 kWh/SWU) or Eurodif plants (2538 kWh/SWU) indicates that real gaseous diffusion plants have an efficiency of about 37%. This represents not only the barrier efficiency, the value of which has not been reported, but also electrical distribution losses, motor and compressor efficiencies, and frictional losses in the process gas flow.

The cost of enriched material from a gaseous diffusion plant depends both on the cost of separative work and of feed material. It can be seen from equation 15 that if the optimum tails concentration from a gaseous diffusion plant is 0.25%, the ratio of the cost of a kg of normal uranium to the cost of a kg of separative work equal to 0.80 is implied. Because the cost of separative work in new gaseous diffusion plants is expected to be about $100/SWU, equation 16 gives the cost per kg of uranium containing 4% ^{235}U as about $1,240.

Pressure Diffusion Processes

The development of the kinetic theory of gases led to the conclusion that a partial separation of the components of a gaseous mixture results when the gas is subjected to a pressure gradient. Thus, a column of gas standing in the earth's gravity field should show a separation effect, the lighter components concentrating at the top of the column, the heavier components at the bottom. This is indeed the case, but the effect is too slight to be utilized in a practical separation process. In the case of the isotopes of uranium, a column about 0.4 km in height would be required to give an enrichment equal to that of a single gaseous diffusion stage. Therefore, in order to utilize the pressure diffusion phenomenon, steeper pressure gradients than are normally available are needed.

Several devices have been developed for the purpose of producing such pressure gradients. The best known is the gas centrifuge (see also CENTRIFUGAL SEPARATION). High-speed centrifuges can develop gravitational fields equal to many thousand times that of the earth. Thus relatively large pressure gradients can exist between the axis and periphery of a centrifuge, giving rise to appreciable separation effects. By moving streams of gas at the periphery and at the axis countercurrently, the centrifuge can be made equivalent to a multistage separating column.

A second type of apparatus based on the pressure diffusion effect is the separation nozzle. Pressure gradients in a curved expanding jet produce an isotopic separation similar to that in a centrifuge. The separation effect obtained with a single jet is relatively small, and separation nozzle stages, similar to gaseous diffusion stages, must be used in a cascade to realize most of the desired separations.

A third device that utilizes pressure diffusion is the vortex chamber. Here, as in the centrifuge, angular acceleration effects in a rapidly rotating gas provide the pressure gradient. The vortex chamber may be considered as a centrifuge with a stationary outer wall. The mechanical difficulties of high-speed rotating machinery are avoided at the expense of friction effects between the gas and the stationary wall. The literature concerning the use of such a device for isotope separation is limited. Results of experimentation indicate the effect of some of the process variables and the separation factors in H_2–CO_2, H_2–HD and ^{36}Ar–^{40}Ar binary gas mixtures have been measured (34,35). A vortex tube has been used for isotope separation (36), and for the separation of gases in nuclear rocket or ramjet engines.

A vortex tube process has been developed in South Africa and is being used there for the enrichment of uranium (37). It appears that cascades of this type are characterized by an extremely high power consumption.

The Gas Centrifuge. The first suggestion that centrifugal gravitational fields might be used to effect separation of isotopes was made in 1919 (38). Then in 1934 the convection-free vacuum ultracentrifuge was developed (39–41). Extensive information on the construction and operation of high-speed centrifuges and experimental data on the separation of the isotopes of argon, xenon, and uranium are available (39,42–47). Early work on centrifuge development and centrifuge theory is discussed in References 48–55. Simplified approximate models of the flow have been developed (56–58), as have more accurate approximations (59–70). Japanese researchers have made an appreciable contribution to the literature in this field (71–78). Other surveys can be found in References 28 and 79–84. An excellent source of information on gas centrifuge development and centrifuge theory can be found in the proceedings of the early workshops on gases in strong rotation and in the proceedings of the workshops on separation phenomena in gases and liquids which followed (85–91).

The Groth and Zippe Centrifuges. A schematic drawing of the ZG 5 gas centrifuge, in Figure 10, is typical of the Groth centrifuge (39,45,46). It is suspended and driven from above directly by an electric motor. The rotor spins in a vacuum-tight casing. Gas is introduced through a central tube and removed through scoop tubes at the ends of the rotor shielded by baffles from the main part of the bowl to prevent disturbance of the internal gas flow pattern. The gas is caused to undergo countercurrent axial flow by maintaining a temperature difference between the ends of the rotor. The top end of the bowl is heated by eddy currents in an aluminum ring at the top end cap; the bottom end cap is cooled by a cooling coil. Thermocouples are used to measure the end cap temperatures and the internal pressure at the centrifuge axis is measured by connections to the center tube. Labyrinth seals are used at the ends to maintain a gas seal.

The short bowl Zippe centrifuge (55) shown in Figure 11 is somewhat simpler. It is supported on a needle bearing at the base and driven by an electric motor, the armature of which is a steel plate rigidly attached to the bottom of the rotor. The stator consists of a flat winding on an iron core positioned so that the poles are separated from the armature by only a small gap (about 6 mm). Power is supplied by an alternator. Damping bearings are used to resist vibrations at both ends of the rotor. The centrifuge is completely closed at the bottom. The other end is connected with the top region of the outer vacuum casing only by a small annular gap around the feed tube. A small amount of gas that leaks from the interior of the bowl at the low pressure near the axis is confined to the region above the top of the rotor by a Holweck-type spiral groove molecular pump surrounding the rotor near the top, and pumped out of this region to maintain the necessary vacuum. Dimensions of the two types of centrifuges are given in Table 1.

558 DIFFUSION SEPARATION METHODS

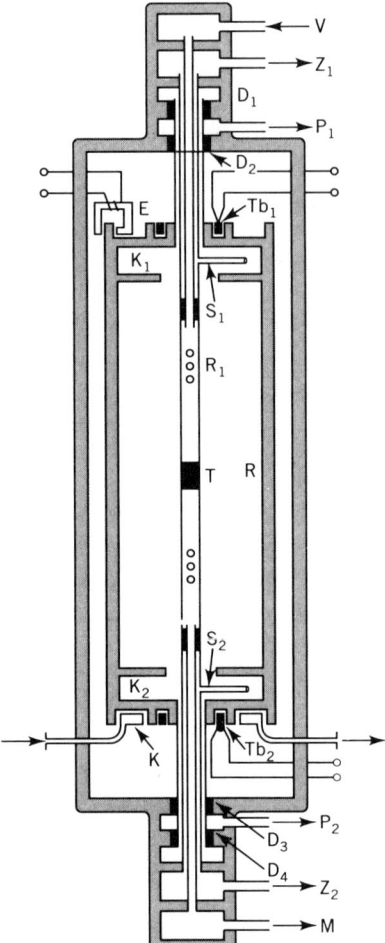

Fig. 10. The Groth ZG 5 centrifuge. R, rotor; R_1, stationary shaft; T, Teflon seal; K_1, K_2, chambers for gas scoops; S_1, S_2, scoops; V, gas supply; M, manometer; Z_1, Z_2, tapping points for enriched and depleted gas; P_1, P_2, vacuum chambers; E, electromagnet for eddy current heating; Tb_1, Tb_2, temperature measuring devices; K, cooling coil; and D_1, D_2, D_3, D_4, labyrinth seals.

The maximum theoretical separative capacity of a centrifuge is proportional to its length and to the fourth power of its peripheral speed, putting a premium particularly on high peripheral speeds and to a lesser extent on long rotors. The allowable peripheral speed of a cylindrical rotor is limited by the ratio of the tensile strength of the material of construction to its density. The maximum peripheral speed, or the burst speed, of a centrifuge rotor is given by:

$$V_{\max} = \left(\frac{\sigma}{\rho}\right)^{1/2} \qquad (61)$$

where σ is the tensile strength of the material and ρ is its density. The rotor

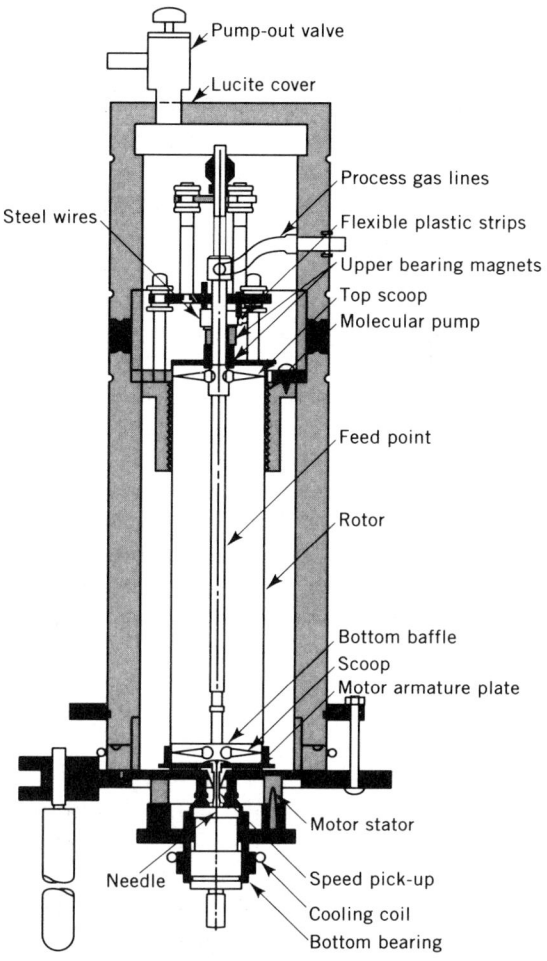

Fig. 11. The Zippe centrifuge.

Table 1. Dimensions of Centrifuge Bowls[a]

Bowl	Length, cm	Radius, cm
UZ I	40	6.0
UZ IIIB	63.5	6.7
ZG 3	66.5	9.25
ZG 5	113.0	9.25
ZG 6[b]	240.0	20.0
ZG 7[b]	316.0	22.5
Zippe	30–38	3.81

[a]See Refs. 45 and 47.
[b]Proposed.

Table 2. Maximum Peripheral Speeds for Various Rotor Materials

Material	Density, ρ, kg/m^3	Tensile strength, σ, GPa[a]	V_{max}, m/s
aluminum[b]	2800	0.448	400
aluminum[c]	2800	0.64	478
steel[b]	7800	1.381	421
maraging steel[b]	7800	1.932	498
maraging steel[c]	8100	3.0	608
glass fiber[b]	1800	0.49	522
carbon fiber[b]	1600	0.829	720
kevlar[c]	1334	2.17	1271
carbon fiber[c]	1560	3.50	1498

[a]To convert from GPa to psi, multiply by 145,000.
[b]From Ref. 82.
[c]From Ref. 90.

must be resistant to the process gas, then only certain materials may be usable. Both Groth and Zippe centrifuges have used an aluminum alloy for rotors for use with UF$_6$. Table 2 gives the values of these properties for several materials that could be used for fabricating centrifuge rotors and the value of the corresponding maximum peripheral speed. The allowable operating speed would be expected to be about 80% of the maximum speed in order to provide a margin of safety.

Mechanical Features. The construction and operation of a precision, high-speed centrifuge in a high vacuum environment presents some formidable mechanical problems. One difficulty with high-speed rotating machinery is the critical-speed phenomena. A long rod, or its equivalent, undergoes resonant vibrations at its fundamental and higher natural frequencies. This can cause large displacements from the axis of rotation unless the rod is properly restrained at high frequencies by damping devices capable of applying sufficient restraining forces. In the Zippe centrifuge the resonance frequency problem is avoided by limiting the ratio of length to diameter to less than four so that the customary operating speeds (300–350 m/s) are below the first fundamental flexural critical frequency, a so-called subcritical centrifuge. On the other hand Groth models, the ZG 6 and ZG 7, are long bowl or supercritical centrifuges. These run at rotational speeds above that corresponding to one or more flexural critical values and operate at a speed not too close to any of the critical values.

The principal power consumption of a centrifuge at operating speed occurs in friction in the bearing systems and in gas drag on the internal parts, particularly the scoops. A poorly balanced rotor results in high power consumption. Wide variations result from variations in the number, length, diameter of tubing, and tip design of the scoops (47). The long-term maintenance and lubrication of the bearing and support systems are a problem.

The scoop system in the Zippe centrifuge is used to control the internal circulation of the gas in the centrifuge, and in common with the Groth machines, must also extract a sufficient volume of gas at a pressure adequate to pump the gas to the feed point of the next centrifuge in a cascade. If this can be accomplished, no intermachine pumps are required in a cascade. This becomes an increasingly difficult problem at higher speeds because the scoop tips must be close to the centrifuge bowl wall in order to have access to the process gas at a higher pressure. The size and length of the piping in the scoops and the feed insertion tubing is critical because of limitations of their conductance for gas flow at low pressures. Other problems include long-term fatigue and creep of structural materials at high speeds and possibly stress corrosion in some systems. The Zippe centrifuge has been taken as the starting point for most of the modern centrifuge research and development programs.

The Urenco/Centec organization, formed in 1971 by British, Dutch, and German companies to carry on centrifuge research and development efforts, is a primary manufacturer and operator of centrifuge cascades for uranium enrichment. The research and development activities pursued by Urenco have succeeded roughly in doubling the separative capacity of individual centrifuges every five years since 1971. This has been accomplished primarily by increases in the length of the centrifuges and by increases in the peripheral speed of the centrifuges by the use of stronger and lighter materials of construction. Urenco currently operates commercial centrifuge enrichment facilities at Almelo in the Netherlands, at Capenhurst in England, and at Gronau in Germany. These plants have a combined separative capacity of 2.5 million SWU per year. A centrifuge plant to be built by Urenco having a separative capacity of 1.5 million SWU per year is planned for Homer, Louisiana.

An aggressive centrifuge development program culminating in the design, construction, and testing of large centrifuges that had a nominal separative capacity of 200 SWU per machine was carried out in the United States. This program was abandoned in 1985 before construction of the centrifuge facility in Portsmouth, Ohio, was completed. No gas centrifuge work has been done in the United States since that date. The former USSR has revealed that it has employed gas centrifuge technology for uranium enrichment since 1960 and that it operates centrifuge plants that have a combined separative capacity of 10 million SWU annually. Japan has also been developing gas centrifuges for a number of years. A 200,000 SWU uranium enrichment demonstration plant was placed in service at Ningyo-Toge in May 1989. A 1.5 million SWU/y gas centrifuge plant is under construction at Rokkashomura.

Design Principles. Although the separation of fluid mixtures can be accomplished using several different types of centrifuges, discussion of the centrifuge separation theory is herein confined to the consideration of the countercurrent gas centrifuge. In order to design separation cascades consisting of countercurrent gas centrifuges, it is necessary to know the separative performance of the individual units. Gas centrifuge theory serves fairly well for predicting the performance of a single centrifuge. However, the separation behavior of a particular gas centrifuge depends on the flow pattern of the gas circulating within it, which in turn depends on the geometry of any baffles and scoops within the centrifuge bowl as well as on any temperature gradients in the gas and on the method used to introduce feed to the centrifuge. Owing to the complexity of the general case, the equations for centrifuge performance are presented for only a few idealized circulation patterns.

Radial Density and Pressure Gradients. Consider a centrifuge of length Z and of radius r_2, the internal dimensions of the centrifuge bowl, that rotates at a constant angular velocity of ω radians per second. If the centrifuge contains a single pure gas rotating at the same angular velocity as the centrifuge bowl, each element of the gas has a force impressed on it by virtue of its angular acceleration. This force is directed outward in a cylindrical coordinate system, and can be expressed as $(\rho\omega^2 r)(r\,dr\,d\phi\,dz)$. At steady state this force must be balanced by a force resulting from the radial pressure gradient established in the centrifuge bowl. The inward force on an element of the gas owing to this pressure gradient is given by $(dp/dr)(r\,dr\,d\phi\,dz)$. Equating these two forces gives:

$$dp/dr = \rho\omega^2 r \qquad (62)$$

where p is the pressure, r is the spatial coordinate in the radial direction, ρ is the density of the gas, and ω is the angular velocity of the centrifuge.

The pressure and the density of a gas are related by an equation of state. If the maximum pressure permitted within the centrifuge bowl is not too high, the equation of state for an ideal gas will suffice. The relationship between the pressure and density of an ideal gas is given by the well-known equation:

$$p = \rho RT/M \qquad (63)$$

where T is the absolute temperature of the gas, K; M is the mol wt of the gas, and R is the gas constant, 8.3147 J/(mol·K). Elimination of the density from equations 62 and 63 yields the differential equation for the pressure gradient in the centrifuge (eq. 64),

$$\frac{dp}{dr} = \frac{Mp}{RT}\omega^2 r \qquad (64)$$

which, for the case of an isothermal centrifuge, is readily integrated to yield

$$p(r) = p(0)\exp(M\omega^2 r^2/2RT) \qquad (65)$$

Equation 65 gives the pressure at any point within the centrifuge, $p(r)$, as a function of the coordinate r, the pressure at the axis $p(0)$, the angular velocity of the centrifuge, and the temperature and mol wt of the gas. Should the centrifuge contain not a single pure gas, but a gas mixture, equations of the above forms could be written for each species present. In particular for the case of a binary gas mixture, consisting of species A and B.

$$p_A(r) = p_A(0)\exp(M_A\omega^2 r^2/2RT) \qquad (66)$$

and

$$p_B(r) = 2p_B(0)\exp(M_B\omega^2 r^2/2RT) \qquad (67)$$

The ratio of these two equations gives the radial separation afforded by the gas centrifuge under equilibrium conditions, that is, for no internal gas circulation. An equilibrium separation factor between gas at the axis of the centrifuge and gas at the periphery is therefore given by:

$$\alpha_0 \equiv \frac{x_A(0)}{x_B(0)} \bigg/ \frac{x_A(r_2)}{x_B(r_2)} = \exp[(M_B - M_A)\omega^2 r_2^2 / 2RT] \qquad (68)$$

It should be noted that the separation factor for the centrifuge process is a function of the difference in the mol wts of the components being separated rather than, as is the case in gaseous diffusion, a function of their ratio. The gas centrifuge process would therefore be expected to be relatively more suitable for the separation of heavy molecules. As an example of the equilibrium separation factor of a gas centrifuge, consider the Zippe centrifuge, operating at 60°C with a peripheral velocity ωr_2 of 350 m/s. From equation 68, α_0 is calculated to be 1.0686 for uranium isotopes in the form of UF_6.

Mass Transport. An expression for the diffusive transport of the light component of a binary gas mixture in the radial direction in the gas centrifuge can be obtained directly from the general diffusion equation and an expression for the radial pressure gradient in the centrifuge. For diffusion in a binary system in the absence of temperature gradients and external forces, the general diffusion equation retains only the pressure diffusion and ordinary diffusion effects and takes the form

$$J_A = -cD_{AB}\left[\frac{M_A x_A}{RT}\left(\frac{\overline{V}_A}{M_A} - \frac{1}{\rho}\right)\frac{dp}{dr} + \frac{dx_A}{dr}\right] \qquad (69)$$

where cD_{AB} is the product of the molar density and binary diffusion coefficient of the process gas and $\overline{V}_a$ is the partial molal volume of component A.

Because the total pressure gradient is the sum of the partial pressure gradients, the following substitution can be made in equation 69

$$\frac{dp}{dr} = \frac{dp_A}{dr} + \frac{dp_B}{dr} = \frac{(M_A p_A + M_B p_B)}{RT}\omega^2 r \qquad (70)$$

and the equation for the radial flux of component A in a mixture of ideal gases is found to be

$$J_A = cD_{AB}\left[\frac{(M_B - M_A)x_A(1 - x_A)}{RT}\omega^2 r + \frac{dx_A}{dr}\right] \qquad (71)$$

Figure 12 is a schematic drawing of a section of a countercurrent gas centrifuge in which an arbitrary axial convective flow pattern is shown. It is assumed that the convective velocity v in the centrifuge can be expressed as a function of r only, and is independent of z. The convective velocity is assumed to be in the z direction only, and the regions at the ends of the centrifuge in which the direction of the flow is changed are neglected.

The net transport of component A in the $+z$ direction in the centrifuge τ_A is equal to the sum of the convective transport and the axial diffusive transport. At

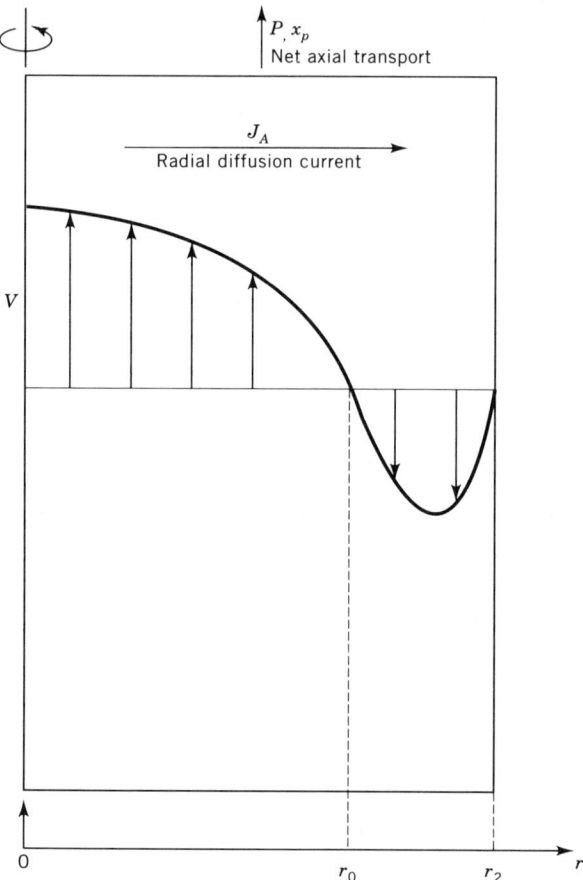

Fig. 12. Axial velocity profile in a countercurrent gas centrifuge.

the steady state the net transport of component A toward the product withdrawal point must be equal to the rate at which component A is being withdrawn from the top of the centrifuge. Thus, the transport of component A is given by equation 72:

$$\tau_A = Px_P = \int_0^{r_2} 2\pi rcvx\,dr - \int_0^{r_2} 2\pi rcD_{AB} \frac{\partial x}{\partial z}\,dr \qquad (72)$$

where x is used in place of x_A for the mol fraction of component A, and the net transport of both components toward the product withdrawal point, τ, is given by

$$\tau = P = \int_0^{r_2} 2\pi rcv\,dr \qquad (73)$$

where P is the product withdrawal rate, mol/s, and x_P is the concentration of component A in the product material. The integrals appearing in equation 72 are evaluated using the flux equation 71 (92). Several approximations are involved and are most satisfactory for the case of relatively long units in which the axial concentration difference is large compared with the concentration differences in the radial direction, and in which the magnitudes of the feed and withdrawal rates are small with respect to the circulation rate of the internal convective flow. The results are not completely satisfactory for application to short-bowl centrifuges with relatively high throughput rates. For the case of the gas centrifuge, application of the method leads to a gradient equation for the enriching section of the centrifuge that can be written in the form

$$S\,dx/dz = (\alpha - 1)x(1 - x) - (P/L)(x_P - x) \tag{74}$$

where S is the stage length in the centrifuge and α is the stage separation factor. The quantity L is a measure of the convective circulation rate of the gas in the centrifuge and may be evaluated from the integral

$$L = \int_0^{r_o} 2\pi r c v\, dr \tag{75}$$

where r_0 is the radius at which the convective velocity is equal to zero (see Fig. 12). The stage length S is the sum of two terms

$$S = \frac{\int_0^{r_2} \dfrac{dr}{rcD_{AB}} \left[\int_0^r r'cv\,dr'\right]^2}{\int_0^{r_0} rcv\,dr} + \frac{\int_0^{r_2} rcD_{AB}\,dr}{\int_0^{r_0} rcv\,dr} \tag{76}$$

The first term may be considered as the contribution of the internal circulation or convective flow to the stage length, the second term as the contribution of the axial diffusion to the stage length. The stage separation factor is given by

$$\alpha - 1 = \frac{(M_B - M_A)\,\omega^2}{RT} \frac{\int_0^{r_2} r\,dr \int_0^r r'cv\,dr'}{\int_0^{r_0} rcv\,dr} \tag{77}$$

From an inspection of the preceding three equations, it is evident that for the case of a given velocity profile in which v retains its functional dependence on r but is permitted to vary in magnitude by a factor, that is, $v = bf(r)$, the convective contribution to the stage length varies directly with the magnitude of L, whereas the diffusive contribution to the stage length varies inversely with the magnitude of the circulation rate L. Thus there exists a value of L for which the stage length for the separation process is a minimum. Designating this value of L by L_0, analysis of the expression for the stage lengths shows that

$$L_0 = 2\pi \int_0^{r_0} rcv\,dr \left[\frac{\int_0^{r_2} rcD_{AB}\,dr}{\int_0^{r_2} \frac{dr}{rcD_{AB}} \left(\int_0^r r'cv\,dr'\right)^2} \right]^{1/2} \qquad (78)$$

The corresponding minimum value of the stage length, designated by S_0, is given by

$$S_0 = \frac{2\left[\int_0^{r_2} \frac{dr}{rcD_{AB}} \left(\int_0^r r'cv\,dr'\right)^2 \int_0^{r_2} rcD_{AB}\,dr\right]^{1/2}}{\int_0^{r_0} rcv\,dr} \qquad (79)$$

The preceding two equations may be used to write the gradient equation for the countercurrent gas centrifuge in an alternative form. If the ratio of the actual gas circulation rate in the centrifuge to the circulation rate that minimizes the stage length L/L_0 is designated by m, then equation 74 may be rewritten

$$\left(\frac{1+m^2}{2m}\right) S_0 \frac{dx}{dz} = (\alpha - 1)x(1-x) - \frac{P}{mL_0}(x_P - x) \qquad (80)$$

For the stripping section of the centrifuge, that is, the section between the point at which the feed is introduced and the end at which the waste stream is withdrawn, the gradient equation has the corresponding form

$$\left(\frac{1+m^2}{2m}\right) S_0 \frac{dx}{dz} = (\alpha - 1)x(1-x) - \frac{W}{mL_0}(x - x_W) \qquad (81)$$

where W is the waste withdrawal rate, mol/s, and x_W is the concentration of component A in the waste material.

Maximum Separative Capacity and the Separative Efficiency. The separative efficiency of a gas centrifuge used for isotope separation is best defined in terms of separative work. Thus, the separative efficiency E is defined by

$$E = \frac{\delta U \text{ (experimental)}}{\delta U \text{ (max)}} \qquad (82)$$

where δU (experimental) is the actual separative work produced per unit time by the centrifuge under consideration and δU (max) is the maximum theoretical separative capacity of the machine. The maximum separative capacity of a gas centrifuge (41) is given by

$$\delta U \text{ (max)} = \frac{\pi Z c D_{AB}}{2}\left(\frac{\Delta M V^2}{2RT}\right)^2 \qquad (83)$$

where δU is the separative capacity in mols per unit time, Z is the length of the rotor, ΔM is the difference in the mol wts of the components being separated, and V is the peripheral velocity of the centrifuge ($V = \omega r_2$). The expression for the maximum separative capacity of a centrifuge indicates a desirability for: (*1*) Low-temperature operation because the theoretical maximum separative capacity of a centrifuge varies inversely as the temperature; (*2*) Long centrifuge bowls because the theoretical maximum separative capacity varies directly as Z and that δU (max) is independent of the radius of the bowl; and (*3*) High peripheral velocity because the theoretical maximum separative capacity varies as the fourth power of the peripheral speed. At the higher speeds the predicted separative capacity increases with increasing peripheral speed much more slowly than the fourth-power relationship. Nevertheless, over the entire range of speeds investigated there is still an appreciable gain in separative capacity to be realized from an increase in speed.

Theoretical Formulation of the Separative Efficiency. The separative efficiency E of a countercurrent gas centrifuge may be considered to be the product of four factors, all but one of which can be evaluated on the basis of theoretical considerations. In this formulation the separative efficiency is defined by

$$E \equiv e_C e_I e_F e_E \qquad (84)$$

where e_C designates the circulation efficiency, e_I designates the ideality efficiency, e_F designates the flow pattern efficiency, and e_E designates the experimental efficiency and includes all phenomena such as turbulence and end effects not taken into account by the preceding terms. The circulation efficiency for a countercurrent gas centrifuge is given by

$$e_C = m^2/(1 + m^2) \qquad (85)$$

As has been previously noted, m is the ratio of the rate at which gas flows upward in a centrifuge to the quantity L_0 that depends on the geometry of the bowl, the physical properties of the gas, and the flow pattern. Thus, m is directly proportional to the upflow rate. It is evident from the definition of e_C that it approaches unity as m takes on increasingly larger values. This is understood when it is realized that the circulation efficiency is representative of the loss in separative capacity owing to axial diffusion against the axial concentration gradient established in the bowl, and is, in fact, equal to the ratio of the convective contribution to the stage length of the sum of the convective and diffusive contributions, that is, to the total stage length. As m increases, the convective contribution of the stage length increases proportionally, and the diffusive contribution decreases as m^{-1}; at high circulation rates the diffusive transport becomes negligible with respect to the convective transport within the centrifuge.

The ideality efficiency takes into account the difference between the shape of a centrifuge that may be regarded as a square cascade and that of an ideal cascade. As has been pointed out in the section on cascade theory, the separative capacity of an element of length in a centrifuge is the greatest when the circulation rate L through the element bears a certain relationship to the withdrawal rate,

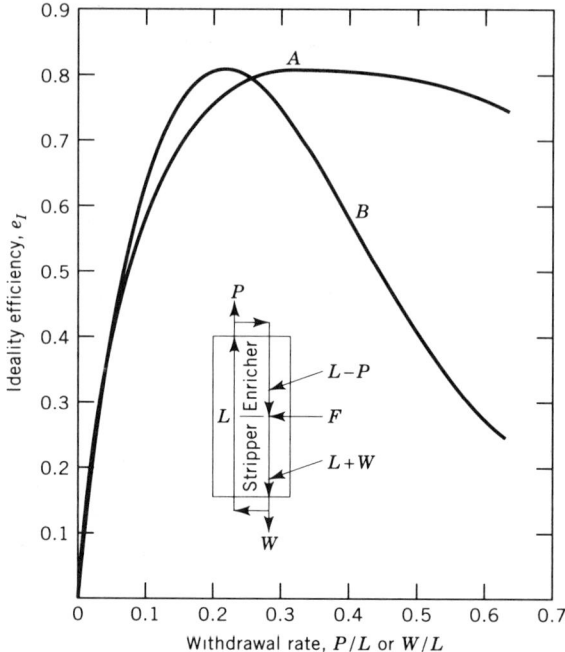

Fig. 13. The ideal efficiency of a five-stage enricher and stripper as a function of the product or waste withdrawal rate, where A represents stripping section efficiency, and B represents enriching section efficiency.

withdrawal concentration, and the concentration in the centrifuge at that point, as has been indicated by equation 28. When this condition is satisfied the cascade is termed ideal. In a square cascade, however, this condition cannot be satisfied at more than a single point in the enricher and in the stripping sections. Thus, one can associate an efficiency with each point in the cascade that is a function of the departure of the actual flow from the ideal flow. The ideality efficiency may be regarded as the average of these point efficiencies over the entire cascade.

Analysis of the gradient equations for a countercurrent gas centrifuge shows that when the withdrawal rates are optimized, the ideal efficiency assumes a maximum value of 81%. Curves of the ideal efficiency for both a stripping and enriching section of five stages ($Z/S = 5$) are shown in Figure 13. The flow model assumed in these calculations is also shown. In order to achieve this maximum value of 0.81 for the ideality efficiency, it is necessary that in addition to operating at the optimum withdrawal rates there be no mixing of gas of unlike concentrations at the feed point. The difference in the behavior of the curves for the efficiency of the enriching and stripping sections results primarily from the fact that in the model considered the feed is assumed to enter the downflowing stream, and therefore the flows in the enriching and stripping sections are not symmetric.

The flow pattern efficiency e_F depends solely upon the shape of the velocity profile in the circulating gas. In terms of the integrals appearing in the gradient equation, the flow pattern efficiency is given by equation 86.

$$e_F = \frac{4 \left(\int_0^{r_2} r\,dr \int_0^r r'cv\,dr' \right)^2}{cD_{AB}r_2^4 \int_0^{r_2} \frac{dr}{rcD_{AB}} \left(\int_0^r r'cv\,dr' \right)^2} \tag{86}$$

To evaluate the flow pattern efficiency, a knowledge of the actual hydrodynamic behavior of the process gas circulating in the centrifuge is necessary. Primarily because of the lack of such knowledge, the flow pattern efficiency has been evaluated for a number of different assumed isothermal centrifuge velocity profiles.

The Optimum Velocity Profile. The optimum velocity profile (41), that is the velocity profile that yields the maximum value for the flow pattern eficiency, is one in which the mass velocity pv is constant over the radius of the centrifuge except for a discontinuity at the wall of the centrifuge ($r = r_2$). This optimum velocity profile is shown in Figure 14a. For this case the following values for the separation parameters of the centrifuge are obtained

$$\alpha - 1 = \frac{1}{2} \frac{\Delta M V^2}{2\,RT} \tag{87}$$

$$L_0 = 2\,(2)^{1/2}\,\pi r_2 c D_{AB} \tag{88}$$

$$S_0 = r_2(2)^{1/2} \tag{89}$$

and $e_F = 1.0$.

The Two-Shell Velocity Profile. A second simple velocity profile (41) is shown in Figure 14b in which the flow consists of two thin streams, one situated at radius r_1, flowing upward, and the other situated at the wall ($r = r_2$), flowing downward. For this case the values of the separation parameters are

$$\alpha - 1 = \left[1 - \left(\frac{r_1}{r_2} \right)^2 \right] \frac{\Delta M V^2}{2\,RT} \tag{90}$$

$$L_0 = \left[\frac{2}{\ln(r_2/r_1)} \right]^{1/2} \pi r_2 c D_{AB} \tag{91}$$

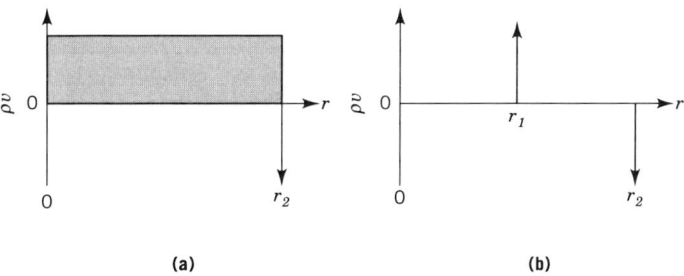

(a)   (b)

Fig. 14. Hypothetical velocity profile models for a countercurrent-flow gas centrifuge. (**a**) The optimum velocity profile in a countercurrent gas centrifuge. (**b**) The two-shell velocity profile.

570 DIFFUSION SEPARATION METHODS

$$S_0 = [2 \ln(r_2/r_1)]^{1/2} r_2 \tag{92}$$

$$e_F = \left[1 - \left(\frac{r_1}{r_2}\right)^2\right]^2 \bigg/ \ln \frac{r_2}{r_1} \tag{93}$$

The value of the flow pattern efficiency is shown as a function of the spacing between the streams in Figure 15a. It can be seen that the flow pattern efficiency is a maximum when the position of the upflowing stream is chosen such that r_1/r_2 is equal to 0.5335. For this particular case the flow pattern efficiency assumes the value $e_F = 0.8145$.

These simple velocity profiles do not indicate directly any dependence of the flow pattern efficiency upon the rotational speed of the centrifuge. A dependence on speed is to be expected on the basis of the argument that at high speeds the gas in the centrifuge is crowded toward the periphery of the rotor and that the effective distance between the countercurrent streams is thereby reduced. It can be seen from the two-shell model that, as the position of upflowing stream approaches the periphery, the flow pattern efficiency drops off from its maximum value.

The Martin Velocity Profile. It has been suggested (50) that the velocity profile in a gas centrifuge in which the countercurrent flow is caused by a temperature difference between the circulating gas and the end caps is given by

$$\int_0^r r'v\,dr' = \frac{\Delta T}{\omega} \left(\frac{\lambda^3}{\eta T}\right)^{1/4} \left(\frac{2\,rp(r_2)}{MRT}\right)^{1/2} \exp{(M\omega^2(r^2 - r_2^2)/4RT)} \tag{94}$$

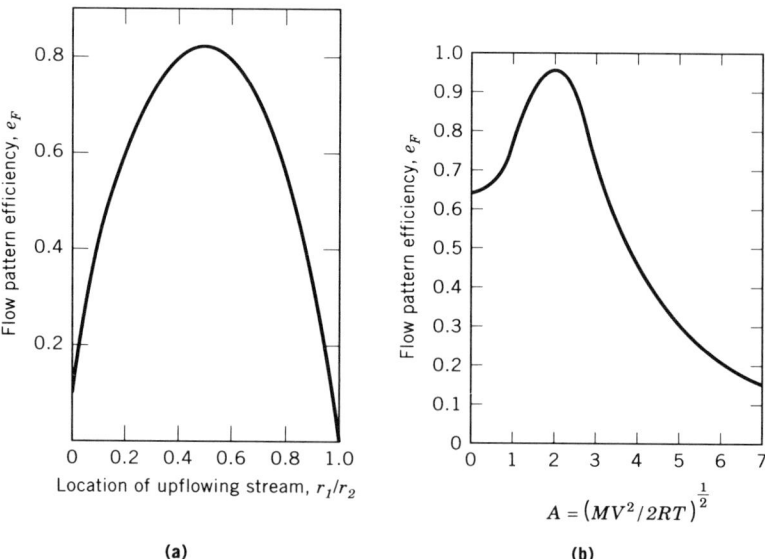

Fig. 15. (a) Values of the flow-pattern efficiency for the two-shell model. (b) The dependence of the flow-pattern efficiency on the dimensionless parameter A for the Martin profile.

where λ is the thermal conductivity of the process gas, η is the viscosity of the process gas, and ΔT is the temperature difference between the gas and the end caps, one warmer, the other cooler than the gas. Equation 94 was derived by considering the flow along a heated plate in a strong gravity field. All other considerations such as coriolis forces and any resistance to the axial flow, were neglected.

The separation parameters have been calculated for a centrifuge in which the behavior of the circulating gas is described by Martin's equation. The flow pattern efficiency is shown in Figure 15(**b**) as a function of the dimensionless parameter A, where A is equal to $(MV^2/2RT)^{1/2}$. In this case the maximum flow pattern efficiency attainable is 0.956.

Cascade Design. The efficiency of a Zippe-type centrifuge, separating uranium isotopes when UF_6 is the process gas, operating at a peripheral speed of 350 m/s and at a temperature of 320 K ($A = 2.85$), would be expected to be

$$E = \frac{m^2}{1 + m^2}(0.81)(0.76) \tag{95}$$

The observed efficiency of 35% could be interpreted to mean that the circulation efficiency of this machine is about 60%, corresponding to an m value of 1.2. According to the theory presented, if the centrifuge could be operated with $m = 3$, it should be possible to obtain a separative efficiency of about 55%. With this assumption, the separative capacity of a single machine would be 1.78×10^{-3} kg/d of uranium. A cascade for uranium isotope separation designed to produce 1 kg/d of enriched uranium containing 90% ^{235}U from natural uranium would therefore require approximately 116,000 Zippe-type gas centrifuges. Table 3 shows the size of a cascade consisting of Zippe-type centrifuges required for the production of 1 kg/d of UF_6. Modern centrifuges have attained much higher separative capacities than the original Zippe machine. Were the cascade described in Table 3 to be constructed using today's centrifuges, the number of centrifuges required would be lower by one to two orders of magnitude.

The Separation Nozzle Process

The separation nozzle process, developed at the Karlsruhe Nuclear Research Center in Germany for the enrichment of the light uranium isotope ^{235}U, is also referred to as the jet diffusion method for the separation of gas mixtures. Isotopes were first separated (93) in a slit-type gas jet (94) in 1946. A device for separating gaseous mixtures by jet diffusion was patented in the United States in 1952. Soon thereafter this separation effect associated with high-speed gas flow through a nozzle was applied to the separation of isotopes (95–97). More recent work by the German research group is described (98–121). Interest in the jet separation process also led to experimental and theoretical work in the United States (94,122–125), and in Japan (126–130).

Apparatus and Method of Operation. The separation nozzle process stage planned for commercial use differs appreciably from the stage used in early investigations. The basic features of the separation nozzle method are illustrated

Table 3. Characteristics of an Ideal Centrifuge Cascade[a]

Characteristics	Value	Source of value
Centrifuge parameters		
length of rotor Z, cm	30.48	given[b]
diameter of rotor $2r_2$, cm	7.62	given[b]
peripheral speed $V = wr_2$, m/s	350	given
operating temperature T, K	320	given
Centrifuge separative capacity		
separative capacity δU, SWU/yr, maximum	1.17	eq. 83
separative capacity, δU, SWU/yr, actual	0.65	E U(max)
circulation efficiency, e_c	0.90	eq. 85
ideality efficiency, e_I	0.81	maximum, Fig. 13
flow pattern efficiency, e_F	0.76	Fig. 15b, $A = 2.85$
Cascade parameters		
product concentration y_P, mol fraction	0.90	given
feed concentration x_F, mol fraction	0.0072	given
waste concentration x_W, mol fraction	0.0025	given
product rate P, kg/d	1.0	given
feed rate F, kg/d	191.0	eqs. 34, 35
separative capacity, δU, SWU/yr	75,380	eq. 36
number of centrifuges required	115,970	

[a] Cascade to yield 1 kg/d UF$_6$ enriched to 0.90 mol fraction ^{235}U.
[b] Predetermined by equipment or operation.

in Figure 16. Gaseous uranium hexafluoride mixed with a light auxiliary gas is expanded through a nozzle into a curved flow channel. At the end of the curved flow path, after turning 180°, the stream is divided by a knife edge into two parts, an interior fraction which is enriched in ^{235}U and a wall fraction which is depleted with respect to the ^{235}U. The light auxiliary gas present in a large molar excess increases the flow velocity of the UF$_6$ and, hence, it increases the centrifugal force determining the separation. In addition, the light gas delays the sedimentation of the two UF$_6$-isotopes in the centrifugal field slightly differently, which also has a favorable effect upon the separation of the isotopes.

Usually, a mixture of 2–5 mol % UF$_6$ and 95–98 mol % H$_2$ is used as a process gas; the expansion ratios range from 1.8–2.5. According to the gas kinetic scaling relations the optimum operating pressure of the nozzle is inversely proportional to its characteristic dimensions; for example, the optimum inlet pressure of a commercial separation nozzle system with a radius of curvature of 0.1 mm is on the order of tens of kPa (tenths of atmospheres). Figure 17 illustrates the design of a commercial separation nozzle element. The ten slit-shaped separation nozzles are mounted on the periphery of an extruded aluminum tube. Feed gas is introduced into the segments marked F and expands through the nozzles. The heavy fraction is pumped off through the segments marked H and the light fraction is pumped off from the space around the element.

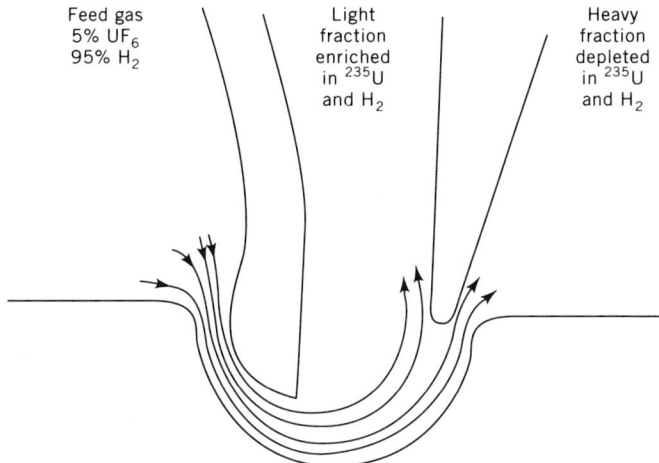

Fig. 16. Cross-section of the separation nozzle system used in the commercial implementation of the separation nozzle process.

Theory. A good understanding of the separation phenomenon of the separation nozzle process is obtained from a very simple model that treats the separation nozzle as a gas centrifuge at steady state. It is assumed that the feed mixture traverses the circular flow path at a constant and uniform angular velocity (wheel flow) and that the peripheral velocity of the flow is equal to the sonic velocity of the entering feed gas. The separation of the isotopes is effected, as in the gas centrifuge, by pressure diffusion in the pressure gradient resulting from the curved streamlines and the associated centrifugal forces. One important function served by the light auxiliary gas in the feed is to increase the flow velocity of the mixture and hence the magnitude of the separation factor attained.

When the gas speed is sufficiently high, the separation factor corresponding to a given value of the cut is essentially independent of the gas velocity and, hence, at high speeds, is given (104) to a good approximation as

$$(\alpha - 1) \cong \left(\frac{1}{1-\theta} \ln \frac{1}{\theta}\right) \frac{\Delta M}{M} \tag{96}$$

where θ is the stage cut defined as the fraction of the uranium in the feed stream to the separation nozzle that is withdrawn in the light or enriched product stream and $\Delta M/M$ is the fractional difference in the mol wts of the isotopic species being separated. In the case of uranium isotopes where UF_6 is the process gas, $\Delta M/M = 3/352 = 0.0085$. The separative work produced by the stage is

$$\delta U = F \frac{\theta(1-\theta)}{2} (\alpha - 1)^2 \tag{97}$$

where F is the feed rate of uranium to the separating unit and δU is the amount of separative work produced by the nozzle system per unit time. The separation

574   DIFFUSION SEPARATION METHODS

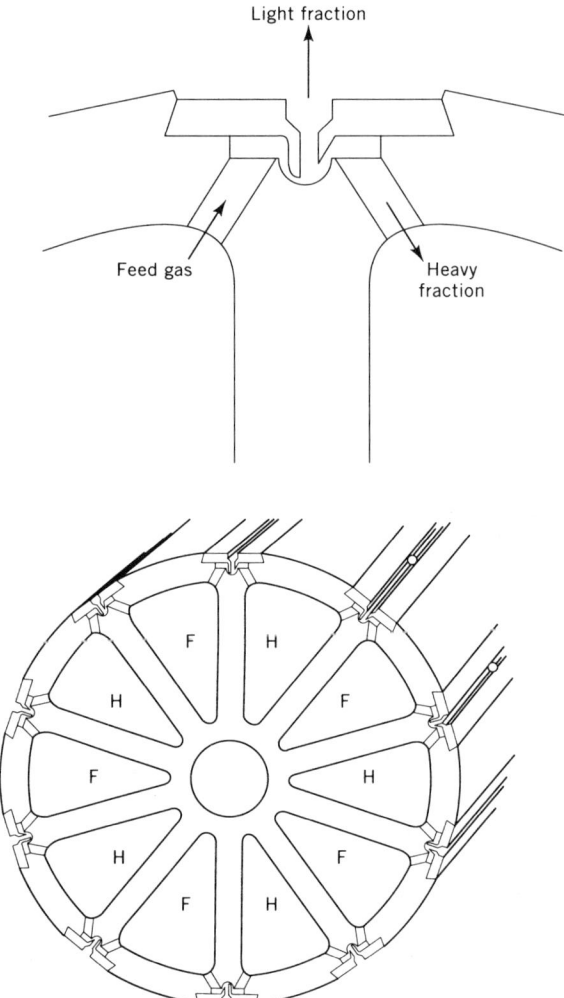

Fig. 17. Schematic representation of a commercial separation element tube manufactured by the Messerschmitt-Bölkow-Blohm Company, Munich. Terms are defined in text.

performance of the separation nozzle in the limit of high gas speed, as described in equations 96 and 97, is shown in graphical form in Figure 18. An equilibrium separation nozzle produces its maximum separative work rate at a cut of about 0.2. The secondary enrichment effects caused by ternary diffusion involving the light auxiliary gas, are treated in some detail in References 131 and 132.

Cascade for Uranium Enrichment. The design for a 5 million SWU/yr plant producing uranium enriched to 3% ^{235}U and stripping the feed to 0.3% ^{235}U in the waste stream is shown in Figure 19. Data for such a plant are given in Table 4. The 570 stages and 2520 MW required by the 5 million SWU/yr nozzle plant should be compared with the 1180 stages (to span the range from 0.25–4.0% ^{235}U) and 2430 MW required by the 8.75 million SWU/yr gaseous diffusion plant shown in Figure 9.

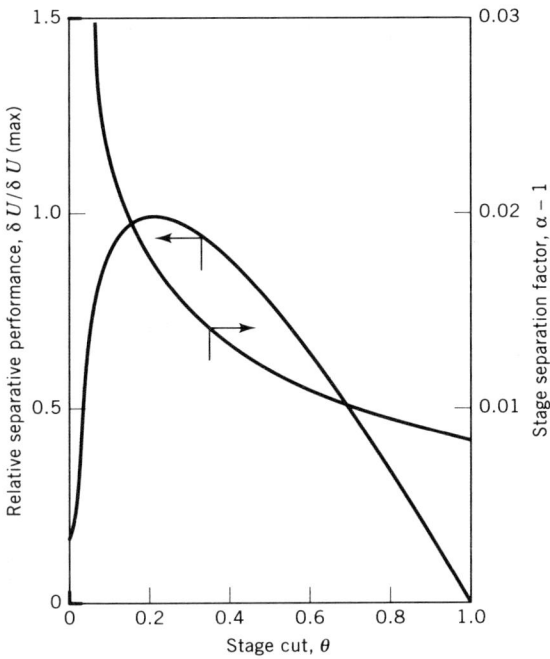

Fig. 18. High speed performance limit of an equilibrium separation nozzle.

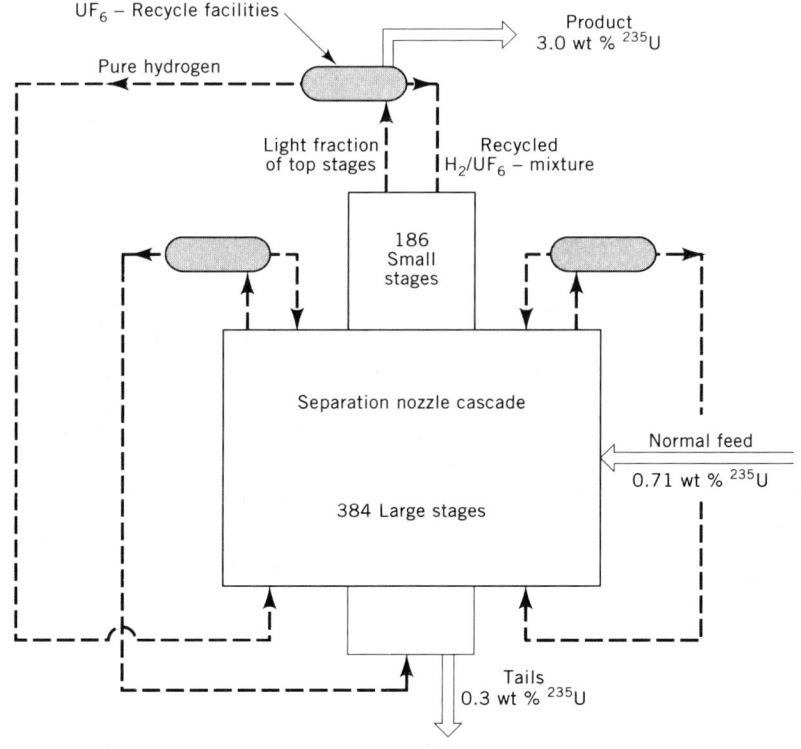

Fig. 19. Separation nozzle cascades of 5 million SWU/yr.

Table 4. Conceptual Design Data For a 5 Million SWU/yr Plant

Characteristics		Value	
separating element[a]			
N_0 (mol % UF_6)		4.2	
expansion ratio, π		2.1	
uranium cut, ϑ_u		¼	
elementary separation effect E_A (%)		1.48	
separation nozzle inlet pressure, kPa[b]		38.7	
separation nozzle outlet pressure, kPa[b]		18.4	
separation stage	*Large*		*Small*
suction flow of compressor, m³/h	1,030,600		312,000
rated power of compressor motor, kW	5,500		1,720
separating element slit length, m	26,660		8,050
separative work capacity, SWU/yr	12,800		3,900
cascade			
mass flow, kg/yr of U			
product[c]		1,585,000	
waste[d]		9,452,000	
feed[e]		11,037,000	
number of large stages		384	
number of small stages		186	
net separative work capacity, kg/yr of U		5,045,000	
total energy requirement of plant, MW		2,520	

[a] Hydrogen is used as a carrier gas.
[b] To convert kPa to mm Hg, multiply by 7.5.
[c] 3% ^{235}U.
[d] 0.337% ^{235}U.
[e] Natural uranium.

Three prototype separation nozzle stages of different sizes have been constructed at the Karlsruhe Nuclear Center. A small stage, designated the SR-33, has a compressor suction volume of 33,000 m³/h, 7.5 m in height and 1.5 m in diameter. A stage of intermediate size, designated the SR-100, has a compressor suction volume of 100,000 cm³/h, 10 m in height and 2.5 m in diameter. A large stage, designated the SR-300, has a compressor suction volume of 300,000 cm³/h, 14 m in height and 4 m in diameter. When separation nozzle systems of an advanced design (double-deflecting nozzles having small radii of curvature of 20 μm and 10 μm) are installed in these stages, the separative capacity of the SR-33 stage is expected to be 2400 SWU annually; the separative capacity of the SR-100, 7200 SWU per year; and the separative capacity of the SR-300, 22,000 SWU per year. It was estimated in 1982 that a 3.8 million SWU per year cascade, designed to enrich uranium to 3.2 percent ^{235}U, would have a separative work cost of $120/SWU (75).

Nomenclature

Some symbols that appear in the text only once and are clearly defined at that time are not listed. Dimensions are given in terms of the mass, M, length, L, time, t, temperature, T, and mols.

C_F = cost of feed material, eq. 14, \$/M or \$/mol.
$C_{\Delta U}$ = cost of separative work, eq. 14, \$/M or \$/mol.
C_{total} = total cost of enriched product, eq. 14, \$/M or \$/mol.
c = total molar concentration, mol/L^3.
D_{AB} = binary diffusion coefficient for the pair $A - B$, L^2/t.
E = overall efficiency of a separation process, eq. 82, dimensionless.
E_b = barrier efficiency in gaseous diffusion, eq. 51, dimensionless.
E_c = cut correction in gaseous diffusion, eq. 54, dimensionless.
E_M = mixing efficiency in gaseous diffusion, eq. 53, dimensionless.
E_p = back-pressure efficiency in gaseous diffusion, eq. 52, dimensionless.
e_C = circulation efficiency of a centrifuge, eq. 85, dimensionless.
e_E = experimental efficiency of a centrifuge, eq. 84, dimensionless.
e_F = flow pattern efficiency of a centrifuge, eq. 86, dimensionless.
e_i = ideality efficiency of a centrifuge, eq. 84, dimensionless.
F = feed flow rate to a unit or cascade, eq. 34, M/t or mol/t.
H = steady-state enriched inventory of desired component in a cascade, eq. 42, M or mols.
h_n = total material holdup of the nth stage, eq. 42, M or mols.
J_A = molar flux of component A in a binary mixture relative to the molar average velocity, eq. 69, mol/tL^2.
L = upflow rate of process gas in a stage, eq. 3, M/t or mol/t.
L_{ideal} = stage upflow of process gas in an ideal cascade, eq. 28, M/t or mol/t.
$L_{\min}$ = minimum stage upflow of process gas, eq. 22, M/t or mol/t.
L_0 = value of interstage flow rate which minimizes the stage length, eq. 78, M/t or mol/t.
l_f = thickness of effective stagnant film on gaseous diffusion barrier, eq. 53, L.
M_i = molecular weight of component i.
m = ratio of actual interstage flow rate to interstage flow rate which minimizes the stage length, eq. 80, dimensionless.
N_A = molar flux of component A with respect to stationary coordinates, mol/L$^2 t$.
N_{ideal} = number of stages required to span a given concentration span in an ideal cascade, eq. 27, dimensionless
$N_{\min}$ = minimum number of stages required to span a given range of concentrations, eq. 21, dimensionless.
N_{sect} = number of stages required to span the concentration range of a square section, eq. 23, dimensionless.
n = stage counting index, eq. 17, dimensionless.
P = product flow rate of a unit or cascade, eq. 34, M/t or mol/t.
p = fluid pressure, $M/t^2 L$.
p_b = fluid pressure on the low pressure side of the gaseous diffusion barrier, eq. 46, $M/t^2 L$.
p_f = fluid pressure on the high pressure side of the gaseous diffusion barrier, eq. 46, $M/t^2 L$.
R = gas constant, $ML^2/t^2 T$ mol.
r = pressure ratio p_b/p_f across the gaseous diffusion barrier, eq. 52, dimensionless.
r_0 = transverse coordinate of the plane of zero axial velocity, eq. 75, L.

r_2 = radius of a centrifuge bowl, eq. 68, L.
S = stage length of a continuous process, L.
T = absolute temperature, T.
T_{eq} = equilibrium time of a cascade or separating unit, eq. 42, t.
δU = separative capacity of a unit. eq. 3, M/t or mol/t.
ΔU = separative capacity of a cascade, M/t or mol/t.
V = peripheral velocity of a centrifuge, eq. 83, L/t.
$\overline{V}_A$ = partial molal volume of a component A in a mixture, eq. 69, L^3/mol.
$v(x)$ = value function, eq. 12, dimensionless.
$v'(x)$ = first derivative of the value function with respect to concentration, eq. 13, dimensionless.
$v''(x)$ = second derivative of the value function with respect to concentration, eq. 11, dimensionless.
W = waste flow rate from a unit or cascade, eq. 34, M/t or mol/t.
x = mol fraction of desired component of a binary mixture in the downflow or depleted stream, dimensionless.
x_F = mol fraction of desired component of a binary mixture in the feed stream of a unit or cascade, dimensionless.
x_i = mol fraction of component i in a mixture, dimensionless.
x_W = mol fraction of desired component of a binary mixture in the waste stream of a unit or cascade, dimensionless.
y = mol fraction of desired component of a binary mixture in the upflow or enriched stream, dimensionless.
y_P = mol fraction of desired component of a binary mixture in the product stream of a unit or cascade, dimensionless.
Z = overall length of the separating column, L.
z = mol fraction of desired component of a binary mixture in the feed stream of a stage, eq. 4, dimensionless.
z = axial distance or length coordinate in a column, L.
α = stage separation factor, eq. 1, dimensionless.
α^* = ideal stage separation factor, eq. 49, dimensionless.
α_0 = equilibrium stage separation factor in a centrifuge, eq. 90, dimensionless.
θ = fraction of stage feed which goes into the stage upflow, "cut," eq. 3, dimensionless.
μ = fluid viscosity, M/Lt.
ρ = mass density, M/L^3.
$\overline{\tau}$ = average net upward transport of desired material in a cascade approaching equilibrium, eq. 42, M/t or mol/t.
ω = angular velocity, eq. 62, radians/t.

BIBLIOGRAPHY

"Diffusion Separation Methods" in the *Encyclopedia of Chemical Technology*, 1st ed., Vol. 5, pp. 76–133, by M. Benedict, Hydrocarbon Research, Inc.; "Diffusion Separation in *ECT* 1st ed., 2nd Suppl., pp. 297–315, by K. B. McAfee, Jr., Bell Telephone Labs.; "Diffusion Separation Methods" in *ECT* 2nd ed., Vol. 7, pp. 91–175, by J. Shacter, E. Von Halle, and R. L. Hoglund, Union Carbide Corporation; in *ECT* 3rd ed., Vol. 7, pp. 639–723, by R. L.

Hoglund, J. Shacter, and E. Von Halle, Union Carbide Nuclear Division; in *ECT* 4th ed., Vol. 8, pp. 149–203, by E. Von Halle, Martin Marietta, Energy Systems, and J. Shacter, Consultant.

1. A. Kanagawa, I. Yamamoto, and Y. Mizuno, *J. Nucl. Sci. Tech.* **14,** 892 (1977).
2. G. Jansen and J. L. Robertson, *Analysis of Nonideal Asymmetric Cascades*, paper presented at American Chemical Society Meeting, Montreal, May 1976.
3. G. A. Garrett and J. Shacter, *Proceedings of the International Symposium on Isotope Separatio*, Amsterdam, 1958, pp. 17–31.
4. G. R. H. Geoghegan in Ref. 3, pp. 518–523.
5. H. Barwich, *Ann. Physik* **20,** 70 (1957).
6. E. Oliveri, *Energia Nucl. Milan* **8,** 453 (1961).
7. J. C. Guais, *BNES Intern. Conference on Uranium Isotope Separation*, Paper 21, London, Mar. 5–7, 1975.
8. N. Ozaki and I. Harada, *BNES Intern. Conference on Uranium Isotope Separation*, Paper 22, London, Mar. 5–7, 1975.
9. H. Harmsen, G. Hertz, and W. Schütze, *Z. Physik* **90,** 703 (1934).
10. D. E. Wooldridge and W. R. Smythe, *Phys. Rev.* **50,** 233 (1936).
11. H. D. Smyth, *Atomic Energy for Military Purposes*, Princeton University Press, Princeton, N.J., 1945.
12. P. C. Keith, *Chem. Eng.* **53,** 112 (1946).
13. J. F. Hogerton, *Chem. Eng.* **52,** 98 (1945).
14. K. E. B. Jay, *Britain's Atomic Factories*, H. M. Stationery Office, London, 1954.
15. M. Molbert in J. R. Merriman and M. Benedict, eds., *Recent Developments in Uranium Enrichment, AIChE Symposium Series, No. 221*, Vol. 78, *The Eurodif Program*, AIChE, New York, 1982.
16. H. Albert, *Proceedings of the U.N. International Conference on Peaceful Uses of Atomic Energy*, 2nd Geneva, Vol. 4, P/1268, 1958, pp. 412–417.
17. *Major Activities in the Atomic Energy Programs*, Jan.–Dec., 1962. U.S. Gov't. Printing Office, Washington, D.C., 1963.
18. D. Massignon in Ref. 16, P/1266, pp. 388–394.
19. J. Charpin, P. Plurien, and S. Mommejac in Ref. 16, P/1265, pp. 380–387.
20. C. Frejacques and co-workers in Ref. 16, P/1262, pp. 418–421.
21. M. Martensson and co-workers in Ref. 16 P/181, pp. 395–404.
22. O. Bilous and G. Counas in Ref. 16, P/1263, pp. 405–411.
23. W. G. Pollard and R. D. Present, *Phys. Rev.* **73,** 762 (1948).
24. P. C. Carman, *Flow of Gases through Porous Media*, Butterworths Publications Ltd., London, 1956.
25. R. D. Present and A. J. de Bethune, *Phys. Rev.* **75,** 1050 (1949).
26. H. T. C. Pratt, *Countercurrent Separation Processes*, American Elsevier Publishing Company, Inc., New York, 1967.
27. C. Boorman in H. London, ed., *Separation of Isotopes*, George Newnes Ltd., London, 1961, Chapt. 8; D. Massignon, *Gaseous Diffusion*, in S. Villani, ed., *Topics in Applied Physics*, Vol. 35, Springer-Verlag, New York, 1979.
28. M. Benedict, T. Pigford, and H. Levi, *Nuclear Chemical Engineering*, 2nd ed., McGraw-Hill Book Co., New York, 1981, Chapt. 14.
29. *AEC Gaseous Diffusion Plant Operations*, USAEC Report No. ORO-684, U.S. Atomic Energy Commission, Washington, D.C., Jan. 1972.
30. *AEC Data on New Gaseous Diffusion Plants*, USAEC Report No. ORO-685, U.S. Atomic Energy Commission, Washington, D.C., Apr. 1972.
31. *CEA Bulletin d'Informations Scientifiques et Techniques*, No. 206, 3-134, Sept. 1975 (in French).
32. G. Besse, *BNES International Conference on Uranium Isotope Separation*, Paper 17, London, Mar. 5–7, 1975.

33. J. P. Gougeau, *Developments in Uranium Enrichment, AIChE Symposium Series 169*, Vol. 73, AIChE, New York, 1977, pp. 12–14.
34. H. J. Mürtz and H. G. Nöller, *Z. Naturforsch.* **16a,** 569 (1961).
35. Ger. Pat. 1,154,793 (Sept. 26, 1963), H. G. Nöller.
36. K. Bornkessel and J. Pilot, *Z. Physik. Chem.* **221,** 177 (1962).
37. A. J. A. Roux, W. L. Grant, R. A. Barbour, R. S. Loubser, and J. J. Wannenburg, *Development and Progress of the South African Enrichment Project, International Conference of Nuclear Power and its Fuel Cycle*, IAEA-CN-36/300, Salzburg, Austria, May 1977.
38. F. A. Lindemann and F. W. Aston, *Phil. Mag.* **37,** 523 (1919).
39. W. Groth in H. London, ed., *Separation of Isotopes*, George Newnes Ltd., London, 1961, Chapt. 6.
40. J. W. Beams, L. B. Snoddy, and A. R. Kuhlthau in Ref. 16, P/723, pp. 428–434.
41. K. Cohen, *The Theory of Isotope Separation as Applied to the Large-Scale Production of U-235*, Natl. Nuclear Energy Ser. Div. III, Vol. 1B, McGraw-Hill Book Co., New York, 1951, Chapt. 6.
42. W. Groth, E. Nann, and K. H. Welge, *Z. Naturforsch.* **12a,** 81 (1957).
43. W. Groth and K. H. Welge, *Z. Physik. Chem.* **19,** 1 (1959).
44. W. Buland and co-workers, *Z. Physik. Chem. Frankfurt* **24,** 249 (1960).
45. W. E. Groth and co-workers in Ref. 16, P/1807, pp. 439–446.
46. K. Beyerle and co-workers in Ref. 3, pp. 667–694.
47. G. Zippe, *The Development of Short Bowl Ultracentrifuges*, Rept. EP-4420-101-60U, Research Laboratories for the Engineering Sciences, University of Virginia, Charlottesville, 1960.
48. A. Bramley, *Science* **92,** 427 (1940).
49. H. Martin and W. Kuhn, *Z. Physik. Chem.* **189,** 219 (1941).
50. H. Martin, *Z. Elektrochem.* **54,** 120 (1950).
51. M. Steenbeck, *Kernenergie* **1,** 921 (1958).
52. J. Los and J. Kistemaker *Proceedings of the International Symposium on Isotope Separation*, Amsterdam, 1958, pp. 695–700.
53. A. Kanagawa and Y. Oyama, *J. At. Energy Soc. Jpn.* **3,** 868 (1961).
54. A. Kanagawa and Y. Oyama, *Nippon Genshiryoku Gakkaishi* **3,** 918 (1961).
55. S. Whitley, *Revs. Modern Physics* **56,** 41 (1984).
56. A. S. Berman, *A Theory of Isotope Separation in a Long Countercurrent Gas Centrifuge*, Rept. K-1536, Union Carbide Corp., Nuclear Div., 1962.
57. A. S. Berman, *A Simplified Model for the Axial Flow in a Long Countercurrent Gas Centrifuge*, Rept. K-1535, Union Carbide Corp., Nuclear Div., 1963.
58. H. M. Parker and T. T. Mayo, IV, *Countercurrent Flow in a Semi-Infinite Gas Centrifuge*, Rept. UVA-279-63R, Research Laboratories for the Engineering Sciences, University of Virginia, Charlottesville, 1963.
59. J. L. Ging, *Countercurrent Flow in a Semi-Infinite Gas Centrifuge: Axially Symmetric Solution in the Limit of High Angular Speed*, Rept. EP-4422-198-62S, Research Laboratories for the Engineering Sciences, University of Virginia, Charlottesville, 1962.
60. J. L. Ging, *The Nonexistence of Pure Imaginary Eigenvalues and the Uniqueness Theorem for the Linearized Gas Flow Equations*, Rept. EP-4422-245-62S, Research Laboratories for the Engineering Sciences, University of Virginia, Charlottesville, 1962.
61. J. L. Ging, *Onsager Minimum Principle for Stationary Flow in Axially Symmetric Rotating Systems*, Rept. EP-3912-321-64U, Research Laboratories for the Engineering Sciences, University of Virginia, Charlottesville, 1964.
62. J. L. Ging, *Eigenvalue Problem—Limit of Low Angular Speed*, Rept. EP-3912-64U, Research Laboratories for the Engineering Sciences, University of Virginia, Charlottesville, 1964.

63. J. L. Ging, *Modified Minimum Principle for Stationary Flow in a Gas Centrifuge*, Rept. EP-3912-325-64U, Research Laboratories for the Engineering Sciences, University of Virginia, Charlottesville, 1964.
64. J. L. Ging, *Onsager Minimum Principle for Axially Decaying Eigenmodes*, Rept. EP-3912-326-64U, Research Laboratories for the Engineering Sciences, University of Virginia, Charlottesville, 1964.
65. G. F. Carrier and S. H. Maslen, *Flow Phenomena in Rapidly Rotating Systems*, Rept. TID 18065, U.S.-D.O.E., 1962.
66. G. F. Carrier in H. Görtler, ed., *Proceedings of the Eleventh International Congress of Applied Mechanics*, Springer, Berlin, 1964.
67. Soubbaramayer in Ref. 27, Chapt. 4, Centrifugation.
68. H. G. Wood, *J. Fluid Mech.* **101,** 1 (1980).
69. R. J. Ribando, *A Finite Difference Solution of Onsager's Model for Flow in a Gas Centrifuge*, Rept. UVA-ER-822-83U, University of Virginia, Charlottesville, 1983.
70. J. J. H. Brouwers, *On the Motion of a Compressible Fluid in a Rotating Cylinder*, Doctoral Dissertation, The Technische Hogeschool, Twente, the Netherlands, June, 1976.
71. T. Sakurai and T. Matsuda, *J. Fluid Mech.* **62,** 727 (1974).
72. T. Sakurai, *J. Fluid Mech.* **72,** 321 (1975).
73. T. Matsuda, K. Hashimoto, and H. Takeda, *J. Fluid Mech.* **73,** 389 (1976).
74. K. Hashimoto, *J. Fluid Mech.* **76,** 289 (1976).
75. T. Matsuda and K. Hashimoto, *J. Fluid Mech.* **78,** 337 (1976).
76. T. Matsuda and K. Hashimoto, *J. Fluid Mech.* **85,** 433 (1978).
77. T. Kai, *J. Nucl. Sci. Technol.* **14,** 267 (1977).
78. T. Kai, *J. Nucl. Sci. Technol.* **14,** 506 (1977).
79. E. Von Halle, *The Countercurrent Gas Centrifuge for the Enrichment of U-235*, K/OA-4058, Union Carbide Nuclear Div., Oak Ridge, Tenn., Nov., 1977.
80. E. Krause and E. H. Hirschel, eds., *DFVLR—Colloquium*, Proz-Wahn, West Germany, 1970.
81. D. R. Olander, *Adv. Nucl. Sci. Technol.* **6,** 105 (1972).
82. D. G. Avery and E. Davies, *Uranium Enrichment by Gas Centrifuge*, Mills & Boon Ltd., London, 1973.
83. S. Villani, *Isotope Separation*, American Nuclear Society, 1976.
84. E. Rätz in *Aerodynamic Separation of Gases and Isotopes, Lecture Series 1978*, Von Karmen Institute for Fluid Dynamics, Belgium, 1978.
85. Soubbaramayer, ed., *Proceedings of the Second Workshop on Gases in Strong Rotation*, Cadarache, France, Apr. 1977.
86. G. B. Scuricini, ed., *Proceedings of the Third Workshop on Gases in Strong Rotation*, Rome, Mar. 1979.
87. E. Rätz, ed., *Proceedings of the Fourth Workshop on Gases in Strong Rotation*, Oxford, UK, Aug. 1981.
88. H. G. Wood, ed., *Proceedings of the Fifth Workshop on Gases in Strong Rotation*, Charlottesville, Va., June 1983.
89. Y. Takashima, ed., *Proceedings of the Sixth Workshop on Gases in Strong Rotation*, Tokyo, Aug. 1985.
90. K. G. Roesner and E. Rätz, eds., *Proceedings of the First Workshop on Separation Phenomena in Liquids and Gases*, Technische Hochschule Darmstadt, Darmstadt, Germany, July 1987.
91. P. Louvet, P. Noe, and Soubbaramayer, eds., *Proceedings of the Second Workshop on Separation Phenomena in Liquids and Gases*, Cite Scientifique Parcs et Technopoles Ile de France Sud, Versailles, France, July 1989.
92. W. H. Furry, R. C. Jones, and L. Onsager, *Phys. Rev.* **55,** 1083 (1939).

93. P. A. Tahourdin, *Final Report on the Jet Separation Methods*, Oxford Rept. No. 36, Br. 694, Clarendon Lab., Oxford, UK, 1946.
94. S. A. Stern, P. C. Waterman, and T. F. Sinclair, *J. Chem. Phys.* **33,** 805 (1960).
95. E. W. Becker and co-workers in Ref. 16, P/1002, pp. 455–457.
96. E. W. Becker in Ref. 3, pp. 560–578.
97. E. W. Becker in H. London, ed., *Separation of Isotopes*, George Newnes Ltd., London, 1961, Chapt. 9.
98. E. W. Becker and co-workers, *Angew. Chemie. Intern. Ed. (Engl.)* **6,** 507 (1967).
99. E. W. Becker and co-workers, *Atomwirtschaft* **18,** 524 (1973).
100. E. W. Becker and co-workers, *International Conference on Uranium Isotope Separation*, London, 1975.
101. E. W. Becker and co-workers, *European Nuclear Conference*, Paris, 1975.
102. E. W. Becker and co-workers, American Nuclear Society Meeting, *KFK-Bericht 2235*, Gesellschaft für Kernforschung, Karlsruhe, 1975.
103. H. Geppert and co-workers, *International Conference on Uranium Isotope Separation*, London, 1975.
104. E. W. Becker and co-workers, *Z. Naturforsch.* **26a,** 1377 (1971).
105. P. Bley and co-workers, *Z. Naturforsch.* **28a,** 1273 (1973).
106. K. Bier and co-workers, *KFK-Bericht 1440*, Gesellschaft für Kernforschung, Karlsruhe, 1971.
107. U. Ehrfeld and W. Ehrfeld, *KFK-Bericht 1634*, Gesellschaft für Kernforschung, Karlsruhe, 1972.
108. W. Ehrfeld and E. Schmid, *KFK-Bericht 2004*, Gesellschaft für Kernforschung, Karlsruhe, 1974.
109. Ger. Pat. 1,096,875 (Jan. 12, 1961), E. W. Becker (to Deutsche Gold-und Silber-Scheideanstalt vorm. Roessler).
110. W. Ehrfeld and U. Knapp, *KFK-Bericht 2138*, Gesellschaft für Kernforschung, Karlsruhe, 1975.
111. E. W. Becker and co-workers, *4th United Nations International Conference on the Peaceful Uses of Atomic Energy*, Geneva, 1971, paper 383.
112. H. J. Fritsch and R. Schütte, *KFK-Bericht 1437*, Gesellschaft für Kernforschung, Karlsruhe, 1971.
113. R. Schütte and co-workers, *Chemie-Ing. Technik* **44,** 1099 (1972).
114. W. Fritz and co-workers, *Chemie-Ing. Technik* **45,** 590 (1973).
115. R. Schütte, *KFK-Bericht 1986*, Gesellschaft für Kernforschung, Karlsruhe, 1974.
116. P. Bley and co-workers, *KFK-Bericht 2092*, Gesellschaft für Kernforschung, Karlsruhe, 1975.
117. W. Ehrfeld in Ref. 84.
118. U. Ehrfeld in Ref. 84.
119. E. W. Becker in Ref. 27.
120. E. W. Becker, P. Noguira Batista, and H. Volcker, *Nucl. Technol.* **52,** 105 (1981).
121. E. W. Becker and co-workers in Ref. 16.
122. P. C. Waterman and S. A. Stern, *J. Chem. Phys.* **31,** 405 (1959).
123. R. R. Chow, *On the Separation Phenomenon of Binary Gas Mixture in an Axisymmetric Jet*, Rept. HE-150-175, Institute of Engineering Research, University of California, Berkeley, 1959.
124. E. E. Gose, *Am. Inst. Chem. Engs. J.* **6,** 168 (1960).
125. V. H. Reis and J. B. Fenn, *J. Chem. Phys.* **39,** 3240 (1963).
126. H. Mikami, *J. Nucl. Sci. Technol.* **6,** 452 (1969).
127. H. Mikami, *I & EC Fundam.* **9,** 121 (1970).
128. H. Mikami and Y. Takashima, *Bull. Tokyo Inst. Technol.* **61,** 67 (1964).

129. H. Mikami and Y. Takashima, *J. Nucl. Sci. Technol.* **5,** 572 (1968).
130. H. Mikami and Y. Takashima, *Int. J. Heat Mass Transfer* **11,** 1597 (1968).
131. W. Berkahn, W. Ehrfeld, and G. Krieg, *Calculations of Uranium Isotope Separation in the Separation Nozzle for Small Mol Fractions of UF_6 in the Auxiliary Gas*, Institut für Kernverfahrenstechnik, Kernforschungszentrum Karlsruhe, West Germany, report KFK-2351, Nov., 1976.
132. G. F. Malling and E. Von Halle, *Aerodynamic Isotope Separation Processes for Uranium Enrichment: Process Requirement*, paper presented at the Symposium on New Advances in Isotope Separation, Div. of Nuclear Chemistry and Technology, American Chemical Society, San Francisco, Calif., Aug. 1976; *UCC-ND Report K/OA-2872*, Oak Ridge Gaseous Diffusion Plant, Oak Ridge, Tenn., Oct. 7, 1976.

General References

M. Benedict, T. Pigford, and H. Levi, *Nuclear Chemical Engineering*, 2nd ed., McGraw-Hill Book Co., New York, 1981, Chapts. 12 and 14.

K. P. Cohen, *The Theory of Isotope Separation as Applied to the Large-Scale Production of U-235, National Nuclear Energy Series Division III*, Vol. 1B, McGraw-Hill Book Co., New York, 1951.

H. London, ed., *Separation of Isotopes*, George Newnes Ltd., London, 1961.

H. R. C. Pratt, *Countercurrent Separation Processes*, American Elsevier Publishing Company, Inc., New York, 1967.

S. Villani, *Isotope Separation*, American Nuclear Society Monograph, ANS Publications, 1976.

S. Villani, ed., *Topics in Applied Physics*, Vol. 35, Springer-Verlag, New York, 1979.

J. Kistemaker, J. Bigeleisen, and A. O. Neir, eds., *Proceedings of the International Symposium on Isotope Separation*, Amsterdam, Apr. 23–27, 1957, North-Holland Publishing Company, Amsterdam, and Interscience Publishers, Inc., New York, 1958.

Proceedings of the Second United Nations International Conference on the Peaceful Uses of Atomic Energy, Geneva, Sept. 1–3, 1958, United Nations publication, Geneva, 1958, particularly Vol. 4: *Production of Nuclear Materials and Isotopes*.

Proceedings of the Third International Conference on the Peaceful Uses of Atomic Energy, Geneva, Aug. 31–Sept. 9, 1964, United Nations publication, New York, 1965, particularly Vol. 12: *Nuclear Fuels—III, Raw Materials*.

Proceedings of the Fourth International Conference on the Peaceful Uses of Atomic Energy, Geneva, Sept. 6–16, 1971, United Nations and the International Atomic Energy Agency, 1972, particularly Vol. 9, *Isotope Enrichment, Fuel Cycles and Safeguards*.

Proceedings of the International Conference on Uranium Isotope Separation, London, Mar. 5–7, 1975, British Nuclear Energy Society, 1975.

M. Benedict, ed., *Development in Uranium Enrichment, AIChE Symposium Series*, Vol. 73, No. 169, American Institute of Chemical Engineers, New York, 1977.

J. R. Merriman and M. Benedict, eds., *Recent Developments in Uranium Enrichment, AIChE Symposium Series*, Vol. 78, no. 221, American Institute of Chemical Engineers, New York, 1982.

E. VON HALLE
Martin Marietta Energy Systems

J. SHACTER
Consultant

DISTILLATION

Distillation is a method of separation that is based on the difference in composition between a liquid mixture and the vapor formed from it. This composition difference arises from the dissimilar effective vapor pressures, or volatilities, of the components of the liquid mixture. When such dissimilarity does not exist, as at an azeotropic point, separation by simple distillation is not possible. Distillation as normally practiced involves condensation of the vaporized material, usually in multiple vaporization/condensation operations, and thus differs from evaporation (qv), which is usually applied to separation of a liquid from a solid but which can be applied to simple liquid concentration operations.

Distillation is the most widely used industrial method of separating liquid mixtures and is at the heart of the separation processes in many chemical and petroleum plants (see SEPARATIONS PROCESS SYNTHESIS). The most elementary form of the method is simple distillation in which the liquid is brought to boiling and the vapor formed is separated and condensed to form a product. If the process is continuous with respect to feed and product flows, it is called flash distillation. If the feed mixture is available as an isolated batch of material the process is a form of batch distillation and the compositions of the collected vapor and residual liquid are thus time dependent. The term fractional distillation, which may be contracted to fractionation, was originally applied to the collection of separate fractions of condensed vapor, each fraction being segregated. In modern practice the term is applied to distillation processes in general, where an effort is made to separate an original mixture into several components by means of distillation. When the vapors are enriched by contact with counterflowing liquid reflux, the process is often called rectification. When fractional distillation is accomplished with a continuous feed of material and continuous removal of product fractions, the process is called continuous distillation. When steam is added to the vapors to reduce the partial pressures of the components to be separated, the term steam distillation is used.

Most distillations conducted commercially operate continuously, with a more volatile fraction recovered as distillate and a less volatile fraction recovered as bottoms or residue. If a portion of the distillate is condensed and returned to the process to enrich the vapors, the liquid is called reflux. The apparatus in which the enrichment occurs is usually a vertical, cylindrical vessel called a still or distillation column. This apparatus normally contains internal devices for effecting vapor–liquid contact. The devices may be categorized as plates or packings.

Distillation has been practiced in one form or another for centuries. It was of fundamental importance to the alchemists and was in use well before the time of Christ. The historical development of distillation has been published (1) as has the history of vapor–liquid contacting devices (2).

Vapor–Liquid Equilibria

The equilibrium distributions of mixture component compositions in the vapor and liquid phases must be different if separation is to be made by distillation. It is important, therefore, that these distributions be known. The compositions at thermodynamic equilibrium are termed vapor–liquid equilibria (VLE) and may be correlated or predicted with the aid of thermodynamic relationships. The driving force for any distillation is a favorable vapor–liquid equilibrium, which provides the needed composition differences. Reliable VLE are essential for distillation column design and for most other operations involving liquid–vapor phase contacting. Many VLE have been measured and reported in the literature, and compilations of such data are available (3,4). Also, bibliographic guides have been published, providing source references for thousands of publications presenting VLE (5–7). If data are not to be found, they may be measured, or estimated by generalized methods (8–10), with some sacrifice in reliability. Even if carefully measured data are available, thermodynamic models are usually required to extrapolate or interpolate the data for conditions not represented by the experiments. Whatever the source and extent of the VLE, some evaluation should be made with regard to accuracy.

The VLE for the system at hand may be simple and easily represented by an equation or, in some systems, may be so complex that they cannot be adequately measured or represented. Excellent treatises are available for selection and implementation of vapor–liquid equilibrium studies (11–14). Typical VLE for binary systems are shown graphically in Figure 1. Figure 1a is a representative boiling point diagram showing equilibrium compositions as functions of temperature at a constant pressure. The lower line is the liquid bubble point line, the locus of points at which a liquid on heating forms the first bubble of vapor. The upper line is the vapor dew point line, representing points at which a vapor on cooling forms the first drop of condensed liquid. The liquid and vapor compositions are conventionally plotted in terms of the low boiling (more volatile) substance, L, in the mixture. The system point A has a vapor composition of y_L^A in equilibrium with a liquid composition of x_L^A at a temperature of T^A. Figure 1b is a typical isobaric phase or y–x diagram. For further discussion, several textbooks are available (15,16).

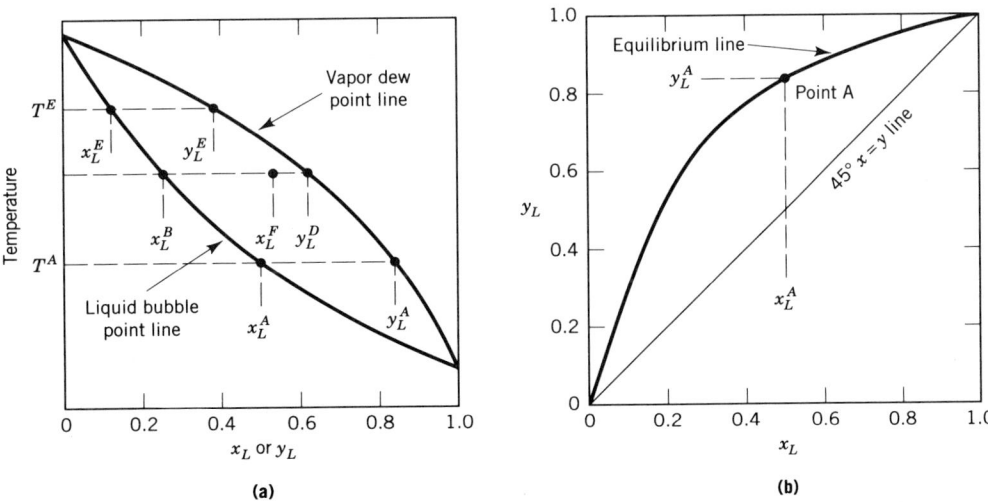

Fig. 1. Isobaric VLE diagrams: (**a**) dew and bubble point; (**b**) vapor–liquid (y–x) equilibrium.

Thermodynamic Relationships. A closed container with vapor and liquid phases at thermodynamic equilibrium may be depicted as in Figure 2, where at least two mixture components are present in each phase. The components distribute themselves between the phases according to their relative volatilities. A distribution ratio for mixture component i may be defined using mole fractions:

$$K_i = y_i^*/x_i \tag{1}$$

where the asterisk is used to denote an equilibrium condition. This K term, known as the vapor–liquid equilibrium ratio, or often the K value, is widely used, especially in the petroleum (qv) and petrochemical industries. For any two mixture components i and j, their relative volatility, often called the alpha value, is defined as

$$\alpha_{ij} = \frac{K_i}{K_j} = \frac{y_i x_j}{x_i y_j} = \frac{y_i (1 - x_i)}{x_i (1 - y_i)} \tag{2}$$

Equation 2 may be rearranged to form an expression for the equilibrium curve in Figure 1b.

$$y_i = \frac{\alpha_{ij} x_i}{1 + (\alpha_{ij} - 1)x_i} \tag{2a}$$

The relative volatility, α, is a direct measure of the ease of separation by distillation. If $\alpha = 1$, then component separation is impossible, because the liquid-and

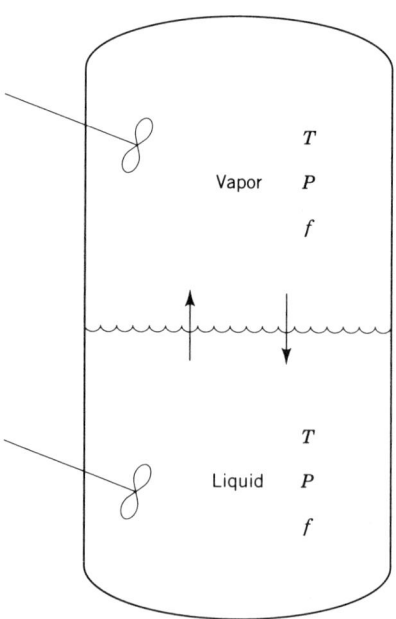

Fig. 2. Equilibrium between vapor and liquid. The conditions for equilibrium are $T^V = T^L$ and $P^V = P^L$. For a given T and P, phase fugacities are equal, ie, $f^V = f^L$ and $f_i^V = f_i^L$.

vapor-phase compositions are identical. Separation by distillation becomes easier as the value of the relative volatility becomes increasingly greater than unity. Distillation separations having α values less than 1.2 are relatively difficult; those which have values above 2 are relatively easy.

When both phases form ideal thermodynamic solutions, ie, no heat of mixing, no volume change on mixing, etc, Raoult's law applies:

$$p_i^V = x_i P_i^0 \tag{3}$$

where P_i^0 is the vapor pressure of i at the equilibrium temperature. Combining this expression with Dalton's law of partial pressures, K values and relative volatilities may be obtained:

$$K_i = P_i^0/P \tag{4}$$

$$\alpha_{ij} = P_i^0/P_j^0 \tag{5}$$

Examples of ideal binary systems are benzene–toluene and ethylbenzene–styrene. The molecules are similar and within the same chemical families. Thermodynamics texts should be consulted before making the assumption that a chosen binary or multicomponent system is ideal. When pressures are low and temperatures are at ambient or above, but the solutions are not ideal, ie, there are dissimilar molecules, corrections to equations 4 and 5 may be made:

$$K_i = \gamma_i^L P_i^0/P \tag{6}$$

$$\alpha_{ij} = \gamma_i^L P_i^0/(\gamma_j^L P_j^0) \tag{7}$$

where the Raoult's law correction factor, γ^L, is a thermodynamically important liquid-phase activity coefficient.

The development and thermodynamic significance of activity coefficients is discussed in most chemical engineering thermodynamics texts. The liquid-phase coefficients are strong functions of liquid composition and temperature and, to a lesser degree, of pressure. A system with positive deviation, ie, the two components having activity coefficients greater than one such that the logarithm of the coefficient is positive, is shown in Figure 3**a**; a system with negative deviation, the coefficients less than unity and logarithms negative, is shown in Figure 3**b**. In a few cases one component of a binary mixture has a positive deviation and the other a negative deviation. Most commonly, however, both coefficients have positive deviations.

Terminal activity coefficients, γ_i^∞, are noted in Figure 3. These are often called infinite dilution coefficients and for some systems are given in Table 1. The hexane–heptane mixture is included as an example of an ideal system. As the molecular species become more dissimilar they are prone to repel each other, tend toward liquid immiscibility, and have large positive activity coefficients, as in the case of hexane–water.

If the molecular species in the liquid tend to form complexes, the system will have negative deviations and activity coefficients less than unity, eg, the system chloroform–ethyl acetate. In azeotropic and extractive distillation (see DISTIL-

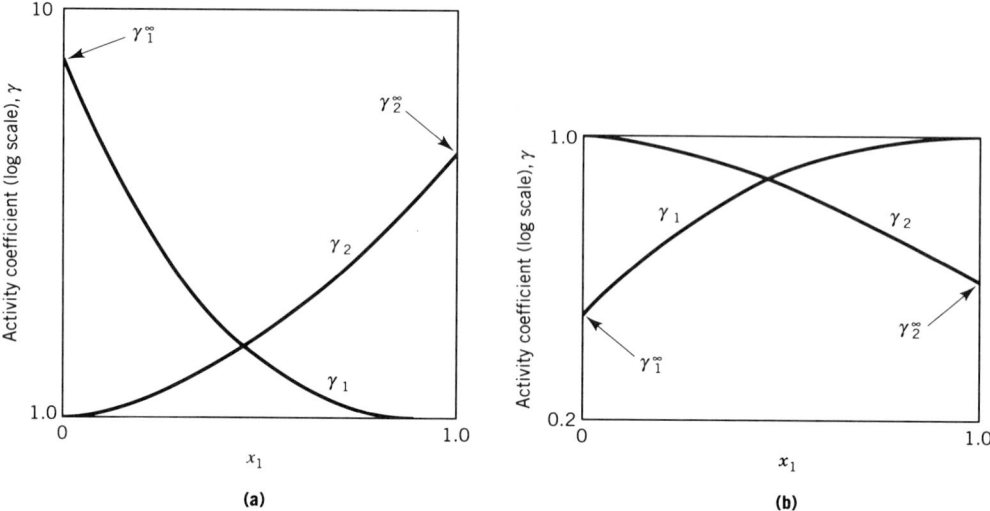

Fig. 3. Binary activity coefficients for two component systems having (**a**) positive and (**b**) negative deviations from Raoult's law. Conditions are either constant pressure or constant temperature and terminal coefficients, γ_i^∞, are noted.

LATION, AZEOTROPIC AND EXTRACTIVE) and in liquid–liquid extraction, nonideal liquid behavior is used to enhance component separation (see EXTRACTION, LIQUID–LIQUID). An extensive discussion on the selection of nonideal addition agents is available (17).

A great deal of study and research has gone into the development of working equations that can represent the curves of Figure 3. These equations are based on solutions of the Gibbs-Duhem equation:

$$x_i \left(\frac{\alpha \ln \gamma_i}{\alpha x_i} \right)_{T,P} + \ldots x_n \left(\frac{\alpha \ln \gamma_n}{\alpha x_i} \right)_{T,P} = 0 \qquad (8)$$

Table 1. Terminal Activity Coefficients at Atmospheric Pressure[a]

Component 1	Component 2	γ_1^∞	γ_2^∞
chloroform	ethyl acetate	0.3	0.3
chloroform	benzene	0.9	0.7
n-hexane	n-heptane	1.0	1.0
ethyl acetate	ethanol	2.5	2.5
ethanol	toluene	6.0	6.0
benzene	methanol	9.0	9.0
ethanol	isooctane	11.0	8.0
methyl acetate	water	20.0	7.0
ethyl acetate	water	100.0	15.0
water	water	>100.0	>100.0

[a]Values are approximate.

One of the simplest and often used equations, or models, is that of Van Laar (18). For a binary system of components 1 and 2, these equations are

$$\ln \gamma = \frac{A_{12}}{\left(1 - \dfrac{A_{12}x_1}{A_{21}x_2}\right)^2} \qquad (9)$$

$$\ln \gamma_2 = \frac{A_{21}}{\left(1 - \dfrac{A_{21}x_2}{A_{12}x_1}\right)^2} \qquad (10)$$

It should be noted that only two parameters are involved. They are directly related to the terminal activity coefficients:

$$\ln \gamma_1^\infty = A_{12} \qquad (11)$$

$$\ln \gamma_2^\infty = A_{21} \qquad (12)$$

A useful and quite popular model is given (19):

$$\ln \gamma_1 = -\ln(x_1 + \Lambda_{12}x_2) + \left(\frac{\Lambda_{12}}{x_1 + \Lambda_{12}x_2} - \frac{\Lambda_{21}}{\Lambda_{21}x_1 + x_2}\right) \qquad (13)$$

$$\ln \gamma_2 = -\ln(x_2 + \Lambda_{21}x_1) - \left(\frac{\Lambda_{12}}{x_1 + \Lambda_{12}x_2} - \frac{\Lambda_{21}}{\Lambda_{21}x_1 + x_2}\right) \qquad (14)$$

This, the Wilson model, is more complex than the Van Laar model, but it does retain the two-parameter feature. The terminal activity coefficients are related to the parameters:

$$\ln \gamma_1^\infty = 1 - \ln \Lambda_{12} - \Lambda_{21} \qquad (15)$$

$$\ln \gamma_1^\infty = 1 - \ln \Lambda_{21} - \Lambda_{12} \qquad (16)$$

Whereas the Wilson model has been found to represent a wide variety of nonideal VLE, it cannot handle the case of partial immiscibility of the liquid phase; for this purpose a three-parameter relationship, the nonrandom, two-liquid (NRTL) model was developed (20).

The most recently developed model is called UNIQUAC (21). Comparisons of measured VLE and predicted values from the Van Laar, Wilson, NRTL, and UNIQUAC models, as well as an older model, are available (3,22). Thousands of comparisons have been made, and Reference 3, which covers the Dortmund Data Base, available for purchase and use with standard computers, should be consulted by anyone considering the measurement or prediction of VLE. The predictive VLE models can be accommodated to multicomponent systems through the use of certain combining rules. These rules require the determination of parameters for all possible binary pairs in the multicomponent mixture. It is possible to use more than one model in determining binary pair data for a given mixture (23).

To estimate VLE when no experimental data or model parameters are available and the cost of special measurements cannot be justified, a group contribution method based on the molecular structures involved called UNIFAC (8) has been developed. Not all possible groups have been evaluated, but regular progress reports are published (24,25). The UNIFAC method, as well as the other models, are critically important in extending limited data to conditions in distillation columns that can cover wide ranges of temperatures, pressures, and compositions. Handling of all these models by computer solution has been described in some detail (26).

The vapor–liquid equilibria of dilute solutions are frequently expressed in terms of Henry's law:

$$p_i^V = H_i^* x_i \tag{17}$$

where, from equation 6, the Henry's law coefficient is

$$H_i^* = \gamma_i^L P_i^0 \tag{18}$$

Henry's law is useful for handling equilibria associated with gas absorption (qv) and stripping problems. Henry's law coefficients are useful for estimating terminal activity coefficients and have been tabulated for many compounds in dilute aqueous solutions (27).

The foregoing discussion has dealt with nonidealities in the liquid phase under conditions where the vapor phase mixes ideally and where pressure–temperature effects do not result in deviations from the ideal gas law. Such conditions are by far the most common in commercial distillation practice. However, it is appropriate here to set forth the completely rigorous thermodynamic expression for the K value:

$$K_i = \frac{\gamma_i^L \phi_i^0 p_i^0 \exp\left(\frac{1}{R'T} \int_{p_i^0}^{P} v_i^L dP\right)}{\hat{\phi}_i P} \tag{19}$$

For nonideal vapor-phase behavior, the fugacity coefficient for component i in the mixture must be determined:

$$\ln \hat{\phi}_1 = \frac{1}{R'T} \int_0^P \left(v_i^v - \frac{R'T}{P}\right) dP \tag{20}$$

If the vapor forms an ideal solution,

$$v_i^v = v^v = zR'T/P \tag{21}$$

where z is the compressibility factor for the mixture. The right side of the numerator in equation 19 is called the Poynting correction, PC:

$$PC = \exp\left(\frac{1}{R'T}\int_{p_i^0}^{p} v_i^L dP\right) \qquad (22)$$

when the liquid is incompressible,

$$PC = \exp\left(\frac{v_i^L(P - p_i^0)}{R'T}\right) \qquad (23)$$

At pressures less than 2 MPa (20 bar) and temperatures greater than 273 K, $PC \sim 1.0$. When the vapor obeys the ideal gas law, $z = 1.0$; then for ideal vapor solutions and for conditions such that $PC = 1.0$, equation 19 reduces to equation 6.

The fugacity coefficient departure from nonideality in the vapor phase can be evaluated from equations of state or, for approximate work, from fugacity/compressibility estimation charts. References 11, 14, and 27 provide valuable insights into this matter.

Journals for the publication of VLE data are available as is a comprehensive tabulation of azeotropic data (28); if the composition and temperature of the azeotrope are known (at a given pressure), then such information may be used to calculate activity coefficients. At the azeotropic point, by definition, $y_i = x_i$; from equation 6,

$$\gamma_i^L = P/P_i^0 \qquad (24)$$

The vapor pressure P_i^0 can be obtained from any of many sources such as handbooks.

The measurement of VLE can be carried out in several ways. A common procedure is to use a recycle still which is designed to ensure equilibrium between the phases. Samples are then taken and analyzed by suitable methods. It is possible in some cases to extract equilibrium data from chromatographic procedures. Discussions of experimental methods are available (5,11). For the more challenging measurements, eg, conditions where one or more components in the mixture can decompose or polymerize, commercial laboratories can be used.

Azeotropic Systems. An azeotropic mixture is one that vaporizes without any change in composition. Figures 4 and 5 represent homogeneous azeotropic systems. Figure 4 depicts a minimum boiling azeotropic system such as ethanol–water; Figure 5 describes a maximum boiling azeotropic system such as acetone–chloroform. The point Z defines the azeotropic composition; this azeotropic point is also called the constant boiling mixture (CBM). Positive activity coefficients tend to produce minimum boiling azeotropes, and negative coefficients tend to produce maximum boiling azeotropes.

Heterogeneous azeotropes are formed when the positive activity coefficients are sufficiently large to produce two liquid phases which exist at the boiling point, and a constant boiling mixture which is formed at some composition, generally within the liquid immiscibility composition range. An example of a heterogeneous azeotropic system is the water/1-butanol system shown in Figure 6. Within the immiscible range, $M-N$, the equilibrium vapor is the heterogeneous azeotrope, Z, of constant composition and the equilibrium temperature is constant. At liquid

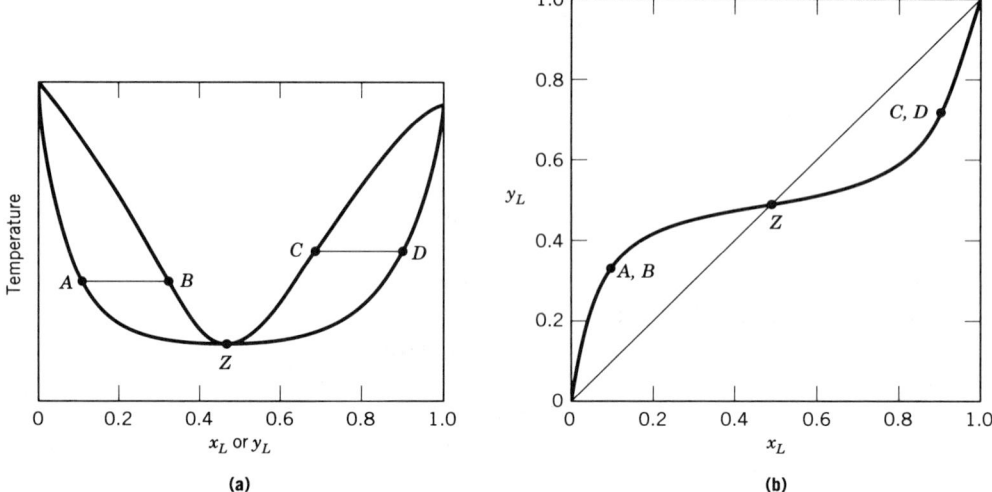

Fig. 4. Boiling point (**a**) and phase diagram (**b**) for a minimum boiling binary azeotropic system at constant pressure. A, B and C, D are representative equilibrium points; Z is the azeotropic point.

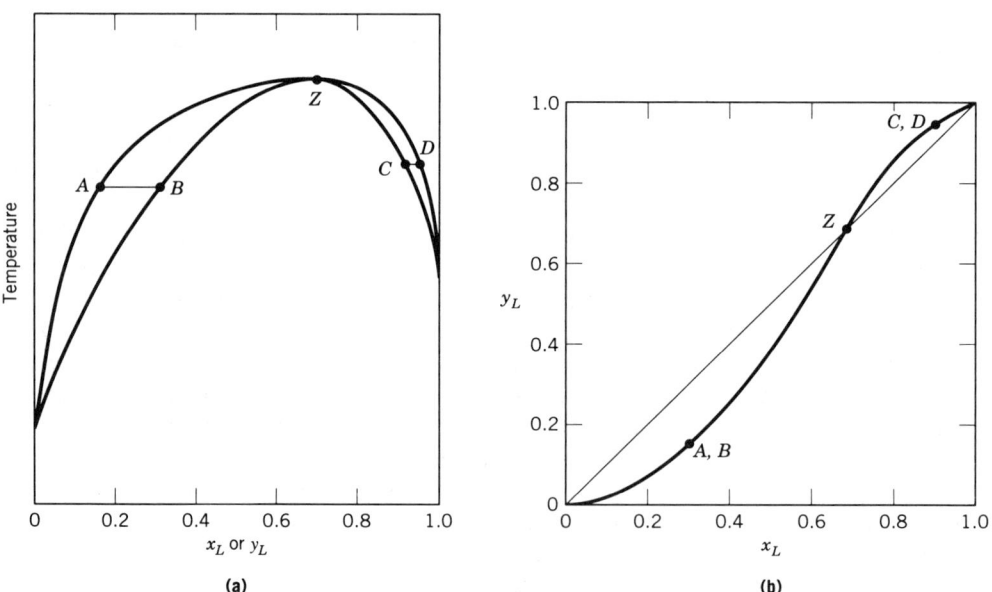

Fig. 5. Boiling point (**a**) and phase diagram (**b**) for a maximum boiling binary azeotropic system at constant pressure. A, B and C, D are representative equilibrium points; Z is the azeotropic point.

compositions lower in water than in the azeotrope, the relative volatility of water/1-butanol is greater than one; at liquid compositions higher in water than in the azeotrope, the relative volatility of water/1-butanol is less than one.

Distillation Processes

Basic distillation involves application of heat to a liquid mixture, vaporization of part of the mixture, and removal of the heat from the vaporized portion. The resultant condensed liquid, the distillate, is richer in the more volatile components and the residual unvaporized bottoms are richer in the less volatile components. Most commercial distillations involve some form of multiple staging in order to obtain a greater enrichment than is possible by a single vaporization and condensation.

For ease of presentation and understanding, the initial discussion of distillation processes involves binary systems. Examining the binary boiling point (Fig. 1a) and phase (Fig. 1b) diagrams, the enrichment from liquid composition x_L to vapor composition y_L represents a theoretical step, or equilibrium stage.

Simple Distillations. Simple distillations utilize a single equilibrium stage to obtain separation. Simple distillation, also called differential distillation, may be either batch or continuous, and may be represented on boiling point or phase diagrams. In Figure 1a, if the batch distillation begins with a liquid of composition $x_L{}^A$ the initial distillate vapor composition is $y_L{}^A$. As the distillate is removed, the remaining liquid becomes less rich in the low boiler, L, and the boiling liquid

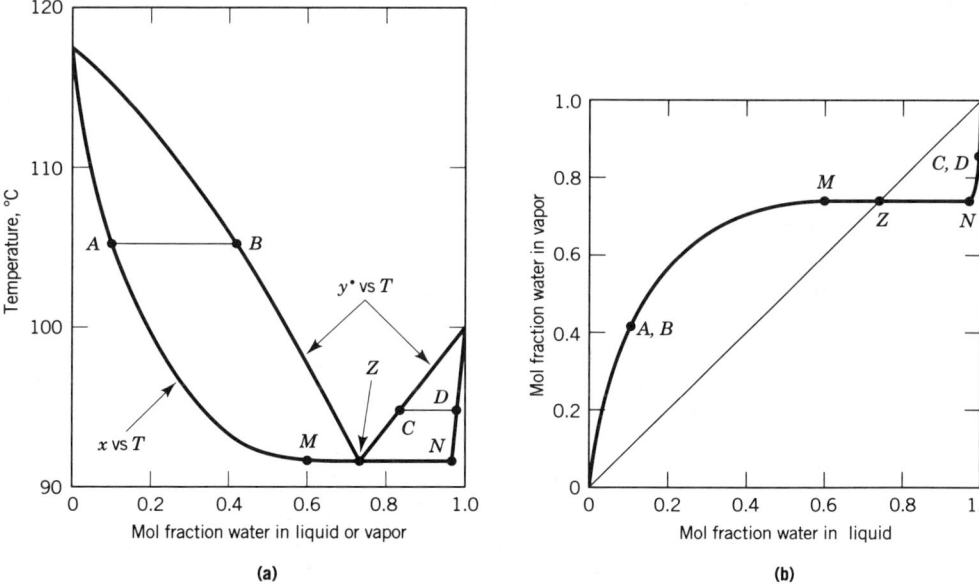

Fig. 6. Boiling point (**a**) and phase diagram (**b**) for the heterogeneous azeotropic system, water/1-butanol at atmospheric pressure. A, B and C, D are representative equilibrium points; Z is the azeotropic point; M and N are liquid miscibility limits.

composition moves to the left along the bubble point line. If the distillation is continued until the liquid has a composition of x_L^E, the last vapor distillate has a composition of y_L^E. Simple batch distillation is not widely used in industry, except for the processing of high valued chemicals in small production quantities, or for distillations requiring regular sanitization. Calculation methods are found in most standard distillation texts, and computer programs, such as BATCHFRAC of Aspen Technology (29), are available for handling the more complex, multicomponent batch distillations.

Simple continuous distillation, also called flash distillation, has a continuous feed to a single equilibrium stage; the liquid and vapor leaving the stage are considered to be in phase equilibrium. On the boiling point diagram (Fig. 1a), the feed is represented by x_L^F, the bottoms liquid by x_L^B, and the equilibrium vapor distillate by y_L^D. The overall mass balance is

$$F = D + B \tag{25}$$

the component L balance is

$$x_L^F F = y_L^D D + x_L^B B \tag{26}$$

Flash distillations are widely used where a crude separation is adequate. Examples of flash multicomponent calculations are given in standard distillation texts (30).

Multiple Equilibrium Staging. The component separation in simple distillation is limited to the composition difference between liquid and vapor in phase equilibrium. To overcome this limitation, multiple equilibrium staging is used to increase the component separation. Figure 7 schematically represents a continuous distillation that employs multiple equilibrium stages stacked one upon another. The feed, F, enters the column at equilibrium stage f. The heat $\bar{q}^s$ required for vaporization is added at the base of the column in a reboiler or calandria. The vapors V^T from the top of the column flow to a condenser from which heat $\bar{q}^c$ is removed. The liquid condensate from the condenser is divided into two streams: the first, a distillate D, which is the overhead product (sometimes called heads or make), is withdrawn from the system, and the second, a reflux R, which is returned to the top of the column. A bottoms stream B is withdrawn from the reboiler. The overall separation is represented by feed F separating into a distillate D and a bottoms B.

Above the feed a typical equilibrium stage is designated as n; the stage above n is $n + 1$ and the stage below n is $n - 1$. The section of column above the feed is called the *rectification section* and the section below the feed is referred to as the *stripping section*.

The mass balance across stage n is (1) vapor (V^{n-1}) from the stage below ($n - 1$) flows up to stage n; (2) liquid (L^{n+1}) from the stage above ($n + 1$) flows down to stage n; (3) on stage n the vapors leaving V^n are in equilibrium with the liquid leaving L^n. The vapors moving up the column from equilibrium stage to equilibrium stage are increasingly enriched in the more volatile components. Similarly, the liquid streams moving down the column are increasingly diminished in the more volatile components.

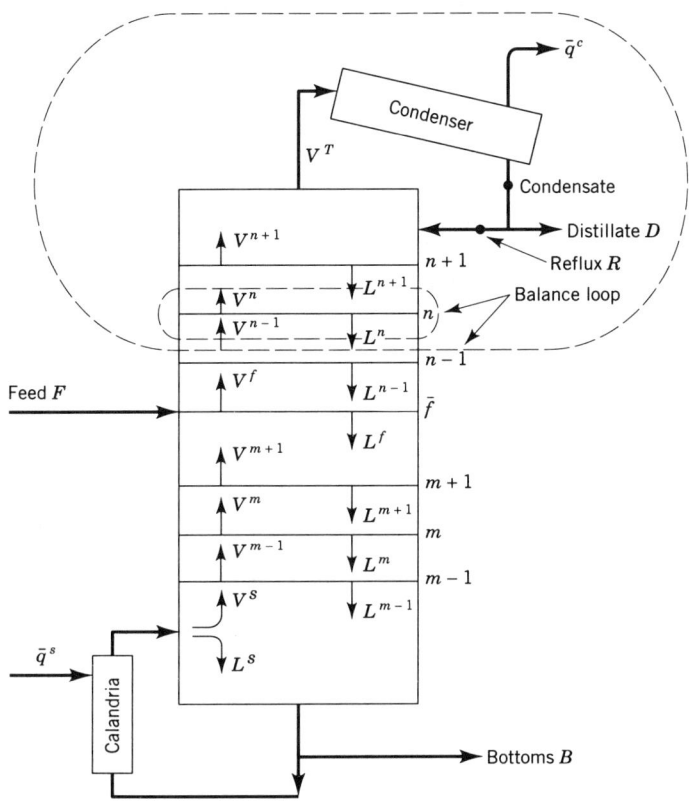

Fig. 7. Distillation column with stacked multiple equilibrium stages. Terms are defined in text.

The overall column mass balances are

$$F = D + B \tag{27}$$

and for any component i,

$$Fx_i^F = Dx_i^D + Bx_i^B \tag{28}$$

The overall enthalpy balance is

$$H^F F + H^S = H^D D + H^B B + H^C \tag{29}$$

A mass balance around plate n and the top of the column gives:

$$V^{n-1} = L^n + D \tag{30}$$

And for any component:

$$V^{n-1} y_i^{n-1} = L^n x_i^n + D x_i^D \tag{31}$$

$$y_i^{n-1} = \left(\frac{L^n}{V^{n-1}}\right) x_i^n + \left(\frac{D}{V^{n-1}}\right) x_i^D \tag{32}$$

Below the feed, a similar balance around plate m and the bottom of the column results in:

$$y_i^{m-1} = \left(\frac{L^m}{V^{m-1}}\right) x_i^m + \left(\frac{B}{V^{m-1}}\right) x_i^B \tag{33}$$

Equation 32 represents the upper (or rectifying) operating line equation and equation 33 represents the lower (or stripping) operating line equation. The slopes L^n/V^{n-1} and L^m/V^{m-1} can vary, depending on heat effects.

Graphical Method. The graphical McCabe-Thiele (31) design method facilitates a visualization of distillation principles while providing a solution to the material balance and equilibrium relationships. Here, the subscripts L and H are not used and x and y refer to the lower boiler, ie, more volatile component, in the binary system. A McCabe-Thiele diagram is given in Figure 8 where P, Q, and S are the x^B, x^F, and x^D compositions on the $y = x$, 45° construction line, respectively. Line OP is the stripping operating line and line OS is the rectifying operating line.

The McCabe-Thiele method employs the simplifying assumption that the molal overflows in the stripping and the rectification sections are constant. This assumption reduces the rectifying and stripping operating line equations to:

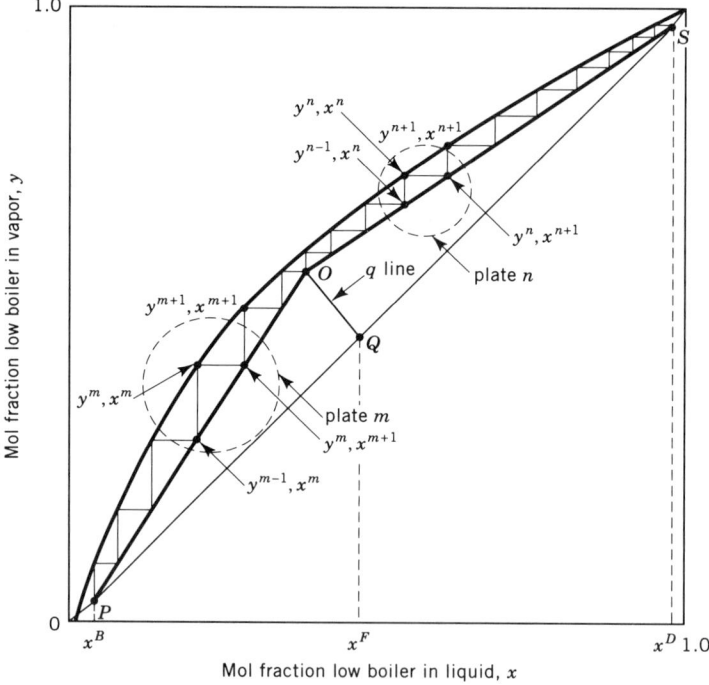

Fig. 8. McCabe-Thiele diagram. Terms are defined in text.

$$y^{n-1} = \left(\frac{\overline{L}}{\overline{V}}\right)_R x^n + \left(\frac{D}{\overline{V}_R}\right) x^D \qquad (34)$$

$$y^{m-1} = \left(\frac{\overline{L}}{\overline{V}}\right)_S x^m + \left(\frac{B}{\overline{V}_S}\right) x^B \qquad (35)$$

The constant molal flows in each section are designated by L and V. The McCabe-Thiele assumption of constant molal overflow implies that the molal latent heats of the two components are identical, the sensible heat effects are negligible, and the heat of mixing and the heat losses are zero. This simplified situation is closely approximated for many distillations. Equation 34 now represents the straight upper operating line OS and equation 35 represents the straight lower operating line OP. The upper operating line has the slope $(L/V)_R$ and the intercept at $x^D (= y^D)$ on the $x = y$ line. Note that this operating line slope is less than one. Similarly, the lower operating line has a slope of $(L/V)_S$ and the intercept is at x^B on the $y = x$ line. This operating line has a slope greater than one. The line QO from the feed intercept Q to the intersection of the operating lines at O is called the q line.

The equilibrium curve gives the vapor–liquid relationships of y^n and x^n above the feed and of y^m and x^m below the feed. The upper operating line gives the relationship between y^{n-1} and x^n and the lower operating line gives the relationship y^{m-1} and x^m, ie, the streams passing each other. The graphical representation of theoretical equilibrium stages n and m is shown. The y^{m-1}, x^m to y^m, x^{L+1} represent the mass balance and phase equilibrium for theoretical stage m. Similarly, y^{n-1}, x^n to y^n, x^{n+1} represent theoretical stage n. The total number of theoretical stages in the column can now be stepped off starting either at the composition x^B and stepping upward or starting at x^D and stepping downward.

Condition of Feed (q Line). The q line, which marks the transition from rectifying to stripping operating lines, is determined by mass and enthalpy balances around the feed plate. These balances are detailed in distillation texts (15).

The slope of the q line is $q/(q-1)$ where:

$$q = \frac{\text{heat needed to vaporize one mole of feed}}{\text{molal latent heat of feed}} \qquad (36)$$

The q line, therefore, depends on the enthalpy condition of the feed. Types of q lines are shown in Figure 9 and are listed below.

Feed enthalpy condition	q	Slope of q line	q Line coordinates
cold liquid	>1	+	Q–E
saturated liquid	1	∞	Q–D
partially vaporized	0–1	−	Q–C
saturated vapor	0	0	Q–B
superheated vapor	<0	+	Q–A

Reflux and Reflux Ratio. The liquid returned to the top of the column is called reflux. The molar ratio R/D is the external reflux ratio. The ratio $(L/V)_R$,

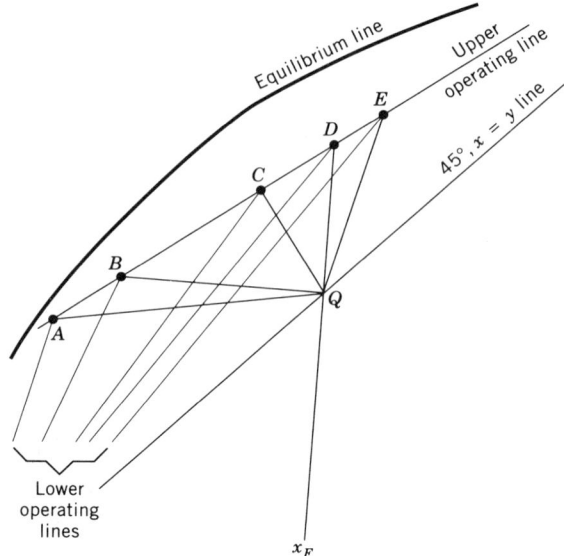

Fig. 9. McCabe-Thiele q lines for various feed enthalpy conditions. Terms are defined in text.

which is the slope of the rectifying operating line, is the rectifying internal reflux ratio. Similarly, the ratio $(L/V)_S$, which is the slope of the stripping operating line, is the stripping internal reflux ratio. As the ratio R/D increases, the rectifying internal reflux ratio increases and numerically approaches unity; similarly, the stripping internal reflux ratio decreases and numerically approaches unity. In the McCabe-Thiele plot the two operating lines move away from the equilibrium line toward the $y = x$ diagonal as the reflux ratio increases, and the individual theoretical stage steps become larger; accordingly, fewer theoretical stages are required to make a given separation.

McCabe-Thiele Example. Assume a binary system $L-H$ that has ideal vapor–liquid equilibria and a relative volatility of 2.0. The feed is 100 mol of $x^F = 0.6$; the required distillate is $x^D = 0.95$, and the bottoms $x^B = 0.05$, with the compositions identified and the lighter component L. The feed is at the boiling point. To calculate the minimum reflux ratio, the minimum number of theoretical stages, the operating reflux ratio, and the number of theoretical stages, assume the operating reflux ratio is 1.5 times the minimum reflux ratio and there is no subcooling of the reflux stream, then:

(*1*) Calculate the vapor composition in equilibrium with the liquid feed. From equation 2a and for $x = 0.60$ mol fraction;

$$y^* = \frac{2(0.6)}{1 + (2.0 - 1)0.6} = 0.75 \text{ mol fraction} \qquad (37)$$

(*2*) Similarly, the entire equilibrium curve is calculated and is plotted in Figure 10. The feed is at the boiling point so the q line is drawn vertically with an infinite slope.

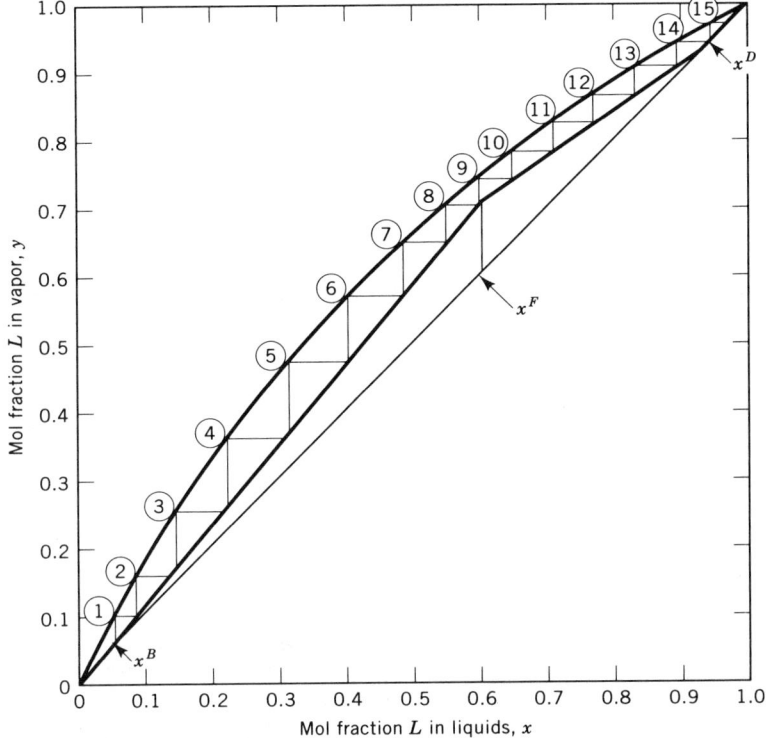

Fig. 10. McCabe-Thiele example. See text.

(3) Calculate mass balances on the basis of 100 mol of feed:

$F = D + B$ (overall column balance)
$0.60F = 0.95D + 0.05B = 60$ (component L balance)
$D = 61.11$ mol distillate
$B = 38.89$ mol bottoms

(4) Calculate reflux ratios. The minimum internal reflux ratio is a line from the intercept of the q line with the equilibrium curve to the x^D point on the 45° line:

slope $= (L/V)_R = (0.95 - 0.75)/(0.95 - 0.60) = 0.5714$ minimum internal reflux ratio
$V = L + D = 0.5714V + 61.11.$
$V = 142.58$ mol (at minimum reflux)
$L = 81.47$ mol $= R$ (at minimum reflux)
$(R/D)_{min} = 81.47/61.11 = 1.333$ (minimum reflux ratio)

(5) Operating reflux ratio = $(R/D)_{operating} = 1.5 \times 1.333 = 2.0$.

(6) Reflux flow = R = 2.0 (61.11) = 122.22 mol = L (at operating reflux ratio).

(7) Rectifying section vapor flow = $V = L + D$ = 122.22 + 61.11 = 183.33 mol.

(8) Upper operating line (eq. 34):

$$y^{n-1} = (122.22/183.33)\,x^n + (61.11/183.33)0.95 = 0.667\,x^n + 0.317$$

(9) Stripping section liquid and vapor flows, because the feed is at the boiling point,

$$L_S = L_R + F = 122.22 + 100 = 222.22 \text{ mol}$$
$$V_S = V_R = 183.33 \text{ mol}$$

(10) Lower operating line (eq. 35):

$$y^{m-1} = (222.22/183.33)\,x^m + (38.89/183.33)(0.05) = 1.212\,x^m - 0.0106$$

(11) Theoretical stages: the complete construction is shown in Figure 10. Stages were stepped off starting at the base. Approximately 14.2 theoretical stages are required. This includes the reboiler, which normally functions as an equilibrium stage. Therefore, a capability of 13.2 theoretical stages in the column is needed. If the condenser were to condense only reflux, with the distillate product leaving the process as a vapor, it could be counted also as an equilibrium stage, making 12.2 stages needed for the column.

Unequal Molal Overflow. The McCabe-Thiele method is based on the simplifying assumption that the molal overflow is constant in both the rectifying and stripping sections. For many problems this assumption is not valid and more precise calculations are necessary. For the more general case, detailed enthalpy balances are made around individual stages or groups of stages. Standard distillation texts discuss the internal enthalpy calculations by algebraic balances or by graphical procedures; eg, Reference 15 details the stage-to-stage mass and enthalpy balances with equilibrium calculations and also by means of the graphical Ponchon-Savarit procedure (32,33). Hand algebraic and graphical methods requiring internal enthalpy calculations have been largely superseded by simulations performed on modern computing devices, including personal computers.

Minimum Number of Theoretical Stages and Minimum Reflux Ratio. There are infinite combinations of reflux ratios and numbers of theoretical stages for any given distillation separation. The larger the reflux ratio, the fewer the theoretical stages required. For any distillation system with its given feed and its required distillate and bottoms compositions, there are two constraints within which the variables of reflux ratio and number of theoretical stages must lie: the minimum number of theoretical stages and the minimum reflux ratio. The minimum reflux ratio occurs when the reflux ratio is reduced so that the upper and lower operating lines and the q line are coincident at a single point on the equilibrium line as

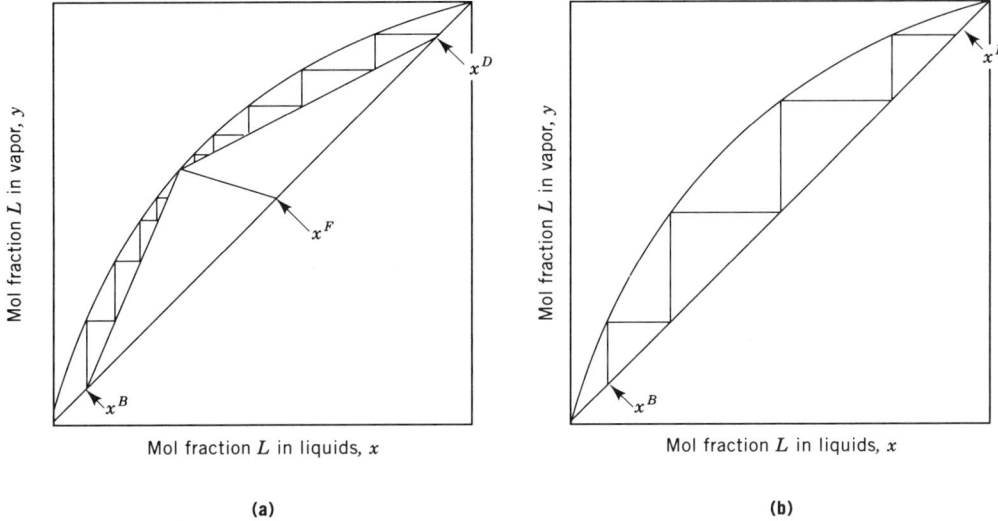

Fig. 11. Limiting conditions in binary distillation. (**a**) Minimum reflux and infinite number of theoretical stages; (**b**) total reflux and minimum number of theoretical stages.

shown in Figure 11**a**. When this condition exists, an infinite number of theoretical stages would be required to make the separation. The minimum number of theoretical stages occurs when the system is at total reflux: no feed, distillate, or bottoms. This is illustrated in Figure 11**b**, where the operating lines are coincident with the 45°, $y = x$ line. For the McCabe-Thiele example presented above, the graphical procedure would give slightly less than nine minimum theoretical stages including the reboiler.

Simple analytical methods are available for determining minimum stages and minimum reflux ratio. Although developed for binary mixtures, they can often be applied to multicomponent mixtures if the two key components are used. These are the components between which the specification separation must be made; frequently the heavy key is the component with a maximum allowable composition in the distillate and the light key is the component with a maximum allowable specification in the bottoms. On this basis, minimum stages may be calculated by means of the Fenske relationship (34):

$$N_{\min} = \frac{\ln\left[(y_i/y_j)_D(x_j/x_i)_B\right]}{\ln \alpha_{ij,\mathrm{avg}}} \qquad (38)$$

where i and j are the light and heavy components of a binary mixture, or the light key and heavy key in a multicomponent mixture. The average relative volatility is often taken as the geometric average of the relative volatilities at the top and bottom of the column. For the McCabe-Thiele example,

$$N_{\min} = \frac{\ln\left[(0.95/0.05)(0.95/0.05)\right]}{\ln 2.0} = 8.50 \text{ stages}$$

For minimum reflux ratio, the following equations (35) may be used:

$$\sum_i \frac{\alpha_i x_{if}}{\alpha_i - \phi} = 1 - q \qquad (39)$$

$$\sum_i \frac{\alpha_i (x_{id})}{\alpha_i - \phi} = R_{min} + 1 \qquad (40)$$

where the value of q is determined as in the McCabe-Thiele procedure. Equation 39 is solved for root ϕ, the value of which must lie between 1.0 and the light key volatility. The root value so determined is then used in equation 40 to obtain the value of R_{min}. Although a trivial example, the McCabe-Thiele problem would yield

$$\frac{2.0(0.6)}{2.0 - \phi} + \frac{1.0(0.4)}{1.0 - \phi} = 1 - q = 0 \text{ (because } q = 1\text{)}$$

solving $\phi = 1.25$. Substituting in equation 40, for the given distillate compositions,

$$\frac{2.0(0.95)}{2.0 - 1.25} + \frac{1.0(0.05)}{1.0 - 1.25} = R_{min} + 1$$

from which $R_{min} = 1.333$.

Both of these limits, the minimum number of stages and the minimum reflux ratio, are impractical for useful operation, but they are valuable guidelines within which the practical distillation must lie. As the reflux ratio decreases toward the

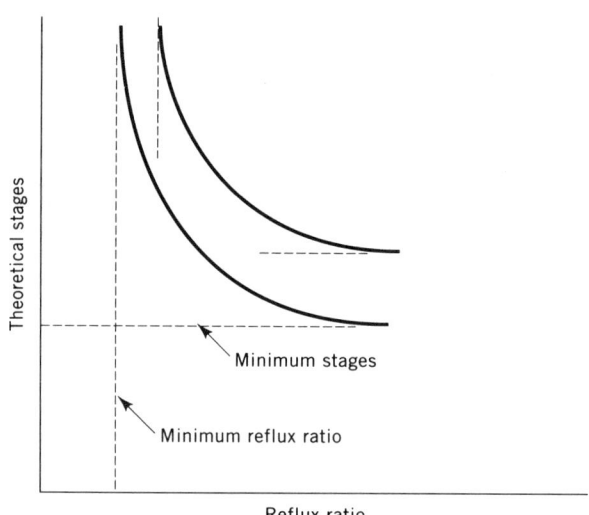

Fig. 12. Representative plot of theoretical stages vs reflux ratio for a given separation. Each curve is the locus of points for a given separation. Note the limiting conditions of minimum reflux and minimum stages.

minimum reflux, the required number of stages increases rapidly. Similarly, as the minimum number of stages is approached, the required reflux ratio increases rapidly. A representative plot of the number of theoretical stages vs reflux ratio for some distillation separation is shown in Figure 12. Both minimum limits may be calculated for any distillation, thereby bracketing the practical design. Actual operating reflux ratios for most commercial columns are in the range of 1.1 to 1.5 times the minimum reflux ratio.

The operating, fixed, and total costs of a distillation system are functions of the relation of operating reflux ratio to minimum reflux ratio. Figure 13 shows a typical plot of costs; as the operating to minimum reflux ratio increases, the operating cost (principally energy cost for the boil-up) increases almost linearly. Similarly, the fixed costs at first decrease from the infinite number of stages, pass through a minimum, and then increase again as the diameter of column increases with increased reflux ratio. These costs for typical distillations have been calculated (36); the ratio of the economic optimum reflux to the minimum reflux is often 1.2 or less.

Minimum Reflux with Pinch Zone. There are some distillations where the minimum reflux does not occur at the intersection of the upper and lower operating lines and the q line. These cases arise when the equilibrium is skewed from positive activity coefficients and when the operating line intersects the equilibrium line in a zone of constant composition, a pinch zone, which is not at the q line intersection. Figure 14 illustrates such a case. An example of such a pinch zone in an ethanol–water column is available (37).

Multicomponent Calculations. The calculations that determine the reflux and stage requirements are more difficult to make for multicomponent systems

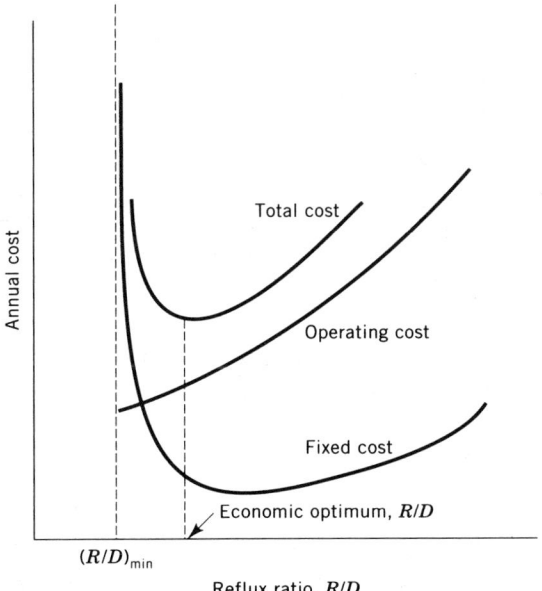

Fig. 13. Fixed, operating, and total costs of a typical distillation, as a function of reflux ratio.

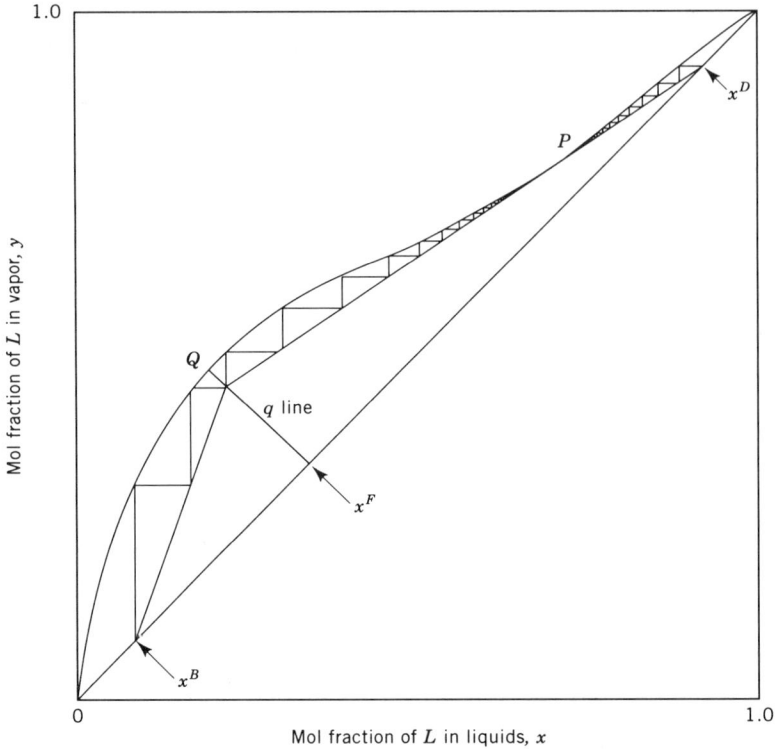

Fig. 14. False minimum reflux for system of skewed equilibria. Minimum reflux occurs at intersection P of operating line and equilibrium line, not at intersection of q line and equilibrium line. Terms are defined in text.

than for binary systems. When the concentration of a component in the distillate and in the bottoms is specified for the overall solution of a binary distillation, the component balance around the column also is completely specified. In the multicomponent case, only a single high boiling key component can be specified in the distillate and a single low boiling key component in the bottoms; the split of other components can be determined only by detailed calculations. These require a series of trial and error computations to obtain the solution at any given reflux ratio and number of stages. As the number of components and number of stages become large, the mathematical problem becomes formidable. Two approaches may be followed: use of approximate, ie, shortcut, methods, or use of a suitable computer program that provides rigorous solutions. The former are used when approximate solutions are adequate or when a computer is not available. For the latter, numerous commercial programs are available and may be used with personal computers.

Most shortcut methods involve: (1) calculating the minimum number of stages; (2) calculating the minimum reflux ratio; and (3) estimating, from empirical correlations, the actual number of stages at an operating reflux. For minimum stages the Fenske relationship (eq. 38) is used, whereas for minimum reflux ratio the Underwood relationships (eqs. 39 and 40) are used. The relationship of operating to minimum reflux ratio and of operating to minimum number of plates is then estimated from the Gilliland correlation (38), or from a more recent correlation such as that of Reference 39.

The Gilliland correlation is in graphical form and the curve has been fitted by several workers (40):

$$\frac{N_t - N_{\min}}{N_t + 1} = 0.75 - 0.75 \left(\frac{R - R_{\min}}{R + 1}\right)^{0.5668} \tag{41}$$

For the McCabe-Thiele example, and using equation 41

$$\frac{N_t - 8.50}{N_t + 1} = 0.75 - 0.75 \left(\frac{2.0 - 1.333}{2.0 + 1}\right)^{0.5668}$$

from which $N_t = 15.7$ stages. The original plot of Gilliland would give $N_t = 14.8$, closer to the McCabe-Thiele value of 14.2 stages.

Discussions of shortcut methods have appeared many times in the literature (16,36), accompanied by the usual admonition to use such methods only for approximate designs or analyses. For multicomponent systems having significant nonidealities, the shortcut methods can be grossly in error.

Rigorous computer solutions are used for complex distillations involving multiple stages, multiple components, nonideal phase equilibria, multiple feeds and drawoffs, and heat addition or removal at intermediate stages. Most calculations are made by computer and the algorithms are generally based on the Thiele-Geddes model (41), which rates a given number of stages and reflux ratio for separation capability. A detailed discussion of computer solutions, including the handling of convergence problems, is available (42).

Computer solutions entail setting up component equilibrium and component mass and enthalpy balances around each theoretical stage and specifying the required design variables as well as solving the large number of simultaneous equations required. The explicit solution to these equations remains too complex for present methods. Studies to solve the mathematical problem by algorithm or iterational methods have been successful and, with a few exceptions, the most complex distillation problems can be solved.

Multiple Products. If each component of a multicomponent distillation is to be essentially pure when recovered, the number of columns required for the distillation system is $N^* - 1$, where N^* is the number of components. Thus, in a five-component system, recovery of all five components as essentially pure products requires four separate columns. However, those four columns can be arranged in 14 different ways (43).

The number of columns in a multicomponent train can be reduced from the $N^* - 1$ relationship if side-stream draw-offs are used for some of the component cuts. The feasibility of multicomponent separation by such draw-offs depends on side-stream purity requirements, feed compositions, and equilibrium relationships. In most cases, side-stream draw-off distillations are economically feasible only if component specifications for the side-stream are not tight. If a single component is to be recovered in an essentially pure state from a mixture containing both lower and higher boiling components, a minimum of two columns is required, one column to separate the lower boilers from the desired component and another column to separate the component from the higher boilers.

The economics of the various methods that are employed to sequence multicomponent columns have been studied. For example, the separation of three-, four-, and five-component mixtures has been considered (44) where the heuristics (rules of thumb) developed by earlier investigators were examined and an economic analysis of various methods of sequencing the columns was made. The study of sequencing of multicomponent columns is part of a broader field, process synthesis, which attempts to formalize and develop strategies for the optimum overall process (45) (see SEPARATIONS PROCESS SYNTHESIS).

Distillation Columns

Distillation columns are vertical, cylindrical vessels containing devices that provide intimate contacting of the rising vapor with the descending liquid. This contacting provides the opportunity for the two streams to achieve some approach to thermodynamic equilibrium. Depending on the type of internal devices used, the contacting may occur in discrete steps, called plates or trays, or in a continuous differential manner on the surface of a packing material. The fundamental requirement of the column is to provide efficient and economic contacting at a required mass-transfer rate. Individual column requirements vary from high vacuum to high pressure, from low to high liquid rates, from clean to dirty systems, and so on. As a result, a large variety of internal devices has been developed to fill these needs. The column devices discussed herein are used for absorption (qv) and stripping as well as distillation. The principal operational difference is that in absorption or stripping, the gas flowing up the column is primarily a noncondensable phase at column conditions, whereas in distillation the gas phase is a condensable vapor.

Plate Columns. There are two general types of plates in use: crossflow and counterflow. These names refer to the direction of the liquid flow relative to the rising vapor flow. On the cross-flow plate the liquid flows across the plate and from plate to plate via downcomers. On the counterflow plate liquid flows downward through the same orifices used by the rising vapor.

Crossflow Plates. As indicated in Figure 15, liquid enters a crossflow plate from the bottom of the downcomer of the plate above and flows across the active or bubbling area where it is aerated by the vapors flowing through orifices from the plate below. It is in this aerated zone where most of the vapor–liquid mass transfer occurs. The aerated mixture flows over the exit weir into a downcomer. A vapor–liquid disengagement takes place in the downcomer and most of the trapped vapor escapes from the liquid and flows back to the interplate vapor space. The liquid, essentially free of entrapped vapor, leaves the plate by flowing under the downcomer to the inlet side of the next lower plate. The vapor, disengaging from the aerated mass on the plate, rises to the next plate above.

The pressure drop incurred by the vapor as it passes through the orifices of

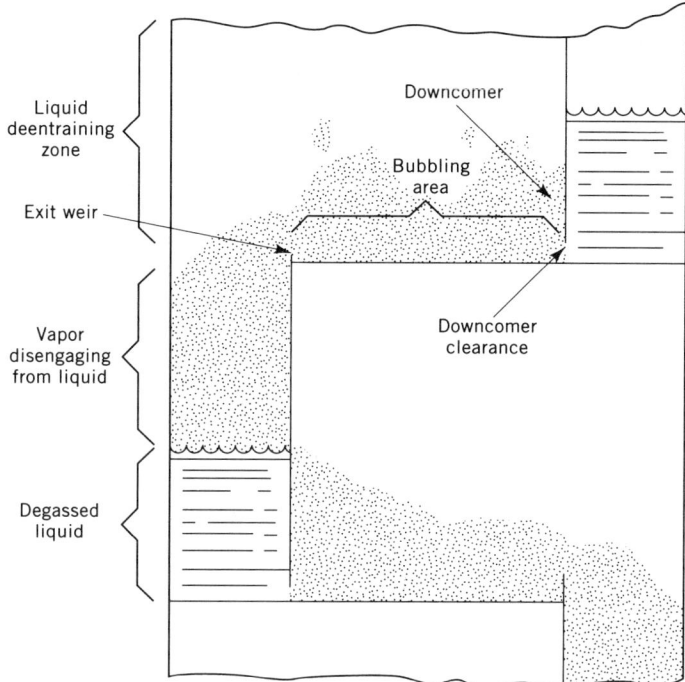

Fig. 15. Flow pattern in a crossflow plate distillation column.

the plate is fundamental to plate operation. In most plate designs, the pressure drop prevents the crossflowing liquid from falling through the plate. The pressure drop also results from the energy consumed to disperse the vapor–liquid mixture, eg, to atomize a portion of the liquid to provide increased interfacial area for mass transfer. Diameters of commercial crossflow plate columns range from 0.3 to 15 m and plate spacings range from 0.15 to 1.2 m. The total pressure drop per plate is often in the range of 0.25 to 1.6 kPa (2–12 mm Hg).

Three principal vapor–liquid contacting devices are used in current crossflow plate design: the sieve plate, the valve plate, and the bubble cap plate. These devices provide the needed intimate contacting of vapor and liquid, requisite to maximizing transfer of mass across the interfacial boundary.

Sieve Plates. The conventional sieve or perforated plate is inexpensive and the simplest of the devices normally used. The contacting orifices in the conventional sieve plate are holes that measure 1 to 12 mm diameter and exhibit ratios of open area to active area ranging from 1:20 to 1:7. If the open area is too small, the pressure drop across the plate is excessive; if the open area is too large, the liquid weeps or dumps through the holes.

Valve Plates. Valve plates are categorized as proprietary and details of design vary from one vendor to another. These represent a variation of the sieve plate in which the holes are large and are fitted with liftable valve units such as those shown in Figure 16. The principal advantage over sieve plates is the ability to maintain efficient operation over a wider operating range through the use of

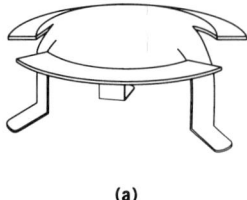 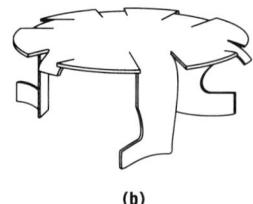 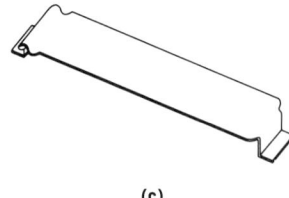

(a) (b) (c)

Fig. 16. Individual valve units used in valve plates: (**a**) Koch Flexitray valve, courtesy of Koch Engineering Co.; (**b**) Glitsch Ballast valve, courtesy of Glitsch, Inc.; and (**c**) Nutter Float Valve, courtesy of Nutter Engineering Co.

variable orifices (valves) which open or close depending on vapor rate. The most common valve units consist of flat disks having attached legs that allow the valve to open or close. Sometimes two weights of valves are used on a single plate to extend operating range and improve vapor distribution. The valve units usually have a tab or indentation that provides a minimum open area of vapor flow, even when the valve is closed, and also prevents the valve from sticking under corrosive or fouling conditions. Details on valve plate geometry, along with methods for valve plate design, are available (46–48).

Bubble Cap Plates. Until the early 1950s, bubble caps were the standard design in the chemical industry. Usage in newer installations is limited to low liquid flow rate applications, or to those cases where the widest possible operating range is desired. A typical bubble cap is shown in Figure 17. The vapor flows through a hole in the plate floor, through the riser, reverses direction in the dome of the cap, flows downward in the annular area between the riser and the cap, and exits through the slots in the cap. Commercial caps range from 50 to 150 mm in diameter and many slot design variations have been used. Bubble cap trays are more expensive and have lower capacity than sieve or valve plates; therefore, use has dropped to a very small percentage of newer column designs.

Multiple Liquid-Path Plates. As the liquid flow rate increases in large diameter crossflow plates (ca 4 m or larger), the crest heads on the overflow weirs and the hydraulic gradient of the liquid flowing across the plate become excessive. To obtain improved overall plate performance, multiple liquid-flow-path plates may be used, with multiple downcomers. These designs are illustrated and discussed in detail in the literature (49).

Counterflow Plates. Counterflow plates are used less frequently than crossflow plates. The liquid flows downward and the vapor upward through the same orifices in a counterflow plate and the plate does not have downcomers. The openings are round holes (dualflow tray) or slots (Turbogrid tray). A variation of the dualflow tray is the ripple tray in which the tray floor is shaped in a corrugated fashion (50). Counterflow plates are used advantageously in fouling services because for each hole vapor and liquid flow alternately, providing a self-cleaning action that is quite effective. The dualflow and Turbogrid plates have similar operating characteristics, and typical operating data have been published (51).

Another important plate which has characteristics similar to a counterflow plate is the multiple downcomer (MD) plate (52). This is a plate where the active area occupies the full column cross section but with a plurality of small downcom-

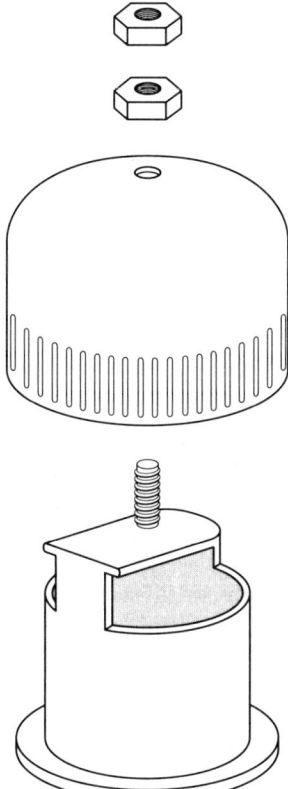

Fig. 17. Expanded view of a bubble cap. Courtesy of Vulcan Manufacturing Co.

ers interspersed among the perforations. The downcomers are specially sealed to prevent upflow of vapor through them. The plate has been used successfully in many high liquid flow cases.

Vapor Capacity Parameters. The diameter of a distillation column is determined by the capacity of the column to handle the required flows of vapor and liquid. The vapor capacity parameter is

$$C_{sb} = V^* \left(\frac{\rho_g}{\rho_L - \rho_g}\right)^{0.5} \tag{42}$$

and its simplification

$$F^* = V^*(\rho_g)^{0.5} \tag{43}$$

The term C_{sb} in equation 42 is called a Souders-Brown capacity parameter and is based on the tendency of the upflowing vapor to entrain liquid with it to the plate above. The term F^* in equation 43 is called an F-factor. For C_{sb} and F^* to be meaningful the cross-sectional area to which they apply must be specified. The capacity parameter is usually based on the total column cross section minus the area blocked for vapor flow by the downcomer(s). For the F-factor, typical operating ranges for sieve plate columns are

	Area basis	$(kg/(m \cdot s^2))^{0.5}$	$(lb/(ft \cdot s^2))^{0.5}$
F_S^*	total cross section	0.6–3.0	0.5–2.5
F_A^*	active area	0.85–4.3	0.7–3.5
F_H^*	hole area	8.5–30	7–25

Entrainment Flooding. The vapor capacity of a column is limited by excessive entrainment, usually called flooding. A flooding condition can be observed when the holdup of liquid becomes excessive, the pressure drop increases dramatically, and the mass-transfer efficiency falls precipitously. Estimates of the vapor velocity for a flooding condition may be made from the chart in Figure 18 (53). The abscissa term $L/G(\rho_g/\rho_L)^{0.5}$ is called a flow parameter and its value can indicate several things about the character of the aerated mass on the plate. For example, a very low value can indicate a phase inversion in which the vapor flow is continuous (spray flow) whereas a high value can indicate a bubbly mass (emulsion flow). The value of the flow parameter is easily determined from the stage calculations (reflux and boilup ratios) and densities of the phases. The ordinate value in Figure 18 leads to a value of the flooding velocity, and prudent design calls for limiting actual flows to 70–80% of this velocity.

Downcomer Flooding. For cases of very high liquid-to-vapor flow ratios the limiting capacity of the column is based on the ability of the downcomers to move the de-aerated liquid from a plate to the next plate below. It is clear that there can be constrictions in the downcomer design or that even with no constrictions there is simply not enough flow area to accommodate the high volume of liquid.

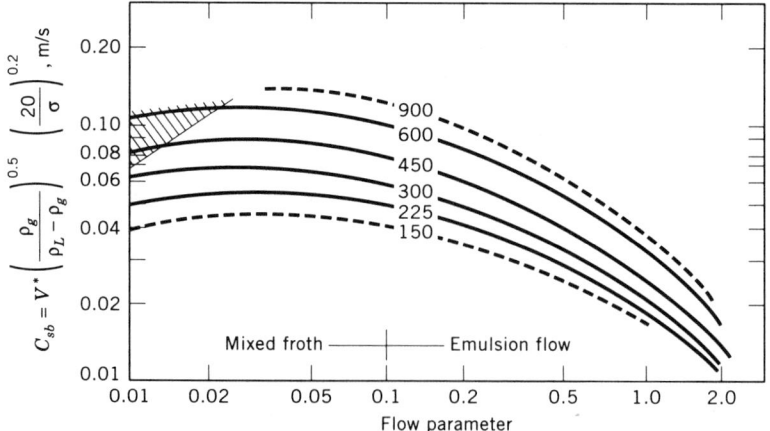

Fig. 18. Flooding correlation for crossflow trays (sieve, valve, bubble-cap) where the numbers represent tray spacing in mm. Also shown are approximate boundaries of the spray zone, ◨, and mixed froth and emulsion flow regimes.

Thus, the downcomer can flood, or choke, when it becomes completely filled with liquid or aerated mass. Typical design heuristics include limiting the downcomer velocity (clear liquid basis) to no more than 0.12 m/s. Also, to allow for complete disengagement of vapor from liquid in the downcomer, a minimum residence time of 4 s is often used. The actual limiting values of these parameters varies somewhat with the properties of the fluids and the exact dimensions of the plate components.

Stable Operating Range. All plates have a stable operating envelope bound by a range of liquid and vapor flow rates as shown in Figure 19. The size and shape of the stable area depends on the plate design and on the system properties. The line AD represents the minimum operable vapor flow rate at various liquid flow rates. Below AD, the vapor rate is too low to maintain the liquid on the plate and, as a result, the liquid weeps excessively or dumps through the plate orifices. Above line BC the column floods by entrainment. To the right of CD the high liquid rate causes downcomer flooding. The area to the left of AB represents high entrainment at low liquid flow rates, with vapor jets at the orifices. Design procedures for bubble caps (49) and sieve trays (53,54) have been published. Additionally, vendors of valve trays make available their design methods.

Plate Efficiencies. Column requirements are calculated in terms of theoretical stages or plates. Actual plates must, however, be specified in the design. Thus the effectiveness of the plate in approaching the equilibrium condition must be predicted. This approach is called the plate efficiency, which is a measure of the rate of mass transfer on the actual plate. This efficiency, expressed either as a fraction or as a percentage, depends on three principal factors: the geometry of the plate (hole arrangement, valve design, etc); the loading of vapor and liquid traffic on the plate; and the diffusional properties of the fluids.

The simplest efficiency is the overall column efficiency which is the number of theoretical plates in a column divided by the number of actual plates:

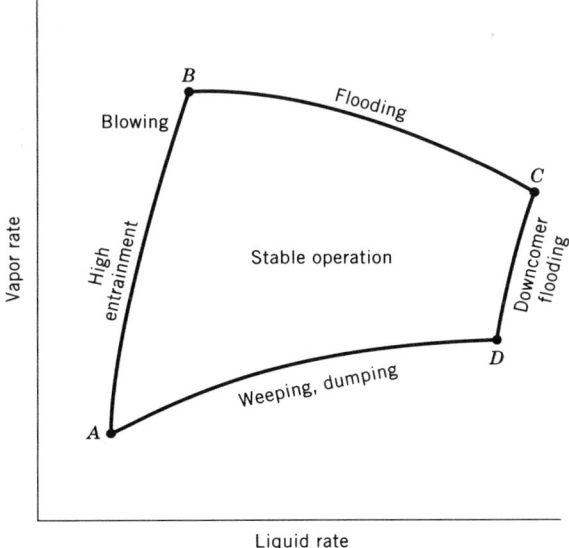

Fig. 19. Stable operating range for crossflow plates.

$$E_o = N_t/N_a \tag{44}$$

Thus the overall efficiency is an averaged efficiency of all the individual plates.

A more useful plate efficiency for theoretical prediction is the Murphree plate efficiency (55):

$$E_{mv} = \frac{y^n - y^{n-1}}{y^{n*} - y^{n-1}} \tag{45}$$

where y^n and y^{n-1} are the vapor compositions from plate n and $n-1$ (the plate below n), and y^{n*} is the vapor composition that would be in equilibrium with the liquid composition leaving plate n. Thus, for a given plate, E_{mv} is a ratio of the actual vapor composition change to the change that would occur if the plate were effective enough to bring the vapor and liquid to thermodynamic equilibrium. This definition is based on the outlet liquid composition, and says nothing about the average liquid composition on the plate. In cases where a significant concentration gradient exists in the liquid composition across the plate, it is possible for E_{mv} to have a value greater than 1.0 (100%). Equation 45 is written in terms of vapor composition. A similar equation can be written in terms of the liquid compositions and is denoted as E_{mL}.

Of still more theoretical importance is the efficiency at some point on the plate:

$$E_{og} = \left(\frac{y^n - y^{n-1}}{y^{n*} - y^{n-1}}\right)_{\text{point}} \tag{46}$$

This parameter is called the point efficiency (or local efficiency). It cannot have a value greater than 1.0, and it has a counterpart term for liquid compositions.

Prediction of Plate Efficiency. As of this writing, the most comprehensive study of plate efficiency known was made in the mid-1950s, based on the then still popular bubble cap plates (56). Unfortunately, the predictive model developed has been shown to be inadequate for many industrial distillations. There has been continuing research effort directed toward a better understanding of the mechanisms that occur in the rather complex aerated mass on the typical plate (57,58). A complicating factor is the lack of uniform liquid flow across the plate, and situations have been found where the liquid actually stagnates in certain zones of larger diameter plates. For larger columns it is possible for the observed Murphree efficiency to exceed 100%. A satisfactory method for predicting plate efficiency does not exist. Most recently there have been studies of the various types of flow regimes that occur on operating plates and of the effect of these regimes on tray performance, including plate efficiency. Pursuit of the flow regime studies (59–62) may lead to improved plate efficiency prediction methods. For example, a newer model (63) takes into account the regime as well as the vapor bubble (froth flow) or liquid drop (spray flow) characteristics in determining mass-transfer coefficients in the aerated zone on the plate.

Empirical Efficiency Prediction Methods. Numerous empirical methods for predicting plate efficiency have been proposed. Probably the most widely used method correlates overall column efficiency as a function of feed viscosity and relative volatility (64). A statistical correlation of efficiency and system variables has been developed from numerous plate efficiency data (65).

General Comments on Plate Efficiency. The plate efficiencies of well-designed commercial bubble cap, sieve, and valve plates are approximately the same when the plates are operated within their normal design range. The plate efficiency decreases both at the low end of the plate's operating range, where the liquid tends to leak through the plate, and at the high end of the operating range, where liquid entrainment becomes substantial.

Most distillation systems in commercial columns have Murphree plate efficiencies of 70% or higher. Lower efficiencies are found under system conditions of a high slope of the equilibrium curve (Fig. 1b), of high liquid viscosity, and of large molecules having characteristically low diffusion coefficients. Finally, most experimental efficiencies have been for binary systems where by definition the efficiency of one component is equal to that of the other component. For multicomponent systems it is possible for each component to have a different efficiency. Practice has been to use a pseudo-binary approach involving the two key components. However, a theory for multicomponent efficiency prediction has been developed (66,67) and is amenable to computational analysis.

Packed Columns. In packed columns, the vapor–liquid contacting takes place in continuous beds of solid packing elements rather than in discrete individual plates. The contacting can be visualized as occurring in differential increments across the height of the packing; thus packings are known as counterflow devices rather than stagewise devices. Mechanically, the packed column is a relatively simple structure. In its simplest form the packed column comprises a vertical shell having dumped or carefully arranged packing elements on an open-type support, together with a suitable liquid distribution device above the packed bed. A packed column having two packed beds and a midcolumn feed is shown in Figure 20. The vapor enters the column below the bottom bed and flows upward through the column. The liquid (reflux or other liquid stream) enters at the top through the liquid distributor and flows downward through the packing countercurrently to the rising vapor. The height of the individual packed beds is limited to 2–9 m by the mechanical strength of the packing or by the need to redistribute the liquid so that good mass-transfer efficiency can be maintained.

Packings. For many years packed columns consisted of randomly dumped packings almost exclusively, with occasional applications of regularly stacked packings or pads of woven or knitted wire. In the late 1960s a partial trend away from random packings began when a special structured packing made of wire gauze was introduced by Sulzer Brothers in Switzerland (68). The indicated advantages of the structured packings were high mass-transfer efficiency and very low pressure drop. These devices appeared to be ideal for high vacuum distillations. However, cost of fabrication was very high and they were considered mainly for the vacuum distillation of specialty chemicals. In 1977, a lower cost sheet metal version was introduced (69), and since that time a large business in structured sheet metal packings has arisen. At the same time, improved random packings have been developed and a comprehensive discussion of their characteristics has been published (70). Some of the common random packings are shown in Figure 21. The Raschig ring, one of the oldest of packings, is an open cylinder of

614 DISTILLATION

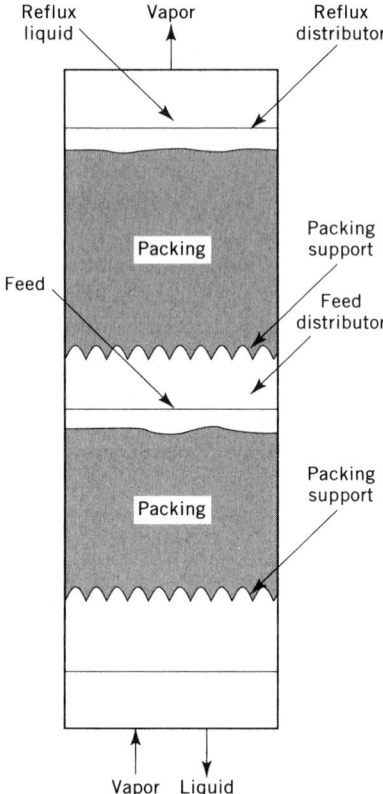

Fig. 20. Packed column shell and internals. Column shown has single packed beds above and below the feed. For separations requiring a large number of stages, additional beds, separated by redistribution devices, are likely to be needed.

equal height and diameter. The Berl saddle and the ceramic Intalox saddle (Norton Co.) have a higher capacity and efficiency than the Raschig ring. The Pall ring is a modification of the Raschig ring which allows through-flow of liquid and vapor, with consequent lower pressure drop and better efficiency. The newer Intalox metal saddle (IMTP) is an example of a random packing having a very high void fraction and low resistance to the flowing phases. Other newer random packings, not shown in Figure 21, include the CMR ring (Glitsch, Inc.) and the Nutter ring

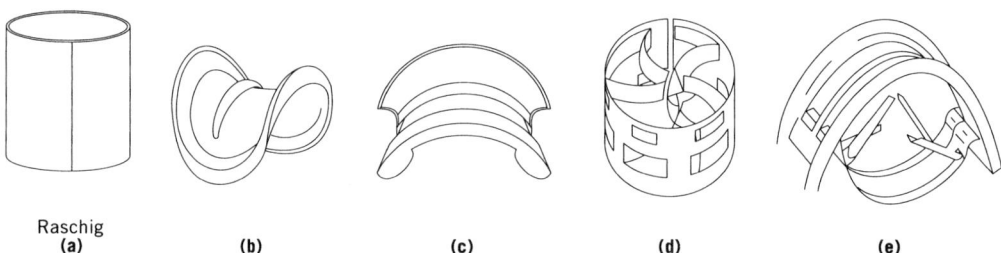

Fig. 21. Random packing elements for distillation columns: (**a**), Raschig ring (metal); (**b**), Berl saddle (ceramic); (**c**), Intalox saddle (ceramic); (**d**), Pall ring (metal); and (**e**), Intalox saddle (metal).

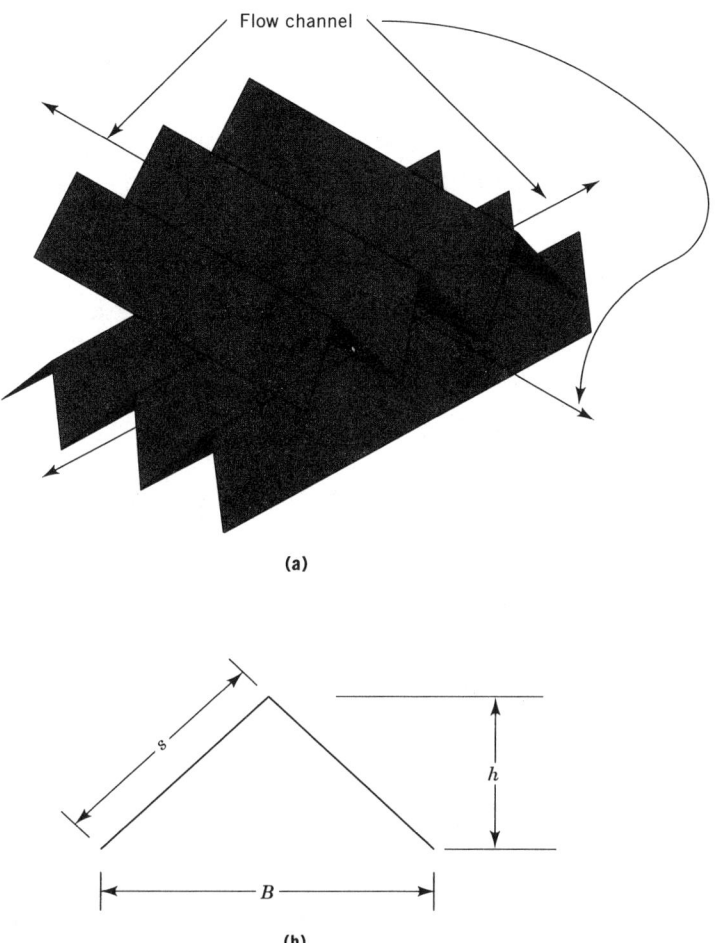

Fig. 22. (a) Flow channel arrangement; (b) flow channel triangular cross section where for angles of 90°, $D_{EQ} = 4 R_H = 4 \left(\dfrac{S \cdot S}{2}\right)\dfrac{1}{2S} = S$.

(Nutter Engineering Co). The random-type packings can generally be made from metal, plastic, or ceramic materials; the approximate nominal size range for the individual elements is 12–75 mm.

Common structured packing geometries are shown in Figure 22. Flat plates of gauze or sheet metal are perforated, or embossed or lanced, and corrugated. Corrugated sheets are then stacked together such that adjacent sheets have opposite corrugation directions. The corrugations have angles with the horizontal of 45 to 60 degrees. Vapor and liquid contact each other in wetted-wall fashion, and the perforations plus other surface enhancements, eg, texturing, serve to promote liquid spreading into thin films. Dimensions, performance characteristics, and design procedures for the structured packings are summarized in Reference 71.

Packed Column Internals. In order to ensure good packed column mass-transfer efficiency, the liquid must be distributed uniformly over the surface of the packing. As a general rule there should be at least 100 pour points per square meter (10 points/ft^2), although fewer points may be used for random packings of the bluff-body type such as Raschig rings and Berl saddles. Although they have capacity and pressure drop limitations, the bluff-body packing elements are able to divide the downflowing liquid and thus improve on an initially marginal distribution. On the other hand, the through-flow type random packings, eg, Pall rings and Intalox metal tower packings, as well as the structured packings, are not able to correct the initial distribution and in fact may allow some deterioration of distribution if the bed heights are greater than about five meters.

Considerable research is in progress on methods for ensuring good liquid distribution in large diameter columns, and the packing manufacturers maintain large test stands where a particular design of distributor can be tested using water before being installed in the column. The distributor design problem becomes more severe at low, ie, <700 cm^3/(s·m^2) (1 gal/(min·ft^2)) liquid rates or in large (>3 m) diameter towers. An example of a more fundamental study of liquid distribution is available (72) as are typical liquid distributor designs and typical packing supports (54).

Packed Column Operation. In the packed column, liquid flows downward in opposition to the upward flow of vapor; both phases flow through the same open space or interstices between the packing elements. At low liquid and gas flow rates, the descending liquid occupies only a small fraction of the interstices and, therefore, offers little hindrance to the rising vapor flow. Figure 23 shows a schematic plot of pressure drop per unit of height as a function of the gas rate at low and high liquid flow rates. At a low rate of gas flow, the log slope of each curve is approximately 2. As the gas flow rate increases, there is an increasing tendency for the liquid to be held up in the void space, thereby decreasing the space available for the gas flow. As the gas flow rate increases further, more liquid is held up until at some high gas rate the packing floods. At this point, the liquid is essentially filling the interstices and can no longer flow downward. At flooding,

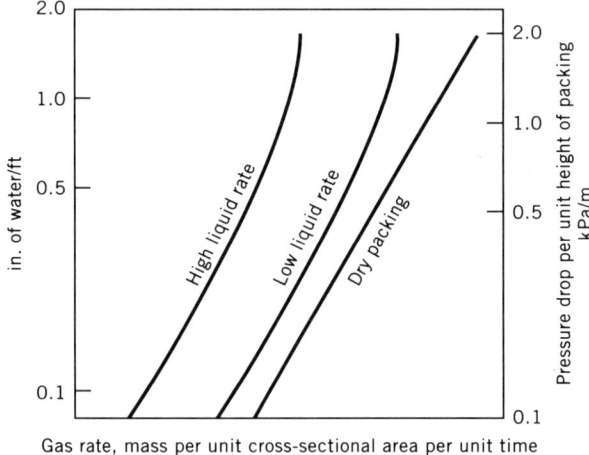

Fig. 23. Log–log plot of pressure drop per unit height of typical packing as a function of gas rate at two liquid rates and for the unirrigated packing.

the log slope is practically infinite. The pressure drop at the inception of flooding ranges from 1.6 to 3.3 kPa/m (2 to 4 in. water/ft) of packing. More comprehensive discussions of packed column hydraulics may be found in distillation texts (15,17,73), monographs (70,74), or handbooks (75,76).

Capacity of Packed Columns. Packed columns are usually designed to operate at some percentage approach to flooding, eg, 60–70%, or at some specified pressure drop per unit height of packing, eg, 0.8 kPa/m (1 in. water/ft) of packing. Flooding correlations have been proposed (77), one revision introducing constant pressure drop lines (78). The most recent revision in these correlations is shown in Figure 24 (70). The idea of flooding has been eliminated from the chart with the stipulation that the topmost curve represents the maximum capacity. Experimentally determined packing factors F_p, presented in Table 2, should be used in the ordinate group. These factors distinguish between the various shapes and sizes of the available packings. The curves are for constant pressure drop and thus the chart enables estimation of both capacity and pressure drop.

Packing Mass-Transfer Characteristics. The contacting for mass transfer (qv) in a packed column occurs differentially along the length of the column. The separation calculations can thus be made on a differential basis along this

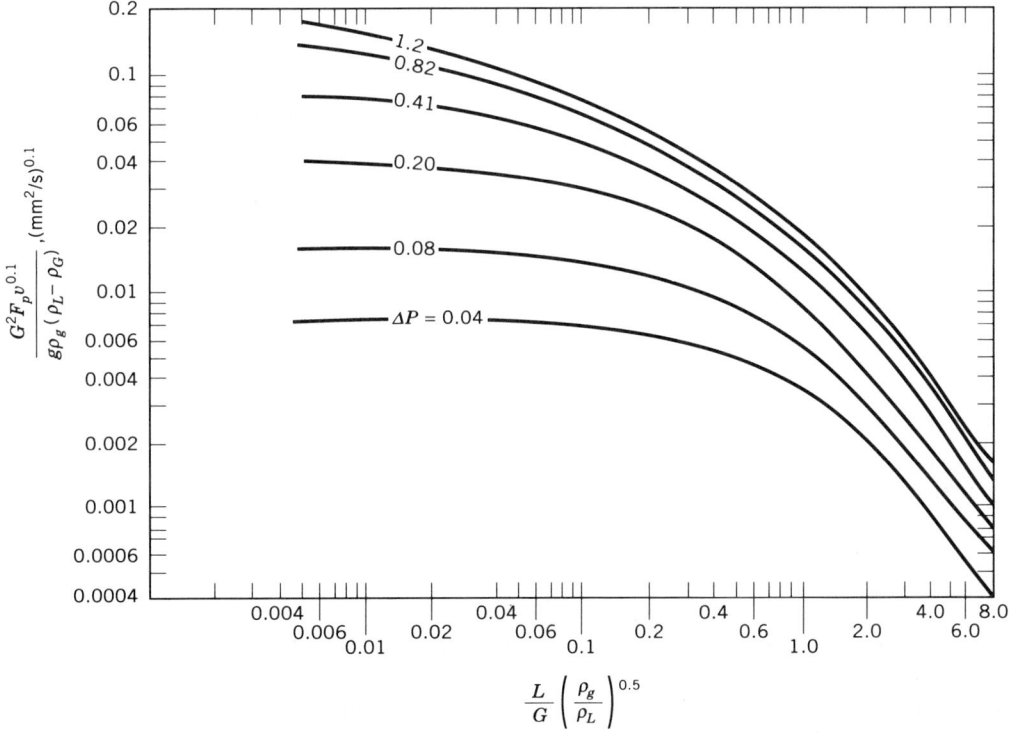

Fig. 24. Generalized method using log scales for estimating packed column flooding and pressure drop, ΔP, in kPa/m; g = gravitational constant, 9.81 m/s^2; v = kinematic viscosity in mm^2/s (= cSt); L, G have units of kg/(m^2·s); ρ_L, ρ_g are in kg/m^3; and the packing factor, F_p, in m^{-1} can be found in Table 2. To convert kPa/m to mm Hg/m, multiply by 7.5 (77).

Table 2. Characteristics of Packing[a]

Packing	Nominal size, mm[b]	Surface area, m^2/m^3	Void fraction	Packing factor, F_p, m^{-1}
\multicolumn{5}{c}{*Dumped (random) packing*}				
Intalox saddles				
ceramic	13	625	0.78	660
	25	255	0.77	300
	50	118	0.79	130
metal (IMTP)	25		0.97	135
	40		0.97	79
	50		0.98	59
plastic	25	206	0.91	130
	50	108	0.93	92
	75	88	0.94	59
Berl saddles, ceramic	13	465	0.62	790
	25	250	0.68	360
	50	105	0.72	150
Pall rings				
metal	16		0.92	265
	25	205	0.94	183
	50	115	0.96	88
plastic	16	341	0.87	310
	25	207	0.90	180
	50	100	0.92	85
Raschig rings, ceramic	13	370	0.64	1902
	25	190	0.74	587
	50	92	0.74	215
\multicolumn{5}{c}{*Structured packing*}				
Flexipac				
1	6	558	0.91	108
2	12	246	0.93	72
3	37	134	0.96	52
4	50	69	0.98	30
Sulzer-BX	6	490	>0.90	66

[a] Ref. 70.
[b] For structured packings, values correspond to crimp height.

length, using mass-transfer coefficients or heights of transfer units. The calculations are somewhat imprecise because of the uncertainty in the fundamental mass-transfer mechanisms in larger-scale columns. Useful models for predicting the mass-transfer efficiency of randomly packed columns have been published (79,80), using the same database of commercial-scale performance data. These models cover the better known packings, eg, metal and ceramic Raschig rings, ceramic Berl saddles, and metal Pall rings, in nominal sizes in the range of 12 to 50 mm. It has been found that to avoid excessive maldistribution of liquid near the wall, a ratio of column diameter to packing element size of at least eight should be maintained. Thus if one wishes to conduct pilot-scale packed column tests, a minimum column diameter of about 100 mm would be used together with 12-mm packing elements. The models would then permit scale-up to large columns containing 50-mm size elements of the same type, eg, Pall rings.

These models provide values of the height of a transfer unit for the liquid-phase H_L and the vapor-phase H_V. These values are combined to form the height of an overall transfer unit, H_{ov}:

$$H_{ov} = H_v + (m'V/L)H_L \qquad (47)$$

where V and L are molar flow rates of vapor and liquid and m' is the slope of the y–x equilibrium curve (Fig. 1b) in the concentration range of interest. The required total height of the packed section is then obtained from the simple relationship,

$$Z_p = (N_{ov})(H_{ov}) \qquad (48)$$

In order to determine the packed height Z_p it is necessary to obtain a value of the overall number of transfer units N_{ov}; methods for doing this are available for binary systems in any standard text covering distillation (73) and, in a more complex way, for multicomponent systems (81). However, it is simpler to calculate the number of required theoretical stages and make the conversion:

$$N_{ov} = N_t(\ln m'V/L)(m'V/L - 1) \qquad (49)$$

An alternative to determining packed height is through the use of an empirical term, height equivalent to a theoretical plate (HETP). This term can be measured in a fashion similar to that used for the overall plate efficiency of a column (eq. 44):

$$\text{HETP} = \frac{\text{total packed height}}{\text{no. of theoretical plates}} = \frac{Z_p}{N_t} \qquad (50)$$

Typical experimental values of HETP for a random packing such as 50-mm Pall rings, and a structured packing, such as Intalox 2T of Norton Co., under the same system conditions, are shown in Figure 25. Many designers of packed columns prefer the use of HETP instead of H_{ov}, but the latter is more fundamental and discriminates between liquid- and vapor-phase resistances. It should be noted that terms such as H_{ov} and N_{ov} are based on vapor-phase concentrations; equivalent terms based on liquid concentrations could be used.

For structured packings, methods for predicting H_v and H_L are somewhat more reliable than those for random packings. Perhaps the most used efficiency correlation for these packings is that in Reference 82; a slightly different model that covers a broader packing size range is found in Reference 83. Methods for predicting pressure drop and flooding in beds of structured packings have been reviewed (71).

Packed vs Plate Columns. Relative to plate towers, packed towers are more useful for multipurpose distillations, usually in small (under 0.5 m) towers

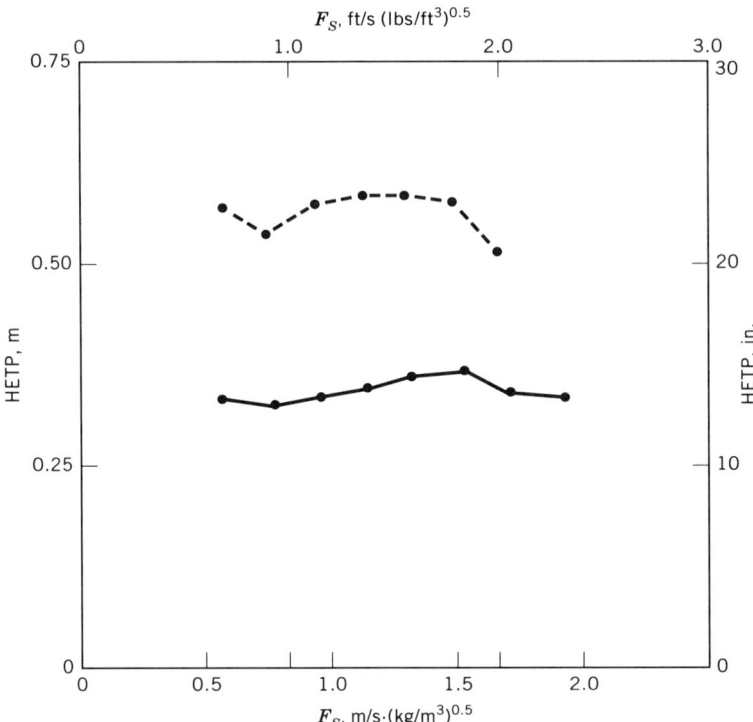

Fig. 25. Values of HETP as a function of throughput for (– – –) 50-mm metal Pall rings and (—) No. 2 structured packing at 12-mm crimp height. Conditions are cyclohexane/n-heptane system, 165 kPa (24 psia) operating pressure, total reflux, 0.43-m diameter column. Courtesy of The University of Texas at Austin.

or for the following specific applications: severe corrosion environment where some corrosion-resistant materials, such as plastics, ceramics, and certain metallics, can easily be fabricated into packing but may be difficult to fabricate into plates; vacuum operation where a low pressure drop per theoretical plate is a critical requirement; high (eg, above 49,000 kg/(h·m²) (~10,000 lb/(h·ft²)) liquid rates; foaming systems; or debottlenecking plate towers having plate spacings that are relatively close, under 0.3 m.

Plate columns have the advantage of lower fabrication cost, less dependence on good liquid and gas distribution, and protection against vapor bypassing the liquid in critical zones, eg, regions of extremely low impurities. Further, methods for the design on plate columns are somewhat more reliable than those for many of the packings, especially those packings of a proprietary nature.

There are notable cases where plate columns have been converted to packed columns to gain advantage of the low pressure drop exacted from the vapor stream. More recently the packings have been largely of the structured type. Illustrative of this is the trend toward the use of structured packing in ethylbenzene–styrene fractionators, some of which have diameters of 10 m or higher.

Steam Distillation

Steam distillation is used to lower the distillation temperatures of high boiling organic compounds that are essentially immiscible with water. If an organic compound is immiscible with water, both liquids exert full vapor pressure upon vaporization from the immiscible two-component liquid. At a system pressure of P, the partial pressures would be:

$$P = p_{\text{water}} + p_{\text{organic}} \tag{51}$$

and because the water and organic compound are immiscible:

$$P = P^0_{\text{water}} + P^0_{\text{organic}} \tag{52}$$

The steam distillation of N-ethylaniline at atmospheric pressure (73) gives the following: the vapor pressures at 99.15°C of water and N-ethylaniline are 98.27 and 3.04 kPa (737 and 22.8 mm Hg), respectively. Thus, according to equation 52,

$$P = 98.27 + 3.04 = 101.3 \text{ kPa} \tag{53}$$

and the concentration of the N-ethylaniline in the vapor is

$$y = 3.04/101.3 = 0.030 \text{ mol fraction} \tag{54}$$

The normal boiling point of N-ethylaniline is 204°C. Therefore, steam distillation makes possible the distillation of N-ethylaniline at atmospheric pressure at a temperature of 99.15°C instead of its normal boiling point of 204°C. Commercial applications of steam distillation include the fractionation of crude tall oil (84), the distilling of turpentine, and certain essential oils. A detailed calculation of steam distillation of turpentine has been reported (85).

Molecular Distillation

Molecular distillation occurs where the vapor path is unobstructed and the condenser is separated from the evaporator by a distance less than the mean-free path of the evaporating molecules (86). This specialized branch of distillation is carried out at extremely low pressures ranging from 13–130 mPa (0.1–1.0 μm Hg). Molecular distillation is confined to applications where it is necessary to minimize component degradation by distilling at the lowest possible temperatures. Commercial usage includes the distillation of vitamins and fatty acid dimers.

Distillation as a Separation Method

Distillation is the most important industrial method of separation and purification of liquid components. Liquid separation methods in less common use include liquid–liquid extraction (see EXTRACTION, LIQUID–LIQUID), membrane diffusion (see DIALYSIS; MEMBRANE TECHNOLOGY), ion exchange (qv), and adsorption (qv). However, distillation does not require a mass separating agent such as a solvent, adsorbent, or membrane, and distillation utilizes energy in a convenient heating medium (often steam). Also, a wealth of experience with design and operations makes distillation column performance prediction more reliable than equivalent predictions for other methods. At times distillation also competes indirectly with methods involving solid–liquid separations such as crystallization (qv). An extensive discussion of the selection of alternative separation methods is available (30) (see SEPARATIONS PROCESS SYNTHESIS).

The suitability and economics of a distillation separation depend on such factors as favorable vapor–liquid equilibria, feed composition, number of components to be separated, product purity requirements, the absolute pressure of the distillation, heat sensitivity, corrosivity, and continuous vs batch requirements. Distillation is somewhat energy-inefficient because in the usual case heat added at the base of the column is largely rejected overhead to an ambient sink. However, the source of energy for distillations is often low pressure steam which characteristically is in long supply and thus relatively inexpensive. Also, schemes have been devised for lowering the energy requirements of distillation and are described in many publications (87).

Favorable Vapor–Liquid Equilibria. The suitability of distillation as a separation method is strongly dependent on favorable vapor–liquid equilibria. The absolute value of the key relative volatilities directly determines the ease and economics of a distillation. The energy requirements and the number of plates required for any given separation increase rapidly as the relative volatility becomes lower and approaches unity. For example: given an ideal binary mixture having a 50 mol % feed and a distillate and bottoms requirement of 99.8% purity each, the minimum reflux and minimum number of theoretical plates for assumed relative volatilities of 1.1, 1.5, and 4, are

Relative volatility	Minimum reflux ratio	Minimum no. of theoretical plates
1.1	20	130
1.5	4.0	31
4	0.66	9

In the example, the minimum reflux ratio and minimum number of theoretical plates decreased 14- to 33-fold, respectively, when the relative volatility increased from 1.1 to 4. Other distillation systems would have different specific reflux ratios and numbers of theoretical plates, but the trend would be the same. As the relative volatility approaches unity, distillation separations rapidly become more costly in terms of both capital and operating costs. The relative volatility can sometimes be improved through the use of an extraneous solvent that modifies the VLE. Binary azeotropic systems are impossible to separate into pure components in a single column, but the azeotrope can often be broken by an extraneous entrainer (see DISTILLATION, AZEOTROPIC AND EXTRACTIVE).

Feed Composition. Feed composition has a substantial effect on the economics of a distillation. Distillations tend to become uneconomical as the feed becomes dilute. There are two types of dilute feed cases, one in which the valuable recovered component is a low boiler and the second when it is a high boiler. When the recovered component is the low boiler, the absolute distillate rate is low but the reflux ratio and the number of plates is high. An example is the recovery of methanol from a dilute solution in water. When the valuable recovered component is a high boiler, the distillate rate, the reflux relative to the high boiler, and the number of plates all are high. An example for this case is the recovery of acetic acid from a dilute solution in water. For the general case of dilute feeds, alternative recovery methods are usually more economical than distillation.

Product Purity. Product purity requirements influence choice of separation methods. For favorable equilibria, distillation energy requirements do not increase significantly as purity specifications become tighter. For example, in an ideal binary distillation of 60 mol % of A in the feed, the minimum and operating reflux ratios would be essentially the same whether the required purity of A was 99 or 99.9999%. The number of plates would increase substantially, however, as the purity requirements became more stringent. The shortcut methods of calculating minimum reflux ratio, minimum number of plates, operating reflux ratio, and number of operating plates allow a rapid evaluation of the effect of changes in purity requirements on the key economic factors in distillation.

Operating Pressure. The absolute pressure of the distillation may have substantial economic impact. The temperatures at which heat is supplied to the reboiler and removed from the condenser determines the unit cost of the energy. The cost of removing heat in the condenser increases rapidly as the condensing temperature drops below the range of air or water cooling capability, eg, the cost of removing a unit quantity of heat at $-25°C$ may be one hundred times as high as removing it at 100°C. Similarly, the cost of the energy required for the reboiler increases rapidly as the boiler temperature increases above some level determined by local conditions. For example, at a particular site low pressure waste steam at 110°C may be essentially without cost, but if a temperature level of 200°C is required, the unit cost of the heat is much higher. The relative cost of the heat being removed and supplied is the controlling factor determining the design of some distillations. The use of multiple interstage reboilers and condensers at different energy levels as well as the use of other operational modes used to optimize the overall economics, has been discussed (88).

The absolute pressure may have a significant effect on the vapor–liquid equilibrium. Generally, the lower the absolute pressure the more favorable the equilibrium. This effect has been discussed for the styrene–ethylbenzene system (30). In a given column, increasing the pressure can increase the column capacity by increasing the capacity parameter (see eqs. 42 and 43). Selection of the economic pressure can be facilitated by guidelines (89) that take into consideration the pressure effects on capacity and relative volatility. Low pressures are required for distillation involving heat-sensitive material.

Heat Sensitivity. The heat sensitivity or polymerization tendencies of the materials being distilled influence the economics of distillation. Many materials cannot be distilled at their atmospheric boiling points because of high thermal degradation, polymerization, or other unfavorable reaction effects that are functions of temperature. These systems are distilled under vacuum in order to lower operating temperatures. For such systems, the pressure drop per theoretical stage is frequently the controlling factor in contactor selection. An excellent discussion of equipment requirements and characteristics of vacuum distillation may be found in Reference 90.

Corrosivity. Corrosivity is an important factor in the economics of distillation. Corrosion rates increase rapidly with temperature, and in distillation the separation is made at boiling temperatures. The boiling temperatures may require distillation equipment of expensive materials of construction; however, some of these corrosion-resistant materials are difficult to fabricate. For some materials, eg, ceramics, random packings may be specified, and this has been a classical application of packings for highly corrosive services. On the other hand, the extensive surface areas of metal packings may make these more susceptible to corrosion than plates. Again, cost may be the final arbiter.

Batch vs Continuous Distillation. The mode of operation also influences the economics of distillation. Batch distillation is generally limited to small-scale operations where the equipment serves several different distillations.

Research. Much of the research on commercial-size distillation equipment is being done by Fractionation Research, Inc. (FRI), a nonprofit, industry-sponsored, research corporation. The industrial sponsors are fabricators, designers, and constructors, or users of distillation equipment. Publications include liquid mixing on sieve plates (91), bubble cap plate efficiency (92), and sieve plate efficiency (93,94). A motion picture of downcomer performance is also available (95). References 96 and 97 cover the literature from 1967 to 1990.

Equipment Costs

A compilation of costs of distillation and related equipment is available (98). Some of the commercial computer-aided process design packages contain equipment cost information. For specialized internals, such as distributors, support plates, packings, crossflow plates, and so on, it is usually necessary to obtain cost information directly from the equipment vendors. It is important to recognize that the cost of a distillation system includes many components in addition to the column itself. For example, an expensive packing may be justified on the basis that it can reduce the cost of the column shell, foundations, piping, and so on. Discussions of economics of distillation systems are available (99,100) (see ENERGY CONSERVATION IN SEPARATION PROCESSES).

Column Control

Distillation columns are controlled by hand or automatically. The parameters that must be controlled are (1) the overall mass balance, (2) the overall enthalpy balance, and (3) the column operating pressure. Modern control systems are designed to control both the static and dynamic column and system variables. For an in-depth discussion, see References 101–104.

NOMENCLATURE

Symbol	Definition	Units
A_{12}, A_{21}	constants in the Van Laar activity coefficient equation	
B	bottoms from column	mol/s
C_{sb}	vapor capacity parameter	m/s
D	distillate from column	mol/s
E_o	overall column plate efficiency (eq. 44)	fractional
E_{mv}	Murphree plate efficiency (eq. 45)	fractional
E_{og}	local, or point, efficiency based on vapor concentrations	fractional
f	fugacity	kPa
F	feed	mol/s
F^*	F-factor (eq. 43)	m/s·(kg/m^3)$^{0.5}$
F_A^*	F-factor based on active (bubbling) area	m/s·(kg/m^3)$^{0.5}$
F_H^*	F-factor based on hole area	m/s·(kg/m^3)$^{0.5}$
F_p	packing factor from Table 2	1/m
G	gas mass rate	kg/s
$\overline{H}$	enthalpy per mole	
H	enthalpy per unit time	
H^*	Henry's law constant (eq. 17)	kPa/mol fraction
HETP	height equivalent of theoretical plate	m
H_L	height of a liquid-phase transfer unit	m
H_{ov}	height of an overall transfer unit, vapor concentrations	m
H_v	height of a vapor-phase transfer unit	m
K	y^*/x, vapor–liquid equilibrium ratio (eq. 1)	
L	liquid rate	mol/s
$\mathbf{L}$	average liquid rate for section	mol/s
$\overline{L}$	liquid mass rate	kg/s
m	an equilibrium stage below the feed	
m'	slope of equilibrium line	
n	an equilibrium stage above the feed	
N	number of stages	
N^*	number of components	
N_a	number of actual stages	
N_{ov}	number of transfer units, vapor concentration basis	
N_t	number of theoretical stages	
P	total pressure of system	kPa
p	partial pressure	kPa
P^0	vapor pressure	kPa
q	heat to vaporize 1 mol feed divided by molal latent heat of feed (eq. 36)	
$\overline{q}$	heat removed or added at column auxiliaries	

Symbol	Definition	Units
R	reflux	mol/s
R'	gas law constant	
T	temperature	K
v	vapor molar volume	m^3/mol
V	vapor molar rate	mol/s
$\overline{V}$	average molar vapor rate for section	mol/s
V^*	vapor velocity	m/s
x	mole fraction in liquid	
y	mole fraction in vapor	
y^*	mole fraction vapor in equilibrium with x	
z	compressibility factor in gas law	
Z_p	height of packed bed	m
α	relative volatility (eq. 2)	
γ^L	liquid-phase activity coefficient (eq. 6)	
γ^∞	terminal activity coefficient, at infinite dilution	
$\Lambda_{12}, \Lambda_{21}$	constant in Wilson activity coefficient model (eq. 13)	
ρ	fluid-phase density	kg/m^3
ϕ	fugacity coefficient (eq. 20)	

Superscripts

B	bottoms	
C	condenser	
D	distillate	
E	end	
F	feed	
f	feed stage	
L	liquid	
m	stage number m	
$m-1$	stage below m	
$m+1$	stage above m	
n	stage number n	
$n-1$	stage below n	
$n+1$	stage above n	
N	Nth component of components i to n	
P	pressure	
S	reboiler	
T	top column	
V	vapor	

Subscripts

$1,2,3\ldots n$	component numbers	
B	bottoms	
D	distillate	
F	feed	
g	gas	
H	component H of binary system L–H, H is the high boiler	
i,j	components of mixture $1\ldots i,j,\ldots n$	

L	component L of binary system L–H, L is the low boiler
L	liquid
min	minimum
P	pressure
R	rectifying section
S	stripping section
T	temperature

BIBLIOGRAPHY

"Distillation" in the *Encyclopedia of Chemical Technology*, 1st ed., Vol. 5, pp. 156–187, by E. G. Scheibel, Hoffmann-LaRoche, Inc.; in *ECT* 2nd ed., Vol. 7, pp. 204–248, by C. D. Holland and J. D. Lindsey, Texas A&M University; in *ECT* 3rd ed., Vol. 7, pp. 849–891, by E. R. Hafslund, E. I. du Pont de Nemours & Co., Inc.; in *ECT* 4th ed., Vol. 8, pp. 311–358, by J. R. Fair, University of Texas at Austin.

1. A. J. V. Underwood, *Trans. I. Chem. E.* **13,** 34 (1935).
2. J. R. Fair, *AIChE Symp. Ser. No. 235* **79,** 1 (1984).
3. J. Gmehling, U. Onken, and W. Arlt, *Vapor–Liquid Equilibrium Collection* (continuing series), DECHEMA, Frankfurt, Germany, 1979.
4. M. Hirata, S. Ohe, and K. Nagahama, *Computer Aided Data Book of Vapor–Liquid Equilibria*, Elsevier, Amsterdam, The Netherlands, 1975.
5. E. Hala, J. Pick, V. Fried, and O. Vilim, *Vapor–Liquid Equilibrium*, 2nd ed., Pergamon Press, Oxford, UK, 1967.
6. E. Hala, I. Wichterle, J. Polak, and T. Boublik, *Vapor–Liquid Equilibrium at Normal Pressures*, Pergamon Press, Oxford, UK, 1968.
7. I. Wichterle, J. Linek, and E. Hala, *Vapor–Liquid Equilibrium Data Bibliography*, Elsevier, Amsterdam, The Netherlands, 1975.
8. A. Fredenslund, J. Gmehling, and P. Rasmussen, *Vapor–Liquid Equilibria Using UNIFAC*, Elsevier, Amsterdam, The Netherlands, 1977.
9. J. H. Hildebrand, J. M. Prausnitz, and R. L. Scott, *Regular and Related Solutions*, Van Nostrand Reinhold Co., Inc., New York, 1970.
10. E. L. Derr and C. H. Deal, *I. Chem. E. Symp. Ser. No. 32* **3**(40), (1969).
11. D. A. Palmer, *Handbook of Applied Thermodynamics*, CRC Press, Inc., Boca Raton, Fla., 1987.
12. J. M. Prausnitz, R. N. Lichtenthaler, and E. G. Azeredo, *Molecular Thermodynamics of Fluid-Phase Equilibria*, 2nd ed., Prentice-Hall, Inc., Englewood Cliffs, N.J., 1986.
13. R. C. Reid, J. M. Prausnitz, and B. Pohling, *The Properties of Gases and Liquids*, 4th ed., McGraw-Hill Book Co., Inc., New York, 1987.
14. S. M. Walas, *Phase Equilibria in Chemical Engineering*, Butterworths, Reading, Mass., 1985.
15. E. J. Henley and J. D. Seader, *Equilibrium-Stage Separation Operations in Chemical Engineering*, John Wiley & Sons, Inc., New York, 1981.
16. P. Wankat, *Equilibrium-Staged Separations*, Elsevier Science Publishing Co., Inc., New York, 1988.
17. M. Van Winkle, *Distillation*, McGraw-Hill Book Co., Inc., New York, 1967.
18. J. J. Van Laar, *Z. Physik. Chem.* **72,** 723 (1910); **83,** 599 (1913).
19. G. M. Wilson, *J. Am. Chem. Soc.* **86,** 127 (1964).
20. H. Renon and J. M. Prausnitz, *AIChE J.* **14,** 135 (1968).
21. D. S. Abrams and J. M. Prausnitz, *AIChE J.* **21,** 116 (1975).

22. Margules, *Sitzber. Math.-Naturw. Kl. Kaiserlichen Akad. Wiss. (Vienna)* **104,** 1243 (1895).
23. H. H. Chien and H. R. Null, *AIChE J.* **18,** 1177 (1972).
24. D. Tiegs, J. Gmehling, P. Rasmussen, and A. Fredenslund, *Ind. Eng. Chem. Res.* **26,** 159 (1987).
25. H. K. Hansen, P. Rasmussen, A. Fredenslund, M. Schiller, and J. Gmehling, *Ind. Eng. Chem. Res.* **30,** 2352 (1991).
26. J. M. Prausnitz and co-workers, *Computer Calculations for Multicomponent Vapor–Liquid and Liquid–Liquid Equilibria*, Prentice-Hall, Inc., Englewood Cliffs, N.J., 1980.
27. *Technical Data Book, Petroleum Refining*, 3rd ed., Vols. I and II, American Petroleum Institute, New York, 1976.
28. L. Horsley, *Azeotropic Data—III*, Advances in Chemistry Series No. 116, American Chemical Society, Washington, D.C., 1973.
29. J. F. Boston, H. I. Britt, S. Jiraphongphan, and V. B. Shah, "An Advanced System for the Simulation of Batch Distillation Operations," in *Foundations of Computer-Aided Chemical Process Design*, Vol. 2, American Institute of Chemical Engineers, New York, 1981.
30. C. J. King, *Separation Processes*, 2nd ed., McGraw-Hill Book Co., Inc., New York, 1980.
31. W. L. McCabe and E. W. Thiele, *Ind. Eng. Chem.* **17,** 605 (1925).
32. M. Ponchon, *Tech. Mod.* **13,** 20, 55 (1921).
33. R. Savarit, *Arts Metiers* **65,** 145, 178, 266, 307 (1922).
34. M. R. Fenske, *Ind. Eng. Chem.* **24,** 482 (1932).
35. A. J. V. Underwood, *Chem. Eng. Progr.* **44,** 603 (1948).
36. J. R. Fair and W. L. Bolles, *Chem. Eng.* **75**(9), 156 (Apr. 22, 1968).
37. G. G. Brown and co-workers, *Unit Operations*, John Wiley & Sons, Inc., New York, 1950.
38. E. R. Gilliland, *Ind. Eng. Chem.* **32,** 918 (1940).
39. J. H. Erbar and R. N. Maddox, *Petrol. Ref.* **40**(5), 183 (1961).
40. H. E. Eduljee, *Hydrocarbon Proc.* **54**(9), 120 (1975).
41. E. W. Thiele and R. L. Geddes, *Ind. Eng. Chem.* **25,** 290 (1933).
42. C. D. Holland, *Fundamentals of Multicomponent Distillation*, McGraw-Hill Book Co., Inc., New York, 1981.
43. R. N. S. Rathore, K. A. Van Wormer, and G. J. Powers, *AIChE J.* **20,** 491 (1974).
44. D. C. Freshwater and B. D. Henry, *Chem. Eng. (London)* (301), 533 (1975).
45. J. E. Hendry, D. F. Rudd, and J. D. Seader, *AIChE J.* **19,** 1 (1973).
46. *Ballast Tray Design Manual*, Bulletin 4900, Glitsch, Inc., Dallas, Tex., 1974.
47. *Flexitray Design Manual*, Bulletin 960, Koch Engineering Co., Wichita, Kans., 1960.
48. *Float Valve Tray Design Manual*, Nutter Engineering Co., Tulsa, Okla., 1976.
49. W. L. Bolles, in B. D. Smith, ed., *Design of Equilibrium Stage Processes*, McGraw-Hill Book Co., Inc., New York, 1963, Chapt. 14.
50. M. H. Hutchinson and R. F. Baddour, *Chem. Eng. Progr.* **52**(12), 503 (1956).
51. F. Kastanek, M. V. Huml, and V. Braun, *I. Chem. E. Symp. Ser. No. 32*, **5**(100), (1969).
52. W. V. Delnicki and J. L. Wagner, *Chem. Eng. Progr.* **52**(1), 28 (1956).
53. J. R. Fair, in Ref. 49, Chapt. 15.
54. J. R. Fair, in R. H. Perry and D. Green, eds., *Perry's Chemical Engineers' Handbook*, 6th ed., McGraw-Hill Book Co., Inc., New York, 1984, section 18.
55. E. V. Murphree, *Ind. Eng. Chem.* **17,** 747, 960 (1925).
56. *Bubble-Tray Design Manual*, American Institute of Chemical Engineers (AIChE), New York, 1958.
57. M. J. Lockett, *Distillation Tray Fundamentals*, Cambridge University Press, Cambridge, Mass., 1986.

58. M. M. Dribika and M. W. Biddulph, *Trans. I. Chem. E.* **70,** Part A, 142 (1992).
59. K. E. Porter, M. J. Lockett, and C. T. Lim, *Trans. I. Chem. E.* **50,** 91 (1972).
60. W. V. Pinczewski, N. D. Benke, and C. J. D. Fell, *AIChE J.* **21,** 1210 (1975).
61. K. E. Porter, A. Safekouri, and M. J. Lockett, *Trans. I. Chem. E.* **51,** 265 (1973).
62. M. Prado, K. L. Johnson, and J. R. Fair, *Chem. Eng. Progr.* **83**(3), 32 (1987).
63. M. Prado and J. R. Fair, *Ind. Eng. Chem. Res.* **29,** 1031 (1990).
64. H. E. O'Connell, *Trans. AIChE* **42,** 741 (1946).
65. G. E. English and M. Van Winkle, *Chem. Eng.* **70**(23), 241 (1963).
66. H. L. Toor and J. K. Burchard, *AIChE J.* **6,** 202 (1960).
67. R. Krishna, H. F. Martinez, R. Sreedhar, and G. L. Standart *Trans. I. Chem. E.* **55,** 178 (1977).
68. A. Sperandio, M. Richard, and M. Huber, *Chem.-Ing.-Tech.* **37,** 22 (1965).
69. W. D. Stoecker and B. Weinstein, *Chem. Eng. Progr.* **73**(11), 71 (1977).
70. R. F. Strigle, *Random Packings and Packed Tower Design*, Gulf Publishing, Houston, Tex., 1987.
71. J. R. Fair and J. L. Bravo, *Chem. Eng. Progr.* **86**(1), 19 (1990).
72. P. J. Hoek, J. A. Wesselingh, and F. J. Zuiderweg, *Chem. Eng. Res. Des.* **64,** 431 (1986).
73. R. E. Treybal, *Mass Transfer Operations*, 3rd ed., McGraw-Hill Book Co., Inc., New York, 1980.
74. W. S. Norman, *Absorption, Distillation and Cooling Towers*, John Wiley & Sons, Inc., New York, 1961.
75. P. A. Schweitzer, ed., *Handbook of Separation Techniques for Chemical Engineers*, 2nd. ed., McGraw-Hill Book Co., Inc., New York, 1988, Chapts. 1.1–1.8.
76. R. W. Rousseau, ed., *Handbook of Separation Process Technology*, John Wiley & Sons, Inc., New York, 1987, Chapt. 5.
77. T. K. Sherwood, G. H. Shipley, and F. A. L. Holloway, *Ind. Eng. Chem.* **30,** 765 (1938).
78. M. Leva, *Chem. Eng. Progr. Symp. Ser. No. 10* **50,** 51 (1954).
79. W. L. Bolles and J. R. Fair, *Chem. Eng.* **89**(14), 109 (July 12, 1982).
80. J. L. Bravo and J. R. Fair, *Ind. Eng. Chem. Proc. Des. Dev.* **21,** 162 (1982).
81. R. Krishnamurthy and R. Taylor, *AIChE J.* **31,** 449, 456 (1985).
82. J. L. Bravo, J. A. Rocha, and J. R. Fair, *Hydrocarbon Proc.* **64**(1), 91 (1985).
83. L. Spiegel and W. Meier, *I. Chem. E. Symp. Ser. No. 104*, A203 (1987).
84. J. Drew and M. Propst, eds., *Tall Oil*, Pulp Chemicals Association, New York, 1981.
85. W. L. McCabe and J. C. Smith, *Unit Operations of Chemical Engineering*, 3rd ed., McGraw-Hill Book Co., Inc., New York, 1976, Chapt. 19.
86. K. C. D. Hickman, in R. H. Perry and C. H. Chilton, eds., *Chemical Engineers' Handbook*, 5th ed., McGraw-Hill Book Co., Inc., New York, 1973, section 13.
87. J. R. Fair, in Y. A. Liu, H. A. McGee, and W. R. Epperly, eds., *Recent Developments in Chemical Process and Plant Design*, John Wiley & Sons, Inc., New York, 1987, Chapt. 3.
88. W. C. Petterson and T. A. Wells, *Chem. Eng.* **84**(20), 79 (1977).
89. H. Z. Kister and I. D. Doig, *Hydrocarbon Proc.* **56**(7), 132 (1977).
90. P. G. Nygren and G. K. S. Connolly, *Chem. Eng. Progr.* **67**(3), 49 (1971).
91. T. Yanagi and B. D. Scott, *Chem. Eng. Progr.* **69**(10), 75 (1973).
92. B. D. Scott and H. S. Myers, in Ref. 91, p. 73.
93. M. Sakata and T. Yanagi, *I. Chem. E. Symp. Ser. No. 56*, **3.2**(21), (1979).
94. T. Yanagi and M. Sakata, *Ind. Eng. Chem. Proc. Des. Dev.* **21,** 712 (1982).
95. T. Yanagi, *Performance of Downcomers in Distillation Columns* (motion picture), AIChE Meeting, Atlanta, Ga., Feb. 1970. Available from Fractionation Research, Inc., Stillwater, Okla.
96. M. S. Ray, *Chemical Engineering Bibliography, 1967–1988*, Noyes Publications, Park Ridge, N.J., 1990.
97. M. S. Ray, *Sepn. Sci. Technol.* **27,** 105 (1992).

98. M. Peters and K. D. Timmerhaus, *Plant Design and Economics for Chemical Engineers*, 4th ed., McGraw-Hill Book Co., Inc., New York, 1991.
99. H. Z. Kister, *Distillation–Operation*, McGraw-Hill Book Co., Inc., New York, 1990.
100. H. Z. Kister, *Distillation—Design*, McGraw-Hill Book Co., Inc., New York, 1992.
101. A. E. Nisenfeld and R. C. Seeman, *Distillation Columns*, Instrument Society of America, Research Triangle Park, N.C., 1981.
102. F. G. Shinskey, *Distillation Control*, 2nd ed., McGraw-Hill Book Co., Inc., New York, 1984.
103. P. B. Deshpande, *Distillation Dynamics and Control*, Instrument Society of America, Research Triangle Park, N.C., 1984.
104. P. S. Buckley, W. L. Luyben, and J. P. Shunta, *Design of Distillation Control Systems*, Instrument Society of America, Research Triangle Park, N.C., 1985.

General References

Fluid Phase Equilibria.
J. Chem. Eng. Data.

JAMES R. FAIR
The University of Texas at Austin

DISTILLATION, AZEOTROPIC AND EXTRACTIVE

Distillation (qv) is the most widely used separation technique in the chemical and petroleum industries. Not all liquid mixtures are amenable to ordinary fractional distillation, however. Close-boiling and low relative volatility mixtures are difficult and often uneconomical to distill, and azeotropic mixtures are impossible to separate by ordinary distillation. Yet such mixtures are quite common (1) and many industrial processes depend on efficient methods for their separation (see also SEPARATIONS PROCESS SYNTHESIS). This article describes special distillation techniques for economically separating low relative volatility and azeotropic mixtures. mixtures.

Whereas there is extensive literature on design methods for azeotropic and extractive distillation, much less has been published on operability and control. It is, however, widely recognized that azeotropic distillation columns are difficult to operate and control because these columns exhibit complex dynamic behavior and parametric sensitivity (2–11). In contrast, extractive distillations do not exhibit such complex behavior and even highly optimized columns are no more difficult to control than ordinary distillation columns producing high purity products (12).

At low to moderate pressures the vapor–liquid phase equilibrium (VLE) of many mixtures can be adequately described by:

$$y_i P = x_i \gamma_i P_i^{\text{sat}} \qquad \text{for } i = 1, ..., c \qquad (1)$$

where y_i is the mole fraction of component i in the vapor phase; x_i is the mole fraction of component i in the liquid phase; P is the system pressure; P_i^{sat} is the vapor pressure of pure component i; c is the number of components in the mixture; and γ_i is the liquid-phase activity coefficient of component i. The activity coefficient is a measure of the nonideality of a mixture and changes both with temperature and composition. When $\gamma_i = 1$ the mixture is said to be ideal and equation 1 simplifies to Raoult's Law. Nonideal mixtures ($\gamma_i \neq 1$) can exhibit either positive ($\gamma_i > 1$) or negative ($\gamma_i < 1$) deviations from Raoult's Law. In many highly nonideal mixtures these deviations become so large that the pressure-composition (P-x,y) and temperature-composition (T-x,y) phase diagrams exhibit a minimum or maximum point (Fig. 1). At these minima and maxima the liquid phase and its equilibrium vapor phase have the same composition, ie,

$$x_i = y_i \qquad \text{for } i = 1, ..., c \qquad (2)$$

and the dew-point (vapor) and bubble-point (liquid) curves are tangent with zero slope. These are the defining conditions for a homogeneous azeotrope where a single liquid phase is in equilibrium with a vapor phase. Mixtures that form two liquid phases are capable of forming heterogeneous azeotropes where the overall liquid composition is identical to the vapor composition, but the vapor and liquid surfaces are not tangent with zero slope. A maximum boiling azeotrope (Fig. 1b) is equivalent to a minimum pressure azeotrope (Fig. 1a) and a minimum boiling azeotrope (Fig. 1d) is also a maximum pressure azeotrope (Fig. 1c). The majority of the known azeotropes are minimum boiling.

Separation by distillation is dependent on the fact that when a liquid is partially vaporized the vapor and liquid compositions differ. The vapor phase becomes enriched in the more volatile components and depleted in the less volatile components with respect to its equilibrium liquid phase. By segregating the phases and repeating the partial vaporization, it is often possible to achieve the desired degree of separation. One measure of the degree of enrichment or the ease of separation is the relative volatility defined as:

$$\alpha_{ij} = \frac{y_i x_j}{x_i y_j} = \frac{\gamma_i P_i^{\text{sat}}}{\gamma_j P_j^{\text{sat}}} \qquad (3)$$

The relative volatility of most mixtures changes with temperature, pressure, and composition. The larger the value of α_{ij}, the easier it is to separate component i from component j. From equation 2, at a c-component, ie, binary, ternary, etc, homogeneous azeotrope, $x_i = y_i$ for all c components in the mixture. Therefore $\alpha_{ij} = 1$ for all components i and j and it is impossible to further enrich the vapor. Thus homogeneous azeotropes can never be separated into pure components by ordinary fractional distillation. Similarly, any mixture, be it ideal, nonideal, close-

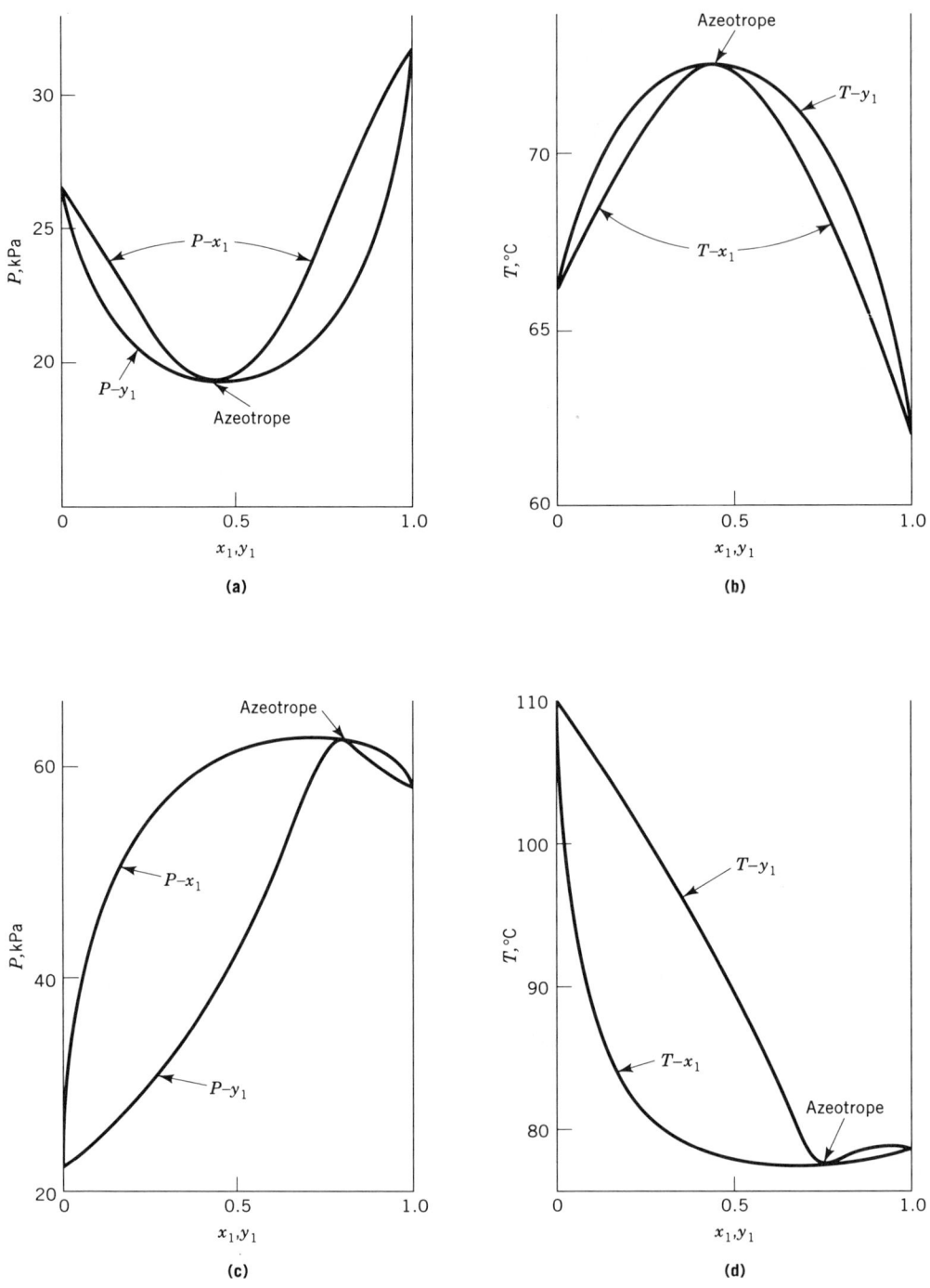

Fig. 1. P-x,y and T-x,y phase diagrams showing maximum and minimum azeotropes. (**a**) Chloroform (1)-tetrahydrofuran (2) at 30°C; (**b**) chloroform (1)-tetrahydrofuran (2) at 101 kPa; (**c**) ethanol (1)-toluene (2) at 65°C; and (**d**) ethanol (1)-toluene (2) at 101 kPa (13). To convert kPa to atm, multiply by 9.87×10^{-3}.

DISTILLATION, AZEOTROPIC AND EXTRACTIVE

boiling, or isomeric, where the relative volatilities are close to unity is difficult to separate by ordinary distillation because little enrichment occurs with each partial vaporization step.

Most methods for distilling azeotropic and low relative volatility mixtures rely on the addition of specially chosen chemicals to facilitate the separation. These separating agents can be divided into distinct classes which define the principal distillation techniques used to separate mixtures containing azeotropes. The five methods for separating azeotropic mixtures are (*1*) extractive distillation and homogeneous azeotropic distillation where the liquid separating agent is completely miscible. For extractive distillation, separating agents are variously known as solvents, extractive agents, entrainers, or extractants. (*2*) Heterogeneous azeotropic distillation or, more commonly, azeotropic distillation where the liquid separating agent, called the entrainer, forms one or more azeotropes with the other components in the mixture and causes two liquid phases to exist over a broad range of compositions. This immiscibility is the key to making the distillation sequence work. (*3*) Distillation in the presence of ionic salts. The salt dissociates in the liquid mixture and alters the relative volatilities sufficiently that the separation becomes possible. (*4*) Pressure-swing distillation where a series of columns operating at different pressures are used to separate binary azeotropes which change appreciably in composition over a moderate pressure range or where a separating agent which forms a pressure-sensitive azeotrope is added to separate a pressure-insensitive azeotrope. (*5*) Reactive distillation where the separating agent reacts preferentially and reversibly with one of the azeotropic constituents. The reaction product is then distilled from the nonreacting components and the reaction is reversed to recover the initial component.

Of these five methods all but pressure-swing distillation can also be used to separate low volatility mixtures and all but reactive distillation are discussed herein. It is also possible to combine distillation and other separation techniques such as liquid–liquid extraction (see EXTRACTION, LIQUID–LIQUID), adsorption (qv), melt crystallization (qv), or pervaporation to complete the separation of azeotropic mixtures.

Residue Curve Maps

The most basic form of distillation, called simple distillation, is a process in which a multicomponent liquid mixture is slowly boiled in an open pot and the vapors are continuously removed as they form. At any instant in time the vapor is in equilibrium with the liquid remaining in the still. Because the vapor is always richer in the more volatile components than the liquid, the liquid composition changes continuously with time, becoming more and more concentrated in the least volatile species. A simple distillation residue curve is a graph showing how the composition of the liquid residue in the pot changes over time. A residue curve map is a collection of residue curves originating from different initial compositions. Residue curve maps contain the same information as phase diagrams, but residue curve maps represent this information in a way that is more useful for understanding how to synthesize a distillation sequence to separate a mixture. The liquid composition profiles in a continuous packed or staged distillation column operating at infinite reflux and reboil are closely approximated by simple distillation residue curves.

Residue curves can only originate from, terminate at, or be deflected by the pure components and azeotropes in a mixture. Pure components and azeotropes that residue curves move away from are called unstable nodes (UN), those where residue curves terminate are called stable nodes (SN), and those that deflect residue curves are called saddles (S).

The simplest residue curve map for a ternary mixture is shown in Figure 2. All ternary nonazeotropic mixtures, including ideal and constant volatility mixtures, are represented qualitatively by this map. All of the residue curves originate at the light (lowest boiling) pure component, move toward the intermediate boiling component, and end at the heavy (highest boiling) pure component. Thus the residue curves point in the direction of increasing temperature. In fact, residue curves must always move in such a way that the boiling temperature of the mixture continuously increases along every curve. From this property and the direction of the arrows in Figure 2, the light component is an unstable node; the intermediate component, which deflects the residue curves, is a saddle; and the heavy component is a stable node. A detailed mathematical treatment of simple distillation residue curve maps can be found in the literature (14,15).

Many different residue curve maps are possible when azeotropes are present. For example, for ternary mixtures containing only one azeotrope there are six possible residue curve maps that differ by the binary pair forming the azeotrope and by whether the azeotrope is minimum or maximum boiling. Figure 3 represents the case where the intermediate and heaviest components form a minimum boiling binary azeotrope. Pure component D is an unstable node, pure components A and B are stable nodes, the minimum boiling binary azeotrope C is a saddle, and the boiling point order from low to high is $D \longrightarrow C \longrightarrow A$ or B. The residue

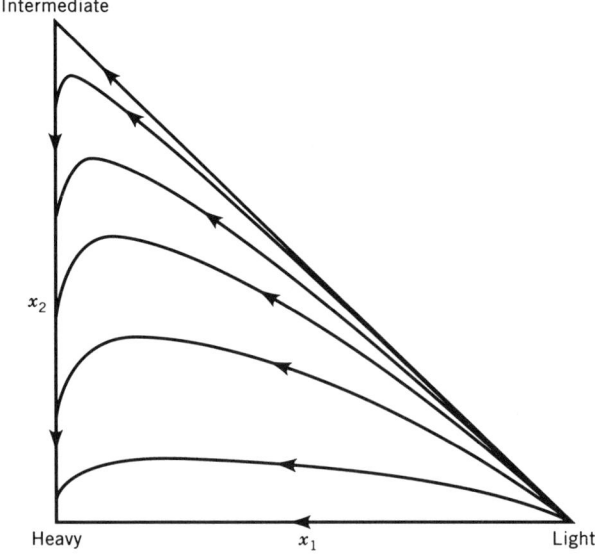

Fig. 2. Residue curve map for a ternary nonazeotropic mixture.

curve connecting component D to the azeotrope C has the special property that it divides the composition triangle into two separate distillation regions. Any initial still-pot composition lying to the left of the curve D–C results in the last drop of liquid being pure B; any initial still-pot composition lying to the right of the curve D–C yields pure A as the last drop of liquid. Residue curves like D–C that divide the composition space into different distillation regions are called simple distillation boundaries or infinite reflux boundaries, often referred to simply as distillation boundaries. There must be a saddle on at least one end of every distillation boundary and each distillation region must contain a stable node, an unstable node, and at least one saddle.

Residue curve maps would be of limited usefulness if they could only be generated experimentally. Fortunately that is not the case. The simple distillation process can be described (14) by the set of equations:

$$\frac{dx_i}{d\xi} = x_i - y_i \qquad \text{for } i = 1, ..., c \qquad (4)$$

where x_i and y_i are the liquid and vapor mole fractions of component i, respectively, ξ is a nonlinear time scale, and c is the number of components in the mixture. Given a method for calculating the vapor–liquid phase equilibrium (VLE) for the mixture of interest (information that is required before a process simulator can be used to design a distillation column), equation 4 can be numerically integrated forward and backward in time from a number of initial conditions to generate a residue curve map. An alternative method for sketching residue curve maps that requires only knowledge of the boiling points of the pure com-

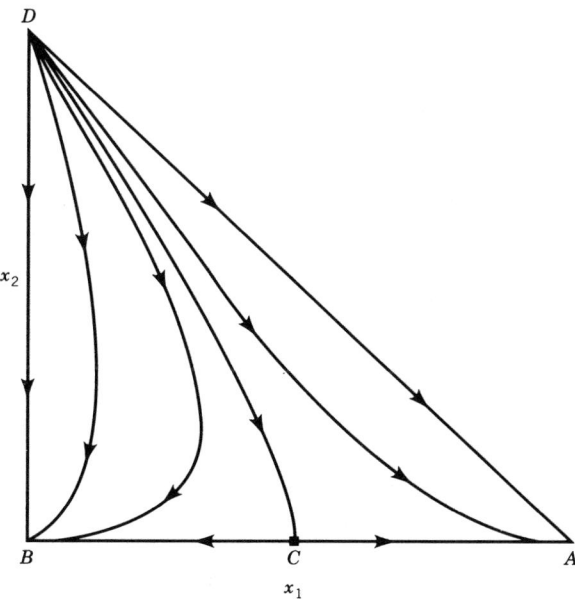

Fig. 3. Residue curve map for a ternary mixture with a distillation boundary running from pure component D to the binary azeotrope C.

ponents and azeotropes in the mixture has also been published (16–18). This latter method is particularly well suited to mixtures for which detailed VLE data is lacking and thus can be useful in entrainer/solvent screening.

Even though the simple distillation process has no practical use as a method for separating mixtures, simple distillation residue curve maps have extremely useful applications. These maps can be used to test the consistency of experimental azeotropic data (16,17,19); to predict the order and content of the cuts in batch distillation (20–22); and, in continuous distillation, to determine whether a given mixture is separable by distillation, identify feasible entrainers/solvents, predict the attainable product compositions, qualitatively predict the composition profile shape, and synthesize the corresponding distillation sequences (16,23–30). By identifying the limited separations achievable by distillation, residue curve maps are also useful in synthesizing separation sequences combining distillation with other methods.

Residue curve maps exist for mixtures having more than three components but cannot be visualized when there are more than four components. However, many mixtures of industrial importance contain only three or four key components and can thus be treated as pseudo-ternary or quaternary mixtures. Quaternary residue curve maps are more complicated than their ternary counterparts but it is still possible to understand these maps using the boiling point temperatures of the pure components and azeotropes (31).

Homogeneous Azeotropic Distillation

The most general definition of homogeneous azeotropic distillation is the separation of any single liquid-phase mixture containing one or more azeotropes into the desired pure component or azeotropic products by continuous distillation. Thus, in addition to azeotropic mixtures which require the addition of a miscible separating agent in order to be separated, homogeneous azeotropic distillation also includes self-entrained mixtures that can be separated without the addition of a separating agent.

The first step in the synthesis of a homogeneous azeotropic distillation sequence is to determine the separation objective. For example, sometimes it is desirable to recover all of the constituents in the mixture as pure components, other times it is sufficient to recover only some of the pure components as products. In other cases an azeotrope may be the desired product. Not every objective is attainable and those that are feasible may require different distillation sequences.

The second step is to sketch the residue curve map for the mixture to be separated. The residue curve map allows one to determine whether the goal can be reached and if so how to reach it, or whether the goal needs to be redefined. The addition of a separating agent to meet a separation objective carries with it the additional responsibility of finding an effective method for its recovery for reuse.

Distillation boundaries for continuous distillation are approximated by simple distillation boundaries. This is a very good approximation for mixtures with nearly linear simple distillation boundaries. Although curved simple distillation boundaries can be crossed to some degree (16,25–30,32,33), the resulting distillation sequences are not normally economical. Mixtures such as nitric acid–water–sulfuric acid, that have extremely curved boundaries, are exceptions. Therefore, a good working assumption is that simple distillation boundaries should not be crossed by continuous distillation. In other words, for a separation to be feasible by distillation it is sufficient that the distillate and bottoms compositions lie in the same distillation region.

An overall material balance for a continuously operated distillation column requires that the feed, distillate, and bottoms compositions lie on a straight line in the composition space (composition triangle for ternary mixtures). Thus feasible distillation sequences for separating homogeneous azeotropic mixtures can first be identified by noting whether the desired products lie in the same distillation region and then can be synthesized by superimposing material balance lines onto simple distillation residue curve maps. When determining the column products for the sequence: (*1*) the distillate composition must have a lower boiling temperature than the bottoms composition, however the component with the lowest (highest) boiling point is not necessarily removed as the distillate (bottoms); (*2*) pure components and azeotropes which are nodes on the residue curve map are easier to obtain as pure products than saddles; and (*3*) double-feed columns are almost always required to obtain saddles as the product, eg, extractive distillations.

As an example, consider the residue curve map for the nonazeotropic mixture shown in Figure 2. It has no distillation boundary so the mixture can be separated into pure components by either the direct or indirect sequence (Fig. 4). In the direct sequence the unstable node (light component, L) is taken overhead in the first column and the bottom stream is essentially a binary mixture of the intermediate, I, and heavy, H, components. In the binary I–H mixture, I has the lowest boiling temperature (an unstable node) so it is recovered as the distillate in the second column and the stable node, H, is the corresponding bottoms stream. The indirect sequence removes the stable node (heavy component) from the bottom of the first column and the overhead stream is an essentially binary L–I mixture. Then in the second column the unstable node, L, is taken overhead and I is recovered in the bottoms.

A second example is the azeotropic mixture of acetone, water, and 2-propanol that arises in the production of acetone by 2-propanol dehydrogenation. The goal is to recover the acetone product, recycle the unreacted 2-propanol to the reactor, and discard the wash water. From the residue curve map (Fig. 5a), it is clear that it is impractical to obtain the three pure components as in the previous example because of the distillation boundary running from the acetone vertex to the 2-propanol–water azeotrope which divides the triangle into two distillation regions. However, by modifying the objective slightly and recycling the 2-propanol–water azeotrope to the reactor instead of pure 2-propanol, a feasible separation sequence is possible (Figs. 5**b** and **c**). The acetone (unstable node) is taken overhead in the first column leaving an essentially binary 2-propanol–water mixture to be distilled in the second column. In this binary mixture the azeotrope is an unstable node so it becomes the distillate and the stable node, water, is removed from the bottom of the second column.

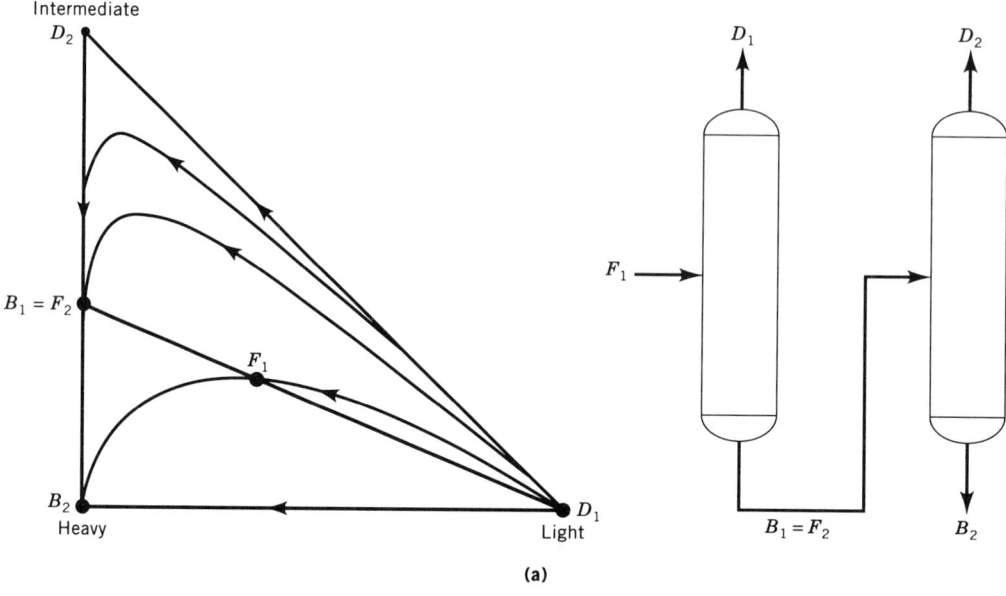

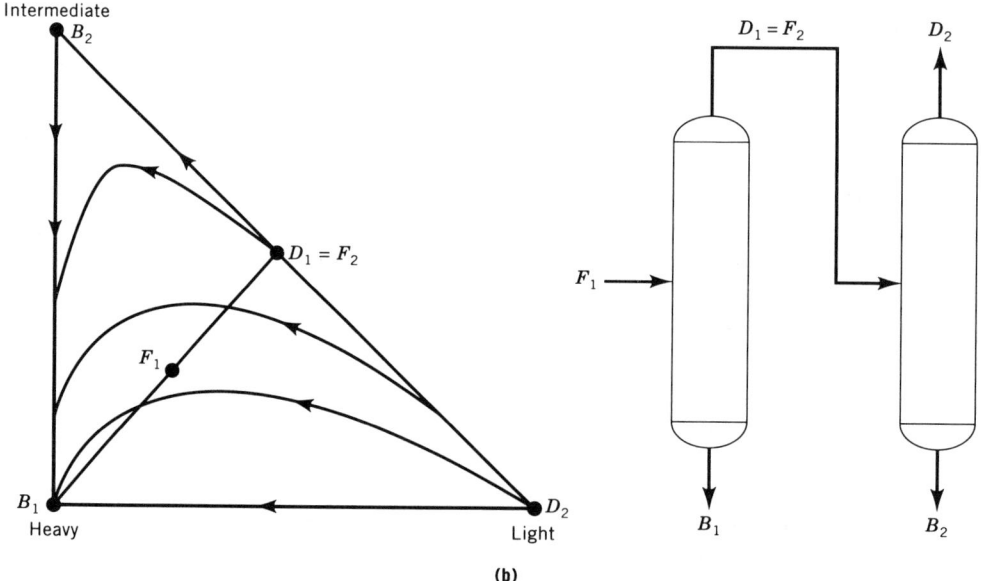

Fig. 4. Column sequences and material balance lines for the (**a**) direct and (**b**) indirect sequences for separating nonazeotropic mixtures.

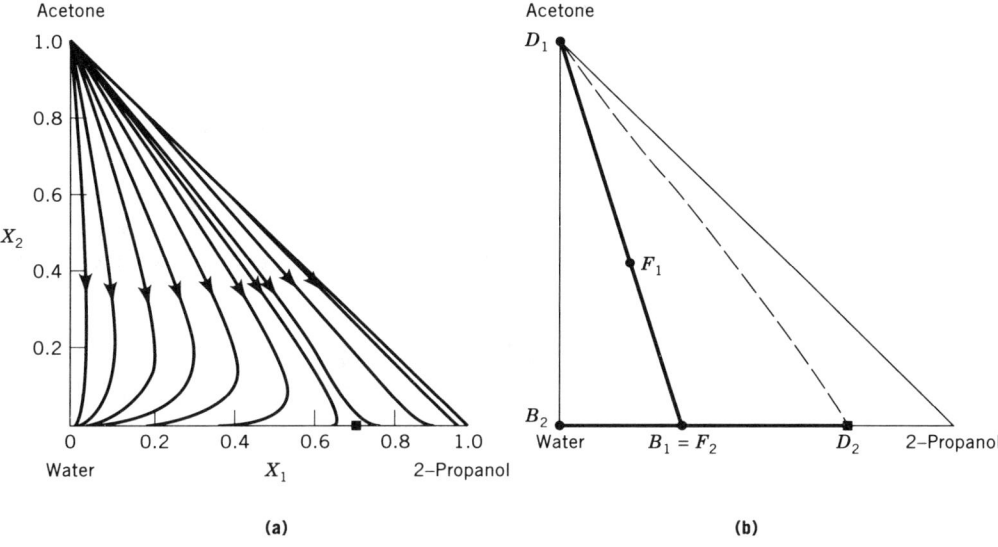

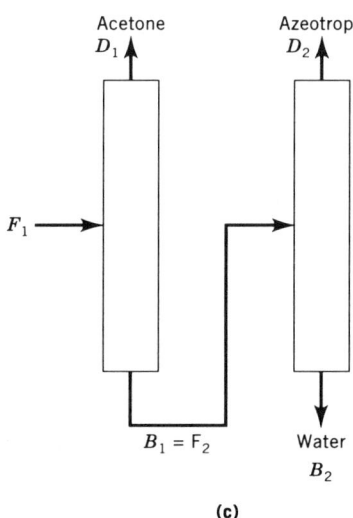

Fig. 5. The acetone–2-propanol–water system where ■ represents the 2-propanol–water azeotrope. (**a**) Residue curve map (34); (**b**) material balance lines showing the (– – –) distillation boundary; and (**c**) column sequence.

The overwhelming majority of all ternary mixtures that can potentially exist are represented by only 113 different residue curve maps (35). Reference 24 contains sketches of 87 of these maps. For each type of separation objective, these 113 maps can be subdivided into those that can potentially meet the objective, ie, residue curve maps where the desired pure component and/or azeotropic products lie in the same distillation region, and those that cannot. Thus knowing the residue curve for the mixture to be separated is sufficient to determine if a given separation objective is feasible, but not whether the objective can be achieved economically.

In the most common situation, a separating agent is added to separate a minimum boiling binary azeotrope into its two constituent pure components by homogeneous azeotropic distillation. The seven most favorable residue curve maps for this task are shown in Figure 6. Other maps have the potential to meet the stated objective (25–30), however they must be studied much more carefully because the design and operation of the resulting sequences are sensitive to the detailed VLE behavior. Thus, for initial screening purposes, only those separating agents that result in one of these seven maps need be considered for meeting the stated objective. Of these seven, the map representing extractive distillation (Fig. 6b) is by far the most common and the most important.

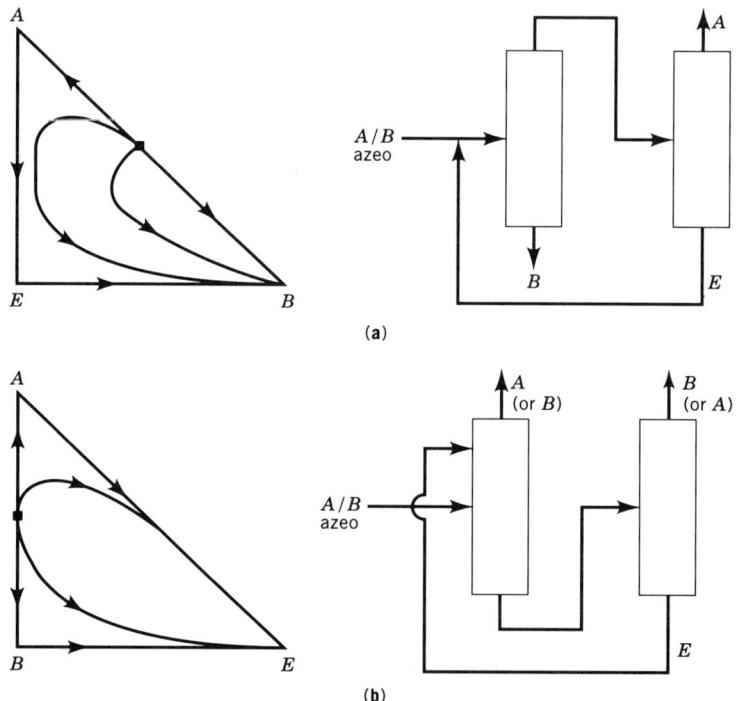

Fig. 6. The seven most favorable residue curve maps and corresponding column sequences for homogeneous azeotropic distillation of a minimum boiling azeotrope where ■ represents an azeotrope (16). (**a**) Case I, where the separating agent is intermediate boiling and does not introduce a new azeotrope; (**b**) Case II, extractive distillation with a heavy solvent which introduces no new azeotrope. In some cases, B can come off the top of the first column; (**c**) Case III, where the separating agent is intermediate boiling and forms a maximum boiling azeotrope with the lighter of the two pure components (ie, A). The agent may or may not form a minimum boiling azeotrope with B, with or without a minimum boiling ternary azeotrope lean in A; and (**d**) the same column configuration as case III, but the separating agent is lower boiling.

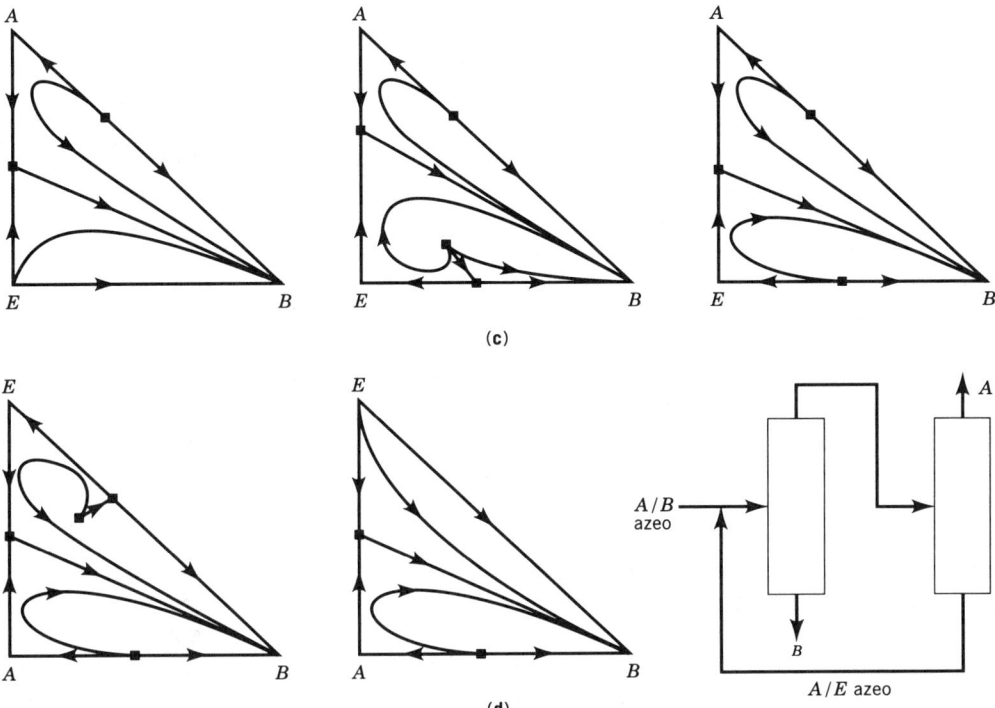

Fig. 6. (*Continued*)

Extractive Distillation

Extractive distillation is defined as distillation in the presence of a miscible, high boiling, relatively nonvolatile component, the solvent, that forms no azeotropes with the other components in the mixture (23). It is widely used in the chemical and petrochemical industries for separating azeotropic, close-boiling, and other low relative volatility mixtures, including those forming severe tangent pinches.

Extractive distillation works because the solvent is specially chosen to interact differently with the components of the original mixture, thereby altering their relative volatilities. Because these interactions occur predominantly in the liquid phase, the solvent is continuously added near the top of the extractive distillation column so that an appreciable amount is present in the liquid phase on all of the trays below. The mixture to be separated is added through a second feed point further down the column (see Fig. 6b). In the extractive column the component having the greater volatility, not necessarily the component having the lowest boiling point, is taken overhead as a relatively pure distillate. The other component leaves with the solvent via the column bottoms. The solvent is separated from the remaining components in a second distillation column and then recycled back to the first column.

A selection of industrial applications of extractive distillation includes: (*1*) the separation of the *n*-butane–butadiene azeotrope in mixed C_4-hydrocarbon streams using furfural [98-01-1], $C_5H_4O_2$, as the solvent (36); (2) the dehydration of ethanol using ethylene glycol [107-21-1] (37–39); (3) the separation of acetone and methanol using water (40); (4) the Ryan-Holmes process for separating the ethane–carbon dioxide azeotrope that arises when carbon dioxide is used for enhanced oil-field recovery (41); (5) the separation of the pyridine–water azeotrope using bisphenol (42); (6) the dehydration of tetrahydrofuran using monopropylene glycol (43); (7) the separation of the cumene–phenol azeotrope using trisubstituted phosphates (44); (8) the separation of the azeotropes formed by alcohols and their esters using aromatic hydrocarbons (45); (9) the separation of the methanol–methylene bromide azeotrope using ethylene bromide (46); (10) the separation of the phenol–cyclohexanone azeotrope using adipic acid diester (47); (11) the removal of close-boiling heptane isomers from cyclohexane, an important raw material for the manufacture of nylon precursors, using a mixture of solvents (48); (12) the separation of the low relative volatility mixture of propylene and propane using acrylonitrile (49); and (13) separating toluene from nonaromatics with phenol (50).

All extractive distillations correspond to one of three possible residue curve maps; one for mixtures containing minimum boiling azeotropes, one for mixtures containing maximum boiling azeotropes, and one for nonazeotropic mixtures. Thus extractive distillations can be divided into these three categories.

Minimum Boiling Azeotropes. All extractive distillations of binary minimum boiling azeotropic mixtures are represented by the residue curve map and column sequence shown in Figure 6b. Typical tray-by-tray composition profiles are shown in Figure 7.

In the distillation of ideal mixtures, the component having the lowest boiling point is always the one recovered as the distillate. This is not true for extractive

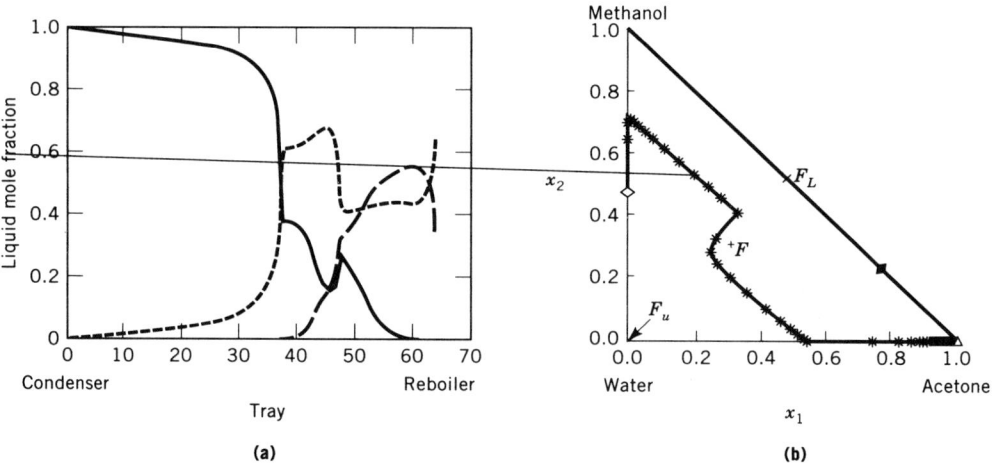

Fig. 7. Extractive distillation column profiles for the acetone–methanol–water separation (40). (**a**) Liquid composition versus theoretical tray location where (—) represents acetone, (– – –) methanol, and (·····) water; and (**b**) liquid composition profiles in mole fraction coordinates where ■ represents the azeotrope; △, the distillate; ×, feed; ◇, bottoms; and *, tray composition (51).

distillations. Neither can the design engineer freely pick which azeotropic component to recover overhead despite the apparent symmetry of the residue curve map (Fig. 6b). For a given solvent, one and only one component can be recovered in the extractive column and it need not be the pure component having the lowest boiling point. For example, the extractive distillation of ethanol and water using gasoline (38), some phenols (53), cyclic ketones, or cyclic alcohols (54) causes water to be removed as the overhead product of the extractive column and the lower boiling ethanol to leave in the bottom stream with the solvent. The higher ketones distill methanol overhead from methanol–acetone mixtures (55,56) and furfural reverses the natural volatility of butane–butadiene mixtures (36). This phenomenon of having the intermediate boiling pure component distill overhead in the extractive column results entirely from the way in which the solvent modifies the volatilities of the other components in the mixture.

Drawing pseudo-binary $y-x$ phase diagrams for the mixture to be separated is the easiest way to identify the distillate product component. A pseudo-binary phase diagram is one in which the VLE data for the azeotropic constituents (components 1 and 2) are plotted on a solvent-free basis. When no solvent is present, the pseudo-binary $y-x$ diagram is the true binary $y-x$ diagram (Fig. 8a). At the azeotrope, where the VLE curve crosses the 45° line, $\alpha_{12} = 1.0$. To determine which component is the distillate, a series of pseudo-binary $y-x$ plots must be drawn at increasing solvent compositions until the pseudo-azeotrope, the point where the solvent-free VLE curve crosses the 45° line, ie, where $\alpha_{12} = 1.0$, disappears into one of the pure component corners. The resulting pseudo-binary phase diagram is either Figure 8b where the solvent increases the volatility of component 1 relative to component 2, making component 1 the distillate, or Figure 8c, where the solvent has the opposite effect, making component 2 the distillate. The expressions to "break" or "negate" an azeotrope originated from the use of these psdudo-binary phase diagrams because that is what appears to happen as the solvent composition increases. For example, aniline increases the volatility of cyclohexane (bp = 80.8°C) relative to benzene (bp = 80.1°C) (Fig. 8d). Thus the higher boiling cyclohexane would be recovered as the distillate in an extractive distillation. Pseudo-binary $y-x$ phase diagrams can also be used to determine the product component for an extractive distillation of a nonazeotropic mixture. For example, phenol enhances the volatility of the nonaromatics relative to toluene so the nonaromatics can be distilled overhead (50).

Because binary homogeneous azeotropic mixtures cannot be separated into pure component products by isobaric distillation without a solvent, but can be separated in the presence of a sufficient amount of solvent, there is clearly some minimum amount of solvent that just makes the separation possible. The minimum solvent flow depends on the solvent used and offers one method for discriminating between solvents. A solvent having a small minimum solvent flow is a better solvent and should result in a lower cost design than a solvent having a large minimum solvent flow. The solvent composition required to just make the pseudo-azeotrope disappear on a pseudo-binary $y-x$ phase diagram, ie, to "break" the azeotrope, can be taken as a qualitative measure of the minimum solvent flow necessary to make a separation feasible. A simple method for estimating minimum solvent flows has been published (57) as has an exact quantitative method (51,58).

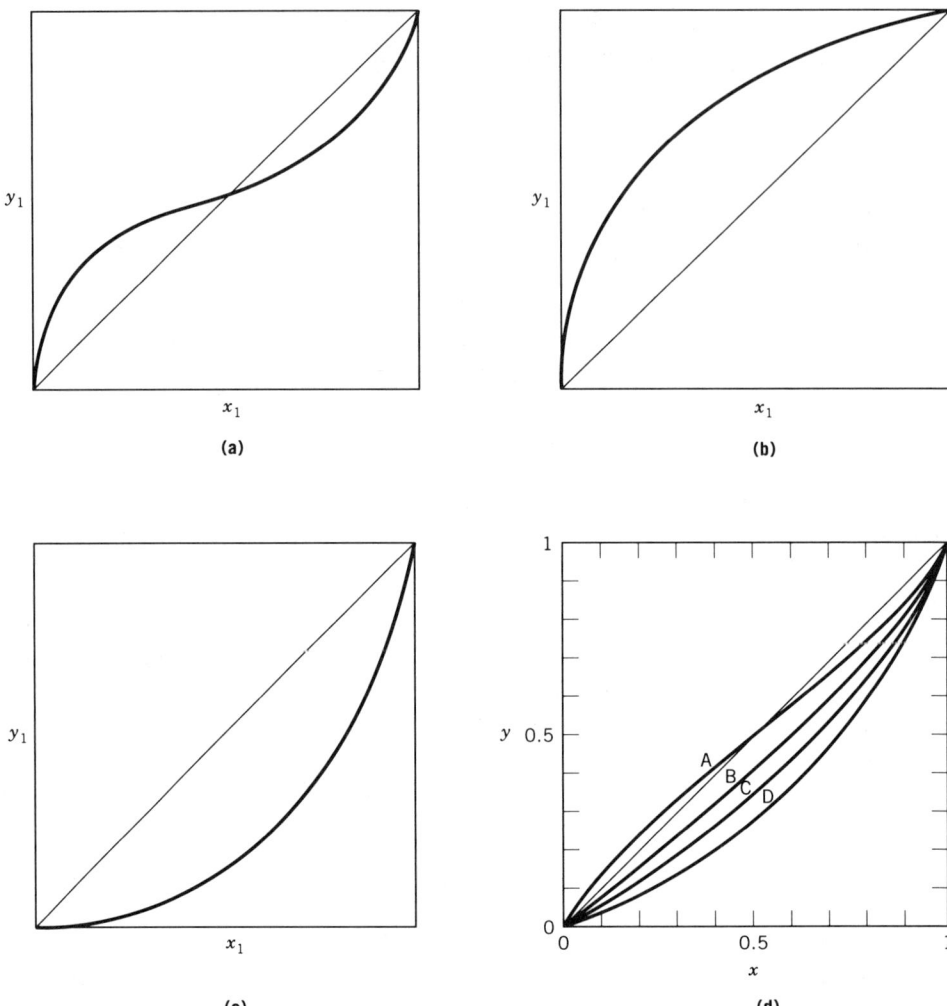

Fig. 8. Pseudo-binary (solvent-free) y–x phase diagrams for determining which component is to be the distillate where (———) is the 45° line. (**a**) No solvent; (**b**) and (**c**) sufficient solvent to eliminate the pseudo-azeotrope where the distillate is component 1 and component 2, respectively (51); and (**d**) experimental VLE data for cyclohexane–benzene where A, B, C, and D represent 0, 30, 50, and 90 mol % aniline, respectively (52).

Optimization. Optimization of the design variables is an important yet often neglected step in the design of extractive distillation sequences. The cost of the solvent recovery (qv) step affects the optimization and thus must also be included. Optimization not only yields the most efficient extractive distillation design, it is also a prerequisite for valid comparisons with other separation sequences and methods.

When several simple heuristics are used the optimization procedure usually reduces to a single-variable optimization of the feed ratio, ie, the molar ratio of the solvent to process feed flow rates, which has the greatest effect on the sequence cost (39). The simple heuristics are for calculating the minimum purity of the

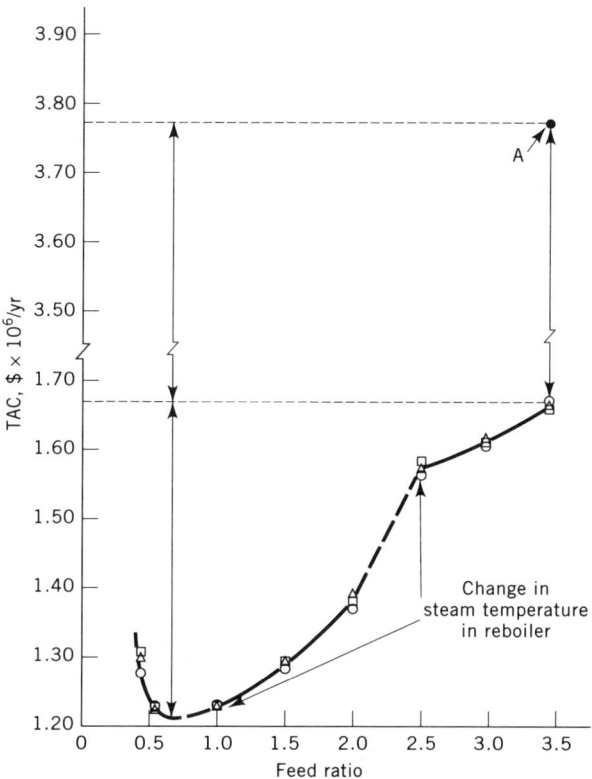

Fig. 9. Extractive distillation sequence cost as a function of the feed ratio for the production of anhydrous ethanol from azeotropic ethanol using ethylene glycol at reflux ratios of △, 1.15 r_{min}; ○, 1.2 r_{min}; and □ 1.3 r_{min} (39). Point A represents a previously published design for the same mixture (37).

recycled solvent (see eqs. 3–6 of Ref. 39) and for setting the reflux ratio in all columns to 1.2–1.5 times the minimum. Minimum reflux ratios for extractive distillations can be calculated using the methods presented in References 59 and 60 to solve the equations given by References 61 and 62. Figure 9 shows how the feed ratio influences the total annualized cost (TAC), defined by Reference 63, for dehydrating ethanol using ethylene glycol. Cost diagrams for all extractive distillations have the same distinctive shape: a very high cost near the minimum feed ratio, ie, the minimum solvent flow, which rapidly decreases to the minimum cost and then slowly increases at higher feed ratios. As a rough rule-of-thumb, the economically optimal feed ratio is often between two and four times the minimum feed ratio (51,58). Many industrial extractive distillations operate at feed ratios between 1 and 4. Figure 9 also shows the significant cost reduction in an extractive distillation sequence published in 1972 (37) that is made possible merely by applying the reflux ratio heuristic.

Maximum Boiling Azeotropes. Maximum boiling azeotropes are far less common than minimum boiling azeotropes. Successful extractive distillations of maximum boiling azeotropes using high boiling solvents are even more rare because the combination of a high boiling solvent and a maximum boiling azeotrope

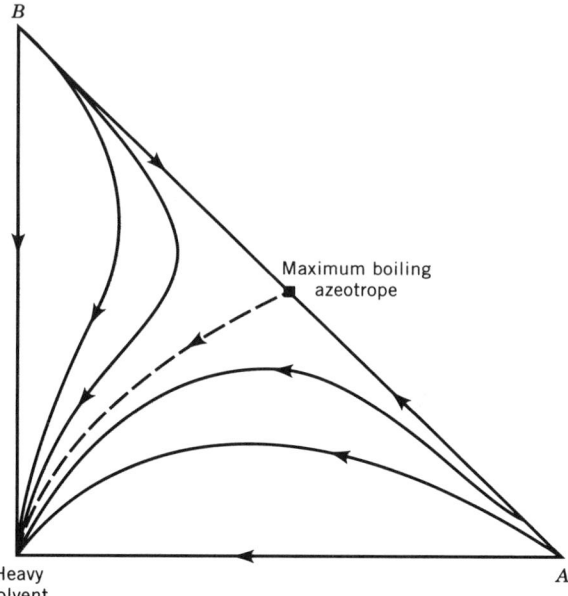

Fig. 10. Residue curve map for separating a maximum boiling azeotrope using a high boiling solvent where (----) represents the distillation boundary and ■, the azeotrope.

gives rise to a distillation boundary running from the maximum boiling azeotrope to the heavy solvent (Fig. 10) that divides the desired pure components into different distillation regions. Consequently, other methods are often used for separating maximum boiling azeotropes. The only way a high boiling solvent can yield an economically viable means for separating a maximum boiling azeotrope is if the resulting distillation boundary is extremely curved. The classic example is the concentration of aqueous nitric acid mixtures using sulfuric acid as the solvent (64–66). Figure 11 shows that the distillation boundary in this mixture is indeed highly curved. Applying the lever arm rule to the material balance lines superimposed on the residue curve map indicates that this process has a relatively high internal recycle rate relative to the product rates (D_1 and D_2).

Nonazeotropic Mixtures. The residue curve map representing the extractive distillation of any close-boiling, isomeric, or other low relative volatility mixtures using a high boiling solvent is represented by Figure 2. Although this map is different from that for extractive distillation of minimum boiling azeotropes, the distillation sequence is identical (see Fig. 6**b**). This process also works because the solvent alters the relative volatilities of the components to be separated via liquid-phase interactions. Depending on the nature of these interactions, the intermediate boiling pure component can become the distillate.

Because there is no azeotrope, these mixtures could be separated without adding a solvent. This, however, would be a difficult and expensive separation. Thus there is no minimum feed ratio (minimum solvent flow) and the only way to determine the optimal solvent-to-process feed ratio is by determining the sequence cost over a range of feed ratios. The best reflux ratios are again 1.2–1.5 times the minimum.

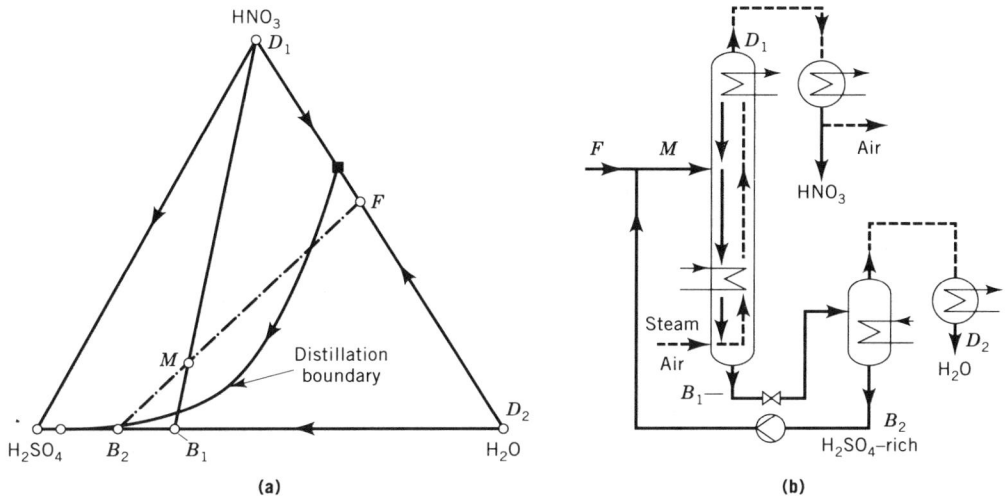

Fig. 11. Separation of nitric acid, HNO_3, and water, H_2O, using sulfuric acid, H_2SO_4, as the solvent. (**a**) Residue curve map and material balance lines where ■ represents the maximum boiling azeotrope and (**b**), column sequence (67).

Solvent Selection. One of the most important steps in developing a successful (economical) extractive distillation sequence is selecting a good solvent. A number of approaches have been proposed. The solvent selection procedure can be thought of as a two-step process. The first step requires techniques for rapidly identifying a group of feasible solvents from among all compounds that could potentially be used and winnowing this group down to a small number of the most promising. In the second step, detailed vapor–liquid equilibrium measurements are made for the top two or three candidates and the final selection is made, preferably by an economic comparison of the optimal sequence for each of the solvents. Residue curve maps are of limited usefulness at the preliminary screening stage because there is usually insufficient information available to sketch them, but they are valuable and should be sketched or calculated as part of the second stage of solvent selection, ie, after the VLE data are available because the screening methods discussed herein are not sufficient. The remaining discussion focuses on the first stage.

As a starting point for identifying candidate solvents, all compounds having boiling points below that of any component in the mixture to be separated should be eliminated. This is necessary to yield the correct residue curve map for extractive distillation, but this process implicitly rules out other forms of homogeneous azeotropic distillation. In fact, compounds which boil as much as 50°C or more above the mixture have been recommended (68) in order to minimize the likelihood of azeotrope formation. On the other hand, the solvent should not boil so high that excessive temperatures are required in the solvent recovery column.

Because extractive solvents work by altering the relative volatility between the components to be distilled, a good solvent causes a substantial change in relative volatility when present at moderate compositions. As seen from equation 3, the solvent modifies the relative volatilities by affecting the ratio of the liquid-phase activity coefficients (γ_i/γ_j). Whereas it is possible to find solvents which increase or decrease this ratio, it is usually preferable to select a solvent which accentuates the natural difference in vapor pressures between the components to be separated; that is, a solvent that increases γ_i relative to γ_j when $P_i^{sat} > P_j^{sat}$ is favored, over one that increases γ_j relative to γ_i. In the latter case, adding small amounts of the solvent actually makes the separation more difficult, and relatively large quantities would be required to completely overcome the natural volatility difference and enhance the separability of the original mixture (65). This second situation corresponds to the intermediate boiling component becoming the distillate. Thus a heuristic is to favor solvents that cause the naturally more volatile component to distill overhead.

To force the naturally more volatile component i overhead, the solvent should either behave essentially ideally with component j and cause positive deviations from Raoult's Law for component i ($\gamma_j \sim 1$ and $\gamma_i > 1$), or behave essentially ideally with component i and cause negative deviations from Raoult's Law for component j ($\gamma_i \sim 1$ and $\gamma_j < 1$). Compounds of similar type and size, eg, pentane–hexane or methanol–ethanol, tend to behave ideally in the liquid phase and thus have activity coefficients close to unity. Dissimilar molecules tend to repel each other causing positive deviations from Raoult's Law and, in the extreme, resulting in liquid-phase immiscibilities. Compounds that tend to associate in the liquid phase exhibit negative deviations from Raoult's Law. Because systems showing positive deviations are more common, the usual approach is to force the lower boiling component overhead by selecting a solvent which is chemically similar to the higher boiling species and dissimilar to the lower boiling species.

Deviations from ideality are often attributed to hydrogen bonding or polarity. General guidelines for predicting the type of deviation that occurs in a mixture based on the hydrogen-bonding tendency of each component are available (69,70). Successful extractive solvents are typically highly hydrogen-bonded liquids such as phenols, aromatic amines, alcohols, glycols, etc (70). Homologues of the component having the smaller natural volatility have also been advocated (54). For example, a higher alcohol can be used to force acetone overhead from the acetone–methanol azeotrope. Polarity arguments suggest using a highly polar solvent to increase the relative volatility of the less polar component in the mixture and a nonpolar solvent to have the opposite effect (65), for example, using water as the solvent to distill the less polar acetone overhead from the more polar methanol. Dipole moment interactions in hydrocarbon systems are discussed in Reference 71.

These qualitative methods are often useful for identifying general classes of compounds that may make good solvents. There are, however, two experimental methods that provide a more quantitative means for screening potential solvents. The first method entails measuring the relative volatility of a fixed composition mixture of the components to be separated (often 50% each) at a constant solvent-to-feed ratio that is typically one to three times the binary mixture on a molar basis (68). The objective is to find the candidate solvent(s) that cause the largest increase in the relative volatility of the components to be separated. Favorable solvents selected by this technique can, however, actually be infeasible solvents. For example, methylethylketone (MEK) has been identified as a solvent for separating acetone and methanol (55,56). Yet in reality methanol and MEK form a minimum boiling azeotrope (1) which results in a distillation boundary that puts acetone and methanol into different distillation regions, making the separation impossible.

The most common method for screening potential extractive solvents is to use gas–liquid chromatography (qv) to determine the infinite-dilution selectivity of the components to be separated in the presence of the various solvent candidates (71,72). The selectivity or separation factor is the relative volatility of the components to be separated (see eq. 3) in the presence of a solvent divided by the relative volatility of the same components at the same composition without the solvent present. A potential solvent can be examined in as little as 1–2 hours using this method. The tested solvents are then ranked in order of infinite-dilution selectivities, the larger values signify the better solvents. Favorable solvents selected by this method may in fact form azeotropes that render the desired separation infeasible.

In addition to its ability to make the separation feasible or easier, the ideal solvent is inexpensive, readily available, nontoxic, noncorrosive, thermally stable, and nonreactive with and easily separated from the other components in the mixture. In reality some compromise in solvent properties is almost always required.

Distillation Using Ionic Salts

Distillation using ionic salts (salt-effect distillation) is analogous to extractive distillation, but, rather than using a high boiling liquid solvent to alter the relative volatility of the mixture to be separated, a nonvolatile and soluble ionic salt is added to modify the volatility. Examples include adding calcium chloride to separate ethanol and water (73), using magnesium nitrate to dehydrate nitric acid, and using ferrous, lithium, or calcium chloride to separate acrylic acid and water (74). As is the case for liquid solvents, salts which have a significant effect on the volatility are preferred. Some salts reverse the natural volatility difference, causing the intermediate boiling pure component to be distilled overhead. See Reference 74 for an example.

For a more complete discussion of salt-effect distilllation see References 75–77.

Pressure-Swing Distillation

It is well known that changing the system pressure can affect the azeotropes in a mixture. This effect can be exploited to separate a binary mixture containing either a minimum or maximum boiling azeotrope which appreciably changes composition over a moderate pressure range in a sequence of two columns operated at different pressures. This process, called pressure-swing distillation, is often used industrially to separate tetrahydrofuran (THF) and water (78–81). Pressure-swing distillation has also been proposed for separating ethanol and water (38,82), a variety of alcohol–ketone azeotropes (83), and the maximum boiling hydrogen chloride–water azeotrope (84). For a binary mixture forming a pressure-sensitive minimum boiling azeotrope, the separation sequence works as shown in Figure 12. The fresh feed, F, is mixed with the recycled stream from the second column

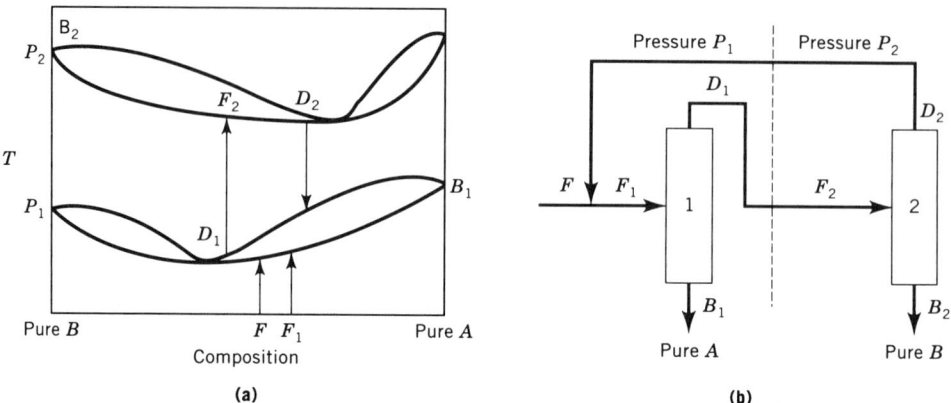

Fig. 12. Pressure-swing distillation of a minimum boiling binary azeotrope. (**a**) Temperature–composition phase diagram showing the effect of pressure on the azeotropic composition; (**b**) column sequence (85).

to form the feed stream, F_1, to the first column, which operates at pressure P_1. Because F_1 lies to the right of the azeotrope at pressure P_1 (Fig. 12**a**), pure A is removed as the bottom product, B_1, and a mixture near the azeotropic composition at pressure P_1 is the distillate, D_1. Stream D_1 is changed to pressure P_2 and fed to the next column as stream F_2. Because F_2 now lies to the left of the azeotropic composition at pressure P_2 (Fig. 12**a**), the other pure component, B, can be recovered in the bottom stream, B_2, and a near azeotropic mixture becomes the distillate, D_2, for recycling to the first column. An analogous procedure is used for binary maximum boiling azeotropes (86).

Only a fraction of the known azeotropes are sufficiently pressure-sensitive for the conventional pressure-swing distillation process to work. However, the concept can be extended to pressure-insensitive azeotropes by adding a separating agent which forms a pressure-sensitive azeotrope and distillation boundary. Then the pressure is varied to shift the location of the distillation boundary (85).

Heterogeneous Azeotropic Distillation

Heterogeneous azeotropic distillation, or simply azeotropic distillation, is widely used for separating nonideal mixtures. The technique uses minimum boiling azeotropes and liquid–liquid immiscibilities in combination to overcome the effect of other azeotropes or tangent pinches in the mixture that would otherwise prevent the desired separation. The azeotropes and liquid heterogeneities that are used to make the desired separation feasible may either be induced by the addition of a separating agent, usually called the entrainer, or they may be intrinsically present, in which case the mixture is sometimes called self-entrained. The most common case is the former; it includes such classic separations as ethanol dehydration using either benzene, heptane, ethyl ether, etc, as the entrainer, and acetic acid recovery from water using either ethyl acetate, 1-propyl acetate, or 1-butyl acetate as the entrainer. In ethanol dehydration the entrainer is used to break the homogeneous minimum boiling azeotrope between ethanol and water; in the acetic acid recovery process the entrainer is used to overcome the tangent pinch between acetic acid and water.

The first successful application of heterogeneous azeotropic distillation was in 1902 (87) and involved using benzene to produce absolute alcohol from a binary mixture of ethanol and water. This batch process was patented in 1903 (88) and later converted to a continuous process (89). Good reviews of the early development and widespread application of continuous azeotropic distillation in the prewar chemical industry are available (90).

Historically azeotropic distillation processes were developed on an individual basis using experimentation to guide the design. The use of residue curve maps as a vehicle to explain the behavior of entire sequences of heterogeneous azeotropic distillation columns as well as the individual columns that make up the sequence provides a unifying framework for design. This process can be applied rapidly, and produces an excellent starting point for detailed simulations and experiments.

Phase Diagrams. For binary mixtures, it is well known that when a liquid–liquid envelope merges with a minimum boiling vapor–liquid-phase envelope the resulting azeotropic phase diagram has the form shown in Figure 13. When the liquid composition $x_1 = x_1^{AZ}$, as in Figure 13**a**, then the vapor composition, y_1, is also equal to x_1^{AZ} and the mixture boils at constant temperature and at constant (and equal) composition in each phase. Thus a homogeneous azeotrope is formed. When the overall liquid composition $x_1^o = (x_1^o)^{AZ}$, as in Figure 13**b**, then y_1 is also equal to $(x_1^o)^{AZ}$ and again the mixture boils at constant temperature and at constant composition in each phase. However, the liquid of composition $(x_1^o)^{AZ}$ splits into two liquid phases so that there are three coexisting equilibrium phases which have different compositions. That is, a heterogeneous azeotrope is formed. Homogeneous and heterogeneous azeotropes share the common property that the overall liquid composition is equal to the vapor composition providing a means for identifying azeotropes experimentally and computationally.

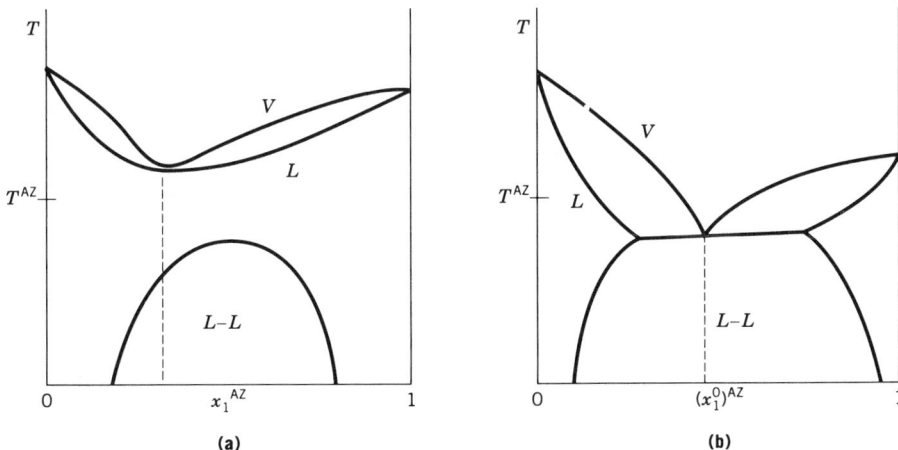

Fig. 13. Schematic isobaric phase diagrams for binary azeotropic mixtures (AZ). (**a**) Homogeneous azeotrope; (**b**) heterogeneous azeotrope.

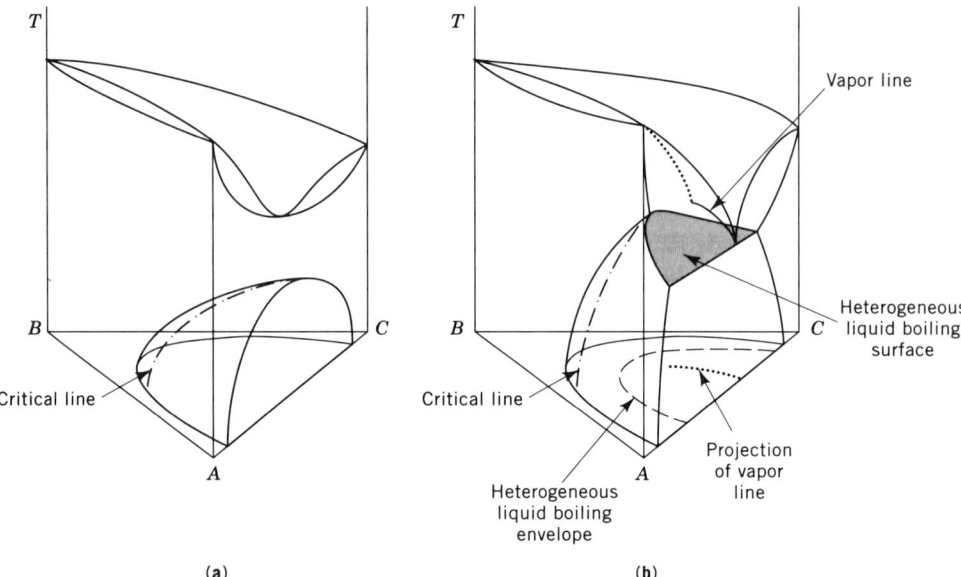

Fig. 14. Schematic isobaric phase diagrams for ternary (A,B,C) azeotropic mixtures. (**a**) Homogeneous liquid phase at all boiling points; (**b**) heterogeneous liquid phase for some boiling points.

The properties of ternary heterogeneous vapor–liquid–liquid equilibrium (VLLE) phase diagrams are not well documented, however, these diagrams are important for understanding azeotropic distillation. The simplest VLLE phase diagram, where a liquid–liquid envelope merges with a vapor–liquid equilibrium (VLE) surface containing a single minimum boiling binary azeotrope, is shown in Figure 14. The characteristic feature is the existence of a heterogeneous liquid boiling surface (Fig. 14**b**). When the overall liquid composition lies inside the heterogeneous boiling envelope, and the temperature lies on the heterogeneous boiling surface, then the liquid boils and splits into two equilibrium liquid phases and one coexisting vapor phase. When the overall liquid composition lies on the boiling surface, the Gibbs phase rule requires that the locus of all the corresponding equilibrium vapor compositions forms a curve in T–y space and *not* a surface, as happens in the homogeneous region.

A convenient way of representing the T–x–y phase diagram (Fig. 14**b**) is by projection onto the composition triangle at the base of the figure. It is understood that the temperature varies from point to point on the projected vapor line and on the projected boiling envelope. The latter looks like an isothermal liquid–liquid binodal envelope, but is not. Each tie line across the boiling envelope is associated with a different boiling temperature (Fig. 15).

Little experimental vapor–liquid–liquid phase equilibrium data for ternary mixtures has been published and thus good model parameters are not available. Nor is it possible to test the reliability of the modeling predictions. Figure 15**a** shows a model-predicted phase diagram for the ethanol–water–benzene mixture at 101.3 kPa (1 atm) pressure (91). In this diagram, the liquids and vapors in equilibrium with each other are signified by a common number. For example, the

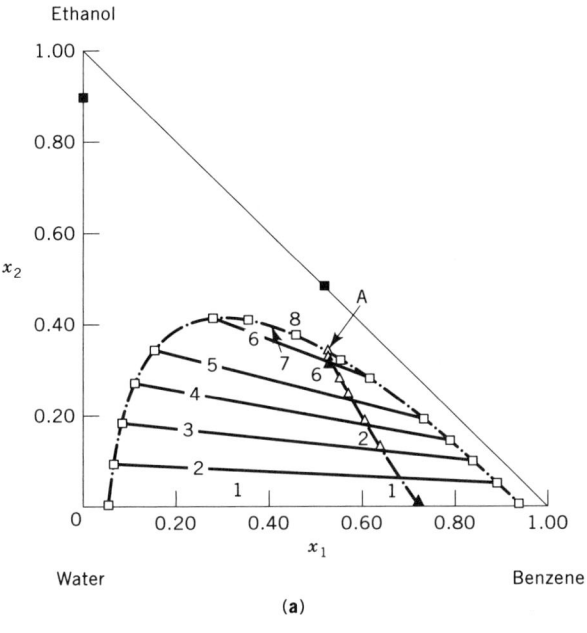
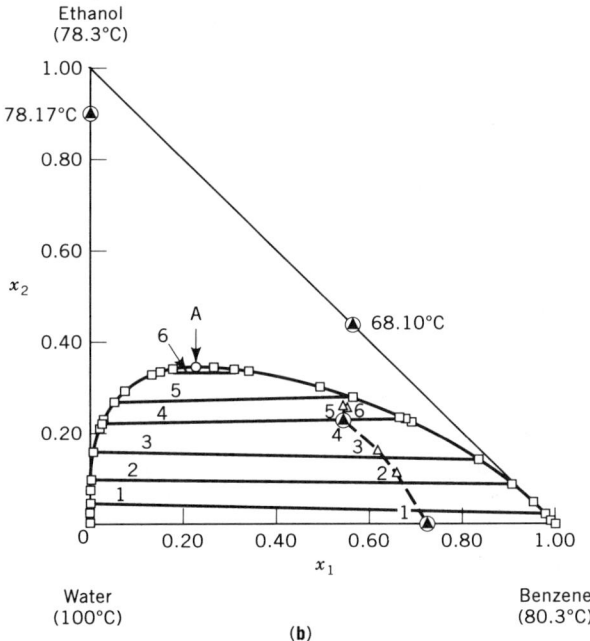

Fig. 15. Isobaric vapor–liquid–liquid (VLLE) phase diagrams for the ethanol–water–benzene system at 101.3 kPa; (□-□) represent liquid–liquid tie-lines; (△-△), the vapor line; ■, homogeneous azeotropes; ▲, heterogeneous azeotropes; ⊛, Horsley's azeotropes. (**a**) Calculated, where A is the end point of the vapor line and the numbers correspond to boiling temperatures in °C of 1, 70.50; 2, 68.55; 3, 67.46; 4, 66.88; 5, 66.59; 6, 66.46; 7, 66.47, and 8, the critical point, 66.48. (**b**) Experimental, where A is the critical point at 64.90°C and the numbers correspond to boiling temperatures in °C of 1, 67; 2, 65.5; 3, 65.0; 4, 64.85; 5, 64.9; and 6, 64.90.

coexisting liquids on tie-line number 2 are in equilibrium with vapor number 2 at a boiling temperature of 68.55°C. The ethanol–water–benzene mixture has a minimum boiling heterogeneous ternary azeotrope. Experimental VLLE data for this mixture is shown in Figure 15**b** (91). Heterogeneous saddle azeotropes are also possible, eg, formic acid–water–*m*-xylene (92), water–acetone–chloroform (93), however, maximum boiling heterogeneous azeotropes cannot exist (94). This differs from homogeneous azeotropes where all three types are found in nature.

The typical phase equilibrium problem encountered in distillation is to calculate the boiling temperature and the vapor composition in equilibrium with a liquid phase of specified composition at a given pressure. If the liquid phase separates, then the problem is to calculate the boiling temperature and the compositions of the two equilibrium liquid phases plus the coexisting vapor phase at the specified overall liquid composition. Robust and practical numerical methods have been devised for solving this problem (95–97) and have become the recommended techniques (98,99).

Thus, using these techniques and a nonideal solution model that is capable of predicting multiple liquid phases, it is possible to produce phase diagrams comparable to those of Figure 15. These predictions are not, however, always quantitatively accurate (2,6,8,91,100).

Residue Curve Maps. Residue curve maps are useful for representing the infinite reflux behavior of continuous distillation columns and for getting quick estimates of the feasibility of carrying out a desired separation. In a heterogeneous simple distillation process, a multicomponent partially miscible liquid mixture is vaporized in a still and the vapor that is boiled off is treated as being in phase equilibrium with all the coexisting liquid phases. The vapor is then withdrawn from the still as distillate. The changing liquid composition is most conveniently described by following the trajectory (or residue curve) of the overall composition of all the coexisting liquid phases. An extensive amount of valuable experimental data for the water–acetone–chloroform mixture, including binary and ternary LLE, VLE, and VLLE data, and both simple distillation and batch distillation residue curves are available (93,101). Experimentally determined simple distillation residue curves have also been reported for the heterogeneous system water–formic acid–1,2-dichloroethane (102).

Using the new generation of VLLE techniques (96), it is possible to calculate residue curve maps for heterogeneous liquid systems (Fig. 16). In this map the heterogeneous liquid boiling envelope has been superimposed on the residue curves in order to distinguish the homogeneous and heterogeneous regions of the triangular diagram.

Systems Having a Ternary Heterogeneous Azeotrope. Binary or ternary heterogeneous azeotropes are restricted to being either unstable nodes or saddles in the residue curve map (94). An example is the computed residue curve map for the ethanol–water–benzene mixture which exhibits a ternary minimum boiling heteroazeotrope (see Fig. 16). This map exhibits three distillation regions. The three distillation boundaries all begin at the minimum boiling ternary heteroazeotrope and end at each of the binary azeotropes. If ethanol is the desired product of the separation, the initial condition for the simple distillation must lie in the upper region of Figure 16. The computed phase diagram is shown in Figure 15**a**. The acetone–chloroform–water system exhibits a ternary heterogeneous saddle azeotrope (93).

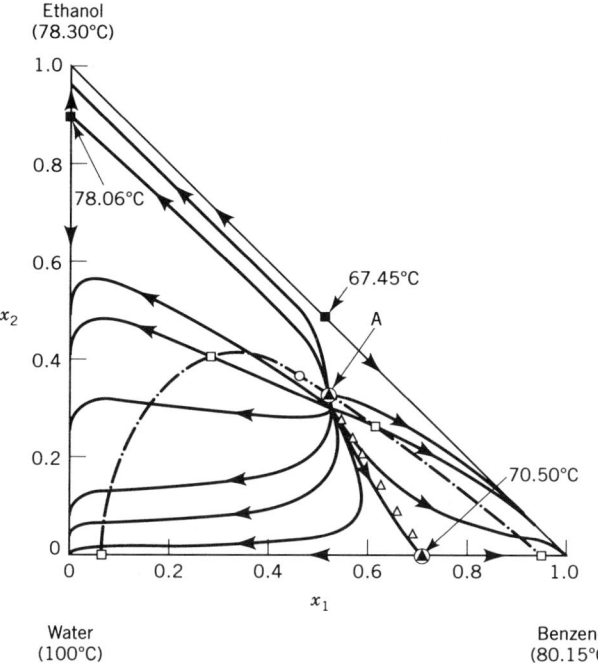

Fig. 16. Residue curve map calculated for the ethanol–water–benzene mixture where A is the end point of the vapor line; ■ represents a homogeneous azeotrope; (▲), heterogeneous azeotropes; (△—△), the vapor line; (−·−), the heterogeneous liquid boiling envelope; ○, the critical point; and □, the end points of tie-lines (103).

Column Sequences. The analysis of residue curve maps and distillation boundaries for homogeneous azeotropic mixtures provides a simple and useful technique for distinguishing between feasible and infeasible sequences of distillation columns. Heterogeneous mixtures also exhibit distillation boundaries. Residue curves cross continuously through the liquid boiling envelope from one side to the other. Thus a simple distillation boundary inside the heterogeneous region does not stop abruptly or exhibit a discontinuity at the liquid boiling envelope but passes continuously through it, becoming a homogeneous distillation boundary thereafter (see Fig. 16). As for homogeneous systems, residue curves cannot cross heterogeneous distillation boundaries. However, if the two individual equilibrium liquid phases resulting from a point x^o on a residue curve inside the heterogeneous region lie in two different distillation regions, then a liquid–liquid phase separation can be exploited to jump across heterogeneous distillation boundaries in a way that is not possible for homogeneous systems. This is the key to devising feasible sequences of columns for separating heterogeneous mixtures.

Binary Mixtures. A binary mixture containing a homogeneous azeotrope can be distilled up to, but not beyond, the azeotropic composition. If the binary azeotrope is heterogeneous, however, the situation is more favorable and a simple sequence of columns, shown in Figure 17, is capable of isolating each pure component. The process feed, x_F, is fed as a saturated liquid to a decanter where it phase-separates into an A-rich phase and a B-rich phase. The A-rich phase, x_{F1}, is then fed to column 1; the B-rich phase, x_{F2}, is fed to column 2. As can be seen

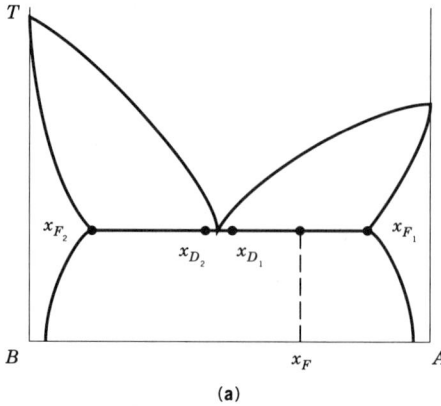

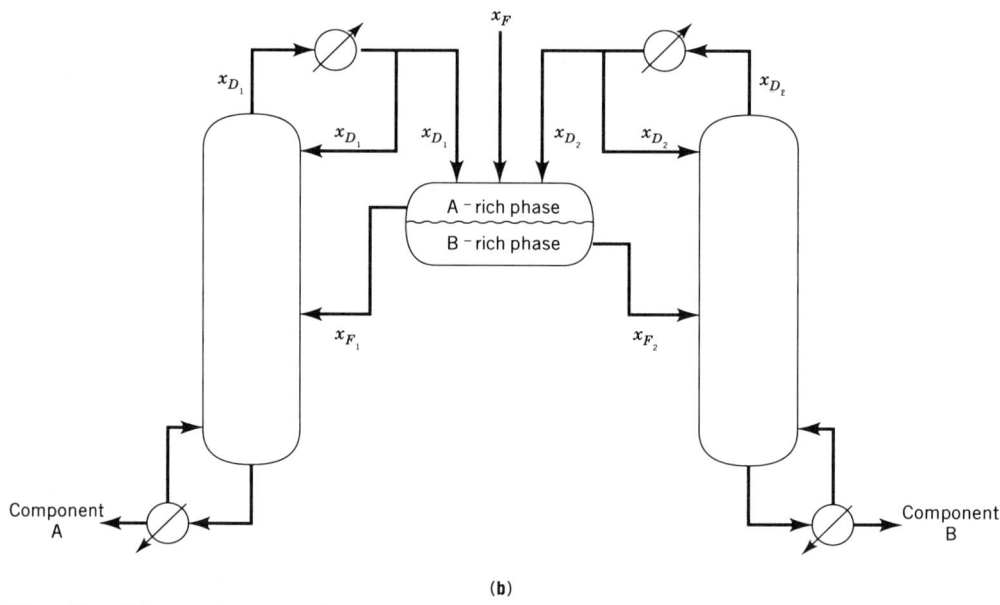

Fig. 17. Column sequence for separating a binary heterogeneous azeotropic mixture, A and B, where x_F represents the process feed mole fraction. (**a**) Phase diagram; (**b**) column sequence.

in Figure 17**a**, because of the liquid–liquid-phase split the compositions of these two feed streams lie on either side of the azeotrope. Therefore, column 1 produces pure A as a bottoms product and the azeotrope as distillate, whereas column 2 produces pure B as a bottoms product and the azeotrope as distillate. The two distillate streams are fed to the decanter along with the process feed to give an overall decanter composition partway between the azeotropic composition and the process feed composition acording to the lever rule. This arrangement is well suited to purifying water–hydrocarbon mixtures, such as a C_4–C_{10} hydrocarbon, benzene, toluene, xylene, etc; water–alcohol mixtures, such as butanol, pentanol, etc; as well as other immiscible systems.

If the process feed does not lie in the liquid–liquid region it can be made to do so by deliberately feeding either pure A or pure B to the decanter, as required. This may only be necessary during start-up or for control purposes because the recycled azeotrope has the beneficial effect of dragging the decanter composition further into the liquid–liquid region.

Ternary Mixtures. When the binary mixture containing the minimum boiling azeotrope is completely homogeneous, ie, the liquid is homogeneous for all compositions, the method given in Figure 17 requires modification. In this case a third component, called the entrainer, is added which induces a liquid–liquid-phase separation over a limited portion of the ternary composition diagram. Many options for sequencing ternary heterogeneous azeotropic distillation systems exist (90). These sequences generally consist of two, three, or four columns using various techniques for handling the entrainer recycle stream. The feasibility of such sequences rests on the use of a liquid–liquid-phase split to provide each column with a feed composition in a different distillation region. In this regard the sequences for ternary mixtures resemble the sequences for binary mixtures. In all cases the heart of the process is the azeotropic column and its decanter.

The classical example is the separation of ethanol from water using benzene as the entrainer. Many other separations fit this mold, eg, ethanol–water–carbon tetrachloride, and 2-propanol–water–benzene. The task is to separate a homogeneous binary mixture of ethanol and water, which contains a minimum boiling binary azeotrope, into its pure components. An entrainer, which has limited miscibility with one of the components, in this case water, is used. In addition, entrainers in this class cause two more minimum boiling binary azeotropes (one homogeneous, the other heterogeneous) to form together with a ternary minimum boiling heterogeneous azeotrope. The resulting residue curve map is similar to that shown in Figure 16, in which there are three distillation regions. Figure 18 shows the column sequence and material balance lines for this system. The map indicates that the components to be separated lie in two different distillation regions, I and II.

The azeo-column must be designed to meet the following target compositions: (*1*) the bottoms composition is specified to be almost pure ethanol, ie, x_B is specified to lie close to the vertex of distillation region II; and (*2*) the overhead vapor composition leaving the last vapor–liquid equilibrium stage of the column, y_N, is specified to lie in the wedge-shaped portion of region II inside the heterogeneous region near the ternary azeotrope. Typical target values for x_B and y_N are shown in Figure 18. The final values of x_B and y_N are subject to optimization and may differ slightly from those shown. All values of y_N inside the wedge-shaped region, except those special values which lie on the vapor line, are in equilibrium with a homogeneous liquid. Therefore, the liquid composition leaving the top tray, x_N, lies in the homogeneous portion of region II as shown (Fig. 18). The azeo-column must therefore be designed so that the steady-state liquid composition profile runs from x_B to x_N, and this can normally be done in such a way that every stage inside the column is homogeneous, ie, has only one liquid phase. A method for doing this is given in Reference 104. Not all azeo-columns have actually been designed this way. Neither do columns that have been designed in this manner

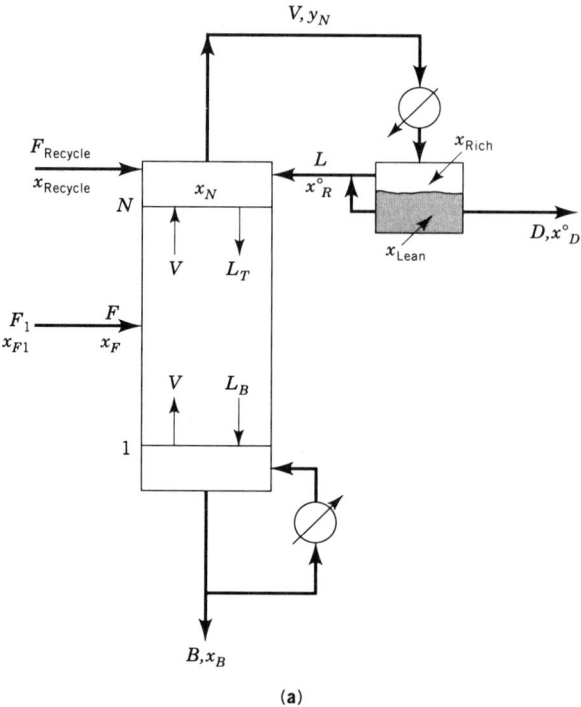

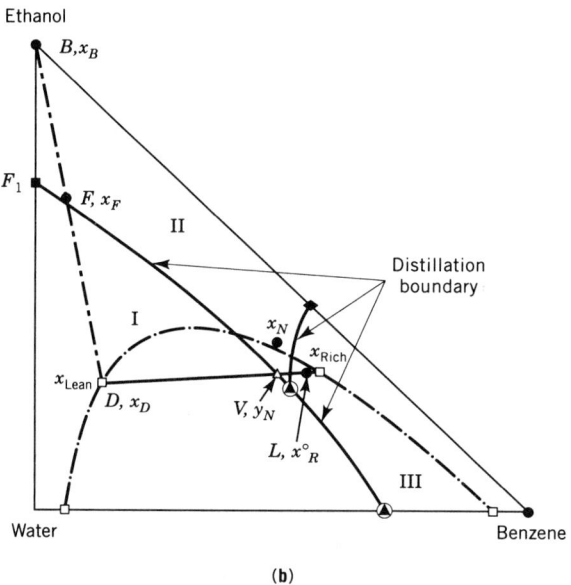

Fig. 18. Separation of ethanol from an ethanol–water–benzene mixture using benzene as the entrainer. (**a**) Schematic representation of the azeo-column; (**b**) material balance lines where ■ denotes the homogeneous and ⊚ the heterogeneous azeotropes; □, the end points of the liquid tie-line and △, the overhead vapor leaving the top of the column. The distillate regions, I, II, and III, and the boundaries are marked. Other terms are defined in text.

necessarily remain homogeneous under the action of disturbances (2–10,99,105).

The overhead vapor of composition y_N is totally condensed into two equilibrium liquid phases, an entrainer-rich phase of composition x_{rich} and an entrainer-lean phase of composition x_{lean}. The relative proportion of these two liquid phases in the condenser, ϕ, is given by the lever rule, where ϕ represents the molar ratio of the entrainer-rich phase to the entrainer-lean phase in the condensate.

The two condensate liquids must be used to provide reflux and distillate streams. Normally, the reflux ratio, r, is chosen so that $r = L/D \geq \phi$. This requires that the reflux rate be greater than the condensation rate of entrainer-rich phase and that the distillate rate be correspondingly less than the condensation rate of entrainer-lean phase. This means that the distillate stream consists of pure entrainer-lean phase, ie, $x_D = x_{\text{lean}}$, and the reflux stream consists of all the entrainer-rich phase plus the balance of the entrainer-lean phase. Thus, the overall composition of the reflux stream, x^o_R, lies on the tie-line between the points x_{rich} and y_N, as shown in Figure 18.

Completing the Separation Sequence. In the remainder of the separation sequence the distillate stream leaving the azeotropic column, column 2 in Fig. 19**a**, must be separated into a product stream and a recycle stream so that the entire sequence is closed with respect to the entrainer.

Kubierschky Three-Column Sequence. If only simple columns are used, ie, no side-streams, side-rectifiers/strippers etc, then the separation sequence can be completed by adding an entrainer recovery column, column 3 in Figure 19**a**, to recycle the entrainer, and a preconcentrator column (column 1) to bring the feed to the azeotropic column up to the composition of the binary azeotrope.

The entrainer recovery column takes the distillate stream, D_2, from the azeocolumn and separates it into a bottoms stream of pure water, B_3, and a ternary distillate stream for recycle to column 2. The overall material balance line for column 3 is shown in Figure 19**b**. This sequence was one of two original continuous processes disclosed in 1915 (106). More recently, it has been applied to other azeotropic separations (38,107,108).

Extensive design and optimization studies have been carried out for this sequence (108). The principal optimization variables, ie, the design variables that have the largest impact on the economics of the process, are the reflux ratio in the azeo-column; the position of the tie-line for the mixture in the decanter, dertermined by the temperature and overall composition of the mixture in the decanter; the position of the decanter composition on the decanter tie-line (see Reference 104 for a discussion of the importance of these variables); and the distillate composition from the entrainer recovery column.

Figure 20 shows material balance lines for three different decanter tie-lines. The process feed to the preconcentrator in each sequence is a binary mixture of 4.2 mol % ethanol and 95.8 mol % water. The product purity from the azeo-column is set at 99.9 mol % ethanol, and the water purity leaving the entrainer recovery column is set at 99.5 mol % water. These specifications are essentially identical to those used for studying the optimal extractive distillation sequences (39,40). For design 1, the decanter tie-line is set at the bubble-point of the mixture leaving the top of the azeo-column, having composition y_N. This temperature is 337.57 K. For design 2, the decanter composition is the same as for design 1 but subcooled

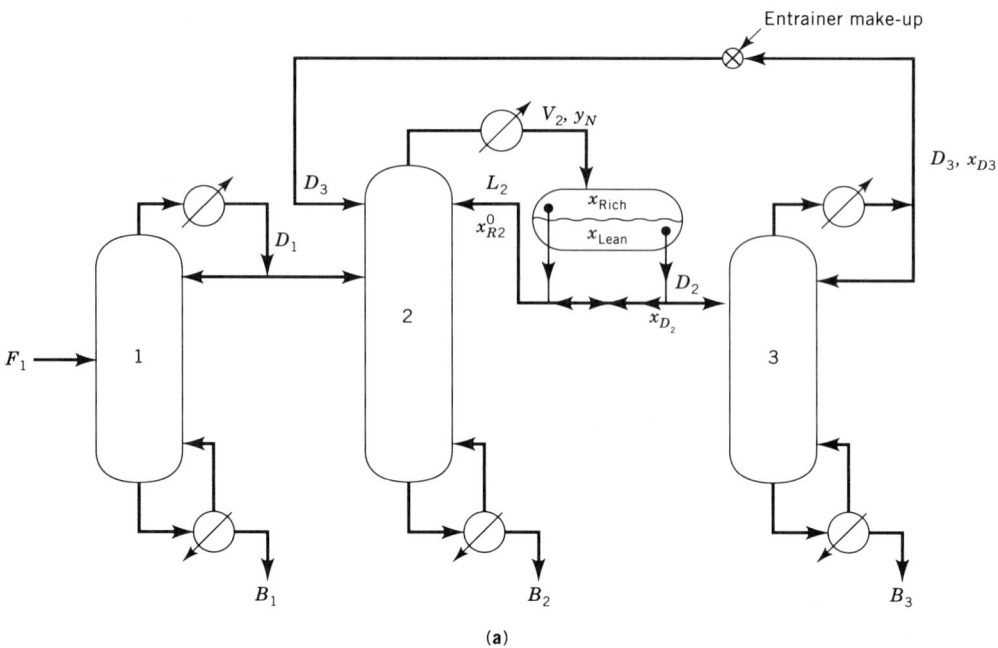

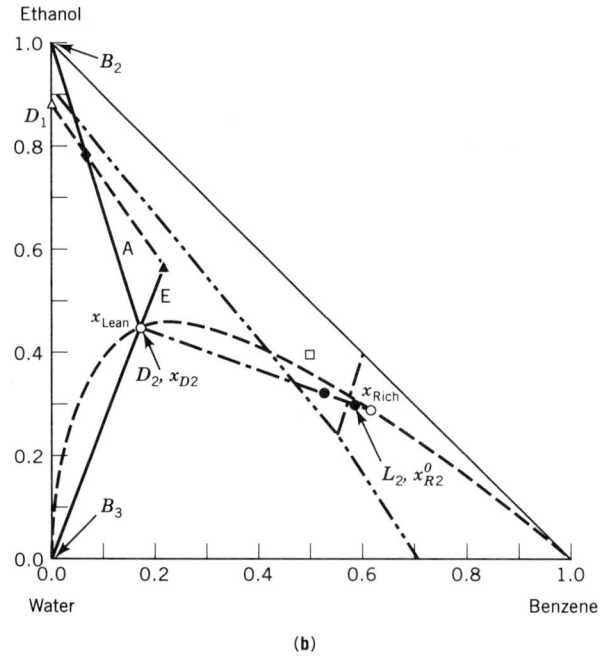

Fig. 19. Separation of ethanol and water from an ethanol–water–benzene mixture. Bottoms B_1 and B_3 are water, B_2 is ethanol. (**a**) Kubierschky three-column sequence where columns 1, 2, and 3 represent the preconcentration, azeotropic, and entrainer recovery columns, respectively. (**b**) Material balance lines from the azeotropic and the entrainer recovery columns, A and E, respectively, where ● represents the overall vapor composition from the azeo-column, V_2, y_N; □, the liquid in equilibrium with overhead vapor composition from the azeo-column, x_N; ▲, distillate composition from entrainer recovery column x_{D3}; and ◆, overall feed composition to the azeo-column, $D_1 + D_3$.

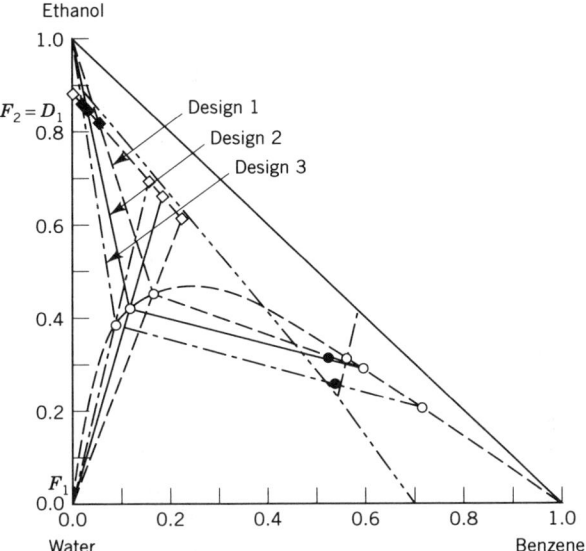

Fig. 20. Three sets of material balance lines for the Kubierschky three-column sequence where design 1 corresponds to the upper tie-line having $r_{min} = 8.78$; design 2, to the subcooled upper tie-line having $r_{min} = 12.23$; and design 3, to the lower tie-line having $r_{min} = 17.31$; ● represents overall decanter composition; ◆, the overall feed composition to the azeo-column; ◇, the distillate composition from the entrainer recovery column; and ○, the end points of the liquid–liquid tie-lines.

to 298.0 K. For design 3 the decanter composition is placed nearer to the ternary azeotrope than for designs 1 and 2, and the decanter temperature is set at the bubble-point of the mixture. In each design, the distillate composition from the entrainer recovery column is placed close to the distillation boundary. The position of the tie-line has a significant influence on the distillate composition from the azeo-column and this influences the position of the material balance line for the entrainer recovery column. The position of the distillate composition from the entrainer recovery column also influences the overall feed composition to the azeo-column.

Optimization studies indicate that the distillate composition from the entrainer recovery column has a strong influence on the process economics and the optimum position is always close to the distillation boundary. This decreases the amount of water being recycled, or equivalently, makes the overall feed to the azeo-column richer in ethanol. For each tie-line, the optimal position of the decanter liquid composition is found by the method proposed in Reference 104. The minimum reflux ratio for a given tie-line can be reduced by as much as 50% by making small changes in the decanter composition. Calculations indicate that the optimal reflux ratio for the azeo-column is normally in the range $1.1–1.5\,r_{min}$ and that the cost of the sequence is insensitive to this factor, leaving the position of the decanter tie-line as the sole remaining optimization variable.

The intrasequence flows, compositions, and reflux ratios are quite sensitive to relatively small changes in the position of the decanter tie-line as evidenced by the r_{min} value for the azeo-column: 8.78 for design 1, 12.23 for design 2, and 17.3 for design 3. Based on knowledge of homogeneous distillations, the vapor rate and total annualized cost for sequence 1 would appear to be the lowest. For homogeneous distillations at the design stage the feed and product flow rates, as well as their compositions, can be held constant as the reflux ratio is changed from one design to another. Thus there is a direct relationship between increased minimum reflux ratio and increased costs. However, such a relationship does not occur for heterogeneous distillations.

A good approximation for the vapor rate leaving the reboiler, V, for any type of distillation is

$$V = (r + 1)D \qquad (5)$$

where r is the reflux ratio and D is the distillate flow rate. For homogenous distillations, D is constant so that V increases as r increases. For azeotropic distillation, however, both r and D change from one tie-line to another. These effects may tend either to reinforce or to cancel each other depending on the mixture. There is no general rule and each mixture must be treated separately. In the ethanol–benzene–water system the reflux ratio increases from one design to another, but the distillate flow rate decreases, as can be seen from the material balance lines for the azeo-column in Figure 20. The net effect is that the vapor rate in the azeo-column hardly changes from one design to another. All the sequences shown in Figure 20 have approximately the same cost. This fortuitous cancellation of effects does not occur in general. It is always worthwhile exploring the economic impact of variations in the position of the decanter tie-line.

Other Sequences. The Kubierschky sequence is not the only way to perform the separation. Alternatives include: (*1*) If the process feed already has a composition at or near the composition of the binary azeotrope then the preconcentrator is not needed. (*2*) Recycling the distillate stream from the entrainer recovery column directly to the decanter analogous to the binary process shown in Figure 17. This recycle alternative causes the reflux ratio in the azeotropic column to be much larger than necessary (109) and should be avoided even though it has been studied extensively in the literature (2,3,110–114). (*3*) Use of the Kubierschky two-column sequence (106). For the ethanol–water–benzene system, this alternative has lower capital costs but higher operating costs than the Kubierschky three-column sequence so that the total annualized cost is about the same for both sequences (108). (*4*) Use of the Steffen three-column sequence, the basic layout of which is the same as the Kubierschky three-column sequence. Reflux to the azeo-column is provided by condensing the overhead vapor and returning part of the ternary liquid mixture before it goes to the decanters. The entrainer-rich phases from the decanters then get returned to the azeo-column as a second feed stream. (*5*) Use of the Ricard-Allenet four-column sequence (90,106, 108,109,115,116). Ricard-Allenet proposed a variation on this sequence (106) in which the overhead vapors from the entrainer recovery column are provided with a separate condenser–decanter system, similar to the one provided for the overhead vapors from the azeo-column. (*6*) Use of the Ricard-Allenet three-column sequence which has good economic possibilities, but there are no available studies of this alternative.

In summary, for systems of the ethanol–water–benzene type, the three most attractive sequences for carrying out azeotropic distillation are the Kubierschky three-column sequence, the Kubierschky two-column sequence, and the Ricard-Allenet three-column sequence. For each of these there is the added possibility of putting a liquid–liquid extraction step after the azeo-column.

Other Classes of Entrainers. Not all azeotropic mixtures are of the ethanol–water–benzene type. The number of azeotropes in the mixture may vary from system to system as may the character, ie, maximum or minimum boiling, heteroegeneous or homogeneous. In addition, the size and shape of the liquid–liquid region varies greatly from system to system. The feasibility and sequencing strategy for each new system is most conveniently established using residue curve maps such as those shown in Figure 21.

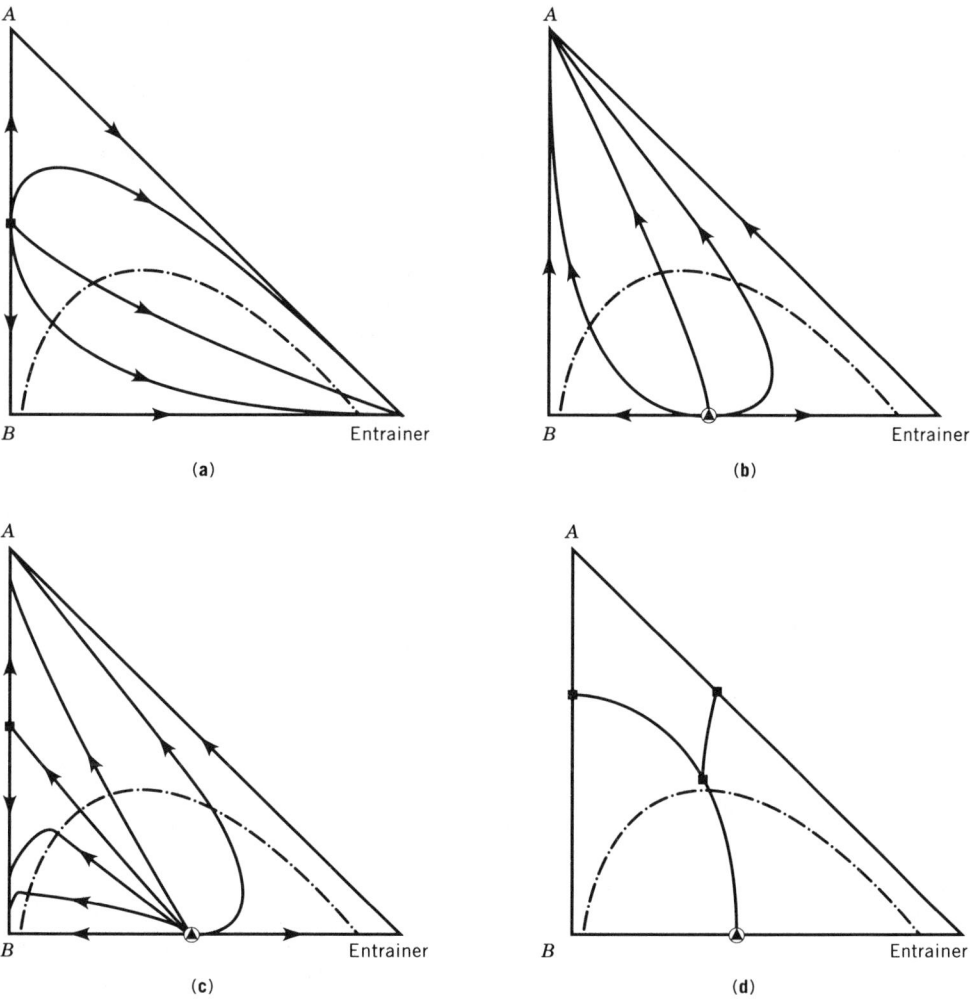

Fig. 21. A selection of feasible residue curve maps for ternary heterogeneous mixtures where ■ represents homogeneous and ⊛ heterogeneous azeotropes. See text.

Any entrainer that induces a liquid-phase heterogeneity over a portion of the composition triangle and which does not divide the components to be separated into different distillation regions is always a feasible entrainer. Examples are shown in Figures 21a and b and it is possible to construct many more such maps by strategically placing heterogeneous regions on the feasible residue curve maps given in Figure 6. Sequences based on these maps normally have multiple liquid phases on some of the stages in the column, which may lower the mass-transfer efficiency on those stages. In practice, mass-transfer efficiency and hydrodynamic performance of three-phase columns does not present a problem. As early as 1938 it was stated that "The efficiency of the plates was apparently undiminished by the heterogeneity of the boiling liquid and that was undoubtedly due to the violent agitation produced on the plates by the rapid bubbling of the vapours through the liquid . . ." (90). A more recent contribution and assessment of the literature from a practitioner's viewpoint is available (117). However, mass-transfer efficiency is of secondary importance to feasibility.

The map shown in Figure 21b is relevant to the separation of acetic acid and water which is of commercial significance. Although this binary mixture does not form an azeotrope, it does have a severe tangent pinch at high aqueous compositions preventing the distillate from being acid free. It is not economical to separate this mixture into pure product streams without the aid of an entrainer. All known commercial entrainers for this separation are heterogeneous and produce residue curve maps similar to the one shown in Figure 21b.

Clarke-Othmer Process for Acetic Acid–Water Separation. Large amounts of dilute acetic acid are purified industrially. The entrainer, eg, ethylene dichloride, ethyl acetate, 1-propyl acetate, or 1-butyl acetate, is charged to the azeocolumn, which has a process feed consisting of acetic acid and water. The bottom stream from the column is pure acetic acid, and the overhead vapor is close to the composition of the minimum boiling heterogeneous binary azeotrope formed by the entrainer and water (see Fig. 21b). The azeotropic vapors are condensed and decanted into a water stream that leaves as distillate, and an entrainer-rich stream that is returned to the column as reflux. Additional reflux, if needed, is achieved by returning some of the water stream. It is typical for these systems to have multiple liquid phases present on many stages in the rectifying section. This process was invented by Clarke and Othmer (118); more detailed operating information and entrainer comparison for this separation is available (119).

Wentworth Process for Ethanol–Water Separation. In the Wentworth process ethyl ether is used as the entrainer for ethanol–water separation, producing a residue curve map similar to the one shown in Figure 21c. Ethyl ether and water form a minimum boiling heterogeneous azeotrope at 34.15°C containing 98.75 wt % ether and 1.25 wt % water at atmospheric pressure. There is no azeotrope between ethyl ether and ethanol and no ternary azeotrope. The dilute ethanol process feed is first preconcentrated up to the composition of the ethanol–water azeotrope. This stream is fed to the azeo-column which produces pure ethanol as bottoms product and an overhead vapor close to the composition of the ethyl ether–water azeotrope. The overhead vapors are condensed and decanted into an ether-rich layer which is returned to the column as reflux, and a water-rich layer which leaves as distillate. The distillate contains no alcohol and very little ether. Pure water may be obtained from this stream by sending it to a stripping column where water is the bottom product and the overhead vapor has a composition near the ethyl ether–water azeotrope. These vapors are condensed and recycled to the decanter. The azeo-column is normally operated at about 700 kPa (7 atm) pressure because this increases the amount of water in the ether–water azeotrope, thereby reducing the amount of ether needed in the system.

This process is described in the literature in more detail (120–122). Its main advantage was a lower energy consumption (about 6,000 kJ/L (20,000 Btu/gal) ethanol product) relative to the benzene process (12,000 kJ/L (43,000 Btu/gal) ethanol) (121). Since the 1940s the gap has been narrowed by better designs for the benzene process which is capable of producing 99.8 mol % ethanol at an energy consumption of 8,400 kJ/L (30,000 Btu/gal) ethanol product (40,108), or using thermally integrated columns for 5,000 kJ/L (18,000 Btu/gal) ethanol product (40,123). In recent years homogeneous separating agents have shown great promise for the ethanol–water separation, and extractive distillation processes using ethylene glycol as the solvent, have been designed having energy consumptions of ca 6,000 kJ/L (22,000 Btu/gal) ethanol product (40), or using thermally integrated columns for 2,200–3,300 kJ/L (8,000–12,000 Btu/gal) ethanol product (40,124).

Rodebush Sequence for Ethanol–Water Separation. When ethyl acetate is used as the entrainer to break the ethanol–water azeotrope the residue curve map is similar to the one shown in Figure 21**d**, ie, the ternary azeotrope is homogeneous. Otherwise the map is the same as for ethanol–water–benzene. In such cases the liquid leaving the condenser from the azeo-column does not separate into two liquid phases, and the sequence is infeasible unless special tricks are employed. In the Rodebush sequence water is continuously added to the decanter in order to shift the overall composition into the two-liquid phase region. Each of the liquid phases from the decanter is fed to a separate distillation column which produces pure water and pure ethyl acetate, respectively. Some of the water is recycled to the decanter and all of the ethyl acetate is recycled to the azeo-column (106). More recently, a clever variation on this sequence was patented (125) for separating a ternary feed consisting of ethanol, water, and diethoxymethane. This variation also has a residue curve map similar to the one shown in Figure 21**d**.

More Complex Mixtures. All the sequences discussed are type I liquid systems, ie, mixtures in which only one of the binary pairs shows liquid–liquid behavior. Many mixtures of commercial interest display liquid–liquid behavior in two of the binary pairs (type II systems), eg, secondary butyl alcohol–water–disecondary butyl ether (SBA–water–DSBE), and water–formic acid–*meta*-xylene (92). Sequences for these separations can be devised on the basis of residue curve maps. The SBA–water–DSBE separation is practiced by ARCO and is considered in detail in the literature (4,5,105,126).

NOMENCLATURE

B	bottom stream or bottoms flow rate
c	number of components in a mixture

D	distillate stream or distillate flow rate
F	feed stream or feed flow rate
P	pressure
P_i^{sat}	vapor pressure of component i
r	reflux ratio
T	temperature
TAC	total annualized cost (63)
V	vapor flow rate
x_i	liquid-phase composition of component i (mol fraction)
y_i	vapor-phase composition of component i (mol fraction)
α_{ij}	volatility of component i relative to component j (defined by eq. 3)
γ_i	liquid-phase activity coefficient of component i
ξ	nonlinear time scale (see eq. 4)

Subscripts

B	bottom
D	distillate
F	feed
lean	entrainer-lean
N	stage N
R	reflux
rich	entrainer-rich

Superscripts

o	overall (composition)

BIBLIOGRAPHY

"Azeotropes" in the *Encyclopedia of Chemical Technology*, 1st ed., under "Distillation," Vol. 5, pp. 176–179, by E. G. Scheibel, Hoffmann-LaRoche Inc.; "Azeotropy and Azeotropic Distillation" in *ECT* 2nd ed., Vol. 2, pp. 839–858, by D. F. Othmer, Polytechnic Institute of Brooklyn; "Azeotropic and Extractive Distillation" in *ECT* 3rd ed., Vol. 3, pp. 352–377, by D. F. Othmer, Polytechnic Institute of New York; in *ECT* 3rd ed., Supl. Vol., pp. 145–158, by G. Prokopakis, Columbia University; in *ECT* 4th ed., Vol. 8, pp. 358–398, by M. F. Doherty, University of Massachusetts at Amherst, and J. P. Knapp, E.I. du Pont de Nemours & Co., Inc.

1. L. H. Horsley, *Azeotropic Data III*, Advances in Chemistry Series No. 116, American Chemical Society, Washington, D.C., 1973.
2. G. J. Prokopakis and W. D. Seider, *AIChE J.*, **29**, 49–60 (1983).
3. *Ibid.*, pp. 1017–1029.
4. J. W. Kovach III and W. D. Seider, *AIChE J.* **33**, 1300–1314 (1987).
5. J. W. Kovach III and W. D. Seider, *Comput. Chem. Eng.* **11**, 593–605 (1987).
6. M. Rovaglio and M. F. Doherty, *AIChE J.*, **36**, 39–52 (1990).
7. B. P. Cairns and I. A. Furzer, *Ind. Eng. Chem. Res.* **29**, 1349–1363 (1990).
8. *Ibid.*, pp. 1383–1395.
9. D. S. H. Wong, S. S. Jang, and C. F. Chang, *Comput. Chem. Eng.* **15**, 325–335 (1991).
10. S. Widagdo, W. D. Seider, and D. H. Sebastian, *AIChE J.* **38**, 1229–1242 (1992).
11. M. Rovaglio, T. Faravelli, G. Biardi, P. Gaffuri, and S. Soccol, *Comput. Chem. Eng.* **17**, in press (1993).
12. E. W. Jacobsen, L. Laroche, M. Morari, S. Skogestad, and H. W. Andersen, *AIChE J.* **37**(12), 1810–1824 (1991).

13. J. M. Smith and H. C. Van Ness, *Introduction to Chemical Engineering Thermodynamics*, 3rd ed., McGraw-Hill Publishing Co., New York, 1975.
14. M. F. Doherty and J. D. Perkins, *Chem. Eng. Sci.* **33**, 281–301 (1978).
15. Ibid., pp. 569–578.
16. E. R. Foucher, M. F. Doherty, and M. F. Malone, *Ind. Eng. Chem. Res.* **30**, 760–772 (1991).
17. M. F. Doherty, *Chem. Eng. Sci.* **40**, 1885–1889 (1985).
18. M. F. Doherty and J. D. Perkins, *Chem. Eng. Sci.* **34**, 1401–1414 (1979).
19. Y. Yamakita, J. Shiozaki, and H. Matsuyama, *J. Chem. Eng. Jpn.* **16**, 145–146 (1983).
20. D. B. Van Dongen and M. F. Doherty, *Chem. Eng. Sci.* **40**, 2087–2093 (1985).
21. C. Bernot, M. F. Doherty, and M. F. Malone, *Chem. Eng. Sci.* **45**, 1207–1221 (1990).
22. C. Bernot, M. F. Doherty, and M. F. Malone, *Chem. Eng. Sci.* **46**, 1311–1326 (1991).
23. M. Benedict and L. C. Rubin, *Trans. AIChE* **41**, 353–370 (1945).
24. M. F. Doherty and G. A. Caldarola, *Ind. Eng. Chem. Fund.* **24**, 474–485 (1985).
25. E. Rev, *Ind. Eng. Chem. Res.* **31**, 893–901 (1992).
26. L. Laroche, N. Bekiaris, H. W. Andersen, and M. Morari, *Ind. Eng. Chem. Res.* **31**, 2190–2209 (1992).
27. L. Laroche, N. Bekiaris, H. W. Andersen, and M. Morari, *AIChE J.* **38**, 1309–1329 (1992).
28. O. M. Wahnschafft, J. W. Koehler, E. Blass, and A. W. Westerberg, *Ind. Eng. Chem. Res.* **31**, 2345–2362 (1992).
29. J. G. Stichlmair and J-R. Herguijuela, *AIChE J.* **38**, 1523–1535 (1992).
30. Z. T. Fidkowski, M. F. Doherty, and M. F. Malone, *AIChE J.* **39**, in press (1993).
31. V. Julka, *A Geometric Theory of Multicomponent Distillation*, Ph.D. dissertation, University of Massachusetts, Amherst, 1992.
32. D. B. Van Dongen, *Distillation of Azeotropic Mixtures: The Application of Simple-Distillation Theory to Design of Continuous Processes*, Ph.D. dissertation, University of Massachusetts, Amherst, 1983.
33. S. G. Levy, *Design of Homogeneous Azeotropic Distillations*, Ph.D. dissertation, University of Massachusetts, Amherst, 1985.
34. S. G. Levy, D. B. Van Dongen, and M. F. Doherty, *Ind. Eng. Chem. Fund.* **24**, 463–474 (1985).
35. H. Matsuyama and H. Nishimura, *J. Chem. Eng. Jpn.* **10**, 181–187 (1977).
36. C. K. Buell and R. G. Boatright, *Ind. Eng. Chem.* **39**(6), 695–705 (1947).
37. C. Black and D. E. Ditsler, in R. F. Gould, ed., *Azeotropic and Extractive Distillation*, Advances in Chemistry Series No. 115, American Chemical Society, Washington, D.C. 1972, pp. 1–15.
38. C. Black, *Chem. Eng. Prog.* **76**(9), 78–85 (1980).
39. J. R. Knight and M. F. Doherty, *Ind. Eng. Chem. Fund.* **28**, 564–572 (1989).
40. J. P. Knapp and M. F. Doherty, *AIChE J.* **36**, 969–984 (1990).
41. R. L. Schendel, *Chem. Eng., Prog.* **80**(5), 39–43 (1984).
42. U.S. Pat. 3,804,722 (Apr. 16, 1974), E. D. Oliver (to Montecatini Edison S. p. A.).
43. U.S. Pat. 4,918,204 (Apr. 17, 1990), T. T. Shih and T. Chang, (to Arco Chemical Technology, Inc.).
44. U.S. Pat. 4,166,772 (Sept. 4, 1979), T. P. Murta (to Phillips Petroleum Co.).
45. U.S. Pat. 4,473,444 (Sept. 25, 1984), J. Feldman and J. M. Hoyt (to National Distillers and Chemical Corp.).
46. U.S. Pat. 3,794,568 (Feb. 26, 1974), G. A. Daniels and J. A. Wingate (to Ethyl Corp.).
47. U.S. Pat. 4,016,049 (Apr. 5, 1977), G. B. Fozzard and R. A. Paul (to Phillips Petroleum Co.).
48. R. E. Brown and F.-M. Lee, *Hydroc. Proc.* **70**(5) 83–86 (1991).
49. E. R. Hafslund, *Chem. Eng. Prog.* **65**(9) 58–64 (1969).

50. H. G. Drickamer and H. H. Hummel, *Trans. AIChE* **41**, 631–644 (1945).
51. J. P. Knapp, *Exploiting Pressure Effects in the Distillation of Homogeneous Azeotropic Mixtures*, Ph.D. dissertation, University of Massachusetts, Amherst, 1991.
52. B. Kolbe, J. Gmehling, and U. Onken, *I. Chem. E. Symp. Ser.* (56), 1.3/23–1.3/40 (1979).
53. U.S. Pat. 4,428,798 (Jan. 31, 1984), D. Zudkevitch, S. E. Belsky, and P. D. Krautheim (to Allied Corp.).
54. U.S. Pat. 4,455,198 (June 19, 1984), D. Zudkevitch, D. K. Preston, and S. E. Belsky (to Allied Corp.).
55. E. G. Scheibel, *Chem. Eng. Prog.* **44**(12), 927–931 (1948); U.S. Pat. 4,501,645 (Feb. 26, 1985), L. Berg, and An-I. Yeh.
56. An-I. Yeh, L. Berg, and K. J. Warren, *Chem. Eng. Comm.* **68**, 69–79 (1988).
57. L. Laroche, N. Bekiaris, H. W. Andersen, and M. Morari, *Can. J. Chem. Eng.* **69**, 1309–1319 (1991).
58. J. P. Knapp and M. F. Doherty, "Minimum Entrainer Flows for Extractive Distillation. A Bifurcation Theoretic Approach," submitted to *AIChE J.* (1993).
59. V. Julka and M. F. Doherty, *Chem. Eng. Sci.* **48**, 1367–1391 (1993).
60. Z. T. Fidkowski, M. F. Malone, and M. F. Doherty, *AIChE J.* **37**, 1761–1779 (1991).
61. S. G. Levy and M. F. Doherty, *Ind. Eng. Chem. Fund.* **25**, 269–279 (1986).
62. J. R. Knight, *Synthesis and Design of Homogeneous Azeotropic Distillation Sequences*, Ph.D. dissertation, University of Masschusetts, Amherst, 1986.
63. J. M. Douglas, *Conceptual Design of Chemical Processes*, McGraw-Hill Publishing Co., New York, 1988.
64. U.S. Pat. 1,074,287 (Sept. 30, 1913), H. Pauling.
65. C. S. Robinson and E. R. Gilliland, *Elements of Fractional Distillation*, 4th ed., McGraw-Hill Publishing Co., New York, 1950.
66. U.S. Pat. 4,966,276 (Oct. 30, 1990), A. Guenkel.
67. J. G. Stichlmair, J. R. Fair, and J. L. Bravo, *Chem. Eng. Prog.* **85**(1), 63–69 (1989).
68. C. S. Carlson and J. Stewart, in E. S. Perry and A. Weissberger, eds., *Techniques of Organic Chemistry*, Vol. IV, *Distillation*, Wiley-Interscience, New York, 1965.
69. R. H. Ewell, J. M. Harrison, and L. Berg. *Ind. Eng. Chem.* **36**(10), 871–875 (1944).
70. L. Berg, *Chem. Eng. Prog.* **65**(9), 52–57 (1969).
71. D. P. Tassios, in Ref. 37, pp. 46–63; D. P. Tassios, *Hydroc. Proc.* **49**(7), 114–118 (1970).
72. D. P. Tassios, *Ind. Eng. Chem. Proc. Des. Dev.* **11**(1), 43–46 (1972).
73. D. Barba, V. Brandini, and G. DiGiacomo, *Chem. Eng. Sci.* **40**, 2287–2292 (1985).
74. U.S. Pat. 4,269,666 (May 26, 1981), C. G. Wysocki (to Standard Oil Co.).
75. W. F. Furter, *Can. J. Chem. Eng.* **55**(6), 229 (1977).
76. W. F. Furter and R. A. Cook, *Int. J. Heat Mass Trans.* **10**, 23–26 (1967).
77. H. R. Galindez and A. Fredenslund, *I. Chem. E. Symp. Ser.* (104), A397–A403 (1987).
78. S. I. Abu-Eishah and W. L. Luyben, *Ind. Eng. Chem. Proc. Des. Dev.* **24**, 132–140 (1985).
79. U.S. Pat. 4,257,961 (May 24, 1981), J. S. Coates (to E. I. du Pont de Nemours & Co., Inc.).
80. U.S. Pat. 4,348,262 (Sept. 7, 1982), A. M. Stock and W. S. Tse (to E. I. du Pont de Nemours & Co., Inc.).
81. U.S. Pat. 4,093,633 (June 6, 1978), Y. Tanabe, J. Toriya, M. Sato, and K. Shiraga (to Mitsubishi Chemical Industries Co., Ltd.).
82. U.S. Pat. 1,676,700 (July 10, 1928), W. K. Lewis.
83. U.S. Pat. 2,324,255 (July 13, 1943), E. C. Britton, H. S. Nutting, and L. H. Horsley, (to The Dow Chemical Company).

84. U.S. Pat. 3,394,056 (July 23, 1968), M. Nadler and co-workers (to Esso Research and Engineering).
85. J. P. Knapp and M. F. Doherty, *Ind. Eng. Chem. Res.* **31,** 346–357 (1992).
86. M. Van Winkle, *Distillation*, McGraw-Hill Publishing Co, New York, 1967.
87. S. Young, *J. Chem. Soc.* **81,** 707–717 (1902).
88. Ger. Pat. 142,502 (June 25, 1903), S. Young.
89. Ger. Pat. 287,897 (Oct. 11, 1915), Kubierschky.
90. H. M. Guinot and F. W. Clark, *Trans. Inst. Chem. Eng.* **16,** 189–199 (1938).
91. H. N. Pham and M. F. Doherty, *Chem. Eng. Sci.* **45,** 1823–1836 (1990).
92. W. Reinders and C. H. De Minjer, *Recl. Trav. Chim.* **66,** 564–572 (1947).
93. *Ibid*, pp. 573–604.
94. H. Matsuyama, *J. Chem. Eng. Jpn.* **11,** 427–431 (1978).
95. L. E. Baker, A. C. Pierce, and K. D. Luks, *Soc. Petrol. Engrs. J.* **22,** 731–742 (1982).
96. M. L. Michelsen, *Fluid Phase Equil.* **9,** 1–19 (1982).
97. *Ibid.*, pp. 21–40.
98. D. J. Swank and J. C. Mullins, *Fluid Phase Equil.* **30,** 101–110 (1986).
99. Ref. 7, pp. 1364–1382.
100. J. M. Prausnitz and co-workers, *Computer Calculations for Multicomponent Vapor–Liquid and Liquid–Liquid Equilibria*, Prentice Hall, Englewood Cliffs, N.J., 1980, Chapt. 4.
101. W. Reinders and C. H. De Minjer, *Recl. Trav. Chim.* **59,** 207–230 (1940).
102. I. N. Bushmakin and P. Ya. Molodenko, *Russ. J. Phys. Chem.* **37,** 2618–2624 (1964).
103. Ref. 91, pp. 1837–1843.
104. H. N. Pham, P. J. Ryan, and M. F. Doherty, *AIChE J.* **35,** 1585–1591 (1989).
105. S. Widagdo, W. D. Seider, and D. H. Sebastian, *AIChE J.* **35,** 1457–1464 (1989).
106. D. B. Keyes, *Ind. Eng. Chem.* **21,** 998–1001 (1929).
107. D. W. Townsend, private communication, 1982.
108. P. J. Ryan and M. F. Doherty, *AIChE J.* **35,** 1592–1601 (1989).
109. Ref. 91, pp. 1845–1854.
110. W. S. Norman, *Trans. Inst. Chem. Eng.* **23,** 66–75 (1945).
111. Ref. 65, pp. 312–324.
112. Zh. A. Bril', A. S. Mozzhukhin, F. B. Petlyuk, and L. A. Serafimov, *Theor. Found. Chem. Eng.* **9,** 761–770 (1975).
113. Zh. A. Bril', A. S. Mozzhukhin, F. B. Petlyuk, and L. A. Serafimov, *Russ. J. Phys. Chem.* **11,** 675–681 (1977).
114. G. J. Prokopakis, W. D. Seider, and B. A. Ross, in R. S. Mah and W. D. Seider, eds., *Foundations of Computer-Aided Chemical Process Design*, Engineering Foundation, New York, 1981.
115. C. D. Holland, S. E. Gallun, and M. J. Lockett, *Chem. Eng.* **88,** 185–200 (1981).
116. C. J. King, *Separation Processes*, 2nd ed., McGraw-Hill Publishing Co., New York, 1980.
117. M. E. Harrison, *Chem. Eng. Prog.* **86**(11), 80–85 (1990).
118. U.S. Pat. 1,804,745 (1931), H. T. Clarke and D. F. Othmer.
119. D. F. Othmer, *Chem. Metall. Eng.* **40,** 91–95 (1941).
120. D. F. Othmer and T. O. Wentworth, *Ind. Eng. Chem.* **32,** 1588–1593 (1940).
121. T. O. Wentworth and D. F. Othmer, *Trans. Am. Inst. Chem. Eng.* **36,** 785–799 (1940).
122. T. O. Wentworth, D. F. Othmer, and G. M. Pohler, *Trans. Am. Inst. Chem. Eng.* **39,** 565–578 (1943).
123. U.S. Pat. 4,217,178 (Aug. 12, 1980), R. Katzen, G. D. Moon, Jr., and J. D. Kumans (to Raphael Katzen Associates).
124. S. Lynn and D. N. Hanson, *Ind. Eng. Chem. Proc. Des. Dev.* **25,** 936–941 (1986).

125. U.S. Pat. 4,740,273 (Apr. 26, 1988), D. L. Martin and P. W. Raynolds (to Eastman Kodak Co.).
126. J. W. Kovach III and W. D. Seider, *J. Chem. Eng. Data* **32,** 16–20 (1988).

General Reference

C. L. Dunn, R. W. Miller, G. J. Pierotti, R. N. Shiras, and M. Souders, Jr., *AIChE J.* **41,** 631–644 (1945).

MICHAEL F. DOHERTY
University of Massachusetts, Amherst

JEFFREY P. KNAPP
E. I. du Pont de Nemours & Co., Inc.

DRYING

Drying is an operation in which volatile liquids are separated by vaporization from solids, slurries, and solutions to yield solid products. In dehydration, vegetable and animal materials are dried to less than their natural moisture contents, or water of crystallization is removed from hydrates. In freeze drying (lyophilization), wet material is cooled to freeze the liquid; vaporization occurs by sublimation. Gas drying is the separation of condensable vapors from noncondensable gases by cooling, adsorption (qv), or absorption (qv) (see also ADSORPTION, GAS SEPARATION). Evaporation differs from drying in that feed and product are both pumpable fluids.

Reasons for drying include user convenience, shipping cost reduction, product stabilization, removal of noxious or toxic volatiles, and waste recycling and disposal. Environmental factors, such as emission control and energy efficiency, increasingly influence equipment choices. Drying operations involving toxic, noxious, or flammable vapors employ gas-tight equipment combined with recirculating inert gas systems having integral dust collectors, vapor condensers, and gas reheaters.

Drying is an applied science; ie, drying theory is based on the laws of physics, physical chemistry, and the principles underlying the transfer processes of chemical and mechanical engineering: heat, mass and momentum transfer (see MASS TRANSFER), vaporization, sublimation, crystallization (qv), fluid mechanics, mixing, and material handling. Drying is one of several unit operations involving simultaneous heat and mass transfer. However, drying is complicated by the presence of solids that interfere with heat, liquid, and vapor flow and retard the transfer processes, at least during the final drying stages or when a solids phase is continuous.

Because all drying operations involve processing of solids, equipment material handling capability is of primary importance. In fact, most industrial dryers are derived from material handling equipment designed to accommodate specific forms of solids. If possible, liquid separation from solids as liquid, by dewatering (qv) in a mechanical separation operation, should precede drying. Solids handling is made easier, and liquid separation without vaporization is less costly (see CENTRIFUGAL SEPARATION; FILTRATION; SEDIMENTATION). Evaporators, which have lower investment and operating costs than dryers, are also used to minimize dryer loads.

Several methods are employed to classify commercial dryers by process application (1–3). Mode of heat transfer is the conventional choice. The principal heat-transfer mechanisms in drying are (*1*) convection from a hot gas that contacts the material, used in direct-heat or convection dryers; (*2*) conduction from a hot surface that contacts the material, used in indirect-heat or contact dryers; (*3*) radiation from a hot gas or hot surface that contacts or is within sight of the material, used in radiant-heat dryers; and (*4*) dielectric and microwave heating in high frequency electric fields that generate heat inside the wet material by molecular friction, used in dielectric, or radio frequency, and microwave dryers. In the last group, high internal vapor pressures develop and the temperature inside the material may be higher than at the surface. Many dryers effect more than one heat-transfer mechanism, but most dryers can be identified by the one that predominates.

In order of priority, the factors that govern the selection of industrial dryers are (*1*) personnel and environmental safety; (*2*) product moisture and quality attainment; (*3*) material handling capability; (*4*) versatility for accommodating process upsets; (*5*) heat- and mass-transfer efficiency; and (*6*) capital, labor, and energy costs.

Costs are determined by energy, labor, capacity, and equipment materials of construction. Continuous dryers are less expensive than batch dryers and drying costs rise significantly if plant size is less than 500 t/yr. Vacuum batch dryers are four times as expensive as atmospheric-pressure batch dryers and freeze dryers are five times as costly as vacuum batch dryers. Once-through air dryers are half as costly as recirculating inert-gas dryers. Per unit of liquid vaporization, freeze and microwave dryers are the most expensive. The cost difference between direct- and indirect-heat dryers is minimal because of the former's large dust recovery requirement. Drying costs for particulate solids at rates of $1 \times 10^3 – 50 \times 10^3$ t/yr are about the same for rotary, fluid-bed, and pneumatic conveyor dryers, although few applications are equally suitable for all three (4,5).

Terminology

Bound moisture is liquid held by a material that exerts a vapor pressure less than that of the pure liquid at the same temperature. Liquid can be bound by solution in cell or fiber walls, by homogeneous solution throughout the material, and by chemical or physical adsorption on solid surfaces.

Capillary flow is liquid flow through the pores, interstices, and over the surfaces of solids which is caused by liquid–solid molecular attraction and liquid surface tension.

Constant rate period is the drying period during which the liquid vaporization rate remains constant per unit of drying surface.

Critical moisture content is that obtained when the constant rate period ends and the falling rate periods begin. Second critical moisture content specifies that remaining in a porous material when capillary flow dominance is replaced by vapor diffusion.

Dry basis describes material moisture content as weight of moisture per unit weight of dry material.

Dryer efficiency is the fraction of total energy consumed which is used to heat and vaporize the liquid.

Equilibrium moisture content is that which a material retains after prolonged exposure to a specific ambient temperature and humidity.

Evaporative efficiency in a direct-heat dryer compares vaporization obtained to that which would be obtained if the drying gas were saturated adiabatically.

Falling rate period is a drying period during which the liquid vaporization rate per unit surface or weight of dry material continuously decreases.

Fiber saturation point is the bound moisture content of cellular materials such as wood.

Free moisture content is the liquid content that is removable at a specific temperature and humidity. Free moisture may include bound and unbound moisture, and is equal to the total average moisture content minus the equilibrium moisture content for the specific drying conditions.

Humidity denotes the amount of condensable vapor present in a gas, expressed as weight of vapor per unit weight of dry gas; ie, dry basis weight.

Internal diffusion occurs during drying when liquid or vapor flow obeys the fundamental diffusion laws.

Moisture is a word used commonly to describe any volatile liquid or vapor involved in drying; ie, it is not used selectively to mean only water.

Moisture gradient is the moisture profile in a material at a specific moment during drying, which usually reveals the mechanisms of moisture movement in the material up to the moment of measurement.

Percent saturation is the ratio of the partial pressure of a condensable vapor in a gas to the vapor pressure of the liquid at the same temperature, expressed as a percentage. For water vapor in air this is called percent relative humidity.

Unaccomplished moisture change is the ratio of free moisture present in a material at any moment during drying to that present initially.

Unbound moisture in a hygroscopic material is moisture that exerts the same vapor pressure as the pure liquid at the same temperature. Unbound moisture behaves as if the material were not present. All moisture in a nonhygroscopic material is unbound.

Wet basis is a material's moisture content expressed as a percentage of the weight of wet material. Although commonly employed, this basis is less satisfactory for drying calculations than the dry basis for which the percentage change of moisture per unit weight of dry material is constant at all moisture contents.

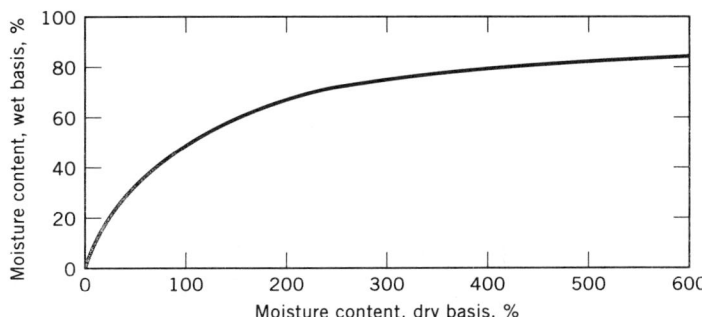

Fig. 1. Relationship between dry and wet weight bases where dry basis = weight moisture per weight of dry material; wet basis = weight moisture per weight of moisture + dry material.

Figure 1 shows the relationship between dry and wet bases. When the wet basis is used to state moisture content, a 2–3% change at moisture contents above 50% may represent a 10–30% change in evaporative load per unit weight of dry material.

Psychrometry

Before drying can begin, a wet material must be heated to such a temperature that the vapor pressure of the contained liquid exceeds the partial pressure of vapor already present in the surrounding atmosphere. The effect of a dryer's atmospheric vapor content and temperature on performance can be studied by construction of a psychrometric chart for the particular gas and vapor. Figure 2 is a standard chart for water vapor in air (6).

The wet bulb or saturation temperature curve indicates the maximum weight of vapor that can be carried by a unit weight of dry gas. For any temperature on the abscissa, saturation humidity is found by reading up to the saturation temperature curve, then across to the ordinate, kg/kg dry air. At saturation, the partial pressure of vapor in the gas is the vapor pressure of the liquid at the specific temperature:

$$H_s = \frac{p_s}{(P - p_s)} \cdot \frac{M_v}{M_g} \tag{1}$$

where H_s = saturation humidity, the weight ratio of moisture/kg dry gas; P = system total pressure; p_s = liquid vapor pressure at the gas temperature; and M_v/M_g = molecular weight ratio of vapor to dry gas. Pressure is in units of kPa. At any condition less than saturation, humidity, H, is expressed similarly:

$$H = \frac{p}{(P - p)} \cdot \frac{M_v}{M_g} \tag{2}$$

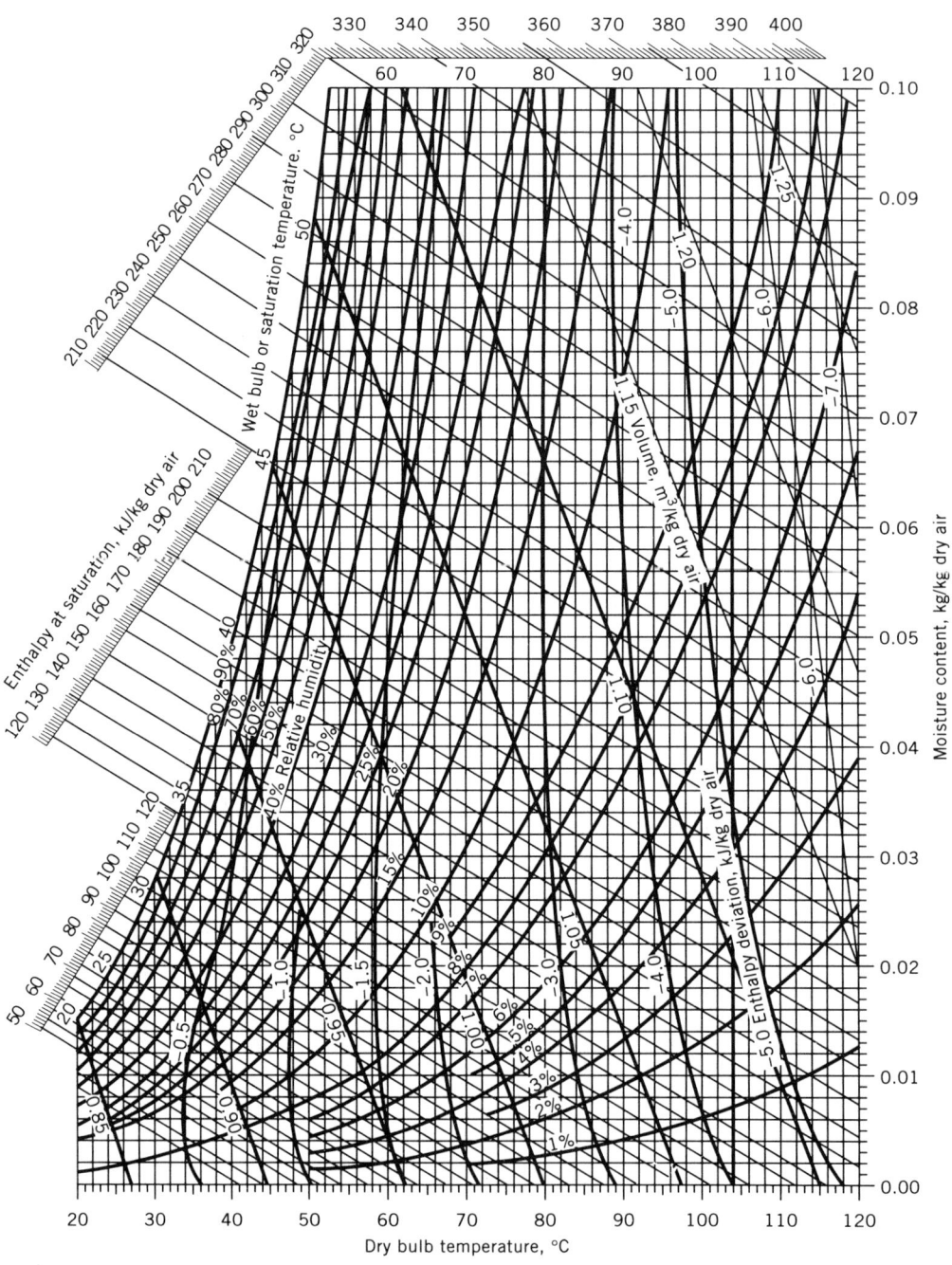

Fig. 2. Carrier psychrometric chart for air and water vapor at 101.325 kPa (1 atmosphere) total pressure. Courtesy of Carrier Corp. (6). To convert J to cal, divide by 4.184.

where p = partial pressure of the vapor in the gas in kPa. Percent relative humidity, % rh, curves indicate percent saturation and are related to vapor pressure:

$$\%(\text{rh}) = 100 \frac{p}{p_s} \qquad (3)$$

In Figure 2 the lines, volume, m³/kg dry air, indicate humid volume, which includes the volume of 1.0 kg of dry gas plus the volume of vapor it carries. Enthalpy at saturation data are accurate only at the saturation temperature and humidity; however, for air–water vapor mixtures, the diagonal wet bulb temperature lines are approximately the same as constant-enthalpy adiabatic cooling lines. The latter are based on the relationship:

$$(H - H_s) = -\frac{C_s}{L_s}(t - t_s) \qquad (4)$$

where H_s and t_s are adiabatic saturation humidity, kg/kg, and temperature, K, respectively, corresponding to gas conditions represented by H and t; C_s = humid heat for humidity H in units of kJ/(kg·K); and L_s = latent heat of vaporization at t_s in kJ/kg. The slope of the constant-enthalpy adiabatic cooling line is $-C_s/L_s$, which is the relationship between temperature and humidity of gas passing through a totally adiabatic direct-heat dryer. The humid heat of a gas–vapor mixture per unit weight of dry gas includes the specific heat of the vapor

$$C_s = c_{pg} + c_{pv}H \qquad (5)$$

where, c_{pg} = specific heat of the dry gas (air) and c_{pv} = specific heat of the vapor (water); both are in units of kJ/(kg·K). The wet bulb temperature is established by a steady-state, nonequilibrium relationship between heat transfer and mass transfer when liquid evaporates from a small mass; eg, the wet bulb of a thermometer, into a sufficiently large mass of flowing gas, so that the latter undergoes no temperature or humidity change. Provided radiant heat transfer is insignificant, steady-state conditions are expressed by the relationship:

$$h_c(t - t_w) = -k'L_w(H - H_w) \qquad (6)$$

where h_c = heat transfer by convection only in units of kW/(m²·K); k' = mass-transfer coefficient in kg/(s·m²)(kg/kg); t = gas dry bulb temperature, K; t_w = gas wet bulb temperature, K; H = humidity at t; H_w = saturation humidity at t_w; and L_w = latent heat of vaporization at t_w, kJ/kg. For air–water vapor mixtures, it happens that $h_c/k' \simeq C_s$. Therefore, because the ratio $(H - H_w)/(t - t_w) = -h_c/k'L_w$, the slope of the wet bulb temperature line in equation 6 is also approximately equal to $-C_s/L_s$, the slope of the constant enthalpy adiabatic cooling line in equation 4, and t_w is approximately the same as t_s. Enthalpy deviation curves in Figure 2 permit enthalpy corrections for humidities less than saturation and reveal the extent to which the wet bulb temperature lines do not coincide with constant enthalpy adiabatic cooling lines. For thorough treatment of wet bulb thermometry, see References 7–9. If system pressure is different from 101.3 kPa (1 atm), the humidity at measured dry bulb and wet bulb temperatures can be corrected (4). A separate chart is preferably constructed for the pressure of interest.

It is a coincidence that for water vapor in air, $h_c/k'C_s$ has a value of approximately one. For most organic vapors encountered in drying, wet bulb temperatures are considerably higher than the adiabatic saturation temperatures. This is because of the higher molecular weights. Larger molecules do not diffuse as easily through air or through most inert gases. For example, values of $h_c/k'C_s$ at 101.3 kPa and 0°C for carbon tetrachloride, benzene, and toluene are 2.17, 1.87, and 1.98, respectively. Psychrometric charts for these vapors in air have been prepared (10). It is necessary to employ such charts for evaluating humidities and particularly material temperatures when vaporizing organic liquids in direct-heat dryers.

To illustrate changes in temperatures and humidities as gas passes through a direct-heat (convection) dryer, such as a pneumatic conveyor, rotary, or fluid-bed, several temperature profiles are drawn on Figure 3. Line AB is an adiabatic saturation line. Gas that enters the dryer at H_1 and t_1 cools and humidifies along line a. Assuming ideal adiabatic operation, the gas could leave at H_2 and t_2. The maximum humidity gain, were the gas cooled to saturation adiabatically, would be $(H_s - H_1)$ at t_s. The ratio, $(H_2 - H_1)/H_s - H_1)$, is the ideal evaporative efficiency of this direct-heat dryer; however, as there are usually heat losses through the dryer enclosure and sensible heat absorbed by the solids, the gas temperature change is rarely accounted for completely by humidity gain; the outlet humidity, H_2', is less than the adiabatic saturation humidity at t_2, and the drying profile traces the path of line b. In dryers that employ internal steam coil gas reheaters, gas may pass several times through the material and heaters. Line c is a characteristic temperature–humidity profile.

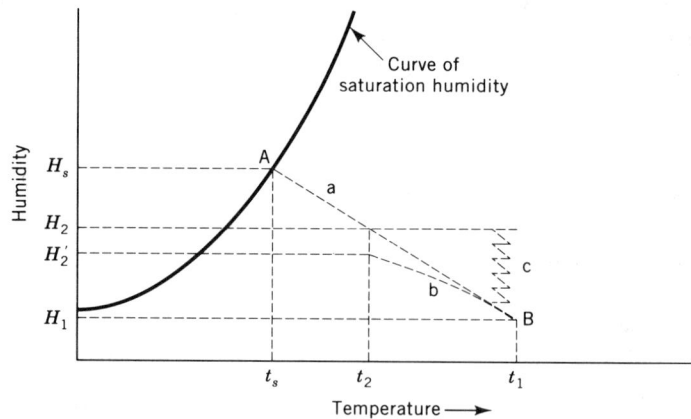

Fig. 3. Humidity chart illustrating changes in air temperature and humidity in adiabatic direct-heat (convection) dryers. AB is an adiabatic saturation line. Terms are defined in text.

Drying Mechanisms

Drying Periods. The goal of most drying operations is not only to separate a volatile liquid, but also to produce a dry solid of a desirable size, shape, porosity, density, texture, color, or flavor. An understanding of liquid and vapor mass-transfer mechanisms is essential for quality control. Mass-transfer mechanisms are best understood by measuring drying behavior under controlled conditions in a prototypic, pilot-plant dryer. No two materials behave alike and a change in material handling method or any operating variable, such as temperature or gas humidity, also affects mass transfer. For example, a layer of sand on a belt conveyor exhibits a different drying profile than sand dried on a vibrating (fluid-bed) conveyor.

Figure 4a shows drying time profiles for one material dried under three conditions. Corresponding rate profiles are in Figure 4b. Three products having uniquely different characteristics were produced by three different kinds of agitation. Other controllable drying conditions were constant. These profiles show that during drying several distinct periods may occur, which depend on how the material is handled. These are (1) an induction period during which wet material

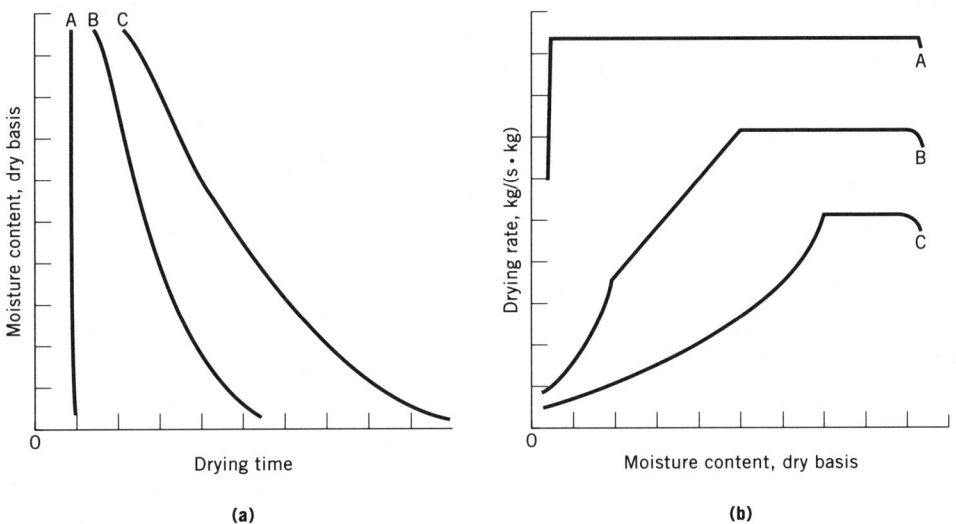

Fig. 4. (a) Profiles of moisture content vs drying time; and (b) drying rate vs moisture content for a slightly soluble, water-wet organic powder centrifuge cake at 4.0 kPa (0.58 psi) absolute pressure using 120°C indirect heat. Profile A was produced in a continuous, high speed agitator dryer provided with scrapers to maintain a clean heating surface. Drying time was 45 s and at an almost entirely constant rate because of high solids surface exposure to the heating surface; the product particle size was 100% less than 150 µm. Profile B was produced in a paddle-agitated batch dryer also having scrapers. Drying time was 70 min, including periods of constant rate, capillary, and diffusion drying. Because of the much slower agitator, the product was a porous, 100–500 µm powder having some dust. Profile C was produced in a double-cone batch dryer using some dry recycle. Drying time was 120 min, and was almost entirely liquid and vapor diffusion-controlled because turning of the double-cone pelletized the wet material early in the cycle; the product was composed of rather dense (200–800 µm) spheres, having negligible dust.

is heated to drying temperature; (2) a constant rate drying period indicated by the horizontal portions of the profiles in Figure 4**b**; (3) a period of decreasing rate shown by the sloping portions of two rate profiles during which the drying rate appears proportional to moisture content; and (4) a period of decreasing rate shown by the curved portions of two rate profiles during which the drying rate is evidently a more complex function of moisture content than simple proportionality.

The moisture content at the end of constant rate drying is the critical moisture content. Drying periods following are falling rate periods. The curved portion of profile B in Figure 4**b** is a second falling rate period; moisture content at the second break is the second critical moisture content. Profile C shows that drying may occur almost entirely in a falling rate period; a slight change in specified product moisture content can have a significant effect on drying time.

Constant Rate Drying. During constant rate drying, vaporization occurs from a liquid surface of constant composition and vapor pressure. Material structure has no influence except moisture movement from within the material must be fast enough to maintain the wet surface. The vaporization rate is controlled by the heat-transfer rate to the surface. The mass-transfer rate adjusts to the heat-transfer rate and the wet surface reaches a steady-state temperature. The drying rate remains constant, therefore, as long as external conditions are constant. If heat is supplied solely by convection, the steady-state temperature is the gas wet bulb temperature. When conduction and radiation contribute, eg, the material contacts and/or receives radiation from a warm surface, a liquid surface temperature between the wet bulb temperature and the liquid's boiling point is obtained. In indirect-heat and radiant-heat dryers, where conduction and radiation predominate, surface liquid may boil regardless of ambient humidity and temperature. During constant rate drying, material temperature is controlled more easily in a direct-heat dryer than in an indirect-heat dryer because in the former the material temperature does not exceed the gas wet bulb temperature as long as all surfaces are wet. For convection, all principles relating to simultaneous heat and mass transfer between gases and liquids apply. The steady-state relationship between heat and mass transfer at the liquid surface is

$$-\frac{dw}{d\theta} = \frac{h_t A}{L'_s}(t - t'_s) = k'_a A(p'_s - p) \qquad (7)$$

where $dw/d\theta$ = moisture loss in kg/s; h_t = sum of all convection, conduction, and radiation components of heat transfer in kW/(m²·K); A = effective surface area for heat and mass transfer in m²; L'_s = latent heat of vaporization at t'_s in kJ/kg; k'_a = mass-transfer coefficient in kg/(s·m²·kPa); t = mean source temperature for all components of heat transfer in K; t'_s = liquid surface temperature in K; p'_s = liquid vapor pressure at t'_s in kPa; p = partial pressure of vapor in the gas environment in kPa. It is often useful to express this relationship in terms of dry basis moisture change. For vaporization from a layer of material:

$$-\frac{dW}{d\theta} = \frac{h_t}{\rho_m d_m L'_s}(t - t'_s) \qquad (8)$$

where $dW/d\theta$ = dry basis drying rate in units of kg/(s·kg); ρ_m = dry material bulk density in kg/m³; and d_m = layer depth in m. A similar equation describes through-circulation drying in which gas flows through a bed of particles:

$$-\frac{dW}{d\theta} = \frac{h_t a}{\rho_m L'_s}(t - t'_s) \tag{9}$$

where a = the effective heat-transfer area per unit of bed volume in units of m²/m³. For a static bed, ρ_m can be measured, but the effective area of a particle bed is difficult to estimate except for uniform shapes such as cylinders and spheres (4). The practice is to conduct drying tests from which a value of the quantity $h_t a$ can be calculated by inserting property and drying data into equation 9. A modification of equation 9 is used to describe dryers in which gas flows among dispersed particles; eg, direct-heat rotary dryers. In dispersed-particle dryers, particle concentration per unit volume of dryer changes continuously and varies from place to place. For these, tests are conducted in a prototype of the commercial dryer and scale-up is based on average dispersion. The designer's concern is that the quality of particle dispersion in the gas in the scaled-up dryer duplicates that in the prototype or that a proper allowance is made for differences. From this procedure comes the concept of the volumetric heat-transfer coefficient:

$$U_a = h_t a \tag{10}$$

where U_a = an average volumetric heat-transfer coefficient having units of kW/(m³·K). For constant rate drying in dispersed-particle dryers, the general relationship is

$$-\frac{dw}{d\theta} = \frac{U_a V \Delta t_m}{L'_s} \tag{11}$$

where V = effective dryer volume in units of m³; Δt_m = log-mean temperature difference between all convection, conduction, and radiation heat sources and the material in K; and L'_s = latent heat of vaporization at the material surface temperature in kJ/kg. For estimating effective areas for various heat-transfer components, methods have been developed for tray dryers that may serve as a guide for other arrangements (11,12).

Convection heat transfer is dependent largely on the relative velocity between the warm gas and the drying surface. Interest in pulse combustion heat sources anticipates that high frequency reversals of gas flow direction relative to wet material in dispersed-particle dryers can maintain higher gas velocities around the particles for longer periods than possible in simple cocurrent dryers. This technique is thus expected to enhance heat- and mass-transfer performance. This is apart from the concept that mechanical stresses induced in material by rapid directional reversals of gas flow promote particle deagglomeration, dispersion, and liquid stream breakup into fine droplets. Commercial applications are needed to confirm the economic value of pulse combustion for drying.

Gas impingement from slots, orifices, and nozzles at 10–100 m/s velocities is used for drying sheets, films, coatings, and thin slabs, and as a secondary heat source on drum dryers and paper machine cans. The general relationship for convection heat transfer is (13,14):

$$h_c = \alpha G^{0.78} \tag{12}$$

where α is a factor dependent on orifice-plate open area, hole, or slot size, and spacing between the plate, slots, or nozzles, and the heat-transfer surface; and G = hole, slot, or nozzle gas velocity in mass flow terms having units of kg/(s·m²). Convection heat- and mass-transfer performance is enhanced by thinning of the laminar gas film immediately above the wet surface caused by direct gas impact on the surface. In a float dryer, cloth, sheet, or film is supported and conveyed on layers of gas which impinge on both sides of the material. This noncontact dryer is an impingement dryer modification.

Contact Drying. Contact drying occurs when wet material contacts a warm surface in an indirect-heat dryer (15–18). A sphere resting on a flat heated surface is a simple model. The heat-transfer mechanisms across the gap between the surface and the sphere are conduction and radiation. Conduction heat transfer is calculated, approximately, by recognizing that the effective conductivity of a gas approaches 0, as the gap width approaches 0. The gas is no longer a continuum and the rarified gas effect is accounted for in a formula that also defines the conduction heat-transfer coefficient:

$$\gamma = \frac{\gamma_g}{[1 + (p/s)]} \quad \text{or} \quad \frac{\gamma}{s} = h_d = \frac{\gamma_g}{(s + p)} \tag{13}$$

where γ_g = the continuum gas heat conductivity; γ = the rarified gas heat conductivity; s = local width of the gas gap; p = the product of the mean-free path of the gas molecules and a function of the accommodation coefficient (19); and h_d = the conduction heat-transfer coefficient based on heating surface. At the contact point, the gap goes to 0, and the coefficient reaches its maximum:

$$h_d(0) = \frac{\gamma_g}{p} \tag{14}$$

Figure 5 shows conduction heat transfer as a function of the projected radius of a 6-mm diameter sphere. Assuming an accommodation coefficient of 0.8, $h_d(0)$ = 3370 W/(m²·K); the average coefficient for the entire sphere is 72 W/(m²·K). This variation in heat transfer over the spherical surface causes extreme nonuniformities in local vaporization rates and if contact time is too long, wet spherical surface near the contact point dries. The temperature profile penetrates the sphere and it becomes a continuum to which Fourier's law of nonsteady-state conduction applies.

If the sphere is one of a mass of wet spherical particles, fastest drying occurs when the specific contact time of each particle approaches 0 in an ideally mixed bed. In general, gas thermal conductivity is independent of pressure between 150

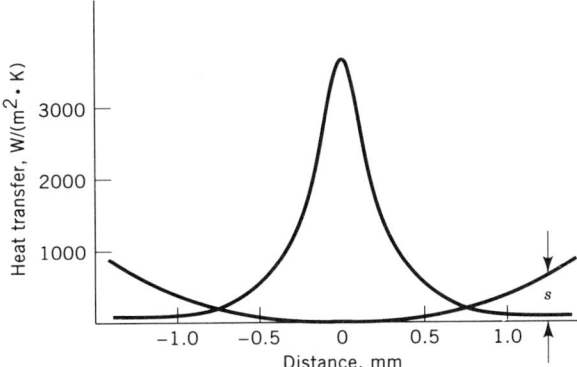

Fig. 5. Profile of conduction heat transfer across the gap between a sphere and a flat plate vs projected radius, R = 3 mm, of the sphere at 40°C and 2.1 kPa (0.30 psi); s is the width of the gas gap (17).

Pa (0.022 psi) and 10^6 Pa (145 psi); below 15 Pa (0.11 mm Hg), conductivity is almost proportional to absolute pressure (20). The mean-free paths of gas molecules are inversely proportional to pressure. Equations 13 and 14 state that conduction heat transfer must decrease as pressure decreases. At very low pressures, conduction heat transfer approaches 0 and radiation alone is effective. Conduction heat transfer is also influenced by particle shape, surface roughness, and probably specific gravity. In agitator-stirred dryers, agitator speed, mixing efficiency, and heating surface clearance are important variables.

Between 1 s and 1 min specific contact time, conduction heat-transfer performance decreases theoretically as the 0.29 power of contact time. This is consistent with empirical data from several forms of indirect-heat dryers which show performance variation as the 0.4 power of rotational speed (21). In agitator-stirred and rotating indirect-heat dryers, specific contact time can be related to rotational speed provided that speed does not affect the physical properties of the material. To describe the mixing efficiency of various devices, the concept of a mixing parameter is employed. An ideal mixer has a parameter of 1.

$$N_{\text{mech}} = \theta_s N \tag{15}$$

where N_{mech} is the mixing parameter, independent of thermal conditions; θ_s = specific particle contact time in seconds; N = agitator or dryer vessel rotational speed in units of 1/s. Values of N_{mech} reported for various dryers lie between 2 and 25. The principle applies to materials other than particulate solids. In the case of drum dryers used for solutions and slurries, capacity often can be increased by increasing drum speed, reducing specific contact time, and laying a thinner film on the drum surface. In the situations of cloth, paper, film, and fiber tow dried on heated cans, an additional resistance to drying is the inability of vapors and entrained gas to escape from between the moving material and the can surface, which also encourages shorter specific contact times.

Critical Moisture Content. Critical moisture content, which is the average material moisture content at the end of constant rate drying, is a function of material properties, the constant drying rate, particle size, or bed depth. Critical moisture content cannot be determined except by a prototypic drying test. For example, while a bed of material is drying during the constant rate period, assume that the drying rate is increased by increasing gas velocity over the material surface. The moisture gradient below the surface which causes liquid flow to replenish the surface becomes steeper and as the surface approaches dryness, the internal moisture content of the material is greater than it would have been had the gas velocity and drying rate not been increased. Critical moisture data are misleading, therefore, unless the exact conditions of drying are known (22,23).

Particle size distribution determines surface-to-mass ratios and the distance internal moisture must travel to reach the surface. Large pieces thus have higher critical moisture contents than fine particles of the same material dried under the same conditions. Pneumatic-conveyor flash dryers work because very fine particles are produced during initial dispersion and these have low critical moisture contents.

Case hardening refers to a circumstance in which a mass of nonporous, soluble, or colloidal material is dried at such a high rate during initial constant rate drying that the surface overheats and shrinks. Because liquid diffusivity decreases with moisture content, the barrier formed by the overdried surface prevents moisture flow from the interior of the mass to the surface. Case hardening of nonporous materials can be minimized by initially maintaining a high relative humidity environment and consequently a high surface equilibrium moisture content until internal moisture has time to escape.

Equilibrium Moisture Content. Equilibrium moisture content is the steady-state equilibrium reached by the gain or loss of moisture when material is exposed to an environment of specific temperature and humidity for a sufficient time. The equilibrium state is independent of drying method or rate. It is a material property. Only hygroscopic materials have equilibrium moisture contents. Clean beach sand is nonhygroscopic and has an equilibrium moisture content of 0. The same rules apply to organic vapors. Hygroscopic material retains a constant fraction of moisture under specific ambient humidity and temperature conditions. At constant temperature, if ambient humidity increases or decreases, an increase or decrease in moisture content follows. This is called equilibrium moisture because it is held in vapor pressure equilibrium with the partial pressure of vapor in the atmosphere. The reason it is retained even when the atmosphere is quite dry is that the retention mechanism reduces effective liquid vapor pressure. It is bound moisture because it is bound to material in solution or by adsorption and bound moisture behaves as if the atmosphere were saturated even when the atmosphere is not saturated relative to the unbound liquid's normal vapor pressure. Chemically combined liquid may behave like bound moisture depending on the nature of the chemical bond. Because equilibrium is influenced by partial vapor pressure in the atmosphere and the effective vapor pressure of the bound liquid, temperature and humidity are both important. For many materials in the 15–50°C temperature range, equilibrium moisture content can be plotted vs relative humidity as an essentially straight line. Equilibrium moisture content appears independent of temperature and relates to Henry's law:

$$p = H(x) \qquad (16)$$

where p = partial vapor pressure in the atmosphere in kPa; H = Henry's constant; and x = dry basis moisture content. Henry's constant is a function of the unbound liquid's vapor pressure:

$$H = f(p_s) \tag{17}$$

where p_s = the unbound liquid's vapor pressure; therefore, $p = f(p_s)(x)$, and since relative humidity = $100(p/p_s)$:

$$100(p/p_s) = 100 f(x) \tag{18}$$

At any given relative humidity, x is constant. In a typical silica gel–air–water vapor system at 5–50°C, $p/p_s = 1.79(x)$, where p = partial pressure of vapor in the air; p_s = water-vapor pressure at the adsorption temperature; x = gel moisture content, kg/kg (24). For many materials as the temperature increases above 50°C, equilibrium moisture content decreases at constant relative humidity. Above 100°C at atmospheric pressure, saturation humidity for water vapor goes to infinity and the concept of relative humidity becomes meaningless; in fact, hygroscopic materials often can be dried in 100% superheated vapor atmospheres.

A profile of equilibrium moisture content vs percent relative humidity (Fig. 6) often is not a perfectly straight line because at high humidities a porous material may retain condensed capillary moisture, whereas at low humidities moisture may be adsorbed in a single-molecular layer on capillary surfaces. The maximum bound moisture a material can hold is identified by the intersection of the equilibrium profile with the 100% relative humidity ordinate. A difference between adsorption and desorption profiles may have several causes. When material initially dries, shrinkage often closes many small capillaries which do not reabsorb moisture when the material is rewet. Also, some capillaries may be cul-desacs that resist vapor reentry once filled with gas. One reason for freeze drying

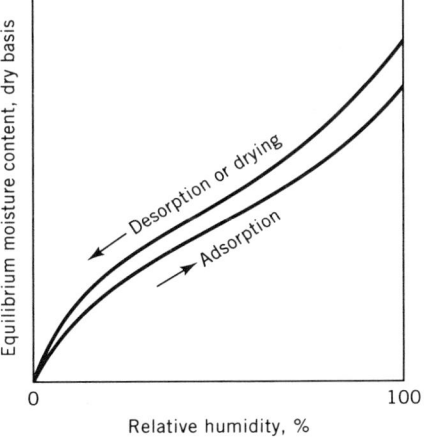

Fig. 6. Equilibrium moisture content profiles vs relative humidity for a hygroscopic material.

solid foods is to minimize shrinkage and capillary closing so the dry residue can be reconstituted to its original moisture content more easily. On the other hand, for laboratory and pilot-plant studies, a once-dried material must never be rewet and reused for a drying test. In fact, once-dried material can never be returned, physically, to its original wet condition.

Falling Rate Drying. Heat transfer is limited by material conductivity, but the drying rate usually is controlled by internal liquid and vapor mass transfer. The principal mass-transfer mechanisms are (1) liquid diffusion in continuous, homogeneous materials; (2) vapor diffusion in porous and granular materials; (3) capillarity in porous and fine granular materials; (4) gravity flow in granular materials; (5) flow caused by shrinkage-induced pressure gradients; and (6) pressure flow of liquid and vapor when porous material is heated on one side, but vapor must escape from the other.

Liquid flows by diffusion through materials in which the liquid is soluble, eg, single-phase systems like soap and gelatin. Movement of other bound moisture by liquid diffusion may occur, but the mechanisms probably are more complicated. Vapor flows by diffusion through the gas phase when liquid vaporizes below the surfaces of porous and granular materials. Wood and other cellular materials at moisture contents less than fiber saturation and the final drying stages of paper, textiles, and hydrophilic solids are examples. Vapor diffusion also occurs in the laminar sublayer of the gas film adjacent to the material surface during all drying periods. Diffusion-controlled mass transfer (qv) is assumed in drying when liquid or vapor flow conforms to Fick's second law of diffusion, which applies to nonsteady-state systems (see also DIFFUSION SEPARATION METHODS):

$$\frac{\delta c_A}{\delta \theta} = D_{AB} \frac{\delta^2 c_A}{\delta z^2} \qquad (19)$$

where c_A = concentration of one component in a two-component phase of A and B; θ = diffusion time; z = distance in the direction of diffusion; and D_{AB} = binary diffusivity of the two components. This equation applies to diffusion in solids, liquids, and gases. The analogous Fourier laws apply to heat conduction. The units of the diffusion coefficient, D_{AB} = $(mol/(s \cdot m^2))(m^3/mol)(m)$ = m^2/s, is the abbreviation usually employed both for this coefficient and the quantity thermal diffusivity in Fourier's heat conduction equations.

In porous and granular materials, liquid movement occurs by capillarity and gravity, provided passages are continuous. Capillary flow depends on the liquid material's wetting property and surface tension. Capillarity applies to liquids that are not adsorbed on capillary walls, moisture content greater than fiber saturation in cellular materials, saturated liquids in soluble materials, and all moisture in nonhygroscopic materials.

When clay or similar material is dried, often a pressure gradient is developed by the forces of repulsion between particles as shrinkage brings the particles close together (25). This gradient forces liquid toward the surface and the resulting moisture profile resembles that characteristic of liquid diffusion.

When a layer of material pervious to gas flow is dried by through-circulation, a drying front usually moves through the layer in the direction of gas flow. In circumstances when material moisture content is sufficiently high, incoming gas is sufficiently hot and the material is of sufficient depth, the gas may cool adiabatically to saturation before it passes fully through the layer. The gas may then be subcooled by material contact with consequent liquid condensation within the layer. Condensed liquid fills the flow passages and is forced through the layer as liquid by the pressure of the blocked gas stream. This phenomenon is known as a vaporization-condensation sequence. It is an ingenious drying process usable most commonly on stationary or moving combination filter-dryers. Thermal efficiency exceeds 100%.

Usually, only one mass-transfer mechanism predominates at any given time during drying, although several may occur together. In most materials, the mechanisms of internal liquid and vapor flow during falling rate drying are complex. Simultaneous heat transfer is a factor and falling rate drying rarely can be described with mathematical precision. Computer models for some materials are published (26), but most employ data from actual drying tests. In the absence of tests, the falling rate drying periods usually are studied on the assumption that internal mass transfer is controlled either by diffusion or capillarity depending on whether the material is porous or nonporous, soluble or not.

Diffusion. Characteristic drying time and rate profiles for liquid diffusion appear as profile C in Figure 4. Figure 7 contains a diffusion drying curve for corn kernels and introduces the complicating factor in solids drying that liquid diffusivity is affected by material moisture content. The diffusion coefficient is not constant, but decreases as material moisture content decreases. Nonetheless, for evaluation of falling rate drying by liquid diffusion an integration of equation 19 is employed using four simplifying assumptions: (1) liquid diffusivity is independent of moisture content; (2) initial moisture distribution is uniform; (3) material size, shape, and density are unchanging; and (4) the material's equilibrium moisture content is constant. For material in the form of a slab:

$$\frac{W_\theta - W_e}{W_c - W_e} = \frac{8}{\pi^2} \left[\sum_{n=0}^{n=\infty} \frac{1}{(2n+1)^2} \exp[-(2n+1)^2 D\theta(\pi/2d)^2] \right] \quad (20)$$

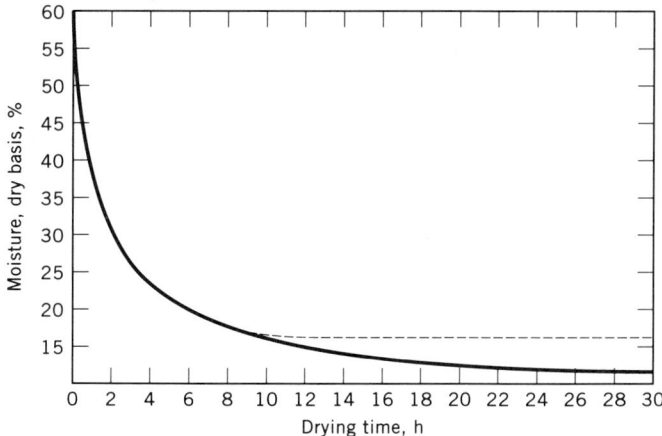

Fig. 7. Drying of corn kernels by liquid diffusion. The dashed line is that predicted by theory based on constant diffusivity. The solid curve shows actual performance at 38°C in air.

where W_θ, W_c, W_e are the moisture content at time θ, the first critical moisture content, and the equilibrium moisture content, respectively, so that the term on the left-hand side is the unaccomplished moisture change as defined; D = liquid diffusivity in units of m^2/s; θ = drying time in s; d = one-half the slab thickness for drying from both sides or total thickness for one-side drying in m. For long drying times, when $D\theta/d^2$ exceeds about 0.1,

$$\frac{W_\theta - W_e}{W_c - W_e} = \frac{8}{\pi^2} \exp[-D\theta(\pi/2d)^2] \qquad (21)$$

from which a drying rate expression can be derived

$$-\frac{dW}{d\theta} = \frac{\pi^2 D}{4d^2}(W_\theta - W_e) \qquad (22)$$

where $dW/d\theta$ = dry basis, drying rate in units of kg/(s·kg). When internal diffusion controls the drying rate, the rate is proportional to the free moisture content and diffusivity, and inversely proportional to the square of material thickness. When equation 20 is plotted semilogarithmically with unaccomplished moisture change as the ordinate and the quantity, $D\theta/d^2$ as the abscissa, a straight line results for $(W_\theta - W_e)/(W_c - W_e)$ values less than 0.6. Equation 21 describes the straight line portion of this plot. An approximate relationship for falling rate drying by liquid diffusion is

$$\theta_f = \frac{4d^2}{\pi^2 D} \ln\left[\frac{8(W_c - W_e)}{\pi^2(W_\theta - W_e)}\right] \qquad (23)$$

where θ_f = falling rate drying time in s. For nonhygroscopic materials, where $W_e = 0$, the unaccomplished moisture change becomes simply W_θ/W_c. Equations 20–23 apply to materials in layers or slabs. Equation 19 is solved for other shapes as well (27), and these equations also may be used to study vapor diffusion in porous and granular materials (28).

An analogy exists between mass transfer by diffusion and heat transfer by conduction. Each involves collisions between molecules and a gradient as the driving force which causes flow. For diffusion, this is a concentration gradient; for conduction, the driving force is an energy gradient. Fourier's nonsteady-state conduction equation is analogous to equation 19, Fick's second law of diffusion,

$$\frac{\delta T}{\delta \theta} = \left(\frac{k}{\rho c_p}\right)\frac{\delta^2 T}{\delta z^2} \qquad (24)$$

where T = material temperature in K; θ = conduction time in s; z = distance in the direction of conduction in m; k = material thermal conductivity in units of kW·m/(m^2·K); ρ = material density in kg/m^3; and c_p = material specific heat in kJ/(kg·K). The quantity, $k/\rho c_p$ is the thermal diffusivity of the material. This term is needed because temperature is a scale and not a quantity. The units of thermal diffusivity are m^2/s, the same as for the diffusion coefficient. Solutions for equation 24 are available for many material shapes (29–32). During falling rate drying, material conductivity is important because liquid in porous materials is vaporized below the surface and heat must be conducted to this liquid by the dry material.

Capillarity. The outer surface of porous material has pore entrances of various sizes. As surface liquid is evaporated during constant rate drying, a meniscus forms across each pore entrance and interfacial forces are set up between the liquid and material. These forces may draw liquid from the interior to the surface. The tendency of liquid to rise in porous material is caused partly by liquid surface tension. Surface tension is defined as the work needed to increase a liquid's surface area by one square meter and has the units J/m^2. The pressure increase caused by surface tension is related to pore size:

$$\Delta p = s_t/r \qquad (25)$$

where Δp = the meniscus pressure increment resulting from surface tension in Pa; s_t = surface tension in J/m^2; and r = pore radius in m. The excess pressure resulting from surface tension is always directed from the concave toward the convex surface of the meniscus.

A second property important to capillarity is surface wetting ability which depends on properties of both the liquid and material. Wetting ability is indicated by the contact angle formed at the liquid–material interface; eg, water and clean glass have a contact angle of 0° and water rises in glass capillaries; mercury has a contact angle with glass of 132° and does not rise. When the contact angle is less than 90°, the force of surface adhesion exceeds liquid cohesive strength. Liquid molecules climb the capillary wall and surface tension causes a liquid column to follow. Liquids rise higher in fine capillaries because adhesive force is a wall effect. In very fine capillaries, meniscus radii become so small that the pressure increase caused by surface tension suppresses liquid vapor pressure. The effect is described in the Kelvin equation

$$\ln(p/p_s) = -2s_t M_1 \cos\phi/(r\rho RT) \qquad (26)$$

where p = effective vapor pressure of the capillary liquid in kPa; p_s = normal liquid vapor pressure in kPa, at temperature T in K; M_1 = liquid molecular weight in kg/mol; ρ = liquid density in kg/m^3; ϕ = liquid contact angle; R = 8314 $J/(mol \cdot K)$, the ideal gas constant; and r = pore radius in m. Calculated effects for water at 50°C and an s_t of 0.06791 J/m^2 give the following:

p/p_s	$r(\mu m)$
0.999	0.910
0.990	0.091
0.960	0.022
0.920	0.011
0.900	0.009
0.800	0.004
0.700	0.003
0.500	0.001

This is the capillary condensation phenomenon, which partly accounts for the hysteresis observed in adsorption profiles of porous materials.

At the critical moisture content, at the end of constant rate drying, dry areas begin to appear on the material surface. Menisci in the larger pores begin to withdraw below the surface. As drying continues, the surface becomes completely dry and liquid withdraws in even the smallest pores. This completes the first falling rate drying period which is represented by the straight line, decreasing rate portions of profiles A and B in Figure 4b. Final drying during the second falling rate period is accomplished by heat conduction to the liquid pockets and vapor diffusion through the pores to the material surface. In most porous materials during this period, the drying rate profile has the concave upward shape of diffusion control as appears in profile B, Figure 4b. In granular materials where pores are large and capillary forces are weak, gravity contributes to retreat of the liquid surface and both falling rate profiles may be straight lines (33). In other capillary porous materials dried from two sides, or in thin layers from one side, complete drying also may occur at a drying rate proportional to residual moisture content. Figure 8 is an example and falling rate drying often may be approximated by assuming the rate proceeds in this manner

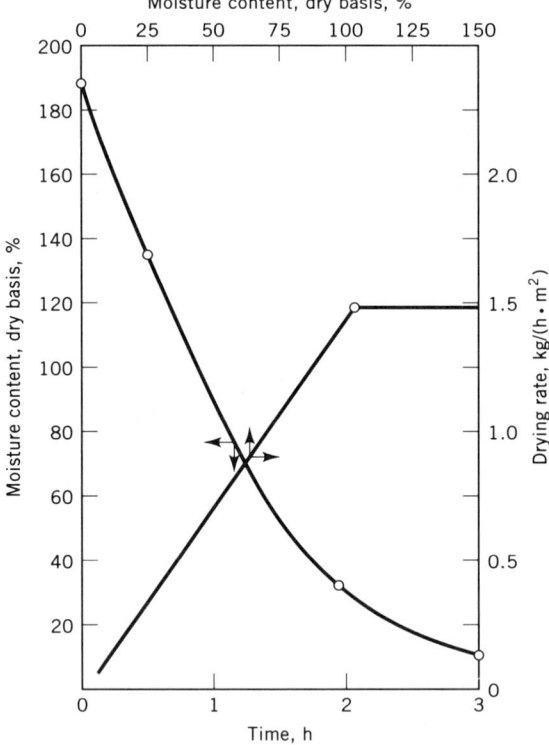

Fig. 8. Drying time and rate profiles for leather pasted on glass plates and dried in two temperature stages. Gas velocity = 5 m/s in parallel flow, 71°C in the first stage, 57°C in the second. The falling rate, drying rate is proportional to residual moisture content.

$$-\frac{dW}{d\theta} = K(W_\theta - W_e) \qquad (27)$$

where K is a function of the constant rate, drying rate at the critical moisture content,

$$-\frac{dW}{d\theta_c} = K(W_c - W_e) \qquad (28)$$

For a layer of wet material, using equation 8,

$$K = \frac{h_t}{\rho_m d_m L'_s}\left[\frac{(t - t'_s)}{(W_c - W_e)}\right] \qquad (29)$$

and

$$-\frac{dW}{d\theta} = \frac{h_t(t - t'_s)}{\rho_m d_m L'_s}\left[\frac{(W_\theta - W_e)}{(W_c - W_e)}\right] \qquad (30)$$

For materials that follow equation 30 the drying rate is inversely proportional to material thickness; falling rate drying time is estimated,

$$\theta_f = \frac{\rho_m d_m L'_s(W_c - W_e)}{h_t(t - t'_s)} \ln\left[\frac{(W_c - W_e)}{(W_\theta - W_e)}\right] \qquad (31)$$

A relationship for through-circulation drying analogous to equation 30 employs equation 9.

$$-\frac{dW}{d\theta} = \frac{h_t a(t - t'_s)}{\rho_m L'_s}\left[\frac{(W_\theta - W_e)}{(W_c - W_e)}\right] \qquad (32)$$

Drying Profiles. An application of diffusion principles to falling rate drying is exemplified in Figure 9 (34). Single drops of whole milk were dried by suspension in a warm air stream. Because of the rapid formation of surface films, drying was mostly by vapor diffusion. Drying times to an unaccomplished moisture change of 0.1 were 190 s and 300 s, respectively, for drops B and A. Based on equations 22 and 23 and employing the square of the initial drop diameter for the d^2 dimension, drying time for the larger drop should have been 317 s. Neglecting the initially 8.0% greater moisture content of the larger drop, drying time and rate varied with the 1.8 power of initial drop diameter. The error in predicting drying time for the larger drop is roughly 6%, which is probably attributable to material shrinkage during drying.

Figure 10 depicts freeze drying data for two milk products (35). Heat and mass transfer involved radiation to the material surface, conduction through the material to the retreating ice phase, and vapor diffusion through ice-free capillaries above the ice phase to the surface. Because both conduction heat transfer

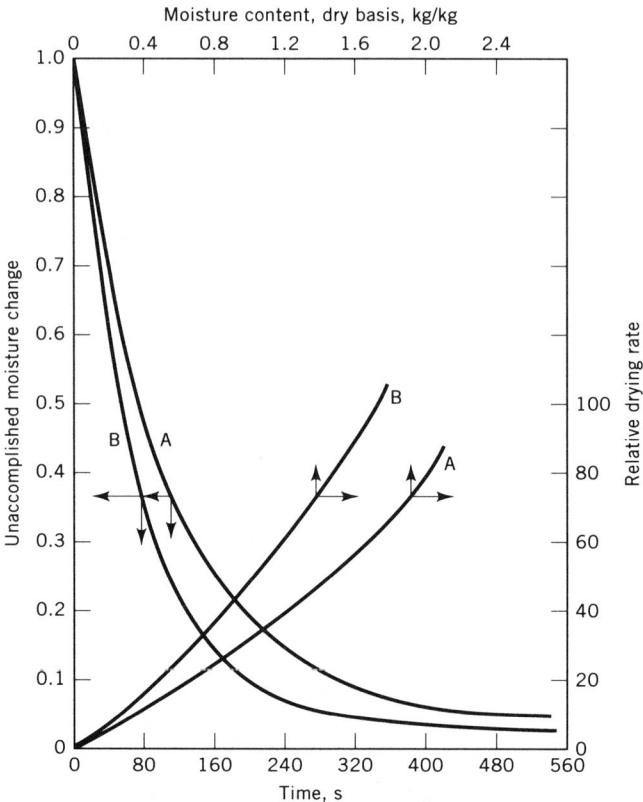

Fig. 9. Drying profiles for single drops of whole milk at 94°C and 0.6 m/s relative air flow. The initial diameter of drop A = 1900 μm, initial moisture content = 2.6 kg/kg, dry basis; drop B = 1470 μm initial diameter and 2.4 kg/kg moisture.

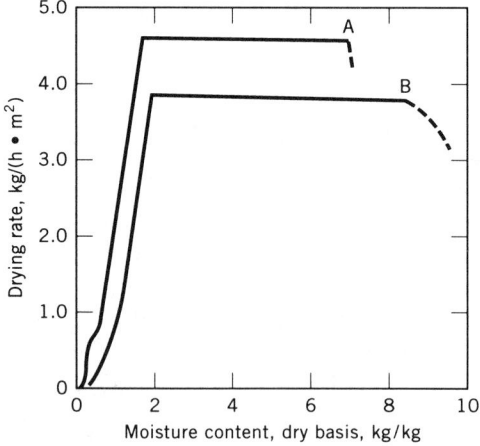

Fig. 10. Freeze drying profiles for A, whole milk, and B, nonfat milk. Heat was transmitted by radiation from heated wires above the frozen milk which rested in a transparent plastic tray. (– – –) is the induction period. Total pressure was 33 Pa (0.25 mm Hg).

and vapor diffusion are distance dependent, a continously falling rate drying profile would be expected. In freeze drying, however, it is necessary to control heat input to prevent temperature rise and melting of the ice. Also, it is necessary to limit vapor transport to condenser capacity so as to prevent drying chamber pressure rise. In this situation, dryer operation is controlled by external factors. The drying profile of necessity is mostly constant rate.

Dryers

Industrial dryers may be broadly classified by heat-transfer method as being either direct or indirect heat. Dryers evolved from material handling equipment and thus most types of industrial dryers are specially suited for certain forms of material. Dryers are also classified as being batch or continuous.

A batch dryer is best suited for small lots and for use in single-product plants. This dryer is one into which a charge is placed; the dryer runs through its cycle, and the charge is removed. In contrast, continuous dryers operate best under steady-state conditions drying continuous feed and product streams. Optimum operation of most continuous dryers is at design rate and steady-state. Periods of low rate operation are energy inefficient; shutdowns and start-ups waste fuel and frequently include periods of offgrade production. Continuous dryers are unsuitable for short operating runs in multiproduct plants.

The material suitability of industrial dryers may be summarized as

Dryer	Material form
spray dryer	pumpable, heat-sensitive pastes, slurries, and solutions; all pumpables at high capacities
indirect-heat drum dryer	pumpable, heat-insensitive pastes, slurries, and solutions
pneumatic conveyor dryer	materials instantly dispersible into discrete particles in the drying gas
fluid-bed dryer	fluidizable particulate materials
spouted-bed dryer	particulate materials too coarse or uniform in size to fluidize adequately
hopper dryer	preheated coarse or uniform materials and beds pervious to gas throughflow
direct-heat rotary dryer	particulate materials too coarse, sticky, or unpredictable to be fluidized or spouted
indirect-heat rotary dryer	fine, dusty materials
double-cone (vacuum) dryer	particulate materials that do not stick together, ball, or pelletize during drying
agitator (vacuum) dryer	particulate materials that may stick together, ball, or pelletize until almost dry
through-circulation or band dryer	materials that can be formed into static beds pervious to gas throughflow
continuous conveyors	continuous webs, paper, fabric, film, and fiber tow
batch, cabinet, and tray dryers	small lots, batch identification, and single-product plants less than 500 t/yr

Direct-Heat Dryers. In direct-heat dryers, steam-heated, extended-surface coils are used for gas heating up to about 200°C. Electric and hot oil or vapor heaters are added for higher temperatures. Diluted combustion products are used for all temperatures. An increasingly popular technique for producing inert gas is to recycle the dryer exit gas and vapor as secondary dilution gas for incoming combustion products. Thereby the oxygen level in the dryer gas stream is reduced to safe levels for organic materials. This is usually less than 10% oxygen, but always material-dependent. These are called self-inerting heaters.

If material must be protected from combustion product contact, gas may be heated indirectly by passing it through tubes in a furnace. A clean, high temperature gas is obtained, but fuel efficiency is 50–70% of direct combustion gas heaters. Unless metal surfaces are protected by insulating or refractory lining, maximum usable gas temperature is about 1000°C. For low temperature operations, gas may be dehumidified. It usually is more economical to recirculate gas back through the dehumidifier after each dryer pass than to continuously dehumidify fresh gas. Polymers dried to very low moisture contents for extrusion or solid-state polymerization require very dry gas regardless of drying temperature. It is more economical to predry these materials as low as possible using ambient humidity gas before final drying in very dry or inert gas.

In most direct-heat dryers, more gas is needed to transport heat than to purge vapor. Larger dust recovery installations are needed than for indirect-heat dryers handling the same vapor load. Strict environmental regulations have eliminated the capital cost advantage of direct-heat dryers, eg, dust recovery investment for a modern spray dryer often exceeds the dryer investment. The greater the gas velocity over, through, or impinging upon a material, the greater the convection heat-transfer coefficient. The more completely material is dispersed, ie, greater surface-to-mass exposure, the faster the drying rate. Gas and material flowing in the same direction in a continuous dryer is called cocurrent flow; gas flow opposing material flow is countercurrent flow. Gas flow across a material is parallel flow or crossflow. Gas flow normal to material flow is impingement flow or through-circulation.

Batch Compartment Dryers. Direct-heat batch compartment dryers are often called tray dryers because of frequent use for drying materials loaded in trays on trucks or shelves. Figure 11 illustrates a two-truck tray dryer. The compartment enclosure comprises insulated panels designed to limit exterior surface temperatures to less than 50°C. Slurries, filter cakes, and particulate solids are placed in stacks of trays; large objects are placed on shelves or stacked in piles. An important design requirement is to assure gas flow uniformity, top-to-bottom of the compartment and back-to-front. This is essential but consumes fan power. Unless the material is dusty, gas is recirculated through an internal heater as shown. Only enough purge is exchanged so as to maintain needed internal humidity. For inert gas operation, purge gas is sent through an external dehumidifier and returned.

These dryers are economical only for a single-product rate less than 500 t/yr, large objects on drying cycles greater than 8 h, multiproduct operations, and batch identification. Tray loading depth and spacing must be uniform throughout the compartment. Tray loading is usually 2–10 cm. Parallel flow gas velocity is 1–10 m/s. Two-speed or variable speed fans are employed to provide higher gas velocity

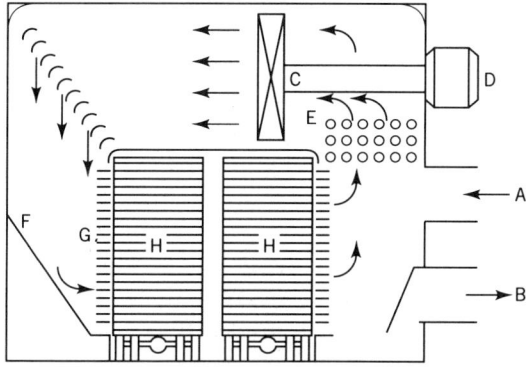

Fig. 11. Two-truck tray dryer. A, air inlet duct; B, air-exhaust duct with damper; C, axial flow fan; D, fan motor, 2–15 kW; E, air heaters; F, air-distribution plenum; G, distribution slots; and H, wheeled trucks and trays. The arrows indicate air and vapor flow pattern.

over the material during early drying stages. To minimize dusting, the fans reduce velocity after constant rate drying when heat transfer at the material surface is no longer the limiting drying mechanism. Deep tray loading reduces labor, but reduces overall capacity because falling rate drying time usually varies with the square of the loading depth. Shallow loading yields faster drying, but care is needed to ensure depth uniformity and labor is increased. Metal trays enhance heat flow through the tray walls and bottom. Screen-bottom trays permit vapor escape through the bottom; ie, two-side drying. Dryer efficiency should be 50–70%. Based on exposed material surface, vaporization rates are 0.2–2.0 kg/(h·m^2).

Through-circulation compartments employ perforated or screen bottom trays and suitable flow baffles so gas is forced through the material. If material is not inherently pervious to gas flow, it may be mechanically shaped into noodles, pellets, or briquettes. These dryers are used in small-scale operations to dry explosives, foods, and pigments. Dryer efficiency is 50–70%. Based on tray area, water vaporization rates are 1–10 kg/(h·m^2).

Continuous Conveyors. Continuous conveyors are characterized by continuous material flow without mixing. Dryer residence time is uniform for all material increments. Tray and tunnel conveyors comprise long insulated compartments through which material is moved on trucks or trays fastened together. Gas flow is usually parallel to material surface and may be cocurrent, countercurrent, or crossflow through recirculation fans and reheaters installed on each side of the compartment. Conveyor movement may be continuous, but usually must be interrupted periodically for introduction to new work at the wet end. Performance is otherwise comparable to batch compartments. Figure 8 data were obtained from a continuous conveyor dryer in which the glass plates were suspended vertically from an overhead chain conveyor.

Turbotray Dryers. The turbotray dryer is a continuous tray dryer comprising a stack of circular trays rotating slowly inside a vertical, insulated, cylindrical housing. Each rotating tray has uniformly spaced radial slots through which material is discharged to the tray below by a stationary plough once per revolution.

Material falling through a slot is leveled to a uniform depth on the tray below by a stationary rake. After another revolution the process repeats. Wet material fed onto the uppermost tray moves down the stack in this manner and exits at the bottom. Circulating fans are carried on the central rotating shaft. Gas reheaters are mounted on the housing walls and gas flows across the trays parallel to material surface. Free flowing, nonsticky, and nondusty materials are dried more rapidly than in static beds. Having a stationary housing, the dryer finds employment in inert gas, solvent recovery, and sublimation processes.

Foam-mat drying is a process in which a suspension, slurry, or solution is transformed into a stable foam by inert gas injection. The foam structure provides porosity and the mat is dried in trays or on a belt in a tunnel compartment, either under vacuum or with circulating gas. A free-flowing powder capable of rapid rehydration results. Fruit juices are dried successfully in this manner.

Continuous Web Dryers. Web Dryers are used for polymer films, paper, cloth, nonwoven fabrics, printed and coated films, and printed fabrics. Gas impinges on or flows parallel to the moving material, called a web, that is supported by various methods. Electric and gas-fired radiant heaters also are usable on some dryer types. On a festoon conveyor the web is draped over sticks that are carried on chains through a heated enclosure. The web is unrestrained and free to shrink or stretch. Gas flow must be comparatively gentle to avoid excessive material movement. On single or multipass roll conveyors web is conveyed either vertically or horizontally over a series of driven rolls while web tension is controlled by differential roll speeds. The rolls are crowned slightly to hold axial alignment, but there is no restriction to lateral shrinkage. Because the web is restrained axially, however, high velocity gas impingement slots or nozzles may be employed on one or both faces. Radiant heaters are used in these dryers. For one-side drying of printed or coated webs, roll conveyors often are installed in ceiling-hung housings to conserve floor space. Tenter frames restrain a web in two directions and are employed to control shrinkage or to stretch a web during drying. It is an ideal setup for two-side gas impingement heating because the nozzles or orifice plates can be mounted close to the stretched web. The closer the spacing between the nozzles and web, the more effective are the impinging jets for heat and mass transfer. Tenter frames are also ideal for electric and gas-fired radiant heaters. Float dryers have closely mounted nozzles both for heat and mass transfer and to support and convey with minimum web tension. For long drying times, festoon or multipass roll conveyors are economical because a long length of web can be contained in a tall enclosure that occupies little floor space.

Through-Circulation Dryers. In through-circulation dryers, permeable materials are conveyed through enclosures on perforated plate or screen conveyors. The enclosures comprise a series of independent compartments, each having its own fans and recirculating gas heaters. Humid gas is removed at the material feed end of the enclosure; fresh dry gas is introduced at the dry end. Conveyor widths are 0.5–5.0 m; length may be 50 m in single or multiple conveyor tiers. Textile fibers, elastomer crumbs, plastic pellets, vegetables, and centrifuge and filter cakes are dried in 1–20-cm deep layers. On a conveyor area basis, drying rates up to 50 kg/(h·m^2) of water are obtained at gas temperatures up to 400°C. Dryer efficiency is 50–70%. Centrifuge and filter cakes are preformed by extrusion into small noodles or by granulation in knife mills. Thin pastes and slurries are predried on indirect-heat drum dryers to form short sticks. Shear-thinning materials are scored and cut in small pieces. Powders are briquetted or pelleted. Pin elevator feeders open fibrous clumps and lay a uniform bed on the conveyor. Fiber tow is distributed by oscillating chutes. Materials that shrink during drying may be redistributed on a second conveyor to prevent gas bypassing by ensuring full conveyor coverage.

Perforated-drum dryers are through-circulation conveyors specially suited for fiber staple, tow, and nonwoven fabrics. Material is continuously supported and conveyed on a series of perforated screen-covered suction drums installed in compartments similar in form to horizontal conveyor compartments. Compared to the former, drying is more uniform because the material is turned over as it passes from one drum to the next. Drying rates are greater than on perforated plate and screen conveyors because greater pressure drop and higher gas velocities can be taken through the drums. Because material is retained by drum suction, edge sealing is less of a problem as well.

Dispersed-Particle Dryers. Through-circulation conveyors realize relatively high drying rates because gas flows through the material and contacts more material surface than do parallel flow, crossflow, and impingement arrangements. Nonetheless, if layer depth or material porosity is not uniform, gas channels through thin areas or larger passages and drying is not uniform. For drying particulate material, a better arrangement is one in which particles are separated completely, so that gas can flow freely among them. The drying rate for all particles of a given size then should be uniform. This is the purpose of all dispersed-particle dryers. Each is intended to provide optimum conditions of particle separation and surface exposure for materials having specific material handling requirements. A disadvantage, compared to continuous conveyors, is the loss of material plug-flow. Particle residence time in dispersed-particle dryers varies around an average and only the average can be calculated from feed rate, dryer fillage, and material density. Variations may be narrowed by various devices, but never eliminated completely. Figure 12 depicts relative particle residence time distributions among four dryers.

Rotary Dryers. A direct-heat rotary dryer is a horizontal rotating cylinder through which gas is blown to dry material that is showered inside. Shell diameters are 0.5–6 m. Batch dryers are usually one or two diameters long. Continuous dryers are at least four and sometimes ten diameters long. At each end, a stationary hood is joined to the cylinder by a rotating seal. These hoods carry the inlet and exit gas connections and the feed and product conveyors. One hood also attaches to the inlet gas heater. For continuous drying, the cylinder may be slightly inclined to the horizontal to control material flow. An array of material showering flights of various shapes is attached to the inside of the cylinder, as shown in Figure 13. Knockers are use to dislodge wet material that sticks to metal surfaces. A flight circle is usually 0.5 diameters long and adjacent circles are offset to minimize gas bypassing. Dry product may be recycled for feed conditioning if material is too fluid or sticky initially for adequate showering. Slurries also may be sprayed into the shell in a manner that the feed strikes and mixes with a moving bed of dry particles. Material fillage in a continuous dryer is 10–18% of cylinder volume. Greater fillage is not showered properly and tends to flush toward the discharge end.

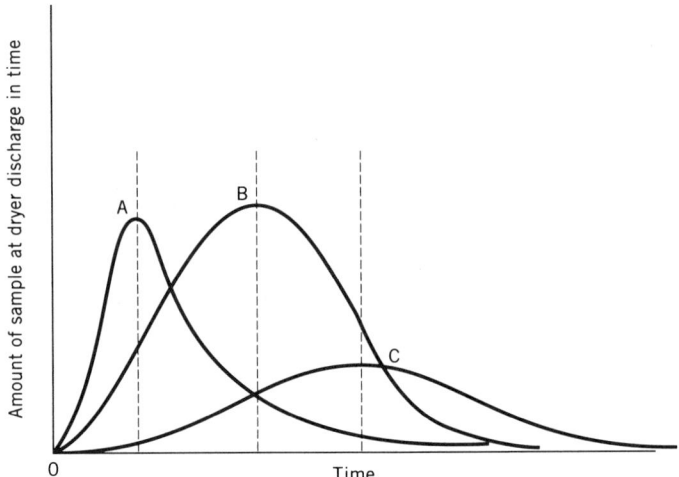

Fig. 12. Material residence time profiles in A, a pneumatic conveyor, B, a spray, and C, a rotary or fluid-bed dryer.

Gas flow in these rotary dryers may be cocurrent or countercurrent. Cocurrent operation is preferred for heat-sensitive materials because gas and product leave at the same temperature. Countercurrent operation allows a product temperature higher than the exit gas temperature and dryer efficiency may be as high as 70%. Some dryers have enlarged cylinder sections at the material exit end to increase material holdup, reduce gas velocity, and minimize dusting. Indirectly heated tubes are installed in some dryers for additional heating capacity. To pre-

Fig. 13. Partial view through a direct-heat rotary dryer.

vent dust and vapor escape at the cylinder seals, most rotary dryers operate at a negative internal pressure of 50–100 Pa (0.5–1.0 cm of water).

Direct-heat rotary dryers are the workhorses of industry. Most particulate materials can somehow be processed through them. These dryers provide reasonably good gas contacting, positive material conveying without serious backmixing, good thermal efficiency in either cocurrent or countercurrent use, and good flexibility for control of gas velocity and material residence time. Usual water drying rates are 10–50 kg/(h·m^3) of cylinder volume.

Fluid and Spouted Beds. A fluid bed of particulate material is produced by introducing gas through a perforated plate, bubble caps, nozzles, or a ceramic grid beneath a static bed of material in such a manner that the solids are lifted uniformly and the particulate material and gas behave together like a boiling fluid. For drying, the upward gas velocity is less than the terminal settling velocity of the particles, so few particles are conveyed out of the bed. At the same time, gas bubbles rise fast enough to lift particles directly above them. Particle motion is violent and a fluid bed exhibits intensive splashing at its surface. A substantial freeboard is included above the bed in the fluidizing vessel to allow particle disentrainment and full-back into the bed. Possibly because of a gas cushion that surrounds each particle as it circulates in the bed, particle attrition is usually moderate. For proper fluidization, it is essential that sufficient pressure drop be taken through the gas distributor so that the gas is distributed uniformly across the entire bed area independently of bed depth or bed behavior.

Spouted beds are used for coarse particles that do not fluidize well. A single, high velocity gas jet is introduced under the center of a static particulate bed. This jet entrains and conveys a stream of particles up through the bed into the vessel freeboard where the jet expands, loses velocity, and allows the particles to be disentrained. The particles fall back into the bed and gradually move downward with the peripheral mass until reentrained. Particle-gas mixing is less uniform than in a fluid bed.

Hopper Dryers. Gas-purged hopper dryers are used for granular materials that need holdup times measured in hours. Gas flow may be upflow or crossflow. Drying of pelleted and extruded animal feeds at less than 100°C are typical operations. Applications also include the continuous final drying of polymer pellets, such as nylon and polyester at temperatures of 150–200°C, prior to melt extrusion or solid-state polymerization. Gas flow is countercurrent to material flow and rarely exceeds 0.25 m/s. Because flow is insufficient to provide needed sensible heat, it is necessary to preheat the polymers to drying temperature before introduction into the hoppers. Drying usually starts at 0.1–0.5% moisture, so evaporative thermal loads are small. The hoppers serve essentially as holding vessels, at temperature, to permit release by diffusion of minute quantities of moisture. This prevents polymer degradation during later processing. A free-flowing character is the principal material requirement. Uniform material holdup is the principal hopper requirement, so all particles are retained the minimum required time. Average holdup usually is 2–3 times the minimum required.

Fluid and spouted beds offer ideal conditions for drying provided the feed material is consistently suitable for fluidization or spouting; however, if the drying operation is preceded by mechanical liquid separation, eg, centrifugation, use of these dryers should be considered with caution. Fluid and spouted beds do not tolerate sticky materials and oversize lumps. Successful applications are particulate and pelleted polymers, grain, sand, coal, and mineral ores, applications wherein the physical size and character of the feed material is known and controllable 100% of the time. Fluid and spouted beds are attractive for inert gas and organic liquid drying because the vessels are stationary. Superheated steam drying is carried out in fluid beds. This is an attractive alternative environmentally, but a process which was stalled for many years because of the lack of a suitable process vessel. The volumetric drying capacity of a fluid bed is many times that of a rotary dryer. The reason is that gas flowing through the latter moves between a series of parallel particle curtains in which the gas must be entrained and mixed to contact particle surfaces. In the former, small bubbles of gas enter through the distributor and immediately penetrate and mix with a cloud of particles. Figure 14 shows that whereas the dryer efficiency of a cocurrent rotary dryer and fluid bed may be comparable, because both are single-stage vessels, the vessel size requirements are quite different. To approach the dryer efficiency of a countercurrent rotary dryer, two or more fluid beds with countercurrent gas flow must be operated in series. Figure 15 shows one form of a two-stage fluid bed.

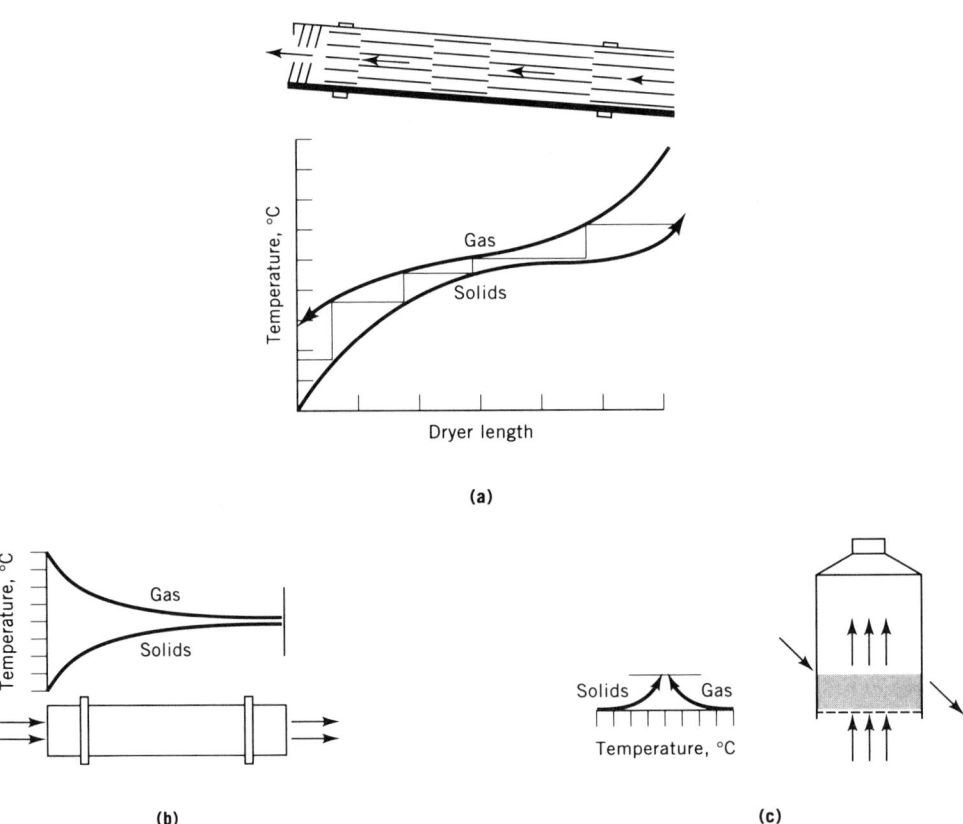

Fig. 14. Temperature profiles of gas and material in direct-heat dryers: (**a**) a countercurrent rotary dryer; (**b**) a cocurrent rotary dryer; and (**c**) a single-stage fluid bed.

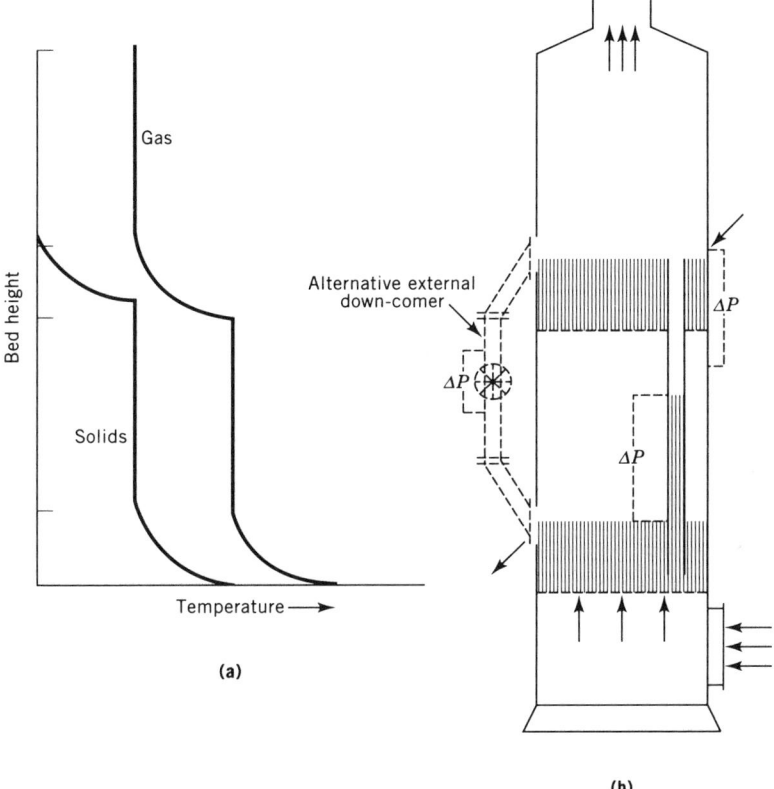

Fig. 15. A two-stage fluid bed dryer: (**a**) gas and material temperature profiles; (**b**) bed arrangements. ΔP = pressure drop through the upper stage distributor and bed.

A vibrating conveyor fluid bed dries while conveying particulate material on a screen-covered perforated deck. Gas is blown up through the material as it is conveyed mechanically and a particle dispersion much like that in a shallow fluid bed may be produced. Both mechanical and fluid energy contribute to fluidization. To minimize dusting, a lower fluidizing velocity is used than in a stationary fluid bed. Bed depth rarely exceeds 50 mm because mechanical energy is not transmitted through deeper beds effectively. Mechanical conveying encourages plug flow and several temperature stages may be incorporated in a single conveyor, but maximum material residence time is about 5 min. As in stationary fluid beds, all feed material must be nonsticky and free-flowing.

Pneumatic Conveyors. Conveyors are adapted for drying by heating the conveying gas, although for drying, gas-to-material ratios must be greater than those sufficient for conveying. Particle residence time is only a few seconds; in fact, most drying takes place near the feed point where the velocity difference between gas and material is the greatest. Conveying tubes rarely need to be over 10 diameters long. For drying accompanied by deagglomeration, two or three pneumatic conveyors may be used in series and dry product may be recycled to the first stage for feed conditioning. Figure 16 shows a simple dryer consisting of a venturi

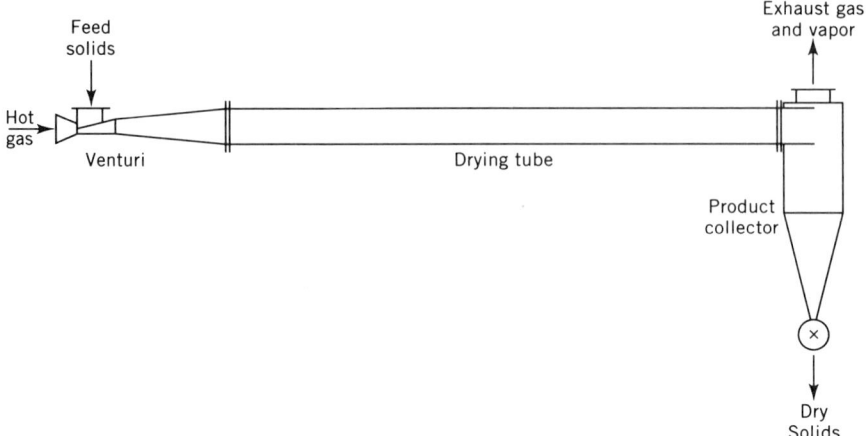

Fig. 16. Single-stage pneumatic conveyor dryer using a venturi for material acceleration and gas–solids mixing.

feeder, tube, and product collector. Knife, hammer, and roller mills are alternative feeding devices. A paddle conveyor, its paddles inclined to retard material flow, may be installed in place of the venturi to increase residence time and enhance dispersion. Conveying tube gas velocity is usually 25–35 m/s; venturi throat velocity is 100–140 m/s. The conveyor is a low power fluid energy mill and particle attrition may be severe. This dryer is single-stage and cocurrent, like the cocurrent rotary, but has much lower residence time. Indirect heat may be combined with direct heat by jacketing the conveying tube.

The principal applications of pneumatic conveyors are for materials that are nonsticky and readily dispersible in the gas stream as drying must be entirely constant rate. Many are employed as predryers ahead of longer residence time fluid-bed and rotary dryers in polymer drying operations.

Spray Dryers. A spray dryer is a large, usually vertical chamber through which hot gas is blown and into which a solution, slurry, or pumpable paste is sprayed by a suitable atomizer. Three atomizers are commonly used: (*1*) two-fluid pneumatic nozzles for very fine particles, between 10 μm and 100 μm, at rates less than 2 t/h; (*2*) single-fluid pressure nozzles for large particles, 125–500 μm, for dust-free products; and (*3*) centrifugal atomizers for various particle sizes at rates up to 150 t/h. The largest spray-dried particle is about 1000 μm; the smallest is about 5 μm. For large capacity dryers, the first two atomizers require multiple-nozzle setups, which may result in spray interference and particle agglomeration. In two-fluid nozzles, particle size is controlled by atomizing fluid pressure, usually compressed air or steam, and atomizing fluid-to-liquid ratio; therefore, very fine particles are obtainable with high fluid-to-liquid ratios and these nozzles are preferred for small, low capacity dryers. Pressure nozzles are limited by the fact that a change in feed rate or atomizing pressure causes a significant change in particle size distribution unless the nozzle orifice size is changed; high pressure feed pumps are required and, with abrasive materials, orifice wear may be rapid with a consequent increase in particle size distribution.

Centrifugal atomizers are characterized by large capacity ranges, simple feed systems, narrow particle size distributions, and ease of particle size control by disk speed changes.

Because all drops must reach a nonsticky state before striking a chamber wall, the largest drop produced determines the size of the drying chamber. Chamber shape is determined by nozzle or disk spray pattern. Nozzle chambers are tall towers, usually having height/diameter ratios of 4–5. Disk chambers are large diameter and short, height being fixed by the fact that the discharge cone slope must be at least 60°, preferably 70° to discourage dry product accumulation on the sloping wall. For any evaporative load, chamber volume may be estimated by assuming a usable inlet gas temperature, exit gas and product at 100°C, and an average gas residence time of 15 s based on total chamber volume and exit gas humid volume, ie, from a simple dryer heat balance. This calculation does not obviate pilot-plant demonstrations, of course.

A spray dryer may be cocurrent, countercurrent, or mixed flow. Cocurrent dryers are used for heat-sensitive materials because relatively high inlet gas temperatures, up to 800°C, may be used while holding the exit gas and product near 100°C. Material temperature usually does not exceed the exit gas temperature provided chamber wall sticking is avoided. At any rate, a maximum dryer inlet gas temperature is about 1100°C because of limits for materials of construction. Countercurrent spray dryers yield higher bulk density products and minimize hollow particle production. Figure 17 shows an open-cycle, cocurrent, disk atomizer chamber with a pneumatic conveyor following for product cooling. Alternative

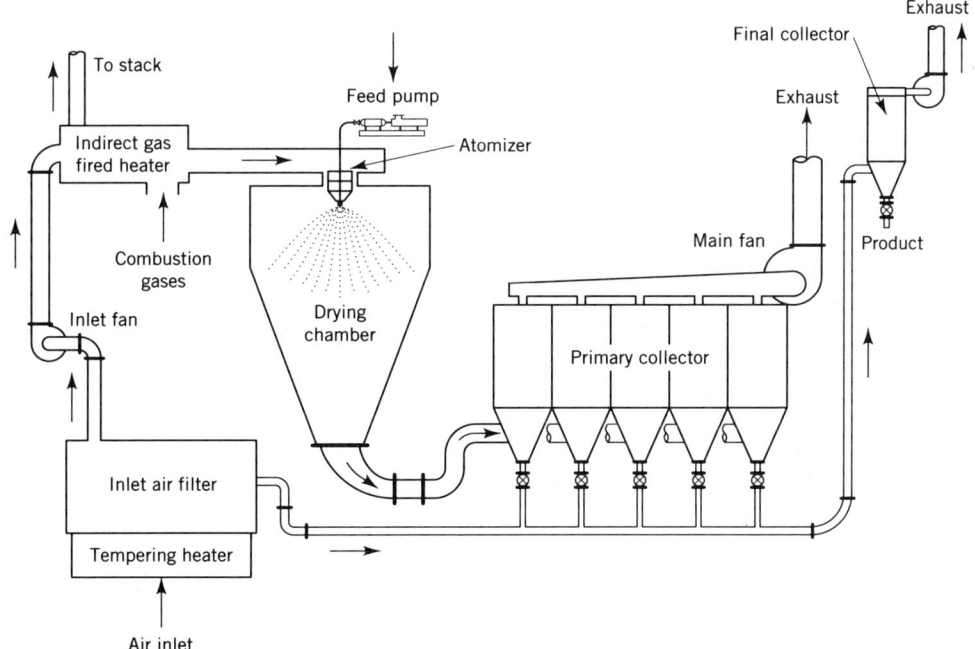

Fig. 17. Open-cycle, cocurrent, disk atomizer spray dryer. Courtesy of Niro Atomizer, Inc., Columbia, Md.

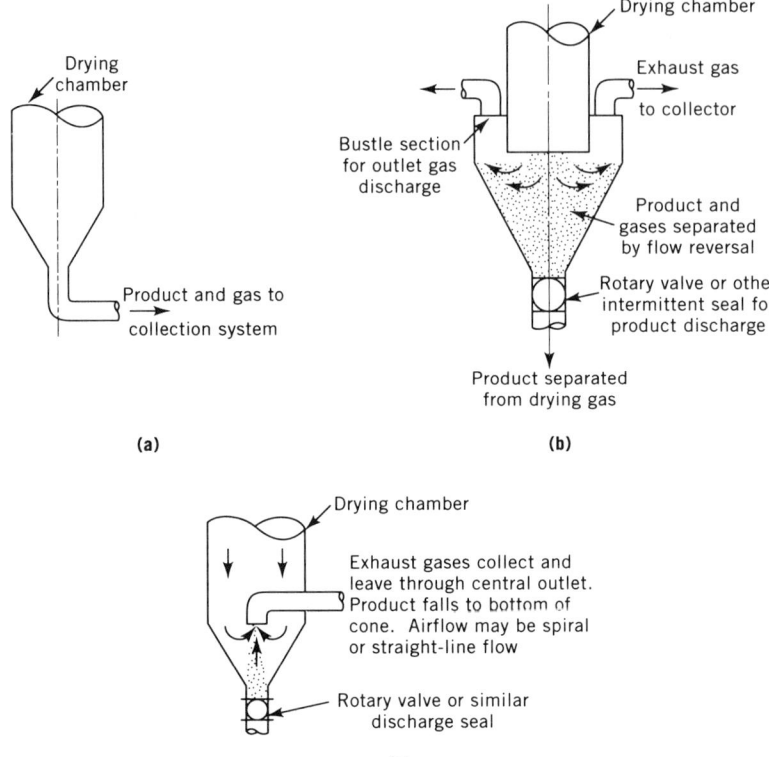

Fig. 18. Product removal arrangements for cocurrent spray dryers: (**a**) simple outlet; (**b**) product separation in an agglomeration chamber; and (**c**) classifying cone to reduce collector load.

cocurrent chamber discharge arrangements, two of which accomplish particle classification, are shown in Figure 18. The scheme of a closed-cycle, gas recirculation dryer, the self-inerting system, is shown in Figure 19. Spray dryers are often followed by fluid beds for second-stage drying or fines agglomeration. Some two-stage dryers make the fluid bed the bottom of the spray chamber, in effect copying the method of slurry and solution drying obtained by spraying directly onto a fluid bed of dry particles (36). Overall, two-stage efficiency is improved. Spray dryer applications include coffee and milk powders, detergents, instant foods, pigments, dyes, and chemical reactions, eg, flue gas desulfurization (see DESULFURIZATION).

Indirect-Heat Dryers. In indirect-heat dryers, heat is transferred mostly by conduction, but heat transfer by radiation is significant when conducting surface temperatures exceed 150°C. For jacketed vessels, steam is the common heating medium because the condensing-side film resistance is insignificant compared to material-side resistance. Hot water is circulated for low temperature heating. Heat-transfer oils or condensing organic vapors are used for high temperatures. Liquid film resistance to heat transfer is much greater than that of condensing vapor; therefore, liquids are better suited for simple heating jobs rather than for

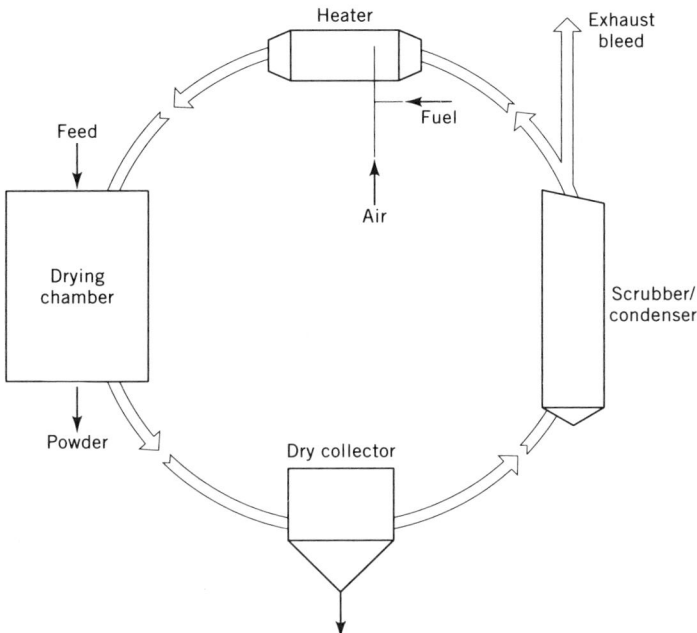

Fig. 19. Self-inerting spray dryer. Courtesy of Niro Atomizer, Inc., Columbia, Md.

drying operations having high evaporative thermal loads. Indirect-heat rotary dryers and calciners operating at temperatures exceeding 200°C usually are furnace-enclosed. The cylinders are heated externally by electric or gas-fired radiant heaters and circulating combustion products. Regardless of heating medium or method, the primary heat-transfer resistance in indirect-heat drying is on the material side. The material-side heat-transfer coefficient is affected by the rapidity of material agitation, particle size, shape, porosity, density, and degree of wetness.

Based on dryer cost alone, indirect-heat dryers are more expensive to build and install than direct-heat dryers designed for the same duty. As environmental concerns and resulting restrictions on process emissions increase, however, indirect-heat dryers are more attractive because they employ purge gas only to remove vapor and not to transport heat as well. Dust and vapor recovery systems for indirect-heat dryers are smaller and less costly: to supply heat for drying, gas throughput in direct-heat dryers is 3–10 kg/kg of water evaporated; indirect-heat dryers require only 1–1.5 kg/kg of vapor removed. System costs vary directly with size, so whereas more money may be spent for the dryer, much more is saved in recovery costs. Wet scrubbers are employed for dust recovery on indirect-heat dryers because dryer exit gas usually is close to saturation. Where dry systems are employed, all external surfaces must be insulated and traced to prevent vapor condensation inside.

Atmospheric Dryers. The rotary steam-tube dryer is a horizontal rotating cylinder in which are installed one or more circumferential rows of steam-heated tubes. These tubes extend axially the length of the cylinder and are connected to a steam and condensate manifold at one end. Figure 20 is a dryer installation

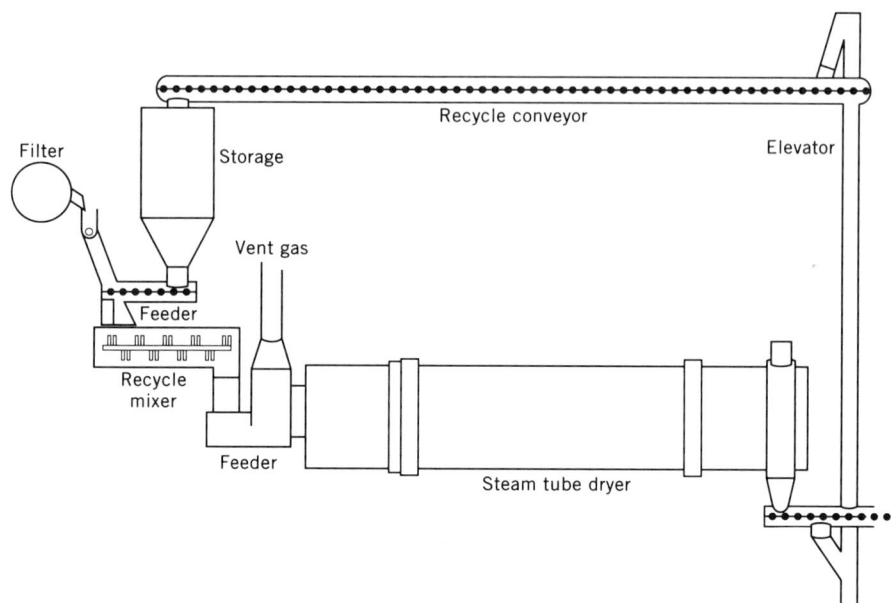

Fig. 20. Steam-tube rotary dryer using a dry product recycle and feed conditioning system.

that incorporates a dry product recycle system for feed conditioning. An essential component of this system is the recycle storage hopper which must be sufficient for one dryer fillage and kept full at all times, so that an empty dryer can be started without immediately fouling tube surfaces with nonconditioned feed. Cylinder diameters are 0.5–4 m; length may be 40 m. A large dryer carrying 1500 m^2 of tube surface in three circumferential rows may evaporate 8 t/h water. A nominal water evaporation rate in a dryer operating with 1.0 MPa (150 psig) saturated steam is 5 kg/(h·m^2). Steam-tube dryers are built for steam-tube pressures up to 3.5 MPa (500 psig). Steam is introduced and condensate is removed through a rotary joint attached to the manifold at the product discharge end. The need for a rotary joint to introduce and remove the heating medium discourages the use of media other than steam and hot water. Feed is introduced and purge gas usually is removed through a stationary throat piece attached to the rotating cylinder by a sliding seal at the end opposite the manifold. The cylinder is slightly inclined to the horizontal to direct material flow and aid condensate drainage from the tubes. The diameter of the throat seal fixes maximum cylinder fillage, which rarely exceeds 20% of cylinder volume, but must be sufficient to ensure that the inner tube row is submerged in material.

To prevent dust and vapor escape at the cylinder seals, a negative internal pressure of 50–100 Pa (0.5–1.0 cm of water column) is maintained. Steam-tube dryers are suitable for any particulate material that can be conditioned so as not to stick to metal when dry. Because of relatively inexpensive heating surface and large capacities, these dryers are probably the most commonly used of the indirect-heat dryers. Gas- and vapor-tight seals sometimes are built for operations involving dangerous vapors and inert gas circulation, but these seals are expensive and high maintenance. Small installations excepted, stationary vessels are preferable.

Calciners. The indirect-heat rotary dryer or calciner resembles the direct-heat type except that the cylinder is enclosed in a furnace and showering flights are replaced by short turning bars attached to the inner cylinder wall, so that material rolls on itself but remains in contact with the wall. Heat is transferred by convection and radiation from the furnace heaters to the cylinder, then by conduction and radiation to the material. Cylinder volume fillage is limited to 5–8% to avoid formation of a core of unheated material in the center of the rolling bed. By choosing suitable metal alloys, cylinders can be fabricated for temperatures up to 1150°C, for drying operations that are too dusty for direct-heat dryers, at temperatures that are too high for steam-tube dryers. Evaporation rates for water are 5–15 kg/(h·m^2), based on total cylinder surface. Cylinders less than 1 m in diameter are employed commonly for calcining and heat-treating operations which require special atmospheres, including hydrogen. Gas-tight rotary seals work successfully on these vessels because cylinder and seal diameters are relatively small.

Fluid-Bed Dryers. Indirect-heat fluid-bed dryers are usually rectangular vessels in which are installed vertical pipe or plate coils. Figure 21 is a diagram of a two-stage, indirect-heat fluid bed incorporating plate coil heaters. This dryer comprises a fully back-mixed section with a centrifugal feed distributor above, followed by a plug-flow section for final drying or cooling. The general design is used for drying several particulate polymers. Vertical plate coils provide heating surface up to 8 m^2/m^3 of bed volume. Excellent heat transfer is obtained in an

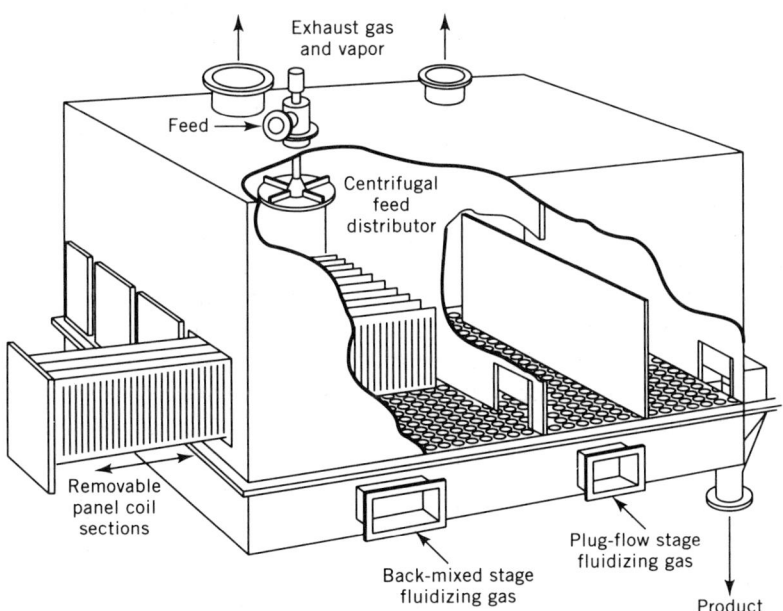

Fig. 21. Indirect-heat, two-stage, back mixed, and plug flow fluid-bed dryer. Courtesy of Niro Atomizer, Inc., Columbia, Md.

environment of intense particle agitation and mixing. Water evaporation rates, based on panel surface, are 10–25 kg/(h·m^2). The fluidizing gas is also the vapor purge gas but contributes little heat to the operation, all of which is transferred indirectly from the coils. Because of its favorable heat- and mass-transfer capabilities, flexibility for staging both temperatures and fluidizing gas velocities and the fact that the vessel is stationary, so rotary seals are not needed except for the feed distributor shaft, an indirect-heat fluid bed is an ideal vessel for vapor recovery and drying in special atmospheres.

Screw Conveyors. Indirect-heat screw conveyor dryers carry hollow, double-wall screws heated with saturated steam or hot oil. The conveyor trough may be jacketed as well, but the trough is only a small fraction of the total heating surface, and because there is a fairly wide clearance between the screw tips and the jacket, transfer coefficients are low. One trough may carry up to four parallel heated screws. A popular arrangement used for both batch and continuous drying consists of two screws that convey in opposite directions in a single trough. This internal recycle arrangement is used for drying slurries and solutions. A bed of dry particles is loaded, circulated, and heated. Feed material is sprayed onto the hot moving bed on one side of the dryer, mixed, and dried by contact with the circulating hot solids. Product is discharged through an overflow weir on the opposite side. Recycle rates as high as 2000 can be obtained. The feed rate is controlled so that the circulating dry solids contain enough sensible heat to fully dry the wet feed falling on and mixing with them before wet feed can contact heating surface. Purge gas is circulated through a vapor hood covering the screws.

Agitator Dryers. Increasing interest in indirect-heat drying has brought out a variety of relatively slow speed, batch, and continuous agitator dryers. These combine the advantages of low purge gas flow, characteristic of all indirect-heat dryers, with a material handling versatility lacking in fluid beds and an ease of gas-tight operation lacking in rotary dryers. Material holdup may be varied from a few minutes to several hours. Figure 22 is an example of this dryer type, called a porcupine dryer. It may have one or two parallel shafts and, as shown, can be provided with stationary, lump-breaker bars that intermesh with the moving paddles. Other types may provide scrapers for both jacket and agitator heating surfaces. Most can be operated under vacuum or well above atmospheric pressure and at temperatures up to 400°C using hot oil, steam, and water. The vessels are stationary and shaft seals are small to minimize leakage.

Paddle dryers are used for drying lumpy materials. Disk dryers contain a series of closely-spaced, parallel, internally-heated plates mounted on the rotating shaft. Disk dryers are suitable for granular, essentially free-flowing materials; these often include adjustable scrapers for continuous disk doctoring to maintain a clean heating surface. Water evaporation capacity of these dryers when operated using 1.0 MPa (145 psig) saturated steam is about 15 kg/(h·m^2). Dryer capacities vary with type, eg, a single, quad-screw conveyor dryer can provide 150 m^2 of heating surface, whereas one disk dryer can carry 400 m^2 of heating surface on a single shaft. Agitator dryers are ideally suited for vapor recovery and drying in special atmospheres.

Both theory and empirical data demonstrate that the faster the movement of particulate material in contact with heated surface, the greater the heat-transfer rate. Nonetheless, agitator speed of the agitator dryers discussed herein

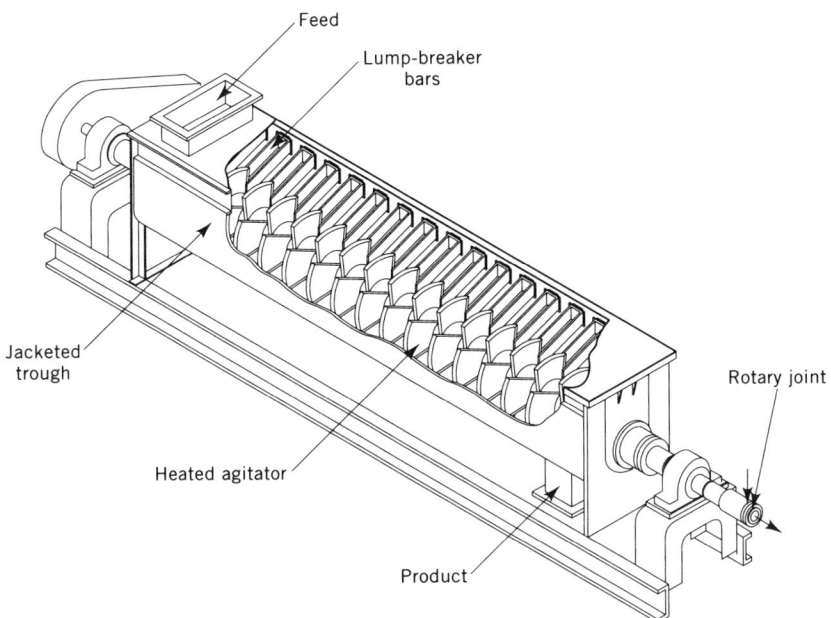

Fig. 22. Indirect-heat, paddle-type agitator dryer. Courtesy of the Bethlehem Corp., Easton, Pa.

rarely exceeds 10 min^{-1}. Because these dryers usually run between 50% and 90% full, mechanical stresses and power demand would become intolerable at higher agitator speeds, especially when drying sticky or sluggish materials. The tradeoff is between heat-transfer efficiency and fillage, power, and mechanical construction being dependent variables. For high fillage, long holdup, and the need to accommodate pasty, sticky, and sluggish materials, the best choice is usually a low speed and as much heating surface as possible built into the agitator.

High speed agitator dryers operate at about 10% fillage and are used for continuous evaporation of easily removable surface moisture. The example in Figure 23 consists of a stationary, horizontal, jacketed cylinder inside of which is carried an array of paddles mounted on a central shaft. Only the cylinder is heated. Paddle tip speed is about 15 m/s. The largest dryer has 100 m^2 of jacket surface. Water evaporation rate when operating using 1.0 MPa (145 psig) saturated steam is 20–25 kg/(h·m^2). To handle fluidlike or sticky feed materials, some high speed agitators include jacket scrapers. These dryers are suitable, however, only for materials that do not stick to metal when dry.

Drum Dryers. Indirect-heat drum dryers, like spray dryers, are usable only for materials that are fluid initially and pumpable. Drying is effected by applying a thin film of material onto the outer surface of a rotating heated drum using applicator rolls, spray nozzles, or by dipping the drum into a reservoir. Usually the drum is cast iron or steel and chrome-plated to provide a smooth surface for ease of product release by doctoring. Drum rotational speed is such that drying occurs in a few seconds. Little thermal damage is experienced and acceptable milk powders have been produced on drum dryers. Nonetheless, material surface di-

708 DRYING

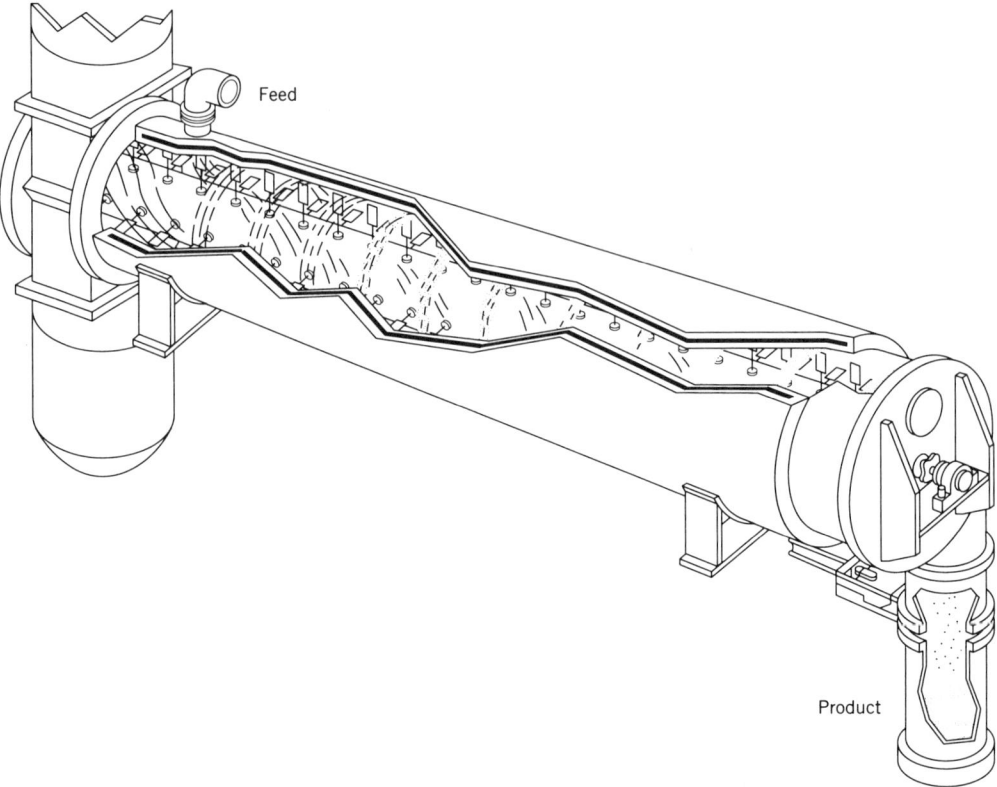

Fig. 23. Indirect-heat, high speed agitator dryer. Courtesy of Bēpex Corp., Minneapolis, Minn.

rectly in contact with the drum reaches drum temperature, so heat effects are less favorable than in spray dryers. Single-, twin-, and double-drum arrangements are used, depending on material properties and feed method.

Drum dryers are operated at atmospheric pressure and enclosed in vacuum housings for heat-sensitive materials. Twin drums are merely two single drums using a common feed system. On double-drum dryers, feed is retained and partially concentrated in a reservoir formed by the nip between two drums. Drum clearance is adjusted to fix film thickness. Material in the nip is drawn through the nip clearance, dried, and released from the back side of each drum by spring-loaded doctor knives. Single and twin drums may be provided with nip applicators by installing auxiliary feed rolls. Water evaporation rates on drum dryers, based on total drum surface, are 50–80 kg/(h·m^2). Greater capacities usually follow higher rotational speeds and thinner films.

The fin-drum dryer is a preforming device for fluid and pastelike materials intended for through-circulation drying. Slurry or paste is forced by a feed roll into circumferential grooves machined in the outer drum surface, partially dried and released in the form of short sticks by finger scrapers. Total drying rarely is attempted. Thin films of solutions, slurries, and pastes also are dried on horizontal belt dryers which are heated by radiant heaters mounted above and below the belt and operate at atmospheric pressure or under vacuum. Thicker films are handled than on drums because residence time can be longer. Temperature staging is feasible.

Can dryers, also called cylinder dryers, are similar in construction to drum dryers and used to dry paper, fiber tow, cloth, and other continuous webs that are insufficiently self-supporting for accommodation by festoon, roll, or tenter-frame equipment. A can dryer may be one 3–5-m diameter can, eg, the Yankee dryer, or it may comprise a number of cans arranged so that material passes over them in series, eg, a paper machine. To enhance conduction heat transfer initially, the web may be forced against the cans by an endless fabric belt, or felt, that also absorbs liquid and is dried separately. Further along, gas impingement nozzles may be mounted close to the web surface to add convection heat transfer. Humid air is removed through hoods above the cans. Paper drying is the largest application.

Vacuum Dryers. The indirect-heat form of batch compartment dryer usually operates under vacuum and is called a vacuum shelf dryer. Wet material is spread on trays that rest on heated shelves in an insulated vacuum chamber. The shelves are heated by steam, hot oil, or water and vacuum is produced by steam jets or pumps. Heat is transferred by conduction and radiation from the supporting shelf and by radiation from the shelf above the material. Conduction heat-transfer rates are low, however, because contact betwen the tray bottom and its supporting shelf rarely is continuous or uniform. In chambers maintained at 1.5 kPa (0.22 psi) pressure, with shelves heated by 200 kPa (29 psig) saturated steam, water evaporation based on exposed tray area is 1–2 kg/(h·m^2). Because of low drying rates, dust losses are negligible, and these dryers are suitable for small lot drying of valuable products. Batch identification is maintained, but if there are alternative choices, shelf dryers rarely are economical for production rates exceeding 200 t/yr.

A rotating vacuum dryer is formed by equipping a double-cone mixer with a jacket and an internal vapor exit tube passing through a rotary joint in one trunnion. Volume capacity is 0.1–30 m^3. Fillage is 50–70% of total volume. Internal operating pressures are 1–10 kPa (0.15–1.5 psi). In dryers operated at 1.5 kPa, having jackets heated by 200 kPa saturated steam, water evaporation based on total heated surface is 4–5 kg/(h·m^2). Rotating vacuum dryers are suitable for materials that do not stick to metal when wet or dry and do not pelletize during drying. Feed conditioning is an option. The ratio of jacket surface to operating volume decreases as dryer size increases, so large dryers often include internal plate or pipe coils to compensate. These internal elements partially destroy a principal attraction of the dryers, however, which is ease of complete emptying and cleaning between batches.

The rotary vacuum dryer is a horizontal stationary jacketed cylinder having an internal rotating ribbon or paddle agitator. If material does not stick to metal when wet or dry, the rotating shaft, ribbon arms, and paddles also may be heated. For materials that are sticky, jacket scrapers can be included. Feed conditioning by wet/dry blending externally and inside the dryer are options. Volume capacity is 0.1–30 m^3. Fillage is 50–90% of total volume. In dryers operated at 1.5 kPa, using jackets and internals heated by 200 kPa saturated steam, water evaporation based on total heated surface is 5–7 kg/(h·m²). Agitator speed is 2–8 min⁻¹, but dust carryover may be severe during initial drying. Vacuum bag-type dust collectors usually are provided to recover dust. Rotary dryers are more versatile than the rotating type, but are difficult to empty completely and are less attractive for multiproduct operations and batch identification.

The vacuum pan dryer, the workhorse of this vacuum group of batch dryers, is a vertical stationary jacketed cylinder having a jacketed dished or flat bottom and a vertical top-driven plough-type agitator to overcome torque loads presented by heavy, sticky, and doughlike materials that would overload or break the ribbons and paddles in rotary vacuum dryers. The agitator stirs these heavy materials at 1–4 min⁻¹ until they are dry enough to break down into particulate form. Power usually peaks just before the material breaks apart. The largest pan is about 4 m diameter. Maximum fillage is about 10 m³. In pans operated at 1.5 kPa pressure, using jackets heated by 200 kPa saturated steam, water evaporation based on total heated surface is 2–4 kg/(h·m²). These dryers also operate at atmospheric pressure; a purge gas is employed for vapor removal.

Freeze Dryers. The original freeze dryer was a vacuum shelf dryer, operated at much lower pressure: 100 Pa (0.8 mm Hg) for seafood, meat, and vegetables, 50 Pa (0.4 mm Hg) for fruits, 20 Pa (150 μm Hg) for concentrated beverages, and 10 Pa (80 μm Hg) for pharmaceuticals. The material is first frozen to effect separation of solutes and solvents by crystallization. Frozen material is placed in trays in a closed compartment that is evacuated and ice is caused to sublime by careful introduction of heat. The purpose is to protect heat-sensitive materials from thermal damage and prevent shrinkage of porous materials, so they can be instantly and fully rehydrated. The sublimation driving force is the difference between the vapor pressure of the ice and the condenser pressure. The drying rate is controlled by heat input and the conductivity of the material.

Most rapid drying and uniform product quality is obtained when all material surfaces are heated uniformly. Use of metal rib trays is helpful because metal conducts heat better than most organic materials and these trays distribute heat more effectively bottom-to-top. Channels for vapor escape are also opened; however, bottom material may still overheat. A suspended rib tray depends on heat transfer entirely by radiation. Higher shelf temperatures may be used without a danger of local material overheating and both top and bottom are heated uniformly. Typical food dryer shelf temperatures of the suspended tray type are 50–150°C, using a refrigeration system at −50°C. Many biological dryers contain movable shelves, so product vials can be stoppered after drying and before compartment venting. The vials are assembled in trays that rest directly on heated shelves. Typical shelf temperatures are −20°C to 50°C, the condenser temperature is −60°C. Based on exposed material surface, sublimation capacity of shelf-type food dryers is 0.2–2.0 kg/(h·m²).

Freeze drying has also been carried out at atmospheric pressure in fluid beds using circulating refrigerated gas. Vacuum-type vibrating conveyors, rotating multishelf dryers and vacuum pans can be used as can dielectric and microwave heating.

Radiant-Heat Dryers. Heat transfer by radiation occurs in all dryers to some degree and is controlled by the temperature and emissivity of the source and the temperature and absorptivity of the receiver. For drying, sources may consist of a number of incandescent lamps, reflector-mounted quartz tubes, electrically heated ceramic surfaces, and ceramic-enclosed gas burners. Usual source temperatures are 800–2500 K. Radiant energy does not penetrate most material surfaces. Heat penetration below the surface is dependent on material conductivity. In situations wherein radiant-heat flow to a surface is high while material thermal conductivity is relatively low, the surface temperature may rise above the liquid boiling point at dryer operating pressure. When drying printed cloth, film, and coatings in continuous conveyor dryers, adjustments to source spacing and temperature often can prevent boiling, skin formation on films, and bubble formation in coatings. For thicker materials, radiant-heat sources are installed alternately with gas impingement and parallel flow zones of more moderate temperature to allow time for liquid diffusion to the material surface, evaporation, and vapor dispersion.

Radiant heaters are most suitable for the drying of thin films, eg, paint films. They are not suitable for large objects and deep material layers in which drying rates are controlled by material internal heat- and mass-transfer mechanisms. When drying heat-sensitive materials, low temperature sources should be used. In continuous dryers, banks of radiant sources are placed above, below, and on both sides of the material in an enclosed tunnel designed to minimize direct and reflection losses to the outside. When materials are dried that may degrade or burn if exposed too long, means are provided to shut off and shutter all radiating surfaces instantly when material flow is interrupted. Purge gas must be provided to remove vapors from atmospheric radiant-heat dryers. A common practice is to pass the incoming purge gas behind the source enclosures to cool the enclosures and preheat the gas. On roll conveyor and tenter frame dryers, water evaporation based on exposed material surface is 10–100 kg/(h·m^2).

Dielectric and Microwave Dryers. Dielectric, also called radio frequency, dryers operate in the frequency range of 1–100 MHz. Microwave dryers in the United States operate at 915 MHz and 2450 MHz. As depicted in Figure 24, a dielectric dryer may consist of two flat metal plates between which material is placed or conveyed. The arrangement forms a capacitor, the plates of which are connected to a high frequency generator. During one-half of a cycle, one plate has a positive charge, the other a negative charge. One-half cycle later, the charges are reversed. Flat plate or platen electrodes are used for bulky objects. Parallel rods of alternating charge, called stray field electrodes, are employed for thin webs and are installed directly in line below or above the moving web. Parallel rods, called staggered-type electrodes, are used for thick webs and boards and are installed alternately above and below the material. The web moves between them.

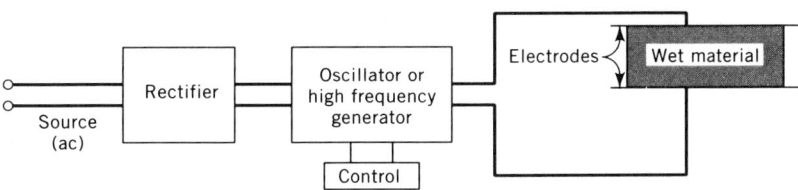

Fig. 24. Diagram of a dielectric (radio frequency) dryer.

Microwave applicators are single, like a microwave oven, or multimode cavities in which material is placed or through which it is conveyed, or rectangular waveguides which in effect surround material as it is conveyed. Rapid reversal of electrode polarity generates heat in the material. In a mechanism called dipole rotation, dipoles, which normally are in random orientation, become ordered in the electrical field. As the field dies, they return to random orientation; as the field reverses, they again become ordered but in the opposite direction. Electrical energy is converted to potential energy, to random kinetic energy, and to heat. In ionic conduction, ions are accelerated by the electrical field. They collide with nonionized molecules in random billiard ball fashion. Electrical energy is converted to kinetic energy and to heat. These are two primary mechanisms of energy conversion.

Industrial applications of dielectric and microwave energy for drying are many; however, response to high frequency electromagnetic radiation depends on a material's dielectric constant and dissipation factor, the product of which is its loss factor. A material having a loss factor greater than 0.05 is a potential drying candidate. Air, glass, some ceramics, and plastics transmit high frequency radiation. Metals may reflect radiation. Water, alcohols, aldehydes, ketones, unsymmetrical halogenated hydrocarbons, and ionic solutions absorb radiation. Hydrocarbons and symmetrical halogenated hydrocarbons do not. Depth of material penetration at which half the energy is absorbed is called a half power depth and is proportional to wave length. For dielectric heating, this may be measured in meters; for microwaves, in centimeters. If the material is large or wide, dielectric heating is preferred. If watt density is high because of a low loss factor, microwaves are preferred; however, if power requirement exceeds 50 kW, economics favor dielectric equipment.

The cost of microwave equipment per kilowatt output is about twice that of the dielectric. For irregular shapes, microwaves are preferable because to avoid hot spots during heating, dielectric electrodes are needed that conform to the material shape. Industrial dielectric dryers are employed for lumber drying, plywood bonding and drying, furniture parts drying, textile skeins and package drying, paper moisture leveling, tire cord drying, and many food products. Dielectric heating frequently is combined with radiant heat and hot air for print and coating drying. Microwave dryers are employed for drying cloth, lumber, and foods. Microwaves are used as an energy source in vacuum and freeze dryers.

Dielectric and microwave heating are generally more costly than alternative methods. Thus many applications involve material preheating and second-stage drying where energy demand is low and cycle times can be reduced significantly. Dielectric and microwave heating are chosen mostly when other methods will not work or are impractical. Material behavior in high frequency electromagnetic fields is frequency-dependent and varies with moisture content, salt concentrations, and other factors. Laboratory testing is necessary. Water evaporation is about 1.0 kg/kWh, and overall power efficiency is about 60%.

BIBLIOGRAPHY

"Drying" in the *Encyclopedia of Chemical Technology*, 1st ed., Vol. 5, pp. 232–265, by W. R. Marshall, Jr., University of Wisconsin; in *ECT* 2nd ed., Vol. 7, pp. 326–378, by W. R.

Marshall, Jr., University of Wisconsin; in *ECT* 3rd ed., Vol. 8, pp. 75–113, by P. Y. McCormick, E. I. du Pont de Nemours & Co., Inc.; in *ECT* 4th ed., Vol. 8, pp. 475–519, by P. Y. McCormick, Drying Unincorporated.

1. D. W. Green, ed., *Perry's Chemical Engineers' Handbook*, 6th ed., McGraw-Hill, Inc., New York, 1984, pp. 20-14.
2. F. W. Dittman, *Chem. Eng.* **84**(2), 106 (1977).
3. J. H. Perry, ed., *Chemical Engineers' Handbook*, 3rd ed., McGraw-Hill, Inc., New York, 1950, pp. 813–817.
4. Ref. 1, Sect. 20.
5. S. F. Sapakie, D. R. Mihalik, and C. H. Hallstrom, *Chem. Eng. Progr.* **75**(4), 44 (1979).
6. *Carrier Psychrometric Chart, Catalog No. 794-005*, copyrighted by Carrier Corp., Syracuse, N.Y., 1975.
7. Ref. 1, Sect. 12.
8. W. L. McCabe, J. C. Smith, and P. Harriott, *Unit Operations of Chemical Engineering*, 4th ed., McGraw-Hill, Inc., New York, 1985.
9. T. K. Sherwood, R. L. Pigford, and C. R. Wilke, *Mass Transfer*, McGraw-Hill, Inc., New York, 1975.
10. Ref. 1, pp. 20-7 and 20-8.
11. Ref. 1, pp. 20-20 and 20-21.
12. Ref. 3, p. 804.
13. Ref. 1, p. 20-20.
14. H. Martin, in J. P. Hartnett and T. F. Irvine, Jr., eds., *Advances in Heat Transfer*, Vol. 13, Academic Press, Inc., New York, 1977.
15. E. U. Schlünder, *Heat Exchangers*, McGraw-Hill, Inc., New York, 1974, pp. 1–19.
16. E. U. Schlünder, *Heat Exchangers*, Hemisphere Publishing Corp., New York, 1981, pp. 177–208.
17. E. U. Schlünder, *Drying '80*, Vol. 1, Hemisphere Publishing Corp., New York, 1980, pp. 184–193.
18. N. Mollekopf and E. U. Schlünder, *Proceedings of the Third International Drying Symposium*, Vol. 2, Drying Research Ltd., Wolverhampton, UK, 1982, pp. 502–513.
19. M. L. Wiedmann and P. R. Trumpler, *Trans. Am. Soc. Mech. Eng.* **68**, 57–64 (1946).
20. Ref. 1, p. 3-282.
21. V. W. Uhl and W. L. Root, *Chem. Eng. Progr.* **58**(6), 37–44 (1962).
22. Ref. 1, p. 20-12.
23. Ref. 3, p. 808.
24. Ref. 3, p. 883.
25. H. H. Macey, *Trans. Br. Ceram. Soc.* **41**, 73 (1942).
26. C. W. Hall and A. S. Mujumdar, eds., *Drying Technology*, Vols. 1–11, Marcel Dekker, Inc., New York, 1983–1993.
27. R. E. Treybal, *Mass Transfer Operations*, 3rd ed., McGraw-Hill, Inc., New York, 1980, p. 91.
28. T. K. Sherwood and R. L. Pigford, *Absorption and Extraction*, McGraw-Hill, New York, 1952, pp. 1–28.
29. Ref. 1, p. 10–10.
30. Ref. 3, p. 462.
31. W. H. McAdams, *Heat Transmission*, 3rd ed., McGraw-Hill, Inc., New York, 1954, pp. 31–54.
32. Ref. 8, pp. 278–285.
33. Ref. 8, p. 796.
34. D. H. Charlesworth and W. R. Marshall, *AIChE J.* **6**(1), 9 (1960).
35. J. Lambert and W. R. Marshall, *Conference on Freeze-Drying of Foods*, National Academy of Sciences, National Research Council, 1962.
36. D. E. Metheny and S. W. Vance, in Ref. 21, pp. 45–48.

General References

D. W. Green, ed., *Perry's Chemical Engineers' Handbook*, 6th ed., McGraw-Hill, Inc., New York, 1984, Sect. 20, pp. 1–74.

J. H. Perry, ed., *Chemical Engineers' Handbook*, 3rd ed., McGraw-Hill, Inc., New York, 1950, Sect. 13, pp. 800–884. This remains the best edition on Drying as a unit operation.

C. M. van'tLand, *Industrial Drying Equipment*, Marcel Dekker, Inc., New York, 1991.

E. M. Cook and H. D. DuMont, *Process Drying Practice*, McGraw-Hill, Inc., New York, 1991.

A. S. Mujumdar, ed., *Handbook of Industrial Drying*, Marcel Dekker, Inc., New York, 1987.

J. L. Ryans and D. L. Roper, *Process Vacuum System Design and Operation*, McGraw-Hill, Inc., New York, 1986.

K. Masters, *Spray Drying Handbook*, 4th ed., Halstead Press, Inc., New York, 1985.

W. R. Marshall, *Chem. Eng. Progr. Monogr. Ser.* **50,** 2(1954). This monograph is still extremely useful.

R. B. Keey, *Drying of Loose and Particulate Materials*, Hemisphere Publishing Corp., New York, 1991.

R. B. Keey, *Introduction to Industrial Drying Operations*, Pergamon press, Elmsford, N.Y., 1978.

R. B. Keey, *Drying, Principles and Practice*, Pergamon Press, New York, 1972.

G. Nonhebel and A. A. H. Moss, *Drying of Solids in the Chemical Industry*, CRC Press, Cleveland, Ohio, 1971.

A. Williams-Gardner, *Industrial Drying*, CRC Press, Cleveland, Ohio, 1971.

PAUL Y. MCCORMICK
Drying Unincorporated

DRYING AGENTS. See DESICCANTS. .

ELECTROSEPARATIONS

Electrodialysis, **715**
Electrophoresis, **728**

ELECTRODIALYSIS

Electrodialysis (ED) is a process for moving ions across a membrane from one solution to another under the influence of a direct electric current (see DIALYSIS). Classically the process was carried out in three-compartment electrolytic cells in which the compartments were separated from each other by essentially nonselective membranes (see MEMBRANE TECHNOLOGY). The end compartments contained electrodes. In 1940, a multicompartment ED process using ion-selective membranes (Fig. 1) was suggested (1) in which membranes A selective to anions alternated with membranes C selective to cations. When a d-c potential is applied, cations M^+ tend to move toward the negatively charged cathode. These ions are able to permeate the cation-selective membranes but not the anion-selective mem-

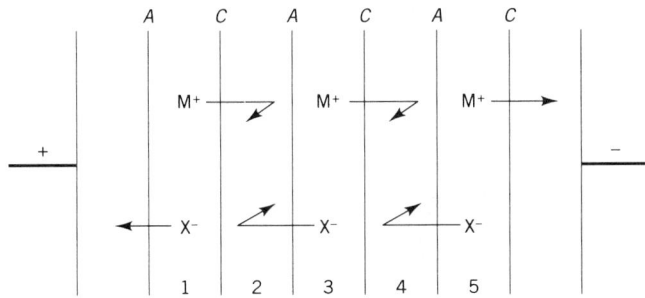

Fig. 1. Principle of multicompartment electrodialysis. See text.

branes if the latter are perfectly selective. Similarly, anions X⁻ tend to move toward the positively charged anode. These ions are able to permeate the anion-selective membranes but not the cation-selective ones. As a result, the odd-numbered compartments in the figure become depleted in electrolyte, the even-numbered compartments enriched.

Membranes

In 1950, ion-selective membranes having high selectivity, low electrical resistance, good mechanical strength, and good chemical stability were described (2). These were essentially insoluble, synthetic, polymeric, organic ion-exchange (IX) resins in sheet form (see ION EXCHANGE). Typical chemical structures for modern membranes of this type are shown schematically in Figure 2. The cation-selective membranes (Fig. 2a) consist of cation-exchange (CX) resin composed of polystyrene having negatively charged sulfonate groups chemically bonded to most of its phenyl groups. The charges of the sulfonate groups are electrically balanced by positively charged cations (counterions). Sulfonated polystyrene swells greatly in water. The amount of swelling is typically controlled by including cross-linking agents in the polymer, represented by divinylbenzene in Figure 2; by incorporating electrically neutral polymers; or by having extensive regions (blocks) in the polymer which lead to substantial microcrystallinity. The positively charged counterions, eg, Na^+, Ca^{2+}, or Mg^{2+}, are appreciably dissociated from the chemically bound negatively charged groups once the membrane is exposed to water. Thus, in water, the counterions are mobile, and may be exchanged for other cations from an ambient solution, maintaining the electrical neutrality of the membrane. This high (typically >1 meq/cm³ of membrane) concentration of counterions in IX resins is responsible for the low electrical resistance of the membrane. The high concentration of bound negatively charged groups tends to exclude mobile negatively charged ions (co-ions) from an ambient solution and is responsible for the high ion selectivity of the membranes.

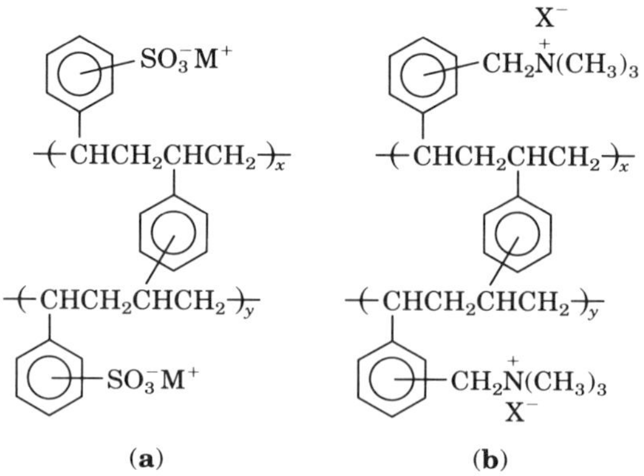

Fig. 2. Schematic representation of (**a**) cation-exchange resin, and (**b**), anion-exchange resin.

The anion-selective (AX) membranes (Fig. 2b) also consist of cross-linked polystyrene but have positively charged quaternary ammonium groups chemically bonded to most of the phenyl groups in the polystyrene instead of the negatively charged sulfonates. In this case the counterions are negatively charged, eg, Cl^-, HCO_3^-, NO_3^-, or SO_4^{2-}.

Commercially available membranes are usually reinforced with woven, synthetic fabrics to improve the mechanical properties. Several hundred thousand square meters of IX membranes are now produced annually, and the mechanical and electrochemical properties are varied by the manufacturers to suit the proposed applications. The electrochemical properties of most importance for ED are (1) the electrical resistance per unit area of membrane; (2) the ion transport number, related to current efficiency; (3) the electrical water transport, related to process efficiency; and (4) the back-diffusion, also related to process efficiency.

Commercial IX membranes have thicknesses of ca 0.15–0.5 mm and electrical resistances of ca 3–20 $\Omega \cdot cm^2$ at 25°C when in equilibrium with 0.5 N sodium chloride. The electrical resistances are somewhat higher in more dilute solutions because co-ions are more effectively excluded from the membrane by the IX resin. The electrical resistance decreases with increasing temperature at a rate of ca $-1.9\%/°C$. The electrical resistance of an ED apparatus depends in large part on the electrical resistances of the membranes when electrolyte solutions are in excess of about 0.1 N. For more dilute solutions the resistance of the apparatus tends to be dominated by the resistance of the solution being demineralized.

The ion transport number is defined as the fraction of current carried through the membrane by counterions. If the concentration of fixed charges in the membrane is high compared to the concentration of the ambient solution, then the mobile ions in the IX membrane are mostly counterions, co-ions are effectively excluded, and the ion transport number then approaches 1. Commercial membranes have ion transport numbers in dilute solutions of ca 0.85–0.95. The relationship between ion transport number and current efficiency is shown in Figure 3 where $\bar{t}_-^A$ is the fraction of current carried by the counterions (anions) through the AX membrane and $\bar{t}_+^C$ is the fraction of current carried by the counterions (cations) through the CX membrane. The remainder of the current $(1 - \bar{t}_-^A) = \bar{t}_+^A$ in the case of the AX membranes and $(1 - \bar{t}_+^C) = \bar{t}_-^C$ in the case of the CX membranes is carried by co-ions and constitutes an electrical inefficiency. The net

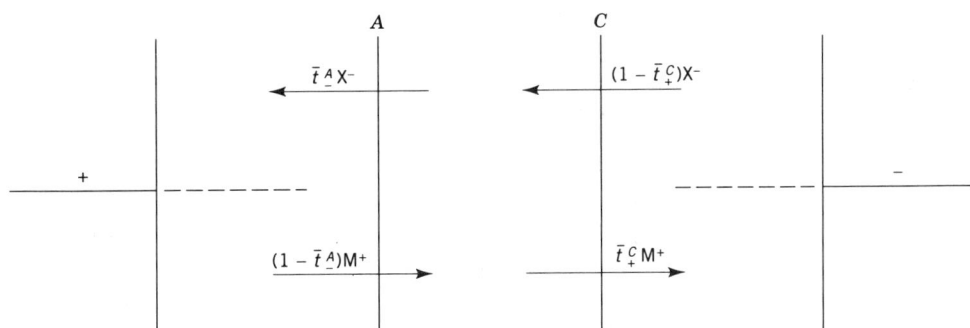

Fig. 3. Relationship between current efficiency and ion transport number.

transport of electrolyte is then given by:

$$\bar{t}^A_- - (1 - \bar{t}^C_+) = \bar{t}^A_- - \bar{t}^C_- = \bar{t}^C_+ - \bar{t}^A_+ \ .$$

The electrical water transport is defined as water that accompanies the electrical transport of ions through the membranes. Owing to the high concentration of counterions in IX membranes, the transport of ions and water are closely coupled, ie, the internal solution tends to move in piston-like flow. The electrical water transport is ca 100–200 cm^3/eq of ions transferred. This value is toward the low end of the range for AX membranes and toward the high end for CX membranes. In the case of dilute solutions, water transport is not of significant engineering importance. However, in more concentrated solutions, such as seawater, the water transport can be a significant fraction of the volume of solution electrodialyzed and, therefore, constitutes a process inefficiency. Generally, membranes having low water contents and high concentrations of fixed, charged groups have low electrical water transport. Such membranes also tend to have high ion transport numbers and are therefore preferred for concentrated electrolytes.

Back-diffusion is the transport of co-ions, and an equivalent number of counterions, under the influence of the concentration gradients developed between enriched and depleted compartments during ED. Such back-diffusion counteracts the electrical transport of ions and hence causes a decrease in process efficiency. Back-diffusion depends on the concentration difference across the membrane and the selectivity of the membrane; the greater the concentration difference and the lower the selectivity, the greater the back-diffusion. Designers of ED apparatus, therefore, try to minimize concentration differences across membranes and utilize highly selective membranes. Back-diffusion between sodium chloride solutions of zero and one normal is generally < ca 2×10^{-6} meq/(s·cm^2).

Bipolar ED. Bipolar ED membranes have one surface consisting of CX resin and the opposite surface of AX resin. When a direct current is passed through such a membrane in a direction to pull anions out of the interface between the AX and CX resins and through the AX resin, the interface rapidly becomes depleted of all ions other than those resulting from the dissociation of water. Dilute alkali can therefore be produced at the outer surface of the AX region and dilute acid at the outer surface of the CX layer (3). An enormous effort has been applied by the AquaTech division of Allied-Signal Corp. to refine and commercialize this technology, which as of this writing is without signal commercial success.

The bipolar membranes are used in a more or less conventional ED stack together with conventional unipolar membranes. Such a stack has many acid–alkali producing membranes between a single pair of end electrodes. The advantages of the process compared to direct electrolysis seem to be that because only end electrodes are required, the cost of the electrodes used in direct electrolysis is avoided, and the energy consumption at such electrodes is also avoided.

The disadvantages appear to be that the bipolar membranes are comparatively expensive, and the economic life is limited to about one year. Such short lifetime appears to result from the very high (~10^6 V/cm) voltage gradients at the interface between the AX and CX regions. Additionally, practical current densities are limited to about 1000 A/m^2 available area.

Apparatus

The apparatus for ED is fundamentally an array of alternating AX and CX membranes terminated by electrodes. The membranes are separated from each other by gaskets which form fluid compartments. Compartments that have AX membranes on the side facing the positively charged anode are electrolyte-depletion compartments. These are also called demineralizing, diluting, diluate, or dilute compartments. The remaining compartments are electrolyte-enrichment compartments, also called concentrating, concentrate, or brine compartments. The enrichment and depletion compartments also alternate through the array. Holes in the gaskets and membranes register with each other to provide two pairs of internal hydraulic manifolds to carry fluid into and out of the compartments. One pair communicates with the depletion compartments, and the other with the enrichment compartments. Much effort has been expended on the design of the entrance and exit channels from the manifolds to the compartments to prevent unwanted cross-leak of fluid intended for one class of compartment into the other class. This effort has been made increasingly difficult by the trend to thinner membranes and gaskets, the latter determining membrane spacing and the thickness of the fluid compartments; such trends are intended to reduce energy consumption. A contiguous group of two membranes and the associated two fluid compartments is called a cell pair. A group of cell pairs and the associated end electrodes is called a stack or a pack. Generally 100–600 cell pairs are arranged in a single stack, the choice being made on the basis of ED capacity desired, the uniformity of flow distribution achieved among the several compartments of the same class in a stack, and the maximum total direct current potential desired. One or more stacks may be arranged in a press, designed to compress the membranes and gaskets against the force of fluid flowing through the compartments thereby preventing fluid leaks to the outside and internal cross-leaks between compartments. For small presses such compression is usually provided by tie-rods; for larger presses hydraulic rams are frequently used.

Commercial membranes have typical thicknesses of ca 0.15–0.5 mm; the compartments between the membranes have typical thicknesses of ca 0.5–2 mm. The thickness of a cell pair is therefore in the 1.3–5.0 mm range, commonly about 3.0 mm. One hundred cell pairs have a combined thickness of about 300 mm. The effective area of a cell pair for current conduction is generally on the order of 0.2–2 m^2.

In concentrated electrolytes the electric current applied to a stack is limited by economic considerations, the higher the current I the greater the power consumption W in accordance with the equation $W = I^2 R_s$, where R_s is the electrical resistance of the stack. In relatively dilute electrolytes the electric current that can be applied is limited by the ability of ions to diffuse to the membranes. This is illustrated in Figure 4 for the case of an AX membrane. When a direct current is passed, a fraction ($\bar{t}^A_- \simeq 0.85$–0.95) is carried by anions passing out of the membrane–solution interface region and through the membrane. In the bulk solution, a fraction t_- of the current is carried by anions passing into the interfacial region. Generally t_- is significantly less than $\bar{t}^A_-$. For example, in the case of sodium chloride, t_- is ca 0.6. As the electric current continues to pass, the inter-

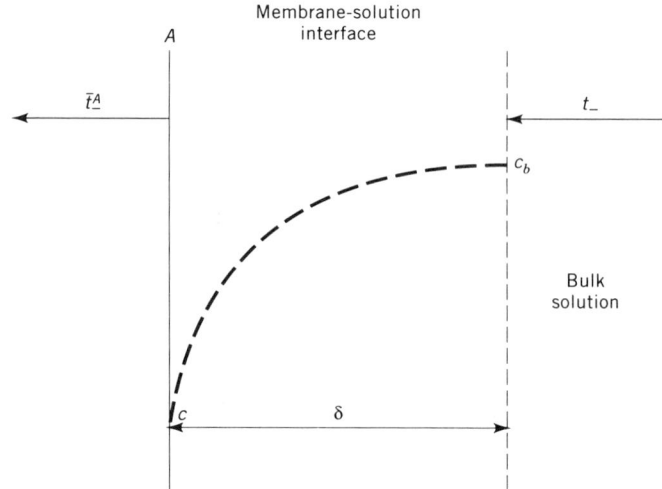

Fig. 4. Current limitation controlled by diffusion to anion membrane.

facial region becomes depleted in electrolyte. The difference between the quantity of anions transferred out of the interfacial region and those transferred in must be made up by convection and diffusion. If the interfacial region is defined as the region of streamline flow, then diffusion is the only mechanism, other than conduction, available to bring anions into the region. At steady state the concentration of electrolyte in the solution at the membrane surface must be reduced sufficiently from the bulk value to provide the concentration gradient necessary to bring in by diffusion the difference between the anions carried out of and into the interfacial regions by the electric current. This may be expressed by:

$$i\bar{t}_-^A = it_- + FD(c_b - c)/\delta$$

where F is Faraday's constant 96,500 C, the quantity of electric current required to transfer one equivalent of ions through a perfectly selective membrane; i is the current density in A/cm^2; D is the diffusion constant for the electrolyte, cm^2/s; c_b, is the concentration of the electrolyte in the bulk (nonstreamline region), g·eq/mL; c, also in g·eq/mL, is the concentration of electrolyte in solution at the membrane surface; and δ in cm is the thickness of the interfacial (streamline flow) region. This expression may be rearranged to:

$$i = \frac{DF(c_b - c)}{\delta(\bar{t}_- - t_-)}$$

Hence in any given situation the maximum current density that can be carried by a combination of electrical conduction and diffusion occurs when c, the concentration in the solution at the membrane surface, approaches zero, ie,

$$i_{max} = \frac{FDc_b}{\delta(\bar{t}_-^A - t_-)}$$

For a typical electrodialysis apparatus, δ appears to be of the order of 5×10^{-3} cm, and $\bar{t}_-^A$ is about 0.9. For dilute sodium chloride solutions, D is ca 1.5×10^{-5} cm^2/s and t_- about 0.6. Under these circumstances the ratio i_{max}/c_b, is ca 1000 A·cm/(g·eq), ie, i_{max} is ca 0.1 A/cm^2 when c_b is ca 0.1 N. Increasing temperature increases D by ca 2.3%/°C, and decreasing δ increases i_{max}. Therefore, designers usually include some kind of structure in the ED compartments to break up the streamline interfacial region and bring electrolyte as close as possible to the membrane surface by convection. The maximum current also increases as the difference $(\bar{t}_-^A - t_-)$ decreases. As $\bar{t}_-^A$ approaches t_-, the diffusion-controlled limitation on current density vanishes and the current density may be increased without limit. However, power consumption increases as i^2R and puts an economic limitation on current density. When $\bar{t}_-^A$ approaches t_-, the anion membrane loses its selectivity and becomes essentially neutral. The net transport of electrolyte is then given by $(t_- - \bar{t}_-^C)$. Thus for many electrolytes, an ED apparatus utilizing alternating neutral and CX membranes exhibits substantial depletion and enrichment (4). For example, in the case of sodium chloride solutions the net transport of electrolyte is 0.8 (current efficiency, 80%) when the ion transport numbers of the AX and CX membranes are each 0.9. If the AX membrane is replaced by a neutral membrane, the net transport is 0.5 (current efficiency, 50%) when the concentration difference between enriched and depleted compartments is small. ED apparatus using neutral membranes has not been commercially successful.

Polarization. When the applied current density equals i_{max}, the AX membrane and the apparatus are said to be concentration polarized or simply polarized. At i_{max} the fluid at the surface of the membrane is essentially depleted of electrolyte and the electrical resistance of the apparatus increases substantially even though the bulk solution in the depletion compartments may still contain an appreciable concentration of electrolyte. This is a sign of polarization.

At the polarization current density, ions resulting from the dissociation of water have concentrations comparable to the concentration of electrolyte at the surface of the membrane. A significant fraction of the current through the AX membrane is then carried by hydroxide ions into the enrichment compartments. Hydrogen ions are carried into the bulk solution in the depletion compartments. Changes in the pH of the enrichment and depletion compartments are another sign of polarization.

If i_{max} is substantially exceeded, electrolytes that are insoluble at high pHs, such as calcium carbonate and magnesium hydroxide, may precipitate at the interface between the AX membrane and the enriched solution. This is a third sign of polarization.

In sodium chloride solutions the ion transport number for Na$^+$ is about 0.4 compared to about 0.6 for Cl$^-$. Thus a CX membrane would be expected to polarize at lower current densities than an AX membrane. Careful measurements show that CX membranes do polarize at lower current densities; however, the effects on pH are not as significant as those found when AX membranes polarize. Such differences in behavior have been satisfactorily explained as resulting from catalysis of water dissociation by weakly basic groups in the AX membrane surfaces and/or by weakly acidic organic compounds absorbed on such surfaces (5).

Many fluids of natural origin contain detectable quantities of high molecular weight organic anions, such as those from humic, fulvic, and tannic acids, which can be carried to and deposited on AX membranes. Such deposits can behave as thin films partially selective to cations (6). The interfaces between such films and the underlying AX membranes then act as very thin stagnant depletion compartments and the AX membranes may exhibit polarization at current densities that are much lower than would be expected for new membranes in the absence of such anions.

Performance. The performance of an ED stack may be estimated by considering the material balance around the stack:

$$\Delta(vc) = v_i c_i - v_o c_o = \frac{\bar{i} E A_p N_p}{F}$$

where v_i and v_o are the flow rates in mL/s into and out of the depletion compartments of the stack, respectively; c_i and c_o are the concentrations in g·eq/mL into and out of the depletion compartments, respectively; $\bar{i}$ is the average current density in A/cm^2; E is the net ion transfer $(\bar{t}_-^A - \bar{t}_-^C)$; A_p is the current-carrying area per cell pair, cm^2; N_p is the number of cell pairs in the stack; and F is Faraday's constant. Typically a stack is designed so that $v_o c_o$ is ca 50% of $v_i c_i$. This is because $\bar{i}_{max}$ is determined by c_o. When c_o is very small compared to c_i, then $\bar{i}$ may be uneconomically low. As a matter of conservative engineering practice, the stack is designed so that $\bar{i}$ is appreciably lower than i_{max}.

The power consumption W may be estimated from $\bar{i}^2 R_p A_p N_p$ where R_p, the electrical resistance of a cell pair per unit area ($\Omega \cdot$cm^2), is given by:

$$R_p = R_d + R_e + R_C + R_A + R_{df} + R_T$$

R_d is the average resistance of a depletion compartment; R_e is the average resistance of an enrichment compartment; R_C is the average resistance of the CX membranes; R_A is the average resistance of the AX membranes; R_{df} is the resistance contributed by depletion in the membrane-solution interfaces; and R_T is the resistance offered by the concentration potential between the enrichment and depletion compartments, ie, a measure of the thermodynamic work required to achieve the concentration difference.

Generally these individual contributions to R_p are not separately measured or calculated. Instead R_p itself is correlated with $\bar{i}, c_o$ and c_i. For dilute solutions (7):

$$R_p = b + \frac{a}{c_a}$$

where a and b are empirically determined constants and c_a is the average concentration defined as:

$$c_a = \frac{2 c_{le} c_{ld}}{c_{le} + c_{ld}}$$

The subscripts have the following significance: ld indicates the log-mean concentration in the depletion compartments; le indicates the log-mean concentration in the enrichment compartments.

Typical values at room temperature are a, 0.001–0.0025 $\Omega \cdot g \cdot eq/cm$; b, ca 20–40 $\Omega \cdot cm^2$. The values of a and b within the ranges depend on electrolyte composition and on design of the electrodialysis apparatus. As stated above, R_p decreases with increasing temperature at ca $-1.9\%/°C$. Typical d-c voltages are ca 0.5–1 volt per cell pair. Assuming an overall current efficiency of 90% this implies a d-c electrical energy requirement of ca 15–30 $W \cdot h/g \cdot eq$ of electrolyte transferred. To this must be added the energy consumption of auxiliary equipment such as pumps and instrumentation, about 500 $W \cdot h/m^3$, and energy losses during conversion of ac to dc. In the case of relatively concentrated electrolytes, for economic reasons, designers tend to choose d-c voltages per cell pair from the low end of the range mentioned above.

Uses

High Purity Water. Electrodialysis is being increasingly used in conjunction with other processes to produce pure and ultrapure water for use in high pressure boilers and in electronics and pharmaceutical manufacture (8,9). A typical application minimally involves six steps: (1) potable or brackish water is treated via cross-flow ultrafiltration (UF) (see ULTRAFILTRATION) to remove particulate matter; (2) the water is treated by reversing-type electrodialysis (EDR) to remove about 90% of the electrolytes; (3) the product of electrodialysis is then subjected to reverse osmosis (qv) to remove a substantial fraction of silica and organics, and about 90% of the remaining electrolytes; (4) the permeate from reverse osmosis is further demineralized by electrodeionization (EDI) to remove a further 90 to 95% of the electrolytes still remaining; (5) the water is finally treated by mixed-bed ion-exchange deionization followed by (6) microfiltration. Activated carbon (qv) may be included to remove organic materials. Chlorine, chlorine dioxide, ozone, hydrogen peroxide, and/or uv irradiation may be used to sterilize the water and/or partially oxidize the organics to enable the latter to be more easily removed.

In any given hybrid process train designed for the production of ultrapure water, any one or more of the process steps may be omitted. These trains have very low chemical and operator requirements. By utilizing EDR, which has almost no requirement for chemical additives or chemical cleaning, before reverse osmosis, the requirement of the latter for chemical additives is essentially eliminated. Usage of EDI reduces the need for chemical regeneration of mixed-bed ion-exchange deionizers by about a factor of 10 to 20. These deionizers are nevertheless needed to produce ultrapure water of the highest possible electrical resistivity.

Brackish Water. Production of potable water from brackish water has been one of the principal applications of ED (see WATER TREATMENT, DESALINATION). The term brackish water is used to define a water that is more saline than potable water but appreciably less so than seawater. Brackish waters readily available worldwide have concentrations of electrolyte of ca 20–200 meq/L. Potable water generally has < 10 meq/L and seawater ca 600 meq/L. A typical application would be demineralization from 0.05 to 0.005 N for which, at one volt per cell pair, d-c energy consumption would be expected to be ca 1.35 kWh/m^3 of potable water. Auxiliaries would add about 0.50 kW·h/m^3. The specific production rates are ca 0.5 m^3/m^2·d of effective cell pair area. In 1993 the installed cost was about $135 per m^2 for large plants, including the immediate electrodialysis auxiliaries such as a-c/d-c conversion equipment, pumps, and instrumentation, but not including the costs of land, buildings, brackish water collection and pretreatment, and potable water storage and distribution. Pretreatment generally consists of filtration to remove suspended particulate matter. Small plants may cost from two to four times as much per square meter.

Direct current enters an ED stack at one end through a positively charged anode, generally a refractory metal such as titanium, coated with a thin film of a noble metal or noble metal oxide. The transition from an electron current in the anode to an ionic current in electrolyte bathing the anode produces hydrogen ions and oxygen from the salts in the brackish water. Because hundreds of cell pairs are usually used in an ED stack, the number of equivalents of hydrogen ions and oxygen is generally <1% of the number of equivalents of electrolyte removed by the membranes. The anodes are usually flushed with some of the brackish water to carry away the anode products. The electric current leaves the stack at the other end through a negatively charged cathode generally also titanium coated with noble metal or noble metal oxide. At the cathode, the electric current produces hydrogen gas and hydroxide ions, again in very small quantities compared to the amount of electrolyte removed through the membranes. However, the hydroxide ions can cause precipitation of calcium carbonate or magnesium hydroxide, or both, resulting from the Ca^{2+} or Mg^{2+} ions in the brackish water.

In some ED plants, strong mineral acids, such as sulfuric or hydrochloric, are added to the brackish water which flushes the cathodes. These acids prevent precipitation but are an additional cost and an inconvenience particularly for small plants.

By using reversing-type or EDR stacks, in which each electrode is alternately anodic and cathodic (10), acid need not be added to the cathode. After passing current through the stack for a period of from about 15 minutes to 24 hours the direction of electric current is reversed for a similar period; that is the electrode that was cathodic in the first half of the cycle and precipitated some alkali-insoluble salts, becomes anodic in the second half. Hydrogen ions generated during the anodic half of the cycle dissolve and separate insoluble salts from the electrode. In the EDR stack, compartments that were enriching during one half of the cycle are depleting during the other half. Appropriately placed automatic valves interchange the entering and exiting fluid streams at the beginning of each half cycle. Several thousand EDR plants have been successfully operated for many years. The periodic reversal also removes foreign substances from the EDR stack per se. Such materials may include alkali-insoluble salts such as calcium carbonate produced by marginal concentration polarization at the interfaces between the depleting fluid and the AX membranes; poorly soluble salts, such as calcium sulfate, which precipitate if the solubility limits are exceeded in the enrichment compartments; and high molecular weight organic anions such as humic and fulvic acids which, if present, may tend to be absorbed on AX membranes.

The success of EDR in water demineralization has apparently resulted from its greater tolerance of particulate and fouling matter compared to reverse osmosis; greater forgivingness of process upsets; greater tolerance for unskilled operators; simplicity in design and construction of EDR stacks compared to reverse osmosis modules; the ability to inspect, clean, or replace one membrane at a time; the existence of a comprehensive global sales and service network; and a vertically integrated manufacture. The economic life of IX membranes in the demineralization of water is generally in excess of 10 years in well-operated plants.

Trends in ED appear to be reduction in pumping and direct ED energy, increase in electric current density, and the use of EDR and hybrid processes in plants in which the manufacturer of the demineralization plant owns and operates the plant, selling water to the user or water distributor.

Concentration of Seawater by ED. In terms of membrane area, concentration of seawater is the second largest use. Warm seawater is concentrated by ED to 18 to 20% dissolved solids using membranes with monovalent-ion-selective skins. The EDR process is not used. The osmotic pressure difference between about 19% NaCl solution and partially depleted seawater is about 20,000 kPa (200 atm) at 25°C, which is well beyond the range of reverse osmosis. Salt is produced from the brine by evaporation and crystallization at seven plants in Japan and one each in South Korea, Taiwan, and Kuwait. A second plant is soon to be built in South Korea. None of the plants are justified on economic grounds compared to imported solar or mined salt.

Industrial Wastes. Closely related to seawater concentration is the simultaneous concentration of industrial effluents and recycle of recovered water. These applications are expected to increase as environmental restrictions increase. Examples are the concentration of blowdown from cooling towers in power plants; concentration of reverse osmosis blowdown; and the processing of metal treatment wastes (11).

Electrodeionization. Electrodeionization (EDI) (12) was developed in the 1960s (13) and has been pursued more recently (14). At least the demineralization compartments of ED apparatus are filled with AX and/or CX beads or fibers or the IX membranes are embossed and in contact with each other. In the reversing type of EDI, both compartments are so filled. At low current densities the filling acts to augment the surface area of the membranes resulting in a very large increase in allowable current density. At high current densities water splitting occurs, as in the case of normal ED, and the IX filling is converted largely to the hydroxide and hydrogen forms, respectively. The apparatus is then essentially IX regenerated in real time electrochemically.

The EDI process has begun to find commercial applications, almost always as part of a hybrid process for producing demineralized water (8). When essentially 18 MΩ water is required, EDI reduces the load on mixed-bed IX by an order of magnitude or more. The need for regeneration and regeneration chemicals is reduced by about the same factor. Applications are for feed water for high pressure boilers and for ultrapure water for the electronics industry. When less than 18 MΩ water is satisfactory, EDI can entirely replace mixed-bed IX. Typically EDI is preceded in a hybrid plant by some combination of EDR, RO, UF or MF, IX water softening and/or absorbent for organics. The tendency of demineralized water producers to reduce chemical requirements seems to assure a future for EDI.

Miscellaneous Applications. The largest miscellaneous application of ED is the demineralization of whey and nonfat milk for food and feed applications. During the manufacture of cheese or of milk casein a serum is often produced which contains much of the albumin, globulin, lactose, and minerals present in the milk plus acid and/or salt added during manufacture. The albumin and globulin are a valuable food or feed sources but use is generally limited by the electrolytes present. Since the 1960s, ED and EDR have been used to remove excessive electrolytes, generally from concentrated serum. There are approximately 70 such ED or EDR plants worldwide having about 35,000 m^2 total installed membrane area and producing about 150,000 t/yr of demineralized whey solids. Some plants remove in excess of 90% of the ash content of whey. Others remove only roughly 50% of such ash, the remainder being removed by strong acid cation exchange followed by weak base anion exchange. Much of the highly demineralized product is used as a component of mothers' milk replacement.

Other applications include (1) recovery of valuable components from metal plating or treating effluents, including the recovery of hydrofluoric and nitric acids by bipolar ED from stainless steel spent pickle liquor; (2) deashing of beet, cane, or other sugar juices and molasses. AX membranes are generally subject to fouling by medium molecular weight organic carboxylic acids in such solutions. Although fouling resistant AX membranes have been proposed, the generally short processing season for such sugar solutions leads to high capital charges. Further the additional sugar which could be recovered through desalting is not justified in view of the agricultural policies of the sugar producing countries. There are few, if any, ED plants desalting sugar solutions; (3) deacidification/acidification of fruit juices. There are a few plants on a commercial scale; (4) desalting of soy sauce, amino acid solutions, fermentation (qv) products, etc. There are a few small ED plants in such applications; and (5) demineralization of blood plasma. There are few such plants.

Economic Aspects

The first commercial ED apparatus was sold in 1954 and installed in Saudi Arabia for desalting brackish water. Since then more than 5000 ED plants have been installed worldwide for the demineralization of brackish and potable water. These range in capacity from a few to more than 10,000 m^3/d.

Ionics Incorporated (Watertown, Massachusetts), the leading ED supplier, has sold more than 2000 ED and EDR plants having a combined capacity of probably more than 600,000 m^3/d. These were originally furnished with probably more than 1.2×10^6 m^2 of membrane. More than 2000 other ED and EDR plants have been built and installed in China. These plants have a combined capacity of more than 600,000 m^3/d and were originally furnished with probably more than 1.2×10^6 m^2 of membrane. Also it is estimated that more than 1000 ED and EDR plants have been built and installed in the CIS. These plants have a combined capacity of probably more than 300,000 m^3/d, probably originally furnished with more than 600,000 m^2 of membrane. Roughly 100 plants have been installed by other companies such as Corning France (formerly S.R.T.I.) and Portals Water Treatment Ltd. (formerly Permutit-Boby).

NOMENCLATURE

a = empirically determined constant representing that portion of cell pair resistance that is inversely proportional to the average concentration in the cell pair, ohm·eq/cm
A_p = current carrying area per membrane, cm^2
b = empirically determined constant representing that portion of cell pair resistance that is independent of the average concentration in the cell pair, ohm·cm^2
c = concentration, eq/mL
c_a = average concentration in the cell pair, eq/mL, harmonic mean
c_b = bulk concentration at any point in a depletion cell, eq/mL
c_i = concentration of solution inlet to depletion compartment, eq/mL
c_{ld} = log-mean concentration in depletion compartment, eq/mL
c_{le} = log-mean concentration in enrichment compartment, eq/mL
c_o = concentration of solution from the outlet of a depletion compartment, eq/mL
D = diffusion constant of an electrolyte, cm^2/s
E = net ion transfer or transport, dimensionless
F = Faraday's constant, about 96,500 C/gram·eq
i = current density, A/cm^2
$\bar{i}$ = average current density, A/cm^2
I = current, A
N_p = number of cell pairs in a stack, dimensionless
R_A = average areal resistance of AX membrane, ohm·cm^2
R_C = average areal resistance of CX membrane, ohm·cm^2
R_d = average areal resistance of depletion compartment, ohm·cm^2
R_{df} = average areal resistance contributed by depletion in the membrane-solution interfaces, ohm·cm^2
R_p = average areal resistance of cell pair, ohm·cm^2
R_s = electrical resistance of stack, ohms
R_T = average areal resistance offered by the concentration potential between the enrichment and depletion compartments, ohm·cm^2
t_- = fraction of current carried by anions in the bulk solution in the depletion compartment, dimensionless
$\bar{t}_+^A$ = fraction of current carried through an AX membrane by cations (co-ions), dimensionless
$\bar{t}_-^A$ = fraction of current carried through an AX membrane by anions (counterions), dimensionless
$\bar{t}_+^C$ = fraction of current carried through CX membrane by cations (counterions), dimensionless
$\bar{t}_-^C$ = fraction of current carried through CX membrane by anions (co-ions), dimensionless
v = volumetric flow rate of solution, mL/s
v_i = volumetric flow rate of solution entering a compartment, mL/s
v_o = volumetric flow rate of solution out of a compartment, mL/s
δ = thickness of interfacial, streamline flow region, cm

BIBLIOGRAPHY

"Electrodialysis" in the *Encyclopedia of Chemical Technology*, 1st ed., Vol. 5, pp. 20–26, by P. Stamberger, Technical Consultant; in *ECT* 2nd ed., Vol. 7, pp. 846–865, by W. K. W.

Chen, Celanese Plastics Co.; in *ECT* 3rd ed., Vol. 8, pp. 726–738, by W. A. McRae, Ionics, Inc.; in *ECT* 4th ed., Vol. 9, pp. 343–356, by W. A. McRae, Consultant.

1. K. H. Meyer and W. Strauss, *Helv. Chim. Acta* **23**, 795 (1940).
2. W. Juda and W. A. McRae, *J. Am. Chem. Soc.* **72**, 1044 (1950).
3. V. J. Frilette, *J. Phys. Chem.* **60**, 435 (1956).
4. U.S. Pat. 2,872,407 (Feb. 3, 1959), P. Kollsman
5. R. Simons, *Electrochimica Acta* **30**, 275 (1985).
6. G. Grossman and A. A. Sonin, *Desalination* **10**, 157 (1972).
7. E. A. Mason and T. A. Kirkham, *Chem. Eng. Prog. Symp. Ser.* **55**, 173 (1959).
8. T. Williamson, G. Coker, K. J. Sims, L. Zhang, and D. Elyanow, *53rd Annual Meeting, International Water Conference*, Pittsburgh, Pa., Oct. 19–21, 1992.
9. H. C. Valcour, Jr., *52nd Annual Meeting, International Water Conference*, Pittsburgh, Pa., Oct. 21–23, 1991.
10. U.S. Pat. 2,863,813 (Dec. 9, 1958), W. Juda and W. A. McRae.
11. L. R. Schmauss, *1988 Biennial Conference, National Water Supply Improvement Association*, San Diego, Calif., Aug. 1988.
12. W. R. Walters, D. W. Weiser, and L. J. Marek, *Indus. Eng. Chem.* **47**, 61 (1955).
13. U.S. Pat. 3,149,061 (Sept. 15, 1964), E. J. Parsi.
14. U.S. Pat. 4,632,745 (Dec. 30, 1986), A. J. Giuffrida, A. D. Jha and G. C. Ganzi.

General References

J. R. Wilson, ed., *Demineralization by Electrodialysis*, Butterworths, London, 1960.

L. H. Shaffer and M. S. Mintz, in K. S. Spiegler, ed., *Principles of Desalination*, Academic Press, Inc., New York, 1966, pp. 199–289.

R. Rautenbach and R. Albrecht, *Membrane Processes,* John Wiley & Sons, Inc., New York, 1989.

K. S. Spiegler, in R. H. Perry and D. Green, eds., *Perry's Chemical Engineers' Handbook*, 6th ed., McGraw-Hill Book Co., New York, 1984, pp. 17–37.

H. Strathmann, *Membrane Separation Systems—A Research Needs Assessment*, U.S. Dept. of Energy, Washington, D.C., p. 8-1.

WAYNE A. MCRAE
Consultant

ELECTROPHORESIS

Electrophoresis is a separation technique most often applied to the analysis of biological or other polymeric samples. It has frequent application to analysis of proteins (qv) and deoxyribonucleic acid (DNA) fragment mixtures. The high resolution of electrophoresis has made it a key tool in the advancement of biotechnology (qv). Variations of this methodology are being used for DNA seqencing, isolating active biological factors associated with diseases such as cystic fibrosis, sickle-cell anemia, myelomas, and leukemia, and establishing immunological reactions between samples on the basis of individual compounds. Electrophoresis is an extremely effective analytical tool because it does not affect a molecule's structure, and it is highly sensitive to small differences in molecular charge and mass.

The term electrophoresis refers to the movement of a solid particle through a stationary fluid under the influence of an electric field. The study of electrophoresis has included the movement of large molecules, colloids, fibers, clay particles, and latex spheres, ie, basically anything that can be said to be distinct from the fluid in which the substance is suspended. This diversity in particle size makes electrophoresis theory very general.

The fundamental principle behind electrophoresis is the existence of charge separation between the surface of a particle and the fluid immediately surrounding it. An applied electric field acts on the resulting charge density, causing the particle to move, the fluid around the particle to move, or both. An applied electric field also generates heat, through resistive heating, and gases, through electrolysis reactions. Each is important in understanding and designing working electrophoresis equipment.

There are three distinct modes of electrophoresis: zone electrophoresis, isoelectric focusing, and isotachophoresis. These three methods may be used alone or in combination to separate molecules on both an analytical (μL of a mixture separated) and preparative (mL of a mixture separated) scale. Separations in these three modes are based on different physical properties of the molecules in the mixture, making at least three different analyses possible on the same mixture.

Distinction is also made among electrophoretic techniques in terms of the type of matrix employed for analysis. Matrices include polymer gels such as agarose and polyacrylamide, paper, capillaries, and flowing buffers. Each matrix is used for different types of mixtures, and each has unique advantages.

There are a variety of techniques for detecting separated sample compounds using chemical stains, photographic media, and immunochemistry. Each detection technique also gives different information about the identity, quantity, and physical properties of the molecules in the mixture. Detection is often the focus of electrophoresis, and usually yields basic information about the mixture being studied.

Principles

Electrophoresis uses the force of an applied electric field to move molecules or particles, often through a polymer matrix. The electric field acts on the intrinsic charge of a substance, and the force on each substance is proportional to the substance's charge or surface potential. The resulting force on the substance results in a distinct velocity for the substance that is proportional to the substance's surface potential. If two different substances have two different velocities, an electric field applied for a fixed amount of time results in different locations on the matrix for these substances.

The application of an electric field to a gel matrix or capillary tube results in heating in the media and gassing at the electrodes. Thus special attention in the design and use of electrophoretic equipment is required.

Theory of Electrophoretic Motion. The study of the mechanics of electrophoresis focuses on the basis of electric potential on the surface of an object, and the relation of the electric potential to the velocity of the particle. Whereas research has been generally limited to nonmolecular particles of well-defined geometry and is not strictly applicable to molecules such as proteins and DNA fragments, this work is useful for understanding the physics of electrophoretic motion.

The Electric Double Layer. Any time an interface between two immiscible phases occurs, an electric potential can be developed at that interface. For example, the walls of a glass beaker have an electric potential when the beaker holds salt water. The potential is generated because some ions preferentially bind to or absorb onto glass, and the glass naturally has silanol groups, SiO, on its surface that ionize in water causing the glass surface to be charged relative to the bulk salt water. Conceptually, there is a separation of charge that exists in a thin section of the glass and in a thin layer of the salt water adjacent to the glass. These two layers of charge are called the electric double layer. Detailed discussions of the properties of ions and the occurrence of electric double layers may be found in the literature (1–4). In electrokinetic phenomena, the primary concern is with the fluid half of the double layer (in the previous example, the layer is in the salt water), because the layer on the glass itself is immobilized. This charged layer is called the diffuse layer. The electric potential in the diffuse layer extends into the fluid phase, and drops off as a Poisson distribution:

$$\phi = \phi_o \exp(-\kappa x) \tag{1}$$

where ϕ_o = potential at interface, mV; ϕ = potential x cm interface, mV; κ = characteristic from the length, cm^{-1}; and x = distance from the interface, cm. The characteristic length that the potential extends into the bulk fluid from the interface is called the double-layer thickness, $1/\kappa$. This length is a function of the concentration and charges of the salts in solution, the temperature, and the permittivity of the solution:

$$\kappa^2 = \left(\frac{F^2 \Sigma c z^2}{\epsilon \epsilon_o R T}\right) \tag{2}$$

where ϵ = relative permittivity; ϵ_o = permittivity of free space = 8.854×10^{-14} C/(V·cm); z = valency of each ion; c = concentration of each ion; R = 8.314 J/(mol·K); T = temperature, K; and $F = 9.65 \times 10^4$ C/mol. The relative permittivity is a measure of the conductance of the pure bulk material relative to a vacuum. In the salt water in the beaker example, the pure bulk material would be water, which has a relative permittivity of about 80.

Not all of the ions in the diffuse layer are necessarily mobile. Sometimes the distinction is made between the location of the true interface, an intermediate interface called the Stern layer (5) where there are immobilized diffuse layer ions, and a surface of shear where the bulk fluid begins to move freely. The potential at the surface of shear is called the zeta potential. The only methods available to measure the zeta potential involve moving the surface relative to the bulk. Because the zeta potential is defined as the potential at the surface where the bulk fluid may move under shear, this is by definition the potential that is measured by these techniques (3).

The physical separation of charge represented allows externally applied electric field forces to act on the solution in the diffuse layer. There are two phenomena associated with the electric double layer that are relevant: electrophoresis when a particle is moved by an electric field relative to the bulk; and electroosmosis, sometimes called electroendosmosis, when bulk fluid migrates with respect to an immobilized charged surface.

Electrokinetics. The first mathematical description of electrophoresis balanced the electrical body force on the charge in the diffuse layer with the viscous forces in the diffuse layer that work against motion (6). Using this force balance, an equation for the velocity, V, of a particle in an electric field is

$$V = \mu E \qquad (3)$$

where E = the electric field strength, V/cm; and $\mu = \dfrac{\epsilon \epsilon_o \zeta}{\eta} f(\kappa a)$ where ζ = zeta potential, mV; η = viscosity, g/(cm·s); and a = particle radius, cm. The function $f(\kappa a)$ is 1 when the particle has a much larger diameter than its double layer (high ionic strengths) and is 1.5 for very low ionic strength solutions where the double layer is much greater than the particle diameter (7). The behavior of a particle is complex when the double layer and particle diameter are close to the same size, which is often the case (8,9).

Electroosmotic flow is also dependent on the zeta potential at the immobilized surface and the strength of the electric field. For electroosmosis, the flow rate generated is

$$F = \dfrac{\epsilon \epsilon_o \zeta}{\eta} \pi r^2 E \qquad (4)$$

where F = flow rate in mL/s and r = radius of flow channel in cm. Electroosmotic flow is generally minimized in polymer networks such as gels, but is important in open channels such as capillaries.

Generation of Heat in Electric Fields. One of the practical problems encountered in electrophoresis is the generation of heat from resistive dissipation of energy in the electrophoretic medium. The generation of heat (Joule heating) is given by

$$W = E I \qquad (5)$$

where W = power in watts and I = current, A. The current and the electric field strength are related by the conductivity of the electrophoretic medium by Ohm's law

$$E = \dfrac{I}{C} \qquad (6)$$

where C = medium conductivity in $(\Omega \cdot cm)^{-1}$. The higher the conductivity of the electrophoretic medium, the more difficult electrophoresis becomes because highly conductive solutions mean a lower field strength per current, and the heat load on the system increases as the current is squared. Electrophoresis is usually done in highly resistive media, using just enough salt to solubilize the compounds of interest. The addition of polymer matrices, such as gels, also serves to increase the resistivity of the media by increasing the viscosity.

Heating in electrophoresis causes changes in the viscosity and density of the electrophoretic medium. High temperatures can also damage electrophoretic equipment by warping cooling blocks, melting plastic, or cracking glass plates. Poorly cooled electrophoretic systems have poor resolution. Usually bands are smeared or appear warped.

When fluids heat unevenly, the hot part of the fluid tends to rise with respect to the cooler part of the fluid because of differences in density. The flow is driven by gravity, and distorts resolution in electrophoretic separations.

The ability to remove heat from electrophoretic systems has severely limited the maximum capacity of these systems in terms of how large or thick the systems can be. Electrophoretic separations have been performed on space flights because the effect of gravity in outer space is small and mixing from heating is negligible. Whereas electrophoresis in outer space has been accomplished (10), the economics for a scaleable process have not.

The heating effect is the limiting factor for all electrophoretic separations. When heat is dissipated rapidly, as in capillary electrophoresis, rapid, high resolution separations are possible. For electrophoretic separations the higher the separating driving force, ie, the electric field strength, the better the resolution. This means that if a way to separate faster can be found, it should also be a more effective separation. This is the opposite of most other separation techniques.

Electrolysis Reactions. The electrodes in electrophoresis equipment are typically constructed from platinum wire, and sodium chloride generally carries the bulk of the current in any electrophoretic medium. This results in the reactions

Cathode
$$2\,H_2O + 2\,e^- \longrightarrow 2\,OH^- + H_2$$
$$HA + OH^- \rightleftharpoons A^- + H_2O$$

Anode
$$H_2O \longrightarrow 2\,H^+ + 0.5\,O_2 + 2\,e^-$$
$$H^+ + A^- \rightleftharpoons HA$$

That is, water is electrolyzed. The hydrogen gas produced at the cathode can be hazardous, especially because it is in the vicinity of an electrode that is also producing heat. For this reason, electrode chambers are usually open to the atmosphere so that gases can vent.

The other reactions at the electrodes produce acid (anode) and base (cathode) so that there is a possibility of a pH gradient throughout the electrophoresis medium unless the system is well buffered. Buffering must take the current load into account because the electrolysis reactions proceed at the rate of the current. Electrophoresis systems sometimes mix and recirculate the buffers from the individual electrode reservoirs to equalize the pH.

Modes of Electrophoretic Separations

Zone electrophoresis, isoelectric focusing, and isotachophoresis are all commonly practiced as analytical techniques. Zone electrophoresis is by far the most commonly practiced and there are several different zone electrophoresis methods and techniques available for different separation goals. Isoelectric focusing is also useful for separation, and possibly more useful for determining charge characteristics of sample proteins. Isotachophoresis takes advantage of the continuity of current across a medium to segregate a sample into contiguous zones of high purity. It is not useful as an analytical technique, but is applied as a potential preparative method.

Electrophoresis Equipment. Most electrophoresis equipment shares a basic design, a diagram of which is shown in Figure 1. Electrophoresis equipment usually consists of two buffer reservoirs, one for the anode and one for the cathode. Very often, electrophoretic buffers have a basic pH and sample compounds migrate to the anode. The equipment includes some sort of electrophoretic medium connecting the two reservoirs, such as a gel, paper, or capillary tubes, to which a sample is applied. A direct current power supply connects the two electrodes that are immersed in the buffer reservoirs. The power supply is usually interlocked to the electrophoresis equipment through a cover, isolating the medium and buffer reservoirs from human interaction because contact can be very dangerous. Small amounts of d-c current through the human heart can cause cardiac arrest or arrhythmia. An electric field is applied between the two electrode reservoirs, causing sample compounds to migrate through the medium and heat to be generated in the medium. The heat is usually removed by a chilled water bath or running tap water. At completion of the electrophoretic run the sample appears resolved into

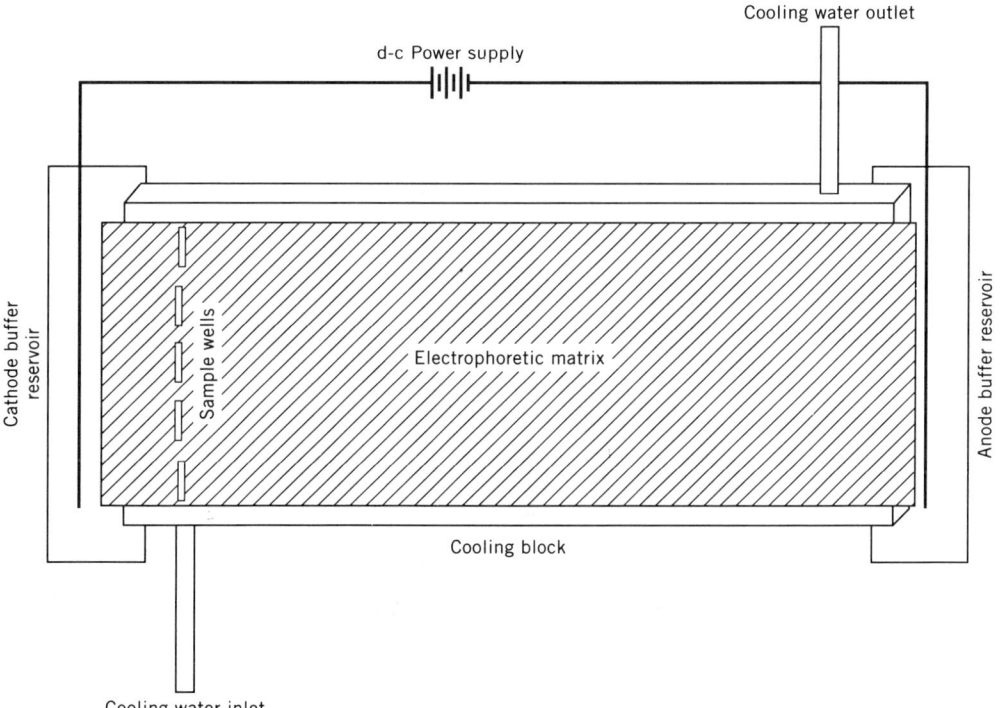

Fig. 1. Electrophoresis equipment.

several different components along the length of the medium. In this respect, gel electrophoresis resembles thin-layer chromatography (see CHROMATOGRAPHY).

Under certain conditions the sample is clearly visible throughout the process. Other times it is necessary to stain the matrix to visualize the components. In cases where a final staining procedure is required, a small amount of dye is often added to the sample before the analysis. The dye typically migrates faster than any sample component. The position of the dye in the matrix indicates the speed of the resolution of the components of the sample. Typically, the electrophoretic medium is discarded after use. Good resolution can be obtained from 1 to 20 hours, using applied voltages of 10 to 2000 V and currents of 5 to 100 mA.

Zone Electrophoresis. In zone electrophoresis multicomponent samples are applied to an electrophoretic medium, most commonly a gel, an electric field is applied, and after a predetermined length of time or after a certain level of power, current, or voltage has been applied, the electrophoretic medium is inspected for resolution of the sample components. A typical zone electrophoresis result is shown in Figure 2, which indicates how closely related molecules can be separated from each other and visualized on a gel. Each band represents a highly enriched substance, has essentially the same shape and width, and is separated from the other bands because each type of substance has a slightly different property. As shown, several samples are typically analyzed side by side on a gel. Bands that migrate the same distance in different sample mixtures represent the same substance and standards are usually run concurrently with samples that are to be compared. In zone electrophoresis, proteins or sample components are com-

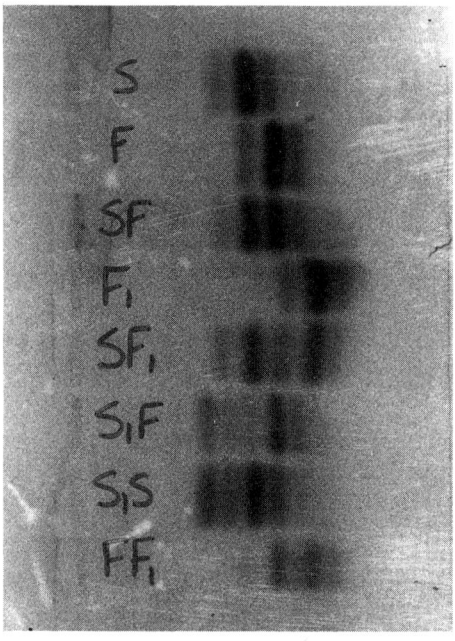

Fig. 2. Zone electrophoresis separation where S, F, S_1, and F_1 are different materials.

pletely separated into discrete zones as they migrate through the media onto which they are applied.

The use of standards with samples makes zone electrophoresis particularly useful as an analytical tool. However, when samples cannot be analyzed on the same gel, differences in the experimental conditions from experiment to experiment make direct comparison more difficult. To make comparisons from experiment to experiment, a relative mobility, R_f, is often measured by measuring the distance a component travels down the gel compared to some reference or standard component.

$$R_f = \frac{\text{distance migrated by component}}{\text{distance migrated by reference}} \qquad (7)$$

Disc Electrophoresis. Resolution in zone electrophoresis depends critically on getting sample components to migrate in a focused band, thus some techniques are employed to concentrate the sample as it migrates through the gel. The most common technique is referred to as discontinuous pH or disc electrophoresis. Disc electrophoresis employs a two-gel system, where the properties of the two gels are different.

Disc electrophoresis was first introduced in the early 1960s (11–13) as various techniques using polyacrylamide gels were being explored and designed. Original work employed several buffer systems and different polyacrylamide gels in order to first concentrate and then separate compounds (14).

The way proteins behave in disc electrophoresis systems depends primarily on differences in the pH in the two gels. The pH of the buffers in both the second gel (the separating gel) and the electrode reservoir are similar, whereas the buffers of the first gel (the stacking gel) and the sample itself are of a lower pH. This difference in pH allows for different sample–gel interactions. The stacking gel is only about 5 cm, including distance for the sample wells, and is stacked on top of the separating gel which is about 20 cm in length. As with most electrophoretic methods, a current is applied and the molecules in the sample wells begin to migrate anodally. Here, the migration is electrochemically different because the buffer in the upper reservoir chamber has a higher pH than that of the samples and stacking gel. This difference in pH allows the molecules in the samples to migrate rapidly through the stacking gel. When the sample compounds enter the separating gel, movement is slowed because of the pH change. This focuses the molecules into narrow bands and allows more bands to be resolved from one another.

Another difference between other types of electrophoresis and disc electrophoresis is that the molecules in a sample do not start to significantly separate until entering the separating gel. A discontinuous gel system may be used with almost any type of zone electrophoresis application.

Native Zone Electrophoresis. In some cases, good resolution between sample species can be obtained with little or no sample pretreatment. In these cases, the gels are said to be native gels. In this method, the charge on the individual sample component is primarily responsible for its differential migration. The relative mobility, R_f, measured is proportional to the charge when the pore size in the electrophoretic medium is large compared to the component. When the pore size and molecular size are about the same, the size of the molecule becomes important in the electrophoretic mobility. Because of this ambiguity, the absolute meaning of R_f from this technique is useful primarily for component identification and comparison, and not for estimating the properties of a molecule.

When separating proteins, native zone electrophoresis leaves the component proteins in folded, globular states, with all subunits intact. This compact size makes for a faster running molecule. Native electrophoresis is useful for resolving components which may differ by a small size and/or a small charge. For example, this method was used to isolate genetic variants of some proteins in cow's milk (15) which differ by one charge unit and less than 0.1% of their total molecular weight.

Reduced SDS Electrophoresis. The combination of sodium dodecyl sulfate (SDS) [151-21-3], $CH_3(CH_2)_{10}CH_2OSO_3Na$, also known as lauryl sulfate, treatment of samples and polyacrylamide gel electrophoresis was first described in the late 1960s (14,16). SDS is an ionic surfactant which solubilizes and denatures proteins. The surfactant coats a protein through hydrophobic interactions with the polypeptide backbone, effectively separating most proteins into their polypeptide subunits. The majority of proteins to which SDS binds then unfold into linear molecules having a similar surface potential. Nonreduced proteins bind approximately 0.9–1.0 grams of SDS per gram of protein (17).

SDS–polyacrylamide gel electrophoresis (SDS–page) allows separation of molecules strictly on the basis of size, ie, molecular weight. When SDS-treated samples migrate into a gel and are electrophoresed, the principal difference is size or length. Smaller molecules travel through the matrix more quickly than those that are larger. The rate at which molecules migrate through a polyacrylamide gel is inversely linear with the logarithm of their molecular weight. Thus denatured samples can be analyzed alongside standards of known molecular weight to aid in the interpretation of a substance's physical size.

Other dissociating agents may be used to further break down a protein. Urea [57-13-6] may be used to disrupt hydrogen bonds. When urea is the only dissociating agent added (no SDS), a protein's intrinsic charge is not affected and separation on the basis of size and charge may be achieved. If a protein contains internal disulfide bonds, a thiol reagent such as β-mercaptoethanol [60-24-2] must be used in order to reduce the sample and break the disulfide bonds. Proteins having reduced disulfide bonds bind approximately 1.4 grams of SDS per gram of protein, compared to about 1 g/g for nonreduced. Typically, both a reduced and nonreduced sample are run in order to evaluate band differences.

Pulsed Field Gel Electrophoresis. Pulsed field gel electrophoresis is a technique that was developed to separate large pieces of DNA in agarose gels. DNA had previously been separated by conventional size sieving electrophoretic techniques. The resolving power of this technique is inversely proportional to the log of the DNA molecular weight. Thus as the molecular weight of the DNA increases, the resolution decreases. In pulsed field electrophoresis, the direction of the field is intermittently changed, either forward and backward or from side to side. A small molecule notices no real difference in its electrophoresis because it can completely reorient in each field direction. However, the redirectioning of the electric field causes larger molecules to travel in a zigzag pattern, putting kinks into the length of the molecule. The longer the molecule, the more kinks in its length, and the slower it travels down the length of the gel. The larger molecule finds itself traveling backward along some sections of its length with respect to the direction of the electric field. This has allowed resolution of megabase size strands of DNA, and has made the Human Genome Project feasible (18–23).

Isoelectric Focusing. Isolectric focusing (ief) is a technique used for protein separation, by driving proteins to a pH where they have no mobility. Resolution depends on the slope of a pH gradient that can be achieved in a gel.

Ampholytes or Zwitterions. An ampholyte is a molecule that can be either positively or negatively charged, depending on the pH. These molecules are also called zwitterions. All amino acids and proteins are ampholytes, or amphoteric. Not only does the sign of the charge of an ampholyte change with pH, but the magnitude of the charge can also vary. The charge on a protein, for example, may vary from +10 or more at low pH, to −10 or more at high pH. For a protein, the pH at which its mobility is zero is called the isoelectric point.

A special class of ampholytes has been synthesized for the purpose of isoelectric focusing (24). These ampholytes have an amino end and a carboxyl end that are separated by varying numbers of methylene groups. The further apart the amino and carboxyl groups, the less one affects the ionization of the other; thus a different isoelectric point (pI) is established for each molecule. These ampholytes, which may be added to an electrophoretic medium, migrate in one direction or another, under the influence of an applied electric field, until they reach a zone in which the pH is the same as that ampholyte's isoelectric point. The ampholyte molecules buffer themselves and establish the local pH as they migrate through the gel. As the ampholytes reach an isoelectric pH, they establish a stationary spatial pH gradient in the electrophoretic medium.

Isoelectric Focusing. Isoelectric focusing (ief) is an electrophoretic technique in which amphoteric samples are separated according to their isoelectric points along a continuous pH gradient. Ief analyses are carried out in various matrices: in acrylamide, agarose, and capillaries. The agarose or acrylamide gels that are used must be prepared with carrier ampholytes bracketing a specific pH range. After some time, the ampholytes separate and there is a pH gradient which covers the range of all the ampholytes' isoelectric points. Initial research on ief (25) was primarily directed toward evaluating the properties of synthetic ampholytes in solution. Later work refined the technique to apply gel matrices (12) and provided the basis for ief methodologies.

Problems associated with gel-to-gel variability have been rectified with the advancement of ampholyte mixtures. One commonly used mixture of ampholytes, called Immobilines, is an improved ampholyte mixture that produces no gradient drift or unequal pH gradient (26) and can be used in a gel matrix reproducibly from one day to the next.

Because protein samples are actually ampholytes, when samples are loaded onto the gel and a current is applied, the compounds migrate through the gel until they come to their isoelectric point where they reach a steady state. This technique measures an intrinsic physicochemical parameter of the protein, the pI, and therefore does not depend on the mode of sample application. The highest sample load of any electrophoretic technique may be used, however, sample load affects the final position of a component band if the load is extremely high, ie, high enough to titrate the gradient ampholytes or distort the local electric field.

Isoelectric focusing takes a long (from ca 3 to 30 h) time to complete because sample compounds move more and more slowly as they approach the pH in the gel that corresponds to their isoelectric points. Because the gradient ampholytes and the samples stop where they have no mobility, the resistivity of the system increases dramatically toward the end of the experiment, and the current decreases dramatically. For this reason, isoelectric focusing is usually run with constant voltage. Constant current application can lead to overheating of the system.

Some forms of agarose are specifically designed to work with large (mol wt >500,000) molecules (27,28). The types of samples for which the agarose ief system are utilized are larger plasma proteins such as immunoglobulins, tissues, and tumors.

Another form of ief is a method called direct tissue isoelectric focusing (dtif) (29) where isoelectric focusing in agarose is used to evaluate tissues. The tissue to be analyzed is placed directly onto the gel. Using the tissue itself and not tissue extracts has advanced the study of proteins that are difficult to extract from tissue, or are damaged by the extraction procedure. Dtif is an important advancement in the area of sample handling and application where direct application of a solid to a gel matrix may actually enhance resolution.

Isotachophoresis. Isotachophoresis takes advantage of the fact that electroneutrality must be maintained in an electrophoretic system in order to support an electric field. If a current passes through a medium, that current must be constant from one electrode to the other, regardless of the local ion concentration or mobility; ie, dilute ions must move faster to keep up with a zone of more concentrated ions. Electric fields compensate for this because the electric field strength does not have to be constant along the length of the medium. The electric field strength is lowest where the ions are most concentrated and most mobile. Isotachophoresis takes advantage of this phenomenon by lining up the ions of interest, fastest (most mobile) to slowest. This is a highly specialized technique that requires detailed knowledge of the properties of the sample to be separated, and is generally not applicable to analytical separations.

An electrophoretic medium, such as a gel or a capillary tube, is cast or filled with an electrolyte that has a higher mobility than any components in the mixture of interest. This electrolyte, called the leading electrolyte, also fills the anode buffer reservoir. A sample is applied on top of the leading electrolyte, and another electrolyte with lower mobility than the sample fills the cathode reservoir. This essentially envelopes the sample. As the electric field is applied, the leading electrolyte moves rapidly toward the anode. The highest mobility component in the sample is drawn toward the lead electrolyte to fill in the conductivity gap left by the quickly migrating lead electrolyte. The electric field driving the first ion in the sample must increase to allow the sample ion to follow at the same speed as the leading electrolyte. The current through the whole media is constant, so the electric field strength increases because the resistivity increases. The resistivity increases because the first ion in the sample to follow the lead electrolyte dilutes until the electric field is balanced to give the two zones the same speed. Progressively, each ion in the sample follows at lower conductivity, just behind an ion of higher mobility and just ahead of an ion of lower mobility. Finally, the "trailing" electrolyte, of lowest mobility, terminates the separation.

This separation technique has been employed primarily for preparative types of separations because detailed knowledge of the properties of the sample is required. Also, because this separation results in discrete zones of sample ions which are virtually pure, it makes sense to use this technique when the sample size is large. This technique is ineffective when the levels of impurities are small with respect to the target compound; small amounts of sample ions do not form zones well and tend to mix with the target compound. Information on this technique is available (30).

Electrophoretic Materials and Matrices

Various support media may be employed in electrophoretic techniques. Separation on agarose, acrylamide, and paper is influenced not only by electrophoretic mobility, but also by sieving of the samples through the polymer mesh. The finer the weave of selected matrix, the slower a molecule travels. Therefore, molecular weight or molecular length, as well as charge, can influence the rate of migration.

In addition to polymeric support media, capillaries and flowing buffers have been used as support media for electrophoresis. Although these are not used as frequently, there are definite advantages for certain types of samples and applications.

Agarose Electrophoresis. Agarose is produced from the processing of red seaweed. When agar is extracted from the seaweed it is in two components, agaropectin and agarose. The agarose portion is nearly uncharged and is therefore the portion desirable for use as an electrophoretic matrix. Charge on the agarose leads to electroosmotic flow through the gel. The chemical composition of agarose is alternatively repeated residues of 1,3-linked-β-D-galactopyranose and 1,4-linked-3,6-anhydro-α-L-galactopyranose (31).

To prepare a gel for electrophoresis a combination of agarose and buffer is heated until the agarose solid is dissolved and boiling. The solution is cooled sufficiently and then poured into a warmed gel casting apparatus which forms the shape of the gel as it cools. After the cooled solution is poured into the casting apparatus, it is allowed to gel and can then be used in an agarose electrophoresis method. Because the composition of agarose is a network of residues that hold water molecules, the extra agarose solution may be stored and used at a later time. The concentration of agarose mixtures typically varies between 0.5 and 2% in a weight-to-volume ratio, although more extreme concentrations have been evaluated. The varying concentrations depend on the desired application. The higher the concentration of agarose, the smaller the pore size.

The use of agarose as an electrophoretic method is widespread (32–35). An example of its use is in the evaluation and typing of DNA both in forensics and to study heritable diseases (36). Agarose electrophoresis is combined with other analytical tools such as Southern blotting, polymerase chain reaction, and fluorescence. The advantages of agarose electrophoresis are that it requires no additives or cross-linkers for polymerization, it is not hazardous, low concentration gels are relatively sturdy, it is inexpensive, and it can be combined with many other analytical methods.

Polyacrylamide Electrophoresis. Polyacrylamide gels are synthesized through the combination of acrylamide [79-60-1], $CH_2=CHCONH_2$, monomer and a cross-linking comonomer. Typically, the cross-linking comonomer of choice is N,N'-methylenebisacrylamide [110-26-9] (bisacrylamide), $(CH_2CHCONH)_2CH_2$, although others are available, such as ethylenediacrylate (EDA) and N,N'-diallyltartardiamide (DATD) [58477-85-3] (37). The cross-linking of polymerized monomer with the comonomer is what controls the pore size of the gel polymer mesh. This level of pore size control makes polyacrylamide gel electrophoresis an effective analytical tool.

The most commonly used combination of chemicals to produce a polyacrylamide gel is acrylamide, bisacrylamide, buffer, ammonium persulfate, and tetramethylenediamine (TEMED). TEMED and ammonium persulfate are catalysts to the polymerization reaction. The TEMED causes the persulfate to produce free radicals, causing polymerization. Because this is a free-radical driven reaction, the mixture of reagents must be degassed before it is used. The mixture polymerizes quickly after TEMED addition, so it should be poured into the gel-casting apparatus as quickly as possible. Once the gel is poured into a prepared form, a comb can be applied to the top portion of the gel before polymerization occurs. This comb sets small indentations permanently into the top portion of the gel which can be used to load samples. If the comb is used, samples are then typically mixed with a heavier solution, such as glycerol, before the sample is applied to the gel, to prevent the sample from dispersing into the reservoir buffer.

Maximum resolution of proteins is the main consideration when determining the acrylamide concentration of a gel. Two parameters define the composition of gel: the wt % of total monomer, % T (acrylamide plus bisacrylamide in grams per 100 mL), and the proportion by weight of monomer that is the cross-linking agent, % C (bisacrylamide) (38). Gels having concentrations between 5 and 15% are typically used to achieve the most desirable separation of the components of interest. Unlike the agarose matrix, once the mixture polymerizes, it cannot be reused.

Polyacrylamide gel electrophoresis is one of the most commonly used electrophoretic methods. Analytical uses of this technique center around protein characterization, for example, purity, size, or molecular weight, and composition of a protein. Polyacrylamide gels can be used in both reduced and nonreduced systems as well as in combination with discontinuous and ief systems (39).

An example of the use of polyacrylamide gels in an ief system in combination with immunoblotting is given in Reference 40 where this method is used to detect low quantities of group specific component (GC) subtypes. Another example of polyacrylamide in combination with ief is given in Reference 41 where the technique is used as a screening tool for inheritance of a certain polymorphic protein.

Both the ease of use of this method for characterization of proteins and nucleic acids, and the ability to analyze many samples simultaneously for comparative purposes, have led to the prevalence of this technique. The drawbacks of a polyacrylamide matrix is that acrylamide is a neurotoxin, the reagents must be combined extremely carefully, and the gels are not as pliable as most agarose gels.

Paper Electrophoresis. Paper (qv) as an electrophoretic matrix was employed in some of the first electrophoretic techniques developed to separate compounds. Paper is easier than a gel matrix because the paper matrix requires no preparation. Besides being easy to obtain, paper is a good medium because it does not contain many of the charges that interfere with the separation of different compounds. Two types of paper employed in this type of electrophoresis are Whatman 3 MM (0.3 mm) and Whatman No. 1 (0.17 mm).

In paper electrophoresis, the sample is placed directly onto chromatographic or filter paper and then exposed to a buffer solution at each end and an electric field is applied. As in most electrophoretic techniques, charged dyes are combined with samples and standards to see the progress of the electrophoresis. The movement of samples on paper is best when the current flow is parallel to the fiber axis in the paper. The paper has high resistance so voltages are typically much higher than in agarose and polyacrylamide matrices. Like agarose and polyacrylamide matrices, paper is combined with other analytical tools to enhance separation and identification of sample components. For example, paper electrophoresis has been combined with chromatography (42).

The difference between paper electrophoresis and paper chromatography is that electrophoresis separates by charge whereas chromatography separates by polarity. This combined technique was used to evaluate polymorphisms of the hemoglobin molecule, ie, normal A-type versus the sickle cell S-type. It has been called peptide fingerprinting. This method was later modified (43) to further evaluate peptides and is one technique known as peptide mapping. The peptide mapping technique also uses high voltages to obtain the desired resolution.

Some advantages of paper in electrophoresis are that paper is readily available, easy to handle, and new methodologies can be developed rapidly. The disadvantages of paper electrophoresis are that the porosity of paper cannot be controlled, the technique is not very sensitive, and it is not easily reproducible.

Capillary Electrophoresis. Capillaries were first applied as a support medium for electrophoresis in the early 1980s (44,45). The glass capillaries used are typically 20 to 200 μm in diameter (46), may be filled with buffer or gel, and are frequently coated on the inside. Capillaries are used because of the high surface-to-volume ratio which allows high voltages without heating effects. The only limitations associated with capillaries are limits of detection and clearance of sample components.

Limits of detection become a problem in capillary electrophoresis because the amounts of analyte that can be loaded into a capillary are extremely small. In a 20 μm capillary, for example, there is 0.03 μL/cm capillary length. This is 1/100 to 1/1000 of the volume typically loaded onto polyacrylamide or agarose gels. For trace analysis, a very small number of molecules may actually exist in the capillary after loading. To detect these small amounts of components, some on-line detectors have been developed which use conductivity, laser Doppler effects, or narrowly focused lasers to detect either absorbance or fluorescence (47,48). The conductivity detector claims detection limits down to 10^6 molecules. The laser absorbance detector has been used to measure some of the components in a single human cell.

Clearance of sample components from a capillary is a problem that has not been as well resolved as detection. Usually, capillary electrophoresis is conducted in a glass capillary that is coated on its inner surface. Some applications call for gel-filled capillaries. Capillaries are difficult to coat or fill with gel, and are too expensive to discard after each use, as is done when using a standard gel. Therefore, all the sample has to be cleared from the capillary before the capillary is used for another sample, much like analytical chromatography. This can be a problem when capillaries are sometimes more than 100 cm long. Another problem arises when analytes absorb to the glass (or coated) walls in the capillary, a frequent occurrence. This latter problem has prevented capillary electrophoresis from becoming a practical method for protein analysis. In coated capillaries, electroosmosis along the capillary walls sometimes adds velocity to the analytes, and clearance is less of an issue. A great deal of research has focused on choosing and optimizing capillary coatings that resist protein fouling and absorb sufficient charge from buffer ions to give electroosmotic flow.

Electroosmotic flow in a capillary also makes it possible to analyze both cations and anions in the same sample. The only requirement is that the electroosmotic flow downstream is of a greater magnitude than electrophoresis of the oppositely charged ions upstream. Electroosmosis is the preferred method of generating flow in the capillary, because the variation in the flow profile occurs within a fraction of κr from the wall (49). When electroosmosis is used for sample injection, differing amounts of analyte can be found between the sample in the capillary and the uninjected sample, because of different electrophoretic mobilities of analytes (50). Two other methods of generating flow are with gravity or with a pump.

Capillary electrophoresis is a commercially available technique, and has been integrated with most automated lab equipment such as autosamplers, computer peak analysis (the charts generated are called electropherograms), temperature control, and recirculating buffers. Capillary electrophoresis separations are rapid (minutes), require high voltage sources (10,000 V), and a small amount of automation, particularly in sample application, to be feasible. The use of capillaries as a support medium for electrophoresis is advantageous because it avoids the effects of heating that occur in a gel, can be very rapid, and produces a chart recording rather than a stained gel for archiving.

Free-Flow Electrophoresis. Free-flow electrophoresis is the most common technique for scaling up electrophoresis for commercial application. In this technique, sample compounds are injected into a curtain of buffer which flows between two flat plates, with electrodes parallel to the flow at each end. The electric field is then applied perpendicularly to the flow direction, so that as compounds flow down between the electrodes they separate horizontally and exit the flow field at different locations. The main challenge for this technique is stabilizing the flow to both heating and electroosmotic forces. Sometimes this is done by dividing the flow curtain into cells using semipermeable membranes, which allow proteins and other sample compounds to migrate from chamber to chamber, but restrict flow (see MEMBRANE TECHNOLOGY). Another method is to apply a very low electric field so that little heating and electroosmosis occur, but then to recycle the material through coolers. The material is then sent back through the separating cell (51–53).

Most electrophoretic methods have been tried in a free-flow format, including isoelectric focusing, native zone electrophoresis, and isotachophoresis. Most free-flow electrophoresis equipment has very low (ca 1 g/(L·h)) capacity, and resolution is reduced by heating and electroosmotic considerations.

Detection Techniques

Most sample components analyzed with electrophoretic techniques are invisible to the naked eye. Thus methods have been developed to visualize and quantify separated compounds. These techniques most commonly involve chemically fixing and then staining the compounds in the gel. Other detection techniques can sometimes yield more information, such as detection using antibodies to specific compounds, which gives positive identification of a sample component either by immunoelectrophoretic or blotting techniques, or enhanced detection by combining two different electrophoresis methods in two-dimensional electrophoretic techniques.

Chemical Staining. Staining techniques that help to visualize banding patterns resulting from electrophoresis vary (54–56). The size and type of the compound as well as the electrophoretic matrix dictate and often limit the variety of stains that can be used. Molecules can be lost during the staining process, so most staining procedures incorporate a "fixing" step either before or in conjunction with staining. With paper electrophoresis, proteins are fixed to the medium by drying the paper. With agarose and acrylamide gels, the fixation of proteins requires additives. Acidic solutions, such as trichloroacetic acid [76-03-9], are commonly used to fix proteins of all sizes. Once proteins are fixed, they can be stained without loss of the separated components.

Amido black is a commonly used stain, but it is not very sensitive. It is often used to visualize concentrated proteins or components that are readily accessible to dyes such as proteins that have been transferred from a gel to nitrocellulose paper. Two of the more sensitive and more frequently used stains are Coomassie Brilliant Blue (R250 and G250) and silver stains. Because these stains interact differently with a variety of protein molecules, optimization of the fixative and staining solutions is necessary. The Coomassie stains are approximately five times more sensitive than amido black and are appropriate for both agarose and polyacrylamide gels. The silver stain is approximately 100 times more sensitive than Coomassie and is typically used for polyacrylamide gels.

A silver stain is used when proteins exist in a very small quantity or when analysis of as many bands as possible created by separation techniques is desired. One positive application of silver stain is its sensitivity. A drawback of the silver stain, however, is that it is more complex and often requires more troubleshooting to obtain the desired results.

To quantitate proteins from staining, a densitometer aided by computer software is used to evaluate band areas of samples compared to band areas of a standard curve. Amido black, Coomassie Brilliant Blue, and silver stains are all applicable for use in quantification of proteins.

Another method to visualize and identify separation products on a gel is through radioactivity. If a sample is radioactive, the bands that form through migration of a sample are subsequently radioactive. This type of gel may be placed against an x-ray until the radiation makes a mirror image of the banding pattern on the x-ray. The x-ray is then developed and the resulting autoradiograph displays the bands.

As an alternative to radiation, a stain such as ethidium bromide is used to visualize DNA. The ethidium may be incorporated into the structure of DNA either before or after electrophoresis. The gel is then visualized under a fluorescent lamp.

Immunoelectrophoretic Techniques. The technique of gel electrophoresis has been successfully combined with immunological techniques in order to further evaluate molecules. Specifically, the concept of double immunodiffusion as described in 1948 (57) and that of single-radial immunodiffusion described in 1963 (58) have been further developed for use with electrophoresis in both the clinical and research setting.

The double-immunodiffusion technique, often referred to as the Ouchterlony technique, uses an agarose gel as the matrix. Holes are made in the agarose where either sample or antisera is placed. The two solutions are allowed to diffuse into the matrix for a predetermined time. If there is a reaction between the antigen in the sample and the antibody, a precipitate is formed. The Ag–Ab reaction line can be visualized after a stain is used (Fig. 3a). Similarly, the single-radial immunodiffusion (Fig. 3b) technique is a reaction of equilibrium between antigen and antibody. The difference is that the antigen is added to a warmed agarose solution before it gels. Holes are cut in the gel as for the Ouchterlony technique where samples are placed. The sample diffuses into the gel and, if reactive with the antibody, forms an area of precipitation around the hole that is visualized after staining.

The Mancini and Ouchterlony techniques are the basis of the techniques employed in immunoelectrophoresis. A technique referred to as rockets (59) is named as such because of the appearance of a rocket-shaped antigen–antibody

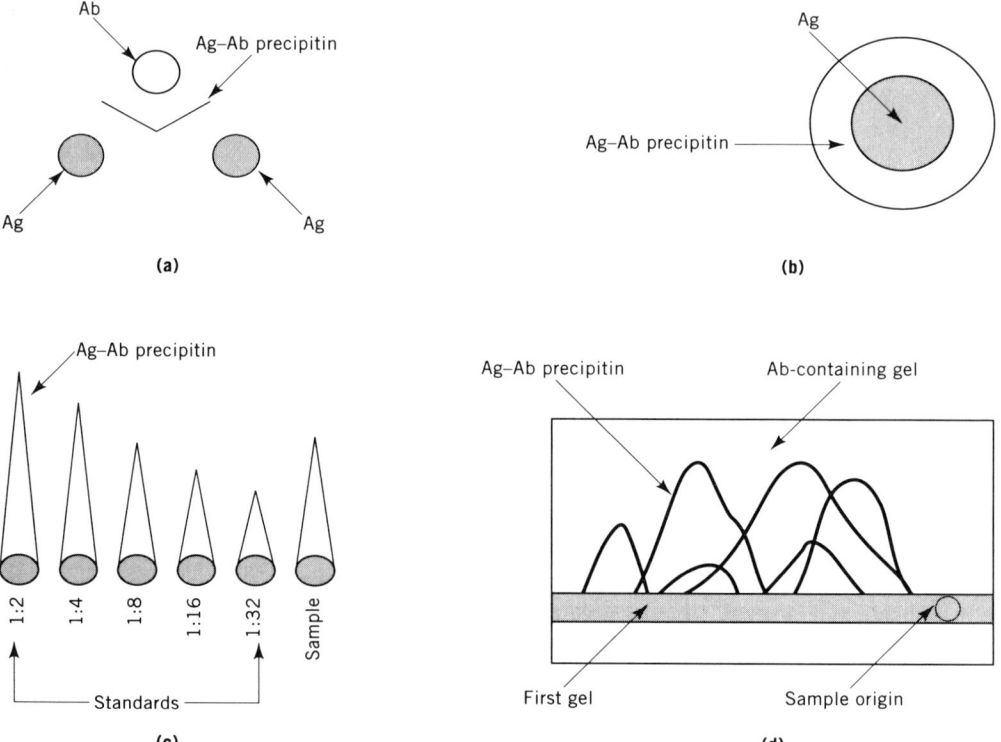

Fig. 3. Immunological reactions, where Ag is antigen and Ab is antibody, for detection in electrophoresis: (**a**) Ouchterlony technique; (**b**) single-radial diffusion; (**c**) rocket immunoelectrophoresis; and (**d**) crossed immunoelectrophoresis.

precipitin formed after an antigenic sample is electrophoresed through a gel-containing antibody (Fig. 3c).

Another frequently used method of immunoelectrophoresis is a technique known as crossed immunoelectrophoresis (Fig. 3d) (60). A sample is first run vertically through an agarose gel for a predetermined time. Secondly, the gel area where the sample was electrophoresed is typically cut out and placed horizontally into a similarly sized area of an antibody-containing gel. As an electrical current is applied to the second gel system, the sample in question electrophoreses through the gel and forms an antigen–antibody precipitin over an area that varies from small to large, depending on the banding pattern of electrophoresis from the first gel system.

At a somewhat more basic level, both agarose and acrylamide gel systems have been used for direct immunofixation. In these gels, samples are electrophoresed and then immunofixed by either using strips of cellulose acetate soaked in an antibody or the antibody is placed directly over the sample area of the gel.

All of these techniques are most often, but not exclusively, used in the clinical setting in order to diagnose abnormalities or to evaluate inheritance patterns of polymorphic proteins. Many applications of these techniques exist (61–65).

Two-Dimensional Electrophoresis. Two-dimensional (2D) electrophoresis is unique, offering an analytical method that is both reproducible and sensitive. It is referred to as 2D because it employs two different methods of electrophoresis, in two different dimensions, to produce one result. Each method separates the sample compounds based on different properties of each compound. The combination of the two methods gives better resolution of the compounds in the sample than could be achieved with either method alone. For example, each method alone may separate up to 100 components of a sample, whereas together they may separate up to 10,000 components.

A pair of electrophoretic techniques commonly employed in 2D analyses are isoelectric focusing (ief) and SDS–polyacrylamide gel electrophoresis (SDS–page). Ief separates sample compounds according to isoelectric point, whereas SDS–page separates the compounds by molecular weight. A 2D analytical technique using ief and SDS–page to separate total protein results in a gel having bands or spots in a random pattern (66). Each spot represents a unique component of a sample. A single charge difference in a component can be identified on the gel by a unique spot. This property of 2D electrophoresis, which allows identification of identical proteins that differ by one charge difference, has made it an invaluable technique for the molecular genetic community. Software is available to identify each spot on a gel and compare abnormalities in samples such as human blood (67), *Escherichia coli*, and yeast proteins.

Blotting Techniques. Problems encountered when trying to analyze resolved components of a sample mixture on a gel, with techniques such as direct immunofixation or application of a ligand, can be circumvented with the use of blotting techniques followed by staining or autoradiography. It is the inability of some compounds, such as antibodies or ligands, to enter the gel matrix of a specific gel system which leads to the development of various blotting techniques which have become widespread since the late 1970s. The nucleic acid and protein blotting techniques have become useful because these combine electrophoretic analyses and sensitive immunological tools. A blotting technique involves the transfer of nucleic acids or proteins, immediately after being separated by electrophoresis, from the gel matrix to another matrix. Typically, the other matrix is nitrocellulose paper, nylon, or other high affinity membrane and the mode of transfer is electrotransfer for proteins and capillary transfer for nucleic acids. Once nucleic acids or proteins are transferred, further analyses can be performed. On nitrocellulose paper or nylon, nucleic acids and proteins are more accessible than in the original gel matrix. This second matrix is then treated with a ligand to identify a specific component of a sample.

For example, the technique of Southern blotting was developed (68) for use with agarose gel electrophoresis of DNA fragments. Southern blots are designed to detect specific sequences of DNA. After electrophoresis is complete, the DNA is denatured and the single stranded DNA transferred to the specially prepared nitrocellulose paper. The nitrocellulose is then incubated with radioactive RNA or DNA complementary to those DNA sequences of interest. After the nitrocellulose has been sufficiently incubated with the radioactive complementary DNA, autoradiography is used to identify the fragments of interest.

The northern blotting technique is similar to the Southern blotting technique with the exception that northern blots detect specific sequences of RNA, not DNA.

The technique of immunoblotting, often referred to as western blotting, is frequently used for a variety of applications where protein concentrations are low and staining of the electrophoretic matrix does not produce adequate resolution. In these instances, as with Southern blotting, proteins are transferred to a second matrix like nitrocellulose or nylon and are then treated with antisera specific to a desired protein (69,70).

Blotting techniques may be used in a variety and combination of electrophoretic systems which makes their use widespread and convenient when protein concentrations are minimal and agarose or polyacrylamide is the matrix choice.

BIBLIOGRAPHY

"Electroseparations, Electrophoresis," in the *Encyclopedia of Chemical Technology*, 4th ed., Vol. 9, pp. 356–376, by S. Rudge and K. Markey, Synergen Inc.

1. J. S. Newman, *Electrochemical Systems*, Prentice Hall, Inc., Englewood Cliffs, N.J. 1973.
2. R. A. Robinson and R. H. Stokes, *Electrolyte Solutions*, Butterworths Scientific Publications, London, 1959
3. R. J. Hunter, in R. H. Ottewill and R. L. Rowell, eds., *Zeta Potential in Colloid Science, Principles and Applications*, Academic Press, Inc., New York, 1981.
4. S. R. Rudge and P. Todd, in M. R. Ladisch, R. C. Willson, C. C. Painton, and S. E. Builder, eds., *Protein Purification, From Molecular Mechanisms to Large-Scale Processes*, ACS Symposium Series 427, Washington, D.C. 1990, p. 244.
5. O. Stern, *Zeitschr. Elektrochem.* **30,** 508 (1924).
6. M. von Smoluchowski, *Bull. Akad. Sci. Cracovie, Classe Sci. Math. Natur.* **1,** 182 (1903).
7. E. Hückel, *Physik. Zeitschr.* **24,** 204 (1924).
8. D. C. Henry, *Proc. Roy. Soc. (London) Ser. A* **133,** 106 (1931).
9. R. W. O'Brien and L. R. White, *J. Chem. Soc. Faraday II* **74,** 1606 (1978).

10. K. Hannig, H. Wirth, and E. Schoen, *Apollo-Soyuz Test Project Summary Science Report*, Vol. 1, NASA SP-412, Washington, D.C., 1977, p. 335.
11. B. J. Davis, *Ann. N.Y. Acad. Sci.* **121,** 404 (1964).
12. L. Ornstein, *Ann. N.Y. Acad. Sci.* **121,** 321 (1964).
13. C. Schafer-Nielsen, in M. J. Dunn, ed., *Gel Electrophoresis of Proteins*, Wright, Bristol, UK, 1986, p. 3.
14. U. K. Laemmli, *Nature* **227,** 680 (1970).
15. R. Aschaffenburg and J. Drewry, *Nature* **176,** 218 (1955).
16. J. V. Maizel, in K. Habel and N. P. Salzman, eds., *Fundamental Techniques in Virology*, Academic Press, Inc., New York, 1969, p. 334.
17. R. Pitt-Rivers and F. S. A. Impiombato, *Biochem. J.* **109,** 825 (1968).
18. D. C. Schwartz and C. R. Cantor, *Cell* **37,** 67 (1984).
19. G. F. Carle, M. Frank, and M. V. Olson, *Science* **232,** 65 (1986).
20. G. F. Carle and M. V. Olson, *Nucleic Acids Research* **12**(14), 5647 (1984).
21. G. Chu, D. Vollrath, and R. W. Davis, *Science* **234,** 1582 (1986).
22. F. S. Collins and co-workers, *Science* **235,** 1046 (1986).
23. S. K. Lawrence, C. L. Smith, R. Srinastava, C. R. Cantor, and S. M. Weissman, *Science* **235,** 1387 (1986).
24. O. Vesterberg, *Acta Chem. Scand.* **23,** 2653 (1969).
25. H. Svensson, *Acta Chem. Scand.* **15,** 325 (1961).
26. A. Chrambach, *The Practice of Quantitative Gel Electrophoresis*, VCH Publishers, Deerfield Beach, Fla., 1985, p. 141.
27. A. Rosen, K. Ek, and P. Aman, *J. Immunol. Methods* **28,** 1 (1979).
28. C. A. Saravis, M. O'Brien, and N. Zamcheck, *J. Immunol. Methods* **29,** 97 (1979).
29. C. A. Saravis and N. Zamcheck, *J. Immunol. Methods* **29,** 91 (1979).
30. F. M. Everaerts, F. E. P. Mikkers, and Th. P. E. M. Verheggen, *Sep. Purif. Meth.* **6,** 287 (1977).
31. C. Araki, *Bull. Chem Soc. Jpn.* **29,** 543–544 (1956).
32. C. A. Alper, in W. J. Williams, E. Beutler, A. J. Erslev, and M. A. Lichtman, eds., *Hematology*, 3rd ed., McGraw-Hill Books, Inc., New York, 1983, p. 1516.
33. E. S. Lander, *Nature* **339,** 501 (1989).
34. T. Maniatis, E. F. Fritsch, and J. Sambrook, *Molecular Cloning*, Cold Spring Harbor Laboratory, Cold Spring Harbor, New York, 1983, pp. 150–171.
35. R. F. Boyer, *Modern Experimental Biochemistry*, The Benjamin/Cummings Publishing Co., Inc., 1986, p. 219.
36. S. Wood and S. Langlois, *J. Chromatog.* **569,** 421 (1991).
37. Ref. 26, p. 64.
38. D. E. Garfin, *Meth. Enzymol.* **182,** 428 (1990).
39. J. F. Robyt and B. J. White, *Biochemical Techniques: Theory and Practice*, Waveland Press, Inc., Prospect Heights, Ill., 1987.
40. S. A. Westwood, *Electrophoresis* **6,** (1985).
41. B. Budlowe and R. S. Murch, *Electrophoresis* **6,** 523 (1985).
42. V. Ingram, *Biochim. Biophys. Acta* **28,** 539 (1958).
43. A. Katz, W. Dreyer, and C. Anfinsen, *J. Biol. Chem.* **234,** 2897 (1959).
44. J. W. Jorgenson and K. D. Lukacs, *Anal. Chem.* **53,** 1298–1302 (1981).
45. J. W. Jorgenson and K. D. Lukacs, *Science* **222,** 266–272 (1983).
46. M. J. Gordon, X. Huang, S. L. Pentoney Jr., and R. N. Zare, *Science* **242,** 224 (1988).
47. X. Huang, T-K. J. Pang, M. J. Gordon, and R. N. Zare, *Anal. Chem.* **59,** 2737–2749 (1987).
48. Y.-F. Cheng and N. J. Dovichi, *Science* **242,** 562 (1988).
49. V. Pretorius, B. J. Hopkins, and J. Schieke, *J. Chromatog.* **99,** 23 (1974).
50. T. Tsuda, K. Nomura, and G. Nakagawa, *J. Chromatog.* **264,** 385 (1983).

51. M. Bier, in J. A. Asenjo and J. Hong, eds., *Separation, Recovery, and Purification in Biotechnology*, ACS Symposium Series No. 314, American Chemical Society, Washing-
52. W. A. Gobie, J. B. Beckwith, and C. G. Ivory, *Biotech. Prog.* **1**(1), 60 (1985).
53. W. A. Gobie and C. F. Ivory, *AIChE. J.* **34,** 474 (1988).
54. I. Syrovy and Z. Hodny, *J. Chromatog.* **569,** 175 (1991).
55. T. Rabilloud, *Electrophoresis* **11,** 785 (1990).
56. C. R. Merril, M. G. Harasewych, and M. G. Harrington, in M. J. Dunn in Ref. 13, p. 323.
57. Ö. Ouchterlony, *Acta Pathol. Microbio. Scand.* **25,** 186 (1948).
58. G. Mancini, J. P. Vaerman, A. O. Carbonara, and J. F. Heremans, *Protides Biol. Fluids* **11,** 370 (1963).
59. C.-B. Laurell, *Anal. Biochem.* **15,** 45 (1966).
60. C.-B. Laurell, *Anal. Biochem.* **10,** 358 (1965).
61. Z. L. Awdeh and C. A. Alper, *Proc. Natl. Acad. Sci.* **77**(6), 3576 (1980).
62. J. P. McCue, R. H. Heinard, and R. Tenold, *Rev. Infect. Dis.* **8**(S4), S374 (1986).
63. R. J. Ziccardi, *Clinical Laboratory Techniques for the 1980's*, Alan R. Liss, Inc., New York, 1980, p. 433.
64. C. A. Alper and R. F. Ritchie, in R. F. Ritchie, ed., *Automated Immunoanalysis*, Marcel Dekker, Inc., New York, 1978, p. 139.
65. C. A. Alper and A. M. Johnson, *Vox Sang.* **17,** 445 (1969).
66. P. H. O'Farrell, *J. Biol. Chem.* **250**(10), 4007 (1975).
67. J. E. Celis, B. Honore, G. Bauw, and J. Vandekerckhove, *BioEssays* **12**(2), 93 (1990).
68. E. Southern, *J. Mol. Biol.* **98,** 503 (1975).
69. D. D. Dykes and H. F. Polesky, *Electrophoresis* **6,** 521 (1985).
70. B. Bountin, S. H. Feng, and P. Arnaud, *Am. J. Hum. Genet.* **37,** 1098 (1985).

General References

R. C. Allen, C. A. Saravis, and H. R. Maurer, *Gel Electrophoresis and Isoelectric Focusing of Proteins*, Walter de Gruyter, New York, 1984.

A. Chrambach, *The Practice of Quantitative Gel Electrophoresis*, VCH Publishers, New York, 1985.

B. A. Baldo and E. R. Tovey, *Protein Blotting: Methodology, Research, and Diagnostic Applications*, Karger, Switzerland, 1989.

J. F. Robyt and B. J. White, *Biochemical Techniques: Theory and Practice*, Waveland Press, Inc., Prospect Heights, Ill., 1987.

M. J. Dunn, ed., *Gel Electrophoresis of Proteins*, Wright, Bristol, UK, 1986.

N. Catsimpoolas, *Methods of Protein Separation*, Plenum Press, New York, 1975.

R. F. Boyer, *Modern Experimental Biochemistry*, The Benjamin/Cummings Publishing Co., Inc., 1986.

R. J. Hunter, in R. H. Ottewill and R. L. Rowell, eds., *Zeta Potential in Colloid Science, Principles and Applications*, Academic Press, Inc., New York, 1981.

SCOTT RUDGE
KATHLEEN MARKEY
Synergen, Inc.

ELECTROSTATIC PRECIPITATION. See AIR POLLUTION CONTROL.

ENERGY CONSERVATION IN SEPARATION PROCESSES

The main driving force for increased energy conservation, which continues in times of both rising and falling energy prices, is broadscale technological process. Advances in technology are responsible for the historical rise in energy efficiency of 1–3% per year achieved by process industries. A wide range of big and little steps have contributed to these advances, such as improved gas turbine efficiency, structured packing in distillation (qv), computer control, variable speed drives, and computer design tools, as well as improved catalysts and synthetic processes for a variety of materials, eg, low density polyethylene, acrylonitrile, ammonia, and acetic acid.

The second force that has driven increased energy conservation is the trade of capital for energy. This trade is optimized within an existing technology and nets large increases when energy prices rise rapidly compared to capital price as in the 1975–1985 time period. The effect of energy usage on total cost is shown in Figure 1. If proper design is used, total costs are relatively tolerant of large deviations from the optimum design. For example, in Figure 1a, if the piping pressure drop is anywhere between one-third and three times the optimal, the penalty in total cost is ≤10%. In piping systems, capital costs dominate over energy costs (1).

Energy Balance

Historically, an energy balance has been prepared for the components of a process primarily to ensure that heat exchangers and utility supply are adequate. Often, an overall process energy balance was not developed. However, beginning in the mid-1980s, the energy balance for the overall process has become a document almost as important as the material balance. The overall energy balance serves as an evergreen framework during design to highlight the areas having greatest potential for improvement. Moreover, this document serves as a tool for plant-operating personnel after start-up, to aid optimization of energy use.

The energy balance should analyze the energy flows by type and amount, ie, present summaries of electricity, fuel gas, steam level, heat rejected to cooling water, etc. It should include realistic loss values for turbine inefficiencies and heat losses through insulation.

Exergy, Lost Work, and Second-Law Analysis. When energy is critically important to process economics, the simple energy balance is sometimes carried into an analysis of lost work. This compares the actual design against the theoretical ideal at each step and defines where the true energy use, or lost work, is occurring. In the discussions herein of reaction, separation, heat exchange, compression, refrigeration, and steam systems, the importance of this concept is illustrated. A few terms are defined below.

Exergy, E, is the potential to do work. It is also sometimes called availability or work potential. Thermodynamically, this is the maximum work a stream can deliver by coming into equilibrium with its surroundings:

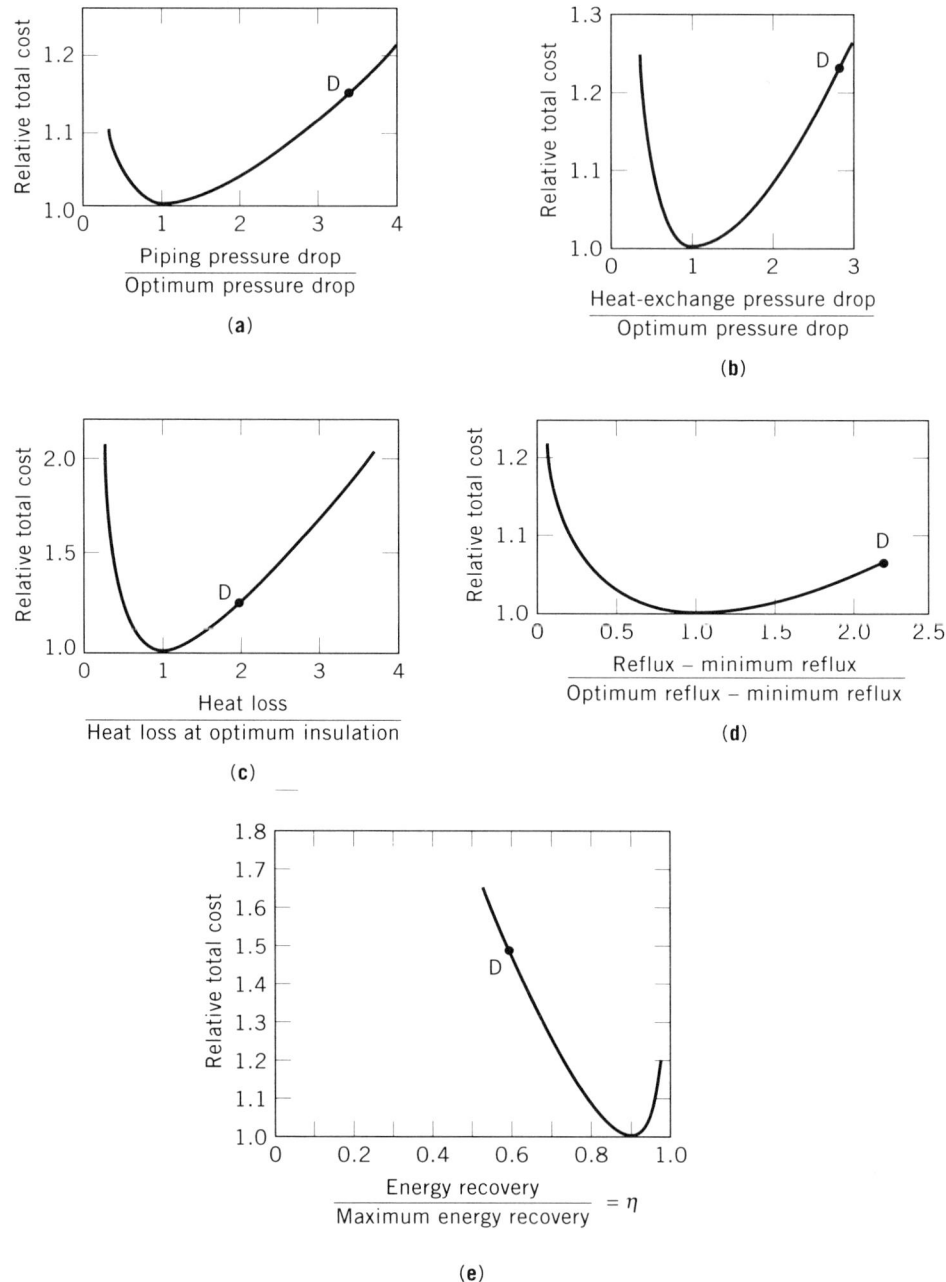

Fig. 1. Effect of energy use on total cost where total cost is the sum of capital and energy costs for the lifetime of the plant, discounted to present value. Point D corresponds to the design point if the designer uses an energy price that is low by a factor of four in projected energy price. Effects on costs of (**a**) pressure drop in piping, (**b**) pressure drop in exchangers, (**c**) heat loss through insulation, (**d**) reflux use, and (**e**) energy recovery through waste-heat boiler use.

$$E = (H - H_0) - T_0(S - S_0)$$

where E = the maximum theoretical work potential; H and S = enthalpy and entropy of the stream at its original conditions; H_0 and S_0 = enthalpy and entropy of the same stream at equilibrium with the surroundings; and T_0 = temperature of the surroundings (sink).

Free energy, G, is a related thermodynamic property. It is most commonly used to define the condition for equilibrium in a processing step. It is identical to ΔE if the processing step occurs at T_0.

$$\Delta G = \Delta H - T \Delta S$$

Lost work, LW, is the irreversible loss in exergy that occurs because a process operates with driving forces or mixes material at different temperatures or compositions.

$$LW = E_{\text{in}} - E_{\text{out}}$$

Second-law analysis looks at the individual components of an overall process to define the causes of lost work. Sometimes it focuses on the efficiency of a step and ratios the theoretical work needed to accomplish a change, eg, a separation, to that actually used.

Sometimes it is more cost-effective simply to compare the design against a second-law violation checklist covering items such as mixing streams at different temperatures and compositions, high pressure drops in control valves, reactions running far from equilibrium, high temperature differentials, and pump-discharge recirculation (2).

Separation

About one-third of the chemical industry's energy is used for separation. A correlation exists between selling price and feed concentration (Fig. 2) as well as between selling price and product purity.

Concentration and purity can both be traced to the minimum work of separation, W, where T_0 is the sink temperature; N_i is the number of moles of a

$$W = RT_0 \Sigma N_i \ln(x_i \gamma_i)$$

species present in the feed; $x_i = N_i/\Sigma N_i$; and γ_i is an activity coefficient. The value of W provides a target that is easily calculated and approachable in practice. For example, work calculated from this expression closely approaches the performance of a real-world distillation after inefficiencies for driving forces are taken into account.

For ideal solutions ($\gamma_i = 1$) of a binary mixture, the equation simplifies to

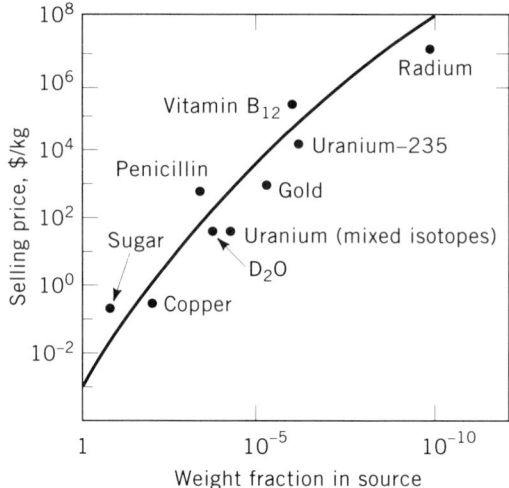

Fig. 2. Commercial selling prices of some separated materials (3). Courtesy of McGraw-Hill Book, Co., Inc.

the following, which applies whether the separation is by distillation or by any other technique.

$$W = RT_0[x_1 \ln x_1 + (1 - x_1) \ln(1 - x_1)]$$

When a separation is not completed, less work is required. For x_1 equal to 0.5,

Product purity, %	Relative work
100	1
99.9	0.99
99.0	0.92
90.0	0.53

This relative work is an important consideration when comparing separation techniques. Some leave much of the work undone, as, for example, in crystallization (qv) involving an unseparated eutectic mixture.

Distillation. Distillation (qv) is by far the most common separation technique because of its inherent advantages. Its phase separation is clean, its equilibrium is closely approached in each stage, and its multistage countercurrent device is relatively easy to build.

Minimum work for an ideal separation at first glance appears unrelated to the slender vertical vessel having a condenser at the top and a reboiler at the bottom. The connection becomes evident when the work embedded in the heat flow that enters the reboiler and leaves at the condenser is calculated. An ideal engine can extract work from this heat.

ENERGY CONSERVATION IN SEPARATION PROCESSES

$$E = QT_0\left(\frac{1}{T_{\text{condenser}}} - \frac{1}{T_{\text{reboiler}}}\right)$$

Comparison of actual use of work potential against the minimum allows calculation of an efficiency relative to the best possible separation:

$$\eta = \frac{RT_0[x_1 \ln x_1 + (1-x_1)\ln(1-x_1)]}{QT_0\left(\frac{1}{T_{\text{condenser}}} - \frac{1}{T_{\text{reboiler}}}\right)}$$

There is still no obvious reason to believe that the efficiency of separating a mixture and an α (relative volatility) of 1.1 is related to that for an α of 2; however, it is known that when α is small, the required reflux and Q are large, but $(T_{\text{condenser}} - T_{\text{reboiler}})$ is small (see DISTILLATION).

The two effects almost cancel one another to yield an approximation for the minimum work potential used in a distillation (3,4).

$$E = RT_0(1 + [\alpha - 1]x_1)$$

When this is combined with the definition of minimum separation work, an approximation for distillation efficiency for an ideal binary can be obtained:

$$\eta = \frac{x_1 \ln x_1 + (1-x_1)\ln(1-x_1)}{1 + (\alpha - 1)x_1}$$

This efficiency is high and shows only minor dependence on α over a broad range of α. For $x_1 = 0.5$:

α	η
1.1	0.66
1.5	0.55
2.0	0.46

The dependence on x_1 is greater:

	$\alpha = 1.05$	$\alpha = 2$
η for $x_1 = 0.1$	0.32	0.30
η for $x_1 = 0.01$	0.056	0.053

These values, which match experience, suggest that distillation should be the preferred separation method for feed concentrations of 10–90%, but is probably a poor choice for feed concentrations of less than 1%. Techniques such as adsorption (qv), chemical reaction, and ion exchange (qv) are chiefly used to remove impurity concentrations of <1%.

The high η values above conflict with the common belief that distillation is always inherently inefficient. This belief arises mainly because past distillation practices utilized such high driving forces for pressure drop, reflux ratio, and temperature differentials in reboilers and condensers. A real example utilizing an ethane–ethylene splitter follows, in which the relative number for the theoretical work of separation is 1.0, and that for the net work potential used before considering driving forces is 1.4.

$$\eta = \frac{\text{theoretical work}}{\text{net work potential used}} = \frac{1.0}{1.4} = 0.7$$

losses for driving forces:
reflux above the minimum	0.1
exchanger ΔT	2.1
ΔP in tower	0.5
ΔP in condenser and tower	0.8
Total	3.5

$$\eta_{\text{including losses}} = \frac{1.0}{1.4 + 3.5} = 0.2$$

These numbers show that, first, the theoretical work can be closely approached by actual work after known inefficiencies are identified and, second, the dominant driving force losses are in pressure drop and temperature difference. This is a characteristic of towers having low relative volatilities.

Optimum Design. Condenser and Reboiler ΔT. The losses for ΔT are typically far greater than those for reflux beyond the minimum. The economic optimum for temperature differential is usually under 15°C, in contrast to the values of over 50°C often used in the past. This is probably the biggest opportunity for improvement in the practice of distillation. A specific example is the replacement of direct-fired reboilers with steam heat.

Adjusting Process to Optimize ΔT. At first glance, there appear to be only three or four utility levels (temperatures), and these can be 50°C apart. Different ways to increase the options include using multieffect distillation, which spreads the ΔT across two or three towers; using waste heat for reboil; and recovering energy from the condenser. To make these options possible, the pressure in a column may have to be raised or lowered.

Reflux Ratio. Generally, the optimum reflux ratio is below 1.15 and often below 1.05 minimum. At this point, excess reflux is a minor contributor to column inefficiency. When designing for this tolerance, correct vapor–liquid equilibrium (VLE) and adequate controls are essential.

State-of-the-Art Control. Computer control using feed-forward capability can save 2–20% of a unit's utilities by reducing the margin of safety (5). Unless the discipline of a controller forces the reduction of the safety margin, operators typically opt for increased safety. Operators are probably correct to do so when a proper set of analyzers and controllers has not been provided and maintained.

Right Feed Enthalpy. Often it is possible to heat the feed with a utility considerably less costly than that used for bottom reboiling. Sometimes the preheating can be directly integrated into the column-heat balance by exchange against the condensing overhead or against the net bottoms from the column. Simulation and careful examination of the overall process are required to assess the value of feed preheating.

A vapor feed is favored when the stream leaves the upstream unit as a vapor or when most of the column feed leaves the tower as overhead product. The use of a vapor feed was a key component in the high efficiency cited previously for the C_2 splitter, where most of the feed goes overhead.

Low Column Pressure Drop. The penalty for column pressure drop is an increase in temperature differential:

$$\Delta T = \left(\frac{dT}{dP}\right)\Delta P$$

$$\frac{dT}{dP} = \frac{R}{\Delta H}\frac{T^2}{P}$$

As this suggests, the penalty becomes large for low vapor pressure materials, ie, for components that are distilled at or below atmospheric pressure. The work penalty associated with this ΔT is approximately defined by the following ratio.

$$\frac{\Delta T_{\text{pressure drop}}}{T_{\text{reboiler}} - T_{\text{condenser}}} = \text{fraction of } W \text{ for } \Delta P$$

This penalty is greatest for close-boiling mixtures. A powerful technique for cutting ΔP is the use of packing. Conventional packings such as 5-cm (2-in.) pall rings can achieve a factor of four reductions over trays, and structured packing can achieve a factor of 10 reduction. Structured packing is more vulnerable to mistakes in detailed engineering and much less tolerant of fouling than trays. Almost 50% of the installations have encountered serious performance problems (6). It is also 2 to 10 times as expensive as the trays it typically replaces. However, despite these obstacles, structured packing is the biggest innovation in energy-saving hardware in the chemical processing industries. The overhead line and condenser pressure drop should be considered as well. (Note the high loss in the C_2 splitter example.)

Intermediate Condenser. As shown in Figure 3, an intermediate condenser forces the operating line closer to the equilibrium line, thus reducing the inherent

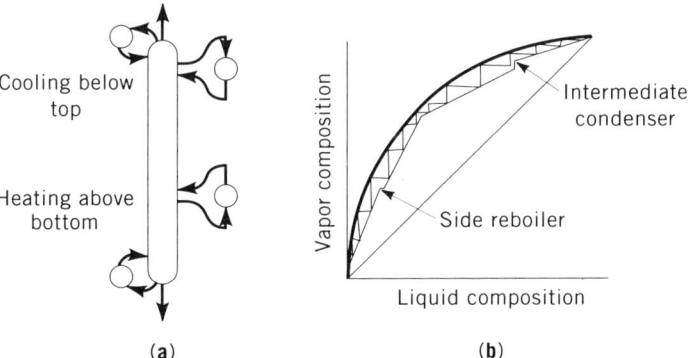

Fig. 3. (a) Schematic of an intermediate condenser and reboiler; (b) the corresponding vapor–liquid equilibrium.

inefficiencies in the tower. Using intermediate condensers and reboilers, it is possible to raise the efficiency above that for a simple reboiler–condenser system, particularly when the feed composition is far from 50:50 in a binary mixture.

	Maximum efficiency of heavy component in feed	
	50%	95%
one condenser, one reboiler	67	20
two condensers, one reboiler	73	47
three condensers, one reboiler	77	62

The intermediate condenser is most effective when a less costly coolant can be substituted for refrigeration.

Intermediate Reboiler. Inclusion of an intermediate reboiler moves the heat-input location up the column to a slightly colder point. It can permit the use of waste heat for reboil when the bottoms temperature is too hot for the waste heat.

Heat Pumps. Because of added capital and complexity, heat pumps are rarely economical, although they were formerly commonly used in ethylene/ethane and propylene/propane splitters. Generally, the former splitters are integrated into the refrigeration system; the latter are driven by low level waste heat, cascading to cooling water.

Lower Pressure. Usually, relative volatility increases as pressure drops. For some systems, a 1% drop in absolute pressure cuts the required reflux by 0.5%. Again, if operating at reduced pressure looks promising, the process can be evaluated by simulation. In a complete study of distillation processes, other questions that need to be asked include, Is the separation necessary? Is the purity necessary? Are there any recycles that could be eliminated? Can the products be sent directly to downstream units, thereby eliminating intermediate heating and cooling?

Other Separation Techniques. Under some circumstances, distillation is not the best method of separation. Among these instances are the following: when relative volatility is <1.05; when <1% of a stream is removed, as in gas drying (adsorption or absorption) or C_2H_2 removal (reaction or absorption); when thermodynamic efficiency of distillation is <5%; and when a high boiling point pushes thermal stability limits. A variety of other techniques may be more applicable in these cases (see SEPARATIONS PROCESS SYNTHESIS).

Reaction. Purification by reaction is relatively common when concentrations are low (ppm) and a high energy but low value molecule is present. Some examples are the hydrogenation of acetylene and the oxidation of waste hydrocarbons:

$$C_2H_2 + H_2 \longrightarrow C_2H_4$$

$$\text{waste hydrocarbon} + O_2 \longrightarrow H_2O + CO_2$$

Absorption. As a separation technique, absorption (qv), also called extractive distillation, starts with an energy deficit because the process mixes in a pure material (solvent) and then separates it again. This process is nevertheless quite common because it shares most of the advantages of distillation. Additionally, because it separates by molecular type, it can be tailored to obtain a high α. The following ratios are suggested for equal costs (7):

$\alpha_{\text{distillation}}$	$\alpha_{\text{extraction distillation}}$	$\alpha_{\text{extraction}}$
1.2	1.4	2.5
1.4	1.9	5
1.6	2.3	8

In practice, most of the applications have come where a small part (<5%) of the feed is removed. Examples include H_2S/CO_2 removal and gas drying with a glycol (see DISTILLATION, AZEOTROPIC AND EXTRACTIVE).

Extraction. The advantage of extraction is that a liquid is purified rather than a vapor, allowing operation at lower temperatures and the removal of a series of similar molecules at the same time, even though these molecules differ widely in boiling point. An example is the extraction of aromatics from hydrocarbon streams (see EXTRACTION, LIQUID–LIQUID).

The disadvantage of extraction relative to extractive distillation is the greater difficulty of getting high efficiency countercurrent processing.

Adsorption. Adsorbents can achieve even more finely tuned selectivity than extraction. The most common application is the fixed bed with thermal regeneration, which is simple, attains essentially 100% removal, and carries little penalty for low feed concentration. An example is gas drying. A variant is pressure-swing adsorption. Here, regeneration is attained by a drop in pressure. By using multiple stages, high impurity rejection can be achieved, but at the expense of losing part of the desired product (see ADSORPTION).

Another approach is the simulated moving-bed system, which has large-volume applications in normal-paraffin separation and *para*-xylene separation. Since its introduction in 1970, the simulated moving-bed system has largely displaced crystallization in xylene separations. The unique feature of the system is that, although the bed is fixed, the feed point shifts to simulate a moving bed (see ADSORPTION, LIQUID SEPARATION).

Melt Crystallization. Crystallization (qv) from a melt is inherently more attractive than distillation because the heat of fusion is much lower than that of evaporation. It also benefits from lower operating temperature. In addition, organic crystals are virtually insoluble in each other so that a pure product is possible in a one-stage operation.

However, crystallization has unique disadvantages that outweigh its virtues and have sharply limited its application. Industry practice suggests the use of a workable alternative, if one exists. The disadvantages of melt crystallization include the following. (*1*) Difficulty of physical separation. Impure liquid is trapped as occlusion, and wets all crystal surfaces. (*2*) Requirement of a second separating process for eutectic mixture. The process thus resembles formation of two liquid phases. Although little energy is required to get the two phases, a great amount of it is required to finish the purification. (*3*) Difficulty of adding or removing heat on account of the thermal resistance of the crystal. (*4*) Difficulty of moving the liquid countercurrent to the crystals.

Thermodynamic efficiency is hurt by the large ΔT between the temperatures of melting and freezing. In an analogy to distillation, the high α comes at the expense of a big spread in reboiler and condenser temperature. From a theoretical standpoint, this penalty is smallest when freezing a high concentration (ca 90%) material.

One process, shown in Figure 4, is a semibatch operation in which liquid falls down the walls of long tubes. This permits both staged operation and sweating of crystals. Sweating is the removal of impurities by melting a small portion of crystals after mother liquor is first drained. The sweating operation washes residual mother liquor off the remaining crystals, and also removes some impurities from within the crystals. Typically, the sweating and staged operations require melting 5 kg of material for each kg of product (8).

Membranes. Liquid separation via membranes, ie, reverse osmosis (qv), is used in production of pure water from seawater. The chief limit to broader use of reverse osmosis is the high pressure required as the concentration of reject rises.

Mole fraction of reject	Minimum ΔP, MPa (psi)
0.05	7.6 (1100)
0.10	15.2 (2200)
0.20	31.8 (4600)

As a result, most systems are limited to achieving a mole fraction reject of 0.1 or less (see MEMBRANE TECHNOLOGY).

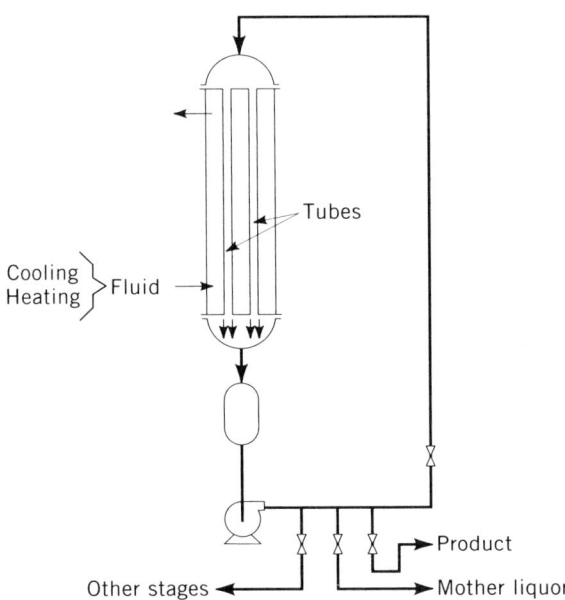

Fig. 4. Falling-film crystallizer, semibatch. Tube length is 12.2 m; tube diameter is 7.6 cm.

Membranes are also used to separate gases, for example, the production of N_2 and O_2 from air and the recovery of hydrogen from ammonia plant purge gas. The working principle is a membrane that is chemically tuned to pass a molecular type (see GAS–SOLID SEPARATIONS).

BIBLIOGRAPHY

"Process Energy Conservation" in the *Encyclopedia of Chemical Technology*, 3rd ed., Suppl. Vol., pp. 669–697, by D. Steinmeyer, Monsanto Co.; in *ECT* 4th ed., Vol. 20, pp. 175–203, by D. Steinmeyer, Monsanto Co.

1. D. E. Steinmeyer, *Chemtech*, **12**(3), 188 (1982).
2. W. F. Kenney, *Proceedings of Industrial Energy Conservation Technology Conference*, Energy Systems Laboratory, Texas A and M University, College Station, Tex., 1981, p. 247.
3. C. J. King, *Separation Processes*, 2nd ed., McGraw-Hill Book Co., Inc., New York, 1980.
4. C. S. Robinson and E. R. Gilliland, *Elements of Fractional Distillation*, 4th ed., McGraw-Hill Book Co., Inc., New York, 1950.
5. T. L. Tolliver, *Chem. Eng.* **93**(22), 99 (1986).
6. A. Sloley, personal communication, Process Consulting Services, Houston, Tex., Aug. 25, 1995.
7. M. Souders, *Chem. Eng. Prog.* **60**(2), 75 (1964).
8. Technical data, D. Carter, Monsanto Corp., St. Louis, Mo., 1982.
9. D. Steinmeyer, *Hydrocarbon Process.* **71**(4), 53 (1992).
10. R. Smith and B. Linnhoff, *Chem. Eng. Res. Des.* **66**, 195 (1988).
11. B. Linnhoff, *Chem. Eng. Prog.* **90**(8), 33 (1994).
12. W. F. Furgerson, *Conserving Energy in Refrigeration*, Massachusetts Institute of Technology Press, Cambridge, Mass., 1982.
13. *Hydrocarbon Process.* **74**(3), 89 (1995).

General References

W. F. Kenney, *Energy Conservation in the Process Industries*, Academic Press, Inc., New York, 1984.
B. Linnhoff and co-workers, *A User's Guide on Process Integration for the Efficient Use of Energy*, Institution of Chemical Engineers, Warwickshire, U.K., 1994.
R. Smith, *Chemical Process Design*, McGraw-Hill Book Co., Inc., New York, 1995.

DAN STEINMEYER
Monsanto Company

EXTRACTION, LIQUID–LIQUID

Liquid–liquid extraction, often loosely referred to as solvent extraction, was carried out as early as Roman times when silver and gold were extracted from molten copper using lead as a solvent. The first significant industrial application of solvent extraction was in the petrochemical industry. This was followed by applications for the recovery of vegetable oils and the purification of penicillin, and since 1945 in the nuclear industry in the refining of uranium, plutonium, and other radioisotopes. Since 1960, solvent extraction has been applied on a large scale in the refining of other nonferrous metals, particularly copper. Most recently it has gained increasing importance as a separation technique in biotechnology.

The physical process of liquid–liquid extraction separates a dissolved component from its solvent by transfer to a second solvent, immiscible with the first but having a higher affinity for the transferred component. The latter is sometimes called the consolute component. Liquid–liquid extraction can purify a consolute component with respect to dissolved components which are not soluble in the second solvent, and often the extract solution contains a higher concentration of the consolute component than the initial solution. In the process of fractional extraction, two or more consolute components can be extracted and also separated if these have different distribution ratios between the two solvents.

The principle of liquid–liquid extraction, and some of the special terminology, are illustrated in Figure 1 which shows a single contacting stage. If equilibrium is fully established after contact, the stage is defined as an ideal or theoretical stage. The two resulting liquid phases are the raffinate from which most of solute C has been removed, and the extract, consisting mainly of solvent B and C.

In the simplest case, the feed solution consists of a solvent A containing a consolute component C, which is brought into contact with a second solvent B. For efficient contact there must be a large interfacial area across which component C can transfer until equilibrium is reached or closely approached. On the laboratory scale this can be achieved in a few minutes simply by hand agitation of the two liquid phases in a stoppered flask or separatory funnel. Under continuous flow conditions it is usually necessary to use mechanical agitation to promote coalescence of the phases. After sufficient time and agitation, the system approaches equilibrium which can be expressed in terms of the extraction factor ϵ for component C:

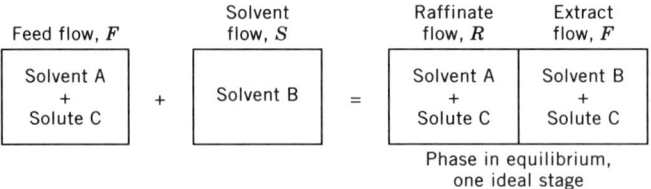

Fig. 1. Single contacting stage.

$$\epsilon = \frac{\text{quantity of C in B-rich phase}}{\text{quantity of C in A-rich phase}} = m\,\frac{B}{A} \qquad (1)$$

where B and A refer to the quantities of the two solvents and m is the distribution coefficient.

The component C in the separated extract from the stage contact shown in Figure 1 may be separated from the solvent B by distillation (qv), evaporation or other means, allowing solvent B to be reused for further extraction. Alternatively, the extract can be subjected to back-extraction (stripping) with solvent A under different conditions, eg, a different temperature; again, the stripped solvent B can be reused for further extraction. Solvent recovery is an important factor in the economics of industrial extraction processes.

Whereas Figure 1 assumes a physical extraction based on different solubilities as expressed by the distribution coefficient, many extractions depend on chemical changes. In such cases the component C in the feed solvent may not itself have any solubility in the extracting solvent B, but can be made to react with an extractant to produce a compound or species which is soluble in B. Many metals can be extracted from aqueous solutions of their salts into organic carrier solvents by using organic extractants which can form organometallic compounds or complexes. Stripping of the metals from the organic to an aqueous phase can be effected by changing a chemical condition such as pH.

Extraction, a unit operation, is a complex and still developing subject area (1,2). The chemistry of extraction and extractants has been comprehensively described (3,4). The main advantage of solvent extraction as an industrial process lies in its versatility because of the enormous potential choice of solvents and extractants. The industrial application of solvent extraction, including equipment design and operation, is a subject in itself (5). The fundamentals and technology of metal extraction processes have been described (6,7), as has the role of solvent extraction in relation to the overall development and feasibility of processes (8). The control of extraction columns has also been discussed (9).

The rapid development of the fundamentals and technology of solvent extraction is reflected in research papers found in chemistry, chemical engineering, and hydrometallurgical journals. Review articles (10–12) provide a critical summary of some of the ongoing research. Leading researchers and practitioners of solvent extraction meet regularly at the International Solvent Extraction Conferences (ISEC) whose proceedings (13–21) provide a useful indication of progress and trends. Papers published in the *ISEC Proceedings* numbered over 400 in 1990 (20), yet these make up less than 20% of all papers published in the solvent extraction area annually (12).

Principles

Physical Equilibria and Solvent Selection. In order for two separate liquid phases to exist in equilibrium, there must be a considerable degree of thermodynamically nonideal behavior. If the Gibbs free energy, G, of a mixture of two solutions exceeds the energies of the initial solutions, mixing does not occur and the system remains in two phases. For the binary system containing only components A and B, the condition (22) for the formation of two phases is

$$\frac{d^2G}{dx_A^2} > 0 \tag{2}$$

The stability criteria for ternary and more complex systems may be obtained from a detailed analysis involving chemical potentials (23). The activity of each component is the same in the two liquid phases at equilibrium, but in general the equilibrium mole fractions are greatly different because of the different activity coefficients. The distribution coefficient m', based on mole fractions, of a consolute component C between solvents B and A can thus be expressed in terms of activity coefficients.

$$m' = \frac{x_{CB}}{x_{CA}} = \frac{\gamma_{CA}}{\gamma_{CB}} \tag{3}$$

If the mutual solubilities of the solvents A and B are small, and the systems are dilute in C, the ratio m' can be estimated from the activity coefficients at infinite dilution. The infinite dilution activity coefficients of many organic systems have been correlated in terms of structural contributions (24), a method recommended by others (5). In the more general case of nondilute systems where there is significant mutual solubility between the two solvents, regular solution theory must be applied. Several methods of correlation and prediction have been reviewed (23). The universal quasichemical (UNIQUAC) equation has been recommended (25), which uses binary parameters to predict multicomponent equilibria.

In addition to thermodynamically based predictions of liquid–liquid equilibria, a great deal of experimental data is to be found in the research literature (26). A liquid–liquid equilibrium data bank is also available (27).

Because of the nonideal nature of liquid–liquid systems, it is common engineering practice to quote data as mass fractions rather than mole fractions. Ternary systems can be represented graphically on a triangular diagram. The triangle can be equilateral or right-angled; the latter has the advantage that ordinary graph paper can be used, and also the vertical and horizontal scales can be changed independently. Figure 2a shows a typical phase diagram for a system where solvents A and B are partially miscible, and the solute C is a liquid. The ordinate scale represents mass fraction x_C and the abscissa represents x_B. The composition x_A can be obtained as $(1 - x_B - x_C)$.

The triangular diagram shows the two-phase envelope which encloses regions of overall composition in which two phases exist in equilibrium. In Figure 2a there is only one two-phase region; other types of systems can have two or more such regions (5). The equilibrium compositions of two liquid phases are connected by tie-lines. Typical tie-lines are shown. It is often convenient for a triangular diagram to be accompanied by a tie-line location curve (Fig. 2b) which allows any tie-line to be drawn by simple geometric construction. The procedure is shown for a given composition x_C of 0.10 in the A-rich phase. A line is drawn across from the triangular diagram to Figure 2b, then reflected from the equilibrium curve to the diagonal line, to give a composition y_C representing the B-rich

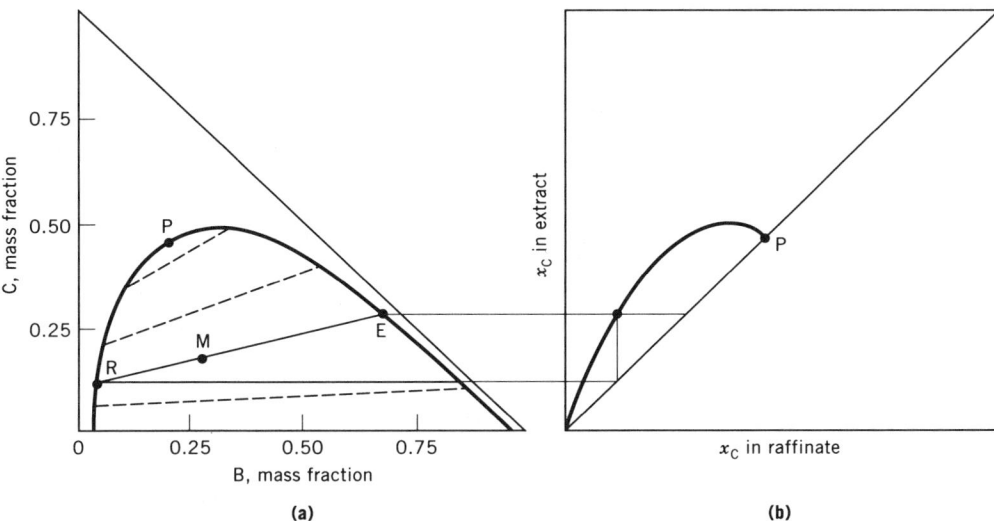

Fig. 2. (a) Triangular diagram, where the dashed lines represent tie-lines, and (b) tie-line location curve. Terms are defined in the text.

phase. A horizontal line is then drawn back to the appropriate point on the two-phase envelope in the triangular diagram, as shown.

It is seen that as the concentration of C is increased, the tie-lines become shorter because of the increased mutual miscibility of the two phases; at the plait point, P, the tie lines vanish. However, P does not necessarily represent the highest possible loading of C which can exist in the system under two-phase conditions. In Figure 2b the plait point lies on the diagonal because the compositions of the two phases approach each other at P.

An important use of the triangular equilibrium diagram is the graphical solution of material balance problems, such as the calculation of the relative amounts of equilibrium phases obtained from a given overall mixture composition. As an example, consider a mixture where the overall composition is represented by point M on Figure 2a. If the A-rich phase is denoted by point R (raffinate) and the B-rich phase is denoted by point E (extract), it can be shown that points R, M, and E are collinear, and also

$$\frac{\text{mass of extract}}{\text{mass of raffinate}} = \frac{\text{distance MR}}{\text{distance ME}} \qquad (4)$$

This is a statement of the inverse lever rule, which is the basis of techniques for graphical multistage calculations in ternary systems. Although triangular diagrams are widely used, there is the disadvantage that at low concentrations of C the tie-lines become almost horizontal and are hard to draw accurately. This problem can be overcome by plotting the compositions on a solvent-free (B-free) basis using a rectangular diagram. Detailed descriptions of this method and its application are given elsewhere (5,28).

Liquid–liquid equilibria having more than three components cannot as a rule be represented on a two-dimensional diagram. Such systems are important in fractional extraction, for example, operations in which two consolute components C and D are separated by means of two solvents A and B. For the special case where A and B are immiscible, the linear distribution law can be applied to components C and D independently:

$$x_{CB} = m_C x_{CA} \tag{5}$$

$$x_{DB} = m_D x_{DA} \tag{6}$$

The selectivity or separation factor between the two solutes is defined as the ratio of the distribution ratios:

$$\beta_{CD} = \frac{m_C}{m_D} = \frac{x_{CB} x_{DA}}{x_{CA} x_{DB}} \tag{7}$$

By convention, the components C and D are assigned so that the ratio β_{CD} exceeds unity. The greater the selectivity, the easier is the separation of C and D using solvents A and B. Selectivity can be defined in terms of mass ratio, mole ratio, or concentration.

The selection of solvents for a given separation depends largely on equilibrium considerations. Other important factors include cost, ease of solvent recovery by distillation (qv) or other means, safety and environmental impact, and physical properties which must permit easy phase dispersion and separation. Solvent selection is therefore a broad-based exercise which is hard to quantify (8). However a useful quantitative approach has been proposed (22) for comparing simplified equilibrium estimations on the basis of regular solution theory. The polar and hydrogen bonding parameters for solutes and possible solvents are plotted as points on a solvent selection diagram. The selectivities can be approximately related to the distances between the points, providing a quick comparison between a large number of potential solvents. Computer-aided solvent selection has been developed using a molecular graphics system (29).

Chemical Equilibria. In many cases, mass transfer between two liquid phases is accompanied by a chemical change. The transferring species can dissociate or polymerize depending on the nature of the solvent, or a reaction may occur between the transferring species and an extractant present in one phase. An example of the former case is the distribution of benzoic acid [65-85-0] between water and benzene. In the aqueous phase, the acid is partially dissociated:

$$C_6H_5COOH \rightleftharpoons C_6H_5COO^- + H^+ \tag{8}$$

In the organic phase the monomer is in equilibrium with the dimer:

$$2\ C_6H_5COOH \rightleftharpoons (C_6H_5COOH)_2 \tag{9}$$

Whereas the linear distribution law can be applied to the undissociated monomer, the interfacial distribution of total benzoic acid, as determined by analysis, is nonlinear.

Many industrial processes involve a chemical reaction between two liquid phases, for example nitration, sulfonation, alkylation, and saponification. These processes are not always considered to be extractions because the main objective is a new chemical product, rather than separation (30). However these processes have many features in common with extraction, for example the need to maintain a high interfacial area with the aid of agitation and the importance of efficient phase separation after the reaction is completed.

In addition to the liquid–liquid reaction processes, there are many cases in both analytical and industrial chemistry where the main objective of separation is achieved by extraction using a chemical extractant. The technique of dissociation extraction is very valuable for separating mixtures of weakly acidic or basic organic compounds such as 2,4-dichlorophenol [120-83-2], $C_6H_4Cl_2O$, and 2,5-dichlorophenol [583-78-8], which are difficult to separate by other means (31). The technique involves the use of a controlled amount of a strong base or acid in the aqueous phase, so that the overall distribution of each species is affected by its dissociation constant as well as the distribution constant of the undissociated form (see eqs. 7 and 8). Dissociation extraction has been applied extensively to mixtures of closely similar phenols, amines, and organic acids (32).

In hydrometallurgical separations (7), a metal ion in aqueous solution can be selectively converted to an organometallic compound or complex which is soluble in an organic carrier solvent. The metal extractants may be classified (33) as: (1) acid and chelating acid extractants, (2) anion exchangers, and (3) solvating extractants. Acid extractants typically contain one or more long hydrocarbon chains, an example being di(2-ethyl-hexyl)phosphoric acid [298-07-7], $[C_4H_9CH(C_2H_5)CH_2]_2HPO_4$. The extractant is dissolved in an organic carrier solvent which may be a pure compound such as hexane, but in industry is commonly a kerosene. When the extractant in the carrier solvent is contacted using an aqueous solution of a metal cation, eg, M^{2+}, equilibria are set up which can be simplified as the overall equation:

$$M^{2+} + \overline{2\,LH} \rightleftharpoons 2\,H^+ + \overline{ML_2} \tag{10}$$

where the overbar denotes species in the organic phase. The control of pH is obviously an important factor in determining the degree of metal extraction and the selectivity of an extractant for two or more cations. Often the logarithm of the distribution coefficient can be plotted as a linear function of pH (34). After a metal has been selectively extracted to the organic phase it can be back-extracted or stripped to a strongly acidic aqueous phase, giving a purer and more concentrated solution than the initial aqueous feed.

Chelating extractants owe effectiveness to the attraction of adjacent groups on the molecule for the metal. Compounds containing long hydrocarbon chains and the hydroxyoxime group, or those based on 8-quinolinol [148-24-3], C_9H_6NOH, form the basis of various commercial formulations (8,33). These extractants often show an amphoteric behavior depending on pH, changing for example from L^- to LH to LH_2^+ as the pH is increased; these agents can be applied in basic as well as acidic conditions (35).

Anionic extractants are commonly based on high molecular weight amines. Metal anions such as MnO_4^- or ReO_4^- can be exchanged selectively with inorganic anions such as Cl^- or SO_4^{2-}. The equilibrium for a quaternary onium compound of organic radicals R for two anion species A^- and B^- might be:

$$\overline{R_4N^+A^-} + B^- \rightleftharpoons \overline{R_4N^+B^-} + A^- \tag{11}$$

Solvating extractants contain one or more electron donor atoms, usually oxygen, which can supplant or partially supplant the water which is attached to the metal ions. Perhaps the best known example of such an extractant is tri-(n-butyl) phosphate [126-73-8] (TBP), $C_{12}H_{27}O_4P$, which forms the basis of the PUREX process (36) for uranium extraction:

$$UO_2^{2+} + 2\,NO_3^- + \overline{2\,TBP} \rightleftharpoons \overline{UO_2(NO_3)_2 \cdot 2\,TBP} \tag{12}$$

TBP and nitric acid also tend to form a complex with each other, but at sufficiently high uranyl nitrate concentrations the nitric acid is mainly displaced into the aqueous phase.

Interfacial Mass-Transfer Coefficients. Whereas equilibrium relationships are important in determining the ultimate degree of extraction attainable, in practice the rate of extraction is of equal importance. Equilibrium is approached asymptotically with increasing contact time in a batch extraction. In continuous extractors the approach to equilibrium is determined primarily by the residence time, defined as the volume of the phase contact region divided by the volume flow rate of the phases.

The rate of mass transfer (qv) depends on the interfacial contact area and on the rate of mass transfer per unit interfacial area, ie, the mass flux. The mass flux very close to the liquid–liquid interface is determined by molecular diffusion in accordance with Fick's first law:

$$N = -D\frac{\partial c}{\partial z} \tag{13}$$

where N refers to the flux in the z direction, c is the concentration of the consolute component, and D is its molecular diffusivity in the solvent. It is accepted practice to use the concentration gradient as the driving force for mass transfer, although it has been suggested that the gradient of chemical potential is more appropriate (37,38). Molecular diffusivities D of solutes are commonly in the range 10^{-10} to 5×10^{-9} m²/s at ambient temperatures. Data for many systems are available in the literature or can be measured experimentally or predicted from correlations (39,40). For dilute solutions of nonelectrolytes, the correlation of Wilke and Chang (41) is recommended.

Although molecular diffusion itself is very slow, its effect is nearly always enhanced by turbulent eddies and convection currents. These provide almost perfect mixing in the bulk of each liquid phase, but the effect is damped out in the vicinity of the interface. Thus the concentration profiles at each side of a liquid–liquid interface have the appearance shown in Figure 3, with essentially uniform bulk composition, and a film on each side of the interface. The films are the regions

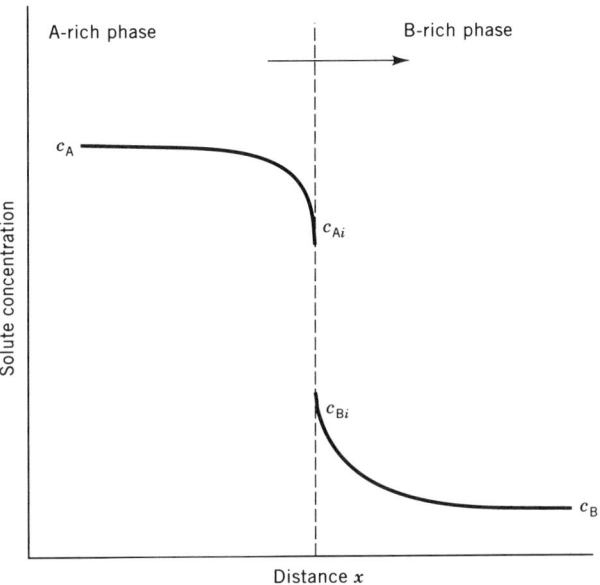

Fig. 3. Concentration profiles near an interface where the arrow represents the direction of mass transfer, c_A = concentration of C in A-rich phase, c_B = concentration of C in B-rich phase, and the subscript i denotes the interface.

in which there are significant concentration gradients. The equivalent film thicknesses depend on hydrodynamic conditions and are in the order of 100 μm. The film thicknesses and concentration profiles are therefore almost impossible to measure directly, and the mass fluxes are expressed by means of mass-transfer coefficients, k_A and k_B, and the concentration differences for each phase.

$$N = k_A (c_A - c_{Ai}) = k_B (c_{Bi} - c_B) \quad (14)$$

The mass-transfer coefficients are typically between 10 to 100 μm/s, depending on hydrodynamic conditions and the values of D.

The solute concentrations very close to the interface, c_{Ai} and c_{Bi}, are assumed to be in equilibrium, in the absence of any slow interfacial reaction. According to the linear distribution law, $c_{Bi} = mc_{Ai}$ and thus from equation 14 the mass-transfer flux can be expressed in terms of an overall mass-transfer coefficient K_A and an overall concentration driving force:

$$N = K_A (c_A - c_A^*) \quad (15)$$

where

$$c_A^* = c_B/m \quad (16)$$

and

$$1/K_A = 1/k_A + 1/(mk_B) \quad (17)$$

This is the important rule of additivity of resistances. In practice, k_A and k_B are often of the same order of magnitude, but the distribution coefficient m can vary considerably. For solutes which preferentially distribute toward solvent B, m is large and the controlling resistance lies in phase A. Conversely, if the distribution favors solvent A the controlling mass-transfer resistance lies in phase B.

Values of the mass-transfer coefficient k have been obtained for single drops rising (or falling) through a continuous immiscible liquid phase. Extensive literature data have been summarized (40,42). The mass-transfer coefficient is often expressed in dimensionless form as the Sherwood number:

$$Sh = kd/D \tag{18}$$

The values of k and hence Sh depend on whether the phase under consideration is the continuous phase, c, surrounding the drop, or the dispersed phase, d, comprising the drop. The notations k_c, k_d, Sh_c, and Sh_d are used for the respective mass-transfer coefficients and Sherwood numbers.

Mass-transfer coefficients are strongly affected by the degree of mobility of the drop surface. When the surface is free of adsorbed molecules it can move in response to surface shear stress, and the drops tend to circulate as they move through the continuous phase. However, even a trace of surface-active material can cause the drop surface to resist shear and the drop circulation is suppressed, resulting in greatly reduced mass transfer (43,44). Table 1 lists some of the proposed equations for mass transfer (expressed as Sh) under various conditions. Nearly a threefold variation in dispersed-phase mass transfer exists between circulating and noncirculating drops. For the continuous phase the effect of circulation is greatest at high Reynolds number. For a more detailed assessment the specialized reviews (11,40,42) are recommended.

Although the adsorption of surfactants tends to reduce mass-transfer coefficients by suppressing drop circulation, a sharp increase in mass transfer can occur if the transferring solute strongly reduces the interfacial tension. This effect, known as the Marangoni effect, results from interfacial turbulence induced by interfacial tension gradients (45). Extensive research in this area has been reviewed (46).

Mass-Transfer Coefficients with Chemical Reaction. Chemical reaction can occur in any of the five regions shown in Figure 3, ie, the bulk of each phase, the film in each phase adjacent to the interface, and at the interface itself. Irreversible homogeneous reaction between the consolute component C and a reactant

Table 1. Equations for Liquid–Liquid Mass Transfer in Single Drops[a]

Type of drop	Sherwood numbers	
	Dispersed phase, Sh_d	Continuous phase, Sh_c
noncirculating	6.58	$2 + 0.6\, Re_c^{0.5} Sc_c^{0.333}$
circulating	17.9[b]	$1.13\, (Re_c \cdot Sc_c)^{0.5}$

[a]Ref. 40.
[b]For laminar conditions.

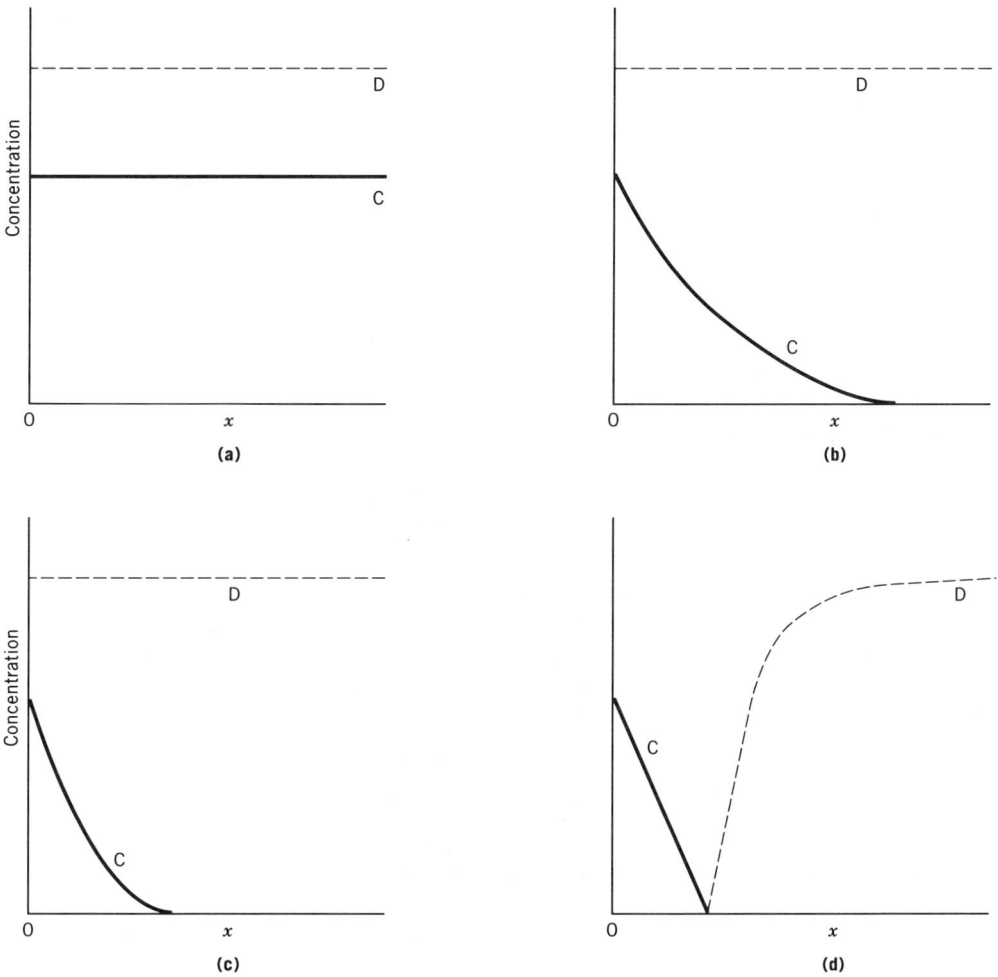

Fig. 4. Concentration profiles for the reaction of equation 19 (43) where (———) is the concentration of C; (– – –) the concentration of D; and x is the distance from the interface. (**a**) Regime 1; (**b**) regime 2; (**c**) regime 3; and (**d**) regime 4, as described in the text.

D in phase B can be described as

$$C + z\,D \rightarrow \text{products} \tag{19}$$

The equations of combined diffusion and reaction, and their solutions, are analogous to those for gas absorption (qv) (47). It has been shown how the concentration profiles and rate-controlling steps change as the rate constant increases (48). When the reaction is very slow and the B-rich phase is essentially saturated with C, the mass-transfer rate is governed by the kinetics within the bulk of the B-rich phase. This is defined as regime 1. Concentration profiles are shown in Figure 4**a**. (48). For a slow reaction defined as regime 2, the consolute component C is almost entirely depleted in the bulk of the B-rich phase (Fig. 4**b**) and the mass transfer of C between the phases controls the rate of the reaction. For a very fast reaction the depletion of C affects the concentration profile in the diffusion film. The steepening of the concentration profile as shown in Figure 4c for regime 3 leads to an enhancement in the film mass-transfer coefficient in the B-rich phase. Finally, the case of an instantaneous reaction (regime 4) leads to the formation of a thin reaction zone to which components C and D diffuse in stoichiometric amounts. This is shown in Figure 4d.

The enhanced rate expressions for regimes 3 and 4 have been presented (48) and can be applied (49,50) when one phase consists of a pure reactant, for example in the saponification of an ester. However, it should be noted that in the more general case where component C in equation 19 is transferred from one inert solvent (A) to another (B), an enhancement of the mass-transfer coefficient in the B-rich phase has the effect of moving the controlling mass-transfer resistance to the A-rich phase, in accordance with equation 17. Resistance in both liquid phases is taken into account in a detailed model (51) which is applicable to the reversible reactions involved in metal extraction. This model, which can accommodate the case of interfacial reaction, has been successfully compared with rate data from the literature (51).

Interfacial Contact Area and Approach to Equilibrium. Experimental extraction cells such as the original Lewis stirred cell (52) are often operated with a flat liquid–liquid interface the area of which can easily be measured. In the single-drop apparatus, a regular sequence of drops of known diameter is released through the continuous phase (42). These units are useful for the direct calculation of the mass flux N and hence the mass-transfer coefficient for a given system.

In industrial equipment, however, it is usually necessary to create a dispersion of drops in order to achieve a large specific interfacial area, a, defined as the interfacial contact area per unit volume of two-phase dispersion. Thus the mass-transfer rate obtainable per unit volume is given as

$$(N \cdot a) = K_A a (c_A - c_A^*) \qquad (20)$$

Drop dispersions are hardly ever uniform, and size distribution must be allowed for in calculating a. This can be done by means of the Sauter mean drop diameter, d_m, based on the average volume-to-area ratio for N drops.

$$d_m = \sum_{i=1}^{N} d_i^3 \Big/ \sum_{i=1}^{N} d_i^2 \qquad (21)$$

The specific interfacial area based on unit volume of two-phase dispersion is given by

$$a = 6h/d_m \qquad (22)$$

where h is the holdup of dispersed phase. The specific rate of interfacial mass transfer can be summarized from equations 20 and 22 as:

$$(N \cdot a) = K_A (6h/d_m)(c_A - c_A^*) \qquad (23)$$

showing that the rate of attainment of equilibrium is proportional to the concentration difference (driving force) and to a physical rate constant ($K_A \cdot 6h/d_m$) which has the units of s^{-1}. The extraction rate constant can be increased operationally by maintaining a high holdup and a small Sauter mean droplet diameter. Typically, in the absence of chemical effects, K_A would be on the order of 10 μm/s, h could be on the order of 0.2, and d_m could be 1 mm; hence ($K_A \cdot 6h/d_m$) is approximately 0.012 s^{-1}. In a batch extraction under these conditions, the deviation from equilibrium would decay exponentially with a half-life of 0.693/0.012 s or just under one minute.

There are certain limits to how far h can be increased and d_m can be reduced; however, if the contact time in a well-mixed extractor can be maintained at several minutes, it can usually be assumed that equilibrium between the exit phases is attained, justifying the use of the equilibrium stage concept represented by Figure 1 and equation 1.

Calculation of Equilibrium Stages. Multistage contacting can be arranged in a cocurrent, crosscurrent, or countercurrent manner as shown in Figure 5. The sequence of stages is sometimes referred to as a cascade, referring to the early use of gravity overflow from stage to stage. Cocurrent stagewise contact (Fig. 5a) is not usually necessary when the stages are ideal, because equilibrium is reached between the streams after stage 1. A crosscurrent cascade (Fig. 5b) in which fresh solvent is added at each stage gives an improvement over the separation obtainable in a single stage for a given ratio of solvent to feed (5,28).

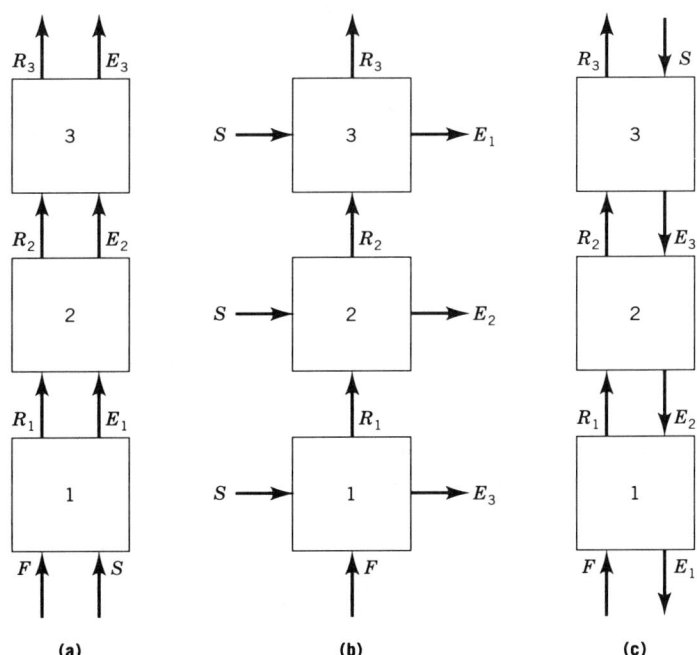

Fig. 5. Arrangement of multistage contactors where F = feed flow (A-rich), R = raffinate flow, S = solvent flow (B-rich), and E = extract flow. (**a**) Cocurrent; (**b**) crosscurrent; and (**c**) countercurrent.

The countercurrent arrangement (Fig. 5c) represents the best compromise between the objectives of high extract concentration and a high degree of extraction of the solute, for a given solvent-to-feed ratio. The feed entering stage 1 is brought into contact with a B-rich stream which has already passed through the other stages, while the raffinate leaving the last stage has been in contact with fresh solvent. Because of the economic advantages, continuous countercurrent extraction is normally preferred for commercial-scale operations. For the case of a partially miscible ternary system, the number of ideal stages in a countercurrent cascade can be estimated graphically on a triangular diagram, using the Hunter-Nash method (53). The feed and solvent compositions and the resulting mixture point M are first located on the diagram as in Figure 2a. If in addition one of the exit stream (extract or raffinate) compositions is given, a point representing the composition of the net flow in the countercurrent cascade can be located. This point, called the delta point, provides the basis for construction of material balance lines and tie-lines representing a sequence of ideal stages for the countercurrent extractor. The Hunter-Nash procedure is well known and useful (5,28). For dilute systems, it is often more convenient to use the delta point construction on a diagram with solvent-free coordinates (5,28). In this case a rectangular diagram is plotted in which the horizontal axis is the mass fraction of the solute C on a B-free basis, and the vertical axis is the mass ratio of B to A + C.

If the feed, solvent, and extract compositions are specified, and the ratio of solvent to feed is gradually reduced, the number of ideal stages required increases. In economic terms, the effect of reducing the solvent-to-feed ratio is to reduce the operating cost, but the capital cost is increased because of the increased number of stages required. At the minimum solvent-to-feed ratio, the number of ideal stages approaches infinity and the specified separation is impossible at any lower solvent-to-feed ratio. In practice the economically optimum solvent-to-feed ratio is usually 1.5 to 2 times the minimum value.

The design of countercurrent contactors is considerably simplified when the solvents A and B are not significantly miscible. The mass flows of A and B then remain constant from one stage to the next, and the material balance at any stage can be written

$$A(X_0 - X) = B(Y_1 - Y) \qquad (24)$$

where A and B are the mass flows of A and B, X is the mass ratio of C to A in the feed, and Y is the ratio of C to B in the extract. The compositions X and Y, expressed as mass ratios, thus vary linearly and equation 24 can be plotted as the operating line on Figure 6. Also shown is the equilibrium curve.

The number of ideal stages can readily be found from Figure 6 by the same type of stepwise construction as used for countercurrent distillation columns. Starting at point (X_0, Y_1), the first horizontal dashed line represents the establishment of equilibrium in stage 1 to give (X_1, Y_1). The second, vertical dashed line represents the solution of the material balance (eq. 24), giving a point on the operating line relating X_1 and Y_2, and so on. The example shown in Figure 6 indicates that between three and four ideal stages are required and in practice the designer would specify four ideal stages.

Although the triangular diagram is normally used when the solvents are

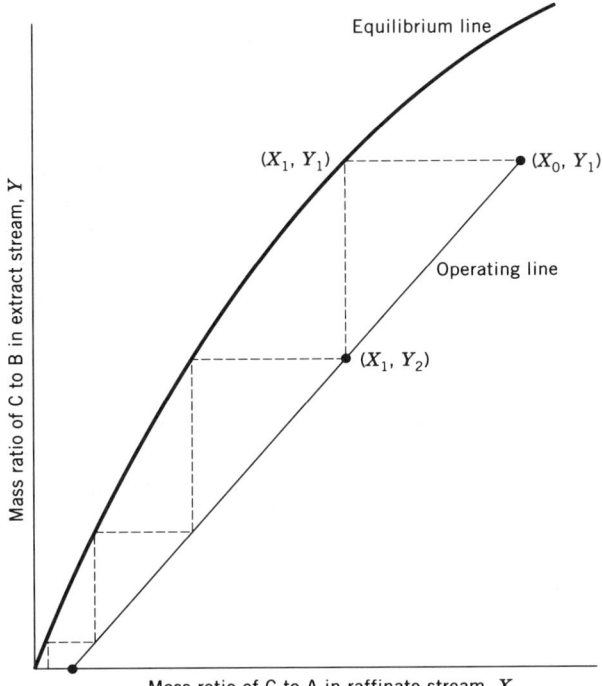

Fig. 6. Countercurrent extraction showing the equilibrium stages (horizontal dashed lines) where A and B are immiscible.

partially miscible, it is also possible to construct an X–Y (mass ratio) or x–y (mass fraction) operating diagram (28) and use the stepwise procedure for calculating stages. When solvents A and B are partially miscible, the operating lines are curved. An important advantage of the stepwise procedure is that it can be adapted to the case where the rate of mass transfer (eq. 23) and contact time are not sufficient to ensure equilibrium between exit streams in each contacting stage. By analogy with distillation, a Murphree stage efficiency can be defined for either phase (28). If the efficiency is less than 1.0, more stages are needed for a given separation than for the ideal case.

When the solvents are substantially immiscible and the equilibrium curve is linear, $Y = mX$, the number of ideal stages can be calculated without the graphical constructions (54,55). When the extraction factor ϵ (eq. 1) is not equal to unity, it can be shown that

$$N_s = \frac{\log\left[\left(\dfrac{X_0 - Y_S/m}{X_N - Y_S/m}\right)\left(1 - \dfrac{1}{\epsilon}\right) + \dfrac{1}{\epsilon}\right]}{\log \epsilon} \qquad (25)$$

and for $\epsilon = 1$, denoting parallel operating and equilibrium lines,

$$N = (X_0 - X_N)/(X_N - Y_S/m) \qquad (26)$$

where X_0 is the ratio C/A in the feed, X_N is the ratio C/A in the raffinate after N ideal stages, and Y_S is the ratio C/B in the entering solvent. For very dilute systems, the mass ratios X and Y in the above equations can be replaced by mass fractions or concentrations.

Fractional Extraction. Fractional extraction is the separation of two or more consolute components by solvent extraction. In single solvent fractional extraction (Fig. 7a) the feed mixture of A and C is added at some point in a countercurrent cascade through which a solvent B is passed. The solvent B preferentially dissolves component C as it passes in the downward direction. The mixture of B and C leaving the cascade is then split into two phases, for example by cooling. Part of the C-rich layer is removed as product and part is returned to the cascade as reflux. The B-rich layer (solvent) is sent to the other end of the cascade where it strips component C from the A + C mixture, allowing an A-rich phase to leave. The two sections of the cascade are analogous to the stripping and enriching sections in a distillation column and the design procedure for estimating the number of stages is somewhat similar (28). Single-solvent fractional extraction has been

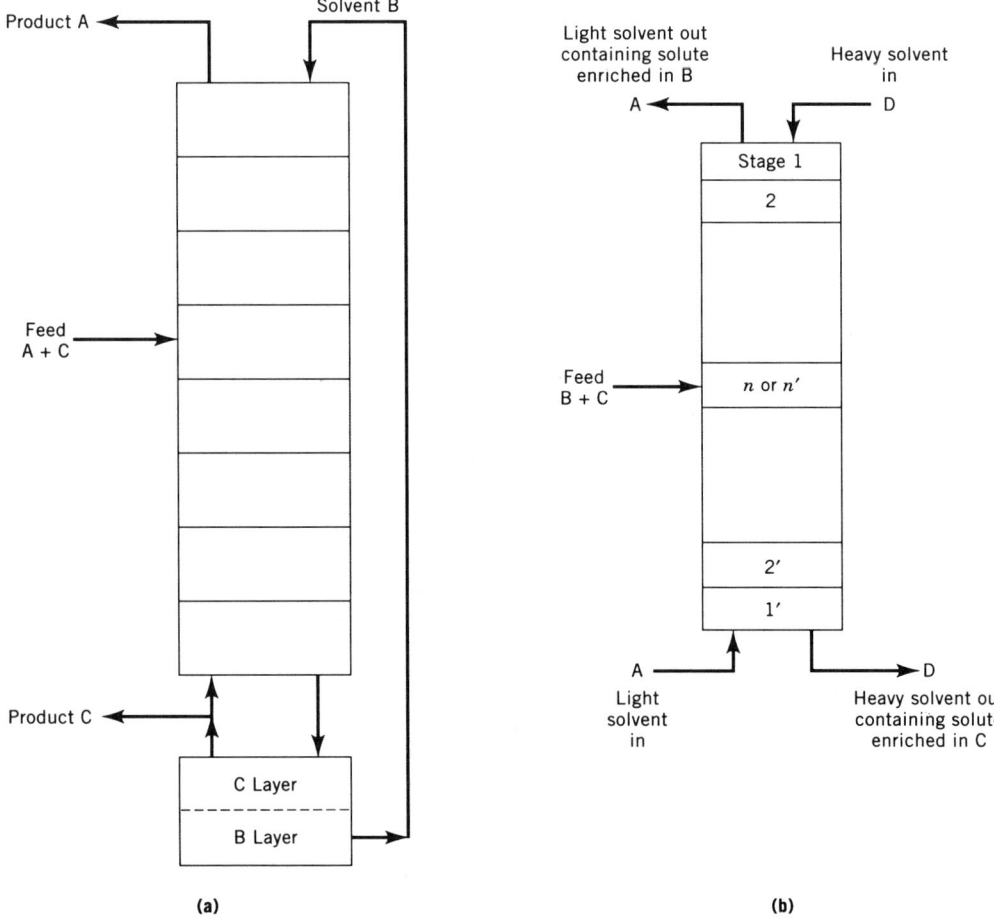

Fig. 7. Fractional extraction: (**a**) one solvent; (**b**) two solvents.

known for many years, but the range of solvents available is limited because of the requirement that the solvents must be sparingly miscible with each of the components A and C.

Dual solvent fractional extraction (Fig. 7b) makes use of the selectivity of two solvents (A and B) with respect to consolute components C and D, as defined in equation 7. The two solvents enter the extractor at opposite ends of the cascade and the two consolute components enter at some point within the cascade. Solvent recovery is usually an important feature of dual solvent fractional extraction and provision may also be made for reflux of part of the product streams containing C or D. Simplified graphical and analytical procedures for calculation of stages for dual solvent extraction are available (5) for the cases where β_{CD} is constant and the two solvents A and B are not significantly miscible. In general, the accurate calculation of stages is time-consuming (28) but a computer technique has been developed (56).

Differential Contacting. Although the equilibrium stage concept has proved extremely useful in describing the performance of mixer-settlers and plate columns having discrete stages, it is not appropriate for spray towers, packed columns, etc, in which no discrete stages can be identified. In such differential types of contactors, equilibrium between phases is never reached and therefore the mass-transfer rate is important in the design procedure.

A differential countercurrent contactor operating with a dilute solution of the consolute component C and immiscible components A and B is shown in Figure 8. Under these conditions, the superficial velocities of the A-rich and B-rich streams can be assumed not to vary significantly with position in the contactor, and are taken to be U_A and U_B, respectively. The concentration of C in the A-rich stream is c_A and that in the B-rich stream is c_B.

A steady-state material balance can be carried out on a small section of length dz and volume dz (on the basis of unit cross-sectional area) in the contactor:

$$U_B dc_B = U_A dc_A = K_A a (c_A - c_A^*) dz \qquad (27)$$

Rearrangement and integration give a relationship for the contactor height in terms of the concentration change:

$$dz = \frac{U_A}{K_A a} \cdot \frac{dc_A}{(c_A - c_A^*)} \qquad (28)$$

$$Z = \left(\frac{U_A}{K_A a}\right) \int_{c_{A0}}^{c_{AZ}} \frac{dc_A}{(c_A - c_A^*)} \qquad (29)$$

The integral can be found graphically if the equilibrium line is curved. An analytical expression for the integral is available for the case where both the equilibrium and operating lines are linear (5):

$$\int_{c_{A0}}^{c_{AZ}} \frac{dc_A}{(c_A - c_A^*)} = \frac{1}{1 - 1/\epsilon} \ln\left[\left(\frac{c_{A0} - c_{AZ}}{c_{AZ} - c_{AZ}^*}\right)(1 - 1/\epsilon) + 1/\epsilon\right] \qquad (30)$$

where $\epsilon = m U_B / U_A$. The integral is unitless and is known as the number of transfer units (NTU) based on the overall A-rich-phase driving force. Obviously the NTU, and hence the contactor length Z required, increase as the difference between c_{AZ} and c_{A0} is increased.

The factor $(U_A/K_A a)$ in equation 29 is known as the height of a transfer unit (HTU). It is a characteristic of the hydrodynamic conditions such as the flow rate A and the specific interfacial area a, but is independent of changes in c_A. It is important that the HTU be specified correctly in regard to the phase and driving force considered; in this case it relates to the overall mass-transfer driving force in the A-rich phase. The HTU may vary with height because of changes in drop size, etc; an average value is usually taken, assuming no variation.

Mass-transfer theory (eq. 17) indicates that the overall mass-transfer resistance $1/K_A$ consists of contributions from each phase, so that the overall HTU is also the sum of two contributions:

$$\begin{aligned}(HTU)_{OA} &= \frac{U_A}{K_A a} + \frac{U_A}{m K_B a} \\ &= \frac{U_A}{K_A a} + \left(\frac{U_B}{K_B a}\right)\left(\frac{U_A}{m U_B}\right) \\ &= (HTU)_A + (HTU)_B/\epsilon \end{aligned} \qquad (31)$$

The heights of a transfer unit in each phase thus contribute to the overall heights of a transfer unit. Data on values of HTU for various types of countercurrent equipment have been reviewed (1,10). In normal operating practice, the extraction factor is chosen to be not greatly different from unity, within the range of 0.5–2.

Although the stagewise model is not physically realistic for differential contactors, it is sometimes used. The number of equivalent theoretical stages N can be determined graphically using the stepwise construction illustrated in Figure 7. For the case where both the equilibrium and operating lines are linear, it can be shown that:

$$\frac{N}{(NTU)_{OA}} = \frac{1 - 1/\epsilon}{\ln \epsilon} \qquad (32)$$

If $\epsilon = 1$, the number of theoretical stages is equal to $(NTU)_{OA}$.

Equations 27–32 are applicable only to dilute, immiscible systems. If the amount of mass transfer is significant in comparison to the total flow rates, more complicated treatments of differential contactors are required (5,28).

Axial Dispersion. The development following equation 27 has assumed that the two phases move countercurrently in plug flow, ie, all the fluid in each phase has the same residence time in the equipment. In practice this is rarely the case, because of axial mixing which arises from the action of turbulent eddies, circulation currents, or the effects of drop wakes (52). The effect is to flatten the axial concentration profiles within each phase. Axial mixing can lead to a reduction in the effective driving force for mass transfer which in turn reduces the NTU below that expected for the plug flow case (eq. 30). An important feature of the profile is the discontinuity or "jump" in concentration which occurs at entry to the contactor when the liquid in the feed line enters the mixed region of the column.

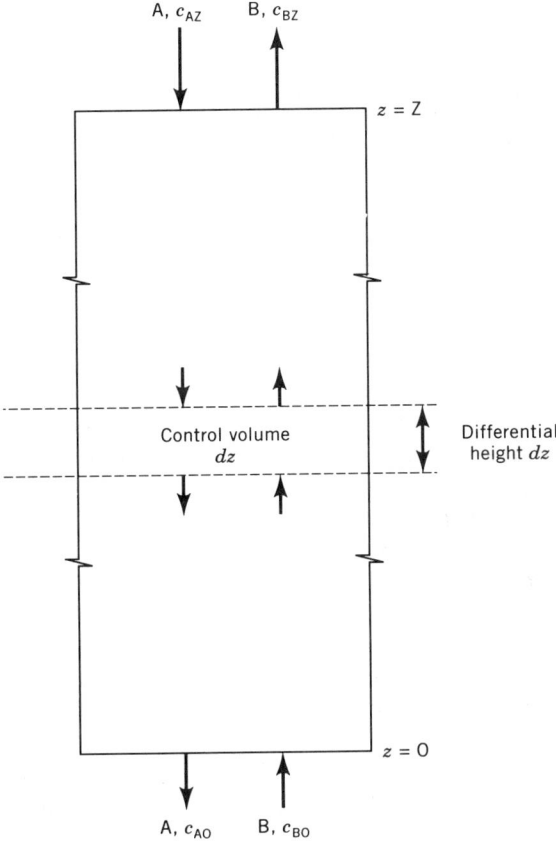

Fig. 8. Mass transfer in a differential contactor. Terms are defined in the text.

Two alternative approaches are used in axial mixing calculations. For differential contactors, the axial dispersion model is used, based on an equation analogous to equation 13:

$$N = -E \frac{\partial c}{\partial z} \qquad (33)$$

Values of E, several orders of magnitude greater than the molecular diffusion coefficient D, are typically in the 10^{-4} to 10^{-3} m²/s range for packed columns, and even larger for spray columns in which circulation currents are unimpeded. For contactors in which discrete well-mixed compartments can be identified, for example sieve-plate columns, axial mixing effects are incorporated into the stagewise model by means of the backflow ratio α which is defined as the fraction of the net interstage flow of one phase which is considered to flow in the reverse direction. For a contactor in which there are many compartments, the axial dispersion coefficient and the backflow ratio, α, are interrelated as follows:

$$E = \frac{UH}{\ln((1 + \alpha)/\alpha)} \qquad (34)$$

where H is the height of one compartment and U is the superficial velocity. The detailed calculations of concentration profiles and mass-transfer rates with axial mixing require the solution of a fourth-order differential equation (dispersion model) or the equivalent difference equation (backflow model) along with appropriate boundary conditions. The methodology was developed in the 1950s and early 1960s and has been concisely reviewed (57). Pratt (58,59) has shown how the profile solutions can be rearranged to give a direct calculation of column height. In the case of the dispersion model, the relative effect of axial mixing is a function of the axial Peclet number, defined as

$$Pe = UZ/E \qquad (35)$$

The Peclet numbers are different for each phase because U and E are different. Axial mixing effects are usually greater in the continuous phase than in the dispersed phase. Plug flow conditions can be assumed only if Pe exceeds about 50. Experimentally measured values of E or α are widely available for laboratory-scale columns with a diameter of up to 15 cm (1). Typically at low agitation rates, circulation effects (hydrodynamic nonuniformity) can lead to large values of E. The circulation effects are mainly a result of the motion of the dispersed-phase droplets, but unstable axial density gradients may also contribute to increased mixing (60,61). Circulation effects are reduced by mechanical agitation which promotes radial uniformity, but at high levels of agitation the increased turbulence leads to an overall increase in E, resulting in a minimum in the plotted curve of E versus agitation.

The effect of increasing column diameter is to increase the tendency for circulation, and hence to increase the axial mixing (62,63). However, extremely few measurements of axial mixing at the industrial scale are available, so large-scale contactor design must still rely quite heavily on empirical experience with the particular type of equipment.

Drop Diameter. In extraction equipment, drops are initially formed at distributor nozzles; in some types of plate column the drops are repeatedly formed at the perforations on each plate. Under such conditions, the diameter is determined primarily by the balance between interfacial forces and buoyancy forces at the orifice or perforation. For an ideal drop detaching as a hemisphere from a circular orifice of diameter d_0 and then becoming spherical:

$$d = (6\sigma d_0/g\Delta\rho)^{1/3} \qquad (36)$$

Equation 36 must be corrected for changes in the drop shape and for the effects of the inertia of liquid flowing through the orifice, viscous drag, etc (64). As the orifice or aperture diameter is increased, d_0 has less effect on the drop diameter and the mean drop size then tends to become a function only of the system properties:

$$d_m \simeq (\sigma/g\Delta\rho)^{1/2} \qquad (37)$$

This type of equation has been found useful in correlating drop diameters in packed columns where the packing size exceeds the drop diameter (65).

In many types of contactors, such as stirred tanks, rotary agitated columns, and pulsed columns, mechanical energy is applied externally in order to reduce the drop size far below the values estimated from equations 36 and 37 and thereby increase the rate of mass transfer. The theory of local isotropic turbulence can be applied to the breakup of a large drop into smaller ones (66), resulting in an expression of the form

$$d_m = K'\sigma^{0.6}\rho_m^{-0.6}\Psi^{-0.4} \tag{38}$$

In this equation, Ψ represents the rate of energy dissipation per unit mass of fluid. In pulsed and reciprocating plate columns the dimensionless proportionality constant K' in equation 38 is on the order of 0.3. In stirred tanks, the proportionality constant has been reported as $0.024(1 + 2.5\,h)$ in the holdup range 0 to 0.35 (67). The increase of drop size with holdup is attributed to the increasing tendency for coalescence between drops as the concentration of drops increases. A detailed survey of drop size correlations is given by the literature (65).

The value of d_m is a mean value, based on a broad distribution of sizes. In a mass-transfer situation the smallest drops, because of the very high specific surface area, quickly come to equilibrium; conversely the largest drops, which typically have a diameter of about $2\,d_m$, are much slower to come to equilibrium with the continuous phase. The effects of drop size distribution on extractor performance are being studied (68–70), although the single parameter d_m is still widely in use for design work.

Holdup and Flooding. The volume fraction of the dispersed phase, commonly known as the holdup h, can be adjusted in a batch extractor by means of the relative volumes of each liquid phase added. In a continuously operated well-mixed tank, the holdup is also in proportion to the volume flow rates because the phases become intimately dispersed as soon as they enter the tank.

$$h = Q_d/(Q_c + Q_d) \tag{39}$$

However, in a countercurrent column contactor as sketched in Figure 8, the holdup of the dispersed phase is considerably less than this, because the dispersed drops travel quite fast through the continuous phase and therefore have a relatively short residence time in the equipment. The holdup is related to the superficial velocities U of each phase, defined as the flow rate per unit cross section of the contactor, and to a slip velocity U_s (71,72):

$$U_s = U_d/h + U_c/(1 - h) \tag{40}$$

In the case of a packed column, the terms on the right-hand side should each be divided by the voidage, ie, the volume fraction not occupied by the solid packing (71). In unpacked columns at low values of h, the slip velocity U_s approximates the terminal velocity of an isolated drop, but the slip velocity decreases with holdup and may also be affected by column internals such as agitators, baffle plates, etc. The slip velocity can generally be represented by (73):

$$U_s = U_k(1 - h)^\beta \tag{41}$$

where the characteristic velocity U_k is a function of drop size and system properties and the exponent β relates to system properties and the degree of flow uniformity in the contactor.

As the throughput in a contactor represented by the superficial velocities U_c and U_d is increased, the holdup h increases in a nonlinear fashion. A flooding point is reached at which the countercurrent flow of the two liquid phases cannot be maintained. The flow rates at which flooding occurs depend on system properties, in particular density difference and interfacial tension, and on the equipment design and the amount of agitation supplied (40,65).

The nonuniformity of drop dispersions can often be important in extraction. This nonuniformity can lead to axial variation of holdup in a column even though the flow rates and other conditions are held constant. For example, there is a tendency for the smallest drops to remain in a column longer than the larger ones, and thereby to accumulate and lead to a localized increase in holdup. This phenomenon has been studied in reciprocating-plate columns (74). In the process of drop breakup, extremely small secondary drops are often formed (64). These drops, which may be only a few micrometers in diameter, can become entrained in the continuous phase when leaving the contactor. Entrainment can occur well below the flooding point.

Coalescence and Phase Separation. Coalescence between adjacent drops and between drops and contactor internals is important for two reasons. It usually plays a part, in combination with breakup, in determining the equilibrium drop size in a dispersion, and it can therefore affect holdup and flooding in a countercurrent extraction column. Secondly, it is an essential step in the disengagement of the phases and the control of entrainment after extraction has been completed.

The role of coalescence within a contactor is not always obvious. Sometimes the effect of coalescence can be inferred when the holdup is a factor in determining the Sauter mean diameter (67). If mass transfer occurs from the dispersed (d) to the continuous (c) phase, the approach of two drops can lead to the formation of a local surface tension gradient which promotes the drainage of the intervening film of the continuous phase (75) and thereby enhances coalescence. It has been observed that d-to-c mass transfer can lead to the formation of much larger drops than for the reverse mass-transfer direction, c to d (76,77).

Phase disengagement occurs at a layer or wedge in which the holdup of the drop phase is very high, providing good opportunities for close contacts between drops (65). Coalescence between drops occurs by a mechanism of drainage of the intervening film of the continuous phase. This process is favored by low viscosity, high interfacial tension, and a relatively large difference in density between the liquid phases. For difficult systems which do not have these properties, various types of mesh-packing coalescence enhancers have been developed (78). These provide a large surface which is preferentially wetted by the drop phase. Another effective technique for enhancing coalescence and phase separation is the application of pulsed electrical fields (79).

Membrane Extraction. An extraction technique which uses a thin liquid membrane or film has been introduced (80,81). The principal advantages of liquid-membrane extraction are that the inventory of solvent and extractant is extremely small and the specific interfacial area can be increased without the problems which accompany fine drop dispersions (see MEMBRANE TECHNOLOGY).

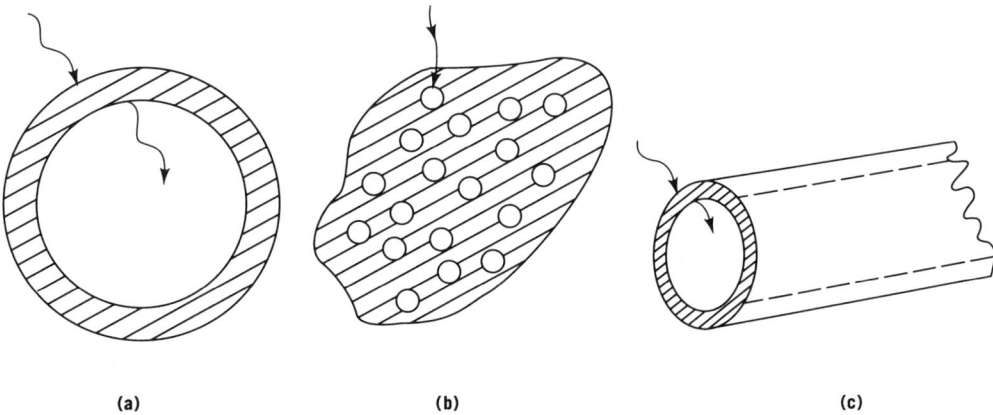

Fig. 9. Membrane extraction where the solvent phase is represented by hatched lines and the arrows show the direction of mass transfer. (**a**) Spherical film; (**b**) emulsion globule where the strip solution is represented by circles; and (**c**) hollow fiber support.

Figure 9a shows an early form of liquid-membrane extraction in which the solute is transferred from the continuous phase to a thin spherical film of an immiscible phase; within this film there is a quantity of strip solution which preferentially removes the solute from the membrane. Thus the membrane is analogous to a selective filter for the diffusional transport of the solute (see FILTRATION). The spherical film can be stabilized by surfactants, but a more convenient arrangement is the emulsion globule as shown in Figure 9b, in which the strip solution is dispersed as very small drops within a globule of solvent. This technique lends itself particularly to chemically driven extraction stripping, for example, hydrometallurgical extractions according to equation 10 (82,83,84). In this case the extractant acts as a carrier to transport the metal complex across the membrane.

In order to maintain a definite contact area, solid supports for the solvent membrane can be introduced (85). Those typically consist of hydrophobic polymeric films having pore sizes between 0.02 and 1 μm. Figure 9c illustrates a hollow fiber membrane where the feed solution flows around the fiber, the solvent–extractant phase is supported on the fiber wall, and the strip solution flows within the fiber. Supported membranes can also be used in conventional extraction where the supported phase is continuously fed and removed. This technique is known as dispersion-free solvent extraction (86,87). The level of research interest in membrane extraction is reflected by the fact that the 1990 International Solvent Extraction Conference (20) featured over 50 papers on this area, mainly as applied to metals extraction. Pilot-scale studies of treatment of metal waste streams by liquid membrane extraction have been reported (88). The developments in membrane technology have been reviewed (89). Despite the research interest and potential, membranes have yet to be applied at an industrial production scale (90).

Supercritical Extraction. The use of a supercritical fluid such as carbon dioxide as extractant is growing in industrial importance, particularly in the food-related industries. The advantages of supercritical fluids (qv) as extractants include favorable solubility and transport properties, and the ability to complete an extraction rapidly at moderate temperature. Whereas most of the supercritical extraction processes are solid–liquid extractions, some liquid–liquid extractions are of commercial interest also. For example, the removal of ethanol from dilute aqueous solutions using liquid carbon dioxide (91) or a supercritical hydrocarbon solvent (92) is under active investigation and several potential applications in food technology have also been reported (92).

Two-Phase Aqueous Extraction. Liquid–liquid extraction usually involves an aqueous phase and an organic phase, but systems having two or more aqueous phases can also be formed from solutions of mutually incompatible polymers such as poly(ethylene glycol) (PEG) or dextran. A system having as many as 18 aqueous phases in equilibrium has been demonstrated (93). Two-phase aqueous extraction, particularly useful in purifying biological species such as proteins and enzymes, can also be carried out in combination with fermentation so that the fermentation product is extracted as it is formed (94).

Because of the growth in biotechnology, two-phase aqueous extraction is becoming more important industrially. Two-phase aqueous systems have low interfacial tension, low interphase density difference, and high viscosity in comparison with most aqueous–organic systems. Although interfacial contact is very efficient, the separation of the phases after contact can be slow, requiring centrifugation. The performance of a spray column for two-phase aqueous extraction has also been reported (95).

Equipment and Processing

The earliest large-scale continuous industrial extraction equipment consisted of mixer–settlers and open-spray columns. The vertical stacking of a series of mixer–settlers was a feature of a patented column in 1935 (96) in which countercurrent flow occurred because of density difference between the phases, avoiding the necessity for interstage pumping. This was a precursor of the agitated column contactors which have been developed and commercialized since the late 1940s. There are several texts (1,2,6,97–98) and reviews (99–100) available that describe the various types of extractors.

The unique ability of solvent extraction to achieve separation according to chemical type rather than physical characteristics, such as vapor pressure, enables a great variety of processes ranging from nuclear-fuel enrichment and reprocessing to fertilizer manufacture, and from petroleum refining to biochemical and food processing. Probably more types of contactors have been developed for solvent extraction than for any other chemical engineering unit operation. Contactors have been developed for specific processes with which they then tend to become associated. As a result, selection of extractors for a new process application is not necessarily simple, and the choice of a contactor remains both an art and a science, largely based on practical experience.

The following criteria should be considered when selecting a contactor for a particular application: (*1*) stability and residence time, (*2*) settling characteristics of the solvent system, (*3*) number of stages required, (*4*) capital cost and maintenance, (*5*) available space and building height, and (*6*) throughput. The prelim-

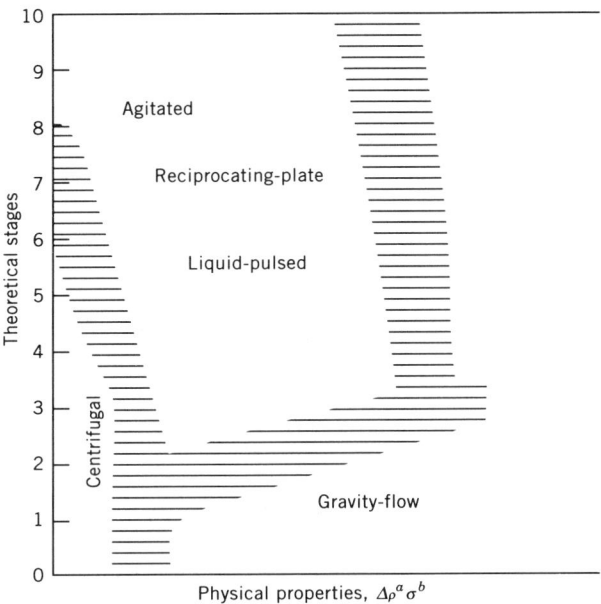

Fig. 10. Economic operating range of extractors. Superscripts a and b are constants. Courtesy of Luwa AG (101).

inary choice of an extractor for a specific process is primarily based on consideration of the system properties and number of stages required for the extraction. A qualitative chart of the economic operating range of various classes of extractors is shown in Figure 10 (101). A useful selection chart is also available (102) (Table 2). The vendor's experience, pilot-testing procedures, scaling-up methods, costs for capital equipment and maintenance, and reliability of operation should be considered and evaluated at an early stage, before the pilot-plant tests are committed. Although cost ought to be a primary balancing consideration, in many cases previous experience and practice are the deciding factors.

An extraction plant should operate at steady state in accordance with the flow-sheet design for the process. However, fluctuation in feed streams can cause changes in product quality unless a sophisticated system of feed-forward control is used (103). Upsets of operation caused by flooding in the column always force shutdowns. Therefore, interface control could be of utmost importance. The plant design should be based on (1) process control decisions made by trained technical personnel, (2) off-line analysis or limited on-line automatic analysis, and (3) control panels equipped with manual and automatic control for motor speed, flow, interface level, pressure, temperature, etc.

Laboratory Extractors, Pilot-Scale Testing, and Scale-Up. Several laboratory units are useful in analysis, process control, and process studies. The AK-UFVE contactor (104,105) incorporates a separate mixer and centrifugal separator. It is an efficient instrument for rapid and accurate measurement of partition coefficients, as well as for obtaining reaction kinetic data. Miniature mixer–settler assemblies set up as continuous, bench-scale, multistage, counter-

Table 2. Extractor Selection Chart[a,b]

	Gravity-separated extractors												Mixer–settlers				Centrifugally separated			
	Continuous contact				Discontinuous contact								Horizontal		Vertical		Continuous contact		Mixer–settler	
	Non-mechanical			Mechanical	Without interstage settling / Mechanical			With settling / Non-mechanical		With settling / Mechanical										
Design requirements	Spray column	Baffle-plate column	Packed column	Pulsed-packed column	Raining-bucket contractor	Rotary-agitated columns	Reciprocating-plate column	Pulsed-plate column	Perforated-plate column	Scheibel column	ARDC column	Rotary film contactor	Pump–settler	Agitated mixer–settler	Pump–settler	Agitated mixer–settler	Perforated plate	Film-flow type (de Laval)	Luwesta	Rotabel
total throughput, m³/h																				
<0.25	3	3	3	3	3	3	3	3	3	3	3	3	0	1	0	1	3	1	0	0
0.25–2.5	3	3	3	3	3	3	3	3	3	3	3	3	1	3	1	3	3	3	1	1
2.5–25	3	3	3	3	3	3	3	3	3	3	3	3	3	3	3	3	3	3	3	3
25–250	3	1	3	3	1	3	1	1	3	1	1	1	3	3	3	1	0[c]	0[c]	0[c]	0[c]
>250	1	0	1	1	0	1	0	1	1	0	0	1	5	5	1	1	0[c]	0[c]	0[c]	0[c]
number of stages																				
<1.0	5[d]	3	3	3	3	3	3	3	3	3	3	3	3	3	3	3	3	3	3	3
1–5	1,*0	3	3	3	3	3	3	3	3	3	3	3	3	3	3	3	3	3	5[f]	3
5–10	0	3	3	3	3	3	3	3	3	3	3	3	3	3	3	3	1[c]	1	0[c]	0[c]
10–15	0	1	1	3	1	1	3	3	1	3	1	1	3	3	1	1	0[c]	0[c]	0[c]	0[c]
>15	0	1	1	1	1	1	1	1	1	1	1	0	3	3	1	1	0[c]	0[c]	0[c]	0[c]
physical properties[g] $(\sigma/\rho g)^{1/2} > 0.60$	1	1	1	3	1	3	3	3	3	3	3	3	3	3	3	3	5	5	5	5
density difference, g/cm³ $0.05 > \Delta\rho > 0.03$	3	3	3	0	3	0	0	0	1	0	1	3	1	1	1	1	1	1	5	5
viscosity[e], mPa·s(=cP) μ_c or $\mu_d > 20$	1	1	1	1	1	1	1	1	1	1	1	1	1	1	1	1	1	1	1	1

slow heterogeneous reaction																
$k_t < 4 \times 10^{-5}$ m/s	0	1	1	3	1	3	3	3	0	0	3	3	3	3	3	3
slow homogeneous reaction																
$t_{1/2}$ = 0.5–5 min	1	1	3	1	1	1	1	1	3	3	3	3	0	0	1	1
> 5 min	0	0	1	0	0	0	0	0	3	3	3	3	0	0	0	0
extreme phase ratio																
F_d/F_f < 0.2 or > 5	1	1	3	1	3	1	3	1	1	1	3	5^h	3	3	3	3
short residence time	0	0	0	1	0	1	1	1	0	0	0	0	0	5	3	3
ability to handle solids																
trace (< 0–0.1% in feed)	3	1	1	5	3	3	3	3	5	3	3	3	3	3	3	3
appreciable (0.1–1% in feed)	1	1	1	3	3	3	3	0	5	1	1	1	1	1	1	1
heavy (> 1% in feed)	1	1	0	1	1	1	1	0	5	1^i	1^i	1^i	0	0	1	1
tendency to emulsify																
slight	3	3	1	3	1	3	1	3	3	1	1	1	5	5	5	5
marked	1	1	0	1	0	1	0	1	1	0	0	0	3	3	3	3
limited space available																
height	0	1	1	5	1	1	1	1	3	3	5	0	5	5	5	5
floor	5	5	5	0	5	5	5	5	0	0	0	5	5	5	5	5
special materials required																
metals (stainless steel, Ti, etc)	5	3	3	3	3	3	3	3	3	3	3	3	3	5	5	5
nonmetals	5	3	5	1	1	0	1	1	0	1	1	5	1	0	0	0
radioactivity present																
weak (mainly α, β)	5	5	3	1	1	1	3	1	1	3	3	5	3	1	1	1
strong γ	5	5	0	0	0	0	0	0	0	3	1	5	1	0	0	0
ease of cleaning	5	3	3	3	3	3	3	3	5	3	3	3	3	3	3	3
low maintenance	5	3	3	3	3	3	3	5	3	3	3	3	3	1	1	1

[a] Ref. 102. [b] Rating of 5, very strongly recommended; 3, satisfactory; 1, may be used; and 0, not suitable. [c] Multiple units in series or parallel can be used. [d] For immeasurably fast homogeneous reaction. [e] For diameters < 15 cm. [f] Two or three stages only in single machine. [g] See text for effect of direction of transfer. [h] With recirculation of separated phases to mixer. [i] Requires provision for solids removal from settler.

current, liquid–liquid contactors (106) are particularly useful for the preliminary laboratory work associated with flow-sheet development and optimization because these give a known number of theoretical stages. Laboratory-scale columns are typified by the 2.5-cm diameter reciprocating-plate extraction column, in which a minimum height of an equivalent theoretical stage (HETS) of 7.1 cm and high volumetric efficiencies were achieved employing a methyl isobutyl ketone (MIBK)–acetic acid–water system (107).

Because the factors relating to mass transfer and fluid dynamics of the systems in an extractor are extremely complex, particularly for mixed solvents and feedstocks of commercial interest, pilot-scale testing remains an almost inevitable preliminary to a full-scale contactor design. These tests provide: (1) total throughput and agitation speed; (2) HETS or HTU; (3) stage efficiency; (4) hydrodynamic conditions, such as droplet dispersion, phase separation, flooding, emulsion layer formation, etc; (5) selection of direction of mass transfer; (6) solvent-to-feed ratio; (7) material of construction and its wetting characteristics; and (8) confirmation of the desired separation in cases where equilibrium data are not available.

For design of a large-scale commercial extractor, the pilot-scale extractor should be of the same type as that to be used on the large scale. Reliable scale-up for industrial-scale extractors still depends on correlations based on extensive performance data collected from both pilot-scale and large-scale extractors covering a wide range of liquid systems. Only limited data for a few types of large commercial extractors are available in the literature.

Commercial Extractors. Extractors can be classified according to the methods applied for interdispersing the phases and producing the countercurrent flow pattern. Figure 11 summarizes the classification of the principal types of commercial extractors; Table 3 summarizes the main characteristics.

Unagitated Columns. Because of the simplicity and low cost, unagitated columns are widely used in industry despite low efficiency, particularly for processes requiring few theoretical stages and for corrosive systems where absence of mechanical moving parts is advantageous (Table 3). Three types of unagitated column extractors are shown in Figure 12. Spray columns (Fig. 12**a**) are the simplest in construction mechanically but have very low efficiency because of poor phase contacting and excessive backmixing in the continuous phase. These generally provide one or, at the most, two equilibrium stages. For example, a baffled spray tower, 2.7 m in diameter and 24 m in height for propane deasphalting of residue was reported to have only 3 to 3.5 theoretical stages (108). Because of the simple construction, however, spray columns are used for industrial operations requiring only a few stages.

Packed columns (Fig. 12**b**) have better efficiency because of improved contacting and reduced backmixing; it is important that the packing material should be wetted by the continuous phase to avoid coalescence of the dispersed phase. To reduce the effects of channeling, redistribution of the liquids at fixed intervals is normally required in the taller columns. Packed columns should not be used if the ratio of the phase-flow rates is beyond the range 0.5 to 2.0 because of probable flooding when suitable holdup and interfacial area are provided (109). Normally, a packed column is preferred over a spray column because the reduced flow capacity is less important than the improved mass transfer. Sulzer static mixers have been reported as a packing in liquid–liquid extraction. The overall values

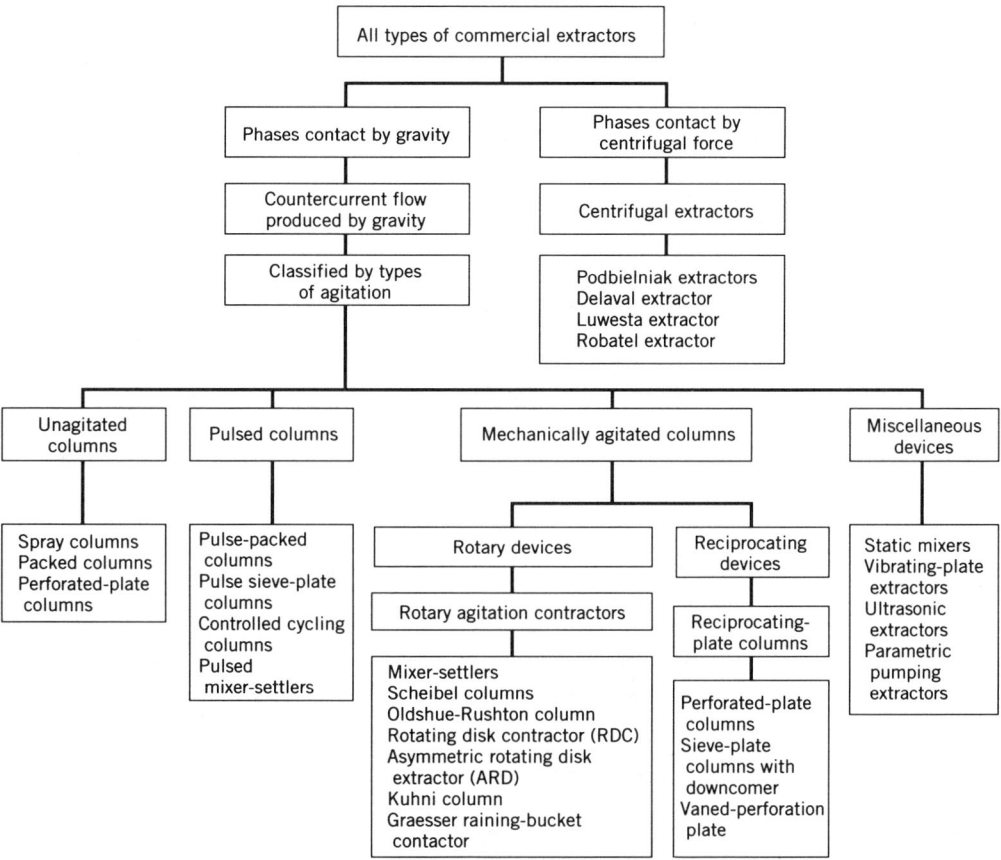

Fig. 11. Classification of commercial extractors.

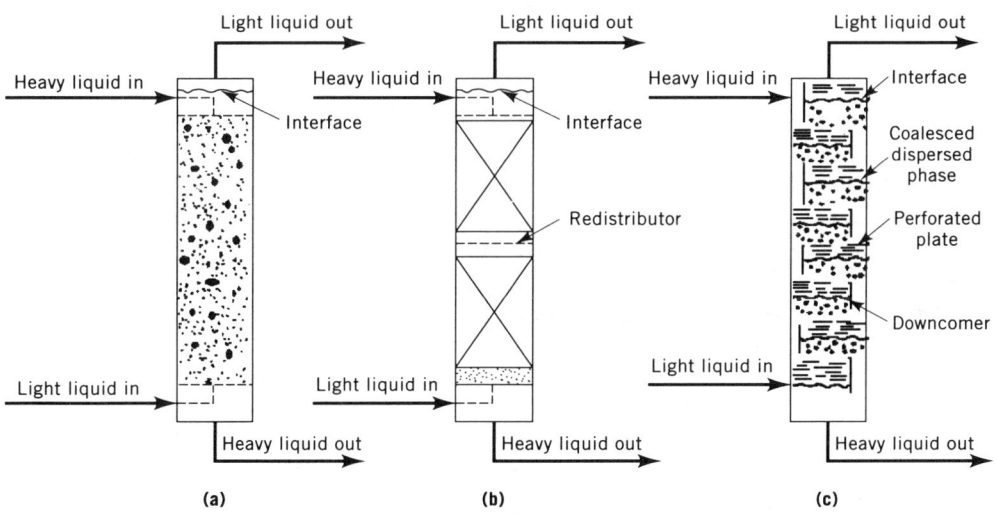

Fig. 12. Unagitated column extractors: (**a**) spray column; (**b**) packed column; and (**c**) perforated-plate column.

Table 3. Summary of Commercial Extractors

Types of extractor	General features[a]	Fields of industrial application
unagitated columns	low capital cost, low operating and maintenance cost, simplicity in construction, handles corrosive material	petrochemical, chemical
mixer–settlers	high stage efficiency, handles wide solvent ratios, high capacity, good flexibility, reliable scale-up, handles liquids with high viscosity	petrochemical, nuclear, fertilizer, metallurgical
pulsed columns	low HETS, no internal moving parts, many stages possible	nuclear, petrochemical, metallurgical
rotary agitation columns	reasonable capacity, reasonable HETS, many stages possible, reasonable construction cost, low operating and maintenance cost	petrochemical, metallurgical, pharmaceutical, fertilizer
reciprocating-plate columns	high throughput, low HETS, great versatility and flexibility, simplicity in construction, handles liquids containing suspended solids, handles mixtures with emulsifying tendencies	pharmaceutical, petrochemical, metallurgical, chemical
centrifugal extractors	short contacting time for unstable material, limited space required, handles easily emulsified material, handles systems with little liquid density difference	pharmaceutical, nuclear, petrochemical

[a] HETS = height of an equivalent theoretical stage.

of height of a transfer unit range from 0.6 to 1.6 m depending on the system and direction of mass transfer (110). Sulzer static mixers have also been used in a column for multistage supercritical fluid extraction. A description of high efficiency packing for liquid–liquid extraction has recently been reported (110).

Perforated-plate columns (Fig. 12c) are operated semistagewise and are reasonably flexible and efficient. If the light phase is dispersed, the light liquid flows through the perforations of each plate and is dispersed into drops which rise through the continuous phase. The continuous phase flows horizontally across each plate and passes to the plate beneath through a downcomer. If the heavy phase is dispersed, the column is reversed and upcomers are used for the continuous phase. A perforated-plate tower 2.13 m in diameter and 24.38 m in height used for extraction of aromatics was reported to have the equivalent of 10 theoretical stages (111). Mass-transfer data obtained in various types of perforated-plate columns up to 225 mm in diameter using different extraction systems have been summarized (112). The data are generally correlated in terms of overall heights of transfer units vs flow velocities of the phases for a specific column and system. There are many variations of basic designs for perforated-plate (sieve-plate) columns and detailed information is given in the literature (108).

Mixer–Settlers. Mixer–settlers are widely used in the chemical process industry because of reliability, flexibility, and high capacity. These extractors are particularly economical for operations that require high throughput and few stages. Mixer–settlers having capacity up to 22.7 m^3/min (6000 gal/min) have been used in the mining industry. The main disadvantages of mixer–settlers are size and the inventory of material held up in the equipment. Considerable development work has been done to improve mixer–settler design, and many newer devices have been reported. Figure 13 shows some of these extractors.

The simple box-type mixer–settler (113) has been used extensively in the UK for the separation and purification of uranium and plutonium (114). In this type of extractor, interstage flow is handled through a partitioned box construction. Interstage pumping is not needed because the driving force is provided by the density difference between solutions in successive stages.

A widely used type of pump–mixer–settler, developed by Israeli Mining Industries (IMI) (115), is shown in Figure 13a. A unit having capacity 8.3 m^3/min (2000 gal/min) has been used in phosphoric acid plants (116). The unique feature of this design is that the pumping device is not required to act as the mixer, and the two phases are dispersed by a separate impeller mounted on a shaft running coaxially with the drive to the pump.

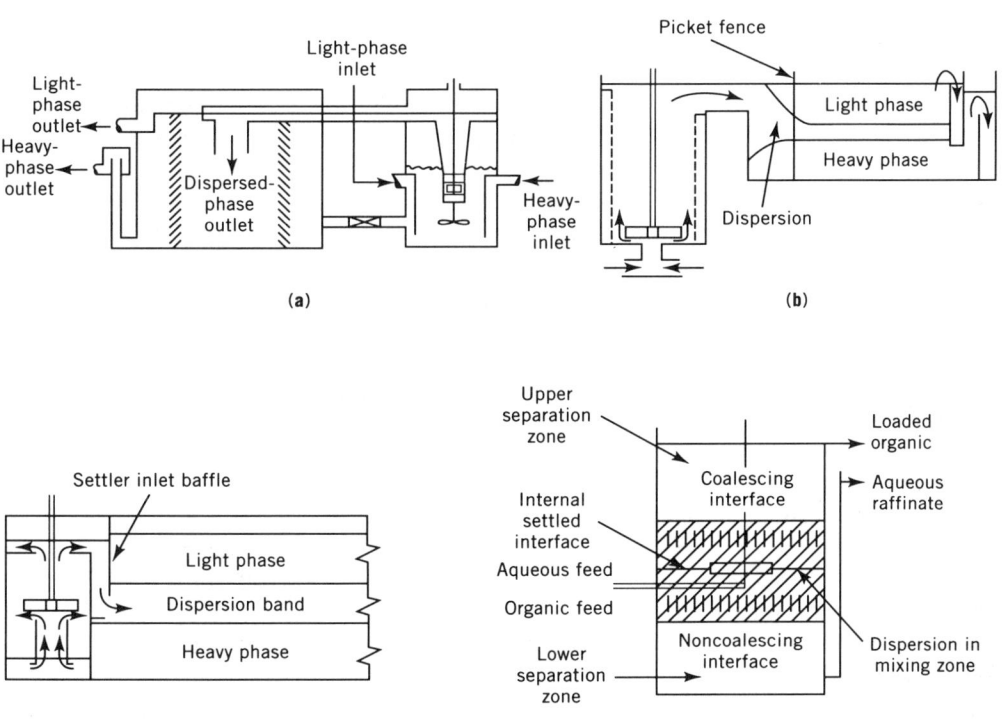

Fig. 13. Mixer–settlers: (**a**) IMI; (**b**) General Mills; (**c**) Davy-McKee; and (**d**) CMS.

The General Mills mixer–settler (117), shown in Figure 13**b**, is a pump–mix unit designed for hydrometallurgical extraction. It has a baffled cylindrical mixer fitted in the base and a turbine that mixes and pumps the incoming liquids. The dispersion leaves from the top of the mixer and flows into a shallow rectangular settler designed for minimum holdup.

In the Davy-Powergas unit (118–120), shown in Figure 13**c**, the liquids run through a draft tube and are pumped by an impeller running directly above the draft tube. The dispersion flows out from the top of the mixer and down through a channel into a rectangular settler. Large units of this type are used for copper extraction (7).

The development of the novel Davy-McKee combined mixer–settler (CMS) has been described (121). It consists of a single vessel (Fig. 13**d**) in which three zones coexist under operating conditions. A detailed description of units used for uranium recovery has been reported (122), and the units have also been studied at the laboratory scale (123). Application of the Davy combined mixer electrostatically assisted settler (CMAS) to copper stripping from an organic solvent extraction solution has been reported (124).

The Lurgi contactor (125), developed in Germany, consists of stacked mixer–settler units. Mixing and interphase transfer take place in pumps attached to the side of the settling column. It has a capacity of 1600 t/h and columns up to 3 m in diameter have been used for aromatic extractions. The Holmes and Narver mixer–settler (126) incorporates a multicompartment mixer and has many other special features. The Kemira mixer–settler (127) developed in Finland also uses the pump–mix concept, in which the phase to be mixed is drawn from a point in line with mixing impeller. Only the heavy phase is pumped into the mixer, and the light phase is allowed to flow freely from the settler. A large auxiliary space is provided between the mixer and the settler. The unit has been successfully used in extraction of the rare earths and nitrophosphate fertilizer processes (128) and found to be particularly flexible when there are great variations in flow rate from stage to stage. A new type of mixer–settler (EC-D) having a delta-type pump–mixing impeller has been developed in China (129). The delta impeller is reported to have the advantages of developing high flow velocities in both the axial and radial directions (130), resulting in high efficiency and relatively low energy consumption. Applications in a large rare-earth extraction plant have been reported (131).

Motionless inline mixers obtain energy for mixing and dispersion from the pressure drops developed as the phases flow at high velocity through an array of baffles or packing in a tube. Performance data on the Kenics (132) and Sulzer (133) types of motionless mixer have been reported.

The scale-up and design of mixer–settlers is relatively reliable because they are practically free of interstage backmixing and stage efficiencies are high, typically 80 to 90%. Various studies (134–136) have shown that (*1*) the rate of extraction is a function of power input, and (*2*) mixers can be reliably scaled up by geometric similitude at constant power input per unit mixer volume, up to a 200-fold factor of throughput (137,138). The processes taking place in the settler are complex. In large industrial mixer–settlers, the settlers usually represent at least 75% of the total volume of the units. The flow capacity of a settler depends on the behavior of a band of dispersion at the interface. The thickness of the band is a measure of the approach to flooding (97). The thickness increases exponentially with increasing flow per unit interfacial area, and settlers can be scaled up by factors of up to 1000 on this basis. A practical means to increase the throughput per unit settler area is needed so that the size of the settler can be reduced and the inventory of solvent lowered. The efficiency of the settler can be enhanced by minimizing turbulence and the formation of small drops, and maintaining low values of the linear velocity along the settler to avoid entrainment of small drops from the dispersion band.

Pulsed Columns. The efficiency of sieve-plate or packed columns is increased by the application of sinusoidal pulsation to the contents of the column. The well-distributed turbulence promotes dispersion and mass transfer while tending to reduce axial dispersion in comparison with the unpulsed column. This leads to a substantial reduction in HETS or HTU values.

The pulsed-plate column is typically fitted with horizontal perforated plates or sieve plates which occupy the entire cross section of the column. The total free area of the plate is about 20–25%. The columns are generally operated at frequencies of 1.5 to 4 Hz with amplitudes 0.63 to 2.5 cm. The energy dissipated by the pulsations increases both the turbulence and the interfacial areas and greatly improves the mass-transfer efficiency compared to that of an unpulsed column. Pulsed-plate columns in diameters of up to 1.0 m or more are widely used in the nuclear industry (139,140).

Figure 14 shows that several regions of operation in the pulsed-plate column can be distinguished, depending on the flow rate and intensity of pulsation (141). At low pulsation velocities (expressed as amplitude × frequency), a discrete layer of liquid appears between plates during each reversal of the pulse cycle. At higher velocities there is little or no layer formation and the column then behaves as a differential contactor. Extensive studies have been made on flooding, mass transfer, and the development of empirical correlations for the column design (142–144), and on the hydrodynamics and performance of columns of various sizes in uranium extraction (139).

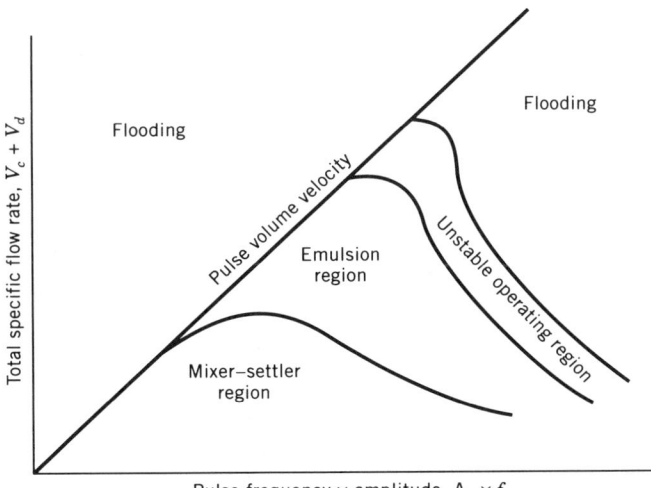

Fig. 14. Regions of operation of a pulsed, perforated-plate column (141).

Pulsed-packed columns consist of vertical cylindrical vessels filled with packing. The light and dense liquids passing countercurrently through the packing are acted on by pulsations transmitted hydraulically to form a dispersion of drops. The pulsation device is connected to the side of the column, usually at the base, through a pulse leg. Mechanical difficulties with the generation of the pulse formerly limited pulsed columns to comparatively small diameters, but the installation of pulsed-packed columns up to 2.74 m in diameter has been reported (145,146). The generation of pulses by compressed air has received increasing attention (147). A detailed model of pulsed-packed column behavior has been developed (147). The controlled cycling column (148) has a high throughput, but no large-scale application has been reported.

Mechanically Agitated Columns. *Rotary Agitated Columns.* Because of the mechanical advantages of rotary agitation, most modern differential contactors employ this method. The best known of the commercial rotary agitated contactors are shown in Figure 15. Features and applications of these columns are given in Table 3.

In the Scheibel column, developed in 1948 (149), every alternate compartment is agitated by an impeller, and the unagitated compartments are packed with open woven wire mesh. Capacity and mass-transfer data are given in the literature (149–151). A newer type of Scheibel column (Fig. 15a) using horizontal baffles with or without wire mesh packing was developed in 1956 (152). Performance data for a 30.5-cm column, with or without wire mesh packing, have shown that the HETS varies as the square root of diameter. A third design (153) is basically similar, but a pumping impeller instead of a turbine is used in the mixing stage.

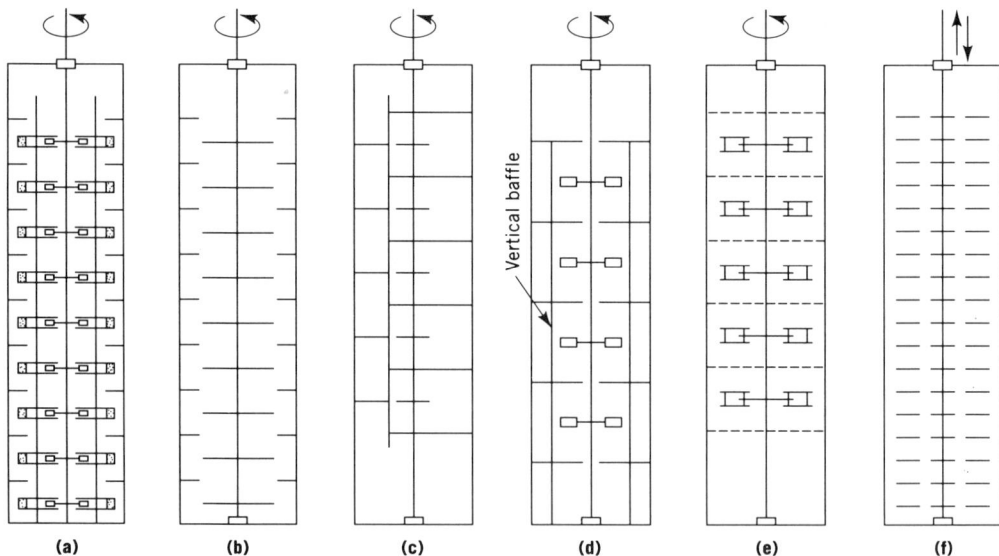

Fig. 15. Mechanically agitated columns: (**a**) Scheibel column; (**b**) rotating-disk contactor (RDC); (**c**) asymmetric rotating-disk (ARD) contactor; (**d**) Oldshue-Rushton multiple-mixer column; (**e**) Kuhni column; and (**f**) reciprocating-plate column.

Scale-up and performance of a 1.47-m Scheibel column have been reported (98,154,155), as have detailed description and design criteria for the Scheibel column (156) and scale-up procedures (157). The same stage efficiency can be maintained on scale-up, and total throughput can be increased by three and one-half times at the expense of higher HETS. As of this writing, Scheibel columns up to 2.75 m in diameter are in service.

The rotating-disk contactor (RDC), developed in the Netherlands (158) in 1951, uses the shearing action of a rapidly rotating disk to interdisperse the phases (Fig. 15b). These contactors have been used widely throughout the world, particularly in the petrochemical industry for furfural [98-01-1] and SO_2 extraction, propane deasphalting, sulfolane [126-33-0] extraction for separation of aromatics, and caprolactam [105-60-2] purification. Columns up to 4.27 m in diameter are in service. An extensive study (159) has provided an excellent theoretical framework for scale-up. A design manual has also been compiled (160). Detailed descriptions and design criteria for the RDC may also be found (161).

The Oldshue-Rushton column (Fig. 15d) was developed (162) in the early 1950s and has been widely used in the chemical industry. It consists essentially of a number of compartments separated by horizontal stator-ring baffles, each fitted with vertical baffles and a turbine-type impeller mounted on a central shaft. Columns up to 2.74 m in diameter have been reported in service (162–167). Scale-up is reported to be reliably predictable (168) although only limited performance data are available (169). A detailed description and review of design criteria are available (170).

The asymmetric rotating disk (ARD) contactor (Fig. 15c) was developed in Czechoslovakia (160,171–174) and has been increasingly used in western Europe. Its design aims at retaining the efficient shearing and dispersing action of the RDC while reducing backmixing by means of the coalescence–redispersion cycle produced in the separate transfer and settling zones. The ARD extractor is used for extraction of petrochemicals, pharmaceuticals, and caprolactam, as well as for propane deasphalting, phenol removal from wastewaster, furfural refining of oils, etc. Columns up to 2.4 m in diameter are in service and a detailed description and review of design procedures are given in the literature (175).

Kuhni contacters (Fig. 15e) have gained considerable commercial application. The principal features are the use of a shrouded impeller to promote radial discharge within the compartments, and a variable hole arrangement to allow flexibility of design for different process applications. Columns up to 5 m in diameter have been constructed (176). Description and design criteria for Kuhni extraction columns have been reported (177,178).

The RTL contactor, formerly known as the Graesser raining bucket contactor (179), is a horizontal design having the phases interdispersed by slowly rotating waterwheel-type impellers. This unit, which has the feature of dispersing each phase into the other, was developed for handling the difficult settling systems found in the coal-tar industry. It is also suitable for solid–liquid systems (180) and data on mass transfer and axial mixing have been reported (181). Units have been built from 100 mm (4 in.) to 1.8 m (6 ft) in diameter.

There are many other types of rotary agitated contactors (182) which have been less widely used.

Reciprocating-Plate Columns. Phase dispersion can also be achieved by reciprocating or vibrating of plates in a column. Improvement of extraction efficiency in a perforated-plate column by pulsing the liquid contents or by reciprocating the plates was proposed in 1935 (183). A reciprocating-plate column (RPC) was later developed (184) and scale-up was shown to be effectively accomplished by adding fixed baffles to the column (185). Many different types of column employing reciprocating plates or packing have been described (186–191) (Fig. 15f).

Reciprocation of plates requires less energy than pulsing the entire volume of liquid in a column, and has the same effect in terms of mixing patterns and uniform dispersion. This is a considerable advantage in large-scale commercial extractors (98). The main difference between the different types of RPC that have been built for industrial use lies in the plate design, as shown in Figure 16. Table 2 outlines the general features and industrial applications of the different types of RPC. Also available are reviews (191,192) and a more detailed description of the design criteria (193).

The open-type (Karr) RPC plate (183) has relatively large (12 to 15 mm) diameter perforations and free area of about 58% (Fig. 16a). It operates only in the emulsion regime. A minimum HETS of 50.8 cm has been measured in a 0.91-m diameter column using the relatively difficult (high interfacial tension) system o-xylene–acetic acid–water. Empirical correlations for scale-up have been proposed (97,107,185,194–196). Hydrodynamics and axial mixing have also been studied (197,198). As well as being operated countercurrently, the Karr RPC can

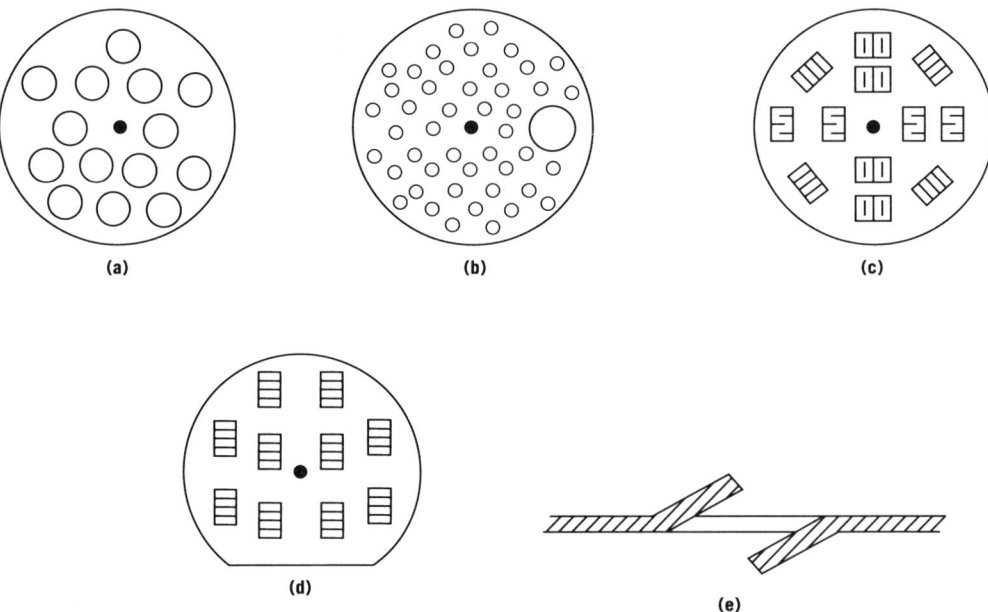

Fig. 16. Types of RPC plate in industrial use: (**a**) Karr RPC plate, ϕ = 0.5–0.6, d_o = 10–16 mm; (**b**) Prochazka RPC plate, ϕ = 0.04–0.3 (excluding (○) downcomer), d_o = 2–5 mm; (**c**) KRIMZ RPC plate, $\phi \approx 0.45$, which has vaned rectangular perforations; (**d**) GIAP II RPC plate, $\phi \approx 0.05–0.15$; and (**e**) sectional view of rectangular perforations for (**d**).

be operated in cocurrent flow. This type of column has gained increasing industrial application in the pharmaceutical, petrochemical, and hydrometallurgical industries, and in wastewater treatment (97,192,194); the Karr RPCs are in service in North America and Europe in diameters up to 1.7 m (199).

RPCs having perforated plates and downcomers (Fig. 16**b**) have been developed industrially in the former Czechoslovakia (200,201) under the trademark VPE (vibrating plate extractor). For large columns a segmental downcomer is used instead of a tubular downcomer. The downcomers permit a much higher throughput than would be possible using perforations alone. The largest units for phenol [108-95-2] extraction have a diameter of 1.2 m and plate stack height 9.1 m. These have a capacity of 80 m^3/h for phenolic wastewater (97,194).

The KRIMZ and GIAP types of RPC (190,191) (Figs. 16**c** and 16**d**, respectively) were developed in the former USSR and the plate designs feature rectangular punched perforations where the displaced metal strips remain attached as inclined vanes. The purpose of the vanes is to deflect the liquid and give it radial motion, which can be beneficial in reducing axial mixing in larger diameter columns. The modeling, design, and scale-up (202) have been based on theoretical principles (203). Industrial applications of KRIMZ and GIAP plates in RPCs up to 1.5 m in diameter have been reported (192,204).

Other types of RPC have been proposed but are not in industrial use as of this writing. These include a reciprocated wire-mesh packing (188), a reciprocating screen-plate (205), and the multistage vibrating disk column (MVDC) developed in Japan (189,206,207). These types of RPC may be useful for gas–liquid contact as well as liquid–liquid contact.

Centrifugal Extractors. In centrifugal extractors, contact time between the phases is reduced and phase separation is accelerated by the application of centrifugal forces which greatly exceed gravitational forces. The units are compact and a relatively high throughput per unit volume can be achieved. Centrifugal extractors are particularly useful for systems which are chemically unstable, eg, extraction of antibiotics, or for systems in which the phases are slow to settle. General features and fields of application are given in Table 3 and a detailed review including design criteria is available (208).

The first differential centrifugal extractor to be used in industry was the Podbielniak extractor which was introduced in the 1950s (209,210) and can be regarded as a perforated-plate column wrapped around a rotor shaft. Rotation creates a centrifugal force which results in a great reduction in the equivalent height and contact time that would be needed in a conventional perforated-plate column.

The behavior of drops in the centrifugal field has been studied (211) and the residence times and mass-transfer rates have been measured (212). Podbielniak extractors have been widely used in the pharmaceutical industry, eg, for the extraction of penicillin, and are increasingly used in other fields as well. Commercial units having throughputs of up to 98 m^3/h (26,000 gal/h) have been reported.

The Alfa-Laval extractor (213) can give up to 20 theoretical stages in one unit. Depending on the system being handled, the capacity of the standard unit ranges between 5.7 and 21.2 m^3/h (1500–5600 gal/h). Antibiotic extractions and petrochemical processing are typical applications.

A countercurrent continuous centrifugal extractor developed in the former

USSR (214) has the feature that mechanical seals are replaced by liquid seals with the result that operation and maintenance are simplified; the mechanical seals are an operating weak point in most centrifugal extractors. The operating units range between 400 and 1200 mm in diameter, and a capacity of 70 m^3/h has been reported in service. The extractors have been applied in coke-oven refining, erythromycin production, lube oil refining, etc.

The class of discrete stage centrifugal extractors includes the Westfalia centrifugal extractor (215,216) which rotates about a vertical axis and is available with up to three contact stages. Its advantage is that the light phase does not have to be introduced under pressure. The capacity of the largest model is reported as ranging from 7.6 m^3/h (2000 gal/h) for three stages, to 49.2 m^3/h (13,000 gal/h) for a single stage. Another important member of this class is the Robatel extractor (220) which consists of a series of mixer–settlers stacked on their sides with the mixing in each stage being provided by a stationary disk attached to the shaft while the chamber rotates. Typical units provide three to eight stages, and throughputs up to 6.2 m^3/h (1600 gal/h) have been reported. Robatel extractors have found general application in the chemical, pharmaceutical, and petrochemical industries, and particularly the nuclear industry. Technical and economic comparisons of the Robatel extractor with mixer–settlers and pulsed columns have been made (217). Research and development on other nondispersive forms of contactors, eg, Hi-Gee solvent extractors that give a high efficiency per unit volume, and contactors effective with very short residence times, eg, improvement on the centrifugal extractor, has been reported (218).

Economics of Extraction. Economic considerations for solvent extraction include both capital and operating costs. Capital cost is made up of the installed cost of equipment and the cost of the inventory of material (including solvent and extractant) held within the plant. Operating costs include the cost of extractor operation, solvent recovery, and solvent losses. Solvent recovery is often the dominant factor because of the high energy consumption involved. Process economy can often be improved by increasing the number of stages, which reduces the solvent recovery despite increasing the capital cost.

Organic Processes

Petroleum and Petrochemical Processes. The first large-scale application of extraction was the removal of aromatics from kerosene [8008-20-6] to improve its burning properties. Jet fuel kerosene and lubricating oil, which require a low aromatics content, are both in demand. Solvent extraction is also extensively used to meet the growing demand for the high purity aromatics such as benzene, toluene, and xylene (BTX) as feedstocks for the petrochemical industry (see BTX SEPARATIONS). Additionally, the separation of aromatics from aliphatics is one of the largest applications of solvent extraction.

Lubricating Oil Extraction. Aromatics are removed from lubricating oils to improve viscosity and chemical stability. The solvents used are furfural, phenol, and liquid sulfur dioxide. The latter two solvents are undesirable owing to concerns over toxicity and the environment and most newer plants are adopting furfural processes. A useful comparison of the various processes is available (219).

Separation of Aromatic and Aliphatic Hydrocarbons. Aromatics extraction for aromatics production, treatment of jet fuel kerosene, and enrichment of gasoline fractions is one of the most important applications of solvent extraction. The various commercial processes are summarized in Table 4.

The Udex process (220) was popular in the United States in the 1970s. The original process produced high purity gasoline by removing aromatics using diethylene glycol as a solvent. The process has also been used for the manufacture of BTX; aqueous tetraethylene glycol [112-60-7] appears to be the best solvent (221). The sulfolane process (222–224), introduced by the Shell Co. in 1962 (Fig. 17), is used in many large units all over the world. Sulfolane [126-33-0] ((tetrahydrothiophene)-1,1-dioxide), is a strongly polar compound that is highly selective for aromatic hydrocarbons and has much greater solvent capacity for hydrocarbons than glycol systems. Additional features in its favor are high density, heat capacity, and chemical stability. The sulfolane process uses the rotating-disk contactor (RDC) (225). The Lurgi Arosolvan process (226) has been used in over a dozen commercial installations. Two process arrangements are available, using as solvent either a mixture of N-methyl-2-pyrrolidinone [872-50-4] (NMP) and water or a mixture of NMP and ethylene glycol [107-21-1]. The polar mixing component (water or ethylene glycol) increases the selectivity of the solvent for aromatics. The Lurgi multistage mixer–settler is used with towers up to 6 m in diameter and 35 m high. The NMP (Arosolvan) process for BTX separation has been described (227). A dimethyl sulfoxide [67-68-5] (DMSO) process which employs two separate extraction steps has been developed (228). The selectivity and low viscosity of the solvent (DMSO plus a few percent water) allow the extraction to take place entirely at ambient temperatures. In addition, DMSO is nontoxic and relatively inexpensive. The process uses the Kuhni column (Fig. 15e) in diameters up to 2.7 m. In the Union Carbide process (229), the solvent (tetraethylene glycol) is recovered by a second extraction step rather than from the extract by distillation. However, it is necessary to distill the raffinate from the first extractor in order to recover the dissolved process solvents. A useful description of the Union Carbide TETRA process is available (230). The Formex process (231) which employs N-formylmorpholine [4394-85-8] and a few percent water as solvent, has the flexibility to handle different feedstocks and product ranges. Either distillation or secondary extraction may be used to regenerate the solvent, depending on the range of aromatics which is to be produced. The Redox process (232) (recycle extract dual extraction) improves the octane number of diesel fuels by extracting an aromatic concentrate. The solvents include furfural–furfuryl alcohol–water mixtures, aqueous tetrahydrofurfuryl alcohol, and aqueous dimethylformamide.

Desulfurization. The sulfur compounds in petroleum oil include hydrogen sulfide, mercaptans, thiophenols, and thioethers in amounts ranging from a few tenths to several percent. Sulfur compounds have objectionable odors and adversely affect the stability of light distillate and the antiknock and oxidation characteristics of gasoline. Sulfurs are generally removed by multistage countercurrent extraction using a relatively large volume of dilute alkali solution.

Butadiene Separation. Solvent extraction is used in the separation of butadiene [106-99-0] from other C-4 hydrocarbons in the manufacture of synthetic rubber. The butadiene is produced by catalytic dehydrogenation of butylene and

Table 4. Extractive Processes for the Separation of Benzene–Toluene–Xylene Mixture from Light Feedstocks[a]

Process	Solvent	Solvent additives and reflux conditions	Operating temperature, °C	Contacting equipment	Comments
Shell process, Universal Oil Products	sulfolane	sulfolane selectivity and capacity insensitive to water content caused by steam-stripping during solvent recovery; heavy paraffinic countersolvent used	120	rotating-disk contactor, up to 4 m in diameter	the high selectivity and capacity of sulfolane leads to low solvent–feed ratios, and thus smaller equipment
Udex process, Universal Oil Products	glycol–water mixture	solvent can be diethylene glycol and water, or mixture of diethylene and dipropylene glycols and water, or tetraethylene glycol and water; light hydrocarbon reflux	150 for diethylene glycol and water	sieve-tray extractor	tetraethylene glycol and water mixtures are claimed to increase capacity by a factor of 4 and also require no antifoaming agent; the extract requires a two-step distillation to recover BTX
Union Carbide Corp.	tetraethylene glycol (TETRA)	the solvent is free of water; a dodecane reflux is used which is later recovered by distillation	100	reciprocating-plate extractor	the extract leaving the primary extractor is essentially free of feed aliphatics, and no further purification is necessary; two-stage extraction uses dodecane as a displacement solvent in the second stage
Institut Français du Pétrole	dimethyl sulfoxide (DMSO)	solvent contains up to 2% water to improve selectivity; reflux consist of aromatics and paraffins	ambient	rotating-blade extractor, typically 10–12 stages	low corrosion allows use of carbon steel equipment; solvent has a low freezing point and is nontoxic; two-stage extraction has displacement solvent in the second stage
Arosolvan process, Lurgi	N-methyl-2-pyrrolidinone (NMP)	a polar mixing component, either water (12–20) or monoethylene glycol (40–50 wt %) must be added to the NMP to increase the selectivity and to decrease the boiling point of the solvent; the NMP–water processes use pentane countersolvent	NMP–glycol, 60; NMP–water, 35	vertical multistage mixer–settler, 24–30 stages, up to 8 m in diameter	the quantity of mixing component required depends on the aromatics content of the feed
Formex process, Snamprogetti	N-formyl-morpholine (FM)	water is added to the FM to increase its selectivity and also to avoid high reboiler temperatures during solvent recovery by distillation	40	perforated-tray extractor, FM density at 1.15 aids phase separation	low corrosion allows use of carbon steel equipment

[a] Ref. 176.

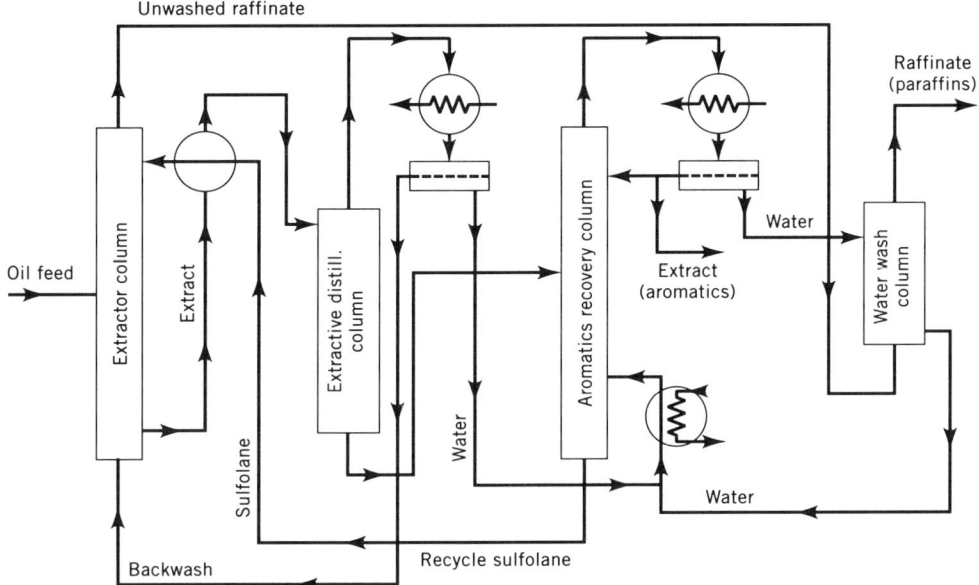

Fig. 17. Aromatic separation, sulfolane process (222–224).

the liquid product is then extracted using an aqueous cuprammonium acetate solution with which the butadiene reacts to form a complex. Butadiene is then recovered by stripping from the extract. Distillation is a competing process.

Caprolactam Extraction. A high degree of purification is necessary for fiber-grade caprolactam, the monomer for nylon-6. Crude aqueous caprolactam is purified by solvent extractions using aromatic hydrocarbons such as toluene as the solvent (233). Many of the well-known types of column contactors have been used; a detailed description of the process is available (234).

Extraction of C-8 Aromatics. The Japan Gas Chemical Co. developed an extraction process for the separation of *p*-xylene [106-42-3] from its isomers using $HF-BF_3$ as an extraction solvent and isomerization catalyst (235). The highly reactive solvent imposes its own restrictions but this approach is claimed to be economically superior to more conventional separation processes.

Anhydrous Acetic Acid. In the manufacture of acetic acid by direct oxidation of a petroleum-based feedstock, solvent extraction has been used to separate acetic acid [64-19-7] from the aqueous reaction liquor containing significant quantities of formic and propionic acids. Isoamyl acetate [123-92-2] is used as solvent to extract nearly all the acetic acid, and some water, from the aqueous feed (236). The extract is then dehydrated by azeotropic distillation using isoamyl acetate as water entrainer (see DISTILLATION, AZEOTROPIC AND EXTRACTIVE). It is claimed that the extraction step in this process affords substantial savings in plant capital investment and operating cost. A detailed description of various extraction processes is available (237).

Synthetic Fuel. Solvent extraction has many applications in synthetic fuel technology such as the extraction of the Athabasca tar sands and Irish peat using *n*-pentane [109-66-0] (238) and a process for treating coal using a solvent under hydrogen (239). In the latter case, coal reacts with a minimum amount of hydrogen so that the solvent extracts valuable feedstock components before the solid residue is burned. Solvent extraction is used in coal liquefaction processes (240) and synthetic fuel refining.

Pharmaceutical Processes. The pharmaceutical industry is a principal user of extraction because many pharmaceutical intermediates and products are heat-sensitive and cannot be processed by methods such as distillation. A useful broad review can be found in the literature (241).

Antibiotics. Solvent extraction is an important step in the recovery of many antibiotics such as penicillin [1406-05-9], streptomycin [57-92-1], novobiocin [303-81-1], bacitracin [1405-87-4], erythromycin, and the cephalosporins. A good example is in the manufacture of penicillin (242) by a batchwise fermentation. Amyl acetate [628-63-7] or n-butyl acetate [123-86-4] is used as the extraction solvent for the filtered fermentation broth. The penicillin is first extracted into the solvent from the broth at pH 2.0 to 2.5 and the extract treated with a buffer solution (pH 6) to obtain a penicillin-rich solution. Then the pH is again lowered and the penicillin is re-extracted into the solvent to yield a pure concentrated solution. Because penicillin degrades rapidly at low pH, it is necessary to perform the initial extraction as rapidly as possible; for this reason centrifugal extractors are generally used.

Fractional extraction has been used in many processes for the purification and isolation of antibiotics from antibiotic complexes or isomers. A 2-propanol–chloroform mixture and an aqueous disodium phosphate buffer solution are the solvents (243). A reciprocating-plate column is employed for the extraction process (154).

Vitamins. The preparation of heat-sensitive natural and synthetic vitamins involves solvent extraction. Natural vitamins A and D are extracted from fish liver oils and vitamin E from vegetable oils; liquid propane [74-98-6] is the solvent. In the synthetic processes for vitamins A, B, C, and E, solvent extraction is generally used either in the separation steps for intermediates or in the final purification.

Miscellaneous Pharmaceutical Processes. Solvent extraction is used for the preparation of many products that are either isolated from naturally occurring materials or purified during synthesis. Among these are sulfa drugs, methaqualone [72-44-6], phenobarbital [50-06-6], antihistamines, cortisone [53-06-5], estrogens and other hormones, and reserpine [50-55-5] and alkaloids. Common solvents for these applications are chloroform, isoamyl alcohol, diethyl ether, and methylene chloride. Distribution coefficient data for drug species are important for the design of solvent extraction procedures. These can be determined with a laboratory continuous extraction system (AKUFVE) (244).

Food Processing. Food processing makes use of solvent extraction in several ways. Industrial refining of fats and oils using propane is known as the Solexol process (245). Vegetable oils are refined by extraction using furfural as solvent (246). Solvent extraction is used in many protein refining processes, for example the extraction of fish protein from ground fish using i-propyl alcohol (247). Recovery of lactic acid by an extractive fermentation has been reported (248). The applications of extraction in the food industry have been reviewed (249).

Other Organic Processes. Solvent extraction has found application in the coal-tar industry for many years, as for example in the recovery of phenols from coal-tar distillates by washing with caustic soda solution. Solvent extraction of fatty and resimic acid from tall oil has been reported (250). Dissociation extraction is used to separate m-cresol from p-cresol (251) and 2,4-xylenol from 2,5-xylenol (252). Solvent extraction can play a role in the direct manufacture of chemicals from coal (253).

Treatment of Industrial Effluents. Solvent extraction appears to have great potential in the field of effluent treatment, both for the economic recovery of valuable materials and for the removal of toxic materials to comply with environmental requirements.

The Phenox process (254) removes phenol from the effluent from catalytic cracking in the petroleum industry. Extraction of phenols from ammoniacal coke-oven liquor may show a small profit. Acetic acid can be recovered by extraction from dilute waste streams (255). Oils are recovered by extraction from oily wastewater from petroluem and petrochemical operations. Solvent extraction is employed commercially for the removal of valuable by-products from wool industry effluents (256) and is applied in the same way in the pharmaceutical industry. A successful extraction process to recover p-nitrophenol [*100-02-7*] from a waste solution containing 8000 ppm has been developed (257). A combination of solvent extraction and wet air oxidation is used in the treatment of toxic pharmaceutical effluent prior to discharge for biological treatment. Several schemes for organic industrial wastewater treatment have been reported (258). Amphiphilic polymer solutions have high capacity for trace organics and can be used with hollow fiber membrane extractors to treat contaminated aqueous streams for environmental applications (258).

Biopolymer Extraction. Research interests involving new techniques for separation of biochemicals from fermentation broth and cell culture media have increased as biotechnology has grown. Most separation methods are limited to small-scale applications but recently solvent extraction has been studied as a potential technique for continuous and large-scale production and the use of two-phase aqueous systems has received increasing attention (259). A range of enzymes have favorable partition properties in a system based on a PEG–dextran–salt solution (97):

Enzyme	Industrial application
α-amylase	glues–food ingredients
glucoamylase	cornstarch–glucose conversion; starch–glucose conversion
α-glucosidase	maltose–glucose conversion
glucose-6-phosphate dehydrogenase	medicinal indicator
formate dehydrogenase	oxalate–formate determination
formaldehyde dehydrogenase	aldehyde–alcohol conversion
catalase	cold milk sterilization
pullunanase	starch–maltose conversion
glucose isomerase	glucose–fructose conversion
β-glucosidase	food processing
interferon	pharmaceutical applications

In many cases rapid and effective removal of contaminants and undesirable products such as nucleic acids and polysaccharides is achieved.

Difficult Separations. Difficult separations, characterized by separation factors in the range 0.95 to 1.05, are frequently expensive because these involve high operating costs. Such processes can be made economically feasible by reducing the solvent recovery load (260); this approach is effective, for example, in the separation of m- and p-cresol, linoleic and abietic components of tall oil, and the production of heavy water.

Inorganic Processes

The first significant application of liquid–liquid extraction in inorganic chemical technology was the separation of uranium and plutonium from nuclear reactor fission products in the late 1940s (261). A few years later, extraction was successfully applied at the front end of the nuclear fuel cycle in separating uranium from ore leach liquors as an alternative to ion exchange (qv). Since then, many other hydrometallurgical applications of liquid–liquid extraction have been developed (1,2,7,262) as well as a number of applications involving nonmetallic inorganic products (263).

Most inorganic compounds are insoluble or sparingly soluble in organic solvents, whereas metal ions in aqueous solution are stable because water has a high dielectric constant and because of the solvation of ions by water molecules. Aqueous affinity must be overcome usually by an extractant which can react with the metal ion to displace the solvated water and form an uncharged species having significant solubility in the organic solvent, as illustrated in equations 10–12. The organic carrier solvent, or diluent, is usually regarded as being chemically inert although the relative aliphatic–aromatic content of the diluent can affect the extraction rates and equilibria (264). Physical properties of the carrier solvent should include a low viscosity, low flash point, and low vapor pressure to minimize evaporative losses (8). The interfacial tension between the extractant–diluent phase and the aqueous phase should preferably be high in order to provide good phase separation and minimize entrainment losses. For reasons of cost, the carrier solvent is usually a cut from petroleum distillation having flash points in the 40–80°C range.

As metal extraction into a diluent–extractant solution proceeds, there is sometimes a tendency for formation of two organic phases in equilibrium with the aqueous phase. A third phase is highly undesirable and its formation can be prevented by adding to the organic phase a few percent of a modifier which is typically a higher alcohol or tri-n-butyl phosphate (TBP) (7).

Nuclear Fuel Reprocessing. Spent fuel from a nuclear reactor contains ^{238}U, ^{235}U, ^{239}Pu, ^{232}Th, and many other radioactive isotopes (fission products). Reprocessing involves the treatment of the spent fuel to separate plutonium and unconsumed uranium from other isotopes so that these can be recycled or safely stored (261,264,265).

The spent fuel is dissolved in nitric acid and the solution is extracted by an appropriate solvent. The Purex process (264–266) uses a 30% solution of tri-n-butyl phosphate (TBP) as extractant, in an aliphatic diluent such as a kerosene. The uranium and plutonium are present in the aqueous phase in the hexavalent state as $UO_2(NO_3)_2$ and $PuO_2(NO_3)_2$ and are selectively extracted (see eq. 12).

Fission products remain in the aqueous raffinate. The organic extract is treated with an aqueous strip solution containing nitric acid and a cationic reducing agent which converts Pu to its trivalent state, which is preferentially stripped to the aqueous phase. Finally the uranium is stripped from the organic phase by contact with a dilute aqueous solution of nitric acid. The aqueous raffinate from the Purex process contains actinides and rare-earth fission products which are long-lived and radiotoxic. Bifunctional extractants of the carbamoylmethylphosphoryl family have been developed with the capability to remove many of these substances and thus improve the economics of safe disposal of the bulk raffinate (265,267). This objective can also be achieved by n,n-dialkyl amides (268). The various alternative approaches to fuel reprocessing have been critically reviewed (268).

Special safety constraints apply to equipment selection, design, and operation in nuclear reprocessing (269). Equipment should be reliable and capable of remote control and operation for long periods with minimal maintenance. Pulsed columns and remotely operated mixer–settlers are commonly used (270). The control of criticality and extensive monitoring of contamination levels must be included in the process design.

Uranium Extraction from Ore Leach Liquors. Liquid–liquid extraction is used as an alternative or as a sequel to ion exchange in the selective removal of uranium [7440-61-1] from ore leach liquors (7,265,271). These liquors differ from reprocessing feeds in that they are relatively dilute in uranium and only slightly radioactive, and contain sulfuric acid rather than nitric acid.

In the Amex process, the feed typically containing 5 g/L uranium and 100 g/L sulfuric acid is first filtered and treated to remove interfering anions such as molybdates and vanadates. The extractant is a commercially available formulation containing tertiary amines having C-8 to C-12 alkyl chains. A kerosene diluent is used, and the extractant is at a concentration of about 5%, plus about 2% of a higher alcohol such as decanol as a modifier. The extracted species is an amine uranyl sulfate. Mixer–settler extractors are commonly used; stripping of the uranium can be carried out under acidic, basic, or neutral conditions. The alternative Dapex process employs alkyl phosphoric acids such as di-(2-ethylhexyl) phosphoric acid [298-07-7] (D2EHPA) as extractants. Although the feed pretreatment requirements are less rigorous than for the Amex process, the extractant is somewhat less selective for uranium (271).

Uranium is present in small (50–200 ppm) amounts in phosphate rock and it can be economically feasible to separate the uranium as a by-product from the crude black acid (30% phosphoric acid) obtained from the leaching of phosphate for fertilizers (qv). The development and design of processes to produce 500 t U_3O_8 per year at Freeport, Louisiana have been detailed (272).

Copper. The recovery of copper [7440-50-8], Cu, from ore leach liquors as a stage in the hydrometallurgical route to the pure metal is one of the largest applications of liquid–liquid extraction (7,198,209). It has been estimated (262) that in 1984 the total copper production capacity of solvent extraction plants was in excess of 700 t per day.

The most common type of copper feed entering a liquid–liquid extraction plant is produced by dilute sulfuric acid leaching and contains between 1 and 10 g/L Cu. This concentration is too low for electrowinning and the purpose of solvent extraction is to raise the concentration as well as to purify the copper solution. A

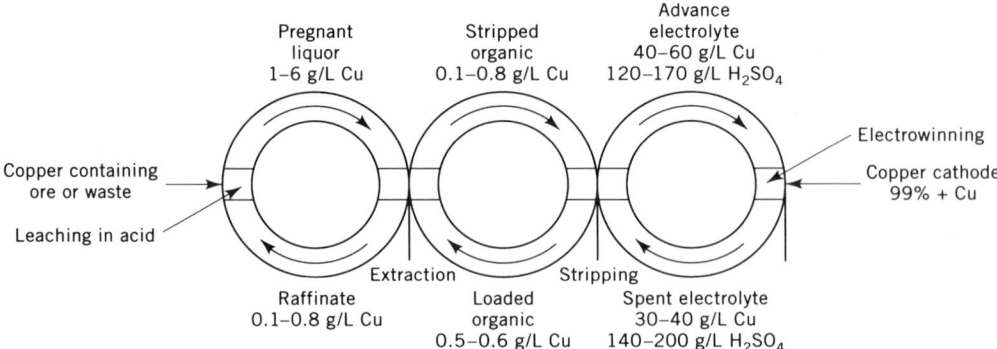

Fig. 18. Diagrammatic representation of copper extraction using solvent extraction (273).

typical extraction circuit for acid leach liquors is shown in Figure 18 (273). Extraction is carried out at a pH of 2 to 4 using an aliphatic kerosene diluent containing chelating extractants based on hydroxyoximes or quinolinol derivatives. These extractants effectively exchange cupric ions and hydrogen ions in the aqueous phase as in the case of acid extractants (eq. 10), so the equilibrium is pH-dependent. Stripping is effected by a strongly acid solution having zero or negative pH, as shown in Figure 18. Only a few stages are needed for extraction and stripping, so mixer–settlers rather than columns are used. Rapid and efficient phase separation in the settlers is an important element of the plant design (7,274).

Nickel and Cobalt. Often present with copper in sulfuric acid leach liquors are nickel [7440-02-0] and cobalt [7440-48-4]. Extraction using an organophosphoric acid such as D2EHPA at a moderate (3 to 4) pH can readily take out the nickel and cobalt together, leaving the copper in the aqueous phase, but the cobalt–nickel separation is more difficult (274). In the case of chloride leach liquors, separation of cobalt from nickel is inherently simpler because cobalt, unlike nickel, has a strong tendency to form anionic chloro-complexes. Thus cobalt can be separated by amine extractants, provided the chloride content of the aqueous phase is carefully controlled. A successful example of this approach is the Falconbridge process developed in Norway (274).

Other Metals. Because of the large number of chemical extractants available, virtually any metal can be extracted from its aqueous solution. In many cases extraction has been developed to form part of a viable process (275). A review of more recent developments in metal extraction including those for precious metals and rare earths is also available (262). In China a complex extraction process employing a cascade of 600 mixer–settlers has been developed to treat leach liquor containing a mixture of rare earths (131).

The depressed prices of most metals in world markets in the 1980s and early 1990s have slowed the development of new metal extraction processes, although the search for improved extractants continues. There is a growing interest in the use of extraction for recovery of metals from effluent streams, for example the wastes from pickling plants and the electroplating plants (276). Recovery of metals from liquid effluent has been reviewed (277), and an AM-MAR concept for metal waste recovery has recently been reported (278). Possible applications exist in this area for liquid membrane extraction (88) as well as conventional extraction. Other schemes proposed for effluent treatment are a wetted fiber extraction process (279) and the use of two-phase aqueous extraction (280).

Extraction of Nonmetallic Inorganic Compounds. Phosphoric acid is usually formed from phosphate rock by treatment with sulfuric acid, which forms sparingly soluble calcium sulfate from which the phosphoric acid is readily separated. However, in special circumstances it may be necessary to use hydrochloric acid:

$$Ca_3(PO_4)_2 + 6\ HCl \rightleftharpoons 3\ CaCl_2 + 2\ H_3PO_4 \tag{42}$$

A process developed in Israel (263) uses solvent extraction using a higher alcohol or other solvating solvent. This removes phosphoric acid and some hydrochloric acid from the system driving the equilibrium of equation 42 to the right. The same principle can be applied in other salt–acid reactions of the form

$$MX + HY \rightleftharpoons MY + HX \tag{43}$$

where M is a metal cation and X and Y are anions. An organic solvent is chosen to remove HX, again driving the equilibrium to the right. Examples of this type of reaction are (1) the production of potassium nitrate from potassium chloride and nitric acid and (2) the production of alkali metal phosphates from the alkali chloride and phosphoric acid (263).

NOMENCLATURE

a	= specific interfacial area	m^{-1} or cm^{-1}
A	= quantity or flow of component A	kg or kg/s
B	= quantity or flow of component B	kg or kg/s
c	= concentration	kg/m^3 or g/cm^3
D	= molecular diffusivity	m^2/s or cm^2/s
d	= drop diameter	m or cm
d_m	= Sauter mean drop diameter	m or cm
d_o	= orifice diameter	m or cm
E	= axial dispersion coefficient	m^2/s or cm^2/s
	Figs. 1 and 5 extract flow	kg/s or g/s
F	= feed flow	kg/s or g/s
g	= acceleration owing to gravity	m/s^2 or cm/s^2
G	= molar Gibbs free energy	J/mol
H	= height of a compartment	m or cm
HTU	= height of a transfer unit	m or cm
h	= holdup of dispersed phase	
k	= mass-transfer coefficient	m/s or cm/s
K	= overall mass-transfer coefficient	m/s or cm/s
K'	= dimensionless constant in equation 38	
m	= distribution ratio based on mass fraction	
m'	= distribution ratio based on mole fraction	
N	= flux of solute (eq. 21) number of drops	$kg/(m^2 \cdot s)$ or $g/(cm^2 \cdot s)$

N_s = number of stages
NTU = number of transfer units
Q = volume flow rate — m³/s or cm³/s
R = raffinate flow — kg/s or g/s
Re = Reynolds number
S = solvent flow — kg/s or g/s
Sc = Schmidt number
Sh = Sherwood number
U = superficial velocity — m/s or cm/s
U_s = slip velocity — m/s or cm/s
U_K = characteristic velocity — m/s or cm/s
X = mass ratio of C to A
x = mole fraction
Y = mass ratio of C to B
Z = contactor height — m or cm
z = distance (eq. 18) stoichiometric factor — m or cm
α = backflow ratio
β = selectivity or (eq. 41) exponent
$\Delta\rho$ = density difference — kg/m³ or g/cm³
γ = activity coefficient
ϵ = extraction factor
σ = interfacial tension — N/m
Φ = fractional open area, perforated plate
Ψ = specific energy dissipation rate — W/kg

SUBSCRIPTS

A = component A
B = component B
C = component C
CA = component C in solvent A
CB = component C in solvent B
c = continuous phase
D = component D
DA = component D in solvent A
DB = component D in solvent B
d = dispersed phase
i = at the interface or (eq. 21) identity of drop
N = exit from stage N
O = overall
0 = feed stream
1 = exit from stage 1
2 = exit from stage 2, etc

BIBLIOGRAPHY

"Liquid–Liquid Extraction" under "Extraction" in the *Encyclopedia of Chemical Technology*, 1st ed., Vol. 6, pp. 122–140, by E. G. Scheibel and A. J. Frey, Hoffmann-La Roche Inc.; "Extraction, Liquid–Liquid" in *ECT* 1st ed., Suppl. 1, pp. 330–365, by Marcel J. P.

Bogart, the Lummus Co.; "Liquid–Liquid Extraction" under "Extraction" in *ECT* 2nd ed., Vol. 8, pp. 719–761, by E. G. Scheibel, Cooper Union School of Engineering and Science; "Extraction, Liquid–Liquid" in *ECT* 3rd ed., Vol. 9, pp. 672–721, by T. C. Lo and M. H. I. Baird; in *ECT* 4th ed., Vol. 10, pp. 125–180, by T. C. Lo, T. C. Lo & Associates, and M. H. I. Baird, McMaster University.

1. T. C. Lo, M. H. I. Baird and C. Hanson, eds., *Handbook of Solvent Extraction,* Wiley-Interscience, New York, 1983.
2. J. D. Thornton, ed., *The Science and Practice of Liquid–Liquid Extraction,* Oxford University Press, Oxford, 1992.
3. T. Sekine and Y. Hasegawa, *Solvent Extraction Chemistry; Fundamentals and Applications,* Marcel Dekker, New York, 1977.
4. S. Alegret, ed., *Developments in Solvent Extraction,* Ellis Horwood, Chichester, UK, 1988.
5. R. E. Treybal, *Liquid Extraction,* 2nd ed., McGraw-Hill, New York, 1963.
6. G. M. Ritcey and A. W. Ashbrook, *Solvent Extraction: Principles and Applications to Process Metallurgy,* Part I, Elsevier, Amsterdam, the Netherlands, 1984.
7. G. M. Ritcey and A. W. Ashbrook, *Solvent Extraction: Principles and Applications to Process Metallurgy,* Part II, Elsevier, Amsterdam, the Netherlands, 1979.
8. R. Blumberg, *Liquid–Liquid Extraction,* Academic Press, London, 1988.
9. K. Najim, *Control of Liquid–Liquid Extraction Columns,* Gordon and Breach, New York, 1988.
10. J. L. Humphrey, J. A. Rocha, and J. R. Fair, *Chem. Eng.,* 76 (Sept. 17, 1984).
11. E. Blass, G. Goldmann, K. Hirschmann, P. Mihailowitsch, and W. Pietzch, *Ger. Chem. Eng.* **9,** 222 (1986).
12. M. H. I. Baird, *Can. J. Chem. Eng.* **69,** 1287 (1991).
13. J. G. Gregory, B. Evans, and P. C. Weston, eds., *Proceedings of the International Solvent Extraction Conference 1971,* Vols. 1 and 2, Society of Chemical Industry, London, 1971.
14. G. V. Jeffreys, ed., *Proceedings of the International Solvent Extraction Conference, 1974,* Vols. 1–3, Society of Chemical Industry, London, 1974.
15. B. H. Lucas, ed., *Proceedings of the International Solvent Extraction Conference, 1977,* Vols. 1–2, Canadian Institute of Mining and Metallurgy, Montreal, 1979.
16. *Proceedings of the International Solvent Extraction Conference 1980,* Vols. 1–3, Association des Ingenieurs Sortis de l'Université de Liege, Liege, Belgium, 1980.
17. *Proceedings of the International Solvent Extraction Conference 1983,* AIChE, New York, 1983.
18. *Preprints of the International Solvent Extraction Conference 1986,* Vols. 1–3, Dechema, Frankfurt, Germany, 1986.
19. *Proceedings of the International Solvent Extraction Conference 1988,* USSR Academy of Sciences, Moscow, Russia, 1990.
20. T. Sekine, ed., *Proceedings of the International Solvent Extraction Conference 1990,* Elsevier, Amsterdam, the Netherlands, 1992.
21. D. H. Logsdail and M. J. Slater, eds., *Proceedings of the International Solvent Extraction Conference, 1993,* Vols. 1–3, Elsevier, Amsterdam, the Netherlands, 1993.
22. J. W. Gibbs, *Collected Works,* Yale University Press, New Haven, Conn., 1928.
23. N. F. Ashton, C. McDermott, and A. Brench, in Ref. 1, Chapt. 1.
24. G. J. Pierotti, C. H. Deal, and E. L. Derr, *Ind. Eng. Chem.* **51,** 95 (1959).
25. D. S. Abrams and J. M. Prausnitz, *AIChE J.* **21,** 116 (1975).
26. J. Wisniak and A. Tamir, *Liquid–Liquid Equilibrium and Extraction,* Physical Science Data Series 7, Elsevier, Amsterdam, the Netherlands, 1981; J. Wisniak and A. Tamir, *Liquid–Liquid Equilibrium and Extraction,* Physical Science Data Series 23, Elsevier, Amsterdam, the Netherlands, 1985.

27. J. M. Sorenson and W. Arlt, *Liquid–Liquid Equilibrium Data Collection,* Dechema Chemistry Data Series, Frankfurt, Germany, Part 1, 1980.
28. H. R. C. Pratt in Ref. 1, Chapt. 5.
29. A. H. Meniai and D. M. T. Newsham, *Chem. Eng. Res. Design* **70,** 78 (1992).
30. C. Hanson in Ref. 1, Chapt. 22.
31. M. M. Milnes in Ref. 11, Vol. 1, p. 983.
32. V. V. Wadekar and M. M. Sharma, *J. Separ. Proc. Tech.* **2,** 1 (1981).
33. M. Cox and D. S. Flett in Ref. 1, Chapt. 2.2.
34. M. Aguilar in Ref. 4, Chapt. 5.
35. A. Leveque and J. Helgorsky in Ref. 12, Vol. 2, p. 439.
36. P. Danesi in Ref. 4, Chapt. 12.
37. G. S. Hartley, *Phil. Mag.* **12,** 473 (1931).
38. Y. Marcus in Ref. 4, Chapt. 2.
39. R. C. Reid and T. K. Sherwood, *The Properties of Gases and Liquids,* McGraw-Hill Book Co., New York, 1958.
40. G. S. Laddha and T. E. Degaleesan, *Transport Phenomena in Liquid Extraction,* Tata-McGraw-Hill, New Delhi, India, 1976.
41. C. R. Wilke and P. C. Chang, *AIChE J.* **1,** 264 (1955).
42. H. R. C. Pratt in Ref. 1, Chapt. 3.
43. K. P. Lindland and S. G. Terjesen, *Chem. Eng. Sci.* **5,** 1 (1956).
44. G. Thorsen and S. G. Terjesen, *Chem. Eng. Sci.* **17,** 137 (1962).
45. J. T. Davies, *Turbulence Phenomena,* Academic Press, Inc., New York, 1972.
46. E. S. Perez de Ortiz in Ref. 2, Chapt. 3.
47. P. V. Danckwerts, *Gas–Liquid Reactions,* McGraw-Hill Book Co., Inc., New York, 1970.
48. M. M. Sharma in Ref. 1, Chapt. 2.1.
49. S. Sarkar, C. J. Mumford, and C. R. Philips, *Ind. Eng. Chem. Proc. Des. Dev.* **10,** 665 (1980).
50. *Ibid.,* p. 672.
51. M. A. Hughes and V. Rod, *Faraday Discuss. Chem. Soc.* **77,** paper 7 (1984).
52. J. B. Lewis, *Chem. Eng. Sci.* **3,** 248, 260 (1954).
53. T. G. Hunter and A. W. Nash, *J. Soc. Chem. Ind. (London)* **53,** 95T (1932).
54. A. Kremser, *Natl. Pet. News* **22**(21), 42 (1930).
55. M. Souders and G. G. Brown, *Ind. Eng. Chem.* **24,** 519 (1932).
56. J. Prochazka and V. Jiricny, *Chem. Eng. Sci.* **31,** 179 (1976).
57. H. R. C. Pratt and M. H. I. Baird in Ref. 1, Chapt. 6.
58. H. R. C. Pratt, *Ind. Eng. Chem. Process Des. Dev.* **14,** 74 (1975).
59. H. R. C. Pratt, *Ind. Eng. Chem. Process Des. Dev.* **15,** 544 (1976).
60. T. L. Holmes, A. E. Karr, and M. H. I. Baird, *AIChE J.* **37,** 360 (1991).
61. M. H. I. Baird and N. V. Rama Rao, *AIChE J.* **37,** 1019 (1991).
62. A. M. Rosen and V. S. Krylov, *Chem. Eng. J.* **7,** 85 (1974).
63. A. E. Karr, S. Ramanujan, T. C. Lo, and M. H. I. Baird, *AIChE J.* **65,** 373 (1987).
64. R. Clift, J. R. Grace, and M. E. Weber, *Bubbles, Drops and Particles,* Academic Press, Inc., New York, 1978.
65. G. S. Laddha and T. E. Degaleesan in Ref. 1, Chapt. 4.
66. R. Shinnar and J. M. Church, *Ind. Eng. Chem.* **52,** 253 (1960).
67. J. W. van Heuven and W. J. Beek in Ref. 13, p. 70.
68. P. M. Bapat, L. L. Tavlarides, and G. W. Smith, *Chem. Eng. Sci.* **38,** 2003 (1983).
69. J. J. C. Cruz-Pinto and W. J. Korchinski, *Chem. Eng. Sci.* **36,** 687 (1981).
70. J. F. Milot, J. Duhamet, C. Gourdon, and G. Cassamatta, *Chem. Eng. J.* **45,** 111 (1991).
71. R. Gayler, N. W. Roberts, and H. R. C. Pratt, *Trans. Inst. Chem. Eng.* **31,** 57 (1953).

72. L. Lapidus and J. C. Elgin, *AIChE J.* **3,** 63 (1957).
73. J. C. Godfrey and M. J. Slater, *Chem. Eng. Res. Design* **69,** 130 (1990).
74. V. Jiriczny and J. Prochazka, *Chem. Eng. Sci.* **35,** 2237 (1980).
75. H. Groothuis and F. J. Zuiderweg, *Chem. Eng. Sci.* **12,** 288 (1960).
76. G. V. Jeffreys and G. B. Lawson, *Trans. Inst. Chem. Engrs.* **43,** 294 (1965).
77. Z. J. Shen, M. H. I. Baird, and N. V. Rama Rao, *Can. J. Chem. Eng.* **63,** 29 (1985).
78. G. A. Davies, G. V. Jeffreys, and M. Azfal, *Br. Chem. Eng.* **17,** 709 (1972).
79. P. J. Bailes and S. K. L. Larkai, *Trans. Inst. Chem. Engrs.* **60,** 115 (1982).
80. N. N. Li, *AIChE J.* **17,** 459 (1971).
81. N. N. Li, *Ind. Eng. Chem. Proc. Des. Dev.* **10,** 215 (1971).
82. N. N. Li, R. P. Cahn, D. Naden, and R. W. M. Lai, *Hydrometallurgy* **9,** 277 (1983).
83. K. Osseo-Asare, *Sepn. Sci. Technol.* **23,** 1269 (1988).
84. J. Draxler, W. Furst, and R. Marr in Ref. 18, Vol. 1, p. 553.
85. P. Danesi in Ref. 4., Chapt. 9.
86. R. Prasad and K. K. Sirkar, *J. Mem. Sci.* **47,** 235 (1989).
87. R. Prasad and K. K. Sirkar, *J. Mem. Sci.* **50,** 153 (1990).
88. J. Draxler and R. Marr in Ref. 20, p. 37.
89. W. S. Ho, *Membrane Handbook,* Van Nostrand Reinhold, New York, 1992.
90. D. S. Flett, Ref. 21, Vol. 1, p. vi.
91. M. A. McHugh and V. Krukonis, *Supercritical Fluid Extraction. Principles and Practice,* Butterworths, Stoneham, Mass., 1986; T. Suzuki, N. Tsuge, and K. Nagahama in Ref. 20 (Area 11).
92. H. Horizoe, T. Tanimoto, I. Yamamoto, K. Ogawa, M. Maki, and Y. Kano in Ref. 20 (Area 11); B. Simandi and co-workers, in Ref. 21, Vol. 2, p. 676; Y. Shibuya, in Ref. 21, Vol. 2, p. 684.
93. P. A. Albertsson, *Partition of Cell Particles and Macromolecules,* 2nd ed., Wiley-Interscience, New York, 1971.
94. I. Kuhn, *Biotech. Bioeng.* **22,** 2393 (1980); C. Weilnhammer and E. Blass, in Ref. 21, Vol. 2, p. 1072.
95. S. B. Sawant, S. K. Sikdar, and J. B. Joshi, *Biotech. and Bioeng.* **36,** 109 (1990).
96. U.S. Pat. 2,000,606 (May 7, 1935), D. F. Othmer.
97. T. C. Lo and M. H. I. Baird, in R. A. Meyers, ed., *Encyclopedia of Physical Science and Technology,* Academic Press, Inc., San Diego, Calif., 1987.
98. T. C. Lo, in P. Schweitzer, ed., *Handbook of Separation Techniques for Chemical Engineers,* 2nd ed., Sec. 1.10, McGraw-Hill Book Co., Inc., New York, 1988.
99. R. W. Cusack and P. Fremeaux, *Chem. Eng.,* 132 (Mar. 1991).
100. J. L. Humphrey, J. A. Rocha, and J. R. Fair, *Chem. Eng.,* 76 (Sept. 17, 1984).
101. J. Marek, technical report, Luwa AG, Zurich, Switzerland, Mar. 1970.
102. H. R. C. Pratt and C. Hanson, in Ref. 1, pp. 476–477, Chapt. 16.
103. S. Ochia, *Automatica* **13,** 435 (1977).
104. J. Rydberg, H. Reinhardt, and J. O. Liljenzin, in A. Marinsky and Y. Marcus, eds., *Ion Exchange and Solvent Extraction,* Vol. 3, Marcel Dekker, Inc., New York, 1973, p. 111.
105. H. Reinhardt and J. Rydberg, *Chem. Ind. (London)* **11,** 488 (1970).
106. M. M. Anwar, C. Hanson, and M. W. T. Pratt, *Chem. Ind. (London)* **9,** 1090 (1969).
107. T. C. Lo and A. E. Karr, *Ind. Eng. Chem. Process Des. Dev.* **11**(4), 495 (1972).
108. S. D. Cavers in Ref. 1, Chapt. 10.
109. R. E. Treybal, *Mass Transfer Operations,* 2nd ed., McGraw-Hill Book Co., Inc., New York, 1968.
110. R. Akell and C. J. King, eds., *New Developments in Liquid–Liquid Extractors: Selected papers from ISEC '83,* Vol. 80, no. 238, AIChE Symposium Series, New York, 1984; W. Y. Fei and co-workers, in Ref. 21, Vol. 1, p. 49.

111. G. H. Reman, *Chem. Eng. Prog.* **62**(9), 56 (1966).
112. R. H. Perry and D. Green, *Chemical Engineers Handbook,* 6th ed., McGraw-Hill Book Co., Inc., New York, 1984, pp. 21–55.
113. L. Lowes and M. J. Larkin, *IChemE Symposium Series No. 26,* Institute of Chemical Engineers, London, 1967, p. 111.
114. B. F. Warner, *Proc. 3rd U.N. Conf. Peaceful Uses Atom. Ener.* **10,** 224 (1964).
115. J. Mizrahi, E. Barnea and D. Meyer, in Ref. 14, Vol. 1, p. 141.
116. IMI staff in Ref. 13, Vol. 2, pp. 1, 386.
117. D. W. Ager and E. R. Dement, *Proceedings of the International Symposium on Solvent Extraction in Metallurgical Processes,* Technologisch Instituut K. VIV, Antwerp, Belgium, 1972, p. 27.
118. G. C. I. Warwick, J. B. Scuffham, and J. D. Lott in Ref. 13, Vol. 2, p. 1373.
119. G. C. I. Warwick and J. B. Scuffham in Ref. 117, p. 36.
120. I. D. Jackson and co-workers, *IChemE Symposium Series No. 26,* Institute of Chemical Engineers, London, 1967, p. 111.
121. J. B. Scuffham, *Chem. Eng.,* 328, July 1981.
122. G. C. I. Warwick and J. B. Scuffham in Ref. 1, Chapt. 9.3.
123. N. V. R. Rao and M. H. I. Baird, *Can. J. Chem. Eng.* **62,** 498 (1984).
124. M. Dilley, M. T. Errington, and D. Nadden, in Ref. 21, Vol. 1, p. 140.
125. W. Mehner, G. Hochfeld, and E. Mueller in Ref. 117, Vol. 2, p. 1265.
126. Exhibition during *International Solvent Extraction Conference ISEC 1974,* Lyon, France, 1974.
127. T. K. Mattile in Ref. 14, Vol. 1, p. 169.
128. L. Niinimaki and J. R. Orjans, *Chem. Eng. Symp. Series* **78**(1), 63 (1971); *Kemira Liquid–Liquid Extraction,* Bulletin, Kemira Oy, Helsinki, Finland.
129. Z. J. Shen, Q. Y. Zhang, B. Y. Sun, and Y. F. Sun in Ref. 17, pp. 24–25.
130. Z. J. Shen and M. H. I. Baird, *Chem. Eng. Res. Design* **69,** 143 (1991).
131. Z. J. Shen, J. Li, and K. G. Song, *Proceedings of the International Conference on Separation Science and Technology,* Vol. 1, Canadian Society for Chemical Engineering, Hamilton, Canada, 1989, p. 244.
132. *Kenics Static Mixers,* Bulletin KTEK-5, Kenics Corp., Danvers, Mass., 1972.
133. *Chem. Eng. (N.Y.)* **80**(7), 111 (1973).
134. R. E. Treybal, *Chem. Eng. Progr.* **62**(9), 67 (1966).
135. S. A. Miller and C. A. Mann, *Trans. Am. Inst. Chem. Eng.* **40,** 709 (1944).
136. A. W. Flynn and R. E. Treybal, *AIChE J.* **1,** 324 (1955).
137. A. D. Ryon, F. L. Daley, and R. S. Lowry, *Chem. Eng. Prog.* **55**(10), 71 (1959).
138. B. F. Warner, joint symposium, *The Scaling-Up of Chemical Plant and Processes,* London, 1957, p. 44.
139. H. Rouyer and co-workers in Ref. 14, Vol. 3, p. 2339.
140. *Liquid–Liquid Extraction in C.E.A. Establishments,* Bulletins 25/74, Commissariat à l'Energie Atomique, Genas, France, 1974.
141. G. Sege and F. W. Woodfield, *Chem. Eng. Prog.* **50**(8), 396 (1954).
142. J. D. Thornton, *Br. Chem. Eng.* **3,** 247 (1958).
143. J. D. Thornton, *Trans. Inst. Chem. Eng.* **35,** 316 (1957).
144. D. H. Logsdail and J. D. Thornton, *Trans. Inst. Chem. Eng.* **35,** 331 (1957).
145. *The Bronswerk Technical Bulletin on Pulsed Packed Column,* Bronswerk, PC ES, Amersfoort, the Netherlands.
146. N. U. Spaay, A. J. F. Simons, and G. P. ten Brink in Ref. 24, Vol. 1, p. 281.
147. M. H. I. Baird and G. M. Ritcey, in Ref. 14, Vol. 2, p. 1571.
148. M. E. Weech and B. E. Knight, *Ind. Eng. Chem. Process Des. Dev.* **6,** 480 (1967); **7,** 157 (1968).
149. U.S. Pat. 2,493,265 (Jan. 3, 1950), E. G. Scheibel (to Hoffmann-La Roche Inc.); *Chem. Eng. Prog.* **44**(9), 681 (1948).

150. A. E. Karr and E. G. Scheibel, *Chem. Eng. Prog. Symp. Ser.* **50**(10), 73 (1954).
151. E. G. Scheibel and A. E. Karr, *Ind. Eng. Chem.* **42**(6), 1048 (1950).
152. U.S. Pat. 2,856,362 (Sept. 2, 1958), E. G. Scheibel (to Hoffmann-La Roche Inc.).
153. E. G. Scheibel, *AIChE J.* **2**, 74 (1956).
154. T. C. Lo in *Engineering Foundation Conference on Mixing Research, Rindge, N.H., 1975*, The Engineering Foundation, New York, 1975.
155. U.S. Pat. 3,389,970 (June 25, 1968), E. G. Scheibel.
156. E. G. Scheibel in Ref. 1, Chapt. 13.3.
157. *Scale-Up Procedures for a Scheibel Extraction Column*, NTIS Report No. DE3-013576, National Technical Information Service, U.S. Department of Commerce, Washington, D.C., 1983.
158. G. H. Reman, *Proceedings of the 3rd World Petroleum Congress*, the Hague, the Netherlands, 1951, Sect. III, P. 121.
159. C. P. Strand, R. Olney, and G. H. Ackerman, *AIChE J.* **8**, 252 (1962).
160. T. Misek, *Rotating Disc Extractors and Their Calculation*, State Publishing House, technical literature, Prague, Czechoslovakia, 1964.
161. W. C. G. Kosters in Ref. 1, Chapt. 13.1.
162. J. Y. Oldshue and J. H. Rushton, *Chem. Eng. Prog.* **48**(6), 297 (1952).
163. J. Y. Oldshue, *Biotech. Bioeng.* **8**(1), 3 (1966).
164. R. Bibaud and R. Treybal, *AIChE J.* **12**, 472 (1966).
165. H. F. Haug, *AIChE J.* **17**, 585 (1971).
166. J. Ingham, *Trans. Inst. Chem. Eng.* **50**, 372 (1972).
167. T. Miyauchi, H. Mitsutake, and I. Harase, *AIChE J.* **12**, 508 (1966).
168. J. Y. Oldshue, private communication, Mixing Equipment Co., Inc., Rochester, N.Y., 1970.
169. J. Y. Oldshue, F. Hodgkinson, and J. C. Pharamond in Ref. 14, Vol. 2, p. 1651.
170. J. Y. Oldshue in Ref. 1, Chapt. 13.4.
171. J. Oldshue in Y. Marcus, ed., *Solvent Extraction Reviews*, Vol. 1, Marcel Dekker, Inc., New York, 1976.
172. T. Misek and J. J. Marek, *Br. Chem. Eng.* **15**, 202 (1970).
173. J. Marek and co-workers, paper presented at the *Society of Chemical Industry Symposium*, Bradford, UK, 1967.
174. B. Seidlova and T. Misek in Ref. 14, Vol. 3, p. 2365.
175. T. Misek and J. Marek in Ref. 1, Chapt. 13.2.
176. P. J. Bailes, C. Hanson, and M. A. Hughes, *Chem. Eng. (N.Y.)* **83**(2), 86 (1976).
177. A. Mogli and U. Buhlman, in Ref. 1, Chapt. 13.5.
178. U. Buhlman, in Ref. 21, Vol. 1, p. 17.
179. UK Pats. 860,880 (Feb. 15, 1961); 972,035 (Oct. 7, 1964); 1,037,573 (July 27, 1966), J. Coleby.
180. J. Hu, Z. J. Shen, and Y. F. Su, paper presented in *International Meeting on Chemical Engineering and Biotechnology*, ACHEMA'88, Frankfurt, Germany, 1988.
181. A. R. Sheikh, C. Hanson, and J. Ingham, *Trans. Inst. Chem. Eng.* **50**, 199 (1972); Z. J. Shen, J. Hu, Y. F. Su, and M. H. I. Baird, *Proceedings of the Second International Conference on Separation Science Technology*, Vol. 1, Canadian Society for Chemical Engineering, Hamilton, Canada, 1989, p. 282.
182. M. H. I. Baird in Ref. 1, Chapt. 14.
183. U.S. Pat. 2,011,186 (Aug. 13, 1935), W. J. D. Van Dijek.
184. A. E. Karr, *AIChE J.* **5**, 446 (1959).
185. A. E. Karr and T. C. Lo in Ref. 15, Vol. 1, p. 229.
186. A. Guyer, A. Guyer, Jr., and K. Mauli, *Helv. Chim. Acta* **38**, 790, 995 (1955); N. Issac and R. L. DeWitte, *AIChE J.* **4**, 498 (1958), *Dechema Monogr.* **32**, 218 (1959); J. Prochazka and co-workers, *Br. Chem. Eng.* **16**, 42 (1971).
187. D. Elenkov and co-workers, *Khim. Inst. Sof.* **4**, 181 (1966).

188. R. Wellek and co-workers, *Ind. Eng. Chem. Process Des. Dev.* **8**, 515 (1969).
189. K. Tojo, T. Miyanami, and T. Yano, *J. Chem. Eng. Jpn.* **7**, 123 (1974).
190. USSR Pat. 175,489 (1965), S. M. Karpacheva, E. I. Zakharov, L. S. Raginski, V. M. Muratov, and A. V. Romanov.
191. I. Y. Gorodetski, A. A. Vasin, V. M. Olevski, and P. A. LuPanov, *Vibratoionnye Massoobmennye Apparaty*, Khimia, Moscow, 1980.
192. T. C. Lo, M. H. I. Baird, and N. V. R. Rao, *Chem. Eng. Comm.* **116**, 67–88 (1992).
193. M. H. I. Baird, N. V. R. Rao, J. Prochazka, and H. Sovova, in M. J. Slater and J. Godfrey, eds. *Solvent Extraction Equipment*, John Wiley & Sons, Inc., New York, 1994 (in press).
194. T. C. Lo and J. Prochazka, in Ref. 1, Chapt. 12.
195. A. E. Karr and T. C. Lo in Ref. 15, paper 8a.
196. A. E. Karr and T. C. Lo, *Chem. Eng. Prog.* **72**(11), 68 (1976).
197. M. H. I. Baird, R. G. McGinnis, and G. C. Tan in Ref. 13, Vol. 1, p. 251.
198. S. D. Kim and M. H. I. Baird, *Can. J. Chem.* **54**, 81 (1976).
199. A. E. Karr and S. Ramanujam, in Ref. 18, Vol. 3, p. 493.
200. J. Prochazka, *Dechema Monogr.* **65**, 325 (1970).
201. U.S. Pat. 3,583,856 (1971), J. Landau, J. Prochazka, and F. Souhrada.
202. I. J. Gorodetskii, A. A. V. M. Olevskii, A. E. Konstanyan, and co-workers, *Khim. Prom.* **8**, 480 (1984).
203. A. M. Rosen, I. G. Martyushin, and co-workers, *Scale Transition in Chemical Industry*, Khimia, Moscow, 1980.
204. I. J. Gorodetskii, and co-workers in Ref. 19, Vol. 2, p. 225.
205. N. S. Yang, B. H. Chen, and A. F. McMillan, *Can. J. Chem. Eng.* **64**, 387 (1986).
206. K. Takeba, preprint of *The 10th General Symposium of the Society of Chemical Engineers*, Japan, 1971, p. 124.
207. K. Miyanami, K. Tojo, and T. Yano, *J. Chem. Eng. (Japan)* **6**, 518 (1973).
208. M. M. Hafez in Ref. 1, Chapt. 15.
209. W. J. Podbielniak, *Chem. Eng. Prog.* **49**(5), 252 (1953).
210. D. B. Todd and G. R. Davis in Ref. 14, Vol. 3, p. 2379.
211. D. B. Todd, G. R. Davis, and H. A. Lange in Ref. 18, Vol. 3, p. 345.
212. F. Otillinger and E. Blass in Ref. 18, Vol. 3, p. 445; in Ref. 19, Vol. 2, p. 235.
213. E. Broadwell, paper presented at *The Society of Chemical Industry Symposium*, Bradford, UK, 1967.
214. Yu. A. Dulatav and I. I. Poniharov in Ref. 18, Vol. 2, p. 259.
215. H. Einsenlohr, *Dechema Monogr.* **19**, 222 (1951).
216. Paper presented at *The Society of Chemical Industry Symposium*, Bradford, UK, 1967.
217. C. Bernard, P. Michel, and M. Amero in Ref. 13, Vol. 2, p. 1282.
218. C. R. Howarth, J. G. M. Lee, and C. Ramshaw, in Ref. 21, Vol. 1, p. 25; C. Judson King, in Ref. 21, Vol. 1, p. 3.
219. B. M. Sankey and D. A. Gudolis in Ref. 1, Chapt. 18.3; W. S. Nogueria and M. F. Moraes, in Ref. 21, Vol. 2, p. 1081.
220. D. Read, paper presented at *American Petroleum Institute Meeting*, San Francisco, Calif., May 1952.
221. T. S. Hoover, *Hydrocarbon Process.* **12**, 69 (1969).
222. H. Voetter and W. C. G. Kosters, *Proceedings of the World Petroleum Congress*, Vol. III, 1963, p. 131.
223. F. S. Beadmore and W. C. G. Kosters, *J. Inst. Pet.* **49**, 469 (1963).
224. W. C. G. Kosters, paper presented at *The Institute for Chemical Engineering Symposium Liquid Extraction*, Newcastle-upon-Tyne, UK, 1967.

225. W. C. G. Kosters in Ref. 1, Chapt. 18.2.3.
226. E. Muller and G. Hoehfeld, *Proceedings of the 7th World Petroleum Congress,* Vol. IV, 1967, p. 13.
227. E. Muller in Ref. 1, Chapt. 18.2.1.
228. *Hydrocarbon Process. Pet. Refiner,* 185 (Sept. 1972).
229. G. S. Somekh in Ref. 1, Vol. 13, p. 323.
230. J. A. Vidueira in Ref. 1, Chapt. 18.2.2.
231. E. Cinelli, S. Noe, and G. Paret, *Hydrocarbon Process. Pet. Refiner,* 141 (Apr. 1972).
232. A. L. Benham and co-workers, *Hydrocarbon Process. Pet. Refiner* **46**(9), 134 (1967).
233. J. Coleby in C. Hanson, ed., *Recent Advances in Liquid–Liquid Extraction,* Pergamon Press, Oxford, UK, 1971.
234. A. J. F. Simon and N. F. Hassen, in Ref. 1, Chapt. 18.4.
235. T. Ueno, in Ref. 1, Chapt. 18.6.
236. E. Lloyd-Jones, *Chem. Ind. (London),* 1590 (1967).
237. C. J. King, in Ref. 1, Chapt. 18.5.
238. F. Panzner, S. R. M. Ellis, and T. R. Bott, in Ref. 15, paper 15g, p. 685.
239. G. H. Beyer, *Proceedings of the International Solvent Extraction Conference ISEC 1977,* The Canadian Institute of Mining and Metallurgy, Ottawa, Canada, 1977, p. 715.
240. J. M. Fox, *Hydrocarbon Process. Pet. Refiner,* 2 (Sept. 1963).
241. K. Ridgeway and E. E. Thorpe, in Ref. 1, Chapt. 19.
242. A. L. Edler, ed., *Chem. Eng. Prog. Symp. Ser.* **66** (1970).
243. U.S. Pat. 3,572,750 (Sept. 8, 1970), A. E. Karr (to Hoffmann-La Roche Inc.); A. E. Karr and co-workers, Hoffmann-La Roche Inc., Nutley, N.J., unpublished report, 1970.
244. S. S. Davis and co-workers, *Chem. Ind. (London),* 677 (Aug. 1976).
245. H. J. Passino, *Ind. Eng. Chem.* **41,** 280 (1949).
246. S. W. Glover, *Ind. Eng. Chem.* **40,** 228 (1948).
247. *Chem. Eng. Prog.* **67**(5), 131 (1971).
248. T. Hano, M. Matsumoto, and T. Ohtake, in Ref. 21, Vol. 2, p. 1025.
249. W. Hamm, in Ref. 1, Chapt. 20.
250. J. M. Nogueria and J. C. Pereira, in Ref. 21, Vol. 2, p. 1088.
251. M. W. T. Pratt and J. Spokes in Ref. 15, paper 31C.
252. J. Coleby, paper presented at *Symposium on Solvent Extraction,* Institute of Chemical Engineers, Newcastle-upon-Tyne, UK, 1967.
253. L. Crainger and W. S. Wise, *Chem. Br.* **4,** 12 (1968).
254. W. L. Lewis and W. L. Martin, *Hydrocarbon Process.* **46**(2), 131 (1967); *Manual on Disposal of Refinery Wastes,* American Petroleum Institute, Washington, D.C., 1969, Chapt. 10.
255. *Chem. Eng. (N.Y.),* 58 (Mar. 15, 1976).
256. P. Ramsden, paper presented at *Institute of Chemical Engineering Research Meeting on Solvent Extraction,* Bradford, UK, 1965.
257. A. E. Karr and S. Ramanujan, *International Solvent Extraction Conference 1986,* Vol. 2, Dechema, Frankfurt, Germany, 1986.
258. D. Mackay and M. Medir in Ref. 1, Chapt. 23; P. N. Hurter and co-workers, in Ref. 21, Vol. 3, p. 1663.
259. M. R. Kula in Ref. 18, Vol. 3, p. 567; T. A. Hatton, in Ref. 19, Part A., p. 23.
260. E. G. Scheibel, *Chem. Eng. Prog.* **62**(9), 66 (1966).
261. J. T. Long, *Engineering for Nuclear Fuel Reprocessing,* Gordon and Breach, Inc., New York, 1967.
262. M. Cox in Ref. 4, Chapt. 11.
263. R. Blumberg in Ref. 1, Chapt. 26.

264. G. R. Choppin and J. Rydberg, *Nuclear Chemistry,* Pergamon Press, New York, 1980.
265. P. Danesi in Ref. 4, Chapt. 12.
266. D. A. Orth in Ref. 18, Vol. 1, p. 75.
267. E. P. Horwitz, H. Diamond, D. Kalina, L. Kaplan, and G. W. Mason in Ref. 17, p. 451.
268. C. Musikas in Ref. 20, p. 297; W. L. Wilkinson, in Ref. 21, Vol. 3, p. 1455.
269. J. A. Williams and W. J. Bowers in Ref. 1, Chapt. 31.
270. A. Naylor and P. D. Wilson in Ref. 1, Chapt. 25.12.
271. P. J. D. Lloyd in Ref. 1, Chapt. 25.11.
272. P. D. Mollere in Ref. 18, Vol. 2, p. 49.
273. J. F. C. Fisher and C. W. Notebaart in Ref. 1, Chapt. 25.1.
274. G. M. Ritcey in Ref. 1, Chapt. 25.2.
275. Ref. 1, Chapt. 25.3–25.14.
276. D. S. Flett in Ref. 20, p. 1.
277. S. O. S. Anderson and H. Reinhardt, in Ref. 1, Chapt. 25.10.
278. H. Reinhardt, in Ref. 21, Vol. 3, p. 1625.
279. G. Angelov and N. Panchev, in Ref. 21, Vol. 3, p. 1649.
280. R. D. Rogers, A. H. Bond, and C. B. Bauer, in Ref. 21, Vol. 3, p. 1641.

TEH C. LO
T. C. Lo & Associates

MALCOLM H. I. BAIRD
McMaster University

EXTRACTION, LIQUID–SOLID

Liquid–solid extraction or leaching is a unit operation that predates large-scale industrial operations, with a history of known uses that goes back to Roman times. The early leaching process was known as lixiviation, a term used to describe the extraction of alkaline slats from wood and other plants. The term leaching, in use in the eighteenth century, was used to describe the process of percolating a liquid through a solid material. It is presumed that a soluble component was removed from the solid phase during the percolation so that the operation was distinguishable from sand filters which were quite widely used by that date, but which served a quite different function. Extraction is used in a wide variety of process industries and traditional terms have evolved in different fields. These include leaching, washing, percolation, digestion, steeping, lixiviation, and infusion, among others.

Many substances used in modern processing industries occur in a mixture of components dispersed through a solid material. To separate the desired solute constituent or to remove an unwanted component from the solid phase, the solid is contacted with a liquid phase in the process called liquid–solid extraction, or simply leaching. In leaching, when an undesirable component is removed from a solid with water, the process is called washing.

In the biological and food processing industries many products are extracted from their original structure by liquid–solid extraction. Sugar is extracted from sugar beets using hot water; instant coffee is leached from ground roasted coffee using water; soluble tea is leached from tea leaves; pharmaceutical components, flavors, and essences are leached from plant roots, leaves and stems; and oil is extracted from peanuts, soybeans, sunflower and cotton seeds, and halibut livers by solvents such as hexane, acetone, or ether. These are all examples of liquid–solid extraction.

Large-scale leaching also occurs in the metal processing industries, where useful metals frequently occur mixed with large quantities of unwanted matter, and leaching is used to remove the metals as soluble salts. For example, gold is leached from its ore using aqueous sodium cyanide solutions; cobalt and nickel by sulfuric acid–ammonia–oxygen mixtures; and copper salts by sulfuric acid or ammoniacal solutions (see METAL SEPARATIONS; MINERALS PROCESSING).

Mechanisms of Extraction

If the solute is uniformly distributed through the solid phase the material near the surface dissolves first to leave a porous structure in the solid residue. In order to reach further solute the solvent has to penetrate this outer porous region; the process becomes progressively more difficult and the rate of extraction decreases. If the solute forms a large proportion of the volume of the original particle, its removal can destroy the structure of the particle which may crumble away, and further solute may be easily accessed by solvent. In such cases the extraction rate does not fall as rapidly.

In general, the following steps can occur in an overall liquid–solid extraction process: solvent transfer from the bulk of the solution to the surface of the solid; penetration or diffusion of the solvent into the pores of the solid; dissolution of the solvent into the solute; solute diffusion to the surface of the particle; and solute transfer to the bulk of the solution. The various fundamental mechanisms and processes involved in these steps make it impracticable or impossible to describe leaching by any rigorous theory.

Any one of the five basic processes may be responsible for limiting the extraction rate. The rate of transfer of solvent from the bulk solution to the solid surface and the rate into the solid are usually rapid and are not rate-limiting steps, and the dissolution is usually so rapid that it has only a small effect on the overall rate. However, knowledge of dissolution rates is sparse and the mechanism may be different in each solid (1).

The overall extraction process is sometimes subdivided into two general categories according to the main mechanisms responsible for the dissolution stage: (*1*) those operations that occur because of the solubility of the solute in or its miscibility with the solvent, eg, oilseed extraction, and (*2*) extractions where the solvent must react with a constituent of the solid material in order to produce a compound soluble in the solvent, eg, the extraction of metals from metalliferous ores. In the former case the rate of extraction is most likely to be controlled by diffusion phenomena, but in the latter the kinetics of the reaction producing the solute may play a dominant role.

Diffusion and Mass Transfer During Leaching. Rates of extraction from individual particles are difficult to assess because it is impossible to define the shapes of the pores or channels through which mass transfer (qv) has to take place. However, the nature of the diffusional process in a porous solid could be illustrated by considering the diffusion of solute through a pore. This is described mathematically by the diffusion equation, the solutions of which indicate that the concentration in the pore would be expected to decrease according to an exponential decay function.

To obtain an indication of the rate of solute transfer from the particle surface to the bulk of the liquid, the concept of a thin film providing the resistance to transfer can be used (2) and the equation for mass transfer written as:

$$\frac{dM}{dt} = \frac{D^* A(c_s - c)}{\delta} \tag{1}$$

where A = the area of the solid–liquid interface, c = the concentration of the solute in the bulk of the liquid at time t, c_s = the concentration of the saturated solution in contact with the particles, D^* = a diffusion coefficient (approximated by the liquid-phase diffusivity), M = the mass of solute transferred in time t, and δ = the effective thickness of the liquid film surrounding the particles. For a batch process where the total volume V of solution is assumed to remain constant, $dM = V\,dc$ and

$$\frac{dc}{dt} = \frac{D^* A(c_s - c)}{\delta V} \tag{2}$$

The time t taken for the concentration of the solution to rise from its initial value c_0 to a value c is obtained by integration of this equation, assuming that both A and δ remain constant, to give:

$$c = c_s - (c_s - c_0)\exp\left(\frac{-D^* At}{\delta V}\right) \tag{3}$$

This simple analysis also shows that the bulk solution approaches a saturated condition exponentially. That A and δ are constant are both significant assumptions rarely met in extraction, although the change with time may be slow. The interfacial area tends to increase as extraction proceeds, and is of course dependent on the extent to which the solid material has been ground prior to extraction. D^* should be treated as an effective mass-transfer coefficient which would be sensitive not only to the composition and properties of the solution surrounding the particle, but also to the hydrodynamic conditions in the bulk of the solution. For larger particles, which are usually present in leaching, equations are available (3,4) to predict the mass-transfer coefficient in agitated vessels.

Process Design

In most leaching operations the maintenance of constant fluid flows, pressures, and temperatures are important. These, together with the need to provide a sufficient contact time between the solvent and the solids, usually indicate a need for continuous, multistage, countercurrent processes in which fresh solvent is fed to the final stage while the solids are fed to the first stage. The objective is to be able to operate at steady conditions, and to be able to avoid extraction of undesirable material while preventing loss of solvent for both economic and safety reasons. This is usually achieved through the use of the usual control equipment, and recording instruments provide a useful means of studying plant performance. There are other factors which must be taken into account in the early stages of a design such as the particle size of the solid and the solvent employed.

Particle Size. The smaller the particle size, the greater is the relative interfacial area between the solid and the liquid and the shorter are diffusional path lengths, and therefore the higher is the rate of transfer of solute. However, smaller particle sizes tend to lead to lower drainage rates from the solid residue and can create problems in the solids flow through countercurrent extraction equipment. A compromise has to be made to select a particle size which offers an acceptable extraction rate but yet does not unduly impede flow of solvent through a percolation process or of solids through a countercurrent process. If the extractable material is a minor proportion of the starting material the disposal of the solids residue may present a problem. If the residues have some commercial value as a product after further processing, excessive grinding of the feed solids may render the residue unsuitable for the product and hence have an adverse effect on the economics of the extraction process.

Solvent. Solvent choice is determined by the chemical structure of the material to be extracted, and the rule that like dissolves like provides useful guidance. Thus vegetable oils consisting of triglycerides of fatty acids are normally extracted with hexane [110-54-3], whereas for free fatty acids, which are more polar than the triglycerides, more polar alcohols are used. Halogenated hydrocarbons and hexane are both widely used as solvents, and liquid carbon dioxide [124-38-9] appears to be suitable for extracting flavor components from plants (5) (see SUPERCRITICAL FLUIDS). Where a choice of solvent other than water exists on the grounds of comparable solubility of the solute, the following criteria are likely to be considered.

Selectivity. Solvent selectivity is intimately linked to the purity of the recovered extract, and obtaining a purer extract can reduce the number and cost of subsequent separation and purification operations. In aqueous extractions pH gives only limited control over selectivity; greater control can be exercised using organic solvents. Use of mixed solvents, for example short-chain alcohols admixed with water to give a wide range of compositions, can be beneficial in this respect (6).

Physical Properties. Low surface tension facilitates wetting of the solids in the first extraction stage, and low viscosity assists diffusion rates in the solvent phase. A low solvent density is desirable to reduce the mass of solvent held up in the solid being extracted, but solvent choice is usually dictated by other factors. A high boiling solvent with a high latent heat of evaporation requires recovery conditions that may be adverse for thermally sensitive extracts and increases the cost of solvent recovery. In chemical leaching the thermodynamics of the leaching reaction must be considered in terms of the redox potential–pH diagram (Pourbaix diagram) which can be constructed from standard free-energy data (7). This provides the basis for choosing equilibrium leaching conditions (acidic vs basic, oxidizing vs reducing), leaving the need for a kinetic study to provide data on leaching rates.

Thermal Stability. At processing temperatures in both the extraction and recovery plants the solvent should be completely stable to avoid expensive solvent losses; contamination of the solvent by any solvent breakdown products must be avoided.

Hazards. The solvent should be nontoxic and nonhazardous; adequate design must take into account flammability and explosivity characteristics of the solvent.

Cost. The cost of fresh solvent is reflected in the operating costs in the form of solvent make-up charges. Avoidance of solvent losses, and hence a reduction of operating costs, may be obtainable through better plant design which is usually associated with increased capital costs.

Temperature. Both the solubility of the material being extracted and its diffusivity usually increase with temperature, and higher extraction rates are obtained. In some cases the upper limit for the operating temperature is determined by factors such as the need to avoid undesirable side reactions.

Agitation of the Fluid. Agitation of the solvent increases local turbulence and the rate of transfer of material from the surface of the particles to the bulk of the solution. Agitation should prevent settling of the solids, to enable most effective use of the interfacial area.

Equilibrium Relationships and Mass Balances

The solid can be contacted with the solvent in a number of different ways but traditionally that part of the solvent retained by the solid is referred to as the underflow or holdup, whereas the solid-free solute-laden solvent separated from the solid after extraction is called the overflow. The holdup of bound liquor plays a vital role in the estimation of separation performance. In practice both static and dynamic holdup are measured in a process study, other parameters of importance being the relationship of holdup to drainage time and percolation rate. The results of such studies permit conclusions to be drawn about the feasibility of extraction by percolation, the holdup of different bed heights of material prepared for extraction, and the relationship between solute content of the liquor and holdup. If the percolation rate is very low (in the case of oilseeds a minimum percolation rate of 3×10^{-3} m/s is normally required), extraction by immersion may be more effective. Percolation rate measurements and the methods of utilizing the data have been reported (8,9); these indicate that the effect of solute concentration on holdup plays an important part in determining the solute concentration in the liquor leaving the extractor.

Single-Stage Leaching. A single-stage leaching process is shown in Figure 1. The solution overflow rate is V kg/h; the mass fraction of solute in the overflow solution is x_A; and the liquid in the slurry is flowing at L kg/h, and has a composition y_A. The mass flow of dry inert solids in the slurry is B kg/h.

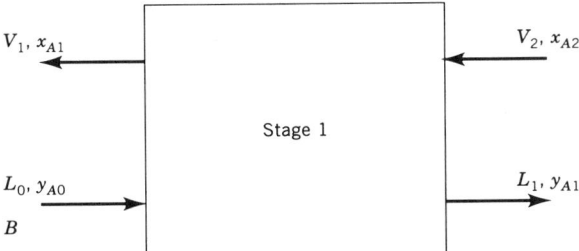

Fig. 1. Flow diagram for single-stage leaching.

The material balance equations are, for the total solution:

$$L_0 + V_2 = L_1 + V_1 = M \qquad (4)$$

where M is the total input flow rate of solution to the unit; for the solute component A:

$$L_0 y_{A0} + V_2 x_{A2} = L_1 y_{A1} + V_1 x_{A1} = M x_{Am} \qquad (5)$$

where $x_1 = y_1 = x_{Am}$; and for the solids:

$$B = L_0 N_0 = L_1 N_1 \qquad (6)$$

where N_i is the mass concentration of inert solids in the ith stream, ie, kg of inert solid per kg solution. From these balances the concentration of the discharged solution can be estimated.

Countercurrent Multistage Leaching. Countercurrent extraction offers the most economical use of solvent, permitting high concentrations in the final extract and high recovery from the initial solid but utilizing the least amount of solvent. In a multistage operation fresh solid enters the first stage and fresh solvent enters the final stage; the latter is gradually enriched in solute until it leaves the extraction battery as overflow from the first stage. The operation is usually discontinuous in that the solvent is pumped from one vessel to the next intermittently and allowed to remain until equilibrium extraction is approached. When the amount of solvent removed with the insoluble solid in the underflow is constant, it is convenient to define the ratio

$$R = \frac{\text{amount of solvent in overflow}}{\text{amount of solvent in underflow}} = \frac{V_n}{L_n} \qquad (7)$$

If perfect mixing occurs in each stage and the solute is not adsorbed preferentially at the surface of the solid, then the concentration of the solution in the underflow is the same as that in the overflow and

$$R = \frac{\text{amount of solute in overflow}}{\text{amount of solute in underflow}} = \frac{V_n x_{An}}{L_n y_{An}} \qquad (8)$$

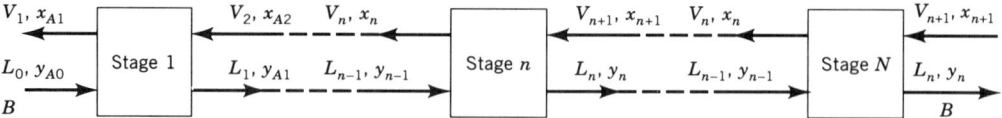

Fig. 2. Flow diagram for countercurrent multistage leaching.

Referring to Figure 2, by considering solute mass balances over n, $(n-1), \ldots 2, 1$ units in turn and eliminating intermediate solute mass fractions and flow rates, the amount of solute associated with the leached solid may be calculated in terms of the composition of the solid and solvent streams fed to the system. The resulting equation is (2)

$$L_0 y_{A0} = \frac{R^{n+1} - 1}{R - 1} L_n y_{An} - \frac{R^n - 1}{R - 1} V_{n+1} x_{An+1} \tag{9}$$

In many cases the amount of solute associated with the leached solid must not exceed a certain value, and it is possible to compute directly the minimum number of units needed by putting $n = N$.

Alternative approaches are to be found in the literature. Derivations of the above equations are given in numerous texts (2,10–12), which also describe graphical or analytical solutions to the problem. Many of these have direct analogues in other separation processes such as distillation (qv) and liquid–liquid extraction, and use plots such as the McCabe-Thiele diagram or Ponchon-Savarit diagram.

Countercurrent Leaching With Variable Underflow. In practice most cases of interest exhibit a variable underflow rate, which is normally greatest at that point in the process where the solute concentration in the solvent is highest. In some leaching operations the viscosity and density of the solution changes appreciably as the solute concentration. Consequently, the operating line derived from mass balance equations for the McCabe-Thiele diagram has a slope which varies from stage to stage, and equation 9 is no longer valid. In the lower numbered stages the solute concentrations are higher and the underflows may retain more solution than the underflows from the higher numbered stages. Clearly, if the underflow rate L_n varies then so does the overflow rate. As in the previous case, each unit is assumed to be well mixed and the solute mass fractions in the overflow and underflow are related by:

$$x_{An} = y_{An} \tag{10}$$

The solute mass fraction in the overflow solution from the first unit ($n = 1$) is

$$x_{A1} = \frac{V_{n+1} x_{An+1} + L_0 y_{A0} - L_n y_{An}}{V_{n+1} + L_0 - L_n} \tag{11}$$

and the solute mass fraction in the solution fed to unit n is

$$x_{An+1} = \frac{V_{n+1} x_{An+1} - L_n y_{An} + L_n y_{An}}{V_{n+1} - L_n + L_n} \tag{12}$$

It is often the case that all quantities in equation 11 except x_{A1} are known. If, instead of V_{n+1}, the concentration of the solution leaving the system is specified, then equation 11 can be used to calculate V_{n+1}.

Extractors

Calculations serve as a guide to the analysis of an extraction plant and, as for the analysis of equipment performance in any other sphere of process engineering, these may be supplemented by empirical correlations or process models pertinent to the particular equipment under consideration. Another process which usually deserves special attention is that needed for solvent regeneration and for solute recovery. Solvent recovery is often energy intensive and a full process energy analysis is recommended to reduce costs. Recovery of organic solvents from the exhausted solids is also important and can be more troublesome than recovery from a liquid, and consideration should be given to the use of superheated solvent vapor for this purpose (13) (see ENERGY CONSERVATION IN SEPARATION PROCESSES).

Extractors rely on either percolation or agitation to ensure intimate contact between the solids and solvent. Percolation and extraction rate data provide guidance on whether extraction should be by percolation or immersion and what extraction time is needed to give an acceptable approach to equilibrium. For a percolation system to be viable the extraction rate needs to be high, as the solvent residence time is often relatively short, although where percolation rates are so high (residence times very short) that extraction becomes inefficient, upward flow through the bed of solids can sometimes be advantageous (14). A percolation process can be carried out either stagewise or in a differential contactor; for an immersion process stagewise contact is often more practicable, particularly where a low extraction rate requires an extended residence time or multiple contact with the solvent. Under these circumstances, or when extraction is accompanied by chemical reaction, a countercurrent multistage operation is often beneficial. When percolation and extraction rates are being measured for equipment specification and sizing purposes, the conditions in the test extractor should match as closely as possible those anticipated in the full-scale extractor in order to provide the most reliable data.

Extractors often contribute substantially to the capital and operating costs of a plant, which provides the impetus to seek ways to reduce the extraction load in order to increase extractor capacity and reduce specific solvent requirements. When the feed material is of plant origin and the solute is contained in cells that can be ruptured by heat or pressure, pre-treatment frequently involves removing part of the solute by pressing. The variety of extractors used in liquid–solid extraction is diverse, ranging from batchwise dump or heap leaching for the extraction of low grade ores to continuous countercurrent extractors to extract materials such as oilseeds and sugar beets where problems of solids transport have dominated equipment development.

Batch Extractors. Coarse solids are leached by percolation in fixed or moving-bed equipment. Both open and closed tanks having false bottoms are used, into which the solids are dumped to a uniform depth and then treated with the solvent by percolation, immersion, or intermittent drainage methods.

The pot extractor is a batch extraction plant in which extraction and solvent recovery from the exhausted solids can be carried out in a single vessel. These extractors are normally agitated vessels having capacities in the range of 2 to 10 m^3, beyond which the battery system becomes a preferred technical alternative.

The diffuser is a closed percolation vessel which is used when the pressure drop is too high for gravity flow of solvent, when evaporative losses of solvent would be too high, or when it is necessary to use elevated temperatures. The solvent is circulated through the tank by pumping, or leaching may be achieved without solvent circulation. A diffuser battery is a semibatch extraction system operating on a cyclical basis. The system comprises a battery of diffusers or vessels, each of which is charged with the solids to be extracted. Fresh solvent is fed to the first one or two vessels, sometimes with the solvent being heated before being fed to the unit. The underflow from one unit is fed to the next, again with the option of interunit heating available. The underflow from the final unit is a solute-rich solution. The actual number of diffusers in the battery depends on extraction and equilibrium conditions, but an additional diffuser above the estimated number is required to permit cyclical operation. When a battery consists of more than four units a close approximation to countercurrent flow is achieved, and owing to the cyclic nature of the operation each unit changes its position in the extraction sequence at each cycle changeover. In the cyclic operation, the most exhausted diffuser is bypassed and emptied, and an empty one is charged with fresh solids. This rather cumbersome plant layout is used, for example, for the extraction of coffee solubles using hot (150–180°C) water, but can be largely replaced by fully continuous devices.

Continuous Extractors. Continuous extractors are available in a variety of forms. The main difference between them is the way by which the solids are transported through the equipment. For convenience the method of solids transport is used as a means of equipment classification.

Moving-bed percolation systems are used for extraction from many types of cellular particles such as seeds, beans, and peanuts. In most of these cases organic solvents are used to extract the oils from the particles. Pre-treatment of the seed or nut is usually necessary to increase the number of cells exposed to the solvent by increasing the specific surface by flaking or rolling. The oil-rich solvent (or miscella) solution often contains a small proportion of fine particles which must be removed, as well as the oil separated from the solvent after leaching.

The Bollman extractor (15) (Fig. 3) is a moving-bed, perforated-basket type of extractor. The solids are loaded into baskets fixed to a chain conveyor in a closed vessel. Solid is fed to the top basket on the downward side of the conveyor and is discharged from the top basket on the upward side. Fresh solvent is sprayed on the solid about to be discharged, leaving some time for drainage from the basket before discharge is effected, and passes downward through the baskets to effect a countercurrent flow. The partially rich solvent (half-miscella) from the bottom of the upward side is pumped to the basket at the top of the downward side, from which solvent flows from basket to basket in cocurrent fashion. The final solvent solution, miscella, is collected from the bottom of the downward side. Control of flake size during pre-treatment is desirable, as is control of the thickness and bulk density of the bed. A typical extractor moves at about 0.3 m/s, each basket contains some 350 kg of seeds, about equal masses of seeds and solvent are used, and the miscella contains about 25% oil by mass (2). Advantages of this design of extractor

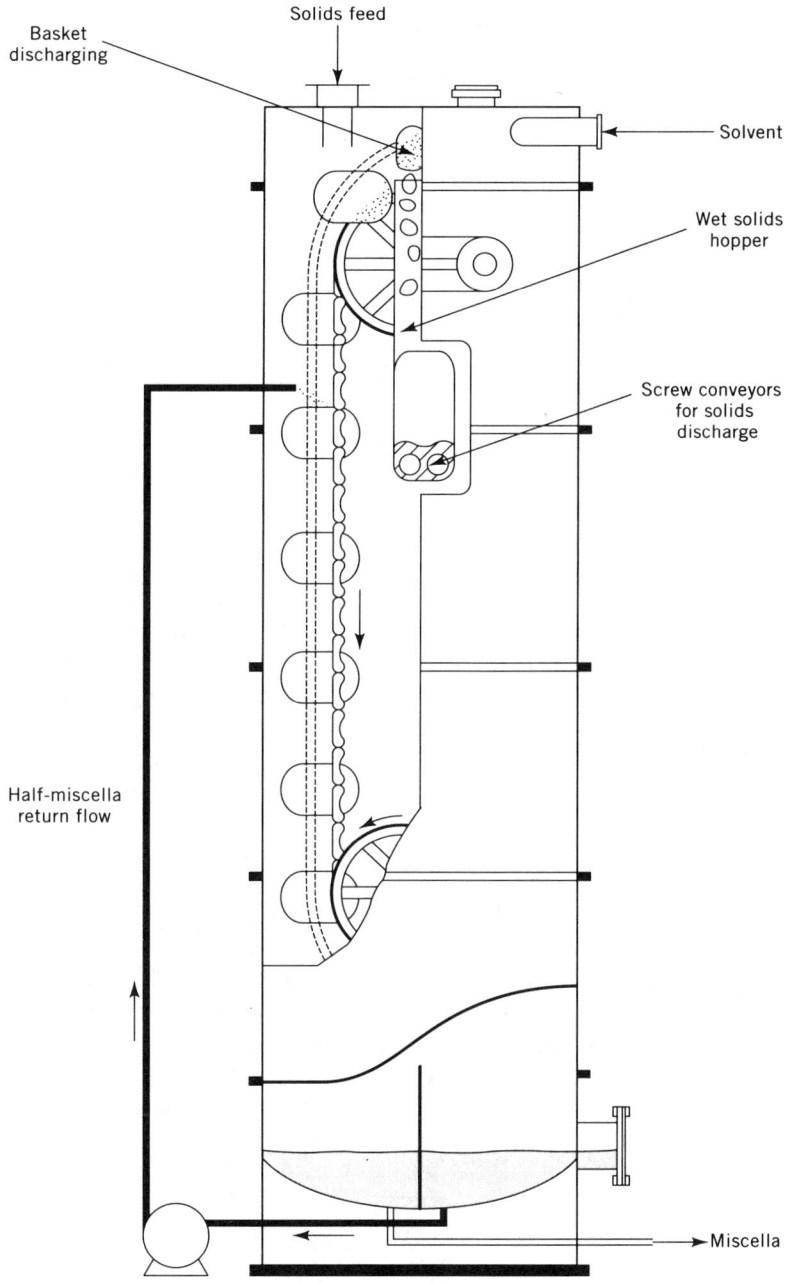

Fig. 3. Bollman moving-bed-type extractor.

are that a solids-free miscella can be obtained, the residue is well drained when the equipment is properly controlled, and large quantities of solids can be extracted continuously.

The Rotocel extractor (16) achieves countercurrent extraction through a sequence of discrete liquid–solid contacts. The solids to be extracted are fed continuously as a dry material or as a slurry to sector-shaped cells arranged around a horizontal rotor. Each cell has a perforated base which allows easy drainage of solvent into a basin at the base of the cell from which the solvent is pumped into the next cell in the countercurrent direction. Fresh solvent is supplied to the last cell, which also occupies a larger sector than the other cells to allow for drainage of the extracted solids prior to discharge. The miscella is filtered by the bed of solids in each cell, and miscella from such rotary-type extractors can be expected to contain less than 5 ppm suspended solids, sometimes effecting a saving on the cost of subsequent solid–liquid separation equipment.

Tipping pan and horizontal filters are also used for leaching: the modus operandi of the Rotocel extractor resembles that of a tipping pan filter, although the details of its design differ slightly.

An alternative tower design, the Bonotto extractor (15) (Fig. 4), is a series of slowly rotating horizontal trays equispaced vertically in a tall cylindrical vessel. The solid is fed continuously close to the outside edge on the top tray and a stationary scraper attached to the vessel causes the solid to cross the tray. The solid then falls through an opening onto the tray beneath, where another scraper moves the solid across the tray in the opposite direction toward a similar opening near the periphery of this tray. This sequence of moving the solid across each plate in opposite directions on alternate plates is continued until the solid reaches the bottom of the tower. It is then transported from the tower by a screw conveyor, although alternative types of solids conveyor could be used. The solvent is fed to the bottom of the vessel and flows upward to give a flow countercurrent to the solids flow direction. Clearly the upward velocity of the solvent should be lower than the fall velocity of the solids to prevent entrainment of the solids, and the density of the solvent may change markedly up the column as the concentration of solute increases.

Endless belt percolation extractors (Fig. 5) such as the uncompartmented de Smet belt extractor and the compartmented Lurgi frame belt extractor are similar in principle and closely resemble a belt filter, and are probably the simplest type of percolation extractor from a mechanical point of view. These are fitted with a slow-moving perforated belt. The belt is made from steel mesh cloths when the solids are fine, or coarser screens when the solids are larger, and is attached to chains which pass over sprockets at each end of the extractor. The solid is fed from a hopper at one end of the extractor to the moving belt, and the bed height is controlled by an adjustable damper at the outlet of the feed hopper. The two side walls of the extractor provide support for the bed on the moving belt. Fresh solvent is fed by spraying it onto the bed close to the discharge end of the belt, but leaving sufficient distance for adequate drainage of the bed prior to discharge. Miscella draining from the bed is collected in a pan below the belt and circulated back to be sprayed onto the bed at a point closer to the solids-feed end of the belt; this process is repeated to achieve extraction operating with a countercurrent flow. The top of the bed is scraped by a hinged rake which has two functions: (*1*)

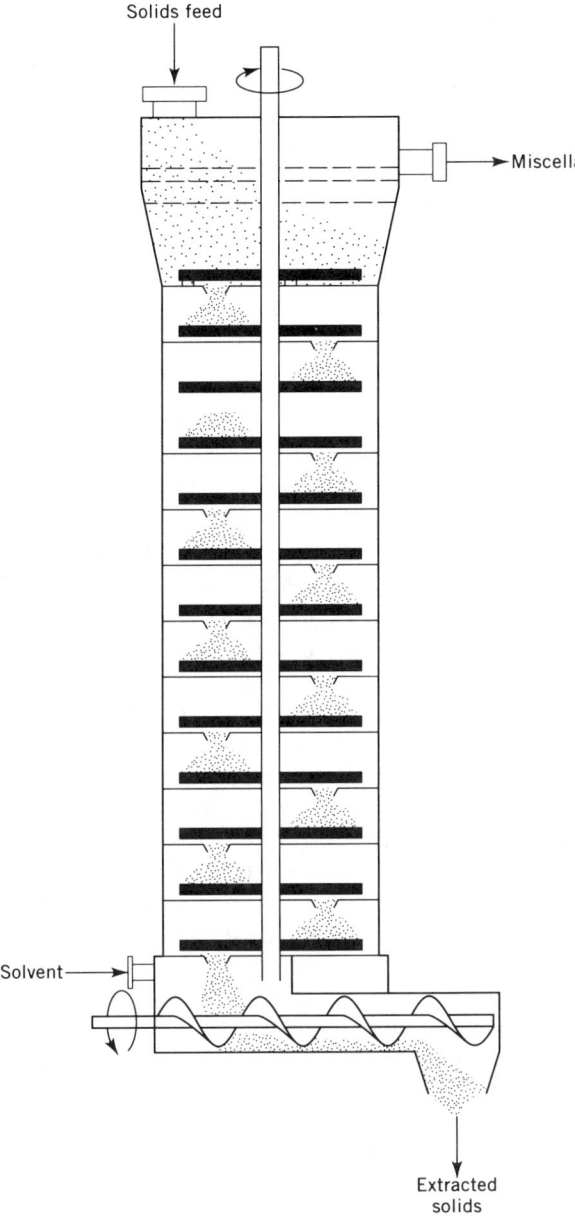

Fig. 4. Principles of the Bonotto-type extractor.

it prevents a layer of fine solids from accumulating at the top of the bed thereby reducing permeability, and (2) it form a solids pile which helps to prevent intermingling of miscella from different feed points at the surface of the bed. The belt is effectively washed twice: once by fresh solvent just after the solids discharge point, and then at the other end of the belt return by miscella. The extraction time and percolation rate determine the belt speed and the amount of drainage

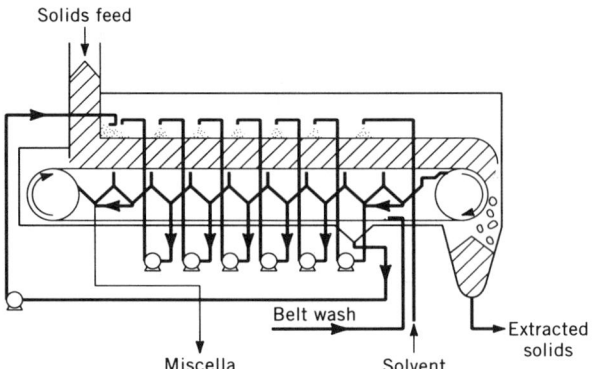

Fig. 5. Principles of the belt-type extractor.

area, and hence linear length of belt, required. These parameters control the plant capacity as the bed height is fixed by the mechanical design of the extractor.

Immersion extraction systems are useful in handling finely ground material or when the percolation rate through the material to be extracted is too rapid to allow effective diffusion from the solids. These systems are applied extensively in the sugar industry, in extraction from oilseeds having a high oil content, and from plant materials. The stepwise extraction by immersion, or continuous countercurrent leaching systems, can be carried out by a number of techniques analogous to the mixer–settler widely used in liquid–liquid extraction. The method is only viable if the solids settle more or less completely so that a clear supernatant liquor remains for decantation and when the slurry formed is pumpable, but it is nonetheless a most important method of leaching. If this is not so, other solid–liquid separation stages, such as filtration (qv) or centrifugation (see CENTRIFUGAL SEPARATION), have to be considered. However, such mixer–settler methods are continuous only by virtue of repeating a sequence of similar stages to achieve a given degree of extraction. More fully continuous methods of extraction were designed as tower systems and later as screw conveyor systems as effective methods of solids transport became reliable.

Continuous countercurrent decantation systems are not uncommon, employing a cascade of thickeners to wash solute from fine particles or to wash the solids formed by chemical reactions. The capacity of each thickener in the cascade is designed so that the residence time of the particles is long enough to allow the reaction to go to completion or to allow an acceptable degree of leaching to be achieved. Interthickener filtration permits use of much smaller volumes of solvent, sometimes also allowing greater removal of solute. The drawback of interstage filtering is that filters tend to be more expensive than equivalent thickeners, but this also has to be set against the smaller amount of space required for the whole installation.

The BMA diffusion tower (17), in common with some other tower systems, employs a central shaft fitted with a series of inclined plates or wings that direct movement of the solids. The tower shell is fitted with staggered guide plates for the same purpose. The solids are fed to the bottom of the tower and transported upward. Such units are found most widely in sugar beet refining. Tower heights

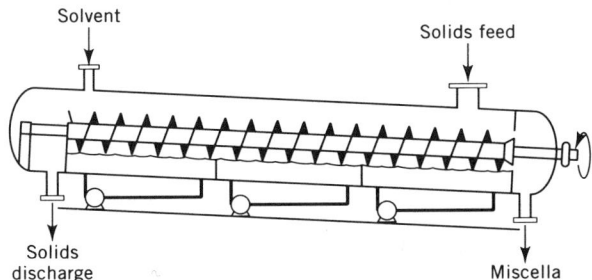

Fig. 6. Principles of the De Danske Sukkerfabriker (DDS) diffuser extractor.

of 10 to 15 m are used, with the diameter being dependent on the solids throughput capacity. For a capacity of 3000 metric tons of beet per day (17), a tower diameter of 5.5 m is required and the power consumption is of the order of 40 kW. In general, immersion extractors take up less space than the percolation types and have a lower power consumption.

Immersion-type extractors have been made continuous through the inclusion of screw conveyors to transport the solids. The Hildebrandt immersion extractor (18) employs a sequence of separate screw conveyors to move solids through three parts of a U-shaped extraction vessel. The helix surface is perforated so that solvent can pass through the unit in the direction countercurrent to the flow of solids. The screw conveyors rotate at different speeds so that the solids are compacted as they travel toward the discharge end of the unit. Alternative designs using fewer screws are also available.

The De Danske Sukkerfabriker (DDS) diffuser extractor (Fig. 6) is a relatively simple version of this family of machines, employing a double screw rotating in a vessel mounted at about 10° to the horizontal. The double screw is used to transport the solids up the gradient of the shell, while solvent flows down the gradient. Equipment using a single screw in a horizontal shell for countercurrent extraction of solids under pressure has been described (19).

Safety and Environmental Considerations

Solvent flammability, the solvent, and dust loading in the atmosphere of the working environment and of the products in the case of edible materials are the main factors that constitute health and safety hazards in extraction plants (20). General safety and environmental standards must therefore be applied and due recognition taken of the most recently published national regulations relating to acceptable threshold limit values (TLVs) for solvents and dusts. The permissible levels of solvent residues and emissions have repeatedly been lowered in recent years, making it important to ensure that the most up-to-date regulations are available to and acted upon by plant designers and engineers.

Disposal of exhausted solids can be easily overlooked at the plant design stage, particularly when these have no intrinsic value; alternative disposal methods might include landfill of inert material or incineration, hydrolysis, or pyrolysis

EXTRACTION, LIQUID–SOLID

of organic materials. Liquid, solid, and gaseous emissions are all subject to the usual environmental considerations.

BIBLIOGRAPHY

"Liquid–Solid Extraction" under "Extraction" in the *Encyclopedia of Chemical Technology*, 1st ed., Vol. 6, pp. 91–122, by F. Lerman, The Vulcan Copper & Supply Co.; in *ECT* 2nd ed., Vol. 8, pp. 761–775, by E. G. Scheibel, The Cooper Union for the Advancement of Science and Art; "Extraction, Liquid–Solid" in *ECT* 3rd ed., Vol. 9, pp. 721–739, by W. Hamm, Unilever Ltd.; in *ECT* 4th ed., Vol. 10, pp. 181–195, by R. J. Wakeman, University of Exeter.

1. G. Karnofsky, *J. Am. Oil Chemists Soc.* **26**, 564 (1949).
2. J. M. Coulson, J. F. Richardson, J. R. Backhurst, and J. H. Harker, *Chemical Engineering*, Vol. 2, 4th ed., Pergamon Press, Oxford, UK, 1991.
3. A. W. Hixson and S. J. Baum, *Ind. Eng. Chem.* **33**, 478 (1941).
4. N. Blakeborough, *Biochemical and Biological Engineering Science*, Vol. 1, Academic Press, Inc., New York, 1968.
5. D. R. J. Laws, *J. Inst. Brew. London* **83**(1), 39 (1977).
6. J. H. Hildebrand, J. M. Prausnitz, and R. L. Scott, *Regular and Related Solutions*, Van Nostrand, New York, 1970.
7. A. R. Burkin, *The Chemistry of Hydrometallurgical Processes*, E. & F. N. Spon, London, 1966.
8. J. D. Keane and C. T. Smith, *J. Am. Oil Chem. Soc.* **35**, 199 (1958).
9. H. Tomschke, M. Meiners, and E. Frohnert, *Tech. Mitt. Krupp Werksber.* **35**(1), 9 (1977).
10. W. L. McCabe and J. C. Smith, *Unit Operations in Chemical Engineering*, 3rd ed., McGraw-Hill Book Co., Inc., London, 1976.
11. R. H. Perry and D. Green, eds., *Perry's Chemical Engineers Handbook*, 50th ed., McGraw-Hill Book Co., Inc., 1984.
12. C. J. Geankoplis, *Transport Processes and Unit Operations*, Allyn and Bacon, Boston, 1978.
13. K. Weber, *Fette Seifen Anstrichmittel* **76**, 495 (1974).
14. *Food Process.* **36**(3), 71 (1975).
15. W. H. Goss, *J. Am. Oil Chem. Soc.* **23**, 348 (1946).
16. K. W. Becker, *AIChE. Symposium Ser.* **64**(86), 60 (1968).
17. F. Schneider, ed., *Technologie des Zuckers*, 2nd ed., M. & M. Schaper, Hannover, Germany, 1968.
18. R. N. Rickles, *Chem. Eng.*, 157 (Mar. 15, 1965).
19. F. A. Cantazini, *Braz. Pedido*, PI BR 80 03788 (Nov. 1981).
20. K. N. Palmer, *Dust Explosions and Fires*, Chapman and Hall, London, 1973.

RICHARD J. WAKEMAN
University of Exeter

EXTRACTIVE METALLURGY. See METAL SEPARATIONS.